MONOGRAPHS OF RARE AND EXOTIC ESSENTIAL OILS AND ABSOLUTES

Exploring the Past to Discover the Future of Medicine

Scott A. Johnson, AMP, CEEOS, CCMA, CPC

Cover design: Scott A. Johnson

Cover Copyright: Scott A. Johnson 2019

Monographs of rare and exotic essential oils and absolutes: Exploring the past to discover the future of medicine/Scott A. Johnson

ISBN-13: 978-0997548761

ISBN-10: 0997548762

Discover more books by Scott A. Johnson at www.authorscott.com.

Published by Scott A. Johnson Professional Writing Services, LLC, Orem, Utah

DISCLAIMERS OF WARRANTY AND LIMITATION OF LIABILITY

The author provides all information on an "as is" and "as available" basis and for informational purposes only. The author makes no representations or warranties of any kind, expressed or implied, as to the information, materials, or products mentioned. Every effort has been made to ensure accuracy and completeness of the information contained; however, it is not intended to replace any medical advice or to halt proper medical treatment, nor diagnose, treat, cure, or prevent any health condition or disease.

Always consult a qualified medical professional before using any dietary supplement or natural product, engaging in physical activity, or modifying your diet; and seek the advice of your physician with any questions you may have regarding any medical condition. Always consult your OB/GYN if you are pregnant or think you may become pregnant before using any dietary supplement or natural product, and to ensure you are healthy enough for exercise or any dietary modifications. The information contained in this book is for educational and informational purposes only, and it is not meant to replace medical advice, diagnosis, or treatment in any manner. Never delay or disregard professional medical advice. Use the information solely at your own risk; the author accepts no responsibility for the use thereof. This book is sold with the understanding that neither the author nor publisher shall be liable for any loss, injury, or harm allegedly arising from any information or suggestion in this book.

The Food and Drug Administration (FDA) has not evaluated the statements contained in this book. The information and materials are not meant to diagnose, prescribe, treat, or cure any disease, condition, illness, or injury.

ACKNOWLEDGEMENTS & DEDICATION

I could not have accomplished this book without the gifts, talents, knowledge, and wisdom I have been blessed with by a loving Heavenly Father. I am humbled by the privilege I have to research plant-based remedies and write about them for others to benefit.

I am grateful to the researchers and colleagues that shared their research to support this project.

I also must acknowledge my family who have been so supportive of my dreams and desire to help others experience greater wellness naturally. I dedicate this book to my wife and four children who mean the world to me. They inspire me to improve and become a better man.

CONTENTS

CHAPTER 1

Plants and herbs have been used for health promotion and disease treatment since the early days of humankind. Plants and humans are intimately connected and have a prolonged relationship. Not only are they used for food, but also plants play a fundamental role in medicine and health. Even most synthetic drugs can trace their origins to a plant molecule. Countless nutraceuticals and many drugs have entered the international market through exploration of traditional uses of medicinal plants. With this deep and ongoing relationship between plants and humans, it's no wonder that thousands of medicinal plants have been employed by all cultures across history and worldwide.

Traditional medicine systems are deeply rooted in oral traditions entrusted to select individuals during the course of history. Healing plants, methods, remedies, and skills developed and refined over hundreds—often thousands—of years have passed from one healer to the next. A reliance on nature and medicinal plants allowed civilizations to thrive and overcome a variety of ailments. These ancient yet living healing traditions are still highly regarded by many civilizations today.

Traditional medicine is the combined knowledge, skills, and practices of cultures founded in their beliefs, theories, and experiences used to prevent, diagnose, and treat disease. Cultural, social, economic, and spiritual beliefs as well as geographic area, environmental, and other prevailing conditions substantially influence the evolution of these traditional healing systems. Although each system is unique, each of these ancient healing systems shares a common holistic philosophy. The emphasis is to promote or restore homeostasis of the mind, body, emotions, and spirit. A greater focus is placed on the overall well-being of the individual as opposed to the disease that he or she is suffering from.

While the era of modern medicine (Western medicine) has brought about remarkable diagnostics and lifesaving techniques not available to indigenous healers of the past, as well as major advances in our understanding of biology, physiology, diseases, and chemistry, it should not and has not completely supplanted traditional medicine. Indeed, it's probable that up to four billion people (more than half of the world's population) still rely on traditional medicine for their primary health-care needs.[1] Estimates suggest that up to 90 percent of people in some areas of Africa, 70 percent of the Indian population, 40 percent of the Chinese population, and 38 percent of US adults use some form of traditional medicine.[2,3] Modern medicine excels at lifesaving treatments and diagnostics, but solely adhering to modern medicine bares the risk of losing not only a vital piece of history but healing methods that can benefit modern man's health.

Increased side effects, lack of curative treatment for chronic conditions, reduced affordability of sick care, high costs of new drugs, emerging microbial resistance to first-line antibiotics, and ideologies that propel individuals to seek more natural approaches are all factors influencing the increased global interest in traditional medicine. An overreliance on synthetic drugs has produced challenges and side effects that consumers are choosing to reject at a rapid pace. A revolution that includes eating better and active lifestyles has triggered a flourishing resurgence in traditional medicine. Frankly, individuals are seeking a more moderate, less invasive, and balanced approach to healing that features preventive measures.

Health-promoting practices—acupuncture, meditation, relaxation, moxibustion (burning of herbs over acupuncture points), energy healing, etc.—long established in cultures across the world are used as complementary or alternatives to modern medicine. Plant-based therapies with sufficient evidence of safety and efficacy are even recommended for the treatment of severe degenerative and chronic conditions where pharmaceuticals have largely failed.

The focus of pharmaceuticals is to inhibit what's perceived as symptoms. Natural medicine considers symptoms (such as vomiting, fever, and inflammation) as corrective actions being taken by the body to restore homeostasis. Using "anti" pharmaceuticals that prevent efforts by the body to heal itself does not address the root cause of the condition and actually inhibits the healing process. Aiding innate and proficient healing functions within the body and providing an environment for cells, organs, and organ systems to thrive is required for full health restoration. In essence, natural medicine aids the body so it can complete the healing process. A greater acceptance of traditional healing solutions can have an enormous impact on the costs of health care and shift sick care to incorporate preventive methods.

Culture values, social norms, and spiritual beliefs affect peoples' attitudes about medical care. Modern health-care practitioners that integrate traditional medicine recognize that any intervention is incomplete without assimilating the social, cultural, and spiritual beliefs of the patient. Failure to do so is a failure to treat the individual rather than the disease.

A resurgence in traditional healing methods is frequently viewed as a failure of modern medicine to satisfy the changing health-care needs of society. It may even be guarded against as a threat to modern medicine. Despite the lack of modern technological advances and experimentation, the ancient healing arts enjoy the advantage of hundreds of years to sharpen and perfect methods. In reality, the two systems don't need to be seen as direct opposites that can't coexist. Instead, they should be considered complements as an integrative approach that emphasizes the least invasive and effective remedies and methods. Merging the ancient healing arts with modern science can meaningfully impact not only the safe and effective use of medicinal plants but also the lives of the millions currently using traditional medicine.

Modern medicine has often failed to address conditions where emotional, behavioral, spiritual, and lifestyle factors play a role. Modern medicine attempts to separate these united and inseparable aspects of health into independent modalities treated by practitioners that specialize in only one aspect of health. With a dramatic increase in the number of conditions caused by these difficult to quantify factors—factors where traditional medicine excels—traditional medicine should be made more available.

The modern lifestyle has ushered in problems associated with a fast-paced, stress-fueled, and environmentally polluted environment. Sleep disturbances, depressed or compromised immune function, infertility, mental health disorders, cancer, and cardiovascular disease are just a few of the conditions linked to modern lifestyles. Plant medicines—particularly when combined with improved dietary and lifestyle choices—are ideally suited to restore balance, homeostasis, and reverse these troubling trends. We can't keep doing what we're doing—seeking solutions in the form of pills filled with synthetic drugs and invasive surgeries—and expect a different outcome. Too many are being duped to believe there's a pill for every ill and we can be rescued from poor eating and lifestyle habits with these magical pills.

We must switch our medical system to incorporate proven ancient remedies and preventive measures that teach people to eat better, be more active, and limit negative environmental factors. Traditional medicines are used in both rural and developed countries, even those where modern medicine dominates. Traditional medicine has been widely embraced as a form of complementary and alternative medicine in developed nations throughout Europe, North America, and Australia.[4,5,6,7] The widespread use of traditional medicine has led to a resurgence and return to the roots of healing—botanical medicine and a holistic approach.

Traditional Healing Systems of India

Ayurveda, Siddha, and Unani are all traditional medicine systems of India found mentioned in ancient Vedas and other texts. Ayurveda is the most well-known of these ancient healing systems, with its origins dating back perhaps to 2500 BC.[8] Its literal translation is "science of life." It's also referred to as the "science of longevity" because its focus is to promote a long and healthy life for man. Three ancient books known as the Great Trilogy—*Caraka Samhita, Sushruta Samhita*, and *Astanga Hridaya*—are considered the main texts of Ayurvedic medicine. Much of India's population uses Ayurvedic medicine alone or in combination with modern medicine.

Relying on key concepts—universal connections with the universe, people, and their health, the body's constitution (*prakriti*), and the life force (*dosha*), practitioners prescribe custom treatments of herbs, proprietary ingredients, nutrition, activity, and lifestyle recommendations. Ayurveda defines optimum health as a balance between body, mind, spirit, and social well-being, which are considered intimately connected. Good health persists as long as balance remains; however, lifestyle choices related to eating, activity, relationships, and career can disrupt this balance. Ultimately, a lack of harmony makes us more susceptible to disease.

Today, Ayurveda remains an integral part of health care in India, and its government supports clinical and laboratory research on its effectiveness. Roughly twenty thousand medicinal plants are found in India, of which, about seven thousand are used by Ayurvedic practitioners to cure disease.[9] This represents a large number of herbal remedies to use for health promotion.

The Siddha system of medicine originates from southernmost areas of India (the ancient Tamil land). Its roots are intertwined with the mythology and culture of the ancient Tamil civilization that predates much of recorded history. The name Siddha is derived from *siddhi*, meaning "attaining perfection in life."

Its guiding principles include that the individual is a microcosm (miniature version) of the universe. Equal emphasis is placed on the physical, mental, and spiritual aspects of health. In addition, the human body consists of five elements: earth, water, fire, air, and space; three humors: vatha, pitta, and kapha; and seven physical constituents: *ooneer* (plasma)—growth, development, and nourishment, *cheneer* (blood)— nourishes the muscles, produces color, and improves intellect, *oon* (muscle)—provides body shape, *koluppu/kozhuppu* (fatty tissue)—lubricates joints and balances oils, *elumbu* (bone)—body structure, posture, and movement, *elumbu majjai* (bone marrow)—forms blood corpuscles, and *sukkilam* (semen)—reproduction and fertility. Food is considered the basic elements from which the body is built and processed into humors, tissues, and wastes. Like Ayurveda, balance is required to maintain good health, and disease occurs when this equilibrium is disturbed.

Eight types of examination are used to determine the appropriate remedy, including the pulse, skin, tongue, complexion, speech, eye, stools, and urine. Pulse is emphasized in the diagnosis. Practitioners prescribe remedies from three different classes: *thaavaram* (herbs), *thaathu* (inorganic substances), and *jangamam* (animal-based remedies).

Presently practiced in India, Asia, and the Middle East, Unani medicine is a Greco-Arabic medicine based on the teachings of the Greek physician Hippocrates and the Roman physician Galen. It was developed into a comprehensive medicine system by Arabian and Persian physicians in the Middle Ages.

Unani focuses on four humors, each of which has its own temperament: blood (hot and wet), phlegm (cold and hot), yellow bile (hot and dry), and black bile (cold and dry). Both the quantity and the quality of the humors determine the individual's state of health. Primary importance is placed on nutrition and digestive health and specific nutritional recommendation are ordered according to temperament during treatment. Good nutrition yields good humors (*Akhlā/Ṣāliḥa*), whereas improper nutrition leads to imbalanced humors (*Akhlā/Radiyya*). Imbalanced humors are corrected with remedies (herbal, animal, and mineral).

Classical diagnosis is based on pulse reading, stool, and urine as well as physical examination. The first lines of treatment include regimental therapy (nutrition, exercise, massage, cupping, climate change, leaching, and balneology—therapeutic bathing) and pharmacotherapy (plant, animal, and mineral remedies). Surgical intervention is reserved as a last resort.

Traditional Chinese Medicine

With a history of more than three thousand years,[10] traditional Chinese medicine (TCM) has accumulated and methodically collected a wealth of information on healing herbs and disease treatments. *The Divine Farmer's Classic of Herbalism* is an ancient Chinese herbal text—and the oldest known herbal pharmacopeia in the world—compiled approximately two thousand years ago. It is rooted in the ancient philosophy of Taoism and possessed a rather sophisticated understanding of anatomy, physiology, biochemistry, and infectious disease.

The Yellow Emperor's Classic, originating between 600 BC and 300 BC, contains the basic concepts of medical philosophy and remarkably accurate information on the human body.[11] The study of disease concentrates on how disorders cause physiological dysfunction. The goal is to restore proper function and physiological balance (*zheng*).

Chinese medicine is largely based on the balance between yin (earth, cold, femininity) and yang (sky, heat, masculinity) and the flow of *Qi* (energy) through twelve meridians. The human body is viewed as a miniature version of the universe and five elements—fire, earth, wood, metal, and water—represent the stages of human life and explain the function of the human body and how it changes during disease. Herbal medicine is a core part of this healing system, and the Chinese *materia medica* describes thousands of medicinal substances, most of which are plants. Different parts of plants (leaves, roots, stems, flowers, and seeds) are used medicinally as teas, capsules, liquid extracts, powders, and granules. TCM also uses acupuncture, moxibustion, tui na (Chinese therapeutic massage), nutrition, tai chi, and qi gong.

Traditional European Herbalism

With a very long and vibrant history, many European citizens still rely on and trust traditional medicine to correct minor and severe conditions. Pharmacopoeias of traditional European healers contain thousands of remedies derived from leaves, roots, bark, herbs, mineral, and animal substances. Traditional European medicine claims its roots in ancient Egypt and with Hippocrates in ancient Greece. Its practice and methodologies were considerably progressed when herbal skills were brought back by crusaders who learned them from Arabians. It continued to flourish for centuries until the advent of chemical medicine caused a slow decline in its use.

Often defined by varied climates, access to imported herbs, local culture, and economic status, medieval European herbalism treated a variety of illness, tended to the wounded in battles, and cared for the home with a variety of modalities and remedies. Many regions would have likely used well-known herbs such as elderberry, sage, nettle, and yarrow. For the most part herbs were used much like today as teas, tinctures, glycerites, infusions, salves, poultices, and ointments. European herbalism is usually overshadowed by other traditional systems, such as Ayurvedic and TCM.

Korean Traditional Medicine

Although influenced by TCM, Korean traditional medicine has distinctive features and has evolved to suit the needs of the Korean people. Its unique identity was fostered during the Goryeo and Joseon dynasties (918–1392 and 1392–1910, respectively). Under the reign of King Sejong, *Hyangyak-Jipseonbang* (Compendium of Aboriginal Medicine), and a comprehensive Korean medical text *Euibang-Yoochui* (Compendium of Aboriginal Medicine) were compiled. In the middle of the Joseon period, the widely popular book on medicine *Dongui-Bogam* (Treasured Mirror of Eastern medicine) was also published. This latter text established external causes of diseases named as the six excesses: wind, cold, summerheat, dampness, dryness, and fire.[12] A balance with the universe is a major concept of Korean traditional medicine.

The primary physiological components of Korean traditional medicine are essence, spirit, qi, blood, fluid, and humor. The essence and spirit are stored in the five viscera (liver, heart, spleen, lungs, and kidney) and represent the core of the body. The qi and blood regulate internal body functions, while the fluid and humor are the body's liquids and become either qi or blood. According to traditional Korean medicine, diseases are caused by insufficiency in healthy qi, invasion by pathogens that interrupt healthy qi flow, or abnormal yin-and-yang activity.

Comprehensive analysis of symptoms and signs and classification of the condition as yin or yang determines the cause, location, and nature of the condition. Pulse is measured to determine the health of qi circulation and auscultation (listening to sounds of the heart, lungs, and other organs) are also used to diagnose illnesses. Herbal remedies, acupuncture, and moxibustion are prescribed based on the diagnosis and according to the *materia medica* to maintain balance of body structures and functions. The priority of Korean traditional medicine is disease prevention.

Traditional African Medicine

African healing traditions have been in existence for centuries and are fixed in indigenous herbalism and spirituality. The use of medicinal plants is still a fundamental element of African health-care systems. In many parts of rural Africa, traditional healers are heavily utilized due to their affordability and availability. Moreover, some ailments (i.e., malaria, HIV/AIDS)—for which no viable treatments exist in modern medicine—disproportionately affect the people of Africa. Given the diversity of cultures on the African continent, it's no surprise that an assortment of therapeutic healing systems exists.

Despite regional differences, each of the systems is holistic in nature, encompassing the physical, spiritual, moral, and social. Traditional African healers typically diagnose and treat the emotional and mental symptoms first because these are considered the root of the illness. Diagnosis may be reached through spiritual means. A range of medicinal herbs (Africa is estimated to contain up to forty-five thousand species of plants, five thousand of which are used medicinally) is prescribed based on the symptoms and according to the patient's local environment. Numerous traditional African herbs have proved beneficial in treating disease and maintaining health.

According to traditional African healing systems, humans experience good health when the physical, spiritual, moral, and social function together harmoniously. If any of these aspects of health are disrupted or out of balance, a person will become ill. Indeed, African healing systems must be viewed as part of a spiritual framework to understand its rationale and treatment of illness.

Persian (Iranian) Traditional Medicine

Persia was a mecca for the study and practice of medicine anciently, attracting scientists from different civilizations to examine early and emerging theories. Continual invasions from nearby empires influenced and sometimes hindered Persian medicine. In more recent history, traditional Persian medicine contributed weakly to the world of research until the last couple of decades when Iranian research facilities increased the study, use, and promotion of Persian medicine. This has led to a significant increase in the number of articles published on Persian medicine, particularly clinical studies. Indeed, traditional Persian medicinal plants—especially essential oils obtained from these plants—have seen a remarkable increase in published clinical research. The knowledge and experiences of multiple Persian scientists and healers has influenced Persian medicinal practices through many centuries.

The medicinal history of Persia details healing performed by divine words (which was considered most effective), surgery, and herbs. The *Venidad ("Given Against the Demons")* is one of the preeminent texts of this age. It details the ways that demons manifest in humans and how to confound them. When the Arabs conquered Iran, they brought with them Islamic religion and destroyed schools, universities, and libraries, burned books, and killed scholars. Medieval healers and scholars strove to preserve and consolidate the medical knowledge and ideas of ancient Persia and expanded on this information with their own observation, experimentation, and skills.

Persian medicine adheres to the concept of personalized medicine (a medical model in which all diagnostic, analytical, and therapeutic aspects of a disease are individualized for each person based on specific testing) and health prevention and preservation.[13] Each person is considered to have a unique temperament that is influenced by age, environment, gender, ethnicity, and season. According to this theory, the same condition (e.g., arthritis) will manifest and progress in each individual differently, which requires an individualized treatment rather than relying on the same herbs to treat the same disease. One of the greatest Persian physicians and scientists, Avicenna, described medicine as "the health maintenance when you are healthy, and its restoration when you are sick."[14]

A large number of ancient tablets outlining traditional Persian medical practices have survived in the library of Ashurbanipal at Nineveh (668 BC), Assyria. One of the oldest (dating to about 1600 BC) and the largest collections—forty total tablets—is known as "Treatise of Medical Diagnosis and Prognoses." Religious leaders (mobeds, magi), surgeons, and herbalists were the healers and physicians of ancient Persia. They believed that health was maintained when all four body fluids—blood, phlegm, yellow bile, and black bile—were in a balanced state. Body-fluid balance was maintained through the digestive system, specifically the digestive processes of the gastrointestinal tract, liver, blood, and cells.

American (Western) Herbalism

When Europeans began voyages to the Americas it opened the possibility of new plants native to the Americas for use in herbalism. They also had the opportunity to learn from the natives of the Americas who had spent centuries exploring, experimenting with, and using the indigenous medicinal herbs. Once settled in the Americas, colonists adapted their medical systems according to the available plants and those that they had brought with them (Mediterranean herbs rich in volatile oils, herbs of Northern Europe, and American herbs).

Natives of the Americas developed a whole medical system that included a variety of holistic treatments employed by indigenous healers for multiple acute and chronic conditions. Healing practices varied according to tribal differences but they shared the common view that good health was obtained through a holistic approach that included biological, psychological, social, and spiritual features. Health and well-being was promoted by nutrition, celebrations of the seasons and harvests, ceremonies that included the sick individual, their family, the healer, and the community, native medicinal plants, and most of all by living in harmony with the earth.

Samuel Thompson, an uneducated and untrained farmer, is considered "the Father of American Herbalism." After treating his two-year-old daughter, who was deathly ill, with steam, he found his neighbors coming to him for treatments that consisted of steam baths and herbs. His methods comprised six basic remedies: lobelia (cleansing through vomiting), cayenne (circulation and raise body temperature), herbs that caused diarrhea (cleansing of the bowels), bitter herbs (to regulate digestion and bile), peach or cherry pits (cure diarrhea and restore strength), and finally, rheumatic drops of myrrh and capsicum (pain relief and promote body temperature equilibrium). Despite a lack of former medical training, he devised a system of healing that swept the American frontier and was adopted by millions of Americans in the early nineteenth century. His success as a healer is largely attributed to his willingness to champion people's medicine and the promotion of self-help health care.

American herbalism prescribed herbs to treat general tissue conditions or states and general body functions or systems (digestion, respiration, circulation, etc.). Herbs were chosen based on the temperature or energetics (hot, cold, damp, dry), affinity for a particular organ function, and the action of the herb in the organism. Rather than attempting to kill foreign pathogens responsible for disease, healers focused on creating an environment inhospitable to these organisms so they would die out and the person's health would be restored.

Western herbalism is a discrete system, distinct from other cultures previously mentioned that used herbs as part of their healing systems and has a unique material medica, principles, and practices. In reality, early American herbalism paved the way for other healing modalities such as Eclectic medicine, homeopathy, and naturopathy.

Australian (Aboriginal) Traditional Medicine

Magic holds a distinctive significance in traditional healing systems of aboriginal Australia, which date back centuries. Aboriginal healers (ngangkari) built a healing system and knowledge base founded on the resources they found on the land. Most often, healers were born into their roles with knowledge and practices being passed down from family member to family member.

Indigenous Australians believed that disease resulted from sorcery or disruption of the spirit, which led to physical symptoms. Tribal societies thought that illness was caused by five major categories: 1) sorcery; 2) loss of the soul; 3) taboo violation; 4) disease-causing objects in the body; or 5) harmful spirits. First and foremost, treatment involved remedying the spirit. Tribal medicine men used magical material such as quartz crystals and stones to converse with the spirit world and gain inspiration for healing practices. In addition, healers used "white" magic to cure and "black" magic to cause illness. Healing rituals included massage and sucking to remove disease-causing objects.

Today, Ngangkari work alongside modern physicians and medical staff in clinics and hospitals. Healers treat everything from pain to childhood illnesses and everything in between. The Commonwealth of Australia has never formerly enacted legislation to legally recognize or regulate Aboriginal traditional medicine.

The Validation of Traditional Healing through Modern Science

Embedded in each of these traditional healing systems is a holistic approach and a reliance upon medicinal botanicals for healing. Polished over centuries, despite the lack of advanced diagnostics methods and a deeper understanding of biology and physiology, these healing systems have proven remarkably effective. Indeed, modern science has validated many of the healing practices of ancient cultures.

In modern times, Western medicine is supplanting traditional healing systems across the globe, placing thousands of years of knowledge and experience at risk. There's an urgent need to validate the basic principles, methods, techniques, and remedies of ancient healing systems to preserve the invaluable wisdom and heritage of botanical medicine. Advancements in the ongoing research, particularly clinical research, will permit the use of evidence-based methods and remedies that are less invasive and less prone to side effects than modern drugs and surgeries.

The prevailing perception is that traditional systems lack a scientific basis and the evidence for safety and efficacy that Western medicine requires. With the renewed interest in traditional remedies, scientists in local areas are beginning the laborious task of evaluating indigenous medicinal plants. Genetic fingerprints of medicinal plants, metabolite profiling, composition standardization, quality control, therapeutic dosing, safety assessments, and risk assessments are several of the factors that require consideration and scientific evaluation. This research is revealing the mechanisms by which many of these remedies work and justifying their use as part of an integrative approach. Merging these ancient healing arts with modern science is elevating medicine as a whole and will largely benefit modern man who is seeking more natural solutions for health maintenance and restoration.

One of the shortcomings of this research is that Western medicine works to antagonize an "abnormal" process that traditional medicine considers a corrective action of the body. Traditional medicine doesn't seek to attack pathogenic targets as much as it seeks to support and restore the self-regulatory ability of the body by adapting its internal and external environment. In other words, help the body thrive in a healthy environment and it's fully capable of correcting illness and disease. So, while scientists can, and have, demonstrated the ability of botanical medicines to inhibit pathogens, its scientific validation is better served by showing by what mechanism it restores normal physiological functions.

While hundreds and thousands of positive studies have been published in support of traditional healing, the negative ones tend to receive the most media coverage. Interestingly, negative reports for synthetic drugs are also available but do not always get media attention. To view the vast amount of supporting evidence for botanical medicines, one may refer to this author's other works *Medicinal Essential Oils: The Science and Practice of Evidence-Based Essential Oil Therapy* and *SuperCritical Essential Oils: A Companion Resource to Medicinal Essential Oils*. Clearly, consumers are seeking for alternatives to modern chemical medicine, so the questions shouldn't be "do traditional remedies have a place in modern medicine?" but "how can we improve their use so consumers enjoy the greatest safety and efficacy?"

Essential Oils

Essential oils are not truly oils but volatile and aromatic molecules obtained from plants. They are complex mixtures of hydrocarbons and oxygenated derivatives produced via two isoprenoid (derived from one of more isoprene units, which is a molecule with five carbon atoms) pathways. Isoprene units are joined together by chemical bonding to form molecules containing five (hemiterpenes; one isoprene unit), ten (monoterpenes; two isoprene units), fifteen (sesquiterpenes; three isoprene units), and twenty (diterpenes; four isoprene units) carbon atoms. Production of essential oils comes at a high metabolic cost to the host plant and plays a pivotal role in plant health and survival.

A large group of enzymes known as terpene synthases form the final terpenoid products—essential oils. Isopentenyl diphosphate (IPP) and dimethylallyl diphosphate (DMAPP) are the universal precursors to produce essential oils. The first pathway for essential oil synthesis is the mevalonic acid pathway. The mevalonic pathway synthesizes mostly sesquiterpene precursors (although in some instances a very small number of monoterpenes are produced). The other route for essential oil production is the 2C-methyl-D-erythrytol 4-phosphate (MEP) pathway. It creates only monoterpene precursors.

Essential oils are produced by glandular trichomes (modified epidermal hairs containing cells that specialize in creation and synthesis of plant volatiles) and other secretory structures in plants. They are stored in glands until they diffuse through the gland wall and evaporate releasing a distinct aroma. The composition of the essential oils continuously adapts to the plant's internal and external environment. Essential oils defend against pathogens, parasites, and predators; aid tissue healing processes in response to damage; and play an important role in plant-to-plant communication, and the attraction of pollinators and animals that disperse seed to create offspring.

Like their herbal cousins, aromatic essences and volatile molecules obtained from plants have a rich tradition of use throughout history. While essential oils may be new to many people, they are not a new therapy. Aromatic essences and volatile molecules from plants have been used for medicinal, religious, and other purposes perhaps as far back as six thousand years ago. Today, this rich tradition of aromatic and volatile plant medicine is being rediscovered.

Much like herbal medicine, it isn't currently possible to say for sure which culture was the first to use volatile compounds from plants. However, archaeological evidence and surviving medical records suggest they were used by the ancient Egyptians, Chinese, Indians, Babylonians, Greeks, Romans, Arabians, and Persians. Plants represented a source of food, feed for domesticated animals, material for clothing, and, of course, medicine during their thousands years' history with humankind. Essential oils have been a part of this history and used in many different traditional healing systems for their biological activities.

Today, scientists in Africa, Asia, Europe, the Middle East, America, and Australia continue to explore essential oils from traditional medicinal plants. The composition and therapeutic properties of plants that are embedded in ancient healing traditions are investigated in laboratory, animal, and human experiments. A growing interest in validating traditional healing systems and the plants they use as medicine has revealed some exciting properties and benefits of essential oils. Further exploration may uncover currently unknown medicinal essential oils and undiscovered human health benefits from traditional medicinal plants.

There are approximately three thousand essential oils produced from more than two thousand plant species, representing about sixty plant families.[15] Of these, about three hundred are considered important for commercial and medicinal use. Many of the commercially important essential oils share similar constituents that are widely distributed in essential oils like alpha-pinene, limonene, 1,8-cineole, alpha-terpineol, beta-caryophyllene, gamma-terpinene, citral (geranial + neral), myrcene, and linalool. However, some essential oils are characterized by unique constituents—massoia lactones, curcumanolides, grandisol, 2-methyloctane, elsholtzia ketone—that may hold unknown medicinal value. Moreover, essential oils with similar major constituents may affect human health differently. Despite both being predominantly limonene, lemon favorably influences genes in the lymphatic and endocrine systems; whereas, orange has an affinity for the muscular and reproductive systems.[16] A synergy and balancing effect of major and minor constituents in essential oils is frequently observed, which may account for some of these differences in therapeutic properties. The continued exploration and scientific evaluation of essential oils is not only valuable but crucial for human health.

Conclusion

Could an undiscovered plant hold the key to reversing diseases (heart disease, diabetes, obesity, cancer) that plague man today? Obviously, the greatest impact will come from people making meaningful dietary and activity changes, but plant medicines are intriguing for their complexity and thousands of years' history with man. Unlike drugs that rely on a single isolated "active" chemical, plants contain mixtures of many different phytochemicals, also known as secondary metabolites, that may act individually, additively, or in a synergistic way to improve health. Regularly, the combined effect of the phytochemical mixtures surpasses the total activity of the primary "active" phytochemical. In addition, these secondary metabolites may increase the satiability of the "active" constituent or act as a buffer to minimize undesirable effects.

Continued exploration of traditional medicinal botanicals is strongly encouraged. An emphasis on essential oils—nature's most potent and versatile remedies—is urged. This text is meant to explore some of the benefits already revealed by science and the composition of essential oils and absolutes not widely commercially available. Understanding these precious botanicals may increase their availability for man's use. The health benefits among rare and exotic essential oils awaiting to be discovered and employed by man is an exciting prospect. Here's to undiscovered territory and the quest to improve health naturally with essential oils.

CHAPTER 2

AFRICAN SANDALWOOD
(East African Sandalwood, Osyris)

Osyris lanceolata Hochst. & Steud., *O. abyssinica* Hochst. ex A. Rich., *O. tenuifolia* Engl.

FAMILY: Santalaceae
NOTE: Base
AROMA INTENSITY: Medium
AROMA: Woody, earthy, slightly balsamic
COMMON EXTRACTION METHOD: Steam distilled or hydrodistilled from the wood
POSSIBLE SUBSTITUTE OILS: Sandalwood
BLENDS WELL WITH: Bergamot, black pepper, cedarwood, chamomile (Roman, German), cypress, frankincense, jasmine, juniper, lavender, myrrh, palo santo, patchouli, petitgrain, rose, rosewood, sandalwood, vetiver, ylang ylang
SUBCELLULAR LOCALIZATION | EPIGENETIC INFLUENCE: Currently unknown | Currently unknown
RECOMMENDED DILUTION RANGE: 3%–33%; neat for some conditions

PRIMARY CONSTITUENTS:[17,18,19]

(Z)-Lanceol	18.0%–56.7%
(Z)-Alpha-Santalol	0.0%–25.1%
(Z)-Beta-Santalol	0.0%–21.4%
(Z)-Alpha-Trans-Bergamotol	0.0%–13.4%
Epi-Cyclosantalal	0.0%–5.9%
2,(7Z,10Z)-Bisabolatrien-13-ol	0.0%–5.6%
(E)-Nuciferol	0.0%–5.5%
Epi-Alpha-Bisabolol	0.0%–5.1%
(E)-Cis,Epi-Beta-Santalol	0.0%–4.7%
Lanceoloxide	0.0%–3.9%
(Z)-Alpha-Bisabolene + Beta-Bisabolene	0.6%–3.7%

OTHER CONSTITUENTS: 7-epi-alpha-cedrene, (Z)-alpha-bergamotene, alpha-cedrene, alpha-santalene, (E)-alpha-bergamotene, epi-beta-santalene, beta-santalene, beta-acoradiene, ar-curcumene, gamma-curcumene, (E)-beta-bergamotene, (Z)-gamma-bisabolene, ar-tenuifolene, (E)-alpha-bisabolene, tenuifolene, para-benzoquinone, (Z)-beta-farnesene, alpha-bisabolol, phthalate

PREFERRED COMPOSITION FOR CLINICAL USE:

Constituent	
(Z)-Lanceol	18%–38%
(Z)-Alpha-Santalol	15%–30%
(Z)-Beta-Santalol	15%–30%
(Z)-Alpha-Trans-Bergamotol	5%–15%
(E)-Nuciferol	2%–10%
(E)-Cis,Epi-Beta-Santalol	1%–7%
Santalenes <α, β, epi-β>	0.1%–5%
Alpha-Bisabolol	0.1%–5%

REPORTED THERAPEUTIC PROPERTIES: Relieves diarrhea, antiarthritic, antirheumatic, relieves headache, analgesic, anti-inflammatory, slows excess bleeding, promotes red blood cell production, aids digestion, wound healing, astringent, antiseptic, soothes skin itching, improves aged skin, reduces the appearance of scars and blemishes, anticancer, sedating, relieves anxiety, stress management, aphrodisiac, enhances meditation, promotes acceptance in times of loss, balancing to the emotions

CAUTIONS:

- None currently known.

SELECTED EVIDENCE:

- None found.

AFRICAN WORMWOOD (Lanyana, Artemisia Afra Herba, Wilde Als)

Artemisia afra Jacq. ex Willd.

FAMILY: Asteraceae
NOTE: Middle
AROMA INTENSITY: Strong
AROMA: Pungent, herbaceous, camphoraceous
COMMON EXTRACTION METHOD: Steam distilled or hydrodistilled from the leaves or aerial parts
POSSIBLE SUBSTITUTE OILS: Sage, tansy, wormwood, mugwort, Spanish sage, camphor, white wormwood
BLENDS WELL WITH: Bay laurel, bergamot, cajeput, camphor, chamomile (German, Roman), coriander, eucalyptus, lavender, lemon, lime, orange, neroli, palo santo, peppermint, petitgrain, ravensara, rosemary, sage, Spanish sage, spikenard, tangerine, tansy, thyme, turmeric, vetiver
SUBCELLULAR LOCALIZATION | EPIGENETIC INFLUENCE: Currently unknown | Currently unknown
RECOMMENDED DILUTION RANGE: 1.5%–20%

PRIMARY CONSTITUENTS:[20,21,22,23,24,25,26,27]

Alpha-Thujone CT	
Alpha-Thujone	78.7%
Beta-Thujone	13.1%
1,8-Cineole	8.2%

Beta-Thujone CT	
Beta-Thujone	30.3%–52.5%
Camphor	14.4%–15.0%
1,8-Cineole	2.4%–13.1%
Hex-3-en-2-one	0.0%–4.4%
Heptane	0.0%–4.1%

Camphor CT	
Camphor	26.8%–46.2%
Davanone	0.0%–16.6%
Alpha-Thujone	0.0%–15.2%
Artemisia Ketone	0.0%–7.4%
1,8-Cineole	0.0%–4.2%
Beta-Thujone	0.0%–4.1%
Bornyl Acetate	0.6%–3.8%
Terpinen-4-ol	1.4%–3.6%
Borneol	0.0%–3.3%
Chamazulene	0.0%–3.2%

Zimbabwe (Wild and Cultivated)

Alpha-Copaene/Camphor	6.2%–50.3%
Artemisia Ketone	tr–41.9%
1,8-Cineole	0.1%–29.3%
Borneol	0.8%–19.1%
Santolina Alcohol	0.0%–10.1%
Alpha-Terpineol	0.1%–8.1%
Camphene	0.3%–8.0%
Sabinene	0.1%–7.9%
Beta-Caryophyllene	0.5%–5.0%
Alpha-Thujone	tr–4.9%
Trans-Sabinene Hydrate	0.0%–4.4%
Bornyl Acetate	0.2%–4.2%

Artemisyl Acetate CT

Artemisyl Acetate	35.5%
Yomogi Alcohol	13.5%
Artemisia Ketone	13.2%
Bornyl Acetate	8.0%
Geraniol	5.4%
1,8-Cineole	3.5%

Thujone CT

Thujone (isomer not identified)	30.3%–32.0%
Camphor	11.7%–13.0%
1,8-Cineole	9.6%–12.9%
Artemisia Ketone	2.9%–3.3%

Note: The composition of African wormwood varies significantly and multiple CTs are reported in the literature. Oils rich in 1,8-cineole (49.5%–67.4%) are also reported in some papers.[28,29,30]

OTHER CONSTITUENTS: *Alpha-Thujone CT*—N/A; *Beta-Thujone CT*—Alpha-pinene, 1,5-heptadien-4-one, borneol, cyclohexene, bornyl formate, carvotanacetone, piperitone, phenol, nonane, cyclohexene, 3-penten-2-one, diepoxy-p-menthane, spathulenol; *Camphor CT*—Camphene, para-cymene, sabina ketone, cuminaldehyde, p-cymen-7-ol, caryophyllene oxide; *Artemisyl Acetate CT*—Sabinene, limonene, beta-phellandrene, artemisia alcohol, camphor, linalool, (Z)-methyl cinnamate, neryl acetate, spathulenol, eudesma-3,5,11-triene; *Zimbabwe (Wild and Cultivated)*—Alpha-pinene, alpha-fenchene, myrcene, alpha-terpinene, dehydro-1,8-cineole, limonene, (E)-beta-ocimene, gamma-terpinene, para-cymene, terpinolene, artemisyl acetate, beta-thujone (< 0.4%), cis-sabinene hydrate, cis-p-menth-2-en-1-ol, terpinen-4-ol, myrtenal, delta-terpineol, bicyclogermacrene, piperitol, delta-cadinene, cuminaldehyde, myrtenol, calamenene, spathulenol, T-muurolol; *Thujone CT*—Camphene, sabinene, yomogi alcohol, para-cymene, 3-p-menthene, terpinen-4-ol, myrtenol, cis-piperitol, piperitol, 2-carene

PREFERRED COMPOSITION FOR CLINICAL USE:
Too great of variance to determine.

REPORTED THERAPEUTIC PROPERTIES: Antiallergic, anti-inflammatory, analgesic, reduces fever, relieves headache, antiarthritic, antirheumatic, antineuralgic, antimicrobial, antibacterial, antifungal, antiviral, antiparasitic, anti-infectious, aids digestion, antioxidant, anticancer, eases cough, decongestant, lowers high blood pressure, insecticidal, aids memory and cognition, reduces mental exhaustion

CAUTIONS:
A full GC-MS should be reviewed before clinical use to determine risks versus benefits and safety due to the great variance in composition. The following are general cautions based on composition and should be observed when meaningful amounts of the listed constituents—or combinations of the constituents—are present, or avoided completely in more vulnerable people (pregnancy, lactation, children, epileptics). See other monographs with similar composition for more details:

- Thujones: Children, pregnancy, lactation, liver disorders, kidney disorders, epilepsy, oral use, drugs metabolized by the CYP450 enzyme pathway
- Camphor: Children, pregnancy, lactation, epilepsy, oral use, broken skin, G6PD, compromised liver
- 1,8-Cineole: Epilepsy, oral use, barbiturates, drugs metabolized by the CYP450 enzyme pathway

Selected Evidence:

- *In vitro* research showed that African wormwood CT camphor essential oil moderately inhibited *S. aureus, S. epidermidis, P. aeruginosa, E. cloacae, K. pneumoniae,* and *E. coli.*[31] The alpha-thujone CT was strongly antimicrobial to multiple pathogenic organisms.[32] Similarly the beta-thujone CT inhibited a panel of pathogenic organisms, with A. *calcoaceticus, B. natriegens, B. linens, B. thermosphacta, C. jiieundii, K. pneumoniae*, and *S. marcescens* being the most susceptible to the oil.[33] Another study reported that the beta-thujone CT displayed broad-spectrum activities against gram-positive and gram-negative bacteria.[34] African wormwood exhibited significant antifungal activity against *A. ochraceus*, *C. albicans*, *A. alternata*, *G. candidum*, *A. niger*, *P. citrium*, and *A. parasiticus*.[35]
- African wormwood CT beta-thujone exerted considerable antioxidant activity in the beta-carotene and linoleic acid test.[36] The camphor CT also exhibited good antioxidant activity in the DPPH and hydroxyl scavenging assays.[37]

ALGERIAN FIR
(Numidian Fir)

Abies numidica de Lannoy ex Carrière

Family: Pinaceae
Note: Top-Middle
Aroma Intensity: Medium
Aroma: Woody, balsamic, piney, fresh
Common Extraction Method: Steam distilled or hydrodistilled from the leaves (needles)
Possible Substitute Oils: Balsam fir, juniper berry, cypress, pine
Blends Well With: Balsam fir, birch, blue spruce, cassia, cedarwood, chamomile (German, Roman), cypress, frankincense, galbanum, lavender, lemon, myrrh, myrtle, palo santo, pine, sandalwood, silver fir, spruce (black), tsuga, white fir, wintergreen
Subcellular Localization | Epigenetic Influence: Currently unknown | Currently unknown
Recommended Dilution Range: 5%–Neat

Primary Constituents:[38,39]

Alpha-Pinene CT	
Alpha-Pinene	22.6%
Limonene	19.7%
Beta-Pinene	12.3%
Camphene	11.2%
Beta-Phellandrene	7.8%
Santene	3.3%

Bornyl Acetate CT	
Bornyl Acetate	29.6%
Alpha-Pinene	13.2%
Delta-3-Carene	12.3%
Limonene	5.6%

Other Constituents: *Bornyl Acetate CT*—Beta-pinene, myrcene, 1,8-cineole, borneol; *Alpha-Pinene CT*—Tricyclene, myrcene, terpinolene, bornyl acetate, citronellyl acetate, geranyl acetate, beta-caryophyllene, 6,9-guaiadiene, alpha-humulene, germacrene D, beta-himachalene, valencene

Preferred Composition for Clinical Use:
Insufficient data/samples to determine.

REPORTED THERAPEUTIC PROPERTIES: Antimicrobial, antifungal, antimicrobial, antioxidant, anti-inflammatory, analgesic, antirheumatic, diuretic, antiseptic, wound healing, relieves chronic skin conditions and itching, reduces the appearance of blemishes, aids respiratory function, ease cough, reduces guilt, relaxing, stress management, reduces anxiety

CAUTIONS:

- None currently known.

SELECTED EVIDENCE:

- *In vitro* research showed that Algerian fir essential oil (unidentified composition but possibly bornyl acetate CT as the study was from the same authors that reported a bornyl acetate CT of oil in another study) inhibited *E. coli, K. pneumoniae, P. aeruginosa, Acinetobacter* sp., and MRSA.[40] The alpha-pinene CT of Algerian fir essential oil demonstrated antimicrobial activity against *S. cerevisiae, M. luteus, S. aureus, S. epidermidis, E. faecalis, K. pneumoniae, E. coli,* and *B. cereus.*[41] Significant activity was observed against *S. cerevisiae, B. cereus, E. coli*, and *E. faecalis*, moderate activity against *M. luteus, S. aureus,* and *S. epidermidis*, and low activity against *K. pneumoniae*. An unidentified CT of Algerian fir essential oil inhibited gram-positive bacteria (*B. subtilis, S. aureus, E. faecalis, S. epidermidis, M. luteus*), gram-negative bacteria (*E. coli, P. aeruginosa, P. syringae, K. pneumoniae*), and yeasts (*C. albicans, C. tropicalis*).[42]
- Algerian fir essential oil exhibited good antioxidant activity in the DPPH assay.[43]

AMERICAN SWEETGUM (American Styrax, Red Gum, Hazel Pine)

Liquidambar styraciflua L., *L. styraciflua* L. var. *macrophylla*

FAMILY: Altingiaceae (Formerly Hamamelidaceae)
NOTE: Middle-Base
AROMA INTENSITY: Medium
AROMA: Balsamic, sweet, slightly floral, spicy, and woody
COMMON EXTRACTION METHOD: Steam distilled or hydrodistilled from the resin; may also be distilled from the leaves or stems
POSSIBLE SUBSTITUTE OILS: Oriental sweetgum, benzoin, frankincense; juniper berry, cypress, pine
BLENDS WELL WITH: Angelica, balsam fir, benzoin, bergamot, cassia, cinnamon, clove, copaiba, frankincense, geranium, grapefruit, lavender, mandarin, neroli, opoponax, orange, palo santo, patchouli, petitgrain, rose, spruce (black), Styrax, tamala, tangerine, tea tree, ylang ylang
SUBCELLULAR LOCALIZATION | EPIGENETIC INFLUENCE: Currently unknown | Currently unknown
RECOMMENDED DILUTION RANGE: Resin (both CTs)—1%–10%; Leaves or Stems—3%–50%; neat from some conditions

PRIMARY CONSTITUENTS:[44,45,46,47]

Resin		*Stems*	
Styrene	30.9%	Alpha-Pinene	11.1%–16.0%
Beta-Caryophyllene	20.2%	Limonene	11.2%–12.9%
Alpha-Pinene	19.6%	Germacrene D	6.7%–10.9%
Beta-Pinene	4.1%	Beta-Pinene	4.4%–8.6%
Acetophenone	3.2%	Beta-Caryophyllene	5.4%–6.9%

var. macrophylla (Resin)	
Cinnamyl Cinnamate	38.0%
3-Phenylpropyl Cinnamate	32.3%
Cinnamic Acid	4.8%
Cinnamyl Alcohol	4.1%
3-Phenylpropyl Alcohol	3.1%

Leaves	
Alpha-Pinene	26.2%–28.0%
Limonene	20.7%–22.3%
Beta-Pinene	10.1%–11.3%
Germacrene D	4.6%–4.9%
Myrcene	3.3%–4.2%
Beta-Caryophyllene	2.5%–3.1%
Gamma-Cadinene	2.1%–3.0%
Alpha-Cadinol	4.8%–5.3%
Delta-Cadinol	4.2%–4.9%
Alpha-Humulene	3.7%–4.9%
Alpha-Terpineol	3.9%–4.6%
Delta-Cadinene	3.9%–4.2%
Cubenol	3.0%–3.2%
Gamma-Cadinene	3.0%–3.1%
Myrcene	2.0%–3.1%

OTHER CONSTITUENTS: *Resin*—Ethyl benzoate, benzaldehyde, camphene, (E)-beta-methylstyrene, p-ethylphenol, phenylpropiolic alcohol, cinnamic alcohol, alpha-cubebene, longifolene, calarene, alpha-humulene; *var. macrophylla (Resin)*—Styrene, beta-caryophyllene, 3-ethylphenol, delta-cadinol, benzyl cinnamate; *Leaves*—n-Octane, alpha-thujene, camphene, sabinene, gamma-terpinene, terpinolene, beta-fenchol, borneol, alpha-terpineol, terpinen-4-ol, alpha-humulene, alloaromadendrene, epizonarene, delta-cadinene, cubenol, delta-cadinol, alpha-cadinol; *Stems*—n-Octane, camphene, gamma-terpinene, terpinolene, n-nonanal, beta-fenchol, trans-pinocarveol, borneol, terpinen-4-ol, verbenone, delta-elemene, eugenol, beta-bourbonene, beta-elemene, beta-gurjunene, alloaromadendrene, gamma-muurolene, caryophyllene oxide, 4-hydroxy germacrene D

PREFERRED COMPOSITION FOR CLINICAL USE:
Insufficient samples/data to determine.

REPORTED THERAPEUTIC PROPERTIES: Soothes skin itching and chronic skin conditions, wound healing, antiseptic, antibacterial, antifungal, antiviral, antiparasitic, antioxidant, reduces high blood pressure, liver protective, expels excess mucus, anti-inflammatory, anticonvulsant, protects stomach against ulcers, anticancer, immune modulatory, insecticidal, improves memory function, antidepressant, relieves anxiety, improves self-confidence

CAUTIONS:

Styrene CT

■ Low to moderate risk of sensitization or skin irritation. Use cautiously on diseased or damaged skin and among people with a compromised immune system.[48]

■ Not recommended for therapy due to the hazards of styrene (largely collected from synthetic styrene hazards): nervous system dysfunction, hearing loss, liver damage, kidney dysfunction, impaired learning, reproductive damage, cancer, irritation (skin, eyes, upper respiratory tract).[49,50]

Cinnamyl Cinnamate CT

■ Moderate risk of sensitization or skin irritation. Use cautiously on diseased or damaged skin and among people with a compromised immune system.[51]

Leaves or Stems

- None currently known.

SELECTED EVIDENCE:

- American sweetgum leaf and stem essential oils were only weakly cytotoxic to cervical cancer cells in laboratory research.[52]
- The stem oil displayed a greater anti-inflammatory activity than the leaf oil as demonstrated by greater inhibition of 5-LOX (5-Lipoxygenase) and PGE2 (prostaglandin E2).[53] Both 5-LOX and PGE2 are involved in pathogenesis of inflammatory diseases.
- Leaf and stem essential oil exhibited antioxidant activity in the DPPH assay and prevented the degradation of the deoxyribose sugar.[54]

ANACHUITA (Arrayán, Anacahuita y Horco Molle, Maria-Preta, Murta)

Blepharocalyx salicifolius (Kunt) O.Berg, *B. acuminatissimus* (Miq.) O.Berg, *B. tweediei* (Hook. et Arn.) Berg, *B. giganteus* Lillo. H

FAMILY: Myrtaceae
NOTE: Middle-Top
AROMA INTENSITY: Medium
AROMA: Medicinal, woody, fresh, sweet
COMMON EXTRACTION METHOD: Steam distilled or hydrodistilled from the leaves; may also be distilled from the seeds
POSSIBLE SUBSTITUTE OILS: Myrtle, eucalyptus, cajeput
BLENDS WELL WITH: Bergamot, black pepper, cedarwood, clary sage, clove, eucalyptus, frankincense, ginger, lavender, lemon, lemongrass, lime, melissa, myrrh, myrtle, neroli, peppermint, rosemary, silver fir, white fir, ylang ylang
SUBCELLULAR LOCALIZATION | EPIGENETIC INFLUENCE: Currently unknown | Currently unknown
RECOMMENDED DILUTION RANGE: 3%–50%; neat for some conditions

PRIMARY CONSTITUENTS:[55,56,57,58,59,60,61,62,63,64]

1,8-Cineole CT

1,8-Cineole	38.5%–55.7%
Alpha-Pinene	0.1%–12.3%
Limonene	0.0%–13.1%
Beta-Pinene	2.5%–9.1%
Linalool	0.2%–4.3%
Caryophyllene Oxide	0.0%–4.1%
Terpinen-4-ol	0.2%–3.9%
Alpha-Terpineol	0.3%–3.2%
Myrcene	0.1%–3.1%

Eudesmol CT

Alpha-Eudesmol	47.7%
Gamma-Eudesmol	31.1%
Elemol	7.3%

Para-Cymene CT

Para-Cymene	25.9%
Gamma-Terpinene	12.5%
Caryophyllene Oxide + Globulol	9.6%
Alpha-Pinene	9.0%

Balanced CT

1,8-Cineole	14.0%–24.5%
Beta-Caryophyllene	0.0%–22.9%
Linalool	7.3%–20.4%
Limonene	3.0%–17.2%
Alpha-Pinene	1.1%–15.8%
Beta-Pinene	0.2%–11.1%
Alpha-Terpineol	2.0%–9.1%
Viridiflorol	0.0%–3.5%
Terpinen-4-ol	0.6%–3.4%
Gamma-Terpinene	0.7%–3.2%
Alpha-Humulene	0.1%–3.0%

Seeds

1,8-Cineole	18.7%–38.4%
Beta-Pinene	13.5%–14.6%
Limonene	3.9%–10.1%
Alpha-Pinene	4.3%–9.9%
Linalool	4.1%–5.9%

Aromadendrene	4.6%
Trans-Calamenene	3.5%
Gamma-Terpinolene	3.3%
Beta-Caryophyllene	3.2%

Bicyclogermacrene CT

Bicyclogermacrene	17.5%
Globulol	14.1%
Viridiflorol	8.8%
Gamma-Eudesmol	7.9%
Alpha-Eudesmol	6.9%
Rosifoliol	4.5%
Beta-Caryophyllene	4.5%
Cubeban-11-ol	4.4%
Palustrol	3.8%
Beta-Eudesmol	3.1%

OTHER CONSTITUENTS: *1,8-Cineole CT*—Para-cymene, (E)-beta-ocimene, beta-caryophyllene, p-cymen-8-ol, (E)-methyl cinnamate, alpha-humulene, gamma-muurolene, beta-selinene, alpha-selinene, delta-cadinene, globulol, humulene epoxide II, 1-epi-cubenol, cubenol, alpha-muurolol, beta-eudesmol, selina-11-en-4alpha-ol; *Balanced CT*—Camphene, myrcene, para-cymene, alpha-terpinene, terpinolene, alpha-fenchol, borneol, beta-selinene, alpha-selinene, delta-cadinene, elina-3,7(11)-diene, caryophyllene oxide, T-cadinol, beta-eudesmol; *Eudesmol CT*—Alpha-pinene, beta-caryophyllene, bicyclogermacrene, spathulenol, globulol, viridiflorol, cubeban-11-ol, rosifoliol, 10-epi-gamma-eudesmol, hinesol; *Para-Cymene CT*—Alpha-thujene, delta-33-carene, alpha-terpinene, limonene, 1,8cineole, alpha-gurjunene, alpha-humulene, alloaromadendrene, spathulenol, viridiflorol, rosifoliol, ledol, humulene epoxide II, 1-epi-cubenol; *Bicyclogermacrene CT*—Alpha-guaiene, alpha-humulene, germacrene D, elemol; *Seeds*—Myrcene, alpha-phellandrene, gamma-terpinene, terpinolene, terpinen-4-ol, alpha-terpineol, alpha-guaiene, alpha-humulene

PREFERRED COMPOSITION FOR CLINICAL USE:
Too great a variance in samples/data to determine.

REPORTED THERAPEUTIC PROPERTIES: Relieves diarrhea, eases bladder and urinary tract infections, aids respiratory function, alleviates cough, analgesic, anti-inflammatory, antispasmodic, antirheumatic, antiarthritic, reduces headache/migraine, antibacterial, antifungal, antiparasitic, astringent, lowers high blood pressure, promotes vaginal health and cleanliness, anticancer, insecticidal, promotes memory and concentration, enhances mental clarity, reduces confusion and anger

CAUTIONS:

1,8-Cineole CT

- Avoid with children under age 3, particularly around the nose and mouth and use cautiously in children under age 5 due to significant 1,8-cineole content. 1,8-cineole may cause seizures, central nervous system problems, or respiratory distress in young children.[65,66,67]

■ Caution is warranted orally due to the significant amounts of 1,8-cineole. Limit it to small doses internally (adults—maximum 10 drops daily). Toxicity has been reported when eucalyptus (also high in 1,8-cineole) was ingested in large doses, and as few as 4 to 5 drops may cause problems in very sensitive individuals.[68,69,70,71,72,73]

■ Avoid with epilepsy and Parkinson's disease due to 1,8-cineole content. May exacerbate or cause seizures/convulsions or reduce seizure-medication efficacy based on 1,8-cineole content.[74,75,76,77]

■ May weakly interfere with the enzymes responsible for metabolizing medications (NSAIDs, proton-pump inhibitors, acetaminophen, antiepileptics, immune modulators, blood-sugar medications, blood pressure medications, antidepressants, antipsychotics, diabetic medications, antihistamines, antibiotics, and anesthetics) due to 1,8-cineole content.[78,79,80]

■ May interfere with pentobarbital and other barbiturates (medications for anxiety and insomnia), based on 1,8-cineole content.[81,82]

Balanced CT, Seeds

■ Avoid with epilepsy and Parkinson's disease due to 1,8-cineole content. May exacerbate or cause seizures/convulsions or reduce seizure-medication efficacy based on 1,8-cineole content.[83,84,85,86]

■ May weakly interfere with the enzymes responsible for metabolizing medications (NSAIDs, proton-pump inhibitors, acetaminophen, antiepileptics, immune modulators, blood-sugar medications, blood pressure medications, antidepressants, antipsychotics, diabetic medications, antihistamines, antibiotics, and anesthetics) due to 1,8-cineole content.[87,88,89]

■ May interfere with pentobarbital and other barbiturates (medications for anxiety and insomnia), based on 1,8-cineole content.[90,91]

Eudesmol, Para-Cymene, and Bicyclogermacrene CTs

■ None currently known.

Selected Evidence:

- □ The bicyclogermacrene CT of anacahuita essential oil was toxic to estrogen-receptor negative breast cancer cells with good selectivity and about seven times less toxicity to normal breast cells, but ineffective against the less aggressive estrogen-receptor positive cancer cells.[92] The anticancer activity of the oil was determined to be due to disruption of cellular metabolic capacity rather than activation of cell death mechanisms.
- □ Anacahuita essential oil CT 1,8-cineole prevented intestinal spasms in isolated rat intestine by noncompetitively inhibiting calcium channels.[93] The oil was also assessed for its ability to reduce coughing and found very effective. The same authors reported that the oil relaxed isolated trachea, suggesting a benefit for asthma and bronchospasm, and provided cardioprotective benefits during a stroke model. The relaxant effect was ten times more potent than papaverine (a well-known phosphodiesterase inhibitor).
- □ *In vitro* research reported that anacahuita essential oil CT bicyclogermacrene was weakly active against *Trichophyton mentagrophytes* and moderately active against *Paracoccidioides brasiliensis* but inactive against *Candida* spp., *Cryptococcus neoformans*, and *Microsporum canis*.[94] The same study reported that this CT is inactive against MRSA, MSSA, *E. coli*, and *P. aeruginosa*.

ANT BUSH
(Negramina)

Siparuna guianensis Aubl.

FAMILY: Siparunaceae
NOTE: Middle
AROMA INTENSITY: Strong
AROMA: Citrusy, lemon-like, sweet, herbaceous, earthy
COMMON EXTRACTION METHOD: Hydrodistilled or steam distilled from leaves or fruits
POSSIBLE SUBSTITUTE OILS: Hemp, hops, helichrysum, copaiba
BLENDS WELL WITH: Bergamot, cedarwood, copaiba, geranium, grapefruit, helichrysum, lemon, lemongrass, lime, melissa, orange, pine, sandalwood, Spanish sage, tangerine
SUBCELLULAR LOCALIZATION | EPIGENETIC INFLUENCE: Currently unknown | Currently unknown
RECOMMENDED DILUTION RANGE: 3%–50%; neat for some conditions

PRIMARY CONSTITUENTS:[95,96,97,98,99,100,101,102,103,104,105]

Leaves (Bicyclogermacrene CT)

Bicyclogermacrene	16.7%
Beta-Myrcene	13.1%
Germacrene D	8.7%
(–)-Spathulenol	4.2%
T-Cadinol	4.1%
Alpha-Bisabolol	3.4%

Leaves (Beta-Myrcene CT)

Beta-Myrcene	28.7%–79.7%
2-Undecanone	8.4%–14.6%
Bicyclogermacrene	0.0%–6.1%
Elixine	0.0%–5.7%
(E)-α-Bisabolene Epoxide	0.0%–4.9%
(Z)-α-Bisabolene Epoxide	0.0%–3.8%
Germacrene D	0.0%–3.6%
Spathulenol	3.3%–3.9%

Leaves (Beta-Myrcene/Epicurzerenone CT)

Beta-Myrcene	24.2%–48.6%
Spathulenol	13.2%–27.2%
Germacrene D	9.9%–14.3%
Gamma-Elemene	4.5%–7.4%
2-Undecanone	5.4%–7.3%
Curzerene	0.0%–7.2%
Elixene	0.0%–4.7%
Spathulenol	1.7%–3.6%

Leaves (Decanoic Acid CT)

Decanoic Acid	46.6%
2-Undecanone	31.7%
Beta-Pinene	6.3%
Nonanol	4.8%

Leaves ((E,E)-Farnesol CT)

(E,E)-Farnesol	18.0%
Beta-Myrcene	16.0%
Spirunone	14.6%
Germacrene D	10.0%
T-Cadinol	3.0%

Leaves ((E)-Nerolidol CT)

(E)-Nerolidol	97.6%–99.3%

Leaves (Gamma-Cadinene CT)

Gamma-Cadinene	47.8%
Gamma-Elemene	12.6%
(E)-Nerolidol	6.0%
Beta-Elemene	4.7%
Alpha-Copaene	3.4%
Gamma-Cadinene	3.3%

Leaves (Valencene CT)

Valencene	27.5%
Beta-Caryophyllene	21.6%
Alpha-Zingiberene	13.0%

Leaves (Alpha-Pinene CT)

Alpha-Pinene	27.6%
1,8-Cineole	22.6%
Para-Cymene	9.8%
Beta-Pinene	7.2%
Delta-3-Carene	6.5%
Gamma-Terpinene	5.1%
Alpha-Terpineol	4.5%

Fruits (2-Undecanone CT)

2-Undecanone	32.5%
Beta-pinene	19.6%
Limonene	13.6%
Delta-3-Carene	8.8%
Nonanol	7.4%
Alpha-Pinene	5.7%
Sabinene	5.6%

Stems

Beta-Myrcene	26.9%
Delta-Elemene	20.9%
2-Undecanone	12.8%
Germacrene D	9.4%
Alpha-Limonene	7.9%
Bicyclogermacrene	7.8%
Spathulenol	6.1%
Alpha-Pinene	7.7%
Beta-Phellandrene	3.5%

Fruits (2-Tridecanone CT)

2-Tridecanone	38.8%
2-Undecanone	26.5%
Beta-Myrcene	16.4%
Nonanol	5.5%
Agarospirol	3.9%
Alpha-Pinene	3.1%

Aerial Parts

Atractylone	0.0%–51.7%
Epi-Alpha-Bisabolol	0.0%–25.1%
Germacrone	0.0%–29.7%
Spathulenol	0.8%–22.0%
Selen-11-en-4α-ol	0.0%–19.4%
Germacrene D	0.0%–14.3%
Germacrene B	0.0%–12.9%
Bicyclogermacrene	0.0%–11.7%
Elemol	0.3%–10.6%
Beta-Eudesmol	0.0%–10.0%
Alpha-Pinene	0.0%–6.4%
Cubebol	1.3%–4.6%
Beta-Bisabolene	0.0%–4.3%
Delta-Cadinene	0.7%–3.7%

Note: The aerial parts could be split into at least three distinct chemotypes but they are combined in this work.

OTHER CONSTITUENTS: *Leaves (Bicyclogermacrene CT)*—Alpha-pinene, 2-undecanone, delta-cadinene; *Leaves (Beta-Myrcene CT)*—Limonene, alpha-pinene, alpha-terpinolene, delta-elemene, beta-cubebene, beta-ocimene, gamma-muurolene, alpha-muurolene, delta-cadinene, delta-cadinol; *Leaves (Beta-Myrcene CT II)— Leaves (Beta-Myrcene/Epicurzerenone CT)*—Pseudolimonene, alpha-cubene, gamma-muurolene, delta-cadinene, alpha-cadinol, germacrene B; *Leaves (Decanoic Acid CT)*—Alpha-pinene, camphene, dodecanoic acid, spathulenol, beta-eudesmol, pentadecanol, octadecanol; *Leaves ((E,E)-Farnesol CT)*—Alpha-pinene, beta-pinene, limonene, para-cymene, 2-undecanone, delta-elemene, beta-elemene, trans-caryophyllene, alpha-humulene, ledol, alpha-muurolene, gamma-cadinene, delta-cadinene, germacrene B, viridiflorol, 1-epi-cubenol, beta-eudesmol, alpha-cadinol, alpha-bisabolol; *Leaves ((E)-Nerolidol CT)*—Gamma-elemene; *Leaves (Gamma-Cadinene CT)*—Alpha-pinene, alpha-cubebene, beta-bourbonene, beta-caryophyllene, alpha-muurolene, spathulenol; *Leaves (Valencene CT)*—Beta-pinene, myrcene, geraniol, beta-elemene, gamma-elemene, alpha-humulene, beta-sesquiphellandrene; *Leaves (Alpha-Pinene CT)*—Camphene, myrcene, linalool, bornylene, (E)-nerolidol terpinolene; *Fruits (2-tridecanone CT)*—Alpha-cadinol, limonene; *Stems*—Delta-3-carene, cadinol, alpha-humulene, beta-elemene; *Fruits (2-Undecanone CT)*—2-nonanone, perillene, nonanal, nonene, spathulenol; *Aerial Parts*—Beta-pinene, myrcene, limonene, alpha-cubebene, undecanol, alpha-copaene, beta-bourbonene, beta-cubebene, beta-elemene, beta-caryophyllene, gamma-elemene, gamma-muurolene, beta-selinene, epi-cubebol, alpha-muurolene, germacrene A, alpha-calacorene, cis-sesquisabinene hydrate, ledol, caryophyllene oxide, guaiol, beta-elemone, humulene epoxide II, 1-epi-cubenol, gamma-eudesmol, alpha-muurolol, cadalene

Preferred Composition for Clinical Use:
Insufficient consistency of samples/data to determine.

Reported Therapeutic Properties: Antiparasitic, antifungal, antibacterial, relieves fever, alleviates cough, diuretic, reduces blood pressure, relieves headache, anti-inflammatory, antispasmodic, aids circulation, wound healing, relieves insect bites, insect repellent, insecticidal, promotes restful sleep, relaxing/calming

Cautions:

Beta-Myrcene CT and Beta-Myrcene/Epicurzerenone CT

- Caution is advised during pregnancy and lactation due to high beta-myrcene content. Extremely high doses (500 mg/kg) of beta-myrcene have been toxic to fetuses according to animal research.[106,107]
- May interact with barbiturates (medications for anxiety and insomnia), antihistamines, benzodiazepines, tricyclic antidepressants, or other central nervous system depressant drugs, increasing their depressant effects.[108,109,110]
- Myrcene CT: May weakly interfere with the enzymes responsible for metabolizing medications (NSAIDs, proton-pump inhibitors, acetaminophen, antiepileptics, immune modulators, blood-sugar medications, blood pressure medications, antidepressants, antipsychotics, diabetic medications, antihistamines, antibiotics, and anesthetics), based on myrcene content.[111,112]

Stems

- Myrcene CT: May weakly interfere with the enzymes responsible for metabolizing medications (NSAIDs, proton-pump inhibitors, acetaminophen, antiepileptics, immune modulators, blood-sugar medications, blood pressure medications, antidepressants, antipsychotics, diabetic medications, antihistamines, antibiotics, and anesthetics), based on myrcene content.[113,114]

Alpha-Pinene CT

- Avoid with epilepsy and Parkinson's disease due to 1,8-cineole content. May exacerbate or cause seizures/convulsions or reduce seizure-medication efficacy based on 1,8-cineole content.[115,116,117]

(E)-Nerolidol CT

- May interfere with antibiotics and increase their effectiveness ((E)-nerolidol sensitizes bacteria to antibiotics).[118]
- May interfere with enzymes (CYP1A, CYP2B, CYP3A) responsible for metabolizing medications based on alpha-bisabolol content (NSAIDs, proton-pump inhibitors, acetaminophen, antiepileptics, immune modulators, blood-sugar medications, blood pressure medications, antidepressants, antipsychotics, diabetic medications, antihistamines, antibiotics, and anesthetics) due to (E)-nerolidol content.[119]

Atractylone CT

- May interfere with the enzymes responsible for metabolizing medications (NSAIDs, proton-pump inhibitors, acetaminophen, antiepileptics, immune modulators, blood-sugar medications, blood pressure medications, antidepressants, antipsychotics, diabetic medications, antihistamines, antibiotics, and anesthetics) due to germacrone content.[120]

Decanoic Acid, (E,E)-Farnesol, Valencene, 2-Tridecanone, 2-Undecanone, and Gamma-Cadinene CTs

- None currently known.

Selected Evidence:

- Ant bush leaf essential oil demonstrated antifungal activity against *C. tropicalis, C. krusei, C. albicans, C. parapsilosis*, and *C. glabrata*.[121] Other research reports that ant bush leaf essential oil (unidentified CT) weakly inhibited bacteria (*S. aureus, L. monocytogenes*) and moderately to weakly inhibited fungi (*A. flavus, A, niger, P. commune, A. carbonarius*).[122] The (E,E)-Farnesol CT showed good activity against *S. mutans* and moderate activity against *S. mitis, S. sanguinis, S. sobrinus, S. salivarius,* and *L. casei*, which caused the authors to conclude it may help prevent cavities.[123] The same research demonstrated that it moderately inhibited *Mycobacterium avium* and weakly inhibited *M. tuberculosis* and *M. kansasii*. Ant bush (beta-myrcene CT, with moderate bisabolenes) inhibited *S. aureus*, but had no effect on *E. coli, P. aeruginosa, I. orientalis,* or *F. oxysporum*.[124]
- An unidentified CT of leaf essential oil weakly inhibited *T. cruzi* (parasite that feeds on blood and lymph and causes Chagas disease).[125] The bicyclogermacrene CT of ant bush was effective against *L. amazonensis* parasites, which was the most effective of sixteen essential oils tested.[126]
- Ant bush CT bicyclogermacrene is considered a weak antioxidant.[127]
- The stem, leaf (beta-myrcene CT), and fruit (2-triecanone CT) essential oils demonstrated strong toxicity against all stages (eggs, larvae, pupae, and adult) of the *Culex quinquefasciatus* and *Aedes aegypti* mosquitoes.[128]

ARABIAN BALSAM TREE (Mecca Myrrh, Balsamon, Apharsemon, Corkwood)

Commiphora gileadensis (L.) C. Chr,
C. opobalsamum (L.) Engl.

Family: Burseraceae
Note: Middle
Aroma Intensity: Medium
Aroma: Piney, peppery, medicinal, herbaceous, slightly minty
Common Extraction Method: Steam distilled from the stems, leaves, fruit, or resin
Possible Substitute Oils: Plai CT sabinene, pink pepper, chaste tree, copaiba, ginger
Blends Well With: Amyris, bergamot, black pepper, cardamom, carrot, cedarwood, clary sage, clove, frankincense, geranium, ginger, grapefruit, lavandin, lavender, lemon, lime, marjoram, may chang, myrrh, neroli, nutmeg, orange, palo santo, patchouli, rosemary, sage, sandalwood, Spanish sage, tangerine, tea tree (melaleuca), vetiver, yarrow, ylang ylang
Subcellular Localization | Epigenetic Influence: Currently unknown | Currently unknown
Recommended Dilution Range: 3%–Neat

Primary Constituents:[129,130]

Stems		*Leaves and Fruit*	
Sabinene	28.5%	Sabinene	21.1%
Alpha-Pinene	15.4%	Beta-Caryophyllene	20.1%
Terpinen-4-ol	11.6%	Germacrene D	19.6%
Gamma-Terpinene	8.1%	Alpha-Pinene	7.2%
Beta-Pinene	7.5%		

Para-Cymene	4.8%	*Fruit*	
Alpha-Terpinene	3.0%	Sabinene	35.9%
		Beta-Pinene	18.0%
Leaves		Alpha-Pinene	15.9%
Terpinen-4-ol	18.5%	Limonene	6.2%
Sabinene	15.8%	Terpinen-4-ol	5.3%
Alpha-Pinene	11.2%	Para-Cymene	5.3%
Para-Cymene	8.4%		
Gamm-Terpinene	5.8%		
Beta-Pinene	5.8%		
Bornyl Acetate	4.4%		

Note: Arabian balsam tree essential oil is very rare.

OTHER CONSTITUENTS: *Stems*—Alpha-thujene, camphene, myrcene, limonene, beta-phellandrene, (Z)-beta-ocimene, terpinolene, cis-p-menth-2-en-1-ol, alpha-terpineol, bornyl acetate; *Leaves*—Alpha-thujene, camphene, myrcene, alpha-terpinene, limonene, (Z)-beta-ocimene, terpinolene, cis-p-menth-2-en-1-ol, alpha-terpineol, trans-piperitol, beta-caryophyllene, germacrene D, caryophyllene oxide, 1,10-di-epi-cubenol; *Fruit*—Alpha-thujene, camphene, myrcene, beta-phellandrene, gamma-terpinene, alpha-terpineol, decanal, alpha-cadinol; *Leaves and Fruit*—Alpha-thujene, beta-pinene, beta-phellandrene, gamma-terpinene, terpinene-4-ol, bornyl acetate, bicyclogermacrene, alpha-ylangene, beta-elemene, beta-copaene, unidentified sesquiterpene hydrocarbons, alpha-humulene, delta-amorphene, gamma-cadinene, germacrene-D-4ol

PREFERRED COMPOSITION FOR CLINICAL USE:
Insufficient data/samples to determine.

REPORTED THERAPEUTIC PROPERTIES: Analgesic, **anti-inflammatory**, anticonvulsive, relieves headache, diuretic, relieves constipation, protects the stomach, alleviates hemorrhoids, antiseptic, wound healing, antimicrobial, antibacterial, anticancer, aids respiratory function, alleviates cough, regulates menstruation, balances female hormones, antidepressant, aphrodisiac, relieves anxiety

CAUTIONS:
- None currently known.

SELECTED EVIDENCE:
- □ *In vitro* research demonstrates that Arabian balsam tree essential oil (leaves and fruits, rich in beta-caryophyllene) kills mouse lymphoma and Epstein-Barr virus-transformed human B lymphocyte cancer cells.[131] The scientists concluded that beta-caryophyllene was responsible for the anticancer activity.

ARABIAN LILAC
(Simpleleaf Chastetree)

Vitex trifolia L. var. *purpurea*; *V. trifolia* L., *V. trifolia* L. var. *simplicifolia* Cham., *V. trifolia* (Willd.) var. *bicolor* Modenke

FAMILY: Lamiaceae (Labiatae)
NOTE: Middle
AROMA INTENSITY: Medium
AROMA: Woody, resinous, herbaceous, slightly spicy and fruity
COMMON EXTRACTION METHOD: Steam distilled or hydrodistilled from the aerial parts or leaves
POSSIBLE SUBSTITUTE OILS: Chaste tree, black pepper, guava leaf, copaiba, clary sage
BLENDS WELL WITH: Black pepper, chaste tree, clary sage, copaiba, geranium, lavender, melissa, rose
SUBCELLULAR LOCALIZATION | EPIGENETIC INFLUENCE: Currently unknown | Currently unknown
RECOMMENDED DILUTION RANGE: 3%–50%; neat for some conditions

PRIMARY CONSTITUENTS:[132,133]

var. purpurea (Aerial Parts)

Constituent	%
Beta-Caryophyllene	22.6%–34.4%
Sabinene	9.2%–18.4%
(6Z,10Z)-Pseudo Phytol	0.0%–15.0%
Manool Oxide	0.0%–8.3%
Caryophyllene Oxide	3.8%–7.3%
Laurenene	2.8%–7.3%
Alpha-Pinene	1.4%–5.4%
Sclareolide	0.0%–5.1%
(5E,9Z)-Farnesyl Acetone	3.0%–5.0%
Dehydroabietal	0.0%–4.9%
Phytol	2.1%–4.6%
Phytane	2.3%–4.4%
7Alpha-Hydroxy Manool	0.0%–3.1%
1-Octen-3-ol	0.0%–3.0%
Manoyl Oxide	0.0%–3.0%
(E,E)-Geranyl Linalool	0.0%–3.0%
1-Octadecanol	0.0%–3.0%

V. trifolia (Leaves) CT Beta-Ocimene

Constituent	%
Beta-Ocimene*	6.9%
1,8-Cineole	6.1%
Alpha-Pinene	4.0%

* Correct isomer not identified.

V. trifolia (Leaves) CT Beta-Caryophyllene

Constituent	%
Beta-Caryophyllene	17.5%
1,8-Cineole	14.9%
Alpha-Terpinyl Acetate	11.1%
Sabinene	8.4%
Alpha-Pinene	6.9%
Bicyclogermacrene	4.7%
Spathulenol	4.6%

var. bicolor (Leaves)

Constituent	%
Beta-Caryophyllene	61.1%
Caryophyllene Oxide	7.8%
Sabinene	9.1%
Octen-1-en-3-ol	4.3%

var. simplicifolia (Leaves)

Constituent	%
1,8-Cineole	38.9%
Alpha-Pinene	14.9%
Alpha-Terpinyl Acetate	12.8%
Sabinene	9.0%
Terpinen-4-ol	3.7%
Beta-Pinene	3.1%

OTHER CONSTITUENTS: *var. purpurea (Aerial Parts)*—Alpha-thujene, beta-pinene, myrcene, alpha-terpinene, para-cymene, beta-phellandrene, gamma-terpinene, terpinen-4-ol, methyl chavicol (< 2.9%), alpha-terpinyl acetate, alpha-humulene, beta-selinene, alpha-selinene, (E)-nerolidol, caryophylladienol II, (Z)-14-hydroxy-caryophyllene, epi-laurenene, laurenan-2-one; *V. trifolia (Leaves) CT Beta-Ocimene*—Delta-3-carene, p-ethyl-toluene, limonene, 3-methoxy-5-methylphenol, beta-caryophyllene; *V. trifolia (Leaves) CT Beta-Caryophyllene*—Beta-pinene, beta-terpinene, limonene, gamma-terpinene, an amyl ester, terpinen-4-ol, alloaromadendrene, alpha-humulene, caryophyllene oxide; *var. simplicifolia (Leaves)*—Myrcene, beta-terpinene, gamma-terpinene, alpha-terpinolene, para-cymene, an amyl ester, alpha-terpineol; *var. bicolor (Leaves)*—Alpha-pinene, beta-pinene, limonene, 1,8-cineole, gamma-terpinene, para-cymene, alpha-humulene

PREFERRED COMPOSITION FOR CLINICAL USE:

Constituent	*var. purpurea*
Beta-Caryophyllene	20%–40%
Sabinene	7%–22%
Caryophyllene Oxide	2%–12%
Laurenene	1%–10%
Alpha-Pinene	0.5%–8%
(5E,9Z)-Farnesyl Acetone	0.5%–7%
Phytol	0.1%–6%
Phytane	0.1%–6%
Beta-Pinene	0.1%–5%
Alpha-Humulene	0.1%–5%

REPORTED THERAPEUTIC PROPERTIES: Balances hormones, balances menstruation, relieves menstrual and menopausal symptoms, **anti-inflammatory**, antirheumatic, soothes sprains/strains, analgesic, reduces fever, antimicrobial, antioxidant, antiallergic, aids respiratory function, eases cough, soothes skin itching and conditions, aids digestion, relieves diarrhea, supports liver function, wound healing, promotes restful sleep, insecticidal, antidepressant, relieves anxiety, stress management, aphrodisiac

CAUTIONS:

Beta-Caryophyllene and Beta-Ocimene CTs

■ None currently known.

1,8-Cineole CT

■ Avoid with children under age 3, particularly around the nose and mouth and use cautiously in children under age 5 due to significant 1,8-cineole content. 1,8-cineole may cause seizures, central nervous system problems, or respiratory distress in young children.[134,135,136]

■ Caution is warranted orally due to the significant amounts of 1,8-cineole. Limit it to small doses internally (adults—maximum 10 drops daily). Toxicity has been reported when eucalyptus (also high in 1,8-cineole) was ingested in large doses, and as few as 4 to 5 drops may cause problems in very sensitive individuals.[137,138,139,140,141,142]

■ Avoid with epilepsy and Parkinson's disease due to 1,8-cineole content. May exacerbate or cause seizures/convulsions or reduce seizure-medication efficacy based on 1,8-cineole content.[143,144,145,146]

■ May weakly interfere with the enzymes responsible for metabolizing medications (NSAIDs, proton-pump inhibitors, acetaminophen, antiepileptics, immune modulators, blood-sugar medications, blood pressure medications, antidepressants, antipsychotics, diabetic medications, antihistamines, antibiotics, and anesthetics) due to 1,8-cineole content.[147,148,149]

■ May interfere with pentobarbital and other barbiturates (medications for anxiety and insomnia), based on 1,8-cineole content.[150,151]

SELECTED EVIDENCE:

- □ *In vitro* research demonstrated that Arabian lilac leaf essential oil CT beta-ocimene inhibited dermatophytes: *C. albicans* (two strains), *C. parapsilosis, C. krusei* (two strains), and *C. kefyr*.[152]
- □ The leaf essential oil CT beta-ocimene exhibited antioxidant activity.[153]
- □ Arabian lilac leaf essential oil (composition not reported) was insecticidal to the jute hairy caterpillar/moth (*Spilosoma obliqua*).[154] A 1,8-cineole CT was also larvicidal to the *Aedes aegypti* mosquito.[155]

ARACA RASTEIRO (Guayabita del Pinar, Araca Preto)

Psidium salutare (Kunth) O.Berg, *Guajava salutaris* (Kunth) Kuntze, *Myrtus salutaris* Kunth

FAMILY: Myrtaceae
NOTE: Top-Middle
AROMA INTENSITY: Medium
AROMA: Green, herbaceous, woody, earthy, slightly floral; or fruity, fresh, earthy
COMMON EXTRACTION METHOD: Steam distilled or hydrodistilled from the leaves or fruits
POSSIBLE SUBSTITUTE OILS: Guava leaf, black pepper, melissa, betony
BLENDS WELL WITH: Balsam fir, cedarwood, copaiba, cypress, eucalyptus, geranium, guava, hemp, hinoki, lavender, peppermint, rose, sandalwood, spruce (black)
SUBCELLULAR LOCALIZATION | EPIGENETIC INFLUENCE: Currently unknown | Currently unknown
RECOMMENDED DILUTION RANGE: 3%–50%; neat for some conditions
PRIMARY CONSTITUENTS:[156,157,158]

Leaves CT Para-Cymene

Para-Cymene	5.1%–17.8%
Gamma-Terpinene	10.3%–17.1%
Terpinolene	6.9%–17.0%
T-Cadinol	10.4%–12.8%
Linalool	4.7%–7.3%
Delta-Cadinene	3.9%–5.3%
Sabinene Hydrate	2.3%–3.9%
Isocaryophyllene	1.2%–3.8%
Delta-Guaiene	0.3%–3.7%
Alpha-Copaene	1.9%–3.5%
Cubenol	0.0%–3.4%
Gamma-Muurolene	2.4%–3.2%
Patchoulane	0.2%–3.1%
Beta-Guaiene	0.0%–3.1%

Fruits

Limonene	17.3%
Myrcene	16.2%
Alpha-Pinene	9.3%
(E)-Nerolidol	6.0%
Alpha-Cadinol	5.9%
(Z)-Beta-Ocimene	5.7%
T-Cadinol	5.6%
(E)-Beta-caryophyllene	4.9%
Gamma-Terpinene	3.3%
Linalool	3.0%

Leaves CT Caryophyllene Oxide

Caryophyllene Oxide	39.8%
ar-Turmerone	17.3%
Beta-Gurjunene	6.7%
Beta-Selinene	6.0%
Alpha-Pinene	5.6%
Beta-Caryophyllene	4.8%
Alpha-Cubebene	4.6%
Viridiflorene	4.4%
Limonene	3.6%

Other Constituents: *Leaves CT Para-Cymene*—1,8-cineole, dimethyl benzylcarbinyl acetate, limonene, myrcene, ocimene, alpha-terpineol, viridiflorol, alpha-humulene, alpha-cubebene, alpha-muurolene, alpha-pinene, beta-cadinene, beta-elemene, delta-cadinol; *Leaves CT Caryophyllene Oxide*—Alloaromadendrene, viridiflorol, guaiol; *Fruits CT Limonene*—Acetal, alpha-thujene, beta-pinene, para-cymene, alpha-phellandrene, (E)-beta-ocimene, terpinen-4-ol, alpha-terpineol, alpha-humulene, gamma-cadinene, caryophyllene oxide, globulol, viridiflorol, cubenol, alpha-eudesmol, T-muurolol

Preferred Composition for Clinical Use:
Insufficient data/samples to determine.

Reported Therapeutic Properties: Antibacterial, antifungal, antioxidant, anti-inflammatory, antirheumatic, analgesic, relieves fever, antispasmodic, aids digestion, heals stomach ulcers, relieves diarrhea, support respiratory function, anticancer, astringent, wound healing, lowers high blood pressure, insect repellent, promotes restful sleep, relieves anxiety, stress management

Cautions:
Leaves CT Para-Cymene
■ May interact with antifungals and enhance their effectiveness based on research with the whole oil.[159]

Leaves CT Caryophyllene Oxide
■ May interact with aspirin, blood pressure, antiplatelet, and anticoagulant medications, and increase the risk of bleeding among people with bleeding disorders, due to caryophyllene-oxide content.[160]

Fruits
■ None currently known.

Selected Evidence:
- □ *In vitro* research showed that araca rastiero leaf essential oil inhibited *C. albicans, C. tropicalis*, and *C. krusei*, and synergized the activity of fluconazole against the fungi.[161]

ARINA
(Iary, Dingadingana)
Psiadia altissima Benth. & J.D. Hook

Family: Asteraceae (Compositae)
Note: Top-Middle
Aroma Intensity: Medium
Aroma: Woody, balsamic, piney, slightly medicinal
Common Extraction Method: Steam distilled or hydrodistilled from the leaves and twigs
Possible Substitute Oils: Galbanum, white fir, gobre salla, juniper berry, frankincense, cypress
Blends Well With: Balsam fir, bay laurel, cajeput, cedarwood, clary sage, copaiba, cypress (Phoenician), grapefruit, lavender, lemon, lime, niaouli, orange, petitgrain, pine, ravintsara, rosemary, spikenard, tangerine, turmeric, silver fir, white fir
Subcellular Localization | Epigenetic Influence: Currently unknown | Currently unknown
Recommended Dilution Range: 3%–33%; 50% for some conditions

PRIMARY CONSTITUENTS:[162,163]

Leaves		*Flower Tops & Leaves*	
Beta-Pinene	39.7%	Beta-Pinene	40.6%–60.0%
(E)-Beta-Ocimene	7.0%	Myrcene	0.6%–12.4%
Terpinen-4-ol	4.9%	Germacrene D	2.3%–10.8%
Alpha-Humulene	4.9%	Alpha-Pinene	3.3%–8.0%
Limonene	3.8%	Limonene	0.8%–7.8%
Delta-Cadinene	3.7%	(E)-Beta-Ocimene	3.8%–5.7%
Beta-Selinene	3.7%		
Alpha-Pinene	3.5%		
Cyperene	3.4%		
Viridiflorene	3.0%		

OTHER CONSTITUENTS: *Flower Tops & Leaves*—(Z)-beta-ocimene, beta-caryophyllene, alpha-humulene; *Leaves*—Myrcene, para-cymene, alpha-ylangene, delta-elemene, terpinolene, beta-elemene, germacrene D, alpha-muurolene, geraniol, (E)-nerolidol, viridiflorol, delta-cadinol, alpha-cadinol, gamma-cadinol, beta-eudesmol, unidentified constituent

PREFERRED COMPOSITION FOR CLINICAL USE:

Constituent	
Beta-Pinene	35%–60%
Myrcene	0.5%–12%
Germacrene D	0.5%–12%
(E)-Beta-Ocimene	2%–10%
Alpha-Pinene	2%–10%
Limonene	0.5%–10%
Alpha-Humulene	0.1%–5%

REPORTED THERAPEUTIC PROPERTIES: Eases cough, aids respiratory function, relieves chronic skin conditions and itching, soothes sore throat, analgesic, anti-inflammatory, aids digestion, relieves diarrhea, expels excess gas, stops excess bleeding, disinfectant, antibacterial, antifungal, insect repellent, energizing, awakens to life opportunities, reduces chaos, grounding, stress management

CAUTIONS:

- None currently known.

SELECTED EVIDENCE:

- *In vitro* research demonstrates that arina essential oil inhibits gram-positive bacteria (*S. aureus, B. subtilis,* and *S. lutea*).[164]

ASPEN
(Quaking Aspen, Common Aspen, Eurasian Aspen, European Aspen)

Populus tremula L., *P. tremuloides* Michx.

FAMILY: Salicaceae
NOTE: Base-Middle
AROMA INTENSITY: Medium-Strong
AROMA: Spicy, smoky, pungent, floral, hint of vanilla
COMMON EXTRACTION METHOD: Hydrodistilled from the aspen buds
POSSIBLE SUBSTITUTE OILS: Black poplar, jasmine, ylang ylang, clove
BLENDS WELL WITH: Balsam fir, black poplar, cedarwood, chamomile (Roman), copaiba, cypress, Douglas fir, elemi, fennel, frankincense, geranium, ginger, grapefruit, ho wood, lavender, lemon, lime, palo santo, pine, rose, sandalwood, spruce (black), vetiver, ylang ylang, araucaria, ginger
SUBCELLULAR LOCALIZATION | EPIGENETIC INFLUENCE: Currently unknown | Currently unknown
RECOMMENDED DILUTION RANGE: 1.5%–20%

PRIMARY CONSTITUENTS:[165]

(E)-Benzyl Cinnamate CT	
(E)-Benzyl Cinnamate	24.7%–26.8%
Benzyl Benzoate	13.5%–26.4%
Salicyl Benzoate	11.7%–16.0%
Unidentified Aliphatic Comp.	4.5%–10.5%
2-Phenylethanol	1.3%–3.8%
Benzyl Alcohol	4.2%–7.0%

Benzyl Benzoate CT	
Benzyl Benzoate	45.5%
Delta-Cadinene	4.8%
Salicyl Benzoate	7.0%
Unidentified Aliphatic Comp.	3.4%

(Z)-Beta-Caryophyllene CT	
(Z)-Beta-Caryophyllene	18.8%–24.0%
Caryophyllene Oxide	12.0%–18.1%
Alpha-Guaiene	6.9%–10.2%
Unidentified Aliphatic Comp.	1.8%–3.4%

2-Phenylethanol CT	
2-Phenylethanol	34.5%
Unidentified Aliphatic Comp.	16.2%
Benzyl Alcohol	15.1%
Eugenol	14.2%

OTHER CONSTITUENTS: *(E)-Benzyl Cinnamate CT*—Delta-cadinene; *Benzyl Benzoate CT*—Benzyl alcohol, eugenol; *(Z)-Beta-Caryophyllene CT*—Eugenol, delta-cadinene, benzyl benzoate, benzyl alcohol; *2-Phenylethanol CT*—Delta-cadinene, benzyl benzoate, (E)-benzyl cinnamate

PREFERRED COMPOSITION FOR CLINICAL USE:
Insufficient samples/data to determine.

REPORTED THERAPEUTIC PROPERTIES: Antibacterial, antifungal, antiviral, antiparasitic, antioxidant, anti-inflammatory, analgesic, stimulates immune function, diuretic, decreases blood pressure, cleanses the urinary tract, wound healing, heals bruises and hemorrhoids, soothes sunburn, relieves skin itching, promotes restful sleep, relieves anxiety, calms emotional chaos, comforting, promotes love and reconciliation

CAUTIONS:

■ Avoid oral use and use cautiously during pregnancy, due to significant amounts of 2-Phenylethanol (PEA). Excessive oral doses of PEA (≥ 140 mg/kg body weight daily; over 10 mL as a human equivalent dose) administered to pregnant rats resulted in reduced body weight of their offspring.[166] Extreme oral doses (≥ 430 mg–1,400 mg/kg daily) caused maternal and embryo-fetal toxicity/abnormalities or an extracervical rib in offspring.

■ Oral caution. Ingestion of benzyl derivatives (benzoic acids, benzoic salts, benzaldehyde, benzyl acetate, benzyl benzoate, and benzyl alcohol) may cause gastrointestinal irritation, vomiting, and diarrhea. The Joint Expert Committee for Food Additives (JECFA) has established an acceptable daily limit for the aforementioned constituents of 0–5 mg/kg body weight for adults to avoid GI discomfort. Based on this limit, it is suggested to limit the oral consumption of orange bird berry essential oil rich in benzyl derivatives to no more than 10 drops daily.

SELECTED EVIDENCE:

- □ None found.

ATRACTYLIS
(Atractylodes, So-jutsu)

Atractylodes lancea (Thunb.) DC;
Korean Atractylis: *A. koreana* Nakai

FAMILY: Asteraceae
NOTE: Middle
AROMA INTENSITY: AI
AROMA: Earthy, woody, green
COMMON EXTRACTION METHOD: Steam distilled or hydrodistilled from the rhizomes
POSSIBLE SUBSTITUTE OILS: Spikenard, Surinam cherry, myrrh
BLENDS WELL WITH: Camphor, cedarwood, frankincense, juniper, lavender, myrrh, myrtle, palo santo, sandalwood, spikenard, Surinam cherry
SUBCELLULAR LOCALIZATION | EPIGENETIC INFLUENCE: Currently unknown | Currently unknown
RECOMMENDED DILUTION RANGE: 3%–33%; 50% for some conditions

PRIMARY CONSTITUENTS:[167,168,169]

Atractylone CT	
Atractylone	48.7%
Longiverbenone	7.3%
Guaia-3,9-Diene	6.2%
Alpha-Elemene	4.8%

Hinesol/Beta-Eudesmol CT	
Hinesol	36.6%–37.2%
Beta-Eudesmol	36.2%–36.4%
4-Trimethyl-3-(1-methylethenyl)-cyclohexanemethanol,	6.8%–9.5%
Guaiol	4.8%–6.5%
Atractylodin	3.9%–5.6%

Beta-Eudesmol CT	
Beta-Eudesmol	26.0%
Beta-Elemene	18.0%
Hinesol	10.0%
Elemol	6.0%

Mixed CT	
Beta-Eudesmol	23.4%–24.9%
Atractylodin	20.3%–23.9%
Atractylone	17.4%–20.1%
Hinesol	6.2%–10.4%
Tetradecanoic Acid	5.0%–5.9%
Propyphenazone	3.0%–3.4%

Korean Atractylis

Eudesma-4(14)-en-11-ol	12.7%
Eudesma-1(14),11-Diene	12.1%
Atractylone	12.0%
Atractylodin	8.9%
Alpha-Thujene	6.6%
Alpha-Elemene	4.6%
Isocomene	3.3%

OTHER CONSTITUENTS: *Atractylone CT*—Alpha-pinene, alpha-phellandrene, undecane, cyclohexene,1-methyl-3-pentyl-, alpha-guaiene, isocomene, calamene, alpha-humulene, 2,2-dimethyl-1-(2,4,6-trimethylphenyl)propan-1-one, eudesma-4(14),11-diene, dehydroaromadendrene, beta-elemene, alpha-vatirenene, 4-methylbiphenyl, bulnesol, germacrene B, atractylenolide I, eudesma-5,11(13)-dien-8,12-olide; *Beta-Eudesmol CT*—Alpha-phellandrene, beta-pinene, sabinene, alpha-terpinene, delta-cadinene, beta-caryophyllene, beta-selinene, muurolene isomers, valencene, alpha-pinene, borneol, camphene, delta-3-carene, para-cymene, myrcene, (Z)-beta-ocimene, (E)-beta-ocimene, terpinolene; *Hinesol/Beta-Eudesmol CT*—4a-Trimethyl-8-methylene-2-naphthalenemethanol, alpha-bisabolol, 2-(1-Methylethenyl)-naphthalene; *Mixed CT*—4-Trimethyl-3-(1-methylethenyl)-cyclohexanemethanol, elemol, guaiol, 4a-Trimethyl-8-methylene-2-naphthalenemethanol, 1,1′-Biphenyl-4-carboxaldehyde, alpha-copaene, 1-Methoxy-2-(1-methyl-2-methylenecyclopentyl)-benzene, N-(3-Chloro-4-fluorophenyl) acetamide, 3-Phenyl-but-2-enoic acid(1-methyl-2-phenyl-ethyl)-amide, (Z)-9-Octadecenamide, 2-(3,7-Dimethyl-octa-2,6-dienyl)-1,4-dimethoxy-benzene, caryophyllene oxide, hexadecanamide; *Korean Atractylis*—Alpha-pinene, alpha-phellandrene, undecane, cyclohexene,1-methyl-3-pentyl-, alpha-guaiene, geranyl acetate, calmene, alpha-humulene, alpha-himachalene, eudesma-4(14),7(11)-diene, alpha-sesquiphellandrene, alpha-cedrene, guaia-3,9-diene, dehydroaromadendrene, beta-elemene, 4-methylbiphenyl, hinesol, bulnesol, alpha-bisabolol, germacrene B, vallereal, atractylenolide I, atractylenolide II

PREFERRED COMPOSITION FOR CLINICAL USE:
Too great of variance in data to determine; however, genuine atractylis essential oil should be predominantly atractylone, hinesol, beta-eudesmol, and atractylodin.[170]

REPORTED THERAPEUTIC PROPERTIES: Aids digestion, relieves stomachache, expels excess gas, relieves diarrhea, eases nausea, increases appetite, diuretic, weight management, anticancer, antiallergic, anti-inflammatory, analgesic, antirheumatic, antiviral, antibacterial, antifungal, liver protective, alleviates night blindness, opens the airways, insecticidal, relaxing, reduces irritability, promotes creative thought

CAUTIONS:

Beta-Eudesmol CT and Mixed CT

- Avoid during pregnancy and lactation (low risk) due to beta-eudesmol content. Beta-eudesmol strongly inhibits the growth of new blood vessels, which is necessary for implantation, and placental and embryonic development.[171]
- May interact with aspirin, blood pressure, antiplatelet, and anticoagulant medications, and increase the risk of bleeding among people with bleeding disorders based on beta-eudesmol content (low risk).[172,173]

Hinesol/Beta-Eudesmol CT

- Avoid during pregnancy and lactation (low risk) due to beta-eudesmol content. Beta-eudesmol strongly inhibits the growth of new blood vessels, which is necessary for implantation, and placental and embryonic development.[174]
- May interact with aspirin, blood pressure, antiplatelet, and anticoagulant medications, and increase the risk of bleeding among people with bleeding disorders based on beta-eudesmol content (low risk).[175,176]

■ May interact with aspirin, blood pressure, antiplatelet, and anticoagulant medications, and increase the risk of bleeding among people with bleeding disorders, due to hinesol content.[177]

Atractylone CT and Korean Atractylis
■ None currently known.

SELECTED EVIDENCE:

- □ Atractylis essential oil restored normal gastric emptying that was delayed due to psychological stress in rats by inhibiting the release of hypothalamic corticotropin-releasing factor (CRF) and activating the vagal pathway.[178]
- □ *In vitro* research demonstrated that atractylis essential oil possesses strong antimicrobial activity against *Rhodotorula glutinis* and *Saprolegnia*, weak activity against *Rhizopus* and *Absidia*, and inhibits *A. niger* and *T. viride*.[179]

AVOCADO
(Alligator Pear)

Persea americana Mill.

FAMILY: Lauraceae
NOTE: Middle-Base
AROMA INTENSITY: Medium to strong
AROMA: Herbaceous, rich or spicy and licorice-like
COMMON EXTRACTION METHOD: Steam distilled from the leaves; may also be steam distilled from the fruit or seeds
POSSIBLE SUBSTITUTE OILS: Copaiba, black pepper, hemp, guava leaf, tropical basil, niaouli CT (E)-nerolidol, cabreuva
BLENDS WELL WITH: Basil, bergamot, black pepper, cardamom, cinnamon, clove, copaiba, fennel, frankincense, ginger, hemp, lavender, lemon, orange, palo santo, tangerine
SUBCELLULAR LOCALIZATION | EPIGENETIC INFLUENCE: Currently unknown | Currently unknown
RECOMMENDED DILUTION RANGE: *Beta-caryophyllene CT*—5%–Neat

PRIMARY CONSTITUENTS:[180,181,182,183,184]

Beta-Caryophyllene CT

Beta-Caryophyllene	43.9%
Valencene	16.0%
Germacrene D	5.9%
Alpha-Humulene	5.0%
Delta-Cadinene	4.8%
Beta-Cubebene	3.7%
Caryophyllene Oxide	3.6%
Spathulenol	3.2%
Cis-Gamma-Cadinene	3.0%

Methyl Eugenol CT

Methyl Eugenol	31.2%
Beta-Caryophyllene	16.9%
Methyl Chavicol	9.0%

Methyl Chavicol CT

Methyl Chavicol	61.9%–78.1%
Sabinene	0.1%–15.2%
Alpha-Pinene	0.3%–14.3%
Alpha-Cubebene	0.0%–3.6%
Methyl Eugenol	0.2%–3.4%
1,8-Cineole	0.3%–3.1%

(E)-Nerolidol CT (Fruits)

(E)-Nerolidol	37.4%–39.8%
Beta-Caryophyllene	2.8%–13.5%
Beta-Pinene	0.7%–13.5%
(E)-Alpha-Bergamotene	1.3%–9.9%
(E,E)-Decadienal	0.0%–9.1%
Beta-Bisabolene	7.0%–7.5%

Delta-Cadinene	4.8%	Germacrene D	4.2%–5.1%
Beta-Pinene	4.2%	Alpha-Copaene	2.7%–4.2%
Alpha-Pinene	3.2%	(E,Z)-2,4-Decadienal	0.8%–4.2%
Alpha-Humulene	3.0%	1-Eicosane	0.1%–3.9%
Germacrene D	3.0%		

Note: A number of varieties of *P. Americana* exist, each of which produces a different CT: *var. flocossa*—Beta-pinene 20.6%, beta-caryophyllene 12.6%, alpha-pinene 10.9%; *var. steyermarkii*—Beta-pinene 10.4%, viridiflorene 10.3%, globulol 9.2%; *var. nubigena*—Alpha-terpineol 18.5%, methyl chavicol 12.4%; *var. drymifolia*—Methyl chavicol 40.2%–95.0%, alpha-terpineol 39.3%; *var guatemalensis*—Beta-caryophyllene 38.3%; *subsp. americana*—Beta-pinene 40.9%, alpha-pinene 27.5%; *var. primatogeia*—Alpha-pinene 24.6%, beta-pinene 20.7%, germacrene D 10.1%; *chemovar. "aguacetale de anis"*—Trans-anethole 67.0%, methyl chavicol 13.1%; *var. tilaranensis*—Germacrene D 26.0%, beta-caryophyllene 21.0%, valencene 14%.

OTHER CONSTITUENTS: *Beta-Caryophyllene CT*—Alpha-pinene, sabinene, beta-pinene, alpha-cubebene; *Methyl Eugenol CT*—Sabinene, alpha-phellandrene, beta-phellandrene, 1,8-cineole, alpha-cubebene, beta-copaene, beta-cubebene, beta-elemene, (Z)-muurola-3,5-diene, (E)-cadina-1(6),4-diene, (E)-cadina-1(2),4-diene, gamma-muurolene, beta-selinene, bicyclogermacrene, alpha-muurolene, (E)-nerolidol, alpha-acorenol, alpha-muurolol, valerianol; *Methyl Chavicol CT*—Camphene, beta-pinene, linalool, undecan-2-one, chavicol, 1-decyl acetate, beta-caryophyllene, delta-cadinene; *(E)-Nerolidol CT (Fruits)*—Alpha-Pinene, limonene, terpinen-4-ol, cyclosativene, beta-cubebene, 1-tetradecene, alpha-humulene, bicyclogermacrene, delta-cadinene, 1,10-di-epi-cubenol, 2,4-tetradecadienal, (E)-nerolidol acetate, pimaradiene

PREFERRED COMPOSITION FOR CLINICAL USE:
Insufficient samples/data to determine.

REPORTED THERAPEUTIC PROPERTIES: Anti-inflammatory, analgesic, antiarthritic, relieves cough, aids respiratory function, balances blood sugar, digestive aid, wound healing, relieves skin itching, nourishes the skin, insecticidal, alleviates mental exhaustion, enhances focus and concentration, encourages emotional healing and release of negative emotions

CAUTIONS:
Due to the significant number of varieties and the wide variation in essential oil composition, avocado essential oil should only be used if you know the exact composition of the oil in your bottle.

Beta-Caryophyllene CT
■ None currently known.

(E)-Nerolidol CT
■ May interfere with antibiotics and increase their effectiveness ((E)-nerolidol sensitizes bacteria to antibiotics).[185]
■ May interfere with enzymes (CYP1A, CYP2B, CYP3A) responsible for metabolizing medications based on alpha-bisabolol content (NSAIDs, proton-pump inhibitors, acetaminophen, antiepileptics, immune modulators, blood-sugar medications, blood pressure medications, antidepressants, antipsychotics, diabetic medications, antihistamines, antibiotics, and anesthetics) due to (E)-nerolidol content.[186]

Methyl Chavicol, Methyl Eugenol Types
■ Not recommended for use in aromatherapy.

SELECTED EVIDENCE:

- ☐ The methyl chavicol (estragole) CT of avocado essential oil demonstrated good larvicidal activity against the *Culex quinquefasciatus* mosquito.[187]

BAKHTIARI SAVORY
(Marze, Marzeh-e-Koohi)

Satureja bachtiarica Bunge

FAMILY: Lamiaceae (Labiatae)
NOTE: Middle
AROMA INTENSITY: Strong
AROMA: Herbaceous, medicinal, sharp
COMMON EXTRACTION METHOD: Steam distilled or hydrodistilled from the aerial parts
POSSIBLE SUBSTITUTE OILS: Thyme, oregano, ajowan, summer savory, mountain savory
BLENDS WELL WITH: Ajowan, bergamot, grapefruit, lavender, lemon, lime, mountain savory, orange, oregano, pine, rosemary, summer savory, tangerine, thyme
SUBCELLULAR LOCALIZATION | EPIGENETIC INFLUENCE: Currently unknown | Currently unknown
RECOMMENDED DILUTION RANGE: 1.5%–20%; 50% for some conditions
PRIMARY CONSTITUENTS:[188,189,190,191,192,193,194,195]

Mixed CT

Carvacrol	15.2%–45.5%
Thymol	15.0%–31.0%
Gamma-Terpinene	0.0%–18.7%
Para-Cymene	3.5%–14.6%
Caryophyllene Oxide	0.5%–12.0%
Borneol	0.0%–11.6%
Linalool	0.0%–9.6%
Carvacrol Acetate	0.0%–4.5%
Beta-Caryophyllene	0.8%–4.4%

Carvacrol CT

Carvacrol	48.6%–63.3%
Para-Cymene	0.0%–16.6%
2-Cyclohene-1-One	0.0%–10.2%
Gamma-Terpinene	0.0%–6.9%
Linalool	0.0%–5.3%
Beta-Caryophyllene	2.1%–4.1%

Thymol CT

Thymol	44.5%–65.1%
Gamma-Terpinene	15.0%–23.9%
Beta-Caryophyllene	4.9%–5.3%
Para-Cymene	4.4%–7.3%
Borneol	3.1%–4.2%
Linalool	2.4%–3.5%
Delta-3-Carene	tr–3.5%

OTHER CONSTITUENTS: *Mixed CT*—Alpha-pinene, camphene, myrcene, alpha-terpinene, limonene, 1,8-cineole, terpinen-4-ol, alpha-thujone (< 1.0%), beta-eudesmol, camphor, trans-sabinene hydrate, cuminaldehyde; *Thymol CT*—Alpha-thujene, alpha-pinene, camphene, myrcene, (E)-beta-ocimene, terpinen-4-ol; *Carvacrol CT*—Alpha-thujene, alpha-pinene, camphene, myrcene, sylvestrene, cis-sabinene hydrate, borneol, terpinene-4-ol, thymol, (E)-anethole, piperitenone, carvacrol acetate, aromadendrene, alpha-curcumene, ledene, cyclohexene, alpha-bisabolene, spathulenol, caryophyllene oxide

PREFERRED COMPOSITION FOR CLINICAL USE:
Too great of variance in samples/data to determine.

Reported Therapeutic Properties: Aids digestion, alleviates diarrhea, relieves stomachache, analgesic, anti-inflammatory, relieves muscle cramps, antiseptic, anti-infectious, antimicrobial, antifungal, antibacterial, antiviral, modulates immune function, antioxidant, antidepressant, reduces fear

Cautions:

■ Avoid during pregnancy and lactation. Oregano (also rich in carvacrol) may negatively affect embryonic development and encourage fetal cell death according to animal research.[196]

■ May interact with diabetic medications and cause low blood-sugar levels due to carvacrol content—carvacrol reduces blood glucose and enhances the antidiabetic activities of rosiglitazone in animals.[197,198,199,200]

■ May interact with aspirin, blood pressure, antiplatelet, and anticoagulant medications, and increase the risk of bleeding among people with bleeding disorders, due to carvacrol and thymol content.[201,202,203,204,205]

■ May interact with anticholinergic (drugs used for asthma, incontinence, gastrointestinal cramps, muscular spasms, depression, and sleep disorders) and cholinergic medications (drugs used to reduce urinary retention, diagnose myasthenia gravis, and for glaucoma), based on thymol and carvacrol content.[206,207]

■ May interact with antibiotics or antifungals and possibly enhance or decrease their effects due to carvacrol or thymol content.[208,209,210,211,212]

■ May irritate mucous membranes (eyes, mouth, nasal passages, vagina, rectum).

■ *Thymol CT*: Oral caution. Thymol has a longer half-life (the time it takes for half of the medication to metabolize or excrete half of the dosage) than most essential oil constituents and should not be administered orally for long periods of time.[213] Thymol is a monoterpene phenol. Reports of fatalities in children who consumed 50 to 200 mg of phenols has been reported.[214] It is recommended that chemotypes with high levels of thymol be limited to 10 drops per day orally for adults with a two- to seven-day break after twenty-one days of use; or limit ingestion to five or six days weekly.

Selected Evidence:

- □ Oral administration of bakhtiari savory CT carvacrol essential oil reversed depression and improved total antioxidant capacity and malondialdehyde (MDA) levels in the brain of rats in a reserpine model of depression (reserpine depletes neurotransmitters important for mood).[215] Resperine increased MDA and reduced brain TAC in rat brains, which were both reversed by bakhtiari essential oil.
- □ Bakhtiari essential oil (mixed CT) exhibited strong antibacterial activity against clinical isolates of *H. pylori*, which was primarily attributed to its carvacrol content.[216] Interestingly, thymol reduced the activity of carvacrol against *H. pylori*.
- □ *In vitro* research demonstrated that bakhtiari (mixed CT) essential oil strongly inhibited *S. agalactiae, S. typhimurium*, and *P. vulgaris*, and also inhibited *B. cereus, S. aureus,* and *L. monocytogenes*.[217] Another study found that the carvacrol CT potently inhibited *Acinetobacter* species bacteria isolated from people being treated in a level-1 trauma center in a hospital.[218]
- □ The mixed CT of bakhtiari essential oil exhibited antioxidant activity that was greater than BHT.[219]

BETLE
(Betle Leaf, Sireh)

Piper betle L.

FAMILY: Piperaceae
NOTE: Base
AROMA INTENSITY: Medium-Strong
AROMA: Spicy, sweet, woody to phenolic, smoky, tar-like
COMMON EXTRACTION METHOD: Hydrodistilled or steam distilled from the leaves; may also be distilled from the stems, fruits, roots (rhizomes), aerial parts, or stalks
POSSIBLE SUBSTITUTE OILS: Clove, mace, tropical basil
BLENDS WELL WITH: Black pepper, cajeput, cardamom, cinnamon, clove, coriander, eucalyptus, ginger, lavender, lemon, orange, patchouli, rosemary, Spanish sage, spikenard, tamala, tangerine, turmeric
SUBCELLULAR LOCALIZATION | EPIGENETIC INFLUENCE: Currently unknown | Currently unknown
RECOMMENDED DILUTION RANGE: 1.5%–20%

PRIMARY CONSTITUENTS:[220,221,222,223,224,225,226,227,228,229,230,231,232,233,234,235,236,237,238]

Safrole CT	
Safrole	22.8%–52.7%
Eugenol	6.4%–27.2%
Allylpyrocatechol Diacetate	0.0%–19.2%
Chavibetol (Chavicol) Acetate	0.0%–12.6%
Isoeugenol	0.0%–8.1%
Beta-Caryophyllene	0.9%–7.8%
Delta-3-Carene	0.0%–7.8%
Beta-Selinene	0.0%–6.4%
Terpinen-4-ol	0.0%–6.4%
Eugenyl Acetate (Acetyleugenol)	0.0%–5.8%
Methyl Chavicol (Estragole)	0.0%–5.8%
Alpha-Cubebene	0.0%–4.4%
Sabinene	0.0%–3.7%
Alpha-Selinene	0.0%–3.6%
Beta-Phellandrene	0.0%–3.4%
Beta-Pinene	0.0%–3.4%
Alpha-Humulene	0.0%–3.1%
Camphene	0.0%–2.8%
Stearaldehyde	0.0%–2.7%

Chavibetol CT	
Chavibetol	22.0%–80.5%
Chavibetol (Chavicol) Acetate	0.0%–15.5%
Methyl Chavicol (Estragole)	0.0%–15.8%
Beta-Cubebene	0.0%–13.6%
Chavicol	0.0%–11.8%
Beta-Caryophyllene	0.4%–11.3%
Eugenyl Acetate (Acetyleugenol)	0.0%–8.3%

Eugenol CT	
Eugenol	33.2%–73.0%
Eugenyl Acetate (Acetyleugenol)	0.0%–18.7%
Isoeugenol	tr–10.6%
Safrole	0.0%–6.5%
Sabinene	0.0%–6.1%
n-Decanal	0.0%–4.2%
Allylpyrocatechol Diacetate	0.0%–3.8%
Terpinen-4-ol	0.0%–3.6%
Methyl Eugenol	0.2%–3.4%

Allylpyrocatechol Diacetate CT	
Allylpyrocatechol Diacetate	34.0%
Beta-Cadinene	5.6%
Alpha-Humulene	4.0%
Gamma-Muurolene	3.8%
Gamma-Selinene	3.2%

Hydroxychavicol CT	
Hydroxychavicol	39.5%–45.3%
Eugenol	11.6%–20.8%
Safrole	0.0%–5.0%
Beta-Caryophyllene	0.0%–4.3%

Trans-Anethole CT	
Trans-Anethole	19.3%
Eugenol	18.9%
Alpha-Humulene	10.6%
Gamma-Cadinene	9.4%

Allylpyrocatechol Diacetate	0.2%–6.2%
Gamma-Muurolene	tr–6.0%
Elixene	0.0%–3.2%

Isoeugenol CT

Isoeugenol	34.6%–59.2%
(E)-Isoeugenol Acetate	0.0%–16.2%
Trans-Isoeugenol	0.0%–12.6%
Eugenyl Acetate (Acetyleugenol)	8.9%–25.3%
Allylpyrocatechol Diacetate	4.8%–11.4%
Gamma-Muurolene	0.4%–4.6%
Germacrene D	1.8%–3.8%

Trans-Isoeugenol CT

Trans-Isoeugenol	28.5%
Isoeugenol	24.2%
Eugenol	18.0%
Allylpyrocatechol Diacetate	6.8%
Germacrene D	6.3%

Stems/Stalks

Safrole	40.5%–41.7%
Eugenol	11.5%–13.2%
Chavibetol (Chavicol) Acetate	10.3%–11.6%
Gamma-Selinene	4.5%–9.6%
Allylpyrocatechol Diacetate	0.0%–5.1%
Beta-Caryophyllene	1.6%–5.0%
Gamma-Muurolene	2.2%–4.8%
Alpha-Humulene	3.0%–4.5%
Alpha-Elemene	3.4%–4.3%

Roots (Rhizomes) CT Safrole

Safrole	31.8%
Camphene	13.5%
Alpha-Pinene	12.9%
Beta-Phellandrene	12.7%
Chavibetol (Chavicol) Acetate	4.7%
Eugenol	4.5%
Allylpyrocatechol Diacetate	4.3%
1,8-Cineole	3.2%
Gamma-Selinene	2.9%

P. betle L. var. *sirugamani*

Germacrene D	16.1%
Bicyclogermacrene	15.0%
Caryophyllene	9.9%
1,3,4-Eugenol	7.2%
Beta-Elemene	5.8%
Gamma-Murrolene	3.2%
Eudesma-7(11)-en-4-ol	3.1%
Methyl Chavicol (Estragole)	7.6%
Alpha-Cubebene	4.7%
Alpha-Cadinol	2.9%

Eugenyl Acetate (Acetyleugenol) CT

Eugenyl Acetate (Acetyleugenol)	54.8%
Caryophyllene	11.2%
Beta-Cubebene	8.3%
Allylpyrocatechol Diacetate	6.6%

Chavicol Acetate CT

Chavicol Acetate	19.3%
Trans-Isoeugenol	18.5%
Acetyl Eugenol	18.4%
Allylpyrocatechol Diacetate	12.6%

Fruits

Beta-Phellandrene	25.0%
Safrole	22.2%
Allylpyrocatechol Diacetate	11.3%
Chavibetol (Chavicol) Acetate	7.2%
Eugenol	6.1%
Limonene	5.3%
Para-Cymene	3.0%
Gamma-Selinene	2.5%

Roots (Rhizomes) CT Cadinol

Alpha-Cadinol	26.2%
T-Cadinol + T-Muurolol	20.7%
Delta-Cadinene	11.7%
Epi-cubenol	4.7%
Delta-Cadinol	3.5%
Gamma-Cadinene	3.5%

Note: The safrole CT from the leaves could be further divided into additional chemotypes but is combined for convenience in this work. Essential oil composition changes according to leaf maturity with the greatest percentage of safrole and chavibetol obtained at harvesting stage.

OTHER CONSTITUENTS: *Safrole CT*—Alpha-pinene, para-cymene, alpha-terpinene, myrcene, gamma-terpinene, gamma-muurolene, beta-cadinene, germacrene D, methyl eugenol (< 2.0%), alpha-copaene, delta-cadinene, caryophyllene oxide, n-decanal, linalool, beta-elemene, gamma-elemene, beta-cubebene, bornylene, 2-methyl-5(1-methylethyl)-bicyclo[3.1.0]hex2-ene, 1-methyl-4-(1-methyl ethyl)benzene, 1-methyl-5-(1-metheyletenyl)cyclohexene, chavicol acetate; *Eugenol CT*—Alpha-thujene, myrcene, (E)-beta-ocimene, terpinolene, allo-ocimene, delta-cadinene, gamma-elemene, linalool, alpha-terpineol, terpinene-1-ol, delta-cadinol, 3,7,11,15-tetramethyl-2-hexadecen-1-ol, geraniol, beta-caryophyllene, alpha-humulene, gamma-muurolene, beta-selinene, alpha-selinene, chavibetol, caryophyllene oxide, methyl benzoate; *Allylpyrocatechol Diacetate CT*—Beta-phellandrene; *Chavibetol CT*—Camphene, methyl chavibetol, allylpyrocatechol, methyl eugenol (< 0.7%), alpha-humulene, gamma-cadinene, cedrol, phytol, 1,8-cineole, alpha-cubebene, beta-elemene, aciphyllene, viridiflorene, delta-cadinene; *Hydroxychavicol CT*—Methyl isoeugenol, methyl eugenol (< 0.9%), isoeugenol; *Trans-Anethole CT*—Camphene, alpha-terpinene, alpha-cadinene, gamma-elemene, aromadendrene, beta-cubebene, 1,8-cineole; *Isoeugenol CT*—Beta-cubebene, beta-elemene, beta-humulene, alpha-humulene, gamma-gurjunene, valencene, delta-cadinene, cada-1,4-diene, alpha-selinene, T-muurolol, ledene, alpha-elemene; *Trans-Isoeugenol CT*—Cis-ocimene, eugenyl acetate, beta-caryophyllene, alpha-humulene, gamma-muurolene, alpha-selinene, bicyclogermacrene, delta-cadinene, alpha-selinene, T-muurolol, alpha-elemene; *Eugenyl Acetate (Acetyleugenol) CT*—Chavicol acetate, chavicol, alpha-humulene, gamma-muurolene, alpha-muurolene, germacrene D, beta-cadinene, cada-1,4-diene, T-muurolol; *Chavicol Acetate CT*—Not listed; *Stems/Stalks*—Alpha-pinene, terpinen-4-ol, beta-cadinene; *Fruits*—Alpha-pinene, camphene, terpinen-4-ol, gamma-muurolene, beta-caryophyllene, alpha-humulene, beta-cadinene; *Roots (Rhizomes) CT Safrole*—Alpha-terpineol, alpha-elemene, beta-caryophyllene; *Roots (Rhizomes) CT Cadinol*—Alpha-pinene, camphene, beta-pinene, 1,8-cineole, o-eugenol, alpha-copaene, alpha-humulene, beta-caryophyllene, gamma-muurolene, alpha-muurolene, cubebene, alpha-cadinene, (E)-nerolidol, ledol; *P. betle* L. var. *sirugamani*—Alpha-pinene, 1,8-cineole, (Z)-beta-ocimene, delta-elemene, copaene, aromadendrene, beta-farnesene, alpha-humulene, beta-selinene, zingiberene, germacrene A, beta-cadinene, alpha-bergamotene, palustrol, globulol, epiglobulol, fonenol, cadr-9-ene, 3-(2-furyl)-N-[2-(4-tert-butyl]phenoxy)ethyl-propenamide

PREFERRED COMPOSITION FOR CLINICAL USE:
Insufficient samples/data to determine.

REPORTED THERAPEUTIC PROPERTIES: Antibacterial, antimicrobial, antifungal, antiparasitic, antioxidant, analgesic, anti-inflammatory, relieves headache, nervine, antineuralgic, aids respiratory function, eases cough, relieves sore throat, anticancer, antiseptic, alleviates constipation, balances blood-sugar levels, reverses loss of appetite, eases nausea, astringent, wound healing, relieves boils, aids scalp and hair health, relieves chronic skin conditions, prevents body odor, **aids oral health**, preserves teeth, improves gum health, combats bad breath, relieves strained voice, insect repellent, **insecticidal**, stimulating, antidepressant, aphrodisiac, promotes self-confidence and courage

CAUTIONS:
Safrole CT, Roots (Rhizomes) CT Safrole, Stems/Stalks, Fruits
Betle CTs high in safrole are generally not recommended for therapy due to their high safrole content.
■ Avoid during pregnancy due to safrole content. Animal research suggests that oral administration of safrole to pregnant mice caused kidney epithelial tumors in 7% of their offspring; and male offspring nursed

by mothers who were administered safrole experienced a high rate of liver tumors.[239] This animal research suggests that safrole increases the risk of cancer in offspring by crossing the placenta and excretion in breast milk. In addition, animal research suggests that the bonding of safrole and its proximate carcinogen (1'-hydroxysafrole) to liver and kidney DNA increases 2.3 to 3.5 fold during pregnancy, which may make them more hazardous and increase the risk of cancer during pregnancy.[240]

■ Oral caution due to moderate to significant safrole content. Animal research suggests that oral administration of safrole may cause liver cancer and is reasonably anticipated to be a human carcinogen.[241]

■ There is a moderate risk that mace essential oil may interact with aspirin, blood pressure, antiplatelet, and anticoagulant medications, and increases the risk of bleeding among people with bleeding disorders (low risk), based on safrole content.[242]

■ May interfere with enzymes responsible for metabolizing medications (NSAIDs, proton-pump inhibitors, acetaminophen, antiepileptics, immune modulators, blood-sugar medications, blood pressure medications, antidepressants, antipsychotics, diabetic medications, antihistamines, antibiotics, and anesthetics) due to safrole content.[243]

■ Isoeugenol may irritate the skin or cause sensitization.[244,245] Dilution strongly recommended.

Eugenol CT

■ Avoid during pregnancy and lactation due to eugenol content. Eugenol is considered strongly toxic to embryos according to animal studies.[246] Animal studies suggest that large doses of clove (also with significant eugenol) may negatively impact embryonic development and encourage fetal cell death.[247,248] Another animal study did not detect any negative influence of clove oil.[249] Based on these conflicting studies it is best to limit betle essential oils rich in eugenol during pregnancy and lactation.

■ May interact with aspirin, blood pressure, antiplatelet, and anticoagulant medications, and increase the risk of bleeding among people with bleeding disorders based on eugenol content.[250,251,252,253]

■ May interact with MAOI antidepressants based on eugenol content. Animal research suggests that eugenol produces antidepressant effects via the monoamine oxidase pathway, which may cause interactions with antidepressants that also interact with this pathway.[254]

■ May interfere with enzymes responsible for metabolizing medications (NSAIDs, proton-pump inhibitors, acetaminophen, antiepileptics, immune modulators, blood-sugar meds, blood pressure meds, antidepressants, antipsychotics, diabetic medications, antihistamines, antibiotics, and anesthetics), based on eugenol content.[255,256]

■ May interact with anticholinergic (drugs used for asthma, incontinence, gastrointestinal cramps, muscular spasms, depression, and sleep disorders) and cholinergic medications (drugs used to reduce urinary retention, diagnose myasthenia gravis, and for glaucoma), based on AChE-inhibitory activity of eugenol.[257]

■ May interact with antibiotics and possibly enhance their effects based on eugenol content.[258]

■ May irritate mucous membranes (eyes, mouth, nasal passages, vagina, rectum) due to eugenol and isoeugenol content.[259,260,261,262]

Chavibetol CT

■ Avoid during pregnancy and lactation due to moderate methyl chavicol (estragole) content.[263] Methyl chavicol (estragole) may cause genetic mutations and oxidative DNA damage.[264,265] The European Medicines Agency recommends exposure to estragole be limited in women who are pregnant or nursing.[266]

■ Use with great caution for children under age six due to animal reports of liver cancer caused by methyl chavicol (estragole) content.[267] The European Medicines Agency recommends exposure to estragole be limited in young children.

■ Avoid oral use due to carcinogenic (cancer-causing) potential of methyl chavicol (estragole).[268,,269,270] The metabolite of estragole (1'-hydroxyestragole) is considered a stronger carcinogen and humans rapidly metabolize estragole to 1'-hydroxyestragole when ingested.[271]

Hydroxychavicol CT

■ Avoid during pregnancy and lactation due to eugenol content. Eugenol is considered strongly toxic to embryos according to animal studies.[272] Animal studies suggest that large doses of clove (also with significant eugenol) may negatively impact embryonic development and encourage fetal cell death.[273,274] Another animal study did not detect any negative influence of clove oil.[275] Based on these conflicting studies it is best to limit betle essential oils rich in eugenol during pregnancy and lactation.

■ May interact with aspirin, blood pressure, antiplatelet, and anticoagulant medications, and increase the risk of bleeding among people with bleeding disorders based on hydroxychavicol content. Hydroxychavicol is a potent inhibitor of cyclooxygenase activity and inhibits platelet calcium signaling, thrombaxan B2 production and aggregation and exhibits anticoagulant activity.[276,277]

Isoeugenol, Trans-Isoeugenol, Chavicol Acetate CTs

■ May interact with aspirin, blood pressure, antiplatelet, and anticoagulant medications, and increase the risk of bleeding among people with bleeding disorders based on isoeugenol content.[278,279]

■ May interact with anticholinergic (drugs used for asthma, incontinence, gastrointestinal cramps, muscular spasms, depression, and sleep disorders) and cholinergic medications (drugs used to reduce urinary retention, diagnose myasthenia gravis, and for glaucoma), based on *in vitro* research.[280]

■ Isoeugenol may irritate the skin or cause sensitization.[281,282] Dilution strongly recommended.

Allylpyrocatechol Diacetate CT, Eugenyl Acetate (Acetyleugenol) CT, Trans-Anethole CT, Chavicol Acetate (4-Allylphenyl Acetate) CT, Roots (Rhizomes) CT Cadinol, and P. betle L. var. *sirugamani*

■ None currently known.

SELECTED EVIDENCE:

- □ Betle essential oil (unknown composition) was highly toxic to cervical cancer cells with low toxicity to leukocytes.[283]
- □ Betle essential oil (unidentified composition) possesses coagulant activity, with higher concentrations progressively increasing time to clotting.[284] The essential oil was able to increase the clotting of blood obtained from people with hemophilia (a condition characterized by excess bleeding from injuries due to abnormal blood clotting).
- □ *In vitro* research shows that betle CT safrole is significantly active against oral pathogens (*Staphylococcus aureus, Streptococcus mutans, Lactobacillus acidophilus, Candida albicans*, and *Saccharomyces cerevisiae*).[285] Another study concluded that an unidentified CT of betle essential oil is effective against periodontal pathogens (*Prevotella intermedia, Porphyromonas gingivalis*, and *Aggregatibacter actinomycetemcomitans*) and possesses antioxidant and anti-inflammatory activity, making it an effective oral health-care product.[286] An unidentified CT of betle essential oil inhibited *Mycobacterium tuberculosis in vitro*.[287] The chavibetol CT of betle essential oil prevented the growth of *Penicillium expansum* on semi-synthetic material in a dose-dependent manner, suggesting it has potential as a natural food preservative.[288] The safrole and allylpyrocatechol diacetate CTs of betle leaf essential oil inhibited bacteria (*Escherichia coli, Streptococcus pyogenes*, and *Staphylococcus aureus*) and fungi (*Colletotrichum* sp., *Fusarium oxysporium* sp., *Corynospora cassicola*, and *Rigidoporous* sp.).[289] Betle leaf essential oil (eugenol CT) inhibited *Aspergillus flavus* and its production of aflatoxin.[290]
- □ Two betle leaf/stalk essential oils with either trans-isoeugenol or 6-ethoxyquinaldine as major constituents significantly inhibited acetylcholinesterase (AChE) activity *in vitro*, with the trans-isoeugenol CT demonstrating greater inhibition.[291] Inhibition of AChE prevents the breakdown of acetylcholine, which is essential for memory and thinking. People with neurodegenerative diseases

make less acetylcholine, and the diseases often break it down at a faster rate, leading to acetylcholine deficits.

- The safrole CT of betle essential oil is insecticidal to the housefly (*Musca domestica*). Isolated safrole and eugenol were also active against the housefly.[292]
- Betle CT chavicol acetate repelled *Aedes aegypti* mosquitoes and showed equivalent repellent activity to 13% DEET when combined with basil (*O. basilicum*) CT citral.[293] An unidentified CT of betle essential oil was also toxic to *Aedes aegypti* mosquitoes.[294]
- The safrole and allylpyrocatechol diacetate CTs showed toxicity to *A. aegypti* mosquito larvae (43% and 100% mortality at 100ppm and 500ppm respectively, within one hour), 100% knockdown rate and mortality against housefly (*Musca domestica*), 100% mortality rate against rice weevil (*Sitophilus oryzae*) at 1% concentration, 100% mortality against the oriental latrine fly (*Chrysomya megacephala*) larvae at 3% and 4% concentration, 100% mortality of the Old World screwworm fly (*Chrysomya bezziana*) larvae—which was better than the control (Asuntol), and good antioxidant activity.[295] Another study also reported betle leaf essential oil (composition not reported) was larvicidal to the oriental latrine fly (*Chrysomya megacephala*).[296]
- Betle essential oil (unidentified CT) significantly inhibited the development of the tobacco cutworm (*Spodoptera litura*) without causing adverse effects to earthworms.[297] The study authors also concluded that twenty constituents of betle essential oil significantly affected the development and behavior of tobacco cutworms.
- A study determined that betle essential oil (unidentified CT) is safer than chemical pesticides for earthworms (nontoxic, no effect on developmental growth, and reduced avoidance of soil compared to soil with synthetic pesticides) and should be considered as an alternative pest control agent.[298]

BIOFLORE ROSEWOOD (Macacaporanga)

Aniba parviflora (Meisn.) Mez

FAMILY: Lauraceae
NOTE: Middle
AROMA INTENSITY: Medium
AROMA: Sweet, woody, slightly fruity, floral
COMMON EXTRACTION METHOD: Hydrodistilled from the leaves or wood
POSSIBLE SUBSTITUTE OILS: Rosewood, basil CT linalool, ho wood, coriander, betony
BLENDS WELL WITH: Bergamot, cedarwood, chamomile (German, Roman), geranium, grapefruit, juniper berry, lavender, lemon, lime, neroli, orange, petitgrain, rose, rosewood, sandalwood
SUBCELLULAR LOCALIZATION | EPIGENETIC INFLUENCE: Currently unknown | Currently unknown
RECOMMENDED DILUTION RANGE: 5%–Neat

PRIMARY CONSTITUENTS:[299,300,301,302,303]

Wood		*Leaves/Stems*	
Linalool	89.5%	Linalool	45.0%
1,8-Cineole	3.4%	Beta-Phellandrene	17.3%
		Alpha-Phellandrene	4.1%
		Beta-Caryophyllene	3.9%

Leaves		*Branches*	
Beta-Phellandrene	15.1%–23.6%	Gamma-Eudesmol	16.8%
Linalool	12.6%–21.3%	Beta-Caryophyllene	15.7%
Gamma-Eudesmol	0.0%–12.9%	Linalool	12.4%
Alpha-Phellandrene	0.5%–7.3%	Beta-Phellandrene	6.7%
Beta-Caryophyllene	2.6%–4.3%	Bicyclogermacrene	6.0%
Alpha-Eudesmol	0.9%–4.0%	Spathulenol	4.8%
Bicyclogermacrene	3.4%–4.3%	Thujopsan-2-Alpha-ol	3.9%
Beta-Eudesmol	0.7%–3.7%	Alpha-Eudesmol	3.6%
Myrcene	0.0%–3.7%	Beta-Eudesmol	3.3%
Alpha-Cubebene	tr–3.4%		
Spathulenol	3.2%–4.7%		
Elemol	0.0%–3.1%		

OTHER CONSTITUENTS: *Wood*—Beta-pinene, alpha-pinene, limonene, geraniol; *Leaves/Stems*—Myrcene, para-cymene, (E)-beta-ocimene, alpha-terpineol, alpha-cubebene, alpha-copaene, alpha-humulene, beta-chamigrene, beta-selinene, alpha-selinene, delta-cadinene, spathulenol, caryophyllene oxide, 10-epi-gamma-eudesmol, gamma-eudesmol, alpha-eudesmol, (E)-nerolidol acetate; *Leaves*—Alpha-pinene, beta-pinene, camphene, (E)-beta-ocimene, gamma-terpinene, borneol, terpinolene, terpinen-4-ol, alpha-terpineol, delta-elemene, alpha-copaene, beta-elemene, (Z)-beta-farnesene, gamma-elemene, alpha-humulene, gamma-muurolene, beta-selinene, guaiol, gamma-cadinene, delta-cadinene, (E)-nerolidol, aromadendrene, (E,E)-alpha-farnesene, alpha-curcumene, thujopsan-2-alpha-ol, globulol, eudesm-4(14)-en-11-ol, guaiol, rosifoliol, humulene epoxide II, 10-epi-gamma-eudesmol, T-muurolol, bulnesol, (Z)-alpha-santalol, benzyl benzoate; *Branches*—(E)-beta-ocimene, alpha-terpineol, delta-elemene, beta-elemene, alpha-guaiene, alpha-humulene, gamma-muurolene, amorpha-4,7(22)-diene, beta-selinene, alpha-bulnesene, elemol, (E)-nerolidol, eudesm-4(14)-en-11-ol, guaiol, rosifoliol, humulene epoxide II, 10-epi-gamma-eudesmol, 7-epi-gamma-eudesmol, bulnesol, 14-hydroxy-9-epi-(E)-caryophyllene

PREFERRED COMPOSITION FOR CLINICAL USE:

Constituent	*Wood*	*Leaves*
Linalool	70%–90%	10%–25%
Cis + Trans-Linalool Oxide	0.5%–4%	–
1,8-Cineole	0.1%–4%	–
Geraniol	0.5%–3%	–
Alpha-Pinene	tr–3%	–
Limonene	tr-3%	–
Beta-Pinene	tr–3%	–
Beta-Phellandrene	–	12%–30%
Eudesmol (alpha, gamma, beta)	–	3%–15%
Alpha-Phellandrene	–	0.5%–10%
Bicyclogermacrene	–	2%–7%
Beta-Caryophyllene	–	1%–7%
Spathulenol	–	2%–6%
Alpha-Cubebene	–	tr–4%

REPORTED THERAPEUTIC PROPERTIES: Analgesic (pain relief), anti-inflammatory, anticonvulsive, anticancer, antiseptic, antibacterial, deodorant, skin disorders, encourages the secretion of hormones, fever reducer, tissue regenerative, helps reduce the appearance of wrinkles and blemishes, wound healing, promotes restful sleep, relaxing, stress management, **antidepressant**, relieves anxiety, aphrodisiac

CAUTIONS:

Wood

■ May interact with antibiotics or antifungals and possibly enhance their effects due to significant linalool content.[304,305]

Leaves and Leaves/Stems

■ May interact with anticholinergic (drugs used for asthma, incontinence, gastrointestinal cramps, muscular spasms, depression, and sleep disorders) and cholinergic medications (drugs used to reduce urinary retention, diagnose myasthenia gravis, and for glaucoma), based on AChE-inhibtory activity of beta-phellandrene.[306]

Branches

■ None currently known.

SELECTED EVIDENCE:

- □ The leaf essential oil of bioflore rosewood (15.1% beta-phellandrene, 14.1% linalool) was active against breast cancer cells.[307]
- □ Intraperitoneal injection of bioflore rosewood leaf/stem essential oil (85 mg/kg body weight) exhibited antidepressant activity in rats without causing sedation or memory impairment in rodents.[308] The study authors attributed this effect to the presence of linalool, which is synergized by the other minor constituents in the oil.
- □ Bioflore rosewood leaf/stem essential oil inhibited *S. pyogenes*, *K. pneumoniae, S. aureus, E. faecalis,* and *S. epidermidis.*[309] Bioflore leaf and branches essential oil demonstrated significant inhibition of *E. coli* (equivalent to the positive control gentamicin) and also inhibited *B. cereus, S. aureus, S. epidermidis,* and *P. aeruginosa.*[310]
- □ Both the leaf and branches essential oils of bioflore rosewood exhibited good antioxidant activity in the DPPH assay.[311]

BITTER GINGER (Lempoyang, Colonia)

Zingiber zerumbet (L.) Sm., *Amomum zerumbet L.*; *Z. zerumbet* (L.) Sm. var. *darcyi*

FAMILY: Zingiberaceae
NOTE: Middle-Base
AROMA INTENSITY: Medium
AROMA: Earthy, warm, woody, slightly spicy
COMMON EXTRACTION METHOD: Hydrodistilled from the rhizomes; may also be distilled from the leaves and flowers
POSSIBLE SUBSTITUTE OILS: Ginger, shell ginger, turmeric
BLENDS WELL WITH: Bay laurel, bergamot, camphor, cardamom, carrot seed, cassia, cedarwood, clove, coriander, eucalyptus, fennel, frankincense, galbanum, geranium, ginger, grapefruit, lemon, lime, neroli, nutmeg, orange, oregano, patchouli, ravensara, rose, sandalwood, tangerine, turmeric, vetiver
SUBCELLULAR LOCALIZATION | EPIGENETIC INFLUENCE: Currently unknown | Currently unknown
RECOMMENDED DILUTION RANGE: 3%–50%; neat for some conditions

PRIMARY CONSTITUENTS:[312,313,314,315,316,317,318,319,320]

Zerumbone CT (Rhizomes)

Constituent	%
Zerumbone	37.0%–84.8%
Alpha-Humulene	2.9%–19.0%
Camphene	1.9%–13.8%
Humulene Epoxide	0.0%–7.3%
Caryophyllene Oxide	tr–5.2%
Alpha-Fenchene	0.0%–4.7%
Beta-Copaen-4-alpha-ol	0.0%–4.3%
1,5,5,8-Tetramethyl-12-oxabicyclo[9.1.0]dodeca-3,7-diene	0.0%–4.3%
(E)-Bisabolol-11-ol	0.0%–4.3%
Beta-Caryophyllene	0.0%–4.0%
Camphor	tr–3.8%
Linalool	0.0%–3.6%
Alpha-Pinene	0.0%–3.5%
1,8-Cineole	0.7%–3.2%

Leaves (Zerumbone CT)

Constituent	%
Zerumbone	37.0%
Alpha-Humulene	16.4%
Camphene	9.2%
1,2-Dihydropyridine,1-(1-oxobutyl),-	5.8%
3-Cyclohexen-1-carboxaldehyde, 3,4-dimethyl-	3.9%
Beta-Caryophyllene	3.3%

Zerumbone/Caryophyllene CT (Rhizomes)

Constituent	%
Zerumbone	39.1%–46.4%
Beta-Caryophyllene	11.4%–28.0%
Terpinen-4-ol	0.0%–20.8%

Curzerenone CT (Rhizomes)

Constituent	%
Curzerenone	14.4%
Camphor	12.8%
Zerumbone	12.6%
Isoborneol	8.9%
1,8-Cineole	7.1%
Borneol	4.9%
Camphene	4.4%
Beta-Elemene	3.3%

Leaves (Beta-Pinene CT)

Constituent	%
Beta-Pinene	31.4%
(E)-Nerolidol	21.4%
Alpha-Pinene	10.3%
Linalool	7.6%
Beta-Caryophyllene	6.9%

Flowers

Constituent	%
(E)-Nerolidol	34.9%
Linalool	17.1%
Beta-Caryophyllene	10.2%
(E)-Beta-Ocimene	3.0%

OTHER CONSTITUENTS: *Zerumbone CT (Rhizomes)*—Alpha-pinene, tricyclene, beta-pinene, myrcene, alpha-phellandrene, delta-3-carene, limonene, (E)-nerolidol, delta-2-carene, para-cymene, isoborneol, beta-caryophyllene, humulenol I, beta-eudesmol, 1H-cycloprop[e]azulen-4-ol, decahydro-1,1,4,7-tetramethyl-, [1ar-(1a.alpha.,4.beta.,4a.beta.,7.alpha.,7a.beta.,7b.alpha.)]-bicyclo[3.1.0]hexane-6-methanol, 2-hydroxy-1,4,4-trimethyl-, kauran-18-al, 17-(acetyloxy)-, (4.beta.)-, carveol, 2-methylenecholestan-3-ol, cycloheptane, 4-methylene-1-methyl-2-(2-methyl-1-propen-1-yl)-1-vinyl-, 2,6,6-trimethylundecan-1,3-dien-9-yn-5-one; *Zerumbone/Caryophyllene CT (Rhizomes)*—Not listed in study; *Curzerenone CT*—Alpha-pinene, beta-pinene, limonene, 2-nonanol, linalool, zingiberene, beta-caryophyllene, terpinen-4-ol, alpha-terpineol, alpha-selinene, beta-selinene, gamma-elemene, caryophyllene oxide, germacrone, spathulenol; *Leaves (Beta-Pinene CT)*—Camphene, myrcene, delta-3-carene, limonene, 1,8-cineole, (E)-beta-ocimene, perillene, terpinen-4-ol, gamma-elemene, (E)-beta-farnesene, beta-chamigrene, zingiberene, (Z)-trans-alpha-bergamotol; *Leaves (Zerumbone CT)*—Tricyclene, alpha-pinene, delta-3-carene, limonene, camphor, borneol, terpinen-4-ol, cycloheptane, 4-methylene-1-methyl-2-(2-methyl-1-propen-1-1-yl)-1-vinyl, 1,6,10-dodecatrien-3-ol, 3,7,11-trimethyl,-[S-(z)], caryophyllene oxide, azulene 1,2,3,4,5,6,7,8-octahydro-1,4-dimethyl-7-(1-methylethylidene), -(IS-cis)-, 2,6-dimethyl bicyclo [3,2,1]octane, 7-octylidenebicyclo [4.1.0] heptan-, 1,5-cycloundecadien, 8,8-dimethyl-9-methylene, 3-isopropyltricyclo [4.3.1.1] (2,5) undec-3-en-l0-ol, beta-eudesmol, agarospirol, trans-longipinene, 3a,9-dimethyldodecahydrocyclohepta[d]inden-3-one; *Flowers*—Alpha-pinene, camphene, beta-pinene, delta-3-carene, 1,8-cineole, beta-elemene, (E)-beta-farnesene, beta-chamigrene, zingiberene, (Z)-trans-alpha-bergamotol

Preferred Composition for Clinical Use:

Constituent	
Zerumbone	50%–80%
Alpha-Humulene	8%–20%
Camphene	4%–14%
Caryophyllene Oxide	tr–6%
1,8-Cineole	0.5%–5%
Camphor	tr–5%

Reported Therapeutic Properties: Reduces cholesterol levels, reduces high blood pressure, antispasmodic, **analgesic**, relieves headache, **anti-inflammatory**, anticonvulsive, antineuralgic, relieves sprains, antiarthritic, antirheumatic, relieves toothache, reduces fever, expels excess gas, relieves diarrhea and constipation, eases stomachache, antiparasitic, antiviral, antimicrobial, diuretic, eases cough, aids respiratory function, soothes skin itching and chronic skin conditions, wound healing, heals bruises, anticancer, antiallergic, modulates immune function, insecticidal, insect repellent, sedating, promotes feelings of stability and courage

Cautions:

Zerumbone CTs

■ May interact with aspirin, blood pressure, antiplatelet, and anticoagulant medications, and increase the risk of bleeding among people with bleeding disorders based on the strong inhibition of human platelet aggregation by zerumbone.[321]

■ May interfere with the enzymes responsible for metabolizing medications (NSAIDs, proton-pump inhibitors, acetaminophen, antiepileptics, immune modulators, blood-sugar medications, blood pressure medications, antidepressants, antipsychotics, diabetic medications, antihistamines, antibiotics, and anesthetics) due to zerumbone's ability to induce CYP450 in rats.[322]

Curzerenone CT

■ Avoid with epilepsy and Parkinson's disease due to camphor content. Camphor may exacerbate or cause seizures/convulsions or reduce seizure-medication efficacy.[323,324,325,326]

■ The camphor content in bitter ginger may negatively impact red blood cells and increase the risk of jaundice in children with glucose-6-phosphate dehydrogenase deficiency (G6PD).[327,328]

Leaves (Beta-Pinene CT) and Flowers

■ May interfere with antibiotics and increase their effectiveness ((E)-nerolidol sensitizes bacteria to antibiotics).[329]

Selected Evidence:

- □ Oral administration of bitter ginger CT zerumbone essential oil (50, 100, 200, or 300 mg/kg body weight) significantly and dose dependently reduced pain perception possibly by inhibiting the glutamatergic system, TRPV1 receptors as well as through activation of l-arginine/nitric oxide/cGMP/protein kinase C/ATP-sensitive K(+) channel pathway. Similar doses injected intraperitoneally produced similar reduction in pain.[330] Another study found that intraperitoneal administration (30, 100, or 300 mg/kg body weight) significantly reduced pain comparable to acetylsalicylic acid, reducing both neuropathic and inflammatory related pain.[331] A third study reported that oral administration (100 and 300 mg/kg body weight) of bitter ginger oil effectively reduces pain possibly by interacting with opioid receptors.[332]

- Intraperitoneal injection of 50, 100, 200, or 300 mg/kg body weight of bitter ginger CT zerumbone essential oil reduced inflammation in both acute and chronic animal models.[333] The same research showed that the oil reduced inflammatory- and noninflammatory-mediated pain.
- *In vitro* research showed that the zerumbone CT of bitter ginger rhizome oil inhibits *Cryptococcus neoformans*.[334] Zerumbone was ineffective against *C. neoformans*.[335] The same oil showed moderate antimalarial activity against chloroquine-sensitive and –resistant *Plasmodium falciparum*; whereas zerumbone showed good activity.
- Immersion in bitter ginger and eucalyptus (*E. globulus*) essential oils prevented hatching of head lice (*Pediculus humanus capitis*) eggs for seven to fourteen days, which significantly exceeded the efficacy of permethrin.[336]
- Inhalation of bitter ginger CT zerumbone essential oil increased food consumption and body weight in rats, which was attributed to a decrease in sympathetic nervous system activity.[337]
- The zerumbone CT of bitter ginger rhizome essential oil was moderately effective against *L. donovani*, with isolated zerumbone being more effective.[338] Another study reported that the zerumbone CT was moderately active against *Leishmania mexicana*.[339]
- Bitter ginger rhizome oil CT zerumbone exhibited antioxidant capacity comparable to ascorbic acid in the DPPH assay.[340] The zerumbone/caryophyllene CT with terpinen-4-ol exhibited significant antioxidant activity in the DPPH assay.[341]
- A 10% dilution of bitter ginger essential oil provided biting protection of 3.1 hours against *Aedes aegypti* and *Culex quinquefasciatus* mosquitoes.[342] A bitter ginger essential oil with equal amounts of zerumbone and alpha-humulene (31.9% and 31.7% respectively) was toxic against both pyrethroid-susceptible and resistant *A. aegypti* mosquitoes.[343]
- Bitter ginger CT zerumbone essential oil displayed contact toxicity against cigar beetle (*L. serricorne*) adults, with alpha-humulene demonstrating stronger contact toxicity and repellent activity than the whole oil.[344] Bitter ginger essential oil also reduced feeding by *Sitophilus zeamais* and *Tribolium castaneum*.[345]

BITTER LEAF (Grawa, Onugbu, Efo Ewuro)

Vernonia amygdalina Del., *V. randii* S.Moore, *V. giorgii* De Wild, *V. vogeliana* Benth., *V. weisseana* Muschl.

FAMILY: Asteraceae
NOTE: Top-Middle
AROMA INTENSITY: Medium
AROMA: Fresh, medicinal, herbaceous
COMMON EXTRACTION METHOD: Steam distilled from the leaves; may also be distilled from the aerial parts or stems
POSSIBLE SUBSTITUTE OILS: Myrtle, helichrysum, thyme
BLENDS WELL WITH: Bergamot, black pepper, cajeput, cedarwood, clary sage, clove, eucalyptus, frankincense, ginger, lavender, lemon, lemongrass, lime, melissa, myrrh, neroli, oregano, peppermint, rosemary, thyme, white fir, ylang ylang
SUBCELLULAR LOCALIZATION | EPIGENETIC INFLUENCE: Currently unknown | Currently unknown
RECOMMENDED DILUTION RANGE: 3%–33%; 50% for some conditions

PRIMARY CONSTITUENTS:[346,347,348,349,350]

Leaves (1,8-Cineole CT)		*Leaves (Thymol CT)*	
1,8-Cineole	25.1%–26.6%	Thymol	27.0%
Beta-Pinene	15.3%–15.5%	(E)-Phytol	15.7%
Myrtenal	6.5%–6.6%	O-Cymene	12.7%
Trans-Pinocarveol	0.0%–6.2%	Beta-Selinene	8.1%
Pinanol	0.0%–6.0%	Gamma-Terpinene	4.4%
Terpinen-4-ol	0.0%–5.2%	Beta-Caryophyllene	3.9%
Alpha-Pinene	4.9%–5.1%	Apiole	3.8%
Unidentified Compound	0.0%–4.9%		
L-Carveol	0.0%–4.4%	*Stems*	
Linalool	0.0%–4.3%	Germacrene D	27.5%
Cryptone	0.0%–3.9%	Eugenol	18.3%
Myrtenol	2.8%–3.5%	1-Octen-3-ol	13.9%
Pinocarvone	3.1%–3.4%	Beta-Caryophyllene	5.9%
		Palmitic Acid	5.9%
Aerial Parts		Cis-Beta-Terpineol	4.0%
Alpha-Muurolol	45.7%	3-Octanol	3.9%
Terpinen-4-ol	3.9%	(E)-Beta-Ocimene	3.8%
Gamma-Muurolene	3.2%	(E,E)-Alpha-Farnesene	3.6%
Isophorone	3.1%	Cedrene	3.0%

OTHER CONSTITUENTS: *Leaves (1,8-Cineole CT)*—Sabinene, beta-phellandrene, gamma-terpinene, verbenene, ascaridol, nopinone, thujene alcohol, cuminal, phellandral; *Leaves (Thymol CT)*—Alpha-pinene, linalool, 2,6,10,14-tetramethyl hexadecane, 6,10,14-trimethyl-2-pentadecanone, isophytol; *Aerial Parts*—Limonene, cis-sabinene hydrate, undecane, linalool, trans-sabinene hydrate, cis-p-menth-2-en-ol, trans-p-menth-2-ol, camphor, dodecane, tetradecane, beta-caryophyllene, beta-selinene, epi-cubebol, cis-cadina-1(2),4-diene, alpha-muurolene, trans-gamma-cadinene, delta-cadinene, trans-calamenene, caryophyllene oxide, cubenol alpha-eudesmol; *Stems*—Nerolidol, phytol, 3.4-dimethyl-1-hexene

PREFERRED COMPOSITION FOR CLINICAL USE:

Insufficient samples/data to determine.

REPORTED THERAPEUTIC PROPERTIES: Antiparasitic, antibacterial, antifungal, antimalarial, antiviral, boosts immune function, anticancer, balances blood-sugar levels, anti-inflammatory, analgesic, relieves headache, reduces fever, laxative, eases nausea, reduces loss of appetite, soothes stomachache, eases cough, enhances female fertility, promotes oral health, insect repellent, insecticidal, aids concentration, antidepressant, relieves anxiety, improves self-confidence

CAUTIONS:

Leaves (1,8-Cineole CT)

■ Avoid with epilepsy and Parkinson's disease due to 1,8-cineole content. May exacerbate or cause seizures/convulsions or reduce seizure-medication efficacy based on 1,8-cineole content.[351,352,353,354]

■ May weakly interfere with the enzymes responsible for metabolizing medications (NSAIDs, proton-pump inhibitors, acetaminophen, antiepileptics, immune modulators, blood-sugar medications, blood pressure medications, antidepressants, antipsychotics, diabetic medications, antihistamines, antibiotics, and anesthetics) due to 1,8-cineole content.[355,356,357]

■ May interfere with pentobarbital and other barbiturates (medications for anxiety and insomnia), based on 1,8-cineole content.[358,359]

Leaves (Thymol CT)

- May interact with aspirin, blood pressure, antiplatelet, and anticoagulant medications, and increase the risk of bleeding among people with bleeding disorders, due to thymol content.[360,361]
- May interact with antibiotics or antifungals and possibly enhance their effects based on thymol content.[362,363,,364,365,366,367,368]
- Avoid in individuals with Refsum disease (a neurological disorder caused by the overaccumulation of phytanic acid in cells and tissues) or who are AMACR-deficient (alpha-methylacyl-CoA racemase-deficient) due to (E)-phytol content. It is believed that humans can convert phytol to phytanic acid.[369] Animal research suggests that phytol can cause liver inflammation, liver failure, and death in AMCAR-deficient mice.[370,371] AMACR is an enzyme found in the mitochondria that plays a role in the breakdown of fatty acids. AMACR deficiency causes neurological problems (seizures, cognitive decline, migraines, encephalopathy, loss of sensation due to nerve damage, muscle stiffness, lack of coordination, and vision problems) that begin in adulthood and progressively get worse.[372] Phytol may also weakly promote tumor formation according to animal research.[373]
- May irritate mucous membranes (eyes, mouth, nasal passages, vagina, rectum).

Stems, Aerial Parts

- None currently known.

SELECTED EVIDENCE:

- □ The stem essential oil from bitter leaf weakly inhibited oral pathogens: periodontopathic *Porphyromonas gingivalis* and *Aggregatibacter actinomycetemcomitans* and cariogenic *Streptococcus mutans*.[374]
- □ Bitter leaf essential oil CT 1,8-cineole was highly toxic to and significantly repelled the maize weevil (*Sitophilus zeamais*) without damaging the corn grains treated with the oil.[375] A follow up study determined that its toxicity was due to the combined effects of its constituents.[376]

BLACKCURRANT | BLACKCURRANT ABSOLUTE (Cassis)

Ribes nigrum L.

FAMILY: Grossulariaceae
NOTE: Base-Middle
AROMA INTENSITY: Strong
AROMA: Green, fruity, woody, slightly earthy and citrusy
COMMON EXTRACTION METHOD: Steam distilled or hydrodistilled from the flower buds, may also be distilled from the leaves or berries; also solvent extracted from the flower buds
POSSIBLE SUBSTITUTE OILS: Pink pepper, black pepper CT delta-3-carene, nutmeg, cypress
BLENDS WELL WITH: Bergamot, cilantro, cistus, coriander, galbanum, grapefruit, jasmine, lavandin, lavender, lemon, lime, orange, neroli, rose, sandalwood, spearmint, tangerine, tarragon, ylang ylang
SUBCELLULAR LOCALIZATION | EPIGENETIC INFLUENCE: Currently unknown | Currently unknown
RECOMMENDED DILUTION RANGE: 3%–Neat

PRIMARY CONSTITUENTS:[377,378,379,380,381,382,383,384,385,386,387]

Buds (HD, Delta-3-Carene CT)

Delta-3-Carene	25.4%–62.9%
Terpinolene	9.2%–25.9%
Beta-Phellandrene	0.3%–17.3%
Sabinene	0.2%–14.6%
Limonene	0.0%–12.8%
Beta-Caryophyllene	0.1%–10.5%
Alpha-Pinene	0.0%–5.4%
Alpha-Humulene	2.2%–5.2%
Beta-Pinene	0.0%–4.6%
Gamma-Terpinene	0.3%–4.1%
(Z)-Beta-Ocimene	0.6%–3.8%
Myrcene	0.0%–3.6%
Cis-Sabinene Hydrate	0.0%–3.4%
Terpinen-4-ol	tr–3.1%

Buds (HD, Sabinene CT)

Sabinene	39.8%–64.2%
(Z)-Beta-Ocimene	1.5%–10.1%
Beta-Caryophyllene	0.0%–9.6%
(E)-Beta-Ocimene	0.8%–8.1%
Germacrene B	0.0%–5.7%
Alpha-Pinene	1.8%–4.3%
Caryophyllene Oxide	0.0%–4.1%
Beta-Phellandrene	0.8%–4.0%
Alpha-Humulene	1.4%–3.8%
Terpinen-4-ol	tr–3.8%
Beta-Pinene	0.0%–3.8%
Myrcene	0.0%–3.7%
Alpha-Terpineol	0.0%–3.5%
Aromadendrene Oxide	0.0%–3.5%

Buds (HD, Sabinene/Delta-3-Carene CT)

Sabinene	17.5%–55.0%
Delta-3-Carene	13.0%–28.5%
Beta-Caryophyllene	tr–9.3%
Terpinolene	4.9%–9.1%
Terpinen-4-ol	1.2%–6.6%
Gamma-Terpinene	0.0%–5.5%
Alpha-Humulene	0.2%–4.1%
Beta-Phellandrene	1.4%–3.7%
Myrcene	0.7%–3.6%
Alpha-Terpinene	0.0%–3.6%
Beta-Pinene	0.5%–3.1%

Buds (SDSE, Delta-3-Carene CT)

Delta-3-Carene	20.2%–35.2%
Beta-Phellandrene	1.0%–11.3%
Terpinolene	10.2%–23.2%
Beta-Caryophyllene	4.5%–8.3%
(Z)-Beta-Ocimene*	2.0%–6.7%
(+)-Limonene	4.9%–5.0%
Germacrene B	0.2%–4.1%
Germacrene D	0.0%–3.3%
Beta-Pinene	0.2%–3.1%

Absolute

Delta-3-Carene	13.0%–19.0%
UI Sesquiterpenes	4.1%–16.5%
Beta-Caryophyllene	9.0%–14.0%
Terpinolene	3.9%–10.9%
Terpinen-4-ol	2.4%–6.3%
Para-Cymene	1.9%–6.3%
Limonene	3.7%–4.9%
Myrcene	2.2%–3.8%
Sabinene	1.8%–3.7%

Buds (SDSE, Sabinene CT)

Sabinene	32.0%–47.5%
Beta-Caryophyllene	4.9%–18.5%
Terpinen-4-ol	5.1%–12.0%
Terpinolene	0.8%–11.5%
(Z)-Beta-Ocimene	0.7%–8.9%
Gamma-Terpinene	2.5%–8.8%
Beta-Phellandrene	0.8%–8.0%
Alpha-Humulene	2.0%–4.9%
Germacrene D	1.2%–3.9%
Beta-Pinene	1.7%–3.2%
Limonene	2.4%–3.0%

Leaves (SDSE)

Delta-3-Carene	25.4%
(Z)-Beta-Ocimene	11.3%
Caryophyllene Oxide	9.8%
Terpinolene	7.7%
(E)-Beta-Ocimene[#]	7.7%
(E)-Beta-Caryophyllene	4.7%

Berries (SDSE)

Delta-3-Carene	20.2%–26.9%
(E)-Beta-Caryophyllene	4.6%–9.3%
Terpinolene	6.8%–8.7%
Beta-Phellandrene	5.4%–6.7%
(Z)-Beta-Ocimene*	2.1%–5.5%
Para-Cymenene	3.5%–4.8%
(+)-Limonene	3.2%–3.9%
Gamma-Elemene	1.1%–3.8%
Germacrene B	1.6%–3.0%

Leaves (HD, Beta-Caryophyllene CT)

Beta-Caryophyllene	12.9%
Hexadecanoic Acid	10.7%
Alpha-Humulene	10.2%
Caryophyllene Oxide	9.6%
Myristic Acid	7.2%
Methyl Oleate	7.2%
Humulene Epoxide II	6.5%
2-Phenethyl Benzoate	5.3%
Caryophyllenol II	3.4%
Clovenol	3.0%

Leaves (HD, Delta-3-Carene CT)

Delta-3-Carene	18.7%
Beta-Caryophyllene	17.7%
Sabinene	11.6%
Terpinolene	10.6%
(Z)-Beta-Ocimene	10.6%
(E)-Beta-Ocimene	6.9%
Germacrene D	4.3%

* Co-eluting with (–)-limonene.

Co-eluting with (–)-beta-phellandrene.

HD=Hydrodistilled. SDSE=Simultaneous distillation/solvent extraction.

OTHER CONSTITUENTS: *Buds (HD, Delta-3-Carene CT)*—Alpha-phellandrene, alpha-terpinene, (E)-beta-ocimene, thujol, alpha-terpineol, germacrene D, alpha-selinene, germacrene-D-4-ol, T-cadinol, caryophyllene oxide; *Buds (HD, Sabinene CT)*—Alpha-thujene, alpha-phellandrene, delta-3-carene, alpha-terpinene, para-cymene, limonene, cis-sabinene hydrate, terpinolene trans-sabinene hydrate, trans-piperitol, bornyl acetate, alpha-selinene, delta-cadinene, T, cadinol, alpha-cadinol; *Buds (SDSE, Delta-3-Carene CT)*—Alpha-pinene, sabinene, myrcene, alpha-phellandrene, (E)-beta-ocimene, alpha-terpinene, citronellyl acetate, alpha-humulene, caryophyllene oxide, spathulenol, cis-1-methyl-4-(1-methethyl)-2-cyclohexen-1-ol, p-mentha-1,4-dien-8-ol, 1,9-decadiyne, alpha-terpinyl acetate, 4-phenyl-1,3-tiazole, aromadendrene oxide, T-cadinol; *Absolute*—Alpha-pinene, alpha-terpinene, (Z)-beta-ocimene, gamma-terpinene, (E)-beta-ocimene, 3-octanone, unidentified monoterpene, alpha-copaene, beta-elemene; *Buds (SDSE, Sabinene CT)*—Alpha-thujene, alpha-pinene, (E)-beta-ocimene, spathulenol, alpha-phellandrene, para-cymene, cis-sabinene hydrate, trans-sabinene hydrate, trans-1-methyl-4-(1-methethyl)-2-cyclohexen-1-ol, cis-1-methyl-4-(1-methethyl)-2-cyclohexen-1-ol, para-cymen-8-ol, bornyl acetate, alpha-terpinyl acetate, citronellyl acetate, beta-elemene, beta-selinene, germacrene B, 4-phenyl-1,3-tiazole, caryophyllene oxide, T-cadinol, alpha-cadinol; *Buds (HD, Sabinene/Delta-3-Carene CT)*—Alpha-thujene, alpha-pinene, para-cymene, (Z)-beta-ocimene, (E)-beta-ocimene, cis-sabinene hydrate, alpha-terpineol, citronellyl acetate, germacrene D, alpha-selinene, aromadendrene oxide; *Berries (SDSE)*—Alpha-pinene, n-heptanal, sabinene, myrcene, alpha-phellandrene, alpha-terpinene, para-cymene, gamma-terpinene, 1,8-cineole, 1-octen-3-ol, methyl chavicol (< 1.3%), terpinen-4-ol, 2-decanol, myrtenal, 1-para-menthen-9-al, alpha-terpineol, isoborneol, bornyl acetate, nerol, citronellyl acetate, perillaldehyde, ethyl decanoate, geranyl acetate, alpha-humulene, bicyclogermacrene, n-hexadecane, decanoic acid, n-heptadecane, caryophyllene oxide, n-octadecane, n-butyl cinnamate, n-nonadecane, n-heneicosane, palmitic acid; *Leaves (HD, Beta-Caryophyllene CT)*—Sabinene, gamma-terpinene, beta-selinene, humulene epoxide I, humulene epoxide II, hexahydrofarnesyl acetone, decanoic acid, caryophyllenol I, pentacosane, dodecanoic acid, ethyl linoleate, tridecanoic acid, phytol, pentadecanoic acid; *Leaves (SDSE)*—Alpha-pinene, sabinene, myrcene, beta-pinene, alpha-terpinene, limonene, para-cymene, beta-phellandrene, para-cymene, bornyl acetate, alpha-humulene, delta-cadinene, germacrene B, n-butyl cinnamate, n-nonadecane; *Leaves (HD, Delta-3-Carene CT)*—1-octen-3-ol, myrcene, alpha-terpinene, beta-phellandrene, geraniol, beta-elemene, alpha-humulene, bicyclogermacrene, caryophyllene oxide, phytol

PREFERRED COMPOSITION FOR CLINICAL USE:
Too great of variance in samples/data to determine.

REPORTED THERAPEUTIC PROPERTIES: Analgesic, anti-inflammatory, antirheumatic, antineuralgic, nervine, relieves headache, antiallergic, diuretic, antiparasitic, **antimicrobial**, antibacterial, **antifungal**, antiviral, aids immune function, promotes balanced menstruation and relieves PMS symptoms, antioxidant, relieves skin itching and chronic skin conditions, strengthens the hair, aids cardiovascular health, supports respiratory function, antiaging, reduces fatigue, promotes mental clarity, antidepressant, relaxing

CAUTIONS:
- May interact with anticholinergic (drugs used for asthma, incontinence, gastrointestinal cramps, muscular spasms, depression, and sleep disorders) and cholinergic medications (drugs used to reduce urinary retention, diagnose myasthenia gravis, and for glaucoma), based on AChE-inhibtory activity of delta-3-carene.[388,389]
- *Absolute:* Avoid oral consumption. Absolutes contain solvent residue and should not be taken orally.

SELECTED EVIDENCE:
- *In vitro* research shows that blackcurrant bud (HD, delta-3-carene CT) exhibits broad-spectrum antibacterial activity against *A. baumanii*, *E. coli*, *P. aeruginosa*, and *S. aureus*.[390] Both the delta-3-carene and the sabinene/delta-3-carene CTs (HD) of blackcurrant essential oil inhibited *E. coli, S. typhimurium, S. enteritidis, P. aeruginosa, P. tolaasii, P. mirabilis, E. cloacae, S. aureus, B. cereus, M. luteus, M. flavus, S. lutea, L. monocytogenes, F. solani, A. flavus, F. verticillioides, A. alternata,* and *T. viride*.[391] Fungi were more sensitive to the oils than bacteria. The leaf oil (HD) inhibited—in order of effectiveness—*E. coli, S. faecalis, S. aureus, C. albicans, T. mentagrophytes, E. floccosum, B. subtilis, S. typhimurium, P. aeruginosa, P. mirabilis, M. luteus, P. tolaasii, M. flavus*, and *L. monocytogenes*.[392]

BLACK POPLAR (Lombardy Poplar)

Populus nigra L., *P. dilatata* Aiton, *P. italica* (Du Roi) Moench, *P. nigra* L. var. *italica* Du Roi, *P. nigra Arn. subsp. pallasiana (Lamb.) Holmboe*

FAMILY: Salicaceae
NOTE: Base
AROMA INTENSITY: Medium-Strong
AROMA: Sweet, spicy, fruity, balsamic, floral, resinous, woody
COMMON EXTRACTION METHOD: Hydrodistilled from the leaf buds or needles; may also be solvent extracted from the leaf buds

POSSIBLE SUBSTITUTE OILS: Copaiba, helichrysum, ylang ylang
BLENDS WELL WITH: Balsam fir, cedarwood, chamomile (Roman), copaiba, cypress, Douglas fir, elemi, fennel, frankincense, geranium, ginger, grapefruit, ho wood, lavender, lemon, lime, palo santo, pine, rose, sandalwood, spruce (black), vetiver, ylang ylang, araucaria, ginger
SUBCELLULAR LOCALIZATION | EPIGENETIC INFLUENCE: Currently unknown | Currently unknown
RECOMMENDED DILUTION RANGE: 3%–50%; neat for some conditions

PRIMARY CONSTITUENTS:[393,394]

Beta-Eudesmol CT (Buds)

Constituent	%
Beta-Eudesmol	19.1%–19.6%
Alpha-Eudesmol	6.7%–9.6%
Gamma-Selinene	7.6%–8.8%
Delta-Cadinene	7.8%–8.6%
Alpha-Copaene-11-ol	1.4%–5.4%
Alpha-Elemene	3.3%–5.2%
Gamma-Cadinene	3.9%–4.2%
Eugenol	1.1%–3.9%
Delta-Cadinol	3.0%–3.6%
(E)-2-Methyl-2-Buten-1-ol	0.9%–3.5%

Ar-Curcumene CT (Buds)

Constituent	%
ar-Curcumene	27.2%
Gamma-Curcumene	9.5%
Delta-Cadinene	7.7%
Benzyl Alcohol	3.2%

Delta-Cadinene CT (Buds)

Constituent	%
Delta-Cadinene	12.0%
Gamma-Curcumene	9.3%
ar-Curcumene	8.9%
2-Phenylethanol	4.3%
Benzyl Alcohol	4.2%

Alpha-Eudesmol CT (Buds)

Constituent	%
Alpha-Eudesmol	22.2%–49.3%
Beta-Eudesmol	16.0%–19.3%
Gamma-Eudesmol	9.9%–14.4%
Bulnezol	0.0%–10.1%
Gamma-Curcumene	0.0%–3.3%

Prenyl Benzoate CT (Buds)

Constituent	%
Prenyl Benzoate	19.8%–21.6%
(E)-Nerolidol	0.3%–17.3%
Salicyl Benzoate	7.2%–12.7%
Gamma-Curcumene	2.3%–11.6%
ar-Curcumene	0.5%–9.1%
Benzyl Benzoate	3.7%–4.5%
Delta-Cadinene	0.3%–4.2%
Bulnezol	0.7%–3.9%

OTHER CONSTITUENTS: *Beta-Eudesmol CT (Buds)*—2-Methyl-3-buten-2-ol, 3-methyl-1-butanol, (E)-2-methyl-2-butenoic acid, geraniol, eugenol, alpha-copaene, alloaromadendrene, alpha-humulene, epizonarene, alpha-muurolene, 1,2,3,4-tetrahydro-1,6-dimethyl-4-isopropylnaphtalene, elemol, T-cadinol, torreyol, benzaldehyde, 1,2-cyclohexadecanedione, 1-phenylethanone, 2-methylhexanoic acid, salicyl aldehyde, benzyl alcohol, 2-H-isoindole, 2-phenyethanol, 1,2-dihydro-1,1,6-trimethylnaphtalene, 1-phenyl-2-propen-1-ol, phenol, octanoic acid, 3-phenyl-2-propen-1-ol acetate, 3-phenyl-2-propen-1-ol, docosane; *Ar-Curcumene CT (Buds)*—2-phenylethanol, unidentified aliphatic compound; *Delta-Cadinene CT (Buds)*—Eugenol, cis-beta-caryophyllene, unidentified aliphatic compound; *Alpha-Eudesmol CT (Buds)*—Delta-cadinene, cis-beta-caryophyllene; *Prenyl Benzoate CT (Buds)*—Benzyl alcohol, 2-phenylethanol, gamma-eudesmol, beta-eudesmol, alpha-eudesmol, unidentified aliphatic compound

PREFERRED COMPOSITION FOR CLINICAL USE:
Insufficient samples/data to determine.

REPORTED THERAPEUTIC PROPERTIES: Antibacterial, antifungal, antiviral, antiparasitic, antioxidant, anti-inflammatory, analgesic, stimulates immune function, diuretic, decreases blood pressure, cleanses the urinary tract, wound healing, heals bruises and hemorrhoids, soothes sunburn, relieves skin itching, promotes restful sleep, relieves anxiety, calms emotional chaos, comforting, promotes love and reconciliation

CAUTIONS:

Beta-Eudesmol CT (Buds) and Alpha-Eudesmol CT (Buds)

■ Avoid during pregnancy and lactation (low risk) due to beta-eudesmol content. Beta-eudesmol strongly inhibits the growth of new blood vessels, which is necessary for implantation, and placental and embryonic development.[395]

■ May interact with aspirin, blood pressure, antiplatelet, and anticoagulant medications, and increase the risk of bleeding among people with bleeding disorders based on beta-eudesmol content (low risk).[396,397]

Ar-Curcumene CT (Buds), Prenyl Benzoate CT (Buds), and Delta-Cadinene CT (Buds)

■ Oral caution. Ingestion of benzyl derivatives (benzoic acids, benzoic salts, benzaldehyde, benzyl acetate, benzyl benzoate, and benzyl alcohol) may cause gastrointestinal irritation, vomiting, and diarrhea. The Joint Expert Committee for Food Additives (JECFA) has established an acceptable daily limit for the aforementioned constituents of 0–5 mg/kg body weight for adults to avoid GI discomfort. Based on this limit, it is suggested to limit the oral consumption of orange bird berry essential oil rich in benzyl derivatives to no more than 10 drops daily.

SELECTED EVIDENCE:

- □ Black polar needle (subsp. *pallasiana*) essential oil was scolicidal to the hydatid worm (Echinococcus granulosus), which causes echinococcosis when the eggs are swallowed or people come in contact with infected animals.[398]

BLACK TEA TREE
(River Tea Tree, Mock Olive)

Melaleuca bracteata F.Muell.

FAMILY: Myrtaceae
NOTE: Middle
AROMA INTENSITY: Strong
AROMA: Spicy, woody, resinous
COMMON EXTRACTION METHOD: Steam distilled or hydrodistilled from the leaves
POSSIBLE SUBSTITUTE OILS: Huon pine, nutmeg,
BLENDS WELL WITH: Basil, bay laurel, bergamot, blue cypress, camphor, cassia, cedarwood, cinnamon, clary sage, coriander, cypress, frankincense, juniper, lavender, lemon, lime, marjoram, orange, palo santo, patchouli, pine, rosemary, sandalwood, Spanish sage, spikenard
SUBCELLULAR LOCALIZATION | EPIGENETIC INFLUENCE: Currently unknown | Currently unknown
RECOMMENDED DILUTION RANGE: *Methyl eugenol CT*—< 3%; *Other CTs*—3%–33%; 50% for some conditions

PRIMARY CONSTITUENTS:[399,400,401,402,403,404,405]

Methyl eugenol CT

Methyl eugenol	50.0%–97.7%
Alpha-Phellandrene	0.0%–12.6%
(E)-Methyl Cinnamate	0.0%–11.4%
Alpha-Pinene	0.0%–10.4%
Para-Cymene	0.0%–5.2%
Beta-Caryophyllene	0.0%–4.3%
Genipin	0.0%–3.2%
Terpinolene	0.0%–3.2%
Eugenol	0.0%–3.0%

(E)-Methyl Isoeugenol CT

(E)-Methyl Isoeugenol	87.7%
Methyl Eugenol	8.7%

Elemicin CT

Elemicin	57.4%
Beta-Caryophyllene	21.4%
(E)-Isoelemicin	5.6%
Alpha-Phellandrene	3.5%

(E)-Isoelemicin CT

(E)-Isoelemicin	76.5%
Elemicin	15.2%

OTHER CONSTITUENTS: *Methyl eugenol CT*—Beta-pinene, alpha-terpinene, limonene, beta-phellandrene, 1,8-cineole, camphor, gamma-terpinene, (E)-beta-ocimene, terpinen-4-ol, linalool, alpha-terpineol, alpha-humulene, germacrene D, carvacrol, undecanol, gamma-eudesmol, decane, 3,4,5-trimethoxybenzoic acid, methyl ester; *Elemicin CT*—Terpinolene, alpha-humulene, germacrene D, bicyclogermacrene; *(E)-Isoelemicin CT*—Germacrene D, (E)-methyl isoeugenol, alpha-eudesmol, beta-eudesmol, (Z)-isoelemicin; *(E)-Methyl Isoeugenol CT*—Linalool, bicyclogermacrene, (Z)-methyl isoeugenol, (E)-isoelemicin

PREFERRED COMPOSITION FOR CLINICAL USE:
The most common CT appears to be the methyl eugenol CT, which is not recommended for therapy.

REPORTED THERAPEUTIC PROPERTIES: Antiseptic, antifungal, antibacterial, antimicrobial, heals stomach ulcers, antioxidant, anti-inflammatory, analgesic, relieves painful menstruation, aids circulation, insecticidal, antidepressant, relieves anxiety, balances the emotions, reduces grief and worrisome thoughts, aphrodisiac

CAUTIONS:
The methyl eugenol CT of black tea tree essential oil is not recommended for essential oil therapy due to significant methyl eugenol content.

- Avoid in children under age 12 due to the multisite cancerous tumors and DNA mutation caused by methyl eugenol in animal studies.[406,407,408,409]
- Avoid during pregnancy and lactation due to changes in embryo form and structure and chromosomal changes in ovary cells caused by methyl eugenol as well as the carcinogenic (cancer-causing) potential of methyl eugenol.[410,411,412,413]
- Avoid oral consumption due to the multisite carcinogenic potential of methyl eugenol.[414,415,416,417]
- May interfere with antibiotics and antifungal medications and increase their effectiveness due to methyl eugenol content.[418]
- May interact with anticholinergic (drugs used for asthma, incontinence, gastrointestinal cramps, muscular spasms, depression, and sleep disorders) and cholinergic medications (drugs used to reduce urinary retention, diagnose myasthenia gravis, and for glaucoma), based on the potent AChE-inhibtory activity of methyl eugenol.[419]

Elemicin CT and (E)-Isoelemicin CT

■ May interact with aspirin, blood pressure, antiplatelet, and anticoagulant medications, and increase the risk of bleeding among people with bleeding disorders, due to the weak platelet aggregation inhibiting effects of elemicin.[420]

(E)-Methyl Isoeugenol CT

■ May interfere with pentobarbital and other barbiturates (medications for anxiety and insomnia), based on (E)-methyl isoeugenol content—(E)-methyl isoeugenol potentiated the hypnotic effect of sodium pentobarbital in mice [421]

SELECTED EVIDENCE:

- *In vitro* research demonstrated that black tea tree CT methyl eugenol essential oil exhibited good activity against *Salmonella typhimurium, Staphylococcus epidermidis*, and moderate activity against *Bacillus subtilis, Staphylococcus aureus, Klebsiella pneumonia,* and *Streptococcus mutans*.[422] Another study reported that the oil (CT methyl eugenol) strongly inhibited *S. marcescens* (two species), *P. aeruginosa, C. violaceum,* and *S. aureus*, sometimes exceeding the efficacy of the positive control ampicillin.[423] The methyl eugenol CT also inhibited common foodborne pathogens, (*B. spizizenii, S. aureus, E. aerogenes, E. coli, K. pneumoniae, P. aeruginosa*, and *S. enterica*).[424]
- Black tea tree CT methyl eugenol essential oil exhibited excellent antioxidant activity in the DPPH (exceeding BHT) assay and good activity in the ABTS assay.[425] Another study reported that the methyl eugenol CT of black tea tree essential oil displayed strong antioxidant activity in the DPPH and reducing power assays.[426]
- Black tea tree CT methyl eugenol essential oil showed good insecticidal activities against the red flour beetle (*Tribolium castaneum*) when compared with pyrethroid as a positive control.[427]

BLACK TURMERIC (Black Zedoary)

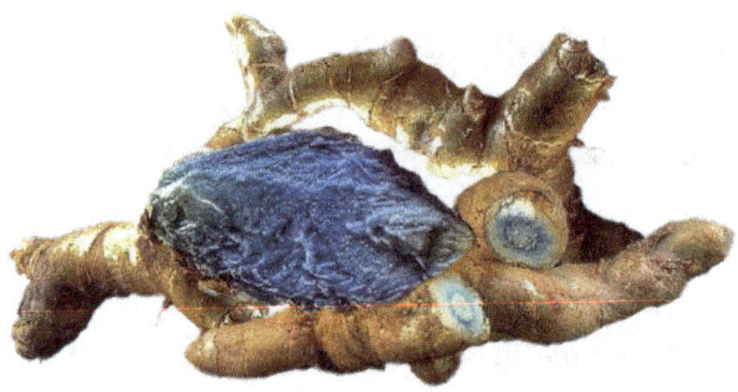

Curcuma caesia Roxb., *C. kuchoor* Royle

FAMILY: Zingiberaceae
NOTE: Base-Middle
AROMA INTENSITY: Medium
AROMA: Earthy, camphoraceous, medicinal, woody, slightly spicy
COMMON EXTRACTION METHOD: Steam distilled or hydrodistilled from the rhizomes; may also be distilled from the leaves
POSSIBLE SUBSTITUTE OILS: Turmeric, white turmeric
BLENDS WELL WITH: Cinnamon, cistus, clary sage, clove, ginger, helichrysum, juniper, lavender, nutmeg, turmeric, ylang ylang
SUBCELLULAR LOCALIZATION | EPIGENETIC INFLUENCE: Currently unknown | Currently unknown
RECOMMENDED DILUTION RANGE: 3%–33%, 50% for some conditions

PRIMARY CONSTITUENTS:[428,429,430,431,432]

Camphor CT		*1,8-Cineole CT*	
Camphor	28.3%	1,8-Cineole	30.1%
ar-Turmerone	12.3%	Camphor	15.2%

(Z)-Beta-Ocimene	8.2%
ar-Curcumene	6.8%
1,8-Cineole	5.3%
Beta-Elemene	4.8%
Borneol	4.4%
Bornyl Acetate	3.3%
Tropolone CT	
Tropolone	15.9%
(Z)-Alpha-Copaene-8-ol	5.5%
6-Methyl-2(1H)-Pteridinone	4.9%
4-(Dimethylamino)-,3,5-Dimethyl-Phenol	4.5%
Bicyclo[3.1.0]Hexane-3one	3.8%
1,2,3,5,6,7,8,8a-octa-hydro-napthalene-2ol Megastigmatrienone	3.4%
4-Dimethyamino-Benzoic Acid	3.4%
Ledol	3.3%
Alpha-Bulnesene	3.0%
Beta-Elemenone	3.0%

ar-Curcumene	14.8%
Camphene	8.2%
Dihydrocarveol	3.6%
Curcumenol	3.1%
Leaves	
1,8-Cineole	27.0%
Camphor	16.8%
Borneol	8.7%
Beta-Pinene	6.3%
Alpha-Terpineol	5.2%

Note: Older and unavailable reports also list delta-linalool (20.4%), camphene (10.5%), and 1,8-cineole (30.4%) with camphor (10.8%), curzerene (8.8%), and curzerenone (5.8%) as major constituents.[433,434,435]

Other Constituents: *Camphor CT*—Camphene, para-cymene, 5-nonanone, alpha-terpinolene, linalool, endo-fenchol, menthone, terpinen-4-ol, endo-fenchyl acetate, beta-cubebene, (Z,E)-alpha-farnesene, beta-caryophyllene, gamma-elemene, gamma-curcumene, elemol, germacrene B; *1,8-Cineole CT*—Alpha-pinene, beta-pinene, limonene, isoborneol, borneol, delta-elemene, beta-farnesene, germacrene D, zingiberene, germacrene B, curzerenone, germacrone; *Tropolone CT*—1,8-cineole, camphor, isoborneol, borneol, alpha-terpineol, isobornyl acetate, beta-elemene, caryophyllene, alpha-selinene, delta-cadinene, caryophyllene oxide, gamma-cadinene, rosifoliol, beta-selinene, beta-guaiene, globulol, ar-turmerone, spathulenol, 2,7-dimethyl oxepine, occidentalol, 1,3,3-trimethyl-2-cyclohexanol, (3-methyl-2-methylene-3-butenylidene)-(2Z)6-isopropenyl-4,8a-dimethyl-, 6-Isopropylidene-bicyclo[3.1.0]hexane, delta1(9)-2-Octalone; *Leaves*—Alpha-pinene, myrcene, limonene, linalool, beta-elemene, beta-caryophyllene, methyl eugenol (< 1.6%), eugenol

Preferred Composition for Clinical Use:
Too great of variance in data/samples composition to determine.

Reported Therapeutic Properties: Aids digestion, expels excess gas, heals stomach ulcers, relieves diarrhea, relieves toothache, analgesic, anti-inflammatory, antirheumatic, relieves sprains, antispasmodic, antiallergic, supports respiratory function, heals bruises, wound healing, soothes chronic skin conditions, reduces the appearance of scars and blemishes, aids circulation, antifungal, antibacterial, antioxidant, anticancer, aids kidney function, stimulating, stabilizes the emotions, improves mental clarity, promotes concentration and focus, aids cognition

CAUTIONS:

Camphor CT

■ Avoid with children under six due to significant camphor content and moderate 1,8-cineole content. Several cases of camphor poisoning and/or seizures from ingestion and topical application have been reported in children.[436,437] Ingestion of camphor-containing products has been lethal in children under age two.[438] Children five years and up may use camphor-containing essential oils topically in dilutions no stronger than 5%. Camphor ingestion by infants and young children may cause cough, vomiting, seizure, burning sensation in the mucous membranes and eyes, or lack of voluntary coordination of muscle movements.[439]

■ Caution is warranted during pregnancy and while lactating, due to significant camphor content. Ingestion of essential oils with significant levels of camphor may lead to abortion because fetuses lack the enzymes to process it.[440] Camphor ingestion by infants and young children may cause cough, vomiting, seizure, burning sensation in the mucous membranes and eyes, or lack of voluntary coordination of muscle movements.[441]

■ Avoid oral consumption due to significant camphor levels. Essential oils with significant levels of camphor may be toxic when taken orally, usually single doses exceeding 2 mL, although the lethal dose for humans is estimated to be greater than 5 mL in a single dose.[442,443,444]

■ Avoid with epilepsy and Parkinson's disease due to the potential of camphor and 1,8-cineole to exacerbate or cause seizures/convulsions or reduce seizure-medication efficacy.[445,446,447,448]

■ The significant camphor content may negatively impact red blood cells and increase the risk of jaundice in children with glucose-6-phosphate dehydrogenase deficiency (G6PD).[449,450]

1,8-Cineole CTs

■ Avoid with epilepsy and Parkinson's disease due to 1,8-cineole and camphor content. May exacerbate or cause seizures/convulsions or reduce seizure-medication efficacy.[451,452,453,454]

■ May weakly interfere with the enzymes responsible for metabolizing medications (NSAIDs, proton-pump inhibitors, acetaminophen, antiepileptics, immune modulators, blood-sugar medications, blood pressure medications, antidepressants, antipsychotics, diabetic medications, antihistamines, antibiotics, and anesthetics) due to 1,8-cineole content.[455,456,457]

■ May interfere with pentobarbital and other barbiturates (medications for anxiety and insomnia), based on 1,8-cineole content.[458,459]

Tropolone CT

■ None currently known.

SELECTED EVIDENCE:

□ *In vitro* research showed that black turmeric essential oil (ar-curcumene as the major constituent) exhibited strong to moderate activity against *Bacillus subtilis, Corynebacterium diphtheriae, Staphylococcus aureus, Salmonella typhi Escherichia coli, Shigella, Aspergillus niger, A. fumigatus, Curvularia lunata, Fusarium psidi*, and *Rhizopus oryzae*.[460] Antifungal activity was also observed in a 1976 study.[461]

BOBONDI KABIR (Mokhaleseh)

Tanacetum polycephalum (L.) Schultz. Bip., *T. polycephalum* Schultz Bip. subsp. p*olycephalum*, *T. polycephalum* Schultz Bip. subsp. *farsicum* Podl., *T. polycephalum* Schultz Bip. subsp. *heterophyllum*, *T. polycephalum* subsp. *duderanum* (Boiss) Podl.

FAMILY: Asteraceae (Compositae)
NOTE: Middle
AROMA INTENSITY: Strong
AROMA: Fresh, medicinal, earthy
COMMON EXTRACTION METHOD: Steam distilled or hydrodistilled from the aerial parts; may also be distilled from the flowers, leaves, or stems
POSSIBLE SUBSTITUTE OILS: Mugwort, tansy, thuja, wormwood
BLENDS WELL WITH: Cedarwood, clary sage, lavandin, lavender, mugwort, patchouli, pine, rosemary, sage, sandalwood, Spanish sage, tansy
SUBCELLULAR LOCALIZATION | EPIGENETIC INFLUENCE: Currently unknown | Currently unknown
RECOMMENDED DILUTION RANGE: 1.5%–20%

PRIMARY CONSTITUENTS:[462,463,464,465,466,467,468,469,470,471,472,473]

Alpha-Thujone CT	
Alpha-Thujone	69.0%–82.3%
Beta-Thujone	5.9%–7.0%
Camphor	3.4%–7.0%
1,8-Cineole	4.5%–9.1%
Borneol	1.0%–3.5%

Camphor/1,8-Cineole CT	
Camphor	7.6%–75.0%
1,8-Cineole	6.6%–33.6%
(E)-Chrysanthenyl Acetate	0.0%–21.8%
Borneol	0.0%–16.7%
Pinocarvone	0.0%–15.3%
Camphene	1.2%–9.1%
(E)-Carveol	0.0%–9.1%
Trans-Isopulegone	0.0%–8.0%
Beta-Thujone	0.0%–7.9%
Alpha-Thujone	0.0%–6.1%
Chrysanthenone	0.0%–5.3%
Alpha-Pinene	0.2%–4.6%
(Z)-Chrysanthenyl Acetate	0.0%–4.5%
Beta-Eudesmol	0.0%–3.8%
Spathulenol	0.0%–3.1%

Beta-Thujone CT	
Beta-Thujone	46.5%–79.8%
Camphor	1.3%–16.3%
1,8-Cineole	8.3%–14.2%
Borneol	1.6%–8.9%

1,8-Cineole CT	
1,8-Cineole	43.4%
Borneol	15.2%
Filifolone	7.9%
Chrysanthenone	6.6%
Para-Cymene	4.0%

Myrtenol CT	
Myrtenol	35.4%
1,8-Cineole	11.9%
Unidentified Ester	8.6%
Alpha-Pinene	4.6%
Pinocarvone	4.0%
Beta-Eudesmol	3.7%
Borneol	3.5%

Borneol CT

Borneol	28.3%
Beta-Pinene	10.1%
Alpha-Pinene	6.5%
Camphene	6.0%
Alpha-Terpineol	5.2%
1,8-Cineole	5.1%
Spathulenol	4.2%
Bornyl Acetate	4.1%

Camphor/Borneol CT

Camphor	14.6%–18.4%
4,5-Epoxy-1-isopropyl-4-methyl-1-cyclohexene	0.0%–18.2%
Borneol	12.5%–16.9%
1,3-Cyclohexadiene, 2-methyl-5-(1-methylethyl)-, monoepoxide	0.0%–11.0%
Chrysanthenone	0.0%–4.9%
Alpha-Pinene	1.7%–3.4%

subsp. farsicum (Aerial Parts)

(E)-Chrysanthenyl Acetate	24.7%
1,8-Cineole	9.2%
(E)-Chrysanthenol	8.9%
(Z)-Chrysanthenyl Acetate	7.1%
(Z)-Chrysanthenol	6.7%
Chrysanthenone	5.6%
(E)-Linalool Oxide Acetate	5.4%
Camphor	4.6%

Flowers (Camphor CT)

Camphor	25.8%
(E)-Chrysanthenyl Acetate	21.4%
Beta-Thujone	7.5%
Alpha-Pinene	5.8%
Santolina Triene	5.4%
Artemisia Ketone	4.8%
(Z)-Chrysanthenol	4.3%
Alpha-Terpinene	3.9%
1,8-Cineole	3.2%

Flowers (Chrysanthenone CT)

Chrysanthenone	39.5%
1,8-Cineole	18.9%
Camphor	7.8%
Alpha-Calacorene	4.2%
Camphene	3.0%

Isomenthol CT

Isomenthol	35.5%
1,8-Cineole	11.9%
Camphor	7.6%
Alpha-Pinene	4.6%
Pinocarvone	4.1%
Beta-Eudesmol	3.8%
Borneol	3.5%

Mixed CT

Borneol	10.5%–36.2%
Camphor	1.7%–27.9%
Bornyl Acetate	7.4%–26.8%
(Z)-Chrysanthenol	0.0%–26.2%
(Z)-Chrysanthenyl Acetate	0.0%–24.5%
1,8-Cineole	14.9%–22.7%
Camphene	1.5%–9.6%
Trans-Linalool Oxide Acetate	0.0%–3.1%

subsp. duderanum (Aerial Parts)

Artemisia Alcohol	42.7%
Camphor	22.0%
Yomogi Alcohol	14.0%
Camphene	4.8%
1,8-Cineole	4.7%

Flowers (Trans-Sabinene Hydrate CT)

Trans-Sabinene Hydrate	56.7%
1,8-Cineole	10.7%
Chrysanthenone	6.3%
Beta-Thujone	4.7%
Camphor	4.1%

Flowers (Dihydro-Eudesmol CT)

Dihydro-Eudesmol	12.0%
Nootkatin	9.6%
Cyclopentadecanolide	9.1%
Germacrene B	5.6%
Germacrone	4.7%
Camphor	4.1%
(Z)-Ethyl Chrysanthemumate	3.6%
Bornyl Acetate	3.6%
Alpha-Himachalene	3.5%
Cis-Calamenene	3.3%

Flower Buds

Camphor	35.5%
Alpha-Terpinene	9.5%
Caryophyllene Oxide	4.9%
1,8-Cineole	4.2%
Alpha-Pinene	4.2%
Cis-Sabinene Hydrate	3.3%

Stems

(E)-Beta-Farnesene	39.7%
Camphor	8.5%
Beta-Bisabolene	7.7%
Caryophyllene Oxide	7.2%
Pinocarvone	3.9%
(E)-Chrysanthenyl Acetate	3.9%
Thymol	3.7%
Alpha-Terpinene	3.1%

Leaves (1,8-Cineole CT)

1,8-Cineole	34.8%–63.5%
Chrysanthenone	5.5%–16.1%
Alpha-Calacorene	6.6%–8.9%
Camphor	2.2%–6.6%
Dihydro-Eudesmol	2.3%–4.1%
Alpha-Terpinene	1.1%–3.9%
Viridiflorene	0.0%–3.4%
Delta-Terpineol-Borneol	0.9%–3.0%

subsp. heterophyllum (Flowers)

Camphor	59.1%
Camphene	14.8%
1,8-Cineole	10.1%
Bornyl Acetate	3.8%
Alpha-Pinene	3.5%

Leaves (Camphor CT)

Camphor	25.4%–29.0%
1,8-Cineole	1.9%–14.3%
Trans-Sabinene Hydrate	0.0%–13.8%
Alpha-Terpinene	0.0%–13.2%
(Z)-Chrysanthenol	0.0%–11.2%
(E)-Chrysanthenyl Acetate	1.0%–9.7%
Trans-Pinene Hydrate	0.0%–9.7%
Gamma-Terpinene	0.0%–9.3%
Alpha-Calacorene	0.0%–7.2%
Cis-Sabinene Hydrate	0.0%–5.8%
Alpha-Thujone	0.0%–4.3%
(Z)-Ethyl Chrysanthemumate	0.0%–4.1%
Beta-Pinene	0.0%–3.7%
Chrysanthenone	0.0%–3.1%
Humulene Epoxide II	0.0%–3.1%

subsp. duderanum (Flowers)

(Z)-Chrysanthenyl Acetate	27.2%
Thuj-3-en-10-al	10.7%
Linalool	9.7%
Linalool Butyrate	8.8%
Carvacrol	5.6%
Geranyl Acetate	5.4%

subsp. heterophyllum (Leaves)

Camphor	53.5%
Bornyl Acetate	12.1%
Camphene	10.9%
1,8-Cineole	7.8%
Borneol	6.1%

OTHER CONSTITUENTS: *Alpha-Thujone CT*—Beta-pinene, para-cymene, pinocarvone; *Beta-Thujone CT*—Camphene, sabinene, para-cymene, cis-sabinene hydrate, alpha-thujone (< 2.6%), gamma-terpinene, chrysanthenone, pinocarvone, bornyl acetate, spathulenol; *Camphor/1,8-Cineole CT*—Beta-pinene, alpha-terpinene, para-cymene, gamma-terpinene, cis-limonene-1,2 epoxide, pinocarveol (isomer not identified), terpinen-4-ol, alpha-terpineol, myrtenol, isopulegol acetate, ethyl 2-methylbutanoate, isovaleric acid, cis-linalool oxide, linalool, carvone, bornyl acetate, trans-linalool oxide acetate, alpha-terpinyl acetate, neryl acetate, cis-jasmone, isobornyl isovalerate, 1,6-dimethylhepta-1,3,5-triene, o-cymene, 6-comphenone, filifolone, isomenthol, thymol, caryophyllene oxide; *Myrtenol CT*—(Z)-epoxy-ocimene, camphene, beta-pinene, o-cymene, beta-thujone (< 1.3%), alpha-thujone (< 0.4%), terpinen-4-ol; *1,8-Cineole CT*—Ethyl 2-methylbutanoate, isovaleric acid, alpha-pinene, camphene, sabinene, beta-pinene, gamma-terpinene, trans-pinocarveol, camphor, pinocarvone, terpinen-4-ol, bornyl acetate, ethyl 2-phenylpropanoate; *Borneol CT*—

Gamma-terpinene, trans-sabinene hydrate, camphor, terpinen-4-ol, myrtenol, trans-carveol, carvone, geraniol, geranyl acetate, bornyl isovalerate, delta-cadinene, (E)-nerolidol, caryophyllene oxide, (E)-sesquilavandulol, octadecane, hexadecanoic acid; *Isomenthol CT*—Camphene, beta-pinene, o-cymene, alpha-thujone, (Z)-epoxy-ocimene, terpinen-4-ol; *Camphor/Borneol CT*—Santolina triene, camphene, sabinene, o-cymene, gamma-terpinene, cis-sabinene hydrate, filifolone, (Z)-beta-terpineol, pinocarvone, terpinen-4-ol, isomenthol, (Z)-pinocarveol, bornyl acetate, thymol, davana ether, davanaone, cis-davanaone, 2,3,4,5-Tetramethylthiophene, 5-epi-neo-intermedeol, isobornyl acetate; *Mixed CT*—Alpha-pinene, beta-pinene, para-cymene, cis-linalool oxide, trans-pinocarveol, pinocarvone, terpinen-4-ol, trans-sabinyl acetate, isobornyl propanoate, isobornyl isovalerate, spathulenol, beta-eudesmol; *subsp. farsicum (Aerial Parts)*—Mesitylene, para-cymene, filifolone, borneol, bornyl acetate, neryl acetate, hexadecanoic acid; *subsp. duderanum (Aerial Parts)*—Alpha-pinene, para-cymene, gamma-terpinene, linalool, chrysanthenone, terpinen-4-ol, spathulenol; *Flowers (Camphor CT)*—Tricyclene, sabinene, myrcene, delta-2-carene, alpha-phellandrene, para-cymene, gamma-terpinene, borneol, (Z)-sabinene hydrate, (Z)-chrysanthenyl acetate, bornyl acetate, spathulenol; *Flowers (Chrysanthenone CT)*—Alpha-pinene, alpha-terpinene, trans-sabinene hydrate, neo-3-thujanol, delta-terpineol-borneol, (Z)-ethyl chrysanthemumate, bornyl acetate, germacrene B, n-tridecanol, gamma-eudesmol, alpha-muurolol, dehydro-eudesmol, (E,Z)-farnesol, cyclopentadecanolide, nootkatin, n-heneicosane, n-tricosane; *Flowers (Trans-Sabinene Hydrate CT)*—2-heptanone, alpha-pinene, camphene, alpha-terpinene, neo-3-thujanol, iso-dihydrocarveol, (Z)-ethyl chrysanthemumate, bornyl acetate, viridiflorene, cis-calamenene, alpha-calacorene, gamma-eudesmol, cyclopentadecanolide; *Flowers (Dihydro-Eudesmol CT)*—Beta-pinene, 1,8-cineole, trans-sabinene hydrate, beta-thujone (< 0.8%), chrysanthenone, delta-terpineol-borneol, terpinen-4-ol, iso-dihydrocarveol, neryl acetate, carvacrol acetate, aromadendrene, n-dodecanol, germacrene D, viridiflorene, beta-sesquiphellandrene, trans-calamenene, alpha-calacorene, n-tridecanol, geranyl isovalerate, humulene epoxide II, beta-cedrene epoxide, gamma-eudesmol, alpha-muurolol, curcumenol, phytol, n-heneicosane, n-tricosane; *Flower Buds*—Tricyclene, beta-pinene, alpha-phellandrene, (Z)-beta-ocimene, gamma-terpinene, Artemisia ketone, alpha-thujone (< 1.3%), trans-pinene hydrate, (Z)-chrysanthenol, pinocarvone, (E)-chrysanthenyl acetate, bornyl acetate, alpha-gurjunene, alpha-himachalene, spathulenol; *Stems*—Alpha-pinene, beta-pinene, gamma-terpinene, (Z)-sabinene hydrate, alpha-thujone (< 1.3%), trans-pinene hydrate, (Z)-chrysanthenol, borneol, bornyl acetate, 3-thujyl acetate, alpha-himachalene, alpha-cadinene, spathulenol; *Leaves (Camphor CT)*—Tricyclene, alpha-pinene, alpha-phellandrene, sabinene, (Z)-beta-ocimene, beta-thujone (< 2.1%), pinocarvone, bornyl acetate, spathulenol, caryophyllene oxide; *Leaves (1,8-Cineole CT)*—Camphene, beta-pinene, (E)-beta-ocimene, Artemisia ketone, trans-linalool oxide, trans-sabinene hydrate, neo-3-thujanol, terpinen-4-ol, iso-dihydrocarveol, (Z)-ethyl chrysanthemumate, bornyl acetate, neryl acetate, carvacrol acetate, aromadendrene, n-dodecanol, germacrene D, beta-sesquiphellandrene, germacrene B, n-tridecanol, geranyl isovalerate, humulene epoxide II, beta-cedrene epoxide, gamma-eudesmol, alpha-muurolol, n-tetradecanol, germacrone, (E,Z)-farnesol, cyclopentadecanolide, iso-bergaptene; *subsp. duderanum (Flowers)*—Alpha-pinene, beta-pinene, 1,8-cineole, trans-linalool oxide, camphor, myrtenal, alpha-terpineol, bornyl acetate, methyl isovalerate, n-docosane; *subsp. heterophyllum (Flowers)*—Sabinene, beta-pinene, para-cymene, borneol, chrysanthenyl acetate; *subsp. heterophyllum (Leaves)*—Alpha-pinene, sabinene, beta-pinene, para-cymene, chrysanthenyl acetate

Preferred Composition for Clinical Use:
Too great of variation in samples/data to determine.

Reported Therapeutic Properties: Antimicrobial, antibacterial, antifungal, anti-infectious, antiseptic, antioxidant, analgesic, anti-inflammatory, anticancer, antiallergic, antiarthritic, soothes chronic skin conditions, reduces the appearance of acne and blemishes, reduces blood pressure, balances heart rate, aids respiratory function, expels excess mucus, combats emotional chaos, promotes courage

CAUTIONS:

Alpha-Thujone, Beta-Thujone CTs

■ Avoid with children under 6 due to thujone content. Cases of seizure have been reported in young children due to exposure to essential oils rich in thujones.[474,475]

■ Avoid during pregnancy and while lactating, due to thujones content. Essential oils high in thujone content may cause abortion.[476,477]

■ Avoid oral consumption. Ingesting plants or products high in thujones can cause vomiting, stomachache, diarrhea, and gastroenteritis followed by absorption disorders, headache, nervous agitation, and chronic convulsions, and symptoms of liver and kidney toxicity extending to yellow liver atrophy (extensive and rapid death of liver cells—parenchymal, possibly by fatty degeneration), arrhythmia, and myocardial bleeding.[478] Thujone is considered significantly neurotoxic and may damage the liver, and it is estimated that as little as 15 mg orally may negatively impact the central nervous system.[479,480,481] Taking 20 drops of thuja oil (also with significant thujones content) twice daily for five days caused an adult woman to have a tonic seizure.[482] As few as 12 drops of sage oil (also with significant thujones content) orally have reportedly caused a seizure and short coma in an adult.[483]

■ Avoid with epilepsy and Parkinson's disease due to thujones, camphor, and 1,8-cineole content.[484,485,486,487,488]

■ Avoid with those who have a compromised liver or kidneys, due to the risk of kidney toxicity, kidney failure, or liver toxicity.[489,490,491] Consuming extremely large amounts of thujone-rich essential oils (10 mL) has caused kidney failure.[492] This would also suggest that those taking medications that could cause liver damage should also use tansy very cautiously or avoid it entirely.

■ May weakly interfere with the enzymes responsible for metabolizing medications (NSAIDs, proton-pump inhibitors, acetaminophen, antiepileptics, immune modulators, blood-sugar medications, blood pressure medications, antidepressants, antipsychotics, diabetic medications, antihistamines, antibiotics, and anesthetics) due to 1,8-cineole and alpha-thujone content.[493,494,495,496]

Camphor CTs

■ Avoid with children under six due to significant camphor content and moderate amounts of 1,8-cineole and thujones. Several cases of camphor poisoning and/or seizures from ingestion and topical application have been reported in children.[497,498] Ingestion of camphor-containing products has been lethal in children under age two.[499] Children five years and up may use camphor-containing essential oils topically in dilutions no stronger than 5%. Camphor ingestion by infants and young children may cause cough, vomiting, seizure, burning sensation in the mucous membranes and eyes, or lack of voluntary coordination of muscle movements.[500] Cases of seizure have been reported in young children due to exposure to essential oils rich in thujones.[501,502]

■ Caution is warranted during pregnancy and while lactating, due to significant camphor content and moderate thujones content. Ingestion of essential oils with significant levels of camphor may lead to abortion because fetuses lack the enzymes to process it.[503] Camphor ingestion by infants and young children may cause cough, vomiting, seizure, burning sensation in the mucous membranes and eyes, or lack of voluntary coordination of muscle movements.[504] Essential oils high in thujone content may cause abortion.[505,506]

■ Avoid oral consumption due to significant camphor and moderate thujones content. Essential oils with significant levels of camphor may be toxic when taken orally, usually single doses exceeding 2 mL, although the lethal dose for humans is estimated to be greater than 5 mL in a single dose.[507,508,509] See oral cautions under *Alpha-Thujone, Beta-Thujone CTs*.

■ Avoid with epilepsy and Parkinson's disease due to camphor, 1,8-cineole, and thujones content, which have the potential to exacerbate or cause seizures/convulsions or reduce seizure-medication efficacy.[510,511,512,513]

■ Avoid with those who have a compromised liver or kidneys, due to the risk of increased liver enzymes and liver or kidney damage based on camphor and thujones content.[514,515,516,517] This would also suggest that those taking medications that could cause liver damage should also use essential oils with significant camphor levels very cautiously or avoid them entirely.

■ The significant camphor content may negatively impact red blood cells and increase the risk of jaundice in children with glucose-6-phosphate dehydrogenase deficiency (G6PD).[518,519]

■ Caution is warranted for the topical application of camphor-containing essential oils on broken or injured skin due to an increased risk of toxicity.[520]

1,8-Cineole CTs

■ Avoid with children under age 3, particularly around the nose and mouth and use cautiously in children under age 5 due to significant 1,8-cineole content. 1,8-cineole may cause seizures, central nervous system problems, or respiratory distress in young children.[521,522,523]

■ Caution is warranted orally due to the significant amounts of 1,8-cineole. Limit it to small doses internally (adults—maximum 10 drops daily). Toxicity has been reported when eucalyptus (also high in 1,8-cineole) was ingested in large doses, and as few as 4 to 5 drops may cause problems in very sensitive individuals.[524,525,526,527,528,529]

■ Avoid with epilepsy and Parkinson's disease due to 1,8-cineole and camphor content. May exacerbate or cause seizures/convulsions or reduce seizure-medication efficacy.[530,531,532,533]

■ May interact with diabetic medications due to significant 1,8-cineole content. 1,8-cineole significantly inhibits alpha-glucosidase activity, particularly when synergized with other constituents (alpha-pinene, beta-pinene, and para-cymene).[534,535,536] Alpha-glucosidase is an enzyme that breaks down carbohydrates by chemical reaction with water. Inhibiting its activity postpones glucose absorption and therefore the impact of carbohydrates on blood-sugar levels. Studies suggest that oils rich in 1,8-cineole may significantly lower glucose in animals, which may increase the action of diabetic medications.[537,538]

■ May interact with aspirin, blood pressure, antiplatelet, and anticoagulant medications, and increase the risk of bleeding among people with bleeding disorders (low risk) due to significant 1,8-cineole content.[539] 1,8-cineole is a weak inhibitor of platelet aggregation.[540]

■ May weakly interfere with the enzymes responsible for metabolizing medications (NSAIDs, proton-pump inhibitors, acetaminophen, antiepileptics, immune modulators, blood-sugar medications, blood pressure medications, antidepressants, antipsychotics, diabetic medications, antihistamines, antibiotics, and anesthetics) due to 1,8-cineole content.[541,542,543]

■ May interfere with pentobarbital and other barbiturates (medications for anxiety and insomnia), based on 1,8-cineole content.[544,545]

Camphor/1,8-Cineole CT, Flowers (Chrysanthenone CT)

■ Follow cautions for both *Camphor* and *1,8-cineole CTs*.

Camphor/Borneol CT

■ Caution is warranted during pregnancy and while lactating, due to significant camphor content and moderate thujones content. Ingestion of essential oils with significant levels of camphor may lead to abortion because fetuses lack the enzymes to process it.[546] Camphor ingestion by infants and young children may cause cough, vomiting, seizure, burning sensation in the mucous membranes and eyes, or lack of voluntary coordination of muscle movements.[547] Essential oils high in thujone content may cause abortion.[548,549]

■ Avoid with epilepsy and Parkinson's disease due to camphor content, which may exacerbate or cause seizures/convulsions or reduce seizure-medication efficacy.[550,551,552,553]

■ The significant camphor content may negatively impact red blood cells and increase the risk of jaundice in children with glucose-6-phosphate dehydrogenase deficiency (G6PD).[554,555]

Isomenthol CT

■ Avoid with epilepsy and Parkinson's disease due to 1,8-cineole and camphor content. May exacerbate or cause seizures/convulsions or reduce seizure-medication efficacy.[556,557,558,559]

Mixed CT
■ Follow cautions for both *Camphor, Borneol,* and *1,8-cineole CTs.*

subsp. duderanum (Aerial Parts)
■ Avoid with children under six due to significant camphor content and moderate amounts of 1,8-cineole. Several cases of camphor poisoning and/or seizures from ingestion and topical application have been reported in children.[560,561] Ingestion of camphor-containing products has been lethal in children under age two.[562] Children five years and up may use camphor-containing essential oils topically in dilutions no stronger than 5%. Camphor ingestion by infants and young children may cause cough, vomiting, seizure, burning sensation in the mucous membranes and eyes, or lack of voluntary coordination of muscle movements.[563]
■ Caution is warranted during pregnancy and while lactating, due to significant camphor content. Ingestion of essential oils with significant levels of camphor may lead to abortion because fetuses lack the enzymes to process it.[564] Camphor ingestion by infants and young children may cause cough, vomiting, seizure, burning sensation in the mucous membranes and eyes, or lack of voluntary coordination of muscle movements.[565]
■ Use cautiously orally due to significant camphor content. Essential oils with significant levels of camphor may be toxic when taken orally, usually single doses exceeding 2 mL, although the lethal dose for humans is estimated to be greater than 5 mL in a single dose.[566,567,568]
■ Avoid with epilepsy and Parkinson's disease due to camphor and 1,8-cineole content. These constituents have the potential to exacerbate or cause seizures/convulsions or reduce seizure-medication efficacy.[569,570,571,572]
■ The significant camphor content may negatively impact red blood cells and increase the risk of jaundice in children with glucose-6-phosphate dehydrogenase deficiency (G6PD).[573,574]

subsp. farsicum (Aerial Parts), Flowers (Trans-Sabinene Hydrate CT)
■ Avoid with epilepsy and Parkinson's disease due to camphor, 1,8-cineole, or thujones content. These constituents have the potential to exacerbate or cause seizures/convulsions or reduce seizure-medication efficacy.[575,576,577,578]

Borneol CT
■ May interact with aspirin, blood pressure, antiplatelet, and anticoagulant medications, and increase the risk of bleeding among people with bleeding disorders, due to the anticoagulant properties of borneol.[579,580]
■ May interfere with enzymes responsible for metabolizing medications (NSAIDs, proton-pump inhibitors, acetaminophen, antiepileptics, immune modulators, blood-sugar medications, blood pressure medications, antidepressants, antipsychotics, diabetic medications, antihistamines, antibiotics, and anesthetics), based on borneol content.[581,582,583,584,585]

Myrtenol CT
■ May interact with barbiturates and other sedative drugs and increase sleeping time based on myrtenol content.[586]

Flowers (Dihydro-Eudesmol CT), Stems, and subsp. duderanum (Flowers)
■ None currently known.

SELECTED EVIDENCE:

- □ *In vitro* evidence demonstrated that bobondi kabir CT camphor/1,8-cineole essential oil inhibited gram-positive (*S. epidermidis, B. subtilis, S. aureus*) and gram-negative bacteria (*E. coli, L. pneumoniae, S. typhi, Sh. dysenteriae*), with the greatest activity against gram-positive bacteria.[587] The borneol CT exhibited remarkable activity against *S. aureus, Sh. flexneri, S. typhi, S. epidermidis, S. saprophyticus, P. aeruginosa,* and *E. coli*.[588]
- □ Bobondi kabir CT borneol essential oil exhibited good antioxidant activity in the DPPH assay.[589]

BORNEO CAMPHOR (Malay Camphor, Sumatra Camphor, Barus Camphor)

Dryobalanops aromatica C.F. Gaertn., *D. camphora* Colebr.

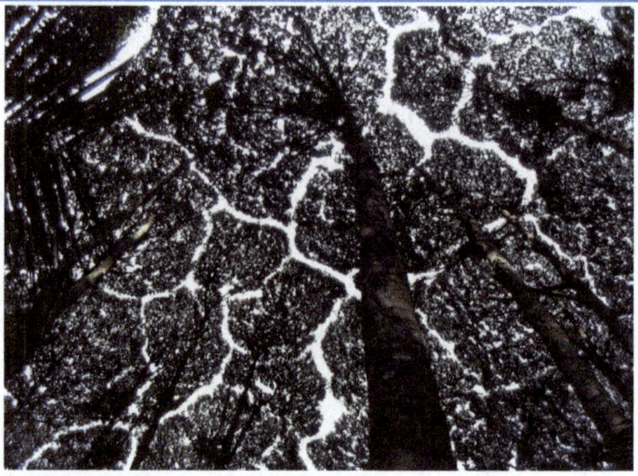

Family: Dipterocarpaceae
Note: Middle-Base
Aroma Intensity: Medium-Strong
Aroma: Camphoraceous and medicinal to woody, resinous, and earthy
Common Extraction Method: Steam distilled from the exudates, leaves, seeds, or wood
Possible Substitute Oils: Frankincense, copaiba, hemp; equal portions of copaiba, cypress, and tea tree
Blends Well With: Balsam fir, bergamot, black spruce, blue spruce, cedarwood, clary sage, copaiba, cypress, galbanum, lavender, lemon, lime, orange, palo santo, pine, ravintsara, tangerine, ylang ylang
Subcellular Localization | Epigenetic Influence: Currently unknown | Currently unknown
Recommended Dilution Range: 3%–Neat

Primary Constituents:[590,591,592]

Exudates (Alpha-Pinene CT)	
Alpha-Pinene	36.7%
Beta-Caryophyllene	21.2%
Terpinen-4-ol	10.6%
Alpha-Humulene	8.4%
Alpha-Terpineol	8.0%
Limonene	4.1%

Leaves	
Alpha-Terpineol	15.8%
Terpinen-4-ol	15.0%
Globulol	8.2%
Alpha-Pinene	6.9%
Viridiflorol	3.8%
Beta-Caryophyllene	3.6%
Alpha-Humulene	3.1%
Limonene	3.0%

Wood	
Alpha-Pinene	54.3%
Beta-Caryophyllene	18.1%
Borneol	8.3%
Alpha-Humulene	4.3%
Alpha-Terpineol	3.0%

Exudates (Beta-Caryophyllene CT)	
Beta-Caryophyllene	36.1%–38.0%
Alpha-Pinene	10.8%–17.0%
Alpha-Humulene	15.8%–16.3%
Terpinen-4-ol	5.8%–9.3%
Alpha-Terpineol	3.8%–5.9%
Beta-Myrcene	0.0%–5.6%
Caryophyllene Oxide	0.4%–4.0%

Seeds	
Alpha-Pinene	40.8%
Alpha-Thujene	12.9%
Beta-Pinene	12.7%
Sabinene	6.4%
Limonene	6.0%
Bicyclogermacrene	5.9%
Myrcene	4.9%
Beta-Caryophyllene	4.3%

OTHER CONSTITUENTS: *Exudates (Alpha-Pinene CT)*—Beta-pinene, 1,4-cineole, 2-carene, camphor, borneol, caryophyllene oxide; *Exudates (Beta-Caryophyllene CT)*—Beta-pinene, alpha-phellandrene, 4-carene, camphor, borneol, bornyl acetate, beta-elemene, curcumene, humulene epoxide; *Leaves*—Alpha-Thujene, beta-pinene, gamma-terpinene, para-cymene, terpinolene, camphor, ledene, borneol, bicyclogermacrene, p-cymen-8-ol, ledol, epiglobulol, caryophyllene oxide, spathulenol, beta-bisabolol, T-muurolol, (E)-alpha-bergamotene, alpha-cadinol, hexadecanoic acid; *Seeds*—Beta-phellandrene, gamma-terpinene, terpinolene, alpha-humulene; *Wood*—Beta-pinene, limonene

PREFERRED COMPOSITION FOR CLINICAL USE:
Insufficient samples/data to determine.

REPORTED THERAPEUTIC PROPERTIES: Aids cardiovascular function, relieves chest pain, antibacterial, antifungal, antiparasitic, analgesic, antiseptic, antispasmodic, anti-inflammatory, antirheumatic, relieves headache, diuretic, supports kidney function, balances menstruation, relieves insect bites, relieves chronic skin conditions, heals boils and abscesses, relieves cold sores, aids respiratory and lung function, enhances sexual function, aphrodisiac, antidepressant, stress management

CAUTIONS:
- None currently known.

SELECTED EVIDENCE:
- □ Animal research showed that borneo camphor essential oil was sedating to mice.[593]

BORONIA ABSOLUTE
(Sweet Boronia)

Boronia megastigma Nees ex Bartling

FAMILY: Rutaceae
NOTE: Top
AROMA INTENSITY: Strong
AROMA: Intensely floral, green, slightly fruity and hay-like
COMMON EXTRACTION METHOD: Solvent extracted from the flowers
POSSIBLE SUBSTITUTE OILS: Rose, jasmine absolute
BLENDS WELL WITH: Bergamot, chamomile (German, Roman), clary sage, clove, coriander, frankincense, geranium, ginger, grapefruit, jasmine, lavender, lemon, lime, Mediterranean mandarin, neroli, orange, palmarosa, patchouli, petitgrain, rose, rosewood, sandalwood, tangerine, vetiver, ylang ylang
SUBCELLULAR LOCALIZATION | EPIGENETIC INFLUENCE: Currently unknown | Currently unknown
RECOMMENDED DILUTION RANGE: 1.0%–3.0%

PRIMARY CONSTITUENTS:[594]

(E)-Beta-Ionone	14.9%
(Z)-Heptadec-8-ene	14.6%
N-[2-(4-prenyloxyphenyl)ethyl]Tiglamide + Methyl(Z)-4-(Geranyloxy)Cinnamate	9.3%
Methyl(Z)-4-(Geranyloxy)Cinnamate	6.7%

(E)-4-(Geranyloxy)Cinnamic Acid	4.3%
8-Hydroxylinalyl Isobutyrate + + Hexadecane	3.8%
(E)-3-Hydroxymegastigm-7-en-9-one	2.4%
5-Acetoxylinalool Isomer B	2.4%

OTHER CONSTITUENTS: Decan-2-one, 7-hydroxyhotrienol, anethole, safrole, nonanoic acid, geranyl acetate, vanillin, 7,8-dihydro-beta-ionone, 7,8-dihydro-beta-ionol, (Z)-retro-alpha-ionone, 8-hydroxylinalyl acetate, dodecanol, 2-phenylethyl 3-methylbutyrate, benzyl tiglate, sesquicineole, spathulenol, (E)-7-hydroxymegastigm-7-en-9-one, (E,E)-pseudo-ionone, 4-oxo-beta-ionone, delta-jasmonic acid lactone, methyl-(Z)-epi-jasmonate, delta-jasmonic acid lactone, methyl cucurbate, heptadecane, methyl (E)-4-hydroxycinnamate, (Z)-4-(geranyloxy)cinnamic acid, 8-hydroxylianlyl octanoate, 8-hydroxylinalyl nonanoate, 8-hydroxylinalyl tridecanoate

PREFERRED COMPOSITION FOR CLINICAL USE:
Insufficient samples/data to determine.

REPORTED THERAPEUTIC PROPERTIES: Soothing to the skin, wound healing, relieves anxiety, deodorant, supports and restore thyroid health, antidepressant, aphrodisiac, stress management, banishes worrisome thoughts, aids meditation

CAUTIONS:
- Avoid oral consumption. Absolutes contain solvent residue and should not be taken orally.

SELECTED EVIDENCE:
- None found.

BRACELET HONEY MYRTLE (Mauve Bracelet Honey Myrtle)

Melaleuca armillaris (Sol. ex Gaertn.) Sm.

FAMILY: Myrtaceae
NOTE: Middle-Top
AROMA INTENSITY: Medium
AROMA: Fresh, medicinal, herbaceous, spicy
COMMON EXTRACTION METHOD: Steam distilled or hydrodistilled from the leaves
POSSIBLE SUBSTITUTE OILS: Tea tree, myrtle, Rosalina, eucalyptus, cajeput, manuka
BLENDS WELL WITH: Bergamot, cajeput, carrot seed, chamomile (German, Roman), cedarwood, clary sage, clove, cypress, eucalyptus, fennel, grapefruit, lavender, lemon, lime, myrrh, orange, patchouli, petitgrain, pine, ravensara, rosemary, sandalwood, tangerine, thyme
SUBCELLULAR LOCALIZATION | EPIGENETIC INFLUENCE: Currently unknown | Currently unknown
RECOMMENDED DILUTION RANGE: 3%–50%, neat for some conditions; *Methyl Eugenol CT*—1%–20%

PRIMARY CONSTITUENTS:[595,596,597,598,599,600]

1,8-Cineole/Terpinen-4-ol CT		*Cis-Calamene CT*	
1,8-Cineole	33.7%–33.9%	Cis-Calamenene	19.0%
Terpinen-4-ol	18.8%–24.8%	Torreyol	15.1%

Gamma-Terpinene	10.4%–12.3%	Dihydrocarveol	9.0%
Sabinene	0.2%–6.6%	Alpha-Terpineol	7.7%
Alpha-Terpinene	5.7%–5.8%	Trans-Pinocarveol	6.9%
Alpha-Pinene	tr–3.6%	Myrtenol	6.4%
Alpha-Terpineol	2.4%–3.5%	Spathulenol	4.4%
Alpha-Terpinyl Acetate	2.9%–3.1%	Aromadendrene	4.1%
		Verbenone	3.7%
1,8-Cineole CT		1,8-Cineole	3.6%
1,8-Cineole	68.9%–85.8%		
Alpha-Pinene	2.0%–7.4%	*Methyl Eugenol CT*	
Borneol	0.4%–6.2%	Methyl Eugenol	80.6%
Camphene	tr–5.1%		

OTHER CONSTITUENTS: *1,8-Cineole/Terpinen-4-ol CT*—Alpha-thujene, beta-pinene, myrcene, limonene, p-cymene, terpinolene, linalool, methyl eugenol (< 1.8%), beta-caryophyllene; *1,8-Cineole CT*—Beta-pinene, myrcene, sabinene, terpinolene, linalool, camphor, alpha-ylangene, beta-elemene, aromadendrene, alloaromadendrene, alpha-cadinol; *Cis-Calamenene CT*—Trans-sabinol, eucarvone, isoborneol, viridene, myrtenal, cis-dihydrocarvone, 9-decen-1ol, citronellyl acetate, cis-piperitone oxide, cis-alpha-ambrinol, gamma-himachalene, germacrene B, 10-undecen-1-ol acetate, caryophyllene oxide, globulol, guaiol, beta-atlantol, 1-epi-cubenol gamma-eudesmol, methyl jasmonate; *Methyl Eugenol CT*—Alpha-phellandrene, linalool, p-cymen-8-ol, alpha-terpineol

PREFERRED COMPOSITION FOR CLINICAL USE:

Constituent	
1,8-Cineole	30%–40%
Terpinen-4-ol	15%–30%
Gamma-Terpinene	7%–15%
Alpha-Terpinene	3%–10%
Sabinene	0.1%–7%
Alpha-Terpineol	1%–6%
Alpha-Terpinyl Acetate	1%–6%
Alpha-Pinene	tr–5%

REPORTED THERAPEUTIC PROPERTIES: Antibacterial, **antifungal**, antimicrobial, antiviral, antiparasitic, aids immune function, antioxidant, stimulates natural antioxidant production and activity, anticancer, reduces high blood pressure, aids respiratory function, insecticidal, encourages creativity, enhances focus

CAUTIONS:

1,8-Cineole/Terpinen-4-ol CT

■ Avoid with children under age 3, particularly around the nose and mouth and use cautiously in children under age 5 due to significant 1,8-cineole content. 1,8-cineole may cause seizures, central nervous system problems, or respiratory distress in young children.[601,602,603]

■ Caution is warranted orally due to the significant amounts of 1,8-cineole. Limit it to small doses internally (adults—maximum 10 drops daily). Toxicity has been reported when eucalyptus (also high in 1,8-cineole) was ingested in large doses, and as few as 4 to 5 drops may cause problems in very sensitive individuals.[604,605,606,607,608,609]

■ Avoid with epilepsy and Parkinson's disease due to 1,8-cineole content. May exacerbate or cause seizures/convulsions or reduce seizure-medication efficacy based on 1,8-cineole content.[610,611,612,613]

■ May weakly interfere with the enzymes responsible for metabolizing medications (NSAIDs, proton-pump inhibitors, acetaminophen, antiepileptics, immune modulators, blood-sugar medications, blood pressure medications, antidepressants, antipsychotics, diabetic medications, antihistamines, antibiotics, and anesthetics) due to 1,8-cineole content.[614,615,616]

■ May interfere with pentobarbital and other barbiturates (medications for anxiety and insomnia), based on 1,8-cineole content.[617,618]

1,8-Cineole CT

■ Avoid with children under age 3, particularly around the nose and mouth and use cautiously in children under age 5 due to significant 1,8-cineole content. 1,8-cineole may cause seizures, central nervous system problems, or respiratory distress in young children.[619,620,621]

■ Caution is warranted orally due to the significant amounts of 1,8-cineole. Limit it to small doses internally (adults—maximum 10 drops daily). Toxicity has been reported when eucalyptus (also high in 1,8-cineole) was ingested in large doses, and as few as 4 to 5 drops may cause problems in very sensitive individuals.[622,623,624,625,626,627]

■ Avoid with epilepsy and Parkinson's disease due to 1,8-cineole content. May exacerbate or cause seizures/convulsions or reduce seizure-medication efficacy based on 1,8-cineole content.[628,629,630,631]

■ May interact with diabetic medications due to significant 1,8-cineole content. 1,8-cineole significantly inhibits alpha-glucosidase activity, particularly when synergized with other constituents (alpha-pinene, beta-pinene, and para-cymene).[632,633,634] Alpha-glucosidase is an enzyme that breaks down carbohydrates by chemical reaction with water. Inhibiting its activity postpones glucose absorption and therefore the impact of carbohydrates on blood-sugar levels. Studies suggest that oils rich in 1,8-cineole may significantly lower glucose in animals, which may increase the action of diabetic medications.[635,636]

■ May interact with aspirin, blood pressure, antiplatelet, and anticoagulant medications, and increase the risk of bleeding among people with bleeding disorders (low risk) due to significant 1,8-cineole content.[637] 1,8-cineole is a weak inhibitor of platelet aggregation.[638]

■ May weakly interfere with the enzymes responsible for metabolizing medications (NSAIDs, proton-pump inhibitors, acetaminophen, antiepileptics, immune modulators, blood-sugar medications, blood pressure medications, antidepressants, antipsychotics, diabetic medications, antihistamines, antibiotics, and anesthetics) due to 1,8-cineole content.[639,640,641]

■ May interfere with pentobarbital and other barbiturates (medications for anxiety and insomnia), based on 1,8-cineole content.[642,643]

Methyl Eugenol CT

Only use the methyl eugenol CT when the benefits outweigh the risks as determined by a qualified professional.

■ Avoid in children under age 12 due to the multisite cancerous tumors and DNA mutation caused by methyl eugenol in animal studies.[644,645,646,647]

■ Avoid during pregnancy and lactation due to changes in embryo form and structure and chromosomal changes in ovary cells caused by methyl eugenol as well as the carcinogenic (cancer-causing) potential of methyl eugenol.[648,649,650,651]

■ Avoid oral consumption due to the multisite carcinogenic potential of methyl eugenol.[652,653,654,655]

■ May interfere with antibiotics and antifungal medications and increase their effectiveness due to methyl eugenol content.[656]

■ May interact with anticholinergic (drugs used for asthma, incontinence, gastrointestinal cramps, muscular spasms, depression, and sleep disorders) and cholinergic medications (drugs used to reduce urinary retention, diagnose myasthenia gravis, and for glaucoma), based on the potent AChE-inhibtory activity of methyl eugenol.[657]

Cis-Calamenene CT

■ None currently known.

SELECTED EVIDENCE:

- □ Bracelet honey myrtle CT 1,8-cineole essential oil displayed significant anticancer activity against breast cancer cells.[658]
- □ A laboratory study concluded that the 1,8-cineole CT of bracelet honey myrtle essential oil moderately relaxed isolated rat aorta, suggesting it may be beneficial for high blood pressure.[659]
- □ *In vitro* research demonstrated that bracelet honey myrtle CT 1,8-Cineole/Terpinen-4-ol essential oil inhibits *B. subtilis, E. coli, A. niger,* and *C. albicans.*[660] The methyl eugenol CT of honey bracelet myrtle exhibited moderate activity against bacteria (*B. spizizenii, S. aureus, E. aerogenes, E. coli, K. pneumoniae, P. aeruginosa, S. enterica*) and excellent activity against fungi (*A. niger, F. oxysporum, F. solani, P. digitatum*).[661]
- □ Oral administration of bracelet honey myrtle CT 1,8-Cineole/Terpinen-4-ol essential oil beginning the second week after infection with *Schistosoma mansoni* significantly ameliorated the disturbed levels of glutathione (GSH), malondialdehyde (MDA) and improved liver and body weights in infected mice.[662]
- □ The 1,8-cineole CT (85.8% 1,8-cineole) of bracelet honey myrtle essential oil exhibited mild antimalarial activity against chloroquine-resistant *Plasmodium falciparum* FcB1 strain.[663]
- □ Honey bracelet myrtle CT 1,8-Cineole/Terpinen-4-ol essential oil was virucidal to herpes simplex virus type 1 (HSV-1).[664]
- □ Injection of bracelet honey myrtle essential oil CT 1,8-Cineole/Terpinen-4-ol (25 mg/kg body weight, three times per week for one month) exhibited marked antioxidant activity and improved vitamin E, vitamin C, and superoxide dismutase parameters in rats.[665] Laboratory research showed that bracelet honey myrtle essential oil (85.8% 1,8-cineole) exhibited weak to moderate antioxidant activity in the ABTS and DPPH assays.[666,667] The methyl eugenol CT showed greater antioxidant activity than BHT in the DPPH assay.[668]

BRAZILIAN GRAPETREE (Jaboticaba, Jabuticaba)

Plinia cauliflora (Mart.) Kausel, *Myrtus cauliflora* (Mart.) O.Berg, *Eugenia cauliflora* DC.; **JABOTICABA-DE-CABINHO**: *P. trunciflora* (O.Berg) Kausel, *Eugenia rabeniana* Kiaersk., *Myrciaria peruviana* (Poir.) Mattos var. *trunciflora* (O.Berg) Mattos

FAMILY: Myrtaceae
NOTE: Middle
AROMA INTENSITY: Medium
AROMA: Herbaceous, earthy, fresh
COMMON EXTRACTION METHOD: Steam distilled or hydrodistilled from the leaves
POSSIBLE SUBSTITUTE OILS: Peta CT spathulenol, betony, kunzea CT spathulenol\globulol, valerian, calendula, hinoki, pink pepper, guggul, helichrysum
BLENDS WELL WITH: Basil, bergamot, copaiba, frankincense, goldenrod, hemp, hyssop, ginger, lavender, lemon, lime, marjoram, orange, patchouli, sage, Spanish sage, valerian, vetiver

SUBCELLULAR LOCALIZATION | EPIGENETIC INFLUENCE: Currently unknown | Currently unknown
RECOMMENDED DILUTION RANGE: *P. cauliflora* and *P. trunciflora (Globulol CT)*—3%–50%, neat for some conditions; *P. trunciflora (Alpha-Cadinol CT)*—1.5–20%

PRIMARY CONSTITUENTS:[669,670]

P. cauliflora

Spathulenol	27.2%
Caryophyllene Oxide	21.6%
Epiglobulol	8.1%
T-Muurolol	5.9%
Globulol	4.1%
Humulene Oxide I	2.9%
Humulene Epoxide II	2.7%
Beta-Elemene	2.4%

P. trunciflora (Globulol CT)

Globulol	10.8%
Bicyclogermacrene	10.6%
Gamma-Muurolene	9.2%
Epiglobulol	8.4%
Beta-Caryophyllene	8.2%
Beta-Pinene	5.1%
Spathulenol	4.2%
Alloaromadendrene	3.7%
Eudesmol*	3.4%
Delta-Cadinene	3.0%
Alpha-Copaene	2.1%

* Correct isomer not identified.

P. trunciflora (Alpha-Cadinol CT)

Alpha-Cadinol	19.2%
Apiole	11.2%
Alpha-Terpineol	7.5%
Cubenol	5.4%
T-Cadinol	3.3%
Alpha-Muurolol	2.8%
Valeranone	2.8%
5-Hydroxyisobornyl Isobutyrate	2.8%
Viridiflorol	2.7%
Vulgarone B	2.7%
Nootkatone	2.1%

OTHER CONSTITUENTS: *P. cauliflora*—Beta-pinene, limonene, 1,8-cineole, alpha-terpineol, alpha-copaene, beta-bourbonene, aromadendrene, alloaromadendrene, gamma-muurolene, 1-epi-cubenol, T-cadinol; *P. trunciflora (Globulol CT)*—n-Hexanol, alpha-pinene, limonene, beta-bourbonene, beta-elemene, alpha-humulene, alpha-muurolene, (E)-nerolidol, 10-epi-gamma-eudesmol, isospathulenol, T-muurolol, alpha-muurolol; *P. trunciflora (Alpha-Cadinol CT)*—Neomenthol, trans-pulegol, ledol, 10-epi-gamma-eudesmol, valerianol, iso-longifolol, (E)-coniferyl alcohol, alpha-sinensal

PREFERRED COMPOSITION FOR CLINICAL USE:
Insufficient samples/data to determine.

REPORTED THERAPEUTIC PROPERTIES: Antifungal, antibacterial, antimicrobial, analgesic, anti-inflammatory, nervine, antirheumatic, aids respiratory function, wound healing, soothes chronic skin conditions, aids circulation, relieves painful menstruation, antioxidant, supports cardiovascular health, reduces high cholesterol, promotes restful sleep, stress management, helps release traumatic emotions

CAUTIONS:

P. trunciflora (Alpha-Cadinol CT)

■ Avoid with children under age twelve. May contain moderate amounts of apiole, which may cause kidney or liver toxicity, abnormal quantities of blood components (blood dyscrasias), dizziness, or hallucinations.[671,672,673,674,675]

■ Avoid during pregnancy and lactation due to the abortive properties (apiole may trigger uterine contractions or necrosis of placental tissue and has been used for decades to end unwanted pregnancies) and high toxicity (several cases of acute poisoning or death have been reported after ingestion of varying amounts of apiole in single or daily doses) of apiole.[676,677,678,679,680,681,682,683]

P. trunciflora (Globulol CT) and P. cauliflora
■ None currently known.

SELECTED EVIDENCE:

- □ *P. trunciflora* CT alpha-cadinol was active against gram-positive bacteria (*Streptococcus equi, Staphylococcus epidermis*) and yeasts (*Candida* spp., *Cryptococcus* spp.).

BRAZILIAN GUAVA (Sour Guava, Guinea Guava, Castilian Guava)

Psidium guineense Sw., *P. molle* Bertol., *P. schiedeanum* Berg., *P. araca* Raddi

FAMILY: Myrtaceae
NOTE: Middle
AROMA INTENSITY: Medium
AROMA: Earthy, herbaceous, woody; or fruity, sweet, slightly musky
COMMON EXTRACTION METHOD: Hydrodistilled or steam distilled from the leaves or fruit
POSSIBLE SUBSTITUTE OILS: Guava leaf, copaiba, black pepper, kunzea
BLENDS WELL WITH: Balsam fir, cedarwood, copaiba, cypress, eucalyptus, geranium, hemp, hinoki, lavender, peppermint, rose, sandalwood, spruce (black)
SUBCELLULAR LOCALIZATION | EPIGENETIC INFLUENCE: Currently unknown | Currently unknown
RECOMMENDED DILUTION RANGE: 5%–Neat

PRIMARY CONSTITUENTS:[684,685,686,687]

Leaves CT Beta-Bisabolene/Alpha-Pinene

Beta-Bisabolene	13.2%
Alpha-Pinene	12.9%
(Z)-Nerolidol	5.5%
Beta-Sesquiphellandrene	5.2%
Limonene	5.1%
(E)-Gamma-Bisabolene	4.3%
Beta-Pinene	4.2%
Beta-Caryophyllene	4.0%
(E)-Beta-Farnesene	3.1%

Fruit

Beta-Caryophyllene	8.6%
Butanol	7.4%
Ethyl Butyrate	7.4%
Selin-11-en-4α-ol	5.9%

Leaves CT Spathulenol

Spathulenol	80.7%
(2Z,6E)-Farnesol	3.7%

Leaves/Stems

Beta-Bisabolol	17.4%
Limonene	6.8%
Epi-Alpha-Bisabolol	6.7%
Alpha-Terpineol	4.5%
ar-Curcumene	3.3%
Alpha-Muurolol	3.1%
3-Phenylpropanal	3.0%

Alpha-Terpineol	4.8%
Alpha-Selinene	4.7%
Beta-Selinene	3.3%
Hexanol	3.3%
Ethyl Acetate	3.0%

OTHER CONSTITUENTS: *Leaves CT Beta-Bisabolene/Alpha-Pinene*—Camphene, myrcene, gamma-terpinene, fenchyl acetate, alpha-ylangene, benzaldehyde, bornyl acetate, (Z)-beta-farnesene, alpha-zingiberene, (Z)-gamma-bisabolene, alpha-curcumene; *Leaves CT Spathulenol*—1,8-cineole, trans-tagetone, gamma-terpineol, alpha-zingiberene, cedrol, T-muurolol, epi-alpha-bisabolol, alpha-bisabolol, (2Z,6Z)-farnesol; *Fruit*—Isobutanol, isoamyl alcohol, 3-methylbutanol, (Z)-3-hexenol, alpha-pinene, beta-pinene, myrcene, ethyl hexanoate, limonene, terpinolene, terpinene-4-ol, linalool, ethyl octanoate, alpha-copaene, beta-bourbonene, beta-elemene, alpha-humulene, beta-bisabolene, alpha-muurolene, germacrene D, gamma-cadinene, caryophyllenol alcohol, spathulenol, caryophyllene oxide, 1-epi-cubenol, gamma-eudesmol, cubenol; *Leaves/Stems*—Benzaldehyde, phenylethyl alcohol, trans-pinocarveol, borneol, endo-fenchyl acetate, trans-pinocarveyl acetate, alpha-copaene, cis-alpha-bergamotene, allo-aromadendrene, alpha-acoradiene, beta-bisabolene, 6,11-oxido-acor-4-ene, (e)-nerolidol, caryophyllene oxide, beta-copaen-4alpha-ol, cedrol, gossonorol, alpha-cadinol, alpha-bisabolol, (Z)-alpha-santalol, (E,E)-farnesol, (E,Z)-farnesol

PREFERRED COMPOSITION FOR CLINICAL USE:
Insufficient samples/data to determine.

REPORTED THERAPEUTIC PROPERTIES: Analgesic, anti-inflammatory, antispasmodic, anticonvulsive, antioxidant, anticancer, diuretic, aids digestion, relieves diarrhea, antimicrobial, antibacterial, antifungal, soothes irritated skin, stress management, relieves anxiety

CAUTIONS:
- None currently known.

SELECTED EVIDENCE:
- The spathulenol CT of Brazilian guava essential oil demonstrated significant antioxidant activity, anti-inflammatory activity (oral administration of 100 or 300 mg/kg in mice), antimycobacterial activity (in a pleurisy model), potently killed ovarian cancer cells, and was active against glioma, breast, multidrug-resistant ovarian, lung, prostate, colon, and leukemia cancer cells.[688]
- Oral administration (400 mg/kg body weight) of an unknown CT of Brazilian guava leaf essential oil reduced pain and inflammation in rats.[689]

BRAZILIAN LEMONGRASS

Pectis brevipedunculata

FAMILY: Asteraceae
NOTE: Top-Middle
AROMA INTENSITY: Strong
AROMA: Lemony, grassy, fresh, earthy
COMMON EXTRACTION METHOD: Steam distilled or hydrodistilled from the aerial parts

POSSIBLE SUBSTITUTE OILS: Lemongrass, lemon verbena, melissa

BLENDS WELL WITH: Basil, bergamot, black pepper, camphor, cardamom, cedarwood, citronella, clary sage, copaiba, cypress, fennel, geranium, ginger, grapefruit, lavender, lavandin, lemon, lemongrass, lemon verbena, lime, melaleuca (tea tree), myrrh, neroli, orange, pine, tangerine, thyme, vetiver, ylang ylang

SUBCELLULAR LOCALIZATION | EPIGENETIC INFLUENCE: Currently unknown | Currently unknown

RECOMMENDED DILUTION RANGE: 1.5%–20%; 50% for some conditions

PRIMARY CONSTITUENTS:[690,691]

Geranial	46.2%–49.2%
Neral	32.5%–35.6%
Geraniol	5.1%–6.6%
Limonene	2.9%–4.5%
Alpha-Pinene	2.6%–3.3%

OTHER CONSTITUENTS: Longipinene epoxide, nerol, 1-tridecene, beta-elemene, neryl acetate, 4-isopropylcyclohexanol

PREFERRED COMPOSITION FOR CLINICAL USE:

Constituent	
Geranial	35%–50%
Neral	25%–40%
Geraniol	3%–10%
Limonene	1%–8%
Alpha-Pinene	1%–7%
Nerol	0.1%–5%

REPORTED THERAPEUTIC PROPERTIES: Antibacterial, antifungal, antiviral, antimicrobial, anti-infectious, antiparasitic, aids cardiovascular function, lowers high blood pressure, digestive aid, expels excess gas, relieves diarrhea, anti-inflammatory, analgesic, anticancer, antioxidant, antiseptic, balances blood sugar, antidepressant, stress management, reduces negative emotions associated with change

CAUTIONS:

- Caution is advised during pregnancy and lactation due to high citral content. Large doses of citral may negatively affect fetal development according to animal studies.[692]
- There is a moderate risk that when wire grass is taken orally it may interfere with enzymes responsible for metabolizing medications (NSAIDs, proton-pump inhibitors, acetaminophen, antiepileptics, immune modulators, blood-sugar medications, blood-pressure medications, antidepressants, antipsychotics, diabetic medications, antihistamines, antibiotics, and anesthetics) due to the citral content.[693,694,695,696]
- Use cautiously with diabetic medications. Large oral doses of citral may improve insulin sensitivity and lower blood-glucose levels according to animal research.[697,698]
- May interact with antibiotics and possibly enhance their effects due to citral content.[699,700]
- There is a moderate risk that wire grass may interfere with pentobarbital and other barbiturates (medications for anxiety and insomnia) based on citral content.[701]
- Dilution is recommended for topical application due to high citral content and the risk for skin irritation or sensitivity.[702,703] May also irritate mucous membranes.

Selected Evidence:

- Brazilian lemongrass essential oil triggered vasorelaxation in isolated rat aorta by affecting the NO/cyclic GMP pathway and calcium influx.[704] Citral also produced vasorelaxation, suggesting both Brazilian lemongrass and citral may lower blood pressure.
- *In vitro* research demonstrated that Brazilian lemongrass essential oil inhibited bacteria (*E. coli, S. aureus, S. epidermidis,* and *T. rubrum*) and fungi (*A. niger, C. albicans, C. neoformans, F. pedrosoi, M. canis,* and *M. gypseum*).[705]

BROOM | BROOM ABSOLUTE (Spanish Broom, Weaver's Broom, Genet)

Spartium junceum L., *Genista juncea* Scop.

Family: Fabaceae
Note: Middle-Base
Aroma Intensity: Medium-Strong
Aroma: Sweet, floral, honey-like, fresh
Common Extraction Method: Steam distilled or hydrodistilled from the flowers or flower buds
Possible Substitute Oils: Jasmine, rose
Blends Well With: Bergamot, chamomile (German, Roman), clary sage, clove, coriander, frankincense, geranium, ginger, grapefruit, lavender, lemon, lime, Mediterranean mandarin, neroli, orange, palmarosa, patchouli, petitgrain, rose, rosewood, sandalwood, tangerine, vetiver, ylang ylang
Subcellular Localization | Epigenetic Influence: Currently unknown | Currently unknown
Recommended Dilution Range: *Essential Oil*—3%–33%; *Absolute*—1.0%–3%

Primary Constituents:[706,707,708,709,710,711]

Hydrodistilled (Tricosane CT)

Tricosane	22.9%
Pentacosane	16.1%
Tetracosane	8.9%
Octadecanyl Acetate	3.7%

Hydrodistilled (Alpha-Thujene CT)

Alpha-Thujene	28.9%
Dimethyl 1-Methyl-1H-Indole-4,6-Dicarboxylate	15.8%
Para-Cymene	13.8%
Gamma-Terpinene	12.9%
Unknown Constituent	11.6%
Nerolidol (isomer not identified)	6.4%

*Hydrodistilled (Camphenol CT)**

Camphenol	28.7%
Ethyl Tetradecanoate	25.0%
Camphor	14.8%
Ethyl Dodecanoate	14.4%
Ethyl Hexadecanoate	4.1%

Hydrodistilled (Flower Buds)

Alpha-Thujene	27.4%
Para-Cymene	19.6%
Gamma-Terpinene	16.4%
Alpha-Terpinene	12.4%
Myrcene	9.2%
Sabinene	6.5%
1-Octen-3-ol	6.5%

Absolute I

Linolenic Acid	15.8%–26.0%
Palmitic Acid	20.9%–24.0%
Ethyl Palmitate	1.9%–14.6%
Linalool	1.7%–10.9%
Methyl Linoleate	0.0%–7.1%
Oleic Acid	5.9%–6.0%
Ethyl Oleate	0.0%–4.9%
Linoleic Acid	4.0%–4.7%
Capric Acid	0.0%–4.1%
Ethyl Stearate	0.0%–3.7%
Myristic Acid	2.9%–3.5%
Stearic Acid	0.0%–3.5%
Linalyl Acetate	0.0%–3.4%
1-Octren-3-ol	1.2%–3.3%

Absolute II	
n-Hexadecanoic Acid	14.3%
9,12,15-Octadecatrien-1-ol	13.1%
Tetradecanoic Acid	6.6%
Octadecanoic Acid	3.7%
Gamma-Sitosterol	3.7%
Pentacosane	3.2%

* Interestingly, the study authors list linalool (26.18%), tetradecanoic acid (22.83%), camphor (13.50%), and dodecanoic acid (13.09%) as the major constituents of the oil but their analysis does not reflect this and instead reflects what is listed here.

OTHER CONSTITUENTS: *Hydrodistilled (Tricosane CT)*—Alpha-thujene, 1-octen-3-ol, para-cymene, hexyl 3-methylbutyrate, 3-dodecanone, (E,E)-alpha-farnesene, ethyl tetradecanoate, ethyl hexadecanoate, methyl octadecanoate; *Hydrodistilled (Alpha-Thujene CT)*—Alpha-pinene, myrcene; *Hydrodistilled (Camphenol CT)*—Limonene, alpha-thujone (< 0.7%), alpha-farnesene, caryophyllene oxide, beta-eudesmol, tetradecanoic acid, methyl ester, methyl linoleate, tricosane; *Hydrodistilled (Flower Buds)*—Carvacrol; *Absolute I*—Ethyl myristate, (E,E)-alpha-farnesene, beta-farnesene, methyl linolenate, 2-phenylethnaol; *Absolute II*—Dodecanoic acid, dodecanoic acid, ethyl ester, tetradecanoic acid methyl ester, octadecane, 2,6,10,14-tetramethyl-hexadecane, 1,2-benzenedicarboxylic acid, bis(2-methylpropyl) ester, hexadecanoic acid, methyl ester, eicosane, cyclic octaatomic sulfur, heneicosane, (Z,Z,Z)-9,12,15-octadecatrienoic acid, ethyl ester, octadecanoic acid, ethyl ester, docosane, tetracosane, 2,4-bis(1-phenylethyl)phenol, dehydroabietic acid, bis(2-ethylhexyl) phthalate, heptacosane, squalene, 4-(benzylamino)-1,3-diphenyl-5,6,7,8-tetrahydroquinolin-2(1H)-one, vitamin E, campesterol, stigmasterol

PREFERRED COMPOSITION FOR CLINICAL USE:
Too great of variance in composition to determine.

REPORTED THERAPEUTIC PROPERTIES: Aids cardiovascular function, diuretic, aids detoxification, relieves constipation, nourishes the skin, wound healing, promotes restful sleep, aphrodisiac, reduces anger, combats grief, reduces traumatic memories

CAUTIONS:

Hydrodistilled (Camphenol CT)

- Avoid with epilepsy and Parkinson's disease due to moderate camphor content. Camphor may exacerbate or cause seizures/convulsions or reduce seizure-medication efficacy.[712,713,714,715]
- The moderate camphor content may negatively impact red blood cells and increase the risk of jaundice in children with glucose-6-phosphate dehydrogenase deficiency (G6PD).[716,717]

Absolutes

- Avoid oral consumption. Absolutes contain solvent residue and should not be taken orally.
- May be mildly irritating when applied topically—dilution below 3% recommended.[718]

Other Hydrodistilled Essential Oils

- None currently known.

SELECTED EVIDENCE:

- None found.

BROOM WEED (Alecrim, Vassoura, Alecrim do Campo)

Baccharis dracunculifolia DC

FAMILY: Asteraceae
NOTE: Middle
AROMA INTENSITY: Medium
AROMA: Herbaceous, green, fresh, woody
COMMON EXTRACTION METHOD: Steam distilled or hydrodistilled from the aerial parts, leaves, or leaf buds; may also be distilled from the propolis
POSSIBLE SUBSTITUTE OILS: Siam wood, niaouli CT (E)-nerolidol, guava leaf CT (E)-nerolidol, helichrysum, lavender, spikenard
BLENDS WELL WITH: Bergamot, cedarwood, coriander, eucalyptus, fennel, helichrysum, juniper, lavandin, lavender, lemon, lime, melaleuca (tea tree), myrtle, niaouli, orange, peppermint, pine, Siam wood, spearmint, tangerine
SUBCELLULAR LOCALIZATION | EPIGENETIC INFLUENCE: Currently unknown | Currently unknown
RECOMMENDED DILUTION RANGE: 3%–50%; neat for some conditions

PRIMARY CONSTITUENTS:[719,720,721,722,723,724,725,726,727]

(E)-Nerolidol CT (Leaves or Aerial Parts)

(E)-Nerolidol	14.8%–33.5%
Beta-Pinene	0.4%–18.3%
Spathulenol	2.9%–17.0%
Limonene	0.4%–12.7%
Bicyclogermacrene	0.0%–11.3%
Germacrene D	0.0%–9.8%
Trans-Caryophyllene	0.4%–6.5%
Alpha-Pinene	0.2%–5.8%
Delta-Cadinene	1.0%–5.5%
Viridiflorol	0.0%–5.0%
Alpha-Muurolol	0.0%–4.7%
Globulol	0.0%–4.2%

Beta-Pinene CT (Aerial Parts)

Beta-Pinene	9.9%–43.4%
(E)-Nerolidol	0.0%–22.0%
Viridiflorol	0.8%–16.4%

Germacrene D CT (Aerial Parts)

Germacrene D	27.4%
Bicyclogermacrene	20.3%
Alpha-Humulene	16.3%
Germacrene A	11.2%
Beta-Pinene	4.7%
Alpha-Copaene	4.2%
Delta-Cadinene	3.4%

Propolis

(E)-Nerolidol	6.6%
Linalool	5.0%
1-Phenyl-Ethanone	4.9%
Trans-Caryophyllene	4.1%
Spathulenol	3.6%
Delta-Cadinene	3.3%
Aromadendrene	3.1%

Limonene	1.2%–13.2%
Delta-Cadinene	0.0%–12.1%
Spathulenol	1.0%–11.7%
T-Cadinol	0.5%–11.6%
Trans-Caryophyllene	0.4%–7.7%
Alpha-Cadinol	0.0%–6.7%
Alpha-Muurolene	0.1%–6.7%
Gamma-Cadinene	0.0%–5.5%
Alpha-Pinene	1.8%–5.3%
1,5,5-Trimethyl-6-Methylene-Cyclohexene	0.0%–5.3%
Germacrene D	0.2%–4.6%
Bicyclogermacrene	0.0%–4.2%
Globulol	0.0%–3.5%
Delta-Amorphene	0.0%–3.0%

OTHER CONSTITUENTS: *(E)-Nerolidol CT (Leaves or Aerial Parts)*—Sabinene, myrcene, (E)-beta-ocimene, alpha-gurjunene, aromadendrene, linalool, beta-elemene, alpha-humulene, alloaromadendrene, gamma-gurjunene, gamma-muurolene, germacrene A, gamma-cadinene, trans-calamenene, guaiol, beta-oplopenone, 1-epi-cubenol, gamma-eudesmol, T-cadinol, T-muurolol, alpha-cadinol, alpha-bisabolol, epi-alpha-bisabolol, bisabolene, benzeneacetonitrile, alpha-terpineol, beta-guaiene; *Beta-Pinene CT (Aerial Parts)*—Sabinene, myrcene, (E)-beta-ocimene, alpha-terpineol, alpha-copaene, alpha-gurjunene, aromadendrene, alpha-humulene, alloaromadendrene, gamma-gurjunene, gamma-muurolene, alpha-amorphene, ar-curcumene, ledene, trans-calamenene, beta-bisabolene, alpha-cadinene, elemol, ledol, guaiol, beta-oplopenone, 1,10-di-epi-cubenol, 1-epi-cubenol, gamma-eudesmol, T-muurolol, alpha-muurolol, beta-bisabolol, alpha-bisabolol, epi-alpha-bisabolol, farnesol, bisabolene, alpha-zingiberene, beta-amorphene, 4-ethylbenzoic acid, but-3-yn-2-yl ester, hotrienol, guaia-9,11-diene, alpha-selinene, patchoulene, cubenene, isoarodendrene epoxide, beta-elemene; *Germacrene D CT (Aerial Parts)*—Alpha-gurjunene, alpha-guaiene, spathulenol, palustrol; *Propolis*—Alpha-pinene, beta-pinene, limonene, benzeneacetonitrile, alpha-terpineol, ethyl 3-phenylpropionate, alpha-humulene, globulol

PREFERRED COMPOSITION FOR CLINICAL USE:
Insufficient samples/data to determine.

REPORTED THERAPEUTIC PROPERTIES: Aids digestion, **relieves heartburn and stomach ulcers**, eases nausea, aids respiratory and sinus function, analgesic, anti-inflammatory, antirheumatic, reduces fever, supports liver and spleen function, liver protective, cleanses the urinary tract, diuretic, antioxidant, **promotes oral hygiene**, antiseptic, antibacterial, antifungal, antiviral, antiparasitic, wound healing, anticancer, insecticidal, stimulating, antidepressant, reduces fear, aids concentration, promotes self-esteem

CAUTIONS:

(E)-Nerolidol CT (Leaves or Aerial Parts)

■ May interfere with antibiotics and increase their effectiveness due to (E)-nerolidol content. (E)-nerolidol sensitizes bacteria to antibiotics.[728]

■ May interfere with enzymes (CYP1A, CYP2B, CYP3A) responsible for metabolizing medications based on alpha-bisabolol content (NSAIDs, proton-pump inhibitors, acetaminophen, antiepileptics, immune modulators, blood-sugar medications, blood pressure medications, antidepressants, antipsychotics, diabetic medications, antihistamines, antibiotics, and anesthetics) due to (E)-nerolidol content.[729]

All Other CTs

- None currently known.

Selected Evidence:

- Elevated concentrations of the (E)-nerolidol CT of broom weed essential oil exerted a cytotoxic activity against human laryngeal epidermoid cancer cells.[730] The propolis was more effective than the essential oil.
- Oral administration (50, 250, and 500 mg/kg body weight) of the (E)-nerolidol CT of broom weed essential oil significantly prevented the formation of stomach ulcers caused by ethanol in rats.[731] Isolated (E)-nerolidol also reduced ulcers caused by stress (the 500 mg/kg was more effective than the drug cimetidine), ethanol (both the 250 and 500 mg/kg doses reduced ulcers better than the drug omeprazole), and indomethacin (the 500 mg/kg was more effective than the drug cimetidine). Another study using the same doses of broom weed essential oil CT (E)-nerolidol found that the oil is antiulcerogenic due to its ability to reduce gastric juice volume and total acidity and increase stomach acid pH after an ulcer was caused by indomethacin (an NSAID).[732]
- The (E)-nerolidol CT of broom weed essential oil moderately inhibited *S. aureus* and *E. coli* in laboratory research.[733] Broom weed essential oil (unidentified CT) inhibited *Candida* spp. (*C. albicans, C. glabrata, C. parapsilosis, C. tropicalis*; drug-sensitive and -resistant strains) isolated from the mouths of breastfeeding infants and the mouths and nipples of breastfeeding women.[734]
- Broom weed essential oil CT (E)-nerolidol exhibited antiviral activity against the poliovirus.[735]
- *In vitro* research showed that an unidentified CT of broom weed essential oil inhibited *S. mutans* (a bacterium associated with dental cavities) and prevented its biofilm formation.[736] A clinical study also reported that rinsing with a broom weed essential oil (unidentified CT) mouthwash four times a day worked as well as two commercial mouthwashes (Listerine and Plax) in reducing dental plaque and promoting oral hygiene.[737]
- The (E)-nerolidol CT of broom weed essential oil was active against *Leishmania donovani* and highly active against *Schistosoma mansoni* adult worms.[738]
- Broom weed essential oil CT germacrene D exhibited weak antioxidant activity.[739]
- Both (E)-nerolidol and broom weed essential oil CT (E)-nerolidol caused greater than 90% mortality of unengorged larvae of *Rhipicephalus microplus* (cattle tick) and significantly reduced the quantity and quality of eggs produced by engorged females.[740]
- The beta-pinene CT of broom weed essential oil was lethal to blowflies (*Cochliomyia macellaria*), causing larvae cuticle abnormalities and adult deformity.[741]

BUGLE
(Lungememen, Hodam)

Ajuga bracteosa Wall. ex Benth.

Family: Lamiaceae (Labiatae)
Note: Middle
Aroma Intensity: Medium
Aroma: Resinous, fresh, herbaceous, slightly sweet, spicy, and fruity
Common Extraction Method: Steam distilled from the aerial parts or leaves
Possible Substitute Oils: Sweet inula, spike lavender, lavandin, elemi

BLENDS WELL WITH: Cassia, cedarwood, chamomile (German, Roman), cinnamon, copaiba, cypress, geranium, hinoki, lavender, myrrh, patchouli, pine, rose, sandalwood, spikenard, spruce (black)
SUBCELLULAR LOCALIZATION | EPIGENETIC INFLUENCE: Currently unknown | Currently unknown
RECOMMENDED DILUTION RANGE: 3%–50%; neat for some conditions

PRIMARY CONSTITUENTS:[742,743]

Aerial Parts		*Leaves*	
Borneol	20.8%	Limonene	18.4%
Hexadecanoic Acid	16.0%	Alpha-Phellandrene	15.5%
(Z,Z)-9,12-Octadecadienoic acid	7.0%	Alpha-Humulene	15.5%
(E)-Phytol	5.6%	Beta-Myrcene	14.6%
Caryophyllene Oxide	5.1%	Elemol	13.0%
1-Octen-3-ol	4.8%	Camphene	12.9%
Camphor	4.4%	Beta-Caryophyllene	8.2%
n-Tetradecanoic Acid	3.3%		
Alpha-Fenchol	3.1%		
6,10,14-Trimethylpentadecane-2-one	3.0%		

OTHER CONSTITUENTS: *Aerial Parts*—Linalool, alpha-terpineol, nonanoic acid, eugenol, decanoic acid, methyl eugenol (< 0.7%), (E)-beta-caryophyllene, trans-alpha-bergamotene, 7-epi-alpha-selinene, n-dodecanoic acid, humulene epoxide II, T-cadinol, beta-eudesmol, eudesm-11-en-4-alpha-ol, bulnesol, pentadecanoic acid, cembrene A, manool; *Leaves*—Not listed

PREFERRED COMPOSITION FOR CLINICAL USE:
Insufficient samples/data to determine.

REPORTED THERAPEUTIC PROPERTIES: Anticancer, balances blood sugar, aids digestion, antiparasitic, antibacterial, antifungal, antiseptic, anti-inflammatory, analgesic, antispasmodic, relieves toothache, supports respiratory function, reduces blood pressure, wound healing, calming/relaxing, enhances meditation, relieves anxiety

CAUTIONS:
- None currently known.

SELECTED EVIDENCE:

- *In vitro* research reported that bugle leaf essential oil potently inhibited *S. aureus*.[744] Bugle aerial parts essential oil potently inhibited *S. aureus* and *B. subtilis*, with an effectiveness greater than amoxicillin.[745]
- Bugle aerial parts essential oil exhibited significant antioxidant activity in the DPPH assay (78% inhibition at 1.0 mg/mL).[746]

BUR CHERVIL
(Bur-Chervil)

Anthriscus caucalis M. Bieb., *A. neglecta* Boiss. & Reut. var. *scandix* (Scop.) Hyl.

FAMILY: Apiaceae (Umbelliferae)
NOTE: Middle
AROMA INTENSITY: Medium-Strong
AROMA: Green, musty, herbaceous
COMMON EXTRACTION METHOD: Hydrodistilled or steam distilled from the aerial parts, stems, leaves, or fruits
POSSIBLE SUBSTITUTE OILS: Wormwood CT (Z)-Chrysanthenyl Acetate, feverfew
BLENDS WELL WITH: Blue tansy, chamomile (German, Roman), clary sage, copaiba, feverfew, frankincense, lavender, mugwort, myrrh, sage, Spanish sage, spruce (black)
SUBCELLULAR LOCALIZATION | EPIGENETIC INFLUENCE: Currently unknown | Currently unknown
RECOMMENDED DILUTION RANGE: 3%–33%; 50% for some conditions

PRIMARY CONSTITUENTS:[747,748]

(Z)-Chrysanthenyl Acetate CT	
(Z)-Chrysanthenyl Acetate	52.8%–73.6%
(Z)-Chrysanthenol	2.1%–16.1%
(E)-Beta-Farnesene	1.6%–9.7%
Caryophyllene Oxide	0.2%–6.7%
Myrcene	0.0%–5.4%
n-Dodecanal	0.9%–5.2%
Beta-Caryophyllene	0.3%–4.0%
Alpha-Pinene	0.0%–3.7%
Limonene	0.0%–2.9%
Geranyl Acetate	0.2%–2.6%
(E)-Nerolidol	0.1%–2.6%

Beta-Bisabolene CT	
Beta-Bisabolene	28.4%
Germacrene D	18.9%
(Z,E)-Alpha-Farnesene	16.7%
Gamma-Muurolene	7.3%
Alpha-Farnesene	3.0%
Beta-Caryophyllene	2.9%
Alpha-Himachalene	2.6%
Delta-Cadinene	2.5%

OTHER CONSTITUENTS: *(Z)-Chrysanthenyl Acetate CT*—Para-cymene, (E)-beta-ocimene, alpha-terpinolene, estragole (< 1.1%), bornyl acetate, neryl acetate, n-decanol, n-dodecanol, n-tetradecanal, germacrene D, (E,E)-alpha-farnesene; *Beta-Bisabolene CT*—Linalool, alpha-copaene, beta-cubebene, beta-elemene, (E)-beta-farnesene, (Z)-beta-farnesene, bicyclogermacrene, beta-elemenone, T-cadinol, alpha-cadinol, hexadecanoic acid

PREFERRED COMPOSITION FOR CLINICAL USE:
Insufficient samples/data to determine.

REPORTED THERAPEUTIC PROPERTIES: Anticancer, antibacterial, antifungal, antioxidant, antiseptic, reduces high blood pressure, wound healing, soothes chronic skin conditions, balances menstruation, reduces feelings of overwhelm, aids concentration and focus, stress management

CAUTIONS:
- None currently known.

Selected Evidence:

- *In vitro* research shows that bur chervil essential oil (beta-bisabolene CT) is potently cytotoxic to liver and breast cancer cells.[749]
- Bur chervil essential oil (beta-bisabolene CT) significantly inhibited *B. subtilis* and *E. coli*.[750]
- Bur chervil essential oil (beta-bisabolene CT) exhibited moderate antioxidant activity in the ferric ion reducing activity (FRAP) assay and weak activity in the DPPH assay.[751]

CABREUVA (Óleo-pardo)

Myrocarpus fastigiatus Allemão

Family: Fabaceae (Leguminosae)
Note: Middle-Base
Aroma Intensity: Medium
Aroma: Woody, sweet, slightly floral
Common Extraction Method: Steam distilled from the wood (sawdust, waste shavings)
Possible Substitute Oils: Niaouli CT (E)-nerolidol, blue mountain sage CT (E)-nerolidol, guava leaf CT (E)-nerolidol
Blends Well With: Bergamot, cassia, cedarwood, cinnamon, lavender, geranium, may chang, palmarosa, tangerine, ylang ylang
Subcellular Localization | Epigenetic Influence: Currently unknown | Currently unknown
Recommended Dilution Range: 3%–33%; 50% for some conditions

Primary Constituents:[752]

(E)-Nerolidol	77.0%
Alpha-Bisabolol	16.0%
(E,E)-Farnesol	2.1%

Other Constituents: Not listed

Preferred Composition for Clinical Use:
Insufficient samples/data to determine.

Reported Therapeutic Properties: Anticancer, antimicrobial, antiviral, anti-inflammatory, antioxidant, aids respiratory function, opens the airways, expels excess mucus, decreases high blood pressure, supports prostate and testis function, soothing to the skin, wound healing, protects kidney and liver function, insecticidal, expands vision, promotes inner strength, relieves anxiety, alleviates obsessions

Cautions:
- May interfere with antibiotics and increase their effectiveness—(E)-nerolidol and alpha-bisabolol sensitize bacteria to antibiotics.[753]

Selected Evidence:

- Higher concentrations of cabreuva essential oil inhibited both gram-negative and gram-positive bacteria when tested against eleven different strains.[754]

CADE
(Prickly Juniper, Cade Juniper, Prickly Cedar)

Juniperus oxycedrus L., *J. oxycedrus* L. ssp. *oxycedrus*, *J. oxycedrus* ssp. *macrocarpa*, *J. oxycedrus* ssp. *Rufescens*, *J. oxycedrus* ssp. *badia*

FAMILY: Cupressaceae
NOTE: Base-Middle
AROMA INTENSITY: Medium-Strong
AROMA: Smoky, rich, burnt, leathery, spicy, piney, fresh, peppery, woody
COMMON EXTRACTION METHOD: Extracted from the branches and wood with dry destructive distillation to produce a crude oil that is rectified to remove benzo[a]pyrene; may also be hydrodistilled or steam distilled from the branches and wood prior to rectification; may also be hydrodistilled or steam distilled from the berries, bark, needles (leaves), or cones
POSSIBLE SUBSTITUTE OILS: Juniper berry, black pepper, dwarf pine, Turkish pine, cypress, pine, copaiba, hops, hemp
BLENDS WELL WITH: Black pepper, cedarwood (Atlas, Himalayan), clove, copaiba, cypress, juniper berry, labdanum, oregano, pine, rosemary, sandalwood, Texas cedarwood, thyme, Virginia cedarwood
SUBCELLULAR LOCALIZATION | EPIGENETIC INFLUENCE: Currently unknown | Currently unknown
RECOMMENDED DILUTION RANGE: 1%–10%

PRIMARY CONSTITUENTS:[755,756,757,758,759,760,761,762,763,764,765,766,767,768,769,770,771]

J. oxycedrus (Wood, SD)

Delta-Cadinene	27.9%
Epi-Cubenol	12.1%
Cubenol	6.3%
Epizonarene	4.4%
14-Hydroxy-9-Epi-Beta-Caryophyllene	3.8%
Trans-Calamenene	3.6%
Beta-Caryophyllene	3.4%
Alpha-Humulene	3.1%

ssp. oxycedrus (Twigs, HD)

Manoyl Oxide	35.4%
Caryophyllene Oxide	16.8%
Caryophyllenol II	4.5%
Dodecanoic Acid	4.2%
Humulene Epoxide II	3.4%

ssp. oxycedrus (Berries, HD)

Alpha-Pinene	27.4%–69.4%
Germacrene D	0.0%–33.0%
Myrcene	2.1%–22.6%

ssp. oxycedrus (Wood, HD)

Delta-Cadinene	14.5%
Cis-Thujopsene	9.2%
Alpha-Muurolene	4.9%
Cadalene	3.7%

J. oxycedrus (Twigs, HD)

Alpha-Pinene	44.7%
T-Cadinol	5.5%
Alpha-Cubebene	4.5%
Germacrene D	3.5%
Alpha-Terpineol	3.1%

ssp. oxycedrus (Bark, HD)

Alpha-Pinene	72.8%
Beta-Pinene	10.7%
Myrcene	5.5%

ssp. oxycedrus (Berries, SD, AP CT)

Alpha-Pinene	62.3%–70.6%
Myrcene	10.2%–17.1%
Germacrene D	12.9%–13.8%

Alpha-Phellandrene	0.0%–7.1%
Limonene	0.0%–6.7%
Sabinene	0.2%–4.5%
Beta-Pinene	1.4%–3.2%

ssp. macrocarpa or rufescens (Berries, HD)

Alpha-Pinene	37.6%–61.1%
Germacrene D	4.0%–31.1%
Myrcene	1.5%–6.4%
Alpha-terpinyl Acetate	0.0%–5.9%
Abietadiene	0.0%–5.8%
Cis-Calamenene	0.8%–3.3%
Beta-Phellandrene	0.0%–3.4%
Beta-Caryophyllene	1.2%–3.0%

J. oxycedrus (Berries, HD)

Myrcene	45.5%–57.0%
Alpha-Pinene	10.2%–36.6%
Limonene	3.6%–13.8%
Germacrene D	1.7%–8.6%
Alpha-Cadinol	1.1%–3.9%

ssp. macrocarpa or rufescens (Needles, HD)

Alpha-Pinene	12.2%–23.8%
Sandaracopimara-8(14)-15-Diene	4.6%–17.6%
Germacrene D	0.0%–11.7%
Abietadiene	2.7%–9.8%
Beta-Phellandrene	1.9%–6.1%
Gamma-Muurolene	1.3%–5.2%
14-Hydroxy-9-Epi-Beta-Caryophyllene	0.0%–5.2%
Beta-Caryophyllene	1.2%–4.7%
Kaurene	4.4%–4.6%
Abieta-8(14),13(15)-Diene	0.0%–4.5%
Cis-Calamenene	3.4%–3.6%
Phyllocladene	0.0%–3.6%

ssp. oxycedrus (Needles, HD, AP/M CT)

Alpha-Pinene	42.0%–55.7%
Myrcene	3.4%–16.6%
Germacrene D	1.3%–8.2%
Delta-3-Carene	tr–6.0%
Para-Cymene	0.0%–5.3%
Limonene	0.4%–4.6%
Beta-Phellandrene	1.8%–3.2%
Gamma-Cadinene	2.5%–3.0%

J. oxycedrus (Berries, SD)

Alpha-Pinene	22.5%–27.1%
Limonene	2.8%–18.1%
Myrcene	11.3%–15.1%
Germacrene D	1.9%–11.5%
Sabinene	1.2%–9.2%
(E)-Beta-Caryophyllene	2.7%–6.2%
Alpha-Humulene	2.7%–4.7%
Delta-Cadinene	2.2%–4.2%
Beta-Pinene	2.6%–3.1%

ssp. oxycedrus (Berries, HD M CT)

Myrcene	44.6%
Alpha-Pinene	19.9%
Germacrene D	15.5%
Beta-Caryophyllene	4.6%
Alpha-Humulene	3.1%

ssp. macrocarpa (Needles, HD, AP CT)

Alpha-Pinene	16.0%–35.5%
Para-Cymene	3.9%–14.5%
Sabinene	9.1%–12.1%
Alpha-Terpinyl Acetate	0.4%–7.3%
13-Epi-Manoyl Oxide	2.5%–6.4%
Delta-3-Carene	0.1%–5.9%
Abietatriene	0.5%–4.5%
Gamma-Terpinene	1.0%–3.1%

ssp. oxycedrus (Needles, HD)

Alpha-Pinene	42.9%–73.3%
Limonene	1.2%–17.8%
Delta-3-Carene	0.1%–8.2%
Caryophyllene Oxide	0.3%–5.1%
Beta-Phellandrene	0.8%–4.2%
Myrcene	2.3%–3.9%
Beta-Pinene	1.5%–3.8%

ssp. oxycedrus (Needles, HD/SD)

Alpha-Pinene	31.6%–86.0%
Germacrene D	0.8%–9.0%
13-Epi-Manoyl Oxide	0.0%–5.6%
Para-Cymene	0.0%–3.6%
Alpha-Copaene	0.0%–3.4%
Cis-Calamenene	0.0%–3.0%

ssp. macrocarpa (Needles, HD, CTII)

Manoyl Oxide	7.7%–21.9%
Alpha-Pinene	7.2%–11.1%
Alpha-Cedrol	2.3%–9.7%
Widdrene	2.1%–5.7%
Alpha-Muurolene	4.1%–4.8%
Trans-Verbenol	1.7%–4.3%
Delta-Cadinene	3.2%–3.8%
Alpha-Campholene Aldehyde	1.7%–3.2%
Trans-Pinocarveol	1.5%–3.0%

ssp. macrocarpa or rufescens (Needles, SFE)

Sandaracopimara-8(14)-15-Diene	4.0%–31.3%
Phyllocladene	0.0%–19.5%
Abieta-8(14),13(15)-Diene	0.0%–15.9%
Abietadiene	4.0%–8.6%
Gamma-Muurolene	0.0%–5.7%
Kaurene	4.4%–5.6%
Cis-Totarol	1.8%–4.7%
Trans-Totarol	2.8%–4.5%
Octadecanol	3.8%–4.0%
Germacrene D	0.0%–4.0%
Alpha-Cadinol	1.3%–3.3%

ssp. badia (Needles, SD)

Alpha-Pinene	39.8%–57.0%
Manoyl Oxide	4.8%–10.2%
Unknown Compound	2.8%–6.8%

ssp. oxycedrus (Needles, HD, MO CT)

Manoyl Oxide	32.8%
Caryophyllene Oxide	11.9%
Germacrene D	5.7%
Humulene Epoxide II	4.7%
Delta-Cadinene	4.1%
Beta-Caryophyllene	3.7%
Alpha-Muurolene	3.2%
Dodecanoic Acid	3.1%

ssp. oxycedrus (Cones, HD)

Myrcene	21.7%
Alpha-Pinene	20.9%
Germacrene D	19.6%
Beta-Caryophyllene	9.8%
Beta-Pinene	8.9%
Limonene	3.4%

ssp. badia (Berries, SD)

Alpha-Pinene	18.6%–65.1%
Germacrene D	0.8%–39.5%
Myrcene	3.7%–6.8%
Delta-Cadinene	0.4%–3.9%
Trans-Pinocarveol	tr–3.5%
Manoyl Oxide	0.8%–3.3%
Alpha-Cadinol	0.0%–3.0%

SD=Steam distilled, HD=hydrodistilled, SFE=Supercritical CO_2 extraction

OTHER CONSTITUENTS: *J. oxycedrus (Wood, SD)*—Beta-elemene, germacrene D, beta-selinene, alpha-muurolene, cubenene, alpha-calacorene, gleenol, T-muurolol, torreyol, alpha-cadinol, selina-11-en-4-alpha-ol, 14-oxo-alpha-muurolene, 14-hydroxy-alpha-humulene, 14-hydroxy-alpha-muurolene, 14-hydroxy-delta-cadinene; *ssp. oxycedrus (Wood, HD)*—Beta-elemene, beta-caryophyllene, alpha-humulene, alloaromadendrene, eremophilene, epi-bicyclosesquiphellandrene, cadina-1,4-diene, alpha-cedrol, methyl palmitate; *ssp. oxycedrus (Twigs, HD)*—Alpha-pinene, sylvestrene, para-cymene, alpha-cubebene, alpha-cubebene, isobornyl acetate, beta-caryophyllene, alpha-terpineol, alpha-muurolene, delta-cadinene, alpha-calacorene, (E)-nerolidol, cedrol, torilenol, cadalene, caryophylladienol I, abietatriene; *J. oxycedrus (Twigs, HD)*—Beta-pinene, myrcene, sylvestrene, alpha-campholenal, trans-pinocarveol, borneol, terpinen-4-ol, trans-verbenone, bornyl acetate, alpha-copaene, beta-caryophyllene, beta-farnesene, alpha-curcumene, delta-cadinene, (E)-nerolidol, caryophyllene oxide; *ssp. oxycedrus (Bark, HD)*—Camphene, limonene, beta-phellandrene; *ssp. oxycedrus (Berries, HD)*—Para-cymene, camphene, sylvestrene, bornyl acetate, alpha-cubebene, beta-caryophyllene, alpha-humulene, beta-selinene, epi-bicyclosesquiphellandrene, alpha-muurolene, terpinolene, gamma-cadinene, delta-cadinene, (E,E)-farnesol, abietadiene; *ssp. oxycedrus (Berries, SD, AP CT)*—Beta-pinene, limonene; *ssp. macrocarpa or rufescens (Berries, HD)*—Camphene, beta-pinene, alpha-phellandrene, para-cymene, terpinolene, trans-pinocarveol, trans-verbenol, para-mentha-1,5-dien-8-ol, meta-cymen-8-ol, meta-alpha-terpineol, myrtenal, verbenone, trans-carveol, isobornyl acetate, carvacrol, alpha-humulene, gamma-muurolene, epi-cubebol, epizonarene, trans-cadinene, germacrene-D-4-ol, caryophyllene oxide, humulene epoxide II, 1,10-di-epi-cubenol, 1,-epi-cubenol, alpha-cadinol, acorenone, cembrene, sandracopimara-8(14)-15-diene, kaurene, octadecanol, sclareol, cis-totarol, 3-alpha-OH-manool; *J. oxycedrus (Berries, SD)*—Alpha-thujene, camphene, alpha-terpinene, gamma-terpinene, terpineolene, linalool, terpinen-4-ol, trans-sabinyl acetate, alpha-cubebene, alpha-copaene, beta-elemene, gamma-elemene, gamma-cadinene, alpha-muurolene, alpha-cadinene, germacrene B, caryophyllene oxide, humulene epoxide II, T-muurolol, alpha-cadinol; *J. oxycedrus (Berries, HD)*—Beta-pinene, terpinolene, linalool, beta-caryophyllene, gamma-cadinene, caryophyllene oxide, sesquilavandulol, torreyol, cembrene, nootkatin, manoyl oxide; *ssp. oxycedrus (Berries, HD M CT)*—Sabinene, limonene, beta-phellandrene, terpinolene, (Z)-beta-farnesene, alpha-muurolene, delta-cadinene, alpha-cadinol; *ssp. macrocarpa or rufescens (Needles, HD)*—Myrcene, delta-3-carene, alpha-humulene, meta-alpha-terpineol, alpha-terpinyl acetate, beta-caryophyllene, cis-muurola-3,5-diene, germacrene D isomer, trans-cadinene, occidentalol, caryophyllene oxide, alpha-cadinol, acorenone, 5-neo-cedranol, cembrene, (Z,E)-farnesol, bifloratriene, trans-totarol, 1,10-di-epi-Cubenol, cis-totarol; *ssp. macrocarpa (Needles, HD, AP CT)*—Beta-pinene, myrcene, alpha-phellandrene, limonene, beta-phellandrene, terpinolene, alpha-campholenal, alpha-terpineol, beta-caryophyllene, alpha-humulene, trans-delta-cadinene, cis-calamenene, occidentalol, caryophyllene oxide, T-muurolol, alpha-eudesmol, cadalene, bisabolene oxide, (E,E)-farnesol; *ssp. oxycedrus (Needles, HD)*—Alpha-fenchene, camphene, alpha-phellandrene, para-cymene, limonene, terpinolene, alpha-campholenal, trans-pinocarveol, p-menta-1,5-dien-8-ol, alpha-terpineol, germacrene D, (E)-nerolidol, (Z)-6-pentadecen-2-one, (E,E)-farnesol, 2,4-decadienal, alpha-cadinol, 1H,naphthol pyran; *ssp. oxycedrus (Needles, HD, AP/M CT)* —Beta-Pinene, terpinolene, linalool, trans-pinocarveol, terpinen-4-ol, p-cymen-8-ol, alpha-terpineol, verbenone, alpha-cubebene, beta-cubebene, beta-caryophyllene, alpha-humulene, gamma-muurolene, alpha-amorphene, alpha-muurolene, (E,E)-alpha-farnesene, delta-cadinene, (E)-nerolidol, caryophyllene oxide, alpha-muurolol, salvial-4(14)-en-1-one, manoyl oxide; *ssp. oxycedrus (Needles, HD/SD)*—Sabinene, beta-pinene, myrcene, alpha-phellandrene, delta-3-carene, gamma-terpinene, limonene, terpinolene, alpha-campholenal, terpinen-4-ol, alpha-terpineol, alpha-terpinyl acetate, beta-bourbonene, beta-caryophyllene, alpha-humulene, delta-muurolene, trans-delta-cadinene, occidentalol, caryophyllene oxide, humulene epoxide II, 1,10-di-epi-cubenol, T-muurolol, alpha-eudesmol, 5-OH-isonbornyl isobutyrate, cadalene, (Z)-alpha-santalol, bisabolene oxide, (E,E)-farnesol, abieta-8,12-diene, abietatriene; *ssp. macrocarpa (Needles, HD, CTII)*—Delta-3-carene, limonene, beta-phellandrene, (E)-beta-ocimene, para-cymene, terpinolene, gamma-campholene aldehyde, alpha-bourbonene, beta-cubebene, octanol, pinocarvone, beta-funebrene, beta-caryophyllene, myrtenal, gamma-cadinene, cubenene, beta-cuprenene, myrtenol, trans-carveol, cuparene, calamenene (isomer not identified), (E)-geranylacetone, epi-cubebol, cubebol, caryophyllene oxide, cinnamaldehyde, humulene epoxide II, 1-epi-cubenol, widdrol, T-cadinol, cis-verbenol, p-menth-1,5-dien-8-ol, epizonarene, alpha-terpineol, alpha-terpinyl acetate, verbenone, cis-muurola-4(15),5-diene, p-menth-1,5-dien-8-ol, alpha-cuprenene, 8,13-abietadiene, dodecanoic acid; *ssp. oxycedrus (Needles, HD, MO CT)*—Alpha-pinene, isobornyl acetate, gamma-cadinene, torilenol, cadalene, abietatriene; *ssp. macrocarpa or rufescens (Needles, SFE)*—Alpha-pinene, para-cymene, beta-phellandrene, alpha-humulene, trans-cadinene, cis-calamenene, 14-hydroxy-9-epi-trans-caryophyllene, meta-alpha-terpineol, alpha-terpinyl acetate, beta-caryophyllene, cis-muurola-3,5-diene, germacrene D isomer, trans-cadinene, cis-calamenene, occidentalol, caryophyllene oxide, acorenone, (Z,E)-farnesol, bifloratriene, sclareol, larixol, 3-alpha-hydroxymanool, acorenone, 5-neo-cedranol; *ssp. oxycedrus (Cones, HD)*—Delta-3-carene, alpha-terpineolene, linalool, cyclohexene,1-mehtyl-4, beta-elemene, alpha-humulene, alpha-amorphene, delta-cadinene, bicyclo[4.4.0]dec-1ene; *ssp. badia (Berries, SD)*—Camphene, beta-pinene, limonene, terpinolene, p-cymenene, camphor, trans-verbenol, borneol, delta-terpineol, alpha-terpienol, terpinen-4-ol, myrtenal, myrtenol, trans-beta-caryophyllene, verbenone, trans-carveol, alpha-humulene, beta-copaene, bornyl iso-butanoate, viridiflorene, alpha-muurolene, gamma-cadinene, caryophyllene oxide, humulene epoxide II, T-muurolol, alpha-cadinol, eudesma-4(15),7-dien-1-beta-ol, abietatriene, abietadiene; *ssp. badia (Needles, SD)*—

Camphene, thuja-2,4-diene, beta-pinene, myrcene, para-cymene, limonene, terpinolene, p-cymenene, alpha-pinene oxide, alpha-campholeal, trans-pinocarveol, camphor, camphene hydrate, delta-terpineol, p-cymen-8-ol, alpha-terpineol, myrtenal, myrtenol, verbenone, trans-carveol, beta-bourbonene, germacrene D, viridiflorene, 2-tridecanone, delta-cadinene, caryophyllene oxide, 10-epi-alpha-muurolol, eudesma-4(15),7-dien-1-beta-ol, (E,E)-farnesol, (E,Z)-farnesol, abietatriene, abietadiene

PREFERRED COMPOSITION FOR CLINICAL USE:

Constituent	*Wood*	*Berries*
Delta-Cadinene	10%–35%	–
Trans-Calamenene	2%–10%	–
Trans-Cadina-1(6),4-diene	1%–10%	–
Beta-Caryophyllene	0.1%–10%	1%–8%
Alpha-Humulene	0.1%–8%	–
Alpha-Cedrene	0.1%–8%	–
Zonarene	0.1%–7%	–
Epi-Cubenol	0.1%–7%	–
Alpha-Muurolene	0.5%–5%	–
Cubenol	0.1%–5%	–
Alpha-Calacorene	0.1%–5%	–
Cadalene	tr–5%	–
Alpha-Pinene	–	20%–50%
Myrcene	–	10%–25%
Limonene	–	2%–15%
Germacrene D	–	2%–15%
Beta-Pinene	–	0.5%–5%
Alpha-Humulene	–	tr–5%

REPORTED THERAPEUTIC PROPERTIES: Anti-inflammatory, wound healing, **reduces dandruff**, **promotes hair health**, **reduces hair loss**, heals chronic skin conditions, reduces skin irritation and itching, antiviral, antiseptic, antibacterial, antifungal, antiviral, antiparasitic, anticancer, reduces high blood pressure, balances blood-sugar levels, insect repellent, insecticidal, antidepressant, combats negative thoughts, aids meditation, stress management

CAUTIONS:

Crude cade oil contains the known carcinogen benzo[a]pyrene and should not be used by any method. Only the rectified oil that has the benzo[a]pyrene removed to ≤ 20 PPB (as opposed to 8 PPM in unrectified oil) or steam distilled or CO_2 extracted oil that you have a GC-MS analysis for indicating the absence of benzo[a]pyrene should be used. Benzo[a]pyrene is one of the carcinogenic components of cigarette smoke and is mutagenic and weakly genotoxic.[772,773,774,775,776,777]

Myrcene CTs (J. oxycedrus (Berries, HD) and ssp. oxycedrus (Berries, HD M CT))

■ Caution is advised during pregnancy and lactation due to high beta-myrcene content. Extremely high doses (500 mg/kg) of beta-myrcene have been toxic to fetuses according to animal research.[778,779]

■ May interact with barbiturates (medications for anxiety and insomnia), antihistamines, benzodiazepines, tricyclic antidepressants, or other central nervous system depressant drugs, increasing their depressant effects.[780,781,782]

■ May weakly interfere with the enzymes responsible for metabolizing medications (NSAIDs, proton-pump inhibitors, acetaminophen, antiepileptics, immune modulators, blood-sugar medications, blood pressure medications, antidepressants, antipsychotics, diabetic medications, antihistamines, antibiotics, and anesthetics), based on myrcene content.[783,784]

All other CTs

- None currently known whether benzo[a]pyrene levels are ≤ 20 PPB.

SELECTED EVIDENCE:

- □ Cade essential oil (ssp. *oxycedrus*, wood, HD) inhibited alpha-amylase activity.[785] Alpha-amylase is the main enzyme in humans responsible for the breakdown of starch into simple sugars. Inhibition of alpha-amylase may improve glucose tolerance and improve diabetes control in humans.
- □ Cade essential oil (ssp. *oxycedrus*, leaves, CT alpha-pinene 76%) inhibited nitric oxide production triggered by nuclear factor-κB activated by interleukin-1β (IL-1).[786] IL-1 is required for the expression of proinflammatory mediators that play an important role in arthritic conditions. A fraction with 93% alpha-pinene also inhibited this activity.
- □ Cade essential oil (ssp. oxycedrus, alpha-pinene and myrcene as major constituents) exhibited antiviral activity against HSV-1 (SI of 5).[787]
- □ *In vitro* research demonstrated that cade (ssp. *macrocarpa* and *rufescens*, leaves, HD and SFE) essential oil significantly inhibited *C. albicans* and inhibited *S. aureus*.[788] Cade leaf oil (ssp. *oxycedrus*, SD) exhibited weak inhibition of *C. albicans* and moderate inhibition of *S. aureus* but was ineffective against *E. coli* and *P. aeruginosa*.[789] The same research showed that the berries oil from the same subspecies was inactive against all organisms. Cade essential oil (*J. oxycedrus, Berries, SD)* showed moderate antimicrobial activity against *Streptococcus pneumoniae, Staphylococcus aureus, Streptococcus agalactiae, Streptococcus pyogenes, Corynebacterium* spp., *Escherichia coli*, and *Campylobacter jejuni*.[790] Cade essential oil (ssp. *macrocarpa* and *oxycedrus*, CT alpha-pinene, HD) both demonstrated moderate inhibition of *S. aureus* but were inactive against *E. coli*.[791] Two berry oils (ssp. *oxycedrus*, HD, CT alpha-pinene) from Morocco inhibited *E. coli, P. aeruginosa*, and *S. aureus* and the oil with the greatest alpha-pinene (69.4%) also inhibited *Enterococcus hirae*.[792] The twig oil CT alpha-pinene inhibited *S. aureus, T. rubrum*, and *C. albicans*.[793]
- □ The alpha-pinene CT of cade oil (*J. oxycedrus*) was active against *Trypanosoma brucei* without being toxic to macrophages.[794] Its activity was attributed to alpha-pinene, which also demonstrated antitrypanosomal effects.
- □ Cade (ssp. *macrocarpa* and *rufescens*, berries, HD) essential oil exhibited high antioxidant activity in the ABTS and DPPH assays.[795] Cade (ssp. *oxycedrus*, wood or berries, HD) essential oil exhibited significant antioxidant activity according to the DPPH method. Two berry oils (ssp. *oxycedrus*, HD, CT alpha-pinene) from Morocco exhibited significant antioxidant activity in the DPPH assay.[796]
- □ Cade (ssp. *oxycedrus*) essential oil in a silica gel was toxic to the rice weevil (*Sitophilus oryzae*).[797]

CAMBUCA

Eugenia edulis Vell., *Marlierea edulis* (Vell.) Nied., *Myrciaria plicatocostata* O. Berg, *M. edulis, Plinia plicatocostata* (O.Berg) Amshoff, *P. edulis* (Vell.) Sobral

FAMILY: Myrtaceae
NOTE: Middle
AROMA INTENSITY: Medium
AROMA: Woody, herbaceous, slightly sweet and fruity
COMMON EXTRACTION METHOD: Steam distilled from the leaves
POSSIBLE SUBSTITUTE OILS: Black pepper, copaiba, guava leaf
BLENDS WELL WITH: Basil, black pepper, cardamom, cassia, cedarwood, clove, copaiba, coriander, frankincense, lavender, lemon, marjoram, nutmeg, orange, palo santo, patchouli, pine, ravensara, rosemary, sandalwood, tangerine, ylang ylang
SUBCELLULAR LOCALIZATION | EPIGENETIC INFLUENCE: Currently unknown | Currently unknown
RECOMMENDED DILUTION RANGE: 5%–Neat

PRIMARY CONSTITUENTS:[798]

Beta-Caryophyllene	21.2%
Caryophyllene Oxide	18.1%
Spathulenol	11.9%
Alloaromadendrene	3.9%
Bicyclogermacrene	2.9%
Humulene Epoxide II	2.7%
Alpha-Humulene	2.6%
Alpha-Cadinol	2.6%
Alpha-Copaene	2.4%
Isospathulenol	2.4%
1,10-di-epi-Cubenol	2.3%
Delta-Cadinene	2.0%

OTHER CONSTITUENTS: Beta-bourbonene, beta-elemene, germacrene D, beta-selinene, beta-bisabolene, gamma-cadinene, globulol, epiglobulol, eudesmol, caryophylladienol II, T-muurolol

PREFERRED COMPOSITION FOR CLINICAL USE:
Insufficient samples/data to determine.

REPORTED THERAPEUTIC PROPERTIES: Anti-inflammatory, analgesic, wound healing, antibacterial, antifungal, relieves diarrhea, aids digestion, relieves stomachache, heals stomach ulcers, supports respiratory function, opens the airways, regulates blood sugar, diuretic, promotes courage and self-confidence, reduces mental fatigue, sharpens focus

CAUTIONS:
- None currently known.

SELECTED EVIDENCE:
- ☐ None found.

CAMPHOR THYME
(Portuguese Thyme)

Thymus camphoratus Hoffm. & Link.

FAMILY: Lamiaceae (Labiatae)
NOTE: Middle
AROMA INTENSITY: Medium
AROMA: Herbaceous, fresh, medicinal, slightly floral, spicy, and woody
COMMON EXTRACTION METHOD: Steam distilled or hydrodistilled from the flowering aerial parts; may also be distilled from the leaf or flower only
POSSIBLE SUBSTITUTE OILS: Rosalina, rosemary, marjoram, myrtle, eucalyptus, lavandin, spike lavender
BLENDS WELL WITH: Balsam fir, bergamot, blue cypress, blue spruce, blue tansy, cajeput, citronella, cypress, Douglas fir, geranium, grapefruit, lavender, lavandin, lemon, lime, Manuka, niaouli, orange, palmarosa, palo santo, Siberian fir, silver fir, spike lavender, spruce (black), sweet inula, tangerine, tea tree (melaleuca), thyme, white fir
SUBCELLULAR LOCALIZATION | EPIGENETIC INFLUENCE: Currently unknown | Currently unknown
RECOMMENDED DILUTION RANGE: 3%–50%; neat for some conditions

PRIMARY CONSTITUENTS:[799,800,801,802]

1,8-Cineole CT	
1,8-Cineole	29.0%–33.0%
Borneol	4.7%–20.0%
Alpha-Pinene	5.4%–8.1%
Camphene	3.6%–6.8%
Linalool	4.7%–6.0%
T-Cadinol	0.0%–5.9%
Camphor	4.6%–4.8%
Geraniol	0.1%–3.8%

Borneol CT	
Borneol	12.7%
Linalool	11.4%
Linalyl Acetate	6.3%
Camphor	6.1%
T-Cadinol	6.1%
Alpha-Pinene	5.4%
Camphene	4.2%
1,8-Cineole	3.8%

Leaf	
1,8-Cineole	62.7%
Limonene	9.9%

Linalool CT	
Linalool	16.6%
Linalyl Acetate	15.2%
1,8-Cineole	10.8%
Intermedeol	6.2%
Alpha-Terpineol	3.4%

Trans-Sabinene Hydrate CT	
Trans-Sabinene Hydrate	19.6%
Terpinen-4-ol	7.8%
Sabinene	5.3%
Borneol	4.0%
1,8-Cineole	3.8%

Flower	
1,8-Cineole	56.7%
Limonene	9.0%
Sabinene	3.7%

OTHER CONSTITUENTS: *1,8-Cineole CT*—Sabinene, beta-pinene, limonene, myrcene, para-cymene, gamma-terpinene, cis-sabinene hydrate, alpha-campholenal, trans-pinocarveol, trans-verbenol, terpinen-4-ol, alpha-terpineol, bornyl acetate, alpha-terpinyl acetate, geranyl acetate, caryophyllene oxide, alloaromadendrene, gamma-cadinene, viridiflorol, epi-cubenol; *Linalool CT*—Alpha-pinene, camphene, myrcene, cis-beta-ocimene, trans-beta-ocimene, cis-linalool oxide, trans-linalool oxide, camphor, trans-verbenol, nerol oxide, borneol, nerol, geraniol, decanol, bornyl acetate, geranyl acetate, beta-caryophyllene, gamma-cadinene, caryophyllene oxide, T-cadinol, delta-cadinol; *Borneol CT*—Sabinene, myrcene, limonene, trans-beta-ocimene, cis-sabinene hydrate, trans-pinocarveol, trans-verbenol, terpinene-4-ol, alpha-terpineol, verbenone, carvone, geraniol, bornyl acetate, geranyl acetate, beta-caryophyllene, delta-cadinene, germacrene D-4-ol, caryophyllene oxide, ledol, intermedeol, alpha-cadinol; *Trans-Sabinene Hydrate CT*—Alpha-pinene, camphene, alpha-terpinene, para-cymene, limonene, trans-beta-ocimene, gamma-terpinene, terpinolene, cis-sabinene hydrate, linalool, alpha-campholenal, trans-p-menthen-1-ol, camphor, cis-verbenol, trans-verbenol, alpha-terpineol, verbenone, cumin alcohol, alpha-terpinyl acetate, beta-caryophyllene, germacrene D, bicyclogermacrene, gamma-cadinene, delta-cadinene, Elemol, germacrene D-4-ol, caryophyllene oxide, ledol, intermedeol, alpha-cadinol; *Leaf*—Alpha-pinene, sabinene, beta-pinene, para-cymene, delta-terpineol, alpha-terpineol, terpinene-4-ol, caryophyllene oxide, T-cadinol, intermedeol; *Flower*—Alpha-pinene, beta-pinene, para-cymene, linalool, delta-terpineol, terpinene-4-ol, alpha-terpineol, caryophyllene oxide, T-cadinol, intermedeol

PREFERRED COMPOSITION FOR CLINICAL USE:
Insufficient data/samples to determine.

REPORTED THERAPEUTIC PROPERTIES: Anti-inflammatory, analgesic, **supports respiratory function**, eases cough, decongestant, antioxidant, antimicrobial, antiallergic, reduces skin itching and chronic skin conditions, reduces the appearance of acne and blemishes, heals blisters and boils, promotes restful sleep, relieves anxiety, stress management, boosts self-confidence

CAUTIONS:

1,8-Cineole CT

- Avoid with epilepsy and Parkinson's disease due to 1,8-cineole and camphor content. May exacerbate or cause seizures/convulsions or reduce seizure-medication efficacy.[803,804,805,806]
- May weakly interfere with the enzymes responsible for metabolizing medications (NSAIDs, proton-pump inhibitors, acetaminophen, antiepileptics, immune modulators, blood-sugar medications, blood pressure medications, antidepressants, antipsychotics, diabetic medications, antihistamines, antibiotics, and anesthetics) due to 1,8-cineole content.[807,808,809]
- May interfere with pentobarbital and other barbiturates (medications for anxiety and insomnia), based on 1,8-cineole content.[810,811]

Linalool CT, Borneol CT, and Trans-Sabinene Hydrate CT

- None currently known.

Leaf or Flower Oil

- Avoid with children under age 3, particularly around the nose and mouth and use cautiously in children under age 5 due to significant 1,8-cineole content. 1,8-cineole may cause seizures, central nervous system problems, or respiratory distress in young children.[812,813,814]
- Caution is warranted orally due to the significant amounts of 1,8-cineole. Limit it to small doses internally (adults—maximum 10 drops daily). Toxicity has been reported when eucalyptus (also high in 1,8-cineole) was ingested in large doses, and as few as 4 to 5 drops may cause problems in very sensitive individuals.[815,816,817,818,819,820]

■ Avoid with epilepsy and Parkinson's disease due to 1,8-cineole content. May exacerbate or cause seizures/convulsions or reduce seizure-medication efficacy based on 1,8-cineole content.[821,822,823,824]

■ May weakly interfere with the enzymes responsible for metabolizing medications (NSAIDs, proton-pump inhibitors, acetaminophen, antiepileptics, immune modulators, blood-sugar medications, blood pressure medications, antidepressants, antipsychotics, diabetic medications, antihistamines, antibiotics, and anesthetics) due to 1,8-cineole content.[825,826,827]

■ May interfere with pentobarbital and other barbiturates (medications for anxiety and insomnia), based on 1,8-cineole content.[828,829]

SELECTED EVIDENCE:

- □ The 1,8-cineole CT of camphor thyme essential oil demonstrated good anti-inflammatory activity by inhibiting excess nitric oxide production and inhibiting proinflammatory enzymes (inducible nitric oxide synthase (iNOS) and cyclooxygenase-2 (COX-2)).[830]
- □ *In vitro* research showed that camphor thyme leaf and flower oils inhibited *C. albicans, E. coli, L. monocytogenes, P. mirabilis*, *Salmonella* spp. and *S. aureus*.[831]
- □ Camphor thyme CT linalool essential oil exhibited weak to moderate antioxidant activity depending on the concentration used in the TBARS assay.[832] Another study found that camphor thyme essential oil exhibited antioxidant activity in the TBARS assay and better antioxidant activity in the micellar model system.[833]

CAPE MAY
(Diosma, White Confetti Bush)

Coleonema album (Thunb.) Bartl. & L.C. Wendl.

FAMILY: Rutaceae
NOTE: Top-Middle
AROMA INTENSITY: Medium-Strong
AROMA: Fresh, green, fruity, floral, herbaceous, slightly peppery
COMMON EXTRACTION METHOD: Hydrodistilled or steam distilled from the leaves and stems or herbal parts
POSSIBLE SUBSTITUTE OILS: Juniper berry, hops, copaiba
BLENDS WELL WITH: Basil, bergamot, blue cypress, cape chamomile, clary sage, fennel, geranium, ginger, juniper, lavender, lemon, neroli, orange, patchouli, petitgrain, vetiver, ylang ylang
SUBCELLULAR LOCALIZATION | EPIGENETIC INFLUENCE: Currently unknown | Currently unknown
RECOMMENDED DILUTION RANGE: 3%–50%; neat for some conditions

PRIMARY CONSTITUENTS:[834,835,836]

Herbal Parts		*Leaves*	
Beta-Phellandrene	5.3%–29.1%	Beta-Caryophyllene	24.9%
Myrcene	10.2%–20.5%	(E)-Beta-Ocimene	9.5%
(Z)-Beta-Ocimene	5.6%–19.0%	Beta-Myrcene	8.0%
Alpha-Pinene	5.9%–8.2%	Beta-Cubebene	7.5%
Beta-Pinene	7.0%–8.0%	Beta-Pinene	7.0%
Limonene	0.2%–7.9%	Delta-3-Carene	6.0%
Alpha-Guaiene	0.0%–7.1%	Bicyclogermacrene	6.0%
Sabinene	1.6%–5.3%	Alpha-Terpineol	3.7%

Linalool	0.5%–4.3%
(E)-Beta-Ocimene	0.8%–4.2%
Isoeugenol	0.0%–3.6%
Germacrene D	tr–3.2%

Other Constituents: *Herbal Parts*—Alpha-thujene, alpha-phellandrene, 1,8-cineole, (E)-beta-ocimene, 6,7-epoxymyrcene, linalool, pinocarvone, alpha-terpineol, delta-cadinene; *Leaves*—1,8-cineole, gamma-elemene, 1,3,6-Octatriene,3,7-dimethyl-, (Z)-beta-ocimene

Preferred Composition for Clinical Use:
Insufficient samples/data to determine.

Reported Therapeutic Properties: Antiseptic, antiparasitic, anti-infectious, anti-inflammatory, analgesic, antispasmodic, relieves sore throat, muscle relaxer, reduces high blood pressure, deodorant, aids detoxification, diuretic, reduces fever, aids circulation, enhances lymph flow, relieves hemorrhoids, reduces the appearance of cellulite and varicose veins, insect repellent, insecticidal, sedating, improves focus, removes emotional obstacles, improves perspective

Cautions:
- None currently known.

Selected Evidence:
- The beta-caryophyllene CT of cape may essential oil inhibited the fungi *Trichophyton rubrum*.[837]

CAPE YELLOWWOOD (Indian Pepper, Indian Ivy-rue, Mullilam, Salai, Trifal, Makaen)

Zanthoxylum rhetsa (Roxb.) DC,
Z. limonella (Dennst.) Alston*

Family: Rutaceae
Note: Middle
Aroma Intensity: Medium
Aroma: Citrusy, fruity, slightly woodsy and spicy
Common Extraction Method: Steam distilled or hydrodistilled from the fresh or dried fruits (seeds), pericarp/seed coat, or leaves
Possible Substitute Oils: Tea tree, nutmeg, plai, pink pepper, chaste tree
Blends Well With: Bergamot, basil (sweet), cassia, cinnamon, citronella, clove, frankincense, grapefruit, lavender, lemon, lemongrass, lime, melissa, nutmeg, orange, tangerine
Subcellular Localization | Epigenetic Influence: Currently unknown | Currently unknown
Recommended Dilution Range: 3%–50%; neat for some conditions

PRIMARY CONSTITUENTS:[838,839,840,841,842,843,844,845]

Sabinene CT (Fruits)

Constituent	Percentage
Sabinene	47.1%–72.7%
Terpinen-4-ol	0.3%–14.1%
Alpha-Pinene	3.9%–10.0%
Alpha-Terpineol	0.7%–7.7%
Beta-Pinene	6.0%–7.5%
Gamma-Terpinene	0.1%–3.7%
Alpha-Terpinene	tr–3.5%
Para-Cymene	tr–3.1%

Terpinen-4-ol CT (Fruits)

Constituent	Percentage
Terpinen-4-ol	32.3%
Sabinene	22.5%
Gamma-Terpinene	8.0%
Decyl Aldehyde	6.0%
Octanal	5.4%
Alpha-Terpinene	4.4%

Pericarp/Seed Coat

Constituent	Percentage
Terpinen-4-ol	25.4%–32.1%
Sabinene	8.1%–16.5%
Beta-Pinene	tr–10.5%
Alpha-Terpineol	7.6%–8.2%
Beta-Phellandrene	4.4%–7.4%
2-Undecanone	0.0%–7.1%
Gamma-Terpinene	4.0%–5.6%
Alpha-Pinene	0.1%–4.3%
Linalool	1.5%–3.3%
Decanal	0.0%–3.3%
Alpha-Phellandrene	0.0%–3.0%

Alpha-Terpinene CT (Fruits)

Constituent	Percentage
Alpha-Terpinene	23.7%
Gamma-Terpinene	23.1%
Terpinolene	5.7%
Limonene	4.7%
p-2,4(8)-Menthadiene	3.4%
Sabinene	3.3%
Terpinen-1-ol	3.0%

Z. limonella (Limonene CT)

Constituent	Percentage
Limonene	57.9%
Alpha-Phellandrene	15.5%
(E)-Beta-Ocimene	8.0%
Linalool	3.2%
Beta-Myrcene	3.1%

Z. limonella (Sabinene CT)

Constituent	Percentage
Sabinene	42.7%
Limonene	39.1%
Terpinen-4-ol	5.4%

* Not all scientists agree that *Z. limonella* is a synonym for *Z. rhetsa* and it may be treated as a separate and distinct species.

OTHER CONSTITUENTS: *Sabinene CT (Fruits)*—Alpha-thujene, myrcene, octanal, decanal, terpinolene, gamma-elemene, beta-elemene, germacrene D, alpha-selinene, delta-cadinol, nonanal, beta-caryophyllene, camphene, gamma-cadinene, alpha-phellandrene; *Terpinen-4-ol CT (Fruits)*—Alpha-thujene, alpha-pinene, beta-pinene, myrcene, para-cymene, octanal, 1-octanol, linalool, cis-sabinene hydrate, terpinolene, 2-cyclohexen-1-ol, 1-terpineol, trans-piperitol, cyclooctane; *Alpha-Terpinene CT (Fruits)*—Alpha-Pinene, beta-pinene, myrcene, alpha-phellandrene, para-cymene, 1,8-cineole, terpinen-4-ol, alpha-terpineol, decanal, alpha-copaene, beta-selinene, alpha-selinene, delta-cadinene; *Z. limonella (Limonene CT)*—Alpha-pinene, (Z)-beta-ocimene, alpha-terpineol, decyl aldehyde, octyl acetate, 2-undecanone, lavandulyl acetate; *Z. limonella (Sabinene CT)*—Alpha-Pinene, alpha-phellandrene, delta-3-carene, beta-cymene, beta-pinene, alpha-terpineol; *Pericarp/Seed Coat*—Alpha-thujene, myrcene, alpha-terpinene, cis-sabinene hydrate, delta-4-carene, para-cymene, terpinolene, (E)-beta-terpineol, 2-propylcyclopentanone, alpha-copaene, alpha-bisabolene, alpha-cadinol, cis-p-menth-2-en-1-ol, trans-p-menth-2-en-1-ol, dodecanal, 2-tridecanone

PREFERRED COMPOSITION FOR CLINICAL USE:

Insufficient samples/data to determine.

REPORTED THERAPEUTIC PROPERTIES: Astringent, aids digestion, relieves diarrhea, diuretic, analgesic, antispasmodic, anti-inflammatory, antirheumatic, relieves toothache, promotes oral health, anticonvulsive, relieves toothache, reduces high cholesterol, balances blood-sugar levels, antibacterial, antimicrobial, antiparasitic, antiseptic, antioxidant, anticancer, relieves chronic skin conditions, **insect repellent**, insecticidal, relieves anxiety, antidepressant

CAUTIONS:

Sabinene and Alpha-Terpinene CTs

■ None currently known.

Terpinen-4-ol CTs

■ Use cautiously orally. Toxicity appears to occur at extreme doses of essential oils rich in terpinen-4-ol; therefore, it is best to limit consumption to 600 mg (about 10 drops with a 60 mg dropper) daily, and avoid long-term use orally. See monograph for Melaleuca (Tea Tree) in *Medicinal Essential Oils* for more information.

■ May interact with antibiotics antifungal drugs and increase or decrease their effectiveness due to terpinen-4-ol content.[846]

SELECTED EVIDENCE:

- □ Both *Z. rhetsa* (terpinen-4-ol CT) and *Z. limonella* (limonene CT) fresh fruit essential oil possessed the ability to inhibit two types of breast cancer cell proliferation and cell viability.[847]
- □ Oral administration of 100 mcL/kg of cape yellowwood pericarp oil significantly reduced diarrhea caused by castor oil in mice.[848]
- □ *Z. limonella* essential oil with limonene (31.1%), terpinen-4-ol (13.9%), and sabinene (9.1%) as major constituents caused dose-dependent, nonspecific contractions of isolated guinea pig ileum and rat thoracic aorta, and in mice.[849] The same researchers reported that oral administration of the essential oil stimulated the gastrointestinal system and increased bowel transit time.
- □ The pericarp essential oil exhibited weak antispasmodic activity in isolated guinea pig ileum.[850]
- □ *In vitro* research shows that the limonene CT (*Z. limonella*) and terpinen-4-ol CT (*Z. rhetsa*) of cape yellowwood essential oil possessed broad-spectrum antibacterial activity against *L. monocytogenes, B. cereus, S. aureus, S. typhi, Sh. enteritidis, P. aeruginosa, V. parahaemolyticus,* and *E. coli.*[851] The study authors directly compared *Z. rhetsa* (terpinen-4-ol CT) and *Z. limonella* (limonene CT) and found that *Z. limonella* was significantly more active against all tested bacteria. The pericarp oil was active against *S. aureus, E. coli*, and *K. pneumonia*e.[852] *Z. limonella* (sabinene CT) potently inhibited *S. aureus* and *E. coli*, and inhibited MRSA and extended-spectrum β-lactamase-producing *E. coli.*[853] *Z. limonella* essential oil inhibited *Clostridium perfringens, Salmonella typhimurium, Campylobacter jejuni, S. enteritidis*, and *E coli.*[854] Other research demonstrates that *Z. limonella* inhibits S. aureus and B. cereus.[855]
- □ *Z. rhetsa* (terpinen-4-ol CT) and *Z. limonella* (limonene CT) fresh fruit essential oil demonstrated similar antioxidant activity in DPPH and TEAC assays.[856] The pericarp oil also showed good antioxidant activity in the DPPH assay.[857]
- □ An unidentified CT of cape yellowwood (*Z. limonella*) essential oil provided 96.2 minutes of complete protection from mosquito (*Aedes albopictus*) bites when used alone, and up to 391 minutes of protection when combined with turmeric and Java patchouli (*Pogostemon heyneanus*) in a 1:1:2 ratio.[858] The same ratio, used as a 20% concentration, repelled blackflies for up to 170.56 minutes.[859] Undiluted *Z. limonella* (0.1 mL of oil was applied per 30 cm2 of exposed skin) provided

two hours of complete protection against *Aedes aegypti* mosquito bites, but 10% and 50% concentrations failed to prevent bites for as long as two hours.[860] Combining clove with *Z. limonella* essential oil in a gel or cream significantly repelled mosquitoes and was more effective than DEET (complete protection for fours and 95.7% repellency versus 82.7% for DEET).[861] *Z. limonella* gave the highest protection time of botanicals tested against *Aedes albopictus* mosquitoes.[862] Protection time was better when combined with mustard oil (296 to 304 minutes) than with coconut oil (223.5 to 245 minutes).

- *Z. limonella* exhibited the most effective larvicidal activity against *Anopheles dirus* (the major malaria vector in Thailand) and *Aedes aegypti* (the main vector of dengue and dengue hemorrhagic fever in urban areas) mosquitoes when compared to caraway, celery, fennel, and turmeric essential oils.[863]

CARNATION ABSOLUTE (Clove Pink)

Dianthus caryophyllus L.

FAMILY: Caryophyllaceae
NOTE: Middle
AROMA INTENSITY: Strong
AROMA: Spicy, sweet, honey-like, floral, slightly minty-herbaceous
COMMON EXTRACTION METHOD: Solvent extracted from the flowers; may also be steam distilled from the flowers
POSSIBLE SUBSTITUTE OILS: Jasmine absolute, ylang ylang, rose
BLENDS WELL WITH: Benzoin, bergamot, black pepper, carrot, cassia, cedarwood, cinnamon, cistus, clary sage, clove, coriander, frankincense, geranium, helichrysum, jasmine, lavender, lemon, neroli, nutmeg, orange, patchouli, Peru balsam, petitgrain, rose, tamala, vetiver, ylang ylang
SUBCELLULAR LOCALIZATION | EPIGENETIC INFLUENCE: Currently unknown | Currently unknown
RECOMMENDED DILUTION RANGE: 1.0%–3.0%

PRIMARY CONSTITUENTS:[864,865,866,867,868]

Absolute	
Benzyl Benzoate	3.9%–45.0%
Hexadecanoic Acid	0.0%–44.3%
(Z,Z)-9,12-Octadecadienoic-Acid	0.0%–31.5%
Eugenol	3.5%–33.5%
Benzyl Salicylate	0.0%–28.6%
Phenyl Ethyl Alcohol	0.0%–23.4%
Nootkatone	0.0%–17.3%
(Z)-3-Hexenyl Benzoate	0.0%–15.0%
Eicosene	0.0%–12.0%
Calamenene*	0.0%–8.7%
Hexyl Benzoate	0.0%–6.3%
Benzoin	0.0%–6.2%
m-Cresyl Phenylacetate	0.0%–5.6%
Elemol	0.0%–5.5%
Limonene	0.0%–4.9%

Gamma-Cadinene	0.0%–4.1%
Phellandrene*	0.0%–3.5%
Para-Cymene	0.0%–3.3%
Bornyl Acetate	0.0%–3.1%
Beta-Pinene	0.0%–3.1%

* Correct isomer not identified.

Essential Oil	
Eugenol	74.0%–88.0%
Beta-Caryophyllene	3.1%–22.4%
Eugenyl Acetate	1.1%–9.5%

OTHER CONSTITUENTS: *Absolute*—Alpha-pinene, camphene, gamma-terpinene, citronellol, methyl eugenol (<1.7%); *Essential Oil*—Alpha-humulene, methyl salicylate (< 0.3%)

PREFERRED COMPOSITION FOR CLINICAL USE:
Insufficient samples/data to determine.

REPORTED THERAPEUTIC PROPERTIES: Anticancer, antiviral, antifungal, antibacterial, antimicrobial, antifungal, antioxidant, kidney protective, analgesic, insecticidal, insect repellent, uplifting, antidepressant, aphrodisiac, stress management, revitalizing to the mind

CAUTIONS:

Absolute

- Caution with pregnancy and lactation. Animal studies suggest that large doses of clove (also rich in eugenol) may negatively impact embryonic development and encourage fetal cell death.[869,870] Another animal study did not detect any negative influence of eugenol and suggested it helped prevent negative effects caused by retinoic acid.[871872] Eugenol is considered strongly toxic to embryos according to animal studies.[873,874] Based on these studies it is best to limit clove oil during pregnancy and lactation.
- Avoid oral consumption. Absolutes contain solvent residue and should not be taken orally.
- May interact with aspirin, blood pressure, antiplatelet, and anticoagulant medications, and increase the risk of bleeding among people with bleeding disorders, due to eugenol content.[875,876,877,878]
- May interact with MAOI antidepressants. Animal research suggests that eugenol produces antidepressant effects via the monoamine oxidase pathway, which may cause interactions with antidepressants that also interact with this pathway.[879]
- May interact with antibiotics and possibly enhance their effects due to eugenol content.[880]
- May interfere with enzymes responsible for metabolizing medications (NSAIDs, proton-pump inhibitors, acetaminophen, antiepileptics, immune modulators, blood-sugar medications, blood pressure medications, antidepressants, antipsychotics, diabetic medications, antihistamines, antibiotics, and anesthetics) due to eugenol content.[881]
- Use cautiously in people with salicylate sensitivities due to the possible presence of benzyl salicylate.
- May irritate mucous membranes (eyes, mouth, nasal passages, vagina, rectum).[882,883]

Essential Oil

- Caution with pregnancy and lactation. Animal studies suggest that large doses of clove (also rich in eugenol) may negatively impact embryonic development and encourage fetal cell death.[884,885] Another animal study did not detect any negative influence of eugenol and suggested it helped prevent negative effects caused by retinoic acid.[886887] Eugenol is considered strongly toxic to embryos according to animal studies.[888,889] Based on these studies it is best to limit clove oil during pregnancy and lactation.

- May interact with aspirin, blood pressure, antiplatelet, and anticoagulant medications, and increase the risk of bleeding among people with bleeding disorders, due to eugenol content.[890,891,892,893]
- May interact with MAOI antidepressants. Animal research suggests that eugenol produces antidepressant effects via the monoamine oxidase pathway, which may cause interactions with antidepressants that also interact with this pathway.[894]
- May interact with antibiotics and possibly enhance their effects due to eugenol content.[895]
- May interfere with enzymes responsible for metabolizing medications (NSAIDs, proton-pump inhibitors, acetaminophen, antiepileptics, immune modulators, blood-sugar medications, blood pressure medications, antidepressants, antipsychotics, diabetic medications, antihistamines, antibiotics, and anesthetics) due to eugenol content.[896]
- May interact with anticholinergic (drugs used for asthma, incontinence, gastrointestinal cramps, muscular spasms, depression, and sleep disorders) and cholinergic medications (drugs used to reduce urinary retention, diagnose myasthenia gravis, and for glaucoma), based on AChE-inhibtory activity of eugenol.[897]
- May irritate mucous membranes (eyes, mouth, nasal passages, vagina, rectum).[898,899]

SELECTED EVIDENCE:

- *In vitro* research demonstrates that carnation essential oil significantly inhibits *S. aureus, S. epidermidis, K. pneumonia,* and *E. coli.*[900] Another study found that carnation essential oil demonstrates antibacterial activity against *E. coli*, which is enhanced when combined with other essential oils.[901]
- Carnation essential oil demonstrated significant antioxidant capacity *in vitro*, which was highest among nineteen tested essential oils and absolutes.[902]
- Carnation essential oil demonstrated moderate larvicidal activity against *Cx. Pipiens* mosquitoes.[903]

CASCARILLA (Sweet Bark)

Croton eluteria (L.) Sw.

FAMILY: Euphorbiaceae
NOTE: Middle
AROMA INTENSITY: Strong
AROMA: Peppery, spicy, sweet, woody
COMMON EXTRACTION METHOD: Steam distilled from the bark
POSSIBLE SUBSTITUTE OILS: Black cumin, pink pepper, thyme
BLENDS WELL WITH: Black pepper, cassia, cinnamon, clove, coriander, ginger, lavender, nutmeg, patchouli, sage, tamala, vetiver
SUBCELLULAR LOCALIZATION | EPIGENETIC INFLUENCE: Currently unknown | Currently unknown
RECOMMENDED DILUTION RANGE: 3%–33%; 50% for some conditions

PRIMARY CONSTITUENTS:[904]

Para-Cymene	25.4%
Geranylacetone	18.2%

Cascarilladiene I	11.8%
Cascarilladienone	10.7%
Alpha-Pinene	6.6%
Cuparene + Alpha-Muurolene	6.3%
Caryophyllene Oxide + Dihydro-ar-Bisabolol	5.6%
Alpha-Thujene	5.4%
Beta-Elemene	4.2%

Other Constituents: Unidentified sesquiterpene hydrocarbons, camphene, 2-carene, myrcene, beta-pinene, limonene, 1,8-cineole, gamma-terpinene, cyclosativene, alpha-copaene, beta-cedrene isomer, thujopsene, alpha-humulene, all-aromadendrene, 1,11-oxidocalamenene, alpha-curcumene, beta-selinene, valencene, gamma-elemene, alpha-chamigrene, delta-cadinene, alpha-calacorene, beta-calacorene, cadalene, nonan-2-one, thujone (isomer not identified; < 1.2%), decan-2-one, carvotanacetone, undeca-5,8-diene-2-one, undecan-2-one, alpha-terpinyl acetate, eugenyl methyl ether, cis-tridec-5-en-2-one, cascarilla epoxide, humulene epoxide isomer, 4-isopropyl-6-methyl-tetralone

Preferred Composition for Clinical Use:
Insufficient samples/data to determine.

Reported Therapeutic Properties: Aids digestion, eases nausea and vomiting, relieves diarrhea, expels excess gas, supports respiratory function, expels excess mucus, relieves fever, soothes hemorrhoids, lowers high blood pressure, antimicrobial, antibacterial, antifungal, combats worrisome and negative thoughts, relives mental exhaustion

Cautions:
- None currently known.

Selected Evidence:
- □ None found.

CAT THYME
(Mint Plant, Germander)

Teucrium marum L., *T. marum* subsp. *marum*

Family: Lamiaceae (Labiatae)
Note: Middle
Aroma Intensity: Strong
Aroma: Musty, pungent, earthy, slightly floral
Common Extraction Method: Steam distilled or hydrodistilled from the aerial parts, calyxes, leaves, stems, flowers, or leaves & flowers; headspace solid-phase microextraction may also be used
Possible Substitute Oils: Melissa, catnip
Blends Well With: Bergamot, big badja gum, cajeput, eucalyptus, grapefruit, gully gum, lavender, lemon, lime, marjoram, myrrh, myrtle, orange, peppermint, rosemary, spearmint
Subcellular Localization | Epigenetic Influence: Currently unknown | Currently unknown
Recommended Dilution Range: 3%–33%; neat for some conditions

PRIMARY CONSTITUENTS:[905,906,907]

Aerial Parts

Dolichodial	20.0%–83.4%
(E)-Beta-Caryophyllene	0.3%–23.1%
Dolicholactone	0.4%–20.9%
(E)-Alpha-Bergamotene	0.2%–20.6%
Methyl chavicol (Estragole)	1.1%–10.5%
Beta-Bisabolene	0.2%–10.0%
Beta-Sesquiphellandrene	0.2%–8.3%
Beta-Barbatene	0.1%–6.7%
2-Methylbutyl Butyrate	0.1%–5.6%
Epi-Dolichodial	0.2%–5.0%
(Z)-Beta-Farnesene	tr–4.6%
Alpha-Humulene	0.4%–4.5%
(E)-2-Hexanal	0.1%–3.4%
Linalool	0.0%–3.1%

Calyxes

Dolichodial	51.5%
(E)-Beta-Caryophyllene	13.7%
Methyl chavicol (Estragole)	6.6%
(E)-Alpha-Bergamotene	5.1%

Leaves

Dolichodial	49.5
(E)-Beta-Caryophyllene	11.1%
(E)-Alpha-Bergamotene	9.1%
Methyl chavicol (Estragole)	4.0%

Stems

(E)-Alpha-Bergamotene	20.7%
(E)-Beta-Caryophyllene	18.2%
Dolicholactone	6.0%
Beta-Bisabolene	5.0%
Methyl chavicol (Estragole)	4.4%
Beta-Sesquiphellandrene	4.2%
Dolichodial	3.8%
Alpha-Humulene	3.4%

Flowers

(E)-Beta-Caryophyllene	21.5%
(E)-Alpha-Bergamotene	9.4%
Nonanal	7.9%
Geranyl Acetate	5.4%
Dolichodial	5.1%
Alpha-Humulene	4.6%
Methyl chavicol (Estragole)	4.0%
Beta-Bisabolene	3.8%
Beta-Sesquiphellandrene	3.1%

Leaves & Flowers (Dolichodial CT)

Dolichodial	28.9%–89.5%
Unidentified (MW=168)*	0.6%–27.8%
Beta-Caryophyllene	0.4%–19.9%
(E)-Alpha-Bergamotene	0.0%–13.9%
Beta-Bisabolene	0.0%–11.9%
Caryophyllene Oxide	0.4%–11.6%
Delta-Cadinene	0.0%–8.5%
Epi-Dolichodial	1.2%–7.8%
Methyl Chavicol (Estragole)	0.7%–7.0%
Alpha-Humulene	0.0%–4.4%

* Possibly dolicholactone?

Leaves & Flowers (BC/CO CT)

Beta-Caryophyllene	14.3%–34.7%
Caryophyllene Oxide	19.3%–30.6%
Delta-Cadinene	1.6%–12.9%
Alpha-Bisabolol	1.5%–10.5%
Beta-Bisabolol	0.0%–10.0%
Alpha-Humulene	3.5%–8.3%
Gamma-Cadinene	0.0%–7.6%
(E)-Alpha-Bergamotene	0.5%–5.9%
Humuladienone	3.2%–5.5%
Dolichodial	0.7%–3.5%

subsp. marum (Aerial Parts)

Isocaryophyllene	20.2%
Beta-Bisabolene	14.7%
Beta-Sesquiphellandrene	11.3%
Alpha-Santalene	11.0%
Dolichodial	9.4%
Alpha-Humulene	7.2%
Methyl Chavicol (Estragole)	3.8%
Caryophyllene Oxide	3.2%

OTHER CONSTITUENTS: *Aerial Parts*—Hexanal, isopentyl acetate, benzaldehyde, 6-methylhept-5-en-2-one, hexyl acetate, nonanal, 1-oct-2-enyl acetate, citronellal, decanal, citronellol, neral, 2-methylhexyl butyrate, (Z)-myrtanol, eugenol, citronellyl acetate, geranyl acetate, (E)-beta-farnesene, (E)-beta-bergamotene, caryophyllene oxide, humulene epoxide II; *Calyxes*—Hexyl acetate, 2-methylbutyl butyrate, linalool, 1-oct-3-enyl acetate, neral, 2-methylhexyl butyrate, epi-dolichodial, geranyl acetate, beta-barbatene, (E)-beta-farnesene, alpha-humulene, (E)-beta-bergamotene, beta-bisabolene, beta-sesquiphellandrene; *Leaves*—(E)-2-hexenal, hexyl acetate, 1-oct-3-enyl acetate, citronellal, 2-methylhexyl butyrate, geranyl acetate, dolicholactone, (E)-beta-farnesene, alpha-humulene, (E)-beta-bergamotene, beta-bisabolene, beta-sesquiphellandrene; *Stems*—Hexanal, 6-methylhept-5-en-2-one, hexyl acetate, linalool, 1-oct-3-enyl acetate, citronellal, 2-methylhexyl butyrate, (Z)-myrtanol, geranyl acetate, (E)-beta-farnesene, (E)-beta-bergamotene, caryophyllene oxide; *Flowers*—Hexanal, 6-,ethylhept-5-en-2-one, hexyl acetate, 2-methylbutyl butyrate, linalool, 1-oct-3-enyl acetate, citronellal, 2-methylhexyl butyrate, (Z)-myrtanol, epi-dolichodial, eugenol, citronellyl acetate, dolicholactone, beta-barbatene, (E)-beta-bergamotene, caryophyllene oxide; *Leaves & Flowers (BC/CO CT)*—Methyl chavicol (< 0.8%), unidentified constituent; *Leaves & Flowers (Dolichodial CT)*—Unidentified constituents, beta-sesquiphellandrene, humuladienone, gamma-cadinene, alpha-bisabolol; *subsp. marum (Aerial Parts)*—Beta-pinene, limonene, n-pentyl butanoate, n-amyl isovalerate, octyl acetate, isopulegol, citronellal, octyl butanoate, ethyl butanoate, isoborneol, epi-dolichodial, citronellyl propionate, geranyl acetate, farnesol

PREFERRED COMPOSITION FOR CLINICAL USE:
Too great a variance in samples/data to determine.

REPORTED THERAPEUTIC PROPERTIES: Diuretic, nervine, aids digestion, supports gallbladder function, astringent, stops excess bleeding, antiparasitic, antibacterial, antifungal, antiviral, anti-inflammatory, analgesic, relieves gout, relieves fever, balances menstruation, ease cough, aids respiratory function, insect repellent, aphrodisiac, relieves anxiety, stress management

CAUTIONS:
- None currently known; however, the monoterpene iridoid dolichodial is not well studied.

SELECTED EVIDENCE:
- *In vitro* research showed that cat thyme CT isocaryophyllene inhibits *Rhizoctonia solani, Fusarium oxysporum, Alternaria solani*, and *Botrytis cinerea.*[908]
- Cat thyme CT isocaryophyllene essential oil exhibited good antioxidant activity in the DPPH, 5-lipoxygenase, and luminol/xanthine/xanthine oxidase chemiluminescence assays when compared to ascorbic acid and BHT.[909]

CATINGA-DE-MULATA (Balanion, Otek, Macassa)

Aeollanthus suaveolens Mart. Ex Spreng

FAMILY: Lamiaceae (Labiatae)
NOTE: Middle
AROMA INTENSITY: Medium-Strong
AROMA: Floral, herbaceous, coconut-like, woody

Common Extraction Method: Steam distilled or hydrodistilled from the aerial parts, flowering or vegetative tops, flowers, or leaves

Possible Substitute Oils: Lavender, German chamomile

Blends Well With: Benzoin, bergamot, cassia, cedarwood, copaiba, eucalyptus, geranium, lavender, neroli, patchouli, Peru balsam, pine, ravintsara, rose, sandalwood, silver fir, white fir

Subcellular Localization | Epigenetic Influence: Currently unknown | Currently unknown

Recommended Dilution Range: 0.5%–5%

Primary Constituents:[910,911,912,913,914,915]

Aerial Parts

Linalool	31.6%–49.3%
(E)-Beta-Farnesene	32.8%–34.9%
Massoia Lactone	2.9%–15.0%
Linalyl Acetate	3.8%–8.5%

Leaves

Linalool	33.3%–49.3%
(E)-Beta-Farnesene	25.4%–37.6%
Massoia Lactone	0.0%–25.9%
Linalyl Acetate	3.2%–5.7%
Alpha-Santalene	0.0%–3.3%

Flowering/Vegetative Tops

Linalool	35.6%–41.9%
(E)-Beta-Farnesene	14.0%–17.1%
Linalyl Acetate	12.5%–15.8%
Massoia Lactone	0.8%–7.1%
Alpha-Terpineol	3.3%–4.4%
(E)-Beta-Ocimene	2.2%–3.1%

Flowers

Linalool	35.9%
(E)-Beta-Farnesene	29.1%
Massoia Lactone	17.0%
Linalyl Acetate	6.8%

Other Constituents: *Aerial Parts*—Myrcene, 1,8-cineole, (E)-beta-ocimene, alpha-terpineol, geraniol, geranyl acetate, alpha-santalene, trans-alpha-bergamotene, delta-decalactone; *Flowering/Vegetative Tops*—Alpha-pinene, camphene, beta-pinene, myrcene, limonene, 1,8-cineole, (Z)-beta-ocimene, terpinolene, trans-alpha-bergamotene, alpha-santalene, cis-alpha-bergamotene, beta-caryophyllene, borneol, neryl acetate, geranyl acetate, nerol, geraniol; *Leaves*—1,8-cineole, (E)-beta-ocimene, alpha-terpineol, geraniol, geranyl formate, trans-alpha-bergamotene, beta-caryophyllene, delta-decalactone; *Flowers*—Alpha-santalene, alpha-bergamotene

Preferred Composition for Clinical Use:

Constituent	
Linalool	30%–55%
(E)-Beta-Farnesene	15%–40%
Linalyl Acetate	3%–16%
Massoia Lactone (Dec-2-en-5-olide)	0.5%–15%
Alpha-Santalene	0.5%–5%
Delta-Decalactone	tr–2%

Reported Therapeutic Properties: Anticonvulsive, ant-seizure, anti-inflammatory, antineuralgic, analgesic, relieves headache, reduces fever, anticancer, aids digestion, relieves diarrhea and stomachache, antibacterial, antifungal, antimicrobial, soothes skin irritation and conditions, insect repellent, relaxing, sedating, antidepressant

Cautions:

■ Massoia lactones are suspected of being strongly irritating to the skin and IFRA (International Fragrance Association) has banned the use of massoia bark essential oil as a fragrance ingredient because of its massoia lactone content. No published data existed as of this writing to determine whether catinga-de-mulata oil causes irritation, sensitization, or photosensitivity reactions.

Selected Evidence:

- Intraperitoneal injection of catinga-de-mulata leaf essential oil (75 mg/kg body weight) exhibited antidepressant activity in rats without causing sedation or memory impairment in rodents.[916] The study authors attributed this effect to the presence of linalool, which is synergized by the other minor constituents in the oil.[917] When isolated major constituents were tested in the same study, massoia lactone was the most active against microorganisms, followed by (E)-beta-farnesene, and linalool.
- *In vitro* research demonstrated that catinga-de-mulata leaf essential oil inhibited *E. coli, Salmonella* sp., and *S. aureus*, with the lowest activity against *S. aureus*.[918] Another study reported that the leaf oil (with 25.9% massoia lactone) inhibited *E. coli, C. neoformans, S. cerevisiae, C. dublinensis, C. albicans, P. aeruginosa, S. aureus, S. epidermidis, K. pneumoniae, B. subtilis*, and *S. Setubal.*
- Catinga-de-mulata leaf essential oil showed weak antioxidant capacity in the DPPH assay.[919]

CHINESE COFFIN TREE (Taiwan Fir)

Taiwania cryptomerioides Hayata

Family: Cupressaceae
Note: Base; Top-Middle
Aroma Intensity: Medium
Aroma: Woody, green, earthy; Piney, fresh, earthy
Common Extraction Method: Steam distilled or hydrodistilled from the heartwood, leaves (needles), or twigs
Possible Substitute Oils: Hinoki (wood), calendula, peta; palo santo (wood), elemi, blue cypress
Blends Well With: Bergamot, blue cypress, cedarwood, cypress, frankincense, geranium, kunzea, lavender, lemon, lime, orange, rose, tagetes, sandalwood, spruce (black), tangerine, vetiver, ylang ylang
Subcellular Localization | Epigenetic Influence: Currently unknown | Currently unknown
Recommended Dilution Range: 3%–50%; Neat for some conditions

Primary Constituents:[920,921,922,923]

Heartwood		*Twigs*	
Alpha-Cadinol	28.0%	Alpha-Cadinol	45.9%
T-Cadinol + T-Muurolol	23.5%	Ferruginol	18.9%
Cedrol	16.3%	Beta-Eudesmol	10.8%
Delta-Cadinene	7.6%	Elemol	4.4%
Delta-Cadinol	4.0%	Cedrol	3.7%
Gamma-Cadinene	3.8%	Gamma-Eudesmol	3.6%

Leaves

Limonene	44.5%
Alpha-Pinene	14.4%
Caryophyllene Oxide	8.5%
Delta-2-Carene	3.9%
Alpha-Terpinyl Acetate	3.6%

OTHER CONSTITUENTS: *Heartwood*—Alpha-cedrene, beta-cedrene, thujopsene, alpha-muurolene, alpha-cadinene, alpha-calacorene, globulol, 1-epi-cubenol, gamma-eudesmol, cadalene, alpha-bisabolol, ferruginol; *Twigs*—Gamma-cadinene, 1,10-di-epi-cubenol, 1-epi-cubenol, cadalene, hinesol acetate, beta-eudesmol acetate; *Leaves*—Myrcene, linalool, camphor, alpha-terpineol, beta-longipinene, beta-caryophyllene, delta-cadinene, trans-calamenene, humulene epoxide, 1-epi-cubenol, alpha-cadinol, 14-hydroxy-9-epi-(E)-caryophyllene

PREFERRED COMPOSITION FOR CLINICAL USE:
Insufficient samples/data to determine.

REPORTED THERAPEUTIC PROPERTIES: Antifungal, antimicrobial, antibacterial, antioxidant, analgesic, anti-inflammatory, soothes skin conditions and itching, wound healing, supports respiratory function, insecticidal, insect repellent, encourages positive thoughts, balances mood, relieves anxiety, calming

CAUTIONS:
■ None currently known.

SELECTED EVIDENCE:

- □ *In vitro* research showed that Chinese coffin tree heartwood essential oil inhibited *B. cereus, S. aureus, S. epidermidis, C. albicans, E. coli, E. aerogenes, K. pneumonia, P. aeruginosa, V. parahaemolyctis,* and *A. niger.*[924] One study reported that Chinese coffin tree heartwood essential oil showed good antifungal, antimicrobial, and antibacterial activity but it appears to be based on studies in Taiwan that used fractions of extracts.[925]
- □ Chinese coffin tree heartwood essential oil exhibited good antioxidant activity in the DPPH assay, with the antioxidant activity primarily being attributed to ferruginol.[926]
- □ Heartwood essential oil demonstrated termiticidal activity against *Coptotermes formosanus* termites.[927] Cedrol, alpha-cadinol, and ferruginol were also active. The heartwood oil also demonstrated insecticidal activity against mites (*Dermatophagoides pteronyssinus*, *D. farinae*).[928] Alpha-cadinol was significantly active, followed by T-muurolol, ferruginol, and T-cadinol.
- □ The heartwood essential oil repelled cat fleas (*Ctenocephalides felis*).[929]

CHINESE NARCISSUS (Chinese Sacred Lily, Cream Narcissus, Bunchflower Daffodil)

Narcissus tazetta L., *N. tazetta* var. *chinensis* Roem, *N. tazetta* L. subsp. *aureus* (Loisel.) Baker, *N. tazetta* L. subsp. *italicus* (Ker. Gawl.) Baker, *N. tazetta* L. subsp. *tazetta*

FAMILY: Amaryllidaceae/Liliaceae
NOTE: Top-Middle
AROMA INTENSITY: Strong
AROMA: Herbaceous, floral, green, fruity
COMMON EXTRACTION METHOD: Headspace Solid-phase microextracted (HS-SPME) from the flowers; may also be solvent extracted
POSSIBLE SUBSTITUTE OILS: Tagetes, ylang ylang, chamomile (Roman, German)
BLENDS WELL WITH: Bergamot, clary sage, frankincense, geranium, grapefruit, juniper berry, juniper (Phoenician), lavender, lemon, lime, myrrh, orange, rose, sandalwood, tangerine, tea tree (melaleuca), ylang ylang
SUBCELLULAR LOCALIZATION | EPIGENETIC INFLUENCE: Currently unknown | Currently unknown
RECOMMENDED DILUTION RANGE: 3%–33%; 50% for some conditions

PRIMARY CONSTITUENTS:[930,931,932]

N. tazetta var. chinensis	
(E)-Beta-Ocimene	62.7%–66.1%
Benzyl Acetate	11.7%–25.0%
(Z)-Beta-Ocimene	1.8%–4.6%
1,8-Cineole	1.5%–4.1%

N. tazetta subsp. tazetta (Absolute)	
Linalool	11.7%
Gamma-Terpinene	10.3%
Benzyl Acetate	9.5%
Para-Cymene	8.5%
Delta-3-Carene	8.4%
Alpha-Pinene	6.1%
Ethyl Cinnamate	5.6%
Limonene	3.3%
1,8-Cineole	3.0%

N. tazetta (Absolute) I	
Gamma-Terpinene	27.2%
Methyl Cinnamate	15.8%
Benzyl Acetate	9.6%
Delta-3-Carene	6.6%
Benzyl Alcohol	4.8%
Benzyl Benzoate	4.0%

N. tazetta (Absolute) II	
Linalool	16.0%
Methyl Anthranilate	15.1%
Benzyl Acetate	12.3%
Benzyl Alcohol	11.3%
Methyl Cinnamate	8.4%
Unidentified Alcohol	6.8%
Cinnamyl Alcohol	5.8%

OTHER CONSTITUENTS: *N. tazetta var. chinensis*—Myrcene, allo-ocimene, phenethyl acetate, 3-phenylpropyl acetate, prenyl acetate; *N. tazetta (Absolute) I*—Alpha-pinene, limonene, 1,8-cineole, para-cymene, linalool, nerol, ethyl cinnamate, cinnamic alcohol; *N. tazetta (Absolute) II*—Trans-Anethole, thymol, eugenol, alpha-terpineol, methyl phenylacetate, nerol, geraniol; *N. tazetta subsp. tazetta (Absolute)*—Alpha-Terpineol, benzyl alcohol, methyl cinnamate, methyl anthranilate, cinnamic alcohol, indole, benzyl benzoate

PREFERRED COMPOSITION FOR CLINICAL USE:
Insufficient samples/data to determine.

REPORTED THERAPEUTIC PROPERTIES: Relieves sore throat, soothes chronic skin conditions, wound healing, antibacterial, antifungal, antiseptic, analgesic, anti-inflammatory, anticancer, aids respiratory function, combats anger and negative thoughts, reduces mental fatigue, antidepressant

CAUTIONS:
HS-SPME
■ Oral caution. Ingestion of benzoic acids, benzoic salts (calcium, potassium, and sodium), benzaldehyde, benzyl acetate, benzyl benzoate, and benzyl alcohol may cause gastrointestinal irritation, vomiting, and diarrhea.[933] The Joint Expert Committee for Food Additives (JECFA) has established an acceptable daily limit for the aforementioned constituents of 0–5 mg/kg body weight for adults to avoid GI discomfort.[934] Based on this limit, it is suggested to limit the oral consumption to no more than 5 drops daily.

Absolute
■ Avoid oral consumption. Absolutes contain solvent residue and should not be taken orally. In addition, the same cautions for benzyl derivatives from the HS-SPME extracted oil would apply.

SELECTED EVIDENCE:
- □ None found.

CHINESE SILK VINE

Periploca sepium Bunge

FAMILY: Apocynaceae
NOTE: Base
AROMA INTENSITY: Medium
AROMA: Sweet, almond-like, floral, fruity, vanilla
COMMON EXTRACTION METHOD: Steam distilled or hydrodistilled from the roots; may also be distilled from the solvent-extracted concrete
POSSIBLE SUBSTITUTE OILS: Peru balsam
BLENDS WELL WITH: Amyris, anise, bergamot, birch, cardamom, cassia, cedarwood, champaca, cinnamon, clove, copaiba, davana, geranium, ginger, helichrysum, jasmine absolute, lavender, lemon, lime, mandarin, may chang, melissa, neroli, orange, Peru balsam, petitgrain, rosewood, rose, sandalwood, tolu balsam, wintergreen, ylang ylang, yuzu, zdravetz
SUBCELLULAR LOCALIZATION | EPIGENETIC INFLUENCE: Currently unknown | Currently unknown
RECOMMENDED DILUTION RANGE: 1%–10%

PRIMARY CONSTITUENTS:[935,936,937]

2-Hydroxy-4-Methoxybenzaldehyde	78.8%–87.8%
Furfural	0.0%–3.1%
Linalool	tr–2.8%
Alpha-Terpineol	tr–2.7%

Other Constituents: Lavender lactone, beta-thujene, myrcene, (Z)-beta-ocimene, borneol, terpinen-4-ol, bornyl acetate, eugenol, n-hexadecanoic acid

Preferred Composition for Clinical Use:

Constituent	
2-Hydroxy-4-Methoxybenzaldehyde	75%–90%
Linalool + Alpha-Terpineol	tr–8%

Reported Therapeutic Properties: Antiarthritic, antirheumatic, anti-inflammatory, analgesic, deodorant, wound healing, antimicrobial, antibacterial, antifungal, insecticidal, combats anger, relieves anxiety, helps release emotional trauma, antidepressant

Cautions:

■ None currently known, although the safety of 2-hydroxy-4-methoxybenzaldehyde has not been fully investigated as of this writing.

Selected Evidence:

- □ *In vitro* research showed that Chinese silk vine essential oil and its main component 2-hydroxy-4-methoxybenzaldehyde possess broad-spectrum antimicrobial activity (*A. tumefaciens, E. coli, P. lachrymans, S. typhimurium, X. vesicatoria, B, subtilis, S. aureus, S. haemolyticus, C. albicans,* and *M. oryzae*).[938]
- □ Chinese silk vine essential oil and 2-hydroxy-4-methoxybenzaldehyde both exhibited moderate antioxidant activity in the DPPH, beta-carotene bleaching, and ferrozine-ferrous ions assays.[939]
- □ Both Chinese silk vine essential oil and its main constituent, 2-hydroxy-4-methoxybenzaldehyde, were significantly toxic to fruit flies (*D. melanogaster*) and moderately toxic to maize weevils (*Sitophilus zeamais*).[940]

CHRYSANTHEMUM

Chrysanthemum boreale (Makino) Makino,
Dendranthema boreale (Makino) Y. Ling ex Kitam.,
D. lavandulifolium var. *seticuspe* (Maxim.) C.Shih,
Ch. indicum L. var. *boreale* Makino

Family: Asteraceae
Note: Middle
Aroma Intensity: Medium-Strong
Aroma: Green, camphorous, earthy, bittersweet
Common Extraction Method: Steam distilled from the flowering aerial parts
Possible Substitute Oils: Sage, tansy, feverfew, Spanish sage, blue tansy
Blends Well With: Blue spruce, blue tansy, camphor, clary sage, copaiba, frankincense, lavender, myrrh, patchouli, peppermint, sage, spearmint, spruce (black), rose, tansy
Subcellular Localization | Epigenetic Influence: Currently unknown | Currently unknown
Recommended Dilution Range: 1.5%–20%; 50% for some conditions

PRIMARY CONSTITUENTS:[941,942,943]

Camphor	15.4%–20.9%
(Z)-Chrysanthenol	0.0%–14.1%
Alpha-Thujone	5.5%–13.3%
Germacrene D	0.0%–10.6%
1,8-Cineole	4.2%–6.7%
Beta-Caryophyllene	3.6%–5.7%
Alpha-Pinene	3.8%–5.3%
Piperitone	0.0%–5.3%
Epi-Sesquiphellandrene	0.0%–5.2%
Beta-Pinene	2.0%–4.5%
Camphene	2.4%–4.2%
Sabinene	0.6%–3.3%

OTHER CONSTITUENTS: Beta-thujone (< 2.7%), alpha-phellandrene, terpinene-4-ol, alpha-terpinolene, borneol, myrcene, cymol, beta-selinene, lepidozene, delta-cadinene, gamma-terpinene, alpha-terpinene, limonene, (Z)-beta-ocimene, para-cymene, trans-chrysanthenyl acetate, bornyl acetate, p-Mentha-1,5-dien-8-ol, carvone, caryophyllene oxide, epiglobulol, alpha-thujene, trans-pinocarveol, umbellulone, (Z)-chrysanthenyl acetate, beta-elemene, zingiberene, beta-sesquiphellandrene, spathulenol, torreyol, T-cadinol, T-muurolol, globulol, alpha-cadinol

PREFERRED COMPOSITION FOR CLINICAL USE:
Insufficient samples/data to determine.

REPORTED THERAPEUTIC PROPERTIES: Anticancer, antibacterial, antifungal, anti-inflammatory, relieves gout, analgesic, antioxidant, balances blood sugar, aids immune function, eases cough, expels excess mucus, **wound healing**, whitens the skin, aids oral health, relieves anxiety, stress management

CAUTIONS:

- Avoid with children under six due to moderate camphor and thujones content. Several cases of camphor poisoning and/or seizures from ingestion and topical application have been reported in children.[944,945] Ingestion of camphor-containing products has been lethal in children under age two.[946] Children 6 years and up may use camphor-containing essential oils topically in dilutions no stronger than 5%. Camphor ingestion by infants and young children may cause cough, vomiting, seizure, burning sensation in the mucous membranes and eyes, or lack of voluntary coordination of muscle movements.[947] Cases of seizure have been reported in young children exposed to essential oils rich in thujones.[948,949,950]
- Avoid during pregnancy and while lactating, due to moderate camphor and thujones content. Ingestion of essential oils with significant levels of camphor may lead to abortion because fetuses lack the enzymes to process it.[951] Essential oils high in thujone content may cause abortion.[952,953] Camphor ingestion by infants and young children may cause cough, vomiting, seizure, burning sensation in the mucous membranes and eyes, or lack of voluntary coordination of muscle movements.[954]
- Oral caution. Essential oils with significant levels of camphor can be toxic when taken orally, usually single doses exceeding 2 mL, although the lethal dose for humans is estimated to be more than 5 mL in a single dose.[955,956,957] Thujone is considered significantly neurotoxic and may damage the liver, and it is estimated that as little as 15 mg orally may negatively impact the central nervous system.[958,959,960] As few as 12 drops of a thujone-rich essential oil (sage) orally have reportedly caused a seizure and short coma in an adult.[961]
- Avoid with epilepsy and Parkinson's due to moderate camphor and thujones content.[962,963,964,965,966]
- The moderate camphor content in chrysanthemum may negatively impact red blood cells and increase the risk of jaundice in children with glucose-6-phosphate dehydrogenase deficiency (G6PD).[967,968]

SELECTED EVIDENCE:

- An *in vitro* study concluded that chrysanthemum essential oil has potential for use in cosmetics due to its antioxidant activity and ability to inhibit tyrosinase activity and melanin synthesis (whitens the skin).[969]
- Chrysanthemum essential oil inhibits the growth, adhesion, acid tolerance, and biofilm formation of *S. Mutans* suggesting it may be useful for the prevention of cavities.[970]
- Animal research suggests that chrysanthemum essential oil may be helpful for skin regeneration and wound healing due to its ability to stimulate keratinocyte proliferation and promote wound closure.[971] Keratinocytes migrate to wound sites and play a role in the initiation, maintenance, and completion of wound healing.
- *In vitro* research showed that chrysanthemum essential oil exhibited varying degrees of antibacterial activity against six gram-positive and eight gram-negative bacteria.[972]

CLEMENTINE

Citrus clementina Hort. Ex Tanaka,
C. reticulata Blanco var. *clementina*;
Nules Clementine: *C. clementina* cv. *nules*

FAMILY: Rutaceae
NOTE: Top
AROMA INTENSITY: Medium
AROMA: Citrusy, fruity, soft, slightly green; floral, sweet, citrusy, fresh
COMMON EXTRACTION METHOD: Cold-pressed or hydrodistilled from the fruit peel (rind); may also be distilled from the leaves or flowers
POSSIBLE SUBSTITUTE OILS: Tangerine, mandarin, orange (sweet); petitgrain, bergamot; neroli
BLENDS WELL WITH: Basil, bergamot, cardamom, carrot seed, cassia, chamomile (German, Roman), cinnamon, clary sage, clove, coriander, frankincense, grapefruit, juniper, lavender, lemon, lemon verbena, lime, nutmeg, neroli, niaouli, orange, patchouli, petitgrain, rose, sage, sandalwood, Spanish sage, ylang ylang
SUBCELLULAR LOCALIZATION | EPIGENETIC INFLUENCE: Currently unknown | Currently unknown
RECOMMENDED DILUTION RANGE: 3%–50%; neat for some conditions

PRIMARY CONSTITUENTS:[973,974,975,976,977,978,979]

Peel (Hydrodistilled)

Limonene	93.8%–96.8%
Myrcene	0.7%–2.1%
Sabinene	0.2%–0.6%
Alpha-Pinene	0.1%–0.6%

Nules Clementine (Peel, Cold-pressed)

Limonene	92.9%
Myrcene	2.5%
Alpha-Pinene	0.6%
Citronellal	0.5%

Peels (Cold-pressed)

Limonene	94.8%–95.0%
Myrcene	1.6%–1.8%
Linalool	0.1%–0.7%

Leaves (Alpha-Sinensal/Linalool CT)

Alpha-Sinensal	5.4%–25.9%
Linalool	19.9%–21.8%
Sabinene	tr–12.8%
Caryophyllene Oxide	0.3%–9.3%
Varamolo	tr–6.7%
Beta-Caryophyllene	1.2%–5.9%

Leaves (Sabinene CT)	
Sabinene	18.5%–36.8%
Linalool	7.5%–21.2%
Terpinen-4-ol	1.8%–10.4%
Beta-Elemene	tr–18.4%
Delta-Selinene	0.0%–8.0%
Beta-Caryophyllene	1.8%–7.5%
(E)-Beta-Ocimene	0.1%–5.9%
Beta-Bisabolene	0.0%–4.3%
Limonene	2.0%–4.2%
Caryophyllene Oxide	tr–4.0%
Alpha-Muurolene	1.4%–3.1%
Beta-Myrcene	1.8%–2.9%
Alpha-Humulene	0.2%–2.3%
Gamma-Terpinene	0.2%–2.3%

Flowers	
Sabinene	34.8%–47.9%
Linalool	16.8%–28.9%
Limonene	6.2%–9.6%
(E)-Nerolidol	3.5%–7.1%
Terpinen-4-ol	3.0%–3.9%
Myrcene	2.4%–3.4%
Para-Cymene	1.7%–2.4%
Beta-Pinene	1.8%–2.2%

Beta-Sinensal	0.5%–5.7%
Delta-3-Carene	tr–5.7%
Citronellol	0.5%–5.6%
Terpinen-4-ol	4.3%–5.3%
(E)-Beta-Ocimene	tr–4.9%
Limonene	tr–4.1%
Alpha-Terpineol	2.2%–3.4%
Geranial	0.6%–3.3%
Bicyclogermacrene	0.2%–2.4%
Myrcene	tr–2.3%
Citronellal	tr–2.3%
Terpinolene	tr–2.1%

Other Constituents: *Peel (Hydrodistilled)*—Beta-pinene, linalool, n-octanal, n-octanol, n-decanal, alpha-sinensal; *Peel (Cold-pressed)*—Alpha-pinene, sabinene, citronellal, n-decanal, alpha-sinensal; *Nules Clementine (Peel, Cold-pressed)*—Sabinene, n-octanal, alpha-phellandrene, terpinolene, linalool, n-decanal, alpha-sinensal; *Leaves (Sabinene CT)*—Alpha-pinene, beta-pinene, alpha-terpinene, o-cymene, para-cymene, alpha-terpinolene, cis-sabinene hydrate, cis-p-menth-2-en-1-ol, trans-p-menth-2-en-1-ol, sabina ketone, alpha-terpineol, trans-piperitol, varamolo, delta-cadinene, beta-eudesmol, alpha-sinensal, spathulenol, cis-Cadin-4-en-7-ol; *Leaves (Alpha-Sinensal/Linalool CT)*—Alpha-pinene, para-cymene, gamma-terpinene, trans-sabinene hydrate, cis-p-menth-2-en-1-ol, neral, geranial, alpha-terpinyl acetate, beta-gurjunene, alpha-humulene, beta-farnesene, elemol, nerolidol, humulene epoxide II, T-muurolol, alpha-muurolol, beta-eudesmol, alpha-cadinol, alpha-bergamotol, nerolidol acetate, beta-sinensal; *Flowers*—Alpha-Pinene, beta-phellandrene, (E)-beta-ocimene, gamma-terpinene, alpha-terpineol, beta-elemene, beta-caryophyllene, caryophyllene oxide, (E,Z)-farnesol

Preferred Composition for Clinical Use:

Constituent	*Cold-Pressed*
Limonene	90%–96%
Myrcene	1%–3%
Linalool	0.1%–1%
Alpha-Pinene	0.1%–1%
Sabinene	0.1%–1%

n-Octanal	0.01%–1%
n-Decenal	0.01%–1%
Alpha-Sinensal	0.01%–1%

REPORTED THERAPEUTIC PROPERTIES: Antimicrobial, antibacterial, antifungal, antiseptic, antispasmodic, anti-inflammatory, expels excess gas, aids digestion, diuretic, relieves constipation, regenerative, wound healing, reduces the appearance of blemishes, stimulates healthy cell cycle, promotes fluid balance and removes excess fluids from the body, purifies the blood, encourages restful sleep, stimulating, stress management, antidepressant, **relieves anxiety**

CAUTIONS:

Peels (Cold-pressed)

▪ Cold-pressed clementine essential oil has not been tested for phototoxicity. It is best to avoid sun and UV exposure for at least 12 hours after topical application until the presence of furanocoumarins is confirmed or ruled out. Hydrodistilled clementine doesn't contain phototoxic furanocoumarins.[980]

Leaves (Sabinene CT)

▪ Avoid during pregnancy and lactation (very low risk) due to moderate beta-elemene content. Beta-elemene may prevent the formation of blood vessels necessary for fetal development according to animal research.[981]

Peels (Hydrodistilled), Flowers, and Leaves (Alpha-Sinensal/Linalool CT)

▪ None currently known.

SELECTED EVIDENCE:

- Clementine (hydrodistilled, peel) essential oil exhibited moderate activity against *B. subtilis, E. coli*, and *C. albicans*.[982] Another study reported that hydrodistilled clementine peel essential oil inhibited C. *albicans, E. coli, Lysteria innocua*, MRSA, and *S. aureus*.[983] Clementine leaf essential oil (sabinene CT) demonstrated strong activity against *S. aureus, B. subtilis, E. coli*, and *C. albicans*.[984]
- Clementine (hydrodistilled, peel) essential oil exhibited moderate antioxidant activity in the DPPH assay.[985]

COGNAC | GRAPE SEED (Green Cognac, White Cognac, Lees)

Vitis vinifera L.

FAMILY: Vitaceae
NOTE: Middle-Base
AROMA INTENSITY: Strong to Medium
AROMA: Complex, green, fruity, tart, wine-like
COMMON EXTRACTION METHOD: Steam distilled or hydrodistilled from the wine lees (leftover yeast deposits) sediment or cold-pressed from the grape seeds; may also be solvent extracted from the seeds
POSSIBLE SUBSTITUTE OILS: Agarwood, coriander

Blends Well With: Bergamot, chamomile (Roman), clary sage, coriander, galbanum, jasmine, lavender, lemon, lime, neroli, osmanthus, rose, tangerine, ylang ylang, yuzu

Subcellular Localization | Epigenetic Influence: Currently unknown | Currently unknown

Recommended Dilution Range: < 5%

Primary Constituents:[986,987,988,989,990]

Wine Lees		*Seeds (Cold-Pressed)*	
Ethyl Decanoate	32.9%–38.2%	Linoleic Acid Ethyl Ester	36.3%
Ethyl Laurate	23.1%–25.0%	Farnesene	33.6%
Ethyl Palmitate	11.9%–15.0%	Docosane	10.0%
Ethyl Octanoate	3.0%–5.2%	Tetracosane	3.1%
Isopentyl Decanoate	4.5%–4.8%		
Isoamyl Laurate	2.9%–3.4%		

Seeds (Solvent Extracted)	
Isoamyl Acetate	16.1%–30.8%
Ethyl Octanoate	10.3%–26.0%
Ethyl Heptanoate	0.9%–12.2%
Phenylethyl Alcohol	0.4%–8.8%
Phenylethyl Acetate	0.0%–8.6%
Hexadecanoic Acid	0.0%–6.9%
Isoamyl Alcohol	0.0%–5.0%
Ethyl Laurate	1.5%–4.3%
Nonanal	1.4%–3.7%
Heptanoic Acid	0.0%–3.1%

Solid-Phase Extraction	
2-Methyl-1-Butanol + 3-Methyl-1-Butanol	59.1%
2-Phenylethanol	16.7%
Ethyl Acetate	7.5%
Ethyl Succinate	4.8%
2-Methyl-1-Propanol	3.6%

Other Constituents: *Wine Lees*—Isoamyl alcohol, decanoic acid, isoamyl octanoate, lauric acid, pentyl decanoate, ethyl myristate, ethyl linoleate, ethyl oleate, ethyl stearate, isopentyl hexanoate, unidentified constituents; *Seeds (Cold-Pressed)*—Octanoate ethyl ester, doconoate ethyl ester; *Seeds (Solvent Extracted)*—Hexyl acetate, 2-nonanon, 1-octen-3-ol, heptyl alcohol, n-decanal, benzaldehyde, nonanal, isovaleric acid, valeric acid, octanoic acid, nonanoic acid; *Solid-Phase Extraction*—3-(methylthio)-1-propanol, 1-hexanol, hexanoic acid, octanoic acid

Preferred Composition for Clinical Use:
Insufficient data/samples to determine.

Reported Therapeutic Properties: Predominantly used for perfumery

Cautions:

- None currently known.

Selected Evidence:

- None found.

COMMON MADDER (Indian Madder, Guamirim, Plinia)

Plinia cordifolia (D. Legrand) Sobral,
Myrciaria cordifolia D. Legrand

FAMILY: Myrtaceae
NOTE: Middle
AROMA INTENSITY: Medium-Strong
AROMA: Sweet, fruity, woody, herbaceous, mildly floral
COMMON EXTRACTION METHOD: Steam distilled from the leaves
POSSIBLE SUBSTITUTE OILS: German chamomile
BLENDS WELL WITH: Bergamot, cassia, cedarwood, eucalyptus, geranium, lavender, neroli, patchouli, pine, ravintsara, rose, silver fir, white fir
SUBCELLULAR LOCALIZATION | EPIGENETIC INFLUENCE: Currently unknown | Currently unknown
RECOMMENDED DILUTION RANGE: 5%–Neat

PRIMARY CONSTITUENTS:[991]

Alpha-Bisabolol Oxide A	28.0%
Beta-Caryophyllene	15.9%
Alpha-Bisabolol Oxide B	7.0%
Alpha-Bisabolol	5.8%
(E)-Beta-Farnesene	5.7%
Alpha-terpineol	4.3%
Delta-Cadinene	3.4%
Alpha-Pinene	2.3%
(E)-Phytol	2.3%

OTHER CONSTITUENTS: Limonene, alpha-humulene, bicyclogermacrene, (E,E)-alpha-farnesene, (E)-gamma-bisabolene, spathulenol, caryophyllene oxide, eudesmol (isomer not identified), 14-hydroxy-9-epi-(E)-caryophyllene

PREFERRED COMPOSITION FOR CLINICAL USE:
Insufficient samples/data to determine.

REPORTED THERAPEUTIC PROPERTIES: Anti-inflammatory, antispasmodic, analgesic, antineuralgic, antirheumatic, wound healing, antiseptic, heals bruises, boils, and blisters, relieves chronic skin conditions, antioxidant, antibacterial, antifungal, calming, relieves anxiety, stress management, reduces tantrums

CAUTIONS:
- None currently known.

SELECTED EVIDENCE:
- None found.

COMMON MUGWORT (Common Wormwood, Crimean Wormwood, St. John's Plant)

Artemisia taurica Willd

FAMILY: Asteraceae (Compositae)
NOTE: Top
AROMA INTENSITY: Medium
AROMA: Fresh, herbaceous, bitter, slightly medicinal
COMMON EXTRACTION METHOD: Steam distilled from the aerial parts
POSSIBLE SUBSTITUTE OILS: Sage, tansy, mugwort, thuja
BLENDS WELL WITH: Bay laurel, bergamot, cajeput, camphor, chamomile (German, Roman), coriander, eucalyptus, lavender, lemon, lime, orange, neroli, palo santo, peppermint, petitgrain, ravensara, rosemary, Spanish sage, spikenard, tangerine, thyme, turmeric
SUBCELLULAR LOCALIZATION | EPIGENETIC INFLUENCE: Currently unknown | Currently unknown
RECOMMENDED DILUTION RANGE: 1.5%–20%

PRIMARY CONSTITUENTS:[992,993]

Alpha-Thujone CT

Alpha-Thujone	69.6%–75.2%
Beta-Thujone	10.9%–16.4%
1,8-Cineole	2.7%–10.6%

Camphor/1,8-Cineole CT

Camphor	37.7%
1,8-Cineole	35.6%
Beta-Caryophyllene	14.0%
Thujone*	7.8%

* Correct isomer not identified.

OTHER CONSTITUENTS: *Alpha-Thujone CT*—Sabinene, myrcene, delta-3-carene, para-cymene, 2(3)-thujon-10-one, trans-pinocarveol, 1,1-dimethyl-3-methylen-2-vinylcyclohexane, camphor, 5(E)-2,6-dimethylocta-1,5,7-trien-3-ol, verbenol, terpinen-4-ol, 1,5,5-trimethyl-6-methylenecyclohex-1-ene, cis-jasmone, germacrene D; *Camphor/1,8-Cineole CT*—Santolina triene, camphene, sabinene, o-cymene

PREFERRED COMPOSITION FOR CLINICAL USE:
Only use if benefits outweigh risks as determined by a qualified professional.

Constituent	*Alpha-Thujone CT*
Alpha-Thujone	65%–80%
Beta-Thujone	7%–20%
1,8-Cineole	2%–12%
Myrcene	0.1%–3%
Borneol	0.1%–3%
Sabinene	0.1%–3%
Camphor	0.1%–3%
Terpinen-4-ol	0.1%–2%

REPORTED THERAPEUTIC PROPERTIES: Promotes menstruation, relieves menstrual cramps, aids digestion, antiseptic, antibacterial, antifungal, antimicrobial, antiparasitic, anticancer, supports respiratory function, eases cough, removes warts and corns, insect repellent, insecticidal, relieves anxiety, aids cognition and memory, promotes focus and concentration

CAUTIONS:

Alpha-Thujone CT

■ Avoid with children under 6 due to thujone content. Cases of seizure have been reported in young children due to exposure to sage essential oil (rich in thujones).[994,995]

■ Avoid during pregnancy and while lactating, due to thujones content. Essential oils high in thujone content may cause abortion.[996,997]

■ Avoid oral consumption. Ingesting plants or products high in thujones can cause vomiting, stomachache, diarrhea, and gastroenteritis followed by absorption disorders, headache, nervous agitation, and chronic convulsions, and symptoms of liver and kidney toxicity extending to yellow liver atrophy (extensive and rapid death of liver cells—parenchymal, possibly by fatty degeneration), arrhythmia, and myocardial bleeding.[998] Thujone is considered significantly neurotoxic and may damage the liver, and it is estimated that as little as 15 mg orally may negatively impact the central nervous system.[999,1000,1001] Taking 20 drops of thuja oil (also with significant thujones content) twice daily for five days caused an adult woman to have a tonic seizure.[1002] As few as 12 drops of sage oil (also with significant thujones content) orally have reportedly caused a seizure and short coma in an adult.[1003]

■ Avoid with epilepsy and Parkinson's disease due to thujone and 1,8-cineole content.[1004,1005,1006,1007,1008]

■ Avoid with those who have a compromised liver or kidneys, due to the risk of kidney toxicity, kidney failure, or liver toxicity.[1009,1010,1011] Consuming extremely large amounts of thujone-rich essential oils (10 mL) has caused kidney failure.[1012] This would also suggest that those taking medications that could cause liver damage should also use tansy very cautiously or avoid it entirely.

■ May weakly interfere with the enzymes responsible for metabolizing medications (NSAIDs, proton-pump inhibitors, acetaminophen, antiepileptics, immune modulators, blood-sugar medications, blood pressure medications, antidepressants, antipsychotics, diabetic medications, antihistamines, antibiotics, and anesthetics) due to 1,8-cineole and alpha-thujone content.[1013,1014,1015,1016]

Camphor/1,8-Cineole CT

■ Avoid with children under six due to significant camphor content. Several cases of camphor poisoning and/or seizures from ingestion and topical application have been reported in children.[1017,1018] Ingestion of camphor-containing products has been lethal in children under age two.[1019] Children five years and up may use camphor-containing essential oils topically in dilutions no stronger than 5%. Camphor ingestion by infants and young children may cause cough, vomiting, seizure, burning sensation in the mucous membranes and eyes, or lack of voluntary coordination of muscle movements.[1020]

■ Caution is warranted during pregnancy and while lactating, due to significant camphor content. Ingestion of essential oils with significant levels of camphor may lead to abortion because fetuses lack the enzymes to process it.[1021] Camphor ingestion by infants and young children may cause cough, vomiting, seizure, burning sensation in the mucous membranes and eyes, or lack of voluntary coordination of muscle movements.[1022]

■ Avoid oral consumption due to significant camphor levels. Essential oils with significant levels of camphor may be toxic when taken orally, usually single doses exceeding 2 mL, although the lethal dose for humans is estimated to be greater than 5 mL in a single dose.[1023,1024,1025]

■ Avoid with epilepsy and Parkinson's disease due to camphor and 1,8-cineole content due to the potential of these constituents to exacerbate or cause seizures/convulsions or reduce seizure-medication efficacy.[1026,1027,1028,1029]

■ Avoid with those who have a compromised liver due to the risk of increased liver enzymes and liver damage based on camphor content.[1030] This would also suggest that those taking medications that could cause liver damage should also use essential oils with significant camphor levels very cautiously or avoid them entirely.

■ May weakly interfere with the enzymes responsible for metabolizing medications (NSAIDs, proton-pump inhibitors, acetaminophen, antiepileptics, immune modulators, blood-sugar medications, blood pressure medications, antidepressants, antipsychotics, diabetic medications, antihistamines, antibiotics, and anesthetics) due to 1,8-cineole content.[1031,1032,1033]
■ May interfere with pentobarbital and other barbiturates (medications for anxiety and insomnia), based on 1,8-cineole content.[1034,1035]
■ The significant camphor content may negatively impact red blood cells and increase the risk of jaundice in children with glucose-6-phosphate dehydrogenase deficiency (G6PD).[1036,1037]
■ Caution is warranted for the topical application of camphor-containing essential oils on broken or injured skin due to an increased risk of toxicity.[1038]

Selected Evidence:

- □ Common mugwort CT 1,8-cineole essential oil exhibited significant antioxidant activity in the DPPH, hydroxyl radical, superoxide radical, and xanthine oxidase assays.[1039]

COPAIBA BALSAM* (Copaiba, Balsam Copaiba)

Copaifera reticulata Ducke, *C. multijuga* Hayne, *C. cearensis* Huber ex Ducke, *C. duckei* Dwyer, *C. paupera* (Herzog) Dwyer, *C. guianensis* Desf., *C. confertiflora* Bth., *C. lucens* Dwyer, *C. coriacea* Mart., *C. martii* Hayne, *C. piresii* Dwyer, *C. pubiflora* Benth.

Family: Fabaceae (Leguminosae)
Note: Base
Aroma Intensity: Medium
Aroma: Woody, mild, slightly balsamic, sweet
Common Extraction Method: Steam distilled or hydrodistilled from the oleoresin
Possible Substitute Oils: Black pepper, guava leaf CT beta-caryophyllene, hemp, frankincense, myrrh, palo santo, German chamomile, Roman chamomile, hairy basil, guava leaf CT beta-bisabolene
Blends Well With: Balsam fir, bergamot, black spruce, blue spruce, cedarwood, clary sage, galbanum, lavender, lemon, lime, orange, palo santo, pine, ravintsara, tangerine, ylang ylang
Subcellular Localization | Epigenetic Influence: Currently unknown | Currently unknown
Recommended Dilution Range: 5%–Neat

Primary Constituents:[1040,1041,1042,1043,1044,1045,1046,1047,1048,1049,1050,1051,1052,1053]

C. reticulata (BC CT)	
Beta-Caryophyllene	27.8%–68.0%
Beta-Bisabolene	0.0%–26.3%
(E)-Alpha-Bergamotene	2.4%–24.0%
Beta-Selinene	0.0%–20.6%
Alpha-Selinene	0.0%–13.2%
Alpha-Humulene	3.9%–9.7%
Beta-Elemene	0.0%–5.6%
Germacrene D	0.0%–5.0%

C. multijuga (BC CT)	
Beta-Caryophyllene	16.0%–85.4%
Trans-Cadina-1(6),4-Diene	0.0%–24.6%
Germacrene D	0.0%–18.9%
Alpha-Copaene	0.3%–15.0%
Germacrene B	0.0%–13.6%
Caryophyllene Oxide	0.0%–11.7%
Alpha-Humulene	0.0%–11.3%

Methyl Kaurenoate	0.0%–3.9%
(Z)-Alpha-Bisabolene	0.0%–3.6%
Methyl Kolavenate	0.0%–3.4%
Alpha-Copaene	0.0%–3.0%
Delta-Cadinene	0.0%–2.6%
Caryophyllene Oxide	0.0%–2.4%
Methyl Copalate	0.0%–2.4%
Methyl Hardwickiate	0.0%–2.3%
Gamma-Amorphene	0.0%–2.2%
Alpha-Bulnesene	0.0%–2.2%
Gamma-Cadinene	0.0%–2.1%

C. cearensis

Beta-Caryophyllene	19.7%
Methyl Clorechinate	11.3%
Alpha-Copaene	8.2%
Beta-Bisabolol	8.2%
Delta-Cadinene	7.2%
Methyl Hardwickiate	6.2%
Alloaromadendrene	3.8%
Alpha-Humulene	3.7%
Germacrene B	3.6%
Beta-Cubebene	3.3%
Beta-Bisabolene	2.8%
(E)-Alpha-Bergamotene	2.1%
Methyl Copalate	2.1%

C. duckei (BC CT)

Beta-Caryophyllene	31.5%–61.8%
Beta-Bisabolene	5.2%–17.4%
(E)-Alpha-Bergamotene	4.1%–10.3%
(E)-Beta-Farnesene	4.1%–7.6%
Beta-Selinene	0.0%–6.7%
Beta-Elemene	0.4%–4.2%
Alpha-Selinene	0.0%–4.2%
(Z)-Alpha-Bisabolene	1.0%–2.5%
Alpha-Bulnesene	0.0%–2.3%

C. reticulata (BB CT)

Beta-Bisabolene	18.2%–54.0%
(E)-Alpha-Bergamotene	8.1%–36.9%
Beta-Caryophyllene	0.0%–20.1%
Beta-Selinene	0.0%–18.1%
Caryophyllene Oxide	0.0%–15.2%
Alpha-Selinene	0.0%–11.4%
Alpha-Guaiene	0.0%–7.9%
Copalic Acid	0.0%–11.0%
Copalic Acid Methyl Ester	0.0%–9.5%
(E)-Alpha-Bergamotene	0.0%–7.5%
Delta-Cadinene	0.0%–6.3%
Methyl Copalate	0.0%–6.2%
3-Acetoxy Copalic Acid	0.0%–6.2%
Beta-Elemene	0.0%–5.3%
Alpha-Amorphene	0.0%–4.6%
Methyl 3-Acetoxy-Copalate	0.0%–3.4%
Alpha-Muurolene	0.0%–3.4%
Delta-Elemene	0.0%–2.9%
Alpha-Cadinol	0.0%–2.7%
Gamma-Muurolene	0.0%–2.7%
Pinifolic Acid	0.0%–2.6%
3Beta-Hydroxy-Copalic acid	0.0%–2.2%
Dimethyl Agathate	0.0%–2.1%

C. pubiflora

Beta-Caryophyllene	65.9%
Beta-Selinene	10.2%
Alpha-Humulene	7.3%
Alpha-Selinene	5.5%
Beta-Elemene	4.7%

C. multijuga (TM CT)

Beta-Caryophyllene	36.2%–60.4%
Trans-Muurola-4(14),5-diene	14.4%–37.0%
Alpha-Copaene	2.4%–18.1%
Delta-Elemene	0.1%–8.5%
Alpha-Humulene	4.2%–6.9%
Delta-Cadinene	0.0%–3.0%
Beta-Elemene	1.1%–2.9%
Trans-Cadina-1(6),4-Diene	0.0%–2.7%

C. duckei (BB CT)

Beta-Bisabolene	15.7%–33.6%
Beta-Caryophyllene	13.0%–25.1%
Beta-Selinene	1.8%–15.4%
(E)-Alpha-Bergamotene	8.3%–12.0%
Alpha-Selinene	1.8%–9.9%
Beta-Elemene	3.0%–9.4%
Alpha-Guaiene	0.0%–4.9%

(Z)-Alpha-Bisabolene	0.0%–5.8%
(E)-Beta-Guaiene	0.0%–5.8%
Alpha-Humulene	0.0%–5.3%
Beta-Elemene	0.0%–5.0%
(E)-Beta-Farnesene	0.0%–4.6%
Delta-Cadinene	0.0%–4.4%
7-Epi-Alpha-Selinene	0.0%–4.0%
(Z)-Beta-Farnesene	0.0%–4.0%
(E)-Gamma-Bisabolene	0.0%–3.8%
Viridiflorene	0.0%–3.4%
Cyclosativene	0.0%–3.3%
Cyclosativene	0.0%–3.2%
Cyperene	0.0%–3.0%
Beta-Sesquiphellandrene	0.0%–2.8%
Beta-Chamigrene	0.0%–2.1%

C. pauper (AC CT)

Alpha-Copaene	42.5%
Beta-Caryophyllene	14.1%
Delta-Cadinene	10.4%
Alpha-Cubebene	5.5%
Alloaromadendrene	3.5%
Beta-Cubebene	3.3%
Beta-Bisabolene	2.6%

C. martii (AC CT)

Alpha-Copaene	36.4%–51.2%
Delta-Cadinene	13.7%–17.2%
Beta-Elemene	4.1%–6.2%
Alloaromadendrene	4.3%–5.0%
Beta-Bisabolene	2.6%–4.5%
Gamma-Muurolene	2.8%–4.4%
Trans-Cadina-1(2),4-Diene	1.9%–2.5%

C. reticulata (AB CT)

(E)-Alpha-Bergamotene	25.6%–39.3%
Beta-Bisabolene	24.2%–32.8%
Beta-Selinene	0.0%–17.1%
Beta-Caryophyllene	1.2%–10.8%
Alpha-Selinene	0.0%–10.4%
Alpha-Guaiene	0.0%–9.5%
(E,E)-Alpha-Farnesene	0.0%–7.6%
Beta-Elemene	0.0%–6.0%
Valencene	0.0%–3.1%
(Z)-Alpha-Bisabolene	0.0%–2.7%
Viridiflorene	0.0%–2.6%
(E)-Beta-Farnesene	0.0%–2.4%
(E)-Gamma-Bisabolene	1.4%–3.3%
Alpha-Bulnesene	1.9%–3.1%
Beta-Sesquiphellandrene	1.1%–2.3%
Aromadendrene	0.8%–3.1%
(E)-Beta-Farnesene	1.4%–2.9%
Alpha-Humulene	1.2%–2.5%

C. paupera (BB CT)

Beta-Bisabolene	20.2%
Alpha-Zingiberene	19.4%
Kaurenoic Acid	13.3%
Copalic Acid	6.1%

C. reticulata (AC CT)

Alpha-Copaene	25.1%
Beta-Caryophyllene	13.1%
Copalic Acid	7.7%
Kaurenoic Acid	7.5%
Hardwickiic Acid	6.9%

C. piresii

Alpha-Coapene	45.5%
Delta-Cadinene	13.7%
Beta-Caryophyllene	10.3%
Alpha-Cubebene	3.5%
Beta-Bisabolene	3.5%
Alloaromadendrene	3.5%
Gamma-Muurolene	2.7%
Alpha-Guaiene	2.0%

C. multijuga (AB CT)

(E)-Alpha-Bergamotene	19.8%–20.9%
Caryophyllene Oxide	11.7%–19.9%
Beta-Caryophyllene	7.6%–16.0%
Beta-Elemene	5.3%–6.1%
Alpha-Copaene	5.5%–6.0%
Ledol	1.9%–3.5%
Alpha-Humulene	2.0%–3.1%
Copalic Acid	2.5%–2.9%
Germacrene B	0.5%–2.8%
3Beta-Hydroxy-Copalic Acid	1.2%–2.2%

C. reticulata (BS CT)

Beta-Selinene	33.0%–35.7%
Beta-Bisabolene	8.1%–32.8%
Alpha-Selinene	21.2%–23.2%
Beta-Elemene	13.1%–13.4%

C. multijuga (CO CT)	
Caryophyllene Oxide	29.2%–44.3%
Beta-Caryophyllene	5.1%–20.8%
Alpha-Copaene	2.0%–15.4%
Beta-Elemene	0.0%–8.6%
Copalic Acid	0.9%–4.9%
Pinifolic Acid	0.8%–4.7%
Ledol	0.0%–4.3%
Alpha-Amorphene	2.0%–4.1%
Alpha-Humulene	1.0%–4.0%
Alpha-Cadinol	0.3%–2.9%
Spathulenol	0.0%–2.1%

C. paupera (AC CT)	
Alpha-Copaene	34.8%
Unidentified	18.9%
Delta-Cadinene	13.9%
Alpha-Cubebene	9.9%
Alloaromadendrene	7.7%
Beta-Caryophyllene	5.4%
Alpha-Humulene	3.2%

(E)-Alpha-Bergamotene	10.1%–11.4%
Beta-Caryophyllene	7.5%–10.8%

C. martii (KA CT)	
Kovalenic Acid	29.0%
Beta-Bisabolene	10.7%
Kaurenoic Acid	7.9%
Alpha-Zingiberene	7.2%

C. lucens	
Polyathic Acid	69.8%
Copalic Acid	11.1%

* This monograph includes additional *Copaifera* spp. not included in *Medicinal Essential Oils* that copaiba essential oil may be extracted from.
Note: Higher viscosity oils indicates the presence of greater levels of diterpenes.

OTHER CONSTITUENTS: *C. reticulata (BC CT)*—Cyclosativene, cyperene, gamma-muurolene, (E,E)-alpha-farnesene, beta-sesquiphellandrene, alpha-gurjunene, aromadendrene, beta-chamigrene, gamma-gurjunene, gamma-curcumene, (E)-gamma-bisabolene, (Z)-beta-farnesene, 7-epi-alpha-selinene, kaurene; *C. reticulata (BB CT)*—Alpha-gurjunene, gamma-muurolene, humulene epoxide II, kaurene, guaia-6,9-diene, gamma-cadinene; *C. reticulata (AC CT)*—Not listed; *C. reticulata (AB CT)*—Guaia-6,9-diene, alpha-humulene, beta-chamigrene, (Z)-gamma-bisabolene, beta-sesquiphellandrene; *C. reticulata (BS CT)*—Cyperene, alpha-humulene, beta-chamigrene, (Z)-alpha-bisabolene, 7-epi-alpha-selinene, delta-cadinene; *C. cearensis*—Alpha-cubebene, longifolene, beta-sesquiphellandrene, aromadendrene, gamma-amorphene, beta-vetivene, ledol, cedrol, alpha-cadinol; *C. multijuga (BC CT)*—Alpha-cubebene, beta-cubebene, bicyclogermacrene, beta-bisabolene, delta-cadinol, (E)-alpha-bergamotene, alloaromadendrene, gamma-cadinene, spathulenol, cis-cadina-1,4-diene, caryophyllene oxide, torreyol, (Z)-beta-ocimene; *C. multijuga (CO CT)*—Alpha-cubebene, gamma-elemene, alloaromadendrene, alpha-muurolene, gamma-cadinene, torreyol, 3beta-hydroxy-copalic acid, 3beta-acetoxy-copalic acid; *C. multijuga (AB CT)*—Delta-elemene, alpha-curcumene, alpha-amorphene, alpha-muurolene, gamma-cadinene, delta-cadinene, torreyol, alpha-cadinol, pinifolic acid; *C. multijuga (TM CT)*—Alpha-cubebene, cyperene, sibirine, alloaromadendrene, gamma-muurolene, gamma-amorphene, gamma-cadinene, germacrene B; *C. duckei (BC CT)*—Alpha-Copaene, alpha-gurjunene, aromadendrene, alpha-humulene, beta-chamigrene, gamma-gurjunene, gamma-curcumene, beta-curcumene, beta-sesquiphellandrene, (E)-gamma-bisabolene; *C. duckei (BB CT)*—(Z)-alpha-bergamotene, beta-chamigrene, gamma-gurjunene, gamma-curcumene, (Z)-alpha-bisabolene, beta-curcumene, epi-alpha-bisabolol, beta-bisabolol; *C. pauper (AC CT)*—(E)-Alpha-bergamotene, alpha-humulene, trans-cadina-1(6),4-diene, epi-cubebol, gamma-cadinene, trans-cadina-1(2),4-diene, T-muurolol; *C. pauper (BB CT)*—Not listed; *C. piresii*—Beta-cubebene, (E)-alpha-bergamotene, alpha-humulene, epi-cubebol, gamma-cadinene, alpha-cadinene, epi-cubenol, cubenol; *C. pubiflora*—(E)-Alpha-bergamotene, germacrene A, caryophyllene oxide; *C. martii (AC CT)*—Alpha-cubebene,

(E)-alpha-bergamotene, aromadendrene, beta-selinene, alpha-selinene, alpha-muurolene, gamma-cadinene, alpha-calacorene, gleenol; *C. martii (KA CT)*—Not listed; *C. lucens*—Beta-caryophyllene; *C. paupera (AC CT)*—Isocaryophyllene, beta-bisabolene, caryophyllene oxide

PREFERRED COMPOSITION FOR CLINICAL USE:

Constituent	*Copaifera sp.*
Beta-Caryophyllene	45%–80%
(E)-Alpha-Bergamotene	2%–12%
Alpha-Copaene	2%–12%
Alpha-Humulene	2%–11%
Germacrene D	0.5%–7%
Beta-Elemene	0.1%–5%
Delta-Elemene	0.1%–5%

REPORTED THERAPEUTIC PROPERTIES: Analgesic (pain relief), **anti-inflammatory**, **antibacterial**, antifungal, antiseptic, expectorant, decongestant, supports immune function, **wound healing**, relieves anxiety, aids digestion, stimulates pancreas and liver function, stress management, stimulating, aids concentration and focus, promotes introspection, encourages emotional healing

CAUTIONS:

C. multijuga (CO CT)

■ May interact with aspirin, blood pressure, antiplatelet, and anticoagulant medications, and increase the risk of bleeding among people with bleeding disorders, due to caryophyllene-oxide content.[1054]

C. reticulata (AB CT), C. reticulata (BC CT), C. reticulata (BB CT), C. duckei (BB CT), and C. pauper (BB CT)

■ May interfere with antibiotics and increase their effectiveness due to significant beta-bisabolene content—beta-bisabolene improves the antimicrobial activity of ampicillin and restores bacterial sensitivity to ampicillin.[1055]

All Other Species/CTs

■ None currently known.

SELECTED EVIDENCE:

- *C. multijuga* essential oil exhibited the strongest anti-inflammatory activity when compared to *C. reticulata* and *C. cearensis*, which isn't surprising since it contained the highest levels of beta-caryophyllene of the three oils (57.5%, 40.9%, and 19.7% respectively) and the lowest diterpenes content (14.5%, 21.85%, and 23.5%).[1056] Some research suggests that the complex mixture of diterpenes and sesquiterpenes may interfere with the active constituents of copaiba oil.[1057] The oil inhibited nitric oxide production *in vitro* and oral administration of 100 mg/kg body weight reduced inflammation in mice (as determined by inhibition of total leukocyte and neutrophil accumulation). Another study found that oral administration (doses of 100 and 200 mg/kg body weight) of *C. multijuga* essential oil reduced lung inflammation *in vitro* and in rats by diminishing leukocyte migration to the pleural cavity.[1058]
- Intraperitoneal injection of 400 mg/kg body weight of copaiba resin (*C. reticulata*) protected against damage to the central nervous system caused by the neurotoxin N-Methyl-D-Aspartate in rats.[1059] Its neuroprotective properties were attributed to a reduction in inflammation.
- Copaiba oil (*C. multijuga* beta-caryophyllene CT) essential oil promoted healing of the cornea after chemical injury in rats.[1060] Eye drops containing 0.1 and 0.5% of copaiba oil reduced corneal ulcer and improved corneal stroma thickness compared to controls (with similar effectiveness to conventional treatments) without causing any damage to the eye.

- Copaiba oil (*C. reticulata*, beta-bisabolene CT) displayed antimicrobial activity against pathogens that cause periodontitis (*L. casei, E. faecalis, S. salivarius, S. sanguinis, S. mitis, S. sobrinus, P. gingivalis, F. nucleatum, A., naeslundii, P. nigrescens*)—both clinical isolates and cultured strains.[1061]
- Varying species of copaiba essential oils exhibited activity against promastigote forms of *Leishmania amazonensis*, with *C. reticulata* (BC CT; 40.9% beta-caryophyllene) being the most active of the species tested.[1062] Another study reported that copaiba resin (beta-caryophyllene CT) was effective against both promastigotes and amastigotes of *Leishmania chagasi*, without any toxicity to monocytes.[1063]
- Interestingly, copaiba essential oil (*C. multijuga*) promoted anxiety in rats administered 100 and 200 mg/kg body weight (orally), with the higher dose provoking greater effects.[1064]

COSTMARY (Balsamite, Alecost)

Tanacetum balsamita L., *Chrysanthemum balsamita* L., *Balsamita major* Desf.

FAMILY: Asteraceae
NOTE: Middle
AROMA INTENSITY: Medium-Strong
AROMA: Herbaceous, minty
COMMON EXTRACTION METHOD: Hydrodistilled or steam distilled from the aerial parts, leaves, or flowers
POSSIBLE SUBSTITUTE OILS: Spearmint, sage, tansy, equal parts spearmint and sage
BLENDS WELL WITH: Chamomile (German, Roman), cajeput, citronella, cypress, eucalyptus, lavender, lemongrass, lemon verbena, peppermint, spearmint, tea tree
SUBCELLULAR LOCALIZATION | EPIGENETIC INFLUENCE: Currently unknown | Currently unknown
RECOMMENDED DILUTION RANGE: 1.5%–20%

PRIMARY CONSTITUENTS:[1065,1066,1067,1068]

Aerial Parts	
Carvone	42.5%–72.9%
Alpha-Thujone	12.4%–21.3%
Beta-Bisabolene	0.5%–10.6%
Beta-Thujone	2.2%–5.6%
Trans-p-Mentha-2,8-Dienol	0.0%–3.0%
Beta-Cubebene	0.0%–2.2%

Leaves	
Carvone	51.8%–68.0%
Alpha-Thujone	10.8%–16.1%
Limonene + 1,8-Cineole	2.9%–5.4%
Beta-Bisabolene	1.3%–4.5%
Beta-Thujone	1.1%–2.2%
T-Muurolol	0.3%–2.2%

Flowers	
Carvone	52.0%–61.7%
Alpha-Thujone	9.0%–11.0%
(E)-Beta-Farnesene	0.0%–3.9%
Limonene + 1,8-Cineole	2.8%–3.8%
T-Muurolol	0.0%–3.8%
Delta-Cadinol	0.0%–2.2%

Other Constituents: *Aerial Parts*—1,8-cineole, trans-p-mentha-2,8-dien-1-ol, pinocarvone, cis-pinocarveol, dihydrocarveol, iso-dihydrocarveol, cis-verbenol, cis-p-mentha-2,8-dien-1-ol, iso-3-thujyl acetate, selin-11-en-4-ol, myrtenol, cis-piperitol, trans-piperitol, viridiflorol; *Leaves*—Sabinene, 2-methylbutyl 2-mehtylbutanoate, trans-p-mentha-2,8-dien-1-ol, cis-p-mentha-2,8-dien-1-ol, pinocarvone, cis-dihydrocarvone, verbenone, trans-piperitol, cis-chrysanthenyl acetate, carvone oxide, trans-carvyl acetate, cis-calamenene, delta-cadinene, methyl isocostate, methyl 4-hydroxy-4,5-dihydroisocostate; *Flowers*—Sabinene, 2-methylbutyl 2-mehtylbutanoate, beta-thujone, trans-p-mentha-2,8-dien-1-ol, cis-p-mentha-2,8-dien-1-ol, cis-dihydrocarvone, cis-chrysanthenyl acetate, beta-bisabolene, cis-calamenene, delta-cadinene, alpha-cadinol, selin-1-en-4-ol, methyl isocostate, methyl 4-hydroxy-4,5-dihydroisocostate, trans-p-Mentha-1(7),8-dien-2-ol, cis-p-Mentha-1(7),8-dien-2-ol, alpha-cedrene, viridiflorol, (Z,E)-farnesol, delta-cadinol

Preferred Composition for Clinical Use:
Insufficient samples/data to determine.

Reported Therapeutic Properties: Antiallergic, liver protective, aids cardiovascular function, insecticidal, antiseptic, antimicrobial, astringent, aids digestion, expels excess gas, sedating, relaxing

Cautions:

All CTs

- Avoid with children under 6 due to thujone content. Cases of seizure have been reported in young children due to exposure to sage essential oil (rich in thujones) exposure.[1069,1070]
- Avoid during pregnancy and while lactating, due to thujones content. Essential oils high in thujone content may cause abortion.[1071,1072]
- Avoid oral consumption. Ingesting plants or products high in thujones can cause vomiting, stomachache, diarrhea, and gastroenteritis followed by absorption disorders, headache, nervous agitation, and chronic convulsions, and symptoms of liver and kidney toxicity extending to yellow liver atrophy (extensive and rapid death of liver cells—parenchymal, possibly by fatty degeneration), arrhythmia, and myocardial bleeding.[1073] Thujone is considered significantly neurotoxic and may damage the liver, and it is estimated that as little as 15 mg orally may negatively impact the central nervous system.[1074,1075,1076] Taking 20 drops of thuja oil (also with significant thujones content) twice daily for five days caused an adult woman to have a tonic seizure.[1077] As few as 12 drops of sage oil (also with significant thujones content) orally have reportedly caused a seizure and short coma in an adult.[1078]
- Avoid with epilepsy and Parkinson's disease due to thujone and 1,8-cineole content.[1079,1080,1081,1082,1083]
- Avoid with those who have a compromised liver or kidneys, due to the risk of kidney toxicity, kidney failure, or liver toxicity.[1084,1085,1086] Consuming extremely large amounts of thujone-rich essential oils (10 mL) has caused kidney failure.[1087] This would also suggest that those taking medications that could cause liver damage should also use tansy very cautiously or avoid it entirely.
- May weakly interfere with the enzymes responsible for metabolizing medications (NSAIDs, proton-pump inhibitors, acetaminophen, antiepileptics, immune modulators, blood-sugar medications, blood pressure medications, antidepressants, antipsychotics, diabetic medications, antihistamines, antibiotics, and anesthetics) due to 1,8-cineole and alpha-thujone content.[1088,1089,1090,1091]
- May interact with diabetes medications and cause low blood sugar based on carvone content.[1092]
- May interact with aspirin, blood pressure, antiplatelet, and anticoagulant medications, and increase the risk of bleeding among people with bleeding disorders based on carvone content. Carvone is a potent calcium channel blocker, which may decrease blood pressure.[1093,1094]
- May interact with barbiturates (medications for anxiety and insomnia), antihistamines, benzodiazepines,

tricyclic antidepressants, or other central nervous system depressant drugs, increasing depressant effects, due to carvone content.[1095]

■ Avoid use in those with iron-deficiency and iron-deficiency anemia. Animal research suggests that spearmint may decrease iron absorption and therefore aggravate iron-deficiency symptoms and anemia.[1096]

SELECTED EVIDENCE:

- □ None found except for research on a subspecies of costmary with dissimilar chemical composition.

CREEK TEA TREE (Tea Tree Oil)

Melaleuca dissitiflora F. Muell.

FAMILY: Myrtaceae
NOTE: Middle
AROMA INTENSITY: Medium
AROMA: Fresh, medicinal, herbaceous, earthy
COMMON EXTRACTION METHOD: Steam distilled or hydrodistilled from the leaves
POSSIBLE SUBSTITUTE OILS: Tea tree, eucalyptus, cajeput
BLENDS WELL WITH: Bergamot, cajeput, carrot seed, chamomile (German, Roman), cedarwood, clary sage, clove, cypress, eucalyptus, fennel, grapefruit, lavender, lemon, lime, melaleuca (tea tree), myrrh, orange, patchouli, petitgrain, pine, ravensara, rosemary, sandalwood, tangerine, thyme
SUBCELLULAR LOCALIZATION | EPIGENETIC INFLUENCE: Currently unknown | Currently unknown
RECOMMENDED DILUTION RANGE: 3%–50%; neat for some conditions

PRIMARY CONSTITUENTS:[1097,1098,1099]

Terpinen-4-ol CT

Terpinen-4-ol	23.1%–55.3%
Gamma-Terpinene	12.0%–18.2%
Beta-Pinene	0.4%–14.7%
Para-Cymene	1.7%–15.8%
1,8-Cineole	1.5%–13.8%
Alpha-Pinene	2.2%–10.7%
Alpha-Terpinene	3.8%–10.1%
Terpinolene	2.0%–3.8%

1,8-Cineole CT

1,8-Cineole	32.6%–65.8%
Terpinen-4-ol	1.5%–32.6%
Alpha-Terpineol	4.4%–8.5%
Limonene	5.3%–6.7%
Para-Cymene	0.7%–6.6%
Gama-Terpinene	0.4%–5.8%
Terpinolene	3.1%–3.3%

OTHER CONSTITUENTS: *Terpinen-4-ol CT*—Sabinene, myrcene, alpha-phellandrene, limonene, linalool, sesquiterpene $C_{15}H_{24}$, aromadendrene, alpha-terpineol, trans-piperitol; *1,8-Cineole CT*—Alpha-pinene, beta-pinene, sabinene, myrcene, alpha-terpinene, monoterpene alcohol $C_{10}H_{18}O$, monoterpene alcohol $C_{10}H_{16}O$, sesquiterpene $C_{15}H_{24}$, trans-piperitol, geraniol, methyl eugenol (< 1.0%)

PREFERRED COMPOSITION FOR CLINICAL USE:

Constituent	*Terpinen-4-ol CT*
Terpinen-4-ol	35%–48%
Gamma-Terpinene	12%–25%

Alpha-Terpinene	5%–12%
Para-Cymene	1%–12%
1,8-Cineole	0.5%–8%
Alpha-Pinene	1%–7%
Alpha-Terpineol	1%–5%
Terpinolene	1%–5%
Sabinene	tr–3%
Aromadendrene	0.1%–3%
Delta-Cadinene	0.1%–3%
Ledene (Viridiflorene)	0.1%–3%
Limonene	0.2%–2%

Reported Therapeutic Properties: Analgesic, anti-inflammatory, **antimicrobial**, antibacterial, antifungal, antiviral, **anti-infectious**, anticancer, aids respiratory function, expectorant, eases cough, antiseptic, helps reduce the appearance of blemishes, helps burns heal, helps boils, blisters, and abscesses heal, helps remove warts, relieves insect bites, helps hemorrhoids heal, **relieves inflammation of the gums**, stimulates immune function, relieves diarrhea, protects against radiation damage and helps radiation burns heal, insecticidal, relieves shock, builds confidence, reduces feelings of oppression

Cautions:

Terpinen-4-ol CT

■ Use cautiously orally. Toxicity appears to occur at extreme doses of essential oils rich in terpinen-4-ol; therefore, it is best to limit tea tree essential oil consumption to 600 mg (about 10 drops with a 60 mg dropper) daily, and avoid long-term use orally. See monograph for Melaleuca (Tea Tree) in *Medicinal Essential Oils* for more information.

■ May interact with antibiotics antifungal drugs and increase or decrease their effectiveness due to terpinen-4-ol content.[1100]

1,8-Cineole CT

■ Avoid with children under age 3, particularly around the nose and mouth and use cautiously in children under age 5 due to significant 1,8-cineole content. 1,8-cineole may cause seizures, central nervous system problems, or respiratory distress in young children.[1101,1102,1103]

■ Caution is warranted orally due to the significant amounts of 1,8-cineole. Limit it to small doses internally (adults—maximum 10 drops daily). Toxicity has been reported when eucalyptus (also high in 1,8-cineole) was ingested in large doses, and as few as 4 to 5 drops may cause problems in very sensitive individuals.[1104,1105,1106,1107,1108,1109]

■ Avoid with epilepsy and Parkinson's disease due to 1,8-cineole content. May exacerbate or cause seizures/convulsions or reduce seizure-medication efficacy based on 1,8-cineole content.[1110,1111,1112,1113]

■ May weakly interfere with the enzymes responsible for metabolizing medications (NSAIDs, proton-pump inhibitors, acetaminophen, antiepileptics, immune modulators, blood-sugar medications, blood pressure medications, antidepressants, antipsychotics, diabetic medications, antihistamines, antibiotics, and anesthetics) due to 1,8-cineole content.[1114,1115,1116]

■ May interfere with pentobarbital and other barbiturates (medications for anxiety and insomnia), based on 1,8-cineole content.[1117,1118]

Selected Evidence:

- □ Although not as well researched as tea tree (*Melaleuca alternifolia*), creek tea tree CT terpinen-4-ol is very similar in composition; therefore, the therapeutic benefits of tea tree may apply. Indeed, the Australian Tea Tree Industry Association (ATTIA) accepts the use of *M. dissitifolia* CT terpinen-4-ol as a quality tea tree oil as long as it meets the established standard.

- Creek tea tree essential oil CT terpinen-4-ol was toxic to mosquito (*Aedes aegypti*) larvae.[1119] Another study reported that creek tea tree essential oil was insecticidal to the Japanese termite (*Reticulitermes speratus*).[1120]

CREEPING THYME (Breckland Thyme, Wild Thyme, Elfin Thyme, Mother of Thyme, Thyme Serpolet)

Thymus serpyllum L.

FAMILY: Lamiaceae
NOTE: Middle
AROMA INTENSITY: Strong
AROMA: Herbaceous, spicy, fresh, woody, pungent
COMMON EXTRACTION METHOD: Hydrodistilled or steam distilled from the aerial parts
POSSIBLE SUBSTITUTE OILS: Oregano, thyme, summer savory
BLENDS WELL WITH: Balsam fir, bergamot, clove, chamomile (German, Roman), cypress, lavandin, lavender, lemon, melissa, orange, oregano, pine, rosemary, sage, tangerine, thyme
SUBCELLULAR LOCALIZATION | EPIGENETIC INFLUENCE: Currently unknown | Currently unknown
RECOMMENDED DILUTION RANGE: 1.5%–20%; 50% for some conditions

PRIMARY CONSTITUENTS:[1121,1122,1123,1124,1125,1126,1127]

Carvacrol CT	
Carvacrol	35.3%–55.2%
Thymol	0.0%–28.6%
Gamma-Terpinene	tr–16.5%
Para-Cymene	11.2%–14.8%
O-Cymene	0.0%–14.0%
Alpha-Terpineol	0.0%–6.5%
Alpha-Pinene	0.0%–6.1%
Beta-Caryophyllene	0.0%–5.3%
1,8-Cineole	0.0%–3.4%

Gamma-Terpinene CT	
Gamma-Terpinene	21.9%–22.7%
Para-Cymene	20.7%–21.1%
Thymol	18.7%–18.7%
Beta-Caryophyllene	0.1%–7.1%
Germacrene D	5.1%–6.0%
Borneol	3.1%–3.9%
Carvacrol Methyl Ether	2.8%–3.2%

Thymol CT	
Thymol	30.5%–58.3%
Carvacrol	3.0%–25.7%
Para-Cymene	0.2%–25.5%
Gamma-Terpinene	0.3%–15.1%
Borneol	2.4%–13.5%
Linalool	0.0%–12.9%
Bornyl Acetate	0.0%–7.0%
Thymol Methyl Ether	0.0%–6.9%
Beta-Caryophyllene	0.0%–6.8%
Limonene	0.0%–4.9%
Trans-Sabinene Hydrate	0.0%–3.8%

Geraniol CT	
Geraniol	26.8%–55.9%
Lavandulyl Acetate	0.0%–28.5%
Thymol	5.1%–17.7%
Geranyl Acetate	0.0%–14.3%
Alpha-Terpinyl Acetate	0.0%–13.9%
Carvacrol	1.2%–7.3%

Para-Cymene CT	
Para-Cymene	16.6%–42.0%
Gamma-Terpinene	4.9%–14.5%
Carvacrol	0.0%–11.5%
(E)-Beta-Caryophyllene	3.8%–10.9%
Thymol Methyl Ether	5.7%–10.0%
Borneol	0.7%–8.3%
Beta-Bisabolene	0.0%–8.0%
Camphene	0.0%–4.3%
(E)-Nerolidol	0.0%–3.8%
Geraniol	0.0%–3.3%
Cis-Sabinene Hydrate	0.0%–3.1%
Thymol	0.0%–3.0%

Borneol CT	
Borneol	22.8%
Caryophyllene Oxide	11.1%
Camphor	10.5%
Camphene	9.4%
Alpha-Terpineol	3.8%
(E)-Nerolidol	3.8%

(E)-Nerolidol CT	
(E)-Nerolidol	20.5%–70.1%
Caryophyllene Oxide	1.1%–24.0%
Germacrene D	1.7%–16.0%
Myrcene	0.0%–15.3%
(E)-Beta-Caryophyllene	1.8%–13.3%
Thymol	1.4%–7.3%
Borneol	0.0%–4.7%
Delta-Cadinene	0.0%–3.7%
Camphor	0.0%–3.6%
Beta-Bisabolene	0.0%–3.3%
Geranyl Acetate	0.0%–3.3%
Linalool	0.0%–3.0%

Myrcene CT	
Myrcene	11.2%–20.2%
(E)-Nerolidol	2.2%–15.2%
(E)-Beta-Caryophyllene	9.4%–13.0%
Germacrene D	3.3%–11.4%
Caryophyllene Oxide	2.2%–10.8%

Not Separated by CT (Estonian)	
(E)-Nerolidol	0.3%–70.1%
Geranyl Acetate	tr–46.4%
Caryophyllene Oxide	0.3%–45.0%
Alpha-Terpineol	0.0%–5.5%
(E)-Nerolidol	0.0%–3.8%

Linalyl Acetate	
Linalyl Acetate	19.6%
Thymol	18.2%
Linalool	11.3%
Geranyl Acetate	5.0%
Geraniol	3.5%
(E)-Beta-Caryophyllene	3.5%
Beta-Bisabolene	3.1%

Alpha-Terpineol CT	
Alpha-Terpineol	21.2%
1,8-Cineole	14.2%
Caryophyllene Oxide	6.5%
Camphor	4.6%
Germacrene D	3.8%
(E)-Beta-Caryophyllene	3.4%

Caryophyllene Oxide CT	
Caryophyllene Oxide	17.7%–45.0%
(E)-Beta-Caryophyllene	4.6%–13.2%
Germacrene D	5.9%–12.4%
(E)-Nerolidol	2.7%–4.8%
Carvacrol	0.9%–3.5%

Germacrene D CT	
Germacrene D	11.2%
(E)-Beta-Caryophyllene	10.5%
Myrcene	10.5%
Caryophyllene Oxide	9.2%
(E)-Nerolidol	6.0%

Gamma-Muurolene CT	
Gamma-Muurolene	10.5%
Alpha-Humulene	7.5%
Alpha-Pinene	6.9%
3-Octanone	6.6%
Thymol	5.6%
Camphor	3.6%

Linalyl Acetate	tr–31.4%
Linalool	0.4%–22.8%
Myrcene	tr–20.2%
(E)-Beta-Caryophyllene	1.4%–13.3%
Germacrene D	0.1%–12.4%
Camphor	tr–10.7%
Camphene	tr–7.3%
Borneol	tr–6.4%
Alpha-Bisabolol	tr–6.4%
Alpha-Terpineol	0.4%–5.4%
1,8-Cineole	0.2%–5.4%
Gamma-Cadinene	tr–5.2%
Spathulenol	tr–5.1%
T-Cadinol	tr–3.9%
(E)-Beta-Ocimene	tr–3.7%
Carvacrol	0.1%–3.5%
Cis-Linalool Oxide	tr–3.4%
Alpha-Farnesol	tr–3.2%

Note: Creeping thyme essential oil composition varies widely according to growing location and the above composition is not inclusive of all CTs but does document most of them.

OTHER CONSTITUENTS: *Carvacrol CT*—Beta-pinene, myrcene, alpha-terpinene, limonene, linalool, borneol, endo-borneol, citronellol, longifolene, beta-bisabolene, alpha-humulene, caryophyllene oxide; *Thymol CT*—Camphene, alpha-pinene, myrcene, sabinene, alpha-terpinene, 1,8-cineole, terpinen-4-ol, cis-sabinene hydrate, camphor, (E)-beta-caryophyllene, beta-bisabolene, (E)-nerolidol, caryophyllene oxide, beta-fenchol, eudesm-3-en-6-ol, thymol acetate, selina-11-en-4α-ol; *Gamma-Terpinene CT*—Alpha-thujene, alpha-pinene, camphene, myrcene, alpha-terpinene, 1,8-cineole, limonene, trans-sabinene hydrate, linalool, thymol methyl ether, carvacrol, thymol acetate, bicyclogermacrene, caryophyllene oxide; *Geraniol CT*—Myrcene, para-cymene, gamma-terpinene, borneol, nerol, geranial, germacrene D, beta-bisabolene, caryophyllene oxide, carvacrol; *Para-Cymene CT*—Alpha-thujene, alpha-pinene, 1-octen-3-ol, alpha-terpinene, thymoquinone, linalool, germacrene D, caryophyllene oxide; *Linalyl Acetate*—Myrcene, para-cymene, 1,8-cineole, camphor, carvacrol, caryophyllene oxide; *Borneol CT*—Myrcene, para-cymene, gamma-terpinene, linalool, (E)-beta-caryophyllene, beta-bisabolene; *Alpha-Terpineol CT*—Myrcene, para-cymene, linalool, borneol, geraniol, geranyl acetate, beta-bisabolene; *(E)-Nerolidol CT*—Alpha-pinene, beta-pinene, para-cymene, limonene, (E)-beta-ocimene, gamma-terpinene, 1,8-cineole, beta-thujone (< 1.9%), alpha-terpineol, carvacrol methyl ether, carvacrol, thymol, geraniol, delta-cadinene, terpinyl acetate, beta-elemene, (E)-beta-farnesene, gamma-muurolene, bicyclogermacrene, trans-calamenene, alpha-muurolene, T-cadinol, viridiflorol, humulene epoxide II, spathulenol, alpha-muurolol, alpha-cadinol, alpha-farnesol, alpha-bisabolol; *Caryophyllene Oxide CT*—Thymol; *Germacrene D CT*—Carvacrol, thymol; *Myrcene CT*—Thymol, carvacrol; *Not Separated by CT (Estonian)*—Alpha-pinene, sabinene, 3-octanol, beta-pinene, para-cymene, gamma-terpinene, beta-thujone (< 2.5%), cis-verbenol, trans-verbenol, terpinen-4-ol, nerol, geraniol, bornyl acetate, thymol, neryl acetate, alpha-copaene, beta-bourbonene, beta-elemene, (E)-beta-farnesene, alpha-humulene, alloaromadendrene, bicyclogermacrene, alpha-farnesene, beta-bisabolene, delta-cadinene, hedycaryol, Selina-3,7(11)-triene, germacrene B, viridiflorol, ledol, alpha-cadinol; *Gamma-Muurolene CT*—Camphene, beta-pinene, alpha-phellandrene, para-cymene, limonene, (E)-beta-ocimene, (Z)-beta-ocimene, gamma-terpinene, 1,8-cineole, linalool, germacrene d-4-ol, alpha-muurolene, beta-bisabolene, beta-bisabolol, delta-cadinene, trans-nerolidol, caryophyllene oxide, T-cadinol

PREFERRED COMPOSITION FOR CLINICAL USE:

Constituent	Carvacrol CT	Thymol CT
Carvacrol	40%–60%	3%–15%
Thymol	10%–25%	40%–60%
Gamma-Terpinene	7%–18%	7%–18%
Para-Cymene	5%–15%	5%–15%
Alpha-Terpinene	0.1%–5%	0.1%–5%
Borneol	–	2%–10%

Note: Most important for both CTs is that there is greater than 40% thymol or carvacrol respectively.

REPORTED THERAPEUTIC PROPERTIES: Antiseptic, antibacterial, antifungal, antimicrobial, antiviral, reviving, relieves skin irritation and chronic skin conditions, aids respiratory function, combats respiratory infections, soothes insect bites/stings, wound healing, enhances circulation, anti-inflammatory, analgesic, relieves headache and backache, antiarthritic, antirheumatic, relieves painful menstruation, antineuralgic, antispasmodic, aids digestion, disinfectant, centers thoughts, brings deep sense of peace

CAUTIONS:

Carvacrol CT

■ Avoid during pregnancy and lactation. Oregano (also rich in carvacrol) may negatively affect embryonic development and encourage fetal cell death according to animal research.[1128]

■ May interact with diabetic medications and cause low blood-sugar levels due to carvacrol content—carvacrol reduces blood glucose and enhances the antidiabetic activities of rosiglitazone in animals.[1129,1130,1131,1132]

■ May interact with aspirin, blood pressure, antiplatelet, and anticoagulant medications, and increase the risk of bleeding among people with bleeding disorders, due to carvacrol and thymol content.[1133,1134,1135,1136,1137]

■ May interact with anticholinergic (drugs used for asthma, incontinence, gastrointestinal cramps, muscular spasms, depression, and sleep disorders) and cholinergic medications (drugs used to reduce urinary retention, diagnose myasthenia gravis, and for glaucoma), based on thymol and carvacrol content.[1138,1139]

■ May interact with antibiotics or antifungals and possibly enhance or decrease their effects due to carvacrol or thymol content.[1140,1141,1142,1143,1144]

■ May irritate mucous membranes (eyes, mouth, nasal passages, vagina, rectum).

Thymol CTs

■ Oral caution. Thymol has a longer half-life (the time it takes for half of the medication to metabolize or excrete half of the dosage) than most essential oil constituents and should not be administered orally for long periods of time.[1145] Thymol is a monoterpene phenol. Reports of fatalities in children who consumed 50 to 200 mg of phenols has been reported.[1146] It is recommended that chemotypes with high levels of thymol be limited to 10 drops per day orally for adults with a two- to seven-day break after twenty-one days of use; or limit ingestion to five or six days weekly.

■ May interact with aspirin, blood pressure, antiplatelet, and anticoagulant medications, and increase the risk of bleeding among people with bleeding disorders, due to thymol and carvacrol content.[1147,1148]

■ May interact with anticholinergic (drugs used for asthma, incontinence, gastrointestinal cramps, muscular spasms, depression, and sleep disorders) and cholinergic medications (drugs used to reduce urinary retention, diagnose myasthenia gravis, and for glaucoma), based on thymol and carvacrol content.[1149,1150]

■ May interact with antibiotics or antifungals and possibly enhance their effects based on thymol and carvacrol content.[1151,1152,1153,1154,1155,1156,1157,1158]

■ May irritate mucous membranes (eyes, mouth, nasal passages, vagina, rectum).

Geraniol CT

■ May moderately interfere with enzymes responsible for metabolizing medications (NSAIDs, proton-pump inhibitors, acetaminophen, antiepileptics, immune modulators, blood-sugar medications, blood pressure medications, antidepressants, antipsychotics, diabetic medications, antihistamines, antibiotics, and anesthetics) due to geraniol content.[1159]

■ May interact with diabetes medications and cause low blood sugar based on significant geraniol content.[1160,1161]

■ May interact with antibiotics and/or antifungals and possibly increase their effectiveness based on significant geraniol content (geraniol inhibits bacterial efflux pumps, which are responsible for antibiotic resistance and resistance to immune system defenses).[1162,1163,1164]

■ Dilution recommended for topical application. Geraniol is metabolized and autoxidized into metabolites like geranial and neral, and may cause sensitization and irritation when applied topically.[1165]

Linalyl Acetate CT

■ May irritate mucous membranes (eyes, mouth, nasal passages, vagina, rectum) due to moderate thymol content.

Borneol CT

■ The moderate camphor content in the borneol CT of creeping thyme may negatively impact red blood cells and increase the risk of jaundice in children with glucose-6-phosphate dehydrogenase deficiency (G6PD).[1166,1167]

■ If borneol content meets or exceeds 25%, caution is warranted for anticoagulants, antiplatelet meds, aspirin, blood pressure meds, bleeding disorders, and drugs metabolized by the CYP450 enzymes.

(E)-Nerolidol CT

■ May interfere with antibiotics and increase their effectiveness ((E)-nerolidol sensitizes bacteria to antibiotics).[1168]

■ May interfere with enzymes (CYP1A, CYP2B, CYP3A) responsible for metabolizing medications based on alpha-bisabolol content (NSAIDs, proton-pump inhibitors, acetaminophen, antiepileptics, immune modulators, blood-sugar medications, blood pressure medications, antidepressants, antipsychotics, diabetic medications, antihistamines, antibiotics, and anesthetics) due to (E)-nerolidol content.[1169]

Alpha-Terpineol CT

■ Avoid with epilepsy and Parkinson's disease due to 1,8-cineole and camphor content. May exacerbate or cause seizures/convulsions or reduce seizure-medication efficacy.[1170,1171,1172,1173,1174,1175]

Caryophyllene Oxide CT

■ May interact with aspirin, blood pressure, antiplatelet, and anticoagulant medications, and increase the risk of bleeding among people with bleeding disorders, due to caryophyllene-oxide content.[1176]

Para-Cymene, Myrcene, Germacrene D, and Gamma-Muurolene CTs

■ None currently known.

SELECTED EVIDENCE:

- □ Creeping thyme essential oil inhibited dermatophytes (clinical animal isolates of *Microsporum canis, Trichophyton mentagrophytes, T. erinacei, T. terrestre,* and *Microsporum gypseum*), which are the main causes of zoonotic or environmental dermatophytes in humans.[1177] An unidentified CT of creeping thyme (possibly thymol) inhibited the growth and mycotoxin production of *Aspergillus ochraceus, A. carbonarius*, and *A. niger*.[1178] An unidentified CT of creeping thyme inhibited *Staphylococcus aureus, Pseudomonas aeruginosa*, and *Escherichia coli*.[1179] Creeping thyme essential oil (near equal carvacrol and para-cymene) showed good antimicrobial activity against *Pseudomonas putida* obtained from meat.[1180] The gamma-terpinene CT of creeping thyme demonstrated bactericidal effects against *E. coli, S. aureus, B. subtilis,* and *K. pneumonia,* but was ineffective against *P. aeruginosa.*[1181] Creeping thyme actively inhibited *H. pylori*, which is associated with peptic ulcers.[1182]

- Supercritical creeping thyme essential oil demonstrated greater antioxidant capacity than the oil extracted by steam distillation.[1183] An unidentified CT of creeping thyme essential oil prevented copper-induced oxidation of LDL cholesterol in animals.[1184] The protective effect was attributed to the presence of thymol and carvacrol.
- Creeping thyme essential oil CT thymol repelled *Dermacentor reticulatus* ticks, providing 82% repellency.[1185] Its repellent effects were even greater (91%) when mixed with citronella at 1.5% each.

CUBEB (Tailed Pepper, Java Pepper, False Pepper)

Piper cubeba L.

FAMILY: Piperaceae
NOTE: Top-Middle
AROMA INTENSITY: Medium
AROMA: Peppery, spicy, woody, camphoraceous
COMMON EXTRACTION METHOD: Hydrodistilled or steam distilled from the fruits (dried berries); may also be distilled from the leaves
POSSIBLE SUBSTITUTE OILS: Pink pepper, savin, nutmeg, plai CT sabinene, muhuhu (leaves), copaiba
BLENDS WELL WITH: Allspice, bergamot, black pepper, camphor, cardamom, clove, frankincense, galbanum, geranium, jasmine, lavender, lemon, myrrh, nutmeg, orange, palo santo, rose, rosemary, sandalwood, turmeric, vetiver, ylang ylang
SUBCELLULAR LOCALIZATION | EPIGENETIC INFLUENCE: Currently unknown | Currently unknown
RECOMMENDED DILUTION RANGE: 3%–33%; 50% for some conditions

PRIMARY CONSTITUENTS:[1186,1187,1188,1189,1190,1191,1192,1193,1194]

Fruits (Sabinene CT)		*Fruits (Cubebol CT)*	
Sabinene	9.1%–46.3%	Cubebol	11.1%–31.0%
Terpinen-4-ol	0.0%–17.0%	Beta-Pinene	0.0%–18.2%
Cubebol	0.0%–15.2%	Gamma-Muurolene	0.0%–11.5%
1,8-Cineole	0.0%–11.9%	Alloaromadendrene	2.3%–11.0%
Beta-Elemene	0.0%–9.4%	Germacrene D	1.5%–11.1%
Alpha-Copaene	0.7%–8.8%	Alpha-Copaene	0.9%–9.8%
Beta-Cubebene	0.0%–8.3%	Beta-Caryophyllene	0.4%–9.5%
Germacrene D	0.0%–7.4%	Beta-Cubebene	3.9%–8.9%
Gamma-Muurolene	0.0%–7.0%	Beta-Elemene	0.0%–7.3%
Beta-Caryophyllene	0.2%–6.7%	Alpha-Cubebene	0.0%–5.1%
Alloaromadendrene	0.0%–6.1%	Sabinene	0.0%–4.9%
Beta-Phellandrene	0.0%–5.9%	Delta-Cadinene	0.0%–4.8%
Beta-Pinene	0.0%–5.9%	(E)-Nerolidol	0.0%–3.5%

Delta-3-Carene	0.0%–5.3%
Linalool	0.2%–5.0%
Epi-Cubebol	0.0%–4.6%
Alpha-Thujene	0.0%–4.5%
Para-Cymene	0.1%–4.4%
Limonene	0.0%–4.4%
Alpha-Terpineol	0.0%–4.4%
Gamma-Terpinene	0.1%–4.2%
T-Muurolene	0.0%–4.2%
Alpha-Pinene	0.0%–4.0%
Alpha-Cubebene	0.0%–3.7%
Isoborneol	0.0%–3.6%
Alpha-Terpinolene	tr–3.0%
Globulol	0.0%–3.5%
Beta-Copaene	0.0%–3.3%
Beta-Bisabolene	0.0%–3.1%

Fruits (Alpha-Copaene CT)

Alpha-Copaene	10.0%–14.0%
Beta-Cubebene	7.0%–11.0%
Delta-Cadinene	9.0%–10.0%
Cubebol	9.0%–10.0%
Alpha-Cubebene	7.0%–9.0%

Leaves

Gamma-Cadinene	16.6%
Trans-Sabinene Hydrate	8.2%
Beta-Caryophyllene	5.0%
Cubebol	4.8%
Epi-Cubebol	4.2%
Sabinene	3.8%
Beta-Pinene	3.8%
Limonene	3.4%
Alpha-Pinene	3.2%

Note: The sabinene CT could be further subdivided into additional chemotypes but is combined in this work.

OTHER CONSTITUENTS: *Fruits (Sabinene CT)*—Alpha-terpinene, myrcene, trans-sabinene, p-cymen-8-ol, linalool, gamma-terpineol, delta-elemene, alpha-humulene, gamma-cadinene, alpha-terpineol, beta-cadinene, cis-sabinene hydrate, trans-sabinene hydrate, terpinen-4-ol, bicyclogermacrene, (E)-gamma-bisabolene, alpha-muurolene, trans-cadina-1(2),4-diene, epiglobulol, guaiol; *Fruits (Cubebol CT)*—Alpha-pinene, limonene, gamma-terpinene, linalool, cubenol, myrcenol, alpha-terpineol, methyl eugenol (< 2.3%), alloaromadendrene, caryophyllene oxide, guaiol, beta-eudesmol, alpha-cadinol, beta-bisabolol; *Fruits (Alpha-Copaene CT)*—Not listed; *Leaves*—Alpha-thujene, myrcene, linalool, alpha-terpineol, 2-undecanone, alpha-cubebene, alpha-copaene, beta-elemene, alpha-humulene, alpha-muurolene, (E)-gamma-bisabolene, (E)-nerolidol, 1-hydroxy-1,7-dimethyl-4-isopropyl-2,7-cyclodecadiene, epi-cubenol, T-cadinol, T-muurolol, alpha-cadinol

PREFERRED COMPOSITION FOR CLINICAL USE:
Significant variance in sample data prohibits creating preferred clinical composition.

REPORTED THERAPEUTIC PROPERTIES: Relieves skin rashes and inflammation, wound healing, astringent, aids oral health, reduces halitosis, eases cough, supports respiratory function, relieves sore throat, expels excess mucus, aids detoxification, diuretic, cleanses the urinary tract, relieves constipation, expels excess gas, digestive aid, antiparasitic, enhances fertility, antiallergic, analgesic, antirheumatic, antiarthritic, anti-inflammatory, reduces fever, aids cardiovascular health, lowers high blood pressure, antimicrobial, antiseptic, aphrodisiac, antidepressant

CAUTIONS:
- None currently known.

SELECTED EVIDENCE:

- □ Oral administration of 300 and 600 mg/kg body weight of cubeb CT sabinene essential oil significantly reduced acute inflammation and inflammatory molecules (myeloperoxidase (MPO), nitric oxide) and proinflammatory cytokines (TNFα, IL-1β) in rats.[1195] Doses of 150, 300, and 600 mg/kg reduced lung inflammation. Another study reported that injection of 150, 300, and 600 mg/kg of cubeb essential oil (unknown CT) reduced pain in animals, with the 600 mg/kg dose exhibiting marked activity.[1196] The same research reported that the oil reduced fever and possessed strong antimicrobial activity.
- □ Cubeb essential oil (sabinene CT) demonstrated significant inhibition of *Penicillium viridicatum* and *Aspergillus flavus*.[1197] A lab-distilled cubeb fruit essential oil (composition not reported) inhibited methicillin- and oxacillin-resistant *S. aureus* by damaging the cell wall and cytoplasmic membrane.[1198]
- □ The sabinene CT of cubeb essential oil showed activity against *Trypanosoma cruzi* (a parasite that causes trypanosomiasis) but was ineffective against *Leishmania amazonensis*.[1199]
- □ The sabinene CT of cubeb essential oil was active against the parasite *Schistosoma mansoni*, exhibiting similar effectiveness as the synthetic control on early stages of the worms.[1200]
- □ Cubeb essential oil CT sabinene exhibited strong antioxidant activity in the DPPH assay.[1201] Another study found that the sabinene CT was only a moderate antioxidant.[1202]
- □ Cubeb essential oil (unknown composition) was toxic against the pulse beetle (*Callosobruchus chinensis*) in both fumigant and contact tests.[1203]

CULANTRILLO
(Maidenhair Fern)

Zanthoxylum limoncello Planch. & Oerst.

FAMILY: Rutaceae
NOTE: Top
AROMA INTENSITY: Medium
AROMA: Floral, fruity, fresh, waxy
COMMON EXTRACTION METHOD: Steam distilled or hydrodistilled from the leaves
POSSIBLE SUBSTITUTE OILS: Rue
BLENDS WELL WITH: Bergamot, chamomile (German, Roman), clary sage, frankincense, geranium, grapefruit, lavender, lemon, lime, myrrh, orange, rose, tangerine, ylang ylang
SUBCELLULAR LOCALIZATION | EPIGENETIC INFLUENCE: Currently unknown | Currently unknown
RECOMMENDED DILUTION RANGE: 3%–33%; 50% for some conditions

PRIMARY CONSTITUENTS:[1204,]

2-Undecanone	31.6%–49.1%
2-Undecenal	25.1%–43.5%
(E)-2-Dodecenal	3.4%–11.2%
Camphene	1.2%–6.3%
1-Decanal	0.7%–6.5%
1-Octanal	1.3%–3.2%

OTHER CONSTITUENTS: Not reported

PREFERRED COMPOSITION FOR CLINICAL USE:

Constituent	
2-Undecanone	30%–50%
2-Undecenal	25%–45%
(E)-2-Dodecenal	2%–12%
Camphene	0.5%–8%
1-Decanal	0.1%–7%
1-Octanal	0.1%–5%

REPORTED THERAPEUTIC PROPERTIES: Anticancer, antibacterial, antifungal, aids digestion, heals stomach ulcers, relieves diarrhea, analgesic, anti-inflammatory, antirheumatic, antiarthritic, supports cardiovascular health, astringent, soothes chronic skin conditions, promotes restful sleep, insecticidal, insect repellent, stress management, relaxing, relieves anxiety, antidepressant

CAUTIONS:

- None currently known.

SELECTED EVIDENCE:

- *In vitro* research demonstrated that culantrillo essential oil moderately inhibited three strains of *H. pylori*, which was partially attributed to the presence of 2-undecanone and 2-undecenal.[1205]
- Culantrillo essential oil exhibited a significant and prolonged repellent effect against the *Aedes aegypti* mosquito.[1206]

CURRY LEAF
(Indian Curry Leaf, Karivepaku)

Murraya koenigii (L.) Sprengel,
Bergera koenigii (L.) Roxb.

FAMILY: Rutaceae
NOTE: Middle-Base
AROMA INTENSITY: Strong
AROMA: Fresh, woody, pungent, minty, herbaceous, peppery, floral
COMMON EXTRACTION METHOD: Steam distilled or hydrodistilled from the leaves (usually fresh)
POSSIBLE SUBSTITUTE OILS: Frankincense, cypress, copaiba, hemp, hops, angelica, lavender
BLENDS WELL WITH: Basil, cedarwood, chamomile (German, Roman), frankincense, geranium, grapefruit, lavender, myrrh, nutmeg, patchouli, sandalwood, tangerine, rose
SUBCELLULAR LOCALIZATION | EPIGENETIC INFLUENCE: Currently unknown | Currently unknown
RECOMMENDED DILUTION RANGE: 3%–33%; neat for some conditions

PRIMARY CONSTITUENTS:[1207,1208,1209,1210,1211,1212,1213,1214,1215,1216,1217,1218,1219,1220,1221]

Alpha-Pinene CT		*Delta-3-Carene CT*	
Alpha-Pinene	19.0%–71.4%	Delta-3-Carene	18.5%–54.2%
Beta-Caryophyllene	3.2%–27.2%	Beta-Pinene	0.0%–13.6%
Beta-Phellandrene	0.0%–27.1%	Beta-Caryophyllene	0.8%–9.5%

Beta-Pinene	0.0%–13.6%
Myrcene	0.0%–13.3%
Globulol	0.0%–10.9%
(E)-Beta-Ocimene	0.1%–5.8%
Limonene	0.0%–5.5%
Bicyclogermacrene	0.0%–5.2%
Alpha-Humulene	0.6%–4.9%

Alpha-Pinene/Sabinene CT

Alpha-Pinene	25.8%–62.6%
Sabinene	3.7%–28.2%
Beta-Caryophyllene	1.6%–16.1%
Beta-Pinene	4.0%–12.1%
Terpinen-4-ol	0.2%–5.5%
Limonene	2.1%–5.4%
Gamma-Terpinene	0.2%–3.4%

Alpha-Pinene/2H-1-Benzopyran CT

Alpha-Pinene	49.3%
2H-1-Benzopyran	24.3%
2-Allyl-4-Methylphenol	16.7%

Beta-Phellandrene CT

Beta-Phellandrene	14.7%–50.2%
Alpha-Pinene	0.0%–17.5%
Beta-Caryophyllene	4.1%–20.1%
Beta-Piperidinoprop-iophenone	0.0%–15.7%
Alpha-Humulene	0.6%–12.6%
Geraniol	0.0%–11.4%
Lavandulol	0.0%–8.9%
Alpha-Phellandrene	0.0%–6.8%
Caryophyllene Oxide	0.0%–6.2%
Limonene	0.0%–6.1%
Terpinen-4-ol	0.0%–6.1%
Linalool	0.0%–4.2%
Sabinene	0.0%–4.1%
Selina-11-en-4α-ol	0.0%–4.0%
Beta-Pinene	0.0%–3.7%
9-Octadecanoic Acid	0.0%–3.5%
Alpha-Selinene	0.0%–3.2%

Beta-Caryophyllene CT

Beta-Caryophyllene	26.5%–38.0%
Naphthalene	0.0%–16.3%
Alpha-Selinene	0.0%–10.4%
Azulene	0.0%–9.7%
(E)-Phytol	0.2%–9.2%
2-Hydroxy-Cyclopenta decanone	0.0%–8.5%

Alpha-Pinene	0.0%–9.4%
Linalool	0.2%–5.4%
Alpha-Eudesmol	0.0%–4.6%
Para-Cymene	0.0%–3.6%
Gamma-Terpinene	2.7%–3.5%
Alpha-Amorphene	0.0%–3.4%
Myrcene	2.4%–3.2%

Alpha-Longipinene CT

Alpha-Longipinene	21.8%
Sativene	6.6%
Spathulenol	5.9%
Bis-1-Methylpropyl-Disulfide	5.2%
Valencene	4.4%
Naphthalene	3.6%

Sabinene CT

Sabinene	30.1%–66.1%
Alpha-Pinene	5.2%–31.3%
Beta-Caryophyllene	2.8%–18.0%
Beta-Pinene	tr–9.9%
Terpinen-4-ol	0.0%–8.4%
Gamma-Terpinene	1.5%–7.4%
Beta-Phellandrene	0.0%–5.3%
Alpha-Terpinene	0.0%–4.5%

Linalool CT

Linalool	32.8%
Linalyl Acetate	16.0%
Elemol	7.4%
Geranyl Acetate	6.2%
Myrcene	6.1%
Allo-Ocimene	5.0%
Alpha-Terpineol	4.9%
(E)-Beta-Ocimene	3.7%
Neryl Acetate	3.5%

Alpha-Copaene CT

Alpha-Copaene	42.0%–44.3%
Beta-Gurjunene	23.3%–25.5%
Isocaryophyllene	11.2%–12.1%
Beta-Caryophyllene	8.6%–8.7%

Alpha-Humulene/Beta-Caryophyllene CT

Alpha-Humulene	15.2%–26.7%
Beta-Caryophyllene	16.6%–26.6%
Para-Cymen-8-ol	0.0%–10.3%
Alpha-Selinene	0.0%–10.2%
(E)-Phytol	0.2%–10.1%
Camphor (Juniper)	1.6%–9.1%

Beta-Phellandrene	1.7%–7.3%	Azulene	0.0%–7.3%
Alpha-Pinene	3.8%–6.5%	Selina-6-en-4-ol	0.0%–7.0%
Alpha-Humulene	0.0%–6.5%	Beta-Selinene	3.8%–6.5%
Selina-11-en-4α-ol	0.0%–6.3%	Naphthalene	0.0%–5.6%
(Z)-Beta-Guaiene	0.0%–5.8%	Viridiflorol	1.5%–5.4%
1,4-Dimethyl-8-Iso-propylidenetricyclo	0.0%–5.6%	2-Naphthalenemethanol	0.7%–4.8%
(E)-Beta-Ocimene	0.0%–3.7%		

Other Constituents: *Delta-3-Carene CT*—Alpha-thujene, camphene, 1,8-cineole, m-cymene, allyl(methoxy)dimethylsilane, alpha-terpinene, limonene, beta-phellandrene, cis-sabinene hydrate, beta-elemene, gamma-elemene, alpha-humulene, caryophyllene oxide, camphor, 3-phenylbutyrophenone, phytol, (Z)-beta-ocimene, (E)-beta-ocimene, alpha-terpineol, carvone, nerol, lavandulyl acetate, myrtenyl acetate, (Z)-jasmone, alpha-gurjunene, germacrene D, delta-cadinene, elemol, viridiflorol, gamma-eudesmol, beta-eudesmol; *Alpha-Pinene CT*—Camphene, terpinolene, alpha-terpinene, linalool, cis-sabinene hydrate, n-nonanal, cis-sabinol, (Z)-beta-terpineol, terpinen-4-ol, bornyl acetate, gamma-cadinene, cedrol, T-cadinol, beta-eudesmol; *Alpha-Pinene/Sabinene CT*—Camphene, myrcene, alpha-terpinene, (E)-beta-ocimene, terpinolene, linalool, neral, bornyl acetate, alpha-copaene, beta-elemene, alpha-humulene, beta-selinene, alpha-selinene, spathulenol, caryophyllene oxide, germacrene D; *Alpha-Pinene/2H-1-Benzopyran CT*—Camphene, elemicin, (E)-beta-farnesene, D-isomenthol, gamma-cadinene; *Beta-Phellandrene CT*—Myrcene, para-cymene, linalool, cis-p-menth-2-en-1-ol, trans-p-menth-2-en-1-ol, alpha-terpineol, citronellol, lavandulyl acetate, geranyl acetate, beta-gurjunene, cis-beta-guaiene, (E)-nerolidol, spathulenol, caryophyllene oxide, humulene epoxide II, alpha-muurolol, phytol, benzaldehyde; *Sabinene CT*—Alpha-thujene, myrcene, limonene, cis-sabinene hydrate, terpinolene, linalool, beta-elemene, alpha-humulene, beta-selinene, alpha-selinene, caryophyllene oxide; *Beta-Caryophyllene CT*—Myrcene, para-cymene, limonene, linalool, cis-p-menth-2-en-1-ol, lavandulol, geraniol, lavandulyl acetate, beta-elemene, (E)-nerolidol, spathulenol, caryophyllene oxide, humulene epoxide I, humulene epoxide II, 1,10-di-epi-cubenol, gamma-eudesmol, alpha-muurolol; *Alpha-Humulene/Beta-Caryophyllene CT*—Linalool, trans-sabinene hydrate, trans-2-cyclohexen-1-ol, cis-2-cyclohexen-1-ol, lavandulol, borneol, terpinen-4-ol, beta-terpineol, alpha-terpineol, lavandulyl acetate, bornyl acetate, alpha-copaene, beta-elemene, trans-alpha-bergamotene, aromadendrene, alloaromadendrene, nopol, beta-farnesene, alpha-amorphene, alpha-gurjunene, butanedioic acid, delta-cadinene, (E)-nerolidol, nerolidol, cycloheptane, beta-humulene, epiglobulol, guaiol, spathulenol, caryophyllene oxide, trivertal, cubenol, eudesma-7(11)-en-4-ol, T-muurolol, ledene, alloaromadendrene oxide, beta-cadin-4-en-10-ol, gamma-cadinene, epi-gamma-eudesmol, alpha-linolenic acid; *Alpha-Longipinene CT*—Dipropyl 4,5-dithiaoctane disulfide, 3-cyclohexen-1-ol, 2-octyn-1-ol, alpha-ylangene, beta-caryophyllene, alloaromadendrene, alpha-bergamotene, aromadendrene, alpha-muurolene, alpha-selinene, caryophyllene oxide, humulene oxide; *Linalool CT*—Limonene, 1,8-cineole, nerol, beta-caryophyllene, alpha-amorphene, viridiflorol, beta-eudesmol; *Alpha-Copaene CT*—Alpha-pinene, beta-pinene, beta-thujene, (E)-beta-ocimene, delta-elemene, germacrene D, spathulenol, caryophyllene, valencene

Preferred Composition for Clinical Use:

Constituent	*BC/AH CT*
Alpha-Humulene	15%–35%
Beta-Caryophyllene	15%–35%
Phytol	0.1%–11%
Beta-Selinene	2%–8%
Viridiflorol	1%–7%

REPORTED THERAPEUTIC PROPERTIES: Relieves nausea or vomiting, aids digestion, expels excess gas, relieves diarrhea, weight management, supports eye health, nourishes the skin, relieves skin itching and irritation, wound healing, analgesic, anti-inflammatory, antispasmodic, reduces fever, antiparasitic, antifungal, antibacterial, immunomodulatory, anticancer, alleviates blood disorders, relieves snake bite, purifies the blood, insect repellent, mentally stimulating, promotes efficiency, encourages cooperation, aids cognition

CAUTIONS:

Alpha-Pinene, Beta-Phellandrene, and Delta-3-Carene CTs

▪ May interact with anticholinergic (drugs used for asthma, incontinence, gastrointestinal cramps, muscular spasms, depression, and sleep disorders) and cholinergic medications (drugs used to reduce urinary retention, diagnose myasthenia gravis, and for glaucoma), based on AChE-inhibtory activity of beta-phellandrene or delta-3-carene.[1222,1223,1224]

All other CTs

▪ None currently known.

SELECTED EVIDENCE:

- ☐ The alpha-humulene/beta-caryophyllene CT of curry leaf essential oil exhibited antiproliferative effects against human breast, human cervical, and murine leukemia cancer cells.[1225]
- ☐ Oral administration of curry leaf essential oil (350 mg/kg body weight per day for seven days) improved the primary immune response and decreased cell-mediated immunity in rats, suggesting curry leaf oil positively modulates immune activity.[1226]
- ☐ Curry leaf CT delta-3-carene essential oil exhibited varying degrees of antifungal activity against *Alternaria alternata, Aspergillus flavus, Aspergillus niger, Aspergillus parasiticus, Fusarium oxysporum, Fusarium moniliforme, Mucor mucedo, Penicillium notatum, Penicillium funiculosum*, and *Trichoderma viride*, with *A. alternate* being the most sensitive fungi.[1227] The alpha-pinene/2H-1-benzopyran CT exhibited strong antimicrobial activity against *B. subtilis, B. cereus, M. luteus, S. aureus, S. epidermidis, E. coli, K. pneumoniae, P. mirabilis, Sh. Flexneri, Sh. dysenteriae.*[1228] Curry leaf CT sabinene essential oil showed pronounced activity against *M. luteus, B. subtilis, E. coli,* and *P. multocida.*[1229] A curry leaf essential oil extracted by supercritical fluid CO_2 significantly inhibited quorum-sensing-controlled virulence factors of *P. aeruginosa.*[1230] The alpha-longipinene CT (supercritical) revealed potent antibiofilm properties against *P. aeruginosa.*[1231] Curry leaf CT linalool essential oil inhibited *C. tuberculosis, P. aeruginosa, S. pyogenes, K. pneumonia, E. aerogenes, S. enterica, P. mirabilis*, and *S. aureus.*[1232] Curry leaf essential oil (alpha-copaene CT) inhibited *Listeria innocua* with complete inhibition observed at the highest concentration.[1233] The alpha-humulene/beta-caryophyllene CT showed potent inhibition of *S. aureus, P. aeruginosa, K. pneumonia, E. coli*, and *S. pneumoniae.*[1234]
- ☐ The delta-3-carene CT of curry leaf essential oil showed good antioxidant activity in the DPPH assay.[1235] The linalool CT also exhibited good antioxidant activity in the DPPH assay.[1236] Curry leaf CT alpha-copaene essential oil exhibited low antioxidant capacity in the DPPH assay.[1237]
- ☐ Both the caryophyllene and beta-phellandrene CTs of curry leaf essential oil repelled cockroaches (*Blattaria* sp.).[1238] The alpha-pinene CT was moderately toxic to the diamondback moth (*Plutella xylostella*) but was a poor repellent.[1239]

CYPRIOL
(Nutgrass, Nagarmotha)

Cyperus scariosus R. Br.

FAMILY: Cyperaceae
NOTE: Base
AROMA INTENSITY: Medium-Strong
AROMA: Woody, earthy, leathery, slightly sweet, peppery, and smoky
COMMON EXTRACTION METHOD: Steam distilled from the roots (rhizomes, tubers)
POSSIBLE SUBSTITUTE OILS: Copaiba, ylang ylang, jointed flatsedge
BLENDS WELL WITH: Agarwood, bergamot, cassia, cedarwood (Atlas), cinnamon, cistus, clary sage, frankincense, galbanum, Himalayan cedarwood, lavender, myrrh, patchouli, sandalwood, Texas cedarwood, vetiver, Virginia cedarwood
SUBCELLULAR LOCALIZATION | EPIGENETIC INFLUENCE: Currently unknown | Currently unknown
RECOMMENDED DILUTION RANGE: 3%–50%

PRIMARY CONSTITUENTS:[1240,1241,1242,1243,1244]

Cyperene CT	
Cyperene	9.4%–30.6%
Cyperotundone	0.0%–13.0%
Caryophyllene Oxide	0.0%–12.5%
Iso-Patchoul-4(5)-en-3-one	0.0%–12.3%
Cis-Pinocarveol	0.0%–11.5%
Rotundene	5.8%–10.7%
Alpha-Humulene	0.0%–10.7%
Zierone	0.0%–10.7%
Trans-Pinocarveol	0.0%–7.2%
Valencene	0.0%–6.2%
Longiverbeone	0.0%–5.9%
Eudesma-4(14)-11-Diene	0.0%–4.6%
Cyperen-6-one	0.0%–4.5%
Corymbolone	0.0%–4.5%
Rotundone	0.0%–4.4%
Epi-Guaipyridine	0.0%–4.2%
Longifolendehyde	0.0%–4.6%
Alpha-Gurjunene	0.0%–3.5%
Alpha-Cyperone	0.0%–3.3%
ar-Curcumene	0.0%–3.3%
Mustakone	0.0%–3.2%
Guaiazulene	0.0%–3.2%

Germacrene D CT	
Germacrene D	27.9%–45.1%
Oxygenated Sesquiterpenes	15.5%–16.0%

Cyperene/Longifolene CT	
Cyperene	11.9%–24.2%
Longifolene	6.0%–17.9%
Valencene	3.3%–13.9%
Longiverbenone	3.7%–9.4%
Caryophyllene Oxide	2.4%–9.0%
Beta-Bulnesene	3.5%–6.8%
Beta-Cadinene	1.9%–5.4%
Ledene Oxide II	2.0%–5.2%
Alpha-Copaene	1.2%–5.2%
Trans-Pinocarvone	2.1%–4.5%
Spathulenol	1.3%–3.0%

Longifolene CT	
Longifolene	14.2%–20.4%
Cyperene	5.8%–15.3%
Longiverbenone	6.9%–12.7%
Caryophyllene Oxide	5.2%–10.4%
Ledene Oxide II	2.6%–5.8%
Beta-Bulnesene	2.1%–5.5%
Valencene	2.9%–4.0%
Trans-Pinocarvone	2.8%–3.8%
Spathulenol	1.7%–3.2%

Camphene CT	
Camphene	11.3%
Alpha-Pinene	8.9%
Alpha-Copaene	7.6%
Caryophyllene Oxide	7.2%

Beta-Pinene	5.3%–11.3%
Alpha-Pinene	3.0%–10.8%
Myrtenol	7.9%–7.9%
3-Cyclopropylthiphene	2.7%–6.6%
Alpha-Copaene	2.0%–5.7%
Caryophyllene Oxide	2.6%–5.4%
Alpha-Selinene	4.6%–5.1%
Caryophylla-3,8(13)-Dien-5-beta-ol	4.2%–4.2%
Patchenol	3.9%–3.9%
Vulgarol B	3.8%–3.8%

Beta-Caryophyllene CT

Beta-Caryophyllene	13.9%–24.4%
Caryophyllene Oxide	0.0%–12.5%
Trans-Pinocarveol	4.0%–7.2%
Longifolene	0.0%–5.8%
Rotundene	0.0%–5.5%
Eudesma-2,4,11-diene	0.0%–4.6%
Alloaromadendrene	0.0%–3.5%
Cyperene	0.0%–3.2%
Guaiajulene	0.0%–3.2%

Myrtenol	7.1%
Myrtenal	6.4%

Iso-Patchoul-4(5)-en-3-one CT

Iso-Patchoul-4(5)-en-3-one	16.5%
Beta-Caryophyllene	15.8%
Trans-Pinocarveol	10.5%
Unidentified Sesquiterpene	7.2%
Spathulenol	6.0%
Rotundene	4.8%
Myrtenol	3.5%
Alpha-Selinene	3.0%

Other Constituents: *Cyperene CT*—guai-9,11-dienpyridine, cananodine, beta-pinene, trans-pinocarveol, cyprotene, cyperadiene, isopatchoula-3,5-diene, isolongifolene, ylanga-2,4(15)-diene, caryophyllene, gamma-gurjunene, beta-selinene, valencene, alpha-selinene, isorotundene, nootkatene, 7-epi-alpha-selinine, delta-cadinene, calamenene, cyperene epoxide, caryophyllene oxide A, brachyloxide, cyperen-6-ol, humulene epoxide II, cyperen-8-one, cubenol, caryophylla-3(15),7(14)-dien-6-ol, cadalene, nor-cyperen-4-one, cis-beta-patchoulenone, rotundan-12-one, cyperenal, nootkatone; *Cyperene/Longifolene CT*—Beta-pinene, trans-pinocarveol, beta-fenchol, myrtenol, beta-copaen-4-alpha-ol, eudesma-4(14),11-diene, alpha-selinene, zicrone, gurjunene epoxide II; *Longifolene CT*—Beta-pinene, trans-pinocarveol, beta-fenchol, myrtenol, copaene, beta-copaene-4-alpha-ol, alpha-selinene, beta-cadinene, zicrone, gurjunene epoxide II, eudesma-4(14),11-diene; *Germacrene D* CT—Camphene, limonene, cyperene, eudesma-2,4,11-triene, beta-selinene, oxo-alpha-ylangene, aromadendrene epoxide, humulene epoxide II, (2R,5E)-caryophyll-5-en-12-al; *Beta-Caryophyllene CT*—Alpha-pinene, alpha-gurjunene, alpha-selinene, 2-cyclopropylthiophene, alpha-copaene, delta-cadinene, valencene, ar-himachalene, spathulenol, globulol, panthenol, vulgarol B; *Camphene CT*—Beta-caryophyllene, alpha-selinene, alpha-gurjunenepoxide; *Iso-Patchoul-4(5)-en-3-one CT*—Iso-patchoul-3-ene, p-mentha-1,5diene-8-ol

Preferred Composition for Clinical Use:
Insufficient samples/data to determine.

Reported Therapeutic Properties: Anti-inflammatory, analgesic, anticonvulsive, antispasmodic, antirheumatic, supports respiratory function, expels excess mucus, aids digestion, relieves stomachache, improves circulation, balances hormones, relieves menstrual discomfort, balances menstrual irregularities, relieves diarrhea, liver protective and restorative, balances blood sugar, reduces blood pressure, promotes scalp health, relieves fever, anti-venom, astringent, diuretic, soothes hemorrhoids, reduces varicose veins, nervine, insect repellent, deodorant (masculine), relaxing, relieves anxiety, antidepressant, aphrodisiac, improves mental clarity, promotes introspection

CAUTIONS:

- None currently known.

SELECTED EVIDENCE:

- Animal research demonstrates that cypriol essential oil reduces inflammation and suppresses adjuvant induced arthritis.[1245]

DAKARA FRANKINCENSE (Black Frankincense)

Boswellia neglecta S. Moore

FAMILY: Burseraceae
NOTE: Base
AROMA INTENSITY: Medium
AROMA: Earthy, musty, balsamic, spicy, subtly sweet
COMMON EXTRACTION METHOD: Steam distilled or hydrodistilled from the oleogum resin
POSSIBLE SUBSTITUTE OILS: Sweet frankincense, frankincense, myrrh
BLENDS WELL WITH: Agarwood, basil, bay laurel, bergamot, black pepper, camphor, cassia, cinnamon, coriander, cypress, frankincense, geranium, grapefruit, lavender, lemon, myrrh, neroli, orange, palo santo, petitgrain, pine, ravensara, ravintsara, rose, sandalwood, spikenard, tangerine, vetiver, ylang ylang
SUBCELLULAR LOCALIZATION | EPIGENETIC INFLUENCE: Currently unknown | Currently unknown
RECOMMENDED DILUTION RANGE: 3%–50%; neat for some conditions

PRIMARY CONSTITUENTS:[1246,1247,1248,1249,1250,1251]

Alpha-Pinene CT	
Alpha-Pinene	32.6%–50.7%
Terpinen-4-ol	0.0%–29.9%
Alpha-Thujene	4.7%–16.5%
Beta-Pinene	1.1%–13.1%
Verbenone	0.0%–8.6%
Para-Cymene	2.0%–5.1%

Alpha-Thujene CT	
Alpha-Thujene	19.2%–26.4%%
Alpha-Pinene	16.7%–21.3%
Terpinen-4-ol	5.3%–15.7%
Para-Cymene	4.5%–11.8%
Delta-3-Carene	0.0%–3.7%

CTs Not Separated in Study	
Alpha-Pinene	1.8%–69.3%
Alpha-Thujene	7.1%–56.4%
Para-Cymene	2.2%–41.8%
Limonene	0.0%–39.9%
Beta-Pinene	0.0%–8.6%
Terpinen-4-ol	0.0%–4.9%
Sabinene	0.0%–3.0%
Delta-3-Carene	0.0%–3.0%

OTHER CONSTITUENTS: *Alpha-Pinene CT*—Sabinene, limonene, alpha-humulene, camphor; *Alpha-Thujene CT*—Camphene, beta-pinene, sabinene, limonene, m-cymene, o-methylanisole, beta-thujone (< 1.9%), trans-sabinene hydrate, linalool, linalyl acetate, beta-elemene, trans-pinocarveol, alpha-terpineol, trans-sabinol, verbenone, cis-sabinol, m-cymen-8-ol, p-cymen-8-ol; *CTs Not Separated in Study*—Camphene, verbenone

PREFERRED COMPOSITION FOR CLINICAL USE:

Constituent	*Alpha-Pinene CT*
Alpha-Pinene	35%–60%
Alpha-Thujene	5%–20%
Terpinen-4-ol	5%–20%
Para-Cymene	2%–10%
Limonene	0.5%–8%
Beta-Pinene	0.5%–5%

REPORTED THERAPEUTIC PROPERTIES: Analgesic, anti-inflammatory, antiarthritic, antirheumatic, anti-inflammatory, antispasmodic, antimicrobial, antifungal, antioxidant, astringent, aids digestion, expels excess gas, diuretic, supports respiratory function, expels excess mucus, immune modulation, would healing, nourishes the skin, restores dry, damaged, and mature skin, soothes chronic skin conditions, reduces the appearance of blemishes and wrinkles, aids liver function, clarifies thoughts, antidepressant, expands consciousness, enhances meditation

CAUTIONS:

- None currently known.

SELECTED EVIDENCE:

- Dakara frankincense CT alpha-pinene inhibited *S. aureus, B. cereus, E. coli, P. vulgaris,* and *C. albicans.*[1252] Another study reported that dakara frankincense (unidentified CT) inhibited *S. aureus, B. cereus, P. aeruginosa, E. coli, C. albicans,* and *C. neoformans*, with synergistic and additive benefits observed when combined with opoponax or myrrh.[1253]

DALZIELII FRANKINCENSE

Boswellia dalzielii Hutch.

FAMILY: Burseraceae
NOTE: Base
AROMA INTENSITY: Medium
AROMA: Piney, balsamic, woody, slightly citrusy, sweet
COMMON EXTRACTION METHOD: Steam distilled or hydrodistilled from the oleoresin; less commonly distilled from the leaves
POSSIBLE SUBSTITUTE OILS: Frankincense (*B. carterii, B. sacra*), myrrh, juniper berry
BLENDS WELL WITH: Basil, bay laurel, bergamot, black pepper, camphor, cassia, cinnamon, coriander, cypress, frankincense, geranium, grapefruit, lavender, lemon, myrrh, neroli, orange, pine, petitgrain, palo santo, ravensara, ravintsara, rose, sandalwood, spikenard, tangerine, vetiver, ylang ylang
SUBCELLULAR LOCALIZATION | EPIGENETIC INFLUENCE: Currently unknown | Currently unknown
RECOMMENDED DILUTION RANGE: 3%–50%; neat for some conditions

PRIMARY CONSTITUENTS:[1254,1255]

Alpha-Pinene CT		*Alpha-Pinene/Myrcene CT*	
Alpha-Pinene	26.3%–72.1%	Alpha-Pinene	21.0%–68.5%
Para-Cymene	0.5%–15.8%	Myrcene	1.7%–19.7%
Alpha-Thujene	0.5%–11.1%	Limonene	0.2%–11.5%
Trans-Verbenol	0.7%–5.6%	Alpha-Thujene	1.2%–10.5%
Carvone	0.0%–5.4%	Carvone	0.1%–4.3%
Alpha-Copaene	0.0%–5.0%	Alpha-Copaene	0.0%–4.1%
Bornyl Acetate	0.6%–3.4%		
Alpha-Cubebene	0.0%–3.4%		

Note: Leaf oils reportedly contain alpha-terpinene (45.7%) and alpha-terpinene (11.5%) or delta-3-carene (27.7%), alpha-pinene (15.2%), para-cymene (9.5%), beta-phellandrene (8.5%), isolongifolene (6.2%), and myrcene (5.7%).[1256,1257]

OTHER CONSTITUENTS: *Alpha-Pinene CT*—Santolina triene, 5,5-Dimethyl-1-vinylbicyclo [2.1.1]hexane, beta-citronellene, camphene, thuja-2,4(10)-diene, sabinene, beta-pinene, myrcene, 1,5,8-p-menthatriene, m-cymene, 1,8-cineole, limonene, gamma-terpinene, cis-sabinene hydrate, terpinolene, para-cymenene, linalool, rans-sabinene hydrate, alpha-thujone (< 1.4%), beta-thujone (< 1.3%), myrcenol, alpha-campholenal, cis-limonene oxide, trans-pinocarveol, camphor, alpha-phellandrene-9-ol, trans-pinocamphone (< 2.1%), terpinen-4-ol, para-cymen-8-ol, alpha-terpineol, myrtenal, verbenone, trans-carveol, carvone, carvotanacetone, 3,5-dimethoxytoluene, bornyl acetate, carvacrol, alpha-terpinyl acetate, alpha-copaene, (3E)-cembrene A, cembrenol, incensole; *Alpha-Pinene/Myrcene CT*—Toluene, 6,6-dimethylhepta-2,4-diene, (Z)-salvene, santolina triene, 5,5-Dimethyl-1-vinylbicyclo [2.1.1]hexane, beta-citronellene, camphene, thuja-2,4(10)-diene, sabinene, beta-pinene, 1,5,8-p-menthatriene, m-cymene, para-cymene, o-cymene, (Z)-beta-ocimene, gamma-terpinene, cis-sabinene hydrate, terpinolene, para-cymenene, trans-sabinene hydrate, alpha-thujone (< 0.8%), beta-thujone (< 1.1%), myrcenol, alpha-campholenal, cis-limonene oxide, trans-pinocarveol, trans-verbenol, alpha-phellandrene-8-ol, trans-pinocamphone (< 1.2%), p-mentha-1,5-dien-8-ol, terpinen-4-ol, para-cymen-8-ol, alpha-terpineol, myrtenal, verbenone, trans-carveol, carvone, carvotanacetone, 3,5-dimethoxytoluene, bornyl acetate, thymol, carvacrol, alpha-terpinyl acetate, (3E)-cembrene A, cembrenol, incensole

PREFERRED COMPOSITION FOR CLINICAL USE:

Constituent	*Alpha-Pinene CT*	*Alpha-Pinene/ Myrcene CT*
Alpha-Pinene	35%–65%	20%–50%
Alpha-Thujene	2%–12%	1%–12%
Para-Cymene	0.5%–8%	0.1%–3%
Limonene	0.5%–5%	1%–12%
Beta-Pinene	0.5%–5%	0.5%–5%
Bornyl Acetate	0.1%–5%	0.1%–4%
Incensoles	tr–5%	tr–5%
Alpha-Copaene	0.0%–5%	tr–5%
Myrcene	0%–5%	5%–20%
Terpinen-4-ol	0.1%–4%	0.1%–3%
Sabinene	0.5%–3%	0.5%–3%
Alpha-Terpineol	0.1%–3%	0.1%–3%

Reported Therapeutic Properties: Analgesic, eases fever, relieves toothache, anti-inflammatory, anti-rheumatic, antiarthritic, relieves headache, eases chest pain, improves liver function, aids digestion, soothes the skin, wound healing, stops excess bleeding, supports respiratory function, anti-infectious, anticancer, supports immune function, aids gum and teeth health, insect repellent, aids cognition, antidepressant, relieves anxiety, aids meditation, promotes emotional healing

Cautions:

■ May interfere with the enzymes responsible for metabolizing medications (NSAIDs, proton-pump inhibitors, acetaminophen, antiepileptics, immune modulators, blood-sugar medications, blood pressure medications, antidepressants, antipsychotics, diabetic medications, antihistamines, antibiotics, and anesthetics).[1258]

Selected Evidence:

- □ None found.

DAUGHTER OF THYME

Thymus bovei Benth.

Family: Labiatae (Lamiaceae)
Note: Middle
Aroma Intensity: Medium-strong
Aroma: Floral, sweet, citrusy
Common Extraction Method: Steam distilled, hydrodistilled, or microwave extracted from the aerial parts
Possible Substitute Oils: Geranium, palmarosa, melissa, citronella, equal parts of palmarosa and lemongrass, thyme CT geraniol
Blends Well With: Bergamot, cedarwood, geranium, lemongrass, lime, palmarosa, rosemary, sandalwood, spikenard, ylang ylang
Subcellular Localization | Epigenetic Influence: Currently unknown | Currently unknown
Recommended Dilution Range: 3%–50%; neat for some conditions
Primary Constituents:[1259,1260]

Geraniol	32.3%–35.4%
Geranial	20.4%–27.7%
Neral	12.4%–14.8%
Nerol	0.0%–7.4%
3-Octanol	2.2%–4.4%
Thymol	1.3%–3.8%

Other Constituents: Camphor, 1,8-cineole, 3-octanone, linalool, farnesene, myrcene, geranyl isobutyrate, borneol, isocaryophyllene, camphene, bergamiol, dihydrocarveol acetate, alpha-cyclocitral, beta-ocimene, (E)-nerolidol, limonene, alpha-terpinene, beta-pinene, alpha-phellandrene

Preferred Composition for Clinical Use:

Constituent	
Geraniol	28%–42%
Geranial	15%–32%

Neral	8%–20%
3-Octanol	0.5%–6%
Thymol	0.5%–5%
Camphor	0.1%–5%
Camphene	0.1%–5%
Linalool	0.1%–5%

Reported Therapeutic Properties: Soothes skin irritation and chronic skin conditions, wound healing, relieves boils and blisters, reduces acne, antiseptic, anti-infectious, antimicrobial, **antibacterial**, **antifungal**, **antiviral**, antiparasitic, antispasmodic, anti-inflammatory, relieves sprains and strains, analgesic, aids respiratory function, anticancer, reduces blood pressure, calming, antidepressant, combats jealous feelings, stress management, balances the emotions

Cautions:

- May moderately interfere with enzymes responsible for metabolizing medications (NSAIDs, proton-pump inhibitors, acetaminophen, antiepileptics, immune modulators, blood-sugar medications, blood pressure medications, antidepressants, antipsychotics, diabetic medications, antihistamines, antibiotics, and anesthetics).[1261]
- May interact with diabetes medications and cause low blood sugar based on significant geraniol content.[1262,1263]
- May interact with antibiotics and/or antifungals and possibly increase their effectiveness based on significant geraniol content.[1264]
- Dilution recommended for topical application. Geraniol is metabolized and autoxidized into metabolites like geranial and neral, and may cause sensitization and irritation when applied topically.[1265]

Selected Evidence:

- Daughter of thyme essential oil exhibited marked anticancer activity against human cervical, colon, and lung cancer cells with very low cytotoxicity against fetal lung fibroblasts—this suggests that the oil is selective against cancer cells, particularly when compared to cisplatin.[1266]
- Both daughter of thyme essential oil and its primary constituent (geraniol) showed prominent inhibition of angiotensin-converting enzyme (ACE), which is a drug target to reduce blood pressure and heart failure.[1267]
- Daughter of thyme essential oil and geraniol displayed pronounced inhibitory properties against herpes simplex virus type 2 (HSV-2) replication. The oil also synergized the antiviral activity of the drug acyclovir.[1268]
- *In vitro* research showed that daughter of thyme essential oil inhibits *S. aureus, C. albicans, P. aeruginosa,* MRSA, and *E. coli.*[1269]
- Powerful anthelmintic activity was observed with daughter of thyme essential oil, which exceeded that of the control drug piperazine citrate.[1270]
- Daughter of thyme essential oil demonstrated excellent antioxidant activity in both the DPPH and trolox assays.[1271]

DENAK
(Den, Moshkorak)

Oliveria decumbens Vent.

Family: Apiaceae (Umbelliferae)
Note: Middle
Aroma Intensity: Strong
Aroma: Spicy, herbaceous, medicinal
Common Extraction Method: Steam distilled or hydrodistilled from the aerial parts
Possible Substitute Oils: Thyme, ajowan, oregano, equal parts of thyme and nutmeg
Blends Well With: Ajowan, bay laurel, bergamot, camphor, clary sage, copaiba, cypress, eucalyptus, geranium, grapefruit, juniper, lavender, lemon, marjoram, melaleuca (tea tree), orange, oregano, petitgrain, pine, ravensara, rose, rosemary, sage, Spanish sage, thyme
Subcellular Localization | Epigenetic Influence: Currently unknown | Currently unknown
Recommended Dilution Range: 1.5%–20%; 50% for some conditions

Primary Constituents:[1272,1273,1274,1275,1276,1277,1278,1279]

Thymol CT		*Gamma-Terpinene CT*	
Thymol	22.0%–47.1%	Gamma-Terpinene	23.3%
Carvacrol	8.5%–32.2%	Myristicin	21.7%
Gamma-Terpinene	7.6%–22.2%	Thymol	20.5%
Para-Cymene	6.1%–19.0%	Para-Cymene	19.4%
Myristicin	0.6%–13.6%	Carvacrol	9.5%

Other Constituents: *Thymol CT*—Beta-pinene, limonene, elemicin; *Gamma-Terpinene CT*—Beta-pinene, limonene

Preferred Composition for Clinical Use:

Constituent	
Thymol	30%–55%
Carvacrol	18%–35%
Gamma-Terpinene	6%–22%
Para-Cymene	5%–20%
Myristicin	0.5%–15%
Limonene	0.1%–5%
Beta-Pinene	0.1%–5%

Reported Therapeutic Properties: Wound healing, aids digestion, protects against and heals stomach ulcers, relieves diarrhea, antibacterial, antifungal, antiviral, **antimicrobial**, immune stimulant, relieves fever, anti-inflammatory, antispasmodic, analgesic, antioxidant, neuroprotective, antidepressant, reduces mental fatigue, reduces anger and guilt

CAUTIONS:

■ Oral caution. Thymol has a longer half-life (the time it takes for half of the medication to metabolize or excrete half of the dosage) than most essential oil constituents and should not be administered orally for long periods of time.[1280] Thymol is a monoterpene phenol. Reports of fatalities in children who consumed 50 to 200 mg of phenols has been reported.[1281] It is recommended that chemotypes with high levels of thymol be limited to 10 drops per day orally for adults with a two- to seven-day break after twenty-one days of use; or limit ingestion to five or six days weekly.

■ May interact with aspirin, blood pressure, antiplatelet, and anticoagulant medications, and increase the risk of bleeding among people with bleeding disorders, due to thymol and carvacrol content.[1282,1283]

■ May interact with anticholinergic (drugs used for asthma, incontinence, gastrointestinal cramps, muscular spasms, depression, and sleep disorders) and cholinergic medications (drugs used to reduce urinary retention, diagnose myasthenia gravis, and for glaucoma), based on thymol and carvacrol content.[1284,1285]

■ May interact with antibiotics or antifungals and possibly enhance their effects based on thymol and carvacrol content.[1286,1287,1288,1289,1290,1291,1292,1293]

■ May irritate mucous membranes (eyes, mouth, nasal passages, vagina, rectum).

SELECTED EVIDENCE:

- □ Diabetic foot ulcers are caused by poor circulation and kidney disease and frequently result in amputation due to improper healing. An herbal cream with geranium and denak CT thymol essential oils (1% of each oil) promoted wound healing of diabetic foot ulcers by reducing inflammation and triggering tissue regeneration.[1294]
- □ *In vitro* evidence shows that denak CT thymol essential oil inhibits both standard and clinical strains of *S. pyogenes, S. epidermidis, E. coli*, and *P. aeruginosa*.[1295] Another study reported that the thymol CT strongly inhibited *S. aureus* and strongly synergized the antimicrobial activity of vancomycin.[1296] Denak essential oil CT thymol inhibited *S. aureus, S. epidermidis, B. cereus, E. coli, P. aeruginosa* (lowest inhibition), *S. marcescens, A. niger,* and *C. albicans*. The gamma-terpinene CT of denak essential oil exhibited significant activity against gram-positive bacteria (*S. aureus, B. subtilis,* MRSA), gram-negative bacteria (*E. coli*), and fungi (*A. niger, C. albicans*), but was weakly active against *P. aeruginosa*.[1297]
- □ Denak essential oil CT thymol acted as a natural food preservative, inhibiting spoilage pathogens, in vegetable oil.[1298] The authors reported that it was the most effective antimicrobial of eight oils tested (carrot seed, galbanum, clove, geranium, calamus, ajowan, and ziziphora).
- □ The thymol CT of denak essential oil demonstrated excellent antifungal activity against *A. niger, A. oryzae, P. chrysogenum, T. harzianum, B. spectabilis*, and *P. variotii*.[1299]
- □ Denak essential oil CT thymol exhibited good antioxidant activity.[1300] Another study reported that the thymol CT exhibited considerable antioxidant activity in the TPC, ABTS, DPPH, and FAR assays.[1301] A study concluded that the antioxidant activity of denak essential oil is attributable to the presence of para-cymene, gamma-terpinene, thymol, carvacrol, and myristicin.[1302]

DITTANY
(Dittany of Crete, Cretan Oregano, Eronda, Diktamo, Hop Marjoram, Wintersweet)

Origanum dictamnus L., *O. dictamnifolium* St.-Lag., *Amaracus dictamnus* (L.) Benth., *A. tomentosus* Moench, *Majorana dictamnus* (L.) Kostel.

FAMILY: Lamiaceae (Labiatae)
NOTE: Middle
AROMA INTENSITY: Strong
AROMA: Herbaceous, spicy, medicinal
COMMON EXTRACTION METHOD: Steam distilled or hydrodistilled from the aerial parts; may also be distilled from the bracts and leaves
POSSIBLE SUBSTITUTE OILS: Oregano
BLENDS WELL WITH: Bergamot, camphor, cedarwood, chamomile (German, Roman), cinnamon, clove, cypress, eucalyptus, frankincense, ginger, lavender, lemon, marjoram, melaleuca (tea tree), myrrh, orange, oregano, pine, rosemary, thyme
SUBCELLULAR LOCALIZATION | EPIGENETIC INFLUENCE: Currently unknown | Currently unknown
RECOMMENDED DILUTION RANGE: 1.5%–20%; 50% for some conditions

PRIMARY CONSTITUENTS:[1303,1304,1305,1306,1307,1308,1309,1310,1311,1312,1313]

Aerial Parts

Constituent	%
Carvacrol	45.3%–82.3%
Para-Cymene	4.3%–22.1%
Linalool	0.0%–13.4%
Gamma-Terpinene	0.4%–12.1%
Thymoquinone	0.0%–6.0%
(E)-Beta-Caryophyllene	0.0%–4.1%

Bracts & Leaves

Constituent	%
Carvacrol	45.6%–89.0%
Gamma-Terpinene	1.2%–11.8%
Thymol	0.1%–9.1%
Thymoquinone	0.0%–7.7%
Para-Cymene	1.2%–6.6%
Alpha-Humulene	0.4%–6.0%
Carvone	0.0%–5.3%
Linalool	0.0%–3.8%

Leaves (Carvacrol CT)

Constituent	%
Carvacrol	58.7%
Thymoquinone	11.2%

Para-Cymene CT

Constituent	%
Para-Cymene	32.7%
Carvacrol	14.7%
Gamma-Terpinene	12.4%
Linalool	7.8%
(E)-Beta-Caryophyllene	3.5%

Leaves (Para-Cymene CT)

Constituent	%
Para-Cymene	41.5%–45.5%
Carvacrol	29.1%–34.8%
Thymoquinone	10.9%–13.6%

Bracts

Constituent	%
Carvacrol	54.5%–85.2%
Para-Cymene	0.0%–26.4%
Thymoquinone	0.0%–10.1%

Other Constituents: *Aerial Parts*—Alpha-thujene, camphene, alpha-terpinene, terpinen-4-ol, beta-caryophyllene, myrcene, thymol, delta-2-carene, cis-sabinene hydrate, trans-sabinene hydrate, borneol, caryophyllene oxide, carvacrol acetate; *Para-Cymene CT*—Alpha-pinene, myrcene, beta-phellandrene, limonene, p-cymenene, trans-sabinene hydrate, terpinen-4-ol, 4-hydroxy-2-methyl acetophenone, alpha-copaene, beta-bisabolene, caryophyllene oxide, caryophylladienol II, T-muurolol, hexadecanoic acid; *Bracts & Leaves*—Tricyclene, alpha-pinene, beta-pinene, myrcene, oct-1-en-3-ol, 3-octanol, alpha-terpinene, limonene, trans-sabinene hydrate, camphor, terpinen-4-ol, beta-bisabolene, delta-cadinene; *Leaves (Para-Cymene CT)*—Alpha-thujene, alpha-pinene, beta-phellandrene, cis-sabinene hydrate, terpinen-4-ol, alpha-copaene, caryophyllene oxide; *Leaves (Carvacrol CT)*—Alpha-pinene, cis-sabinene hydrate, alpha-copaene, beta-cubebene, beta-caryophyllene, alpha-humulene, gamma-muurolene, beta-bisabolene, delta-cadinene, caryophyllene oxide; *Bracts*—Alpha-thujene, alpha-pinene, cis-sabinene hydrate, caryophyllene oxide

Preferred Composition for Clinical Use:

Constituent	
Carvacrol	45%–75%
Para-Cymene	5%–20%
Gamma-Terpinene	2%–12%
Thymoquinone*	0%–8%
Beta-Caryophyllene	0.5%–6%

* The presence of thymoquinone is strongly preferred for medicinal value. Thymoquinone may be more abundant in large-leaved dittany plants as opposed to narrow-leaved plants, and concentrate in the leaves.[1314,1315]

Reported Therapeutic Properties: Wound healing, heals scars, relieves burns, antiaging, soothes insect bites, relieves skin itching, antioxidant, anticancer, **antibacterial**, **antifungal**, **antimicrobial**, antiparasitic, analgesic, anti-inflammatory, antirheumatic, relieves sore throat, aids digestion, heals gastric ulcer, enhances uterine tone, eases childbirth, aids circulation, supports respiratory health and combats respiratory infections, eases cough, anticonvulsive, insecticidal, insect repellent, antidepressant, promotes a sense of security

Cautions:

- Avoid during pregnancy and lactation. Oregano (also rich in carvacrol) may negatively affect embryonic development and encourage fetal cell death according to animal research.[1316]
- May interact with diabetic medications and cause low blood-sugar levels due to carvacrol content—carvacrol reduces blood glucose and enhances the antidiabetic activities of rosiglitazone in animals.[1317,1318,1319,1320]
- May interact with aspirin, blood pressure, antiplatelet, and anticoagulant medications, and increase the risk of bleeding among people with bleeding disorders, due to carvacrol and thymol content.[1321,1322,1323,1324,1325]
- May interact with anticholinergic (drugs used for asthma, incontinence, gastrointestinal cramps, muscular spasms, depression, and sleep disorders) and cholinergic medications (drugs used to reduce urinary retention, diagnose myasthenia gravis, and for glaucoma), based on thymol and carvacrol content.[1326,1327]
- May interact with antibiotics or antifungals and possibly enhance or decrease their effects due to carvacrol or thymol content.[1328,1329,1330,1331,1332]
- May irritate mucous membranes (eyes, mouth, nasal passages, vagina, rectum).

Selected Evidence:

- Dittany essential oil (aerial parts) demonstrated excellent cytotoxicity to liver cancer cells.[1333] Another study found that dittany (aerial parts; para-cymene as the main constituent) oil was active against liver and colon (supraclavicular lymph node metastasis) cells.[1334]

- Intraperitoneal administration of dittany essential oil (aerial parts) showed increased seizure latency and reduced severity of seizures (ranging from simple twitches to complete seizures) in mice.[1335]
- *In vitro* research showed that dittany (aerial parts) essential oil effectively inhibits *Staphylococcus aureus, Staphylococcus epidermidis, Escherichia coli, Listeria monocytogenes, Salmonella enteritidis, Salmonella typhimurium, Saccharomyces cerevisiae*, and *Aspergillus niger*.[1336] Another study found that dittany oil (aerial parts) inhibited *E. coli, S. enteritidis, S. aureus, L. monocytogenes*, and *B. cereus*, sometimes exceeding the effectiveness of the control drugs (gentamycin, tetracycline).[1337] The aerial parts oil inhibited *E. coli, P. aeruginosa, S. typhimurium, S. aureus, Rh. Leguminosarum*, and *B subtilis*. A dittany oil with para-cymene as the major constituent inhibited *B. cereus, B. subtilis, S. aureus, S. epidermidis, S. faecalis, E. coli, P. mirabilis*, and *S. typhi*.[1338] Dittany (aerial parts) oil also suppressed the development of grey mold rot (*Botrytis cinerea*) in a laboratory and in eggplant fruit.[1339]
- Dittany (aerial parts) essential oil exhibited significant antioxidant activity in the DPPH assay.[1340] Another study found that dittany (para-cymene CT) exhibited weak antioxidant activity.[1341]
- Dittany essential oil displayed high nematicidal activity against the southern root-knot nematode (*M. incognita*).[1342]
- Dittany (aerial parts) essential oil demonstrated potent repellent properties against *Ae. albopictus* adult mosquitoes.[1343]

DRAGONHEAD (Moldavian Balm, Moldavian Dragonhead)

Dracocephalum moldavica L.

FAMILY: Lamiaceae
NOTE: Top-Middle
AROMA INTENSITY: Medium
AROMA: Lemony, floral, fruity
COMMON EXTRACTION METHOD: Steam distilled from the flowering aerial parts
POSSIBLE SUBSTITUTE OILS: Melissa, lemongrass
BLENDS WELL WITH: Citronella, geranium, lavender, lemon, lemon verbena, lemongrass, melissa, myrtle, neroli, orange, palo santo, rose, ylang ylang
SUBCELLULAR LOCALIZATION | EPIGENETIC INFLUENCE: Currently unknown | Currently unknown
RECOMMENDED DILUTION RANGE: 3%–50%; neat for some applications

PRIMARY CONSTITUENTS:[1344,1345,1346,1347,1348]

Geranyl Acetate	18.9%–50.4%
Geraniol	12.7%–43.3%
Geranial	11.2%–35.2%
Neral	0.0%–18.8%
Carvacrol	0.0%–8.9%
Neryl Acetate	0.7%–5.4%
Nerol	0.0%–3.3%
Sulcatone	0.0%–3.2%

Other Constituents: Beta-ocimene, beta-pinene, para-cymene, myrcene, limonene, tridecane, 1-octen-3-ol, copaene, (-)-beta-bourbonene, 2,2-dimethylocta-3,4-dienal, trans-3-pinanone, isopinocamphone (< 1.0%), linalool, beta-caryophyllene, bicyclogermacrene, caryophyllene oxide, linalool oxide, cis-chrysanthenol, carvone, thymol, germacrene D, phellandrene (isomer not identified), trancelimonene oxide, 1,8-cineole

Preferred Composition for Clinical Use:
Insufficient samples/data to determine.

Reported Therapeutic Properties: Analgesic, anti-inflammatory, antirheumatic, relieves headache, soothes toothache, aids cardiovascular function, antibacterial, antiviral, antifungal, anticancer, antioxidant, supports liver function, relieves stomachache, aids digestion, alleviates nausea, expels excess gas, relieves kidney infections, relives insect bites/stings, relaxing, sedating, antidepressant, insecticidal, insect repellent

Cautions:

■ Caution is advised during pregnancy and lactation due to citral (geranial + neral) content. Large doses of citral may negatively affect fetal development according to animal studies.[1349]

■ May moderately interfere with enzymes responsible for metabolizing medications (NSAIDs, proton-pump inhibitors, acetaminophen, antiepileptics, immune modulators, blood-sugar medications, blood pressure medications, antidepressants, antipsychotics, diabetic medications, antihistamines, antibiotics, and anesthetics) due to geraniol and citral content.[1350,1351,1352,1353,1354]

■ May interact with diabetes medications and cause low blood sugar based on significant geraniol content.[1355,1356] Large doses of citral may also improve insulin sensitivity and lower blood-glucose levels according to animal research.[1357,1358]

■ May interact with antibiotics and/or antifungals and possibly increase their effectiveness based on significant geraniol (geraniol inhibits bacterial efflux pumps, which are responsible for antibiotic resistance and resistance to immune system defenses) and citral content.[1359,1360,1361,1362]

■ Dilution recommended for topical application due to geraniol and citral content. Geraniol is metabolized and autoxidized into metabolites like geranial and neral, and may cause sensitization and irritation when applied topically.[1363] Citral may cause sensitization.[1364,1365]

■ There is a moderate risk that dragonhead may interfere with pentobarbital and other barbiturates (medications for anxiety and insomnia), based on citral content.[1366]

Selected Evidence:

- □ An interesting chemotype of dragonhead essential oil (geranial and neral as major constituents and poor in geranyl acetate) significantly reduced pain when injected intraperitoneally at a dose of 5 to 20 mg/kg body weight.[1367]

DWARF EVERLAST (Sandy Everlasting, Yellow Chaste Weed, Xeranthemum)

Helichrysum arenarium (L.) Moench, *Helichrysi flos*

FAMILY: Asteraceae (Compositae)
NOTE: Middle-Base
AROMA INTENSITY: Medium
AROMA: Burnt sugar and ham, sweet, rich, earthy
COMMON EXTRACTION METHOD: Hydrodistilled or steam distilled from the inflorescences, aerial parts, or leaves
POSSIBLE SUBSTITUTE OILS: Copaiba, helichrysum
BLENDS WELL WITH: Bergamot, chamomile (German, Roman), clary sage, clove, cypress, geranium, grapefruit, helichrysum, lavender, lemon, lime, melaleuca (tea tree), neroli, orange, oregano, rosemary, tangerine, vetiver, yarrow, ylang ylang
SUBCELLULAR LOCALIZATION | EPIGENETIC INFLUENCE: Currently unknown | Currently unknown
RECOMMENDED DILUTION RANGE: 3%–50%, neat for some conditions

PRIMARY CONSTITUENTS:[1368,1369,1370,1371,1372,1373]

Inflorescences (Heneicosane/β-Caryophyllene CT)

Heneicosane	1.5%–32.1%
Beta-Caryophyllene	4.4%–25.6%
n-Nonadecane	0.0%–12.2%
Delta-Cadinene	0.6%–11.2%
n-Octadecane	tr–8.9%
Tetradecanoic Acid	0.0%–7.8%
Alpha-Copaene	1.5%–7.2%
1,8-Cineole	0.8%–7.0%
Gamma-Cadinene	0.8%–5.8%
Caryophyllene Oxide	0.2%–5.7%
Gamma-Muurolene	0.2%–5.4%
Selina-3,7(11)-Diene	0.0%–4.7%
Phthalide (MZ=149)	0.0%–4.3%
(E)-Muurola-4(14),5-Diene	0.0%–4.1%
(E)-Citronellyl Tiglate	0.0%–4.1%
T-Cadinol	0.2%–4.0%
Alpha-Muurolene	0.0%–4.0%
Delta-Selinene	0.0%–3.8%
Alpha-Cadinol	0.2%–3.2%

Leaves

Beta-Caryophyllene	5.8%–22.3%
Delta-Cadinene	6.6%–14.4%
Selina-3,7(11)-Diene	0.0%–8.6%
Alpha-Copaene	1.7%–7.0%

Inflorescences (Methyl Palmitate CT)

Methyl Palmitate	8.0%–28.5%
Capric Acid	8.5%–19.8%
Lauric Acid	6.5%–14.6%
Dodecanoic Acid	0.0%–11.9%
Delta-Cadinene	0.7%–7.9%
Thymol	0.0%–7.7%
Pelargonic Acid	5.2%–6.9%
Nonanoic Acid	0.0%–6.9%
Carvacrol	2.0%–6.2%
Caprylic Acid	0.6%–6.0%
Octanoic Acid	0.0%–6.0%
Trans-Cadina-1(6),4-Diene	tr–4.5%
Delta-Cadinol	1.3%–3.5%
(E)-Anethole	0.1%–3.2%
Beta-Caryophyllene	0.3%–3.1%

Aerial Parts (Alpha-Pinene CT)

Alpha-Pinene	32.0%
1,8-Cineole	16.0%
Alpha-Humulene	15.0%
Beta-Caryophyllene	8.0%

Di-Epi-Alpha-Cedrene CT (Part not identified)

Di-Epi-Alpha-Cedrene	17.9%
Alpha-Ylangene	14.0%

T-Cadinol	0.9%–6.8%
Alpha-Cadinol	0.2%–5.8%
Gamma-Cadinene	3.4%–5.5%
Alpha-Cadinene	tr–5.4%
(E)-Beta-Ionone	2.5%–4.9%
Germacrene D-4-ol	0.0%–4.9%
Tricosane	0.0%–4.8%
Alpha-Pinene	0.6%–4.7%
Heneicosane	0.3%–4.6%
Isophytol	0.0%–4.3%
1,8-Cineole	0.8%–4.2%
Alpha-Muurolene	0.6%–4.2%
Delta-Selinene	0.0%–4.1%
Gamma-Muurolene	0.0%–3.9%
(E)-Muurola-4(14),5-Diene	tr–3.1%
Alpha-Calacorene	0.6%–3.0%

Cyclosativene	11.9%
Limonene	11.4%
Alpha-Selinene	8.2%
Para-Cymene	3.3%

Other Constituents: *Inflorescences (Heneicosane/β-Caryophyllene CT)*—Alpha-pinene, beta-pinene, n-nonanal, nonen-1-al, beta-trans-terpineol, borneol, alpha-terpineol, safranal, n-decanal, lavandulyl acetate, alpha-ylangene, n-tetradecane, methyl eugenol (< 0.6%), dodecanal, alpha-humulene, beta-farnesene, trans-cadina-1(6),4-diene, beta-ionone, gamma-amorphene, alpha-cadinene, alpha-calacorene, (E)-nerolidol, gleenol, n-hexadecane, 1,10-di-epi-cubenol, 1-epi-cubenol, alpha-muurolol, n-tetradecanol, chamazulene, methyl tetradecanoate, hexadecanol, eicosane, n-tricosane, terpinolene, trans-sabinyl acetate, alpha-humulene, viridiflorene, trans-cadina-1(6),4-diene, trans-cadina-1(2),4-diene, caryophyllene oxide, neryl isovalerate, octadecane; *Inflorescences (Methyl Palmitate CT)*—Linalool, alpha-terpineol, carvone, alpha-copaene, 4-anisaldehyde, alpha-humulene, eugenol, gamma-muurolene, alpha-gurjunene, anisole, caryophyllenol, beta-asarone, globulol; *Aerial Parts (Alpha-Pinene CT)*—Not reported; *Leaves*—Alpha-pinene, terpinolene, nonanal, alpha-terpineol, decanal, tridecane, alpha-ylangene, tetradecane, alpha-humulene, (E)-beta-farnesene, germacrene D, alpha-amorphene, gamma-amorphene, zonarene, trans-cadina-1(2),4-diene, alpha-cadinene, alpha-calacorene, caryophyllene oxide, neryl isovalerate, hexadecane, 1,10-di-epi-cubenol, 1-epi-cubenol, alpha-muurolol, octadecane, hexadecanol, nonadecane, eicosane; *Di-Epi-Alpha-Cedrene CT (Part not identified)*—Tricyclene, artemisiatriene, sabinene, beta-caryophyllene, alpha-cedrene, (Z)-alpha-bergamotene, epizonarene, gamma-selinene, delta-cadinene

Preferred Composition for Clinical Use:
Insufficient samples/data to determine.

Reported Therapeutic Properties: Aids digestion, stimulates gallbladder function and the release of bile, liver protective, aids pancreas function, diuretic, antispasmodic, aids detoxification, antibacterial, antifungal, anti-inflammatory, nervine, antineuralgic, analgesic, antispasmodic, antioxidant, aids circulation, wound healing, relieves skin itching and irritation, heals boils and bruises, relieves hemorrhoids, astringent, balances the emotions, antidepressant, promotes creativity and imagination

Cautions:

- None currently known.

SELECTED EVIDENCE:

- *In vitro* research showed that dwarf everlast CT di-epi-alpha-cedrene essential oil exhibits strong antimicrobial activity against gram-positive (*Micrococcus luteus, Staphylococcus aureus, Staphylococcus epidermidis*) and gram-negative bacteria (*Escherichia coli, Pseudomonas tolaasii, Salmonella enteritidis, Salmonella typhimurium*) and the yeast *Candida albicans*.[1374] The alpha-pinene CT of dwarf everlast essential oil inhibited the growth of three foodborne pathogens including *Serratia marcescens, Streptococcus agalactiae*, and *Staphylococcus aureus*.[1375] Dwarf everlast essential oil (unidentified composition) inhibited the foodborne pathogens *A. flavus, A. parasiticus, C. albicans, S. aureus, E. coli*, and *B. subtilis*.[1376]

EASTERN ST. PAUL'S-WORT (Common St. Paul's Wort, Herba Siegesbeckiae)

Siegesbeckia orientalis L.

FAMILY: Asteraceae
NOTE: Middle
AROMA INTENSITY: Strong
AROMA: Herbaceous, earthy, floral, sweet
COMMON EXTRACTION METHOD: Steam distilled or hydrodistilled from the aerial parts
POSSIBLE SUBSTITUTE OILS: Melissa, German chamomile
BLENDS WELL WITH: Bergamot, cassia, cedarwood, chamomile (German, Roman), eucalyptus, geranium, lavender, neroli, patchouli, pine, ravintsara, rose, silver fir, white fir
SUBCELLULAR LOCALIZATION | EPIGENETIC INFLUENCE: Currently unknown | Currently unknown
RECOMMENDED DILUTION RANGE: 5%–Neat

PRIMARY CONSTITUENTS:[1377]

(E)-Alpha-Bisabolene	24.4%
Caryophyllene Oxide	16.9%
Beta-Caryophyllene	14.1%
Spathulenol	12.5%

OTHER CONSTITUENTS: Nerol, alpha-humulene, gamma-elemene, (Z)-alpha-bisabolene, acoradiene, gamma-muurolene, curcumene, elemol, alpha-bisabolene, (Z)-alpha-bisabolene epoxide, globulol, himbaccol, trans-longipinocarveol, pentadecanol, 4a,5-Dimethyl-3-(prop-1-en-2-yl)-1,2,3,4,4a,5,6,7-octahydronaphthalen-1-ol, aristol-1(10)-en-9-ol, beta-copaen-4alpha-ol, (+)-cycloisolonifol-5-ol, 2,15-hexadecanedione, 6-Methyl-2-(4-methylcyclohex-3-en-1-yl)hepta-1,5-dien-4-ol, geranyl-alpha-terpinene

PREFERRED COMPOSITION FOR CLINICAL USE:
Insufficient data/samples to determine.

REPORTED THERAPEUTIC PROPERTIES: Antirheumatic, anti-inflammatory, analgesic, antimalarial, antiviral, antibacterial, antifungal, relieves snakebites, antiallergic, balances immune activity, prevents blood clots, anticancer, wound healing, relieves skin itching, relieves nightmares, relaxing, uplifting

CAUTIONS:

- None currently known.

SELECTED EVIDENCE:

- Eastern St. Paul's wort essential oil exhibited antitumor activities against cervical and liver cancer cells.[1378] The study authors found that caryophyllene oxide significantly contributed to the oils cytotoxicity against the tested cancer cells.
- *In vitro* research demonstrates that Eastern St. Paul's wort essential oil possesses anti-inflammatory activity as evidenced by reduced nitric oxide production and release of the inflammatory cytokine (IL-6) by LPS-stimulated macrophages.[1379]

ECHINACEA (Purple Coneflower)

Echinacea purpurea (L.) Moench, *Brauneria purpurea* (L.) Britton, *E. angustifolia* DC, *E. pallida* (Nutt.) Nutt., *E. atrorubens* Nutt.

FAMILY: Asteraceae (Compositae)
NOTE: Middle-Base
AROMA INTENSITY: Medium
AROMA: Herbaceous, floral, earthy
COMMON EXTRACTION METHOD: Steam distilled or hydrodistilled from the flowers, aerial parts, roots, leaves, stems, or achenes (fruits)
POSSIBLE SUBSTITUTE OILS: Ylang ylang, feverfew, helichrysum, hops, hemp CT myrcene
BLENDS WELL WITH: Bay laurel, bergamot, cajeput, cardamom, chamomile (German, Roman), clary sage, clove, copaiba, coriander, cypress, eucalyptus, geranium, ginger, helichrysum, juniper, lavender, melissa, neroli, orange, palmarosa, patchouli, rose, sandalwood, turmeric, vetiver, ylang ylang
SUBCELLULAR LOCALIZATION | EPIGENETIC INFLUENCE: Currently unknown | Currently unknown
RECOMMENDED DILUTION RANGE: 3%–33%; 50% for some conditions

PRIMARY CONSTITUENTS:[1380,1381,1382,1383,1384]

Flowers (E. purpurea, Germacrene D CT)	
Germacrene D	11.3%
Caryophyllene Oxide	8.7%
Beta-Caryophyllene	7.2%
Alpha-Cadinol	6.3%
Naphthalene,1,2,3,5,6,8a-Hexahydro-4,7-dimethyl-1-(1-ethylethyl)-,(1S-cis)-	3.3%
1,5 Epoxysalvia-4(14)ene	3.3%

Flowers (E. purpurea, Beta-Caryophyllene CT)	
Beta-Caryophyllene	25.0%
Palmitic, Linoleic Acid	17.2%

Flowers (E. angustifolia)	
Myrcene	32.8%
Alpha-Pinene	13.5%
Limonene	9.6%
(E)-Beta-Ocimene	8.3%
Camphene	7.0%
Sabinene + Beta-Thujene	6.0%
Beta-Pinene	5.3%

Flowers (E. pallida)	
Myrcene	60.1%
Beta-Pinene	6.9%
Alpha-Pinene	5.0%

Unidentified	14.2%
Germacrene D	13.8%
Cyclodeca-1.5-Diene	9.5%
Alpha-Pinene + Beta-Pinene	9.5%
Nerolidol*	6.6%
Alpha-Phellandrene	4.2%

Flowers (E. purpurea, Myrcene CT)

Myrcene	42.7%
Alpha-Pinene	22.7%
Beta-Pinene	7.4%
Limonene	4.0%

Roots (E. purpurea)

Myrcene	16.7%
Dimethyl Sulfide	14.7%
2-Methylnutanal	12.9%
3-Methylbutanal	11.6%
2-Methylpropanal + Acetone	11.5%
Para-Cymene	11.0%
Trichloroacetic Acid	3.4%

Roots (E. angustifolia)

2-Methylpropanal + Acetone	35.2%
Dimethyl Sulfide	13.2%
3-Methylbutanal	12.8%
2-Methylbutanal	8.3%
Camphene	6.5%
7-Dodecenol	6.3%
Para-Cymene	5.5%
Acetaldehyde	5.3%
Myrcene	4.9%

Roots (E. pallida, Dimethyl Sulfide CT)

Dimethyl Sulfide	34.4%
2-Methylpropanal + Acetone	20.3%
7-Dodecenol	11.8%
2-Methylbutanal	11.0%
3-Methylbutanal	9.3%
Limonene	7.6%
Acetaldehyde	4.9%
Camphene	3.9%

Leaves (E. angustifolia)

Myrcene	37.9%
Limonene	9.8%
(Z)-3-Hexen-1-ol	9.5%
2-Methyl-4-Pentenal	6.2%
Camphene	3.1%
(Z)-3-Hexen-1-ol	3.0%

Aerial Parts (E. purpurea)

Germacrene D	52.6%–57.8%
Myrcene	1.7%–8.8%
Alpha-Pinene	2.3%–7.5%
Beta-Caryophyllene	2.1%–3.1%
Valencene	1.9%–3.0%

Roots (E. pallida, 8-Pentadecene-2-on CT)

8-Pentadecene-2-one	26.1%
Cyclopentadienone	18.3%
Unidentified	15.8%
Ledol	14.4%
2.5-Octadecadiionoic Acid	11.7%
1.4-Decahydroetanonaphtalene	7.3%
2-Pentadecadone	3.9%

Roots (E. atrorubens)

8-Pentadecene-2-one	24.3%
Cyclopentadecanone	21.3%
2.5-Octadecadiinoic Acid	17.9%
Unidentified	13.2%
Ledol	11.9%
1.4-Decahydroetanonaphtalene	6.3%
Hexadecene-4-in (Isomer I+II)	3.9%

Leaves (E. purpurea)

Myrcene	27.2%
3-Hexen-1-ol Acetate	12.7%
Alpha-Pinene	12.0%
2-Methyl-4-Pentenal	11.7%
(Z)-3-Hexen-1-ol	9.2%
(E)-2-Hexenal	7.0%
Limonene	3.1%

Leaves (E. pallida)

Myrcene	31.4%
3-Hexen-1-ol Acetate	13.8%
Beta-Pinene	9.0%
(Z)-3-Hexen-1-ol	8.3%
2-Methyl-4-Pentenal	7.1%
Alpha-Pinene	5.4%
Limonene	4.1%

3-Hexen-1-ol Acetate	5.1%
Camphene	4.4%
Beta-Pinene	4.0%
Alpha-Pinene	3.4%
(E)-2-Hexenal	3.4%

Stems (E. purpurea)

Myrcene	45.3%
Alpha-Pinene	33.8%
Limonene	4.6%
Beta-Pinene	4.4%

Stems (E. pallida)

Myrcene	45.3%
Alpha-Pinene	33.8%
Limonene	4.6%
Beta-Pinene	4.4%

Echinacea spp. (Leaves)

Germacrene D	18.1%
Caryophyllene Oxide	11.3%
Alpha-Cadinol	9.1%
Naphthalene	7.8%
Cedrol	7.2%
Alpha-Phellandrene	6.9%
Trans-Carveol	4.8%
Beta-Caryophyllene	4.5%
Para-Cymene	3.7%

* Correct isomer not identified.

(E)-2-Hexenal	3.7%
Camphene	3.6%

Stems (E. angustifolia)

Myrcene	44.7%
Limonene	11.3%
Alpha-Pinene	9.0%
Camphene	8.9%
Sabinene + Beta-Thujene	4.9%
Beta-Pinene	3.6%

Echinacea spp. (Roots)

Germacrene D	20.3%
Caryophyllene Oxide	12.2%
Cedrol	10.5%
Alpha-Phellandrene	6.6%
Naphthalene	6.4%
Alpha-Cadinol	5.9%
Beta-Caryophyllene	4.0%
Carvone	3.8%
Alpha-Pinene	3.7%
Camphor	3.4%
Methyl Chavicol (Estragole)	3.3%
Ledol	3.1%

OTHER CONSTITUENTS: *Flowers (E. purpurea, Germacrene D CT)*—Beta-pinene, alpha-phellandrene, para-cymene, beta-elemene, alpha-cadinene, alpha-farnesene, naphthalene,1,2,3,4,4a,7-hexahydro-1,6-dimethyl-4-(1-methylethyl)-, alpha-bisabolene, gamma-cadinene, decatone, ethyl oleate, vulgarol B, aromadendrene oxide, 3,4-difloro-4 methoxybiphenyl, iso-aromadendrene epoxide, trans-(Z)-alpha-bisabolene epoxide, benzenepropanoic acid, octadecyl ester, diepi-alpha-cedrene epoxide; *Flowers (E. purpurea, Beta-Caryophyllene CT)*—Not listed; *Flowers (E. purpurea, Myrcene CT)*—Dimethyl sulfide, 2-methylpropanal, acetone, camphene, hexanal, sabinene, beta-thujene, trans-2-hexenal, ocimene, gamma-terpinene, para-cymene, 1-hexanol, allo-ocimene, (Z)-3-hexen-1-ol, beta-caryophyllene; *Flowers (E. angustifolia)*—Dimethyl sulfide, 2-methylpropanal + acetone, 2-butanone, alpha-thujene, hexanal, alpha-terpinene, 1,8-cineole, sabinene, beta-thujene, gamma-terpinene, alpha-terpineolene, 1-hexanol, allo-ocimene, (Z)-3-hexen-1-ol; *Aerial Parts (E. purpurea)*—Camphene, sabinene, beta-pinene, alpha-phellandrene, limonene, menthol, beta-cubebene, beta-elemene, alpha-humulene, alpha-muurolene, beta-bisabolene, (Z)-gamma-bisabolene, germacrene B, germacrene D-4-ol, guaiol, gamma-eudesmol, alpha-muurolol, alpha-cadinol, dehydro-eudesmol, 7-epi-alpha-eudesmol, unidentified oxygenated sesquiterpene; *Flowers (E. pallida)*—Dimethyl sulfide, 2-methylpropanaol + acetone, alpha-thujene, geranyl acetate, sabinene, beta-thujene, limonene, (E)-2-hexenal, (E)-beta-ocimene, gamma-terpinene, 3-hexen-1-ol acetate, 1-hexanol, (E)-2-hexen-1-ol; *Roots (E. purpurea)*—Acetaldehyde, 2-propenal, camphene, hexanal, 2-methyl-1-propanol, alpha-terpinene, 3-methyl-1-butanol, ocimene; *Roots (E. angustifolia)*—2-propenal, alpha-pinene, hexenal, beta-pinene, heptanal, limonene, 6-methyl-hepten-2-one; *Roots (E. pallida, Dimethyl Sulfide CT)*—2-propenal, myrcene; *Roots (E. pallida, 8-Pentadecene-2-on CT)*—Beta-caryophyllene; *Roots (E. atrorubens)*—Beta-caryophyllene; *Leaves (E. purpurea)*—Acetaldehyde, pentanal, hexanal, beta-pinene, 2-pentanal, (Z)-2-hexenal, trans-ocimene, hexyl acetate, allo-ocimene, alpha-ylangene, 5-ethyl-2(5H)-furanone; *Leaves (E. angustifolia)*—Pentanal, alpha-thujene, hexanal, 3-hexenal, alpha-terpinene, gamma-terpinene, (E)-beta-ocimene, 1-hexanol, 5-ethyl-2(5H)-furanone; *Leaves (E. pallida)*—Pentanal, alpha-thujene, hexanal, sabinene + beta-thujene, 3-hexenal, (Z)-2-hexenal, sabinene + beta-thujene, gamma-terpinene, (E)-beta-ocimene, 1-hexanol, 5-ethyl-2(5H)-furanone; *Stems (E. purpurea)*—Geranyl acetate, camphene, sabinene, beta-thujene, trans-3-hexenal, ocimene, gamma-terpinene, allo-ocimene, (Z)-3-hexen-1-ol; *Stems (E. angustifolia)*—Dimethyl sulfide, alpha-thujene, hexanal,2-methyl-4-pentenal, alpha-terpinene, sabinene, beta-thujene, (e)-2-hexenal, gamma-terpinene, (E)-beta-ocimene, terpinolene, endo-bornyl acetate; *Stems (E. pallida)*—Alpha-thujene, hexanal, camphene, sabinene + beta-thujene, 2-methyl-4-pentenal, (E)-2-hexenal, (E)-beta-ocimene, 3-hexen-1-ol-acetate, endo-bornyl acetate; *Echinacea spp. (Roots)*—Alpha-fenchene, camphene, beta-pinene, limonene, linalool, isomenthone, borneol, lavandulol, p-cymen-8-ol, trans-carveol, cis-carveol, alpha-cubebene, methyl eugenol (< 0.7%); *Echinacea spp. (Leaves)*—2-Hexenal, alpha-pinene, beta-pinene, limonene, linalool, methyl chavicol (< 2.3%), cis-carveol, carvone, piperitone, alpha-cubebene, beta-cubebene, alpha-humulene, gamma-muurolene, viridiflorol

PREFERRED COMPOSITION FOR CLINICAL USE:
Too great of variance to determine.

REPORTED THERAPEUTIC PROPERTIES: Anti-inflammatory, analgesic, antiarthritic, wound healing, stimulates immune function, aids respiratory function, eases cough, antifungal, antibacterial, anticancer, antiaging, promotes restful sleep, relieves anxiety, antidepressant, promotes creativity, boosts self-confidence

CAUTIONS:

Myrcene CTs

■ Caution is advised during pregnancy and lactation due to high beta-myrcene content. Extremely high doses (500 mg/kg) of beta-myrcene have been toxic to fetuses according to animal research.[1385,1386]

■ May interact with barbiturates (medications for anxiety and insomnia), antihistamines, benzodiazepines, tricyclic antidepressants, or other central nervous system depressant drugs, increasing their depressant effects.[1387,1388,1389]

■ May weakly interfere with the enzymes responsible for metabolizing medications (NSAIDs, proton-pump inhibitors, acetaminophen, antiepileptics, immune modulators, blood-sugar medications, blood pressure medications, antidepressants, antipsychotics, diabetic medications, antihistamines, antibiotics, and anesthetics), based on myrcene content.[1390,1391]

Roots (All CTs)

■ The root oils are not recommended for essential oil therapy due to potentially troubling constituents.

All others

■ None currently known.

SELECTED EVIDENCE:

- Oral administration (10 g/kg body weight) of an unidentified CT of Echinacea (*E. purpurea*) flower essential oil reduced inflammation by reducing the levels of proinflammatory cytokines (IL-2, IL-6 and TNF-α) when used at the highest dose.[1392]

ELEPHANT TREE
(Torote, Mexican Copal)

Bursera microphylla A. Gray

FAMILY: Burseraceae
NOTE: Base
AROMA INTENSITY: Medium
AROMA: Woody, piney, earthy, balsamic, hint of citrus
COMMON EXTRACTION METHOD: Steam distilled or hydrodistilled from the oleoresin
POSSIBLE SUBSTITUTE OILS: Copaiba, frankincense, elemi
BLENDS WELL WITH: Basil, bay laurel, bergamot, black pepper, camphor, cassia, cinnamon, copaiba, coriander, cypress, frankincense, geranium, grapefruit, lavender, lemon, myrrh, neroli, orange, pine, petitgrain, palo santo, ravensara, ravintsara, rose, sandalwood, spikenard, tangerine, vetiver, ylang ylang
SUBCELLULAR LOCALIZATION | EPIGENETIC INFLUENCE: Currently unknown | Currently unknown
RECOMMENDED DILUTION RANGE: 3%–50%; neat for some conditions

PRIMARY CONSTITUENTS:[1393]

Beta-Caryophyllene	35.7%–72.9%
Myrcene	0.4%–14.4%
Caryophyllene Oxide	4.8%–8.8%
Delta-3-Carene	0.1%–8.3%
Alpha-Humulene	2.1%–4.0%

OTHER CONSTITUENTS: Beta-pinene, alpha-phellandrene, para-cymene, 6-methyl-5-hepten-2-one, delta-cadinene, octanoic acid, cembrene, neocembrene A, caryophylla-2(12),6-dien-5-ol

PREFERRED COMPOSITION FOR CLINICAL USE:
Insufficient samples/data to determine.

REPORTED THERAPEUTIC PROPERTIES: Analgesic, anti-inflammatory, antirheumatic, antispasmodic, relives headache, soothes sore throat, anti-infectious, anti-cancer, astringent, wound healing, relieves chronic skin conditions and skin itching, aids respiratory function, cleanses the urinary tract, antimicrobial, antiviral, relieves venereal diseases, stimulates white blood cell production, aids immune function, decreases high blood pressure, balances menstruation, diuretic, balances mood, combats anxiety and worrisome thoughts

CAUTIONS:
- None currently known.

SELECTED EVIDENCE:
- □ None found.

ENGLISH WALNUT (Persian Walnut, Carpathian Walnut, Madeira Walnut)

Juglans regia L.; **Eastern Black Walnut**: *J. nigra* L.

FAMILY: Juglandaceae
NOTE: Middle
AROMA INTENSITY: Medium
AROMA: Earthy, piney, spicy, sweet, woody, herbaceous, balsamic, hint of citrus and floral
COMMON EXTRACTION METHOD: Steam distilled or hydrodistilled from the leaves
POSSIBLE SUBSTITUTE OILS: Copaiba, ylang ylang, helichrysum, melissa, clove
BLENDS WELL WITH: Balsam fir, bergamot, black spruce, blue spruce, cedarwood, copaiba, clary sage, clove, galbanum, lavender, lemon, lime, orange, palo santo, pine, ravintsara, tangerine, ylang ylang
SUBCELLULAR LOCALIZATION | EPIGENETIC INFLUENCE: Currently unknown | Currently unknown
RECOMMENDED DILUTION RANGE: 3%–50%; neat for some conditions

PRIMARY CONSTITUENTS:[1394,1395,1396,1397,1398,1399]

Caryophyllene Oxide CT

Caryophyllene Oxide	16.9%–33.8%
Alpha-Eudesmol	1.9%–22.6%
Beta-Caryophyllene	0.8%–22.5%
Beta-Pinene	tr–9.5%
Germacrene D	0.0%–9.4%
Alpha-Pinene	tr–6.5%
Valencene	0.0%–5.0%
Beta-Bourbonene	0.2%–3.9%
Alpha-Guaiol	0.0%–4.2%
Hexahydrofarnesyl Acetone	0.0%–4.0%
(E)-Geranyl Acetone	0.0%–3.7%

Beta-Pinene CT

Beta-Pinene	22.5%–39.5%
Germacrene D	5.0%–19.8%
Alpha-Pinene	12.7%–18.1%
Beta-Caryophyllene	6.4%–15.5%
Alpha-Zingiberene	0.0%–11.3%
Limonene	2.8%–8.6%
ar-Curcumene	0.0%–7.2%
(E)-Beta-Ocimene	1.2%–3.9%
(E)-Myrtanol Acetate	0.0%–3.8%
Alpha-Humulene	0.1%–3.7%
1,8-Cineole	tr–3.7%
Caryophyllene Oxide	0.1%–3.0%

Beta-Caryophyllene CT

Beta-Caryophyllene	18.5%–24.5%
Germacrene D	14.0%–22.2%
Beta-Pinene	8.5%–15.2%
Alpha-Humulene	2.8%–11.8%
Alpha-Pinene	3.1%–8.5%
Alpha-Copaene	0.7%–5.6%
Delta-Cadinene	1.0%–5.5%
ar-Curcumene	0.0%–4.6%
Alpha-Zingiberene	0.1%–3.9%
1,8-Cineole	tr–3.8%
(E)-Muurola-3,5-diene	0.0%–3.8%
(E)-Beta-Farnesene	0.0%–3.1%

Germacrene D CT

Germacrene D	17.1%–23.3%
Beta-Caryophyllene	13.0%–16.7%
Beta-Pinene	5.4%–15.9%
Alpha-Humulene	1.2%–9.3%
Alpha-Pinene	2.4%–8.9%
Alpha-Copaene	2.9%–6.5%
Delta-Cadinene	3.8%–6.2%
Gamma-Curcumene	0.7%–4.2%

Eugenol CT

Eugenol	27.5%
Germacrene D	21.4%
Methyl Salicylate	16.2%
(E)-Beta-Farnesene	8.2%

9-Octadecene CT	
9-Octadecene	17.7%
Hexadecanoic Acid	11.3%
Cembrene	9.5%
Octadecanoic Acid	7.0%
5-Octadecane	6.2%
2-Dodecene	4.8%
Octadecanal	3.3%

Thymol CT	
Thymol	23.1%
Unknown Compound	11.0%
Beta-Eudesmol	9.5%
(E)-Geranyl Acetone	5.8%
Caryophyllene Oxide	3.7%
Isothymol	3.6%
Farnesyl Acetate	3.3%

Eastern Black Walnut (J. nigra)	
Germacrene D	7.1%–22.5%
Beta-Caryophyllene	17.3%–20.4%
Alpha-Pinene	6.3%–11.4%
(2E)-Hexenal	tr–10.0%
Alpha-Hydrojuglone	1.0%–9.5%
Juglone	1.0%–8.8%
Delta-Cadinene	3.8%–8.7%
(3Z)-Hexanol	0.0%–4.5%
Alpha-Humulene	1.1%–4.2%
Alpha-Cadinol	2.9%–4.0%
(E)-Phytol	1.0%–3.6%
Beta-Pinene	1.5%–3.3%

OTHER CONSTITUENTS: *Caryophyllene Oxide CT*—Limonene, myrtenal, bornyl acetate, alpha-humulene, geranyl acetate, beta-farnesene, gamma-muurolene, gamma-curcumene, gamma-amorphene, alpha-muurolene, gamma-cadinene, manool, eremoligenol, 4,8-dimethyl-1,3,7-nonatirene, caryophylladienol I; *Beta-Caryophyllene CT*—Ethyl isovalerate, alpha-phellandrene, limonene, (E)-beta-ocimene, camphor, terpinen-4-ol, methyl salicylate (< 0.6%), alpha-longipinene, 7-epi-sesquithujene, beta-copaene, trans-alpha-bergamotene, gamma-curcumene, gamma-amorphene, alpha-muurolene, (Z)-alpha-bisabolene, caryophyllene oxide, (Z)-cadina-4-en-7-ol, alpha-cadinol, agarospirol; *Beta-Pinene CT*—Ethyl isovalerate, myrcene, alpha-phellandrene, linalool, terpinen-4-ol, methyl salicylate (< 0.7%), bornyl acetate, alpha-copaene, beta-cubebene, 7-epi-sesquithujene,(Z)-beta-farnesene, (E)-muurola-3,5-diene, (E)-beta-farnesene, gamma-curcumene, alpha-muurolene, (Z)-alpha-bisabolene, delta-cadinene, (Z)-cadina-4-en-7-ol, alpha-cadinol; *Germacrene D CT*—Myrcene, limonene, 1,8-cineole, (E)-beta-ocimene, 7-epi-sesquithujene, (Z)-beta-farnesene, alpha-zingiberene, gamma-amorphene, alpha-muurolene, germacrene B, spathulenol, caryophyllene oxide, (Z)-cadina-4-en-7-ol, alpha-cadinol; *9-Octadecene CT*—Hexadec-7-en-16-olide, methanone, oxalic aicd, octadecane, tetradecanoic acid, 8-hydroxy-1-(2-hydroxyethyl)-1,2,5,5-tetramethyl-cisdecalin; *Eugenol CT*—1-hexanol, hexanoic acid, benzyl alcohol, methyl benzoate, trans-pinocarveol, terpinen-4-ol, trans-verbenol, alpha-terpineol, myrtenal, isobornyl acetate, geranyl acetate, alpha-zingiberene, (E,E)-alpha-farnesene, delta-cadinene, (3Z)-hexenyl benzoate, hexyl benzoate, alpha-cadinol, germacra-4(15),5,10(14)-trien-1alpha-ol; *Thymol CT*—Bet-pinene, para-cymene, alpha-bourbonene, linalool, nopinone, pinocarvone, myrtenal, trans-pinocarveol, carvone, (E)-anethole, (E)-beta-ionone, (E)-nerolidol, (E,Z)-pseudoionone, 3,4-dimethyl-5-pentylidene-2(5H)-furanone, eremoligenol, hinesol, alpha-eudesmol, caryophylladienol I, dodecanoic acid, tetradecanoic acid, phytol, hexadecanoic acid; *Eastern Black Walnut*—2-Furaldehyde, myrcene, (3Z)-hexenyl acetate, limonene, nonanal, para-vinylguaiacol, delta-elemene, alpha-cubebene, eugenol, alpha-copaene, beta-bourbonene, beta-elemene, beta-copaene, 6,9-guaiadiene, cis-cadina-1(6),4-diene, gamma-muurolene, trans-muurola-4(14),5-diene, cis-cadina-1,4-diene, gamma-amorphene, alpha-muurolene, delta-amorphene, gamma-cadinene, trans-cadina-1,4-diene, alpha-cadinene, globulol, viridiflorol, junenol, T-cadinol, T-muurolol, alpha-muurolol (torreyol), germacra-4(15),5,10(14)-trien-1alpha-ol, naphthoquinone, palmitic acid

PREFERRED COMPOSITION FOR CLINICAL USE:

Constituent	*β-Caryophyllene CT*	*Beta-Pinene CT*
Beta-Caryophyllene	15%–30%	5%–20%
Germacrene D	10%–25%	5%–20%
Beta-Pinene	8%–18%	20%–45%
Alpha-Humulene	2%–14%	0.1%–5%
Alpha-Pinene	2%–10%	10%–25%
Delta-Cadinene	0.5%–7%	–
Alpha-Copaene	0.5%–7%	–
Limonene	–	2%–10%

Note: This preferred composition is very preliminary based on the available data.

REPORTED THERAPEUTIC PROPERTIES: Eases cough, aids respiratory health, anti-cancer, anti-rheumatic, antiarthritic, anti-inflammatory, analgesic, relieves toothache, antineuralgic, antifungal, antibacterial, antiparasitic, wound healing, heals hemorrhoids, relieves skin infections and itching, astringent, aids circulation, relieves diarrhea, antioxidant, anticancer, insecticidal, relieves anxiety, promotes self-confidence, encourages emotional healing, combats fear and frustration

CAUTIONS:

Eugenol CT

■ Avoid during pregnancy and lactation due to eugenol and methyl salicylate content. Methyl salicylate may cause congenital abnormalities and fetal malformations.[1400,1401] Eugenol is considered strongly toxic to embryos according to animal studies.[1402] Animal studies suggest that large doses of clove (also with significant eugenol) may negatively impact embryonic development and encourage fetal cell death.[1403,1404] Another animal study did not detect any negative influence of clove oil.[1405] Based on these conflicting studies it is best to limit use of eugenol during pregnancy and lactation.

■ Avoid with children under age 12. The moderate methyl salicylate content may increase the risk of a very serious and life-threatening illness called Reye syndrome in children under 12. Reye syndrome is sudden brain damage and liver dysfunction that most commonly occurs in children ages 4 to 12. It has an unknown cause, but it typically occurs in children who were given aspirin when they are experiencing a fever or recovering from flu-like symptoms or chickenpocks.[1406]

■ Oral caution—As little as 2.5 mL can cause toxicity and 4 mL may be fatal in children. As little as 5 mL could result in methyl salicylate poisoning in adults, and more than 5 mL can be fatal.[1407,1408,1409,1410,1411] Do not exceed 5 drops per day orally for an adult.

■ May interact with aspirin, blood pressure, antiplatelet, and anticoagulant medications, and increase the risk of bleeding among people with bleeding disorders based on methyl salicylate (both topically and orally) and eugenol content.[1412,1413,1414,1415,1416,1417,1418]

■ Avoid with epilepsy and Parkinson's disease due to methyl salicylate content.[1419,1420]

■ May interact with MAOI antidepressants based on eugenol content. Animal research suggests that eugenol produces antidepressant effects via the monoamine oxidase pathway, which may cause interactions with antidepressants that also interact with this pathway.[1421]

■ May interfere with enzymes responsible for metabolizing medications (NSAIDs, proton-pump inhibitors, acetaminophen, antiepileptics, immune modulators, blood sugar medications, blood pressure medications, antidepressants, antipsychotics, diabetic medications, antihistamines, antibiotics, and anesthetics) based on eugenol content.[1422,1423]
■ May interact with antibiotics and possibly enhance their effects based on eugenol content.[1424]
■ Toxicity may occur with overuse topically and the absorption of methyl salicylate increases with repeated applications.[1425] Not intended for long-term use.
■ Avoid in individuals allergic to aspirin, methyl salicylate, or other NSAIDs; or those with salicylate sensitivities. Methyl salicylate may be metabolized to the known NSAID, salicylic acid, by the liver.[1426]
■ May irritate mucous membranes (eyes, mouth, nasal passages, vagina, rectum) due to eugenol content.[1427,1428,1429,1430]

Thymol CT
■ May interact with aspirin, blood pressure, antiplatelet, and anticoagulant medications, and increase the risk of bleeding among people with bleeding disorders due to thymol content.[1431,1432]
■ May interact with antibiotics or antifungals and possibly enhance their effects based on thymol content.[1433,1434,,1435,1436,1437,1438,]
■ Avoid in individuals with Refsum disease (a neurological disorder caused by the overaccumulation of phytanic acid in cells and tissues) or who are AMACR-deficient (alphamethylacyl-CoA racemase-deficient) due to (E)-phytol content. It is believed that humans can convert phytol to phytanic acid.[1439] Animal research suggests that phytol can cause liver inflammation, liver failure, and death in AMCAR-deficient mice.[1440,1441] AMACR is an enzyme found in the mitochondria that plays a role in the breakdown of fatty acids. AMACR deficiency causes neurological problems (seizures, cognitive decline, migraines, encephalopathy, loss of sensation due to nerve damage, muscle stiffness, lack of coordination, and vision problems) that begin in adulthood and progressively get worse.[1442] Phytol may also weakly promote tumor formation according to animal research.[1443]
■ May irritate mucous membranes (eyes, mouth, nasal passages, vagina, rectum).

Caryophyllene Oxide CT
■ May interact with aspirin, blood pressure, antiplatelet, and anticoagulant medications, and increase the risk of bleeding among people with bleeding disorders due to caryophyllene oxide content.[1444]

All Other CTs and Eastern Black Walnut
■ None currently known.

SELECTED EVIDENCE:

- □ *In vitro* research demonstrated that English walnut CT 9-octadecene essential oil inhibited *S. aureus* subsp. *Aureus, P. vulgaris, B. cereus, E. aerogenes, B. subtilis, S. enterica serovar typimuririum,* and *Y. pseudotuberculosis.*[1445] The beta-pinene CT inhibited *B. subtilis, S. epidermidis, S. aureus, P. vulgaris, P. aeruginosa, S. dyssenteriae, S. typhi, E. coli,* and *K. pneumoniae.*[1446]
- □ English walnut CT beta-pinene essential oil exhibited high antioxidant activity in the DPPH activity, which was comparable to ascorbic acid and BHT.[1447]

ENGLISH YEW
(Common Yew, European Yew)

Taxus baccata L.

FAMILY: Taxaceae
NOTE: Middle
AROMA INTENSITY: Medium
AROMA: Earthy, green, musty, woody, herbaceous
COMMON EXTRACTION METHOD: Steam distilled or hydrodistilled from the needles (leaves) and twigs
POSSIBLE SUBSTITUTE OILS: Monarda, fragonia, Formosan cypress
BLENDS WELL WITH: Bergamot, cajeput, copaiba, geranium, grapefruit, frankincense, lavender, lemon, lemon myrtle, myrrh, orange, palmarosa, rhododendron, rosalina, rose, tangerine, tea tree (melaleuca)
SUBCELLULAR LOCALIZATION | EPIGENETIC INFLUENCE: Currently unknown | Currently unknown
RECOMMENDED DILUTION RANGE: *Not recommended for use currently*

PRIMARY CONSTITUENTS:[1448,1449,1450]

1-Octen-3-ol CT

1-Octen-3-ol	15.6%–27.6%
(3Z)-Hex-3-en-1-ol	4.8%–19.8%
Myrtenol	9.2%–13.3%
Hexahydrofarnesyl Acetone	5.3%–5.7%
Octan-1-ol	3.4%–4.2%
Nonanal	1.7%–4.1%
Hexan-1-ol	2.5%–3.9%
(2E)-Hex-2-enal	0.8%–3.6%
Heptacosane	2.0%–3.5%
Hexadecanoic Acid	0.2%–3.3%
Junicedranol	0.7%–3.1%

Myrtenol CT

Myrtenol	18.3%
Hexahydrofarnesyl Acetone	18.3%
(3Z)-Hex-3-en-1-ol	6.0%
3-Methyl-2-Butenoic Acid	5.9%
Tricosane	5.5%

Leaves Only

1-Octen-3-ol	20.7%–32.4%
Hexan-1-ol	2.7%–10.9%
(2E)-2-Hexen-1-ol	7.3%–8.2%
Myrtenol	0.0%–7.8%
Caryophyllene Oxide	0.0%–7.2%
Hexahydrofarnesyl Acetone	3.2%–6.8%
Sabinene	2.5%–3.2%

OTHER CONSTITUENTS: *1-Octen-3-ol CT*—Oct-1-en-3-one, octan-3-ol, limonene, decanal, nerol, geraniol, (2E)-dec-2-enal, decan-1-ol, (2E)-undec-2-enal, (E)-beta-ionone, caryophyllene oxide, tetra decanal, heptadecane; *Myrtenol CT*—Hexanol, 1-octen-3-ol, octabol, geraniol, 3,5-dimethoxyphenol, pentadecanal, hexadecanal, methyl hexadecanoate, isophytol, hexadecanoic acid, eicosane, ar-abiatriene, 1-heneicosaene, (E)-phytol, (Z)-9-octadecenoic acid, octadecanoic acid, 1-docosene, docosane delta-octadecalactone, tetracosane; *Leaves Only*—1-Penten-3-ol, (Z)-3-penten-1-ol, (E)-2-pentanol, hexenal, (3Z)-hex-3-en-1-ol, alpha-pinene, benzaldehyde, 1-heptanol, hexanoic acid, beta-pinene, 3-octanol, alpha-phellandrene, para-cymene, (Z)-2-octenal, heptanoic acid, alpha-terpinolene, nonanal, (E)-2-nonenal, benzyl acetate, octanoic acid, ethyl benzoate, geraniol, nonanoic acid, menthyl acetate, alpha-terpinyl acetate, eugenol, decanoic acid, geranyl acetate, beta-caryophyllene, alpha-humulene, (Z)-3-hexyl benzoate, (E,E)-2.6-farnesyl acetate, eicosane, nonadecanal, 1-nonadecanol, docosane, tricosane

PREFERRED COMPOSITION FOR CLINICAL USE:
Not recommended for use currently.

REPORTED THERAPEUTIC PROPERTIES: Anticancer, anti-infectious, anticonvulsant, antispasmodic, antirheumatic, analgesic, relieves fever, aids cardiovascular function, stimulates menstruation, supports respiratory function, insect repellent, promotes restful sleep, deeply relaxing

CAUTIONS:
▪ *Not recommended for essential oil therapy until full safety studies can be performed.* The volatile oil of English yew is considered irritating to skin and mucous membranes and may trigger gastrointestinal problems if ingested.[1451] Cases of horses collapsing after ingesting the needles have been reported—although this is likely be due to the alkaloids and not the oil. Human ingestion of needles has also caused severe multiorgan dysfunction syndrome and coma, which led to death twelve hours after ingestion.[1452] The death was attributed to the taxine-derived alkaloids (e.g., taxine A and B, isotaxine B, paclitaxel), taxane-derived substances (e.g., taxol A and B), and glycosides (e.g., taxicatine).

SELECTED EVIDENCE:
- ☐ None found.

FERN-LEAF LAVENDER (Egyptian Lavender, California Lavender, Pterostachys Lavender)

Lavandula multifidi L.

FAMILY: Lamiaceae (Labiatae)
NOTE: Middle
AROMA INTENSITY: Strong
AROMA: Herbaceous, medicinal, slightly spicy
COMMON EXTRACTION METHOD: Steam distilled or hydrodistilled from the aerial parts, inflorescences or leaves; also hexane extracted, which produces an absolute
POSSIBLE SUBSTITUTE OILS: Oregano, summer savory, melissa, lavender
BLENDS WELL WITH: Bergamot, eucalyptus, grapefruit, lavandin, lavender, lemon, lime, marjoram, orange, oregano, pine, rosemary, tangerine, summer savory, thyme
SUBCELLULAR LOCALIZATION | EPIGENETIC INFLUENCE: Currently unknown | Currently unknown
RECOMMENDED DILUTION RANGE: 1.5%–20%; 50% for some conditions

PRIMARY CONSTITUENTS:[1453,1454,1455,1456,1457,1458,1459,1460,1461]

Carvacrol/Beta-Bisabolene CT (Aerial parts)		*Inflorescences*	
Carvacrol	29.7%–65.1%	Carvacrol	61.7%
Beta-Bisabolene	9.0%–38.4%	Linalool	5.7%
Caryophyllene Oxide	tr–7.5%		

Linalool	tr–7.4%
Spathulenol	0.0%–6.2%
Menthone	0.0%–5.0%
Myrcene	0.3%–4.9%
Para-Cymene	tr–4.1%
Beta-Caryophyllene	2.1%–3.9%
1,8-Cineole	0.0%–3.5%
Beta-Pinene	0.0%–3.2%

Carvacrol CT (Aerial parts)

Carvacrol	66.2%
Spathulenol	4.9%
p-Cymen-8-ol	4.2%

Linalool CT (Aerial parts)

Linalool	50.1%
Camphene	10.1%
Linalyl Acetate	7.3%
Alpha-Thujene	3.8%
Bornyl Acetate	3.0%

Durenol CT (Aerial parts)

Durenol	90.0%

Carvacrol/(Z)-Beta-Ocimene CT

Carvacrol	41.5%
(Z)-Beta-Ocimene	27.0%
Myrcene	5.5%
Beta-Bisabolene	5.0%

Leaves

Carvacrol	50.1%
(E)-Anethole	17.4%
Beta-Bisabolene	5.8%
Linalool	3.4%

Carvacrol CT (Hexane, leaves)

Carvacrol	10.7%–51.6%
Lauryl Acrylate	2.3%–25.9%
Beta-Bisabolene	12.2%–21.0%
Caryophyllenol	tr–9.3%
Alpha-Phellandrene	1.3%–7.4%
Para-Cymene	tr–6.7%
Myrcene	tr–5.7%
Terpinen-4-ol	tr–3.9%
Alpha-Thujene	tr–3.8%
Caryophyllene Oxide	tr–3.6%

Caryophyllene Oxide CT (Hexane, leaves)

Caryophyllene Oxide	43.1%
Carvacrol	25.0%
Beta-Bisabolene	17.3%
Beta-Phellandrene	5.6%

Thymol CT (Aerial Parts)

Thymol	32.0%
Carvacrol	27.8%
Para-Cymene	15.7%
Gamma-Terpinene	9.5%

OTHER CONSTITUENTS: *Carvacrol/Beta-Bisabolene CT*—Alpha-pinene, sabinene, (Z)-beta-ocimene, (E)-beta-ocimene, terpinolene, linalool, alpha-terpineol, camphor, carvone, methyl eugenol (< 0.7%)bicyclogermacrene; *Linalool CT*—Alpha-pinene, sabinene, beta-pinene, terpinolene, para-cymene, myrcene, lavandulol, delta-terpineol, carvacrol, neral, cis-linalool oxide; *Carvacrol CT*—Alpha-pinene, octanol, p-cymene, 1,8-cineole, beta-ocimene, terpinolene, alpha-thujone (< 0.9%), beta-caryophyllene, alpha-sesquisabinene, caryophyllene oxide, beta-eudesmol; *Durenol CT*—Aromadendrene, carvacrol methyl ether, spathulenol, caryophyllene oxide; *Inflorescences*—Furfural, benzaldehyde, (Z)-beta-ocimene, alpha-terpinolene, exo-fenchol, terpinen-4-ol, durenol, 2,3-dihydrobenzofuran, p-menth-1-en-9-al, 6-mwthyl-3-cyclohexene-1-carboxaldehyde, (Z)-geraniol, geraniol, 2-ethenyl-1,3,3-trimethyl-cyclohexene, 2-methoxy-4-vinylphenol, beta-caryophyllene, caryophyllene oxide, epizonarene, m-hexadecanoic acid, (Z)-9-octadecanoic acid; *Leaves*—Furfural, (Z)-3-hexen-1-ol, (Z)-beta-ocimene, terpinolene, durenol, alpha-terpineol, (+/–)-1-methyl-3-(1-methylethenyl)-cyclohexene, 2,3-dihydrobenzofuran, p-menth-1-en-9-al, (1R,5-cis,6-cis)-bicycle[3.2.2]non-8-en-6-ol, 1-methoxy-4-methyl-2-(1-methylethyl)-benzene, geraniol, p-menth-1-en-9-ol, 2-methoxy-4-vinylphenol, beta-caryophyllene, cis-(–)-2,4a,5,6,9a Hexahydro-3,5,5,9-tetramethyl (1H) benzocycloheptene, epizonarene, (Z)-9-octadecenamide; *Carvacrol CT (Hexane, leaves)*—Alpha-pinene, camphene, sabinene, limonene, 1,8-cineole, gamma-terpinene, terpinolene, camphor, beta-bourbonene, alpha-gurjunene, alpha-cubebene, alpha-fenchol, trans-beta-caryophyllene, alloaromadendrene, geraniol, cis-alpha-bisabolene, (E)-nerolidol, spathulenol; *Caryophyllene Oxide CT (Hexane, leaves)*—Terpinolene, lauryl acrylate; *Carvacrol/(Z)-Beta-Ocimene CT*—Alpha-pinene, 1-octen-3-ol, alpha-terpinene, (E)-beta-ocimene, terpinolene, beta-caryophyllene, alpha-muurolene, (E,E)-alpha-farnesene, spathulenol; *Thymol CT*—Alpha-pinene, myrcene, alpha-terpinene, 1,8-cineole, trans-thujan-4-ol, beta-caryophyllene, linalool, terpinen-4-ol, pulegone (< 1.4%), caryophyllene oxide

PREFERRED COMPOSITION FOR CLINICAL USE:
Too great of variation in data/samples to determine.

REPORTED THERAPEUTIC PROPERTIES: Analgesic, anti-inflammatory, antirheumatic, antiarthritic, alleviates colds, lowers blood sugar, antioxidant, antifungal, antibacterial, **antimicrobial**, reduces high cholesterol, aids digestion, improves circulation, reduces shock, antidepressant

CAUTIONS:
Carvacrol CTs
■ Avoid during pregnancy and lactation. Oregano (also rich in carvacrol) may negatively affect embryonic development and encourage fetal cell death according to animal research.[1462]
■ May interact with diabetic medications and cause low blood-sugar levels due to carvacrol content—carvacrol reduces blood glucose and enhances the antidiabetic activities of rosiglitazone in animals.[1463,1464,1465,1466]
■ May interact with aspirin, blood pressure, antiplatelet, and anticoagulant medications, and increase the risk of bleeding among people with bleeding disorders, due to carvacrol and thymol content.[1467,1468,1469,1470,1471]
■ May interact with anticholinergic (drugs used for asthma, incontinence, gastrointestinal cramps, muscular spasms, depression, and sleep disorders) and cholinergic medications (drugs used to reduce urinary retention, diagnose myasthenia gravis, and for glaucoma), based on thymol and carvacrol content.[1472,1473]
■ May interact with antibiotics or antifungals and possibly enhance or decrease their effects due to carvacrol or thymol content.[1474,1475,1476,1477,1478]
■ May irritate mucous membranes (eyes, mouth, nasal passages, vagina, rectum).

Thymol CT
■ Oral caution. Thymol has a longer half-life (the time it takes for half of the medication to metabolize or excrete half of the dosage) than most essential oil constituents and should not be administered orally for long periods of time.[1479] Thymol is a monoterpene phenol. Reports of fatalities in children who consumed 50 to 200 mg of phenols has been reported.[1480] It is recommended that chemotypes with high levels of thymol be limited to 10 drops per day orally for adults with a two- to seven-day break after twenty-one days of use; or limit ingestion to five or six days weekly.
■ May interact with aspirin, blood pressure, antiplatelet, and anticoagulant medications, and increase the risk of bleeding among people with bleeding disorders, due to thymol and carvacrol content.[1481,1482]
■ May interact with anticholinergic (drugs used for asthma, incontinence, gastrointestinal cramps, muscular spasms, depression, and sleep disorders) and cholinergic medications (drugs used to reduce urinary retention, diagnose myasthenia gravis, and for glaucoma), based on thymol and carvacrol content.[1483,1484]
■ May interact with antibiotics or antifungals and possibly enhance their effects based on thymol and carvacrol content.[1485,1486,1487,1488,1489,1490,1491,1492]
■ May irritate mucous membranes (eyes, mouth, nasal passages, vagina, rectum).

Linalool CT
■ None currently known.

Durenol CT

■ None currently known, but the phenol durenol (2,3,5,6-Tetramethylphenol) is not well studied for safety.

Absolutes (Hexane extracts)

■ Not recommended for essential oil therapy. Provided for informational purposes only.

SELECTED EVIDENCE:

- *In vitro* research showed that the fern-leaf lavender carvacrol CT essential oil significantly inhibits *S. aureus, E. coli, P. aeruginosa , B. cereus, Alternaria* sp., *P. expansum, R. stolonifer*, and *B. cinerea.*[1493] The carvacrol/beta-bisabolene CT of fern-leaf lavender oil inhibited clinical isolates and reference strains of MRSA, *S. Aureus, B. subtilis, P. mirabilis, P. vulgaris, L. innocua, L. monocytogenes*, and *E. coli* (two strains).[1494,1495] A fern-leaf lavender CT with carvacrol and (Z)-beta-ocimene as the major constituents inhibited dermatophytes and *C. neoformans* causing the study authors to conclude it may be a viable complementary therapy for candidiasis.[1496] The durenol CT of fern-leaf lavender oil inhibited *E. coli, P. aeruginosa*, and *K. pneumoniae.*[1497] A fern-leaf lavender essential oil (composition not reported) inhibited *E. faecalis, E. coli, P. aeruginosa, K. pneumoniae, S. aureus, C. albicans, L. monocytogenes, B. cereus, B. subtilis.*[1498] The thymol CT of fern-leaf lavender essential oil showed strong activity against three bacteria: *E. coli, S. aureus*, and *P. aeruginosa.*[1499] Fern-leaf lavender CT carvacrol/(Z)-beta-ocimene inhibited several pathogenic fungi responsible for candidosis, meningitis, dermatophytosis, and aspergillosis (*C. albicans, C. tropicalis, C. krusei, C. guilliermondii, C. parapsilosis, C. neoformans, T. mentagrophytes, T. mentagrophytes* var. *interdigitale, T. rubrum, M. canis, M. gypseum, R. floccosum, A. niger, A. fumigatus*, and *A. flavus*).[1500]
- Fern-leaf lavender carvacrol/beta-bisabolene CT exhibited weak antioxidant activity in the DPPH, FRAP, and chelating ability assays.[1501]

FIG LEAF | FIG LEAF ABSOLUTE (Common Fig)

Ficus carica L.

FAMILY: Moraceae
NOTE: Base
AROMA INTENSITY: Medium
AROMA: Green, earthy, herbaceous, sweet, fruity
COMMON EXTRACTION METHOD: Steam distilled (SD), hydrodistilled (HD), solvent extracted (Absolute), or solid-phase micro extracted (SPME) from the leaves; may also be extracted from the fruits
POSSIBLE SUBSTITUTE OILS: Blackcurrant absolute, boronia absolute, ylang ylang
BLENDS WELL WITH: Other absolutes, floral oils
SUBCELLULAR LOCALIZATION | EPIGENETIC INFLUENCE: Currently unknown | Currently unknown
RECOMMENDED DILUTION RANGE: Not recommended for topical application

PRIMARY CONSTITUENTS:[1502,1503,1504,1505,1506]

Leaves (HDI)

Beta-Damascenone	10.1%
Psoralen	10.1%
[1S-(1α,7α,8aβ)]-1,2,3,5,6,7,8,8aoctahydro-1,4-dimethyl-7-(1-methylethenyl)-azulene	5.9%
n-Docosanoic Acid (Behenic Acid)	4.8%
Benzyl Alcohol	4.6%
m-Tert-Butyl-Phenol	4.3%
Benzaldehyde	4.0%
2-Furancarboxaldehyde	3.9%

Fruits (HD)

Palmitic Acid	15.7%
Furfural	10.6%
5-Methyl-2-Furaldehyde	10.1%
Ethyl Palmitate	8.8%

Absolute (Leaves, n-Octacosane CT)

n-Octacosane	22.8%
n-Hexadecanoic Acid (Palmitic Acid)	14.5%
Lupeol Acetate	8.3%
n-Eicosanoic Acid (Arachidic Acid)	7.2%
n-Tridecane	6.6%
n-Docosanoic Acid (Behenic Acid)	6.3%
Germanicol	4.9%
n-Tetradecane	4.6%
Stearic Acid	4.2%
Tetracosan-1-ol (Lignoceryl Alcohol)	3.9%
Tetracosanoic Acid (Lignoceric Acid)	3.4%
Octacosane-1-ol (Montanyl Alcohol)	3.3%
n-Dodecane	3.1%

Leaves (HDII)

(Z)-3-Hexenyl benzoate	19.8%
n-Tetracosane	11.6%
n-Hexadecanoic Acid (Palmitic Acid)	9.2%
n-Docosane	7.7%
(E)-2-Hexenal	7.2%
Phytol	6.7%
n-Nonanal	3.9%

Fruits (SPME)[#]

Hexanal	6.2%–56.3%
Benzaldehyde	4.5%–47.0%
2-Pentylfuran	0.9%–7.6%
Ethyl Alcohol	0.5%–5.6%
Nonanal	1.9%–4.7%
Pentanal	0.2%–4.3%
Limonene	0.2%–3.6%
Octanal	3.0%–3.2%
1-Octen-3-ol	0.8%–3.0%

Absolute (Leaves, Phytol CT)

Phytol	21.3%–32.5%
Beta-Sitosterol	11.2%–14.4%
Alpha-Linoleic Acid	5.2%–13.8%
n-Hexadecanoic Acid (Palmitic Acid)	4.3%–10.1%
Cycloartenol Acetate	4.0%–9.0%
n-Octacosane	0.0%–7.5%
n-Tridecane	0.8%–7.3%
Lupeol Acetate	4.0%–6.7%
Alpha-Amyrin	3.2%–5.8%
n-Tetradecane	0.5%–4.6%
Linoleic Acid	1.8%–4.5%
n-Dodecane	0.3%–4.5%

[#] Could be split into a benzaldehyde and a hexanal CT.

OTHER CONSTITUENTS: *Leaves (HDI)*—4-methyl-3-penten-2-one, (E)-2-hexenal, 6-methyl-5-hepten-2-one, pelargonic aldehyde, alpha-lonene, cis-tagetone, (E,E)-2,4-heptadienal, trans-tagetone, 1-octanol, hotrienol, [1S-(1α,4α,7α)]-1,2,3,4,5,6,7,8-octahydro-1-4-dimethyl-7-(1-methylethenyl)-azulene, 2,6,6-trimethyl-1-cyclohexene-1-carboxaldehyde, (E)-2-decenal, 2,6,6-trimethyl-1,3-cyclohexadiene-1-carboxaldehyde, [3aS-(3aα,3bβ,4β,7α)]-octahydro-7-methyl-3-methylene-4-(1-methylethyl)-1H-cyclopenta [1,3] cyclopropa [1,2]benzene, 4,6,6-trimethyl-bicyclo [3.1.1]hept-3-en-2-one, 1,2-dihydro-

1,1,6-trimethyl-naphthalene, 3-ethyl-4-methyl-2,5-furandione, (1-methoxyethyl)-benzene, methyl salicylate (< 0.6%), geranylacetone, mequinol, beta-ionone, hexahydrofarnesyl acetone, tricosane, (E,E)-6,10,14-trimethyl-5,9,13-pentadecatrien-2-one, pentacosane, phthalic acid diisobutyl ester, phytol, dibutyl phthalate, 17-pentatriacontene, hexacosane, butyl 2-methylpropyl-1,2-benzenedicarboxylate, phenanthrene, heptacosane, bergapten (2%); *Leaves (HDII)*—Alpha-pinene, 1-octanol, methyl salicylate (< 0.7%), dill ether, (E)-beta-damascenone, nerylacetone, beta-ionene, cis-muurola-4(15),5-diene, (E,E)-alpha-farnesene, n-pentadecane, n-heptadecane, n-tetradecanoic acid, n-octadecane, 6,10,14-trimethylpentadecan-2-one, nonadecane, heptacosane, n-pentacosane, phytane, fanesene, n-heneicosane; *Fruits (HD)*—Dihydro-2-methyl-3 (2H)-furanone, phenylethane, 1,4-dimethyl benzene, 2-furylmethyl ketone, benzenemethanol, α-methyl-α-[4-methyl-3-pentenyl]oxiranemethanol, trans-5-ethenyltetrahydro-α, α, 5-trimethyl-2-furanmethanol, beta-linalool, hotrienol, prehnitene, isodurene, tetrahydro-2,2,6-trimethyl-6-vinyl-2H-pyran-3-ol, alpha-terpineol, α,4-dimethyl-3-cyclohexene-1-acetaldehyde, cis-geraniol, copaene, cadina-1(10),4-diene, alpha-calacorene, 1R,trans-calamenene, myristaldehyde, Ω-benzyl acetophenone, diisobutyl phthalate, methyl palmitate, dibutyl palmitate, methyl-(Z, Z)-9,12-octadecadienoate, linolenic acid methyl ester, cis-9,cis-12-octadecadienoic acid, ethyl linoleate, linolenic acid ethyl ester; *Fruits (SPME)*—Methyl alcohol, isopentyl alcohol, 1-hexanol, 1-octanol, 1-nonanol, 1-dodecanol, methyl acetate, ethyl acetate, ethyl-2-butanoate, ethyl caproate, 2-heptanone, 3-octanone, 2-octanone, 3-hydroxy-2-butanoate, 6-methyl-5-hepten-2-one, 2-nonanone, acetaldehyde, 2-methyl-butanal, 3-methyl-butanal, heptanal, (E)-2-hexenal, (E)-2-octenal, decanal, styrene; *Absolute (Phytol CT)*—Stearic acid, n-eicosanoic acid (arachidic acid), n-docosanoic acid (behenic acid), tetracosan-1-ol (lignoceryl alcohol), squalene, tetracosanoic acid (lignoceric acid), 1-hexacosanol (ceryl alcohol), octacosane-1-ol (montanyl alcohol), stigmasterol, triacontan-1-ol (melissyl alcohol), germanicol, lanosterol, beta-amyrin, beta-amyrin acetate; *Absolute (n-Octacosane CT)*—n-Heptadecanoic acid (palmitic acid), 1-octadecanol (stearyl alcohol), 1-docosanol (behenyl alcohol), 1-hexacosanol (ceryl alcohol), hexacosanoic acid (cerotic acid), triacontan-1-ol (melissyl alcohol)

PREFERRED COMPOSITION FOR CLINICAL USE:
Not recommended for essential oil therapy. Fragrance use only.

REPORTED THERAPEUTIC PROPERTIES: Nourishes the skin, wound healing, lowers high blood sugar, aids metabolism, reduces colic, relieves diarrhea, analgesic, soothes sore throat, anti-inflammatory, reduces fever, antispasmodic, supports respiratory function, ease cough, expels excess mucus, aids cardiovascular function, reduces high cholesterol, antioxidant, antibacterial, antifungal, anticancer, reduces anger and negative thinking, stress management

CAUTIONS:
Not recommended for essential oil therapy. Fragrance use only. Do not ingest or apply topically. Monograph is for informational purposes only.
■ Strongly phototoxic due to the presence of significant amounts of psoralen and bergapten.[1507] Avoid UV rays for a minimum of 12 hours after topical application. The absolute is also strongly irritating to the skin.[1508]

SELECTED EVIDENCE:
- □ None found.

FOREST SPUR-FLOWER
(Bitterbos, Asbossie)

Plectranthus fruticosus L'Hér.

FAMILY: Lamiaceae (Labiatae)
NOTE: Middle
AROMA INTENSITY: Strong
AROMA: Pungent, herbaceous, spicy, minty
COMMON EXTRACTION METHOD: Steam distilled or hydrodistilled from the leaves
POSSIBLE SUBSTITUTE OILS: Savin
BLENDS WELL WITH: Doesn't blend well with other oils due to strong unpleasant and pungent odor
SUBCELLULAR LOCALIZATION | EPIGENETIC INFLUENCE: Currently unknown | Currently unknown
RECOMMENDED DILUTION RANGE: Not recommended.

PRIMARY CONSTITUENTS:[1509,]

Trans-Sabinyl Acetate	60.7%
Alpha-Cubebene	7.7%
Sabinene	6.1%
Beta-Bisabolene	4.5%
Trans-Farnesol	3.3%

OTHER CONSTITUENTS: Alpha-thujene, gamma-terpinene, linalool, terpinen-4-ol, alpha-humulene, gamma-cadinene, alpha-elemene

PREFERRED COMPOSITION FOR CLINICAL USE:
Not recommended for therapy.

REPORTED THERAPEUTIC PROPERTIES: Wound healing, relieves burns, removes warts, antimicrobial, antiparasitic, anti-inflammatory, analgesic, insect repellent, grounding, calming

CAUTIONS:
Forest spur-flower is not recommended for use in essential oil therapy in any form of administration, but the profile is included for informational purposes and safety information.

- Avoid in children due to the toxicity of trans-sabinyl acetate.[1510]
- Avoid during pregnancy. Essential oils rich in sabinyl acetate, including forest spur-flower, are toxic to developing embryos, cause fetal malformations, and prevent implantation during days 0–4 of gestation according to animal research.[1511,1512,1513,1514] Other research using a Spanish sage essential oil (containing 50% trans-sabinyl acetate) demonstrated trans-sabinyl acetate can cause abortions and maternal toxicity (which indicated a greater susceptibility to trans-sabinyl acetate toxicity during pregnancy) in animals.[1515] The researchers concluded that trans-sabinyl acetate possesses significant antifertility activity during the preimplantation period, and attribute the fetotoxic and abortifacient activity to this constituent.
- Avoid oral consumption. Essential oils rich in sabinyl acetate irritate the mucosal lining of the gastrointestinal tract, damage the kidneys, cause blood in the urine, obstruct blood flow in the abdominal organs, and cause abnormally heavy menstrual bleeding, which the study authors attributed to the combined

effects of sabinene, sabinol, and trans-sabinyl acetate.[1516] However, more recent research suggests it is the trans-sabinyl acetate that causes toxicity and other adverse effects.

■ Avoid with epilepsy and Parkinson's disease due to thujone and trans-sabinyl acetate content.[1517]

■ Avoid in those who have a compromised liver due to the increased risk of liver toxicity and liver damage.[1518] This would also suggest that those taking medications that could cause liver damage should also use it very cautiously or avoid it entirely.

SELECTED EVIDENCE:

- □ None found.

FORSKOHLII
(Coleus)

Plectranthus barbatus Andrews, *P. forskohlii* auct., *Coleus forskohlii* auct., *C. barbatus* (Andrews) Benth.

FAMILY: Lamiaceae (Labiatae)
NOTE: Middle
AROMA INTENSITY: Medium-Strong
AROMA: Spicy to woody/earthy to unpleasant
COMMON EXTRACTION METHOD: Hydrodistilled or steam distilled from the roots; may also be distilled from the aerial parts, leaves, or stems
POSSIBLE SUBSTITUTE OILS: Spruce (black), clove, juniper berry
BLENDS WELL WITH: Basil, black pepper, cardamom, cassia, clove, coriander, frankincense, lavender, lemon, marjoram, nutmeg, palo santo, patchouli, pine, ravensara, rosemary, sandalwood, ylang ylang
SUBCELLULAR LOCALIZATION | EPIGENETIC INFLUENCE: Currently unknown | Currently unknown
RECOMMENDED DILUTION RANGE: 3%–33%; 50% for some conditions

PRIMARY CONSTITUENTS:[1519,1520,1521,1522,1523,1524,1525,1526,1527,1528,1529]

Roots (Bornyl Acetate CT)	
Bornyl Acetate	15.0%–23.5%
2-Decen-1-ol	0.0%–19.3%
Beta-Sesquiphellandrene	0.0%–13.2%
Gamma-Eudesmol	0.0%–12.5%
3-Decanone	0.0%–7.0%
Nonylcyclopropane	0.0%–6.7%
Beta-Farnesene	0.0%–6.1%
Alpha-Pinene	1.9%–5.3%
Beta-Pinene	1.5%–5.1%
Alpha-Gurjunene	0.0%–5.0%
Alpha-Selinene	0.0%–4.4%
Borneol	0.0%–4.0%
1-Napthalol	0.0%–3.9%
Camphene	2.2%–3.6%

Roots (Decenal CT)	
Decenal	30.7%–39.7%
Bornyl Acetate	9.2%–13.0%
N-Decyl Acetate	6.1%–7.5%
1-Decanol	0.0%–4.8%
(E,E,Z)-1,3,12-Nonadecatriene-5,14-diol	2.7%–4.4%
Beta-Sesquiphellandrene	0.0%–3.8%
Guaia-1-10,(11)-diene	3.2%–3.4%
Beta-Pinene	1.9%–3.4%

Aerial Parts (Eugenol CT)	
Eugenol	25.1%
Thymol	15.3%
Carvacrol	12.1%
Beta-Caryophyllene	10.7%

Roots ((Z)-Beta-Ocimene CT)

(Z)-Beta-Ocimene	37.6%
Alpha-Santalene	8.1%
Beta-Caryophyllene	6.8%
Bornyl Acetate	6.2%
Sabinene	6.0%
(E,E)-Farnesol	5.6%
Alpha-Pinene	4.7%

Aerial Parts (Beta-Farnesene CT)

Beta-Farnesene	24.9%
Germacrene D	21.4%
Alpha-Copaene	12.7%
(Z)-3-Hexen-1-ol	7.8%
(Z)-3-Hexen-1-ol Acetate	4.5%
Alpha-Zingiberene	4.4%
Beta-Myrcene	3.2%

Leaves (Valencene CT)

Valencene	22.9%
Beta-Caryophyllene	21.0%
Germacrene D	15.5%
Alpha-Copaene	5.4%
Beta-Bourbonene	4.8%
Caryophyllene Oxide	3.9%
Beta-Bisabolene	3.3%
Delta-Cadinene	3.1%

Leaves (Eugenol CT)

Eugenol	31.1%
Alpha-Pinene	19.4%
Beta-Caryophyllene	18.4%
Beta-Cubebene	3.7%
Alpha-Copaene	8.9%
Beta-Cubebene	3.7%

Aerial Parts (Caryophyllene Oxide CT)

Caryophyllene Oxide	36.4%
Alpha-Copaene	17.1%
Alpha-Pinene	10.9%
Beta-Farnesene	6.9%
Beta-Myrcene	4.3%

Leaves (Alpha-Pinene CT)

Alpha-Pinene	22.2%
Eremophilene	13.3%
Beta-Myrcene	12.3%
Humulenone	10.0%
Beta-Caryophyllene	8.0%
(Z)-Beta-Ocimene	6.5%
Alpha-Selinene	3.6%

Leaves (Beta-Caryophyllene CT)

Beta-Caryophyllene	18.0%
Germacrene D	17.4%
Viridiflorene	14.1%
Cyclosativene	9.9%
Alpha-Pinene	8.9%
Beta-Cedrene	5.7%

Stems (Beta-Phellandrene CT)

Beta-Phellandrene	26.1%
Alpha-Pinene	13.5%
Alpha-Copaene	9.8%
Sabinene	8.7%
Caryophyllene Oxide	6.2%
Limonene	5.4%
Beta-Caryophyllene	5.0%
Alpha-Humulene	4.3%
Humulene Epoxide II	3.6%

Flowers

Alpha-Pinene	67.0%
Beta-Caryophyllene	10.4%
Germacrene D	6.7%

Roots (Beta-Caryophyllene CT)

Beta-Caryophyllene	15.1%
Beta-Phellandrene	10.6%

Leaves + Stems (CTs not separated)

1-Octen-3-ol	1.2%–28.2%
Alpha-Pinene	0.8%–23.1%
Beta-Pinene	1.8%–21.7%
Beta-Bourbonene	0.0%–12.0%
Beta-Caryophyllene	7.2%–11.9%
Limonene	0.7%–11.2%
Alpha-Humulene	0.8%–10.8%
Sabinene	0.4%–8.6%
Decanal	tr–8.3%
Germacrene D	0.6%–6.8%

Alpha-Copaene	0.5%–6.8%
Caryophyllene Oxide	0.3%–6.6%
Beta-Copaene	0.8%–6.0%
Beta-Selinene	0.0%–5.8%
Linalool	0.3%–5.7%
Ledol	tr–5.5%
Beta-Phellandrene	tr–4.3%
Terpinen-4-ol	0.4%–4.0%
1,8-Cineole	tr–4.0%
Gamma-Cadinene	tr–3.8%
Beta-Sesquiphellandrene	0.2%–3.5%
Alpha-Muurolene	0.1%–3.2%
Alpha-Cubebene	tr–3.2%
Alpha-Terpinyl Acetate	0.0%–3.0%

Beta-Gurjunene	10.6%
Alpha-Humulene	6.3%

Stems (Linalool CT)

Linalool	10.4%
Alpha-Copaene	9.4%
Beta-Caryophyllene	8.3%
Delta-Cadinene	8.1%

Leaves (Anethole CT)

Anethole (isomer not identified)	18.8%
Beta-Caryophyllene	12.2%
Germacrene D	9.4%
1-Octen-3-ol	8.1%

OTHER CONSTITUENTS: *Roots (Bornyl Acetate CT)*—Alpha-phellandrene, beta-thujene, terpinene-4-ol, alpha-terpineol, bicyclo [5.3.0.]decane,2-methylene-5-(1-methylvinyl)-8-methyl, 1-tridecene, benzoic acid,2,4-dimethyl-,methyl ester, delta-elemene, cis-alpha-bergamotene, bicyclogermacrene, isolongifolene, gamma-elemene, cedran-8-ol, abietatriene, beta-himachalene, 7-Isopropyl-1,4-dimethyl-1,2,3,4,5,6,7,8-octahydroazulene, (E)-nerolidol, dehydroaromadendrene, cubenol, T-muurolol; *Roots ((Z)-Beta-Ocimene CT)*—Camphene, alpha-phellandrene, delta-elemene, alpha-copaene, aromadendrene, alpha-humulene, trans-alpha-bergamotene, humulenone, beta-sesquiphellandrene, abietatriene; *Roots (Decenal CT)*—Alpha-pinene, camphene, limonene, borneol, cyclosativene, alpha-copaene, gamma-maaliene, cis-alpha-bergamotene, trans-alpha-bergamotene, cubebol, beta-bisabolene, germacrene B, beta-maaliene, selina-3,7(11)-diene, (Z)-valerenyl acetate, 8-propoxycedrane, alpha-muurolol, juniper camphor, dibutyl phthalate, manoyl oxide, (11E,13Z)11813-labdadie-8-ol, 1,5,9,9-tetramethyl-spirol[3.5]nonan-5-ol, pentacosane, ferruginol, hexacosane, triacontane, sabinene, 1,8-cineole, camphor, sclareol, cedran-8-ol, n-tricosane; *Aerial Parts (Eugenol CT)*—Sabinene, myrcene, (Z)-beta-ocimene, (E)-beta-ocimene, (E)-alpha-bergamotene, alpha-humulene, caryophyllene oxide, T-cadinol; *Aerial Parts (Beta-Farnesene CT)*—Alpha-pinene, 1-octen-3-ol, (Z)-beta-ocimene, delta-cadinene, beta-sesquiphellandrene, caryophyllene oxide; *Aerial Parts (Caryophyllene Oxide CT)*—(Z)-beta-ocimene, alpha-cubebene, beta-bourbonene, alpha-zingiberene, delta-cadinene, beta-sesquiphellandrene; *Leaves (Valencene CT)*—Alpha-humulene; *Leaves (Alpha-Pinene CT)*—Sabinene, limonene, (E)-beta-ocimene, beta-bourbonene, aromadendrene, beta-selinene, delta-cadinene, caryophyllene oxide, manool; *Leaves (Eugenol CT)*—Sabinene, myrcene, (Z)-beta-ocimene, (E)-beta-ocimene, alpha-copaene, alpha-humulene, germacrene D, beta-selinene, viridiflorene, germacrene A, 7-epi-alpha-selinene, delta-cadinene, germacrene B, caryophyllene oxide, T-cadinol; *Stems (Beta-Phellandrene CT)*—Camphene, alpha-phellandrene, para-cymene, borneol, bornyl acetate, cyclosativene, alpha-santalene, delta-cadinene; *Leaves (Beta-Caryophyllene CT)*—Tetrahydrocitronellene, alpha-cubebene, alpha-copaene, beta-bourbonene, beta-cubebene, beta-longipinene, alpha-humulene, beta-selinene, gamma-muurolene, germacrene A, 7-epi-alpha-selinene, delta-cadinene, caryophyllene oxide, (E)-epi-14-hydroxy-9-caryophyllene; *Leaves + Stems (CTs not separated)*—Camphene, sabinene, octanal, myrcene, para-cymene, (Z)-beta-ocimene, (E)-beta-ocimene, gamma-terpinene, octan-1-ol, nonanal, trans-verbenol, borneol, naphthalene, alpha-terpineol, octyl acetate, decan-1-ol, bornyl acetate, cyclosativene, dodecanal, isocaryophyllene, delta-cadinene, T-cadinol, delta-cadinol, beta-eudesmol, alpha-cadinol; *Flowers*—Octanal, myrcene, limonene, gamma-terpinene, linalool, trans-verbenol, alpha-copaene, alpha-humulene, gamma-cadinene, delta-cadinene, T-cadinol; *Roots (Beta-Caryophyllene CT)*—Not listed; *Stems (Linalool CT)*—Not listed; *Leaves (Anethole CT)*—Not listed

PREFERRED COMPOSITION FOR CLINICAL USE:
Too great of variance in samples/data to determine.

REPORTED THERAPEUTIC PROPERTIES: Aids digestion, relieves intestinal disturbances, reduces nausea, supports liver function, nervine, antineuralgic, analgesic, antispasmodic, anticonvulsive, anti-inflammatory, antirheumatic, decreases high blood pressure, stops excess bleeding, decreases chest pain, aids oral health, supports respiratory function, relieves cough, wound healing, heals bruises, soothes chronic skin conditions, reduces the appearance of blemishes, improves oily skin, antiviral, antibacterial, antifungal, promotes restful sleep, relieves anxiety, aphrodisiac

CAUTIONS:
Eugenol CTs (Leaves, Aerial parts)
■ Caution with pregnancy and lactation. Animal studies suggest that large doses of clove (also rich in eugenol) may negatively impact embryonic development and encourage fetal cell death.[1530,1531] Another animal study did not detect any negative influence of eugenol and suggested it helped prevent negative effects caused by retinoic acid.[15321533] Eugenol is considered strongly toxic to embryos according to animal studies.[1534,1535] Based on these studies it is best to limit clove oil during pregnancy and lactation.
■ May interact with aspirin, blood pressure, antiplatelet, and anticoagulant medications, and increase the risk of bleeding among people with bleeding disorders, due to eugenol content.[1536,1537,1538,1539]
■ May interact with MAOI antidepressants. Animal research suggests that eugenol produces antidepressant effects via the monoamine oxidase pathway, which may cause interactions with antidepressants that also interact with this pathway.[1540]
■ May interact with antibiotics and possibly enhance their effects due to eugenol content and research that the eugenol CT of forskohlii essential oil enhances the activity of aminoglycoside antibiotics.[1541,1542]
■ May interfere with enzymes responsible for metabolizing medications (NSAIDs, proton-pump inhibitors, acetaminophen, antiepileptics, immune modulators, blood-sugar medications, blood pressure medications, antidepressants, antipsychotics, diabetic medications, antihistamines, antibiotics, and anesthetics) due to eugenol content.[1543]
■ May irritate mucous membranes (eyes, mouth, nasal passages, vagina, rectum).[1544,1545]

Stems (Beta-Phellandrene CT)
■ May interact with anticholinergic (drugs used for asthma, incontinence, gastrointestinal cramps, muscular spasms, depression, and sleep disorders) and cholinergic medications (drugs used to reduce urinary retention, diagnose myasthenia gravis, and for glaucoma), based on AChE-inhibtory activity of beta-phellandrene.[1546]

All other CTs
■ None currently known.

SELECTED EVIDENCE:
- □ Preclinical research shows that the alpha-pinene CT of forskohlii essential oil relaxes the intestines and reduces intestinal spams by modulating calcium balance in cells.[1547] The study authors concluded that the main active constituent responsible for the antispasmodic activity of forskohlii essential oil was alpha-pinene.
- □ *In vitro* research demonstrates that forskohlii essential oil (varied composition) inhibited *B. cereus, B. subtilis, S. aureus, L. monocytogenes, E. coli, P. aeruginosa, H. pylori*, and *S. cerevisiae*.[1548] The aerial parts (eugenol CT) inhibited *E. coli, P. vulgaris, B. cereus, P. aeruginosa*, and *S. aureus*, and

enhanced the antibacterial activity of aminoglycoside antibiotics (Amikacin, Kanamycin, and Gentamicin).[1549] Forskohlii leaf essential oil CT beta-caryophyllene inhibited *C. neoformans* and *C. gatti*.[1550] A leaf oil with anethole as the primary constituent inhibited *E. coli* and *C. albicans*.[1551] The same research reported that forskohlii roots (beta-caryophyllene CT) and stems (linalool CT) inhibited *C. albicans*.

- Forskohlii essential oil (varied composition) exhibited weak antioxidant activity in the DPPH assay and moderate activity in the TBARS assay at high concentrations.[1552] Forskohlii roots (Beta-caryophyllene CT) and stems (linalool CT) exhibited good antioxidant activity (DPPH and ABTS assays), but the leaves (anethole as the primary constituent) was inactive.[1553]
- The leaf oil (Eugenol CT) exhibited strong toxicity against early third-instar larvae of *An. subpictus, Cx. tritaeniorhynchus*, and *Ae. albopictus* mosquitoes.[1554] Its major constituents eugenol, alpha-pinene, and beta-caryophyllene were also effective against the mosquito larvae.

FRINGED LAVENDER (French Lavender)

Lavandula dentata L., *L. santolinifolia* Spach., *Stoechas dentata* (L.) Mill

FAMILY: Lamiaceae
NOTE: Top-Middle
AROMA INTENSITY: Medium to strong
AROMA: Sweet, floral, herbaceous to fresh, medicinal, and camphoraceous
COMMON EXTRACTION METHOD: Hydrodistilled or steam distilled from the flowers, aerial parts (leaves, twigs, flowers), inflorescences, shoot tips, or leaves
POSSIBLE SUBSTITUTE OILS: Lavandin, spike lavender, myrtle, rosemary CT camphor
BLENDS WELL WITH: Bay laurel, bergamot, camphor, chamomile (German, Roman), citronella, clary sage, geranium, jasmine, lavender, lemon, orange, palmarosa, patchouli, pine, tangerine, thyme, rosemary, rosewood, ylang ylang
SUBCELLULAR LOCALIZATION | EPIGENETIC INFLUENCE: Currently unknown | Currently unknown
RECOMMENDED DILUTION RANGE: Linalool CT: 5%–Neat; Other CTs: 1.5%–20%; 50% for some conditions

PRIMARY CONSTITUENTS:[1555,1556,1557,1558,1559,1560,1561,1562,1563,1564]

Flowers	
Camphor	23.3%
1,8-Cineole	19.9%
L-Fenchone	7.1%
D-Fenchol	4.0%
Beta-Eudesmol	3.5%
Beta-Selinene	3.0%

Aerial Parts (Camphor CT)	
Camphor	12.4%
Trans-Pinocarveol	7.5%
Beta-Eudesmol	7.1%
Alpha-Guaiol	6.1%

Aerial Parts (Linalool CT)	
Linalool	47.3%
Linalyl Acetate	28.7%
Bicyclogermacrene	3.4%

Aerial Parts (Terpinolene CT)	
Alpha-Terpinolene	51.1%
Camphor	13.4%
Borneol	7.9%
2-Furanmethanol,5-Ethenyltetra-Hydro-a,a,5-Trimethyl-,cis-	7.1%
Cis-Linalool Oxide	7.0%
Terpinen-4-ol	5.5%

Beta-Selinene	4.5%
Myrtenol	3.9%
Linalool	3.7%
Pinocarvone	3.1%
Caryophyllene Oxide	3.1%
Alpha-Fenchol	3.0%
Beta-Pinene	3.0%
Myrtenal	3.0%
Cryptone	3.0%

Aerial Parts (1,8-Cineole CT)

1,8-Cineole	18.3%–48.0%
Beta-Pinene	0.0%–18.3%
Sabinene	0.1%–13.9%
Trans-Pinocarveol	0.0%–7.6%
Bicyclo [3.1.0] Hexan-3-Ol, 4-Methylene-1-(1-Methylethyl)-, (1à,3à,5à)-	0.0%–6.8%
Linalool	2.8%–5.4%
Myrtenal	2.5%–5.1%
Alpha-Pinene	2.1%–4.1%
Pinocarvone	1.8%–3.5%
Myrtenol	2.0%–3.4%

Leaves

1,8-Cineole	33.5%
Camphor	18.9%
L-Fenchone	8.4%
D-Fenchol	4.1%
Beta-Eudesmol	3.5%
Beta-Selinene	3.0%

Flowers/Leaves

Camphor	61.4%
Fenchone	24.3%
3,7-Octadiene-2,6-diol,2,6-Dimethyl	4.3%
1,8-Cineole	3.6%

Aerial Parts (p-Mentha-1,5-Dien-8-ol CT)

p-Mentha-1,5-Dien-8-ol CT	26.8%
Caryophyllene Oxide	16.4%
Alpha-Guaiol	15.6%
1,8-Cineole	5.0%
Geraniol	3.9%
Nonanoic Acid	3.7%
Alpha-Citronellol	3.0%
Carvone	3.0%

Inflorescences

1,8-Cineole	17.9%–27.0%
Camphor	15.1%–23.5%
Fenchone	10.4%–13.2%
Alpha-Fenchol	5.5%–7.1%
Beta-Pinene	1.9%–5.3%
Limonene	1.7%–4.4%
Linalool	1.0%–3.2%

Shoot Tips

1,8-Cineole	12.2%–25.6%
Fenchol	15.9%–19.4%
Camphor	15.2%–19.1%
Borneol	12.8%–16.3%
Beta-Pinene	2.0%–4.4%

Other Constituents: *Flowers*—Beta-pinene, limonene, linalool, trans-pinocarveol, trans-verbenol, borneol, alpha-terpineol, p-cymen-8-ol, alpha-terpineol, myrtenal, myrtenol, caryophyllene oxide, beta-eudesmol, bisabolol oxide B; *Aerial Parts (Camphor CT)*—Alpha-pinene, camphene, 1-hepten-3-ol, para-cymene, trans-linalool oxide, limonene, para-cymene, fenchone, exo-fenchol, trans-verbenol, borneol, terpinen-4-ol, trans-carveol, carvone, (E)-beta-caryophyllene, alpha-humulene, beta-ionene, gamma-muurolene, beta-curcumene, alpha-alaskene, alpha-calacorene, humulene epoxide II, gamma-eudesmol, bulnesol, beta-bisabolol; *Aerial Parts (1,8-Cineole CT)*—Camphene, para-cymene, limonene, cis-linalool oxide, fenchone, trans-linalool oxide, trans-p-menth-2,8-dien-1ol, alpha-campholenal, nopinone, camphor, trans-verbenol, borneol, terpinen-4-ol, cryptone, trans-carveol, carvone, cuminaldehyde, undecanol, perillyl alcohol, beta-selinene, verbenone, alpha-selinene, caryophyllene oxide, beta-eudesmol, alpha-bisabolol, 1,1,1,3,5,7,9,9,9-nonamethylpentasiloxane, mesityl acetyl propynylphosphine, 3-methyl-5-propyl-4-butylidene-cyclohex-2-ene-1-one; *Aerial Parts (Linalool CT)*—Myrcene, 3-octanone, terpinen-4-ol, alpha-terpineol, delta-terpineol, thymol, camphor, 14-norcadin-5-en-4-one isomer B, alpha-bisabolol, (Z)-beta-santalol; *Aerial Parts (Terpinolene CT)*—N/A; *Aerial Parts (p-Mentha-1,5-Dien-8-ol CT)*—Tricyclene, alpha-pinene, gamma-terpinene, cuminyl alcohol, decanoic acid, beta-caryophyllene, alpha-humulene, alloaromadendrene, germacrene D, beta-bisabolene, gamma-cadinene, farnesol, octadecatrienoic acid methyl ester, 6-dodecanone, docosane, 17-octadecynoic acid; *Leaves*—Alpha-pinene, beta-pinene, para-cymene, limonene, linalool, trans-pinocarveol, trans-verbenol, pinocarvone, borneol, p-cymen-8-ol, alpha-terpineol, myrtenal, myrtenol, caryophyllene oxide, 8-oxoneoisolongifolene, 14-norcadin-5-en-4-one isomer B, (Z)-beta-santalol; *Inflorescences*—Alpha-pinene; *Flowers/Leaves*—Linalool, fenchol, myrtenol, allo-ocimene; *Shoot Tips*—Alpha-pinene, camphene, cis-verbenol, myrtenol, campholenal, myrtenal, pinocarvone, fenchone, gamma-gurjunene, cuparene, cadinene, epoxycaryophyllene, alpha-eudesmol

PREFERRED COMPOSITION FOR CLINICAL USE:
Insufficient samples/data to determine.

REPORTED THERAPEUTIC PROPERTIES: Aids respiratory function, expectorant, eases cough, analgesic, antispasmodic, anti-inflammatory, antirheumatic, relieves headache, wound healing, helps boils and blisters heal, relieves insect bites, antioxidant, anticancer, antibacterial, antimicrobial, brain protective, insecticidal, promotes restful sleep, relieves anxiety, relieves mental exhaustion and burnout, promotes peaceful feelings

CAUTIONS:

Flowers

■ Avoid with children under six due to significant camphor content. Several cases of camphor poisoning and/or seizures from ingestion and topical application have been reported in children.[1565,1566] Ingestion of camphor-containing products has been lethal in children under age two.[1567] Children five years and up may use camphor-containing essential oils topically in dilutions no stronger than 5%. Camphor ingestion by infants and young children may cause cough, vomiting, seizure, burning sensation in the mucous membranes and eyes, or lack of voluntary coordination of muscle movements.[1568]

■ Caution is warranted during pregnancy and while lactating, due to significant camphor content. Ingestion of essential oils with significant levels of camphor may lead to abortion because fetuses lack the enzymes to process it.[1569] Camphor ingestion by infants and young children may cause cough, vomiting, seizure, burning sensation in the mucous membranes and eyes, or lack of voluntary coordination of muscle movements.[1570]

■ Exercise caution or avoid oral consumption due to significant camphor levels. Essential oils with significant levels of camphor may be toxic when taken orally, usually single doses exceeding 2 mL, although the lethal dose for humans is estimated to be greater than 5 mL in a single dose.[1571,1572,1573]

■ Avoid with epilepsy and Parkinson's disease due to 1,8-cineole, fenchone, or camphor content due to the potential of these constituents to exacerbate or cause seizures/convulsions or reduce seizure-medication efficacy.[1574,1575,1576,1577,1578,1579,1580]

■ The significant camphor content in fringed lavender may negatively impact red blood cells and increase the risk of jaundice in children with glucose-6-phosphate dehydrogenase deficiency (G6PD).[1581,1582]

Aerial Parts (Camphor CT)

■ Avoid with epilepsy and Parkinson's disease due to camphor content. Camphor may exacerbate or cause seizures/convulsions or reduce seizure-medication efficacy.[1583,1584,1585,1586]

■ The camphor content in fringed lavender may negatively impact red blood cells and increase the risk of jaundice in children with glucose-6-phosphate dehydrogenase deficiency (G6PD).[1587,1588]

Aerial Parts (1,8-Cineole CT)

■ Caution is warranted orally due to the significant amounts of 1,8-cineole. Limit it to small doses internally (adults—maximum 10 drops daily). Toxicity has been reported when eucalyptus (also high in 1,8-cineole) was ingested in large doses, and as few as 4 to 5 drops may cause problems in very sensitive individuals.[1589,1590,1591,1592,1593,1594]

■ Avoid with epilepsy and Parkinson's disease due to 1,8-cineole content. May exacerbate or cause seizures/convulsions or reduce seizure-medication efficacy based on 1,8-cineole content.[1595,1596,1597,1598]

■ May weakly interfere with the enzymes responsible for metabolizing medications (NSAIDs, proton-pump inhibitors, acetaminophen, antiepileptics, immune modulators, blood-sugar medications, blood pressure medications, antidepressants, antipsychotics, diabetic medications, antihistamines, antibiotics, and anesthetics) due to 1,8-cineole content.[1599,1600,1601]

■ May interfere with pentobarbital and other barbiturates (medications for anxiety and insomnia), based on 1,8-cineole content.[1602,1603]

Aerial Parts (Terpinolene CT), Aerial Parts (p-Mentha-1,5-Dien-8-ol CT)

■ Avoid with epilepsy and Parkinson's disease due to 1,8-cineole, fenchone, or camphor content due to the potential of these constituents to exacerbate or cause seizures/convulsions or reduce seizure-medication efficacy.[1604,1605,1606,1607,]

■ The significant camphor content in fringed lavender may negatively impact red blood cells and increase the risk of jaundice in children with glucose-6-phosphate dehydrogenase deficiency (G6PD).[1608,1609]

Flowers/Leaves (Camphor CT)

■ Avoid with children under six due to significant camphor content. Several cases of camphor poisoning and/or seizures from ingestion and topical application have been reported in children.[1610,1611] Ingestion of camphor-containing products has been lethal in children under age two.[1612] Children five years and up may use camphor-containing essential oils topically in dilutions no stronger than 5%. Camphor ingestion by infants and young children may cause cough, vomiting, seizure, burning sensation in the mucous membranes and eyes, or lack of voluntary coordination of muscle movements.[1613]

■ Caution is warranted during pregnancy and while lactating, due to significant camphor content. Ingestion of essential oils with significant levels of camphor may lead to abortion because fetuses lack the enzymes to process it.[1614] Camphor ingestion by infants and young children may cause cough, vomiting, seizure, burning sensation in the mucous membranes and eyes, or lack of voluntary coordination of muscle movements.[1615]

■ Avoid oral consumption due to significant camphor levels. Essential oils with significant levels of camphor may be toxic when taken orally, usually single doses exceeding 2 mL, although the lethal dose for humans is estimated to be greater than 5 mL in a single dose.[1616,1617,1618]

■ Avoid with epilepsy and Parkinson's disease due to camphor and fenchone content due to the potential of these constituents to exacerbate or cause seizures/convulsions or reduce seizure-medication efficacy.[1619,1620,1621,1622]

■ Avoid with those who have a compromised liver due to the risk of increased liver enzymes and liver damage based on camphor content.[1623] This would also suggest that those taking medications that could cause liver damage should also use essential oils with significant camphor levels very cautiously or avoid them entirely.

■ The significant camphor content in fringed lavender may negatively impact red blood cells and increase the risk of jaundice in children with glucose-6-phosphate dehydrogenase deficiency (G6PD).[1624,1625]

■ Caution is warranted for the topical application of camphor-containing essential oils on broken or injured skin due to an increased risk of toxicity.[1626]

Leaves, Inflorescences, and Shoot Tips (1,8-Cineole CTs)

■ Avoid with children under six due to significant camphor content. Several cases of camphor poisoning and/or seizures from ingestion and topical application have been reported in children.[1627,1628] Ingestion of camphor-containing products has been lethal in children under age two.[1629] Children five years and up may use camphor-containing essential oils topically in dilutions no stronger than 5%. Camphor ingestion by infants and young children may cause cough, vomiting, seizure, burning sensation in the mucous membranes and eyes, or lack of voluntary coordination of muscle movements.[1630]

■ Caution is warranted during pregnancy and while lactating, due to significant camphor content. Ingestion of essential oils with significant levels of camphor may lead to abortion because fetuses lack the enzymes to process it.[1631] Camphor ingestion by infants and young children may cause cough, vomiting, seizure, burning sensation in the mucous membranes and eyes, or lack of voluntary coordination of muscle movements.[1632]

■ Avoid oral consumption due to significant camphor levels. Essential oils with significant levels of camphor may be toxic when taken orally, usually single doses exceeding 2 mL, although the lethal dose for humans is estimated to be greater than 5 mL in a single dose.[1633,1634,1635]

■ Avoid with epilepsy and Parkinson's disease due to 1,8-cineole, fenchone, or camphor content due to the potential of these constituents to exacerbate or cause seizures/convulsions or reduce seizure-medication efficacy.[1636,1637,1638,1639,1640,1641,1642]

■ Avoid with those who have a compromised liver due to the risk of increased liver enzymes and liver damage based on camphor content.[1643] This would also suggest that those taking medications that could cause liver damage should also use essential oils with significant camphor levels very cautiously or avoid them entirely.

■ The significant camphor content in fringed lavender may negatively impact red blood cells and increase the risk of jaundice in children with glucose-6-phosphate dehydrogenase deficiency (G6PD).[1644,1645]

■ May weakly interfere with the enzymes responsible for metabolizing medications (NSAIDs, proton-pump inhibitors, acetaminophen, antiepileptics, immune modulators, blood-sugar medications, blood pressure medications, antidepressants, antipsychotics, diabetic medications, antihistamines, antibiotics, and anesthetics) due to 1,8-cineole content.[1646,1647,1648]

■ May interfere with pentobarbital and other barbiturates (medications for anxiety and insomnia), based on 1,8-cineole content.[1649,1650]

■ Caution is warranted for the topical application of camphor-containing essential oils on broken or injured skin due to an increased risk of toxicity.[1651]

Aerial Parts (Linalool CT)

■ None currently known.

SELECTED EVIDENCE:

- □ Fringed lavender essential oil demonstrated a neuroprotective effect by reducing excessive activation of AMPA (a subtype of the excitatory glutamate receptor) receptors.[1652] Specifically, the oil decreased the desensitization rate of GluA1 and GluA2 homomers, as well as GluA1/2 heteromers. Excessive activation of glutamate receptors can cause brain damage and neurological disorders like seizures, ALS, and stroke.
- □ *In vitro* research showed that fringed lavender CT 1,8-cineole strongly inhibits *L. monocytogens, S. pneumoniae, H. influenzae (2), Streptococcus* sp., and *Neisseria meningitides (1)*, modestly inhibited *S. aureus, Salmonella* sp., *Neisseria meningitidis (2), Klebsiella pneumoniae (1* and *2), E. coli, Pantoea sp., P. mirabilis,* and *Salmonella* sp. *(2)*, but was ineffective against *P. aeruginosa* and only weakly inhibited some strains of *E. coli*.[1653] The camphor CT (aerial parts) showed good antimicrobial activity against *S. aureus, B. subtilis,* and *E. coli*, which was greater than the tested antibiotics (amoxicillin, gentamicin).[1654]
- □ Oral administration of the p-menth-1,5-Dien-8-ol CT of fringed lavender showed that it enhances antioxidant activity against streptozotocin-induced oxidative stress in diabetic rats.[1655]
- □ The camphor CT (aerial parts) of fringed lavender exhibited weak antioxidant capacity.[1656]
- □ Fringed lavender CT terpinolene demonstrated larvicidal activity against two species of mosquito (*Culiseta longiareolata* and *Culex pipiens*).[1657] The camphor CT (flowers/leaves) was insecticidal

to the cowpea seed beetle (*Callosobruchus maculatus*).[1658] Another study reported that fringed lavender CT 1,8-cineole was insecticidal to houseflies (*Musca domestica*) and blowflies (*Chrysomya albiceps*).[1659]

- An alpha-thujone CT of fringed lavender was aca ricidal to ticks (*Hyalomma scupense*).[1660]

GARDENIA | GARDENIA ABSOLUTE (Cape Jasmine, Common Gardenia)

Gardenia jasminoides J.Ellis, *G. grandiflora* Lour., *G. angustifolia Lodd., G. florida* L., *G. radicans* (Thunb.) Baill.

FAMILY: Rubiaceae
NOTE: Middle-Base
AROMA INTENSITY: Strong
AROMA: Sweet, floral, green, slightly fruity
COMMON EXTRACTION METHOD: Hydrodistilled, steam distilled, or solvent extracted from the flower petals; less commonly extracted by enfleurage
POSSIBLE SUBSTITUTE OILS: Ylang ylang, osmanthus, jasmine, honeysuckle, carnation
BLENDS WELL WITH: Bergamot, cassia, cinnamon, clove, coriander, grapefruit, jasmine, lavender, lemon, lime, mandarin, neroli, rose, sandalwood, tangerine, tuberose, ylang ylang
SUBCELLULAR LOCALIZATION | EPIGENETIC INFLUENCE: Currently unknown | Currently unknown
RECOMMENDED DILUTION RANGE: *EOs*—3%–33%; 50% for some conditions; *Absolutes*—1%–5%

PRIMARY CONSTITUENTS:[1661,1662,1663,1664,1665]

Essential Oil (Alpha-Farnesene CT)

Alpha-Farnesene	17.5%–28.4%
Linalool	21.5%–22.1%
(Z)-3-Hexenyl Tiglate	5.5%–12.7%
(E)-Beta-Ocimene	8.6%–10.6%
Alpha-Terpineol	3.1%–9.0%
Guaiol	5.9%–6.4%
Unidentified (RT=25.182)	0.0%–5.9%
Tetracosane	1.0%–5.6%
Hexyl Tiglate	4.8%–5.5%
Bulnesol	4.3%–5.0%
(Z)-3-Hexenyl Benzoate	1.8%–4.2%

Absolute (Alpha-Farnesene CT)

Alpha-Farnesene	18.0%–23.4%
Linalool	12.0%–22.5%
(Z)-3-Hexenyl Tiglate	10.0%–15.3%
Cyclopenta[c]pyran-4-Carboxylic acid, 7-methyl-, methyl ester	0.0%–8.4%
Jasmin Lactone	0.0%–8.0%
Palmitic Acid	0.0%–7.2%

Essential Oil ((Z)-3-Hexenyl Tiglate CT)

(Z)-3-Hexenyl Tiglate	54.8%–65.7%
Linalool	18.9%–27.0%
6-Tridecanol	4.8%–5.6%
(Z)-3-Hexenyl Benzoate	2.7%–4.7%
Alpha-Terpineol	3.6%–4.4%
Methyl Benzoate	1.5%–3.2%

Absolute ((Z)-3-Hexenyl Tiglate CT)

(Z)-3-Hexenyl Tiglate	36.3%–80.5%
Methyl Benzoate	12.4%–21.5%
Nonadecanol	0.0%–9.6%
Linalool	0.0%–8.8%
2,4-di-tert-Butylphenol	0.0%–7.4%
Beta-Farnesene	0.0%–7.1%
Octadecyl Vinyl Ether	0.0%–6.8%
7-Deceno-5-Olide	0.0%–3.8%

Absolute (Enfleurage I)

(E,E)-Alpha-Farnesene	0.6%
(Z)-3-Hexenyl Tiglate	0.4%

(Z)-3-Hexenyl Benzoate	0.8%–5.0%
Hexyl Tiglate	1.0%–3.5%
Ethyl 5-Hydroxy-(Z)-7-Decenoate	0.0%–3.0%
Squalene	0.0%–3.0%
*Concrete**	
Bicyclo[4.3.1]dec-1(9)-ene	0.0%–2.8%
Gamma-Sitosterol	0.0%–2.5%
Linalool	0.0%–2.3%
(E)-5,10-Secocholest-1(10)-en-3,5-dione	0.0%–1.8%
Stigmasta-5,22-dien-3-ol	0.0%–1.8%
Guaiol	0.0%–1.7%
(Z)-3-Hexenyl Tiglate	0.0%–1.6%
Palmitic Acid	0.0%–1.2%
Ethyl Linoleate	0.0%–1.1%
Linoleic Acid	0.0%–0.8%
6-(Pent-2'-enyl)-Tetrahydro-Pyran-2-one	0.0%–0.8%
2-Methoxy-4-Vinylphenol	0.0%–0.7%
2,4-di-tert-Butylphenol	0.0%–0.5%
(E)-3,7-Dimetyl-1,3,6-octatriene	0.0%–0.4%

* The semi-solid, waxy mixture left over after the solvent is removed. The absolute is then extracted from the concrete (usually with ethyl alcohol).

(E)-3,7-Dimetyl-1,3,6-octatriene	0.3%
(Z)-3-Hexenyl Benzoate	0.2%
1H-Indole	0.2%
(E)-9-Octadecene	0.1%

Absolute (Enfleurage Hot)	
Methyl Benzoate	65.6%
(E)-Beta-Ocimene	34.4%

Absolute (Enfleurage Cold)	
Methyl Benzoate	46.4%
(Z)-3-Hexenyl Tiglate	29.3%
(E)-Beta-Ocimene	21.1%
Alpha-Farnesene	3.2%

OTHER CONSTITUENTS: *Essential Oil (Alpha-Farnesene CT)*—(Z)-3-hexenol, hotrienol, nerol, geraniol, calarene, beta-caryophyllene; *Essential Oil ((Z)-3-Hexenyl Tiglate CT)*—(Z)-6-octen-2-one, 2-isononenal, unidentified (RT=37.746); *Absolute (Alpha-Farnesene CT)*—Methyl benzoate, (E)-beta-ocimene, phenyl ethyl alcohol, ethyl benzoate, borneol, methyl salicylate (< 1.1%), calarene, beta-caryophyllene, isoeugenol, (Z)-3-hexenyl benzoate, guaiol, bulnesol, tricosane; *Absolute ((Z)-3-Hexenyl Tiglate CT)*—6-Tridecanol, tetradecanol, (Z)-3-hexenyl benzoate; *Absolute (Enfleurage I)*—N/A; *Absolute (Enfleurage Hot)*—N/A; *Absolute (Enfleurage Cold)*—N/A; *Concrete*—N/A

PREFERRED COMPOSITION FOR CLINICAL USE:

Constituent	
Alpha-Farnesene	15%–32%
Linalool	12%–28%
(Z)-3-Hexenyl Tiglate	5%–15%
(E)-Beta-Ocimene	5%–15%
Alpha-Terpineol	2%–12%
Guaiol	2%–10%
Hexyl Tiglate	1%–8%
Bulnesol	1%–8%
Tetracosane	0.5%–7%
(Z)-3-Hexenyl Benzoate	0.5%–7%

REPORTED THERAPEUTIC PROPERTIES: Antiseptic, deodorant, aids digestion, relives constipation, eases stomachache, balances menstruation, relieves menopausal symptoms, enhances lactation, aids liver function and the release of bile, analgesic, anti-inflammatory, relieves headaches, supports respiratory function, aids circulation, soothes chronic skin conditions and skin itching, wound healing, anticancer, antioxidant, aids immune function, antibacterial, anti-infectious, cleanses the urinary system, diuretic, promotes restful sleep, relieves anxiety, antidepressant

CAUTIONS:

Essential Oils

- None currently known.

Absolutes/Concretes

- Avoid oral consumption. Absolutes contain solvent residue and should not be taken orally. In addition, the possibility of high levels of benzyl derivatives requires caution for ingestion. Ingestion of benzyl derivatives (benzoic acids, benzoic salts, benzaldehyde, benzyl acetate, benzyl benzoate, and benzyl alcohol) may cause gastrointestinal irritation, vomiting, and diarrhea.[1666] The Joint Expert Committee for Food Additives (JECFA) has established an acceptable daily limit for the aforementioned constituents of 0–5 mg/kg body weight for adults to avoid GI discomfort.[1667] Based on this limit, it is suggested to limit the oral consumption of orange bird berry essential oil rich in benzyl derivatives to no more than 10 drops daily.

SELECTED EVIDENCE:

- Both gardenia essential oil CT alpha-farnesene and absolute CT alpha-farnesene inhibited *S. aureus* and *S. epidermidis*, only the oil was active against *E. coli*, and only the absolute was active against *C. albicans*.[1668]
- Gardenia absolute CT alpha-farnesene and essential oil CT alpha-farnesene exhibited antioxidant activity in the TEAC and FRAP assays, with the absolute being more active.[1669]

GARDEN CHERVIL
(French Parsley, Chervil)

Anthriscus cerefolium (L.) Hoffm., *A. longirostris* Bertol., *Scandix cerefolium* L.

FAMILY: Apiaceae (Umbelliferae)
NOTE: Middle
AROMA INTENSITY: Medium
AROMA: Licorice-like, sweet, spicy, herbaceous
COMMON EXTRACTION METHOD: Hydrodistilled or steam distilled from the aerial parts; may also be distilled from the stem, leaves, or fruits alone
POSSIBLE SUBSTITUTE OILS: Tarragon, tropical basil, fennel CT methyl chavicol (estragole)
BLENDS WELL WITH: Balsam fir, basil, bergamot, black spruce, blue spruce, camphor, copaiba, cypress, eucalyptus, fennel, geranium, lavender, lemongrass, lime, marjoram, neroli, pine, rosemary, tarragon
SUBCELLULAR LOCALIZATION | EPIGENETIC INFLUENCE: Currently unknown | Currently unknown
RECOMMENDED DILUTION RANGE: *Estragole CT*—3%–20%; *Caryophyllene CT*—5%–Neat

PRIMARY CONSTITUENTS:[1670,1671,1672,1673,16741675]

Caryophyllene CT		*Estragole CT*	
Caryophyllene	16.9%	Methyl Chavicol (Estragole)	58.0%–95.9%
Delta-Cadinene	16.4%	1-Allyl-2,4-Dimethoxybenzene	1.4%–33.0%
Trans-Pinocarveol	12.5%	n-Undecane	0.8%–10.0%
Spathulenol	7.5%		
Caryophyllene Oxide	6.8%	*1-Allyl-2,4-Dimethoxybenzene CT*	
Germacrene D	5.8%	1-Allyl-2,4-Dimethoxybenzene	52.5%–77.8%
Beta-Elemene	5.1%	Methyl Chavicol (Estragole)	12.6%–40.7%
Beta-Pinene	4.3%	n-Undecane	2.4%–2.6%
Beta-Myrcene	4.1%		
Alpha-Terpinene	3.1%		

Note: The estragole CT is the most common CT reported.

OTHER CONSTITUENTS: *Estragole CT*—1-allyl-3,4-dimethoxybenzene; *1-Allyl-2,4-Dimethoxybenzene CT*—2-(E)-nonenal, beta-caryophyllene, germacrene D; *Caryophyllene CT*—Beta-ocimene, terpinen-4-ol

PREFERRED COMPOSITION FOR CLINICAL USE:
Insufficient samples/data to determine.

REPORTED THERAPEUTIC PROPERTIES: Reduces high blood pressure, blood purifier, antioxidant, antiseptic, reduces perspiration, diuretic, aids digestion, expels excess gas, wound healing, soothes chronic skin conditions, antiarthritic, analgesic, expels excess mucus, aids circulation, balances menstruation, insecticidal, insect repellent, antidepressant, mentally stimulating

CAUTIONS:
Estragole CT and 1-Allyl-2,4-Dimethoxybenzene CT

- Use with great caution for children under age six due to animal reports of liver cancer caused by methyl chavicol (estragole) content.[1676] The European Medicines Agency considers estragole a possible human carcinogen and recommends exposure to estragole be limited in young children.
- Avoid during pregnancy and lactation due to high methyl chavicol (estragole) content.[1677] Methyl chavicol (estragole) may cause genetic mutations and oxidative DNA damage.[1678,1679] The European Medicines Agency recommends exposure to estragole be limited in women who are pregnant or nursing.[1680]
- Avoid oral use due to carcinogenic (cancer-causing) potential of methyl chavicol (estragole).[1681,1682,1683,1684] The metabolite of estragole (1'-hydroxyestragole) is considered a stronger carcinogen and humans rapidly metabolize estragole to 1'-hydroxyestragole when ingested.[1685]
- May interact with aspirin, blood pressure, antiplatelet, and anticoagulant medications, and increase the risk of bleeding among people with bleeding disorders based on methyl chavicol content. Methyl chavicol is a more potent inhibitor of platelet aggregation than aspirin.[1686]
- May interfere with enzymes responsible for metabolizing medications (NSAIDs, proton-pump inhibitors, acetaminophen, antiepileptics, immune modulators, blood-sugar medications, blood pressure medications, antidepressants, antipsychotics, diabetic medications, antihistamines, antibiotics, and anesthetics) due to methyl chavicol (estragole) content.[1687]
- May interact with barbiturates (medications for anxiety and insomnia), antihistamines, benzodiazepines, tricyclic antidepressants, or other central nervous system depressant drugs.[1688]
- May interact with antibiotics or antifungal drugs and possibly enhance or decrease their effectiveness based on estragole content.[1689,1690,1691]

Caryophyllene CT

■ None currently known.

SELECTED EVIDENCE:

□ None found.

GARDENCRESS
(Pepperweed, Cress)

Lepidium sativum L.

FAMILY: Brassicaceae (Cruciferae)
NOTE: Middle
AROMA INTENSITY: Medium
AROMA: Spicy, fresh, balsamic
COMMON EXTRACTION METHOD: Steam distilled from the seeds; may also be distilled from the leaves or aerial parts
POSSIBLE SUBSTITUTE OILS: Cardamom
BLENDS WELL WITH: Bergamot, black pepper, cardamom, cedarwood, cinnamon, clove, eucalyptus, fennel, ginger, grapefruit, lavender, lemon, lemongrass, neroli, orange, patchouli, rose, sandalwood, tangerine, vetiver, ylang ylang
SUBCELLULAR LOCALIZATION | EPIGENETIC INFLUENCE: Currently unknown | Currently unknown
RECOMMENDED DILUTION RANGE: 3%–33%; 50% for some conditions

PRIMARY CONSTITUENTS:[1692,]

Alpha-Terpinyl Acetate	44.1%
1,8-Cineole	35.8%
Linalool	4.3%
Alpha-terpineol	4.3%

OTHER CONSTITUENTS: Alpha-pinene, sabinene, alpha-phellandrene, gamma-terpinene, terpinen-4-ol, linalool propanoate, (E)-nerolidol

PREFERRED COMPOSITION FOR CLINICAL USE:
Insufficient samples/data to determine.

REPORTED THERAPEUTIC PROPERTIES: Antimicrobial, antiparasitic, antiseptic, antioxidant, anticancer, supports respiratory and lung function, eases chronic respiratory disorders (bronchitis, asthma), expels excess mucus, relieves cough, antiallergic, aids digestion, expels excess gas, soothes insect bites, relieves skin conditions and itching, diuretic, increases lactation, reduces worrisome thoughts, shame, and guilt, stress management

CAUTIONS:

■ Avoid with epilepsy and Parkinson's disease due to 1,8-cineole content. May exacerbate or cause seizures/convulsions or reduce seizure-medication efficacy based on 1,8-cineole content.[1693,1694,1695,1696]

■ May weakly interfere with the enzymes responsible for metabolizing medications (NSAIDs, proton-pump inhibitors, acetaminophen, antiepileptics, immune modulators, blood-sugar medications, blood pressure medications, antidepressants, antipsychotics, diabetic medications, antihistamines, antibiotics, and anesthetics) due to 1,8-cineole content.[1697,1698,1699]

■ May interfere with pentobarbital and other barbiturates (medications for anxiety and insomnia), based on 1,8-cineole content.[1700,1701]

SELECTED EVIDENCE:

- Garden cress essential oil (aerial parts, major constituent alpha-thujene—88.9%) demonstrated significant antiparasitic activity against *Echinococcus granulosus*.[1702]

GERMANDER

Teucrium pruinosum Boiss., *T. pruinosum* var. *aksarayense* M. Dinç & S. Doğu, *T. flavum* L., *T. flavum* L. subsp. *flavum, T. flavum* L. subsp. *hellenicum, T. polium* L., *T. polium* subsp. *capitatum* (L.), *T. capitatum* L., *T. polium* L. subsp. *gabesianum* (L.H.), *T. polium* L. *ssp. aurasiacum*, T. *brevifolium* Schreber, *T. montbretii Benth. ssp. heliotropiifolium (Barbey) Davis*

FAMILY: Lamiaceae (Labiatae)
NOTE: Middle-Base
AROMA INTENSITY: Medium
AROMA: Woody, resinous, earthy, sweet
COMMON EXTRACTION METHOD: Microwave-ultrasonic isolated from the leaves; steam distilled from the aerial parts; hydrodistilled from the aerial parts or leaves, flowers, and fruits; hydrodistilled from the aerial parts
POSSIBLE SUBSTITUTE OILS: Agarwood, copaiba, black pepper, hemp, hops, ylang ylang, German chamomile, melissa
BLENDS WELL WITH: Agarwood, cassia, cedarwood, chamomile (German, Roman), cinnamon, copaiba, cypress, geranium, hinoki, lavender, myrrh, patchouli, pine, rose, sandalwood, spikenard, spruce (black)
SUBCELLULAR LOCALIZATION | EPIGENETIC INFLUENCE: Currently unknown | Currently unknown
RECOMMENDED DILUTION RANGE: 5%–Neat

PRIMARY CONSTITUENTS:[1703,1704,1705,1706,1707,1708,1709,1710,1711,1712,1713,1714,1715]

T. pruinosum	
Agarospirol	45.5%
Beta-Caryophyllene	19.4%
Limonene	3.7%
Alpha-Humulene	3.4%
Caryophyllene Oxide	3.1%
Seychellene	3.0%

T. flavum CT Beta-Bisabolene	
Beta-Bisabolene	31.5%–48.7%
Germacrene D	4.3%–14.0%

T. flavum subsp. flavum	
Beta-Caryophyllene	32.5%
Alpha-Humulene	17.8%
Germacrene D	6.0%
Caryophyllene Oxide	4.9%
(Z)-Gamma-Bisabolene	4.0%

T. flavum subsp. hellenicum (Leaf)	
Germacrene D	21.9%
Beta-Caryophyllene	18.1%
Spathulenol	17.9%

Viridiflorol	1.9%–5.0%
Beta-Caryophyllene	3.1%–4.7%
Delta-Cadinene	2.3%–4.3%
Phytol	0.0%–4.0%

T. flavum CT Caryophyllene

Caryophyllene	12.2%
4-Vinyl Guaiacol	9.7%
Caryophyllene Oxide	7.9%
Alpha-Humulene	6.0%
Linalool	3.4%

T. flavum subsp. hellenicum (Flower)

Germacrene D	22.3%
Beta-Caryophyllene	22.2%
Alpha-Humulene	11.8%
Beta-Bisabolene	7.6%
Alpha-Pinene	6.8%
Beta-Pinene	5.2%

T. polium CT Germacrene D

Germacrene D	3.8%–19.0%
Beta-Pinene	15.0%–18.1%
Alpha-Pinene	14.1%–18.0%
Myrcene	8.2%–10.4%
Limonene	5.3%–8.7%
Delta-Elemene	0.0%–4.5%
Sabinene	2.7%–4.3%
Alpha-Humulene	0.8%–4.2%
Alpha-Thujopsene 2-ol	0.0%–3.5%
Alpha-Cadinol	0.0%–3.2%

T. polium CT Limonene

Limonene	37.7%
2,4di-Tetr-Butylphenol	10.8%
Para-Cymene	8.2%
Alpha-Pinene	4.3%
Valerianol	3.9%
Beta-Pinene	3.7%
Spathulenol	3.6%

T. polium CT Beta-Bisabolol

Beta-Bisabolol	37.2%–45.6%
Beta-Caryophyllene	9.3%–9.4%
11-Acetoxyeudesman-4-alpha-ol	1.9%–8.5%
(E)-Beta-Farnesene	7.3%
Beta-Bisabolene	7.1%
Beta-Cubebene	5.2%
Myrcene	3.7%
Alpha-Pinene	3.4%
Beta-Pinene	3.1%

T. polium CT 8-Cedren-13-ol

8-Cedren-13-ol	24.8%
Beta-Caryophyllene	8.7%
Unidentified Sesquiterpene Hydrocarbon	6.9%
Germacrene D	6.8%
Sabinene	5.2%
Alloaromadendrene	4.5%
Alpha-Humulene	4.3%
Delta-Cadinene	3.5%

T. polium CT Beta-Caryophyllene

Beta-Caryophyllene	18.0%
Germacrene D	13.2%
Spathulenol	10.4%
Beta-Pinene	9.7%
Alpha-Pinene	9.0%
Bicyclogermacrene	9.0%
Caryophyllene Oxide	7.1%
Limonene	4.6%
(Z)-Beta-Farnesene	3.1%

T. polium CT Mixed

11-Acetoxyeudesman-4-alpha-ol	16.7%–30.2%
Alpha-Bisabolol	9.3%–25.9%
Beta-Caryophyllene	8.0%–12.9%
Bisabolol Oxide B	4.4%–7.4%
Caryophyllene Oxide	3.5%–6.7%
Beta-Pinene	2.0%–4.0%
Dehydro-Sesquicineol	2.1%–3.6%
Alpha-Humulene	2.2%–3.4%
Limonene	1.2%–3.4%
Bicyclogermacrene	0.0%–3.3%

T. polium CT Carvacrol

Carvacrol	10.1%
Caryophyllene	9.8%

Alpha-Bisabolol	2.0%–7.2%
Germacrene B	0.0%–4.8%
Dihydro-Sesquicineol	1.0%–4.4%
Caryophyllene Oxide	1.7%–4.2%
Beta-Pinene	2.6%–3.8%
Gossonerol	0.0%–3.5%

T. polium subsp. capitatum

Alpha-Pinene	28.8%
Beta-Pinene	7.2%
Para-Cymene	7.0%
Alpha-Thujene	5.0%
Terpinen-4-ol	4.6%
Limonene	3.0%
Para-Cymene-8-ol	3.0%

T. polium ssp. aurasiacum

Alpha-Cadinol	46.8%
3Beta-Hydroxy-Alpha-Muurolene	22.5%
Alpha-Pinene	9.5%
Beta-Pinene	8.3%

T. brevifolium

Spathulenol	9.0%
Delta-Cadinene	4.2%
Trans-Pinocarveol	3.8%
Caryophyllene Oxide	3.8%
Beta-Pinene	3.6%
Viridiflorol	3.4%
Beta-Eudesmol	3.4%
Cadalene	3.4%

Torreyol	7.6%
Alpha-Cadinol	4.5%
Alpha-Humulene	3.8%
Cis-Verbenone	3.7%
Germacrene D	3.1%
Alpha-Amorphene	3.0%

T. polium subsp. gabesianum

Beta-Pinene	36.0%
Alpha-Pinene	13.3%
Alpha-Thujene	8.5%
Para-Cymene	5.3%
Verbenone	5.1%
Myrcene	4.7%

T. montbretii ssp. heliotropiifolium

Carvacrol	13.9%
Caryophyllene Oxide	12.7%
Hexadecanoic Acid	9.8%
Caryophyllene	7.8%
Caryophylladienol I	3.7%
Linalool	3.5%
Beta-Bourbonene	3.5%
Alloaromadendrene	3.0%

Note: *T. polium* essential oil composition varies widely and may be related to variety type (tetraploid or diploid).[1716] Other CTs reported in the literature include myrcene, ledene oxide II, and spathulenol.

Other Constituents: *T. pruinosum*—Di-epi-alpha-cedrene, guaiol, isoledene, beta-humulene, alpha-bourbonene, spathulenol, alpha-cubebene, trans-alpha-bergamotene, thujopsene, butylidene-cyclohexene; *T. flavum CT Beta-Bisabolene*—Limonene, isoamyl hexanoate, alpha-copaene, beta-bourbonene, beta-copaene, (E)-beta-farnesene, alpha-humulene, gamma-muurolene, bicyclogermacrene, alpha-muurolene, ledol, germacrene d-4-ol, caryophyllene oxide, copaborneol, cadin-4-en-10-ol, alpha-bisabolol, phytone, hexadecanol; *T. flavum CT Caryophyllene*—Terpinolene, isoamyl isovalerate, cis-pinocarveol, trans-pinocarveol, cis-verbenol, veratrole, myrtenol, isoamyl hexanoate, eugenol, (Z)-alpha-trans-bergamotene, alloaromadendrene, germacrene D, epi-bicyclosesquiphellandrene, (E,E)-alpha-farnesene, beta-bisabolene, delta-cadinene, guaiol, viridiflorol, caryophylladienol I, alpha-cadinol, benzyl benzoate, hexahydrofarnesylacetone, hexadecanoic acid methyl ester, (Z)-phytol, (Z)-9-octadecenoic acid methyl ester, tricosane, pentacosane, heptacosane, squalene; *T. flavum* subsp. *flavum*—Isopentyl hexanoate, carvacrol, beta-bourbonene, 1-tetradecane, (E)-beta-farnesene, gamma-muurolene, beta-curcumene, delta-cadinene, viridiflorol, ledol, humulene epoxide II, caryophylla-4(14),8(15)-dien-5-a-ol, alpha-muurolol, alpha-cadinol, 1-octadecene, 1-eicosane, phytol, 1-docosane, 1-tetracosane, 1-hexacosane; *T. flavum* subsp. *hellenicum (Leaf)*—Limonene, alpha-copaene, beta-bourbonene, linalool, alloaromadendrene, bicyclogermacrene, calacorene, caryophyllene oxide, (Z)-nerolidol, viridiflorol; *T. flavum* subsp. *hellenicum (Flower)*—Myrcene, limonene, amyl valerate, alpha-copaene, linalool, beta-cubebene, beta-bourbonene, beta-selinene, bicyclogermacrene, caryophyllene oxide, (Z)-nerolidol, viridiflorol, T-cadinol, carvacrol; *T. polium CT Germacrene D*—Alpha-thujene, (E)-beta-ocimene, gamma-terpinene, trans-pinocarveol, terpinen-4-ol, myrtenal, beta-bourbonene, beta-elemene, beta-caryophyllene, beta-copaene, alpha-guaiene(E)-9-epi-caryophyllene, gamma-cadinene, bourbonanone-1-nor, spathulenol, caryophyllene oxide, humulene epoxide II, dauca-5,8-diene, T-cadinol, T-muurolol, beta-bisabolol-ep; *T. polium CT 8-Cedren-13-ol*—Alpha-thujene, alpha-pinene, beta-pinene, myrcene, para-cymene, limonene, gamma-terpinene, beta-cubebene, cis-beta-guaiene, valencene, (Z)-alpha-bisabolene, gamma-cadinene, (Z)-gamma-bisabolene, spathulenol, caryophyllene oxide, guaiol, beta-oplopenone, T-cadinol, alpha-cadinol, 14-hydroxy-9-epi-trans-caryophyllene, 5-iso-cedrenal, cadalene; *T. polium CT Beta-Caryophyllene*—Myrcene, linalool, alpha-campholenal, trans-pinocarveol, trans-verbenol, myrtenal, verbenone, beta-bourbonene, alpha-humulene, delta-cadinene; *T. polium CT Limonene*—Delta-2-carene, allo-ocimene, verbenone, dodecane, tetradecane, beta-caryophyllene, alloaromadendrene, pentadecane, hexadecane, hexadecene; *T. polium CT Mixed*—Alpha-pinene, 6-methyl-5-hepten-2-one, beta-bisabolene, delta-cadinene, (E)-gamma-bisabolene, germacrene B, spathulenol, gossonerol; *T. polium CT Beta-Bisabolol*—Alpha-pinene, 6-methyl-5-heptene-2-one, limonene, alpha-humulene, bicyclogermacrene, beta-bisabolene, delta-cadinene, (E)-gamma-bisabolene, spathulenol; *T. polium CT Carvacrol*—Sabinene, linalool, cis-sabinol, nopinone, trans-pinocarveol, cis-verbenone, sabina ketone, umbellulone, terpinen-4-ol, alpha-terpineol, myrtenol, cis-carveol, p-cymen-7-ol, p-methoxyacetophenone, alpha-copaene, alpha-gurjunene, ar-curcumene, cis-beta-guaiene, epi-bicyclosesquiphellandrene, delta-cadinene, germacrene B, germacrene D-4-ol, humulene epoxide II, caryophylladienol I, cadalene, pentacosane, heptacosane, nonacosane, hentriacontane; *T. polium* subsp. *capitatum*—Camphene, sabinene, myrcene, linalool, beta-thujone (< 0.6%), alpha-campholenal, nopinone, trans-pinocarveol, trans-verbenol, sabina ketone, pinocarvone, myrtenal, alpha-terpineol, myrtenol, verbenone, trans-carveol, carvone, bornyl acetate, alpha-terpinyl acetate, delta-cadinene, guaiol, bulnesol, cadalene; *T. polium subsp. gabesianum*—Camphene, limonene, terpinolene, alpha-thujone (< 0.6%), linalool, fenchol, alpha-campholenol, trans-verbenol, myrtenol, alpha-cubenol, terpinen-4-ol, p-cymen-8-ol, 4-metoxyacetophenone, pinocarvone, borneol; *T. polium ssp. aurasiacum*—Myrcene, limonene, trans-pinocarveol, myrtenal; *T. brevifolium*—Alpha-pinene, 1-octen-3-ol, myrcene, pinocarvone, myrtenal, myrtenol, cis-verbenone, trans-carveol, phellandral, p-methoxyacetophenone, delta-elemene, eugenol, alpha-copaene, (E)-beta-damascenone, beta-bourbonene, alpha-gurjunene, beta-gurjunene, aromadendrene, beta-selinene, gamma-muurolene, epizonarene, gamma-cadinene, alpha-calacorene, beta-calacorene, cis-alpha-copaen-8-ol, ledol, cedrenol, alpha-eudesmol, ledene alcohol, alpha-bisabolol, (Z)-alpha-bisabolene epoxide, aristolone, hexadecanal, hexahydrofarnesylacetone, phenanthrene, octadecanal; *T. montbretii ssp. heliotropiifolium*—Geraniol, thymol, eugenol, beta-gurjunene, alpha-humulene, germacrene D, dihydro actinidiolide, gamma-muurolene, ledol, globulol, epi-bicyclosesquiphellandrene, delta-cadinene, T-muurolol, alpha-cadinol, vulgarol B, (Z,Z)-9,12-octadecadienoic acid, hexahydrofarnesylacetone, pentacosane, heptacosane, octacosane, nonacosane, hentriacontane

Preferred Composition for Clinical Use:

Too great of variance in data/samples composition to determine.

Reported Therapeutic Properties: Antioxidant, antirheumatic, antispasmodic, analgesic, anti-inflammatory, reduces fever, anticonvulsive, antiparasitic, antibacterial, diuretic, antiseptic, astringent, wound healing, reduces blood sugar, anticancer, supports respiratory function, reduces appetite, aids digestion, heals stomach ulcer, reduces blood pressure, balances heart function, reduces post-partum bleeding, insecticidal, insect repellent, aphrodisiac, relieves anxiety, aids meditation, stress management

Cautions:

- None currently known.

Selected Evidence:

- Germander essential oil (*T. pruinosum*) demonstrated anticancer activity against cervical cancer cells, significantly disrupting mitochondrial activity (cell viability) in cancer cells.[1717] *Teucrium* species essential oils were cytotoxic to colon, melanoma, and large lung cancer cells.[1718]
- Germander essential oil (*T. pruinosum*) inhibited both the COX-1 and COX-2 enzymes, demonstrating anti-inflammatory activity.[1719] The oil was more selective toward COX-2 than COX-1. Macrophages produce a large number of cytotoxic molecules when activated by inflammatory stimulation. Several germander essential oil species and CTs significantly inhibited inflammation by reducing macrophage production of cytotoxic molecules.[1720] The most effective oil was *t. brevifolium*, followed by *t. montbretii* ssp. *helitropiifolium*.
- *T. polium* essential oil reduced pain in mice when administered intraperitoneally, with 37.5 and 75 mg/kg body weight being the most effective doses.[1721]
- *In vitro* research reported that germander (*T. polium* subsp. *gabesianum*) inhibited bacteria (*E. coli, E. faecalis, S. aureus, P. aeruginosa, C. freundeii, P. mirabilis*) and fungi (*R. rubrum, M. canis, T. soudanense, S. brevicaulis, C. neoformans, C. albicans, A. fumigatus*).[1722] *T. polium* CT Germacrene D essential oil inhibited *M. luteus, B. cereus, E. coli,* and *S. aureus*.[1723] Germander (*T. polium* CT limonene) inhibited *S. pneumoniae, E. coli, K. pneumoniae, S. aureus*, and *S. epidermidis*.[1724]
- Germander (*T. polium*; CT (E)-3-caren-2-ol) exhibited promising activities against *Leishmania donovani* promastigotes.[1725]
- The DPPH assay showed that germander essential oil (*T. pruinosum*) is a good antioxidant.[1726] *T. flavum* L. subsp. *flavum* exhibited moderate antioxidant activity in the DPPH assay.[1727] Germander (T. polium; 26.3% 11-acetoxyeudesman-4-α-ol, 24.6% alpha-bisabolol) essential oil exhibited good antioxidant activity in the DPPH and beta-carotene-linoleic acid assays.[1728]
- A germander essential oil (*T. polium* subsp. *capitatum*) rich in alpha-cadinol was insecticidal to the cowpea weevil (*Callosobruchus maculatus*).[1729]

GRAND FIR

Abies grandis (Douglas ex D. Don) Lindl.

Family: Pinaceae
Note: Top-Middle
Aroma Intensity: Medium
Aroma: Woody, piney, balsamic, sweet, fresh
Common Extraction Method: Steam distilled or hydrodistilled from the needles (leaves)
Possible Substitute Oils: Balsam fir, spruce (black)

Blends Well With: Balsam fir, birch, cassia, blue spruce, cedarwood, chamomile (German, Roman), cypress, frankincense, galbanum, lavender, lemon, myrrh, myrtle, palo santo, pine, sandalwood, silver fir, spruce (black), tsuga, white fir, wintergreen
Subcellular Localization | Epigenetic Influence: Currently unknown | Currently unknown
Recommended Dilution Range: 5%–Neat

Primary Constituents:[1730,1731]

Beta-Pinene CT		*Bornyl Acetate CT*	
Beta-Pinene	20.3%–31.1%	Bornyl Acetate	19.8%–24.2%
Bornyl Acetate	12.7%–26.2%	Beta-Phellandrene	19.8%–21.5%
Beta-Phellandrene	9.9%–25.2%	Camphene	13.7%–19.2%
Camphene	8.3%–13.5%	Alpha-Pinene	7.1%–8.9%
Alpha-Pinene	4.4%–7.4%		
Alpha-Terpineol	0.9%–3.6%		

Other Constituents: *Beta-Pinene CT*—Tricyclene, myrcene, alpha-phellandrene, delta-3-carene, alpha-terpinene, limonene, gamma-terpinene, terpinolene, camphor, citronellal, borneol, citronellol, citronellyl acetate, geranyl acetate, trans-muurola-4(14),5-diene, cubebol, delta-cadinene, 1-epi-cubenol, cubenol; *Bornyl Acetate CT*—Tricyclene, myrcene, limonene, camphor, borneol, alpha-terpineol, citronellyl acetate, geranyl acetate

Preferred Composition for Clinical Use:

Constituent	
Beta-Pinene	20%–35%
Bornyl Acetate	10%–25%
Beta-Phellandrene	8%–25%
Camphene	7%–16%
Limonene	3%–15%
Alpha-Pinene	3%–10%
Delta-Cadinene	0.5%–5%
Alpha-Terpineol	0.1%–5%

Reported Therapeutic Properties: Analgesic, reduces fever, anti-inflammatory, antiarthritic, antirheumatic, relieves sprains and strains, antibacterial, antifungal, anticancer, antiseptic, antioxidant, aids circulation, increases localized circulation, antispasmodic, disinfectant, wound healing, removes excess mucus, relieves cough, supports respiratory function, calming, stress management, relieves anxiety, enhances focus, encourages feelings of stability and empowerment, combats negative emotions and memories

Cautions:

- None currently known.

Selected Evidence:

- □ None found.

GRASSHOPPER'S GRASS (Ahibalala)

Helichrysum faradifani Sc. Ell.

FAMILY: Asteraceae (Compositae)
NOTE: Middle
AROMA INTENSITY: Medium
AROMA: Herbaceous, earthy, slightly floral
COMMON EXTRACTION METHOD: Steam distilled or hydrodistilled from the aerial parts
POSSIBLE SUBSTITUTE OILS: Helichrysum, copaiba, cedarwood (Atlas)
BLENDS WELL WITH: Bergamot, chamomile (German, Roman), clary sage, clove, cypress, geranium, grapefruit, helichrysum, lavender, lemon, lime, melaleuca (tea tree), neroli, orange, oregano, rosemary, tangerine, vetiver, ylang ylang
SUBCELLULAR LOCALIZATION | EPIGENETIC INFLUENCE: Currently unknown | Currently unknown
RECOMMENDED DILUTION RANGE: 3%–50%; Neat for some conditions

PRIMARY CONSTITUENTS:[1732,1733,1734,1735]

Alpha-Fenchene CT	
Alpha-Fenchene	32.3%–35.6%
Gamma-Curcumene	17.7%–19.4%
Beta-Caryophyllene	4.8%–14.2%
Limonene	2.8%–8.3%
Lavandulyl Acetate	2.1%–4.7%
ar-Curcumene	2.9%–3.4%

Mixed CT (Fresh)	
Beta-Himachalene	15.7%–32.8%
Alpha-Fenchene	13.1%–27.3%
Beta-Caryophyllene	14.8%–20.5%
ar-Curcumene	2.1%–7.9%
Bornyl Acetate	2.5%–5.2%
Limonene	1.9%–4.5%
Unknown Compound	2.7%–3.9%
Linalool	1.2%–3.3%
Alpha-Terpineol	0.9%–3.2%

Beta-Himachalene/Beta-Caryophyllene CT (Dried)	
Beta-Himachalene	30.0%–35.5%
Beta-Caryophyllene	15.4%–29.2%
Alpha-Terpineol	1.0%–6.1%
Bornyl Acetate	0.3%–5.1%
Unknown Compound	1.2%–4.8%
Eudesmol (isomer NI)	2.1%–4.2%
Bisabolol (isomer NI)	1.4%–4.0%
ar-Curcumene	2.5%–3.7%
Beta-Bisabolene	1.8%–3.7%

Beta-Caryophyllene CT	
Beta-Caryophyllene	34.6%
Linalool	16.1%
Alpha-Humulene	5.4%
Limonene	4.6%
Caryophyllene Oxide	4.3%
Delta-Cadinene	3.0%

OTHER CONSTITUENTS: *Alpha-Fenchene CT*—2-bornene, alpha-pinene, beta-pinene, alpha-terpinene, para-cymene, linalool, exo-fenchol, alpha-fenchyl hydrate, endo-fenchyl acetate, iso-italicene, italicene, alpha-humulene, alpha-guaiene, alpha-bulnesene, beta-bisabolene, beta-curcumene, delta-cadinene, caryophyllene alcohol, epi-alpha-bisabolol; *Mixed CT (Fresh)*—Tricyclene, alpha-pinene, camphene, beta-pinene, myrcene, 1,8-cineole, gamma-terpinene, para-cymene, terpinolene, alpha-fenchyl acetate, alpha-copaene, alpha-cedrene, alpha-bergamotene, alpha-fenchol, terpinene-4-ol, alpha-humulene, (Z)-beta-farnesene, gamma-muurolene, beta-bisabolene, geranial, delta-cadinene, caryophyllene oxide, eudesmol, bisabolol; *Beta-Himachalene/Beta-Caryophyllene CT (Dried)*—Alpha-pinene, limonene, terpinolene, alpha-copaene, alpha-cedrene, linalool, alpha-bergamotene, alpha-fenchol, terpinen-4-ol, alpha-humulene, (Z)-beta-farnesene, gamma-muurolene, borneol, beta-selinene, gamma-cadinene, geranial, delta-cadinene, (Z,Z)-farnesol; *Beta-Caryophyllene CT*—Alpha-pinene, beta-pinene, myrcene, para-cymene, 1,8-cineole, (E)-beta-ocimene, borneol, alpha-terpineol, alpha-copaene, alloaromadendrene, gamma-muurolene, ar-curcumene, beta-selinene, gamma-cadinene, calamenene, humulene 6,7-epoxide, T-cadinol, beta-eudesmol

PREFERRED COMPOSITION FOR CLINICAL USE:
Too great of variance in samples/data to determine.

REPORTED THERAPEUTIC PROPERTIES: Wound healing, relieves skin itching and chronic skin conditions, reduces the appearance of scars and blemishes, diuretic, aids and protects liver function, balances blood sugar, aids digestion, stops excess bleeding, analgesic, anti-inflammatory, anti-cancer, antiallergic, supports respiratory function, antifungal, insecticidal, relieves anxiety, calming, helps relieve emotional trauma

CAUTIONS:
- None currently known.

SELECTED EVIDENCE:
- ☐ Grasshopper's grass CT alpha-fenchene essential oil was larvicidal to *Culex quinquefasciatus* mosquitoes with the 2nd instar being most sensitive to the oil.[1736]

GREEK SAGE
(Turkish Sage, Greek Oregano)

Salvia fruticosa Mill., *S. libanotica* Boiss. & Gaill., *S. triloba* L., *S. cypria, S. lobryana*

FAMILY: Family
NOTE: Top-Middle
AROMA INTENSITY: Strong
AROMA: Herbaceous, woody, earthy, camphoraceous, fresh
COMMON EXTRACTION METHOD: Steam distilled or hydrodistilled from the leaves; may also be distilled from the stem, inflorescences, and aerial parts
POSSIBLE SUBSTITUTE OILS: White sage, rosemary, Spanish sage, spike lavender
BLENDS WELL WITH: Bergamot, camphor, cedarwood, frankincense, lavender, lemon, marjoram, orange, pine, rosemary, sandalwood, Spanish sage, tangerine, thyme, vetiver, white sage, yuzu
SUBCELLULAR LOCALIZATION | EPIGENETIC INFLUENCE: Currently unknown | Currently unknown
RECOMMENDED DILUTION RANGE: 3%–20%; 50% for some condition

PRIMARY CONSTITUENTS:[1737,1738,1739,1740,1741,1742,1743]

		Stems, Aerial Parts, or Inflorescences	
1,8-Cineole	17.2%–64.2%	1,8-Cineole	22.5%–56.9%
Camphor	tr–30.3%	Alpha-Pinene	2.6%–37.3%
Beta-Thujone	0.0%–25.6%	Viridiflorol	0.0%–15.7%
Alpha-Pinene	1.8%–18.6%	Manool	0.0%–13.2%
Beta-Caryophyllene	0.2%–16.0%	Beta-Caryophyllene	0.3%–12.2%
Alpha-Thujone	0.0%–19.2%		

Globulol	0.0%–10.0%
Camphene	tr–9.9%
Beta-Pinene	tr–9.4%
Manool	0.0%–6.4%
Alpha-Campholenal	0.0%–6.1%
Alpha-Terpineol	0.0%–7.5%
Viridiflorol	0.0%–5.5%
Myrcene	0.2%–5.3%
Alpha-Terpinyl Acetate	0.0%–5.3%
Cis-Sabinene Hydrate	0.0%–4.6%
Eugenol	0.0%–4.5%
Linalyl Propionate	0.0%–4.0%
Alpha-Humulene	0.0%–3.9%
Isoborneol	0.0%–3.8%
Caryophyllene Oxide	0.0%–3.0%
Bornyl Acetate	0.0%–3.0%

Myrcene	0.1%–7.0%
Camphor	0.4%–6.8%
Beta-Pinene	2.0%–6.6%
Aromadendrene	0.0%–4.5%
Alpha-Terpineol	0.0%–4.3%
Alpha-Terpinyl Acetate	0.0%–3.4%
Camphene	0.3%–3.1%

Aerial Parts CT Viridiflorol

Viridiflorol	25.4%–37.9%
Manool	14.4%–25.7%
Beta-Caryophyllene	9.9%–11.5%
Alpha-Humulene	8.5%–10.2%
Alpha-Terpineol	6.2%–6.6%
Caryophyllene Oxide	3.3%–3.9%

OTHER CONSTITUENTS: *OC*—Tricyclene, para-cymene, o-cymene, limonene, gamma-terpinene, linalool, trans-pinocamphone, borneol, terpinen-4-ol, linalyl formate, carvacrol, aromadendrene, 9-epi-(E)-caryophyllene, gamma-muurolene, ar-curcumene, viridiflorene, trans-calamenene, delta-cadinene, elemol, spathulenol, humulene epoxide, neo-intermedeol; *Stems, Aerial Parts, or Inflorescences*—Limonene, borneol, para-cymene, alpha-thujone (< 1.5%), beta-thujone (< 1.2%), alpha-humulene, caryophyllene oxide, delta-terpineol, alpha-gurjunene, alloaromadendrene, ledene, delta-cadinene, labd-7,13-dien-15-ol, trans-ferruginol; *Aerial Parts CT Viridiflorol*—Delta-terpineol, carvacrol, alpha-gurjunene, aromadendrene, alpha-amorphene, gamma-cadinene, delta-cadinene, epiglobulol, spathulenol, caryophylla-4(14),8(15)dien-5-beta-ol, labd-7,13-dien-15-ol, trans-ferruginol

PREFERRED COMPOSITION FOR CLINICAL USE:

Constituent	
1,8-Cineole	25%–55%
Camphor*	5%–20%
Alpha-Pinene	1.5%–10%
Beta-Caryophyllene	0.5%–10%
Thujone < alpha + beta >	< 10%
Camphene	0.1%–10%
Beta-Pinene	0.1%–8%
Myrcene	0.5%–6%
Alpha-Humulene	0.1%–5%

* Camphor content varies widely but the most therapeutic oils will contain around 10%–20%.

REPORTED THERAPEUTIC PROPERTIES: Antimicrobial, antibacterial, antifungal, antiviral, anti-infectious, antiseptic, astringent, anti-inflammatory, analgesic, antioxidant, anticancer, protects brain function, **supports memory and cognitive function**, supports respiratory function, **eases cough**, diuretic, aids digestion, aids circulation, reduces high blood sugar, soothes skin itching, balances menstruation, insecticidal, aids concentration, boosts self-confidence, reduces mood swings

CAUTIONS:

Leaves

■ Avoid with children under 6 due to thujone and camphor content. Cases of seizure have been reported in young children due to exposure to essential oils rich in thujones.[1744,1745] Several cases of camphor poisoning and/or seizures from ingestion and topical application have been reported in children.[1746,1747] Ingestion of camphor-containing products has been lethal in children under age two.[1748] Children five years and up may use camphor-containing essential oils topically in dilutions no stronger than 5%. Camphor ingestion by infants and young children may cause cough, vomiting, seizure, burning sensation in the mucous membranes and eyes, or lack of voluntary coordination of muscle movements.[1749] The significant 1,8-cineole content warrants the avoidance of Greek sage with children under age 3, particularly around the nose and mouth, due to the potential to cause seizures, central nervous system problems, or respiratory distress in young children.[1750,1751,1752]

■ Avoid during pregnancy and while lactating, due to thujones and camphor content. Essential oils high in thujone content may cause abortion.[1753,1754] Ingestion of essential oils with significant levels of camphor may lead to abortion because fetuses lack the enzymes to process it.[1755] Camphor ingestion by infants and young children may cause cough, vomiting, seizure, burning sensation in the mucous membranes and eyes, or lack of voluntary coordination of muscle movements.[1756]

■ Avoid oral consumption. Ingesting plants or products high in thujones can cause vomiting, stomachache, diarrhea, and gastroenteritis followed by absorption disorders, headache, nervous agitation, and chronic convulsions, and symptoms of liver and kidney toxicity extending to yellow liver atrophy (extensive and rapid death of liver cells—parenchymal, possibly by fatty degeneration), arrhythmia, and myocardial bleeding.[1757] Thujone is considered significantly neurotoxic and may damage the liver, and it is estimated that as little as 15 mg orally may negatively impact the central nervous system.[1758,1759,1760] Taking 20 drops of thuja oil (also with significant thujones content) twice daily for five days caused an adult woman to have a tonic seizure.[1761] As few as 12 drops of sage oil (also with significant thujones content) orally have reportedly caused a seizure and short coma in an adult.[1762] Essential oils with significant levels of camphor may be toxic when taken orally, usually single doses exceeding 2 mL, although the lethal dose for humans is estimated to be greater than 5 mL in a single dose.[1763,1764,1765]

■ Avoid with epilepsy and Parkinson's disease due to thujones, camphor, and 1,8-cineole content.[1766,1767,1768,1769,1770]

■ May weakly interfere with the enzymes responsible for metabolizing medications (NSAIDs, proton-pump inhibitors, acetaminophen, antiepileptics, immune modulators, blood-sugar medications, blood pressure medications, antidepressants, antipsychotics, diabetic medications, antihistamines, antibiotics, and anesthetics) due to 1,8-cineole and alpha-thujone content.[1771,1772,1773,1774]

Stems, Aerial Parts, or Inflorescences

■ Avoid with epilepsy and Parkinson's disease due to 1,8-cineole and camphor content. May exacerbate or cause seizures/convulsions or reduce seizure-medication efficacy.[1775,1776,1777,1778]

■ Caution is warranted orally due to the significant amounts of 1,8-cineole. Limit it to small doses internally (adults—maximum 10 drops daily). Toxicity has been reported when eucalyptus (also high in 1,8-cineole) was ingested in large doses, and as few as 4 to 5 drops may cause problems in very sensitive individuals.[1779,1780,1781,1782,1783,1784]

■ May weakly interfere with the enzymes responsible for metabolizing medications (NSAIDs, proton-pump inhibitors, acetaminophen, antiepileptics, immune modulators, blood-sugar medications, blood pressure medications, antidepressants, antipsychotics, diabetic medications, antihistamines, antibiotics, and anesthetics) due to 1,8-cineole content.[1785,1786,1787]

■ May interfere with pentobarbital and other barbiturates (medications for anxiety and insomnia), based on 1,8-cineole content.[1788,1789]

Aerial Parts CT Viridiflorol

■None currently known.

SELECTED EVIDENCE:

- Greek sage essential oil (26.4% 1,8-cineole, 18.9% camphor, 9.5% camphene) demonstrated remarkable neuroprotective (brain astrocytes) effects against hydrogen peroxide induced (oxidative stress) cell death, which suggests it may have potential as a therapy against neurodegenerative diseases.[1790]
- *In vitro* research demonstrated that Greek sage essential oil significantly inhibits *C. albicans*, gram-positive, and gram-negative bacteria, including multi-antibiotic-resistant strains.[1791] Greek sage essential oil with an unidentified composition reduced antibiotic resistance of tetracycline-resistant clinical isolates of *Staphylococcus epidermidis* by inhibition of efflux pumps.[1792] Inhibition of efflux pumps reduces the ability of antibiotics to resist antibiotic therapy. The aerial parts oil was inactive against *M. tuberculosis*.[1793] Although it was the least effective of four oils tested, Greek sage essential oil inhibited thirteen fungi (food poisoning, plant, animals and human pathogenic species).[1794] Greek sage essential oil weakly inhibited the citrus fungi *Penicillium digitatum*.[1795]
- The expression of matrix metalloproteinases (MMPs) is increased in inflammatory conditions (rheumatoid arthritis, osteoarthritis, Alzheimer's disease, multiple sclerosis, etc.), cancers (including metastasis), autoimmune disorders of the skin, and metabolic bone diseases. Activation of MMP-2 has been associated with arthritic conditions, chronic airway diseases, cancer progression (cause disorganized membranes by degrading type IV collagen), and excess activity may lead to immune system disruption that allows pathogens to persist and spread. Greek sage essential oil (49.5% 1,8-cineole, 13.3% camphor, 7.2% beta-pinene) inhibited MMP-2 to reduce periodontal pathogen virulence and tissue destruction.[1796]
- Greek sage aerial parts essential oil exhibited antioxidant activity with a greater effect observed when more oxygenated sesquiterpenes and diterpenes were present.[1797]

GUACO
(Huaco, Snake-Herb)

Mikania glomerata Spreng., *M. laevigata* Schultz Bip. ex Baker

FAMILY: Asteraceae
NOTE: Middle
AROMA INTENSITY: Medium
AROMA: Spicy, sweet, slightly floral to fruity with a hint of vanilla
COMMON EXTRACTION METHOD: Steam distilled or hydrodistilled from the leaves; may also be distilled from the seeds or flowers (inflorescences)
POSSIBLE SUBSTITUTE OILS: Ylang ylang, helichrysum, copaiba, melissa
BLENDS WELL WITH: Bergamot, chamomile (German, Roman), clary sage, clove, cypress, geranium, grapefruit, helichrysum, lavender, lemon, lime, melaleuca (tea tree), neroli, orange, oregano, rosemary, tangerine, vetiver, ylang ylang
SUBCELLULAR LOCALIZATION | EPIGENETIC INFLUENCE: Currently unknown | Currently unknown
RECOMMENDED DILUTION RANGE: 3%–50%; neat for some conditions

PRIMARY CONSTITUENTS:[1798,1799,1800,1801]

M. glomerata (Leaves, CO CT)

Constituent	%
Caryophyllene Oxide	24.8%
Spathulenol	23.7%
Germacrene D	7.5%
Calamenenol (isomer not identified)	5.2%
Beta-Caryophyllene	4.2%

M. glomerata (Leaves, GD CT)

Constituent	%
Germacrene D	23.4%–41.5%
Beta-Caryophyllene	14.5%–21.3%
Bicyclogermacrene	9.8%–11.4%
Delta-Elemene	0.0%–9.9%
Caryophyllene Oxide	3.5%–5.9%
Germacrene B	0.0%–4.1%
Elemol	2.0%–3.3%
Spathulenol	3.0%–3.3%
Alpha-Cadinol	0.0%–3.0%

M. glomerata (Flowers)

Constituent	%
Germacrene D	25.4%
Beta-Caryophyllene	16.0%
Globulol	5.2%
Epi-Alpha-Bisabolol	5.0%
Spathulenol	3.9%
Beta-Elemene	3.8%
Alpha-Cadinol	3.0%

M. glomerata (Seeds)

Constituent	%
Spathulenol	30.8%
Epi-Alpha-Bisabolol	9.9%
Caryophyllene Oxide	8.0%
Beta-Caryophyllene	6.1%
Germacrene D	4.0%

M. laevigata (Leaves, GD CT)

Constituent	%
Germacrene D	23.6%–29.8%
Beta-Caryophyllene	7.6%–20.9%
Bicyclogermacrene	11.2%–13.4%
Myrcene	0.0%–9.2%
Delta-Elemene	0.0%–8.7%
Spathulenol	3.3%–7.3%
Coumarin	0.0%–5.4%
Beta-Elemene	0.0%–3.6%

M. laevigata (Flowers)

Constituent	%
Germacrene D	18.7%
Myrcene	17.6%
Bicyclogermacrene	11.9%
Beta-Caryophyllene	9.8%
Delta-Elemene	4.1%
Spathulenol	3.0%

M. laevigata (Seeds)

Constituent	%
Spathulenol	21.7%
Myrcene	14.9%
Germacrene D	11.2%
Beta-Caryophyllene	8.8%
Bicyclogermacrene	7.4%

Note: Although distinct species, *M. laevigata* and *M. glomerata* are very similar morphologically. Their primary difference is their flowering period (*M. laevigata*—September; *M. glomerata*—January), but their leaves also vary slightly with the lobes more prominent in *M. glomerata*.

OTHER CONSTITUENTS: *M. glomerata (Leaves, CO CT)*—Alpha-pinene, beta-pinene, alpha-ylangene, beta-elemene, bicyclogermacrene, gamma-cadinene, delta-cadinene, alpha-calacorene, occidentalol, (E)-nerolidol, beta-copaen-4-alpha-ol, humulene epoxide I, humulene epoxide II, gamma-eudesmol, alpha-muurolol, 7-epi-alpha-eudesmol, 14-hydroxy-9-epi-beta-caryophyllene, alpha-bisabolol; *M. glomerata (Leaves, GD CT)*—Limonene, sabinene, gamma-elemene, alpha-humulene, alpha-elemene, alpha-copaene, beta-elemene, delta-cadinene, T-muurolol; *M. glomerata (Flowers)*—Sylvestrene, nonanal, delta-elemene, gamma-elemene, alpha-humulene, germacrene B, (E)-nerolidol, T-muurolol; *M. glomerata (Seeds)*—Sylvestrene, terpinen-4-ol, beta-elemene, alpha-humulene, germacrene B, (E)-nerolidol, T-muurolol, alpha-cadinol; *M. laevigata (Leaves, GD CT)*—Alpha-Pinene, sabinene, beta-pinene, gamma-elemene, alpha-humulene, elemol, caryophyllene oxide, beta-bourbonene, delta-cadinene, germacrene B, globulol, guaiol, humulene oxide (isomer not identified), beta-eudesmol, T-cadinol, alpha-cadinol; *M. laevigata (Flowers)*—Alpha-pinene, beta-pinene, sylvestrene, (E)-beta-ocimene, terpinen-4-ol, nonanal, beta-elemene, gamma-elemene, alpha-humulene, elemol, germacrene B, (E)-nerolidol, globulol, alpha-cadinol; *M. laevigata (Seeds)*—Alpha-pinene, beta-pinene, (E)-beta-ocimene, nonanal, delta-elemene, alpha-copaene, beta-elemene, gamma-elemene, alpha-humulene, germacrene B, (E)-nerolidol, alpha-cadinol

PREFERRED COMPOSITION FOR CLINICAL USE:
Insufficient samples/data to determine.

REPORTED THERAPEUTIC PROPERTIES: Antiallergic, aids respiratory function, relieves asthma, bronchodilator, eases cough, expels excess mucus, analgesic, reduces fever, anti-inflammatory, soothes sore throat, antineuralgic, antispasmodic, antiarthritic, antirheumatic, antibacterial, antifungal, antiparasitic, anti-infectious, cleanses the blood, wound healing, relieves chronic skin conditions, soothes itching, relieves snakebites, eases stomachache, anticancer, stimulates appetite, antioxidant, helps release emotional trauma, stress management, encourages self-confidence, reduces hysteria and rage

CAUTIONS:
M. glomerata (Leaves, CO CT)
■ May interact with aspirin, blood pressure, antiplatelet, and anticoagulant medications, and increase the risk of bleeding among people with bleeding disorders, due to caryophyllene-oxide content.[1802]

All other CTs
■ None currently known. Exact GC-MS composition should be known before use to rule out the presence of high amounts of coumarin. Only 5.4% was reported in one of the samples in the literature, suggesting it is not likely to be present in sufficient quantities to warrant oral and drug interaction cautions.

SELECTED EVIDENCE:
- □ The germacrene D CT of guaco (*M. glomerata*) was ineffective against *C. albicans*.[1803]

HAIRY BEGGARTICKS
(Blackjack, Spanish Needles)

Bidens pilosa L., *B. odorata* Cav., *B. pilosa* L. var. *bimucronata* (Turcz.) Sch. Bip., *B. pilosa* L. var. *minor*, *B. pilosa* L. var. *pilosa*, *B. pilosa* L. var. *radiata*

FAMILY: Asteraceae
NOTE: Middle
AROMA INTENSITY: Medium
AROMA: Earthy, woody, mildly sweet and floral
COMMON EXTRACTION METHOD: Steam distilled from the leaves, stems, flowers, or roots
POSSIBLE SUBSTITUTE OILS: Copaiba, ylang ylang, spruce (black), black pepper
BLENDS WELL WITH: Basil, cardamom, cassia, clove, coriander, frankincense, lavender, lemon, marjoram, nutmeg, palo santo, patchouli, pine, ravensara, rosemary, sandalwood, ylang ylang

SUBCELLULAR LOCALIZATION | EPIGENETIC INFLUENCE: Currently unknown | Currently unknown
RECOMMENDED DILUTION RANGE: 2%–10%

PRIMARY CONSTITUENTS:[1804,1805,1806,1807,1808]

B. pilosa (Leaves)	
Unidentified Constituent	24.4%
Beta-Caryophyllene	12.6%
Beta-Cubebene	11.7%
Elixene	5.1%
Thujopsene	4.6%
Cadinene*	3.8%

Var. Minor (Flowers, Stems, or Leaves)	
1-Phenylhepta-1,3,5-Triyne	30.7%–48.0%
Germacrene D	7.2%–21.5%
1-Phenylhept-5-Ena-1,3-Diyne	13.3%–20.9%
Beta-Copaene	1.1%–10.2%
Beta-Caryophyllene	6.1%–7.5%
Beta-Phellandrene	tr–3.5%
Alpha-Pinene	0.1%–3.2%

Var. Radiata (Leaves, AP CT)	
Alpha-Pinene	14.7%
Beta-Caryophyllene	13.5%
Beta-Ocimene*	12.8%
Cadinene*	10.1%
Megastigmatrienone	7.1%
Cubebene*	4.8%
Linalool	4.4%
Bourbonene*	3.9%
Para-Cymen-8-ol	3.6%
Caryophyllene Oxide	3.5%

* Correct isomer not identified.

Var. Minor (Roots)	
Bornyl Acetate	24.1%
Octadecadienol	23.8%
7-Phenylhept-2-Ena-4,6-Diynyl Acetate	22.5%
Hexadecanol	7.1%
Alpha-Cubebene	3.0%

Var. Radiata (Leaves, BC CT)	
Beta-Caryophyllene	10.9%
τ-Cadinene	7.8%
Megastigmatrienone	5.4%

Var. Radiata (Flowers)	
τ-Cadinene	6.1%
Alpha-Pinene	6.0%
Beta-Caryophyllene	5.1%

OTHER CONSTITUENTS: *B. pilosa (Leaves)*—Linalool, cedrene, alpha-cadinol, globulol, nerolidol, 1-hexanol, 2-hexen-1-ol, 2hexenal; *Var. Minor (Flowers, Stems, or Leaves)*—(E)-Beta-ocimene, linalool, delta-elemene, alpha-humulene, bicyclogermacrene, delta-cadinene, gamma-cadinene, (E)-nerolidol, caryophyllene oxide, alpha-cadinol, 7-phenylhept-2-ena-4,6-diynyl acetate; *Var. Minor (Roots)*—Sabinene, cis-chrysanthenyl acetate, eugenol isobutyrate, hexadecyl acetate; *Var. Radiata (Leaves, BC CT)*—Alpha-Pinene, cis-hexenyl acetate, (E)-beta-ocimene, (Z)-beta-ocimene, T-muurolene, beta-bourbonene, beta-elemene, beta-cubebene, beta-farnesene, alpha-humulene, isoledene, (–)-beta-cadinene, 2,5,9-trimethylcycloundeca-4,8-dienone, caryophyllene oxide, diphenylenemethane; *Var. Radiata (Flowers)*—Myrcene, delta-3-carene, limonene, (E)-beta-ocimene, (Z)-beta-ocimene, (4E,6Z)-2,6-dimethyl-2,4,6-octatriene, T-muurolene, beta-bourbonene, beta-elemene, beta-cubebene, alpha-humulene, isoledene, caryophyllene oxide, megastigmatrienone, diphenylenemethane; *Var. Radiata (Leaves, AP CT)*—Limonene, alpha-terpinolene, delta-elemene, 1-hexanol, myrcene, delta-3-carene, trans-linalool oxide, cis-linalool oxide, farnesene, isoledene, (E)-nerolidol

PREFERRED COMPOSITION FOR CLINICAL USE:
Insufficient samples/data to determine.

REPORTED THERAPEUTIC PROPERTIES: Antifungal, antibacterial, antimicrobial, anti-inflammatory, analgesic, relieves toothache, antioxidant, antiseptic, liver protective, lowers blood pressure, balances blood-sugar levels, aids oral health and cleanliness, insecticidal, emotionally balancing, releases traumatic emotional memories, fosters courage, reduces mental fatigue

CAUTIONS:
▪ Strongly phototoxic due to the presence of polyacetylenes (i.e., 1-phenylhepta-1,3,5-triyne); do not expose skin to UV rays for up to twenty-four hours following topical application.[1809,1810,1811]

SELECTED EVIDENCE:
- □ *In vitro* research showed that hairy beggarticks leaf and flower essential oil (var. *radiata*; beta-caryophyllene and T-cadinene CTs) inhibited bacteria (*M. flavus, B. subtilis, B. cereus, B. pumilus, E. coli,* and *P. ovalis*) and fungi (*Corticium rolfsii, F. solani,* and *F. oxysporum*).[1812]
- □ A CT of hairy beggarticks leaf essential oil with an unidentified constituent as the primary constituent inhibited periodontopathic *A. actinomycetemcomitans* and *P. gingivalis*, moderately inhibited cariogenic *S. mutans*, and was least effective against *L. acidophilus*.[1813]
- □ Hairy beggarticks (var. *radiata*; CT alpha-pinene) effectively reduced the formation of fungi (Aspergillus spp., Penicillium sp., Fusarium sp., and Mucor spp.) on bambara groundnut and deterred cowpea weevils (*Callosobruchus maculatus*) from infesting the stored food.[1814]
- □ The alpha-pinene CT of hairy beggarticks (var. *radiata*) exhibited good antioxidant capacity in the DPPH, RP, and beta-carotene methods.[1815]

HAIRY LADY'S MANTLE (Lady's Mantle)

Alchemilla xanthochlora Rothm., *A. mollis* Rothm., *A. speciosa* Rothm.

FAMILY: Rosaceae
NOTE: Middle
AROMA INTENSITY: Medium
AROMA: Green, floral, woody
COMMON EXTRACTION METHOD: Steam distilled or hydrodistilled from the aerial parts
POSSIBLE SUBSTITUTE OILS: Lavender, helichrysum
BLENDS WELL WITH: Bergamot, chamomile (German, Roman), clary sage, clove, cypress, geranium, grapefruit, lavandin, lavender, lemon, lime, melaleuca (tea tree), neroli, orange, oregano, rosemary, tangerine, vetiver, ylang ylang
SUBCELLULAR LOCALIZATION | EPIGENETIC INFLUENCE: Currently unknown | Currently unknown
RECOMMENDED DILUTION RANGE: 3%–50%; neat for some conditions

PRIMARY CONSTITUENTS:[1816]

(Z)-3-Hexenol	11.2%
Linalool	10.4%
Oct-1-en-3-ol	9.0%
Nonanal	7.8%
Myrtenol	4.7%
Hexadecanoic Acid	4.6%
(Z)-3-Hexenyl Acetate	3.9%
Alpha-Terpineol	3.1%

OTHER CONSTITUENTS: Not listed in study.

PREFERRED COMPOSITION FOR CLINICAL USE:
Insufficient data/samples to determine.

REPORTED THERAPEUTIC PROPERTIES: Analgesic, anti-inflammatory, antirheumatic, astringent, diuretic, antiparasitic, balances menstruation, relieves painful menstruation, relieves menopausal symptoms, reduces excess bleeding, astringent, wound healing, reduces the appearance of blemishes, relieves diarrhea, relaxing and calming, stress management, relieves anxiety

CAUTIONS:
- None currently known.

SELECTED EVIDENCE:
- None found.

HEADACHE TREE (California Laurel, Oregon Myrtle)

Umbellularia Californica (Hook. & Arn.) Nutt.

FAMILY: Lauraceae
NOTE: Middle
AROMA INTENSITY: Strong
AROMA: Pungent, medicinal, spicy, herbaceous
COMMON EXTRACTION METHOD: Steam distilled from the leaves; may also be distilled from the bark
POSSIBLE SUBSTITUTE OILS: Bay laurel
BLENDS WELL WITH: Bay laurel, cinnamon, clary sage, clove, coriander, eucalyptus, fennel, ginger, lavender, patchouli, rosemary, tamala, thyme
SUBCELLULAR LOCALIZATION | EPIGENETIC INFLUENCE: Currently unknown | Currently unknown
RECOMMENDED DILUTION RANGE: < 0.5%

PRIMARY CONSTITUENTS:[1817,1818]

Umbellulone	36.7%–41.0%
1,8-Cineole	19.5%–22.0%
Sabinene	0.1%–10.2%
Methyl Eugenol	0.0%–8.4%
Thymol	0.0%–7.8%
Terpinen-4-ol	0.0%–6.6%
Alpha-Terpineol	0.0%–6.5%

OTHER CONSTITUENTS: Para-cymene, delta-terpineol, beta-bisabolene, (E)-alpha-bisabolene, cumin alcohol

PREFERRED COMPOSITION FOR CLINICAL USE:
Insufficient samples/data to determine.

REPORTED THERAPEUTIC PROPERTIES: Analgesic, antirheumatic, nervine, antineuralgic, aids digestion, relieves intestinal cramps, antiseptic, antibacterial, kills lice, insecticidal, insect repellent, stimulating, encourages attention and focus

CAUTIONS:

- Not recommended for essential oil therapy due to significant umbellulone (a monoterpene ketone) content. Inhalation of umbellulone has been associated with cluster headaches or migraines by diffusing to nerve terminals and triggering a cascade of chemical events that activate the Transient receptor potential ankyrin (TRPA1) and the trigeminovascular system.[1819,1820]

SELECTED EVIDENCE:

- □ Topical application of headache tree essential oil to the cutaneous microvasculature of frogs caused sustained, but reversible constriction of blood vessels.[1821]

HIBAWOOD (ood, Hiba Arborvitae, False Arborvitae, Aomori Hiba)

Thujopsis dolabrata (Thunb. ex L. f.) Siebold & Zucc., *T. dolobrata* Sieb. et Zucc. var. *hondai Makino*

FAMILY: Cupressaceae
NOTE: Top
AROMA INTENSITY: Strong
AROMA: Woody, fresh, sweet, cedar-like
COMMON EXTRACTION METHOD: Steam distilled or hydrodistilled from the wood; may also be distilled from the leaves (needles)
POSSIBLE SUBSTITUTE OILS: Texas cedarwood, Virginia cedarwood, Formosan cypress
BLENDS WELL WITH: Bergamot, cassia, cinnamon, clary sage, cypress, frankincense, juniper berry, labdanum, pine, rosewood, rosemary, Texas cedarwood, vetiver, Virginia cedarwood, ylang ylang
SUBCELLULAR LOCALIZATION | EPIGENETIC INFLUENCE: Currently unknown | Currently unknown

RECOMMENDED DILUTION RANGE: 3%–33%; 50% for some conditions

PRIMARY CONSTITUENTS:[1822,1823,1824]

Wood	
(Z)-Thujopsene	16.5%–52.8%
Widdrol	2.1%–13.8%
Cedrol	5.5%–6.1%
Valerenal	0.0%–5.5%
Beta-Costol	0.0%–5.3%
Cuparene	1.7%–5.0%
Palustrol	0.0%–4.8%
Beta-Herbertenol	0.0%–4.7%
Beta-Himachalene	0.0%–4.1%
Hongkui<121,1682>	0.0%–3.6%
Alpha-Bisabolol	0.0%–3.1%
Unidentified Compound	0.0%–3.0%

Leaves (var. hondai)	
Sabinene	35.7%
Gamma-Terpinene	13.5%
Alpha-Pinene	7.7%
Myrcene	7.7%
Alpha-Terpinene	6.7%
Alpha-Thujene	5.2%
Para-Cymene	5.1%
Limonene	4.8%
Terpinen-4-ol	4.5%

OTHER CONSTITUENTS: *Wood*—Terpinen-4-ol, carvacrol, beta-elemene, 2-epi-beta-funebrene, alpha-cedrene, beta-cedrene, beta-chamigrene, alpha-chamigrene, gamma-cadinene, delta-cadinene, cuprenene isomer, alpha-cuprenene, gamma-cuprenene, delta-cuprenene, hiba<81,1571>, allo-cedrol, widdrol, unidentified sesquiterpene, cedr-8-en-15-ol, Ucypress, hongjui<203,1685>, Ujuniper, 4-cuprenene-1-ol, (Z)-thujopsenal, gamma-costol, alpha-costol, hongkui<105,1765>, hiba<148,1795>, hinokitiol, ylanga-2,4(15)-diene, alpha-microbiotene, alpha-longipinene, beta-acorenol, pogostol, clareolide, cyperena210rticulcan-1-ene; *Leaves (var. hondai) Leaves (var. hondai)*—Delta-3-carene, terpinolene

PREFERRED COMPOSITION FOR CLINICAL USE:

Constituent	*Wood*
(Z)-Thujopsene	45%–65%
Cedrol	5%–12%
Widdrol	2%–12%
Cuprenene Isomers	2%–10%
Cuparene	0.5%–8%
Hinokitiol (Beta-Thujaplicin)	0.1%–5%

REPORTED THERAPEUTIC PROPERTIES: Anticancer, antibacterial, antifungal, antiviral, anti-inflammatory, antiarthritic, odor elimination, reduces the appearance of skin blemishes, soothes chronic skin conditions and itching, wound healing, **antiallergic**, encourages hair growth, insect repellent, insecticidal, promotes restful sleep, relaxing, relieves anxiety, stress management, emotionally balancing, reduces fear

CAUTIONS:

Wood

- May weakly interfere with the enzymes responsible for metabolizing medications (NSAIDs, proton-pump inhibitors, acetaminophen, antiepileptics, immune modulators, blood-sugar medications, blood pressure medications, antidepressants, antipsychotics, diabetic medications, antihistamines, antibiotics, and anesthetics) due to thujopsene, cedrol, and beta-cedrene content.[1825]
- Dilute to less than 1% if hinokitiol reaches or exceeds 10% total content. Hinokitiol can cause skin irritation and allergic contact dermatitis.[1826,1827]

Leaves

- None currently known.

SELECTED EVIDENCE:

- Hibawood essential oil inhibited human gastric cancer tumor formation in mice.[1828] The study authors also tested isolated hinokitiol and found that it exhibited weaker activity than the whole oil suggesting synergy between multiple constituents provides the anticancer activity of hiba wood.
- Topical application of 1.5%, 3%, and 6% hibawood essential oil (var. *hondai*; with hinokitiol as the major constituent) for every two days for two weeks reduced eczema-like skin lesions in mice by reducing the infiltration of mast cells into the lesion area and inhibiting the release of beta-hexosaminidase, histamine, and IgE (immunoglobulin E).[1829]
- Inhalation of hibawood essential oil (50 mcL placed on filter paper for 30 minutes following a stressor) reduced chronic stress in rats.[1830]
- Hiba leaf oil (var. *hondai*) essential oil exhibited good antioxidant activity in the DPPH assay.[1831] The study authors also reported that gamma-terpinene exhibited significant antioxidant activity.

HIMALAYAN ANGELICA (Smooth Angelica)

Angelica glauca Edgew.

FAMILY: Apiaceae
NOTE: Middle
AROMA INTENSITY: Medium
AROMA: Green, fresh, herbaceous, slightly sweet and spicy
COMMON EXTRACTION METHOD: Steam distilled from the roots or aerial parts
POSSIBLE SUBSTITUTE OILS: Lovage, celery, angelica
BLENDS WELL WITH: Angelica, basil, bay laurel, cardamom, cassia, chamomile (German, Roman), cinnamon, clove, frankincense, geranium, lavender, myrrh, patchouli, sandalwood, tamala
SUBCELLULAR LOCALIZATION | EPIGENETIC INFLUENCE: Currently unknown | Currently unknown
RECOMMENDED DILUTION RANGE: 3%–20%; 50% for some conditions

PRIMARY CONSTITUENTS:[1832,1833,1834,1835,1836,1837]

Roots		*Aerial Parts*	
(Z)-Ligustilide	5.2%–74.1%	Alpha-Phellandrene	13.0%–13.5%
(Z)-3-Butylidenephthalide	1.0%–32.8%	Trans-Carveol	12.0%–12.0%
Beta-Phellandrene	0.0%–15.3%	Beta-Pinene	11.0%–11.7%
(Z)-3-Methyl-2-Octene	0.0%–13.8%	Alpha-Thujene	0.0%–7.5%
(E)-Ligustilide	0.5%–12.4%	Caryophyllene Oxide	7.0%–7.2%
Methyl Eugenol	0.0%–10.7%	Beta-Caryophyllene	7.0%–7.0%
Limonene	0.0%–10.2%	Gamma-Terpinene	6.5%–6.7%
(Z)-Beta-Ocimene	0.0%–9.3%	(E)-Nerolidol	0.0%–6.5%
Myrcene	0.0%–8.3%	Beta-Bisabolene	5.2%–5.7%
Alpha-Phellandrene	0.0%–6.9%	Germacrene D	4.5%–4.5%
Methyl Chavicol	0.0%–6.3%	(E,E)-Farnesol	0.0%–3.5%

Camphene	0.0%–6.0%
(E)-3-Butylidenephthalide	0.2%–5.9%
Beta-Pinene	0.0%–5.3%
Beta-Phellandrene	0.0%–4.6%
Caryophyllene Oxide	0.0%–4.0%
Citronellyl Acetate	0.0%–3.4%
Beta-Caryophyllene	0.0%–3.4%
Citronellol	0.0%–3.0%

Note: The root oil could be subdivided into multiple chemotypes but it is combined in this work.

OTHER CONSTITUENTS: *Roots*—Para-cymene, alloaromadendrene, germacrene D, beta-eudesmol; *Aerial Parts*—Alpha-pinene, sabinene, delta-3-carene, cis-ocimene, camphor, cis-piperitol, p-cymene-8-acetate, beta-elemene, alloaromadendrene, beta-eudesmol

PREFERRED COMPOSITION FOR CLINICAL USE:
Insufficient samples/data to determine.

REPORTED THERAPEUTIC PROPERTIES: Anticancer, brain protective, antiallergic, aids respiratory function, eases asthma and respiratory disorders, expels excess mucus, stimulates appetite, expels excess gas, aids digestion, relieves stomachache, relieves constipation, aids gallbladder and liver function, aids cardiovascular function, improves circulation, analgesic, relieves headache, antiarthritic, reduces fever, eases chronic skin conditions, antibacterial, balances menstruation, encourages restful sleep, aids memory function, combats hysteria and irrationality, releases traumatic memories

CAUTIONS:

- Avoid during pregnancy and lactation due to the estrogenic activity of (Z)-ligustilide and *A. sinensis* (also with significant ligustilide content). (Z)-ligustilide may prolong menstruation and stimulate or relax the uterus.[1838,1839,1840]
- May interact with diabetic medications and cause low blood-sugar levels based on (Z)-ligustilide content.[1841]
- May interact with aspirin, blood pressure, antiplatelet, and anticoagulant medications, and increase the risk of bleeding among people with bleeding disorders, due to ligustilide content. (Z)-ligustilide inhibits platelet aggregation without affecting coagulation time of peripheral blood.[1842]
- May interact with levodopa (L-dopa) and increase its cytotoxic effects (causes neuronal cell death) due to (Z)-ligustilide content.[1843]

SELECTED EVIDENCE:

- Oral administration of Himalayan angelica (200 mcL/kg body weight) with alpha-pinene as the major constituent significantly decreased blood eosinophil count, serum IgE (Immunoglobulin E), the number of eosinophils and neutrophils in respiratory fluids (bronchoalveolar lavage fluid), and relaxed airways constricted (bronchorelaxation) by histamine and ovalbumin in guinea pigs and mice, suggesting it reverses airway changes caused by allergies.[1844] The number of eosinophils increases during inflammatory and allergic reactions (including asthma) involving the lungs. Neutrophils can produce large amounts of histamine (the inflammatory compound released during allergic reactions that causes runny nose, watery eyes, and wheezing). IgE is an antibody produced when your immune system overreacts to a foreign substance and plays a key role in asthma.

HIMALAYAN CYPRESS (Bhutan Cypress)

Cupressus torulosa D.Don

FAMILY: Cupressaceae
NOTE: Middle-Base
AROMA INTENSITY: Medium
AROMA: Woody, balsamic, fresh, sweet
COMMON EXTRACTION METHOD: Hydrodistilled or steam distilled from the wood or needles; may also be steam distilled from the cones
POSSIBLE SUBSTITUTE OILS: Japanese cedarwood, cypress, cedarwood (Atlas), tea tree, plai, pink pepper, neroli, black pepper, Turkish pine
BLENDS WELL WITH: Amyris, balsam fir, cardamom, cedarwood, frankincense, geranium, lavandin, lavender, marjoram, myrrh, neroli, palmarosa, patchouli, petitgrain, rosewood, sandalwood, vetiver
SUBCELLULAR LOCALIZATION | EPIGENETIC INFLUENCE: Currently unknown | Currently unknown
RECOMMENDED DILUTION RANGE: 3%–Neat (unless umbellulone is present, then < 0.5%)

PRIMARY CONSTITUENTS:[1845,1846,1847,1848,1849,1850]

Needles (Alpha-Pinene CT)	
Alpha-Pinene	16.0%–45.4%
Delta-3-Carene	6.5%–38.3%
Limonene	3.1%–23.8%
Sabinene	tr–19.2%
Cedrol	0.0%–12.5%
Umbellulone	0.0%–10.4%
Terpinen-4-ol	tr–8.4%
Alpha-Terpinolene	0.6%–6.3%
Aromadendrene	0.0%–6.3%
Alpha-Terpinyl Acetate	0.0%–5.9%
Beta-Myrcene	2.1%–4.6%
Gamma-Terpinene	0.0%–4.5%
Thujene-2-one	0.0%–3.6%
Alpha-Terpinene	tr–3.1%
Beta-Cubebene	0.0%–3.1%

Needles (Sabinene CT)	
Sabinene	16.5%–23.4%
Terpinen-4-ol	13.0%–15.0%
Umbellulone	0.0%–11.7%
Alpha-Pinene	8.0%–9.8%
Delta-3-Carene	0.0%–7.5%
Cis-Muurola-4(14),5-Diene	0.0%–6.8%
Limonene	2.7%–6.0%
Gamma-Terpinene	4.7%–5.9%
Myrcene	3.5%–3.9%

Leaves (Umbellulone CT)	
Umbellulone	16.3%–21.6%
Terpinen-4-ol	0.1%–19.6%
Alpha-Pinene	9.6%–17.8%
Limonene	13.3%–14.6%
Sabinene	5.9%–9.3%
Gamma-Terpinene	2.0%–6.9%
Alpha-Terpinene	2.0%–5.3%
Delta-3-Carene	0.2%–4.2%
Myrcene	3.3%–4.1%
Alpha-Terpinolene	0.2%–3.4%
Abietatriene	0.2%–3.3%

Leaves (Terpinen-4-ol CT)	
Terpinen-4-ol	13.7%–22.9%
Alpha-Pinene	9.7%–19.0%
Limonene	5.0%–15.1%
Umbellulone	0.0%–9.6%
Gamma-Terpinene	4.3%–7.4%
Alpha-Terpinene	3.2%–5.5%
Delta-3-Carene	0.9%–5.4%
Myrcene	3.5%–3.9%
Alpha-Terpinolene	2.1%–3.3%

Wood	
Alpha-Copaene	29.9%
Cyperene	13.1%

Alpha-Terpinene	3.0%–3.6%	Trans-Caryophyllene	12.1%
Cis-Muurola-4(14),5-diene	0.0%–3.4%	Delta-Cadinene	6.9%
Alpha-Cadinol	0.0%–3.2%	Cyperol	3.6%
		Cyperotundone	3.2%

Leaves (Limonene CT)		*Cones*	
Limonene	18.4%–35.9%	Umbellulone	26.7%
Alpha-Pinene	10.9%–13.8%	Alpha-Pinene	14.7%
Terpinen-4-ol	3.4%–11.9%	Sabinene	14.3%
Delta-3-Carene	4.0%–11.1%	Limonene	13.9%
Beta-Cubebene	0.0%–7.1%	Gamma-Terpinene	5.5%
Sabinene	2.1%–4.1%	Myrcene	4.6%
Alpha-Terpinyl Acetate	2.0%–3.8%		
Myrcene	2.4%–3.2%		
Terpinen-4-ol	2.4%–3.0%		

OTHER CONSTITUENTS: *Needles (Alpha-Pinene CT)*—Alpha-thujene, alpha-fenchene, camphene, beta-pinene, para-cymene, thujene-2-one, alpha-terpineol, bornyl acetate, beta-terpinyl acetate, alpha-cubebene, ar-curcumene, germacrene D, pimaradiene, manool oxide, abienol, sempervirol, trans-p-menth-2-en-1-ol, cis-p-menth-2-en-1-ol; *Needles (Sabinene CT)*—Alpha-thujene, beta-pinene, terpinolene, alpha-terpineol, bornyl acetate, alpha-terpinyl acetate, epizonarene, cis-calamenene, cis-muurola-5-en-4-alpha-ol, gamma-cadinol, phyllocladene, 13(16),14-labdien-8-ol, totarol; *Leaves (Umbellulone CT)*—Alpha-thujene, alpha-phellandrene, para-cymene, fenchone, linalool, trans-p-menth-2-en-1-ol, p-cymen-8-ol, alpha-terpineol, thymol, alpha-terpinyl acetate, alpha-cadinol, (Z)-nuciferol, isohibaene, isophyllocladene, abietadiene; *Leaves (Terpinen-4-ol CT)*—Alpha-thujene, para-cymene, 2-nonanone, trans-p-menth-2-en-1-ol, cis-p-menth-2-en-1-ol, camphor, alpha-terpineol, bornyl acetate, alpha-terpinyl acetate, alpha-cubebene, beta-cubebene, epizonarene; *Leaves (Limonene CT)*—Alpha-thujene, beta-pinene, alpha-terpinene, para-cymene, gamma-terpinene, 2-nonanone, alpha-terpinolene, cis-p-menth-2-en-1-ol, camphor, alpha-terpineol, bornyl acetate, alpha-cubebene, epizonarene; *Wood*—Alpha-cubebene, beta-cubebene, beta-elemene, alpha-gurjunene, alpha-humulene, alloaromadendrene, gamma-gurjunene, trans-cadina-1(6),4-diene, gamma-selinene, beta-selinene, valencene, alpha-selinene, alpha-panasinensen, isospathulenol, caryophyllene oxide, 5-oxo-isolongifolene, zierone, valerenal, alpha-cyperone; *Cones*—Alpha-thujene, alpha-terpinene, para-cymene, cis-sabinene hydrate, linalool, fenchone, terpinen-4-ol, (E)-isoamyl cinnamate, (E)-nuciferol, abietatriene

PREFERRED COMPOSITION FOR CLINICAL USE:
Insufficient samples/data to determine.

REPORTED THERAPEUTIC PROPERTIES: Analgesic, anti-inflammatory, antispasmodic, nervine, antineuralgic, antiarthritic, antirheumatic, anticancer, relieves painful menstruation, alleviates symptoms of menopause, aids circulation, antibacterial, diuretic, relieves cough, reduces the appearance of scars, wound healing, insecticidal, insect repellent, relaxing, promotes restful sleep, stress management, promotes feelings of security

CAUTIONS:

■ Due to the potential for umbellulone to be present, Himalayan cypress should not be used unless a full GC-MS is available to determine if it is present and in what percentage. Leaf and cone samples from India,

Greece, and the Ukraine had varying levels of umbellulone, whereas no umbellulone was detected in the wood sample from Nepal. If the sample contains greater than or equal to 10% of umbellulone, it would not be recommend for essential oil therapy. Inhalation of umbellulone has been associated with cluster headaches or migraines by diffusing to nerve terminals and triggering a cascade of chemical events that activate the Transient receptor potential ankyrin (TRPA1) and the trigeminovascular system.[1851,1852]

SELECTED EVIDENCE:

- *In vitro* research shows that the alpha-pinene CT of Himalayan cypress essential oil inhibits *M. luteus, S. aureus, B. subtilis, B. cereus, Alcaligenes denitrificans, P. aeruginosa,* and *P. alcaligenes.*[1853]
- Himalayan cypress essential oil CT alpha-pinene was moderately toxic to the *Aedes albopictus* mosquito, but was an inadequate repellent.[1854]

HIMALAYAN SILVER FIR (Abies Oil, East Himalayan Fir)

Abies spectabilis (D. Don) Spach, *A. spectabilis* subsp. *densa* (Griff.) Silba, *A. spectabilis* subsp. *langtangensis* (Silba) Silba

FAMILY: Pinaceae
NOTE: Top-Middle
AROMA INTENSITY: Medium
AROMA: Piney, balsamic, fresh, sweet
COMMON EXTRACTION METHOD: Steam distilled from the needles or twigs
POSSIBLE SUBSTITUTE OILS: Balsam fir, spruce (black), pine
BLENDS WELL WITH: Balsam fir, birch, blue spruce, cassia, cedarwood, chamomile (German, Roman), cypress, frankincense, galbanum, lavender, lemon, myrtle, palo santo, pine, sandalwood, spruce (black), tsuga, white fir, wintergreen
SUBCELLULAR LOCALIZATION | EPIGENETIC INFLUENCE: Currently unknown | Currently unknown
RECOMMENDED DILUTION RANGE: 5%–Neat

PRIMARY CONSTITUENTS:[1855]

Langtangensis (Needles)	
Alpha-Pinene	33.4%
Bornyl Acetate	10.9%
Selen-11-en-4β-ol	9.4%
Alpha-Bisabolol	4.0%
Delta-3-Carene	3.3%
Limonene	3.3%
Caryophyllene Oxide	3.2%
Camphene	3.0%
Beta-Pinene	3.0%

Densa (Needles)	
Limonene	26.9%
Alpha-Pinene	22.2%
Bornyl Acetate	10.3%
Beta-Phellandrene	9.7%
Camphene	7.1%
Beta-Pinene	6.5%
Delta-3-Carene	3.1%

OTHER CONSTITUENTS: *Densa*—Tricyclene, myrcene, alpha-phellandrene, terpinolene, alpha-terpineol, thymol methyl ether, alpha-bourbonene, trans-caryophyllene, delta-cadinene, selin-11-en-4beta-ol; *Langtangensis*—Myrcene, para-cymene, trans-pinocarveol, trans-verbenol, borneol, m-cymen-8-ol, p-cymen-8-ol, alpha-terpineol, verbenone, alpha-bourbonene, trans-cadina-1(6),4-diene, beta-chamigrene, trans-beta-guaiene, beta-bisabolene, trans-nerolidol, humulene epoxide II, alpha-cadinol

PREFERRED COMPOSITION FOR CLINICAL USE:
Insufficient samples/data to determine.

REPORTED THERAPEUTIC PROPERTIES: Antirheumatic, relieves arthritis, analgesic, anti-inflammatory, alleviates cough, aids respiratory function, relieves sinusitis, expels excess mucus, aids circulation, reduces anxiety, grounding, emotionally stabilizing, antidepressant

CAUTIONS:
- None currently known.

SELECTED EVIDENCE:
- □ None found.

HIMALAYAN THYME (Wild Thyme, Mother of Thyme)

Thymus linearis Benth.

FAMILY: Lamiaceae
NOTE: Middle-Top
AROMA INTENSITY: Strong
AROMA: Herbaceous, medicinal, fresh
COMMON EXTRACTION METHOD: Steam distilled from the aerial parts
POSSIBLE SUBSTITUTE OILS: Thyme, ajowan
BLENDS WELL WITH: Ajowan, bay laurel, bergamot, camphor, clary sage, copaiba, cypress, eucalyptus, geranium, grapefruit, lavender, lemon, juniper, marjoram, melaleuca (tea tree), orange, oregano, petitgrain, pin216rticulatara, rose, rosemary, sage, Spanish sage, thyme
SUBCELLULAR LOCALIZATION | EPIGENETIC INFLUENCE: Currently unknown | Currently unknown
RECOMMENDED DILUTION RANGE: 1.5%–20%; 50% for some conditions

PRIMARY CONSTITUENTS:[1856,1857,1858]

Thymol CT	
Thymol	34.3%–67.3%
Para-Cymene	1.8%–27.4%
Gamma-Terpinene	1.9%–24.4%
Para-Cymen-8-ol	0.0%–6.7%
Borneol	0.0%–5.8%
Carvacrol	0.9%–5.4%
Beta-Bisabolene	0.0%–5.1%
Thymol Acetate	0.0%–4.6%
Thymol Methyl Ether	0.0%–3.7%
Alpha-Terpinene	0.0%–3.6%
Bicyclogermacrene	0.0%–3.0%

Germacrene D CT	
Germacrene D	65.1%
Gamma-Terpinene	19.4%
Beta-Caryophyllene	4.4%

OTHER CONSTITUENTS: *Thymol CT*—Alpha-thujene, alpha-pinene, camphene, 3-octanone, myrcene, limonene, cis-sabinene hydrate, terpinen-4-ol, beta-caryophyllene; *Germacrene D CT*—Alpha-humulene, gamma-cadinene

PREFERRED COMPOSITION FOR CLINICAL USE:

Constituent	*Thymol CT*
Thymol	40%–65%
Gamma-Terpinene	10%–25%
Para-Cymene	5%–20%
Carvacrol	0.5%–6%
Borneol	0.1%–6%
Thymol Methyl Ether	0.5%–5%
Alpha-Terpinene	tr–5%
Beta-Caryophyllene	tr–5%
Thymol Acetate	tr–5%
Alpha-Thujene	tr–4%

REPORTED THERAPEUTIC PROPERTIES: Relieves fever, eases cough, aids respiratory function, antiseptic, antibacterial, antifungal, anti-infectious, antiviral, aids digestion, promotes oral health and cleanliness, relieves painful menstruation, regulates menstruation, relieves chronic skin conditions and itching, antioxidant, anticancer, analgesic, anti-inflammatory, antiarthritic, insecticidal, reduces mental fatigue, aids concentration and focus, promotes self-confidence

CAUTIONS:

Thymol CT

▪ Oral caution. Thymol has a longer half-life (the time it takes for half of the medication to metabolize or excrete half of the dosage) than most essential oil constituents and should not be administered orally for long periods of time.[1859] Thymol is a monoterpene phenol. Reports of fatalities in children who consumed 50 to 200 mg of phenols has been reported.[1860] It is recommended that chemotypes with high levels of thymol be limited to 10 drops per day orally for adults with a two- to seven-day break after twenty-one days of use; or limit ingestion to five or six days weekly.

▪ May interact with aspirin, blood pressure, antiplatelet, and anticoagulant medications, and increase the risk of bleeding among people with bleeding disorders, due to thymol and carvacrol content.[1861,1862]

▪ May interact with anticholinergic (drugs used for asthma, incontinence, gastrointestinal cramps, muscular spasms, depression, and sleep disorders) and cholinergic medications (drugs used to reduce urinary retention, diagnose myasthenia gravis, and for glaucoma), based on thymol and carvacrol content.[1863,1864]

▪ May interact with antibiotics or antifungals and possibly enhance their effects based on thymol and carvacrol content.[1865,1866,1867,1868,1869,1870,1871,1872]

▪ May irritate mucous membranes (eyes, mouth, nasal passages, vagina, rectum).

Germacrene D CT

▪ None currently known.

SELECTED EVIDENCE:

- None found.

HONEYSUCKLE | HONEYSUCKLE ABSOLUTE (European Honeysuckle, Woodbine)

Lonicera periclymenum L., *Lonicera* spp.

FAMILY: Caprifoliaceae
NOTE: Middle
AROMA INTENSITY: Medium-Strong
AROMA: Intensely sweet, rich, floral
COMMON EXTRACTION METHOD: Solvent extracted (Absolute) or hydrodistilled (HD) from the flowers; uncommonly distilled from the leaves or stems or cold macerated from the flowers
POSSIBLE SUBSTITUTE OILS: Jasmine, tuberose, lavender, ylang ylang, sandalwood
BLENDS WELL WITH: Bergamot, chamomile (German, Roman), clary sage, clove, coriander, frankincense, geranium, grapefruit, jasmine, lavender, lemon, lime, mandarin, neroli, orange, palmarosa, patchouli, petitgrain, rose, rosewood, sandalwood, tangerine, tuberose, vetiver, ylang ylang
SUBCELLULAR LOCALIZATION | EPIGENETIC INFLUENCE: Currently unknown | Currently unknown
RECOMMENDED DILUTION RANGE: *Absolute*—1%–3%; *EO*—3%–33%; 50% for some conditions

PRIMARY CONSTITUENTS:[1873,1874,1875,1876,1877,1878,1879]

L. caprifolium (Absolute)

Linalool	75.0%
Germacrene D	5.8%
Indole	3.3%
Alpha-Farnesene	3.0%

L. caprifolium (HD)

Germacrene D	33.1%
Farnesol*	26.2%
(Z)-3-Hexenyl Tiglate	5.7%
3-Hexyl Benzoate	5.2%
Farnesyl Acetate	5.2%

L. japonica (HD)

Farnesol*	41.5%–51.0%
Germacrene D	11.3%–19.7%
Linalool	9.6%–11.6%
(Z)-3-Hexenyl Tiglate	2.6%–6.8%
Farnesyl Acetate	0.1%–3.4%

L. japonica (HD, (Z,Z)-Farnesol CT)

(Z,Z)-Farnesol	16.2%
Linalool	11.0%
Methyl Linoleate	9.0%
Hexadecanoic Acid	7.1%

L. japonica (Absolute, CT Linalool)

Linalool	40.4%
Germacrene D	24.1%
Alpha-Farnesene	10.3%
(Z)-3-Hexenyl Tiglate	3.8%
(Z)-Jasmone	3.6%

L. japonica (Absolute, CT Germacrene D)

Germacrene D	31.1%–53.2%
Alpha-Farnesene	0.9%–29.4%
Linalool	2.6%–19.4%
Benzyl Benzoate	0.2%–8.3%
Hexenyl Benzoate	0.5%–7.7%
Methyl Palmitate	2.7%–6.1%
(Z)-3-Hexenyl Tiglate	1.5%–5.3%
Linalool Oxide II	2.1%–5.0%
Phenylacetaldehyde	1.4%–3.8%

L. fragrantissima (Absolute)

10-Nonacosanol	18.1%–24.1%
Loganetin	2.1%–19.1%
Nonacosane	14.4%–15.0%
Hexyl Tetracosanoate	5.1%–7.6%
Geraniol	5.9%–7.1%
Hexyl Triacontanoate	5.0%–6.8%

(Z)-Hexenyl Tiglate	6.9%
Geraniol	4.8%
Germacrene D	4.8%
Fenchyl Acetate	4.1%
(E,Z)-Farnesol	3.9%
Methyl Hexadecanoate	3.6%
Benzyl Benzoate	3.4%
Nerol	3.2%

L. reticulaica (HD, Nerolidol CT)

(E)-Nerolidol	16.3%
Caryophyllene Oxide	11.2%
Linalool	8.6%
Para-Cymene	7.4%
Hexadecanoic Acid	6.4%
Eugenol	6.1%
Geraniol	5.0%
Trans-Linalool Oxide	3.8%

L. macranthoides (HD)

Dibutyl Phthalate	23.1%
Dimethyl Phthalate	14.1%
Hexadecanoic Acid, Ethyl Ester	12.7%
Tetradecanoic Acid	11.9%
Hexadecanoic Acid	4.6%
Geraniol	4.2%
Cis-(–)-Linalool Oxide Pyranoid	3.7%
Nonadecane	3.1%

L. macranthoides (Absolute)

9,12-Octadecanoic Acid	0.0%–27.5%
Hexadecanoic Acid	17.9%–25.7%
9,12-Octadecadien-1-ol	0.0%–24.6%
Dimethyl Phthalate	13.3%–19.3%
Dibutyl Phthalate	16.1%–18.4%
1,2,3-Propanetriol, Monoacetate	0.0%–8.4%
1-Heptacosanol	0.0%–7.4%
9,12,15-Octadecatrienoic acid, methyl ester	0.0%–7.2%
Pentatriacontane	0.0%–4.6%
10-Nonadecanol	0.0%–4.3%
Octadecanoic Acid	2.8%–3.5%
Heneicosane	0.0%–3.2%

* Correct isomer not identified.

Hexyl Docosanoate	5.2%–5.6%
Geranial	1.4%–4.6%
Citronellol	0.5%–3.3%
Unknown	2.6%–3.1%

L. macranthoides (Cold Maceration)

9,12-Octadecadien-1-ol	22.4%
Octadecanoic Acid	18.2%
Hexadecanoic Acid	12.3%
9,12,15-Octadecatrienoic acid, methyl ester	8.0%
Dimethyl Phthalate	7.7%
Dibutyl Phthalate	7.1%
10-Nonadecanol	3.6%

L. japonica (HD, Leaves)

Hexadecanoic Acid	16.0%
Linalool	8.7%
(E,E)-Alpha-Farnesene	7.1%
(Z,Z)-Farnesol	5.5%
Bis(2-Ethylhexyl) Phthalate	4.6%
Nonacosane	3.5%

L. japonica (HD, Stems)

Hexadecanoic Acid	31.4%
Bis(2-Ethylhexyl) Phthalate	5.8%
(E)-Beta-Damascenone	5.5%
Linalool	4.2%
Germacrene D	3.9%
(Z,E)-Alpha-Farnesene	3.6%
m-Xylene	3.6%
(E,E)-Alpha-Farnesene	3.4%
Eugenol	3.1%
2,6-Diisopropylnaphthalene	3.0%

Other Constituents: *L. caprifolium (Absolute)*—Nerol220rticulasmine lactone; *L. caprifolium (HD)*—Linalool, beta-ele220rtic alpha-gurjunene, bicyclogermacrene, delta-cadinene, elemol, nerol220rtic alpha-cadinol, benzyl benzoate, 1-octen-3-ol benzoat220rticul*aponica (HD)*—3-hexen-1-ol, methyl benzoate, nonanal, hexyl tiglate, (Z)-jasmone, geranyl acetone, bicyclogermacrone, delta-cadinene, elemol, nerolidol, hexyl benz220rtic alpha-cadinol, farnesal, benzyl benzoat220rticul*aponica (HD, (Z,Z)-Farnesol CT)*—Heptacosane, nonacosane, 2-pentadecanone, (E,E)-farnesyl acetate, tetradecanoic acid, butyl hexadecanoate, methyl docosanoate, methyl tetracosan220rtic alpha-terpineol, geranyl avetone, delta-cadinene, germacre(E)B, (E)-nerolidol, ledol, 10-epi-alpha-muurolol, T-m220rticol, alpha-cad220rticula *japonica (HD, Nerolidol CT)*—(Z)-4-heptenal, 1-octen-3-ol, 1,8-cineole, acetophenone, benzyl alcohol, cis-linalool oxide, phenylethyl 220rticol, alpha-terpineol, citronellyl acetate, decanal, dodecane, undecanol, geranyl acetate, cis-jasmone, dodecanal, beta-caryophyllene, geranylacetone, aromadendrene, beta-ionone, elemol, spathulenol, globulol, viridiflorol, epiglobulol, juniper 220rticor, alpha-cadinol, (Z,E)-farnesol, tetradecanoic acid, hexahydrofarnesylacetone, pentadecanoic acid, octadecanoic 220rticula *japonica (Absolute, CT Linalool)*—Alpha-terpineol, linalool oxide I, trans-geraniol, germacrene B, hexenyl benzoate, nerolidol, methyl palmi220rticula *japonica (Absolute, CT Germacrene D)*—Methyl phenylacetate, methyl salicylate (< 1.0%), trans-geraniol, cis-jasmone, beta-bourbonene, beta-caryophyllene, geranylacetone, germacrene B, benzyl sal220rticue, indole; *L. fragrantissima (A(E)lute)*—(E)-2-Hexenal, 1-hexanol, l220rticul, indole, geranic acid, hexahydrofarnesyl acetone, hexadecano220rticulates220220leic acid, (E,Z)-9,12-octadecadienoic acid, methyl tetracosanoate, hexyl eicosanoate, triacontane, hentriacontane, unknown compounds; *L. macranthoides (HD)*—Octadecanoic acid, 9,12,15-octadecatrienoic acid, methyl ester, hexadecanoic acid, butyl ester, phenylethyl alcohol, 1-hexadecanol, heptadecane,2,6,10,15-tetra220rticulates220zoic acid,4-formyl-,methyl ester, undecanoic acid, hexadecanoic acid, methyl ester, 12,15-octadecadienoic acid, methyl ester, 9,12,15-octadecatrienoic acid, methyl ester, 9,12,15-octadecatrien-1-ol, hexadecane; *L. macranthoides (Absolute)*—Tetradecanoic acid, 3,7,11,15-tetramethyl-2-hexadecen-1-ol, phenylethyl alcohol, heptadecane,2,6,10,15-tetramethyl-, pentanoic acid, ethyl ester, hexanoic acid, 1,2-benzenedicarboxylic acid, bis(2-methylpropyl) ester, 2-nonadecanone, tetracosane, octadecanoic acid, butyl ester; *L. macranthoides (Cold Maceration)*—Tetradecanoic acid, 1,2,3-propanetriol, monoacetate, heneicosane, acetic acid, octadecyl ester, citronellyl isobutyrate, eicosanoic 220rticula *japonica (HD, Leaves)*—o-Xylene, 2,6-diisopropylnaphthalene, 1-undecaane, 1,19-eicosadiene, docosane, tricosane, tetracosane, heptacosane, octacosane, (Z)-3-hexenol, (Z,Z,Z)-9,12,15-octadecatrien-1-ol, dodecanoic acid, tetradecano220rticulates220220leic acid, octadecanoic acid, hotrienol, p-menth-1-en-9-al, nerol, g(E)niol, (E)-beta-damascenone, beta-caryophyllene, aromadendrene, germacrene D, beta-bisabolene, germacrene B, beta-elemol; *L. japonica (HD, Stems)*—Para-xylene, pentacosane, heptacosane, 3-hexanol, hexanal, 2-vinyl-crotonaldehyde, 3-hexanone, tetradecanoic acid, pentadecanoic acid, but220rticulates220220inoleic acid, geraniol, beta-caryophyllene, thujopsene, aromadendrene, squalene

Preferred Composition for Clinical Use:
Insufficient and inconsistent data/samples to determine.

Reported Therapeutic Properties: Antioxidant, anticancer, antibacterial, antifungal, antimicrobial, antiviral, liver protective, antiallergic, eases cough, supports respiratory function, astringent, wound healing, moisturizes the skin, soothes rashes and skin itching, digestive aid, nervine, antispasmodic, anti-inflammatory, reduces fever, aids immune function, balances female hormones, protects cells, heals traumatic memories, relaxing, sedating, antidepressant, relieves anxiety

CAUTIONS:

Absolutes

■ Avoid oral consumption. Absolutes contain solvent residue and should not be taken orally. In addition, ingestion of benzoic acids, benzoic salts (calcium, potassium, and sodium), benzaldehyde, benzyl acetate, benzyl benzoate, and benzyl alcohol may cause gastrointestinal irritation, vomiting, and diarrhea. The Joint Expert Committee for Food Additives (JECFA) has established an acceptable daily limit for the aforementioned constituents of 0–5 mg/kg body weight for adults to avoid GI discomfort. Based on this limit, it is suggested to limit the oral consumption to no more than 5 drops daily.

L. japonica (HD), L. japonica (HD, (Z,Z)-Farnesol CT), and L. caprifolium (HD)

■ May interfere (low risk) with enzymes responsible for metabolizing medications based on alpha-bisabolol content (NSAIDs, proton-pump inhibitors, acetaminophen, antiepileptics, immune modulators, blood-sugar medications, blood pressure medications, antidepressants, antipsychotics, diabetic medications, antihistamines, antibiotics, and anesthetics) due to farnesol content.[1880,1881,1882]

■ May interact with antibiotics and/or antifungals and possibly increase their effectiveness based on farnesol content.[1883]

L. macranthoides (HD), L. macranthoides (Absolute), and L. macranthoides (Cold Maceration)

■ Not recommended due to the association of synthetic dibutyl phthalate to reproductive and developmental toxicity.[1884]

L. japonica (HD, Nerolidol CT), L. japonica (HD, Leaves), and L. japonica (HD, Stems)

■ None currently known.

SELECTED EVIDENCE:

- □ Honeysuckle essential oil (*L(E)aponica,* (E)-nerolidol CT) inhibited *L. monocytogenes, B. subtilis, B. cereus, S. aureus* (two strains), *S. enteritidis, S. typhimurium, E. aerogenes,* and *E. coli* ATCC 8739, but was ineffective against *E. coli* ATCC 43888 and *P. aeruginosa* KCTC 2004.[1885] The same oil was later tested against fungi and found to inhibit *M. canis* (three strains), *T. rubrum* (three strains), and *T. mentagrophytes.*[1886]

HORSERADISH

Armoracia rusticana G. Gaertn. B. M & Scherb., *A. lapathifolia* Gilib., *Cochlearia armoracia* L., *Nasturtium armoracia* (L.) Fr., *Radicula armoracia* (L.) B.L. Rob., *Rorippa armoracia* (L.) Hitchc.

FAMILY: Brassicaceae (Cruciferae)
NOTE: Top
AROMA INTENSITY: Strong
AROMA: Pungent, spicy, mustard-like
COMMON EXTRACTION METHOD: Steam distilled or hydrodistilled from the roots; may also be solvent extracted or solid-phase microextracted
POSSIBLE SUBSTITUTE OILS: Garlic, onion
BLENDS WELL WITH: Basil, cumin, garlic, lavender, onion
SUBCELLULAR LOCALIZATION | EPIGENETIC INFLUENCE: Currently unknown | Currently unknown
RECOMMENDED DILUTION RANGE: Not recommended for topical use

Primary Constituents:[1887,1888,1889,1890,1891,1892]

Allyl Isothiocyanate	44.3%–97.8%
2-Phenethyl Isothiocyanate	tr–51.3%
4-Pentenyl Isothiocyanate	0.0%–4.0%

Solid-Phase Microextraction (SPME)

Allyl Isothiocyanate	63.6%–81.8%
2-Phenethyl Isothiocyanate	4.2%–25.7%
3-Methylbutyl Isothiocyanate	2.6%–14.6%
Cyclopentyl Isothiocyanate	0.3%–4.4%

Absolute (Solvent Extracted)

Allyl Isothiocyanate	21.8%–55.7%
2-Phenethyl Isothiocyanate	10.3%–47.8%
4-Pentenyl Isothiocyanate	0.3%–4.0%
Unidentified Constituent	0.6%–3.7%
Allyl Thiocyanate	0.7%–3.2%

Other Constituents: 3-Butenyl isothiocyanate, 2-phenyl isothiocyanate, allyl thiocyanate, 2-butyl isothiocyanate; *Solid-Phase Microextraction (SPME)*—2-butyl isothiocyanate, 3-butenyl isothiocyanate, unidentified constituent, benzyl isothiocyanate; *Absolute (Solvent Extracted)*—2-butyl isothiocyanate, unidentified constituent

Preferred Composition for Clinical Use:
Not recommended for essential oil therapy due to strong skin irritant concerns.

Reported Therapeutic Properties: Anticancer, diuretic, reduces kidney stones, aids gallbladder function, analgesic, antineuralgic, antirheumatic, relieves painful menstruation, eases cough, aids respiratory function, antimicrobial, antifungal, antibacterial, treats urinary tract infections, **insecticidal**, aphrodisiac, mentally stimulating

Cautions:
Horseradish essential oil is not recommended for essential oil therapy and listed for informational Purposes only.

- Allyl isothiocyanate (AITC) was not carcinogenic in mice but was associated with urinary cancer in a long-term study in rats; however, it was not genotoxic in mice or rats.[1893] The lowest-observed-adverse-effect level (LOAEL)—the lowest amount of a substance found by experiment or observation that causes an adverse effect— for AITC was determined to be 9mg/kg of body weight, which led to the establishment of an acceptable daily intake of 0.02 mg/kg body weight.[1894] Based on this, it is recommended to avoid oral administration of horseradish Essential oil.
- May interact with aspirin, blood pressure, antiplatelet, and anticoagulant medications, and increase the risk of bleeding among people with bleeding disorders based on AiTC content.[1895]
- May interact with diabetic medications and cause low blood sugar based on the ability of AITC to increase the utilization of excess blood glucose by increasing insulin secretion.[1896,1897]
- The vapor can be irritating to the eyes and possibly cause temporary vision impairment.[1898]
- Horseradish essential oil is strongly irritating to the skin and mucus membranes and should not be applied topically or on sensitive mucous membranes.[1899,1900,1901,1902]

Selected Evidence:

- □ The vapor phase of horseradish essential oil exhibited antifungal activity against the cause of Chalkbrood disease (*Ascosphaera apis*).[1903]
- □ Horseradish essential oil achieved similar insecticidal efficacy against Indian meal moth (*Plodia interpunctella*) and maize weevil (*Sitophilus zeamais*) compared to those of standard chemical fumigants.[1904] Horseradish essential oil also demonstrated potent larvicidal activity against the

mushroom fly (*Lycoriella ingenua*).[1905] Of isolated constituents tested from anise, garlic, and horseradish oils (*m*-anisaldehyde, *o*-anisaldehyde, *p*-anisaldehyde, trans-anethole, 4-allylanisole, diallyl sulfide, diallyl disulfide, diallyl trisulfide, and allyl isothiocyanate), allyl isothiocyanate was the most toxic. Horseradish essential oil controlled the house dust mite (*De223rticuagoides farinae*) and allyl isothiocyanate was considered highly toxic.[1906]

HUON PINE (Macquarie Pine)

Lagarostrobos franklinii (Hook.f.) Quinn, *Dacrydium franklinii* J.D. Hook

FAMILY: Podocarpaceae
NOTE: Middle-Top
AROMA INTENSITY: Strong; medium
AROMA: Herbaceous, spicy; fresh, piney, woody, fruity
COMMON EXTRACTION METHOD: Steam distilled or hydrodistilled from the wood or leaves (needles)
POSSIBLE SUBSTITUTE OILS: T223rticulatasil, ravensara CT methyl eugenol, Rosalina CT methyl eugenol, holy basil CT methyl eugenol; pine, cypress
BLENDS WELL WITH: Basil, bergamot, cassia, cinnamon, lavender, marjoram, patchouli, rosemary; bay laurel, bergamot, blue cypress, camphor, cedarwood, clary sage, coriander, cypress, frankincense, juniper, lavender, lemon, lime, marjoram, orange, palo santo, pine, rosemary, sandalwood, Spanish sage, spikenard
SUBCELLULAR LOCALIZATION | EPIGENETIC INFLUENCE: Currently unknown | Currently unknown
RECOMMENDED DILUTION RANGE: *Wood*—< 3%; *Leaves*—3%–neat

PRIMARY CONSTITUENTS:[1907,1908,1909]

Leaves		*Wood*	
Alpha-Pinene	25.6%	Methyl eugenol	95.0%–98.0%
Limonene	24.1%	Eugenol	0.5%–1.0%
Delta-3-Carene	16.7%		
16-Kaurene	6.0%		
Phyllocladene	4.2%		

OTHER CONSTITUENTS: *Wood*—N/A; *Leaves*—Alpha-fenchene, beta-pinene, myrcene, beta-phellandrene, terpinolene, (E)-beta-farnesene, bicyclogermacrene, methyl eugenol (< 0.9%), (E)-nerolidol, viridiflorol, sclarene, 15-isoprimaradiene, (E,E)-farnesol

PREFERRED COMPOSITION FOR CLINICAL USE:
Wood oil not recommended for therapy. Insufficient data/samples to determine for leaf oil.

REPORTED THERAPEUTIC PROPERTIES: Antibacterial, antifungal, anti-inflammatory, analgesic, antiseptic, astringent, circulatory stimulant, relieves erectile dysfunction, diuretic, insecticidal, promotes confidence, reduces emotional trauma and grief, antidepressant

CAUTIONS:

Leaves

- None currently known.

Wood

Not recommended for essential oil therapy.

- Avoid in children under age 12 due to the multisite cancerous tumors and DNA mutation caused by methyl eugenol in animal studies.[1910,1911,1912,1913]
- Avoid during pregnancy and lactation due to changes in embryo form and structure and chromosomal changes in ovary cells caused by methyl eugenol as well as the carcinogenic (cancer-causing) potential of methyl eugenol.[1914,1915,1916,1917]
- Avoid oral consumption due to the multisite carcinogenic potential of methyl eugenol.[1918,1919,1920,1921]
- May interfere with antibiotics and antifungal medications and increase their effectiveness due to methyl eugenol content.[1922]
- May interact with anticholinergic (drugs used for asthma, incontinence, gastrointestinal cramps, muscular spasms, depression, and sleep disorders) and cholinergic medications (drugs used to reduce urinary retention, diagnose myasthenia gravis, and for glaucoma), based on the potent AChE-inhibtory activity of methyl eugenol.[1923]

SELECTED EVIDENCE:

- □ None found.

HYACINTH | HYACINTH ABSOLUTE (Dutch Hyacinth, Garden Hyacinth)

Hyacinthus orientalis L.

FAMILY: Asparagaceae
NOTE: Middle-Base
AROMA INTENSITY: Strong
AROMA: Intensely sweet, green, and floral with subtle foliage notes
COMMON EXTRACTION METHOD: Solvent extracted, steam distilled, carbon dioxide (Co2) or hydrodistilled (HD) from the flowers
POSSIBLE SUBSTITUTE OILS: Benzoin absolute, tuberose absolute, carnation absolute, ylang ylang
BLENDS WELL WITH: Cassia, champaca, cinnamon, coriander, frangipani, galbanum, jasmine, nerol224rticulatroot, osmanthus, Peru balsam, saffron, styrax, tuberose, ylang ylang
SUBCELLULAR LOCALIZATION | EPIGENETIC INFLUENCE: Currently unknown | Currently unknown
RECOMMENDED DILUTION RANGE: *Absolute—%–; Essential Oil—*

PRIMARY CONSTITUENTS:[1924,1925,1926,1927]

Absolute (Phenylethyl Alcohol CT)		*HD*	
Phenylethyl Alcohol	48.1%–75.4%	Benzyl Benzoa(E)	4.4%–44.3%
(E)-Cinnamyl Alcohol	0.6%–28.9%	Beta-Phenylethyl Benzoate	6.1%–34.3%

1-Hepten-3-ol	1.7%–12.6%
Benzyl Benzoate	0.0%–11.2%
Unknown Compound	0.0%–9.8%
Benzyl Alcohol	1.6%–4.9%

Absolute (Benzyl Benzoate CT)

Benzyl Benzoate	34.1%
(E)-Cinnamyl Alcohol	24.6%
1,2,4-Trimethoxy Benzene	7.3%
Beta-Phenylethyl Benzoate	5.1%
Dimethyl Salicylate	3.1%

Absolute ((E,E)-Alpha-Farnesene CT)

(E,E)-Alpha-Farnesene	16.2%–31.7%
Benzyl Acetate	16.4%–31.2%
Beta-Phenylethyl Alcohol	5.0%–19.8%
Benzyl Benzoate	5.1%–14.6%
1,2,4-Trimethoxy Benzene	2.8%–6.8%
Hydroquinone Dimethyl Ether	tr–5.1%
(E)-Beta-Ocimene	3.7%–4.9%
(E)-Cinnamyl Alcohol	2.1%–4.1%
Benzyl Alcohol	0.5%–3.1%

Benzyl Acetate	4.0%–17.2%
(E,E)-Alpha-Farnesene	8.1%–11.7%
1-Octen-3-ol	0.3%–4.5%

Co2

Benzyl Benzoate	23.0%
Cinnamyl Alcohol	12.8%
Benzyl Acetate	6.4%
Cuminaldehyde	5.4%
Para-Cymene	5.0%
Alpha-Terpinene	4.7%
Beta-Pinene	3.7%
1,2,4-Trimethoxy Benzene	3.6%
Phenylethyl Alcohol	3.1%

Note: Today, hyacinth absolute is most likely a blend of synthetic and isolated aromatic compounds rather than a genuine extraction from the flowers.

OTHER CONSTITUENTS: *Absolute (Phenylethyl Alcohol CT)*—1-Pentanol, 4-hexen-1-ol, cinnamyl acetate, benzaldehyde, phenylacetaldehyde, benzyl acetate, 3-phenyl propanol; *Absolute (Benzyl Benzoate CT)*—Benzyl acetate, phenylethyl alcohol, methyl eugenol (< 1.4%), phenylpropyl alcohol, eugenol, elemicin; *Absolute ((E,E)-Alpha-Farnesene CT)*—Myrcene, oct-1-en-3-ol, 2,(E)ethylocta-1,3(E),5(E),7-tetrene, benzaldehyde, phenylacetaldehyde, (Z,E)-alpha-farnes(E), 2,6-dimethylocta-3(E),5(E),7-trien-3-ol, beta-phenylethyl acetate, methyl eugenol (< 0.6%), elemicin, 2-(4-,ethoxyphenyl) ethyl alcohol, beta-phenylethyl benzoate; *HD*—Not listed; *Co2*—Ethyl alcohol, alpha-pinene, alpha-thujene, beta-caryophyllene, alpha-farnesene, p-metha-1,4-dien-7-al, benzyl alcohol, methyl eugenol (< 1.0%), phenylpropyl alcohol, dimethyl salicylate, eugenol, elemicin, pentacosane, n-octacosane, beta-phenylethyl benzoate, squalene, n-butyl benzenesulfonamide

PREFERRED COMPOSITION FOR CLINICAL USE:
Insufficient samples and too great of variance in data to determine.

REPORTED THERAPEUTIC PROPERTIES: Deodorant, moisturizes the skin, reduces the appearance of blemishes, soothes chronic skin conditions, treats snakebites, anti-inflammatory, improves lactation, stress management, relieves anxiety, reduces sadness

CAUTIONS:

■ Avoid oral consumption. Absolutes contain solvent residue and should not be taken orally. In addition, ingestion of benzyl derivatives (benzoic acids, benzoic salts, benzaldehyde, benzyl acetate, benzyl benzoate, and benzyl alcohol; present in HD, Co2, and absolutes) may cause gastrointestinal irritation, vomiting, and diarrhea. The Joint Expert Committee for Food Additives (JECFA) has established an acceptable daily limit for the aforementioned constituents of 0–5 mg/kg body weight for adults to avoid GI discomfort. Based on this limit, it is suggested to limit the oral consumption of orange bird berry essential oil rich in benzyl derivatives to no more than 10 drops daily.

SELECTED EVIDENCE:

- None found.

INCENSE CEDAR (California River Cedar)

Calocedrus decurrens Torr. Florin, *Heyderia decurrens* (Torr.) K. Koch, *Libocedrus decurrens* Torr.

FAMILY: Cupressaceae
NOTE: Middle
AROMA INTENSITY: Medium
AROMA: Citrusy, piney, woody; woody, spicy, herbaceous
COMMON EXTRACTION METHOD: Steam distilled or hydrodistilled from the heartwood, leaves (needles), or branches
POSSIBLE SUBSTITUTE OILS: Pink pepper, black pepper, cypress; black cumin
BLENDS WELL WITH: Amyris, bergamot, black pepper, cedarwood, cardamom, carrot, clary sage, clove, frankincense, geranium, ginger, grapefruit, lavandin, lavender, lemon, lime, marjoram, may chang, myrrh, neroli, nutmeg, orange, palo santo, patchouli, rosemary, sage, sandalwood, Spanish sage, tangerine, tea tree (melaleuca), vetiver, yarrow, ylang ylang
SUBCELLULAR LOCALIZATION | EPIGENETIC INFLUENCE: Currently unknown | Currently unknown
RECOMMENDED DILUTION RANGE: 3%–50%; Neat for some conditions

PRIMARY CONSTITUENTS:[1928,1929,1930,1931,1932]

Heartwood (Thymoquinone CT)

Thymoquinone	35.9%
Carvacrol	29.2%
p-Methoxythymol	11.0%
p-Methoxycarvacrol	3.2%

Branches (Delta-3-Carene CT)

Delta-3-Carene	43.1%
Alpha-Fenchyl Acetate	14.2%
Alpha-Fenchol	13.1%
Cedrol	4.5%
Linalool	3.9%
Terpinolene	3.7%

Leaves (Delta-3-Carene CT)

Delta-3-Carene	31.3%
Myrcene	13.4%

Branches (Alpha-Pinene CT)

Alpha-Pinene	22.3%–56.6%
Delta-3-Carene	5.2%–11.1%
Pin-2-en-8-ol	4.5%–10.3%
Myrcene	8.4%–9.6%
Limonene	5.1%–5.5%
Methyl Pin-2-en-8-oate	1.8%–3.6%
Methyl Myrtenate	1.6%–3.5%
Terpinolene	1.5%–3.2%

Leaves (Limonene CT)

Limonene	18.2%–31.3%
Delta-3-Carene	15.2%–21.0%
Alpha-Pinene	8.7%–15.8%
Alpha-Fenchyl Acetate	0.0%–9.7%
Myrcene	6.2%–8.2%
Terpinolene	3.9%–8.0%

Alpha-Pinene	11.2%	Alpha-Terpinyl Acetate	0.2%–5.7%
Terpinolene	6.9%	Methyl Myrtenate	0.0%–4.3%
Limonene	6.4%		
Pin-2-en-8-ol	4.2%		
Methyl Pin-2-en-8-oate	3.0%		

OTHER CONSTITUENTS: *Heartwood (Thymoquinone CT)*—Para-cymene, terpinen-4-ol, alpha-terpineol, piperitone, carvenone, heyderiol, libocedrol; *Branches (Delta-3-Carene CT)*—Alpha-pinene, para-cymene, limonene, gamma-terpinene, alpha-thujone (< 1.9%), thymol, beta-caryophyllene, alloaromadendrene, alpha-cadinol, palmitic acid; *Branches (Alpha-Pinene CT)*—Beta-pinene, beta-phellandrene, pin-2-en-8-al, terpinen-4-ol, (E)-dec-4-enal, verbenone, carvone, piperitone, bornyl acetate, pin-2-en-8-yl acetate, alpha-terpinyl acetate, beta-elemene, cedrol; *Leaves (Delta-3-Carene CT)*—1-(2-methylene cyclopropyl) cyclopentene, alpha-fenchene, sylvestrene, beta-phellandrene, pin-2-en-8-al, terpinen-4ol, (E)-dec-4-enal,methyl myrtenate, alpha-terpinyl acetate; *Leaves (Limonene CT)*—Hex-2-en-1-al, 3-methylene-bicyclo-[3.2.1]oct-2-ene, alpha-fenchene, verbenene, beta-pinene, gamma-terpinene, linalool, terpinen-4-ol, alpha-terpineol, carvone, bornyl acetate, piperitone, verbenone, piperityl acetate, cedrol

PREFERRED COMPOSITION FOR CLINICAL USE:
Too great of variance in samples/data to determine.

REPORTED THERAPEUTIC PROPERTIES: Aids digestion, expels excess mucus, supports respiratory function, antibacterial, antifungal, antioxidant, anticancer, wound healing, anti-inflammatory, insecticidal, promotes restful sleep, relieves anxiety, improves concentration and focus

CAUTIONS:
Heartwood (Thymoquinone CT)

- Avoid during pregnancy and lactation. Oregano (also rich in carvacrol) may negatively affect embryonic development and encourage fetal cell death according to animal research.[1933] In addition, CTs with significant thymoquinone would require a caution because thymoquinone prevents the formation of new blood vessels (angiogenesis), which occurs during pregnancy beginning twenty-one days after conception.[1934,1935,1936] The growth of a dense network of blood vessels within the placenta is necessary for the exchange of respiratory gases, nutrients, and wastes between the mother and fetus. If angiogenesis is inhibited during pregnancy it may lead to insufficient blood supply to the placenta and pregnancy complications (fetal growth restriction, preeclampsia, and fetal death). Thymoquinone also negatively effects embryonic development if administered during the second trimester (fetal malformations).[1937]
- May interact with diabetic medications and cause low blood-sugar levels due to carvacrol and thymoquinone content.[1938,1939,1940,1941,1942,1943,1944,1945,1946,1947,1948] Higher thymoquinone content will increase the effect on blood sugar levels.[1949]
- May interact with aspirin, blood pressure, antiplatelet, and anticoagulant medications, and increase the risk of bleeding among people with bleeding disorders due to carvacrol, thymoquinone content.[1950,1951,1952,1953] Thymoquinone is known to cause platelet apoptosis (cell death).[1954,1955]
- May interact with anticholinergic (drugs used for asthma, incontinence, gastrointestinal cramps, muscular spasms, depression, and sleep disorders) and cholinergic medications (drugs used to reduce urinary retention, diagnose myasthenia gravis, and for glaucoma) based on thymoquinone, thymol, and carvacrol content.[1956]
- May interfere with the enzymes responsible for metabolizing medications based on trans-anethole and thymoquinone content (NSAIDs, proton-pump inhibitors, acetaminophen, antiepileptics, immune modulators, blood-sugar medications, blood pressure medications, antidepressants, antipsychotics, diabetic medications, antihistamines, antibiotics, and anesthetics).[1957,1958,1959,1960]

■ May interact with antibiotics and possibly enhance their effects due to thymol, thymoquinone, or carvacrol content.[1961,1962]
■ May interfere with the antiseizure medication valproate (possibly other antiseizure drugs) and enhance its anticonvulsive activity due to thymoquinone content.[1963]
■ May irritate mucous membranes (eyes, mouth, nasal passages, vagina, rectum).

All Other CTs
■ None currently known.

SELECTED EVIDENCE:

- □ The branches (delta-3-carene CT) of incense cedar essential oil demonstrated strong antimicrobial properties against gram-positive (*S. aureus, S. epidermidis, S. pyogenes*) and gram-negative (*E. coli, K. pneumonia, P. vulgaris, P. aeruginosa, S. boydii*) bacteria.[1964]
- □ Incense cedar heartwood essential oil was toxic to mosquitoes (*Aedes aegypti*), the deer tick (*Ixodes scapularis*), and the Oriental rat flea (*Xenopsylla cheopis*).

INDIAN DILL
(Sowa)

Anethum sowa Roxb. Ex Flem.

FAMILY: Apiaceae (Umbelliferae)
NOTE: Top-Middle
AROMA INTENSITY: Medium
AROMA: Fresh, spicy, herbaceous, minty
COMMON EXTRACTION METHOD: Steam distilled or hydrodistilled from the seeds (fruits) or aerial parts (leaves and stems); may also be distilled from the stems or roots
POSSIBLE SUBSTITUTE OILS: Spearmint, caraway, dill
BLENDS WELL WITH: Anise, basil, bergamot, cajeput, cardamom, eucalyptus, ginger, grapefruit, frankincense, lavender, lemon, lemongrass, marjoram, niaouli, nutmeg, orange, peppermint, petitgrain, pine, rosemary, spearmint, tangerine, thyme
SUBCELLULAR LOCALIZATION | EPIGENETIC INFLUENCE: Currently unknown | Currently unknown
RECOMMENDED DILUTION RANGE: 1.5%–50%

PRIMARY CONSTITUENTS:[1965,1966,1967,1968,1969,1970,1971,1972,1973]

Seeds (Dill Apiole CT)	
Dill Apiole	20.7%–52.5%
Limonene	5.9%–45.0%
Carvone	20.9%–23.1%
(E)-Dihydrocarvone	4.2%–16.6%
Alpha-Phellandrene	tr–6.5%
(Z)-Dihydrocarvone	0.8%–5.2%

Aerial Parts	
Dill Apiole	25.4%–26.2%
Alpha-Thujene	14.8%–16.5%
O-Cymene	13.8%–15.3%
Beta-Phellandrene	7.1%–7.2%
6,6-Dimethyl-2-(3-Oxobutyl)-Bicycle[3.1.1]Heptan-3-one	6.9%–7.0%
Exo-2-Hydroxycineole	5.0%–5.1%
Limonene	4.1%–4.2%

Seeds (Limonene CT)

Limonene	19.7%–83.0%
(E)-Dihydrocarvone	0.0%–31.6%
Carvone	1.5%–30.0%
Dill Apiole	0.0%–19.0%
Grandisol	0.0%–7.4%
Bis-1,2-Benzenedicarb-(E)ylic Acid	0.0%–5.7%

Seeds (Carvone CT)

Carvone	40.0%–60.0%
Limonene	0.0%–28.7%
Dill Apiole	2.5%–15.6%
(E)-Dihydrocarvone	1.8%–7.2%

Stems

Dill Apiole	45.6%
5-Isopropyl-6-Methyl-Hepta-3,5-dien-2-ol	7.8%
Limonene	7.3%
(E)-6-Nonenal	6.7%
m-Cymene	5.8%
1,3-Cyclohexadiene, 2-Methyl-5-(1-Methylethyl)-	5.0%
3,9-Epoxy-1-p-Menth-1-ene	4.6%
2-cyclohexene-1-4-Hydroxy-3-Methyl-6'	3.2%

OTHER CONSTITUENTS: *Seeds (Dill Apiole CT)*—Not listed; *Seeds (Limonene CT)*—Myrcene, alpha-phellandrene, gamma-terpinene, thymol, cyclohexane, beta-phellandrene, thujyl alcohol; *Seeds (Carvone CT)*—Not listed; *Aerial Parts*—Alpha-pinene, terpinolene, undecane, dill ether, 1,3-cyclohexadiene,2-methyl-5-(1-ethylethyl)-monoepoxide, 6-camphenol, lilac alcohol B, 3-isopropyl-4-methyl-1-pentyn-3-ol, alpha-selinene, phytol; *Stems*—6,6-dimethyl-2-(3-oxobutyl)Bicyclo[3.1.1]heptan-3-one, 6-camphenol, exo-2-hydroxycineole, 3-isopropyl-4-methyl-1-pentyn-3-ol, 3-cyclohexene-1-one,-3:hydryoxymethyl-6-1-methylethyl, myristicin

PREFERRED COMPOSITION FOR CLINICAL USE:
Too great of variance in composition to determine.

REPORTED THERAPEUTIC PROPERTIES: Antimicrobial, antifungal, anticancer, antispasmodic, anti-inflammatory, analgesic, relieves headache, antiallergic, antioxidant, aids digestion, expels excess gas, disinfectant, reduces high cholesterol and lipid levels, diuretic, aids cardiovascular function, relieves skin itching, increases lactation, promotes restful sleep, sedating, reduces mental fatigue, antidepressant, stress management, reduces fear

CAUTIONS:

■ Avoid with children under age twelve. May contain moderate to significant amounts of dill apiole, which may cause kidney or liver toxicity, abnormal quantities of blood components (blood dyscrasias), dizziness, or hallucinations.[1974,1975,1976,1977,1978]

■ Avoid during pregnancy and lactation due to the abortive properties (dill apiole may trigger uterine contractions or necrosis of placental tissue and has been used for decades to end unwanted pregnancies) and high toxicity (several cases of acute poisoning or death have been reported after ingestion of varying amounts of apiole in single or daily doses) of apiole.[1979,1980,1981,1982,1983,1984,1985,1986]

■ Avoid oral consumption due to high toxicity of dill apiole, reported poisonings by apiol, and deaths reported after ingestion of significant amounts of apiole.[1987,1988,1989,1990,1991,1992]

■ May interact with diabetes medications and cause low blood sugar based on potential significant carvone content.[1993]

■ May interact with aspirin, blood pressure, antiplatelet, and anticoagulant medications, and increase the risk of bleeding among people with bleeding disorders based on potential significant carvone content. Carvone is a potent calcium channel blocker, which may decrease blood pressure.[1994,1995]

■ May interact with barbiturates (medications for anxiety and insomnia), antihistamines, benzodiazepines, tricyclic antidepressants, or other central nervous system depressant drugs, increasing depressant effects, due to potential significant carvone content.[1996]

■ Avoid with compromised liver or kidney due to potential to cause liver and kidney toxicity based on dill apiole content.[1997,1998,1999,2000,2001]

SELECTED EVIDENCE:

□ Indian dill essential oil exhibited antifungal activity depending on the concentration used.[2002]

IRONWOOD TREE
(Sri Lankan Ironwood, Ceylon Ironwood, Indian Rose Chestnut, Cobra's Saffron)

Mesua ferrea L.

FAMILY: Calophyllaceae
NOTE: Middle
AROMA INTENSITY: Medium
AROMA: Woody, earthy, green, slightly sweet; floral, spicy, green, citrusy, slightly pungent
COMMON EXTRACTION METHOD: Steam distilled or hydrodistilled from the bark, leaves, buds, or flowers; may also be macerated from the flowers or petals
POSSIBLE SUBSTITUTE OILS: Copaiba, guava leaf, black pepper, Helichrysum, German chamomile
BLENDS WELL WITH: Balsam fir, bergamot, black pepper, black spruce, blue spruce, cedarwood, clary sage, copaiba, frankincense, galbanum, Helichrysum, lavender, lemon, lime, orange, palo santo, pine, ravintsara, tangerine, ylang ylang
SUBCELLULAR LOCALIZATION | EPIGENETIC INFLUENCE: Currently unknown | Currently unknown
RECOMMENDED DILUTION RANGE: 5%–Neat

PRIMARY CONSTITUENTS:[2003,2004]

Leaves (Beta-Caryophyllene CT)

Beta-Caryophyllene	26.0%–30.9%
Caryophyllene Oxide	11.4%–17.9%
Beta-Selinene	1.9%–10.1%
Spathulenol	0.0%–10.0%
Alpha-Copaene	1.1%–9.9%
Alpha-Humulene	3.4%–6.0%
Alpha-Selinene	1.1%–5.7%
Delta-Cadinene	tr–4.1%
Beta-Elemene	0.5%–3.8%
Gamma-Muurolene	0.0%–3.5%

Bark

(E)-Alpha-Bisabolene	31.3%
Unidentified Compounds	17.9%

Leaves (Alpha-Copaene CT)

Alpha-Copaene	19.3%
Beta-Caryophyllene	18.8%
Caryophyllene Oxide	15.8%
Unidentified Compounds	15.5%
Spathulenol	10.4%
Alpha-Humulene	3.1%

Buds

Alpha-Copaene	28.7%
Germacrene D	19.0%
(Z)-Gamma-Bisabolene	10.6%
(Z)-Alpha-Bisabolene	5.0%
Alpha-Selinene	4.0%
Elema-1,3,11(15)-Trien-12-al	3.9%

Alpha-Selinene	12.2%
Beta-Selinene	7.0%
(Z)-Alpha-Bisabolene	5.5%
ar-Curcumene	5.0%
(Z)-Gamma-Bisabolene	4.0%
Caryophyllene Oxide	3.4%
Beta-Bisabolene	3.0%

Flowers

Alpha-Copaene	20.2%
Germacrene D	16.1%
(Z)-Alpha-Bisabolene	8.2%
(Z)-Gamma-Bisabolene	7.6%
Beta-Caryophyllene	7.0%
Beta-Selinene	5.9%
Alpha-Selinene	5.7%
Beta-Elemene	3.9%
Delta-Cadinene	3.4%
Unidentified Compounds	3.1%
Delta-Cadinene	3.6%
Beta-Elemene	3.2%

Flowers (Maceration)

Dioctyl Phthalate	13.9%
Alpha-Copaene	13.5%
Geranyl Benzoate 1	9.8%
Trans-Alpha-Bergamotene	9.1%
Beta-Caryophyllene	4.5%
Beta-Selinene	4.5%
Alpha-Selinene	4.0%
Beta-Elemene	3.6%

Petals (Maceration)

Alpha-Copaene	22.7%
Trans-Alpha-Bergamotene	17.4%
Dioctyl Terephthalate	13.9%
Beta-Selinene	8.9%
Alpha-Selinene	6.8%

Note: One study reported that mature leaves produce the beta-caryophyllene CT of ironwood tree essential oil, whereas the young leaves are rich in alpha-copaene.[2005]

Other Constituents: *Leaves (Beta-Caryophyllene CT)*—Beta-bourbonene, aromadendrene, alpha-acoradiene, alloaromadendrene, germacrene D, valencene, alpha-selinene, alpha-muurolene, beta-bisabolenee, gamma-cadinene, (Z)-calamenene, caryophyllene alcohol, T-muurolol, n-hexadecanoic acid, globulol, humulene oxide; *Leaves (Alpha-Copaene CT)*—nonanal, beta-bourbonene, beta-elemene, alpha-acoradiene, beta-selinene, alpha-selinene, (Z)-alpha-bisabolene, (Z)-3-hexenyl benzoate, globulol, humulene oxide; *Bark*— (E)-6-methyl-3,5-heptadien-2-one, 4-acetyl-1-methylcyclohexene, alpha-copaene, beta-caryophyllene, trans-alpha-bergamotene, alpha-humulene, alpha-acoradiene, (E)-gamma-bisabolene; *Buds*—Alpha-cubebene, aromadendrene, alpha-humulene, alloaromadendrene, alpha-acoradiene, beta-selinene, (E)-gamma-bisabolene, (E)-alpha-bisabolene, spathulenol, caryophyllene oxide, junenol, 1-epi-cubenol, T-muurolol, T-cadinol. alpha-cadinol; *Flowers*—Linalool, alpha-cubebene, aromadendrene, alpha-humulene, alloaromadendrene, alpha-calacorene, beta-bisabolene, (E)-gamma-bisabolene, (E)-nerolidol, T-muurolol, T-cadinol, alpha-cadinol; *Flowers (Maceration)*—Beta-ionone, alpha-farnesene; *Petals (Maceration)*—Beta-elemene, beta-caryophyllene, trans-beta-farnesene, gamma-selinene, alpha-farnesene, delta-cadinene, alpha-bisabolene, geranyl benzoate 1, dioctyl phthalate

Preferred Composition for Clinical Use:
Too great of variance in samples/data to determine.

Reported Therapeutic Properties: Anti-inflammatory, analgesic, relieves sore throat, antirheumatic, antispasmodic, antiarthritic, relieves fever, antiseptic, antibacterial, antifungal, antiparasitic, blood purification, expectorant, aids respiratory function, eases cough, supports cardiovascular function, aids digestion, expels excess gas, snake bite antidote, wound healing, eases skin itching, astringent, liver protective, antioxidant, anticancer, relieves anxiety, encourages emotional healing, combats worrisome thoughts

CAUTIONS:

Leaves, Buds, and Bark

- None currently known.

Flowers (Maceration) and Petals (Maceration)

- Not recommended for aromatherapy due to cold maceration with hexane (a neurotoxin) and presence of phthalates that are skin irritants.

SELECTED EVIDENCE:

- The leaf oil (beta-caryophyllene CT) of ironwood tree exhibited anticancer activities against breast, lung (small-cell), and cervical cancer cells but nontoxic to normal cells.[2006]
- *In vitro* research demonstrated that ironwood tree leaf essential oil (beta-caryophyllene CT) significantly inhibited *E. coli* and *S. aureus*.[2007]
- Ironwood tree leaf essential oil (beta-caryophyllene CT) showed good antioxidant activity in the DPPH assay.[2008]

JABORANDI (Indian Hemp)

Pilocarpus jaborandi Holmes, *P. microphyllus* Stapf.

FAMILY: Rutaceae
NOTE: Middle
AROMA INTENSITY: Medium
AROMA: Balsamic, earthy, herbaceous, pungent
COMMON EXTRACTION METHOD: Steam distilled or hydrodistilled from the leaves
POSSIBLE SUBSTITUTE OILS: Ant bush, copaiba, bergamot
BLENDS WELL WITH: Bergamot, cedarwood, copaiba, geranium, grapefruit, helichrys232rticulon, lemongrass, lime, melissa, orange, pine, sandalwood, Spanish sage, tangerine
SUBCELLULAR LOCALIZATION | EPIGENETIC INFLUENCE: Currently unknown | Currently unknown
RECOMMENDED DILUTION RANGE: 3%–33%; neat for some conditions

PRIMARY CONSTITUENTS:[2009,2010,2011]

P. microphyllus (Limonene CT)

Limonene	47.0%
2-Undecanone	22.5%

P. microphyllus (Mixed CT)

Beta-Caryophyllene	7.8%–40.6%
2-Tridecanone	2.4%–36.0%
2-Pentadecanone	6.7%–28.1%
Caryophyllene Oxide	2.7%–15.4%
Germacrene D	3.2%–10.5%
Unidentified Hydrocarbon	0.0%–6.9%
Unidentified Sesquiterpenes	0.0%–6.6%
Alpha-Humulene	1.1%–5.4%

P. jaborandi (Limonene CT)

Limonene	67.1%
2-Tridecanone	18.6%

P. jaborandi (2-Tridecanone CT)

2-Tridecanone	22.0%–86.2%
Beta-Caryophyllene	0.0%–19.0%
2-Pentadecanone	0.0%–18.8%
Sandaracopimara-8(14),15-Diene	0.0%–13.1%
3,7,7-Trimethylbicyclo [3.1.1]-2-Heptene	0.0%–12.0%
Limonene	0.0%–9.6%

Alpha-Copaene	0.7%–4.7%	Spathulenol	0.0%–7.6%
(E)-Alpha-Bergamotene	0.0%–4.2%	Delta-Cadinene	0.0%–4.7%
Beta-Sesquiphellandrene	0.5%–3.4%	Alpha-Guaiene	0.0%–4.4%
Beta-Bourbonene	0.0%–3.4%	Alpha-Humulene	0.0%–4.0%
		Alpha-Copaene	0.0%–3.0%

P. microphyllus (2-Tridecanone CT)

2-Tridecanone	64.1%–88.1%
2-Pentadecanone	6.9%–18.9%
Beta-Caryophyllene	2.5%–7.5%

OTHER CONSTITUENTS: *P. microphyllus (Limonene CT)*—Myrcene, alpha-pinene, vinyl dodecanoate, sandaracopimara-8(14),15-diene; *P. microphyllus (Mixed CT)*—Delta-elemene, alpha-cubebene, beta-cubebene, beta-elemene, alpha-gurjunene, cis-alpha-bergamotene, aromadendrene, (E)-beta-farnesene, beta-selinene, beta-bisabolene, (E)-nerolidol, spathulenol, 2-tetradecanone, carotol, humulene epoxide II; *P. microphyllus (2-Tridecanone CT)*—Alpha-copaene, alpha-humulene, germacrene D, beta-sesquiphellandrene, 2-tetradecanone, unidentified hydrocarbon, unidentified sesquiterpenes; *P. jaborandi (2-Tridecanone CT)*—Beta-ocimene, alpha-pinene, myrcene, beta-elemene, caryophyllene oxide alpha-cadinol; *P. jaborandi (Limonene CT)*—Alpha-pinene, myrcene, sandaracopimara-8(14),15-diene

PREFERRED COMPOSITION FOR CLINICAL USE:
Insufficient samples/data to determine.

REPORTED THERAPEUTIC PROPERTIES: Antibacterial, antifungal, anti-infectious, aids kidney function, reduces kidney stones, poison antidote, diuretic, aids respiratory function, eases cough, soothes sore throat, anticonvulsive, anti-inflammatory, antirheumatic, reduces fever, increases sweating, relieves stomachache, soothes chronic skin conditions, reduces hysteria, relieves anxiety

CAUTIONS:
- None currently known.

SELECTED EVIDENCE:
- □ None found.

JACK PINE
(Grey Pine)

Pinus banksiana Lamb., *P. divaricata* (Aiton) Sudw.

FAMILY: Family
NOTE: Top
AROMA INTENSITY: Medium
AROMA: Piney, fresh, green, woody, fruity
COMMON EXTRACTION METHOD: Steam distilled from the branches and needles
POSSIBLE SUBSTITUTE OILS: Black spruce, pine
BLENDS WELL WITH: Anise, balsam fir, bergamot, blue cypress, cajeput, cedarwood, cypress, frankincense, grapefruit, lavender, lemon, peppermint, pine, rose, rosemary, silver fir, white fir

SUBCELLULAR LOCALIZATION | EPIGENETIC INFLUENCE: Currently unknown | Currently unknown
RECOMMENDED DILUTION RANGE: 3%–50%

PRIMARY CONSTITUENTS:[2012,2013,2014]

Alpha-Pinene	23.1%–53.5%
Beta-Pinene	7.3%–15.4%
Delta-3-Carene	6.4%–12.7%
Bornyl Acetate	6.1%–12.6%
Limonene	5.3%–10.9%
Camphene	3.4%–7.4%
Myrcene	3.8%–6.1%
Beta-Phellandrene	1.0%–4.8%
Alpha-Terpineol	1.3%–4.0%

OTHER CONSTITUENTS: Tricyclene, santene, terpinolene, borneol, delta-cadinene

PREFERRED COMPOSITION FOR CLINICAL USE:

Constituent	
Alpha-Pinene	30%–55%
Beta-Pinene	5%–20%
Delta-3-Carene	5%–15%
Bornyl Acetate	5%–15%
Limonene	4%–12%
Camphene	2%–9%
Myrcene	3%–7%

REPORTED THERAPEUTIC PROPERTIES: Anti-inflammatory, analgesic, antineuralgic, nervine, diuretic, antibacterial, antifungal, antiseptic, antioxidant, supports respiratory function, eases cough, expectorant, aids circulation, focuses the mind, relieves anxiety, promotes self-confidence, enhances feelings of security

CAUTIONS:

- None currently known.

SELECTED EVIDENCE:

- None found.

JADE CYPRESS
(White Cypress-Pine, Australian White Cypress, Murray Pine Cypress)

Callitris glaucophylla Joy Thomps. & L.A.S. Johnson

FAMILY: Cupressaceae
NOTE: Base-middle
AROMA INTENSITY: Medium
AROMA: Earthy, woody, sweet; Balsamic, woody, spicy, sweet, slightly fruity
COMMON EXTRACTION METHOD: Steam distilled or hydrodistilled from the heartwood or needles (leaves); may also be distilled from the needles and twigs or solvent extracted from the wood
POSSIBLE SUBSTITUTE OILS: White cypress, blue cypress
BLENDS WELL WITH: Balsam fir, bergamot, black pepper, blue cypress, cardamom, cedarwood, cistus, frankincense, juniper berry, lavandin, lavender, lemon, lime, myrtle, orange, pine, sandalwood, silver fir, white fir, ylang ylang
SUBCELLULAR LOCALIZATION | EPIGENETIC INFLUENCE: Currently unknown | Currently unknown
RECOMMENDED DILUTION RANGE: 5%–Neat

PRIMARY CONSTITUENTS:[2015,2016,2017]

Wood (Guaiol CT)

Guaiol	28.5%–79.9%
Alpha-Eudesmol + Beta-Eudesmol	2.8%–21.3%
Bulnesol	0.3%–16.0%
Columellarin	1.0%–13.2%
Dihydrocolumellarin	1.4%–11.8%
Eudesma-1,4(15),11-Triene	0.0%–6.8%
Beta-Selinene	0.0%–6.6%
Alpha-Selinene	0.0%–5.6%
10-Epi-Gamma-Eudesmol	0.0%–3.7%

Wood (Sandaracopimarinal CT)

Sandaracopimarinal	60.1%
Unknown (MW 273)	4.7%
Unknown (MW 281)	3.0%

Wood (Citronellic Acid CT)

Citronellic Acid	40.3%
Beta-Eudesmol	10.0%
Guaiol	6.2%
Geranic Acid	3.9%
Cadalene	3.3%
Alpha-Eudesmol	3.0%
Dihydrocolumellarin	3.0%

Wood (Dihydrocolumellarin CT)

Dihydrocolumellarin	20.9%
Unknown (MW 258)	8.5%
Sandaracopimarinal	8.2%
Unknown (MW 256)	8.0%
Gamma-Lactone Methyl Ester	6.3%
Beta-Costol	5.4%
Gamma-Costol	4.8%
Columellarin	4.0%
Unknown (MW 207)	3.8%
Beta-Eudesmol	3.4%

Needles

Limonene	22.2%–26.2%
Alpha-Pinene	17.4%–22.8%
Fenchyl Acetate	11.7%–19.7%
Bornyl Acetate	16.1%–16.1%
Myrcene	6.9%–9.2%
Camphor	2.1%–3.2%
Camphene	2.7%–3.0%

OTHER CONSTITUENTS: *Wood (Guaiol CT)*—Sandaracopimarinal, gamma-eudesmol, methyl citronellate, citronellic acid, neric acid, geranic acid, Selina-4,11-diene, alpha-guaioxide, callitrin, callitrisin; *Wood (Dihydrocolumellarin CT)*—Alpha-pinene, methyl myrtenate, citronellic acid, beta-selinene, alpha-selinene; *Wood (Citronellic Acid CT)*—(E,E)-methyl dehydrogeranate, alpha-guaioxide, elemol, gamma-eudesmol, bulnesol, (–)-alpha-costol, (+)-gamma-costol, (+)-beta-costol, elemanolide, callitrin, eudesmanolide isomer, isokhusenic acid isomer 1, callitrisin, columellarin, germacranolide, dihydrocallitrisin; *Needles*—Tricyclene, para-cymene, gamma-terpinene, terpinolene, borneol, nerolic acid, alpha-terpinyl acetate, beta-caryophyllene, caryophyllene oxide

PREFERRED COMPOSITION FOR CLINICAL USE:
Insufficient samples/data to determine.

REPORTED THERAPEUTIC PROPERTIES: Anti-inflammatory, analgesic, reduces fever, antibacterial, antifungal, anti-infectious, anticancer, soothes skin irritation and itching, wound healing, heals boils and blisters, relieves stomachache, eases cough, supports respiratory function, grounding, calming/relaxing, centering, relieves anxiety

CAUTIONS:
- None currently known.

SELECTED EVIDENCE:
- □ Jade cypress essential oil (various CTs) demonstrated cytotoxicity toward liver cancer cells.[2018]

JAMROSA

Cymbopogon khasans Sobti, *C. jwarancusa* x *C. nardus* var. *confertiflorus*

FAMILY: Poaceae (Gramineae)
NOTE: Middle
AROMA INTENSITY: Medium-strong
AROMA: Floral, green, fresh, slightly sweet and rose-like
COMMON EXTRACTION METHOD: Steam distilled or hydrodistilled from the roots
POSSIBLE SUBSTITUTE OILS: Palmarosa, geranium
BLENDS WELL WITH: Bergamot, cedarwood, geranium, lemongrass, lime, neroli, palmarosa, petitgrain, rose, rosemary, sandalwood, spikenard, ylang ylang
SUBCELLULAR LOCALIZATION | EPIGENETIC INFLUENCE: Currently unknown | Currently unknown
RECOMMENDED DILUTION RANGE: 3%–50%; neat for some conditions

PRIMARY CONSTITUENTS:[2019,2020,2021,2022,2023,2024]

Geraniol CT	
Geraniol	29.4%–75.5%
Geranyl Acetate	18.0%–38.3%
(Z)-Beta-Ocimene	3.8%–7.8%
Piperitone	1.6%–3.4%
Linalool	1.0%–3.4%

Geranyl Acetate CT	
Geranyl Acetate	20.3%–56.2%
Geraniol	13.0%–40.5%
Citral (Geranial/Neral)	1.2%–5.1%

Neral CT	
Neral	59.7%
Linalyl Acetate	35.0%

OTHER CONSTITUENTS: *OC*—(E)-Beta-Ocimene, isopulegol; *Geranyl Acetate CT*—Not listed; *Neral CT*—Beta-ocimene (isomer not identified), patchulane, beta-caryophyllene, farnesol

PREFERRED COMPOSITION FOR CLINICAL USE:

Constituent	*Geraniol CTI*	*Geraniol CTII*
Geraniol	50%–65%	60%–80%
Geranyl Acetate	15%–35%	5%–18%
Beta-Ocimene Isomers	1%–10%	–
Piperitone	0.5%–5%	–
Linalool	0.1%–5%	–

REPORTED THERAPEUTIC PROPERTIES: Antifungal, antibacterial, antiviral, anti-infectious, antiseptic, aids digestion, wound healing, reduces the appearance of scars, heals boils and blisters, relieves chronic skin conditions, anti-inflammatory, relieves fever, aids circulation, insecticidal, insect repellent, promotes restful sleep, antidepressant, relieves anxiety, calming, heals hurt emotions, promotes unconditional love

CAUTIONS:

Geraniol and Geranyl Acetate CTs

■ May moderately interfere with enzymes responsible for metabolizing medications (NSAIDs, proton-pump inhibitors, acetaminophen, antiepileptics, immune modulators, blood-sugar medications, blood pressure medications, antidepressants, antipsychotics, diabetic medications, antihistamines, antibiotics, and anesthetics) due to geraniol content.[2025]

■ May interact with diabetes medications and cause low blood sugar based on significant geraniol content.[2026,2027]

■ May interact with antibiotics and/or antifungals and possibly increase their effectiveness based on significant geraniol content (geraniol inhibits bacterial efflux pumps, which are responsible for antibiotic resistance and resistance to immune system defenses).[2028,2029,2030]

■ Dilution recommended for topical application. Geraniol is metabolized and autoxidized into metabolites like geranial and neral, and may cause sensitization and irritation when applied topically.[2031]

Neral CT

■ Caution is advised during pregnancy and lactation due to moderate neral. Large doses of citral (neral + geranial) may negatively affect fetal development according to animal studies.[2032]

■ There is a moderate risk that when the neral CT is taken orally, it may interfere with enzymes responsible for metabolizing medications (NSAIDs, proton-pump inhibitors, acetaminophen, antiepileptics, immune modulators, blood-sugar medications, blood pressure medications, antidepressants, antipsychotics, diabetic medications, antihistamines, antibiotics, and anesthetics).[2033,2034,2035]

■ Use cautiously with diabetic medications. Large oral doses of citral (neral + geranial) may improve insulin sensitivity and lower blood-glucose levels according to animal research.[2036,2037]

■ May interact with antibiotics and possibly enhance their effects based on neral content.[2038,2039]

■ Dilution is recommended for topical application due to high citral content and the risk for skin irritation or sensitivity.[2040] May also irritate mucous membranes.

SELECTED EVIDENCE:

- ☐ *In vitro* evidence showed that jamrosa essential oil CT neral inhibited the highest aflatoxin B1 producing strain, *A. flavus*, and suppressed its production of aflatoxin B1.[2041,2042]
- ☐ The neral CT of jamrosa essential oil exhibited antioxidant activity in the DPPH assay.[2043]

JAPANESE CATNIP
(Jing Jie)

Nepeta tenuifolia Benth., *Schizonepeta tenuifolia* (Benth.) Briq.

FAMILY: Lamiaceae (Labiatae)
NOTE: Middle
AROMA INTENSITY: Strong
AROMA: Minty, fresh, herbaceous
COMMON EXTRACTION METHOD: Hydrodistilled or steam distilled from the flower spikes, leaves, or stems
POSSIBLE SUBSTITUTE OILS: Pennyroyal, buchu, spearmint CT pulegone, apple mint CT pulegone
BLENDS WELL WITH: Buchu, camphor, catnip, citronella, eucalyptus, geranium, lavender, lemongrass, myrtle, pennyroyal, peppermint, rosemary, sage, Spanish sage, spearmint
SUBCELLULAR LOCALIZATION | EPIGENETIC INFLUENCE: Currently unknown | Currently unknown
RECOMMENDED DILUTION RANGE: 1.0%–5.0%

PRIMARY CONSTITUENTS:[2044,2045,2046,2047,2048,2049,2050,2051]

*Pulegone CT**

Pulegone	24.3%–91.0%
Menthone	0.0%–68.1%
2-Ethylhexyl Hexyl Sulfite	0.0%–26.7%
Anisic Acid,Tridec-2-ynyl Ester	0.0%–13.3%
Heptane,2,2,4,6,6-Pentamethyl	0.0%–9.6%
Limonene	0.0%–7.0%
2-Methylcyclohexanol	0.0%–6.0%
Shizonal	0.0%–5.4%
2,5-Dimethyl-3-Hexanone	0.0%–4.8%
Cis-Pulegone Oxide	0.0%–4.1%
(Z)- p-Mentha-6,8-dien-2-ol	0.0%–3.5%
Germacrene D	0.0%–3.0%
Isomenthone	0.0%–3.0%

2-Isopropyl-5-Methylcyclohexanone CT

2-Isopropyl-5-Methylcyclohexanone	50.2%
2-Isopropylidene-5-Methylcyclohexanone	33.9%
Limonene	3.2%

* The pulegone CT could be further divided into additional chemotypes—including CTs with menthone as the major constituent and pulegone secondary—but is combined for simplicity in the work.

OTHER CONSTITUENTS: *Pulegone CT*—1-Octen-3-ol, octen-1-ol, cis-pulegone, 3-octanone, myrcene, menthofuran, beta-caryophyllene, 1,4-butanediol, carvone, trans-pulegone oxide, Ethanone,1-cyclopropyl-2(1-pyrrolidinyl), 3-Heptanone,5-ethyl-4-methyl, 2,2′-methylenebis(6-238rticulates238-methyl-phenol), hexadecanoic acid, bis(2-ethylhexyl) ester, glycine, N-(4-butylbenzoyl)-,butyl ester, 8-hydroxy-p-menthan3-one, piperitone, piperitenone, mint furanone 1, caryophyllene oxide; *2-Isopropyl-5-Methylcyclohexanone CT*—1-Octen-3-ol, cis-verbenol, piperitenone, beta-caryophyllene, germacrene D, caryophyllene oxide

PREFERRED COMPOSITION FOR CLINICAL USE:

Constituent	
Pulegone	55%–75%
Menthone	15%–30%
Limonene	2%–8%

Beta-Caryophyllene	0.1%–5%
Myrcene	0.1%–5%
1-Octen-3-ol	tr–3%
3-Octanone	tr–3%
Menthofuran	≤ 1.0%

Reported Therapeutic Properties: Relieves chronic skin conditions, antiallergic, wound healing, analgesic, anti-inflammatory, antiarthritic, antispasmodic, soothes sore throat, relieves fever, antibacterial, antifungal, antiviral, anti-infectious, aids respiratory function, eases cough, antioxidant, insecticidal, insect repellent, reduces jealousy, uplifting

Cautions:

Japanese catnip essential oil should only be used in essential oil therapy when the benefits of use outweigh the risks as determined by a health professional trained in medical essential oil therapy.

▪ Avoid with children under age twelve. Multiple organ failure, liver failure, brain swelling (cerebral edema), severe epileptic encephalopathy (severe brain disorder with seizures), organ cell death (necrosis), and death has been reported in infants who were administered mint tea containing pennyroyal essential oil (also rich in pulegone).[2052] Pulegone and its metabolite menthofuran (or metabolites of menthofuran) are believed to cause the toxic effects of pennyroyal, which include liver and kidney toxicity, bronchiolar cell destruction, depletion of liver glutathione levels (the primary protective antioxidant of the liver), and direct cellular damage.[2053,2054,2055] Children are more likely to experience these effects.[2056]

▪ Avoid during pregnancy and lactation due to significant amounts of pulegone and potential toxicity to multiple organs.[2057,2058] Pennyroyal (also rich in pulegone) is believed to trigger uterine contractions by irritating the genito-urinary system. Pennyroyal is commonly used to induce abortions.[2059,2060] While pennyroyal herb is considered abortifacient, the essential oil is not.[2061] When pennyroyal essential oil is used alone to induce abortion, it generally is unsuccessful but causes serious health consequences to the mother. Carachipita is an abortifacient sold over the counter in South America that contains pennyroyal, yerba de la perdiz, oregano, and marsh rosemary. The literature reports successful use of Carachipita to induce abortion, but not without serious consequences to the mother (multiple organ failure, vaginal bleeding, destruction of red blood cells, numbness, dizziness, and death).[2062,2063,2064] It is prudent to avoid Japanese catnip oil while pregnant or nursing due to the toxicity of pulegone alone, despite a lack of evidence that it is a direct abortifacient.

▪ Avoid oral consumption due to the toxic potential of pulegone, menthofuran, and their metabolites to multiple organs. Menthofuran is generally a very minor constituent of Japanese catnip essential oil, but it is considered more toxic than pulegone.[2065] In addition, pulegone is metabolized by human liver cytochrome P-450s (CYP2E1, CYP1A2, and CYP2C19) to menthofuran after administration, which actually increases its toxic effect to the liver.[2066,2067] Pulegone is also metabolized to p-cresol, which depletes glutathione levels and can increase liver damage.[2068] Animal research suggests oral administration of pulegone is significantly toxic to the liver and causes death at 300 mg/kg (over 23 mL for a human adult equivalent dose) for two weeks.[2069] The same research concluded that lower doses of 75 and 150 mg/kg for three months resulted in liver, kidney, bone marrow, heart, stomach, lung, ovarian, and thymus damage, and one death in animals. A two-year animal study reported urinary cancer and papillomas, liver cancer, osteosarcoma, and multiple organ damage or enlargement (liver, kidney, olfactory epithelium, lung, bile duct).[2070] Ingestion of pulegone-rich essential oils can cause abdominal pain and tenderness, nausea, bloody vomiting, burning of the throat, fever, lethargy alternating with agitation, confusion, delirium, restlessness, seizures, dizziness, auditory and visual hallucinations, increased blood pressure, rapid heartbeat, swollen blood vessels in the lungs and filling of the alveoli with blood, excessive acidity of the blood, blood clots, abortion, liver failure, kidney failure, respiratory failure, shock, coma, and death.[2071] Neurological damage has been reported in adults who consumed 2.5 to 5.0 mL of pennyroyal essential oil, moderate to severe poisoning with single doses of 10 mL, and deaths occurred at over 15 mL of ingestion.[2072,2073,2074,2075,2076,2077,2078,2079,2080,2081,2082]

■ Avoid with epilepsy and Parkinson's disease due to pulegone content, which may exacerbate or cause seizures or convulsions.[2083,2084,2085]

■ Avoid in people with compromised liver or kidney due to significant potential to cause liver and kidney toxicity or damage caused by pulegone.[2086,2087,2088,2089,2090] This warning includes using Japanese catnip with medications known to harm the liver.

SELECTED EVIDENCE:

- □ Oral administration of Japanese catnip essential oil CT pulegone (once daily for four days) reduced inflammation in a rat pleurisy model as indicated by reduced exudate volume, leukocytes, protein level, myeloperoxidase (MPO), malondialdehyde (MDA), tumor necrosis factor-α (TNF-α) and interleukine-1β (IL-1β).[2091] Another study found that injecting Japanese catnip essential oil significantly inhibited the expression of intercellular adhesion molecule CD54 (a major regulator of pulmonary inflammation), the activation of nuclear factor-kappa B p65 (NF-kappaB p65; a protein complex involved in lung inflammation and disease), and the transcription of NF-kappaB p65 mRNA in lung cells.[2092]
- □ *In vitro* research showed that a menthone-rich (68.1%) Japanese catnip essential oil stron240rticulaibits fungi (*Botrytis cinerea* and *Colletotrichum gloeosporioides*) associated with postharvest diseases.[2093] The pulegone CT demonstrates broad-spectrum antimicrobial activity against antibiotic-resistant and -sensitive pathogens including significant inhibition of *Salmonella enterica, Citrobacter freundii*, and *Escherichia coli*, and good activity against *B. cereus, L. monocytogenes, S. aureus,* and *C. freundii.*[2094] The oil was least active against strains of *P. aeruginosa.*
- □ Japanese catnip essential oil CT pulegone exhibited antioxidant activity, with the authors concluding that the antioxidant potential of the oil correlates with pulegone and menthone content.[2095]
- □ Japanese catnip essential oil (49.7% menthone, 43.3% pulegone) caused 96.6% mortality against the mushroom fly (*Lycoriella ingenua*) larvae at 12.5 mcg/mL. Isolated pulegone was also highly toxic to larvae.[2096] Japanese catnip essential oil exhibited acaricidal activity against house dust and stored food mites, which was attributed to the presence of 2-isopropyl-5-methylcyclohexanone.[2097]
- □ Japanese catnip essential oil strongly repelled the booklouse (*Liposcelis bostrychophila*) and red flour beetle (*Tribolium castaneum*).[2098]

JASHIR
(Joshire Kotolei)

Prangos acaulis (DC.) Bornm.,
Cachrys acaulis DC.

FAMILY: Apiaceae (Umbelliferae)
NOTE: Middle-Strong
AROMA INTENSITY: Medium
AROMA: Earthy, hay-like, balsamic, woody

Common Extraction Method: Hydrodistilled or steam distilled from the aerial parts; may also be distilled from the stems, leaves, or flowers
Possible Substitute Oils: Katrafay, galbanum, helichrysum, pine
Blends Well With: Basil, bay laurel, bergamot, black pepper, camphor, cassia, cinnamon, coriander, cypress, geranium, grapefruit, lavender, lemon, myrrh, neroli, orange, p241rticulattgrain, palo santo, ravensara, ravintsara, rose, sandalwood, spikenard, tangerine, vetiver, ylang ylang
Subcellular Localization | Epigenetic Influence: Currently unknown | Currently unknown
Recommended Dilution Range: 3%–50%

Primary Constituents:[2099,2100,2101,2102]

Aerial Parts (Cis-Sesquisabinene Hydrate CT)	
Cis-Sesquisabinene Hydrate	27.1%
Alpha-Pinene	12.5%
Gamma-Eudesmol	7.4%
Spathulenol	7.2%
Beta-Elemene	6.8%
(E)-Phytol	5.6%
Beta-Eudesmol	3.4%

Aerial Parts (Alpha-Pinene CT)	
Alpha-Pinene	22.9%
3-Ethylidene-2-Methyl-1-Hexen-4-yne	21.4%
Delta-3-Carene	18.2%
Limonene	17.6%
Alpha-Phellandrene	6.9%
Myrcene	5.3%
Alpha-Terpinolene	3.2%

Leaves	
Alpha-Pinene	39.5%
3-Ethylidene-2-Methyl-1-Hexen-4-yne	37.9%
Alpha-Terpinene	10.9%
Limonene	5.2%
Alpha-Phellandrene	3.3%

Aerial Parts (Delta-3-Carene CT)	
Delta-3-Carene	25.5%
Alpha-Terpinolene	14.8%
Alpha-Pinene	13.6%
Limonene	12.9%
Myrcene	8.1%
Beta-Pinene	5.4%
Caryophyllene	3.0%

Flowers	
Alpha-Pinene	25.0%
3-Ethylidene-2-Methyl-1-Hexen-4-yne	23.5%
Alpha-Terpinene	17.3%
Limonene	13.6%
Alpha-Phellandrene	6.9%
Myrcene	5.4%

Stems	
3-Ethylidene-2-Methyl-1-Hexen-4-yne	56.8%
Alpha-Pinene	34.2%
Alpha-Terpinene	3.4%

Other Constituents: *Aerial Parts (Cis-Sesquisabinene Hydrate CT)*—Camphene, sabinene, beta-pinene, myrcene, limonene, alpha-terpinyl acetate, beta-caryophyllene, gamma-elemene, alpha-humulene, germacrene D, bicyclogermacrene, delta-cadinene, caryophyllene oxide, (Z,E)-farnesol; *Aerial Parts (Delta-3-Carene CT)*—Para-cymene, trans-ocimene, alpha-methylstyrene, linalool, nerol, alpha-humulene, gamma-curcumene; *Aerial Parts (Alpha-Pinene CT)*—Sabinene, beta-pinene; *Flowers*—Beta-pinene; *Leaves*—Myrcene, alpha-terpinolene; *Stems*—(Z)-2-nonen-4-yne, limonene, alpha-terpinolene, m-methyl styrene

Preferred Composition for Clinical Use:
Insufficient samples/data to determine.

Reported Therapeutic Properties: Moisturizes the skin, antiseptic, aids digestion, expels excess gas, antifungal, antibacterial, antiviral, antiparasitic, anticancer, antioxidant, anti-inflammatory, relieves kidney stones, relieves mental fatigue, aids concentration, combats inflexibility

Cautions:

- None currently known.

Selected Evidence:

- None found.

JAVA PATCHOULI (Indian Patchouli, Kadlum, False Patchouli, Lagumtum)

Pogostemon heyneanus Benth., *P. suavis* Ten.

Family: Lamiaceae
Note: Middle
Aroma Intensity: Medium
Aroma: Sweet, spicy, woodsy
Common Extraction Method: Steam distilled or hydrodistilled from the leaves
Possible Substitute Oils: Patchouli
Blends Well With: Bay laurel, bergamot, black pepper, cardamom, cedarwood, cassia, cinnamon, clove, frankincense, ginger, lavender, lemon, myrrh, orange, sandalwood, spikenard, vetiver
Subcellular Localization | Epigenetic Influence: Currently unknown | Currently unknown
Recommended Dilution Range: 3%–50%; neat for some conditions

Primary Constituents:[2103,2104,2105]

Beta-Pinene CT	
Beta-Pinene	25.4%–40.1%
Alpha-Pinene	12.8%–21.4%
Patchoulol	6.3%–20.2%
Acetophenone	5.1%–17.3%
(E)-Nerolidol	6.7%–15.5%
Beta-Caryophyllene	2.2%–8.2%
Alpha-Guaiene	1.5%–3.6%

Patchoulol CT	
Patchoulol	29.0%–57.1%
Beta-Caryophyllene	2.2%–22.9%
Seychellene	5.7%–14.8%
Alpha-Bulnesene (Delta-Guaiene)	6.1%–14.8%
Alpha-Guaiene	0.0%–9.7%
Alpha-Patchoulene	0.0%–4.5%
Beta-Pinene	0.0%–4.0%
Beta-Patchoulene	0.0%–3.2%

Acetophenone CT	
Acetophenone	51.0%
Patchoulol	14.0%
Beta-Pinene	5.3%
(E)-Nerolidol	5.4%
Limonene	4.0%
Benzyl Acetate	3.1%

OTHER CONSTITUENTS: *Beta-Pinene CT*—Limonene, beta-patchoulene, seychellene, alpha-guaiene, alpha-humulene, alpha-patchoulene, alpha-bulnesene (delta-guaiene), caryophyllene oxide; *Patchoulol CT*—Alpha-pinene, beta-elemene, beta-caryophyllene, acetophenone, alpha-humulene, caryophyllene oxide, (E)-nerolidol, pogostol; *Acetophenone CT*—Alpha-pinene, alpha-terpineol, alpha-guaiene, seychellene, alpha-patchoulene, alpha-bulnesene (delta-guaiene), pogostol

PREFERRED COMPOSITION FOR CLINICAL USE:
Insufficient samples/data to determine.

REPORTED THERAPEUTIC PROPERTIES: Diuretic, cleanses the urinary tract, gum health, reduces bad breath, aids digestion, expels excess gas, relieves diarrhea, eases nausea and vomiting, wound healing, relieves boils and abscesses, reduces the appearance of blemishes, supports cardiovascular function, insect repellent, antidepressant, stress management, mentally stimulating

CAUTIONS:
- None currently known.

SELECTED EVIDENCE:
- *In vitro* research reported that Java patchouli essential oil prevented biofilm formation and reduced virulence factors like staphylo xanthin and hemolysin in MRSA.[2106]
- An unidentified CT of Java patchouli essential oil provided 123.4 minutes of complete protection from mosquito (*Aedes albopictus*) bites when used alone (the highest of the three oils tested), and up to 391 minutes of protection when combined with turmeric and cape yellowwood essential oils in a 2:1:1 ratio.[2107] The same mixture of essential oils, used as a 20% concentration, repelled blackflies for up to 170.56 minutes.

JOINTED FLATSEDGE
(Piprioca, Piri Piri)

Cyperus articulatus L., *C. articulatus* L. var. *articulatus*, *C. articulatus* L. var. *nodosus*

FAMILY: Cyperaceae
NOTE: Middle
AROMA INTENSITY: Medium
AROMA: Woody, floral, spicy
COMMON EXTRACTION METHOD: Steam distilled or hydrodistilled from the rhizomes (roots); may also be distilled from the stems
POSSIBLE SUBSTITUTE OILS: Helichrysum, ginger, cypriol
BLENDS WELL WITH: Cassia, cinnamon, clove, cypress, geranium, ginger, helichrysum, lavandin, lavender, patchouli, ylang ylang
SUBCELLULAR LOCALIZATION | EPIGENETIC INFLUENCE: Currently unknown | Currently unknown
RECOMMENDED DILUTION RANGE: 3%–33%; 50% for some conditions

Primary Constituents:[2108,2109,2110,2111,2112,2113,2114,2115]

Cyperene CT

Cyperene	33.6%
Cedryl Methyl Ketone	7.1%
Beta-Santalene	5.5%
3-Ethyl Hexane	4.6%
2-Nonanone	3.9%
Mandassidione	3.1%
Himachalene-Gamma-Dehydrone	3.1%
Cis-Calamenene	3.0%

Caryophyllene Oxide CT

Caryophyllene Oxide	10.1%–23.4%
Mustakone	8.0%–14.5%
Gamma-Patchoulene	0.0%–12.8%
Humulene Epoxide II	2.4%–11.2%
Alpha-Pinene	0.6%–7.7%
Beta-Pinene	0.6%–7.4%
Alpha-Cadinol	0.0%–6.6%
Trans-Pinocarveol	3.8%–6.0%
Myrtenal + Myrtenol	4.1%–5.9%
Cyperotundone	0.6%–5.4%
Cedrol*	0.0%–5.3%
Alpha-Cyperone	1.3%–5.2%
Ledol	2.1%–4.6%
Cis-Carveol	0.0%–4.6%
Cyperene	0.5%–4.3%

Cyperotundone CT

Cyperotundone	42.3%
Piperitone	10.1%
Beta-Maaliene	8.4%
Germacrone	5.3%
1-Epi-Cubenol	3.3%
Germacrene D	3.0%

Verbenone CT

Verbenone	19.6%
Trans-Pinocarveol	17.4%
Myrtenal	8.2%
P-Mentha-1,5-Dien-8-ol	7.1%
Myrtenol	4.6%
P-Cymen-8-ol	4.4%
Pinocarvone	3.7%
Alpha-Terpineol	3.4%

Alpha-Pinene CT

Alpha-Pinene	5.7%–12.9%
Mustakone	5.7%–9.8%
Myrtenal + Myrtenol	4.5%–7.7%
Trans-Pinocarveol	4.4%–7.5%
Beta-Pinene	3.5%–7.4%
Caryophyllene Oxide	3.4%–7.0%
Alpha-Cyperone	0.0%–5.9%
Alpha-Copaene	2.0%–5.0%
Beta-Selinene	1.9%–5.0%
Alpha-Bulnesene	1.0%–5.0%
Cyperotundone	0.0%–4.5%
Ledol	0.0%–4.3%
Verbenone	1.7%–3.9%
P-Mentha-1,5-Dien-8-ol	0.0%–3.0%

Cedrol CT

Cedrol*	19.0%
Guaia-5-en-11-ol	14.9%
Cyperotundone	9.6%
Sabinene	5.2%
Alpha-Pinene	5.2%
Trans-Pinocarveol	5.2%
Cis-Carveol	4.7%
Trans-Carveol	4.0%
Alpha-Cadinol	3.4%

Stems

Caryophyllene Oxide	18.2%–27.4%
Gamma-Patchoulene	0.0%–12.1%
Mustakone	0.0%–10.0%
Ledol	2.1%–8.8%
Cyperotundone	0.6%–6.7%
Humulene Epoxide II	0.0%–6.5%
Alpha-Copaene	2.1%–6.5%
Dill Apiole	0.0%–5.3%
Alpha-Cadinol	0.0%–5.2%
Limonene	0.4%–4.1%
Beta-Caryophyllene	0.0%–3.9%
Myrtenol	0.3%–3.8%
Trans-Pinocarveol	0.2%–3.8%
Cyperene	1.1%–3.4%
Cedrol	0.0%–3.1%

Alpha-Cadinol CT

Alpha-Cadinol	12.1%
Trans-Pinocarveol	9.9%
Cyperenone	7.3%
Cyperene	6.3%
Myrtenol	5.5%
Myrtenal	4.9%
Cis-Carveol	4.4%
Nopinone	3.6%
Caryophyllene Oxide	3.2%
Alpha-Pinene	3.2%

var. articulatus

Mustakone	9.8%–14.5%
Alpha-Pinene	5.7%–12.3%
Caryophyllene Oxide	4.6%–10.8%
Beta-Pinene	4.2%–6.6%
Myrtenol + Myrtenal	5.8%–6.3%
Trans-Pinocarveol	5.5%–6.0%
Alpha-Cyperone	1.4%–5.9%
Cyperotundone	3.7%–5.4%
Ledol	3.2%–4.6%
Unidentified M218	2.1%–4.4%
Cyperene	1.8%–4.3%
Verbenone	1.5%–3.9%
Alpha-Copaene	2.2%–3.3%

* Correct isomer not identified.

var. nodosus

Mustakone	6.6%–14.5%
Caryophyllene Oxide	5.4%–13.7%
Alpha-Pinene	4.7%–11.9%
Humulene Epoxide II	1.5%–11.2%
Cyperene	0.6%–8.9%
Myrtenol + Myrtenal	3.8%–8.3%
Ledol	2.5%–8.3%
Trans-Pinocarveol	3.8%–8.1%
Beta-Pinene	3.1%–7.4%
Alpha-Copaene	2.1%–7.3%
Cyperotundone	2.6%–5.3%
Alpha-Cyperone	1.4%–4.6%
Unidentifed M218	0.7%–4.0%
Eudesma-3,11-dien-5-ol	0.9%–3.6%
Beta-Selinene	0.0%–3.3%
Alpha-Bulnesene	0.8%–3.0%
Verbenone	0.1%–3.0%

OTHER CONSTITUENTS: *Cyperene CT*—2-Octene, 1-octene(Z), 2-methyl heptane, trans-1-Ethyl-3-methyl cyclopentane, 3-methyl-4-heptanone, alpha-thujene, isopropyl tiglate, 1-octen-3-ol, n-octanal, 1,2,4-trimethyl benzene, 2,6-fimethyl nonane, bergamal, 2-phenyl propanal, p-1,3,8-menthatriene, decanone, methyl chavicol (< 0.7%), isoledene, thujopsadiene, 1,4-Methano-1H-indene, octahydro-4-methyl-8-methylene-7-(1-methylethyl), gamma-gurjunene, alpha-copaene-8-ol, 3-thujopsanone, corymbolone; *Alpha-Pinene CT*—Thuja-2,4(10)-diene, para-cymene, limonene, alpha-campholenal, trans-verbenol, cis-verbenol, pinocarvone, terpinen-4-ol, p-cymen-8-ol, trans-carveol, carvone, cyperene, rotundene, germacrene D, eudesma-2,4,11-triene, alpha-selinene, delta-cadinene, trans-calamenene, alpha-calacorene, humulene epoxide II, dill apiole (< 1.1%), pathchoulenone, eudesma-3,11-dien-5-ol, aristolone, ishwarane, alpha-guaiene; *Caryophyllene Oxide CT*—Thuja-2,4(10)-diene, para-cymene, para-cymenene, alpha-campholenal, trans-verbenol, cis-verbenol, pinocarvone, p-mentha-1,5-dien-8-ol, terpinen-4-ol, p-cymen-8-ol, trans-carveol, carvone, p-cymen-7-ol, alpha-copaene, beta-gurjunene, beta-caryophyllene, alpha-guaiene, alpha-humulene, rotundene, germacrene D, eudesma-2,4,11-triene, beta-selinene, alpha-selinene, delta-cadinene, trans-calamenene, cis-calamenene, ledene oxide, alpha-amorphene, globulol, alpha-bulnesene, germacrone, cyclocolorenone, alpha-calacorene, beta-copaen-4alpha-ol, patchoulenone, caryophylla-4(14),8(15)-dien-5alpha-ol, eudesma-3,11-dien-5-ol, aristolone; *Cyperotundone CT*—Alpha-pinene, sabinene, myrcene, 1,8-cineole, trans-pinocarveol, borneol, alpha-terpineol, neral, cyperadiene, alloaromadendrene, rotundene, gamma-cadinene, cyperene oxide, cyclocolorenone; *Cedrol CT*—1,8-cineole, para-cymene, linalool, camphor, terpinen-4-ol, alpha-terpineol, beta-cyclocitral, geranyl formate, cyperene, aromadendrene, cubeb-11-ene, alpha-bulnesene, cis-calamenene, caryophyllene oxide, beta-bisabolol; *Verbenone CT*—1,8-cineole, alpha-campholenal, nopinone, trans-verbenol, sabinene ketone, terpinen-4ol, trans-carveol, carvone; *Alpha-Cadinol CT*—Diacetone alcohol, thuya-2,4(10)-diene, beta-pinene, 1,8-cineole, p-mentha-1,5-dien-8-ol, alpha-terpineol, p-cymen-8-ol, verbenone, methyl thymol, carvone, beta-maaliene, eudesma-2,4-11-triene, beta-calacorene, ledol, globulol, beta-copaene-4alpha-ol, humulene epoxide II, cedrol, patchoulenone, caryophylla-4(14),8(15)-dien-5alpha-ol, mustakone, cyperotundone; *Stems*—Alpha-pinene, sabinene, alpha-campholenal, myrtenal, myrtenol, carvone, p-cymen-7-ol, alpha-humulene, germacrene D, beta-selinene, alpha-selinene, alpha-calacorene, beta-copaene-4-alpha-ol, patchoulenone, caryophylla-4(14),8(15)-dien-5alpha-ol, alpha-cyperone, aristolone, cis-verbenol, beta-cyclocitral, beta-elemene, beta-gurjunene, gamma-muurolene, cis-calamenene, ledene oxide, amorphene, longipinocarvone, ledol, 1-epi-cubenol, germacrone, beta-bisabolol, alpha-cyp246rticulatesocolorenone; *var. articulatus*—Thuja-2,4(10)-diene, para-cymene, limonene, cis-verbenol, pinocarvone, terpinen-4-ol, para-cymen-8-ol, trans-carveol, carvone, rotundene, germacrene D, eudesma-2,4,11-triene, beta-selinene, alpha-selinene, alpha-bulnesene, delta-cadinene, trans-calamenene, alpha-calacorene, beta-copaen-4alpha-ol, humulene epoxide II, dill apiole (< 1.5%), patchoulenone, caryophylla-4(14),8(15)-dien-5-alpha-ol, eudesma-3,11-dien-5-ol, aristolone; *var. nodosus*—Thuja-2,4(10)-diene, para-cymene, limonene, para-cymenene, 1,8-cineole, alpha-campholenal, trans-verbenol, cis-verbenol, pinocarvone, p-mentha-1,5-dien-8-ol, terpinen-4-ol, para-cymen-9-ol, trans-carveol, carvone, beta-caryophyllene, alpha-guaiene, alpha-humulene, rotundene, germacrene D, eudesma-2,4,11-triene, alpha-selinene, delta-cadinene, trans-calamenene, alpha-calacorene, beta-copaen-4alpha-ol, dill apiole (< 1.7%), patchoulenone, caryophylla-4(14),8(15)-dien-5-alpha-ol, aristolone

PREFERRED COMPOSITION FOR CLINICAL USE:
Insufficient samples/data to determine.

REPORTED THERAPEUTIC PROPERTIES: Antibacterial, antifungal, anti-infectious, antiparasitic, relieves migraine, analgesic, reduces fever, anti-inflammatory, anticonvulsant, alleviates diarrhea, aids digestion, reduces nausea and vomiting, expels excess gas, anticancer, improves circulation, wound healing, soothes upset stomach, calming/relaxing, promotes vivid dreams, aphrodisiac

CAUTIONS:

Caryophyllene Oxide CTs

- May interact with aspirin, blood pressure, antiplatelet, and anticoagulant medications, and increase the risk of bleeding among people with bleeding disorders, due to caryophyllene-oxide content.[2116]

All Other CTs

- None currently known.

SELECTED EVIDENCE:

- *In vitro* research reported that jointed flatsedge CT cyperene essential oil inhibited *S. aureus, B. subtilis, E. coli,* and *C. albicans*.[2117] The verbenone CT of jointed flatsedge essential oil inhibited *Candida albicans*.[2118] The alpha-pinene CT of jointed flatsedge essential oil inhibited and prevented biolfilm formation by oral pathogens (*Candida albicans, S. sanguis, S. mitis, P. gingivalis*, and *F. nucleatum*) demonstrating broad-spectrum antimicrobial activity.[2119]

KARO KAROUNDE ABSOLUTE
(Karo-Karoundé, Kaulathi)

Leptactina senegambica Hook.f., *L. senegambica* Hook

FAMILY: Rubiaceae
NOTE: Base
AROMA INTENSITY: Medium
AROMA: Fruity, musky, floral, green, herbaceous, woody
COMMON EXTRACTION METHOD: Solvent extracted from the flowers
POSSIBLE SUBSTITUTE OILS: Tuberose, gardenia, jasmine absolute
BLENDS WELL WITH: Bergamot, cassia, cinnamon, clove, coriander, grapefruit, jasmine, lavender, lemon, lime, mandarin, neroli, rose, sandalwood, tangerine, tuberose, ylang ylang
SUBCELLULAR LOCALIZATION | EPIGENETIC INFLUENCE: Currently unknown | Currently unknown
RECOMMENDED DILUTION RANGE: 1%

PRIMARY CONSTITUENTS:[2120,2121]

Benzyl Cyanide	4.8%
2-Phenylnitroethane	1.1%

Note: The bulk of the absolute is saturated and unsaturated fatty acids, together with their ethyl-, 2-methylbutyl-, benzyl, and phenylethyl esters.

OTHER CONSTITUENTS: Isoeugenol

PREFERRED COMPOSITION FOR CLINICAL USE:
Insufficient data/samples to determine.

REPORTED THERAPEUTIC PROPERTIES: Deodorant

CAUTIONS:
- Karo karounde absolute is not generally recommended for therapy unless the benzyl cyanide content is determined to be less than 5%.

SELECTED EVIDENCE:
- □ None found.

KESUM
(Small Water Pepper, Kesom, Pygmy Smartweed)

Persicaria minor (Huds.) Opiz,
Polygonum minus Huds.

FAMILY: Polygonaceae
NOTE: Middle
AROMA INTENSITY: Strong
AROMA: Pungent, fresh, herbaceous, green, citrusy and slightly floral
COMMON EXTRACTION METHOD: Steam distilled or hydrodistilled from the leaves; may also be distilled form the stems
POSSIBLE SUBSTITUTE OILS: Vietnamese cilantro, cinnamon
BLENDS WELL WITH: Bergamot, chamomile (German, Roman), cinnamon, lavender, lemon, marjoram, orange, oregano, tea tree, thyme, vetiver
SUBCELLULAR LOCALIZATION | EPIGENETIC INFLUENCE: Currently unknown | Currently unknown
RECOMMENDED DILUTION RANGE: 3%–33%; 50% for some conditions

PRIMARY CONSTITUENTS:[2122,2123,2124,2125]

Dodecanal CT		*Stems*	
Dodecanal	43.5%–54.3%	Dodecanal	47.3%
Decanal	16.3%–24.4%	Decanal	5.7%
1-Decanol	1.6%–12.7%	1-Dodecanol	5.5%
Beta-Caryophyllene	0.2%–3.8%	1-Decanol	4.0%
8-Bromo-Neoisolongifolene	0.0%–3.1%		

Caryophyllene Oxide CT	
Caryophyllene Oxide	42.5%
Alpha-Panasinsene	6.8%
Pinocarveol	4.2%

Note: Some consider Vietnamese coriander (*Persicaria odorata*) and kesum (*Persicaria minor*) to be the same plant.

OTHER CONSTITUENTS: *Dodecanal CT*—Tetradecanal, geraniol, 1-dodecanol, nerolidol, alpha-bisabolol, bornyl acetate, isobornyl acetate, n-decanoic acid, beta-himachalene, drimenol, alpha-bergamotene, seychellene, alpha-humulene, beta-farnesene, beta-selinene, alpha-zingiberene, valencene, germacrene D, alpha-panasinsene, delta-cadinene, caryophyllene oxide; *Caryophyllene Oxide CT*—Decanal, undecanol, dodecanal, tetradecanol, alpha-sinensal, 1-nonanol, myrtenol, 1-decanol, geraniol, farnesol, nerolidol, globulol, derminol, bornyl acetate, beta-farnesene, alpha-curcumene, beta-selinene, alpha-zingiberene, valencene, beta-bisabolene, alpha-sesquiphellandrene, delta-cadinene, alpha-humulene, alloaromadendrene, isocaryophyllene, n-decanoic acid, dodecanoic acid; *Stems*—1-Nonanol, undecanol, 1-undecanol, tetradecanal

PREFERRED COMPOSITION FOR CLINICAL USE:

Constituent	
Dodecanal	40%–55%
Decanal	12%–27%
1-Decanol	1%–13%
Beta-Caryophyllene	0.1%–7%

REPORTED THERAPEUTIC PROPERTIES: Aids digestion, heals stomach ulcers, relieves diarrhea, nourishes the scalp and hair, analgesic, anti-inflammatory, antimicrobial, **antibacterial**, antifungal, anticancer, antioxidant, antiaging, relaxing, stress management

CAUTIONS:

Dodecanal CTs

- None currently known.

Caryophyllene Oxide CT

- May interact with aspirin, blood pressure, antiplatelet, and anticoagulant medications, and increase the risk of bleeding among people with bleeding disorders, due to caryophyllene-oxide content.[2126]

SELECTED EVIDENCE:

- ☐ None found.

KEWDA (Screw-Pine, Keora, Umbrella Tree, Ruh Kewda, Kewda Attar)

Pandanus odorifer (Forssk.) Kuntze, *P. fascicularis* Lam., *P. odoratissimus* L.f.

FAMILY: Pandanaceae
NOTE: Top
AROMA INTENSITY: Medium-Strong
AROMA: Sweet, floral, fruity, honey-like
COMMON EXTRACTION METHOD: Steam distilled or hydrodistilled from the flowers or inflorescences (the complete flower head, including stems, stalks, bracts, and flowers)
POSSIBLE SUBSTITUTE OILS: Ylang ylang, rose
BLENDS WELL WITH: Agarwood, allspice, ambrette, angelica, anise, balsam fir, benzoin, Buddha wood, cassia, cedarwood, cinnamon, clove, galbanum, ginger, jasmine, helichrysum, opoponax, neroli, patchouli, Peru balsam, sandalwood, silver fir, white fir, ylang ylang
SUBCELLULAR LOCALIZATION | EPIGENETIC INFLUENCE: Currently unknown | Currently unknown
RECOMMENDED DILUTION RANGE: 3%–33%; 50% for some conditions

PRIMARY CONSTITUENTS:[2127,2128,2129,2130,2131]

2-Phenylethyl Methyl Ether	16.1%–75.4%
2-Phenethyl Alcohol	0.0%–33.2%

Terpinen-4-ol	0.0%–21.0%
Ortho-Cymene	0.0$–11.3%
Benzyl Benzoate	0.0%–11.0%
Viridine	0.0%–8.7%
Germacrene B	0.0%–8.3%
Alpha-Terpineol	0.0%–8.3%
2-Phenylethyl Isobutyrate	0.0%–4.6%
Para-Cymene	0.0%–3.1%

OTHER CONSTITUENTS: Alpha-thujene, alpha-pinene, sabinene, beta-pinene, alpha-terpinene, limonene, 1,8-cineole, gamma-terpinene, delta-2-carene, cis-linalool oxide, trans-linalool oxide, linalool, terpinolene, trans-p-menth-2-en-1-ol, piperitone, 2-phenylethyl benzoate, geranyl formate, benzaldehyde, benzeneacetaldehyde, isoborneol, para-cymenol, piperitone, beta-phenethyl acetate

PREFERRED COMPOSITION FOR CLINICAL USE:

Constituent	
2-Phenylethyl Methyl Ether	60%–82%
Terpinen-4-ol	5%–20%
Alpha-Terpineol	0.1%–7%

REPORTED THERAPEUTIC PROPERTIES: Analgesic, anti-inflammatory, antirheumatic, relieves headache, antispasmodic, anti-infectious, anticonvulsant, would healing, heals boils and scabies, aids digestion, heals stomach ulcers, supports liver function, anticancer, aids cardiovascular function, antioxidant, deodorant, aphrodisiac, promotes joyful feelings, emotionally balancing

CAUTIONS:

- None currently known.

SELECTED EVIDENCE:

- ☐ None found.

KHENJUK
(Kolkhoung, Khinjuk)

Pistacia khinjuk Stocks

FAMILY: Anacardiaceae
NOTE: Middle
AROMA INTENSITY: Medium-Strong
AROMA: Herbaceous, woody, slightly spicy
COMMON EXTRACTION METHOD: Steam distilled or hydrodistilled from the aerial parts, fruits, or leaves
POSSIBLE SUBSTITUTE OILS: Summer savory CT gamma-terpinene, fingerroot CT gamma-terpinene, lovage, betony, dill CT alpha-phellandrene, pink pepper, vetiver
BLENDS WELL WITH: Bergamot, betony, copaiba, frankincense, goldenrod, hemp, hyssop, ginger, lavender, lemon, lime, marjoram, orange, patchouli, sage, Spanish sage, vetiver

SUBCELLULAR LOCALIZATION | EPIGENETIC INFLUENCE: Currently unknown | Currently unknown
RECOMMENDED DILUTION RANGE: 3%–33%; 50% for some applications

PRIMARY CONSTITUENTS:[2132,2133,2134,2135,2136,2137,2138]

Fruits (Phellandrene CT)

Phellandrene*	52.3%–53.1%
Alpha-Pinene	15.3%–15.6%
Octadecanoic Acid	6.0%–6.3%
Limonene	4.1%–4.6%

Leaves (Spathulenol CT)

Spathulenol	20.9%
Germacrene B	9.5%
Dehydroaromadendrene	8.8%
Alpha-Eudesmol	6.8%
Zizanol	6.5%
Thujopsanone	3.8%
Delta-Occidentalol Cadinene	3.8%
Occidentalol Acetate	3.7%
Beta-Caryophyllene	3.7%

* Correct isomer not identified in either study.

Aerial Parts (Gamma-Terpinene CT)

Gamma-Terpinene	81.1%
Beta-Pinene	3.9%

Leaves (Alpha-Pinene CT)

Alpha-Pinene	18.4%
Sabinene	13.5%
Terpinen-4-ol	12.1%
Beta-Pinene	9.4%
Para-Cymene	6.4%
Caryophyllene Oxide	4.0%
Beta-Caryophyllene	3.3%

OTHER CONSTITUENTS: *Aerial Parts (Gamma-Terpinene CT)*—Camphene, sabinene, myrcene, delta-3-carene, limonene, terpinolene, 3-cyclohexene-1-carboxaldehyde, bornyl acetate; *Fruits (Phellandrene CT)*—Sabinene, 1,3,6-octatriene, gamma-terpinene, terpinolene, (Z)-4,8-dimetyl-1,3,7-nonatriene, linalool, thujopsene, caryophyllene oxide, hexadecanoic acid, 9-octadecanoic acid, ethyl oleate; *Leaves (Spathulenol CT)*—Tricyclene, alpha-pinene, camphene, sabinene, beta-pinene, myrcene, limonene, nonanal, camphenol, p-menth-1,5-dien-8-ol, alpha-terpineol, bornyl acetate, delta-elemene, alpha-guaiene, luparene, beta-eudesmol, khusimol, aristolone, pentacosane; *Leaves (Alpha-Pinene CT)*—Tricyclene, alpha-thujene, camphene, myrcene, alpha-terpinene, 1,8-cineole, beta-phellandrene, limonene, gamma-terpinene, linalool, cis-sabinene hydrate, alpha-terpineol, alpha-humulene, alpha-farnesene, (Z)-3-hexenyl benzoate

PREFERRED COMPOSITION FOR CLINICAL USE:
Insufficient samples/data to determine.

REPORTED THERAPEUTIC PROPERTIES: Anticancer, antioxidant, antibacterial, antifungal, antiviral, aids digestion, reduces nausea, vomiting, and motion sickness, heals stomach ulcer, relieves diarrhea, reduces fever, analgesic, anti-inflammatory, soothes sore throat, relieves toothache, astringent, soothes skin itching and chronic skin conditions, helps eliminate kidney stones, grounding, aids cognitive function, enhances focus and concentration, reduces excitability

CAUTIONS:
- None currently known.

SELECTED EVIDENCE:
- Khenjuk leaf CT spathulenol essential oil showed moderate anthelmintic activity against the hydatid worm (*Echinococcus granulosus*).[2139]

- *In vitro* research showed that khenjuk essential CT gamma-terpinene oil inhibited *P. aeruginosa* and *B. subtilis*.[2140] Other research reported that the gamma-terpinene CT strongly inhibited *E. coli, S. aureus*, *P. aeruginosa,* and *C. albicans*.[2141,2142,2143]
- Higher concentrations of the phellandrene CT of khenjuk essential oil prevented lipid peroxidation and exhibited antioxidant capacity in the DPPH assay.[2144]

LANTANA (Big-Sage, Wild Sage, Red Sage Red Sedge, Tickberry)

Lantana camara L., *L. aculeata* L., *L. strigocamara* R.W. Sanders

FAMILY: Verbenaceae
NOTE: Middle-Top
AROMA INTENSITY: Medium-Strong
AROMA: Herbaceous, fresh, fruity, floral, slightly balsamic and resinous
COMMON EXTRACTION METHOD: Steam distilled or hydrodistilled from the leaves, flowers, or aerial parts; may also be distilled from the fruits or stems
POSSIBLE SUBSTITUTE OILS: Threeleaf shrubverbena, goldenrod, copaiba, davana, ylang ylang, helichrysum, lavender
BLENDS WELL WITH: Bergamot, blue tansy, chamomile (Roman, German), copaiba, davana, goldenrod, lavender, lemon, orange, neroli, tangerine, ylang ylang
SUBCELLULAR LOCALIZATION | EPIGENETIC INFLUENCE: Currently unknown | Currently unknown
RECOMMENDED DILUTION RANGE: 3%–33%; 50% for some conditions

PRIMARY CONSTITUENTS:[2145,2146,2147,2148,2149,2150,2151,2152,2153,2154,2155,2156,2157,2158,2159,2160]

Leaves ((E)-Caryophyllene CT)

(E)-Caryophyllene	13.6%–23.8%
Bicyclogermacrene	0.0%–15.8%
Alpha-Humulene	2.9%–11.8%
Germacrene D	2.0%–11.7%
Isocaryophyllene	0.0%–9.6%
Sabinene	0.0%–9.2%
ar-Curcumene	0.0%–8.6%
Valencene	0.0%–8.3%
Davanone	0.0%–7.3%
Gamma-Muurolene	0.0%–6.9%
Geranial	0.0%–6.9%
Gamma-Curcumene	0.0%–6.3%
Alpha-Terpinolene	0.0%–6.0%
Gamma-Elemene	0.0%–5.7%
Squalene	0.0%–5.6%
1,8-Cineole	0.0%–5.5%
(E)-Nerolidol	0.0%–4.7%
Alpha-Pinene	0.0%–4.0%

Leaves (Germacrene D CT)

Germacrene D	19.8%–46.1%
(E)-Caryophyllene	9.4%–19.7%
Bicyclogermacrene	0.0%–11.7%
Gamma-Elemene	0.0%–10.3%
Alpha-Humulene	0.0%–9.3%
Beta-Elemene	0.0%–7.3%
Alpha-Copaene	1.1%–4.9%
T-Muurolol	0.0%–4.8%
Germacrene B	0.0%–4.2%
(E)-Phytol	0.0%–4.0%
Delta-Cadinene	0.0%–3.6%
Alpha-Cadinene	0.0%–3.3%

Leaves (Bicyclogermacrene CT)

Bicyclogermacrene	19.4%
Isocaryophyllene	16.7%
Valencene	12.9%
Germacrene D	12.3%

Caryophyllene Oxide	0.0%–3.8%
Gamma-Terpinene	0.0%–4.0%
Neral	0.0%–3.5%
Limonene	0.0%–3.4%
Spathulenol	0.0%–3.1%
Beta-Pinene	0.5%–3.0%

Flowers (Germacrene D CT)

Germacrene D	32.4%
(E)-Caryophyllene	13.6%
Limonene	7.8%
Alpha-Copaene	4.7%
Bicyclogermacrene	5.8%
Germacrene B	5.0%
Delta-Cadinene	3.3%
Gamma-Terpinene	3.2%

Aerial Parts (Germacrene D CT)

Germacrene D	15.9%
Germacrene B	6.2%
Cycloheptatriene	5.4%
Bicyclogermacrene	4.7%
1,8-Cineole	4.6%
(E)-Caryophyllene	3.9%
Gamma-Terpinene	3.4%

Aerial Parts (Davanone CT)

Davanone	44.4%
(E)-Nerolidol	13.0%
Octen-1-en-3-ol	5.3%
Alpha-Humulene	3.4%
Unidentified	3.3%

Aerial Parts (Sabinene CT)

Sabinene	16.9%
(E)-Caryophyllene	13.8%
1,8-Cineole	9.0%
Alpha-Humulene	5.9%
Bicyclogermacrene	5.8%
Unidentified	3.5%

Aerial Parts (Not Separated in Study)

(E)-Caryophyllene	3.8%–43.6%
Davanone	0.0%–35.1%
Beta-Bisabolene	0.4%–23.7%
ar-Curcumene	0.0%–22.6%
Sabinene	1.8%–20.4%
Epi-Cubenol	0.0%–12.0%
Azulenol	5.6%
Cis-p-Menth-2-en-ol	4.2%
Sabinene	4.8%

Flowers (Beta-Elemene CT)

Beta-Elemene	14.5%
Germacrene D	10.6%
Beta-Cadinene	7.2%
(E)-Caryophyllene	7.0%
Gamma-Elemene	6.8%

Aerial Parts ((E)-Caryophyllene CT)

(E)-Caryophyllene	34.6%
Caryophyllene Oxide	10.8%
Beta-Elemene	7.3%
Germacrene D	4.4%
ar-Curcumene	4.3%
Delta-Cadinene	4.1%
Bicyclosesquiphellandrene	3.8%
Alpha-Humulene	3.7%
Unidentified	3.4%

Aerial Parts ((E)-Nerolidol CT)

(E)-Nerolidol	16.6%
(E)-Beta-Farnesene	11.3%
Unidentified	7.8%
Germacrene D	7.3%
1,8-Cineole	6.9%
Alpha-Pinene	3.2%

Aerial Parts (Caryophyllene Oxide)

Caryophyllene Oxide	50.3%
(E)-Nerolidol	13.1%
Alpha-Copaene	10.6%
(E)-Caryophyllene	7.5%
Spathulenol	3.2%

Fruits or Stems

Palmitic Acid	22.8%–32.7%
Stearic Acid	12.8%–23.9%
Germacrene D	3.6%–7.1%
(–)-Italicene Ether	0.6%–4.6%
Ethyl Linolenate	2.0%–4.1%
Gamma-Curcumenal	0.5%–3.4%
Sesquiphellandrene	2.3%–3.1%

Alpha-Humulene	0.0%–11.7%
1,8-Cineole	0.0%–10.6%
Alpha-Pinene	0.0%–9.0%
Alpha-Terpineol	0.0%–8.7%
Linalool	0.0%–8.6%
Germacrene D	0.9%–7.3%
Gamma-Cadinene	0.0%–7.3%
Beta-Pinene	0.0%–6.6%
Humulene Oxide	0.0%–6.0%
(Z)-Nerolidol	0.0%–5.3%
Spathulenol	0.0%–4.7%
Delta-Cadinene	0.0%–4.4%
(E)-Beta-Ocimene	0.0%–4.1%
Alpha-Cadinol	0.0%–3.3%
Delta-Elemene	0.0%–3.3%
Delta-3-Carene	0.0%–3.0%

L. camara var. moritziana	
Germacrene D	31.0%
(E)-Caryophyllene	14.8%
Alpha-Phellandrene	6.7%
Limonene	5.7%
1,8-Cineole	5.2%
Alpha-Chamigrene	5.1%
L. camara var. aculeata	
Sabinene	19.6%–21.5%
1,8-Cineole	12.6%–14.8%
Beta-Caryophyllene	12.7%–13.4%
Alpha-Humulene	5.8%–6.3%
Alpha-Pinene	2.2%–4.3%
Delta-3-Carene	2.1%–3.2%
Humulene Epoxide III	2.3%–3.0%

OTHER CONSTITUENTS: *Leaves (Germacrene D CT)*—Alpha-pinene, sabinene, beta-pinene, myrcene, n-decane, delta-3-carene, para-cymene, (Z)-beta-ocimene, (E)-beta-ocimene, gamma-terpinene, linalool, tetradecane, beta-gurjunene, (E)-beta-farnesene, alpha-muurolene, germacrene A, cubenol, germacrene-d-4-ol, davanone, humuladyenone, terpinen-4-ol, eugenol, alpha-bergamotene, alloaromadendrene, alpha-zingiberene, sesquiphellandrene, (E)-nerolidol, spathulenol, caryophyllene oxide, helifolen-12-al, 1-epi-cubenol, T-cadinol, alpha-cadinol, epi-alpha-bisabolol, 1-nonadecanol, delta-elemene, beta-bourbonene; *Leaves ((E)-Beta-Caryophyllene CT)*—Alpha-thujene, camphene, myrcene, delta-3-carene, alpha-terpinene, alpha-phellandrene, beta-phellandrene, para-cymene, (Z)-beta-ocimene, (E)-beta-ocimene, alpha-terpineol, (Z)-beta-terpineol, 4-nonanone, linalool, piperitone, trans-geraniol, epi-bicyclosesquiphellandrene, alpha-muurolene, Selina-3,7(11)-diene, alpha-cubebene, alpha-copaene, beta-elemene, alloaromadendrene, cubebol, delta-cadinene, (Z)-nerolidol, o-cymene, germacrene-d-4-ol, borneol, davana ether 1, davana ether 3, cis-alpha-bisabolene oxide, 8-alpha-acetoxyelemol, phytol, 1-octen-3-ol, gamma-cadinene; *Leaves (Bicyclogermacrene CT)*—Para-cymene, (Z)-beta-ocimene, (E)-beta-ocimene, gamma-terpinene, terpinolene, terpinen-4-ol, alpha-elemene, beta-elemene, (E)-beta-caryophyllene, alloaromadendrene, gamma-elemene, alpha-humulene, gamma-cadinene, spathulenol, caryophyllene oxide; *Flowers (Germacrene D CT)*—Alpha-thujene, alpha-pinene, sabinene, beta-pinene, (E)-beta-ocimene, delta-elemene, beta-elemene, alpha-humulene, (E)-beta-farnesene, alloaromadendrene, germacrene A, germacrene-d-4-ol, alpha-cadinol, nonane; *Flowers (Beta-Elemene D CT)*—Eugenol, alpha-bergamotene, alpha-zingiberene, alpha-humulene, (E)-beta-farnesene, alloaromadendrene, sesquiphellandrene, calacorene, (E)-nerolidol, spathulenol, caryophyllene oxide, helifolen-12-al, cubenol, 1-epi-cubenol, T-cadinol, alpha-cadinol, alpha-bisabolol, epi-alpha-bisabolol, trans-farnesol, epi-alpha-bisabolol-1-one, tetradecanoic acid, n-nonadecane; *Aerial Parts (Germacrene D CT)*—Aromadendrene epoxide, phytol isomer, caryophyllene oxide, 3-eicosane, para-cymene, sabinene, gamma-muurolene, cymol, alpha-myrcene, linalool, gamma-cadinene, phytol, alpha-ylangene, gamma-elemene, hexadecanal, camphene, alpha-copaene, camphor, delta-3-carene, delta-elemene, methyl salicylate (<0.3%), borneol, delta-cadinene, alpha-phellandrene, alpha-pinene, xylene; *Aerial Parts ((E)-Caryophyllene CT)*—Beta-pinene, linalool, alpha-muurolene, alpha-cadinol; *Aerial Parts (Davanone CT)*—Methyl salicylate (< 0.8%), dodecane, eugenol, alpha-copaene, beta-caryophyllene, germacrene D, caryophyllene oxide, humulene epoxide II, T-cadinol, (E)-14-hydroxy-9-epicaryophyllene, decane, linalool, camphor; *Aerial Parts ((E)-Nerolidol CT)*—Camphene, para-cymene, limonene, linalool, camphor, isoborneol, terpinen-4-

ol, alpha-terpineol, alpha-copaene, beta-elemene, alloaromadendrene, beta-selinene, alpha-selinene, alpha-muurolene, cubebol, delta-cadinene, germacrene B, spathulenol, alpha-cadinol; *Aerial Parts (Sabinene CT)*—Camphene, beta-pinene, oct-1-en-3-ol, myrcene, limonene, (Z)-beta-ocimene, (E)-beta-ocimene, gamma-terpinene, cis-sabinene hydrate, trans-sabinene hydrate, camphor, terpinen-4-ol, alpha-copaene, beta-elemene, beta-copaene, germacrene D, alpha-muurolene, germacrene A, cubebol, germacrene B, isodavanone, spathulenol, caryophyllene oxide, T-cadinol; *Aerial Parts (Caryophyllene Oxide)*—Beta-elemene, beta-gurjunene, alpha-muurolene; *Aerial Parts (Not Separated in Study)*—Camphene, myrcene, alpha-phellandrene, limonene, beta-phellandrene, gamma-terpinene, para-cymene, alpha-cubebene, alpha-copaene, camphor, beta-elemene, aromadendrene, (E)-beta-farnesene, gamma-muurolene, alpha-selinene, caryophyllene oxide, cubenol, viridiflorol, zingiberenol, T-cadinol, delta-cadinol; *Fruits or Stems*—Linalool, borneol, terpinen-4-ol, alpha-terpineol, geraniol, eugenol, alpha-copaene, beta-elemene, tetradecane, alpha-humulene, alloaromadendrene, alpha-zingiberene, gamma-elemene, beta-curcumene, (–)-italicene ether, (E)-nerolidol, spathulenol, caryophyllene oxide, humulene epoxide II, cubenol, 1-epi-cubenol, T-cadinol, alpha-cadinol, alpha-bisabolol, epi-alpha-bisabolol, (Z)-alpha-trans-bergamotol, n-nuciferyl acetate, phytol; *L. camara var. moritziana*—(Z)-3-Hexenol, n-hexanol, alpha-pinene, sabinene, beta-pinene, myrcene, delta-3-carene, o-cymene, (E)-beta-ocimene, terpinolene, linalool, camphor, alpha-copaene, beta-bourbonene, beta-cubebene, beta-elemene, alpha-humulene, (E)-beta-farnesene, alloaromadendrene, T-muurolene, alpha-farnesene, delta-cadinene, germacrene B, spathulenol, caryophyllene oxide; *L. camara var. aculeata*—Alpha-thujene, camphene, beta-pinene, myrcene, alpha-phellandrene, alpha-terpinene, (E)-beta-ocimene, gamma-terpinene, cis-sabinene hydrate, terpinolene, linalool, camphor, borneol, terpinen-4-ol, alpha-terpineol, germacrene D, bicyclogermacrene, (E)-nerolidol, germacrene B, 8-hydroxy bicyclogermacrene

PREFERRED COMPOSITION FOR CLINICAL USE:
Too great of variance in composition to determine.

REPORTED THERAPEUTIC PROPERTIES: Wound healing, relieves skin rash and itching, soothes chronic skin conditions, heals boils, antiseptic, anticancer, antiviral, antibacterial, antifungal, antimicrobial, antiparasitic, expels excess mucus, aids respiratory function, supports liver and gallbladder function, balances blood-sugar levels, stops excess bleeding, eases cough, reduces high blood pressure, relieves fever, anti-inflammatory, analgesic, antiarthritic, antirheumatic, antispasmodic, balances menstruation, insecticidal, connects the mind, body, emotions, and spirit, enhances motivation, combats anger, frustration, and fear, antidepressant, relieves anxiety

CAUTIONS:

Leaves (Bicyclogermacrene CT)

▪ May interact with antibiotics and enhance their effectiveness.[2161]

Aerial Parts (Caryophyllene Oxide)

▪ May interact with aspirin, blood pressure, antiplatelet, and anticoagulant medications, and increase the risk of bleeding among people with bleeding disorders, due to caryophyllene-oxide content.[2162]

All other CTs

▪ None currently known.

Selected Evidence:

- *In vitro* research shows that lantana leaf essential oil CT beta-caryophyllene inhibits *M. luteus, E. coli, S. aureus, B. cereus,* and *P. aeruginosa.*[2163] The germacrene D CT of lantana leaf essential oil moderately inhibited the phytopathogenic fungi *Corynespora cassiicola.*[2164] An unidentified CT of lantana leaf essential oil inhibited *B. subtilis, S. aureus, E. coli,* and *S. gallinarum.*[2165] Lantana (var. *moritziana*) essential oil inhibited the growth of *E. faecalis* and *S. aureus.*[2166] The bicyclogermacrene CT of lantana leaf essential oil inhibited multidrug-resistant *E. coli* and *S. aureus* and synergized the activity of aminoglycoside antibiotics (neomycin, amikacin, kanamycin, gentamisin) against both bacteria.[2167] An odd CT of lantana essential oil with 3,7,11-trimethyl-1,6,10-dodecatriene as the primary constituent inhibited *A. protophormiae, M. luteus, R. rhodochrous*, and *S. aureus.*[2168] An unknown CT of lantana essential oil demonstrated broad-spectrum antibacterial and antifungal activity against seven bacteria (*B. megaterium, B. subtilis, E. coli, S. aureus, P. aeruginosa, P. vulgaris, X. campestris*) and eight fungi (*A. niger, A. parasiticus, R. oryzae, R. oryzae-sativae, F. solani, C. albicans, C. musae, A. brassicicola*).[2169]
- Lantana aerial parts essential oil (CT (E)-caryophyllene) was larvicidal to the *Culex pipiens* mosquito.[2170] The beta-caryophyllene CT also killed adult mosquitoes (*Ae. aegypti, Cx. quinquefasciatus, An. culicifacies, An. fluviatilis, An. stephensi*).[2171]
- An unidentified CT of lantana leaf essential oil proved insecticidal to the maize grain weevil (*Sitophilus zeamais*).[2172] The (E)-caryophyllene CT effectively killed *Sitophilus granarius* adults.[2173]
- The beta-caryophyllene CT of lantana leaf essential oil exhibited leishmanicidal and trypanocidal activity against *Leishmania braziliensis* and *Trypanosoma cruzi.*[2174]
- Lantana aerial parts essential oil (caryophyllene oxide CT) inhibited larval development of the gastrointestinal parasite *Haemonchus contortus in vitro.*[2175]

LARGE CARDAMOM (Greater Cardamom, Black Cardamom)

Amomum subulatum Roxb.

Family: Family
Note: Top
Aroma Intensity: Strong
Aroma: Fresh, medicinal, clean, harsh, slightly woody
Common Extraction Method: Steam distilled or hydrodistilled from the seeds (fruit); may also be distilled from the rind/pericarp, whole capsules, or leaves
Possible Substitute Oils: Eucalyptus, cajeput, ravintsara CT 1,8-cineole, myrtle, rosemary CT 1,8-cineole, niaouli CT 1,8-cineole
Blends Well With: Bay laurel, bergamot, blue cypress, cajeput, camphor, cedarwood, chamomile (German, Roman), eucalyptus, geranium, ginger, grapefruit, juniper, lavender, lemon, lemongrass, melaleuca (tea tree), myrtle, peppermint, petitgrain, pine, ravensara, ravintsara, rosemary, sage, tangerine, thyme
Subcellular Localization | Epigenetic Influence: Currently unknown | Currently unknown
Recommended Dilution Range: 3%–33%; neat for some conditions

PRIMARY CONSTITUENTS:[2176,2177,2178,2179,2180,2181,2182,2183]

Seeds

1,8-Cineole	51.9%–86.0%
Alpha-Terpineol	0.0%–16.5%
Limonene	0.0%–11.8%
Beta-Pinene	0.0%–11.1%
Alpha-Pinene	0.0%–6.5%
(E)-Nerolidol	0.0%–6.0%
Terpinen-4-ol	0.0%–5.4%
Delta-Terpineol	0.0%–3.4%
Linalool Oxide	0.0%–3.2%
Alloaromadendrene	0.0%–3.2%

Pericarp/Rind

1,8-Cineole	38.7%–39.0%
Beta-Pinene	13.6%–17.7%
Alpha-Terpineol	12.3%–12.6%
Spathulenol	3.4%–8.7%
Terpinen-4-ol	3.2%–4.5%
Germacrene D	0.2%–3.0%

Whole Capsules

1,8-Cineole	68.5%–78.0%
4-Trimehtylcyclohexene	2.1%–4.8%
Beta-Terpineol	1.2%–3.8%
Limonene	1.0%–3.8%
Alpha-Terpineol	0.1%–3.7%
Beta-Pinene	2.3%–3.6%
Terpinen-4-ol	1.1%–3.5%

Leaves

1,8-Cineole	39.8%
Alpha-Terpineol	11.5%
Beta-Pinene	4.2%
(E)-Nerolidol	4.2%
Terpinen-4-ol	3.9%
Spathulenol	3.4%

OTHER CONSTITUENTS: *Seeds*—Alpha-thujene, delta-3-carene, sabinene, myrcene, beta-phellandrene, alpha-phellandrene, alpha-terpinene, gamma-terpinene, (E)-beta-ocimene, trans-sabinene hydrate, cis-beta-terpineol, linalool, trans-pinocarveol, delta-elemene, germacrene D, spathulenol, longifolenaldehyde, tetracosane; *Pericarp*—Alpha-pinene, limonene, gamma-terpinene, alpha-terpinolene, trans-pinocarveol, myrcenol, beta-elemene, beta-selinene, germacrene B, gamma-cadinene, globulol, alpha-eudesmol, alpha-bisabolol, delta-terpineol, myrtenol, (E)-nerolidol, caryophyllene oxide; *Whole Capsules*—Beta-phellandrene, benzyl acetate, carvone, cis-sabinene hydrate, (E)-nerolidol, alpha-pinene, alpha-terpinyl acetate, gamma-terpineol; *Leaves*—Alpha-thujene, alpha-pinene, camphene, myrcene, alpha-terpinene, gamma-terpinene, borneol, carvone, cinnamyl alcohol, anethole, carvacrol, eugenol, beta-elemene, methyl eugenol (< 1.0%), trans-beta-caryophyllene, alpha-humulene, germacrene D, viridiflorene, alpha-muurolene, gamma-cadinene, alpha-cadinene, viridiflorol, humulene epoxide II, alpha-cadinol

PREFERRED COMPOSITION FOR CLINICAL USE:

Constituent	***Seeds***
1,8-Cineole	60%–85%
Alpha-Terpineol	2%–12%
Beta-Pinene	3%–11%
Alpha-Pinene	1%–%
(E)-Nerolidol	tr–6%
Terpinen-4-ol	tr–5%

REPORTED THERAPEUTIC PROPERTIES: Aids digestion, relieves stomach upset, heals stomach ulcers, eases nausea and vomiting, astringent, soothes sore throat, promotes oral cleanliness, analgesic, anti-inflammatory, antirheumatic, antiarthritic, relieves headache, supports lung and respiratory function, decongestant, stimulates appetite, aids cardiovascular function, lowers high cholesterol, antioxidant, antibacterial, antifungal antimicrobial, aids liver and gallbladder function (promotes elimination of bile), alleviates jaundice, treats gonorrhea, relieves mental fatigue, promotes focus and concentration, aids memory, stimulating

CAUTIONS:

■ Avoid with children under age 3, particularly around the nose and mouth and use cautiously in children under age 5 due to significant 1,8-cineole content. 1,8-cineole may cause seizures, central nervous system problems, or respiratory distress in young children.[2184,2185,2186]

■ Caution is warranted orally due to the significant amounts of 1,8-cineole. Limit it to small doses internally (adults—maximum 10 drops daily). Toxicity has been reported when eucalyptus (also high in 1,8-cineole) was ingested in large doses, and as few as 4 to 5 drops may cause problems in very sensitive individuals.[2187,2188,2189,2190,2191,2192]

■ Avoid with epilepsy and Parkinson's disease due to 1,8-cineole content. May exacerbate or cause seizures/convulsions or reduce seizure-medication efficacy based on 1,8-cineole content.[2193,2194,2195,2196]

■ May interact with diabetic medications due to the ability of 1,8-cineole to significantly inhibit alpha-glucosidase activity, particularly when synergized with other constituents (alpha-pinene, beta-pinene).[2197,2198,2199] Alpha-glucosidase is an enzyme that breaks down carbohydrates by chemical reaction with water. Inhibiting its activity postpones glucose absorption and therefore the impact of carbohydrates on blood-sugar levels. Studies suggest that oils rich in 1,8-cineole significantly lower glucose in animals, which may increase the action of diabetic medications.[2200,2201]

■ May interact with aspirin, blood pressure, antiplatelet, and anticoagulant medications, and increase the risk of bleeding among people with bleeding disorders (low risk) due to the weak inhibition of platelet aggregation by 1,8-cineole.[2202]

■ May weakly interfere with the enzymes responsible for metabolizing medications (NSAIDs, proton-pump inhibitors, acetaminophen, antiepileptics, immune modulators, blood-sugar medications, blood pressure medications, antidepressants, antipsychotics, diabetic medications, antihistamines, antibiotics, and anesthetics) due to 1,8-cineole content.[2203,2204,2205]

■ May interfere with pentobarbital and other barbiturates (medications for anxiety and insomnia), based on 1,8-cineole content.[2206,2207]

SELECTED EVIDENCE:

- □ Large cardamom seed and rind essential oil were marginally toxic to breast cancer cells.[2208]
- □ Oral administration of large cardamom seed essential oil (200 mg/kg body weight; composition not reported) significantly prevented stomach ulcers caused by ethanol but not pylorus ligation and aspirin.[2209]
- □ An interesting CT of large cardamom seed essential oil (2-methylheptane (9.1%), 2-methylnonane (7.2%), n-hexane (6.5%), n-non-2-ene (6.5%) and n-heptane (6.2%) as major constituents) inhibited edema and inflammation caused by xylene (a chemical that causes instant irritation of the skin and edema characteristic of acute inflammation).[2210]
- □ *In vitro* research shows that large cardamom seed and rind essential oil inhibit *A. niger*, but are weakly antimicrobial against *B. cereus* and *S. aureus*. The seed oil also exhibited activity against drug-resistant *S. typhi, S. paratyphi, E. coli, S. aureus, B. licheniformis,* and *P. fluorescens*.[2211] An interesting large cardamom seed oil (2-methylheptane (9.1%), 2-methylnonane (7.2%), n-hexane (6.5%), n-non-2-ene (6.5%) and n-heptane (6.2%) as major constituents) inhibited *B. pumilus, B. subtilis, S. aureus, M. luteus, S. epidermidis, P. aeruginosa, C. albicans, A. niger,* and *S. cervisiae*.[2212] It was more active than standard drugs (for bacteria: ciprofloxacin; for fungi: fluconazole) against *B. pumilus, B. subtilis, S. aureus, S. epidermidis, S. cerevisiae*, and *P. aeruginosa*. Large cardamom seed essential oil (composition not reported) inhibited gram-positive bacteria (*B. pumilus, B. subtilis, S. aureus, M. luteus, S. epidermidis*), gram-negative bacteria (*E. coli, P. aeruginosa*), and fungi (*C. albicans, A. niger, S. cerevisiae*).[2213]
- □ The addition of large cardamom essential oil to sweet orange juice effectively controlled the growth of spoilage microorganisms.[2214]
- □ Large cardamom seed and rind essential oil exhibited moderate insecticidal activity against the fruit fly (*Drosophila melanogaster*) and weak activity against the red imported fire ant (*Solenopsis*

invicta × *richteri*).[2215] The leaf oil was significantly toxic to *An. subpictus, Ae. albopictus,* and *Cx. tritaeniorhynchus* mosquitoes, while demonstrating low toxicity (hundredfold higher than against the mosquitoes) against nontarget biocontrol insects (*Anisops bouvieri, Diplonychus indicus, Poecilia reticulata*, and *Gambusia affinis*).[2216]

- The seed essential oil was moderately nematocidal to *Caenorhabditis elegans*.[2217]

LAVENDER COTTON (Santolina, Santolin, Ground Cypress, Cotton Lavender)

Santolina chamaecyparissus L., *S. chamaecyparissus* L. subsp. *incana, S. chamaecyparissus* L. subsp. *squarrosa, S. chamaecyparissus* L. subsp. *tomentosa*

FAMILY: Asteraceae (Compositae)
NOTE: Middle-Top
AROMA INTENSITY: Medium
AROMA: Balsamic, floral, herbaceous, reminiscent of Roman chamomile
COMMON EXTRACTION METHOD: Steam distilled or hydrodistilled from the aerial parts; may also be distilled from the flowers or roots
POSSIBLE SUBSTITUTE OILS: Sweet wormwood, sweet nancy, African bluegrass, camphor
BLENDS WELL WITH: Bay laurel, bergamot, black pepper, camphor, cedarwood, chamomile (German, Roman), clary sage, copaiba, cypress, feverfew, grapefruit, lavender, lemon, lime, neroli, orange, pine, sandalwood, sweet wormwood, tangerine, vetiver, ylang ylang
SUBCELLULAR LOCALIZATION | EPIGENETIC INFLUENCE: Currently unknown | Currently unknown
RECOMMENDED DILUTION RANGE: 1.5%–20%; 50% for some conditions

PRIMARY CONSTITUENTS:[2218,2219,2220,2221,2222,2223,2224,2225,2226,2227,2228,2229,2230,2231,2232,2233,2234,2235,2236]

French

Artemisia Ketone	25.0%–45.0%
Beta-Phellandrene	5.0%–17.9%
Myrcene	3.6%–15.0%
Unidentifieds	0.0%–10.7%
Sabinene	3.0%–5.5%
Beta-Pinene	2.9%–5.0%
Camphene	0.5%–3.0%

Turkish

Artemisia Ketone	38.1%
Camphor	11.7%
Beta-Phellandrene	9.2%
Alpha-Bisabolol	6.6%
Myrcene	4.3%

Spanish (AK/TC CT)

Artemisia Ketone	25.6%–27.8%
T-Cadinol	4.8%–23.6%

Spanish (AK/DA CT)

Artemisia Ketone	27.2%
Dihydroaromadendrene	18.2%
Beta-Phellandrene	7.5%
Myrcene	6.9%
Camphor	3.9%

Algerian (AK CT)

Artemisia Ketone	40.3%
Beta-Thujone	9.8%
(2Z,6E)-Farnesol	7.3%
Limonene	6.9%
Camphene	3.1%

Syrian

Artemisia Ketone	15.7%
Germacrene D	12.1%
Beta-Phellandrene	10.6%
Myrcene	7.4%

Myrcene	0.9%–16.1%
1,8-Cineole	2.8%–13.2%
Cubenol	1.7%–8.6%
Beta-Pinene + Sabinene	2.0%–5.7%
Camphor	2.0%–4.7%
(E)-Nerolidol + Spathulenol	1.9%–3.7%
Alpha-Cadinol	0.9%–3.0%

Spanish (Camphor CT)

Camphor	24.9%–42.9%
Cubenol	1.7%–17.3%
Borneol	8.4%–12.8%
Camphene	0.5%–8.0%
1,8-Cineole	0.3%–7.2%
(E)-Nerolidol + Spathulenol	0.7%–6.8%
Beta-Eudesmol	1.8%–4.6%
Artemisia Ketone	0.1%–4.5%
Gamma-Terpineol	tr–3.9%
Terpiene-4-ol	1.0%–3.5%

Spanish (Copaenol CT)

Copaenol	14.6%–15.2%
Borneol	11.3%–11.6%
(E)-Nerolidol + Spathulenol	8.6%–9.3%
1,8-Cineole	6.0%–8.7%
Beta-Eudesmol	1.7%–7.5%
Cubenol	1.6%–6.7%
Ledol	4.1%–5.2%
Elemol	0.9%–5.0%
Terpinen-4-ol	3.4%–4.4%
Unidentified	2.6%–3.3%

Spanish (Borneol CT)

Borneol	26.0%–28.4%
Camphor	18.9%–22.5%
1,8-Cineole	2.3%–17.6%
(E)-Nerolidol + Spathulenol	3.0%–7.7%
Copaenol	tr–5.8%
Terpinen-4-ol	3.1%–5.1%
Beta-Eudesmol	0.6%–3.9%
Ledol	0.7%–3.5%
Camphene	tr–3.0%

Nootkatone	7.0%
Germacrene A	6.4%
Alpha-Pinene	4.0%
Sabinene	3.5%
Beta-Pinene	3.4%

Indian (AK CT)

Artemisia Ketone	31.8%
1,8-Cineole	15.6%
Myrcene	14.2%
Germacrene D	8.8%
Beta-Pinene	4.7%
Sabinene	4.5%

Algerian (Camphor CT)

Camphor	31.1%
Cubenol	17.0%
Para-Cymene	8.3%
Sabinene	4.0%
Limonene	3.5%

Spanish (Cubenol CT)

Cubenol	12.9%
Caryophylladienol	10.2%
Camphor	8.3%
(E)-Nerolidol + Spathulenol	6.1%
T-Cadinol	6.0%
Beta-Eudesmol	5.7%
Copaenol	4.9%
ar-Curcumene	4.6%
Delta-Cadinene	3.2%

Egyptian

1,8-Cineole	17.8%–19.5%
Santolina Alcohol	18.0%–18.4%
Artemisia Ketone	15.6%–16.6%
Borneol	7.4%–14.3%
Limonene	1.1%–6.4%
Unidentified	1.1%–5.8%
Beta-Pinene	4.3%–4.5%
Alpha-Phellandrene	4.1%–4.5%
Dimethyl Styrene	3.6%–3.8%
Sabinene	1.8%–3.1%

Spanish (1,8-Cineole CT)

1,8-Cineole	20.0%–24.8%
Beta-Pinene	1.3%–17.5%
Linalool	0.2%–9.3%
Borneol	0.0%–8.3%
Myrcene	0.0%–7.9%
Camphor	5.0%–7.4%
Terpinen-4-ol	1.3%–7.4%
Linalyl Acetate	0.0%–6.8%
Alpha-Pinene	0.1%–4.5%
Limonene	0.0%–3.8%
Germacrene D	2.7%–3.2%
Trans-Sabinene Hydrate	0.0%–3.1%
Spathulenol	0.0%–3.1%
Sabinene	0.0%–3.0%

Spanish (Balanced)

Unidentifieds	10.1%
1,8-Cineole	9.8%
8-methylene-3-oxatricyclo [5.2.0.0(2,4)]nonane	8.2%
Viridiflorol	7.5%
Neoallo-ocimene	4.8%
Alpha-Gurjunene	4.0%
Santolina Triene	3.8%
Myrtenol	3.7%
Beta-Pinene	3.0%

Flowers (Tunisia)

1,8-Cineole	12.9%
Beta-Eudesmol	10.5%
Terpinen-4-ol	7.0%
Gamma-Cadinene	6.6%
Spathulenol	5.8%
Camphor	5.3%
Germacrene D	5.0%
Myrtenol	4.3%
Borneol	3.7%
T-Cadinol	3.2%

Indian (1,8-Cineole CT)

1,8-Cineole	13.5%
(E)-Beta-Ocimene	11.7%
Limonene	7.0%
Linalool	5.4%

Indian (P-Cymene CT)

Para-Cymene	33.0%
Beta-Thujone	24.0%
Linalool	10.0%
Alpha-Thujone	7.5%
Methyl Heptenone	3.6%
(Z)-Heptanol	3.4%
Unidentified	3.0%

*Saudi Arabian**

Caryophyllene Oxide	1.5%–3.6%

Spanish (AK/BP CT)

Artemisia Ketone	26.6%
Beta-Phellandrene	21.7%
Myrcene	9.7%
Sabinene	7.2%
Beta-Pinene	6.9%
Longiverbenone	5.9%

Roots (Tunisia)

Unidentified (MW=200)	48.1%
Unidentified (MW=200)	37.3%
Cuminaldehyde	3.7%

Indian (Linalool CT)

Linalool	11.5%–26.6%
(Z)-Beta-Ocimene	1.9%–10.2%
Limonene	9.2%–9.6%
Myrcene	1.8%–7.9%
Borneol	2.1%–3.2%

Indian (ZBO CT)

(Z)-Beta-Ocimene	13.5%
Limonene	11.7%
Alpha-Terpinene	8.8%
Myrcene	7.0%
(E)-Beta-Ocimene	5.4%

* Data appears to be incomplete with only about 20% of the oil's composition reported.

Other Constituents: *French*—Alpha-pinene, yomogi alcohol, limonene, artemisia alcohol, camphor, 2-propylfuran, linalyl acetate, 3,3,5-trimethyl-cyclohexanone; *Turkish*—Santolina triene, camphene, beta-pinene, sabinene, limonene, terpinolene, yomogi alcohol, artemisia alcohol, terpinen-4-ol, cryptone, borneol, (E)-nerolidol; *Spanish (AK/TC CT)*—Camphene, yomogi alcohol, terpinolene, borneol, terpinen-4-ol, gamma-terpineol, linalyl acetate, gamma-terpinyl acetate, ar-curcumene, elemene, delta-cadinene, elemol, copaenol, cadinenol, beta-eudesmol, 3beta-hydroxymuurolene, 3beta-hydroxy-delta-cadinene, germacrone, beta-bisabolol; *Spanish (AK/DA CT)*—Sabinene, alpha-zingiberene, ar-curcumene, spathulenol; *Spanish (Camphor CT)*—Alpha-pinene, beta-pinene, sabinene, yomogi alcohol, para-cymene, linalool, isoborneol, alpha-terpineol, myrtenol, dihydrocarveol, verbenone, carveol, irregular monoterpene acetate, alloaromadendrene, ar-curcumene, delta-cadinene, elemol, copaenol, ledol, caryophylladienol, alpha-cadinol, 3beta-hydroxymuurolene, alpha-bisabolol; *Spanish (Copaenol CT)*—Alpha-pinene, camphene, beta-pinene, sabinene, para-cymene, artemisia ketone, beta-terpineol, isoborneol, alpha-terpineol, gamma-terpineol, dihydrocarveol, carveol, irregular monoterpene acetate, alloaromadendrene, ar-curcumene, elemene, delta-cadinene, caryophylladienol, T-cadinol, alpha-cadinol, 3beta-hydroxymuurolene, 3beta-hydroxy-delta-cadinene, germacrone, beta-bisabolol; *Spanish (Borneol CT* —Alpha-pinene, beta-pinene, sabinene, para-cymene, artemisia ketone, linalool, isoborneol, alpha-terpineol, gamma-terpineol, myrtenol, bornyl acetate, isobornyl acetate, irregular monoterpene acetate, ar-curcumene, caryophylladienol, alpha-cadinol, cubenol, oplopenone, alpha-bisabolol; *Spanish (Cubenol CT)*—Beta-pinene, sabinene, para-cymene, 1,8-cineole, linalool, borneol, terpinen-4-ol, alpha-terpineol, elemene, gamma-cadinene, elemol, alpha-cadinol, oplopenone, germacrone, alpha-bisabolol, beta-bisabolol, 3-oxo-T-muurolol; *Spanish (1,8-Cineole CT)*—Camphene, alpha-terpinene, para-cymene, terpinolene, cis-sabinene hydrate, trans-p-menthen-1-ol, thuj-3-en-10-al, myrtenol, isobornyl acetate, alloaromadendrene, bicyclogermacrene, caryophyllene oxide, anhydro-oplopenone, T-cadinol, ar-curcumene, longipinene, verbenone, trans-ocimene, beta-caryophyllene, lavandulyl acetate, delta-cadinene, alpha-thujene; *Algerian (AK CT)*—Alpha-pinene, beta-pinene, yomogi alcohol, 4-dycene, p-cymenene, camphor, alpha-terpineol, beta-longipinene, alpha-gurjunene, (Z)-nerolidol, 1-nor-bourbounanone; *Algerian (Camphor CT)*—Santolina triene, alpha-pinene, myrcene, 1,8-cineole, santolina alcohol, undecane, borneol, alpha-terpineol, carvone, linalyl acetate, thymol, alpha-ylangene, beta-elemene, beta-caryophyllene, alloaromadendrene, beta-ionone, germacrene D, beta-bisabolene, dodecanoic acid, caryophyllene oxide; *Syrian*—Camphene, terpinolene, yomogi alcohol, 1,8-cineole, artemisia alcohol, camphor, trans-chrysanthenol, borneol, terpinen-4-ol, thymol, delta-elemene, alpha-longipinene, beta-caryophyllene, alpha-cadinene, trans-alpha-bisabolene, spathulenol, alpha-sinensal; *Egyptian*—Myrcene, alpha-terpinene, myrtenol, thujone (isomer not identified); *Indian (AK CT)*—Alpha-pinene, camphene, limonene, gamma-terpinene, artemisia alcohol, terpinolene, camphor, isopulegol, isoborneol, terpinen-4-ol, alpha-terpineol gamma-gurjunene, spathulenol; *Indian (P-Cymene CT)*—Sabinene, beta-pinene, gamma-terpinene; *Saudi Arabian*—1,8-Cineole, alpha-terpineol, camphor, camphanone, limonene oxide, limonene, borneol, beta-caryophyllene, carveol, cis-sabinene hydrate, trans-sabinene hydrate, alpha-bisabolene oxide, thujol, limonene diepoxide, zingiberene, oxazolidin, alpha-bisabolol, curcumene, p-cymen-8-ol; *Spanish (Balanced)*—Para-cymene, artemisia ketone, linalool, trans-pinocarveol, borneol, isopinocamphone (< 2.1%), terpinen-4-ol, aromadendrene, gamma-cadinene, delta-cadinene, spathulenol, caryophyllene oxide; *Spanish (AK/BP CT)*—Longipinene, ar-curcumene; *Indian (Linalool CT)*—Santolina triene, delta-3-carene, para-cymene, 1,8-cineole, (E)-beta-ocimene, alpha-terpinene, artemisia alcohol, alpha-thujone (<2.9%), cryptone, terpinen-4-ol, beta-elemene, beta-caryophyllene, beta-gurjunene, alpha-humulene, caryophyllene oxide, globulol, beta-oplopenone; *Indian (1,8-Cineole CT)*—Borneol, terpinen-4-ol, alpha- terpineol, beta-elemene; *Indian (ZBO CT)*—Santolina triene, 1,8-cineole, linalool, cryptone, borneol, terpinen-4ol, alpha-terpineol, beta-caryophyllene, beta-gurjunene, alpha-humulene, caryophyllene oxide, globulol, beta-oplopenone; *Flowers (Tunisia)*—Beta-pinene, delta-terpinene, para-cymene, linalyl acetate, isobornyl acetate, myrtenal, alloaromadendrene, trans-pinocarveol, isoborneol, beta-gurjunene, alpha-terpineol, delta-cadinene, cuminaldehyde, ar-curcumene, carveol, beta-bisabolol, elemol, ledol, viridiflorol, patchoulene, alpha-cadinol, carvacrol, alpha-bisabolol, phytol, hexadecanoic acid; *Roots (Tunisia)*—Terpinen-4-ol, trans-pinocarveol, borneol, gamma-cadinene, ar-curcumene, ledol, hexadecanoic acid

PREFERRED COMPOSITION FOR CLINICAL USE:
Too great of variance in composition to determine.

REPORTED THERAPEUTIC PROPERTIES: Aids digestion, reduces stomachache, promotes balanced menstruation, antiparasitic, antiseptic, wound healing, relieves skin itching and chronic skin conditions, soothes insect bites and snakebites, aids kidney function, antispasmodic, anti-inflammatory, analgesic, antimicrobial, antibacterial, **antifungal**, insecticidal, promotes concentration, aids memory function, promotes courage, stress management

CAUTIONS:
French, Spanish (AK/DA CT), Syrian, Spanish (Copaenol CT), Saudi Arabian, Indian (AK CT), Spanish (Cubenol CT), Spanish (Balanced CT), Indian (Linalool CT), Indian (1,8-Cineole CT), Indian (ZBO CT), and Roots (Tunisia)
■ May interact with antifungals and possibly enhance their effects.[2237]

Turkish
■ The camphor content in cotton lavender may negatively impact red blood cells and increase the risk of jaundice in children with glucose-6-phosphate dehydrogenase deficiency (G6PD).[2238,2239]
■ May interact with antifungals and possibly enhance their effects.[2240]

Spanish (AK/TC CT) and Flowers (Tunisia)
■ Avoid with epilepsy and Parkinson's disease due to 1,8-cineole and camphor content. May exacerbate or cause seizures/convulsions or reduce seizure-medication efficacy.[2241,2242,2243,2244]
■ May interact with antifungals and possibly enhance their effects.[2245]

Algerian and Indian (Para-Cymene CT)
■ Avoid with children under 6 due to thujone content. Cases of seizure have been reported in young children due to exposure to essential oils rich in thujones.[2246,2247]
■ Avoid during pregnancy and while lactating, due to thujones content. Essential oils high in thujones may cause abortion.[2248,2249]
■ Avoid oral consumption. Ingesting plants or products high in thujones can cause vomiting, stomachache, diarrhea, and gastroenteritis followed by absorption disorders, headache, nervous agitation, and chronic convulsions, and symptoms of liver and kidney toxicity extending to yellow liver atrophy (extensive and rapid death of liver cells—parenchymal, possibly by fatty degeneration), arrhythmia, and myocardial bleeding.[2250] Thujone is considered significantly neurotoxic and may damage the liver, and it is estimated that as little as 15 mg orally may negatively impact the central nervous system.[2251,2252,2253] Taking 20 drops of thuja oil (also with significant thujones content) twice daily for five days caused an adult woman to have a tonic seizure.[2254] As few as 12 drops of sage oil (also with significant thujones content) orally have reportedly caused a seizure and short coma in an adult.[2255]
■ Avoid with epilepsy and Parkinson's disease due to the presence of thujones.[2256,2257,2258,2259,2260]
■ Avoid with those who have a compromised liver or kidneys, due to the risk of kidney toxicity, kidney failure, or liver toxicity.[2261,2262,2263] Consuming extremely large amounts of thujone-rich essential oils (10 mL) has caused kidney failure.[2264] This would also suggest that those taking medications that could cause liver damage should also use tansy very cautiously or avoid it entirely.
■ May interact with antifungals and possibly enhance their effects.[2265]

Spanish (Camphor CT), Algerian (Camphor CT)

■ Avoid with children under six due to significant camphor content and moderate amounts of 1,8-cineole. Several cases of camphor poisoning and/or seizures from ingestion and topical application have been reported in children.[2266,2267] Ingestion of camphor-containing products has been lethal in children under age two.[2268] Children five years and up may use camphor-containing essential oils topically in dilutions no stronger than 5%. Camphor ingestion by infants and young children may cause cough, vomiting, seizure, burning sensation in the mucous membranes and eyes, or lack of voluntary coordination of muscle movements.[2269]

■ Caution is warranted during pregnancy and while lactating, due to significant camphor content. Ingestion of essential oils with significant levels of camphor may lead to abortion because fetuses lack the enzymes to process it.[2270] Camphor ingestion by infants and young children may cause cough, vomiting, seizure, burning sensation in the mucous membranes and eyes, or lack of voluntary coordination of muscle movements.[2271]

■ Avoid oral consumption due to significant camphor content. Essential oils with significant levels of camphor may be toxic when taken orally, usually single doses exceeding 2 mL, although the lethal dose for humans is estimated to be greater than 5 mL in a single dose.[2272,2273,2274]

■ Avoid with epilepsy and Parkinson's disease due to camphor and 1,8-cineole content, which have the potential to exacerbate or cause seizures/convulsions or reduce seizure-medication efficacy.[2275,2276,2277,2278]

■ Avoid with those who have a compromised liver or kidneys, due to the risk of increased liver enzymes and liver or kidney damage based on camphor.[2279,2280,2281,2282] This would also suggest that those taking medications that could cause liver damage should also use essential oils with significant camphor levels very cautiously or avoid them entirely.

■ The significant camphor content may negatively impact red blood cells and increase the risk of jaundice in children with glucose-6-phosphate dehydrogenase deficiency (G6PD).[2283,2284]

■ May interact with antifungals and possibly enhance their effects.[2285]

Spanish (Borneol CT)

■ Avoid with children under six due to significant camphor content and moderate amounts of 1,8-cineole. Several cases of camphor poisoning and/or seizures from ingestion and topical application have been reported in children.[2286,2287] Ingestion of camphor-containing products has been lethal in children under age two.[2288] Children five years and up may use camphor-containing essential oils topically in dilutions no stronger than 5%. Camphor ingestion by infants and young children may cause cough, vomiting, seizure, burning sensation in the mucous membranes and eyes, or lack of voluntary coordination of muscle movements.[2289]

■ Caution is warranted during pregnancy and while lactating, due to significant camphor content. Ingestion of essential oils with significant levels of camphor may lead to abortion because fetuses lack the enzymes to process it.[2290] Camphor ingestion by infants and young children may cause cough, vomiting, seizure, burning sensation in the mucous membranes and eyes, or lack of voluntary coordination of muscle movements.[2291]

■ Avoid oral consumption due to significant camphor content. Essential oils with significant levels of camphor may be toxic when taken orally, usually single doses exceeding 2 mL, although the lethal dose for humans is estimated to be greater than 5 mL in a single dose.[2292,2293,2294]

■ Avoid with epilepsy and Parkinson's disease due to camphor and 1,8-cineole content, which have the potential to exacerbate or cause seizures/convulsions or reduce seizure-medication efficacy.[2295,2296,2297,2298]

■ The significant camphor content may negatively impact red blood cells and increase the risk of jaundice in children with glucose-6-phosphate dehydrogenase deficiency (G6PD).[2299,2300]

■ May interact with aspirin, blood pressure, antiplatelet, and anticoagulant medications, and increase the risk of bleeding among people with bleeding disorders, due to the anticoagulant properties of borneol.[2301,2302]

■ May interfere with enzymes responsible for metabolizing medications (NSAIDs, proton-pump inhibitors, acetaminophen, antiepileptics, immune modulators, blood-sugar medications, blood pressure medications, antidepressants, antipsychotics, diabetic medications, antihistamines, antibiotics, and anesthetics), based on borneol content.[2303,2304,2305,2306,2307]

Spanish (1,8-Cineole CT) and Egyptian

▪ Avoid with epilepsy and Parkinson's disease due to camphor or 1,8-cineole content, which have the potential to exacerbate or cause seizures/convulsions or reduce seizure-medication efficacy.[2308,2309,2310,2311]

▪ May weakly interfere with the enzymes responsible for metabolizing medications (NSAIDs, proton-pump inhibitors, acetaminophen, antiepileptics, immune modulators, blood-sugar medications, blood pressure medications, antidepressants, antipsychotics, diabetic medications, antihistamines, antibiotics, and anesthetics) due to 1,8-cineole content.[2312,2313,2314]

▪ May interfere with pentobarbital and other barbiturates (medications for anxiety and insomnia), based on 1,8-cineole content.[2315,2316]

▪ May interact with antifungals and possibly enhance their effects.[2317]

SELECTED EVIDENCE:

- Lavender cotton (Saudi Arabian) essential oil was highly cytotoxic to liver cancer cells.[2318]
- An unidentified CT of lavender cotton essential oil significantly controlled vaginal candidiasis (VC) and systemic candidiasis (SC).[2319] A 4% solution of the oil was squirted on the vagina of mice to control VC; whereas, the oil was orally administered to mice in olive oil at a dosis of 25 mg/kg body weight (twice daily for seven days) to control SC. The same research reported that topical application of the oil controlled fungal skin infections in guinea pigs.
- *In vitro* research demonstrated that Algerian lavender cotton CT camphor essential oil moderately to strongly inhibited *P. aeruginosa, E. coli, K. pneumonia, B. bronchiseptica* (clinical strains isolated directly from patients), *S. aureus, S. epidermis, E. faecalis, M. luteus, C. albicans*, and *S. cerevisiae*.[2320] Syrian lavender cotton essential oil inhibited gram-positive bacteria (*B. subtilis, S. aureus, M. luteus*) and gram-negative bacteria (*E. coli, P. aeruginosa, K. pneumonia*), and fungi (*F. solani, Mucor* sp., *Rhyzopus* sp., *C. albicans*).[2321] Spanish (AK/DA CT) lavender cotton essential oil inhibited *L. innocua, S. marcescens, A. hydrophila, A. faecalis*, and *A. denitrificans*.[2322] The flower essential oil showed strong antifungal properties against *E. floccosum, S. brevicaulis, T. rubrum, S. dimidiatum, A. fumigatus*, and *M. canis*; whereas the root oil was only moderately antifungal against most of the same species (strong activity was observed against *T. rubrum*).[2323] The flower oil also inhibited *E. coli, P. aeruginosa, E. faecalis, S. aureus, C. freundeii*, and *P. mirabilis*, but the root oil was only moderately active and inactive against *P. mirabilis*. Algerian lavender cotton CT AK showed broad-spectrum antimicrobial activity against *E. coli, B. bronchiseptica, K. pneumoniae, B. cereus, B. subtillis, M. luteus, S. aureus, E. faecium, Penicillum* sp., and *F. culmorum*.[2324] Lavender cotton essential oil proved a good inhibitor of the molds *A. flavus* and *M. circinelloides*.[2325] Lavender cotton essential oil (Spanish, 1,8-cineole CT) was antifungal to pathogenic fungi of mushrooms and vegetables (*Botrytis cinerea, Sclerotinia sclerotiorum, Fusarium oxysporum, Phytophthora parasitica, Pythium aphanidermatum, Alternaria brassicae, Cladobotryum mycophilum*, and *Trichoderma aggressivum f.sp. europaeum*).[2326]
- Lavender cotton (Algerian, AK CT) exhibited weak antioxidant activity in the DPPH and beta-carotene/linoleic acid assays.[2327] Another study reported that lavender cotton displayed good antioxidant capacity in the ferrous iron-chelating assay.[2328]
- The Spanish balanced CT of lavender cotton essential oil was a strong antifeedant against the aphid *Rhopalosiphum padi*, toxic to *Hyalomma lusitanicum* tick and with moderate effects against the Colorado potato beetle (*Leptinotarsa decemlineata*), Egyptian cotton leafworm (*Spodoptera*

littoralis), and perennial ryegrass (*Lolium perenne*).[2329] An unidentified CT of lavender cotton reduced the survival of the brown planthopper (*Nilaparvata lugens*) insect.[2330] Lavender cotton (Spanish AK/BP CT) was also toxic to Japanese termites (*Reticulitermes speratus*).[2331]

- Lavender cotton (Spanish, 1,8-cineole CT) demonstrated promising herbicidal activity against two weeds (*Portulaca oleracea* L. and *Vicia sativa* L.) with low harm to seedlings of four crops (*Zea mays* L., *Triticum durum* L., *Pisum sativum* L., and *Lactuca sativa* L.).[2332]

LEEK (Wild Leek, Broadleaf Wild Leek, Egyptian Leek, Elephant Garlic, Persian Shallot)

Allium ampeloprasum L., *A. ampeloprasum* var. *atroviolaceum*

Family: Alliaceae
Note: Top-Middle
Aroma Intensity: Strong
Aroma: Woody, fresh, fruity; pungent, sulfur-like
Common Extraction Method: Steam distilled or hydrodistilled from the aerial parts or bulbs
Possible Substitute Oils: Onion, garlic, wild garlic
Blends Well With: Basil (sweet), cumin, garlic, lavender, onion, peppermint, rosemary, wild garlic
Subcellular Localization | Epigenetic Influence: Currently unknown | Currently unknown
Recommended Dilution Range: *Sulfides CT*—1.5%–5.0%; *Other CTs*—3%–50%

Primary Constituents:[2333,2334,2335]

Borneol/Alpha-Muurolol CT	
Borneol	4.8%–16.4%
Alpha-Muurolol	6.3%–9.3%
Tepinen-4-ol	3.6%–9.3%
Pentacosane	5.3%–7.1%
Alpha-Cadinol	1.2%–3.9%
Gamma-Elemene	2.5%–3.5%
Nonacosane	2.4%–3.4%
Alpha-Phellandrene	2.0%–3.2%
Alpha-Copaene	2.8%–3.1%

Sulfides CT (Bulbs)	
Dipropyl Disulfide	47.7%
Dipropyl Trisulfide	15.0%
Methyl Propyl Disulfide	4.5%
1-Propenyl Propyl Disulfide	3.8%
Methyl Propyl Trisulfide	3.2%

var. atroviolaceum	
Limonene	27.0%
Beta-Pinene	25.3%
9-Octadecanoic Acid	17.3%
Hexadecanoic Acid	14.9%
Beta-Caryophyllene	7.4%

OTHER CONSTITUENTS: *Borneol/Alpha-Muurolol CT*—Alpha-pinene, beta-pinene, myrcene, alpha-terpinene, limonene, gamma-terpinene, 1,8-cineole, linalool, alpha-terpineol, thymol, carvacrol, eugenol, beta-caryophyllene, gamma-muurolene, delta-cadinene, palmitic acid; *Sulfides CT (Bulbs)*—Allyl propyl disulfide, n-nonanol, 3-ethyl-5-methyl-1,2,4-trithiolane, geranyl acetate, 2-methyl-3,4-dithiaheptane, palmitic acid, ethyl palmitate; *var. atroviolaceum*—Alpha-pinene, dimethyl trisulfide, gamma-terpinene, dimethyl tetrasulfide, elemene, caryophyllene oxide, 1-heptadecane

PREFERRED COMPOSITION FOR CLINICAL USE:
Insufficient data/samples to determine.

REPORTED THERAPEUTIC PROPERTIES: Antimicrobial, antibacterial, antifungal, anti-inflammatory, antispasmodic, analgesic, aids digestion, lowers blood sugar, cleanses the blood, aids detoxification, digestive aid, mentally stimulating, combats anger and rage

CAUTIONS:
Sulfides CT

■ Caution is advised during pregnancy and nursing due to the presence of sulfides. Methanolic extracts of onion have caused temporary (400 mg/kg body weight) to permanent (800 mg/kg) antifertility effects in mice due to the antifertility effects of some of the sulfur compounds and other compounds in onions.[2336] Sulfur compounds from other sulfurous vegetables (garlic) have been detected in the amniotic fluid of women who consumed a single capsule of garlic herb, and a review concluded that garlic herb has the potential to cause abortion, trigger menstruation, or stimulate the uterus.[2337,2338] However, no adverse outcomes were reported in pregnant women who took 800 mg of garlic herb daily during the third trimester of pregnancy, and diallyl sulfide prevents embryo-feto toxicity.[2339,2340,2341] Animal research also reports that garlic herbal oil—containing diallyl trisulfide—reduces pregnancy complications (decreased newborn body weight, abortion, and musculoskeletal abnormalities) due to its ability to reduce cholesterol in pregnant rats with high cholesterol.[2342] Based on conflicting information it is best to use onion cautiously during pregnancy until definitive research determines its safety. In addition, ingestion of garlic by lactating mothers reportedly altered the flavor of breast milk because garlic constituents are secreted in the breast milk and made infants prone to extended nursing.[2343,2344]

■ May interact with aspirin, blood pressure, antiplatelet, and anticoagulant medications, and increase the risk of bleeding among people with bleeding disorders, due to the sulfur compounds.[2345,2346,2347,2348,2349,2350,2351,2352,2353,2354,2355,2356] The study authors frequently attributed this effect to the presence of volatile organosulfur constituents (diallyl trisulfide, methyl allyl trisulfide).

■ May interfere with enzymes responsible for metabolizing medications (NSAIDs, proton-pump inhibitors, acetaminophen, antiepileptics, immune modulators, blood-sugar medications, blood pressure medications, antidepressants, antipsychotics, diabetic medications, antihistamines, antibiotics, and anesthetics) due to diallyl disulfide, diallyl trisulfide, allyl methyl sulfide, and allyl diallyl sulfide content.[2357,2358,2359,2360,2361,2362] Diallyl disulfide and diallyl trisulfide also interfere with protein transporters that move drugs across cell membranes, which can increase or decrease drug effectiveness.[2363]

■ May interact with antifungal medications (ketoconazole) and increase their effectiveness based on presence of sulfur-containing compounds similar to garlic essential oil.[2364]

■ Topical application may cause skin irritation or skin sensitivity reactions. Cases of contact dermatitis or irritation have been reported with onions and other sulfurous vegetables.[2365,2366,2367,2368,2369] Onion compounds are both irritating and sensitizing. Dilutions under 2% are strongly recommended.

Borneol CT and var. atroviolaceum

■ None currently known.

SELECTED EVIDENCE:

- Oral administration of leek essential oil (150 mg/kg body weight) with higher borneol content (16.4%) decreased blood-glucose levels and prevented damage to pancreas beta-cells in diabetic rats.[2370]
- Leek essential oil diminished skin damage caused by trichothecene when administered prior to and after exposure to the toxin in mice.[2371]
- *In vitro* evidence showed that leek essential oil inhibited *S. cerevisiae, A. tamarii, P. griseofulvum,* and *E. amstelodami.*[2372]

LEMON-SCENTED IRON BARK (Lemon Ironbark)

Eucalyptus staigeriana F. Muell. ex Bailey

FAMILY: Myrtaceae
NOTE: Middle
AROMA INTENSITY: Medium
AROMA: Sweet, lemony, citrusy, fresh
COMMON EXTRACTION METHOD: Steam or hydrodistilled from the leaves
POSSIBLE SUBSTITUTE OILS: Equal parts of lemongrass and lemon
BLENDS WELL WITH: Bergamot, cedarwood, lavender, lemon, lemongrass, lime, marjoram, pine, rosemary, thyme
SUBCELLULAR LOCALIZATION | EPIGENETIC INFLUENCE: Currently unknown | Currently unknown
RECOMMENDED DILUTION RANGE: 3%–33%; 50% for some conditions

PRIMARY CONSTITUENTS:[2373,2374,2375]

Limonene/Citral CT		*Limonene CT*	
Limonene	7.0%–28.8%	Limonene	72.9%
Geranial	14.2%–16.0%	1,8-Cineole	9.5%
Geraniol	4.2%–14.8%	O-Cymene	4.6%
Methyl Geranate	3.7%–11.0%		
Neral	9.7%–10.8%		
Alpha-Terpinolene	0.0%–9.4%		
Geranyl Acetate	3.9%–9.2%		
Citronellol	0.0%–6.3%		
Citronellal	0.8%–5.5%		
1,8-Cineole	5.0%–5.4%		
Alpha-Pinene	1.8%–3.3%		

Note: Commercial lemon-scented iron bark essential oils also shows a beta-phellandrene CT.

OTHER CONSTITUENTS: *Limonene/Citral CT*—Beta-pinene, o-cymene, delta-3-carene, linalool, menth-3-en-8-ol, terpinene-4-ol, cymen-8-ol (meta), alpha-terpineol, citronellol acetate, neryl acetate; *Limonene CT*—Myrcene, alpha-thujene, para-cymene, alpha-pinene, alpha-terpinolene

PREFERRED COMPOSITION FOR CLINICAL USE:

Constituent	*Limonene/ Citral CT*	*Beta-Phellandrene CT*
Limonene	7%–32%	4%–14%
Geranial	8%–18%	6%–19%

Geraniol	4%–15%	2%–12%
Neral	7%–14%	6%–15%
Methyl Geranate	3%–14%	3%–15%
Geranyl Acetate	2%–12%	0.5%–12%
Citronellal	tr–7%	–
1,8-Cineole	3%–7%	0.5%–7%
Alpha-Pinene	0.5%–5%	0.1%–6%
Beta-Phellandrene	–	12%–34%
Linalool	–	1%–19%
Para-Cymene	–	tr–4%
Nerol	–	tr–3%

REPORTED THERAPEUTIC PROPERTIES: Antiviral, anti-infectious, antibacterial, antifungal, antiseptic, antiparasitic, anti-inflammatory, analgesic, antispasmodic, relieves sprains/strains, antirheumatic, antineuralgic, relieves headache, soothes sore throat, relieves fever, wound healing, insecticidal, insect repellent, sedating, calming/relaxing, stress management, antidepressant, reduces mental fatigue

CAUTIONS:

Limonene/Citral CT

■ There is a moderate risk that when lemongrass is taken orally, it may interfere with enzymes responsible for metabolizing medications (NSAIDs, proton-pump inhibitors, acetaminophen, antiepileptics, immune modulators, blood-sugar medications, blood pressure medications, antidepressants, antipsychotics, diabetic medications, antihistamines, antibiotics, and anesthetics).[1232–1235]

■ Use cautiously with diabetic medications. Large oral doses of citral may improve insulin sensitivity and lower blood-glucose levels according to animal research.[1236,1237]

■ May interact with antibiotics and possibly enhance their effects.[1238]

■ Dilution is recommended for topical application due to high citral content and the risk for skin irritation or sensitivity.[4185,4193] May also irritate mucous membranes.

Limonene CT

■ None currently known.

SELECTED EVIDENCE:

- □ Both lemon-scented iron bark essential oil CT limonene and the chitosan-encapsulated essential oil significantly inhibited *H. contortus* (a common parasite that infects the gut of humans and animals) larval hatching and killed larvae.[2376] It also proved effective against sheep and goat gastrointestinal nematodes.[2377,2378]
- □ Lemon-scented iron bark essential oil CT limonene/citral was insecticidal to the *Lutzomyia longipalpis* sand fly.[2379]
- □ Lemon-scented iron bark was toxic to the corn pest *Spodoptera frugiperda*.[2380]

LESSER CALAMINT
(Calamint, Basil-Thyme)

Calamintha nepeta (L.) Savi, *C. nepeta* (L.) Savi subsp. nepeta, *C. Nepetoides*, *C. nepeta* (L.) Savi subsp. *glandulosa* (Riquien) P.W. Ball, *Satureja calamintha* (L.) Scheele

FAMILY: Lamiaceae (Labiateae)
NOTE: Middle
AROMA INTENSITY: Strong
AROMA: Minty, herbaceous, fresh
COMMON EXTRACTION METHOD: Hydrodistilled or steam distilled from the dried aerial parts
POSSIBLE SUBSTITUTE OILS: Applemint, pennyroyal, buchu, peppermint
BLENDS WELL WITH: Bergamot, buchu, camphor, catnip, citronella, eucalyptus, geranium, Japanese catnip, lavender, lemon, lemongrass, myrtle, orange, pennyroyal, peppermint, rosemary, sage, Spanish sage, spearmint, tangerine
SUBCELLULAR LOCALIZATION | EPIGENETIC INFLUENCE: Currently unknown | Currently unknown
RECOMMENDED DILUTION RANGE: 1.0%–5.0%

PRIMARY CONSTITUENTS:[2381,2382,2383,2384,2385,2386,2387,2388,2389,2390,2391,2392,2393,2394,2395,2396,2397,2398,2399,2400,2401]

ssp. nepeta (Pulegone CT)	
Pulegone	12.0%–61.7%
Piperitenone	2.1%–36.0%
Menthone	2.6%–20.0%
Carvacrol	0.0%–10.0%
Limonene	0.4%–7.8%
Gamma-Terpinene	0.0%–7.0%
Isomenthone	0.0%–5.3%
Trans-Piperitone Oxide	0.0%–5.3%
Beta-Caryophyllene	0.0%–4.5%
Thymol	0.0%–4.2%
Para-Cymene	0.0%–3.3%
Trans-Sabinene Hydrate	0.0%–3.2%

C. nepeta (Pulegone CT)	
Pulegone	21.4%–58.8%
Menthone	4.8%–33.2%
Piperitenone	0.0%–16.4%
Neomenthone	0.0%–14.5%
Piperitone	0.0%–13.1%
Isomenthone	0.2%–12.0%
Limonene	0.5%–7.0%
Cis-Sabinene Hydrate	0.0%–5.7%
Menthol	0.0%–4.6%
Piperitenone Oxide	0.0%–4.6%
Chrysanthenone	0.0%–4.1%
Piperitone Oxide	0.0%–3.9%

ssp. grandulosa (Pulegone CT)	
Pulegone	19.5%–76.1%
Menthone	7.0%–35.7%
Piperitenone	0.0%–15.0%
Piperitone	0.0%–10.2%
Caryophyllene Oxide	0.0%–7.9%
Limonene	0.3%–6.0%
Trans-Piperitone Oxide	0.0%–5.7%
Piperitenone Oxide	0.0%–5.4%
Menthol	0.0%–5.0%

ssp. grandulosa (Pulegone/Chrysanthenone CT)	
Pulegone	37.7%–84.7%
Chrysanthenone	1.3%–33.9%
Menthone	0.0%–20.3%
Limonene	0.0%–13.6%
Methylisopulegone	0.0%–12.6%
Isopulegone	0.0%–9.4%
Cinerolone	0.0%–5.8%
Piperitenone	0.0%–4.5%

Trans-Sabinene Hydrate	0.0%–3.5%
Isomenthone	0.0%–3.0%

ssp. grandulosa var. subisodonda

Pulegone	75.5%
Piperitenone Oxide	6.0%
Menthone	5.3%
Menthol	4.3%

ssp. nepeta (Menthone CT)

Menthone	36.5%–59.8%
Pulegone	10.9%–26.2%
Piperitone	3.1%–8.8%
Piperitone Oxide*	0.0%–8.3%
Piperitenone	0.0%–8.2%
Limonene	2.8%–5.2%
Isomenthone	2.1%–4.2%
Neomenthol	0.0%–3.2%
Alpha-Pinene	0.0%–3.2%

ssp. nepeta (Trans-Piperitone Oxide CT)

Trans-Piperitone Oxide	30.9%–68.0%
Piperitenone Oxide	3.6%–11.7%
Caryophyllene Oxide	0.2%–7.8%
Limonene	0.1%–7.1%
(E)-Phytol	0.0%–4.7%
Beta-Caryophyllene	1.3%–4.6%
Neo-Isopulegol + Terpinen-4-ol	0.0%–4.0%
Octadecanol	0.0%–3.3%
Menthone	0.0%–3.1%

C. nepeta (Cis-Piperitone Oxide)

Cis-Piperitone Oxide	48.7%
Piperitenone Oxide	22.1%
Limonene	13.5%
Terpinen-4-ol	4.6%

C. nepeta (Isopulegol CT)

Isopulegol	22.6%–51.6%
Isopulegone	17.2%–20.5%
1,8-Cineole	11.9%–18.3%
Pulegone	4.5%–8.6%
Terpinen-4-ol	1.1%–6.7%
Menthone	1.6%–6.5%
Menthol	0.0%–5.1%
Beta-Caryophyllene	2.5%–3.6%

Isocaryophyllene	0.0%–3.8%
3-Octanol	0.0%–3.0%

ssp. grandulosa (Menthone CT)

Menthone	40.1%–55.8%
Pulegone	17.6%–36.8%
Trans-Piperitone Oxide	0.0%–12.4%
Terpinen-4-ol	0.0%–6.0%

ssp. nepeta (Piperitenone Oxide CT)*

Piperitenone Oxide	47.8%–59.1%
Piperitone Oxide	0.0%–19.1%
Limonene	5.1%–9.1%
Bornyl Acetate	0.0%–8.3%
Terpinen-4-ol	0.2%–4.8%
Cis-Sabinene Hydrate	0.0%–4.5%
Beta-Caryophyllene	1.5%–3.1%

ssp. grandulosa (Trans-Piperitone Oxide CT)

Trans-Piperitone Oxide	34.2%
Limonene	17.3%
Piperitenone Oxide	10.8%

C. nepeta (Trans-Piperitone Oxide)

Trans-Piperitone Oxide	30.5%
Limonene	12.8%
Piperitenone Oxide	12.5%
Pulegone	12.4%
Menthone	9.3%

ssp. grandulosa (Piperitenone Oxide CT)

Piperitenone Oxide	43.8%
Trans-Piperitone Oxide	25.2%
Limonene	13.0%

ssp. nepeta (Piperitone CT)

Piperitone	28.8%
Pulegone	17.4%
Piperitenone	15.5%
Menthone	10.3%
Limonene	4.9%

ssp. nepeta (Piperitone Oxide CT)*

Piperitone Oxide	35.7%
Limonene	17.5%
Piperitenone Oxide	14.9%
Menthone	4.6%
Thymol	3.4%

* Correct isomer not identified.

Other Constituents: *ssp. nepeta (Pulegone CT)*—Alpha-pinene, beta-pinene, myrcene, alpha-terpinene, 3-octanol, trans-sabinene hydrate, carvacrol methyl ether, alpha-terpineol, germacrene D, cis-piperitone oxide, beta-bisabolene, piperitenone oxide, caryophyllene oxide, spathulenol; *ssp. grandulosa (Pulegone CT)*—3-Octanol, cis-isopulegone, trans-isopulegone, terpinen-4ol, menthol, Germacrene D, cis-piperitone oxide, nementhol, alpha-pinene, sabinene, beta-pinene, 1,8-cineole, bornyl acetate, alpha-terpineol, alpha-thujene, beta-caryophyllene; *C. nepeta (Pulegone CT)*—Alpha-pinene, sabinene, beta-pinene, delta-2-carene, gamma-terpinene, trans-sabinene hydrate, terpinen-4-ol, caryophyllene, 3-octanol; *ssp. grandulosa var. subisodonda*—Isomenthone, dihydroedulan; *ssp. grandulosa (Pulegone/Chrysanthenone CT)*—Para-cymene, terpinen-4-ol, delta-cadinene, germacrene D, isopiperitenone, isopulegone, linalool, p-mentha-1,8-dien-3-one, sabinene, alpha-terpineol, p-menth-1-en-8-ol; *ssp. nepeta (Menthone CT)*—Sabinene, beta-pinene, 3-octanol, myrcene, trans-sabinene hydrate, linalool, piperitone oxide I, dihydrocarvyl acetate, alpha-terpinyl acetate ; *ssp. grandulosa (Menthone CT)*—3-octanol; *ssp. nepeta (Trans-Piperitone Oxide CT)*—Isopulegol acetate, thymol, 4-hydroxypiperitone, diosphenol, germacrene D; *ssp. grandulosa (Piperitone Oxide CT)*—Alpha-pinene, beta-pinene, myrcene, sabinene, 3-octanol, trans-sabinene hydrate, beta-bourbonene, linalool, bornyl acetate, terpinen-4-ol, alpha-terpineol, borneol, isopiperitenone, piperitenone, caryophyllene oxide, thymol, hexadecanoic acid; *C. nepeta (Trans-Piperitone Oxide)*—Beta-pinene, 3-octanol, myrcene, 1,8-cineole, linalool, neomenthol, trans-piperitone oxide, piperitone, dihydrocarvyl acetate, piperitenone, beta-caryophyllene, germacrene D; *C. nepeta (Cis-Piperitone Oxide)*—3-Octanol, linalool, carvacrol, dihydrocarvacryl acetate, beta-caryophyllene, caryophyllene oxide; *ssp. nepeta (Piperitenone Oxide CT)*—Alpha-pinene, beta-pinene, myrcene, (Z)-beta-ocimene, sabinene, alpha-terpinene, gamma-terpinene, alpha-terpinolene, alpha-terpineol, dyhydroneocarveol, diosphenol, alpha-humulene, thymol, germacrene D, 4-octanol; *ssp. grandulosa (Piperitenone Oxide CT)*—Beta-pinene, myrcene, gamma-terpinene, 6-methyl-3-heptanol, linalool, terpinen-4-ol, alpha-terpineol, borneol, cis-piperitone oxide, isopiperitenone, piperitenone; *C. nepeta (Isopulegol CT)*—Limonene, trans-pinocarveol, alpha-pinene, camphene, beta-pinene, alpha-phellandrene, sabinene hydrate, estragole (< 0.8%); *ssp. nepeta (Piperitone CT)*—3-Octanol, trans-sabinene hydrate, isopiperitenol, thymol, beta-caryophyllene, beta-cubebene; *ssp. nepeta (Piperitone Oxide* CT)* —3-Octanol, gamma-terpinene, trans-sabinene hydrate, terpinen-4-ol, isopiperitenol, pulegone, piperitenone, bornyl acetate, thymol, isopulegyl acetate, beta-caryophyllene, beta-cubebene

Preferred Composition for Clinical Use:
Insufficient samples/data to determine.

Reported Therapeutic Properties: Anti-inflammatory, analgesic, antirheumatic, relieves painful menstruation, reduces fever, aids respiratory function, expels excess mucus, eases cough, combats respiratory infections, **antibacterial**, antifungal, antiviral, anti-infectious, antiseptic, reduces the appearance of blemishes, relieves chronic skin conditions and itching, aids gallbladder function, reduces stomachache, aids digestion, promotes restful sleep, insecticidal, insect repellent, relieves anxiety, reduces mental exhaustion, stress management

Cautions:

Pulegone CTs

Lesser calamint essential oil should only be used in essential oil therapy when the benefits of use outweigh the risks as determined by a health professional trained in medical essential oil therapy.

■ Avoid with children under age twelve. Multiple organ failure, liver failure, brain swelling (cerebral edema), severe epileptic encephalopathy (severe brain disorder with seizures), organ cell death (necrosis), and death has been reported in infants who were administered mint tea containing pennyroyal essential oil (also rich in pulegone).[2402] Pulegone and its metabolite menthofuran (or metabolites of menthofuran) are believed to cause the toxic effects of pennyroyal, which include liver and kidney toxicity, bronchiolar cell destruction, depletion of liver glutathione levels (the primary protective antioxidant of the liver), and direct cellular damage.[2403,2404,2405] Children are more likely to experience these effects.[2406]

■ Avoid during pregnancy and lactation due to significant amounts of pulegone and potential toxicity to multiple organs.[2407,2408] Pennyroyal (also rich in pulegone) is believed to trigger uterine contractions by irritating the genito-urinary system. Pennyroyal is commonly used to induce abortions.[2409,2410] While pennyroyal herb is considered abortifacient, the essential oil is not.[2411] When pennyroyal essential oil is used alone to induce abortion, it generally is unsuccessful but causes serious health consequences to the mother. Carachipita is an abortant sold over the counter in South America that contains pennyroyal, yerba de la perdiz, oregano, and marsh rosemary. The literature reports successful use of Carachipita to induce abortion, but not without serious consequences to the mother (multiple organ failure, vaginal bleeding, destruction of red blood cells, numbness, dizziness, and death).[2412,2413,2414] It is prudent to avoid lesser calamint oil while pregnant or nursing due to the toxicity of pulegone alone, despite a lack of evidence that it is a direct abortifacient.

■ Avoid oral consumption due to the toxic potential of pulegone, menthofuran, and their metabolites to multiple organs. Menthofuran is generally a very minor constituent of lesser calamint essential oil, but it is considered more toxic than pulegone.[2415] In addition, pulegone is metabolized by human liver cytochrome P-450s (CYP2E1, CYP1A2, and CYP2C19) to menthofuran after administration, which actually increases its toxic effect to the liver.[2416,2417] Pulegone is also metabolized to p-cresol, which depletes glutathione levels and can increase liver damage.[2418] Animal research suggests oral administration of pulegone is significantly toxic to the liver and causes death at 300 mg/kg (over 23 mL for a human adult equivalent dose) for two weeks.[2419] The same research concluded that lower doses of 75 and 150 mg/kg for three months resulted in liver, kidney, bone marrow, heart, stomach, lung, ovarian, and thymus damage, and one death in animals. A two-year animal study reported urinary cancer and papillomas, liver cancer, osteosarcoma, and multiple organ damage or enlargement (liver, kidney, olfactory epithelium, lung, bile duct).[2420] Ingestion of pulegone-rich essential oils can cause abdominal pain and tenderness, nausea, bloody vomiting, burning of the throat, fever, lethargy alternating with agitation, confusion, delirium, restlessness, seizures, dizziness, auditory and visual hallucinations, increased blood pressure, rapid heartbeat, swollen blood vessels in the lungs and filling of the alveoli with blood, excessive acidity of the blood, blood clots, abortion, liver failure, kidney failure, respiratory failure, shock, coma, and death.[2421] Neurological damage has been reported in adults who consumed 2.5 to 5.0 mL of pennyroyal essential oil, moderate to severe poisoning with single doses of 10 mL, and deaths occurred at over 15 mL of ingestion.[2422,2423,2424,2425,2426,2427,2428,2429,2430,2431,2432]

■ Avoid with epilepsy and Parkinson's disease due to pulegone content, which may exacerbate or cause seizures or convulsions.[2433,2434,2435]

■ Avoid in people with compromised liver or kidney due to significant potential to cause liver and kidney toxicity or damage caused by pulegone.[2436,2437,2438,2439,2440] This warning includes using lesser calamint with medications known to harm the liver.

Menthone CTs

■ May interact with aspirin, blood pressure, antiplatelet, and anticoagulant medications, and increase the risk of bleeding among people with bleeding disorders, due to significant menthone content. Menthone is a weak inhibitor of platelet aggregation.[2441]

■ Follow cautions for pulegone CTs also due to the potential for menthone CTs to contain significant pulegone levels.

Piperitenone Oxide CTs, Piperitenone Oxide CT

■ May interact with antibiotics, antifungals, or antiviral drugs and increase their effectiveness based on piperitenone oxide and piperitone content.[2442,2443,2444,2445,2446,2447]

■ A GC-MS with major constituents should be reviewed to determine pulegone levels. If moderate to significant pulegone (≥ 15%) is present, follow the cautions for pulegone as well.

Isopulegol CT

■ Do not take orally due to slight toxicity of isopulegol.[2448,2449]

SELECTED EVIDENCE:

- □ The pulegone CT of lesser calamint (ssp. *grandulosa*) strongly inhibited lipoxygenase (an important target in the treatment of multiple inflammatory disorders) *in vitro*. Interestingly, the same study reported that pulegone was inactive against lipoxygenase.[2450]
- □ The isopulegol CT of lesser calamint showed good antioxidant and AChE and BChE inhibitory activity *in vitro*.[2451]
- □ *In vitro* research showed that lesser calamint (ssp. *grandulosa*, pulegone CT) inhibited *A. niger, E. coli, S. aureus, S. enteritidis, C. subtilis,* and *P. aeruginosa*.[2452] The isopulegol CT also demonstrated broad-spectrum antimicrobial activity against *S. aureus, S. pneumoniae, E. coli, B. subtilis,* and *P. aeruginosa*.[2453] Remarkably the isopulegol CT exhibited greater inhibition of *S. pneumonia* than ampicillin and ciprofloxacin, and equal inhibition of *S. aureus* when compared to ciprofloxacin. The pulegone CT of lesser calamint (ssp. *glanduosa*) demonstrated anti-candida activity.[2454] The piperitenone oxide CT of lesser calamint inhibited *P. aeruginosa, S. enteritidis, S. aureus, B. cereus,* and *E. coli*.[2455] Lesser calamint (ssp. *nepeta* var. *subisodonda*) inhibited *A. niger, E. coli, S. aureus, S. enteritidis, B. subtilis,* and *P. aeruginosa*.[2456] Lesser calamint (*C. nepeta* CT pulegone) inhibited *S. venexiana, S. paratyphi* B, *S. typhimulium, P. oryzae, A. niger, B. cinerea, B. cersus*, and *L. monocytogenes*.[2457] Only pulegone showed antimicrobial activity when the main constituents (pulegone, menthone, menthol, limonene) of the essential oil were tested against the same organisms. The pulegone CT inhibited *S. aureus, E. coli, P. aeruginosa, B. subtilis,* and *C. albicans*.[2458]
- □ Lesser calamint CT piperitone oxide exhibited significant antimicrobial activity against twenty phytopathogens, which often exceeded standard antibiotics.[2459]

MACE

Myristica fragrans Houtt., M. officinalis L. f.

FAMILY: Myristicaceae
NOTE: Middle
AROMA INTENSITY: Medium
AROMA: Sweet, spicy, rich
COMMON EXTRACTION METHOD: Steam distilled from the pericarp (aril; outer shell/layer)
POSSIBLE SUBSTITUTE OILS: Nutmeg, ravensara (sabinene CT)
BLENDS WELL WITH: Camphor, cassia, cinnamon, clary sage, clove, copaiba, coriander, eucalyptus, fennel, geranium, ginger, lavender, lime, nutmeg, orange, oregano, patchouli, pine, rosemary, turmeric, vetiver, ylang ylang
SUBCELLULAR LOCALIZATION | EPIGENETIC INFLUENCE: Currently unknown | Currently unknown
RECOMMENDED DILUTION RANGE: 3%–25%

PRIMARY CONSTITUENTS:[2460,2461,2462,2463,2464,2465]

Malaysian

Terpinen-4-ol	13.5%–19.1%
Alpha-Terpineol	8.2%–15.5%
Alpha-Pinene	14.1%–15.2%
Gamma-Terpinene	5.2%–9.1%
Beta-Pinene	6.8%–9.0%
Alpha-Terpinolene	5.1%–6.5%
Alpha-Terpinene	3.4%–6.1%
Limonene	4.6%–5.7%
Beta-Phellandrene	2.8%–5.4%
Isoeugenol	3.1%–4.5%
Beta-Myrcene	3.3%–4.1%
Delta-3-Carene	1.7%–3.7%
Myristicin	1.0%–3.0%

Southern Indian

Sabinene	20.2%
Terpinen-4-ol	12.1%
Safrole	10.3%
Alpha-Pinene	9.7%
Beta-Phellandrene	6.6%
Gamma-Terpinene	5.9%
Beta-Pinene	5.6%
Alpha-Terpinene	3.7%

Pakistani

Gamma-Terpinene	19.1%
Safrole	18.2%
Terpinen-4-ol	12.7%
Alpha-Pinene	11.6%
Sabinene	11.2%
Myrysticin	7.5%
Methyl Eugenol	3.5%
Alpha-Terpinolene	3.4%

East Indian/Indian

Beta-Pinene + Sabinene	23.1%–46.7%
Alpha-Pinene	15.2%–26.7%
1,8-cineole + Limonene*	7.0%–21.0%
Terpinen-4-ol	4.0%–14.0%
Gamma-Terpinene	1.8%–11.6%
Myristicin	1.3%–5.9%
Alpha-Terpinene + Para-Cymene	3.5%–7.5%
Alpha-Phellandrene	0.0%–3.2%
Elemicin	0.2%–3.1%

* Co-eluted with beta-phellandrene

OTHER CONSTITUENTS: *Pakistani*—Trans-p-2-Menthen-1-ol, 2-pentyl-anisole, eugenol, isoeugenol methyl ether; *Malaysian*—Sabinene, alpha-phellandrene, para-cymene, linalool, nerol, safrole; *Southern Indian*—Alpha-thujene, myrcene, alpha-phellandrene, beta-3-carene, para-cymene, terpinolene, cis-p-menth-2-en-1-ol, alpha-terpineol, alpha-copaene, trans-isocroweacin, methyl eugenol (< 2.0%); *East Indian/Indian*—Beta-phellandrene, alpha-terpineol, isoeugenol, safrole (< 2.0%), para-cymen-8-ol, methyl eugenol (< 0.3%)

PREFERRED COMPOSITION FOR CLINICAL USE:
Insufficient samples/data to determine.

REPORTED THERAPEUTIC PROPERTIES: Analgesic, anti-inflammatory, antispasmodic, antirheumatic, relieves headache, antioxidant, aids oral health, relieves toothache, expels excess gas, enhances circulation, reduces male sexual dysfunction, alleviates painful menstruation, expels excess mucus, supports respiratory function, antidepressant, relieves anxiety, encourage self-esteem

CAUTIONS:

Pakistani, Southern Indian

- Avoid mace use in children under 6, and use cautiously through age 12. Children are more susceptible to nutmeg toxicity (similar in composition to mace), which can cause dry mouth, flushing, irregular heart rhythm, tremors, agitation, and delirium.[2466,2467,2468]
- Avoid during pregnancy and lactation due to myristicin and safrole content, CTS with higher levels of myristicin and safrole. Animal research suggests that higher doses of safrole or myristicin (0.01 mL/kg every

two days while nursing, and 0.001 to 10 mg while pregnant) taken orally during pregnancy and while nursing may cause liver toxicity in offspring born to the mother and increase the offspring's risk of liver cancer.[2469,2470]

■ Oral caution. Animal research suggests that oral administration of safrole may cause liver cancer and is reasonably anticipated to be a human carcinogen.[2471] Long-term oral use of small doses (1 to 6 drops) for more than twenty-one days without at least a seven-day break and single large doses (5 g or more of the whole herb) should be avoided due to the psychotropic potential of mace, as well as its constituents methyl eugenol and safrole. Safrole and methyl eugenol can be toxic in large doses.[2472,2473,2474] Myristicin is believed to be at least partially responsible for the psychotropic (a substance capable of affecting the mind, emotions, and behavior) effects of mace; although it is likely a combination of more than one component of the oil that produces this effect. Large doses of mace with high myristicin content (5 g in a single dose), may produce a psychotropic (hallucinations, anxiety, fear, and delusions) effect in adults.[2475,2476] In addition, scientists suspect constituents in mace may be chemically modified once ingested to amphetamine-like constituents, though this has not been proven conclusively.[2477]

■ There is a moderate risk that mace essential oil may interact with aspirin, blood pressure, antiplatelet, and anticoagulant medications, and increase the risk of bleeding among people with bleeding disorders (low risk), based on safrole content.[2478]

■ May interact with psychotropic medications—medications prescribed for anxiety, depression, schizophrenia, mania, obsessive-compulsive disorder, ADHD, and bipolar disorder.[2479,2480]

■ May interact with monoamine oxidase inhibitors (MAOIs)—medications prescribed for the treatment of depression.[2481]

■ May interfere with enzymes responsible for metabolizing medications (NSAIDs, proton-pump inhibitors, acetaminophen, antiepileptics, immune modulators, blood-sugar medications, blood pressure medications, antidepressants, antipsychotics, diabetic medications, antihistamines, antibiotics, and anesthetics).[2482,2483,2484,2485,2486,2487,2488]

Malaysian, East Indian/Indian

■ Avoid mace use in children under 6, and use cautiously through age 12. Children are more susceptible to nutmeg toxicity (similar in composition to mace), which can cause dry mouth, flushing, irregular heart rhythm, tremors, agitation, and delirium.[2489,2490,2491]

■ Oral caution. Long-term oral use of small doses (1 to 6 drops) for more than twenty-one days without at least a seven-day break and single large doses (5 g or more of the whole herb) should be avoided due to the psychotropic potential of mace. Myristicin is believed to be at least partially responsible for the psychotropic (a substance capable of affecting the mind, emotions, and behavior) effects of mace; although it is likely a combination of more than one component of the oil that produces this effect. Large doses of mace with high myristicin content (5 g in a single dose), may produce a psychotropic (hallucinations, anxiety, fear, and delusions) effect in adults.[2492,2493] In addition, scientists suspect constituents in mace may be chemically modified once ingested to amphetamine-like constituents, though this has not been proven conclusively.[2494]

■ May interact with psychotropic medications—medications prescribed for anxiety, depression, schizophrenia, mania, obsessive-compulsive disorder, ADHD, and bipolar disorder.[2495,2496]

■ May interact with monoamine oxidase inhibitors (MAOIs)—medications prescribed for the treatment of depression.[2497]

■ May interfere with enzymes responsible for metabolizing medications (NSAIDs, proton-pump inhibitors, acetaminophen, antiepileptics, immune modulators, blood-sugar medications, blood pressure medications, antidepressants, antipsychotics, diabetic medications, antihistamines, antibiotics, and anesthetics).[2498,2499,2500,2501,2502,2503,2504]

SELECTED EVIDENCE:

- ◻ Southern Indian mace essential oil demonstrated moderate to good antioxidant activity (radical scavenging and reducing power) when compared to synthetic antioxidants (BHA, BHT).[2505] The same researchers concluded that mace oil showed strong antibacterial activity against *B. subtilis*, and good activity against *S. aureus* and *B. cereus*. It was ineffective against *E. coli* and *P. aeruginosa*.
- ◻ Malaysian mace essential oil inhibited seven bacterial strains including two gram-positive (*B. subtilis* subsp. *spizizenii* and *S. aureus*) and five gram-negative strains (S. *enterica, E. coli, E. aerogenes, K. pneumonia,* and *P. aeruginosa*), and six fungal strains (*A. niger, A. flavus, F. solani, F. oxysporum, P. digitatum,* and *A. alternata*).[2506] The same study reported that mace oil possesses significant antioxidant activity. Mace oil was a more effective antimicrobial than nutmeg oil, but not as strong in antioxidant activity in this study.
- ◻ Mace demonstrated moderate antioxidant activity *in vitro*.[2507]

MALLEE (Kochii Eucalyptus)

Eucalyptus kochii Maiden & Blakely, *E. kochii* subsp. *kochii, E. kochii* subsp. *plenissima, E. kochii* subsp. *borealis*

FAMILY: Myrtaceae
NOTE: Top
AROMA INTENSITY: Strong
AROMA: Fresh, medicinal, clean, slightly woody
COMMON EXTRACTION METHOD: Steam distilled from the leaves
POSSIBLE SUBSTITUTE OILS: Eucalyptus, cajeput, myrtle
BLENDS WELL WITH: Bay laurel, bergamot, blue cypress, cajeput, camphor, cedarwood, eucalyptus, geranium, ginger, grapefruit, juniper, lavender, lemon, lemongrass, melaleuca (tea tree), myrtle, petitgrain, pine, ravensara, ravintsara, rosemary, sage, tangerine, thyme
SUBCELLULAR LOCALIZATION | EPIGENETIC INFLUENCE: Currently unknown | Currently unknown
RECOMMENDED DILUTION RANGE: 3%–33%; neat for some conditions

PRIMARY CONSTITUENTS:[2508,2509,2510,2511]

1,8-Cineole 77.3%–97.3%

OTHER CONSTITUENTS: Alpha-pinene, limonene, para-cymene, terpinen-4-ol, sabinene, beta-terpineol, cuminal

PREFERRED COMPOSITION FOR CLINICAL USE:

Constituent	
1,8-Cineole	75%–95%
Alpha-Pinene	0.5%–3%
Para-Cymene	0.5%–3%
Terpinen-4-ol	0.1%–3%
Beta-Pinene	0.1%–3%

REPORTED THERAPEUTIC PROPERTIES: Analgesic, antibacterial, antifungal, antirheumatic, antiseptic, antispasmodic, anesthetic, antiviral, antineuralgic, nervine, **decongestant**, diuretic, expectorant, eases cough, **supports respiratory function**, reduces fevers, promotes oral cleanliness, helps balance blood-sugar levels, wound healing, insect repellent, invigorating, aids concentration and memory, combats negative emotions, relieves mental exhaustion

CAUTIONS:

■ Avoid with children under age 3, particularly around the nose and mouth and use cautiously in children under age 5 due to significant 1,8-cineole content. 1,8-cineole may cause seizures, central nervous system problems, or respiratory distress in young children.[2512,2513,2514]

■ Caution is warranted orally due to the significant amounts of 1,8-cineole. Limit it to small doses internally (adults—maximum 10 drops daily). Toxicity has been reported when eucalyptus (also high in 1,8-cineole) was ingested in large doses, and as few as 4 to 5 drops may cause problems in very sensitive individuals.[2515,2516,2517,2518,2519,2520]

■ Avoid with epilepsy and Parkinson's disease due to 1,8-cineole content. May exacerbate or cause seizures/convulsions or reduce seizure-medication efficacy based on 1,8-cineole content.[2521,2522,2523,2524]

■ May interact with diabetic medications due to significant 1,8-cineole content. 1,8-cineole significantly inhibits alpha-glucosidase activity, particularly when synergized with other constituents (alpha-pinene, beta-pinene, and para-cymene).[2525,2526,2527] Alpha-glucosidase is an enzyme that breaks down carbohydrates by chemical reaction with water. Inhibiting its activity postpones glucose absorption and therefore the impact of carbohydrates on blood-sugar levels. Studies suggest that oils rich in 1,8-cineole may significantly lower glucose in animals, which may increase the action of diabetic medications.[2528,2529]

■ May interact with aspirin, blood pressure, antiplatelet, and anticoagulant medications, and increase the risk of bleeding among people with bleeding disorders (low risk) due to significant 1,8-cineole content.[2530] 1,8-cineole is a weak inhibitor of platelet aggregation.[2531]

■ May weakly interfere with the enzymes responsible for metabolizing medications (NSAIDs, proton-pump inhibitors, acetaminophen, antiepileptics, immune modulators, blood-sugar medications, blood pressure medications, antidepressants, antipsychotics, diabetic medications, antihistamines, antibiotics, and anesthetics) due to 1,8-cineole content.[2532,2533,2534]

■ May interfere with pentobarbital and other barbiturates (medications for anxiety and insomnia), based on 1,8-cineole content.[2535,2536]

SELECTED EVIDENCE:

- □ *In vitro* research demonstrated that mallee essential oil inhibited *E. coli. S. aureus, E. faecalis, S. typhimurium, A. baumannii, C. albicans, S. epidermidis*, and MRSA, with *E. faecalis* and *C. albicans* being the most susceptible microbes and *A. baumannii* the least susceptible.[2537]

MANGO GINGER
(Amahaldi)

Curcuma amada Roxb.

FAMILY: Zingiberaceae
NOTE: Middle-Base
AROMA INTENSITY: Medium to Strong

AROMA: Earthy, fruity, woody, camphoraceous, spicy, hint of vanilla
COMMON EXTRACTION METHOD: Steam distilled or hydrodistilled from the rhizomes; may also be extracted by Co2 (supercritical); less commonly extracted from the leaves
POSSIBLE SUBSTITUTE OILS: Hemp, hops, turmeric, ginger, myrrh
BLENDS WELL WITH: Bay laurel, bergamot, camphor, cardamom, carrot seed, cassia, cedarwood, clove, coriander, eucalyptus, fennel, frankincense, galbanum, geranium, ginger, grapefruit, lemon, lime, neroli, nutmeg, orange, oregano, patchouli, ravensara, rose, sandalwood, tangerine, turmeric, vetiver
SUBCELLULAR LOCALIZATION | EPIGENETIC INFLUENCE: Currently unknown | Currently unknown
RECOMMENDED DILUTION RANGE: 3%–50%; neat for some conditions

PRIMARY CONSTITUENTS:[2538,2539,2540,2541,2542,2543,2544,2545,2546]

Myrcene CT	
Myrcene	40.2%–88.8%
Beta-Pinene	3.7%–15.2%
ar-Curcumene	0.0%–10.0%
2,6,6-Trimethyl-bicyclohept-2-ene	0.0%–5.0%
Camphor	0.0%–3.2%
(E)-Beta-Ocimene	0.0%–2.6%

ar-Curcumene CT	
ar-Curcumene	28.1%
Camphor	11.2%
Beta-Curcumene	11.2%
Curzerenone	7.1%
Isoborneol	4.5%
1,8-Cineole	6.0%
Beta-Elemene	2.8%

(Z)-Beta-Farnesene CT	
(Z)-Beta-Farnesene	21.9%
Guaia-6,9-Diene	19.8%
Alpha-Longipinene	14.8%
Alpha-Guaiene	14.5%
Camphor	5.5%
Thymol	4.9%

Myrcene/Alpha-Asarone CT	
Myrcene	63.9%
Alpha-Asarone	30.3%

Beta-Ocimene CT	
(E & Z)-Beta-Ocimene	47.2%
Alpha-Pinene	18.0%
Linalool	11.2%
Safrole	9.3%
Linalyl Acetate	9.1%

Leaves	
Camphor	17.9%
Epi-Curzerenone	10.8%
Curzerenone	9.5%
Furanosesquiterpenoid	8.6%
Isoborneol	7.3%
Curzerene	4.0%
Camphene	3.6%
Beta-Caryophyllene	3.3%
6-Methoxy-2-(1-Buten-3-yl)-Naphthalene	3.3%

Note: The myrcene CT is the most common reported in the literature.

OTHER CONSTITUENTS: *Myrcene CT*—Alpha-pinene, beta-thujone (< 0.8%), perillene, beta-caryophyllene, (Z)-beta-farnesene, alpha-farnesene, (Z)-beta-ocimene; *ar-Curcumene CT*—Alpha-pinene, camphene, beta-pinene, borneol, gamma-terpinene, beta-elemene, alpha-selinene, beta-selinene, alpha-zingiberene, beta-caryophyllene; *Myrcene/Alpha-Asarone CT*—Beta-Pinene, myrophine, 1-nitroguanidine; *Beta-Ocimene CT*—N/A; *(Z)-Beta-Farnesene CT*—N/A; *Leaves*—Alpha-pinene, beta-pinene, 1,8-cineone, linalool, camphene hydrate, borneol, beta-cubebene, beta-elemene, gamma-elemene, 1,3,5-bisanolatrien-7-ol, T-cadinol, (2E,6Z)-farnesol, 14-hydroxy α-muurolene, furanogermenone

PREFERRED COMPOSITION FOR CLINICAL USE:

Constituent	
Myrcene	65%–90%
Beta-Pinene	3%–15%
Alpha-Pinene	0.1%–5%
Perillene	tr–5%

REPORTED THERAPEUTIC PROPERTIES: Aids digestion, improves appetite, expels excess gas, relieves constipation, **analgesic**, reduces fever, anti-inflammatory, antispasmodic, antiarthritic, antirheumatic, antineuralgic, antimicrobial, antibacterial, antifungal, aids detoxification, anticancer, antioxidant, antiallergic, astringent, moisturizes the skin, relieves skin conditions and itching, wound healing, heals bruises, nourishes the scalp, aids respiratory function, opens the airways, eases cough, insect repellent, aphrodisiac, relaxing

CAUTIONS:

Myrcene CT

■ Caution is advised during pregnancy and lactation due to high beta-myrcene content. Extremely high doses (500 mg/kg) of beta-myrcene have been toxic to fetuses according to animal research.[2547,2548]

■ May interact with barbiturates (medications for anxiety and insomnia), antihistamines, benzodiazepines, tricyclic antidepressants, or other central nervous system depressant drugs, increasing their depressant effects.[2549,2550,2551]

■ May weakly interfere with the enzymes responsible for metabolizing medications (NSAIDs, proton-pump inhibitors, acetaminophen, antiepileptics, immune modulators, blood-sugar medications, blood pressure medications, antidepressants, antipsychotics, diabetic medications, antihistamines, antibiotics, and anesthetics), based on myrcene content.[2552,2553]

Myrcene/Alpha-Asarone CT

This CT is not recommended for therapy unless the benefits outweigh the risks as determined by a qualified professional.

■ Avoid in children under age 12 due to liver toxicity concerns of asarone content.[2554,2555,2556,2557]

■ Avoid during pregnancy and lactation reduced maternal weight gain, embryo toxicity, abortion, and fetal malformations (hydrocephaly, extra-ribs, clubfeet, and cleft lips) reportedly caused by alpha-asarone in animal studies.[2558] However, one animal study did not report toxicity to embryos or fetuses nor birth defects.[2559]

■ Avoid oral consumption due to asarone content. Beta-asarone is genotoxic (destructive to a cell's genetic material—DNA, RNA—and integrity), and may cause damage to organs (liver, heart, and kidneys), abnormal fluid accumulation in the abdominal (peritoneal cavity), and malignant intestinal tumors at large to extreme doses according to animal research.[2560,2561,2562,2563,2564,2565] Long-term administration increases the damage caused. The exposure limit for total asarones (alpha- and beta-) was set to 115 mcg (or 2 mcg/kg/day) from herbal medicinal products, which would equate to less than one drop of essential oil.[2566] A case report involving a middle-aged woman reports tachycardia, dizziness, tremor, irregular breathing, pallor, anxiety, nausea and vomiting reportedly due to asarone intoxication after consuming an unknown quantity of asarone-containing herbal medicines.[2567]

■ May interfere with pentobarbital and other barbiturates (medications for anxiety and insomnia) due to asarone content.[2568,2569,2570,2571]

■ Avoid with compromised liver or kidney due to significant potential to cause liver and kidney toxicity.[2572,2573,2574,2575,2576]

Leaves

■ Avoid with epilepsy and Parkinson's disease due to the potential of camphor to exacerbate or cause seizures/convulsions or reduce seizure-medication efficacy.[2577,2578,2579,2580]

■ The significant camphor content may negatively impact red blood cells and increase the risk of jaundice in children with glucose-6-phosphate dehydrogenase deficiency (G6PD) due to camphor content.[2581,2582]

ar-Curcumene and (Z)-Beta-Farnesene

■ None currently known.

Beta-Ocimene CTs

■ Not recommended for therapy due to the possible presence of significant safrole.

SELECTED EVIDENCE:

- *In vitro* research demonstrated that mango ginger CT myrcene inhibited *S. aureus, E. coli, K. pneumoniae, P. aeruginosa, S. paratyphi, V. cholera, E. aerogenes, S. pneumoniae, B. subtilis, B. cereus, P. mirabilis,* and *P. vulgaris*.[2583] Mango ginger essential oil inhibited fungi (*R. solani* and *S. solani*) more effectively than its extracts.[2584]
- Mango ginger CT myrcene (with 30.3% alpha-asarone) essential oil exhibited comparable antioxidant activity to BHT in the DPPH, ABTS, and FRAP assays.[2585] The myrcene CT also exhibited moderate to good antioxidant activity in the DPPH, FRAP, and nitric oxide scavenging assays.[2586]
- Mango ginger essential oil showed 100% repellency against the housefly (*Musca domestica*) even when used at a 0.5% concentration.[2587]

MASSOIA BARK
(Massoy Bark, Masoyi, Massoi)

Cryptocarya massoy (Oken) Kosterm., *C. massoia* (Oken) Kosterm., *Massoia aromatica* Becc.

FAMILY: Lauraceae
NOTE: Base
AROMA INTENSITY: Medium
AROMA: Rich, woody, balsamic, toasted coconut, slightly fruity
COMMON EXTRACTION METHOD: Steam distilled or hydrodistilled from the bark; may also be distilled from the heartwood or fruits
POSSIBLE SUBSTITUTE OILS: Ylang ylang, jasmine
BLENDS WELL WITH: Anise, balsam fir, benzoin, bergamot, carrot seed, cassia, cedarwood, cinnamon, clove, copaiba, coriander, davana, helichrysum, lavender, lemon, lime, nutmeg, orange, patchouli, Peru balsam, sandalwood, spearmint, tangerine, ylang ylang
SUBCELLULAR LOCALIZATION | EPIGENETIC INFLUENCE: Currently unknown | Currently unknown
RECOMMENDED DILUTION RANGE: Not currently recommended for topical use

PRIMARY CONSTITUENTS:[2588,2589,2590]

Bark

C-10 Massoia Lactone	56.2%–92.1%
C-12 Massoia Lactone	6.2%–17.4%
Benzyl Benzoate	2.7%–13.4%
C-8 Massoia Lactone	0.0%–3.4%
4-Propylcyclohexene	0.0%–2.2%

Heartwood

C-10 Massoia Lactone	67.8%
C-12 Massoia Lactone	27.7%
Delta-Decalactone	2.5%

Fruits

Benzyl Benzoate	68.3%
Beta-Caryophyllene	12.9%
Alpha-Humulene	2.2%
Caryophyllene Oxide	2.2%
Alpha-Copaene	2.1%

Note: Massoia bark essential oil may be available in grades according to its C-10 lactone content. C-10 lactone content also changes its color from pale yellow to orange.

OTHER CONSTITUENTS: *Bark*—Benzyl salicylate (< 1.9%), beta-bisabolene, borneol, linalool, delta-decalactone, 1phenylcyclohexa-2,4-diene, tuberolactone, delta-dodecalactone, C-14 massoia lactone; *Heartwood*—C-14 massoia lactone; *Fruits*—Beta-elemene, C-10 massoia lactone, delta-cadinene, benzyl salicylate (< 1.9%)

PREFERRED COMPOSITION FOR CLINICAL USE:

Constituent	
C-10 Massoia Lactone (5,6-dihydro-6-pentyl-2H-pyran-2-one)	60%–90%
C-12 Massoia Lactone (5,6-dihydro-6-heptyl-2H-pyran-2-one)	3%–18%
Benzyl Benzoate	0.5%–15%

REPORTED THERAPEUTIC PROPERTIES: Aids circulation, anticancer, antibacterial, antifungal, anti-infectious, modulates immune system activity, supports respiratory function, eases cough, antispasmodic, relieves sore throat, soothes the digestive tract, calming and relaxing, revitalizing, relieves anxiety

CAUTIONS:

Bark, Heartwood

■ Massoia lactones are suspected of being strongly irritating to the skin, and IFRA (International Fragrance Association) has banned the use of massoia bark essential oil as a fragrance ingredient. No published data existed as of this writing to determine whether massoia bark oil causes irritation, sensitization, or photosensitivity reactions. Based on this, massoia bark is not recommended for aromatic use only until safety evaluations can be performed.

Fruit

■ Oral caution. Ingestion of benzyl derivatives (benzoic acids, benzoic salts, benzaldehyde, benzyl acetate, benzyl benzoate, and benzyl alcohol) may cause gastrointestinal irritation, vomiting, and diarrhea.[2591] The Joint Expert Committee for Food Additives (JECFA) has established an acceptable daily limit for the aforementioned constituents of 0–5 mg/kg body weight for adults to avoid GI discomfort.[2592] Based on this limit, it is suggested to limit the oral consumption of massoia bark essential oil rich in benzyl derivatives to no more than 10 drops daily.

- Use cautiously in people with salicylate sensitivities due to the possible presence of benzyl salicylate.

SELECTED EVIDENCE:

- Both massoia bark essential oil and C-10 massoia lactone significantly activated macrophage phagocytosis, suggesting an immunomodulatory effect.[2593] No effect on lymphocyte proliferation was observed in the same study. A second study found that massoia bark essential oil acted as an anti-infective in an immunosuppressed condition by stimulating macrophage phagocytosis.[2594] The study authors reported that C-10 massoia lactone played a major role in the immunomodulatory activity of the essential oil.
- *In vitro* research found that massoia bark essential oil partially inhibits *C. albicans* biofilm formation and synergized the antibiofilm activity of basil CT citral and padang cassia CT cinnamaldehyde essential oils.[2595] Another study also reported anti-candida activity.[2596]

MEXICAN OREGANO (Redbrush Lippia, Scented Matgrass, Scented Lippia, Orégano Cimmaron)

Lippia graveolens Kunth

FAMILY: Verbenaceae
NOTE: Middle
AROMA INTENSITY: Strong
AROMA: Herbaceous, medicinal, fresh, sharp
COMMON EXTRACTION METHOD: Steam distilled or hydrodistilled from the leaves or aerial parts
POSSIBLE SUBSTITUTE OILS: Oregano, thyme, ajowan
BLENDS WELL WITH: Ajowan, bergamot, grapefruit, lavender, lemon, lime, mountain savory, orange, oregano, pine, rosemary, summer savory, tangerine, thyme
SUBCELLULAR LOCALIZATION | EPIGENETIC INFLUENCE: Currently unknown | Currently unknown
RECOMMENDED DILUTION RANGE: 1.5%–20%; 50% for some conditions

PRIMARY CONSTITUENTS:[2597,2598,2599,2600,2601,2602,2603,2604,2605,2606]

Carvacrol CT	
Carvacrol	30.9%–83.5%
O-Cymene	0.0%–32.2%
Thymol	2.7%–17.2%
Para-Cymene	0.2%–10.1%
Beta-Caryophyllene	0.4%–5.4%
Gamma-Terpinene	tr–5.4%
3-(Tert-Butyl)-4-Hydroxyanisole	0.0%–4.9%
Alpha-Humulene	0.2%–3.3%
Caryophyllene Oxide	0.4%–3.2%

Balanced CT	
Thymol	18.4%–24.6%
Carvacrol	10.3%–24.5%
Para-Cymene	12.3%–19.7%
1,8-Cineole	3.4%–13.2%
Gamma-Terpinene	4.7%–10.0%
Beta-Caryophyllene	0.0%–6.8%
Myrcene	0.0%–3.0%

Carvacrol/Alpha-Terpinyl Acetate CT	
Carvacrol	37.8%
Alpha-Terpinyl Acetate	22.4%

Thymol CT

Thymol	42.2%–91.8%
Para-Cymene	0.4%–17.7%
Gamma-Terpinene	0.1%–4.7%
Thymol Methyl Ether	0.7%–3.8%

Caryophylla-4(12),8(13)-dien-5a-ol CT

Caryophylla-4(12),-8(13)-Dien-5a-ol	12.7%–25.2%
Alpha-Eudesmol + Alpha-Cadinol	7.4%–22.5%
Beta-Eudesmol	2.8%–21.7%
T-Cadinol + T-Muurolol	3.1%–15.0%
Viridiflorol	1.0%–12.6%
Caryophyllene Oxide	4.9%–11.3%
Thymol	1.1%–11.3%
(Z)-Calamen-10-ol	1.7%–8.9%
Beta-Epi-Bisabolol	4.5%–7.6%
Humulene Epoxide II	3.0%–6.0%
1,8-Cineole	0.1%–5.1%
Spathulenol	0.3%–4.0%
1,10-Diepicubenol	1.3%–3.7%
Muuroloa-4,10(14)-Dien-1Beta-ol	1.4%–3.5%
Beta-Caryophyllene	1.3%–3.3%

M-Cymene	20.4%
Thymol	6.7%

1,8-Cineole CT

1,8-Cineole	10.2%–16.8%
Thymol	0.0%–13.7%
Carvacrol	12.6%–13.5%
Alpha-Terpineol	1.8%–7.6%
Para-Cymene	5.7%–7.5%
Caryophyllene Oxide	0.0%–7.1%
Beta-Caryophyllene	6.1%–6.3%
Gamma-Octanolide	0.0%–5.7%
Alpha-Terpinene	3.9%–4.5%
Gamma-Terpinene	0.8%–3.9%

Para-Cymene CT

Para-Cymene	34.7%–42.4%
Carvacrol	7.3%–23.0%
Thymol	1.5%–19.4%

Note: Other plants may also be called Mexican oregano, such as *Lippia berlandieri* Schauer and *Poliomintha longiflora* Gray, but *L. graveolens* is the most common species called Mexican oregano. Mexican oregano can contain a variety of cymene derivatives (para-cymene, o-cymene, m-cymene) depending on multiple factors.

OTHER CONSTITUENTS: *Carvacrol CT*—1,8-cineole, gamma-terpinene, linalool, terpinen-4-ol, thymol methyl ether, carvacryl acetate, humulene epoxide II; *Thymol CT*—Para-cymene, 1,8-cineole, linalool, terpinen-4-ol, carvacrol, thymol acetate, beta-caryophyllene, alpha-humulene, caryophyllene oxide, humulene epoxide II; *Balanced CT*—Alpha-thujene, alpha-pinene, limonene, linalool, terpinen-4-ol, alpha-humulene, caryophyllene oxide; *Carvacrol/Alpha-Terpinyl Acetate CT*—Alpha-thujene, beta-pinene, isocaryophyllene, alpha-humulene; *1,8-Cineole CT*—Alpha-pinene, sabinene, beta-pinene, alpha-myrcene, alpha-phellandrene, m-cymene, (Z)-beta-ocimene, (E)-beta-farnesene, germacrene D, thymol methyl ether, carvacryl acetate, spathulenol, humulene oxide; *Caryophylla-4(12),8(13)-dien-5a-ol CT*—Trans-sabinol, borneol, terpinen-4-ol, alpha-terpineol, trans-piperitol, trans-carveol, thymol methyl ether, carvacrol, alpha-copaene, alpha-humulene, gamma-muurolene, beta-selinene, alpha-selinene, trans-calamenene, cis-sesquisabinene hydrate, trans-sesquisabinene hydrate, trans-nerolidol, guaiol; *Para-Cymene CT*—Not listed

PREFERRED COMPOSITION FOR CLINICAL USE:

Constituent	
Carvacrol	50%–70%
Para-Cymene	2%–12%
Thymol	2%–10%
Beta-Caryophyllene	2%–8%

Gamma-Terpinene	0.5%–8%
Alpha-Humulene	0.1%–5%

REPORTED THERAPEUTIC PROPERTIES: Aids digestion, relieves diarrhea, eases stomachache, supports respiratory function, opens the airways, expectorant, antibacterial, antifungal, antiviral, anti-infectious, antiparasitic, antioxidant, analgesic, reduces fever, anti-inflammatory, relieves menstrual discomfort, balances blood sugar, anticancer, insecticidal, antidepressant, reduces anger and frustration

CAUTIONS:

Carvacrol CT, Carvacrol/Alpha-Terpinyl Acetate CT, Balanced CT, Thymol CT, 1,8-Cineole CT

■ Avoid during pregnancy and lactation. Oregano (also rich in carvacrol) may negatively affect embryonic development and encourage fetal cell death according to animal research.[2607]

■ May interact with diabetic medications and cause low blood-sugar levels due to carvacrol content—carvacrol reduces blood glucose and enhances the antidiabetic activities of rosiglitazone in animals.[2608,2609,2610,2611]

■ May interact with aspirin, blood pressure, antiplatelet, and anticoagulant medications, and increase the risk of bleeding among people with bleeding disorders, due to carvacrol and thymol content.[2612,2613,2614,2615,2616]

■ May interact with anticholinergic (drugs used for asthma, incontinence, gastrointestinal cramps, muscular spasms, depression, and sleep disorders) and cholinergic medications (drugs used to reduce urinary retention, diagnose myasthenia gravis, and for glaucoma), based on thymol and carvacrol content.[2617,2618]

■ May interact with antibiotics or antifungals and possibly enhance or decrease their effects due to carvacrol or thymol content.[2619,2620,2621,2622,2623]

■ May irritate mucous membranes (eyes, mouth, nasal passages, vagina, rectum).

■ *Thymol CT*: Oral caution. Thymol has a longer half-life (the time it takes for half of the medication to metabolize or excrete half of the dosage) than most essential oil constituents and should not be administered orally for long periods of time.[2624] Thymol is a monoterpene phenol. Reports of fatalities in children who consumed 50 to 200 mg of phenols has been reported.[2625] It is recommended that chemotypes with high levels of thymol be limited to 10 drops per day orally for adults with a two- to seven-day break after twenty-one days of use; or limit ingestion to five or six days weekly.

■ *1,8-Cineole CT*: Avoid with epilepsy and Parkinson's disease due to 1,8-cineole content. May exacerbate or cause seizures/convulsions or reduce seizure-medication efficacy.[2626,2627,2628,2629]

Caryophylla-4(12),8(13)-dien-5a-ol CT

■ Avoid during pregnancy and lactation (low risk) due to beta-eudesmol content. Beta-eudesmol strongly inhibits the growth of new blood vessels, which is necessary for implantation, and placental and embryonic development.[2630]

■ May interact with aspirin, blood pressure, antiplatelet, and anticoagulant medications, and increase the risk of bleeding among people with bleeding disorders based on beta-eudesmol content (low risk).[2631,2632]

SELECTED EVIDENCE:

- □ The thymol CT (70% thymol, 9% para-cymene) of Mexican oregano essential oil demonstrated antigenotoxic (prevents mutations caused by damage to genetic information within the cell, which may lead to cancer) activity caused by UV rays.[2633]
- □ Mexican oregano essential oil (19.8% thymol, 16.9% para-cymene) was active against *Giardia lamblia*, a parasite that attaches to the epithelium of the small intestine and causes giardiasis.[2634]
- □ Mexican oregano essential oil CT 1,8-cineole showed marginal antispasmodic activity by inhibiting contractions caused by carbachol and histamine in isolated guinea pig ileum.[2635] Another study

reported that the same essential oil was less active than clove and thyme essential oils against *Giardia lamblia* parasites.[2636]

- *In vitro* evidence demonstrated that Mexican oregano essential oil CT 1,8-cineole weakly inhibited *B. subtilis, S. aureus, E. coli,* and *S. typhi*, with carvacrol showing the best antibacterial activity of isolated major constituents (carvacrol, para-cymene, thymol).[2637] The carvacrol/thymol CT of Mexican oregano exhibited antifungal activity against *Fusarium verticillioides*.[2638] Both the carvacrol and para-cymene CTs, as well as their microencapsulated forms, inhibited *E. coli, P. aeruginosa,* and *S. aureus*.[2639] An interesting CT with carvacrol and alpha-terpinyl acetate as the major constituents inhibited gram-positive (*S. aureus, S. epidermidis, B. subtilis, Sarcina lutea*) and gram-negative (*Sh. boydii, V. cholerae, V. cholerae* (clinical strain), *V. cholerae* (isolated from water), *V. cholerae* Tor, *E. coli, E. agglomerans, E. aerogenes, Y. enterocolitica, S. typhi*) bacteria.[2640] The carvacrol CT showed high levels of antifungal activity against fluconazole-susceptible *C. glabrata*.[2641] Mexican oregano essential oil (composition not reported) exhibited significant antimicrobial activity against pathogenic microbes (*S. aureus, E. coli, S. mutans,* and *C. albicans*), which exceeded the reference drugs nyastatin and gentamycin.[2642]
- Mexican oregano essential oil CT carvacrol (with o-cymene) exhibited antiviral activity against acyclovir-resistant herpes simplex virus type 1 (ACVR-HSV-1), acyclovir-sensitive HSV-1, human respiratory syncytial virus (HRSV), bovine herpesvirus type 2 (BoHV-2), and bovine viral diarrhea virus (BVDV), but was inactive against human rotovirus (RV) and BoHV-1 and 5.[2643]
- Microencapsulation of Mexican oregano essential oil with beta-cyclodextrin significantly increased its antioxidant activity.[2644]
- Mexican oregano essential oil (balanced CT) was acaricidal to the Asian blue tick (*Rhipicephalus microplus*).[2645] The carvacrol/thymol CT of Mexican oregano was not insecticidal to the maize weevil (*Sitophilus zeamais*) but powerfully inhibited acetylcholinesterase (AChE) activity in the weevil.[2646] Para-cymene showed fumigant toxicity against the weevil in the same research.

MICROMELUM

Micromelum integerrimum (Buch.-Ham. ex DC.) Wight & Arn. ex M.

FAMILY: Rutaceae
NOTE: Middle-Top
AROMA INTENSITY: Medium
AROMA: Woody, sweet, fresh, piney, herbaceous, green
COMMON EXTRACTION METHOD: Steam distilled or hydrodistilled from the fruit or flowers
POSSIBLE SUBSTITUTE OILS: Douglas fir, pine, tagetes
BLENDS WELL WITH: Balsam fir, bergamot, cedarwood, frankincense, grapefruit, lavender, lemon, lime, marjoram, neroli, orange, petitgrain, pine, rosemary, sandalwood, silver fir, tangerine, vetiver, white fir
SUBCELLULAR LOCALIZATION | EPIGENETIC INFLUENCE: Currently unknown | Currently unknown
RECOMMENDED DILUTION RANGE: 5%–Neat

PRIMARY CONSTITUENTS:[2647]

Fruit

Alpha-Terpinolene	32.2%
Alpha-Pinene	17.2%

Beta-Pinene	17.2%
Camphene	4.1%
(Z)-Beta-Ocimene	3.2%
7-Diethoxyethyl Bicylo [3.2.0] Heptan-2-one	3.1%
Neral	3.0%
Alpha-Terpinene	3.0%

Note: A Chinese study reported the primary composition of flowers oil as gamma-muurolene (53.5%), gamma-elemene (10.6%), β-elemene(9.6%), and benzyl salicylate (5.0%); and the fruit oil as decanal (47.5%), lauric aldehyde (35.9%), gamma-muurolene (6.4%), and decanol (2.0%), but no full text was available to see the full composition reported in the study.[2648]

OTHER CONSTITUENTS: 6-Methyl-heptan-2-ol, myrcene, limonene, 1,8-cineole, cis-p-mentha-2,8-dien-1-ol, iso-pulegyl acetate, 9-octadecenoic acid, neo-iso-isopulegyl acetate, methyl geranate, neryl acetate, alpha-copaene, dodecanal, gamma-muurolene, alpha-acoradiene

PREFERRED COMPOSITION FOR CLINICAL USE:
Insufficient data/samples to determine.

REPORTED THERAPEUTIC PROPERTIES: Anti-inflammatory, analgesic, antiarthritic, reduces fever, anticancer, antioxidant, antibacterial, antifungal, wound healing, stress management, relieves anxiety

CAUTIONS:
- None currently known.

SELECTED EVIDENCE:
- Micromelum essential oil exhibited broad-spectrum antibacterial activity against *B. subtilis, B. spizizenii, C. diphtheria, E. coli, K. pneumonia, M. luteus, P. aeruginosa, P. putida,* and *S. aureus*.[2649]

MIMOSA ABSOLUTE (Silver Wattle, Blue Wattle)

Acacia dealbata Link, *A. decurrens* (Wendl. f.) Willd. var. *dealbata* (Link) F. Muell.

FAMILY: Fabaceae
NOTE: Base
AROMA INTENSITY: Strong
AROMA: Sweet, floral, spicy, woody, waxy, slightly green and fruity
COMMON EXTRACTION METHOD: Solvent extracted from the flowers and twig ends with hexane or petroleum ether
POSSIBLE SUBSTITUTE OILS: Jasmine, frankincense
BLENDS WELL WITH: Balsam, benzoin, bergamot, chamomile (German, Roman), citronella, copaiba, geranium, jasmine, lavender, lemon, neroli, orange, Peru balsam, pine, rose, tangerine, ylang ylang
SUBCELLULAR LOCALIZATION | EPIGENETIC INFLUENCE: Currently unknown | Currently unknown
RECOMMENDED DILUTION RANGE: 1%–10%

PRIMARY CONSTITUENTS:[2650]

Lupenone	20.0%
Lupeol	7.8%
(Z)-Heptadec-8-ene	6.0%

Note: The absolute is very thick and may need to be warmed to be used. It consists of approximately 20% volatile (alkanes, fatty acids), 50% semivolatile (triterpene ketones and alcohols), and nonvolatile (aromatic and fatty acid triterpene esters) compounds.

OTHER CONSTITUENTS: Butanone, pent-1-en-3-ol, pentanal, ethyl propanoate, (E)-pent-2-en-1ol, hexanal, ethyl butanoate, butyl acetate, (E)-hex-2-enal, (Z)-hex-3-enol, (E)-hex-2-enol, hexan-1-ol, heptanal, ethyl pentanoate, isobutyl pentanoate, isobutyl butanoate, oct-1-en-3-ol, 6-methylhept-5-en-2-one, ethyl hexanoate, 1,8-cineole, (E)-oct-2-en-1-ol, guaiacol, methyl benzoate, nonanal, 2-phenylethyl alcohol, (E,Z)-nona-2,6-dienal, benzyl acetate, (E)-non-2-enal, ethyl benzoate, decanal, (E)-dec-2-enal, methyl p-anisate, vanillin, (Z)-jasmone, ethyl p-anisate

PREFERRED COMPOSITION FOR CLINICAL USE:
Insufficient data/samples to determine.

REPORTED THERAPEUTIC PROPERTIES: Nourishes and soothes the skin, helps remove excess oil and dirt from pores, improves the appearance of the skin, wound healing, astringent, antiseptic, nourishes the scalp and hair, deodorant, relieves diarrhea, antidepressant, relieves anxiety, stress management, reduces nervousness, combats overreactions, sedating

CAUTIONS:

- Avoid oral consumption. Absolutes contain solvent residue and should not be taken orally.

SELECTED EVIDENCE:

- None found.

MOHAR MADON (Simple-Leaved Frankincense, Emerald Frankincense, Frankincense Occulta)

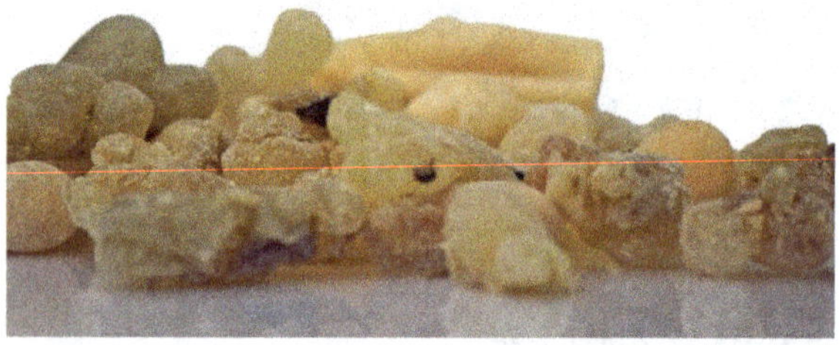

Boswellia occulta Thulin DeCarlo & S.P.Johnson sp. nov

FAMILY: Burseraceae
NOTE: Base
AROMA INTENSITY: Medium
AROMA: Balsamic, medicinal, woody
COMMON EXTRACTION METHOD: Steam distilled or hydrodistilled from the
POSSIBLE SUBSTITUTE OILS: Frankincense, myrrh
BLENDS WELL WITH: Agarwood, basil, bay laurel, bergamot, black pepper, camphor, cassia, cinnamon, coriander, cypress, frankincense, geranium, grapefruit, lavender, lemon, myrrh, neroli, orange, palo santo, petitgrain, pine, ravensara, ravintsara, rose, sandalwood, spikenard, tangerine, vetiver, ylang ylang

SUBCELLULAR LOCALIZATION | EPIGENETIC INFLUENCE: Currently unknown | Currently unknown
RECOMMENDED DILUTION RANGE: 3%–50%; neat for some conditions

PRIMARY CONSTITUENTS:[2651]

1-Methoxydecane	26.6%–47.9%
Serratol	2.7%–31.8%
Unidentified Guaiol	1.2%–15.1%
1-Methoxyoctane	3.6%–9.2%
Sabinene	0.1%–8.3%

OTHER CONSTITUENTS: Alpha-pinene, alpha-thujene, para-cymene

PREFERRED COMPOSITION FOR CLINICAL USE:
Insufficient data/samples to determine.

REPORTED THERAPEUTIC PROPERTIES: Analgesic, anti-inflammatory, antiarthritic, antirheumatic, anti-inflammatory, antispasmodic, antimicrobial, antifungal, antioxidant, astringent, aids digestion, expels excess gas, diuretic, supports respiratory function, expels excess mucus, immune modulation, would healing, nourishes the skin, restores dry, damaged, and mature skin, soothes chronic skin conditions, reduces the appearance of blemishes and wrinkles, aids liver function, clarifies thoughts, antidepressant, expands consciousness, enhances meditation

CAUTIONS:
- None currently known.

SELECTED EVIDENCE:
- □ None found.

MOJITO MINT (Cubin Mint, Common Mint Egyptian Mint)

Mentha × *villosa* Huds. (pro sp.) [*spicata* × *suaveolens*], *M. alopecuroides* Hull, *M. nemorosa* Willd.

FAMILY: Lamiaceae (Labiatae)
NOTE: Top-Middle
AROMA INTENSITY: Medium-Strong
AROMA: Minty, sweet, slightly fruity
COMMON EXTRACTION METHOD: Steam distilled or hydrodistilled from the leaves
POSSIBLE SUBSTITUTE OILS: Applemint, spearmint, caraway
BLENDS WELL WITH: Applemint, bergamot, copaiba, eucalyptus, frankincense, grapefruit, lavandin, lavender, lemon, lime, neroli, orange, pennyroyal, peppermint, petitgrain, rosemary, spearmint, tangerine
SUBCELLULAR LOCALIZATION | EPIGENETIC INFLUENCE: Currently unknown | Currently unknown
RECOMMENDED DILUTION RANGE: Piperitenone Oxide CT—5% to neat; Carvone CT—3.0%–50%, neat for some conditions

PRIMARY CONSTITUENTS:[2652,2653,2654,2655,2656,2657,2658,2659,2660,2661,2662]

Piperitenone Oxide CT (Brazil, Greece)	
Piperitenone Oxide	51.0%–95.9%
Gamma-Muurolene	0.0%–13.1%
Germacrene D	0.0%–9.7%
Gamma-Cadinene	0.0%–9.7%
Limonene	0.0%–8.8%
Trans-Sabinene Hydrate	0.0%–5.8%
1,8-Cineole	0.0%–4.5%
(Z)-Beta-Ocimene	0.0%–4.0%
Myrcene	0.0%–3.7%
Beta-Caryophyllene	0.0%–3.6%

Carvone CT (USA, Czech Republic)	
Carvone	44.2%–72.1%
Dihydrocarvyl Acetate	0.0%–18.0%
Limonene	1.0%–14.2%
Cis-Dihydrocarvone	2.3%–14.1%
1-Octen-3-ol	0.0%–9.1%
Neodihydrocarveol	0.0%–6.2%
Piperitone Oxide	0.0%–4.5%
Germacrene D	0.0%–4.5%
1,8-Cineole	0.2%–4.3%

OTHER CONSTITUENTS: *Piperitenone Oxide CT* —Alpha-thujene, alpha-pinene, beta-pinene, sabinene, octyl acetate, beta-elemene, cis-muurola-4(14),5-diene, alpha-humulene, trans-muurola-4(14),5-diene, (E)-beta-farnesene, bicyclogermacrene, germacrene A, 1,10-di-epi-cubenol, alpha-muurolol, (E)-beta-farnesene, alpha-gurjunene, trans-calamenene, alpha-cadinol, alpha-bisabolol, caryophyllene oxide, cis-calamenene; *Carvone CT*—Alpha-pinene, beta-pinene, myrcene, (Z)-beta-ocimene, cis-sabinene hydrate, terpinolene, linalool, trans-sabinene hydrate, menthone, terpinen-4-ol, trans-dihydrocarvone, alpha-terpineol, trans-carveol, dihydrocarveol, pulegone (< 1.2%), cis-carvone oxide, beta-bourbonene, beta-caryophyllene

PREFERRED COMPOSITION FOR CLINICAL USE:

Constituent	
Piperitenone Oxide	55%–80%
Limonene	0.5%–8%
1,8-Cineole	tr–8%
Beta-Caryophyllene	1%–5%
Myrcene	tr–4%

REPORTED THERAPEUTIC PROPERTIES: Analgesic, antispasmodic, antineuralgic, anti-inflammatory, relieves painful menstruation, anticancer, **reduces blood pressure**, aids digestion, relieves diarrhea, soothes stomachache, antiparasitic, antifungal, antibacterial, supports respiratory function, eases cough, insecticidal, relieves anxiety, stress management

CAUTIONS:

Piperitenone Oxide CT

- Caution during pregnancy. Mojito mint CT piperitenone oxide did not impair pregnancy or gestation in rats but did cause hemorrhaging in the liver, kidney, brain, and blood vessels of some fetuses when their mothers were given 10, 25, or 50 mcg/kg orally daily.[2663]
- May interact with blood pressure medications when taken orally and cause low blood pressure due to the effects on blood pressure of the whole oil and piperitenone oxide.[2664,2665,2666,2667,2668]
- May interact with antibiotics, antifungals, or antiviral drugs and increase their effectiveness based on piperitenone oxide content.[2669,2670,2671,2672]

Carvone CT

- May interact with diabetes medications and cause low blood sugar based on carvone content.[2673]

■ May interact with aspirin, blood pressure, antiplatelet, and anticoagulant medications, and increase the risk of bleeding among people with bleeding disorders based on carvone content. Carvone is a potent calcium channel blocker, which may decrease blood pressure.[2674,2675]

■ May interact with barbiturates (medications for anxiety and insomnia), antihistamines, benzodiazepines, tricyclic antidepressants, or other central nervous system depressant drugs, increasing depressant effects, due to carvone content.[2676]

Selected Evidence:

- □ Laboratory and *in vivo* (in mice) research concluded that mojito mint CT piperitenone oxide possesses significant antitumor activity (colorectal, ovarian, and glioblastoma cancer) with low systemic toxicity (no adverse effects on body weight or organs).[2677] Piperitenone oxide was only weakly cytotoxic, which caused the study authors to conclude the antitumor effect may be due to the synergistic action of mojito mint's minor constituents.
- □ Oral administration of 200 mg/kg body weight of mojito mint CT piperitenone oxide produced a significant antinociceptive effect, which was attributed to an indirect anti-inflammatory activity rather than direct activity on the central nervous system.[2678]
- □ Mojito mint CT piperitenone oxide prevented muscle spasms caused by high potassium and caffeine in toad muscle by influencing calcium channels.[2679]
- □ IV administration of mojito mint CT piperitenone oxide (1 to 20 mg/kg body weight) decreased blood pressure in a dose-dependent manner in rats likely by triggering smooth muscle relaxation rather than affecting sympathetic nervous system activity.[2680] Other research found that the piperitenone oxide CT of mojito mint reduces blood pressure and heart rate in rats without high blood pressure probably due to a direct cardiodepressant action and peripheral vasodilation (influenced nitric oxide, prostacyclin, and calcium channels).[2681,2682,2683] Piperitenone oxide appears to be the active constituent responsible for mojito mint's cardiovascular effects.[2684,2685]
- □ Mojito mint (unidentified CT but likely piperitenone oxide because the plant materials were harvested in Brazil) demonstrated schistosomicidal effects against the water-borne parasite of humans *S. mansoni*.[2686] Another study reported that mojito mint reduced motility of adult worms of *S. masoni* at low concentrations and caused death at higher concentrations.[2687] Piperitenone oxide and limonene also decreased worm motility at low concentrations and piperitenone oxide caused death at higher concentrations.
- □ *In vitro* research reported that mojito mint CT piperitenone oxide (and its constituents piperitenone oxide, limonene oxide, and (+)- pulegone) inhibited *S. aureus* and *C. albicans*, but were ineffective against *E. coli*.[2688] Mojito mint CT piperitenone oxide essential oils inactivated spoilage yeasts (*C. albicans, P. anomala*, and *S. cerevisiae*) in nut and fruit juices.[2689] Mojito mint essential oil delayed the growth of postharvest mold in cherry tomato fruit during storage without affecting the quality of the fruit.[2690]
- □ Mojito mint CT piperitenone oxide essential oil exhibited excellent toxicity against *Ae. aegypti* mosquito larvae, which exceeded its isolated piperitenone oxide.[2691] Interestingly, limonene was the most toxic isolated constituent and more toxic than the whole oil. Another study reported that the piperitenone oxide CT was only weakly larvicidal against third instar of *C. quinquefasciatus* mosquitoes.[2692]

MOMI FIR
(Japanese Fir)

Abies firma Siebold & Zucc.

FAMILY: Pinaceae
NOTE: Top
AROMA INTENSITY: Medium-Strong
AROMA: Green, piney, balsamic, resinous, woody
COMMON EXTRACTION METHOD: Steam distilled from the needles (leaves); may also be distilled from the shoots or wood
POSSIBLE SUBSTITUTE OILS: Cypress, dwarf pine, Siberian fir, pink pepper, black pepper CT delta-3-carene
BLENDS WELL WITH: Agarwood, bergamot, cajeput, camphor, cedarwood, clary sage, coriander, Douglas fir, Engelmann spruce, eucalyptus, frankincense, galbanum, grapefruit, juniper, lavender, lemon, lime, marjoram, melaleuca (tea tree), neroli, niaouli, peppermint, ravensara, ravintsara, rosemary, Spanish sage, Siberian fir, silver fir, spikenard, thyme, vetiver, white fir, white pine, white spruce, ylang ylang
SUBCELLULAR LOCALIZATION | EPIGENETIC INFLUENCE: Currently unknown | Currently unknown
RECOMMENDED DILUTION RANGE: 3%–50%; neat for some conditions

PRIMARY CONSTITUENTS:[2693,2694]

Needles	
Delta-3-Carene	16.3%–54.1%
Bornyl Acetate	2.6%–27.1%
Alpha-Pinene	14.0%–22.3%
Beta-Caryophyllene	1.8%–9.0%
Camphene	2.1%–6.1%
Levomenol	0.0%–6.0%
Myrcene	2.6%–5.2%
Delta-Cadinene	0.4%–4.9%
Borneol	tr–3.9%
Longifolene	0.0%–3.6%
Limonene	0.0%–3.0%

Shoots	
Delta-3-Carene	47.6%
Alpha-Pinene	31.7%
Myrcene	9.8%

OTHER CONSTITUENTS: *Needles*—Beta-pinene, terpinolene, alpha-humulene, beta-bisabolene, isoledene, alpha-muurolene, caryophyllene oxide, germacrene D; *Shoots*—Camphene, beta-pinene, limonene, alpha-terpineol, longifolene, beta-caryophyllene, alpha-humulene

PREFERRED COMPOSITION FOR CLINICAL USE:
Insufficient samples/data to determine.

REPORTED THERAPEUTIC PROPERTIES: Antibacterial, antiparasitic, antiseptic, diuretic, astringent, analgesic, anti-inflammatory, antiarthritic, supports respiratory function, expels excess mucus, antioxidant, aids immune function, aids circulation, supports/protects cardiovascular health, cleanses and supports urinary tract function, wound healing, relieves blisters/boils, relieves oily skin, stress management, relieves anxiety, promotes a sense of security

CAUTIONS:

■ May interact with anticholinergic (drugs used for asthma, incontinence, gastrointestinal cramps, muscular spasms, depression, and sleep disorders) and cholinergic medications (drugs used to reduce urinary retention, diagnose myasthenia gravis, and for glaucoma), based on AChE-inhibtory activity of delta-3-carene.[2695,2696]

SELECTED EVIDENCE:

□ None found.

MONTEREY CYPRESS
(Monterey Cypress 'Gold Crest')

Hesperocyparis macrocarpa (Hartw.) Bartel, *Cupressus macrocarpa* Hartw. ex Gord., *Callitropsis macrocarpa* (Hartw.) D.P. Little, *Neocupressus macrocarpa* (Hartw.) de Laub.

FAMILY: Cupressaceae
NOTE: Middle
AROMA INTENSITY: Medium
AROMA: Fresh, evergreen, slightly woody
COMMON EXTRACTION METHOD: Hydrodistilled or steam distilled from the needles (leaves); may also be distilled from the branches
POSSIBLE SUBSTITUTE OILS: Plai, tea tree, shell ginger, lemongrass
BLENDS WELL WITH: Bay laurel, bergamot, blue cypress, camphor, cedarwood, clary sage, coriander, frankincense, juniper, lavender, lemon, lime, marjoram, orange, palo santo, pine, rosemary, sandalwood, Spanish sage, spikenard
SUBCELLULAR LOCALIZATION | EPIGENETIC INFLUENCE: Currently unknown | Currently unknown
RECOMMENDED DILUTION RANGE: 3%–Neat

PRIMARY CONSTITUENTS:[2697,2698,2699,2700,2701,2702]

Sabinene/Terpinen-4-ol CT		*Alpha-Terpineol CT*	
Sabinene	7.1%–24.9%	Alpha-Terpineol	19.0%
Terpinen-4-ol	18.9%–24.0%	Unknown (RT27.0)	10.0%
Alpha-Pinene	1.6%–19.5%	Camphenilone	9.8%
Delta-3-Carene	0.0%–11.6%	Elemol	8.9%
Citronellol	1.3%–10.3%	2-Tridecanone	8.8%
Phyllocladene	0.0%–9.9%	Alpha-Terpinyl Acetate	8.6%
Gamma-Terpinene	3.9%–7.9%	2-Pentyl-2-Cyclopenten-1-one	6.9%
Alpha-Terpinene	1.6%–5.7%	Beta-Bisabolol	5.8%
Citronellal	0.0%–4.9%	(E)-Santaline	3.7%
Myrcene	1.7%–4.1%	Unknown (RT59.25)	3.6%
Camphor	0.0%–3.1%	Dendolasin	3.5%
Cis-Muurola-4(14),5-Diene	0.0%–3.0%	Delta-Elemene	3.4%

Neral CT	
Neral	31.4%–35.8%
Hydroxy Citronellal	12.6%–16.6%
Trans-Piperitol	7.9%–8.3%
Isobornyl Isobutyrate	0.2%–6.6%
Nezukol	0.2%–5.6%
Linalool	0.2%–5.2%
Trans-Ferruginol	0.4%–4.9%
Alpha-Terpinyl Acetate	0.1%–4.3%
Para-Cymenene	0.3%–4.0%
Geraniol	3.1%–4.0%
Stemone	0.2%–3.5%
Furfuryl Hexanoate	0.3%–3.0%

Branches	
Terpinen-4-ol	23.7%
Alpha-Phellandrene	19.2%
Alpha-Citronellol	17.3%
Citronellal	6.6%
Camphor	5.4%
Gamma-Terpinene	5.3%
Alpha-Terpinene	3.0%

OTHER CONSTITUENTS: *Sabinene/Terpinen-4-ol CT*—Alpha-thujene, para-cymene, cis-sabinene hydrate, limonene, beta-pinene, 1,8-cineole, terpinolene, isobutyl butyrate, alpha-terpineol, terpinene-4-ol acetate, epizonarene, cis-calamenene, nerolidol, linalool, alpha-cadinol, cypressene, kaur-15-ene, epi-manoiloksid, dihydro abietane, cis-p-mentha-2-en-ol, trans-p-mentha-2-en-ol, trans-piperitol, citronellic acid, isophyllocladene, phyllocladanol; *Alpha-Terpineol CT*—Camphene, para-cymene, alpha-phellandrene, alpha-terpinene, isophyllocladene; *Neral CT*—O-vanillin, limonene aldehyde, 4-methoxy acetophenone, 2,6-dimethoxy phenol, dihydroeugenol, cyclosativene, carvacryl acetate, alpha-copaene, isopimarol, abietol; *Branches*—Alpha-thujene, alpha-myrcene, m-cymene, limonene, terpinolene, alpha-linalool, trans-2-menthenol, alpha-terpineol, piperitol, citronellic acid

PREFERRED COMPOSITION FOR CLINICAL USE:
Insufficient samples/data to determine.

REPORTED THERAPEUTIC PROPERTIES: Antibacterial, antifungal, antimicrobial, anti-inflammatory, analgesic, antirheumatic, eases cough, aids respiratory function, antioxidant, wound healing, insect repellent, grounding, promotes self-confidence

CAUTIONS:
Sabinene/Terpinen-4-ol and Alpha-Terpineol CTs, Branches
- None currently known.

Neral CT
- Caution is advised during pregnancy and lactation due to moderate neral. Large doses of citral (neral + geranial) may negatively affect fetal development according to animal studies.[2703]
- There is a moderate risk that when the neral CT is taken orally, it may interfere with enzymes responsible for metabolizing medications (NSAIDs, proton-pump inhibitors, acetaminophen, antiepileptics, immune modulators, blood-sugar medications, blood pressure medications, antidepressants, antipsychotics, diabetic medications, antihistamines, antibiotics, and anesthetics).[2704,2705,2706]
- Use cautiously with diabetic medications. Large oral doses of citral (neral + geranial) may improve insulin sensitivity and lower blood-glucose levels according to animal research.[2707,2708]
- May interact with antibiotics and possibly enhance their effects based on neral content.[2709,2710]
- Dilution is recommended for topical application due to high citral content and the risk for skin irritation or sensitivity.[2711] May also irritate mucous membranes.

SELECTED EVIDENCE:

- *In vitro* research demonstrates that Monterey cypress essential oil (Alpha-terpineol CT) is antimicrobial against fungi (*Fusarium solani, F. oxysporum, A. niger*), yeast (*C. albicans*), gram-negative bacteria (*P. aeruginosa, E. coli*), and gram-positive bacteria (*S. aureus, B. subtillis*).[2712] The sabinene/terpinene-4-ol CT demonstrated good activity against *Trichophyton mentagrophytes, T. soudanense, T. violaceum,* and *T. tonsurans*.[2713] When the researchers tested individual constituents of the oil against *T. rubrum*, they found that terpinen-4-ol showed the greatest activity. The neral CT also showed considerable antimicrobial activity against *S. aureus*, *P. aeruginosa, E. coli, A. niger, A. parasiticus,* and *C. albicans*.[2714] The branches oil (CT Terpinen-4-ol) inhibited *Agrobacterium tumefaciens, B. cereus, Dickeya solani, E. coli, Pectobacterium atrosepticum, Pectobacterium carotovorum,* and *S. aureus*.[2715] An unidentified CT of Monterey cypress demonstrated broad-spectrum antimicrobial activity against gram-positive (*B. subtillis, S. aureus*), gram-negative (*E.coli, P. aeruginosa*) and fungi (*A. niger, A. flavus, C. albicans, A. fumigatus*).[2716]
- Monterey cypress essential oil CT neral demonstrated remarkable antioxidant activity when compared to BHA.[2717]
- The sabinene/terpinene-4-ol CT of Monterey cypress essential oil demonstrated excellent mosquito repellency comparable to DEET.[2718]

MOSQUITO FLATSEDGE
(Piprioca Verdadeira, Pipiriocao)

Cyperus prolixus Kunth

FAMILY: Cyperaceae
NOTE: Middle-Base
AROMA INTENSITY: Medium-strong
AROMA: Earthy, woody, slightly sweet, peppery, and smoky
COMMON EXTRACTION METHOD: Steam distilled from the roots (tubers)
POSSIBLE SUBSTITUTE OILS: Jointed flatsedge, poorland flatsedge, melissa, copaiba, ginger
BLENDS WELL WITH: Agarwood, bergamot, cassia, cedarwood (Atlas), cinnamon, cistus, clary sage, frankincense, galbanum, Himalayan cedarwood, lavender, myrrh, patchouli, sandalwood, Texas cedarwood, vetiver, Virginia cedarwood
SUBCELLULAR LOCALIZATION | EPIGENETIC INFLUENCE: Currently unknown | Currently unknown
RECOMMENDED DILUTION RANGE: 3%–50%

PRIMARY CONSTITUENTS:[2719]

Caryophyllene Oxide	6.9%–26.8%
Alpha-Cyperone	13.5%–20.6%
Beta-Caryophyllene	9.6%–16.7%
Beta-Selinene	3.1%–6.5%
Humulene Epoxide II	4.4%–4.9%
Caryophylla-4(14),8(15)-Dien-5-Alpha-ol	3.6%–4.6%
Eudesma-3,11-Dien-5-ol	0.8%–4.5%

Other Constituents: Alpha-pinene, beta-pinene, limonene, 1,8-cineole, trans-pinocarveol, pinocarvone, myrtenal, myrtenol, verbenone, alpha-copaene, thujopsadiene, eudesma-2,4,11-triene, alpha-selinene, alpha-calacorene, Selina-3,11-dien-6alpha-ol

Preferred Composition for Clinical Use:
Insufficient samples/data to determine.

Reported Therapeutic Properties: Antibacterial, antifungal, antiviral, anti-infectious, antiparasitic, relieves migraine, analgesic, reduces fever, anti-inflammatory, anticonvulsant, alleviates diarrhea, aids digestion, reduces nausea and vomiting, expels excess gas, anticancer, improves circulation, wound healing, soothes upset stomach, calming/relaxing, promotes vivid dreams, aphrodisiac

Cautions:
- May interact with aspirin, blood pressure, antiplatelet, and anticoagulant medications, and increase the risk of bleeding among people with bleeding disorders, due to caryophyllene-oxide content.[2720]

Selected Evidence:
- None found.

MOUNTAIN SALVIA
(Wild Mountain Sage, Purple Sage, Gaia)

Hemizygia petiolata Ashby

Family: Lamiaceae (Labiatae)
Note: Middle
Aroma Intensity: Strong
Aroma: Sweet, herbaceous, earthy, slightly bitter
Common Extraction Method: Steam distilled or hydrodistilled from the leaves and flowers
Possible Substitute Oils: Katrafay, German chamomile CT (E)-beta-farnesene, lavender, copaiba
Blends Well With: Agarwood, cedarwood, copaiba, frankincense, ginger, lavender, lemon, marjoram, sandalwood, tea tree, vetiver
Subcellular Localization | Epigenetic Influence: Currently unknown | Currently unknown
Recommended Dilution Range: 3%–50%; neat for some conditions

Primary Constituents:[2721]
(E)-Beta-Farnesene > 70.0%

Other Constituents: Linalool, bicyclogermacrene, germacrene D

Preferred Composition for Clinical Use:
Insufficient data/samples to determine.

Reported Therapeutic Properties: Reduces the appearance of acne and blemishes, relieves chronic skin conditions and itching, analgesic, anti-inflammatory, antineuralgic, antirheumatic, insect repellent, reduces mental fatigue, stress management, relieves anxiety

CAUTIONS:

- None currently known.

SELECTED EVIDENCE:

- None found.

MUSTARD (Black Mustard, Brown Mustard, White/Yellow Mustard)

Brown: *Brassica juncea* (L.) Czern., *Sinapis juncea* L., *B. Integrifolia* Rupr., *B. japonica* Thunb., *B. willdenowii* Boiss.; **Black**: *B. nigra* (L.) W.D.J. Koch, *S. nigra* L.; **White/Yellow**: *B. alba* (L.) Rabenh., *S. alba* L., *B. hirta* Moench

FAMILY: Brassicaceae
NOTE: Top
AROMA INTENSITY: Strong
AROMA: Pungent, spicy
COMMON EXTRACTION METHOD: Steam distilled or hydrodistilled from the black or brown seeds
POSSIBLE SUBSTITUTE OILS: Horseradish, garlic, onion
BLENDS WELL WITH: Basil, cumin, garlic, horseradish, lavender, onion
SUBCELLULAR LOCALIZATION | EPIGENETIC INFLUENCE: Currently unknown | Currently unknown
RECOMMENDED DILUTION RANGE: Not recommended for topical use

PRIMARY CONSTITUENTS:[2722,2723,2724,2725,2726,2727]

B. juncea (Brown)	
Allyl Isothiocyanate	43.8%–73.4%
1-Butene-4-Isothiocyanate	0.0%–19.1%
Diallyl Trisulfide	0.0%–9.7%
3-(Methylthio)propyl Isothiocyanate	0.7%–9.4%
3-Butenyl Isothiocyanate	0.0%–5.9%
Diallyl Sulfide	0.0%–5.5%
Butyl Isothiocyanate	0.0%–4.7%
Phenyl Isothiocyanate	0.0%–3.7%
2-Phenylethyl Isothiocyanate	1.2%–3.7%

B. nigra (Black)	
Allyl Isothiocyanate	98.4%–98.9%

B. alba (White/Yellow 1B4I CT)	
1-Butene-4-Isothiocyanate	96.9%

B. alba (White/Yellow AITC CT)	
Allyl Isothiocyanate	71.1%
Cyclopropyl Isothiocyanate	12.2%
2-Phenylethyl Cyanide	4.2%
Furfural	3.4%

OTHER CONSTITUENTS: *B. juncea*—Thymol, 2-phenethyl isothiocyanate; *B. nigra*—Allylsulfide; *B. alba (White/Yellow 1B4I CT)*—2-phenethyl isothiocyanate; *B. alba (White/Yellow AITC CT)*—3-Pentenenitrile, 2-furanmethanol, isobutyl isothiocyanate, 3-butenyl isothiocyanate, benzeneacetaldehyde, 3-(methylthio)propyl isothiocyanate, 2-phenethyl isothiocyanate

PREFERRED COMPOSITION FOR CLINICAL USE:
Not recommended for clinical use.

REPORTED THERAPEUTIC PROPERTIES: Aids detoxification, improves appetite, aids digestion, antimicrobial, antibacterial, antifungal, antioxidant, increases localized circulation, promotes sweating, antiarthritic, antirheumatic, decongestant, insect repellent, mentally stimulating

CAUTIONS:
Mustard oil is not recommended for therapy unless advised to use by a qualified practitioner that knows the possible adverse effects.

▪ Allyl isothiocyanate (AITC) was not carcinogenic in mice but was associated with urinary cancer in a long-term study in rats; however, it was not genotoxic in mice or rats.[2728] The lowest-observed-adverse-effect level (LOAEL)—the lowest amount of a substance found by experiment or observation that causes an adverse effect— for AITC was determined to be 9mg/kg of body weight, which led to the establishment of an acceptable daily intake of 0.02 mg/kg body weight.[2729] Based on this, it is recommended to avoid oral administration of mustard essential oil.

▪ May interact with aspirin, blood pressure, antiplatelet, and anticoagulant medications, and increase the risk of bleeding among people with bleeding disorders based on AITC content.[2730]

▪ May interact with diabetic medications and cause low blood sugar based on the ability of AITC to increase the utilization of excess blood glucose by increasing insulin secretion.[2731,2732]

▪ Not for inhalation. Mustard oil produces an unpleasant sensation in the occipital regions (back) of the head and can irritate and inflame the eyes and mucus membranes of the respiratory system.[2733,2734]

▪ Not for topical use. Mustard oil is strongly irritating to the skin, resulting in an almost immediate burning sensation due to its allyl isothiocyanate content.[2735,2736,2737,2738] Other isothiocyanates are similarly irritating to the skin and strong to extreme sensitizers.[2739,2740,2741]

▪ See cautions for Leek CT sulfides for additional cautions if mustard essential oil has significant sulfides (diallyl trisulfide, diallyl sulfide, etc.) present.

SELECTED EVIDENCE:

- □ *In vitro* research demonstrated that brown and white mustard essential oil considerably inhibit fungi (*A. niger, P. virisicatum, Fusarium* spp.), strongly inhibited gram-positive bacteria (*S. aureus*) and one gram-negative bacterium (*B. subtilis*), and only weakly inhibited other gram-negative bacteria (*E. coli, P. aeruginosa*).[2742] White/Yellow mustard CT AITC essential oil inhibited gram-positive bacteria (*S. aureus, M. luteus, S. epidermidis*) and gram-negative bacteria (*E.coli, P.aeruginosa, B. subtilis, Sh. sonnei, S. lignieres. P. fluorescens*).[2743] Black mustard essential oil was the most effective of three oils tested (thyme, Mexican oregano, and mustard) against foodborne pathogenic bacteria (*L. monocytogenes, S. aureus,* and *S. enteritidis*), and synergized the antibacterial activity of the other oils.[2744]
- □ Both white and brown mustard essential oil exhibited good to moderate antioxidant activity and moderate to high chelating activity.[2745]

MYRTLE-LEAVED ORANGE (Myrtle-Leaf Orange, Myrtifolia Sour Orange, Chinotto, Sour Orange)

Citrus myrtifolia Rafinesque, *C. aurantium* var. *myrtifolia* Guill.

FAMILY: Rutaceae
NOTE: *Peels*—Top; *Leaves*—Middle
AROMA INTENSITY: *Peels*—Medium; *Leaves*—Strong
AROMA: *Peels*—Citrusy, sweet, slightly floral; *Leaves*—Sweet floral, fresh, citrusy
COMMON EXTRACTION METHOD: Cold-pressed or steam distilled from the peels (rinds) or hydrodistilled or steam distilled from the leaves
POSSIBLE SUBSTITUTE OILS: *Peels*—Bergamot, bitter orange, orange (sweet), tangerine; *Leaves*—neroli, petitgrain
BLENDS WELL WITH: Balsam fir, bay leaf, bergamot, bitter orange, black pepper, camphor, cardamom, carrot seed, cassia, cedarwood, chamomile (Roman, German), coriander, clary sage, cypress, fennel, frankincense, geranium, helichrysum, kaffir lime, lavender, lemon, lemon verbena, lime, neroli, niaouli, nutmeg, orange (sweet), palmarosa, petitgrain, ravensara, sage, Spanish sage, vetiver, ylang ylang
SUBCELLULAR LOCALIZATION | EPIGENETIC INFLUENCE: Currently unknown | Currently unknown
RECOMMENDED DILUTION RANGE: 3%–50%; neat for some conditions

PRIMARY CONSTITUENTS:[2746,2747,2748,2749]

Peel (Balanced CT)

Limonene	26.9%–54.3%
Linalyl Acetate	12.0%–47.5%
Linalool	12.1%–32.4%
Gamma-Terpinene	tr–3.7%

Peel (Limonene CT)

Limonene	48.8%–94.9%
Beta-Pinene	1.0%–19.3%
Gamma-Terpinene	tr–7.8%
Para-Cymene	0.0%–6.5%
Sabinene	0.3%–3.2%

Peel (Steam Distilled)

Limonene	80.1%
Linalool	5.5%

Leaves

Limonene	0.7%–44.0%
Linalool	1.2%–37.7%
Linalyl Acetate	0.0%–29.2%
Beta-Pinene	1.6%–16.6%
Alpha-Terpineol	0.6%–9.8%
Geranial	tr–8.6%
Neral	0.0%–6.1%
(E)-Beta-Ocimene	2.5%–5.8%
Geraniol	0.9%–5.5%
Geranyl Acetate	0.2%–5.4%
Sabinene	0.2%–3.1%

OTHER CONSTITUENTS: *Peel (Balanced CT)*—Beta-Pinene, myrcene, neral, geranial, geranyl acetate; *Peel (Limonene CT)*—Alpha-pinene, myrcene, beta-phellandrene, terpinolene, linalool, neral, alpha-terpineol, beta-bisabolene, geranial, geranyl acetate; *Peel (Steam Distilled)*—Linalyl acetate, nootkatone, farnesol; *Leaves*—Alpha-thujene, alpha-pinene, myrcene, 1,8-cineole, (Z)-beta-ocimene, terpinolene, 6-methylhept-5-en-2-one, citronellal, neryl acetate, nerol

PREFERRED COMPOSITION FOR CLINICAL USE:

Constituent	*Peel*	*Leaves*
Limonene	30%–55%	0.1%–5%
Linalyl Acetate	20%–35%	15%–35%
Linalool	5%–20%	25%–50%
Gamma-Terpinene	2%–10%	–
Geranyl Acetate	0.1%–3%	2%–7%
Myrcene	tr–3%	0.1%–5%
Beta-Pinene	tr–3%	0.5%–5%
Alpha-Terpineol	–	5%–15%
Geraniol	–	2%–8%
(E)-Beta-Ocimene	–	0.5%–5%
Neryl Acetate	–	0.1%–5%

REPORTED THERAPEUTIC PROPERTIES: Anti-inflammatory, analgesic, antispasmodic, antioxidant, anticancer, antibacterial, antiseptic, aids cardiovascular function, balances female hormones, digestive aid, wound healing, reduces the appearance of scars and blemishes, promotes restful sleep, antidepressant, relieves anxiety, combats negative emotions and thinking, reduces mood swings

CAUTIONS:

Peel (Cold-pressed)

■ Avoid exposure of application site to UV rays for up to twelve hours after topical application due to the presence of furanocoumarins (bergapten and epoxybergamottin).[2750]

Leaves

■ None currently known.

SELECTED EVIDENCE:

- □ Laboratory research showed that myrtle-leaved orange peel essential oil (half ripe) reduces inflammation by reducing nitric oxide production, expression of inflammatory genes, COX-2 (cyclooxygenase-2), inducible nitric oxide synthase (iNOS), and cytokines (IL-1β, IL-6, MCP-1).[2751]
- □ Myrtle-leaved orange peel essential oil exhibited antioxidant activity in both the DPPH and ABTS assays, with half ripe and fully ripe fruits performing better than green fruits.[2752]

NASTURTIUM
(Indian Cress, Garden Nasturtium)

Tropaeolum majus L.

FAMILY: Tropaeolaceae
NOTE: Middle
AROMA INTENSITY: Strong
AROMA: Rich, earthy, herbaceous, green, peppery, musky, floral, sweet, slightly spicy
COMMON EXTRACTION METHOD: Steam distilled or hydrodistilled from the flowering aerial parts; may also be solvent extracted as an absolute

POSSIBLE SUBSTITUTE OILS: Copaiba
BLENDS WELL WITH: Angelica, basil, cedarwood, chamomile (German, Roman), frankincense, geranium, grapefruit, lavender, myrrh, patchouli, sandalwood, tangerine
SUBCELLULAR LOCALIZATION | EPIGENETIC INFLUENCE: Currently unknown | Currently unknown
RECOMMENDED DILUTION RANGE: Not for topical application.

PRIMARY CONSTITUENTS:[2753,2754,2755]

Benzyl Isothiocyanate	82.5%
Benzene Acetonitrile	3.9%
2-Phenyl Ethyl Isovalerate	2.9%

OTHER CONSTITUENTS: Alpha-thujene, isocaryophyllene

PREFERRED COMPOSITION FOR CLINICAL USE:
Insufficient samples/data to determine.

REPORTED THERAPEUTIC PROPERTIES: Supports cardiovascular function, anticancer, antimicrobial, antibacterial, antiparasitic, treats urinary tract infections, modulates immune function, aids respiratory function, antiallergic, relieves constipation, antioxidant, **anti-inflammatory**, antiarthritic, reduces skin irritation and itching, improves blemishes, relieves cracked heels, balances blood sugar, weight management, reduces blood pressure, relieves anxiety, stress management

CAUTIONS:

■ Do not use during pregnancy or while lactating, due to minor fetal toxicity and reproductive toxicity reported in animal studies.[2756]

■ Do not take orally due to benzyl derivatives. Ingestion of benzyl derivatives (benzoic acids, benzoic salts, benzaldehyde, benzyl acetate, benzyl benzoate, and benzyl alcohol) may cause gastrointestinal irritation, vomiting, and diarrhea. The Joint Expert Committee for Food Additives (JECFA) has established an acceptable daily limit for the aforementioned constituents of 0–5 mg/kg body weight for adults to avoid GI discomfort. Based on this limit, it is suggested to limit the oral consumption of orange bird berry essential oil rich in benzyl derivatives to no more than 10 drops daily. Although benzyl isothiocyanate (BITC) is commonly ingested because it is found in cruciferous vegetables, the levels are very high in nasturtium essential oil. The metabolism and excretion of BITC has been investigated—metabolized to N-acetyl-S-(N-benzylthiocarbamoyl)-L-cysteine, 53.7% of the BITC excreted via the kidneys, maximum concentrations two to six hours after administration, and complete excretion/metabolism by twelve hours—in man and doesn't appear to be harmful.[2757] It also inhibits carcinogenesis by chemicals.[2758,2759]

■ May interfere with the enzymes (CYP2E1, CYP2B1) responsible for metabolizing medications (NSAIDs, proton-pump inhibitors, acetaminophen, antiepileptics, immune modulators, blood-sugar medications, blood pressure medications, antidepressants, antipsychotics, diabetic medications, antihistamines, antibiotics, and anesthetics) due to benzyl isothiocyanate content.[2760,2761]

■ Not for topical use. Isothiocyantes are very irritating to the skin and mucous membranes and strong to extreme sensitizers.[2762,2763,2764, 2765,2766,2767,2768] They may also be irritating to the eyes and cause headaches or pressure when inhaled.[2769,2770]

SELECTED EVIDENCE:

- □ Topical application of nasturtium essential oil significantly reduced skin irritation and inflammation (up to 85%) caused by SLS. Remarkably, the beneficial effects of the oil were observed immediately after application (significantly improved the integrity of the hydric layer of the skin) and produced better results than pharmaceutical compounds.[2771]

- Nasturtium selectively inhibited the 5-lipoxygenase (5-LOX) enzyme, which is an inflammatory enzyme involved in the pathology of inflammatory conditions such as cancer, arthritis, asthma, and atherosclerosis. Its primary constituent benzyl isothiocyanate also exhibited potent anti-inflammatory effects against 5-LOX.[2772]
- *In vitro* research demonstrated that nasturtium essential oil inhibited gram-negative (*E. coli, Salmonella* sp., *P. aeruginosa*), gram-positive (*L. monocytogenes, S. aureus, B, subtilis*), and yeasts (*C. albicans*).[2773]
- Nasturtium essential oil exhibited good antioxidant activity in the DPPH (near equal to ascorbic acid) and FRAP assays.[2774]

NOOTKA (Nootka Cypress, Alaska Cypress, Yellow Cypress, Nootka Cedar, Alaska Cedar)

Cupressus nootkatensis D. Don, *Callitropsis nootkatensis* (D. Don) Oerst. ex D.P. Little, *Chamaecyparis nootkatensis* (D. Don) Spach, *Xanthocyparis nootkatensis* (D.Don) Farjon & D.K. Harder

FAMILY: Chamaecyparis
NOTE: Middle-Base (wood) or Top (stems and needles)
AROMA INTENSITY: Medium-Strong
AROMA: Woody, fresh, balsamic, resinous, citrusy, earthy, slightly smoky
COMMON EXTRACTION METHOD: Steam distilled or hydrodistilled from the wood (bark or heartwood); may also be distilled from the stems and leaves (needles)
POSSIBLE SUBSTITUTE OILS: Equal mixture of oregano and grapefruit; frankincense, cypress, palo santo
BLENDS WELL WITH: Balsam fir, bergamot, black pepper, cardamom, cedarwood (Atlas), cistus, copaiba, cypress, grapefruit, lavender, lavandin, lemon, mandarin, patchouli, pine, Siberian fir, silver fir, Texas cedarwood, Virginia cedarwood, white fir, ylang ylang
SUBCELLULAR LOCALIZATION | EPIGENETIC INFLUENCE: Currently unknown | Currently unknown
RECOMMENDED DILUTION RANGE: 3%–33%; 50% for some conditions

PRIMARY CONSTITUENTS:[2775,2776,2777]

Carvacrol CT (Heartwood)		*Needles*	
Carvacrol	27.2%–35.4%	Limonene	35.4%–42.4%
Nootkatene	13.3%–20.1%	Delta-3-Carene	11.5%–23.4%
Nootkatone	12.9%–17.4%	Alpha-Pinene	8.7%–16.3%
Valencene-13-ol	5.6%–6.4%	Sabinene	0.2%–10.5%
Nootkatol	5.2%–5.7%	Terpinen-4-ol	0.2%–6.2%
Epi-Nootkatol	0.0%–3.9%	Terpinolene	3.1%–3.3%
Nootkatin	3.1%–3.5%	Myrcene	2.6%–3.2%

Stems/Needles

Limonene	53.2%
Delta-3-Carene	21.0%
Alpha-Pinene	12.2%
Myrcene	6.1%

Note: Commercial samples tend to have lower carvacrol content (< 10%) than those reported in the literature.

OTHER CONSTITUENTS: *Carvacrol CT*—Terpinen-4-ol, alpha-terpineol, carvacrol methyl ether, valencene, 7-epi-alpha-selinene, T-muurolol, alpha-cadinol; *Stems/Needles*—Alpha-fenchene, terpinolene, terpinen-4-ol; *Needles*—(E)-2-hexenal, alpha-fenchene, 1-octen-3-ol, beta-pinene, alpha-terpinene, gamma-terpinene, cis-sabinene hydrate

PREFERRED COMPOSITION FOR CLINICAL USE:

Constituent	*Wood*
Nootkatene	25%–55%
Nootkatone	2%–12%
Valencene	3%–10%
Carvacrol	1%–10%
Valencene-13-ol	2%–8%
Carvacrol Methyl Ether	1%–5

REPORTED THERAPEUTIC PROPERTIES: Antioxidant, analgesic, anti-inflammatory, anti-infectious, antibacterial, antifungal, antimicrobial, antiparasitic, soothes the skin, aids respiratory function, bronchodilator, insecticidal, insect repellent, antidepressant, stress management, relieves anxiety

CAUTIONS:

Carvacrol CT (if carvacrol exceeds 25%)

■ Avoid during pregnancy and lactation. Oregano (also rich in carvacrol) may negatively affect embryonic development and encourage fetal cell death according to animal research.[2778]

■ May interact with diabetic medications and cause low blood-sugar levels due to carvacrol content—carvacrol reduces blood glucose and enhances the antidiabetic activities of rosiglitazone in animals.[2779,2780,2781,2782]

■ May interact with aspirin, blood pressure, antiplatelet, and anticoagulant medications, and increase the risk of bleeding among people with bleeding disorders, due to carvacrol and thymol content.[2783,2784,2785,2786,2787]

■ May interact with antibiotics or antifungals and possibly enhance or decrease their effects due to carvacrol or thymol content.[2788,2789,2790,2791,2792]

■ May irritate mucous membranes (eyes, mouth, nasal passages, vagina, rectum).

Needles and Stems/Needles

■ None currently known.

SELECTED EVIDENCE:

- □ *In vitro* evidence showed that nootka stem/needle essential oil inhibits *B. subtilis* and *C. albicans*.[2793] Another study found that nootka essential oil inhibited *Fusobacterium necrophorum, Clostridium perfringens, Actinomyces bovis*, and *Candida albicans*.[2794]

- □ Nootka essential oil repelled, suppressed, and killed hybrid imported fire ant workers (*Solenopsis invicta* Buren × *Solenopsis richteri* Forel).[2795] Other studies demonstrating the repellent or insecticidal activity of isolated components from nootka essential oil also exist in the literature.

NORDMANN FIR (Caucasian Fir)

Abies nordmanniana (Steven) Spach,
A. nordmanniana subsp. *bornmuelleriana* Mattf.,
A. nordmanniana subsp. *nordmanniana* Mattf.

FAMILY: Pinaceae
NOTE: Top-Middle
AROMA INTENSITY: Medium
AROMA: Fresh, woody, piney
COMMON EXTRACTION METHOD: Steam distilled or hydrodistilled from the needles or twigs
POSSIBLE SUBSTITUTE OILS: Balsam fir, Siberian fir, spruce (black)
BLENDS WELL WITH: Balsam fir, birch, blue spruce, cassia, cedarwood, chamomile (German, Roman), cypress, frankincense, galbanum, lavender, lemon, myrtle, palo santo, peppermint, pine, sandalwood, Siberian fir, silver fir, spruce (black), tsuga, white fir, wintergreen
SUBCELLULAR LOCALIZATION | EPIGENETIC INFLUENCE: Currently unknown | Currently unknown
RECOMMENDED DILUTION RANGE: 5%–Neat

PRIMARY CONSTITUENTS:[2796,2797,2798,2799]

Alpha-Pinene CT	
Alpha-Pinene	65.7%
Beta-Pinene	9.6%
Limonene	7.2%

Delta-3-Carene CT	
Delta-3-Carene	30.5%–33.8%
Beta-Phellandrene	7.0%–15.4%
Alpha-Pinene	12.2%–13.7%
Beta-Pinene	7.4%–9.6%
Camphene	6.7%–9.0%
Limonene	4.6%–6.3%
Terpinolene	3.2%–3.3%
Myrcene	2.7%–3.1%

Limonene CT	
Limonene	24.5%
Bornyl Acetate	20.3%
Delta-3-Carene	11.4%
Linalool	9.1%
Camphene	9.0%
Alpha-Pinene	6.9%
Beta-Pinene	4.1%
Borneol	3.5%

OTHER CONSTITUENTS: *Alpha-Pinene CT*—Camphene, 2,4(10)-thujadiene, myrcene, delta-3-carene, alpha-campholene aldehyde, trans-pinocarveol, trans-verbenol, p-mentha-1,5-dien-8-ol, alpha-terpineol, myrtenal, myrtenol, verbenone, beta-caryophyllene, alpha-humulene, cis-alpha-bisabolene, caryophyllene oxide; *Delta-3-Carene CT*—Santene, tricyclene, alpha-phellandrene, gamma-terpinene, terpinen-4-ol, alpha-terpineol, bornyl acetate, beta-caryophyllene, alpha-humulene; *Limonene CT*—Beta-phellandrene, myrcene, 1,8-cineole, terpinolene, camphor, alpha-terpineol, alpha-cedrol

Preferred Composition for Clinical Use:
Too great of variance in limited samples to determine.

Reported Therapeutic Properties: Analgesic (pain relief), reduces fever, anti-inflammatory, antiarthritic, antirheumatic, antibacterial, antifungal, antiseptic, antioxidant, aids circulation, antispasmodic, expectorant, decongestant, supports immune system function, disinfectant, removes excess mucus, relieves cough, supports respiratory function, anticancer, wound healing, relieves anxiety, grounding, stress management, encourages feelings of stability and empowerment

Cautions:
- None currently known.

Selected Evidence:
- Animal research demonstrated that topical application of Nordmann fir cone essential oil (once daily for nine days; composition not reported) reduced inflammation and accelerated wound healing activities.[2800]
- *In vitro* research demonstrated that Nordmann fir essential oil (composition not reported) inhibited bacteria (*B. megaterium, B. cereus, B, brevis, B. subtilis, S. aureus, L. monocytogenes, E. aerogenes,* and *P. aeruginosa*) and yeasts (*S. cerevisiae* and *C. albicans*).

NORWAY SPRUCE
(Shoots Spruce, Spruce Abies, Excelsa Spruce, Himalayan Blue Pine)

Picea abies (L.) Karst, *P. excelsa* (Lam.) Link

Family: Pinaceae
Note: Top-Middle
Aroma Intensity: Medium-Strong
Aroma: Fresh, piney, green, woody
Common Extraction Method: Steam distilled or hydrodistilled from the needles (leaves) or shoots; may also be distilled from the cones
Possible Substitute Oils: Hinoki, frankincense, white spruce, blue spruce, spruce (black)
Blends Well With: Balsam fir, bergamot, cedarwood, cypress, frankincense, galbanum, grapefruit, hinoki, juniper, lavender, lemon, myrrh, orange, palo santo, peppermint, pine, rosemary, sandalwood, Siberian fir, spearmint, spruce (black), tangerine, white fir, ylang ylang
Subcellular Localization | Epigenetic Influence: Currently unknown | Currently unknown
Recommended Dilution Range: 3%–50%; neat for some conditions

Primary Constituents:[2801,2802,2803,2804,2805]

Alpha-Cadinol CT (Shoots)		*Limonene CT (Shoots)*	
Alpha-Cadinol	11.2%–25.3%	Limonene	39.5%–48.3%
Bornyl Acetate	11.8%–15.0%	Alpha-Pinene	3.0%–14.4%
Camphene	3.9%–14.1%	Bornyl Acetate	9.4%–13.2%

Limonene	6.3%–13.0%	Germacrene D	tr–6.2%
Alpha-Muurolol	6.4%–11.0%	Alpha-Humulene	0.6%–5.4%
Alpha-Pinene	2.4%–10.4%	Alpha-Terpineol	2.9%–4.3%
Beta-Myrcene	0.8%–10.1%	Camphene	1.9%–4.3%
Delta-Cadinene	6.1%–9.8%	Beta-Caryophyllene	0.5%–4.2%
Manool	2.5%–6.2%	Beta-Pinene	2.3%–3.1%
Epi-Alpha-Cadinol	tr–5.1%		
Santene	0.8%–3.1%		

Needles (Leaves)

Alpha-Pinene	17.5%
Myrcene	16.7%
Beta-Caryophyllene	11.4%
Beta-Pinene	10.6%
Gamma-Eudesmol	8.3%
Elemol	6.6%
Kaur-16-ene	5.4%
Germacrene D	4.8%
Beta-Fenchyl Alcohol	4.5%

Cones

Limonene	31.0%
Beta-Pinene	30.0%
Delta-3-Carene	15.0%
Alpha-Pinene	11.0%
Myrcene	11.0%

OTHER CONSTITUENTS: *Alpha-Cadinol CT*—Tricyclene, alpha-thujene, beta-pinene, 1,8-cineole, camphor, borneol, alpha-terpineol, beta-caryophyllene, alpha-humulene, gamma-muurolene, beta-selinene, alpha-muurolene, gamma-cadinene, (E)-nerolidol, 1-epi-cubenol, delta-cadinol; *Limonene CT*—Myrcene, terpinolene, camphor, linalool; *Needles (Leaves)*—Alpha-Terpinolene, limonene, bornyl acetate, terpinen-4-ol, alpha-copaene, alpha-humulene, alpha-muurolene, delta-cadinene; *Cones*—Not listed in study

PREFERRED COMPOSITION FOR CLINICAL USE:

Constituent	
Limonene	35%–50%
Bornyl Acetate	5%–15%
Alpha-Pinene	3%–15%
Alpha-terpineol	1.5%–8%
Alpha-Humulene	0.1%–7%
Beta-Pinene	1%–6%
Camphene	1%–6%
Beta-Caryophyllene	0.1%–6%

REPORTED THERAPEUTIC PROPERTIES: Antibacterial, antifungal, antiviral, anti-infectious, antioxidant, anticancer, disinfectant, **aids respiratory function**, **decongestant**, eases cough, diuretic, antiarthritic, antirheumatic, relieves gout, antineuralgic, analgesic, antispasmodic, anti-inflammatory, relieves fever, soothes chronic skin conditions, relieves sore throat, balances thyroid and adrenal function, promotes restful sleep, antidepressant, combats negative thoughts and emotions, relieves anxiety and nervousness

CAUTIONS:

- None currently known.

Selected Evidence:

- *In vitro* research demonstrates that Norway spruce CT alpha-cadinol essential oil inhibits *S. aureus, B. cereus,* and *P. vulgaris*, with the gram-positive bacteria mores susceptible to the oil than gram-negative.[2806] The limonene CT inhibited *S. aureus* (three strains), *S. faecalis, C. albicans, Candida* sp., *B. subtilis,* Beta-haemolyzing *Streptococcus* A, *S. pneumoniae, S. milleri, Staphylococcus* coag. Negative I and II, and *C. diptheriae*.[2807]

NUT GRASS (Nut Sedge, Musta, Piprioquinha, Purple Nutsedge)

Cyperus rotundus Linn.

Family: Cyperaceae
Note: Base
Aroma Intensity: Medium-Strong
Aroma: Green, earthy, spicy, musky, herbaceous, mildly woody, fruity, and camphoraceous
Common Extraction Method: Steam distilled or hydrodistilled from the rhizomes (tubers, roots); may also be distilled from the aerial parts
Possible Substitute Oils: Cypriol, jointed flatsedge, turmeric, ginger, Himalayan cypress,
Blends Well With: Agarwood, bergamot, cassia, cedarwood (Atlas), cinnamon, cistus, clary sage, cypriol, frankincense, galbanum, Himalayan cedarwood, lavender, myrrh, patchouli, sandalwood, Texas cedarwood, vetiver, Virginia cedarwood
Subcellular Localization | Epigenetic Influence: Currently unknown | Currently unknown
Recommended Dilution Range: 3%–50%

Primary Constituents:[2808,2809,2810,2811,2812,2813,2814,2815,2816,2817,2818,2819]

Alpha-Cyperone CT		*Cyperene CT*	
Alpha-Cyperone	11.0%–38.5%	Cyperene	9.8%–30.9%
Cyperene	0.3%–20.4%	Alpha-Cyperone	0.0%–17.7%
4-Oxo-Alpha-Ylangene	0.0%–12.8%	Cyperotundene	0.0%–8.8%
Alpha-Selinene	0.0%–11.7%	Humulen-(v1)	0.0%–8.0%
Myrtenol	0.0%–7.9%	Beta-Selinene	0.0%–7.9%
Isolongifolen-5-one	0.0%–7.0%	Rotundene	0.0%–7.6%
Caryophyllene Oxide	2.3%–6.9%	Zierone	0.0%–4.6%
Isocyperol	0.0%–6.7%	Alpha-Cyperone	0.0%–4.5%
Eudesma-3,11-Dien-5-ol	0.0%–6.2%	Cyperol	0.0%–4.0%
Cyperol	0.0%–5.6%	Beta-Pinene	1.0%–3.9%
Beta-Pinene	0.0%–5.3%	Mustakone	0.0%–3.8%
Beta-Selinene	0.0%–5.1%	Campholenic Aldehyde	0.0%–3.8%
Rotundone	0.0%–4.8%	Isorotundene	0.0%–3.6%
Caryophyllenol 11	0.0%–4.8%	Alpha-Cubebene	0.0%–3.4%
Cyperorotundene	0.0%–4.7%	Caryophyllene-8,9-Epoxide	0.0%–3.2%
Rotundene	0.0%–4.6%		
Patchoulenone	0.0%–4.3%		

Caryophylla-3,8(13)-Dien-5-Beta-ol	0.0%–4.2%
Trans-Pinocarveol	0.8%–4.0%
Patchenol	0.0%–3.9%
Vulgarol B	0.0%–3.8%
Limonene	0.0%–3.6%
Gamma-Gurjunene Epoxide	0.0%–3.6%
Oplopenone	0.0%–3.4%
Alpha-Calacorene	0.0%–3.1%
Gamma-Muurolene	0.0%–3.1%
Mustakone	0.0%–3.1%
Alpha-Pinene	0.0%–3.0%

Cyperotundone CT

Cyperotundone	19.7%
Cyperene	15.2%
Mustakone	5.8%
Nootkatone	3.8%
Rotundene	3.6%

Alpha-Copaene CT

Alpha-Copaene	11.4%–12.1%
Alpha-Cyperene	8.4%–11.7%
Valerenal	8.7%–9.8%
Caryophyllene Oxide	7.8%–9.7%
Trans-Pinocarveol	5.2%–7.4%
Nootkatone	1.4%–6.7%
Valencene	3.6%–5.1%
Alpha-Gurjunene	3.2%–3.8%
Humulene Oxide	2.9%–3.5%

Aerial Parts

Cyperene	37.9%
Cyperotundone	11.2%
Isorotundene	9.5%
Cyperol	6.4%
Alpha-Cyperone	4.3%
Beta-Pinene	3.9%
Alpha-Cubebene	3.7%
Mustakone	3.7%

Beta-Pinene CT

Beta-Pinene	11.3%
Alpha-Pinene	10.8%
Alpha-Cyperone	7.9%
Myrtenol	7.1%
Alpha-Selinene	6.6%
Limonene	5.7%
Beta-Selinene	4.6%
Trans-Pinocarveol	4.0%

2,5,9-Trimethylcycloundeca-4,8-Dienone CT

2,5,9-trimethylcyclo-undeca-4, 8-dienone	13.4%
Alloaromadendrene Oxide-(1)	8.5%
Piperitone	7.4%
Methyl (Z)-5,11,14,17-Eicosatetraenoate	7.3%
Beta-Elemol	7.1%
Selina-6-en-4-ol	7.0%
Longiverbenone	4.3%
Selina-3,7(11)-Diene	3.7%
Isoaromadendrene Epoxide	3.3%

OTHER CONSTITUENTS: *Alpha-Cyperone CT*—Beta-cubebene, alpha-cubebene, beta-elemene, beta-copaene, trans-calamenene, delta-cadinene, cyperene oxide, alpha-copaene oxide, spathulenol, nardol, humulene epoxide, delta-cadinol, alpha-muurolol, intermediol, myrtenal, trans-verbenol, alpha-terpineol, trans-carveol, cubebol, pinocarvone, eudesma-2,4,11-triene, epi-cubebol, epoxy-cyperene, humulene epoxide II, eudesma-3,11-dien-2-one, alpha-terpinolene, myrcene, para-cymene, terpinen-4-ol, beta-elemene, beta-caryophyllene, alloaromadendrene, (2R,5E)-caryophyll-5-en-12-al, caryophylla-3,8(13)-dien-5-α-ol, aristolone; *Cyperene CT*—Alpha-pinene, 1,8-cineole, limonene, para-cymene, beta-elemene, trans-pinocarveol, gamma-muurolene, alpha-selinene, delta-cadinene, ar-curcumene, humulene epoxide I, caryophylladienol I, caryophylladienol II, isocyperol, T-cadinol, trans-calamenene, alpha-cadinol, (E,E)-farnesol, caryophyllene oxide, cypera-2,4-diene, alpha-copaene, alpha-humulene, alpha-muurolene, alpha-muurolol, delta-cadinene, cadina-1,4-diene, alpha-calacorene, artemiseole, aristola-1(10),8-diene, myrtenal, chamigrene, alpha-terpineol, verbenone, azulene, isolongifolene, beta-vatirenene, myrtenol, ermophila-1(10), isogermacrene D, 4-boraperhydroindane, 4-ethylguaiacol, alloaromadendrene oxide-(2), duvatriendiol, 2(3h)-naphthalenone, viridiflorol, fokienol, kaur-16-ene, (Z)-alpha-bisabolene epoxide, 1-heptatriacontanol, 1,3,6,10-cyclotetradecatetraene, kaur-18-al; *Cyperotundone CT*—Trans-pinocarveol, pinocarvone, myrtenal, myrtenol, cyprotene, alpha-cubebene, cyperadiene, alpha-ylangene, alpha-copaene, beta-copaene, gamma-muurolene, trans-calamenene, delta-cadinene, alpha-calacorene, alpha-copaene oxide, gamma-calacorene, spathulenol, nardol, caryophyllene oxide, humulene epoxide, eudesma-5-en-11-alpha-ol, epi-cubenol, delta-cadinol, alpha-muurolol, intermediol; *Alpha-Copaene CT*—Hexanol, alpha-pinene, beta-pinene, 1-octen-3-ol, para-cymene, limonene, 1,8-cineole, octanol, alpha-campholenal, trans-verbenol, pinocamphone (< 1.5%), pinocarvone, isopinocamphone (< 0.9%), cis-verbenol, p-cymen-8-ol, myrtenal, alpha-terpineol, myrtenol, verbenone, trans-carveol, beta-elemene, beta-caryophyllene, aromadendrene, alpha-amorphene, alpha-guaiene, rotundene, alpha-muurolene, beta-selinene, caryophyllene alcohol, ledol, guaiol, widdrol, cedrol, valeranone, mustakone, cyperenone, eudesm-7(11)-en-4-ol, phytol; *Beta-Pinene CT*—Camphene, para-cymene, pinocarvone, alloaromadendrene, verbenone, alpha-copaene, beta-elemene, caryophyllene oxide, aristolone, caryophyllene-2-6-beta-oxide, vulgarol B, caryophyllenol; *2,5,9-Trimethylcycloundeca-4,8-Dienone CT*—Alpha-pinene, beta-pinene, para-cymene, alpha-terpineol, myrtenol, verbenone, alpha-cubebene, alpha-copaene, alloaromadendrene, patchoulene, beta-H,5alpha-eremophila, ledol, palustrol, 1,5-cycloundecadien,8,8-dimethyl-1-9-methylene, tertacyclo(6,3,2,o(2,5),o(1,8)tridecan-9-ol, gamma-eudesmol, androstan-17-one,3ethyl3-hydroxy-,(5-a), 4,8,13-duvatriene-1,3-diol, tricycle(20,8,o,o(7,16)-tricontane,1(22),7(16)diepoxy; *Aerial Parts*—Alpha-pinene, beta-selinene, trans-calamenene, isocyperol, T-cadinol, alpha-muurolol, alpha-cadinol

PREFERRED COMPOSITION FOR CLINICAL USE:
Too great of variance in samples/data to determine.

REPORTED THERAPEUTIC PROPERTIES: **Anticancer**, antibacterial, antifungal, antimicrobial, anti-infectious, antiparasitic, antioxidant, wound healing, heals bruises, anti-inflammatory, analgesic, antispasmodic, relieves fever, regulates menstruation, relieves painful menstruation, balances hormones, aids digestion, relieves diarrhea, eases nausea and vomiting, expels excess gas, balances blood sugar, supports respiratory function, decongestant, **reduces hair growth**, insect repellent, insecticidal, antidepressant, sedating, relieves anxiety, improves focus and concentration

CAUTIONS:
- None currently known.

SELECTED EVIDENCE:
- The alpha-cyperone CT of nut grass essential oil was potently cytotoxic to colon, liver, and breast cancer cells.[2820] Another study reported that the alpha-cyperone CT was cytotoxic to neuroblastoma cancer cells.[2821] The 2,5,9-Trimethylcycloundeca-4,8-Dienone CT of nut grass essential oil exhibited anti-proliferative activity against two types of colon (HT29 and HCT116) cancer cells.[2822] The cyperene CT was very effective against leukemia cancer cells.[2823]

- A pilot clinical study found that topical application of nut grass essential oil decreased both dark and white underarm hair growth as effectively as Alexandrite laser therapy.[2824] Hair was removed by their usual method every three weeks followed by the topical application of 0.25 mL of oil that was rubbed in for two minutes until absorbed. In addition, 0.25 mL of oil was applied twice daily after washing and cleansing for six months. A second study found that topical application of the oil (0.25 mL after hair removal, massaged into the area for two minutes; then 0.25 mL applied twice daily, after washing the area for six months) reduced hirsutism (unwanted, male-pattern hair growth in women) and underarm hair growth.[2825]
- Oral administration of 0.1 g/kg of the cyperene CT of nut grass essential oil showed good analgesic effects in mice.[2826]
- *In vitro* research demonstrated that nut grass essential CT cyperotundone oil inhibited *S. aureus, streptococcus A, S. emteritidis,* and *K. pneumoniae* in lower concentrations and *P. aeruginosa* and *E. coli* in higher concentrations, with *S. aureus* being the most sensitive organism.[2827] The alpha-cyperone CT showed inhibition of gram-positive bacteria (*S. epidermidis*) and the gram-negative bacteria (*K. pneumoniae, Sh. Flexneri, S. enteritidis*) comparable to standard drugs (ampicillin, gentamycin).[2828] The same research reported that the oil demonstrated broad-spectrum inhibition of fungi (*A. corymbifera, T. mentagrophytes, G. candidum, C. albicans*), gram-positive bacteria (*S. aureus, S. pyogenes, L. innocua*), and the gram-negative bacterium (*P. vulgaris*). Nut grass CT alpha-cyperone inhibited several foodborne pathogens (*S. aureus, S. epidermidis, B. subtilis, S. typhimurium, E. coli*, and *S. dysenteriae*).[2829] The 2,5,9-Trimethylcycloundeca-4,8-Dienone CT of nut grass essential oil inhibited *E. coli, P. aeruginosa, B. subtilis*, and *S. aureus*.[2830] A nut grass essential oil with unidentified sesquiterpenes as the major constituents (MW 218, RI 1676; MW 218, RI 1687; MW 218, RI 1692) inhibited *C. albicans*.[2831]
- Nut grass CT cyperotundone essential oil exhibited moderate antioxidant activity in the DPPH assay, which was attributed to the presence of its sesquiterpenes (cyperene, cyperotundone, caryophyllene oxide and mustakone).[2832] Another study stated that the alpha-cyperone CT of nut grass possesses excellent antioxidant activity, protecting against DNA damage.[2833] The 2,5,9-Trimethylcycloundeca-4,8-Dienone CT of nut grass essential oil was inactive as an antioxidant in the DPPH assay.[2834] The cyperene CT of nut grass essential oil exhibited low antioxidant capacity as determined by the DPPH assay, and weakly inhibits superoxide according to the XOD-inhibition assay.[2835] The same researchers reported that the cyperene CT decreased the genotoxic (antimutagenic) activity of H_2O_2 UV photolysis.
- Nut grass CT alpha-cyperone essential oil was insecticidal to first instar *Aedes aegypti* mosquito larvae.[2836] Nut grass CT cyperene synergized the insecticidal activity of permethrin against mosquitoes (*Ae. aegypti*).[2837]

OCOTEA, COMOROS

Ocotea comoriensis Kosterm.

FAMILY: Lauraceae
NOTE: Middle-Top
AROMA INTENSITY: Medium
AROMA: Woody, piney, fresh, balsamic
COMMON EXTRACTION METHOD: Steam distilled or hydrodistilled from the bark
POSSIBLE SUBSTITUTE OILS: Spruce (black), Siberian fir, white spruce, blue spruce, tsuga

Blends Well With: Balsam fir, bergamot, cedarwood, Douglas fir, frankincense, grapefruit, lavender, lemon, lime, orange, palo santo, rosemary, sandalwood, Siberian fir, silver fir, spruce (black), tangerine, tsuga, vetiver, white fir
Subcellular Localization | Epigenetic Influence: Currently unknown | Currently unknown
Recommended Dilution Range: 3%–Neat

Primary Constituents:[2838]

Camphene	18.1%
Bornyl Acetate	13.8%
Alpha-Pinene	13.7%
Beta-Pinene	8.4%
Limonene	5.6%
Alpha-Cubebene	4.5%
T-Muurolol + T-Cadinol	3.0%
Alpha-Cadinol	3.0%

Other Constituents: Myrcene, 1,8-cineole, linalool, alpha-copaene, beta-cubebene, beta-caryophyllene, alpha-humulene, germacrene D, alpha-muurolene, cubebol I, gamma-cadinene, cubebol II, calamenene, delta-cadinene, germacrene D-4-ol, caryophyllene oxide, humulene epoxide II, 1-epi-cubenol, 3,4,5-trimethoxyphenyl isopropylketone

Preferred Composition for Clinical Use:
Insufficient data/samples to determine.

Reported Therapeutic Properties: Analgesic, relieves headache, anti-inflammatory, aids digestion, eases stomachache, antibacterial, antiviral, relieves urinary conditions, improves circulation, supports respiratory function, eases cough, aids cognition, improves concentration, relieves anxiety

Cautions:
- None currently known.

Selected Evidence:
- □ Comoros ocotea exhibited antimalarial activity *in vitro*.[2839] The same study reported that the oil lacks antioxidant activity and is weakly larvicidal to *Ae. aegypti* mosquito larvae.

OCOTEA, COSTA RICA

Ocotea brenesii Standl, *Nectandra brenesii* (Standl.) C.K. Allen, *N. caparrapi* Sand.-Groot ex Nates

Family: Lauraceae
Note: Middle
Aroma Intensity: Medium
Aroma: Woody, sweet, slightly balsamic and herbaceous
Common Extraction Method: Steam distilled or hydrodistilled from the leaves or wood (twigs)
Possible Substitute Oils: Copaiba, katafray
Blends Well With: Balsam fir, bergamot, black spruce, blue spruce, cedarwood, clary sage, copaiba, frankincense, galbanum, lavender, lemon, lime, myrrh, orange, palo santo, pine, tangerine, ylang ylang

SUBCELLULAR LOCALIZATION | EPIGENETIC INFLUENCE: Currently unknown | Currently unknown
RECOMMENDED DILUTION RANGE: 5%–Neat

PRIMARY CONSTITUENTS:[2840]

Leaves		*Wood (Twigs)*	
Alpha-Copaene	21.1%	Alpha-Copaene	6.6%
Delta-Cadinene	9.2%	Caryophyllene Oxide	6.3%
Spathulenol	7.3%	Beta-Caryophyllene	6.1%
Globulol	5.6%	Globulol	5.0%
Beta-Caryophyllene	5.2%	Humulene Epoxide II	4.6%
Alpha-Calacorene	3.8%	Delta-Cadinene	3.8%
Germacrene D	3.2%	14-Oxy-α-Muurolene	3.2%
eudesma-4(15),7-Dien-1-β-ol	3.0%	Cadalene	3.1%

OTHER CONSTITUENTS: *Wood (Twigs)*—Linalool, alpha-cubebene, beta-cubebene, beta-elemene, alpha-humulene, gamma-muurolene, germacrene D, gamma-amorphene, cubebol, alpha-calacorene, (E)-nerolidol, salvial-4(14)-en-1-one, 1-epi-cubenol, cubenol, 3-thujopsanone, cis-calamenen-10-ol, trans-calamenen-10-ol, pentadecanal, cyclocolorenone, 14-hydroxy-α-muurolene, hexadecanoic acid; *Leaves*—Alpha-cubebene, isoledene, beta-cubebene, beta-copaene, alpha-humulene, gamma-muurolene, gamma-amorphene, alpha-muurolene, beta-calacorene, salvia-4(14)-en-1-one, 1-epi-cubenol, cubenol, alpha-muurolol, 3-thujopsanone, cadalene, cyclocolorenone, hexadecanoic acid

PREFERRED COMPOSITION FOR CLINICAL USE:
Insufficient data/samples to determine.

REPORTED THERAPEUTIC PROPERTIES: Anti-inflammatory, analgesic, reduces intestinal inflammation, digestive aid, anticancer, aids respiratory function, decongestant, supports immune function, wound healing, reduces the appearance of blemishes and acne, antibacterial, antifungal, promotes emotional healing, aids meditation, improves concentration

CAUTIONS:
- None currently known.

SELECTED EVIDENCE:
- □ None found.

OCOTEA, MADAGASCAR

Ocotea auriculiformis Kost.

FAMILY: Lauraceae
NOTE: Middle
AROMA INTENSITY: Medium
AROMA: Earthy, herbaceous, spicy
COMMON EXTRACTION METHOD: Steam distilled or hydrodistilled from the leaves
POSSIBLE SUBSTITUTE OILS: Hops, copaiba

Blends Well With: Bergamot, cassia, cinnamon, clary sage, clove, copaiba, grapefruit, hemp, lavender, lemon, lime, Mediterranean mandarin, nutmeg, orange, patchouli, pine, tangerine, valerian, yuzu
Subcellular Localization | Epigenetic Influence: Currently unknown | Currently unknown
Recommended Dilution Range: 3%–33%; neat for some conditions

Primary Constituents:[2841]

Alpha-Humulene	42.6%
Beta-Pinene	8.5%
(E)-Beta-Caryophyllene	7.8%
Alpha-Pinene	6.4%
Humulene Oxide II	4.8%
Beta-Eudesmol	3.1%

Other Constituents: Alpha-ylangene, alpha-copaene, beta-bourbonene, germacrene D, alpha-selinene, delta-cadinene, caryophyllene oxide, humulol, sequithuriferol, alpha-cadinol bulnesol

Preferred Composition for Clinical Use:
Insufficient data/samples to determine.

Reported Therapeutic Properties: Antibacterial, antifungal, antioxidant, anticancer, analgesic, anti-inflammatory, antispasmodic, aids respiratory function, ease cough, aids digestion, antiallergic, modulates immune function, insect repellent, insecticidal, protects memory function, relieves anxiety, antidepressant

Cautions:
- None currently known.

Selected Evidence:
- *In vitro* research demonstrated that Madagascar ocotea essential oil inhibited *S. pneumoniae* (exceeding the control drugs: netilmycine, spectinomycine) and *B. cereus*.[2842]
- Madagascar ocotea essential oil exhibited antioxidant activity comparable to ascorbic acid in the DPPH assay.[2843]

ONION
(Garden Onion)

Allium cepa L., *A. cepa* L. var. *aggregatum* G. Don. (formerly *A. ascalonicum*), *A. cepa* L. var. *cepa*, *A. cepa* L. var. *multiplicans* L.H., *A. cepa* L. var. *solaninum* Alef

Family: Alliaceae
Note: Top-Middle
Aroma Intensity: Strong
Aroma: Sulfurous, onion-like
Common Extraction Method: Steam distilled or hydrodistilled from the bulbs
Possible Substitute Oils: Garlic, wild garlic
Blends Well With: Garlic, wild garlic

SUBCELLULAR LOCALIZATION | EPIGENETIC INFLUENCE: Currently unknown | Currently unknown
RECOMMENDED DILUTION RANGE: 0.1%–2.0%

PRIMARY CONSTITUENTS:[2844,2845,2846,2847,2848,2849]

Red 1

Constituent	%
3,5-Diethyl -1,2,4-Trithiolane	10.2%
1,3,5-Trithiolane	7.8%
3(2H-Furanone,2-Hexyl-5-Methyl)	7.7%
Dodecane	6.8%
4-Dibutylaminobut-2-yn-1-ol	5.9%
3(2H)-Furanone,5-Methyl-2-Octyl	5.3%
3,5-Diethyl-1,2,4-Tritholane	3.1%
2,6-Dimethyl–Undecane	3.0%
4-Chloro-Benzo(1,2,5Trithiazol-5ol)	3.0%

White

Constituent	%
Dodecane	28.7%
2-Carboxylic Acid,3-Methyl(Methyl-Sulfonylamino)Thiophene	9.0%
Undecane	8.1%
Undecane,2,6-Dimethyl	6.9%
1,2,4-Trithiolane,3,5-Dimethyl	5.8%
Undecane,2-Methyl	3.6%

Kupusinski jabuˇcar Variety

Constituent	%
Dimethyl Trisulfide	16.6%
Dietil-1,2,4-Tritiolan Isomers	15.7%
Methyl Propyl Trisulfide	14.2%
Methyl-(1-Propenyl)-Disulfide*	7.5%
Dimethyl Tetrasulfide	7.2%
Methyl-(1-Propenyl)-Trisulfide*	7.2%
Methyl-(1-Propenyl)-Trisulfide*	5.9%
Methyl-(1-Propenyl)-Disulfide*	5.6%
Dipropyl Tetrasulfide	3.1%

Shallot (var. aggregatum)

Constituent	%
Dipropyl Disulfide	15.2%
Dipropyl Trisulfide	11.1%
Methyl Propyl Trisulfide	9.2%
2-Hexyl-5-Methyl 3(2H)-Furanone	5.4%
1-Propenyl Propyl Disulfide	4.6%
2-Methyl-3,4-Dithiaheptane	4.4%
Methyl Propyl Disulfide	3.3%

* Authors were unable to distinguish (Z) and (E) isomers.

Red 2

Constituent	%
Diallyl Trisulfide	22.2%
Dipropyl Trisulfide	11.1%
2-Methyl-3,4-Dithiaheptane	9.9%
Dipropyl Tetrasulfide	8.1%
Methyl Propyl Trisulfide	8.1%
2-Propenyl Propyl Disulfide	5.2%
2,5-Dimethylthiazole	4.6%
1-Propenyl Propyl Disulfide	3.8%
p-Methoxybenzylisothiocyanate	3.6%

Unidentified Variety I

Constituent	%
Diallyl Trisulfide	58.8%
Methyl Allyl Trisulfide	18.5%
Diallyl Disulfide	17.7%
Diallyl Tetrasulfide	3.1%

Unidentified Variety II

Constituent	%
Dipropyl Disulfide	30.9%
Dipropyl Trisulfide	17.1%
1-Propenyl Propyl Disulfide	7.3%
2-Methyl-3,4-Dithiaheptane	6.5%
Methyl Propyl Trisulfide	5.2%
Di-1-Propenyl Trisulfide	3.1%

OTHER CONSTITUENTS: *Red 1*—3,4–dimethyl-thiophene, disulfide-methyl-propyl, disulfide-methyl-1-propenyl, dimethyl trisulfide, undecane, 2-ethyl-5-chloro-1,3,4-triazole, 4,4,4-trifloro-1-(3-pyridinyl)-1,3-butadienone, 2,6,10-trimethyl–tetradecane, 1-ethenyloxy–octadecane, dimethyl tetrasulfide, 4,6-dimethyl-1,2,3-trithiolane, 2,2-bis(methylthiol)-propane, 2,2-bis(methyl thio)-propane, 1,1-thiobis(3-methylthiol)-propane, 1,1dioxide-1,2-dithiolane, N,(1-Cyclohexylrthyl)-2-methoxy-propanamide, 4,5-decanediol-6-ethyl, cyclohpetane carboxylic acid-pentadecyl ester, trans-3,5-diethyl-1,2,4-trithiolane, 3-(methylsulfonyl amino)-thiphene-2-carboxyli acid, 1,5-dithiaspiro (5,6) dodecan-7-ol, 2,2-bis(methylthiol) propane, 2,2,4,6-tetramethyl-trans-1,3-dioxane, 2-quinolinecarboxylic acid, methyl ester, 3-methyl,3-nonanol, 3,5-diethyl-1,2,4-trithiolane; *Red 2*—Diallyl sulfide, 2,5-dimethyl thiophene, octyl aldehyde, phenylacetaldehyde, trans-propenyl propyl disulfide, dipropyl disulfide, 3,5-dimethyl-1,2,4-trithiolane, benzylcyanide (< 2.6%), dimethyl tetrasulfide, 1-propenyl propyl sulfide, di-1-propenyl sulfide, 2-methyl-3-isothiozolone; *White*—3,4-Dimethyl thiophene, n-butyl-pentyldisulfide, 1,1,2,3-tetramethyl cyclohexane, dimethyl trisulfide, 1-methyl-2-propyl cyclohexane, decane, 4-methyl decane, 11,11-dimethyl bicyclo(8,2)dodecane, 2-methyl decane, undecane,3-methyl, norleucine,2-butyl-N-N-dimethylmethylester, thiophene-2-carboxamide,N-(2-furfuryl), 1,2,4-benzenetricarboxilic acid, 1,2-dimethyl, propane,1.1`-thiobis (3-methylthio), pentasiloxane,dodecamethyl, 3-butoxy-1,1,1,5,5,5-hexamethyl-3-(trimethylsiloxyl)trisiloxane, 1,2,4-trithiolane,3,5-dimethyl, 3-oxa-6-thia-2,7-disilaoctane,2,2,7,7-tetramethyl, 1,3,5-trithiolane, 1,2-dithiolane,1,1-dioxide, 2-methyl-3-decanol; *Kupusinski jabu˘car Variety*—Dimethyl disulfide, 2-etil-2-butenal/2-methyl-2-pentenal, dimethyl thiphenes, (Z)-(1-propenyl)-propyl-disulfide, dipropyl disulfide, (E)-(1-propenyl)-propyl-disulfide, etil-methyl-1,2,4-tritiolan, (1-propenyl)-propyl-trisulfide, 1,2,4,6-tetrathiepane, butane-1,4-dithiol; *Unidentified Variety I*—Methyl allyl disulfide, 2-vinyl-4H-1,3-dithin; *Unidentified Variety II*—Methyl propyl disulfide, methyl 1-propenyl disulfide, methyl-1-(methylthio)ethyl-disulfide, 2-undecanone, tridecane, allyl propyl trisulfide, 2-hexyl-5-methyl 3(2H)-furanone, dipropyl tetrasulfide, methyl palmitate, methyl linoleate; *Shallot (var. aggregatum)*—Methyl 1-propenyl trisulfide, methyl-1-(methylthio)ethyl-disulfide, dimethyl tetrasulfide, 2-undecanone, 3-methoxyoctane, 1-propenyl propyl trisulfide, allyl propyl trisulfide, 2-tridecanone, methyl palmitate

PREFERRED COMPOSITION FOR CLINICAL USE:
Too great of variance in samples to determine.

REPORTED THERAPEUTIC PROPERTIES: Antibacterial, antifungal, antiviral, anti-infectious, antiseptic, antiparasitic, aids immune function, lower blood sugar, anticancer, relieves earache, analgesic, anti-inflammatory, antirheumatic, antispasmodic, relieves fever, promotes respiratory function, expels excess mucus, aids digestion, expels excess gas, aids detoxification, cleanses the blood, diuretic, decreases blood pressure, reduces cholesterol levels, promotes hair growth, aphrodisiac, increases awareness, mentally stimulating

CAUTIONS:

■ Caution is advised during pregnancy and nursing. Methanolic extracts of onion have caused temporary (400 mg/kg body weight) to permanent (800 mg/kg) antifertility effects in mice due to the antifertility effects of some of the sulfur compounds and other compounds in onions.[2850] Sulfur compounds from other sulfurous vegetables (garlic) have been detected in the amniotic fluid of women who consumed a single capsule of garlic herb, and a review concluded that garlic herb has the potential to cause abortion, trigger menstruation, or stimulate the uterus.[2851,2852] However, no adverse outcomes were reported in pregnant women who took 800 mg of garlic herb daily during the third trimester of pregnancy, and diallyl sulfide prevents embryo-feto toxicity.[2853,2854,2855] Animal research also reports that garlic herbal oil—containing diallyl trisulfide—reduces pregnancy complications (decreased newborn body weight, abortion, and musculoskeletal abnormalities) due to its ability to reduce cholesterol in pregnant rats with high cholesterol.[2856] Based on conflicting information it is best to use onion cautiously during pregnancy until definitive research determines its safety. In addition, ingestion of garlic by lactating mothers reportedly altered the flavor of breast milk because garlic constituents are secreted in the breast milk and made infants prone to extended nursing.[2857,2858]

■ May interact with aspirin, blood pressure, antiplatelet, and anticoagulant medications, and increase the risk of bleeding among people with bleeding disorders, due to the sulfur compounds.[2859,2860,2861,2862,2863,2864,2865,2866,2867,2868,2869,2870] The study authors frequently attributed this effect to the presence of volatile organosulfur constituents (diallyl trisulfide, methyl allyl trisulfide).

■ May interfere with enzymes responsible for metabolizing medications (NSAIDs, proton-pump inhibitors, acetaminophen, antiepileptics, immune modulators, blood-sugar medications, blood pressure medications, antidepressants, antipsychotics, diabetic medications, antihistamines, antibiotics, and anesthetics) due to diallyl disulfide, diallyl trisulfide, allyl methyl sulfide, and allyl diallyl sulfide content.[2871,2872,2873,2874,2875,2876] Diallyl disulfide and diallyl trisulfide also interfere with protein transporters that move drugs across cell membranes, which can increase or decrease drug effectiveness.[2877]

■ May interact with antifungal medications (ketoconazole) and increase their effectiveness based on presence of sulfur-containing compounds similar to garlic essential oil.[2878]

■ May be photosensitizing due to reports that diallyl disulfide causes photo contact dermatitis.[2879,2880] Avoid sun exposure to area of application for at least twelve hours after topical application.

■ Topical application may cause skin irritation or skin sensitivity reactions. Cases of contact dermatitis or irritation have been reported with onions and other sulfurous vegetables.[2881,2882,2883,2884,2885] Onion compounds are both irritating and sensitizing. Dilutions under 2% are strongly recommended.

Selected Evidence:

- □ Red onion (red 2) essential oil showed significant anti-inflammatory activity, which exceeded that of sodium diclofenac (the reference NSAID) *in vitro*.[2886]
- □ Red onion essential (red 1) oil significantly killed adult *Schistosoma mansoni* worms.[2887]
- □ *In vitro* research demonstrated that onion essential oil (Kupusinski jabuˇcar variety) inhibits *A. versicolor* mycelial growth and its production of sterigmatocystin (a cancer-causing mycotoxin produced by common molds).[2888] The Junski srebrnjak and Kupusinski jabuåar varieties of onion essential oil (compositions not reported) inhibited *E. amstelodami, A. tamarii, P. griseofulvum, C. tropicalis,* and *S. cerevisiae*.[2889] An onion essential oil (unidentified composition) exhibited weak antibacterial activity against *E. coli*.[2890]
- □ The butanolic fraction of white onion essential oil showed the highest antioxidant activity in the DPPH and phosphomolybdnum assays.[2891] Red (red 2) onion essential oil exhibited high antioxidant and DPPH radical scavenging effect but low ferric reducing power activity.[2892]

ORANGE BIRD BERRY
(Butter-Berry, Kamynye)

Hoslundia opposita Vahl.

FAMILY: Lamiaceae
NOTE: Middle or Top depending on CT
AROMA INTENSITY: Medium
AROMA: Fresh, herbaceous, medicinal, slightly floral, spicy, and vanilla-like
COMMON EXTRACTION METHOD: Hydrodistilled or steam distilled from the leaves; may also be distilled from the fruit or root bark
POSSIBLE SUBSTITUTE OILS: Eucalyptus, cajeput, copaiba, guava leaf, black pepper, clove, ylang ylang, camphor, rosemary
BLENDS WELL WITH: Black pepper, cajeput, camphor, cardamom, cinnamon, clove, copaiba, coriander, eucalyptus, ginger, lavandin, lavender, patchouli, rosemary, turmeric, ylang ylang
SUBCELLULAR LOCALIZATION | EPIGENETIC INFLUENCE: Currently unknown | Currently unknown
RECOMMENDED DILUTION RANGE: 1.5%–20%; 50% for some conditions

PRIMARY CONSTITUENTS:[2893,2894,2895,2896,2897,2898,2899]

1,8-Cineole CT	
1,8-Cineole	61.2%–72.3%
Alpha-Terpineol	7.2%–16.8%
Beta-Phellandrene	0.0%–13.3%
Sabinene	0.0%–4.5%
Thymol	0.0%–4.2%
Delta-3-Carene	0.0%–3.7%
Beta-Farnesene	0.0%–3.6%

Beta-Caryophyllene CT	
Beta-Caryophyllene	20.5%–40.9%
Germacrene D	10.7%–26.1%
Alpha-Humulene	0.0%–18.3%
Thymol	0.0%–13.9%
Benzyl Benzoate	0.0%–11.6%
Menthol	0.0%–10.9%
Caryophyllene Oxide	0.0%–10.5%
Menthone	0.0%–7.0%
Para-Cymene	0.0%–6.1%
(E)-Phytol	0.0%–5.9%

Eugenol CT	
Eugenol	55.1%–76.2%
Beta-Caryophyllene	6.5%–8.5%
Alpha-Humulene	0.0%–7.6%
Germacrene D	0.4%–4.6%
(Z)-Beta-Ocimene	0.2%–4.5%
1,8-Cineole	0.2%–4.4%

Germacrene D CT	
Germacrene D	26.8%–31.1%
Alpha-Humulene	0.0%–24.4%
Beta-Caryophyllene	9.5%–17.1%
Delta-Cadinene	0.0%–11.4%
Palmitic Acid	0.0%–10.2%
Benzyl Benzoate	0.0%–8.0%
Alpha-Phellandrene	0.0%–4.6%
Alpha-Copaene	0.0%–4.6%
Cadina-1,4-Diene	0.0%–4.4%
Guaiol	0.0%–4.4%
(E)-Phytol	0.0%–4.3%
Beta-Cadinene	0.0%–3.9%
Ethyl Linolenate	0.0%–3.7%

Benzyl Benzoate CT	
Benzyl Benzoate	30.1%–40.4%
Germacrene D	5.5%–19.0%
Beta-Caryophyllene	10.4%–13.3%
Thymol	7.2%–7.5%
(E)-Phytol	0.0%–4.6%
Alpha-Copaene	0.0%–4.2%

Limonene CT	
Limonene	35.8%
Beta-Caryophyllene	18.8%
Germacrene D	14.0%

Alpha-Humulene-14-Hydroxy CT

Alpha-Humulene-14-Hydroxy	19.9%–26.5%
Beta-Caryophyllene	0.0%–15.7%
(E)-Phytol	0.0%–15.0%
Caryophyllene Oxide	0.0%–9.0%
Germacrene D	0.0%–8.6%
Alpha-Humulene	0.0%–8.2%

Phytol CT

(E)-Phytol	18.8%–20.4%
Beta-Caryophyllene	11.3%–17.4%
Germacrene D	11.0%–17.3%
Alpha-Humulene	0.0%–8.6%
Alpha-Humulene-14-Hydroxy	0.0%–7.5%

Caryophyllene Oxide CT

Caryophyllene Oxide	26.3%
Germacrene D	22.3%
Beta-Caryophyllene	11.5%
(E)-Phytol	7.4%

Fruit

Camphor	69.5%
Linalool	5.4%
Beta-Pinene	3.4%

Other Constituents: *1,8-Cineole CT*—Alpha-pinene, (Z)-beta-ocimene, terpinen-4-ol, carvacrol, beta-cubebene, caryophyllene, (Z)-alpha-farnesene, germacrene D, cis-sabinene-hydrate; *Beta-Caryophyllene CT*—Alpha-pinene, limonene, 1,8-cineole, gamma-terpinene, linalool, camphor, alpha-terpineol, pulegone (< 1.4%), menthyl acetate, alpha-copaene, bicyclogermacrene, beta-elemene, gamma-cadinene, germacrene B, caryophyllene oxide, salvia-4(14)-en-1-one, alpha-cadinol; *Germacrene D CT*—Alpha-pinene, myrcene, limonene, 1,8-cineole, (Z)-beta-ocimene, (E)-beta-ocimene, alpha-cubebene, beta-copaene, beta-cubebene, trans-beta-bergamotene, alloaromadendrene, cis-muurola-4,14-5-diene, bicyclogermacrene, beta-sesquiphellandrene, elemol, spathulenol, nerolidol, caryophyllene oxide, T-muurolol, alpha-cadinol, bulnesol, para-cymene, thymol; *Eugenol CT*—(Z)-Hex-3-en-1-ol, alpha-pinene, beta-pinene, myrcene, (E)-beta-ocimene, linalool, camphor, cryptone, artemisyl acetate, 2-isopropylen-3-methyl-cyclopent-2-en-4-ol, neral, 2-methyl-5(1-methylethenyl) cyclohex-2-ene-1-ol + t-carvone, geranial, alpha-copaene, beta-elemene, trans-alpha-bergamotene, cinnamyl acetate, geraniol, linalyl acetate, (Z)-beta-farnesene, beta-bisabolene, beta-sesquiphellandrene, caryophyllene oxide; *Benzyl Benzoate CT*—Limonene, gamma-terpinene, beta-bourbonene, beta-cubebene, alpha-humulene, germacrene A, gamma-cadinene, germacrene B, para-cymene; *Alpha-Humulene-14-Hydroxy CT*—Not listed; *Phytol CT*—Not listed; *Limonene CT*—Not listed; *Caryophyllene Oxide CT*—Not listed; *Fruit*—Alpha-thujene, alpha-pinene, sabinene, limonene, (E)-beta-ocimene, gamma-terpinene, borneol, terpinen-4-ol, alpha-terpineol, linalyl acetate, neral, geranial, eugenol, germacrene D, gossonorol

Preferred Composition for Clinical Use:
Insufficient samples and excess variability in composition data make it unable to determine.

Reported Therapeutic Properties: Balances blood sugar, reduces cholesterol levels, reduces fever, analgesic, anti-inflammatory, anticonvulsant, relieves sore throat, diuretic, anti-venom (snakes), relieves chronic skin conditions, wound healing, heals oral wounds, anti-infectious, antibacterial, antimalarial, relieves abdominal pain, promotes mental and emotional balance, aids memory, antidepressant

Cautions:

1,8-Cineole CT

■ Avoid with children under age 3, particularly around the nose and mouth and use cautiously in children under age 5 due to significant 1,8-cineole content. 1,8-cineole may cause seizures, central nervous system problems, or respiratory distress in young children.[2900,2901,2902]

■ Caution is warranted orally due to the significant amounts of 1,8-cineole. Limit it to small doses internally (adults—maximum 10 drops daily). Toxicity has been reported when eucalyptus (also high in 1,8-cineole) was ingested in large doses, and as few as 4 to 5 drops may cause problems in very sensitive individuals.[2903,2904,2905,2906,2907,2908]

■ Avoid with epilepsy and Parkinson's disease due to 1,8-cineole content. May exacerbate or cause seizures/convulsions or reduce seizure-medication efficacy based on 1,8-cineole content.[2909,2910,2911,2912]

■ May weakly interfere with the enzymes responsible for metabolizing medications (NSAIDs, proton-pump inhibitors, acetaminophen, antiepileptics, immune modulators, blood-sugar medications, blood pressure medications, antidepressants, antipsychotics, diabetic medications, antihistamines, antibiotics, and anesthetics) due to 1,8-cineole content.[2913,2914,2915]

■ May interfere with pentobarbital and other barbiturates (medications for anxiety and insomnia), based on 1,8-cineole content.[2916,2917]

Eugenol CT

■ Avoid during pregnancy and lactation due to eugenol content. Eugenol is considered strongly toxic to embryos according to animal studies.[2918] Animal studies suggest that large doses of clove (also with significant eugenol) may negatively impact embryonic development and encourage fetal cell death.[2919,2920] Another animal study did not detect any negative influence of clove oil.[2921] Based on these conflicting studies it is best to limit orange bird berry essential oils rich in eugenol during pregnancy and lactation.

■ May interact with aspirin, blood pressure, antiplatelet, and anticoagulant medications, and increase the risk of bleeding among people with bleeding disorders based on eugenol content.[2922,2923,2924,2925]

■ May interact with MAOI antidepressants based on eugenol content. Animal research suggests that eugenol produces antidepressant effects via the monoamine oxidase pathway, which may cause interactions with antidepressants that also interact with this pathway.[2926]

■ May interfere with enzymes responsible for metabolizing medications (NSAIDs, proton-pump inhibitors, acetaminophen, antiepileptics, immune modulators, blood-sugar medications, blood pressure medications, antidepressants, antipsychotics, diabetic medications, antihistamines, antibiotics, and anesthetics), based on eugenol content.[2927,2928]

■ May interact with anticholinergic (drugs used for asthma, incontinence, gastrointestinal cramps, muscular spasms, depression, and sleep disorders) and cholinergic medications (drugs used to reduce urinary retention, diagnose myasthenia gravis, and for glaucoma), based on AChE-inhibtory activity of eugenol.[2929]

■ May interact with antibiotics and possibly enhance their effects based on eugenol content.[2930]

■ May irritate mucous membranes (eyes, mouth, nasal passages, vagina, rectum) due to eugenol content.[2931,2932]

Benzyl Benzoate CT

■ Oral caution. Ingestion of benzyl derivatives (benzoic acids, benzoic salts, benzaldehyde, benzyl acetate, benzyl benzoate, and benzyl alcohol) may cause gastrointestinal irritation, vomiting, and diarrhea.[2933] The Joint Expert Committee for Food Additives (JECFA) has established an acceptable daily limit for the aforementioned constituents of 0–5 mg/kg body weight for adults to avoid GI discomfort.[2934] Based on this limit, it is suggested to limit the oral consumption of orange bird berry essential oil rich in benzyl derivatives to no more than 10 drops daily.

Phytol CT and Alpha-Humulene-14-Hydroxy CT

■ Orange bird berry CTs rich in phytol should be avoided by individuals with Refsum disease (a neurological disorder caused by the overaccumulation of phytanic acid in cells and tissues) or who are AMACR-deficient (alphamethylacyl-CoA racemase-deficient) due to phytol content. It is believed that humans can convert phytol to phytanic acid.[2935] Animal research suggests that phytol can cause liver inflammation, liver failure, and death in AMACR-deficient mice.[2936,2937] AMACR is an enzyme found in the mitochondria that plays a role in the breakdown of fatty acids. AMACR deficiency causes neurological problems (seizures, cognitive decline, migraines, encephalopathy, loss of sensation due to nerve damage, muscle stiffness, lack of coordination, and vision problems) that begin in adulthood and progressively get worse.[2938] Phytol may also weakly promote tumor formation according to animal research.[2939]

Fruit

■ Avoid with children under six due to significant camphor content. Several cases of camphor poisoning and/or seizures from ingestion and topical application have been reported in children.[2940,2941] Ingestion of camphor-containing products has been lethal in children under age two.[2942] Children five years and up may use camphor-containing essential oils topically in dilutions no stronger than 5%. Camphor ingestion by infants and young children may cause cough, vomiting, seizure, burning sensation in the mucous membranes and eyes, or lack of voluntary coordination of muscle movements.[2943]

■ Caution is warranted during pregnancy and while lactating, due to significant camphor content. Ingestion of essential oils with significant levels of camphor may lead to abortion because fetuses lack the enzymes to process it.[2944] Camphor ingestion by infants and young children may cause cough, vomiting, seizure, burning sensation in the mucous membranes and eyes, or lack of voluntary coordination of muscle movements.[2945]

■ Avoid oral consumption due to significant camphor levels. Essential oils with significant levels of camphor may be toxic when taken orally, usually single doses exceeding 2 mL, although the lethal dose for humans is estimated to be greater than 5 mL in a single dose.[2946,2947,2948]

■ Avoid with epilepsy and Parkinson's disease due to camphor content due to the potential of these constituents to exacerbate or cause seizures/convulsions or reduce seizure-medication efficacy.[2949,,2950,2951]

■ Avoid with those who have a compromised liver due to the risk of increased liver enzymes and liver damage based on camphor content.[2952] This would also suggest that those taking medications that could cause liver damage should also use essential oils with significant camphor levels very cautiously or avoid them entirely.

■ The significant camphor content may negatively impact red blood cells and increase the risk of jaundice in children with glucose-6-phosphate dehydrogenase deficiency (G6PD).[2953,2954]

■ Caution is warranted for the topical application of camphor-containing essential oils on broken or injured skin due to an increased risk of toxicity.[2955]

Caryophyllene Oxide CT

■ May interact with aspirin, blood pressure, antiplatelet, and anticoagulant medications, and increase the risk of bleeding among people with bleeding disorders, due to caryophyllene-oxide content.[2956]

Beta-Caryophyllene CT, Germacrene D CT, and Limonene CT

■ None currently known.

SELECTED EVIDENCE:

- □ Intraperitoneal administration of orange bird berry essential oil (110 and 220 mg/kg body weight, once daily, for four days) normalized dyslipidemia caused by diabetes.[2957] Total cholesterol, total triglycerides, and low density cholesterol (LDL) were all significantly reduced and HDL cholesterol was significantly elevated. Remarkably, the atherogenic index (LDL/HDL ratio) returned to levels that were not significantly different than the control group. The oil also reduced cholesterol levels in nondiabetic rats. The composition of the oil was not identified but the same researchers identified 1,8-cineole as the primary constituent in a subsequent study. Another study using the same administration method found that 110 mg/kg of the oil reduced fasting blood-glucose levels in diabetic rats to near normal levels by day four.[2958]

- Intraperitoneal administration of orange bird berry essential oil (110 and 220 mg/kg body weight, once daily, for four days) increased white blood cell (WBC) count and reduced lymphocyte percentages in nondiabetic rats administered the higher dose.[2959] Diabetic rats experienced a significant increase in red blood cells (RBCs) at the lower dose, whereas the nondiabetic rats experienced reduced RBCs at the lower dose. The study authors concluded that the oil significantly improved anemia caused by alloxan in diabetic rats, which made it promising for the management of type 2 diabetes and its associated hematological complications. The composition of the oil was not identified but the same researchers identified 1,8-cineole as the primary constituent in a subsequent study.
- The germacrene D CT of orange bird berry essential oil inhibited oral pathogens: periodontopathic *Porphyromonas gingivalis* and *Aggregatibacter actinomycetemcomitans* and cariogenic *Streptococcus mutans*.[2960] Orange bird berry essential oil is significantly active against *Aspergillus niger, Acinetobacter calcoacetica, Brochothrix thermosphacta*, and *Flavobacterium suaveolens*.
- Leaf essential oil of orange bird berry (1,8-cineole CT) was insecticidal to and repelled the cowpea seed bruchid (*Callosbruchus maculatus*).[2961]
- Orange bird berry essential oil consisting of mainly sesquiterpenes and sesquiterpene alcohols killed and repelled the rust-red flour beetle (*Tribolium castaneum*).[2962]

ORIENTAL SWEET GUM (Asian Styrax, Storax, Turkish Sweetgum, Styrax Liquidus, Levantine Styrax)

Liquidambar orientalis Mill., *L. orientalis* Mill. var. *Orientalis*, *L. orientalis* Mill. var. *integriloba*

FAMILY: Altingiaceae (Formerly Hamamelidaceae)
NOTE: Middle-Base
AROMA INTENSITY: Medium-Strong
AROMA: Balsamic, sweet, slightly floral, spicy, and woody (may be very metallic or gasoline-like initially); or medicinal, fresh, herbaceous, earthy, woody
COMMON EXTRACTION METHOD: Steam distilled or hydrodistilled from the gum resin; less commonly distilled from the leaves
POSSIBLE SUBSTITUTE OILS: Benzoin, American sweetgum, frankincense; or tea tree, plai, shell ginger
BLENDS WELL WITH: Angelica, balsam fir, benzoin, bergamot, cassia, cinnamon, clove, copaiba, frankincense, geranium, grapefruit, lavender, mandarin, neroli, opoponax, orange, palo santo, patchouli, petitgrain, rose, spruce (black), styrax, tamala, tangerine, tea tree, ylang ylang
SUBCELLULAR LOCALIZATION | EPIGENETIC INFLUENCE: Currently unknown | Currently unknown
RECOMMENDED DILUTION RANGE: Resin—1%–10%; Leaves—3%–50%; neat from some conditions

PRIMARY CONSTITUENTS:[2963,2964,2965,2966,2967,2968,2969]

Gum Resin CT Styrene		*Leaves*	
Styrene	70.4%–81.9%	Terpinen-4-ol	9.4%–35.0%
Alpha-Pinene	3.5%–19.0%	Alpha-Terpineol	1.9%–29.0%

Cinnamyl Alcohol	0.3%–6.9%
Beta-Pinene	0.6%–4.3%
Benzene Propanol	0.1%–3.4%

Gum Resin CT Cinnamyl Alcohol

Cinnamyl Alcohol	45.1%
Hydrocinnamyl Alcohol	41.1%
Beta-Caryophyllene	3.6%

Gum Resin CT Cinnamyl Acetate

Cinnamyl Cinnamate	21.0%
3-Phenylpropyl Cinnamate	7.5%
Cinnamic Acid	4.0%

Sabinene	1.7%–15.0%
Gamma-Terpinene	tr–14.4%
Germacrene D	0.4%–11.0%
Viridiflorene	1.7%–9.7%
Alpha-Terpinene	tr–9.0%
Alpha-Pinene	2.6%–8.6%
Para-Cymene	tr–6.5%
Cis-Carveol	0.0%–5.2%
Beta-Gurjunene	0.0%–4.4%
Delta-Cadinene	0.1%–4.3%
Trans-Carvyl Acetate	0.2%–4.2%
Terpinolene	tr–3.6%
Myrtenol	0.0%–3.2%

Other Constituents: *Gum Resin CT Styrene*—Camphene, beta-pinene, limonene, 4-ethyl phenol; *Gum Resin CT Cinnamyl Alcohol*—Styrene, alpha-pinene, benzaldehyde, benzyl alcohol; *Gum Resin CT Cinnamyl Acetate*—Cinnamyl alcohol; *Leaves*—Beta-pinene, myrcene, alpha-phellandrene, para-cymene, limonene, beta-phellandrene, n-octanol, 4-terpinyl acetate, n-nonanol, menthol, isomenthol, trans-piperitol, trans-carveol, trans-myrtanol, alpha-longipinene, beta-caryophyllene, aromadendrene, alpha-farnesene, spathulenol, caryophyllene oxide, humulene epoxide, globulol, T-cadinol, farnesol (isomer not identified), farnesyl acetate (isomer not identified)

Preferred Composition for Clinical Use:
Too great of variance in samples/data to determine.

Reported Therapeutic Properties: Soothes skin itching and chronic skin conditions, wound healing, antiseptic, antibacterial, antifungal, antiviral, antiparasitic, antioxidant, reduces high blood pressure, liver protective, expels excess mucus, anti-inflammatory, anticonvulsant, protects stomach against ulcers, anticancer, immune modulatory, insecticidal, improves memory function, antidepressant, relieves anxiety, improves self-confidence

Cautions:

Gum Resin EO

- Low to moderate risk of sensitization or skin irritation. Use cautiously on diseased or damaged skin and among people with a compromised immune system.[2970]
- *Styrene CT*: Not recommended for therapy due to the hazards of styrene (largely collected from synthetic styrene hazards): nervous system dysfunction, hearing loss, liver damage, kidney dysfunction, impaired learning, reproductive damage, cancer, irritation (skin, eyes, upper respiratory tract).[2971,2972]

Leaf EO

- Use cautiously orally. Toxicity appears to occur at extreme doses of essential oils rich in terpinen-4-ol; therefore, it is best to limit consumption to 600 mg (about 10 drops with a 60 mg dropper) daily, and avoid long-term use orally. See monograph for Melaleuca (Tea Tree) in *Medicinal Essential Oils* for more information.
- May interact with antibiotics antifungal drugs and increase or decrease their effectiveness due to terpinen-4-ol content.[2973]

SELECTED EVIDENCE:

- Oral administration of 150 and 300 mg/kg body weight of Oriental sweet gum essential oil significantly protected against stomach ulcers in rats.[2974]
- Feeding chickens Oriental sweet gum essential oil resulted in reduced abdominal fat, blood cholesterol levels, and *E. coli* counts in the intestines.[2975]
- Oriental sweet gum CT cinnamyl alcohol essential oil was toxic to the Japanese termite (*Reticulitermes speratus*).[2976] Another study reported that the same CT was nematicidal to the pine wood nematode (*Bursaphelenchus xylophilus*).[2977]

OSTOKHODUS (Persian Lavender)

Nepeta menthoides Boiss. & Bohse

FAMILY: Labiatae (Lamiaceae)
NOTE: Top-Middle
AROMA INTENSITY: Strong
AROMA: Fresh, medicinal, earthy
COMMON EXTRACTION METHOD: Steam distilled from the leaves and flowers
POSSIBLE SUBSTITUTE OILS: Catnip, lemon catnip, eucalyptus, rosalina, melaleuca (tea tree), fragonia
BLENDS WELL WITH: Bergamot, cajeput, copaiba, geranium, grapefruit, frankincense, lavender, lemon, lemon myrtle, myrrh, orange, palmarosa, rhododendron, rosalina, rose, tangerine, tea tree (melaleuca)
SUBCELLULAR LOCALIZATION | EPIGENETIC INFLUENCE: Currently unknown | Currently unknown
RECOMMENDED DILUTION RANGE: 3%–33%; neat for some conditions

PRIMARY CONSTITUENTS:[2978,2979,2980,2981,2982,2983,2984]

1,8-Cineole	16.6%–70.1%
Nepetalactones	0.0%–36.0%
Dihydromyrcen-1-ol	0.0%–9.2%
Beta-Pinene	3.4%–8.8%
Geranyl Acetate	0.0%–8.1%
Terpinen-4-ol	1.8%–7.1%
Alpha-Terpineol	0.0%–5.7%
Neryl Acetate	0.0%–3.5%
Sabinene	0.0%–3.3%

Note: Ostokhodus could be split into at least two chemotypes (nepetalactones and 1,8-cineole) but the CTs are combined in this work.

OTHER CONSTITUENTS: Alpha-thujene, myrcene, alpha-pinene, gamma-terpinene, n-amyl-isovalerate, isobutyl isovalerate, isoamyl isovalerate, alpha-terpinene, para-cymene, cis-sabinene hydrate, linalool, terpinolene, pinocarvone, trans-dihydrocarvone, linalool, alpha-fenchol, menth-3-en-1-ol, d-terpineol, pinocamphone, myrthanol, trans-anethole, benzyl propanoate, (E)-alpha-bisabolene, (Z)-alpha-bisabolene, neryl acetate, 2-methyl butyl isovalerate, 2-methyl butyl 2-methyl butanoate, linalyl propionate, n-undecane, cis-dihydro-α-terpineol

PREFERRED COMPOSITION FOR CLINICAL USE:

Constituent	*Nepetalactone CT*	*1,8-Cineole CT*
Nepetalactones	20%–40%	–
1,8-Cineole	15%–40%	40%–70%
Beta-Pinene	2%–10%	3%–13%
Terpinen-4-ol	2%–10%	0.5%–10%
Alpha-Terpineol	2%–10%	2%–10%
Geranyl Acetate + Neryl Acetate	2%–10%	2%–10%
Sabinene	0.1%–5%	0%–5%
Alpha-Pinene	0.1%–5%	0.1%–5%

REPORTED THERAPEUTIC PROPERTIES: Reduces high blood pressure, analgesic, anti-inflammatory, antirheumatic, antineuralgic, nervine, antispasmodic, aids respiratory function, bronchodilator, antioxidant, antibacterial, antiseptic, insecticidal, promotes restful sleep, protects the brain and neurons, aids cognition and memory, antidepressant, promotes balanced thinking, aphrodisiac

CAUTIONS:

1,8-Cineole CT

▪ Avoid with children under age 3, particularly around the nose and mouth and use cautiously in children under age 5 due to significant 1,8-cineole content. 1,8-cineole may cause seizures, central nervous system problems, or respiratory distress in young children.[2985,2986,2987]

▪ Caution is warranted orally due to the significant amounts of 1,8-cineole. Limit it to small doses internally (adults—maximum 10 drops daily). Toxicity has been reported when eucalyptus (also high in 1,8-cineole) was ingested in large doses, and as few as 4 to 5 drops may cause problems in very sensitive individuals.[2988,2989,2990,2991,2992,2993]

▪ Avoid with epilepsy and Parkinson's disease due to 1,8-cineole content. May exacerbate or cause seizures/convulsions or reduce seizure-medication efficacy based on 1,8-cineole content.[2994,2995,2996,2997]

▪ May interact with anticholinergic (drugs used for asthma, incontinence, gastrointestinal cramps, muscular spasms, depression, and sleep disorders) and cholinergic medications (drugs used to reduce urinary retention, diagnose myasthenia gravis, and for glaucoma), based on AChE-inhibtory activity of both CTs of essential oil and the potent AChE inhibition by 1,8-cineole.[2998,2999]

▪ May weakly interfere with the enzymes responsible for metabolizing medications (NSAIDs, proton-pump inhibitors, acetaminophen, antiepileptics, immune modulators, blood-sugar medications, blood pressure medications, antidepressants, antipsychotics, diabetic medications, antihistamines, antibiotics, and anesthetics) due to 1,8-cineole content.[3000,3001,3002]

▪ May interfere with pentobarbital and other barbiturates (medications for anxiety and insomnia), based on 1,8-cineole content.[3003,3004]

Nepetalactones CT

▪ Avoid with epilepsy and Parkinson's disease due to 1,8-cineole content. May exacerbate or cause seizures/convulsions or reduce seizure-medication efficacy based on 1,8-cineole content.[3005,3006,3007,3008]

▪ May interact with anticholinergic (drugs used for asthma, incontinence, gastrointestinal cramps, muscular spasms, depression, and sleep disorders) and cholinergic medications (drugs used to reduce urinary retention, diagnose myasthenia gravis, and for glaucoma), based on AChE-inhibtory activity of both CTs of essential oil and the potent AChE inhibition by 1,8-cineole.[3009,3010]

■ May weakly interfere with the enzymes responsible for metabolizing medications (NSAIDs, proton-pump inhibitors, acetaminophen, antiepileptics, immune modulators, blood-sugar medications, blood pressure medications, antidepressants, antipsychotics, diabetic medications, antihistamines, antibiotics, and anesthetics) due to 1,8-cineole content.[3011,3012,3013]

■ May interfere with pentobarbital and other barbiturates (medications for anxiety and insomnia), based on 1,8-cineole content.[3014,3015]

SELECTED EVIDENCE:

- *In vitro* research shows that ostokhodus (1,8-cineole CT) potently inhibits three different breast cancer cells.[3016] A nepetalactone-rich CT of ostokhodus inhibited breast ductal, colon, and colorectal cancer cells without harming normal cells.[3017] The greatest effect was observed against breast ductal cancer cells.
- Ostokhodus (1,8-cineole rich) potently inhibited acetylcholinesterase (AChE) activity, demonstrating greater potency than the standard drug rivastigmine and etoposide.[3018] 1,8-cineole was more potent than the whole essential oil. Inhibition of AChE prevents the breakdown of acetylcholine, which is essential for memory and thinking. People with neurodegenerative diseases make less acetylcholine, and the diseases often break it down at a faster rate, leading to acetylcholine deficits. A nepetalactone-rich CT (36.0% nepetalactones and 16.7% 1,8-cineole) also demonstrated AChE inhibition.[3019] Based on these two reports, the 1,8-cineole CT is a more potent inhibitor of AChE than the nepetalactone CT.
- Ostokhodus (1,8-cineole CT) demonstrated moderate antimicrobial activity against bacteria (*B. subtilis, S. epidermidis, S. aureus, K. pneumonia, Sh. Dysenteriae, P. vulgaris, S. paratyphi*) and fungi (*C. albicans, A. niger, A. brasiliensis*).[3020] Another research reported that ostokhodus rich in nepetalactones inhibited *Bacillus subtilis, B. cereus, Enterococcus faecalis, Staphylococcus aureus, S. epidermidis, Escherichia coli*, and *Klebsiella pneumoniae*.[3021] The researchers noted that 1,8-cineole and 4aα-7α-7aα-Nepetalactone exhibited similar antibacterial activity as the whole oil, suggesting they are responsible for the antibacterial activity.
- A nepetalactone-rich CT of ostokhodus demonstrated moderate antioxidant activity, which was attributed to myrcene, 1,8-cineole, limonene, α-terpinene, and α-terpinolene.

PADANG CASSIA (Indonesian Cinnamon, Korintje, Batavia Cassia)

Cinnamomum burmannii (Nees & Th Nees) Nees ex Blume

FAMILY: Lauraceae
NOTE: Middle
AROMA INTENSITY: Strong
AROMA: Spicy, sweet, fresh, slightly citrus
COMMON EXTRACTION METHOD: Steam distilled or hydrodistilled from the bark; may also be distilled from the leaves or branches
POSSIBLE SUBSTITUTE OILS: Cinnamon, cassia, tamala, eucalyptus (*E. radiata*), rosemary
BLENDS WELL WITH: Bergamot, black pepper, cardamom, copaiba, eucalyptus, frankincense, geranium, lavandin, lavender, lemon, lime, orange, rosemary, tangerine, vetiver
SUBCELLULAR LOCALIZATION | EPIGENETIC INFLUENCE: Currently unknown | Currently unknown
RECOMMENDED DILUTION RANGE: 1.5%–20%

PRIMARY CONSTITUENTS:[3022,3023,3024]

Bark (Cinnamaldehyde CT)

(E)-Cinnamaldehyde	92.0%
Alpha-Copaene	4.1%

Bark (1,8-Cineole CT)

1,8-Cineole	49.8%–52.9%
Alpha-Terpineol	12.0%–13.0%
Camphor	8.1%–9.8%
Terpinen-4-ol	7.7%–9.3%

Leaves (Borneol CT)

Borneol	78.5%
Bornyl Acetate	3.3%

Branches

Borneol	34.2%
Alpha-Terpineol	10.1%
Bornyl Acetate	9.6%
Terpinen-4-ol	7.2%
1,8-Cineole	6.9%
Para-Cymene	5.8%
Beta-Caryophyllene	5.1%

Leaves (Cinnamaldehyde CT)

(E)-Cinnamaldehyde	60.2%
Eugenol	17.6%
Coumarin	13.4%
Borneol	6.8%

Leaves (1,8-Cineole CT)

1,8-Cineole	28.5%
Borneol	16.5%
Alpha-Terpineol	6.4%
Para-Cymene	6.1%
Spathulenol	5.8%
Terpinen-4-ol	4.1%
Bornyl Acetate	3.1%

OTHER CONSTITUENTS: *Bark (Cinnamaldehyde CT)*—3-phenyl-2-prophenyl acetate; *Bark (1,8-Cineole CT)*—Alpha-pinene, camphene, para-cymene, beta-caryophyllene, elemol, beta-eudesmol; *Leaves (Borneol CT)*—1,8-cineole, 1,7,7-trimethyl-, 3-Cyclohexen-1-o, Bicyclo[2.2.1]hept-2-ene, caryophyllene, gamma-elemene, spathulenol, caryophyllene oxide, guaiol, docosahexaenoic acid methyl ester; *Leaves (1,8-Cineole CT)*—Alpha-pinene, beta-pinene, cinnamyl acetate, beta-caryophyllene, myristicin, elemol; *Leaves (Cinnamaldehyde CT)*—Alpha-terpineol, benzylidenemalonaldehyde; *Branches*—Camphor, alpha-humulene, myristicin, elemol, spathulenol, beta-eudesmol

PREFERRED COMPOSITION FOR CLINICAL USE:
Insufficient samples/data to determine.

REPORTED THERAPEUTIC PROPERTIES: Antibacterial, antimicrobial, antifungal, anti-infectious, antiparasitic, antioxidant, antiseptic, aids immune function, antispasmodic, anti-inflammatory, promotes oral health, aids digestion, eases nausea, expels excess gas, relieves diarrhea, eases cough, aids respiratory function, aphrodisiac, promotes mental clarity, aids concentration, emotionally uplifting

CAUTIONS:

Cinnamaldehyde CTs

■ Avoid during pregnancy and lactation. Several studies have tested the major constituents (e.g., cinnamaldehyde) of cinnamon and cassia essential oils with pregnant animals, and the results have been inconsistent.[3025,3026,3027] Therefore, it is advised to avoid essential oils with significant quantities of cinnamaldehyde during pregnancy until further research is conclusive.

■ May interact with aspirin, blood pressure, antiplatelet, and anticoagulant medications, and increase the risk of bleeding among people with bleeding disorders, due to cinnamaldehyde content.[3028,3029,3030]

■ May interfere with enzymes responsible for metabolizing medications (NSAIDs, proton-pump inhibitors, acetaminophen, antiepileptics, immune modulators, blood-sugar medications, blood pressure medications, antidepressants, antipsychotics, diabetic medications, antihistamines, antibiotics, and anesthetics) due to cinnamaldehyde or eugenol content.[3031,3032,3033]

■ May interact with diabetes medications and cause low blood-sugar levels due to cinnamaldehyde content.[3034,3035,3036]

■ May interact with antibiotics and possibly enhance their effects due to cinnamaldehyde content.[3037,3038]

■ May irritate mucous membranes (eyes, mouth, nasal passages, vagina, rectum) due to cinnamaldehyde content.[3039]

■ Topical application (significant dilution strongly encouraged). Cinnamaldehyde is very irritating to the skin and prone to cause severe reactions.[3040]

■ *Leaves with coumarin*: Use cautiously orally—maximum of 5 drops daily for adults due to potentially significant coumarin content. Animal research suggests that coumarin may be toxic to the liver and lungs when taken orally in very high doses.[3041] The total safe daily intake was established at up to 0.1 mg/kg body weight daily by the European Food Safety Authority.[3042] In addition, the coumarin content may interact with aspirin, blood pressure, antiplatelet, anticoagulant, and antidepressant medications, and increase the risk of bleeding among people with bleeding disorders.[3043]

1,8-Cineole CTs

■ Avoid with children under age 3, particularly around the nose and mouth and use cautiously in children under age 5 due to potentially significant 1,8-cineole content. 1,8-cineole may cause seizures, central nervous system problems, or respiratory distress in young children.[3044,3045,3046]

■ Caution is warranted orally due to the significant amounts of 1,8-cineole. Limit it to small doses internally (adults—maximum 10 drops daily). Toxicity has been reported when eucalyptus (also high in 1,8-cineole) was ingested in large doses, and as few as 4 to 5 drops may cause problems in very sensitive individuals.[3047,3048,3049,3050,3051,3052]

■ Avoid with epilepsy and Parkinson's disease due to 1,8-cineole content. May exacerbate or cause seizures/convulsions or reduce seizure-medication efficacy based on 1,8-cineole content.[3053,3054,3055,3056]

■ May weakly interfere with the enzymes responsible for metabolizing medications (NSAIDs, proton-pump inhibitors, acetaminophen, antiepileptics, immune modulators, blood-sugar medications, blood pressure medications, antidepressants, antipsychotics, diabetic medications, antihistamines, antibiotics, and anesthetics) due to 1,8-cineole content.[3057,3058,3059]

■ May interfere with pentobarbital and other barbiturates (medications for anxiety and insomnia), based on 1,8-cineole content.[3060,3061]

Borneol CTs

■ May interact with aspirin, blood pressure, antiplatelet, and anticoagulant medications, and increase the risk of bleeding among people with bleeding disorders, due to the anticoagulant properties of borneol.[3062,3063]

■ May interfere with enzymes responsible for metabolizing medications (NSAIDs, proton-pump inhibitors, acetaminophen, antiepileptics, immune modulators, blood-sugar medications, blood pressure medications, antidepressants, antipsychotics, diabetic medications, antihistamines, antibiotics, and anesthetics), based on borneol content.[3064,3065,3066,3067,3068]

SELECTED EVIDENCE:

- □ *In vitro* research demonstrates that the cinnamaldehyde CT of padang cassia essential oil inhibits biofilm formation and breaks down existing biofilms of *P. aeruginosa* and *S. aureus*.[3069]
- □ The cinnamaldehyde CT of padang cassia inhibited *E. coli, S. aureus,* and *C. albicans.*[3070] This antimicrobial activity was attributed to the presence of (E)-cinnamaldehyde.

PARSNIP
(Wild Parsnip)

Pastinaca sativa L., *P. sativa* L. var. *pratensis* Pers., *P. sativa* L. ssp. *sativa* L. var. *hortensis* Ehrh, *P. sativa* L. subsp. *urens* (Req. ex Godron) Celak

FAMILY: Apiaceae (Umbelliferae)
NOTE: Middle
AROMA INTENSITY: Medium
AROMA: Fresh, sweet, herbaceous, slightly spicy
COMMON EXTRACTION METHOD: Steam distilled or hydrodistilled from the roots; may also be distilled from the seeds or aerial parts
POSSIBLE SUBSTITUTE OILS: Nutmeg, ginger, safrole
BLENDS WELL WITH: Camphor, cassia, cinnamon, clary sage, copaiba, coriander, clove, eucalyptus, fennel, geranium, ginger, lavender, lime, nutmeg, orange, oregano, patchouli, pine, rosemary, turmeric, vetiver, ylang ylang
SUBCELLULAR LOCALIZATION | EPIGENETIC INFLUENCE: Currently unknown | Currently unknown
RECOMMENDED DILUTION RANGE: 3%–33%; 50% for some conditions

PRIMARY CONSTITUENTS:[3071,3072,3073,3074,3075]

Roots (Terpinolene CT)

Terpinolene	37.1%–69.0%
Myristicin	17.2%–44.4%
Beta-Pinene	2.4%–8.6%
(Z)-Beta-Ocimene	0.7%–3.7%
Limonene	1.7%–3.2%

Roots (Myristicin CT)

Myristicin	44.6%–66.2%
Terpinolene	25.3%–44.1%

Seeds

Octyl Butyrate	79.5%
Octyl Hexanoate	5.3%

Aerial Parts

(Z)-Beta-Ocimene	10.8%
Hexyl Butanoate	10.4%
(E)-Beta-Farnesene	6.1%
Lavandulyl Acetate	5.2%

Note: The composition of root essential oil changes during the course of plant development: seedling stage (0–18 days)—(E)-beta-farnesene and myristicin; secondary stage (23–30 days)—terpinolene and myristicin content increases as the sesquiterpene content decreases; and adult stage (38–160 days)—primarily terpinolene and myristicin.[3076]

OTHER CONSTITUENTS: *Roots (Terpinolene CT)*—Myrcene, gamma-terpinene, (E)-beta-ocimene; *Roots (Myristicin CT)*—Beta-bisabolene, beta-sesquiphellandrene, gamma-palmitolactone, (E)-beta-farnesene, monoterpene hydrocarbons; *Seeds (subsp. urens)*—Decanal, octanol, decyl butyrate, phenylethyl butyrate; *Aerial Parts*—Not listed in study

PREFERRED COMPOSITION FOR CLINICAL USE:

Constituent	*Roots*
Terpinolene	40%–65%
Myristicin	20%–40%
Beta-Pinene	2%–10%
(E)-Beta-Ocimene	0.1%–7%
Limonene	0.1%–7%

Reported Therapeutic Properties: Antimicrobial, antibacterial, antifungal, soothes chronic skin conditions, weight management, reduces blood pressure, aids cardiovascular function, supports respiratory function, eases cough, antiallergic, improves red blood cell production, aids circulation, analgesic, anti-inflammatory, reduces fever, insecticidal, relieves anxiety, reduces worrisome thoughts, combats negative emotions, antidepressant

Cautions:

Roots

- Use cautiously during pregnancy and lactation due to myristicin content. Animal research suggests that higher doses of myristicin (0.01 mL/kg every two days while nursing, and 0.001 to 10 mg while pregnant) taken orally during pregnancy and while nursing may cause liver toxicity in offspring born to the mother and increase the offspring's risk of liver cancer.[3077,3078]
- May interact with psychotropic medications—medications prescribed for anxiety, depression, schizophrenia, mania, obsessive-compulsive disorder, ADHD, and bipolar disorder due to myristicin content.[3079,3080]
- May interact with monoamine oxidase inhibitors (MAOIs)—medications prescribed for the treatment of depression—due to MAO inhibition by myristicin.[3081]
- May interfere with enzymes responsible for metabolizing medications (NSAIDs, proton-pump inhibitors, acetaminophen, antiepileptics, immune modulators, blood-sugar medications, blood pressure medications, antidepressants, antipsychotics, diabetic medications, antihistamines, antibiotics, and anesthetics).[3082,3083]

Seeds, Aerial Parts

- None currently known.

Selected Evidence:

- Parsnip aerial parts essential oil inhibited *B. cereus, E. coli, C. albicans, S. aureus, P. aeruginosa, S. enteritidis,* and *L. monocytogenes*.[3084]

PERILLA
(Korean Perilla, Beefsteak Plant)

Perilla frutescens (L.) Britton, *P. frutescens* (L.) Britton var. *crispa* f. *viridis, P. frutescens* Britton var. *japonica*

Family: Lamiaceae (Labiatae)
Note: Top-Middle
Aroma Intensity: Strong
Aroma: Spicy, green, herbaceous, slightly woody and fruity
Common Extraction Method: Steam distilled or hydrodistilled from the aerial parts; may also be distilled from the inflorescences or leaves
Possible Substitute Oils: Mugwort, cumin, nutmeg
Blends Well With: Angelica root, black pepper, caraway, chamomile (German, Roman), coriander, cumin, lavender, nutmeg, rosemary, turmeric
Subcellular Localization | Epigenetic Influence: Currently unknown | Currently unknown
Recommended Dilution Range: 1%–10%

PRIMARY CONSTITUENTS:[3085,3086,3087,3088,3089,3090,3091,3092,3093]

Perillaldehyde CT

Perillaldehyde	36.7%–86.8%
Limonene	1.1%–23.8%
Caryophyllene Oxide	0.0%–12.6%
(Z,E)-Alpha-Farnesene	0.0%–11.5%
(Z)-Dihydrocarveol	0.0%–9.5%
Beta-Caryophyllene	0.0%–7.8%
Perill Alcohol	0.0%–5.4%
Thymol	0.0%–4.7%
Linalool	1.1%–4.6%
Phytol	0.0%–4.6%

Perillaketone CT

Perillaketone	17.5%–69.2%
Isoegomaketone	1.0%–45.7%
Egomaketone	0.0%–29.8%
Beta-Caryophyllene	0.1%–18.0%
Linalool	0.2%–5.0%
Germacrene D	0.0%–4.3%
(Z,E)-Alpha-Farnesene	0.0%–3.9%
Alpha-Humulene	0.0%–3.7%
Caryophyllene Oxide	0.0%–3.4%
1,8-Cineole	0.0%–3.3%
Isobornyl Formate	0.0%–3.2%

var. japonica CT Beta-Caryophyllene

Beta-Caryophyllene	24.1%
Unidentified (RT69.755)	16.1%
Unidentified (RT47.737)	16.0%
Unidentified (RT80.487)	12.4%
Beta-Farnesene	11.4%
Unidentified (RT70.805)	6.8%

var. crispa f. viridis

Perrilaldehyde	49.5%–66.7%
Limonene	11.8%–15.4%
Cis-Limonene Oxide	0.0%–6.8%
Trans-Limonene Oxide	0.0%–3.0%

var. japonica CT Perillaketone

Perillaketone	17.6%–54.2%
Unidentified Compound	4.2%–49.2%
Beta-Caryophyllene	7.8%–24.3%
Elemicin	0.3%–21.3%
Myristicin	5.7%–20.2%
Apiole	0.0%–14.7%
(E)-Alpha-Bergamotene	1.1%–3.0%

var. japonica CT Myristicin

Myristicin	33.6%–37.8%
Perillaketone	21.0%–37.0%
Beta-Caryophyllene	7.3%–23.4%
(E)-Alpha-Bergamotene	2.3%–6.6%
Unidentified Compound	3.4%–6.4%
Elemicin	0.4%–3.0%
Perrilaldehyde	0.2%–3.0%

Note: *P. frutescens* var. *japonica* composition varies widely and several CTs with various unidentified compounds have been reported.[3094] Isoegomaketone may be the dominant compound in perillaketone CTs.

OTHER CONSTITUENTS: *Perilladehyde CT*—Alpha-pinene, beta-pinene, trans-linalool oxide, cis-linalool oxide, alpha-humulene, (E,E)-alpha-farnesene, (E)-beta-ionone, perilla alcohol, (E)-nerolidol, humulene epoxide II, hexahydrofarnesyl acetone, spathulenol; *Perillaketone CT*—Beta-muurolene, perillene, 1-octen-3-ol, alpha-copaene, benzaldehyde, humulene epoxide II, hexahydrofarnesylacetone, spathulenol, 1-(2-furanyl)-2-butanone; *var. japonica CT Perillaketone*—Trans-2-hexenal, 6-methyl-5-hepten-2-one, linalool, 1-octen-3-ol, alpha-humulene, beta-farnesene, germacrene D, perillaldehyde, phytol, eugenol; *var. japonica CT Myristicin*—Alpha-humulene, beta-farnesene, delta-elemene, germacrene D, phytol; *var. japonica CT Beta-Caryophyllene*—Alpha-humulene, unidentified constituents; *var. crispa f. viridis*—Myrcene, trans-furanoid, cis-pyranoid, terpinolene

PREFERRED COMPOSITION FOR CLINICAL USE:

Constituent	
Perillaldehyde	40%–75%
Limonene	5%–25%
Beta-Caryophyllene	0.5%–10%
Linalool	0.5%–7%

REPORTED THERAPEUTIC PROPERTIES: Antioxidant, antibacterial, antifungal, antiparasitic, analgesic, anti-inflammatory, antiarthritic, eases cough, relieves asthma, aids digestion, eases nausea/vomiting, relieves constipation, eases stomachache, anticancer, aids cardiovascular health, reduces high cholesterol, insecticidal, insect repellent, antidepressant, relieves anxiety, stress management

CAUTIONS:

Perrilaldehyde CTs

■ Dilution strongly recommended, and avoidance by people with sensitive skin is recommended. Perilla workers are highly prone to skin irritation by perilla and those with dermatitis are highly likely to experience allergic reactions.[3095]

Perillaketone CTs

■ This CT is not recommended for therapy due to the presence of ketones (perillaketone, isoegomaketone, and egomaketone) that have caused pulmonary toxicity and deaths in animals.[3096,3097,3098,3099]

var. japonica CT Myristicin

■ This CT is not recommended due to the presence of perillaketone, which is associated with pulmonary toxicity and deaths in animals.[3100,3101,3102,3103] The possible presence of apiole is also concerning and requires additional cautions during pregnancy, while lactating, in children under age 12, for oral use, and with medications.

■ Use cautiously during pregnancy and lactation due to myristicin content. Animal research suggests that higher doses of myristicin (0.01 mL/kg every two days while nursing, and 0.001 to 10 mg while pregnant) taken orally during pregnancy and while nursing may cause liver toxicity in offspring born to the mother and increase the offspring's risk of liver cancer.[3104,3105]

■ May interact with psychotropic medications—medications prescribed for anxiety, depression, schizophrenia, mania, obsessive-compulsive disorder, ADHD, and bipolar disorder due to myristicin content.[3106,3107]

■ May interact with monoamine oxidase inhibitors (MAOIs)—medications prescribed for the treatment of depression—due to MAO inhibition by myristicin.[3108]

■ May interfere with enzymes responsible for metabolizing medications (NSAIDs, proton-pump inhibitors, acetaminophen, antiepileptics, immune modulators, blood-sugar medications, blood pressure medications, antidepressants, antipsychotics, diabetic medications, antihistamines, antibiotics, and anesthetics).[3109,3110]

var. japonica CT Beta-Caryophyllene

■ This CT is uncommon and not recommended for therapy due to a significant amount of its composition not identified.

SELECTED EVIDENCE:

- ☐ Administration of perilla oil once daily for five consecutive weeks reduced mild stress-induced depression by reversing alterations in serotonin and inflammation in mice.[3111] Another study concluded that perillaldehyde was responsible for the antidepressant effects of perilla essential oil.[3112]

- Perilla essential oil restored brain-derived neurotrophic factor (BDNF) expression that were reduced after stress in mice.[3113]
- *In vitro* research demonstrated that perilla essential oil reduced exotoxin production by *S. aureus* even when used at subinhibitory concentrations.[3114] Microencapsulated perilla essential oil demonstrated antibacterial activity to preserve strawberries.[3115]
- The perillaketone CT of perilla essential oil exhibited good antioxidant activity in the DPPH assay.[3116]
- Both green and red perilla essential oil (perillaldehyde CT) deterred biting by the *Aedes aegypti* mosquito, with green being more effective.[3117]
- Application of 1 mg/mL of perilla essential oil in pear orchards controlled insect (*Cacopsylla chinensis*) infestations.[3118] Perilla oil was the most effective of twenty-six essential oils screened in the laboratory, which spurred the researchers to conduct the field test.

PHOEBE (Imbuia, Brazilian Walnut)

Ocotea porosa (Nees & Martius) Barroso, *Phoebe porosa* (Nees & Mart.) Mez, *Oreodaphne porosa* Nees & Mart.

FAMILY: Lauraceae
NOTE: Middle-Base
AROMA INTENSITY: Medium
AROMA: Smoky, woody
COMMON EXTRACTION METHOD: Steam distilled or hydrodistilled from the wood chips
POSSIBLE SUBSTITUTE OILS: Agarwood, equal parts valerian and German chamomile
BLENDS WELL WITH: Agarwood, balsam fir, cassia, cedarwood, chamomile (German, Roman), cinnamon, copaiba, cypress, frankincense, geranium, hinoki, lavender, myrrh, patchouli, pine, rose, sandalwood, spikenard, spruce (black), white fir
SUBCELLULAR LOCALIZATION | EPIGENETIC INFLUENCE: Currently unknown | Currently unknown
RECOMMENDED DILUTION RANGE: 3%–20%

PRIMARY CONSTITUENTS:[3119,3120]

Eremoligenol	8.4%–8.8%
Beta-Eudesmol	6.9%–8.4%
Valerianol	5.0%–7.6%
Alpha-Copaene	5.6%–6.3%
Alpha-Bisabolol	3.3%–3.6%
Gamma-Cadinene	1.2%–3.6%
Delta-Cadinene	3.1%–3.3%
Hinesol	1.9%–3.2%
Beta-Bisabolol	2.5%–2.9%
Alloaromadendrene	2.4%–2.6%
Eremophila-9,10-en-11-ol	1.6%–2.5%
(E)-Alpha-Bergamotene	1.5%–2.5%
Beta-Curcumene	0.0%–2.5%

OTHER CONSTITUENTS: Alpha-eudesmol, 7-epi-delta-eudesmol, agarospirol, 10-epi-gamma-eudesmol, carquejyl acetate, sesquithujene, epi-sesquithujene, viridiflorol, alpha-clacorene, alpha-humulene, gamma-muurolene, alpha-pinene, calamenene, (E)-beta-farnesene, 7-epi-alpha-eudesmol, beta-selinene, beta-caryophyllene, caryophyllene oxide, ledol, 1,10-aristolene, beta-bisabolene, alpha-cadinol, humulene epoxide I, alpha-selinene, safrole, beta-pinene, para-cymene, sesquisabinene, germacrene D, beta-elemol, gamma-amorphene, alpha-agarofuran, humulene epoxide II, 1-epi-cubenol, porosadienone

PREFERRED COMPOSITION FOR CLINICAL USE:
Insufficient data/samples to determine.

REPORTED THERAPEUTIC PROPERTIES: Wound healing, antioxidant, improves circulation, antimicrobial, antibacterial, antifungal, anti-inflammatory, antirheumatic, analgesic, antispasmodic, aphrodisiac, reduces hysteria, balances the emotions, relieves anxiety

CAUTIONS:
- None currently known.

SELECTED EVIDENCE:
- None found.

PINK AND BLUE GINGER (Neelakua, Kamindam)

Curcuma aeruginosa Roxb.

FAMILY: Zingiberaceae
NOTE: Base
AROMA INTENSITY: Medium
AROMA: Earthy, slightly spicy
COMMON EXTRACTION METHOD: Hydrodistilled or steam distilled from the rhizomes
POSSIBLE SUBSTITUTE OILS: Turmeric, myrrh, fingerroot, ginger
BLENDS WELL WITH: Cinnamon, cistus, clary sage, clove, ginger, helichrysum, juniper, lavender, nutmeg, ylang ylang
SUBCELLULAR LOCALIZATION | EPIGENETIC INFLUENCE: Currently unknown | Currently unknown
RECOMMENDED DILUTION RANGE: 5%–50%; neat for some conditions

PRIMARY CONSTITUENTS:[3121,3122,3123,3124,3125,3126,3127]

Germacrone CT	
Germacrone	23.2%–23.5%
Furanodiene	0.0%–15.3%
1,8-Cineole	10.9%–14.4%
Curzerenone	0.0%–11.8%
Beta-Pinene	4.0%–9.1%
Delta-Cadinene	0.0%–9.1%
Ar-Curcumene	0.0%–6.7%
Dehydrocurdione	0.0%–3.8%
Cis-Beta-Guaiene	0.0%–3.4%

Cycloisolongifolene, 8,9-dehydro-9-formyl- CT	
Cycloisolongifolene, 8,9-dehydro-9-formyl-	35.3%
Dihydrocostunolides	22.5%
Valleral	10.0%
Beta-Elemene	4.8%
Alloaromadendrene Oxide-(2)	4.1%
1,8-Cineole	4.0%

Camphor CT	
Camphor	29.4%
Germacrone	21.2%
Germacrene B	5.2%
Curzerene	4.9%
Borneol	7.3%

Curzerenone CT	
Curzerenone	24.6%
1,8-Cineole	11.2%
Camphor	10.5%
Zedoarol	6.4%
Isocurcumenol	5.8%
Curcumenol	5.6%
Furanogermenone	5.5%
Isoborneol	3.2%

Curcumanolides A,B CT	
Curcumanolides	11.4%
Curcumenol	9.9%
Dehydrocurdione	9.4%
Isocurcumenol	8.5%
Beta-Eudesmol	6.5%
Limonene	6.0%
1,8-Cineole	6.0%
Camphor	5.9%
Curdione	3.6%

Curcumenol CT	
Curcumenol	38.7%
Beta-Pinene	27.5%
Beta-Eudesmol	3.6%

OTHER CONSTITUENTS: *Germacrone CT*—Alpha-pinene, beta-elemene, trans-caryophyllene, gamma-muurolene, beta-selinene, alpha-muurolene, beta-bisabolene, alpha-bulnesene, gamma-cadinene, germacrene B, spathulenol, beta-atlantol, T-cadinol, alpha-cadinol, 8-hydroxy-eremophilone; *Camphor CT*—Camphene, 1,8-cineole, endo-borneol, beta-elemene, alpha-humulene, germacrene A; *Cycloisolongifolene, 8,9-dehydro-9-formyl- CT*—Camphor, caryophyllene, beta-farnesene, alpha-bulnesene, beta-cubebene, eudesma-4(14),11-diene, aromadendrene oxide-(2), xanthinin; *Curcumanolides A,B CT*—Camphene, nonan-2-ol, borneol, isoborneol, beta-elemene, alpha-guaiene, ar-curcumene, terpinene-4-ol, Selina-3,7(11)-diene, beta-caryophyllene oxide, curzerenone, humuladienone, germacrone, curcumenone; *Curcumenol CT*—Limonene, beta-farnesene, zingiberene, (Z,Z)-farnesol; *Curzerenone CT*—Borneol, terpinene-4-ol, beta-caryophyllene, alpha-selinene, beta-selinene, furanodienone, germacrone

PREFERRED COMPOSITION FOR CLINICAL USE:
Insufficient samples/data to determine.

REPORTED THERAPEUTIC PROPERTIES: Anticancer, antimicrobial, antibacterial, antifungal, antioxidant, antiparasitic, wound healing, relieves chronic skin conditions, relieves diarrhea, expels excess gas, alleviates heartburn, purifies the blood, relieves painful menstruation, supports respiratory function, eases cough, aids detoxification, promotes mental clarity, aids cognitive function, balances the emotions

CAUTIONS:
Camphor CT

- Avoid with children under six due to significant camphor content. Several cases of camphor poisoning and/or seizures from ingestion and topical application have been reported in children.[3128,3129] Ingestion of camphor-containing products has been lethal in children under age two.[3130] Children five years and up may use camphor-containing essential oils topically in dilutions no stronger than 5%. Camphor ingestion by infants and young children may cause cough, vomiting, seizure, burning sensation in the mucous membranes and eyes, or lack of voluntary coordination of muscle movements.[3131]
- Caution is warranted during pregnancy and while lactating, due to significant camphor content. Ingestion of essential oils with significant levels of camphor may lead to abortion because fetuses lack the enzymes to process it.[3132] Camphor ingestion by infants and young children may cause cough, vomiting, seizure, burning sensation in the mucous membranes and eyes, or lack of voluntary coordination of muscle movements.[3133]

■ Avoid oral consumption due to significant camphor levels. Essential oils with significant levels of camphor may be toxic when taken orally, usually single doses exceeding 2 mL, although the lethal dose for humans is estimated to be greater than 5 mL in a single dose.[3134,3135,3136]
■ Avoid with epilepsy and Parkinson's disease due to the potential of camphor and thujones to exacerbate or cause seizures/convulsions or reduce seizure-medication efficacy.[3137,3138,3139,3140]
■ Avoid with those who have a compromised liver due to the risk of increased liver enzymes and liver damage based on camphor content.[3141] This would also suggest that those taking medications that could cause liver damage should also use essential oils with significant camphor levels very cautiously or avoid them entirely.
■ The significant camphor content may negatively impact red blood cells and increase the risk of jaundice in children with glucose-6-phosphate dehydrogenase deficiency (G6PD).[3142,3143]
■ Caution is warranted for the topical application of camphor-containing essential oils on broken or injured skin due to an increased risk of toxicity.[3144]

Germacrone CT
■ May interfere with the enzymes responsible for metabolizing medications (NSAIDs, proton-pump inhibitors, acetaminophen, antiepileptics, immune modulators, blood-sugar medications, blood pressure medications, antidepressants, antipsychotics, diabetic medications, antihistamines, antibiotics, and anesthetics) due to germacrone content.[3145]

All other CTs
■ None currently known.

SELECTED EVIDENCE:

- □ Both a 1% and 5% pink and blue ginger essential oil lotion (applied twice daily for 5 to 11 weeks) prevented unwanted hair growth and brightened the skin in women.[3146] The skin brightening effect occurred within three weeks of use and lasted two weeks after discontinuation of use.
- □ Pink and blue ginger (Camphor CT) essential oil demonstrated greater antimicrobial activity against gram-positive bacteria (*Staphylococcus aureus, Bacillus cereus*) than gram-negative bacteria (*Escherichia coli, Pseudomonas aeruginosa*).[3147] The Cycloisolongifolene, 8,9-dehydro-9-formyl- CT inhibited gram-positive bacteria (*S. aureus, B. cereus, P. aeruginosa*) and fungi (*C. neoformans, C. albicans*), but was ineffective against gram-negative bacteria (*E. coli*).[3148]
- □ The germacrone CT of pink and blue ginger essential oil demonstrated antioxidant activity comparable to vitamin C (ascorbic acid) in the DPPH and OH radicals test, but poorer activity in the H2O2 test.[3149] The same researchers reported that it was moderately active against *Enterococcus faecalis* and weakly active against *Mycobacterium tuberculosis*.

PINK SAVORY (Thyme-Leaved Savory, Savory of Crete)

Satureja thymbra L., *S. collina* Salisb., *S. biroi* Jav., *S. tragoriganum* (L.), Tausch, *Clinopodium thymbra* (L.) Kuntze, *Micromeria thymbra* (L.) Kostel., *Thymbra hirsuta* Pers., *T. hirsutissima Vent. ex Pers., Thymus hirsutissimus Poir., Th. tragoriganum L.*

Family: Lamiaceae (Labiatae)
Note: Middle
Aroma Intensity: Strong
Aroma: Herbaceous, spicy, medicinal
Common Extraction Method: Steam distilled or hydrodistilled from the aerial parts; may also be distilled from the branches or leaves
Possible Substitute Oils: Oregano, thyme
Blends Well With: Bergamot, camphor, cedarwood, chamomile (German, Roman), cinnamon, clove, cypress, eucalyptus, frankincense, ginger, lavender, lemon, marjoram, melaleuca (tea tree), mountain savory, myrrh, orange, oregano, pine, rosemary, summer savory, thyme, savory (wild)
Subcellular Localization | Epigenetic Influence: Currently unknown | Currently unknown
Recommended Dilution Range: 1.5%–20%; 50% for some conditions

Primary Constituents:[3150,3151,3152,3153,3154,3155,3156,3157,3158,3159,3160,3161,3162,3163,3164,3165]

Carvacrol CT

Carvacrol	27.6%–66.5%
Gamma-Terpinene	9.7%–32.4%
Thymol	0.0%–19.9%
Para-Cymene	5.3%–16.4%
Beta-Caryophyllene	tr–10.3%
Thymol Methyl Ether	0.0%–4.4%
Borneol	0.0%–3.3%
Delta-3-Carene	0.0%–3.3%
Myrcene	0.0%–3.2%

Gamma-Terpinene CT

Gamma-Terpinene	25.1%–41.0%
Thymol	13.2%–25.2%
Carvacrol	4.2%–23.1%
Para-Cymene	7.2%–14.7%
Beta-Caryophyllene	2.8%–6.8%
Alpha-Terpinene	0.0%–3.5%
Caryophyllene Oxide	0.0%–3.5%
Carvacrol Methyl Ether	0.0%–3.3%

Leaves

Carvacrol	40.2%
Gamma-Terpinene	26.6%
Para-Cymene	16.4%
Thymol	13.2%

Thymol CT

Thymol	20.7%–65.6%
Gamma-Terpinene	4.4%–31.3%
Para-Cymene	7.2%–18.2%
Carvacrol	1.4%–12.8%
Beta-Caryophyllene	1.7%–8.1%
Caryophyllene Oxide	0.0%–4.9%
Terpinen-4-ol	tr–4.6%
Thymol Methyl Ether	0.0%–4.3%
Borneol	tr–4.0%
Limonene	0.0%–3.7%
Myrcene	0.8%–3.6%
Alpha-Terpinene	0.0%–3.1%

Branches

Carvacrol	24.3%
Gamma-Terpinene	22.2%
Para-Cymene	17.2%
Caryophyllene Oxide	14.6%
Thymol	8.2%
Beta-Caryophyllene	5.2%
Carvacrol Methyl Ether	4.2%

Other Constituents: *Carvacrol CT*—Alpha-thujene, alpha-pinene, beta-pinene, limonene, trans-sabinene hydrate, terpinen-4-ol, alpha-terpineol, caryophyllene oxide, (Z)-beta-ocimene, (E)-beta-ocimene, germacrene D, beta-bisabolene, aromadendrene, alpha-humulene; *Thymol CT*—Alpha-thujene, alpha-pinene, camphene, beta-pinene, trans-sabinene hydrate, cis-sabinene hydrate, alpha-terpinolene, linalool,

1,8-cineole, (Z)-beta-ocimene, eugenol, germacrene D, bicyclogermacrene, beta-bisabolene; *Gamma-Terpinene CT*—Alpha-thujene, alpha-pinene, camphene, myrcene, beta-pinene, limonene, ocimene, linalool, terpinen-4-ol, thymol methyl ether, alpha-bisabolene, 1,8-cineole, cis-sabinene hydrate, borneol, eugenol, germacrene D, bicyclogermacrene, beta-bisabolene; *Leaves*—Beta-pinene, alpha-terpinene, beta-caryophyllene; *Branches*—Alpha-pinene, beta-pinene

PREFERRED COMPOSITION FOR CLINICAL USE:

Constituent	*Carvacrol CT*	*Thymol CT*
Carvacrol	35%–55%	4%–20%
Gamma-Terpinene	10%–30%	10%–30%
Thymol	5%–20%	35%–60%
Para-Cymene	5%–17%	5%–18%
Beta-Caryophyllene	tr–10%	2%–10%
Myrcene	–	0.1%–5%
Terpinen-4-ol	–	tr–5%
Borneol	–	tr–5%

REPORTED THERAPEUTIC PROPERTIES: Antibacterial, antifungal, **antimicrobial**, antiviral, anti-infectious, antiseptic, antiparasitic, enhances circulation, aids immune function, anti-inflammatory, analgesic, antispasmodic, antirheumatic, eases muscle cramps, relieves diarrhea, expels excess gas, relieves nausea and indigestion, aids respiratory function, relieves asthma and bronchitis, reduces high cholesterol, relieves gingivitis, insecticidal, insect repellent, aphrodisiac, reduces fear, promotes a sense of security, antidepressant

CAUTIONS:

Carvacrol CT

■ Avoid during pregnancy and lactation. Oregano (also rich in carvacrol) may negatively affect embryonic development and encourage fetal cell death according to animal research.[3166]

■ May interact with diabetic medications and cause low blood-sugar levels due to carvacrol content—carvacrol reduces blood glucose and enhances the antidiabetic activities of rosiglitazone in animals.[3167,3168,3169,3170]

■ May interact with aspirin, blood pressure, antiplatelet, and anticoagulant medications, and increase the risk of bleeding among people with bleeding disorders, due to carvacrol and thymol content.[3171,3172,3173,3174,3175]

■ May interact with anticholinergic (drugs used for asthma, incontinence, gastrointestinal cramps, muscular spasms, depression, and sleep disorders) and cholinergic medications (drugs used to reduce urinary retention, diagnose myasthenia gravis, and for glaucoma), based on thymol and carvacrol content.[3176,3177]

■ May interact with antibiotics or antifungals and possibly enhance or decrease their effects due to carvacrol or thymol content.[3178,3179,3180,3181,3182]

■ May irritate mucous membranes (eyes, mouth, nasal passages, vagina, rectum).

Thymol CTs

■ Oral caution. Thymol has a longer half-life (the time it takes for half of the medication to metabolize or excrete half of the dosage) than most essential oil constituents and should not be administered orally for long periods of time.[3183] Thymol is a monoterpene phenol. Reports of fatalities in children who consumed 50 to 200 mg of phenols has been reported.[3184] It is recommended that chemotypes with high levels of thymol be limited to 10 drops per day orally for adults with a two- to seven-day break after twenty-one days of use; or limit ingestion to five or six days weekly.

■ May interact with aspirin, blood pressure, antiplatelet, and anticoagulant medications, and increase the risk of bleeding among people with bleeding disorders, due to thymol and carvacrol content.[3185,3186]
■ May interact with anticholinergic (drugs used for asthma, incontinence, gastrointestinal cramps, muscular spasms, depression, and sleep disorders) and cholinergic medications (drugs used to reduce urinary retention, diagnose myasthenia gravis, and for glaucoma), based on thymol and carvacrol content.[3187,3188]
■ May interact with antibiotics or antifungals and possibly enhance their effects based on thymol and carvacrol content.[3189,3190,3191,3192,3193,3194,3195,3196]
■ May irritate mucous membranes (eyes, mouth, nasal passages, vagina, rectum).

Gamma-Terpinene CT
■ Follow cautions for both carvacrol and thymol CTs.

SELECTED EVIDENCE:

- □ The carvacrol CT of pink savory essential oil showed antiproliferative capacity against breast cancer cells. Carvacrol also exhibited significant activity against non-small cell lung cancer cells.[3197]
- □ Pink savory CT carvacrol inhibited both acetylcholinesterase (AChE) and butylcholinesterase (BChE) activity, with thymol, carvacrol, and gamma-terpinene also inhibiting AChE and BChE.[3198]
- □ Animal research showed that pink savory CT gamma-terpinene essential oil exhibits central analgesic activity but does not exert anti-inflammatory activity in mice and rats.[3199]
- □ *In vitro* research demonstrated that pink savory CT carvacrol essential oil inhibited *E. coli, S. aureus, P. aeruginosa, E. aerogenes,* and *C. albicans*.[3200] The carvacrol CT also inhibited *E. coli, P. aeruginosa, S. typhimurium, Sh. sonnei, S. aureus*, and *C. albicans* in other research.[3201] Another study found that pink savory essential oil (unidentified CT at 1%) presented sufficient bactericidal effect against bacterial (*S. simulans, L. fermentum, P. putida, S. enterica, P. putida*) biofilms formed on stainless steel surfaces, which exceeded the activity of acid-based disinfectants.[3202] Pink savory essential oil (possibly thymol CT) was also highly effective against food spoilage pathogens (*Ps. fragi, E. coli*) grow in in biofilm.[3203] The thymol CT inhibited *S. aureus, B. cereus, M. flavus, L. monocyotgenes, P. aeruginosa, E. faecalis, S. typhimurium*, and *E. coli*.[3204] Carvacrol exhibited greater activity when carvacrol and thymol were tested individually. The thymol CT demonstrated greater activity than the fungicide bifonazole, and was fungistatic, bacteriostatic, and bactericidal.[3205] Pink savory essential oil was more effective than *Thymus* and *Origanum* species against *E. coli, S. enteritidis, S. aureus, L. monocytogenes,* and *B. cereus*.[3206] The thymol CT (supercritical and hydrodistilled oils) demonstrated broad-spectrum antifungal activity when tested against (*Candida albicans, C. tropicalis, C. krusei, C. guillermondii, C. parapsilosis, Cryptococcus neoformans, Trichophyton rubrum, T. mentagrophytes, T. mentagrophytes var. interdigitale, Trichophyton rubrum, T. verrucosum, Microsporum canis, M. gypseum, Epidermophyton floccosum, Aspergillus niger, A. fumigatus*, and *A. flavus*).[3207] Pink savory essential oil (unknown composition) inhibited the growth of *L. monocytogenes* and *S. enteritidis* at 0.1%.[3208] An unidentified CT of pink savory essential oil inhibited gram-positive (*B. subtilis, M. luteus, S. mutans, S. aureus, S. epidermidis*) and gram-negative (*E. coli, P. aeruginosa, P, fluorescens, P. stutzeri, S. maltophilia, C. luteola*) bacteria, including multidrug-resistant strains, and the yeast *C. albicans*.[3209]
- □ Pink savory CT thymol was effective against the protozoan parasite *Plasmodium falciparum*, which is a deadly species that causes malaria in humans.[3210]
- □ The carvacrol CT of pink savory essential oil exhibited antioxidant activity in the beta-carotene, ABTS, and DPPH assays (in descending order of activity).[3211] Carvacrol, thymol, and gamma-

terpinene also exhibited good antioxidant activity. Pink savory CT carvacrol exhibited strong antioxidant activity in the DPPH assay.[3212] The study authors also noted that thymol exhibited a twofold greater antioxidant capacity when compared to carvacrol.

- Of 14 Lamiaceae essential oils tested, pink savory (equal carvacrol and gamma-terpinene) was one of the most effective repellents against the *Aedes albopictus* mosquito (providing total protection against bites), and weakly killed larvae as well.[3213] Another study reported that the carvacrol CT of pink savory essential oil was significantly larvicidal to *Culex pipiens* mosquito larvae.[3214] Pink savory CT carvacrol was more toxic to the Vine Mealybug (*Planococcus ficus*) than the reference product tested.[3215] The thymol CT was toxic to larvae and adults of *Anopheles gambiae* mosquitoes.[3216] Pink savory CT carvacrol was highly effective against the Indianmeal moth (*P. interpunctella*) and the Mediterranean flour moth (*E. kuehniella*).[3217] Pink savory CT carvacrol essential oil applied to cotton wicks completely killed adult ticks (*Hyalomma marginatum*).[3218]

PINNATA LAVENDER (Jagged Lavender)

Lavandula pinnata Lundmark,
L. pinnata L. fil. var. *pinnata*

FAMILY: Lamiaceae (Labiatae)
NOTE: Middle
AROMA INTENSITY: Medium
AROMA: Herbaceous, floral, pungent, slightly spicy and minty
COMMON EXTRACTION METHOD: Hydrodistilled from the flowering tops, leaves and stems, or flowers and leaves
POSSIBLE SUBSTITUTE OILS: Betany, lovage, goldenrod; oregano, summer savory CT carvacrol
BLENDS WELL WITH: Bergamot, betony, copaiba, frankincense, goldenrod, ginger, hemp, hyssop, lavender, lemon, lime, marjoram, orange, patchouli, sage, Spanish sage, vetiver
SUBCELLULAR LOCALIZATION | EPIGENETIC INFLUENCE: Currently unknown | Currently unknown
RECOMMENDED DILUTION RANGE: *Beta-Phellandrene CT*—3%–33%, neat for some conditions; *Carvacrol CT*—1.5%–20%, 50% for some conditions

PRIMARY CONSTITUENTS:[3219,3220,3221]

Carvacrol CT		*Beta-Phellandrene CT*	
Carvacrol	27.0%–83.7%	Beta-Phellandrene	12.2%–31.7%
Caryophyllene Oxide	0.0%–14.6%	Alpha-Phellandrene	6.3%–15.9%
Bisabolene Isomers	0.0%–13.6%	Beta-Caryophyllene	10.5%–11.4%
Alpha-Terpinolene	0.0%–12.8%	Phenylacetaldehyde	1.3%–9.2%
Spathulenol	0.0%–12.2%	(Z)-Beta-Ocimene	4.2%–8.9%
(Z)-Beta-Ocimene	0.0%–11.0%	Alpha-Pinene	3.5%–6.5%
Farnesene Isomers	0.0%–7.0%	Octen-1-en-3-ol	0.7%–6.0%
Benzeneacetaldehyde	0.0%–6.4%	Sabinene	3.1%–5.7%
Myrcene	0.0%–5.8%	Limonene	1.8%–4.0%
Beta-Caryophyllene	0.0%–3.7%		

Note: Pinnata lavender is a rare species native to southern Madeira and the Canary Islands.

OTHER CONSTITUENTS: *Beta-Phellandrene CT*—Alpha-thujene, benzaldehyde, camphene, beta-pinene, beta-myrcene, delta-3-carene, para-cymene, (E)-beta-ocimene, Artemisia ketone, gamma-terpinene, n-octanol, n-nonanal, linalool, camphor, trans-p-menth-2-en-ol, allo-ocimene, naphthalene, terpinen-4-ol, alpha-terpineol, n-decanal, n-dodecane, alpha-humulene, myristicin, alpha-farnesene, delta-cadinene, alpha-calacorene, caryophyllene oxide, beta-eudesmol, alpha-bisabolol; *Carvacrol CT*—Benzaldehyde, cis-linalool oxide, camphor, borneol, iso-neomenthol, alpha-terpineol, carvacrol methyl ether, thymol methyl ether, p-7-cymenol, beta-damascenone

PREFERRED COMPOSITION FOR CLINICAL USE:
Insufficient samples/data to determine.

REPORTED THERAPEUTIC PROPERTIES: Antibacterial, antifungal, antimicrobial, antiviral, antiparasitic, analgesic, anti-inflammatory, antirheumatic, anticancer, supports respiratory function, wound healing, antioxidant, aids circulation, balances female hormones, reduces painful menstruation, aids digestion, expels excess gas, nervine, antineuralgic, relieves headache/migraine, purification, relieves insect bites/stings, insecticidal, insect repellent, stress management, relieves anxiety, combats nightmares, calming/relaxing

CAUTIONS:

Beta-Phellandrene CT

▪ May interact with anticholinergic (drugs used for asthma, incontinence, gastrointestinal cramps, muscular spasms, depression, and sleep disorders) and cholinergic medications (drugs used to reduce urinary retention, diagnose myasthenia gravis, and for glaucoma), based on AChE-inhibitory activity of beta-phellandrene.[3222]

Carvacrol CT

▪ Avoid during pregnancy and lactation. Oregano (also rich in carvacrol) may negatively affect embryonic development and encourage fetal cell death according to animal research.[3223]

▪ May interact with diabetic medications and cause low blood-sugar levels due to carvacrol content—carvacrol reduces blood glucose and enhances the antidiabetic activities of rosiglitazone in animals.[3224,3225,3226,3227]

▪ May interact with aspirin, blood pressure, antiplatelet, and anticoagulant medications, and increase the risk of bleeding among people with bleeding disorders, due to carvacrol and thymol content.[3228,3229,3230,3231,3232]

▪ May interact with anticholinergic (drugs used for asthma, incontinence, gastrointestinal cramps, muscular spasms, depression, and sleep disorders) and cholinergic medications (drugs used to reduce urinary retention, diagnose myasthenia gravis, and for glaucoma), based on thymol and carvacrol content.[3233,3234]

▪ May interact with antibiotics or antifungals and possibly enhance or decrease their effects due to carvacrol or thymol content.[3235,3236,3237,3238,3239]

▪ May irritate mucous membranes (eyes, mouth, nasal passages, vagina, rectum).

SELECTED EVIDENCE:

- None found.

POET'S NARCISSUS ABSOLUTE
(Narcissus Absolute)

Narcissus poeticus L.

FAMILY: Amaryllidaceae/Liliaceae
NOTE: Middle-Base
AROMA INTENSITY: Strong
AROMA: Floral, sweet, green, herbaceous
COMMON EXTRACTION METHOD: Solvent extracted from the flowers
POSSIBLE SUBSTITUTE OILS: Jasmine, ylang ylang
BLENDS WELL WITH: Balsam fir, bergamot, blue tansy, cassia, cedarwood, chamomile (Roman, German), chastetree, cinnamon, clary sage, clove, copaiba, davana, frankincense, geranium, jasmine absolute, juniper berry, lavender, lemon, myrrh, patchouli, rose, rosewood, sandalwood, tamala, tangerine, ylang ylang
SUBCELLULAR LOCALIZATION | EPIGENETIC INFLUENCE: Currently unknown | Currently unknown
RECOMMENDED DILUTION RANGE: 1.0%–3.0%

PRIMARY CONSTITUENTS:[3240,3241,3242]

Gamma-Terpinene CT	
Gamma-Terpinene	10.3%–27.2%
(E)-Methyl Cinnamate	1.1%–15.8%
Linalool	0.6%–11.7%
Benzyl Acetate	9.5%–9.6%
Benzyl Benzoate	0.7%–8.9%
Para-Cymene	1.3%–8.5%
Delta-3-Carene	6.6%–8.4%
Alpha-Terpineol	0.3%–7.0%
Alpha-Pinene	2.4%–6.1%
(E)-Ethyl Cinnamate	0.5%–5.6%
Benzyl Alcohol	1.7%–4.8%
Limonene	2.1%–3.3%
1,8-Cineole	1.5%–3.0%

Alpha-Terpineol CT	
Alpha-Terpineol	23.7%
(E)-Methyl Isoeugenol	20.0%
Benzyl Benzoate	19.4%
Coumarin	6.9%
Benzyl Alcohol	5.0%
Delta-3-Carene	3.4%

Benzyl Benzoate CT	
Benzyl Benzoate	19.0%
Phenylethyl Alcohol	17.5%
Benzyl Alcohol	11.0%

OTHER CONSTITUENTS: *Gamma-Terpinene CT*—Cinnamyl alcohol, (E)-beta-ocimene, 2-phenylethanol, (E)-methyl isoeugenol, indole, coumarin, (Z)-beta-ocimene; *Alpha-Terpineol CT*—Phenylethyl alcohol, ethyl palmitate, cinnamyl alcohol, phenylpropyl acetate, 1,8-cineole, beta-caryophyllene, benzyl acetate, isoeugenol, (Z)-3-hexenyl acetate, cinnamyl cinnamate; *Benzyl Benzoate CT*—Benzyl acetate, 2-phenylethyl acetate, 3-methoxy-2-methyl-phenol, (Z)-α-pentylcinnamaldehyde, astaric acid, palmitic acid

PREFERRED COMPOSITION FOR CLINICAL USE:
Insufficient samples/data to determine.

REPORTED THERAPEUTIC PROPERTIES: Nervine, analgesic, relieves muscle cramps, anti-inflammatory, antispasmodic, deodorant, expectorant, relaxing/calming, antidepressant, relieves anxiety, combats mental exhaustion, stress management

CAUTIONS:

■ Avoid oral consumption. Absolutes contain solvent residue and should not be taken orally. In addition, the benzyl derivatives (benzoate, alcohol, acetate) are known to cause gastrointestinal irritation.

SELECTED EVIDENCE:

- □ The benzyl alcohol CT of Poet's narcissus absolute inhibited BChE (butyrylcholinesterase) but only weakly inhibited AChE (acetylcholinesterase).[3243] This suggests it may be useful for cognitive function as inhibition of BChE has been shown to reduce Aß-amyloid levels (a hallmark of dementia and Alzheimer's disease). Interestingly, two of its primary constituents (benzyl alcohol and phenylethyl alcohol) were weak inhibitors of AChE and BChE suggesting the BChE inhibition is likely a result of synergy between multiple components.

POORLAND FLATSEDGE

(Annual Sedge)

Cyperus compressus L.

FAMILY: Cyperaceae
NOTE: Base-Middle
AROMA INTENSITY: Medium-Strong
AROMA: Earthy, woody-sweet, slightly smoky and peppery
COMMON EXTRACTION METHOD: Hydrodistilled or steam distilled from the rhizomes (roots)
POSSIBLE SUBSTITUTE OILS: Cypriol, jointed flatsedge, helichrysum, copaiba
BLENDS WELL WITH: Agarwood, bergamot, cassia, cedarwood (Atlas), cinnamon, cistus, clary sage, frankincense, galbanum, Himalayan cedarwood, lavender, myrrh, patchouli, sandalwood, Texas cedarwood, vetiver, Virginia cedarwood
SUBCELLULAR LOCALIZATION | EPIGENETIC INFLUENCE: Currently unknown | Currently unknown
RECOMMENDED DILUTION RANGE: 3%–50%

PRIMARY CONSTITUENTS:[3244]

Caryophyllene Oxide	34.0%
Cyperene	25.6%
Cyperotundone	6.4%
Beta-Caryophyllene	4.3%
Rotundene	3.2%

OTHER CONSTITUENTS: Beta-pinene, alpha-copaene, alpha-humulene, aristolochene, humulene epoxide II, beta-biotone, caryophylla-4(12),8(13) dien 5-ol, caryophylla-3,8(13)-dien-5beta-ol, alpha-cyperone

PREFERRED COMPOSITION FOR CLINICAL USE:
Insufficient samples/data to determine.

REPORTED THERAPEUTIC PROPERTIES: Wound healing, astringent, relieves chronic skin conditions, analgesic, anti-inflammatory, reduces fever, antibacterial, aids digestion, reduces nausea and vomiting, insect repellent, calming, enhances meditation, relieves mental exhaustion

CAUTIONS:

■ May interact with aspirin, blood pressure, antiplatelet, and anticoagulant medications, and increase the risk of bleeding among people with bleeding disorders, due to caryophyllene-oxide content.[3245]

SELECTED EVIDENCE:

- □ None found.

PORT ORFORD CEDARWOOD (Incense Cedar, Oregon Cedar, Rose of Cedar, Lawson Cypress)

Chamaecyparis lawsoniana (A. Murray bis) Parl., *Cupressus lawsoniana* A. Murray bis, *Retinispora lawsoniana* (A. Murray bis) A.V. Bobrov & Melikyan

FAMILY: Cupressaceae
NOTE: Base (Wood) or Middle (Aerial Parts)
AROMA INTENSITY: Medium
AROMA: Coniferous, green, woody, herbaceous, green, slightly sweet
COMMON EXTRACTION METHOD: Steam distilled from the heartwood or wood shavings; may also be distilled from the aerial parts, needles, or twigs
POSSIBLE SUBSTITUTE OILS: Cedarwood, sandalwood, balsam fir
BLENDS WELL WITH: Agarwood, balsam fir, bergamot, cedarwood, clary sage, cypress, frankincense, jasmine, juniper, lavender, myrrh, neroli, orange, pine, petitgrain, rose, rosemary, sandalwood, silver fir, spearmint, vetiver, white fir, ylang ylang
SUBCELLULAR LOCALIZATION | EPIGENETIC INFLUENCE: Currently unknown | Currently unknown
RECOMMENDED DILUTION RANGE: 3%–33%; 50% for some conditions

PRIMARY CONSTITUENTS:[3246,3247,3248,3249,3250,3251]

Wood		*Aerial Parts (Terpinen-4-ol CT)*	
Alpha-Terpineol	14.3%–29.5%	Terpinen-4-ol	22.0%
Delta-Cadinene	8.2%–15.0%	Sabinene	21.0%
Borneol	1.3%–12.2%	Camphor	7.8%
Alpha-Fenchol	5.5%–11.1%	Citronellol	7.3%
Camphor	5.9%–9.9%	Gamma-Terpinene	7.0%
Fenchone	4.7%–7.4%	Alpha-Terpinene	4.0%
Alpha-Pinene	5.0%–6.5%	Limonene	3.5%
Alpha-Cadinol	0.0%–5.3%	(2E,6E)-Farnesol	3.5%
Beta-Terpineol	3.3%–4.8%		
Alpha-Muurolene	0.0%–4.2%		
Citronellyl Acetate	0.6%–3.5%		
T-Cadinol	0.0%–3.4%		

Aerial Parts (Limonene CT)

Limonene	18.5%–77.7%
Beyerene	0.0%–17.1%
Oplopanonyl Acetate	1.6%–15.9%
Methyl Myrtenate	0.0%–9.7%
Terpinen-4-ol	1.0%–6.9%
Sabinene	0.0%–5.4%
p-Cymen-7-ol	0.0%–7.0%

OTHER CONSTITUENTS: *Wood*—Camphene, 1,4-cineole, limonene, 1,8-cineole, para-cymene, terpinolene, para-cymenene, isopulegol, bornyl acetate, beta-elemene, terpinen-4-ol, alpha-amorphene, citronellol, myrtenol, T-muurolol; *Aerial Parts (Terpinen-4-ol CT)*—Alpha-thujene, alpha-pinene, camphene, myrcene, para-cymene, beta-phellandrene, terpinolene, linalool, cis-p-menth-2-en-1-ol, cis-beta-terpineol, citronellal, alpha-terpineol, 2-undecanone, germacrene D, 2-tridecanone; *Aerial Parts (Limonene CT)*—Alpha-thujene, myrcene, gamma-terpinene, terpinolene, trans-sabinene hydrate, cis-p-menth-2-en-1-ol, alpha-terpineol, (4Z)-decenal, (4Z)-decen-1-ol, perillaldehyde, bornyl acetate, alpha-terpinyl acetate, cis-muurola-3,5-diene, cis-cadina-1(6),4-diene, epizonarene, delta-cadinene, 10-epi-cubenol, beta-oplopenone, alpha-cadinol, manool oxide

PREFERRED COMPOSITION FOR CLINICAL USE:

Constituent	
Alpha-Terpineol	15%–30%
Delta-Cadinene	7%–18%
Borneol	5%–15%
Alpha-Fenchol	3%–12%
Alpha-Pinene	3%–10%
Camphor	2%–10%
Fenchone	2%–8%
Alpha-Muurolene	0.5%–7%
Beta-Terpineol	0.5%–6%
Alpha-Cadinol	0.1%–6%
Citronellyl Acetate	0.5%–5%

REPORTED THERAPEUTIC PROPERTIES: Antibacterial, antifungal, antiviral, antimicrobial, anticancer, wound healing, relieves chronic skin conditions and itching, anti-inflammatory, analgesic, antispasmodic, insect repellent, **insecticidal**, grounding, relaxing, increases awareness, sharpens focus, improves cognition

CAUTIONS:

Wood and Aerial Parts (Terpinen-4-ol CT)

- Avoid with epilepsy and Parkinson's disease due to camphor or fenchone content. [3252,3253,3254,3255] These constituents exacerbate or trigger seizures/convulsions or reduce medication efficacy.

Aerial Parts (Limonene CT)

- None currently known.

SELECTED EVIDENCE:

- The limonene CT of Port Orford cedarwood essential oil inhibited *C. albicans, S. aureus, B. subtilis,* and *M. luteus*.[3256]
- Port Orford cedarwood essential oil (aerial parts; limonene CT) provided equivalent mosquito (*Aedes albopictus*) protection to the standard repellent DEET, and was moderately toxic to mosquito larvae.[3257]
- Port Orford cedarwood (wood) essential oil was toxic to deer ticks (*Ixodes scapularis*), fleas (*Xenopsylla cheopis*), and mosquitoes (*Aedes aegypti*).[3258]

PORTUGUESE THYME
(Tomillo Carnoso)

Thymus carnosus Boiss.

FAMILY: Lamiaceae (Labiatae)
NOTE: Middle
AROMA INTENSITY: Medium
AROMA: Fresh, camphoraceous, woody, medicinal
COMMON EXTRACTION METHOD: Steam distilled or hydrodistilled from the aerial parts
POSSIBLE SUBSTITUTE OILS: Marjoram, tea tree, sweet inula, thyme, ajowan, rosemary, lavender
BLENDS WELL WITH: Bay laurel, bergamot, cajeput, eucalyptus, frankincense, geranium, grapefruit, helichrysum, hyssop, lavender, lavandin, lemon, lemon myrtle, lime, marjoram, myrtle, orange, rosalina, rose, rosemary, sweet inula, thyme
SUBCELLULAR LOCALIZATION | EPIGENETIC INFLUENCE: Currently unknown | Currently unknown
RECOMMENDED DILUTION RANGE: 3%–33%; neat for some conditions

PRIMARY CONSTITUENTS:[3259,3260,3261,3262,32633264]

Borneol CT	
Borneol	17.6%–43.7%
Camphene	5.5%–19.5%
Cis-Sabinene Hydrate	0.0%–16.4%
Terpinen-4-ol	0.1%–12.2%
Trans-Sabinene Hydrate	0.0%–10.1%
Alpha-Pinene	2.0%–8.9%
Bornyl Acetate	1.5%–8.6%
Camphor	0.9%–6.1%
Gamma-Terpinene	0.6%–5.3%
Beta-Pinene	0.1%–3.6%
Para-Cymene	0.4%–3.5%
Alpha-Thujene	1.1%–3.4%
Elemol	0.0%–3.4%
Limonene	0.8%–3.1%
(E)-Beta-Ocimene	tr–3.0%

Thymol CT	
Thymol	27.2%–36.1%
Para-Cymene	21.3%–26.2%
Gamma-Terpinene	19.1%–19.6%

Linalool CT	
Linalool	9.5%–25.5%
Borneol	15.0%–20.1%
Trans-Sabinene Hydrate	8.5%–17.0%
Cis-Sabinene Hydrate	3.8%–11.2%
Terpinen-4-ol	5.9%–7.2%
Camphene	2.5%–6.8%
Bornyl Acetate	1.5%–5.0%
Alpha-Pinene	1.5%–3.1%

Camphene CT

Camphene	23.1%
Borneol	14.4%
Alpha-Pinene	10.7%
Gamma-Terpinene	8.5%
Alpha-Thujene	6.0%
Alpha-Terpinene	5.1%
Beta-Pinene	4.6%
(E)-Beta-Ocimene	3.1%
Bornyl Acetate	3.0%

Note: Portuguese thyme is considered a threatened species, making its commercial availability unlikely.

OTHER CONSTITUENTS: *Borneol CT*—Tricyclene, sabinene, myrcene, alpha-terpinene, beta-phellandrene, (Z)-beta-ocimene, terpinolene, (Z)-dihydrocarvone, cis-verbenol, trans-verbenol, alpha-terpineol, beta-caryophyllene, viridiflorol; *Thymol CT*—Alpha-thujene, alpha-pinene, camphene, sabinene, myrcene, alpha-terpinene, 1,8-cineole, limonene, trans-sabinene hydrate, linalool, borneol, carvacrol, beta-caryophyllene, caryophyllene oxide; *Linalool CT*—Alpha-thujene, limonene, gamma-terpinene, (E)-beta-ocimene, para-cymene, camphor, trans-verbenol, caryophyllene oxide, elemol, viridiflorol; *Camphene CT*—Tricyclene, sabinene, para-cymene, limonene, cis-sabinene hydrate

PREFERRED COMPOSITION FOR CLINICAL USE:

Constituent	***Borneol CT***
Borneol	20%–40%
Camphene	5%–20%
Cis-Sabinene Hydrate	2%–15%
Terpinen-4-ol	4%–14%
Alpha-Pinene	2%–10%
Bornyl Acetate	1.5%–10%
Trans-Sabinene Hydrate	0.1%–7%
Camphor	0.5%–7%
Gamma-Terpinene	0.1%–7%
Beta-Pinene	0.1%–5%
Para-Cymene	0.1%–5%
Alpha-Thujene	0.1%–5%
Limonene	0.1%–5%
(E)-Beta-Ocimene	tr–5%
Beta-Caryophyllene	0.1%–4%
Terpinolene	0.1%–4%
Sabinene	0.1%–4%

REPORTED THERAPEUTIC PROPERTIES: Anti-inflammatory, analgesic, aids respiratory function, eases cough, anticancer, antimicrobial, antibacterial, antifungal, antiviral, antiparasitic, preserves brain and memory function, antioxidant, supports digestion, heals stomach ulcers, wound healing, insecticidal, relieves anxiety, promotes a sense of peace, calming

CAUTIONS:

Borneol CT

■ May interact with aspirin, blood pressure, antiplatelet, and anticoagulant medications, and increase the risk of bleeding among people with bleeding disorders, due to the anticoagulant properties of borneol.[3265,3266]

■ May interfere with enzymes responsible for metabolizing medications (NSAIDs, proton-pump inhibitors, acetaminophen, antiepileptics, immune modulators, blood-sugar medications, blood pressure medications, antidepressants, antipsychotics, diabetic medications, antihistamines, antibiotics, and anesthetics), based on borneol content.[3267,3268,3269,3270,3271]

Thymol CT

■ May interact with aspirin, blood pressure, antiplatelet, and anticoagulant medications, and increase the risk of bleeding among people with bleeding disorders, due to thymol and carvacrol content.[3272,3273]

■ May interact with anticholinergic (drugs used for asthma, incontinence, gastrointestinal cramps, muscular spasms, depression, and sleep disorders) and cholinergic medications (drugs used to reduce urinary retention, diagnose myasthenia gravis, and for glaucoma), based on thymol and carvacrol content.[3274,3275]

■ May interact with antibiotics or antifungals and possibly enhance their effects based on thymol and carvacrol content.[3276,3277,3278,3279,3280,3281,3282,3283]

■ May irritate mucous membranes (eyes, mouth, nasal passages, vagina, rectum).

Linalool and Camphene CTs

■ None currently known.

SELECTED EVIDENCE:

- □ Portuguese thyme CT borneol displayed anti-inflammatory activity by reducing the expression of iNOS and COX-2 enzymes (two enzymes involved in the inflammatory process).[3284] Another study found that the camphene CT inhibited 5-LOX activity, another enzyme involved in the inflammatory process.[3285]
- □ The borneol CT exhibited good antioxidant activity in the TBARS assay.[3286] Another study reported that the borneol CT possesses better antioxidant capacity than BHT in sunflower oil.[3287] The camphene CT also demonstrated antioxidant activity in the ORAC, metal chelating, and FE^{3+} assays, but was only weakly active in the lipid-peroxidation test.[3288]

PROTIUM COPAL (Amazonia Resintree)

Protium amazonicum (Cuatrec.) Daly

FAMILY: Burseraceae
NOTE: Base
AROMA INTENSITY: Medium
AROMA: Balsamic, rich, sweet, woody, slightly citrusy
COMMON EXTRACTION METHOD: Hydrodistilled from the oleoresin
POSSIBLE SUBSTITUTE OILS: Pink pepper, pine, black pepper CT delta-3-carene, cypress, frankincense
BLENDS WELL WITH: Basil, bay laurel, bergamot, black pepper, camphor, cassia, cinnamon, coriander, cypress, frankincense, geranium, grapefruit, lavender, lemon, myrrh, neroli, orange, palo santo, petitgrain, pine, ravensara, ravintsara, rose, sandalwood, spikenard, tangerine, vetiver, ylang ylang
SUBCELLULAR LOCALIZATION | EPIGENETIC INFLUENCE: Currently unknown | Currently unknown
RECOMMENDED DILUTION RANGE: 3%–Neat

PRIMARY CONSTITUENTS:[3289]

Delta-3-Carene	47.9%
Alpha-Terpineol	5.5%
Limonene	5.1%
m-Cymen-8-ol	4.8%
Para-Cymene	4.1%
Alpha-Pinene	4.0%
Para-Cymenene	3.2%
4-Methyleneisophorone	3.0%

OTHER CONSTITUENTS: Alpha-thujene, 3,7,7-trimethyl-1,3,5-cycloheptatriene, beta-pinene, alpha-phellandrene, 1,8-cineole, terpinolene, 2-isobutylnorbornane, trans-dihydro-alpha-terpineol, alpha-phellandrene-8-ol, beta-phellandrene-8-ol, borneol, p-cymen-8-ol, 2-carone, alpha-trans-bergamotene

PREFERRED COMPOSITION FOR CLINICAL USE:
Insufficient samples/data to determine.

REPORTED THERAPEUTIC PROPERTIES: Analgesic, anti-inflammatory, antirheumatic, relieves headache, alleviates toothache, wound healing, stomach protective, aids respiratory function, expels excess mucus, antifungal, antimicrobial, antiseptic, anticancer, supports circulation, moisturizes the skin, insect repellent, antidepressant, relaxing, aids mental clarity, enhances meditation

CAUTIONS:
- None currently known.

SELECTED EVIDENCE:
- *In vitro* research inhibited the *Cryptococcus neoformans* fungus.[3290]

PTERONIA
(Karoo Gumbush)

Pteronia incana (Burm.) DC., *P. xantholepis* DC., *Chrysocoma incana* Burm.

FAMILY: Asteraceae (Compositae)
NOTE: Middle
AROMA INTENSITY: Medium
AROMA: Piney, woody, fresh, sweet, green, balsamic
COMMON EXTRACTION METHOD: Steam distilled or hydrodistilled from the leaves or leaves and stalks
POSSIBLE SUBSTITUTE OILS: Galbanum, white fir, gobre salla, black pepper CT beta-pinene, Turkish pine; pink pepper, black cumin, shell ginger CT para-cymene
BLENDS WELL WITH: Balsam fir, bergamot, cedarwood, cypress, Douglas fir, frankincense, grapefruit, lavender, lemon, lime, orange, palo santo, pine, rosemary, sandalwood, silver fir, spruce (black), tangerine, tsuga, vetiver, white fir
SUBCELLULAR LOCALIZATION | EPIGENETIC INFLUENCE: Currently unknown | Currently unknown
RECOMMENDED DILUTION RANGE: 3%–33%, 50% for some conditions

PRIMARY CONSTITUENTS:[3291,3292,3293,3294,3295]

Beta-Pinene CT (Leaves)

Beta-Pinene	22.2%–36.4%
Epi-Bicyclosesquiphellandrene	0.0%–30.4%
Benzene, 1-Methyl-2-(1-Methylethyl)	0.0%–27.8%
1,8-Cineole + Limonene	12.1%–24.8%
Alpha-Pinene	12.6%–22.3%
Para-Cymene	0.0%–19.1%
Myrcene	0.0%–17.0%
1,3-Dioxolo[4,5-b] Acridin-10(5H)-O	0.0%–8.6%
Ortho-Cymene	0.0%–4.5%
2-(3'-(2',’-Dimethylpyridyl)Benzene,1,2-dimethoxy-4(2-propen)	0.0%–3.3%

Leaves/Stalks

Beta-Pinene	32.5%
Alpha-Pinene	18.6%
Para-Cymene	11.3%
Myrcene	10.3%
1,8-Cineole	9.0%
Methyl eugenol	7.2%
Limonene	7.0%
Terpinen-4-ol	5.3%
Beta-Caryophyllene	3.2%
Ledene	3.1%

Para-Cymene CT (Leaves)

Para-Cymene	4.2%–36.7%
Spathulenol	2.9%–22.9%
1,8-Cineole	0.4%–18.2%
Isoelemicine	0.0%–14.8%
Alpha-Bisabolol	0.0%–13.5%
Limonene	0.8%–10.9%
Methyl eugenol	0.0%–10.9%
Beta-Pinene	0.3%–10.2%
Globulol	0.0%–8.3%
Myrcene	0.2%–8.0%
Terpinen-4-ol	0.0%–7.1%
Alpha-pinene	0.0%–6.7%
Bicyclogermacrene	0.4%–4.5%
Pentadecane	0.0%–4.1%
Beta-Caryophyllene	0.0%–4.1%
Isoledene	0.0%–3.7%
Caryophyllene Oxide	0.0%–3.0%

Para-Cymene/Camphor CT (Part Unknown)

Para-Cymene	19.2%
Camphor	7.5%
Myrtenol	6.9%
Spathulenol	4.4%
Heptan-3-one	3.9%

OTHER CONSTITUENTS: *Beta-Pinene CT (Leaves)*—Alpha-phellandrene, (E)-beta-farnesene, delta-4-carene, linalyl propionate, delta-cadinene, 2H-3,9a-methanol-1-benzoxepin, beta-bisabolene, beta-farnesene, germacrene B, Phenol, 4-(methylthio)-8–acetyl-3,3-; *Leaves/Stalks*—(E)-beta-ocimene, linalool, trans-pinocarveol, alpha-terpinyl acetate, muurolene, bicyclogermacrene, ar-curcumene, delta-cadinene, myrtenol, p-cymen-8-ol, alpha-calacorene, (E)-nerolidol, ledol, epiglobulol, T-muurolol, carvacrol, alpha-bisabolol, isospathulenol, alpha-cadinol, unidentified sesquiterpenes; *Para-Cymene CT (Leaves)*—Sabinene, alpha-phellandrene, alpha-terpinene, gamma-terpinene, 1-hexanol, tetradecane, presilhiperfol-7-ene, perillen, pinguisene, dill ether, modhephene, isocomene, cis-sabinene hydrate, pinocarvone, myrtenal, trans-pinocarveol, delta-terpineol, alpha-terpineol, heptadecane, germacrene D, carvone, gamma-cadinene, octadecane, myrtenal, ar-curcumene, p-cymen-8-ol, geranyl acetone, (E)-nerolidol, viridiflorol, carvacrol, zingiberenol; *Para-Cymene/Camphor CT (Part Unknown)*—Cis-sabinene hydrate, 1-undecene, furan, terpinen-4-ol, verbenone, trans-carveol, 2-cyclohexen-1-one, bornyl acetate, junipene, eugenol, alpha-humulene, ar-curcumene, naphthalene, farnesene

PREFERRED COMPOSITION FOR CLINICAL USE:
Too great of variance in samples/data to determine.

REPORTED THERAPEUTIC PROPERTIES: Antimicrobial, antibacterial, antifungal, anti-infectious, reduces fever, analgesic, anti-inflammatory, antirheumatic, treats kidney conditions, anticancer, weight management, aids memory function, insecticidal, combats anger, relieves anxiety, stress management

CAUTIONS:

■ Avoid with epilepsy and Parkinson's disease due to potentially moderate 1,8-cineole and camphor content. May exacerbate or cause seizures/convulsions or reduce seizure-medication efficacy.[3296,3297,3298,3299]

■ May weakly interfere with the enzymes responsible for metabolizing medications (NSAIDs, proton-pump inhibitors, acetaminophen, antiepileptics, immune modulators, blood-sugar medications, blood pressure medications, antidepressants, antipsychotics, diabetic medications, antihistamines, antibiotics, and anesthetics) due to 1,8-cineole content.[3300,3301,3302]

■ May interfere with pentobarbital and other barbiturates (medications for anxiety and insomnia), based on 1,8-cineole content.[3303,3304]

■ Not recommended in children, during pregnancy/lactation, or orally if methyl eugenol meets or exceeds 15%.

SELECTED EVIDENCE:

- □ *In vitro* research demonstrated that pteronia essential oil CT para-cymene inhibited *C. neoformans, M. smegmatis, K. pneumoniae,* and *M. catarrhalis*, with the greatest inhibition of *C. neoformans*.[3305] The beta-pinene CT of pteronia essential oil exhibited varying degrees of antimicrobial activity against more than twenty bacteria and yeast strains.[3306] Pteronia CT para-cymene/camphor inhibited *L. monocytogenes, B. cereus, S. pyogenes, B. subtilis, P. vulgaris, E. coli, K. pneumoniae, A. faecalis, Sh. flexneri,* and *P. flourescense.*[3307]

QUEENSLAND SANDALWOOD (Northern Sandalwood, Plumbush)

Santalum lanceolatum R.Br.

FAMILY: Santalaceae
NOTE: Base
AROMA INTENSITY: Medium
AROMA: Fresh, woody, slightly sweet and spicy
COMMON EXTRACTION METHOD: Steam distilled from the wood
POSSIBLE SUBSTITUTE OILS: Sandalwood, cedarwood
BLENDS WELL WITH: Basil, bergamot, black pepper, blue cypress, cardamom, cedarwood, chamomile (German, Roman), coriander, cypress, frankincense, juniper, lavender, myrrh, neroli, palmarosa, palo santo, patchouli, petitgrain, ravensara, rose, sandalwood, Spanish sage, vetiver, ylang ylang
SUBCELLULAR LOCALIZATION | EPIGENETIC INFLUENCE: Currently unknown | Currently unknown
RECOMMENDED DILUTION RANGE: 5%–Neat

PRIMARY CONSTITUENTS:[3308,3309]

(Z)-Lanceol CT	
(Z)-Lanceol	22.1%
(Z)-Beta-Curumen-12-ol	6.8%
(Z)-Nuciferol	4.6%
(Z)-Alpha-Santalol	4.5%
Alpha-Sinensal	3.6%
(Z)-Farnesol	3.3%
Epi-Beta-Bisabolol	3.0%

Santalols CT (Cape York)	
(Z)-Alpha-Santalol	30.5%–50.0%
(Z)-Beta-Santalol	12.5%–18.5%

OTHER CONSTITUENTS: *(Z)-Lanceol CT*—Alpha-santalene, ar-curcumene, beta-bisabolene, nerolidol, epi-alpha-santalol; *Santalols CT (Cape York)*—(Z)-nuciferol, (Z)-beta-curumen-12-ol (percentages not listed)

PREFERRED COMPOSITION FOR CLINICAL USE:
Insufficient data/samples to determine.

REPORTED THERAPEUTIC PROPERTIES: Reduces fever, supports respiratory function, eases cough, expels excess mucus, aids liver and gallbladder function, aids digestion, relieves diarrhea, analgesic, antispasmodic, anti-inflammatory, stress management, relieves anxiety

CAUTIONS:
- None currently known.

SELECTED EVIDENCE:
- □ None found.

QUINA
(Bálsamo, Sándalo)

Myroxylon peruiferum L.f., *Myrospermum pedicellatum* Lam.

FAMILY: Fabaceae (Leguminosae)
NOTE: Middle-Base
AROMA INTENSITY: Medium
AROMA: Balsamic, spicy, sweet, hint of vanilla
COMMON EXTRACTION METHOD: Steam distilled from the leaves or bark
POSSIBLE SUBSTITUTE OILS: Peta, helichrysum, blue cypress, kunzea, copaiba, guggul
BLENDS WELL WITH: Balsam fir, black pepper, cardamom, cedarwood, cinnamon, clove, copaiba, frankincense, ginger, lemon, lime, neroli, orange, patchouli, petitgrain, rose, sandalwood, silver fir, spruce (black), tamala, tangerine, white fir, ylang ylang
SUBCELLULAR LOCALIZATION | EPIGENETIC INFLUENCE: Currently unknown | Currently unknown
RECOMMENDED DILUTION RANGE: 3%–Neat

PRIMARY CONSTITUENTS:[3310,3311,3312]

Leaves CT Spathulenol	
Spathulenol	13.3%
Alpha-Pinene	9.6%
Caryophyllene Oxide	7.3%
Limonene	7.2%
Beta-Elemene	6.8%
Bicyclogermacrene	5.5%
Beta-Bourbonene	5.0%
Alpha-Muurolene	4.2%
Beta-Caryophyllene	3.5%
Delta-Cadinene	2.8%
Germacrene A	2.7%
Alpha-Copaene	2.6%

Bark	
Alpha-Copaene	13.4%
Guaiol	9.4%
Safrole	8.4%
Delta-Cadinene	7.1%
Alpha-Amorphene	6.6%
(–)-Sinularene	5.8%
Gamma-Gurjunene	5.3%
Alloaromadendrene	5.2%
Ar-Curcumene	5.1%
(E)-Alpha-Bergamotene	3.5%
Beta-Selinene	3.3%

Leaves CT Germacrene D

Germacrene D	17.1%
Alpha-Pinene	14.8%
Beta-Caryophyllene	10.8%
Gamma-Elemene	6.5%
Limonene	5.8%
Beta-Pinene	4.8%
Germacrene B	3.5%
Alpha-Humulene	3.4%

OTHER CONSTITUENTS: *Leaves CT Spathulenol*–Alpha-campholenal, trans-verbenol, alpha-cubebene, ylanga-2,4 (15)-diene, beta-cubebene, alpha-humulene, germacrene D, beta-selinene, alpha-selinene, gamma-cadinene; *Bark*—(+)-aromadendrene, beta-bisabolene, beta-eudesmol; *Leaves CT Germacrene D*—Alpha-thujene, delta-2-carene, terpinen-4-ol, alpha-copaene, delta-cadinene, spathulenol, caryophyllene oxide

PREFERRED COMPOSITION FOR CLINICAL USE:
Insufficient samples/data to determine.

REPORTED THERAPEUTIC PROPERTIES: Analgesic, **anti-inflammatory**, relieves headache/migraine, supports respiratory function, expels excess mucus, wound healing, antiseptic, anti-infectious, antifungal, antibacterial, antiparasitic, soothes chronic skin conditions, aid circulation, relieves diarrhea, antiparasitic, energizing, uplifting, aids mental clarity

CAUTIONS:
- None currently known.

SELECTED EVIDENCE:
- Quina leaf essential oil demonstrated antifungal activity against *C. tropicalis, C. krusei, C. albicans, C. parapsilosis*, and *C. glabrata.*[3313]
- Quina bark essential oil demonstrated antileishmanial properties against *L. amazonensis.*[3314]
- The germacrene D CT of quina leaf essential oil inhibited *S. aureus, S. epidermidis,* and *T. rubrum.*[3315] The same study reported good antioxidant activity.

RED BOTTLEBRUSH (Crimson Bottlebrush, Lemon Bottlebrush)

Callistemon citrinus (Curtis) Skeels

FAMILY: Myrtaceae
NOTE: Top
AROMA INTENSITY: Medium
AROMA: Fresh, medicinal, slightly woody
COMMON EXTRACTION METHOD: Steam distilled or hydrodistilled from the leaves; may also be distilled from the flowers or stems

POSSIBLE SUBSTITUTE OILS: *Eucalyptus radiata*, *Eucalyptus globulus*, myrtle, cajeput

BLENDS WELL WITH: Bay laurel, bergamot, blue cypress, cajeput, camphor, cedarwood, eucalyptus, geranium, ginger, grapefruit, juniper, lavender, lemon, lemongrass, melaleuca (tea tree), myrtle, petitgrain, pine, ravensara, ravintsara, rosemary, sage, tangerine, thyme

SUBCELLULAR LOCALIZATION | EPIGENETIC INFLUENCE: Currently unknown | Currently unknown

RECOMMENDED DILUTION RANGE: 3%–33%; neat for some conditions

PRIMARY CONSTITUENTS:[3316,3317,3318,3319,3320,3321,3322,3323]

Leaves CT 1,8-Cineole

1,8-Cineole	52.1%–87.2%
Alpha-Terpineol	0.0%–14.7%
Eugenol	tr–14.2%
Linalool	0.0%–4.5%
Myrcene	0.0%–3.7%
Alpha-Phellandrene	0.0%–3.6%
Terpinolene	0.1%–3.1%
Terpinen-4-ol	0.1%–3.0%

Leaves CT Alpha-Pinene

Alpha-Pinene	32.3%
Alpha-Terpineol	14.6%
Limonene	13.1%
1,8-Cineole	9.8%

Flowers CT Alpha-Eudesmol

Alpha-Eudesmol	12.9%
Beta-Caryophyllene	11.9%
Bornyl Acetate	10.0%
1,8-Cineole	8.1%
Bicyclogermacrene	5.7%
Gamma-Eudesmol	5.2%
Alpha-Pinene	4.6%
Camphor	4.4%
Camphene	3.8%
Aromadendrene	3.8%
Spathulenol	3.2%

Leaves CT 1,8-Cineole/Alpha-Pinene

1,8-Cineole	34.1%–61.2%
Alpha-Pinene	13.4%–29.0%
Alpha-Terpineol	0.0%–10.7%
Alpha-Phellandrene	0.0%–8.9%
Trans-Pinocarveol	0.0%–5.8%
Beta-Pinene	0.9%–4.7%
Limonene	0.0%–3.4%

Flowers CT 1,8-Cineole

1,8-Cineole	36.6%
Alpha-Pinene	29.7%
Alpha-Terpineol	5.6%
Alpha-Phellandrene	4.4%
Limonene	4.0%

Stems

1,8-Cineole	56.0%
Alpha-Pinene	31.0%
Limonene	4.0%
Alpha-Terpineol	3.7%

OTHER CONSTITUENTS: *Leaves CT 1,8-Cineole*—Alpha-pinene, (Z)—beta-ocimene, trans-pinocarveol, geraniol, trans-carveol, beta-caryophyllene, alpha-humulene, dihydroxydurene, flavesone, geranyl acetate, spathulenol, caryophyllene oxide; *Leaves CT 1,8-Cineole/Alpha-Pinene*—Alpha-thujene, para-cymene, alpha-phellandrene, gamma-terpinene, terpinolene, linalool, geraniol, linalyl acetate, spathulenol, tridecanoic acid, benzoic acid,3,4-dimethoxy methyl ester; *Leaves CT Alpha-Pinene*—Beta-pinene, myrcene, para-cymene, 1,8-cineole, gamma-terpinene, terpinen-4-ol, beta-caryophyllene, aromadendrene, beta-sesquiphellandrene, ledol, spathulenol, viridiflorol, globulol; *Flowers CT 1,8-Cineole*—Isoamyl acetate, isobutyl isobutyrate, alpha-thujene, para-cymene, alpha-terpinolene, linalool, terpinen-4-ol, ledene, beta-asarone; *Flowers CT Alpha-Eudesmol*—Beta-thujene, beta-pinene, o-cymene, gamma-terpinene, borneol, terpinen-4-ol, alpha-gurjunene, alpha-humulene, alloaromadendrene, alpha-elemene, delta-cadinene, elemol, ledene, viridiflorol, rosifoliol, cis-1(7),8-p-mentha-diene-2-ol, 5-amino-1-phenylpyrazole; *Stems*—Beta-pinene, linalool, pinocarveol, terpinen-4-ol

PREFERRED COMPOSITION FOR CLINICAL USE:
Insufficient and too great of variance in data/samples to determine.

REPORTED THERAPEUTIC PROPERTIES: Anti-inflammatory, analgesic, antispasmodic, aids digestion, relieves stomachache, antibacterial, antifungal, antiviral, antiparasitic, anticancer, eases cough, aids respiratory function, opens the airways, deodorant, insecticidal, aids memory function, stimulating

CAUTIONS:
1,8-Cineole CTs
■ Avoid with children under age 3, particularly around the nose and mouth and use cautiously in children under age 5 due to significant 1,8-cineole content. 1,8-cineole may cause seizures, central nervous system problems, or respiratory distress in young children.[3324,3325,3326]
■ Caution is warranted orally due to the significant amounts of 1,8-cineole. Limit it to small doses internally (adults—maximum 10 drops daily). Toxicity has been reported when eucalyptus (also high in 1,8-cineole) was ingested in large doses, and as few as 4 to 5 drops may cause problems in very sensitive individuals.[3327,3328,3329,3330,3331,3332]
■ Avoid with epilepsy and Parkinson's disease due to 1,8-cineole content. May exacerbate or cause seizures/convulsions or reduce seizure-medication efficacy based on 1,8-cineole content.[3333,3334,3335,3336]
■ May interact with diabetic medications due to significant 1,8-cineole content. 1,8-cineole significantly inhibits alpha-glucosidase activity, particularly when synergized with other constituents (alpha-pinene, beta-pinene, and para-cymene).[3337,3338,3339] Alpha-glucosidase is an enzyme that breaks down carbohydrates by chemical reaction with water. Inhibiting its activity postpones glucose absorption and therefore the impact of carbohydrates on blood-sugar levels. Studies suggest that oils rich in 1,8-cineole may significantly lower glucose in animals, which may increase the action of diabetic medications.[3340,3341]
■ May interact with aspirin, blood pressure, antiplatelet, and anticoagulant medications, and increase the risk of bleeding among people with bleeding disorders (low risk) due to significant 1,8-cineole content.[3342] 1,8-cineole is a weak inhibitor of platelet aggregation.[3343]
■ May weakly interfere with the enzymes responsible for metabolizing medications (NSAIDs, proton-pump inhibitors, acetaminophen, antiepileptics, immune modulators, blood-sugar medications, blood pressure medications, antidepressants, antipsychotics, diabetic medications, antihistamines, antibiotics, and anesthetics) due to 1,8-cineole content.[3344,3345,3346]
■ May interfere with pentobarbital and other barbiturates (medications for anxiety and insomnia), based on 1,8-cineole content.[3347,3348]

Alpha-Pinene and Alpha-Eudesmol CTs
■ None currently known.

SELECTED EVIDENCE:
- □ Red bottlebrush flower and leaf (alpha-pinene CT) essential oils showed cytotoxicity against human lung carcinoma and rat glioma cancer cells while demonstrating nontoxicity against human peripheral blood mononuclear cells (PBMCs).[3349] The flower oil demonstrated stronger activity.
- □ Laboratory research showed that red bottlebrush essential oil (composition not reported) reduced contractions caused by acetylcholine in isolated rat ileum.[3350]
- □ *In vitro* research demonstrated that red bottlebrush leaf essential oil (1,8-cineole/alpha-pinene CT) inhibited bacteria (*Staphylococcus aureus, S. epidermidis, Streptococcus mutans, Enterococcus faecalis, Pseudomonas aeruginosa, Enterobacter aerogenes*, and *Escherichia coli*) and fungi (*Candida albicans, Cryptococcus neoformans, Trichophyton rubrum, Aspergillus flavus, A. niger*, and *Sporothrix schenckii*).[3351] The 1,8-cineole/alpha-pinene CT also exhibited strong activity

against *S. faecalis*, two strains of S. *aureus, B. cereus*, and *S. macrcesens* when compared to standard antibiotics gentamycin and tetracycline used as controls.[3352] The leaf oil (1,8-cineole CT) inhibited *B. cinerea*, *A. flavus*, *F. verticillioides*, *A. alternate*, and *A. parasiticus*, with the greatest activity against *B. cinerea*.[3353] The leaf (1,8-cineole/alpha-pinene CT) and flower (alpha-eudesmol CT) essential oils demonstrated broad-spectrum antibacterial activity against gram-negative (*A. hydrophilia, E. coli, V. alginolyticus, S. typhi*) and gram-positive (*S. enteritidis, S. aureus, L. monocytogenes*) bacteria, with the leaf and flower oil showing the greatest activity against gram-negative bacteria.[3354]

- Red bottlebrush leaf (1,8-cineole/alpha-pinene CT) and flower (alpha-eudesmol CT) essential oils exhibited moderate to good antioxidant activity in the DPPH assay.[3355]
- Red bottlebrush leaf (1,8-cineole CT) essential oil was insecticidal to the fruit fly (*Reticulitermes virginicus*) and worker termites (*Reticulitermes virginicus*).[3356]
- The 1,8-cineole/alpha-pinene CT of red bottlebrush essential oil repelled and killed the cowpea weevil (*Callosobruchus maculatus*), which is a storage pest.[3357]

RED GUM
(Red Stringybark)

Eucalyptus macrorhyncha F.Muell. ex Bent

FAMILY: Myrtaceae
NOTE: Top
AROMA INTENSITY: Strong
AROMA: Fresh, medicinal, slightly green and woody
COMMON EXTRACTION METHOD: Steam distilled from the leaves
POSSIBLE SUBSTITUTE OILS: Myrtle, eucalyptus, cajeput, blue cypress
BLENDS WELL WITH: Bay laurel, bergamot, blue cypress, cajeput, camphor, chamomile (German, Roman), cedarwood, geranium, ginger, grapefruit, juniper, lavender, lemon, lemongrass, melaleuca (tea tree), peppermint, petitgrain, pine, ravensara, ravintsara, rosemary, sage, tangerine, thyme
SUBCELLULAR LOCALIZATION | EPIGENETIC INFLUENCE: Currently unknown | Currently unknown
RECOMMENDED DILUTION RANGE: 3%–33%; neat for some conditions

PRIMARY CONSTITUENTS:[3358,3359]

1,8-Cineole	29.8%–50.2%
Piperitone	0.0%–12.7%
Beta-Eudesmol	9.3%–11.0%
Alpha-Eudesmol	4.4%–7.3%
Limonene	tr–6.3%
Alpha-Terpinene	0.0%–5.6%
Caryophyllene Oxide	0.0%–5.5%
Para-Cymene	0.6%–5.4%
Gamma-Eudesmol	0.0%–5.0%
Geraniol	0.0%–4.9%
Beta-Caryophyllene	0.0%–3.7%
Alpha-Terpineol	1.5%–3.7%
Geranyl Acetate	0.0%–3.2%

OTHER CONSTITUENTS: Alpha-pinene, gamma-terpinene, linalool, alpha-terpinyl acetate, terpinen-4-ol, globulol

PREFERRED COMPOSITION FOR CLINICAL USE:
Insufficient samples/data to determine.

REPORTED THERAPEUTIC PROPERTIES: Antimicrobial, antibacterial, antifungal, anti-infectious, analgesic, anti-inflammatory, aids bone health, antioxidant, heals stomach ulcers, supports respiratory function, eases cough, insect repellent, aids concentration, relieves mental exhaustion, inspires confidence

CAUTIONS:

- Avoid during pregnancy and lactation (low risk) due to beta-eudesmol content. Beta-eudesmol strongly inhibits the growth of new blood vessels, which is necessary for implantation, and placental and embryonic development.[3360]
- Avoid with children under age 3, particularly around the nose and mouth and use cautiously in children under age 5 due to significant 1,8-cineole content. 1,8-cineole may cause seizures, central nervous system problems, or respiratory distress in young children.[3361,3362,3363]
- Caution is warranted orally due to the significant amounts of 1,8-cineole. Limit it to small doses internally (adults—maximum 10 drops daily). Toxicity has been reported when eucalyptus (also high in 1,8-cineole) was ingested in large doses, and as few as 4 to 5 drops may cause problems in very sensitive individuals.[3364,3365,3366,3367,3368,3369]
- Avoid with epilepsy and Parkinson's disease due to 1,8-cineole content. May exacerbate or cause seizures/convulsions or reduce seizure-medication efficacy based on 1,8-cineole content.[3370,3371,3372,3373]
- May weakly interfere with the enzymes responsible for metabolizing medications (NSAIDs, proton-pump inhibitors, acetaminophen, antiepileptics, immune modulators, blood-sugar medications, blood pressure medications, antidepressants, antipsychotics, diabetic medications, antihistamines, antibiotics, and anesthetics) due to 1,8-cineole content.[3374,3375,3376]
- May interfere with pentobarbital and other barbiturates (medications for anxiety and insomnia), based on 1,8-cineole content.[3377,3378]

SELECTED EVIDENCE:

- None found.

RED SPRUCE
(Yellow Spruce, Eastern Spruce, West Virginia Spruce, He-Balsam)

Picea rubens Sarg., P. australis Small

FAMILY: Pinaceae
NOTE: Top-Middle
AROMA INTENSITY: Medium
AROMA: Sweet, fresh, woody, green, resinous, slightly fruity
COMMON EXTRACTION METHOD: Steam distilled or hydrodistilled from the leaves (needles)
POSSIBLE SUBSTITUTE OILS: Spruce (black), sweet inula CT bornyl acetate, tsuga, Siberian fir
BLENDS WELL WITH: Balsam fir, bergamot, copaiba, cypress, eucalyptus, frankincense, grapefruit, juniper, lavender, lemon, myrtle, neroli, palo santo, petitgrain, rose, silver fir, tsuga, vetiver, white fir

SUBCELLULAR LOCALIZATION | EPIGENETIC INFLUENCE: Currently unknown | Currently unknown
RECOMMENDED DILUTION RANGE: 5%–50%; neat for some conditions

PRIMARY CONSTITUENTS:[3379,3380]

HS-SPME		*Distilled*	
Borneol	12.4%	Bornyl Acetate	54.0%–56.0%
Alpha-Pinene	10.4%	Camphene	6.3%–6.8%
Germacrene D	9.9%	Borneol	4.8%–5.5%
Delta-Cadinene	8.3%	Piperitone	2.5%–3.6%
Alpha-Amorphene	6.6%	Alpha-Pinene	2.5%–3.4%
Bornyl Acetate	5.7%		
Naphthalene	5.7%		
Camphor	5.4%		
1.3-Benzenediamine	4.4%		
Bicyclo[4.4.0]dec-1-ene	3.8%		
Beta-Pinene	3.6%		
Beta-Caryophyllene	3.1%		

OTHER CONSTITUENTS: *OC*—Santene, tricyclene, beta-pinene, delta-3-carene, myrcene, limonene, beta-phellandrene, para-cymene, thujones (< 0.4%), linalool, terpinen-4-ol, isoborneol, alpha-terpineol, gamma-cadinene, delta-cadinene; *HS-SPME*—Camphene, beta-phellandrene, alpha-copaene, beta-elemene, gamma-muurolene, isobornyl acetate, gamma-cadinene, delta-4-carene, phenol

PREFERRED COMPOSITION FOR CLINICAL USE:
Insufficient samples/data to determine.

REPORTED THERAPEUTIC PROPERTIES: Analgesic, anti-inflammatory, antispasmodic, astringent, supports respiratory function, eases cough, antibacterial, antifungal, aids circulation, improves focus and concentration, reduces mood swings, relieves anxiety, promotes self-acceptance

CAUTIONS:

- None currently known.

SELECTED EVIDENCE:

- □ None found.

REDFLOWER RAGLEAF (Ebolo, Thickhead, Fireweed)

Crassocephalum crepidioides (Benth.) S. Moore, *Gynura crepidioides* Benth., *G. sarcobensis* DC., *Senecio diversifolius* A. Rich

FAMILY: Asteraceae
NOTE: Middle or Base
AROMA INTENSITY: Medium
AROMA: Herbaceous, earthy, floral, slightly sweet
COMMON EXTRACTION METHOD: Steam distilled or hydrodistilled from the flowers, leaves, stems, or roots, or combinations of plant parts

POSSIBLE SUBSTITUTE OILS: Bergamot, cassia, cinnamon, clary sage, clove, copaiba, grapefruit, hemp, hops, lavandin, lavender, lemon, lime, Mediterranean mandarin, nutmeg, orange, patchouli, pine, tangerine, valerian, yuzu
BLENDS WELL WITH: Hemp, thyme, copaiba
SUBCELLULAR LOCALIZATION | EPIGENETIC INFLUENCE: Currently unknown | Currently unknown
RECOMMENDED DILUTION RANGE: 3%–33%; neat for some conditions

PRIMARY CONSTITUENTS:[3381,3382,3383,3384]

Leaves CT 4-Cyclohexylbutyramide

4-Cyclohexylbutyramide	20.9%
Beta-Cubebene	13.8%
Alpha-Humulene	10.3%
Phytol	6.3%
Beta-Caryophyllene	6.1%
(Z)-1-[1-butenyl]aziridine	3.7%
Alpha-Farnesene	3.3%
Beta-Elemene	3.2%
(Z)-Beta-Farnesene	3.1%

Stems CT Myrcene

Myrcene	26.1%
Alpha-Pinene	10.7%
Alpha-Humulene	5.9%
(E)-Beta-Farnesene	5.2%
Caryophyllene Oxide	5.0%
Beta-Caryophyllene	4.4%
Humulene Epoxide II	3.8%
Alpha-Copaene	3.2%
Bornyl Acetate	3.0%

Flowers

Myrcene	43.3%
Beta-Phellandrene	10.7%
Cryptone	8.1%
Alpha-Copaene	3.3%
Alpha-Humulene	3.1%

Leaves CT Myrcene

Myrcene	59.3%
Beta-Phellandrene	11.9%
Cryptone	6.4%

Stems CT Thymol

Thymol	43.9%
Alpha-Humulene	15.2%
Beta-Caryophyllene	10.7%
6-Phenyl Dodecane	9.3%

Aerial Parts

Myrcene	65.9%
Beta-Phellandrene	8.8%
Alpha-Pinene	3.1%

Roots

(E)-Beta-Farnesene	30.6%
Alpha-Humulene	10.3%
Beta-Caryophyllene	7.2%
(Z)-Beta-Guaiene	6.1%
Alpha-Bulnesene	5.3%
Cyperene	3.7%
Bornyl Acetate	3.5%
Alpha-Copaene	3.5%
Beta-Elemene	3.4%
Beta-Chamigrene	3.0%

OTHER CONSTITUENTS: *Leaves CT 4-Cyclohexylbutyramide*—m-Tert-butyl-phenol, p-myrecene, 1-methyl-2-(1-methylethyl)-3-(1-methylethylidene)cyclopropane, myrtenol, 2,6-dimethyl-3,5,7-octatriene-2-ol, 2-isopropylidene-3-methylhexa-3,5dienal, 3-methylene-p-menth-8-ene, artemisia triene, alpha-copaene, 2-dodecanone, beta-elemene, 2-tridecanone, delta-cadinene, gamma-elemene, 3,7,11-trimethyl-1,6,10-dodecatrien-3-ol, caryophyllene oxide, humulene epoxide II, alpha-santalol, (Z)-beta-ocimene, 2-methylene-6,8,8-trimethyltricycle[5.2.2.0(1,6)]undecan-3-ol, 2,6,8-trimethyl decanoic acid methylester, 1,19-eicosadione, n-hexadecanoic acid; *Leaves CT Myrcene*—Alpha-pinene, para-cymene, limonene, (E)-beta-ocimene, 2,6-dimethyl-2,6-octadiene, perillene, neryl formate, para-cymen-7-ol, 3-oxo-p-menth-1-en-7-al, alpha-copaene, beta-caryophyllene, (3E)-4,8-dimethyl-3,7-nonadien2-ol, alpha-humulene, neryl isobutanoate, phytol; *Stems CT Thymol*—N/A; *Aerial Parts*—Para-cymene, limonene, perillone, cis-para-mentha-2-en-1-ol, crytptone, alpha-copaene, alpha-humulene, germacrene D; *Stems CT Myrcene*—Sabinene, beta-pinene, para-cymene, beta-phellandrene, 2,6-dimethyl-2,6-octadiene, perillene, linalool, neryl formate, p-cymen-7-ol, beta-elemene, cyperene, (3E)-4,8-dimethyl-3,7-nonadien2-ol, germacrene D, beta-selinene, neryl isobutanoate, cyperotundone A, T-cadinol, alpha-cadinol, selin-11-en-4alpha-ol, phytol; *Roots*—Myrcene, beta-phellandrene, carvacrol methyl ether, 2,5-dimethoxy-p-cymene, trans-alpha-bergamotene, alpha-selinene, cubebol, delta-cadinene, caryophyllene oxide, guaiol, hinesol, valerianol; *Flowers*—Alpha-pinene, para-cymene, limonene, 2,6-dimethyl-2,6-octadiene, perillene, linalool, cuminal, neryl formate, para-cymen-7-ol, 3-hydroxycineole, 4-hydroxycryptone, 3-oxo-p-menth-1-en-7-al, beta-caryophyllene, (3E)-4,8-dimethyl-3,7-nonadien2-ol, (E)-beta-farnesene, caryophyllene oxide, humulene epoxide II, phytol

PREFERRED COMPOSITION FOR CLINICAL USE:
Insufficient data to determine.

REPORTED THERAPEUTIC PROPERTIES: Aids digestion, relieves constipation, supports liver function, antioxidant, analgesic, anti-inflammatory, reduces fever, relieves headache, anticancer, antibacterial, antiallergic, wound healing, heals burns, stops excess bleeding, insecticidal, relieves anxiety, calming, antidepressant, stress management

CAUTIONS:
Thymol CT
■ Oral caution. Thymol has a longer half-life (the time it takes for half of the medication to metabolize or excrete half of the dosage) than most essential oil constituents and should not be administered orally for long periods of time.[3385] Thymol is a monoterpene phenol. Reports of fatalities in children who consumed 50 to 200 mg of phenols has been reported.[3386] It is recommended that chemotypes with high levels of thymol be limited to 10 drops per day orally for adults with a two- to seven-day break after twenty-one days of use; or limit ingestion to five or six days weekly.
■ May interact with aspirin, blood pressure, antiplatelet, and anticoagulant medications, and increase the risk of bleeding among people with bleeding disorders due to thymol and carvacrol content.[3387,3388]
■ May interact with anticholinergic (drugs used for asthma, incontinence, gastrointestinal cramps, muscular spasms, depression, and sleep disorders) and cholinergic medications (drugs used to reduce urinary retention, diagnose myasthenia gravis, and for glaucoma) based on thymol and carvacrol content.[3389,3390]
■ May interact with antibiotics or antifungals and possibly enhance their effects based on thymol and carvacrol content.[3391,3392,3393,3394,3395,3396,3397,3398]
■ May irritate mucous membranes (eyes, mouth, nasal passages, vagina, rectum).
Myrcene CT
■ Caution is advised during pregnancy and lactation due to high beta-myrcene content. Extremely high doses (500 mg/kg) of beta-myrcene have been toxic to fetuses according to animal research.[3399,3400]
■ May interact with barbiturates (medications for anxiety and insomnia), antihistamines, benzodiazepines, tricyclic antidepressants, or other central nervous system depressant drugs, increasing their depressant effects.[3401,3402,3403]
■ May weakly interfere with the enzymes responsible for metabolizing medications (NSAIDs, proton-pump inhibitors, acetaminophen, antiepileptics, immune modulators, blood-sugar medications, blood pressure medications, antidepressants, antipsychotics, diabetic medications, antihistamines, antibiotics, and anesthetics) based on myrcene content.[3404,3405]

Roots and 4-Cyclohexylbutyramide CT
■ None currently known; although, the safety of 4-Cyclohexylbutyramide has not been investigated.

Selected Evidence:

- Redflower ragleaf stem essential oil moderately inhibited AChE (acetylcholinesterase) activity, suggesting it may be useful to support memory function.[3406] This may be due to its high thymol content since thymol is a known AChE inhibitor.
- *In vitro* research demonstrated that redflower ragleaf essential oils (leaves CT 4-cyclohexylbutyramide and stems CT thymol) inhibited gram-positive (*S. aureus, B. subtilis, S. typhi*) and gram-negative bacteria (*E. coli*, *P. aeruginosa*).[3407]
- Myrcene CTs of redflower ragleaf essential oil were larvicidal to mosquitoes (*Aedes aegypti, Aedes albopictus*, and *Culex quinquefasciatus*).[3408]

REDWOOD (Coastal Redwood, California Redwood, Coast Redwood)

Sequoia sempervirens (Lamb. ex D. Don) Endl.

Family: Cupressaceae
Note: Top-Middle
Aroma Intensity: Medium-Strong
Aroma: Woody, piney, earthy
Common Extraction Method: Steam distilled or hydrodistilled from the leaves (needles)
Possible Substitute Oils: Cypress, pine, pink pepper
Blends Well With: Bergamot, cajeput, camphor, cedarwood, clary sage, coriander, eucalyptus, frankincense, galbanum, grapefruit, juniper, lavender, lemon, lime, melaleuca (tea tree), marjoram, neroli, niaouli, peppermint, ravensara, ravintsara, rosemary, Spanish sage, spikenard, thyme, silver fir, white fir
Subcellular Localization | Epigenetic Influence: Currently unknown | Currently unknown
Recommended Dilution Range: 5%–50%; Neat for some conditions

Primary Constituents:[3409,3410]

Alpha-Pinene CT		*Alpha-Phellandrene CT*	
Alpha-Pinene	5.8%–29.5%	Alpha-Phellandrene	29.6%
Gamma-Terpinene	3.6%–16.4%	Limonene	15.6%
Beta-Phellandrene	1.0%–11.4%	Alpha-Pinene	8.7%
Myrcene	1.1%–8.1%	Germacrene B	4.9%
Limonene	1.0%–6.6%	Terpinen-4-ol	4.4%
Alpha-Terpinyl Acetate	1.1%–5.9%	Alpha-Teprinyl Acetate	3.7%
Sabinene	0.9%–4.9%		
Germacrene D	0.8%–4.9%		
Terpinen-4-ol	1.3%–3.1%		

Other Constituents: *Alpha-Pinene CT*—Beta-pinene, alpha-phellandrene, alpha-terpinene, (Z)-beta-ocimene, para-cymene, terpinolene, alpha-copaene, beta-elemene, trans-2-menthen-1-ol, alpha-muurolene, cis-piperitol, delta-cadinene; *Alpha-Phellandrene CT*—Sabinene, myrcene, para-cymene, (Z)-beta-ocimene, (E)-beta-ocimene, m-cymene, gamma-terpinene, borneol, fenchone, terpinolene, para-cymen-8-

ol, cis-piperitol, trans-piperitol, citronellol, thymol, carvacrol, thymol acetate, beta-caryophyllene, germacrene D, alpha-humulene, trans-beta-farnesene, (E,E)-alpha-farnesene, spathulenol, citronellyl pentanoate, alpha-cadinol, phytol

PREFERRED COMPOSITION FOR CLINICAL USE:
Insufficient data/samples to determine.

REPORTED THERAPEUTIC PROPERTIES: Anticancer, antibacterial, antioxidant, anticancer, relieves earache, anti-inflammatory, analgesic, antiseptic, purifies the blood, reduces fatigue, relaxing, stress management, reduces anger and resentment

CAUTIONS:
- None currently known.

SELECTED EVIDENCE:
- Redwood CT alpha-phellandrene essential oil exhibited potent anticancer activity against liver, colon, and breast cancer cells.
- *In vitro* research showed that redwood CT alpha-phellandrene essential oil inhibited gram-positive (*S. aureus, B. subtilis, L. acidophilus*) and gram-negative (*S. typhimurium, E. coli, K. pneumonia, S. flexneri, P. aeruginosa, P. vulgaris*) bacteria and fungi (*A. flavus, A. niger, G. candidum, T. mentagrophytes*).[3411] The strongest activity was observed against *B. subtilis, T. mentagrophytes*, and *S. aureus*.
- The alpha-phellandrene CT of redwood essential oil displayed good antioxidant activity in the DPPH assay.[3412]

RIBBED CACHRYS (Jashir*)

Prangos ferulacea (L.) Lindl.,
Cachrys ferulacea (L.) Calestani

FAMILY: Apiaceae (Umbelliferae)
NOTE: Middle-Top
AROMA INTENSITY: Medium
AROMA: Green, woody, fresh, piney, slightly citrusy and sweet
COMMON EXTRACTION METHOD: Steam distilled or hydrodistilled from the aerial parts; may also be distilled from the roots
POSSIBLE SUBSTITUTE OILS: Douglas fir (needles), tagetes (*T. erecta*, leaves), hinoki, pink pepper, black pepper CT delta-3-carene, Turkish pine, dwarf pine, cypress, juniper berry, palo santo
BLENDS WELL WITH: Balsam fir, bergamot, black pepper, cedarwood, cypress, Douglas fir, frankincense, grapefruit, juniper berry, lavender, lemon, lime, marjoram, neroli, orange, palo santo, petitgrain, pine, rosemary, sandalwood, silver fir, tangerine, vetiver, white fir
SUBCELLULAR LOCALIZATION | EPIGENETIC INFLUENCE: Currently unknown | Currently unknown
RECOMMENDED DILUTION RANGE: 3%–50%; neat for some conditions

PRIMARY CONSTITUENTS:[3413,3414,3415,3416,3417]

Alpha-Terpinolene CT

Alpha-Terpinolene	38.1%–56.3%
Gamma-Terpinene	6.1%–13.3%
Alpha-Phellandrene	0.0%–6.2%
Beta-Phellandrene	0.0%–5.8%
Trans-Caryophyllene	3.6%–4.7%
Isopentyl Isovalerate	0.0%–4.7%
(Z)-Beta-Ocimene	0.0%–3.5%
Bornyl Acetate	1.8%–3.0%

Delta-3-Carene CT

Delta-3-Carene	28.6%–45.9%
Alpha-Pinene	0.0%–26.7%
Myrcene	0.0%–15.2%
Beta-Pinene	tr–15.0%
Indole	0.0%–11.6%
Alpha-Terpinolene	2.4%–9.6%
O-Cymene	tr–8.0%
Alpha-Terpinene	0.0%–7.6%
Para-Cymen-8-ol	0.0%–6.2%
n-Pentadecanol	0.0%–5.5%
Sabinene	0.0%–5.4%

Roots (Delta-3-Carene CT)

Delta-3-Carene	22.5%
Beta-Phellandrene	11.8%
Alpha-Pinene	8.6%
Alpha-Terpinolene	7.2%
Para-Cymene	6.3%
Alpha-Phellandrene	6.2%
Myrcene	4.5%
Sabinene	3.6%
Bornyl Acetate	3.2%
Gamma-Terpinene	3.0%

Alpha-Pinene CT

Alpha-Pinene	24.2%–41.4%
Delta-3-Carene	7.7%–34.6%
Limonene	0.0%–14.6%
Beta-Pinene	8.6%–9.5%
Alpha-Terpinolene	3.8%–8.1%
Myrcene	0.0%–7.4%
Sabinene	1.5%–4.7%
Beta-Phellandrene	0.0%–4.4%
Alpha-Phellandrene	2.0%–4.2%

Limonene CT

Limonene	55.1%
Gamma-Terpinene	10.7%
Bornyl Acetate	8.5%
Trans-Caryophyllene	3.0%

Beta-Pinene CT

Beta-Pinene	43.1%
Alpha-Pinene	22.1%
Delta-3-Carene	16.9%
Alpha-terpinolene	3.9%

Roots (Beta-Phellandrene CT)

Beta-Phellandrene	32.1%
m-Tolualdehyde	26.2%
Delta-3-Carene	25.8%
Alpha-Pinene	4.7%

* Shared name with *Prangos acaulis*.

Note: Two additional studies reported beta-pinene, alpha-pinene, and delta-3-carene as the primary constituents of ribbed cachrys essential oil (aerial parts), but no full text was available to see full composition.[3418,3419]

OTHER CONSTITUENTS: *Alpha-Terpinolene CT*—Alpha-Terpinene, (E)-beta-ocimene, 1-terpineol, allo-ocimene, alpha-terpineol, o-cymene, terpinen-4-ol, eugenol, 1-octen-3-yl acetate, cis-limonene oxide, (E)-nerolidol, endo-1-bourbonanol; *Delta-3-Carene CT*—Gamma-Terpinene, (E)-beta-ocimene, alpha-phellandrene, camphene, limonene, alpha-humulene, bornyl acetate, acetophenone; *Alpha-Pinene CT*—Gamma-Terpinene, o-cymene, trans-caryophyllene, bornyl acetate, camphene, sylvestrene, isopentyl2-methyl butanoate, methyl eugenol (< 1.1%); *Limonene CT*—(E)-Beta-ocimene, alpha-terpinolene, alpha-pinene, methyl eugenol (< 0.6%), (E)-nerolidol, beta-bisabolene, allo-ocimene; *Beta-Pinene CT*—Sabinene, myrcene, o-cymene, limonene, beta-phellandrene, trans-verbenol, p-cymen-8-ol, beta-elemene, alpha-bisabolol, osthole; *Roots (Delta-3-Carene CT)*—Alpha-Thujene, camphene, beta-pinene, p-mentha-1,5-diene-8-ol, terpinen-4-ol, m-cymen-8-ol, p—cymen-8-ol, beta-cedrene, trans-beta-farnesene, cis-muurola-4(14),5-diene, beta-bisabolene, cyclopentadecanolide, oxacycloheptadecan-2-one; *Roots (Beta-Phellandrene CT)*—Camphene, sabinene, myrcene, gamma-terpinene, coumarone, sabinyl acetate

PREFERRED COMPOSITION FOR CLINICAL USE:
Insufficient samples/data to determine.

REPORTED THERAPEUTIC PROPERTIES: Wound healing, moisturizes the skin, antiseptic, aids digestion, expels excess gas, antifungal, antibacterial, antiviral, antiparasitic, anticancer, antioxidant, anti-inflammatory, relieves kidney stones, relieves mental fatigue, aids concentration, combats inflexibility

CAUTIONS:
Delta-3-Carene CT
■ May interact with anticholinergic (drugs used for asthma, incontinence, gastrointestinal cramps, muscular spasms, depression, and sleep disorders) and cholinergic medications (drugs used to reduce urinary retention, diagnose myasthenia gravis, and for glaucoma), based on AChE-inhibtory activity of delta-3-carene.[3420,3421]

Roots (Beta-Phellandrene CT)
■ May interact with anticholinergic (drugs used for asthma, incontinence, gastrointestinal cramps, muscular spasms, depression, and sleep disorders) and cholinergic medications (drugs used to reduce urinary retention, diagnose myasthenia gravis, and for glaucoma), based on AChE-inhibtory activity of beta-phellandrene.[3422]

All Other CTs
■ None currently known.

SELECTED EVIDENCE:

- Ribbed cachrys root essential oil (beta-phellandrene CT) promoted wound healing by significantly enhancing the migration of fibroblast (L929) cells to the wound site and increasing their production of collagen type I.[3423] The study authors also reported that fibroblasts remained viable (near control levels) even at high concentrations.
- *In vitro* research demonstrates that ribbed cachrys essential oil (composition not reported) inhibits the oral pathogen *E. faecalis*.[3424] Another study found that the root oil (beta-phellandrene CT) inhibited the growth of *S. aureus* and *P. aeruginosa*.[3425]

ROMERILLO
(Butterfly Needles)

Bidens alba (L.) DC.

FAMILY: Asteraceae
NOTE: Middle
AROMA INTENSITY: Medium
AROMA: Earthy, woody, mildly sweet and floral
COMMON EXTRACTION METHOD: Steam distilled from the flowers, leaves, seeds, or roots
POSSIBLE SUBSTITUTE OILS: Hairy beggarticks
BLENDS WELL WITH: Basil, cardamom, cassia, clove, coriander, frankincense, lavender, lemon, marjoram, nutmeg, palo santo, patchouli, pine, ravensara, rosemary, sandalwood, ylang ylang

SUBCELLULAR LOCALIZATION | EPIGENETIC INFLUENCE: Currently unknown | Currently unknown
RECOMMENDED DILUTION RANGE: 2%–10%

PRIMARY CONSTITUENTS:[3426]

Flowers, Leaves, Seeds		*Roots*	
1-Phenylhepta-1,3,5-Triyne	39.1%–54.0%	7-Phenylhept-2-Ena-4,6-Diynyl Acetate	15.4%
Alpha-Pinene	10.1%–33.5%	1-Isobutyl-Eugnol Isobutyrate	14.1%
Germacrene D	tr–8.0%	Hexadecanol	12.3%
1-Phenylhept-5-Ena-1,3-Diyne	1.0%–6.4%	Octadecadienol	11.6%
Unidentified Constituent	0.9%–6.0%	1-Phenylhept-5-Ena-1,3-Diyne	10.4%
(Z)-Beta-Ocimene	1.0%–5.3%	1-Phenylhepta-1,3,5-Triyne	7.8%
Beta-Caryophyllene	0.8%–4.9%	1'-Pentanoyloxy-Eugenol Isobutyrate	6.8%
Beta-Pinene	0.3%–3.7%		

OTHER CONSTITUENTS: *Flowers, Leaves, Seeds*—Camphene, sabinene, myrcene, delta-2-carene, limonene, (E)-beta-ocimene, bornyl acetate, beta-copaene, alpha-humulene, bicyclogermacrene, spathulenol, caryophyllene oxide; *Roots*—Alpha-pinene, camphene, beta-pinene, delta-2-carene, (Z)-beta-ocimene, bornyl acetate, eugenol isobutyrate, hexadecyl acetate, octadecadienyl acetate

PREFERRED COMPOSITION FOR CLINICAL USE:
Insufficient samples/data to determine.

REPORTED THERAPEUTIC PROPERTIES: Antifungal, antibacterial, antimicrobial, anti-inflammatory, analgesic, relieves toothache, antioxidant, antiseptic, liver protective, lowers blood pressure, balances blood-sugar levels, aids oral health and cleanliness, insecticidal, emotionally balancing, releases traumatic emotional memories, fosters courage, reduces mental fatigue

CAUTIONS:

- Strongly phototoxic due to the presence of polyacetylenes (i.e., 1-phenylhepta-1,3,5-triyne); do not expose skin to UV rays for up to twenty-four hours following topical application.[3427,3428,3429]

SELECTED EVIDENCE:

- ☐ None found.

ROSE GERANIUM
(Bourbon Geranium)

Pelargonium roseum (Andrews) DC.,
P. graveolens L'Hér. ex Aiton var. *roseum*

FAMILY: Geraniaceae
NOTE: Middle
AROMA INTENSITY: Strong
AROMA: Floral, rose-like (slightly more than geranium), green
COMMON EXTRACTION METHOD: Steam distilled or hydrodistilled from the leaves

POSSIBLE SUBSTITUTE OILS: Geranium, rose

BLENDS WELL WITH: Basil, bay laurel, bergamot, carrot seed, cassia, chamomile (German, Roman), citronella, clary sage, coriander, cypress, fennel, galbanum, ginger, grapefruit, juniper, lavender, lemon, lemongrass, lime, melissa, neroli, orange, palmarosa, patchouli, peppermint, ravensara, rose, rosemary, spikenard, tangerine, ylang ylang

SUBCELLULAR LOCALIZATION | EPIGENETIC INFLUENCE: Currently unknown | Currently unknown

RECOMMENDED DILUTION RANGE: 3%–50%; neat for some conditions

PRIMARY CONSTITUENTS:[3430,3431,3432,3433]

Citronellol	27.3%–44.6%
Geraniol	0.0%–18.5%
Citronellyl Formate	0.0%–17.2%
Beta-Caryophyllene	2.6%–10.3%
Isomenthone	0.0%–6.3%
Gamma-Muurolene	0.0%–6.1%
Linalool	0.5%–5.7%
Delta-Selinene	0.0%–5.5%
Menthone	0.9%–5.0%
Caryophyllene Oxide	0.0%–4.2%
Citronellyl Acetate	0.7%–3.5%
Delta-Cadinene	1.0%–3.4%

OTHER CONSTITUENTS: Alpha-pinene, limonene, cis-rose oxide, linalool oxide, trans-rose oxide, terpinen-4-ol, alpha-terpineol, citronellal, geranyl formate, alpha-ylangene, alpha-copaene, beta-bourbonene, beta-cubebene, alpha-gurjunene, beta-gurjunene, alpha-humulene, alpha-cadinene, germacrene D, bicyclogermacrene, valencene, gamma-cadinene, spathulenol, globulol, 1,10-diepi-cubenol, alpha-cadinol, cubenol, citronellyl butanoate, phenylethyl tiglate, geranyl tiglate, alpha-agarofuran, beta-eudesmol

PREFERRED COMPOSITION FOR CLINICAL USE:

Constituent	*w/ Citronellyl Formate*	*w/o Citronellyl Formate*
Citronellol	25%–50%	30%–50%
Citronellyl Formate	12%–22%	–
Geraniol	5%–15%	5%–20%
Menthone	0.1%–7%	0.1%–7%
Cis- + Trans-Rose Oxide	1%–5%	1%–5%
Linalool	0.1%–5%	2%–10%
Delta-Cadinene	0.1%–5%	0.5%–5%
Citronellyl Acetate	0.1%–3%	1%–7%

REPORTED THERAPEUTIC PROPERTIES: Analgesic (pain relief), antibacterial, antidepressant, supports blood-sugar balance, stimulates lymph flow and drainage, anti-inflammatory, antiseptic, astringent, antineuralgic, antioxidant, diuretic, **relieves menopausal and menstrual symptoms**, aids liver function, balances menstruation, stops excess bleeding, regenerative, helps bruises heal, nervine (calms and soothes the nerves), **wound healing, reduces the appearance of scars and blemishes**, insecticidal, antidepressant, relieves anxiety, promotes a sense of security, stimulates imagination, promotes balanced emotions

CAUTIONS:

■ Use cautiously with diabetic medications (low risk). Large oral doses of citronellol and geraniol may lower blood-glucose levels.[3434,3435]

■ May interact with antibiotics and possibly enhance their effects due to citronellol and geraniol content.[3436,3437]

SELECTED EVIDENCE:

- Intraperitoneal administration of rose geranium essential oil significantly reduced anxiety and depression in mice by influencing the serotonergic system.[3438]
- Rose geranium essential oil inhibited gram-negative (*Pseudomonas aeruginosa, Proteus mirabilis, Escherichia coli*), and gram-positive bacteria (*Staphylococcus aureus, Enterococcus faecalis*) and fungi (*Candida albicans*).[3439] The oil completely inhibited the growth of *C. albicans* and exhibited antibacterial activity against *S. aureus* and *P. aeruginosa* comparable to antibacterial controls. Another study reported that rose geranium essential oil inhibited *B. cereus, S. faecium, S. aureus, E. coli, S. typhi, S. dysenteria, P. aeruginosa,* and *S. cervisiae*.[3440]
- *In vitro* and *in vivo* (in pigeons) showed that rose geranium essential oil inhibits the parasite *Trichomonas gallinae*, which infects pigeons and doves.[3441]
- Rose geranium essential oil significantly reduced populations of the greenhouse-crop-pest sweet potato whitefly (*Bemisia tabaci*).[3442] Rose geranium essential oil also demonstrated acaricidal effects against ticks (*Rhipicephalus annulatus*).[3443]

RUGOSA ROSE (Japanese Rose, Beach Rose, Ramanas Rose, Chinese Rose)

Rosa rugosa Thunb., *R. rugosa* Thunb. var. *plena* Regel

FAMILY: Rosaceae
NOTE: Middle
AROMA INTENSITY: Strong
AROMA: Floral, sweet, honey-spicy
COMMON EXTRACTION METHOD: Steam distilled or hydrodistilled from the flower petals
POSSIBLE SUBSTITUTE OILS: Rose, geranium, jasmine
BLENDS WELL WITH: Bay laurel, bergamot, black pepper, cardamom, cedarwood, chamomile (German, Roman), citronella, clary sage, copaiba, eucalyptus, geranium, helichrysum, jasmine, lavender, lemongrass, melissa, neroli, palo santo, patchouli, pine, rose, sandalwood, spikenard, vetiver, ylang ylang
SUBCELLULAR LOCALIZATION | EPIGENETIC INFLUENCE: Currently unknown | Currently unknown
RECOMMENDED DILUTION RANGE: 3%–33%; neat for some conditions

PRIMARY CONSTITUENTS:[3444,3445]

R. rugosa		*var. plena*	
Citronellol	44.5%	Citronellol	60.0%
Geraniol	12.8%	Geraniol	8.6%
Farnesol	3.8%	Nerol	2.8%

2-Tridecanone	3.3%	Citronellyl Acetate	2.7%
Aromadendrene	2.5%	(E,E)-Farnesol	2.4%
2-Dodecanone	2.5%		

Other Constituents: *R. rugosa*—2-pentadecanone, linalool, 2-tridecanol, beta-cubebene, eugenol, methyl eugenol (< 0.2%); *var. plena*—Heptanal, alpha-terpineol, geranial, (E,Z)-farnesol, methyl eugenol (< 1.6%), tridecan-2-One, linalool

Preferred Composition for Clinical Use:

Constituent	
Citronellol	35%–60%
Geraniol	5%–20%

Reported Therapeutic Properties: Aids circulation, aids digestion, relieves diarrhea, increases appetite, supports spleen and liver function, promotes hormone release by the endocrine system, diuretic, reduces the appearance of acne and other blemishes, astringent, antiarthritic, analgesic, balances menstruation, relieves menopausal symptoms, reduces fever, aids cardiovascular function, promotes restful sleep, antidepressant, relieves anxiety, relieves despair and grief, reduces loneliness

Cautions:

▪ Use cautiously with diabetic medications (low risk). Large oral doses of citronellol and geraniol may lower blood-glucose levels.[3446,3447]

▪ May interact with antibiotics and possibly enhance their effects due to citronellol and geraniol content.[3448,3449]

▪ Dilution recommended for topical application of rose oil with high levels of geraniol. Geraniol is metabolized via the CYP450 pathway into metabolites such as geranial and neral and may cause sensitization and irritation when applied topically.[3450]

Selected Evidence:

- □ None found.

RUSSIAN WORMWOOD

Artemisia vestita Wall. ex Besser,
A. vestita var. *vestita*

Family: Asteraceae
Note: Note
Aroma Intensity: Medium
Aroma: Fresh, herbaceous, slightly sweet, woody, and balsamic
Common Extraction Method: Steam distilled from the aerial parts; may also be distilled from the leaves
Possible Substitute Oils: Eucalyptus, Atlas cedarwood, rosemary
Blends Well With: Bay laurel, bergamot, cajeput, camphor, chamomile (German, Roman), coriander, eucalyptus, lavandin, lavender, lemon, orange, oregano, peppermint, ravensara, rosemary, sage, Spanish sage, tea tree, thyme
Subcellular Localization | Epigenetic Influence: Currently unknown | Currently unknown
Recommended Dilution Range: 3%–20%; 50% for some conditions

PRIMARY CONSTITUENTS:[3451,3452,3453,3454]

Beta-Himachalene CT (Indian)		*Grandisol CT (Chinese)*	
Beta-Himachalene	10.1%	Grandisol	40.3%–45.1%
(E)-Alpha-Atlantone	6.9%	1,8-Cineole	14.9%–22.5%
1,8-Cineole	5.3%	Camphor	5.2%–11.4%
Himacholol	5.0%	Germacrene D	5.1%–7.6%
Artemisia Alcohol	4.5%	Gamma-Elemene	0.0%–3.9%
Alpha-Himachalene	3.5%	Beta-Caryophyllene	2.5%–3.4%
Santolina Alcohol	3.1%	Alpha-Myrcene	0.0%–3.2%
Artemisyl Acetate	3.0%		
Allohimachalol	3.0%		

Leaves	
1,8-Cineole	46.8%
Geranial	13.7%
Limonene	9.8%
Alpha-Phellandrene	6.4%
Camphor	5.0%
Alpha-Thujone	3.0%
Beta-Thujone	3.0%

OTHER CONSTITUENTS: *Beta-Himachalene CT (Indian)*—Alpha-pinene, camphene, sabinene, yomogi alcohol, para-cymene, Artemisia ketone, alpha-thujone (< 1.9%), beta-thujone (< 0.7%), cis-p-menthen-1-ol, trans-thujanol, sabinol, borneol, santolinyl acetate, trans-piperitol, fragranol; *Grandisol CT (Chinese)*—Alpha-pinene, beta-pinene, camphene, delta-3-carene, artemiseole, pinocamphone (< 2.2%), pinocarvone, terpinen-4-ol, alpha-humulene, caryophyllene oxide, alpha-bisabolol, gamma-pyronene, alpha-copaene, gamma-himachalene, alpha-zingiberene, delta-cadinene, (Z)-alpha-santalol; *Leaves*—Pinocamphone (< 1.5%), beta-cubebene, bisabolol oxide B

PREFERRED COMPOSITION FOR CLINICAL USE:
Insufficient samples/data to determine.

REPORTED THERAPEUTIC PROPERTIES: Anti-inflammatory, antiarthritic, antirheumatic, reduces fever, antibacterial, antifungal, antioxidant, insecticidal, aids mental clarity, promotes concentration

CAUTIONS:

Beta-Himachalene CT (Indian)

- None currently known.

Grandisol CT (Chinese)

- Avoid with children under six due to significant camphor content. Several cases of camphor poisoning and/or seizures from ingestion and topical application have been reported in children.[3455,3456] Ingestion of camphor-containing products has been lethal in children under age two.[3457] Children five years and up may use camphor-containing essential oils topically in dilutions no stronger than 5%. Camphor ingestion by infants and young children may cause cough, vomiting, seizure, burning sensation in the mucous membranes and eyes, or lack of voluntary coordination of muscle movements.[3458]

■ Caution is warranted during pregnancy and while lactating, due to significant camphor content. Ingestion of essential oils with significant levels of camphor may lead to abortion because fetuses lack the enzymes to process it.[3459] Camphor ingestion by infants and young children may cause cough, vomiting, seizure, burning sensation in the mucous membranes and eyes, or lack of voluntary coordination of muscle movements.[3460]
■ Oral caution—Essential oils with significant levels of camphor can be toxic when taken orally, usually single doses exceeding 2 mL, although the lethal dose for humans is estimated to be more than 5 mL in a single dose.[3461,3462,3463]
■ Avoid with epilepsy and Parkinson's disease due to 1,8-cineole and camphor content. May exacerbate or cause seizures/convulsions or reduce seizure-medication efficacy.[3464,3465,3466,3467,3468,3469]
■ The significant camphor content in feverfew may negatively impact red blood cells and increase the risk of jaundice in children with glucose-6-phosphate dehydrogenase deficiency (G6PD).[3470,3471]

Leaves
■ Avoid with children under 6 due to thujone, 1,8-cineole, and camphor content. Cases of seizure have been reported in young children due to sage essential oil (rich in thujones) exposure.[3472,3473] Several cases of camphor poisoning and/or seizures from ingestion and topical application have been reported in children.[3474,3475] Ingestion of camphor-containing products has been lethal in children under age 2.[3476] 1,8-cineole may cause seizures, central nervous system problems, or respiratory distress in young children.[3477,3478,3479] Children 6 years and up may use toothache tree essential oil topically in dilutions no stronger than 5%.
■ Avoid during pregnancy and while lactating, due to thujone and camphor content. Camphor ingestion may lead to abortion because fetuses lack the enzymes to process it.[3480] Essential oils high in thujone content may cause abortion.[3481,3482] Camphor ingestion by infants and young children may cause cough, vomiting, seizure, burning sensation in the mucous membranes and eyes, or lack of voluntary coordination of muscle movements.[3483]
■ Caution is warranted orally due to the significant amounts of 1,8-cineole. Limit it to small doses internally (adults—maximum 10 drops daily). One text recommends a maximum daily dose of 6 drops, whereas the European Medicines Agency recommends 3.5 to 7 drops (100–200 mg) of high 1,8-cineole essential oils orally, 2 to 5 times daily for adolescents, adults, and the elderly.[3484,3485] Toxicity has been reported when eucalyptus (also high in 1,8-cineole) was ingested in large doses, and as few as 4 to 5 drops may cause problems in very sensitive individuals.[3486,3487,3488,3489,3490] In humans, 3.5 to 5 mL has proven fatal orally.[3491] The small amounts of neurotoxic thujones in conjunction with the 1,8-cineole further warrants caution.
■ Avoid with epilepsy and Parkinson's disease due to 1,8-cineole, thujones, and camphor content. May exacerbate or cause seizures/convulsions or reduce seizure-medication efficacy based on 1,8-cineole content.[3492,3493,3494,3495]
■ May weakly interfere with the enzymes responsible for metabolizing medications (NSAIDs, proton-pump inhibitors, acetaminophen, antiepileptics, immune modulators, blood-sugar medications, blood pressure medications, antidepressants, antipsychotics, diabetic medications, antihistamines, antibiotics, and anesthetics) due to 1,8-cineole content.[3496,3497,3498]
■ May interfere with pentobarbital and other barbiturates (medications for anxiety and insomnia), based on 1,8-cineole content.[3499,3500]

SELECTED EVIDENCE:

- □ *In vitro* research demonstrates that Russian wormwood essential oil CT grandisol inhibits *Streptococcus pyogenes* and its biofilm formation.[3501] The same study concluded that oral administration of Russian wormwood essential oil (100 mcg/kg body weight) significantly reduced the number of viable bacterial cells in the lungs of mice.

- ☐ The grandisol CT of Russian wormwood essential oil demonstrated contact toxicity and strong fumigant toxicity against adult maize weevils (*Sitophilus zeamais*).[3502]

SAKHALIN FIR
(Japanese Fir, Japanese Pine)

Abies sachalinensis F.Schmidt

FAMILY: Pinaceae
NOTE: Top-Middle
AROMA INTENSITY: Medium
AROMA: Woody, piney, fresh, balsamic
COMMON EXTRACTION METHOD: Steam distilled or hydrodistilled from the leaves (needles) or leaves and twigs; may also be distilled from the cones
POSSIBLE SUBSTITUTE OILS: Balsam fir, white fir, silver fir, spruce (black), cypress, pine
BLENDS WELL WITH: Balsam fir, birch, blue spruce, cassia, cedarwood, chamomile (German, Roman), cypress, frankincense, galbanum, lavender, lemon, myrrh, myrtle, palo santo, pine, sandalwood, silver fir, spruce (black), tsuga, white fir, wintergreen
SUBCELLULAR LOCALIZATION | EPIGENETIC INFLUENCE: Currently unknown | Currently unknown
RECOMMENDED DILUTION RANGE: 5%–Neat

PRIMARY CONSTITUENTS:[3503,3504,3505]

Alpha-Pinene CT	
Alpha-Pinene	17.4%–26.6%
Camphene	13.7%–20.0%
Bornyl Acetate	7.8%–18.5%
Borneol	tr–14.0%
Beta-Phellandrene	1.2%–10.7%
Beta-Pinene	2.7%–9.5%
Limonene	4.4%–8.9%
Myrcene	0.5%–6.1%
Unidentified [MW=222]	0.0%–6.0%
Beta-Caryophyllene	tr–4.4%

Bornyl Acetate CT	
Bornyl Acetate	17.0%–30.3%
Alpha-Pinene	12.2%–22.3%
Camphene	16.8%–21.2%
Beta-Pinene	4.1%–8.2%
Borneol	0.7%–8.2%
Beta-Phellandrene	2.6%–7.5%
Limonene	3.5%–6.5%
Unidentified [MW=222]	3.0%–5.0%
Beta-Caryophyllene	1.4%–4.2%
Myrcene	1.3%–3.9%

Limonene CT	
Limonene	29.6%–33.1%
Alpha-Pinene	14.0%–16.6%
Bornyl Acetate	7.6%–11.5%
Beta-Phellandrene	5.7%–11.2%
Beta-Pinene	3.6%–10.2%
Camphene	5.9%–8.8%
Unidentified [MW=222]	1.8%–6.4%
Myrcene	2.0%–5.1%
Borneol	1.1%–3.9%

Note: Higher bornyl acetate and reduced alpha-pinene content are typically obtained from needles harvested during the colder months, like December and January.[3506]

OTHER CONSTITUENTS: *Alpha-Pinene CT*—Santene, tricyclene, terpinolene, alpha-fenchyl acetate, linalool, linalyl acetate, alpha-humulene, gamma-selinene, alpha-terpinyl acetate, unknown [MW=204]; *Bornyl Acetate CT*—Santene, tricyclene, sabinene, terpinolene, alpha-fenchyl acetate, alpha-humulene, gamma-selinene, alpha-terpinyl acetate, delta-cadinene, unknown [MW=204], (E)-nerolidol; *Limonene CT*—Tricyclene, delta-3-carene, terpinolene, linalool, linalyl acetate, beta-caryophyllene, alpha-humulene, gamma-selinene, unknown [MW=204], tricosane

PREFERRED COMPOSITION FOR CLINICAL USE:

Constituent	*Alpha-Pinene CT*	*Bornyl Acetate CT*
Alpha-Pinene	15%–35%	10%–27%
Camphene	10%–25%	12%–25%
Bornyl Acetate	5%–20%	15%–35%
Beta-Pinene	2%–12%	3%–15%
Limonene	2%–10%	2%–10%
Myrcene	0.1%–8%	0.5%–8%
Beta-Caryophyllene	tr–7%	0.5%–8%

REPORTED THERAPEUTIC PROPERTIES: Analgesic, reduces fever, anti-inflammatory, antiarthritic, antirheumatic, relieves sprains and strains, antibacterial, antifungal, anticancer, antiseptic, antioxidant, aids circulation, increases localized circulation, antispasmodic, disinfectant, wound healing, removes excess mucus, relieves cough, supports respiratory function, calming, stress management, **relieves anxiety**, enhances focus, encourages feelings of stability and empowerment, combats negative emotions and memories

CAUTIONS:

- None currently known.

SELECTED EVIDENCE:

- Inhalation of Sakhalin fir CT alpha-pinene essential oil significantly reduced anxiety in mice.[3507] Interestingly, intraperitoneal administration of the oil did not produce any effect possibly because inadequate levels of its major constituent were achieved in the brain with this method. The study authors also observed that alpha-pinene was present in the brain to a greater degree after inhalation when compared to intraperitoneal administration (19.7 mg/kg vs. 5.7 mg/kg).

SALVA-DE MARAJÓ (Salve of Happiness, Alecrim-d'Angola, Mexican Oregano)

Lippia origanoides H.B.K.

FAMILY: Verbenaceae
NOTE: Middle
AROMA INTENSITY: Strong
AROMA: Herbaceous, spicy, medicinal
COMMON EXTRACTION METHOD: Steam distilled from the leaves; may also be distilled from the inflorescences or aerial parts

Possible Substitute Oils: Oregano, thyme; copaiba, hemp; hops; eucalyptus
Blends Well With: Bergamot, camphor, cedarwood, chamomile (German, Roman), cinnamon, clove, cypress, eucalyptus, frankincense, ginger, lavender, lemon, marjoram, melaleuca (tea tree), myrrh, orange, oregano, pine, rosemary, summer savory, thyme, winter savory
Subcellular Localization | Epigenetic Influence: Currently unknown | Currently unknown
Recommended Dilution Range: 1.5%–20%; 50% for some conditions

Primary Constituents:[3508,3509,3510,3511,3512,3513,3514,3515,3516,3517,3518,3519,3520,3521]

Carvacrol CT

Carvacrol	35.9%–63.8%
Para-Cymene	6.3%–18.4%
Gamma-Terpinene	0.1%–17.0%
Thymol	4.9%–15.0%
p-Methoxythymol	0.0%–14.1%
(E)-Beta-Caryophyllene	0.8%–4.6%
Linalool	0.0%–3.9%
p-Methyoxycarvacrol	0.0%–3.7%
Caryophyllene Oxide	0.0%–3.6%
Beta-Myrcene	1.0%–3.3%

1,8-Cineole CT

1,8-Cineole	21.9%–58.2%
Alpha-Phellandrene	0.4%–17.8%
Alpha-Terpineol	0.5%–8.6%
Carvacrol	0.0%–7.4%
(E)-Beta-Caryophyllene	2.5%–7.3%
Germacrene D	0.0%–6.9%
Gamma-Terpinene	1.3%–5.8%
Sabinene	3.0%–4.2%
Alpha-Pinene	1.6%–4.2%
Limonene	1.8%–4.1%
Alpha-Humulene	tr–3.7%
Para-Cymene	1.0%–3.4%

Beta-Caryophyllene CT

Beta-Caryophyllene	33.7%–60.6%
Delta-Cadinene	2.9%–10.3%
Alpha-Copaene	8.5%–8.5%
Caryophyllene Oxide	6.0%–6.2%
Alpha-Muurolol	0.7%–5.5%
(E)-Beta-Ocimene	0.0%–4.5%
1-Epi-Cubenol	0.0%–4.5%
1,8-Cineole	0.0%–4.3%
Alpha-Humulene	2.7%–3.7%

Thymol CT

Thymol	29.6%–68.2%
Para-Cymene	0.0%–18.0%
Benzyl Benzoate	0.0%–17.2%
Gamma-Terpinene	0.0%–10.9%
(E)-Beta-Caryophyllene	1.9%–6.2%
Beta-Selinene	0.0%–4.6%
Viridiflorene	0.0%–3.9%
Caryophyllene Oxide	0.0%–3.8%
Salicylic Acid	0.0%–3.6%
Alpha-Humulene	0.0%–3.3%
Thymol Methyl Ether	0.4%–3.1%

Para-Cymene CT

Para-Cymene	18.4%
1,8-Cineole	17.4%
Thymol Methyl Ether	10.7%
Gamma-Terpinene	8.5%
Thymol	8.2%
(E)-Beta-Caryophyllene	6.2%
Alpha-Phellandrene	4.1%
Limonene	4.0%
Alpha-Humulene	3.1%

Thymol Methyl Ether CT

Thymol Methyl Ether	21.6%
Para-Cymene	19.6%
Thymol	13.5%
Gamma-Terpinene	12.0%
(E)-Beta-Caryophyllene	6.4%
1,8-Cineole	6.0%

Gamma-Selinene CT

Gamma-Selinene	16.7%
7-Epi-Alpha-Selinene	13.5%

Alpha-Humulene CT

Alpha-Humulene	40.0%–66.3%
Beta-Caryophyllene	9.5%–24.9%
Humulene Epoxide II	4.6%–5.0%
Alpha-Copaene	4.2%–4.6%
Caryophyllene Oxide	0.7%–3.9%

Aerial Parts (Carvacrol CT)

Carvacrol	33.5%–46.2%
Thymol	0.0%–18.5%
Para-Cymene	10.3%–15.8%
Gamma-Terpinene	4.1%–13.2%
Thymol Methyl Ether	1.0%–8.7%
1,8-Cineole	0.0%–6.2%
Beta-Caryophyllene	1.6%–5.9%
Alpha-Terpinene	1.5%–3.7%

Aerial Parts (Thymol CT)

Thymol	53.6%–88.2%
Para-Cymene	6.4%–11.5%
Gamma-Terpinene	0.0%–6.3%
Myrcene	0.0%–4.5%
(E)-Beta-Caryophyllene	0.3%–3.2%

Aerial Parts (Beta-Caryophyllene CT)

Beta-Caryophyllene	22.1%
Beta-Phellandrene	17.7%
Alpha-Phellandrene	17.6%
Caryophyllene Oxide	4.9%
Trans-Calamenene	4.3%
Linalool	4.2%

Aerial Parts ((E)-Nerolidol CT)

(E)-Nerolidol	27.9%
Beta-Caryophyllene	12.7%
1,8-Cineole	7.7%
Carvacrol	5.3%
Caryophyllene Oxide	3.2%
Alpha-Pinene	3.1%
Globulol	3.0%

Inflorescences (α-Humulene CT)

Alpha-Humulene	58.7%
Beta-Caryophyllene	12.7%
Alpha-Pinene	8.4%
Alpha-Copaene	3.4%
Limonene	3.0%
Alpha-Humulene	10.3%
Beta-Caryophyllene	9.5%
Alpha-Cadinol	8.6%
Selina-11-en-4-alpha-ol	8.1%
Alpha-Selinene	6.0%
Linalool	4.1%
Valencene	3.6%
Beta-Selinene	3.1%

Aerial Parts (1,8-Cineole CT)

1,8-Cineole	64.1%–70.5%
Alpha-Terpineol	5.0%–12.0%
Sabinene	2.2%–4.9%
Alpha-Pinene	2.5%–4.6%
Beta-Caryophyllene	1.0%–3.8%
Gamma-Terpinene	1.7%–3.7%
Terpinen-4-ol	3.0%–3.1%

Aerial Parts (Para-Cymene CT)

Para-Cymene	11.2%–15.7%
(E)-Beta-Caryophyllene	9.4%–11.3%
Alpha-Phellandrene	8.7%–9.9%
Limonene	6.9%–7.2%
Beta-Phellandrene	6.5%–6.7%
Alpha-Humulene	5.7%–6.0%
Caryophyllene Oxide	2.2%–3.7%
Alpha-Eudesmol	1.4%–3.2%
Alpha-Pinene	2.3%–3.1%
Borneol	1.9%–3.1%

Inflorescences (β-Caryophyllene CT)

Beta-Caryophyllene	34.2%–63.4%
Alpha-Humulene	2.2%–35.0%
Alpha-Pinene	0.0%–27.2%
1,8-Cineole	0.2%–8.7%
(E)-Beta-Ocimene	0.0%–7.9%
Alpha-Copaene	3.7%–5.4%
Delta-Cadinene	0.8%–5.4%
Linalool	1.2%–5.2%
Gamma-Terpinene	0.0%–5.1%
Beta-Elemene	1.6%–4.2%
Bicyclogermacrene	tr–4.1%

Inflorescences (γ-Selinene CT)

Gamma-Selinene	14.9%
Beta-Caryophyllene	11.2%
7-Epi-Alpha-Selinene	10.5%
Beta-Elemene	10.2%

Inflorescences (1,8-Cineole CT)

1,8-Cineole	52.2%
Alpha-Terpineol	6.7%
Sabinene	5.8%
Bicyclogermacrene	5.3%
Alpha-Pinene	4.6%
Beta-Caryophyllene	3.9%
Spathulenol	3.4%

Leaves & Inflorescences

Carvacrol	33.2%–56.5%
Linalool	2.7%–15.9%
Bicyclogermacrene	4.0%–8.5%
Gamma-Terpinene	4.5%–8.0%
Para-Cymene	4.3%–7.8%
Beta-Caryophyllene	4.0%–5.3%
Thymol	2.3%–4.0%
Camphor	0.9%–3.2%

Aerial Parts ((E)-Methyl Cinnamate CT)

(E)-Methyl Cinnamate	40.0%
Hedycaryol	8.0%
Alpha-Eudesmol	7.6%
Beta-Eudesmol	7.3%
Beta-Caryophyllene	4.2%
Gamma-Eudesmol	3.5%

Alpha-Humulene	8.3%
Alpha-Selinene	6.0%
Alpha-Cadinol	5.9%
Selin-11-en-4-alpha-ol	5.8%
Linalool	5.7%
Beta-Selinene	3.7%
Valencene	3.0%

Leaves & Flowers

Thymol	34.5%–59.6%
Carvacrol	0.0%–25.8%
Para-Cymene	8.4%–9.3%
Gamma-Terpinene	7.3%–9.1%
Beta-Caryophyllene	3.4%–4.3%
Myrcene	3.0%–3.6%

OTHER CONSTITUENTS: *Carvacrol CT*—Alpha-thujene, alpha-pinene, alpha-terpinene, limonene, 1,8-cineole, terpinen-4-ol, thymol methyl ether, thymol acetate, carvacrol acetate, geranyl acetate, alpha-humulene, beta-bisabolene, (Z)-alpha-bisabolene, trans-gamma-bisabolene, germacrene D, viridiflorol; *Thymol CT*—Alpha-thujene, myrcene, alpha-terpinene, limonene, 1,8-cineole, terpinen-4-ol, thymol acetate, cis-alpha-bergamotene, beta-cubebene, 9-epi-(E)-caryophyllene, germacrene D, gamma-muurolene, bicyclogermacrene, germacrene D-4-ol, palustrol, viridiflorol, delta-amorphene, humulene epoxide II, beta-bisabolene; *1,8-Cineole CT*—Alpha-pinene, alpha-terpinene, borneol, thymol methyl ether, carvacrol acetate, cis-alpha-bergamotene, 9-epi-(E)-caryophyllene, bicyclogermacrene, germacrene D-4-ol, viridiflorol, spathulenol, caryophyllene oxide; *Para-Cymene CT*—Alpha-pinene, camphene, beta-pinene, myrcene, alpha—3-carene, linalool, borneol, terpinen-4-ol, alpha-terpineol, thymol acetate, germacrene D, bicyclogermacrene, caryophyllene oxide, humulene epoxide II, alpha-eudesmol; *Thymol Methyl Ether CT*—Alpha-thujene, myrcene, alpha-phellandrene, limonene, beta-phellandrene, alpha-terpinene, thymol acetate, alpha-humulene, gamma-muurolene, bicyclogermacrene, caryophyllene oxide; *Beta-Caryophyllene CT*—Alpha-pinene, camphene, para-cymene, (Z)-beta-ocimene, linalool, alpha-gurjunene, alpha-guaiene, alloaromadendrene, germacrene D, alpha-selinene, alpha-muurolene, humulene epoxide II, cubenol, alpha-cadinol; *Alpha-Humulene CT*—Alpha-pinene, 1,8-cineole, linalool, alloaromadendrene, germacrene D, alpha-muurolene, delta-cadinene, 1-epi-cubenol, cubenol, alpha-muurolol, alpha-cadinol; *Gamma-Selinene CT*—Sabinene, alpha-copaene, aromadendrene, gamma-muurolene, germacrene B, caryophyllene oxide, humulene epoxide II, 1-epi-cubenol, cubenol, alpha-muurolol; *Aerial Parts (Carvacrol CT)*—Alpha-thujene, alpha-pinene, camphene, sabinene, myrcene, alpha-phellandrene, beta-phellandrene, (E)-beta-ocimene, limonene, 1,8-cineole, terpinen-4-ol, linalool, thymol methyl ether, carvacrol methyl ether, carvacrol acetate, alpha-humulene, alloaromadendrene, beta-selinene, gamma-bisabolene, 2-phenyl ethyl tiglate, alpha-cadinol, butyl hydroxy anisole, (Z)-gamma-bisabolene, butyl hydroxyl anisole acetate, caryophyllene oxide; *Aerial Parts (1,8-Cineole CT)*—Beta-pinene, delta-3-carene, alpha-terpinene, para-cymene, cis-sabinene hydrate, terpinolene, trans-sabinene hydrate, delta-terpineol, thymol methyl ether, caryophyllene oxide; *Aerial Parts (Thymol CT)*–Linalool, alpha-humulene; *Aerial Parts (Beta-Caryophyllene CT)*—Terpinen-4-ol, alpha-copaene, (Z)-beta-farnesene, geranyl acetate, gamma-curcumene, globulol, alpha-acorenol, caryophyllne-4(12),8(13)-dien-5α-ol, alpha-cadinol, epi-alpha-bisabolol, alpha-bisabolol, (2E,6Z)-farnesal, (2E,6E)-farnesal, isoamyl (E)-cinnamate; *Aerial Parts (Para-Cymene CT)*—Myrcene, alpha-terpinene, (E)-beta-ocimene, gamma-terpinene, linalool, thymol, delta-cadinene; *Aerial Parts ((E)-Nerolidol CT)*—Alpha-phellandrene, para-cymene, gamma-terpinene, linalool, p-menth-1,5-dien-8-ol, terpinene-4ol, alpha-terpineol, thymol methyl ether, thymol, alpha-cubebene, alpha-humulene, trans-muurola-4(14),5-diene, caryophyllne-4(12),8(13)-dien-5α-ol; *Inflorescences (β-Caryophyllene CT*—Myrcene, alpha-terpinene, para-cymene, limonene, 1-octen-3-ol, borneol, terpinen-4-ol, alpha-terpineol, alpha-guaiene, alloaromadendrene, germacrene D, beta-selinene, alpha-muurolene, germacrene A, 1-epi-cubenol, cubenol, alpha-muurolol, alpha-cadinol; *Inflorescences (α-Humulene CT)*—Camphene, myrcene, (E)-beta-ocimene, gamma-terpinene, 1-octen-3-ol, linalool, beta-elemene, gamma-muurolene, beta-selinene, delta-cadinene, caryophyllene oxide, 1-epi-cubenol, alpha-cadinol; *Inflorescences (1,8-Cineole CT)*—Alpha-thujene, beta-pinene, para-cymene, gamma-terpinene, linalool, borneol, terpinen-4-ol, beta-elemene, germacrene D, caryophyllene oxide, cubenol; *Inflorescences (γ-Selinene CT)*—(E)-Beta-ocimene, alpha-copaene, aromadendrene, gamma-muurolene, germacrene D, germacrene A, germacrene B, caryophyllene oxide, humulene epoxide II, 1-epi-cubenol, alpha-muurolol; *Leaves & Flowers*—Alpha-thujene, alpha-terpinene, limonene, cis-sabinene hydrate, terpinene-4-ol, thymol methyl ether, thymol acetate, trans-alpha-bergamotene, alpha-humulene; *Leaves & Inflorescences*—Alpha-thujene, camphene, myrcene, alpha-terpinene, camphor, borneol, terpinen-4-ol, thymol methyl ether, carvacrol acetate, alpha-bergamotene, aromadendrene, alpha-humulene, spathulenol, caryophyllene oxide; *Aerial Parts ((E)-Methyl Cinnamate CT*—Alpha-pinene, limonene, (E)-beta-ocimene, (Z)-methyl cinnamate, alpha-copaene, alpha-humulene, germacrene D, delta-cadinene, spathulenol, guaiol, ermoligenol, hinesol

PREFERRED COMPOSITION FOR CLINICAL USE:

Constituent	*Carvacrol CT*
Carvacrol	40%–65%
Para-Cymene	5%–18%
Thymol	5%–15%
Gamma-Terpinene	0.1%–10%
(E)-Beta-Caryophyllene	0.5%–5%
Beta-Myrcene	0.5%–5%

REPORTED THERAPEUTIC PROPERTIES: Aids digestion, combats nausea, relieves stomachache, eases heartburn, expels excess gas, relieves diarrhea, antiseptic, anticancer, **antimicrobial**, **antibacterial**, **antifungal**, antiviral, anti-infectious, antiparasitic, antioxidant, anti-inflammatory, analgesic, antispasmodic, relieves menstrual pain, aids respiratory function, opens airways, expectorant, relieves skin conditions, insecticidal, antidepressant, promotes a sense of security

CAUTIONS:

Carvacrol CTs

■ Avoid during pregnancy and lactation. Oregano (also rich in carvacrol) may negatively affect embryonic development and encourage fetal cell death according to animal research.[3522]

■ May interact with diabetic medications and cause low blood-sugar levels due to carvacrol content—carvacrol reduces blood glucose and enhances the antidiabetic activities of rosiglitazone in animals.[3523,3524,3525,3526]

■ May interact with aspirin, blood pressure, antiplatelet, and anticoagulant medications, and increase the risk of bleeding among people with bleeding disorders, due to carvacrol and thymol content.[3527,3528,3529,3530,3531]

■ May interact with anticholinergic (drugs used for asthma, incontinence, gastrointestinal cramps, muscular spasms, depression, and sleep disorders) and cholinergic medications (drugs used to reduce urinary retention, diagnose myasthenia gravis, and for glaucoma), based on thymol and carvacrol content.[3532,3533]

■ May interact with antibiotics or antifungals and possibly enhance or decrease their effects due to carvacrol or thymol content, and based on research with the whole essential oil.[3534,3535,3536,3537,3538]

■ May irritate mucous membranes (eyes, mouth, nasal passages, vagina, rectum).

Thymol CTs

■ Oral caution. Thymol has a longer half-life (the time it takes for half of the medication to metabolize or excrete half of the dosage) than most essential oil constituents and should not be administered orally for long periods of time.[3539] Thymol is a monoterpene phenol. Reports of fatalities in children who consumed 50 to 200 mg of phenols has been reported.[3540] It is recommended that chemotypes with high levels of thymol be limited to 10 drops per day orally for adults with a two- to seven-day break after twenty-one days of use; or limit ingestion to five or six days weekly.

■ May interact with aspirin, blood pressure, antiplatelet, and anticoagulant medications, and increase the risk of bleeding among people with bleeding disorders.[3541,3542]

■ May interact with anticholinergic (drugs used for asthma, incontinence, gastrointestinal cramps, muscular spasms, depression, and sleep disorders) and cholinergic medications (drugs used to reduce urinary retention, diagnose myasthenia gravis, and for glaucoma), based on thymol and carvacrol content.[3543,3544]

■ May interact with antibiotics or antifungals and possibly enhance their effects based on thymol, carvacrol, or linalool content.[3545,3546,3547,3548,3549,3550,3551,3552]

■ May irritate mucous membranes (eyes, mouth, nasal passages, vagina, rectum).

1,8-Cineole CTs

■ Avoid with children under age 3, particularly around the nose and mouth and use cautiously in children under age 5 due to significant 1,8-cineole content. 1,8-cineole may cause seizures, central nervous system problems, or respiratory distress in young children.[3553,3554,3555]

■ Caution is warranted orally due to the significant amounts of 1,8-cineole. Limit it to small doses internally (adults—maximum 10 drops daily). Toxicity has been reported when eucalyptus (also high in 1,8-cineole) was ingested in large doses, and as few as 4 to 5 drops may cause problems in very sensitive individuals.[3556,3557,3558,3559,3560,3561]

■ Avoid with epilepsy and Parkinson's disease due to 1,8-cineole content. May exacerbate or cause seizures/convulsions or reduce seizure-medication efficacy based on 1,8-cineole content.[3562,3563,3564,3565]

■ May weakly interfere with the enzymes responsible for metabolizing medications (NSAIDs, proton-pump inhibitors, acetaminophen, antiepileptics, immune modulators, blood-sugar medications, blood pressure medications, antidepressants, antipsychotics, diabetic medications, antihistamines, antibiotics, and anesthetics) due to 1,8-cineole content.[3566,3567,3568]

■ May interfere with pentobarbital and other barbiturates (medications for anxiety and insomnia), based on 1,8-cineole content.[3569,3570]

Para-Cymene CT

■ Avoid with epilepsy and Parkinson's disease due to 1,8-cineole content. May exacerbate or cause seizures/convulsions or reduce seizure-medication efficacy based on 1,8-cineole content.[3571,3572,3573,3574]

(E)-Nerolidol CT

■ May interfere with antibiotics and increase their effectiveness due to (E)-nerolidol content. ((E)-nerolidol sensitizes bacteria to antibiotics).[3575]

■ May interfere with enzymes (CYP1A, CYP2B, CYP3A) responsible for metabolizing medications based on alpha-bisabolol content (NSAIDs, proton-pump inhibitors, acetaminophen, antiepileptics, immune modulators, blood-sugar medications, blood pressure medications, antidepressants, antipsychotics, diabetic medications, antihistamines, antibiotics, and anesthetics) due to (E)-nerolidol content.[3576]

Alpha-Humulene CTs, Beta-Caryophyllene CTs, Thymol Methyl Ether CT, (E)-Methyl Cinnamate, Gamma-Selinene CTs

■ None currently known.

SELECTED EVIDENCE:

- □ Salva-de marajó CT thymol (leaves & flowers) essential oil protected DNA against bleomycin-induced damage, suggesting a chemopreventive (cancer preventing) effect.[3577] The study authors attributed the chemopreventive effect to thymol and carvacrol, which also protected DNA.
- □ The carvacrol CT of salva-de marajó relaxed isolated guinea pig tracheal smooth muscle possibly by stimulating soluble guanylyl cyclase and subsequent activation of calcium and potassium channels.[3578]
- □ Salva-de marajó CT 1,8-cineole (aerial parts) inhibited tyrosinase activity, suggesting it may lighten the skin.[3579] Tyrosinase is an enzyme involved in the production of melanin (a dark brown to black pigment that is responsible for tanning of skin when exposed to sunlight) by oxidation of tyrosine.
- □ Salva-de marajó CT carvacrol reduced yellow fever virus yield 100% *in vitro*.[3580]
- □ *In vitro* research showed that salva-de marajó CT carvacrol essential oil *S. aureus* and *E. coli*.[3581] Another study showed that the carvacrol CT of salva-de marajó aerial parts significantly inhibited *C. albicans* (two types), *c. guilliermondii, C. parapsilosis, Cryptococcus neoformans, T. rubrum, F. pedrosi, S. aureus,* MRSA, *S. mutans,* and *L. casei*, with inhibition zones greater than vancomycin, methicillin, and amphotericin B for some organisms.[3582] Other researchers also reported strong inhibition of *E. coli, S. aureus, MRSA, C. albicans*, and *C tropicalis* by the carvacrol CT (aerial parts). The carvacrol CT demonstrated bacteriostatic and bactericidal action against *A. hydrophila* (a gram-negative bacterium found in fresh to brackish water associated with stomach upset, diarrhea, and vomiting).[3583] A study determined that the carvacrol CT of salva-de marajó essential oil is a safe and effective food, cosmetic, and pharmaceutical preservative due to its ability to inhibit *E. coli, P. aeruginosa, S. aureus, A. basiliensis,* and *C. albicans*.[3584] The aerial parts oil with (E)-methyl cinnamate as the primary constituent inhibited bacteria (*E. coli*) and fungi (*Fonsecea pedrosoi, A. niger, T. rubrum, Cryptococcus neoformans* Serotype A).[3585]
- □ Salva-de marajó CT carvacrol essential oil demonstrated good antioxidant activity, but its activity was lower than BHT.[3586] One study compared the antioxidant activity of salva-de marajó CT carvacrol essential oil extracted from plants at different maturity stages and grown with different fertilizers. The study authors concluded that young plants fertilized with organic fertilizer had the best antioxidant activity.[3587]
- □ Salva-de marajó CT carvacrol essential oil was insecticidal against the green peach aphid (Myzus persicae).[3588] An unknown CT of salva-de marajó essential oil effectively killed cattle ticks (Rhipicephalus microplus).[3589]
- □ An unknown CT of salva-de marajó exhibited cytotoxic activity against trypomastigote and amastigote forms of *T. cruzi* (a parasite that causes Chagas disease, which is an important cause of mortality in Latin America).[3590]

- The thymol and carvacrol CTs of salva-de marajó aerial parts showed activity against *Leishmania chagasi* (*L. infantum*) promastigotes—a parasite that causes infantile visceral leishmaniasis in the Mediterranean region and in Latin America—without harming normal mammalian cells.[3591] They were also moderately toxic to epimastigotes and amasitgotes of *T. cruzi*. The para-cymene CT was inactive.

SANDALWOOD, PNG/INDONESIAN | SANDALWOOD, PACIFIC (Yasi, Ahi) | SANDALWOOD, POLYNESIAN (A'i)

Santalum macgregorii F.Muell. | *S. yasi* Seem. | *S. insulare* (Bertero ex A. DC.)

FAMILY: Santalaceae
NOTE: Base
AROMA INTENSITY: Medium
AROMA: Woody, earthy, slightly sweet
COMMON EXTRACTION METHOD: Hydrodistilled from the heartwood
POSSIBLE SUBSTITUTE OILS: Sandalwood, cedarwood
BLENDS WELL WITH: Basil, bergamot, black pepper, blue cypress, cardamom, cedarwood, chamomile (German, Roman), coriander, cypress, frankincense, juniper, lavender, myrrh, neroli, palo santo, palmarosa, patchouli, petitgrain, ravensara, rose, sandalwood, Spanish sage, vetiver, ylang ylang
SUBCELLULAR LOCALIZATION | EPIGENETIC INFLUENCE: Currently unknown | Currently unknown
RECOMMENDED DILUTION RANGE: 3%–33%; neat for some conditions

PRIMARY CONSTITUENTS:[3592,3593,3594,3595,3596]

PNG/Ind. Sandalwood ((Z)-α-Santalol CT)	
(Z)-Alpha-Santalol	30.2%–47.0%
(Z)-Beta-Santalol	13.5%–23.6%
(Z)-Lanceol	1.7%–12.1%
Alpha-Trans-Bergamotol	5.7%–12.0%
(E)-Beta-Santalol	1.3%–4.5%
(Z)-Nuciferol	0.7%–4.0%
Epi-Beta-Santalol	3.0%–3.9%

Pacific Sandalwood	
(Z)-Alpha-Santalol	36.9%–47.4%
(Z)-Beta-Santalol	23.2%–26.5%
Alpha-Trans-Bergamotol	5.2%
(Z)-Nuciferol	4.2%
Epi-Beta-Santalol	4.0%

PNG/Ind. Sandalwood ((Z)-β-Curcumen-12-ol CT)	
(Z)-Beta-Curcumen-12-ol	21.4%–30.6%
(Z)-Nuciferol	14.2%–30.3%
(Z)-Gamma-Curcumen-12-ol	5.3%–17.5%
(Z)-Alpha-Santalol	0.0%–8.8%
(E,E)-Farnesol	0.0%–8.1%
Spirosantalol	0.0%–4.5%
(Z)-Lanceol	0.2%–4.4%

PNG/Ind. Sandalwood ((Z)-Lanceol CT)	
(Z)-Lanceol	31.3%–74.8%
(Z)-Alpha-Santalol	2.6%–24.8%
(Z)-Nuciferol	1.7%–17.8%
(Z)-Beta-Santalol	1.1%–11.2%
Alpha-Trans-Bergamotol	0.6%–5.7%
(E,E)-Farnesol	0.0%–5.1%

Polynesian Sandalwood	
(Z)-Alpha-Santalol	39.1%–48.7%
(Z)-Beta-Santalol	16.9%–19.9%

Polynesian Sandalwood (Concrete)	
(Z)-Alpha-Santalol	3.5%–53.2%
(Z)-Nuciferol	tr–29.3%

(Z)-Trans-Alpha-Beramotol	3.7%–6.4%	(Z)-Beta-Santalol	2.7%–23.6%
(Z)-Nuciferol	0.9%–5.7%	6,13-Dihydroxybisbola-2,10-diene	tr–20.5%
Epi-Beta-Santalol	3.9%–4.7%	Alpha-Santaldiol	tr–13.0%
		(Z)-Alpha-Trans-Bergamotol	3.1%–11.8%
		Epi-Beta-Santalol	0.5%–4.7%
		Beta-Santaldiol	tr–3.7%

Note: *S. macgregorii* is found in Papua New Guinea (PNG) and Indonesia; *S. yasi* is found in Fiji, Niue, and Tonga; and *S. insulare* is found in Southeast Polynesia.

OTHER CONSTITUENTS: *PNG/Ind. Sandalwood ((Z)-α-Santalol CT)*—Epi-beta-santalene, gamma-curcumene, beta-bisabolene, ar-curcumene, (E)-nerolidol, beta-bisabolol, alpha-bisabolol; *PNG/Ind. Sandalwood ((Z)-Lanceol CT)*—Beta-bisabolol, alpha-bisabolol, epi-beta-santalol, (Z)-gamma-curcemen-12-ol, (E)-beta-santalol, (Z)-beta-curcuem-12-ol; *PNG/Ind. Sandalwood ((Z)-β-Curcumen-12-ol CT)*—Dendrolasin, beta-bisabolol, alpha-bisabolol, epi-beta-santalol, (Z)-beta-santalol; *Pacific Sandalwood*—Alpha-santalene, epi-beta-santalene, beta-santalene, (E)-nuciferol, (Z)-lanceol; *Polynesian Sandalwood*—Alpha-santalene, epi-beta-santalene, beta-santalene, cyclosantalol, alpha-santalal, beta-bisabolol, (E)-alpha-santalol, (Z)-gamma-curcumen-12-ol, (Z)-gamma-bisabolen-12-ol, (Z)-beta-curcumen-12-ol, (Z)-lanceol, spirosantalol; *Polynesian Sandalwood (Concrete)*—Beta-bisabolol

PREFERRED COMPOSITION FOR CLINICAL USE:

Constituent	*PNG/Ind.*	*Pacific*	*Polynesian*
(Z)-Alpha-Santalol	30%–50%	35%–55%	35%–55%
(Z)-Beta-Santalol	10%–25%	15%–30%	12%–27%
Alpha-Trans-Bergamotol	4%–13%	2%–8%	2%–10%
(Z)-Lanceol	1%–13%	0.5%–6%	0.1%–5%
Epi-Beta-Santalol	2%–8%	2%–8%	2%–8%
(E)-Beta-Santalol	0.5%–8%	–	0.1%–3%
(Z)-Nuciferol	0.1%–6%	2%–8%	0.5%–7%

REPORTED THERAPEUTIC PROPERTIES: Analgesic (pain relief), antiseptic, antispasmodic, astringent, antibacterial, anti-inflammatory, antitumor, stimulates lymph drainage, expels excess gas, antineuralgic, reduces fever, tones and strengthens the cardiovascular system, disinfectant, reduces the appearance of scars and blemishes, relieves skin conditions (especially those that involve inflammation), **anti-aging**, **nourishes and softens skin**, decongestant, eases cough, diuretic, helps lower high blood pressure, antifungal, sedating, aids concentration and memory, aphrodisiac, grounding, reduces aggressive behavior, encourages acceptance, stress management, reduces irritability and despondency, diminishes guilt

CAUTIONS:

- None currently known.

SELECTED EVIDENCE:

- ☐ None found.

SASSAFRAS

North American Sassafras: *Sassafras albidum* (Nutt.) Nees, *S. officinale* Nees & Eberm., *S. variifolium* (Salisb.) Kuntze, *Laurus albidus* Nutt., *L. sassafras* L.; **Brazillian Sassafras**: *Nectandra sanguinea* Rol Ex. Rottb., *Netandra globosa* Mez, *Ocotea odorifera* (Vell.) Rohwer; **Chinese Sassafras**: *Cinnamomum porrectum* (Roxb.) Kosterm, *C. rigidissimum* HT Chang

FAMILY: Lauraceae
NOTE: Top-Middle
AROMA INTENSITY: Medium to strong
AROMA: Spicy, sweet, woody, floral
COMMON EXTRACTION METHOD: Steam distilled or hydrodistilled from the wood, bark, or roots; may also be distilled from the leaves
POSSIBLE SUBSTITUTE OILS: Betle CT safrole, nutmeg, mace
BLENDS WELL WITH: Black pepper, cajeput, cardamom, cinnamon, clove, coriander, eucalyptus, ginger, lavender, lemon, orange, patchouli, rosemary, Spanish sage, spikenard, tamala, tangerine, turmeric
SUBCELLULAR LOCALIZATION | EPIGENETIC INFLUENCE: Currently unknown | Currently unknown
RECOMMENDED DILUTION RANGE: 1%–5%

PRIMARY CONSTITUENTS:[3597,3598,3599,3600,3601,3602,3603,3604,3605,3606]

North American Sassafras

Safrole	75.0%–92.0%
Camphor	0.6%–6.8%
Methyl eugenol	1.1%–3.0%

Brazillian Sassafras

Safrole	71.2%–92.9%

Chinese Sassafras

Safrole	61.7%–99.8%
Methyl eugenol	0.0%–28.6%
Elemicin	0.0%–4.3%

Brazillian Sassafras (Leaves)

Camphor	6.5%–43.0%
Safrole	36.3%–42.0%
Gamma-Cadinene	0.0%–6.6%
Alpha-Copaene	0.0%–6.0%

Chinese Sassafras (Leaves)

Safrole	93.2%

Note: Sassafras oil rich in safrole has reportedly been used to manufacture the illicit drug MDMA (also known as ecstasy). Sassafras may also be used to make illicit natural root beer.

OTHER CONSTITUENTS: *North American Sassafras*—1,8-cineole, limonene, carvone, alpha-cubebene, eugenol, beta-cubebene, cis-calamenene, alpha-calacorene, ledene oxide; *Brazillian Sassafras*—Camphor, methyl eugenol, alpha-pinene; *Chinese Sassafras*—N/A; *Brazillian Sassafras (Leaves)*—Not listed; *Chinese Sassafras (Leaves)*—Beta-caryophyllene, gamma-cadinene, tetradecanal

PREFERRED COMPOSITION FOR CLINICAL USE:
Not recommended for therapy.

Reported Therapeutic Properties: Relieves skin itching and chronic skin conditions, treats head lice, antifungal, antibacterial, aids digestion, expels excess gas, anti-inflammatory, antirheumatic, antiarthritic, reduces high blood pressure, cleanses the urinary tract, insecticidal, antidepressant, promotes self-esteem

Cautions:

Sassafras varieties high in safrole are generally not recommended for therapy due to the risks of their high safrole content. In addition, potentially high methyl eugenol or camphor content warrant additional cautions.
Methyl eugenol: Children, pregnancy, lactation, ingestion, antibiotics, antifungals, cholinergic drugs
Camphor: Children, pregnancy, lactation, ingestion, epilepsy, G6PD

■ Avoid during pregnancy due to safrole content. Animal research suggests that oral administration of safrole to pregnant mice caused kidney epithelial tumors in 7% of their offspring; and male offspring nursed by mothers who were administered safrole experienced a high rate of liver tumors.[3607] This animal research suggests that safrole increases the risk of cancer in offspring by crossing the placenta and excretion in breast milk. In addition, animal research suggests that the bonding of safrole and its proximate carcinogen (1'-hydroxysafrole) to liver and kidney DNA increases 2.3-to-3.5-fold during pregnancy, which may make them more hazardous and increase the risk of cancer during pregnancy.[3608]

■ Oral caution due to moderate to significant safrole content. Animal research suggests that oral administration of safrole may cause liver cancer and is reasonably anticipated to be a human carcinogen.[3609]

■ There is a moderate risk that mace essential oil may interact with aspirin, blood pressure, antiplatelet, and anticoagulant medications, and increase the risk of bleeding among people with bleeding disorders (low risk), based on safrole content.[3610]

■ May interfere with enzymes responsible for metabolizing medications (NSAIDs, proton-pump inhibitors, acetaminophen, antiepileptics, immune modulators, blood-sugar medications, blood pressure medications, antidepressants, antipsychotics, diabetic medications, antihistamines, antibiotics, and anesthetics) due to safrole content.[3611]

Selected Evidence:

- □ Brazillian sassafras leaf essential oil displayed significant activity against the promastigote forms of *Leishmania amazonensi*.[3612]
- □ *In vitro* evidence demonstrated that North American sassafras essential oil inhibited several fungi (*Aspergillus niger, A. ochraceus, A. versicolor, A. flavus, A. terreus, Alternaria alternate, Aureobasidium pullulans, Penicillium ochrochloron, P. funiculosum, Cladosporium cladosporioides, C. fulvium, Trichoderma viride, Fusarium tricinctum, F. sporotrichoides, Phoma macdonaldii, Phomopsis helianthi*, and *Mucor mucedo*).[3613] Chinese sassafras essential oil showed an inhibitory effect on mycelial growth, sporulation, and aflatoxin B1 production of *A. parasiticus* and *A. flavus*.[3614] Chinese sassafras also inhibited *C. albicans*.[3615] Chinese sassafras inhibited bacteria (*S. aureus*, MRSA, *B. subtilis, S. mutans, E. coli*), yeasts (*C. albicans, C. neoformans*), and dermatophytes (*T. rubrum, T. mentagrophytes, M. gypseum*).[3616]
- □ Brazillian sassafras leaf essential oil was insecticidal to the maize weevil (*Sitophilus zeamais*).[3617]

SATSUMA MANDARIN (Satsuma Orange, Satsuma Tangerine, Unshu Mikan)

Citrus unshiu (Yu.Tanaka ex Swingle); **Unshu Orange**: *C. reticulata Blanco* ssp. *unshiu* (Marcow.) D.Rivera Núñez

FAMILY: Rutaceae
NOTE: Top
AROMA INTENSITY: Medium
AROMA: Citrusy, sweet, fruity, fresh
COMMON EXTRACTION METHOD: Cold-pressed (expressed) or steam distilled/hydrodistilled from the fruit rind (peel)
POSSIBLE SUBSTITUTE OILS: Yuzu, tangerine, mandarin
BLENDS WELL WITH: Basil, bergamot, black pepper, chamomile (Roman), cinnamon, clove, grapefruit, kumquat, lavandin, lavender, lemon, lime, mandarin, marjoram, neroli, nutmeg, orange, palmarosa, petitgrain, rose, tangerine, yuzu, ylang ylang
SUBCELLULAR LOCALIZATION | EPIGENETIC INFLUENCE: Currently unknown | Currently unknown
RECOMMENDED DILUTION RANGE: 3%–50%; neat for some conditions

PRIMARY CONSTITUENTS:[3618,3619,3620,3621,3622,3623]

Cold-Pressed		*Distilled*	
Limonene	37.5%–93.5%	Limonene	67.4%–88.1%
o-Cymene	0.0%–14.3%	Para-Cymene	0.0%–7.7%
Gamma-Terpinene	0.0%–13.9%	Myrcene	0.9%–7.2%
Alpha-Pinene	0.8%–9.1%	Gamma-Terpinene	0.0%–4.7%
Myrcene	2.0%–5.4%	Delta-3-Carene	0.0%–4.4%
Sabinene	0.0%–4.1%	Alpha-Pinene	0.0%–2.5%
Cis-Carveol	0.0%–3.6%		
Alpha-Copaene	0.0%–2.6%		
Beta-Elemene	0.0%–2.2%		
Trans-p-Mentha-2,8-Dienol	0.0%–2.0%		

OTHER CONSTITUENTS: *Cold-Pressed*—Delta-terpinene, dehydro-para-cymene, alpha-humulene, alpha-farnesene, linalool, cis-limonene oxide, trans-limonene oxide, carvone, 1-octanol, alpha-terpineol, limonene glycol; *Distilled*—O-Cymene, terpinolene, linalool, cyclohexane,2,4-diisopropenyl-1-methyl-1-vinyl, alpha-farnesene, diethyl phthalate, beta-elemene, beta-sesquiphellandrene

PREFERRED COMPOSITION FOR CLINICAL USE:
Insufficient samples/data to determine.

REPORTED THERAPEUTIC PROPERTIES: Aids digestion, relieves hiccup, soothes skin conditions and itching, reduces fatigue, weight management, digestive aid, promotes skin health, anti-inflammatory, anticancer, antibacterial, antiseptic, antioxidant, promotes restful sleep, antidepressant, relieves anxiety

CAUTIONS:

■ None currently known. Neither the distilled or cold-pressed essential oils are considered phototoxic due to the negligible presence of furanocoumarins.[3624]

SELECTED EVIDENCE:

- □ Distilled Satsuma mandarin reduced the secretion of nitric oxide triggered by lipopolysaccharide, demonstrating that it has anti-inflammatory effects.[3625]
- □ *In vitro* research reported that Satsuma mandarin essential oil (hydrodistilled) inhibited the bacteria *S. aureus, B. cereus,* and *B. subtilis.*[3626] Another study reported that distilled Satsuma mandarin essential oil inhibited antibiotic-resistant and -susceptible skin pathogens (*S. epidermidis, P, acnes, M. furfur, C. albicans*).[3627]
- □ Satsuma mandarin essential oil exhibited good antioxidant activity in the DPPH assay.[3628]

SAWTOOTH CANDYLEAF (Sawtooth Stevia)

Stevia serrata Cav.

FAMILY: Asteraceae
NOTE: Middle
AROMA INTENSITY: Strong
AROMA: Herbaceous, sweet, fruity
COMMON EXTRACTION METHOD: Steam distilled from the aerial parts
POSSIBLE SUBSTITUTE OILS: German chamomile, blue tansy, yarrow
BLENDS WELL WITH: Balsam fir, bergamot, blue tansy, cassia, cedarwood, eucalyptus, geranium, lavender, neroli, patchouli, pine, ravintsara, rose, silver fir, white fir, yarrow
SUBCELLULAR LOCALIZATION | EPIGENETIC INFLUENCE: Currently unknown | Currently unknown
RECOMMENDED DILUTION RANGE: 5%–Neat

PRIMARY CONSTITUENTS:[3629]

Chamazulene	60.1%
(E)-Nerolidol	7.3%
Caryophyllene Oxide	6.3%
Germacrene D	5.4%
Spathulenol	4.3%

OTHER CONSTITUENTS: Linalool, caryophyllene, bicyclogermacrene, 14-hydroxy-9-epi-(E)-caryophyllene

PREFERRED COMPOSITION FOR CLINICAL USE:
Insufficient samples/data to determine.

REPORTED THERAPEUTIC PROPERTIES: Anti-inflammatory, analgesic, antispasmodic, antimicrobial, antibacterial, antifungal, antioxidant, improves blisters, bruises, and cuts, relieves skin conditions, calming, stress management

CAUTIONS:

■ May interfere with the enzymes responsible for metabolizing medications (NSAIDs, proton-pump inhibitors, acetaminophen, antiepileptics, immune modulators, blood-sugar medications, blood pressure medications, antidepressants, antipsychotics, diabetic medications, antihistamines, antibiotics, and anesthetics).[895]

SELECTED EVIDENCE:

- □ Oral administration of sawtooth candyleaf essential oil markedly reduced pain in mice.[3630]

SEA WORMWOOD (Seaworm Wood, Sea Sagebrush)

Artemisia maritima L., *A. pseudogallica* (Rouy) A.W.Hill, *A. salina* Willd.

FAMILY: Family
NOTE: Top-Middle
AROMA INTENSITY: Medium-strong
AROMA: Camphoraceous, herbaceous, green, woody, piney
COMMON EXTRACTION METHOD: Steam distilled or hydrodistilled from the aerial parts
POSSIBLE SUBSTITUTE OILS: Feverfew, sweet wormwood, yarrow, camphor, Spanish rosemary
BLENDS WELL WITH: Blue spruce, blue tansy, camphor, copaiba, eucalyptus, feverfew, frankincense, lavandin, lavender, myrrh, myrtle, peppermint, Spanish sage, spearmint, spruce (black), rose
SUBCELLULAR LOCALIZATION | EPIGENETIC INFLUENCE: Currently unknown | Currently unknown
RECOMMENDED DILUTION RANGE: 1.5%–20%; 33% for some conditions

PRIMARY CONSTITUENTS:[3631,3632,3633,3634,3635]

Camphor/1,8-Cineole CT	
Camphor	20.3%–44.4%
1,8-Cineole	27.0%–41.1%
Camphene	4.6%–5.9%
Beta-Caryophyllene	0.1%–4.2%

1,8-Cineole/Chrysanthenone CT	
Chrysanthenone	23.6%–25.7%
1,8-Cineole	23.6%–25.0%
Germacrene D	6.7%–7.4%
Isoborneol	0.0%–4.3%
Borneol	0.0%–3.4%

Alpha-Thujone CT	
Alpha-Thujone	63.3%
Sabinene	7.8%
1,8-Cineole	6.5%

OTHER CONSTITUENTS: *Camphor/1,8-Cineole CT*—Alpha-pinene, beta-pinene, sabinene, 2,3-dehydro-1,8-cineole, alpha-terpinene, gamma-terpinene, para-cymene, beta-thujone (< 1.2%), chrysanthenone, borneol, trans-pinocarveol, pinocarvone, myrtenol, myrtenal, carvone, alpha-terpineol, (Z)-chrysanthenyl acetate, (E)-geranyl acetate, bornyl acetatecis-davanone; *1,8-Cineole/Chrysanthenone CT*—Alpha-pinene, camphene, sabinene, para-cymene, gamma-terpinene, filifolone, trans-sabinene hydrate, trans-pinocarveol, camphor, trans-verbenol, pinocarvone, terpinen-4-ol, piperitone, chrysanthenyl acetate, bornyl acetate, beta-caryophyllene, trans-beta-farnesene, alpha-copaene, gamma-elemene, delta-cadinene; *Alpha-Thujone*

CT—Para-cymene, trans-sabinene hydrate, isopinocamphone (< 0.6%), beta-caryophyllene, terpinen-4-ol, sabinaketone, germacrene D, cuminaldehyde

PREFERRED COMPOSITION FOR CLINICAL USE:

Constituent	
1,8-Cineole	25%–45%
Camphor	20%–45%
Camphene	3%–10%
Beta-Caryophyllene	0.1%–5%

REPORTED THERAPEUTIC PROPERTIES: Aids respiratory function, opens the airways, relieves asthma, decongestant, expectorant, anti-inflammatory, antispasmodic, analgesic, eases stomachache, antineuralgic, aids digestion, increases appetite, relieves diarrhea, supports cardiovascular function, relieves stings, insect repellent, aphrodisiac, relieves anxiety, stress management, combats anger and irritability

CAUTIONS:

Camphor/1,8-Cineole CT

■ Avoid with children under six due to significant camphor content and moderate amounts of 1,8-cineole and thujones. Several cases of camphor poisoning and/or seizures from ingestion and topical application have been reported in children.[3636,3637] Ingestion of camphor-containing products has been lethal in children under age two.[3638] Children five years and up may use camphor-containing essential oils topically in dilutions no stronger than 5%. Camphor ingestion by infants and young children may cause cough, vomiting, seizure, burning sensation in the mucous membranes and eyes, or lack of voluntary coordination of muscle movements.[3639] Cases of seizure have been reported in young children due to exposure to essential oils rich in thujones.[3640,3641]

■ Caution is warranted during pregnancy and while lactating, due to significant camphor content and moderate thujones content. Ingestion of essential oils with significant levels of camphor may lead to abortion because fetuses lack the enzymes to process it.[3642] Camphor ingestion by infants and young children may cause cough, vomiting, seizure, burning sensation in the mucous membranes and eyes, or lack of voluntary coordination of muscle movements.[3643] Essential oils high in thujone content may cause abortion.[3644,3645]

■ Avoid oral consumption due to significant camphor and moderate thujones content. Essential oils with significant levels of camphor may be toxic when taken orally, usually single doses exceeding 2 mL, although the lethal dose for humans is estimated to be greater than 5 mL in a single dose.[3646,3647,3648] See oral cautions under *Alpha-Thujone, Beta-Thujone CTs*.

■ Avoid with epilepsy and Parkinson's disease due to camphor, 1,8-cineole, and thujones content, which have the potential to exacerbate or cause seizures/convulsions or reduce seizure-medication efficacy.[3649,3650,3651,3652]

■ Avoid with those who have a compromised liver or kidneys, due to the risk of increased liver enzymes and liver or kidney damage based on camphor and thujones content.[3653,3654,3655,3656] This would also suggest that those taking medications that could cause liver damage should also use essential oils with significant camphor levels very cautiously or avoid them entirely.

■ The significant camphor content may negatively impact red blood cells and increase the risk of jaundice in children with glucose-6-phosphate dehydrogenase deficiency (G6PD).[3657,3658]

■ Caution is warranted for the topical application of camphor-containing essential oils on broken or injured skin due to an increased risk of toxicity.[3659]

1,8-Cineole/Chrysanthenone CT

■ Avoid with epilepsy and Parkinson's disease due to 1,8-cineole content, which may exacerbate or cause seizures/convulsions or reduce seizure-medication efficacy.[3660,3661,3662,3663]

■ May weakly interfere with the enzymes responsible for metabolizing medications (NSAIDs, proton-pump inhibitors, acetaminophen, antiepileptics, immune modulators, blood-sugar medications, blood pressure medications, antidepressants, antipsychotics, diabetic medications, antihistamines, antibiotics, and anesthetics) due to 1,8-cineole content.[3664,3665,3666]

■ May interfere with pentobarbital and other barbiturates (medications for anxiety and insomnia), based on 1,8-cineole content.[3667,3668]

Alpha-Thujone CT

■ Avoid with children under 6 due to thujone content. Cases of seizure have been reported in young children due to exposure to essential oils rich in thujones.[3669,3670]

■ Avoid during pregnancy and while lactating, due to thujones content. Essential oils high in thujone content may cause abortion.[3671,3672]

■ Avoid oral consumption. Ingesting plants or products high in thujones can cause vomiting, stomachache, diarrhea, and gastroenteritis followed by absorption disorders, headache, nervous agitation, and chronic convulsions, and symptoms of liver and kidney toxicity extending to yellow liver atrophy (extensive and rapid death of liver cells—parenchymal, possibly by fatty degeneration), arrhythmia, and myocardial bleeding.[3673] Thujone is considered significantly neurotoxic and may damage the liver, and it is estimated that as little as 15 mg orally may negatively impact the central nervous system.[3674,3675,3676] Taking 20 drops of thuja oil (also with significant thujones content) twice daily for five days caused an adult woman to have a tonic seizure.[3677] As few as 12 drops of sage oil (also with significant thujones content) orally have reportedly caused a seizure and short coma in an adult.[3678]

■ Avoid with epilepsy and Parkinson's disease due to thujones, camphor, and 1,8-cineole content.[3679,3680,3681,3682,3683]

■ Avoid with those who have a compromised liver or kidneys, due to the risk of kidney toxicity, kidney failure, or liver toxicity.[3684,3685,3686] Consuming extremely large amounts of thujone-rich essential oils (10 mL) has caused kidney failure.[3687] This would also suggest that those taking medications that could cause liver damage should also use tansy very cautiously or avoid it entirely.

■ May weakly interfere with the enzymes responsible for metabolizing medications (NSAIDs, proton-pump inhibitors, acetaminophen, antiepileptics, immune modulators, blood-sugar medications, blood pressure medications, antidepressants, antipsychotics, diabetic medications, antihistamines, antibiotics, and anesthetics) due to 1,8-cineole and alpha-thujone content.[3688,3689,3690,3691]

SELECTED EVIDENCE:

- □ Sea wormwood essential oil CT camphor/1,8-cineole demonstrated antispasmodic activity in isolated rabbit jejunum by suppressing spontaneous and high potassium-induced contractions and modifying calcium concentrations.[3692] The oil also reduced contractions in isolated trachea caused by carbachol, suggesting it acts as a bronchodilator.
- □ *In vitro* research showed that the 1,8-cineole/chrysanthenone CT of sea wormwood essential oil possesses potent antibacterial activity against *B. cereus, E. coli, K. pneumonia, P. vulgaris, S. typhimurium, S. pyogenes, S. aureus*, and effectively inhibits the fungi *A. niger, A. ochraceus*, and *A. terreus*.[3693]
- □ The camphor/1,8-cineole CT of sea wormwood essential oil repelled *Aedes aegypti* mosquitoes and significantly deterred biting.[3694]

SHRUBBY HARE'S EAR
(Thorow-Wax)

Bupleurum fruticosum L., *Tenoria fruiticosa* Spreng.

FAMILY: Apiaceae (Umbelliferae)
NOTE: Top
AROMA INTENSITY: Strong
AROMA: Pungent, herbaceous, woody, green, slightly minty
COMMON EXTRACTION METHOD: Steam distilled or hydrodistilled from the flowers (umbels), leaves, branches, twigs, aerial parts, or fruit; may also be extracted by Co2 (supercritical)
POSSIBLE SUBSTITUTE OILS: Bettany, lovage, German chamomile
BLENDS WELL WITH: Cedarwood, copaiba, cypress, Douglas fir, juniper berry, lavender, Phoenician juniper, sage, sandalwood, Spanish sage, Texas cedarwood, vetiver
SUBCELLULAR LOCALIZATION | EPIGENETIC INFLUENCE: Currently unknown | Currently unknown
RECOMMENDED DILUTION RANGE: 5%–Neat

PRIMARY CONSTITUENTS:[3695,3696,3697,3698,3699]

Beta-Phellandrene CT

Beta-Phellandrene	39.1%–71.4%
Sabinene	9.3%–22.0%
Para-Cymene	0.7%–9.3%
Limonene	0.0%–5.6%
Alpha-Phellandrene	0.1%–5.4%
Cryptone	0.0%–4.3%
Alpha-Terpinene	0.0%–3.4%
Geranyl Acetate	0.0%–3.0%

Gamma-Terpinene CT

Gamma-Terpinene	48.4%
Alpha-Phellandrene	18.6%
Sabinene	7.6%
Bicyclogermacrene	3.2%

Supercritical Beta-Phellandrene CT

Beta-Phellandrene	53.0%–65.1%
Sabinene	12.0%–13.9%
Methyl eugenol	1.2%–9.8%

Alpha-Pinene/Beta-Pinene CT

Alpha-Pinene	35.4%–42.7%
Beta-Pinene	33.0%–39.1%
Beta-Phellandrene	6.2%–7.8%
(E)-Beta-Ocimene	0.2%–4.7%

Sabinene CT

Sabinene	39.7%
Beta-Phellandrene	38.7%
Limonene	5.0%

Supercritical AP/GD CT

Alpha-Pinene	30.3%–33.3%
Germacrene D	18.2%–24.6%
Beta-Pinene	19.1%–21.9%
Beta-Phellandrene	5.7%–6.9%

OTHER CONSTITUENTS: *Beta-Phellandrene CT*—Alpha-pinene, myrcene, (Z)-beta-ocimene, (E)-beta-ocimene, gamma-terpinene, terpinen-4-ol, estragole (< 0.6%), octyl acetate, decanol, trans-carveol, cuminal, geraniol, carvacrol, methyl eugenol (<2.9%), alpha-copaene; *Alpha-Pinene/Beta-Pinene CT*—Alpha-thujene, camphene, (Z)-beta-ocimene, myrcene, gamma-terpinene, terpinen-4-ol, hexyl-2-methyl butyrate, carvacrol methyl ether, germacrene D, beta-bisabolene, beta-caryophyllene, beta-cedrene; *Gamma-Terpinene CT*—Alpha-pinene, alpha-terpinene, para-cymene, beta-phellandrene, (Z)-beta-ocimene, terpinen-4-ol, citronellyl acetate, geranyl acetate, alpha-copaene, germacrene D; *Sabinene CT*—Alpha-pinene, myrcene, elemicin, terpinen-4-ol, decyl acetate; *Supercritical Beta-Phellandrene CT*—Alpha-pinene, beta-pinene, myrcene, alpha-phellandrene, (Z)-beta-ocimene; *Supercritical AP/GD CT*—Sabinene, myrcene, para-cymene, (E)-beta-ocimene, gamma-terpinene, bicyclogermacrene

PREFERRED COMPOSITION FOR CLINICAL USE:

Constituent	*Beta-Phellandrene CT*	*Alpha-Pinene/Beta-Pinene CT*
Beta-Phellandrene	50%–75%	4%–12%
Sabinene	10%–25%	0.1%–5%
Para-Cymene	0.5%–10%	0.1%–5%
Alpha-Phellandrene	0.1%–8%	tr–5%
Alpha-Pinene	0.1%–5%	30%–50%
Beta-Pinene	–	25%–45%
(E)-Beta-Ocimene	–	0.1%–7%

REPORTED THERAPEUTIC PROPERTIES: Antibacterial, antifungal, anti-infectious, reduces fever, anti-inflammatory, antispasmodic, antiarthritic, decreases blood pressure, insecticidal, aids respiratory function, eases cough, aids digestion, relieves diarrhea, soothes skin itching and irritation, relieves anxiety, aids mental clarity, stress management

CAUTIONS:

- None currently known.

SELECTED EVIDENCE:

- □ The alpha-pinene/beta-pinene CT of shrubby hare's ear essential oil demonstrated anti-inflammatory and antispasmodic activity, which was attributed to the presence of alpha-pinene and beta-pine with thymol and carvacrol potentiating the effects.[3700]
- □ *In vitro* evidence showed that shrubby hare's ear essential oil inhibits Candida albicans, *C. tropicalis, C. krusei, C. guillermondii, C. parapsilosis, Cryptococcus neoformans, Trichophyton rubrum, T. mentagrophytes, Microsporum canis, M. gypseum, Epidermophyton floccosum, Aspergillus niger,* and *A. fumigatus*.[3701] In general, the beta-phellandrene CT was more active than the alpha-pinene CTs.

SICHUAN PEPPER (Japanese Pepper, Korean Pepper, Fagara, Chopi)

Zanthoxylum piperitum DC

FAMILY: Rutaceae
NOTE: Middle
AROMA INTENSITY: Medium
AROMA: Fresh, green, citrusy

Common Extraction Method: Steam distilled or hydrodistilled from the green or ripe fruits or aerial parts; may also be solvent extracted or distilled from the leaves
Possible Substitute Oils: Xanthoxylum, pink pepper, black pepper
Blends Well With: Amyris, bergamot, black pepper, cardamom, carrot, cedarwood, clary sage, clove, frankincense, geranium, ginger, grapefruit, lavandin, lavender, lemon, lime, marjoram, may chang, myrrh, neroli, nutmeg, orange, palo santo, patchouli, rosemary, sage, sandalwood, Spanish sage, tangerine, tea tree (melaleuca), vetiver, yarrow, ylang ylang
Subcellular Localization | Epigenetic Influence: Currently unknown | Currently unknown
Recommended Dilution Range: 3%–Neat

Primary Constituents:[3702,3703,3704,3705,3706]

Fruits CT Beta-Phellandrene

Beta-Phellandrene	45.0%
Limonene	24.2%
Beta-Pinene	11.4%
Geranyl Acetate	5.5%
Citronellol	3.8%

Fruits Isocarvestrene CT

Isocarvestrene	29.0%
Geranyl Acetate	22.7%
Citronellal	14.9%
Citronellyl Acetate	9.5%
4-Isopropyl-2-Cyclohexenone	5.5%

Leaves

Beta-Phellandrene	28.2%
Alpha-Pinene	23.3%
Limonene	12.0%
Isopulegol	7.1%
Beta-Pinene	5.5%
Alpha-Terpineol	5.2%
Citronellal	3.1%

Aerial Parts

Beta-Phellandrene	29.4%
(E,E)-Farnesyl Acetate	14.6%
Citronellol	10.3%
Citronellal	6.8%
Beta-Caryophyllene	5.6%
Limonene	5.5%
(E,E)-Farnesol	4.0%
Myrcene	3.3%
(E)-Nerolidol	3.0%

Fruits CT Limonene

Limonene	11.7%–20.6%
Citronellal	0.0%–16.1%
Isopulegol	tr–15.5%
Myrcene	7.9%–14.2%
Beta-Phellandrene	7.1%–14.1%
Acetic Acid	1.7%–8.0%
Gamma-Terpinene	0.0%–7.4%
4(10)-Thujene	0.0%–6.5%
Terpinen-4-ol	0.0%–5.9%
(E)-Beta-Ocimene	1.6%–5.8%
Geranyl Acetate	0.0%–5.5%
Alpha-Terpinene	0.0%–5.3%
Terpinolene	0.0%–3.6%
Linalool	0.9%–3.1%
Citronellyl Acetate	0.4%–3.1%

Leaves (Solvent Extracted)

Beta-Phellandrene	15.0%–37.5%
Citronella	1.8%–25.1%
Farnesyl Acetate	1.1%–8.8%
Myrcene	1.5%–4.5%
Alpha-Pinene	1.3%–4.0%

OTHER CONSTITUENTS: *Fruits CT Beta-Phellandrene*—Citronellal, linalool, isopulegol, gamma-terpineol; *Fruits Isocarvestrene CT*—Myrcene, citronellol; *Fruits CT Limonene*—(Z)-Beta-ocimene, alpha-cubebene, copaene, alpha-humulene, alpha-terpineol, geraniol, piperitone, geranial, carvone, cuminaldehyde; *Leaves*—Toluene, 3-methylfuran, (Z)-beta-ocimene, 1,4-cyclohexadiene, (E)-beta-ocimene, alpha-humulene, geranial, piperitone, geranyl acetate, citronellol; *Aerial Parts*—(Z)-Beta-ocimene, alpha-terpineol, geraniol, citronellyl acetate, neryl acetate, alpha-humulene, germacrene D; *Leaves (Solvent Extracted)*—1,8-cineole, 1,3,5-octatriene, terpinolene, linalool, citronellol, 2-undecanone, citronellyl acetate, tridecane, geranyl acetate, beta-caryophyllene, 2-tridecane, nerolidol, tetradecanoic acid, 2-pentadecanone, farnesol

PREFERRED COMPOSITION FOR CLINICAL USE:
Too great of variance and insufficient data to determine.

REPORTED THERAPEUTIC PROPERTIES: Aids digestion, relieves nausea/vomiting, antibacterial, antifungal, antiparasitic, anti-inflammatory, analgesic, antirheumatic, antioxidant, wound healing, relieves skin itching, modulates immune function, insecticidal, insect repellent, aids cognition, mentally stimulating

CAUTIONS:
■ May interact with anticholinergic (drugs used for asthma, incontinence, gastrointestinal cramps, muscular spasms, depression, and sleep disorders) and cholinergic medications (drugs used to reduce urinary retention, diagnose myasthenia gravis, and for glaucoma), based on AChE-inhibtory activity of beta-phellandrene or delta-3-carene.[3707,3708,3709]

SELECTED EVIDENCE:

- □ Sichuan pepper essential oil reduces inflammation by suppressing inducible nitric oxide, COX-2 expression, and modulating immune activity.[3710,3711]
- □ Animal research showed that oral administration Sichuan pepper essential oil reduced pain by reducing the excitatory transmission induced by glutamate.[3712]
- □ *In vitro* research demonstrated that Sichuan pepper essential oil inhibited foodborne pathogens (*Listeria monocytogenes, Staphylococcus aureus, Aeromonas hydrophila, Vibrio parahaemolyticus, Salmonella choleraesuis, Salmonella enterica*, and *Vibrio vulnificus*).[3713]
- □ Laboratory research reported that Sichuan pepper essential oil repelled female *Ae. aegypti* mosquitoes for 1.5 hours, which improved to 2.5 hours when 5% vanillin was added.[3714] The same authors found that the oil provided better protection against mosquitoes (*Ae. gardnerii, An. barbirostris, Ar. subalbatus, C. tritaeniorhynchus, C. gelidus, C. vishnui group*, and *M. uniformis*) than DEET in field tests when combined with vanillin. Another study reported that the limonene CT repelled *Ae. aegypti* mosquitoes, with better activity observed when 10% vanillin was added.[3715]
- □ Sichuan pepper essential oil was toxic to the mushroom fly (*Lycoriella ingenua*).[3716]
- □ The isocarvestrene CT of Sichuan pepper essential oil was insecticidal to mites (*Dermatophagoides farinae, D. pteronyssinus*, and *Tyrophagus putrescentiae*).[3717]

SICILIAN SUMAC (Elm-Leaved Sumach, Tanner's Sumach)

Rhus coriaria L.

FAMILY: Anacardiaceae
NOTE: Middle
AROMA INTENSITY: Medium
AROMA: Green, waxy, spicy, woody, fruity
COMMON EXTRACTION METHOD: Steam distilled or hydrodistilled from the fruits (seeds), pericarp, leaves, or branches
POSSIBLE SUBSTITUTE OILS: Copaiba, black pepper, melissa, cilantro
BLENDS WELL WITH: Balsam fir, basil, bergamot, black pepper, carrot, cassia, chamomile (German, Roman), cinnamon, copaiba, galbanum, ginger, grapefruit, lavender, lemon, lime, marjoram, melissa, orange, oregano, tangerine
SUBCELLULAR LOCALIZATION | EPIGENETIC INFLUENCE: Currently unknown | Currently unknown
RECOMMENDED DILUTION RANGE: 3%–50%; neat for some conditions

PRIMARY CONSTITUENTS:[3718,3719,3720,3721,3722,3723]

Italian (Fruits)

p-Anisaldehyde	0.0%–28.4%
Alpha-Pinene	0.7%–26.0%
n-Nonanal	9.8%–25.1%
Fenchone	0.4%–20.7%
Camphor	0.3%–12.5%
Decanal	1.4%–11.9%
Beta-Caryophyllene	tr–6.8%
Alpha-Muurolene	0.0%–6.8%
(Z)-2-Heptenal	0.0%–5.3%
1,8-Cineole	tr–5.2%
Delta-Cadinene	0.0%–4.2%
Gamma-Muurolene	0.0%–4.1%
Valencene	0.0%–3.9%
Isobornyl Acetate	0.3%–3.5%
Limonene	0.4%–3.4%
(E)-2-Nonenal	0.0%–3.3%
(E)-Geranylacetone	tr–3.0%

Iranian (Fruits)

Beta-Caryophyllene	5.9%–50.3%
Cembrene	1.0%–23.8%
n-Nonanal	0.0%–23.3%
Alpha-Pinene	tr–19.7%
Nonanoic Acid	0.0%–15.8%
(3E)-Cembrene A	0.0%–7.0%

Turkish (Fruits)

(E)-2-Decenal	2.9%–22.2%
Cembrene	1.5%–15.4%
Beta-Caryophyllene	4.0%–13.8%
n-Nonanal	2.4%–11.5%
Carvacrol	0.0%–10.4%
(2E,4E)-Decadienal	1.3%–5.8%
Heptacosane	0.9%–5.7%
Palmitic Acid	0.8%–5.1%
(E)-2-Heptenal	1.3%–4.4%
Nonacosane	0.0%–4.0%
Naphthalene	tr–3.6%
Alpha-Pinene	1.1%–3.5%
Caryophyllene Oxide	0.4%–3.1%

Egyptian (Fruits)

O-Cymene	7.7%
Beta-Ocimene	7.5%
Limonene	7.4%
1,8-Cineole	7.3%
(E)-Beta-Farnesene	6.1%
(Z,Z)-Alpha-Farnesene	6.0%
Beta-Caryophyllene	4.5%
Allyl p-Methylbenzyl Ether	4.5%
Isocaryophyllene	4.4%

Caryophyllene Oxide	0.0%–6.6%
(2E,4Z)-Decadienal	0.0%–6.5%
(2E)-Undecenal	0.0%–6.4%
Delta-Cadinene	0.0%–6.1%
Linalool	0.0%–5.8%
(2E)-Decenal	0.0%–5.2%
Para-Cymene	0.0%–5.2%
Alpha-Terpinene	0.0%–4.9%
Trans-Pinocamphone	0.0%–4.5%
Carvacrol	0.0%–4.5%
Alpha-Humulene	0.4%–3.9%

Jordanian (Fruits)

Naphthalene	15.9%
(Z,Z)-Alpha-Farnesene	7.6%
(E)-Beta-Farnesene	6.6%
(E)-Beta-Farnesene Isomer	6.5%
Isocaryophyllene	5.3%
Beta-Caryophyllene	5.3%
Longifolene	5.2%
O-Cymene	4.3%
Beta-Ocimene	3.8%
Furfural	3.7%
Aromadendrene	3.7%
Limonene	3.4%
1,8-Cineole	3.3%

Turkish (Pericarp)

(Z)-2-Decenal	9.9%–42.4%
n-Nonanal	10.8%–13.1%
Limonene	0.2%–9.5%
(2E,4E)-Decadienal	1.3%–7.0%
Cembrene	0.0%–6.4%
Nonanoic Acid	0.0%–4.0%

Turkish (Leaves)

Caryophyllene Oxide	3.1%–23.9%
Beta-Caryophyllene	0.3%–17.0%
Cembrene	1.2%–7.8%
(E)-2-Hexenal	tr–4.4%
Limonene	2.1%–3.7%
Alpha-Pinene	0.3%–3.6%
Farnesyl Acetone	2.6%–3.2%

Longifolene	4.4%
Alpha-Pinene	4.1%
(E)-Beta-Farnesene Isomer	3.7%
Linalool	3.4%

Palestinian (Fruits)

Furfural	0.0%–34.4%
Alpha-Pinene	2.0%–16.7%
n-Nonanal	0.9%–12.2%
(Z,Z)-Alpha-Farnesene	6.8%–8.6%
(E)-Beta-Farnesene	0.1%–8.3%
Beta-Caryophyllene	4.2%–6.2%
Isocaryophyllene	0.5%–5.8%
Longifolene	4.8%–5.7%
(E)-Beta-Farnesene Isomer	0.6%–5.7%
3-Thujanone	0.8%–5.2%
Itaconic Anhydride	1.8%–4.3%
Limonene	0.7%–3.9%
5-methyl-Fufural	0.0%–3.9%
Beta-Ocimene	0.9%–3.2%

Turkish (Branches)

Cembrene	10.7%–26.5%
Beta-Caryophyllene	12.4%–21.9%
Alpha-Pinene	3.8%–17.7%
Caryophyllene Oxide	4.9%–8.4%
Delta-Cadinene	1.4%–6.7%
Farnesyl Acetone	1.8%–4.9%
Gamma-Muurolene	0.7%–3.5%

Note: The Sicilian sumac is considered a vulnerable population with decreasing numbers. This work has differentiated the essential oils based on country of origin, but the great variance in composition among samples—even among plants growing in the same country—suggests there are multiple chemotypes.

OTHER CONSTITUENTS: *Italian (Fruits)*—Hexanal, heptanal, beta-pinene, 1-octen-3-ol, 6-methyl-5-hepten-2-one, myrcene, 2-pentylfuran, octanal, para-cymene, (Z)-beta-ocimene, 3-octen-2-one, (E)-beta-ocimene, gamma-terpinene, (E)-2-octenal, linalool, (Z)-6-nonenal, isomenthone, neomenthol, pulegone (< 2.8%), trans-2-hydroxypinocamphone, linalyl acetate, 1-decanol, citronellyl formate, nonanoic acid, undecanol, cis-2,3-pinanediol, myrtenyl acetate, cyclosativene, alpha-copaene, n-tetradecane, dodecanal, alpha-humulene, alpha-amorphene, trans-gamma-cadinene, eugenyl acetate, alpha-cadinene, cis-dracunculifolol, caryophyllene oxide, n-hexadecane, n-heptadecane, (Z)-beta-santalol-cis-santalol, n-octadecane, 6,10,14-trimethyl-2-pentadecanone; *Turkish (Fruits)*—Hexanal, alpha-terpinene, limonene, 1,8-cineole, 2-pentylfuran, gamma-terpinene, (E)-beta-ocimene, para-cymene, terpinolene, octanal, 6-methylhept-5-en-2-one, (E)-oct-2-enal, oct-1-en-3-ol, cis-linalool oxide, (E)-non-2-enal, octanol, fenchol, bornyl acetate, terpinen-4-ol, oct-2-enol, alpha-humulene, alpha-terpineol, (2E,4E)-nonadienal, borneol, beta-bisabolene, carvone, (E)-undec-2-enal, (2E,4Z)-nonadienal, ar-curcumene, cuminaldehyde, trans-anethole, geranyl acetone, methyl eugenol (< 0.6%), caryophyllene alcohol, hexahydro-farnesylacetone, pelargonic acid, eugenol, thymol, carypophylla-1(12),8(15)-dienol-(9), farnesyl acetone, nonadecna-2-one, methyl oleate, pentacosane, myristic acid; *Iranian (Fruits)*—Camphene, beta-pinene, myrcene, n-decane, alpha-phellandrene, (Z)-beta-ocimene, (E)-beta-ocimene, gamma-terpinene, alpha-terpineol, n-decanal, isobornyl acetate, undecanol, alpha-terpinyl acetate, alpha-ylangene, alpha-copaene, isoamyl benzoate, geranyl acetone, beta-selinene, alpha-selinene, alpha-muurolene, gamma-cadinene, trans-cadina-1,4-diene, Selina-3,7(11)-diene, caryophyllene alcohol, n-hexadecane, 5-epi-7-epi-eudesmol, 1-epi-cubenol, gamma-eudesmol, alpha-eudesmol, 14-hydroxy-9-epi-(E)-caryophyllene, n-octadecane, hexahydrofarnesyl acetone, n-nonadecane, (5E,9E)-farnesyl acetone, (3Z)-cembrene A; *Egyptian (Fruits)*—Alpha-terpineol, methyl nonanoate, precocene I, itaconic anhydride, octanal, nonanal, 6-camphenone, 3-thujanone, beta-pinene, myrcene, beta-thujene, 1,3,8-p-menthatriene, beta-phellandrene, gamma-terpinene, beta-terpinene, alpha-cubebene, aromadendrene, farnesene isomer, alpha-humulene, (Z)-murrola-4(14),5-diene, germacrene D, beta-bisabolene; *Jordanian (Fruits)*—Linalool, terpinen-4-ol, styrene, allyl p-methylbenzyl ether, octanal, nonanal, (Z)-2-decenal, 3-thujanone, alpha-pinene, beta-pinene, beta-thujene, 1,3,8-p-menthatriene, alpha-longipinene, alpha-cubebene, farnesene isomer, alpha-humulene, (Z)-muurola-4(14),5-diene, germacrene D, beta-bisabolene; *Palestinian (Fruits)*—Linalool, styrene, allyl p-methylbenzyl ether, maleic anhydride, octanal, 6-camphenone, alpha-fenchene, beta-pinene, beta-thujene, O-cymene, 1,8-cineole, alpha-longipinene, alpha-cubebene, aromadendrene, farnesene isomer, alpha-humulene, (Z)-muurola-4(14),5-diene, germacrene D, gamma-muurolene, alpha-muurolene, gamma-cadinene, delta-cadinene, calamenene, naphthalene,1,2,3,4,4a,7-hexahydro-1,6-dimethyl-4-(1-methylethyl)-; *Turkish (Pericarp)*—Alpha-pinene, hexanal, isoamyl acetate, myrcene, 1,8-cineole, amyl furan, gamma-terpinene, para-cymene, 2-octanone, octanal, (Z)-2-heptenal, 6-methyl-5-hepten-2-one, (E)-2-octenal, 1-octen-3-ol, decanal, 1,1-diethoxynonane, (Z)-2-nonenal, octanol, beta-caryophyllene, hexadecane, alpha-terpineol, naphthalene, (E)-2-undecenal, delta-cadinene, hexanoic acid, geranyl acetone, caryophyllene oxide, octanoic acid, hexahydrofarnesyl acetone, methyl palmitate, farnesyl acetone; *Turkish (Leaves)*—Para-cymene, 6-methyl-5-hepten-2-one, alpha-copaene, linalool, bornyl acetate, alpha-humulene, gamma-muurolene, alpha-terpineol, delta-cadinene, geranyl acetone, heptanoic acid, nerolidol, carvacrol; *Turkish (Branches)*—Camphene, hexanal, beta-pinene, limonene, beta-phellandrene, (Z)-beta-ocimene, (E)-beta-ocimene, alpha-copaene, bornyl acetate, alpha-humulene, germacrene D, alpha-muurolene, pentyl benzoate, T-cadinol

PREFERRED COMPOSITION FOR CLINICAL USE:
Too great of variance in data/samples.

Reported Therapeutic Properties: Aids digestion, stimulates appetite, relieves diarrhea, heals stomach ulcers, soothes hemorrhoids, wound healing, stops excess bleeding, anti-inflammatory, analgesic, relieves sore throat, supports vaginal cleanliness and health, aids and protects liver function, antimicrobial, antibacterial, antifungal, antioxidant, lowers high blood pressure, aids and protects cardiovascular function, supports bone strength, balances blood sugar, DNA protective, anticancer, aids eye health, promotes courage, reduces anxiety, stress management

Cautions:

■ Avoid with epilepsy and Parkinson's disease due to potentially moderate 1,8-cineole and camphor content, which may exacerbate or cause seizures/convulsions or reduce seizure-medication efficacy.[3724,3725,3726,3727,3728,3729]

Selected Evidence:

- Sicilian sumac essential oil with beta-caryophyllene and cembrene as the major constituents inhibited *Escherichia coli, Pseudomonas aeruginosa, Staphylococcus aureus*, and *Bacillus subtilis*.[3730]

SILVER WORMWOOD (Owyhee, Western Mugwort, White Sagebrush, Gray Sagewort, Luisianna Wormwood, Estafiate, Iztauyat)

Artemisia ludoviciana Nutt, *A. ludoviciana* var. *latiloba* Nutt

Family: Asteraceae
Note: Middle
Aroma Intensity: Medium-Strong
Aroma: Fruity, green, floral, fresh, resinous, honey-like, herbaceous, hint of peach
Common Extraction Method: Steam distilled or hydrodistilled from the aerial parts
Possible Substitute Oils: Spanish sage, camphor, rosemary (Spanish); Roman chamomile
Blends Well With: Balsam fir, bergamot, chamomile (Roman), clary sage, geranium, grapefruit, lavender, lemon, lime, mandarin, neroli, Peru balsam, petitgrain, rose, ylang ylang
Subcellular Localization | Epigenetic Influence: Currently unknown | Currently unknown
Recommended Dilution Range: *Camphor/1,8-Cineole CT*—1.5%–20%; *Esters CT*—5%–neat

Primary Constituents:[3731,3732,3733,3734,3735]

Camphor/1,8-Cineole CT	
Camphor	15.9%–40.6%
1,8-Cineole + Beta-Phellandrene	10.9%–26.2%
Borneol	2.3%–16.0%
Davanone	0.0%–11.5%
Linalool	0.0%–4.1%
Camphene	0.6%–4.7%
Terpinen-4-ol	1.4%–3.5%
Artemisia Alcohol	0.0%–3.0%

Commercial Oils (CT Esters)	
Beta-Artemisia Acetate	48.1%–62.5%
Yomogi Alcohol	11.1%–16.2%
Camphor	0.0%–9.4%
Artemisia Alcohol	6.5%–8.2%
1,8-Cineole	2.2%–5.1%
Camphene	1.7%–3.5%

Note: Information from most suppliers of commercial silver wormwood essential oil (owyhee) report the composition as predominantly esters and similar to Roman chamomile, but this CT was not found in the published literature. Silver wormwood essential oil is very rare and often called Owyhee (perhaps to distance it from the neurotoxic wormwood essential oils). The only CT commercially found was the esters CT.

OTHER CONSTITUENTS: *Camphor/1,8-Cineole CT*—Alpha-pinene, sabinene, beta-pinene, 1-octen-3-ol, yomogi alcohol, alpha-terpinene, para-cymene, gamma-terpinene, cis-sabinene hydrate, trans-sabinene hydrate, chrysanthenone, trans-pinocarveol, trans-verbenol, artemisyl acetate, bornyl acetate, trans-sabinyl acetate (< 0.7%), alpha-copaene, beta-selinene, gamma-cadinene, spathulenol, hexadecanoic acid, 2-methyl butyl 2-methyl butyrate, trans-p-menth-2-en-1-ol, pinocarvone, myrtenal, alpha-terpinyl acetate, verbenone, davana ether isomers, nor-davanone, myrtenol, alpha-terpineol, hexyl isobutyrate, filifolone; *Commercial Oils (CT Esters)*—Artemisia triene isomer, artemiseole, 1,8-cineole, linalool, owyhee MZ=41,69,96,153, trans-chrysanthemal, borneol, owyhee MZ=43,121,81,93,67, bornyl acetate

PREFERRED COMPOSITION FOR CLINICAL USE:

Constituent	*Esters CT*	*Camphor/1,8-Cineole CT*
Beta-Artemisia Acetate	40%–65%	–
Yomogi Alcohol	8%–20%	–
Artemisia Alcohol	3%–12%	–
1,8-Cineole	1%–7%	10%–35%
Camphene	0.5%–5%	0.1%–7%
Linalool	tr–5%	–
Camphor	–	15%–40%
Borneol	–	3%–15%
Terpinen-4-ol	–	0.1%–5%

REPORTED THERAPEUTIC PROPERTIES: Analgesic, reduces fever, relieves menstrual pain, nervine, antineuralgic, anti-inflammatory, antiarthritic, antirheumatic, antispasmodic, antibacterial, antifungal, anti-infectious, antiparasitic, diuretic, aids digestion, expels excess gas, relieves nausea, expectorant, eases cough, supports respiratory function, nourishes the skin, eases skin itching and irritation, rejuvenates the skin, wound healing, aids liver and gallbladder function, promotes restful sleep, dispels moodiness, combats anger, stress management, antidepressant, reduces resentment, aids meditation

CAUTIONS:

Camphor/1,8-Cineole CT

■ Avoid with children under six due to significant camphor content. Several cases of camphor poisoning and/or seizures from ingestion and topical application have been reported in children.[3736,3737] Ingestion of camphor-containing products has been lethal in children under age two.[3738] Children five years and up may use camphor-containing essential oils topically in dilutions no stronger than 5%. Camphor ingestion by infants and young children may cause cough, vomiting, seizure, burning sensation in the mucous membranes and eyes, or lack of voluntary coordination of muscle movements.[3739]

■ Caution is warranted during pregnancy and while lactating, due to significant camphor content. Ingestion of essential oils with significant levels of camphor may lead to abortion because fetuses lack the enzymes to process it.[3740] Camphor ingestion by infants and young children may cause cough, vomiting, seizure, burning sensation in the mucous membranes and eyes, or lack of voluntary coordination of muscle movements.[3741]

■ Avoid oral consumption due to significant camphor levels. Essential oils with significant levels of camphor may be toxic when taken orally, usually single doses exceeding 2 mL, although the lethal dose for humans is estimated to be greater than 5 mL in a single dose.[3742,3743,3744]

■ Avoid with epilepsy and Parkinson's disease due to camphor and 1,8-cineole content due to the potential of these constituents to exacerbate or cause seizures/convulsions or reduce seizure-medication efficacy.[3745,3746,3747,3748]

■ Avoid with those who have a compromised liver due to the risk of increased liver enzymes and liver damage based on camphor content.[3749] This would also suggest that those taking medications that could cause liver damage should also use essential oils with significant camphor levels very cautiously or avoid them entirely.

■ May weakly interfere with the enzymes responsible for metabolizing medications (NSAIDs, proton-pump inhibitors, acetaminophen, antiepileptics, immune modulators, blood-sugar medications, blood pressure medications, antidepressants, antipsychotics, diabetic medications, antihistamines, antibiotics, and anesthetics) due to 1,8-cineole content.[3750,3751,3752]

■ May interfere with pentobarbital and other barbiturates (medications for anxiety and insomnia), based on 1,8-cineole content.[3753,3754]

■ The significant camphor content may negatively impact red blood cells and increase the risk of jaundice in children with glucose-6-phosphate dehydrogenase deficiency (G6PD).[3755,3756]

■ Caution is warranted for the topical application of camphor-containing essential oils on broken or injured skin due to an increased risk of toxicity.[3757]

CT Esters

■ Caution is advised with epilepsy due to the potential moderate amounts of camphor and 1,8-cineole, which may exacerbate or cause seizures/convulsions or reduce seizure-medication efficacy.[3758,3759,3760,3761]

SELECTED EVIDENCE:

- □ Intravenous administration of silver wormwood essential oil significantly reduced nerve-related pain at least partially by influencing the opioid system.[3762]
- □ Silver wormwood CT camphor/1,8-cineole essential oil was effective against three parasites: *Acanthamoeba castellanii, Leishmania infantum*, and *Trichomonas vaginalis* in laboratory research.[3763]
- □ *In vitro* research demonstrated that silver wormwood CT camphor/1,8-cineole essential oil inhibited *S. aureus, S. epidermidis, E. coli, C. albicans, C. neoformans, F. pedrosoi, T. rubrum, M. canis, M. gypseum,* and *A. niger*.[3764]
- □ Silver wormwood CT camphor/1,8-cineole essential oil exhibited weak antioxidant activity.[3765]

SKIMMIA
(Nair, Nairpati, Kedarpati)

Skimmia laureola (DC.) Sieb. & Zucc. ex Walp.

FAMILY: Rutaceae
NOTE: Top-Middle
AROMA INTENSITY: Medium
AROMA: Herbaceous, floral, sweet, fresh
COMMON EXTRACTION METHOD: Steam distilled or hydrodistilled from the leaves; less commonly extracted from the stems
POSSIBLE SUBSTITUTE OILS: Lavender, lavandin, spike lavender, basil CT linalool
BLENDS WELL WITH: Bay laurel, bergamot, camphor, chamomile (German, Roman), citronella, clary sage, geranium, jasmine, lavandin, lavender, lemon, orange, palmarosa, patchouli, pine, tangerine, thyme, rosemary, rosewood, spike lavender, ylang ylang

SUBCELLULAR LOCALIZATION | EPIGENETIC INFLUENCE: Currently unknown | Currently unknown
RECOMMENDED DILUTION RANGE: 5%–Neat

PRIMARY CONSTITUENTS:[3766,3767,3768,3769,3770]

Leaves		*Stems*	
Linalyl Acetate	33.0%–51.6%	Linalool	22.0%
Linalool	12.5%–25.0%	Linalyl Acetate	17.9%
Beta-Phellandrene	0.0%–11.7%	Dictamnol	13.1%
Geranyl Acetate	2.8%–8.5%	Alpha-Terpineol	11.1%
Limonene	0.0%–8.1%	Alpha-Terpinyl Acetate	9.3%
Alpha-Terpinyl Acetate	0.0%–6.2%	Neryl Acetate	7.1%
Cis-p-Menth-2-en-1-ol	0.0%–6.2%		
Alpha-Terpineol	0.0%–5.9%		
Dictamnol	0.0%–3.9%		
Myrcene	0.0%–3.4%		

Note: CTs with limonene (32.3%) and linalyl butyrate (35.0%) are also reported in the literature.[3771,3772]

OTHER CONSTITUENTS: *Leaves*—Para-cymene, (Z)-beta-ocimene, alpha-pinene, (E)-beta-ocimene, sabinene, geraniol, 2-heptyl acetate, trans-linalool oxide, geyrene isomer, geyrene, neoiso-dihydro carveol acetate, neryl acetate, spathulenol, elemol acetate, (2E,6E)-farnesyl acetate; *Stems*—Myrcene, beta-thujone (< 0.8%), geyrene

PREFERRED COMPOSITION FOR CLINICAL USE:

Constituent	
Linalyl Acetate	30%–55%
Linalool	12%–30%
Geranyl Acetate	2%–10%
Alpha-Terpineol	3%–7%
Alpha-Terpinyl Acetate	0.5%–7%
Neryl Acetate	0.1%–5%

REPORTED THERAPEUTIC PROPERTIES: Antimicrobial, antibacterial, antifungal, anti-infectious, antiparasitic, aids respiratory function, decongestant, eases cough, anti-inflammatory, antispasmodic, analgesic, reduces fever, relieves headache, antioxidant, wound healing, soothes skin itching and chronic skin conditions, antiallergic, insecticidal, promotes restful sleep, relieves anxiety, antidepressant, stress management, reduces shock and fear

CAUTIONS:

- None currently known. Some reports suggest it may be phototoxic because bergaten has been reported in the leaves of skimmia, but its presence in the essential oil was not reported in the cited studies.[3773]

SELECTED EVIDENCE:

- Injection of skimmia essential oil (20, 100 and 200 mg/kg body weight) significantly reduced pain and reversed fever in mice, with the 200mg dose being the most effective.[3774]
- Skimmia essential oil exhibited remarkable antispasmodic activity in isolated rabbit jejunum, possibly by blocking the release of stored calcium or by blocking calcium channels.[3775]

- *In vitro* evidence showed that skimmia essential oil strongly inhibits gram-positive bacteria (*B. subtilis, S. epidermidis, S. aureus*), gram-negative bacteria (*P. aeruginosa, P. vulgaris*), and fungi (*P. chrysogenum, A. niger*), with its activity exceeding that of drug standards (ampicillin, vancomycin, fluconazole).[3776] Another study reported that it inhibited *M. luteus, E. coli, S. aureus, S. viridans, P. multocida, T. longifusis, C. albicans, F. solani, M. canis, A. flavus,* and *C. glabrata*, but was inactive against *P. aeruginosa* and *B. subtilis*.[3777] A skimmia essential oil with linalyl butyrate as the primary constituent also showed appreciable antimicrobial activity against several bacteria and fungi.[3778]
- An odd CT of skimmia essential oil (35.0% linalyl butyrate, 9.5% geraniol, 8.8% nerol, 7.7% citronellol, 6.5% alpha-phellandrene) exhibited good antioxidant activity in the DPPH assay.[3779]
- Skimmia CT limonene (32.3% limonene, 23.5% linalyl acetate, 16.7% alpha-terpineol) essential oil was insecticidal to and repelled the black garden ant (*Lasius niger*).[3780]

SMOKETREE (Sumac, Venetian Sumach, Smoch Bush)

Cotinus coggygria Scop., *Rhus cotinus* L.

FAMILY: Anacardiaceae
NOTE: Top-Middle
AROMA INTENSITY: Medium
AROMA: Green, woody, balsamic, sweet, fruity, resinous, slightly spicy
COMMON EXTRACTION METHOD: Steam distilled or hydrodistilled from the leaves or leaves and twigs; less commonly distilled from the inflorescences and infructescences (fruiting head)
POSSIBLE SUBSTITUTE OILS: Elemi, black pepper, cypress, frankincense, pink pepper, Douglas fir, tagetes
BLENDS WELL WITH: Balsam fir, bergamot, cedarwood, copaiba, Douglas fir, frankincense, grapefruit, lavender, lemon, lime, marjoram, myrrh, neroli, orange, petitgrain, pine, rosemary, sandalwood, silver fir, tangerine, vetiver, white fir
SUBCELLULAR LOCALIZATION | EPIGENETIC INFLUENCE: Currently unknown | Currently unknown
RECOMMENDED DILUTION RANGE: 3%–50%; neat for some conditions

PRIMARY CONSTITUENTS:[3781,3782,3783,3784,3785,3786]

Leaves/Twigs CT Limonene

Limonene	39.2%–47.1%
(Z)-Beta-Ocimene	15.2%–26.3%
(E)-Beta-Ocimene	4.6%–9.0%
Alpha-Pinene	8.2%–8.8%
(E)-2-Hexenal	1.2%–7.6%
Terpinolene	4.1%–6.8%
Beta-Pinene	2.1%–3.5%

Leaves/Twigs CT Alpha-Pinene

Alpha-Pinene	44.0%
Limonene	20.0%

Leaves CT Limonene

Limonene	48.5%–67.4%
(Z)-Beta-Ocimene	tr–27.9%
Alpha-Pinene	1.6%–14.7%
(E)-Beta-Ocimene	tr–9.7%
Terpinolene	0.1%–8.6%
Beta-Caryophyllene	4.5%–4.6%

Leaves CT Camphor

Camphor	14.5%
Gamma-Elemene	9.6%
Globulol	8.0%

Beta-pinene	11.4%
Beta-Phellandrene	4.1%

Leaves CT Myrcene

Myrcene	32.0%
Sabinene	18.0%
Alpha-Pinene	15.9%
Germacrene D	6.0%
Terpinen-4-ol	4.0%

Leaves CT Sabinene

Sabinene	24.2%
Myrcene	14.0%
Limonene	10.9%
Terpinen-4-ol	10.9%
Alpha-Terpinene	9.7%
Gamma-Terpinene	8.0%
Alpha-Pinene	5.5%
Beta-Caryophyllene	4.9%
Germacrene D	4.1%

Infructescences CT Sabinene

Sabinene	41.4%
Alpha-Terpinene	9.5%
Myrcene	9.3%
Terpinen-4-ol	8.8%
Gamma-Terpinene	6.8%
Limonene	6.5%
Alpha-Pinene	4.4%
Beta-Phellandrene	7.8%
Beta-Caryophyllene	7.7%
Epi-Bicyclosesqui-Phellandrene	5.6%
Myrcene	5.4%
Limonene	4.6%
Terpinen-4-ol	4.3%
Delta-2-Carene	4.1%
Gamma-Terpinene	3.8%
Stigmasterol	3.7%

Inflorescences

Limonene	41.4%
Alpha-Pinene	32.4%
Terpinolene	14.3%
Beta-Caryophyllene	9.3%

Infructescences CT Myrcene

Myrcene	29.5%
Sabinene	15.5%
Alpha-Pinene	10.5%
Terpinen-4-ol	7.1%
Alpha-Terpinene	6.4%
Beta-Caryophyllene	5.9%
Germacrene D	5.6%
Gamma-Terpinene	5.5%
Delta-3-Carene	3.2%

OTHER CONSTITUENTS: *Leaves/Twigs CT Limonene*—(Z)-3-hexenal, myrcene, alpha-phellandrene, camphene, allo-ocimene, beta-caryophyllene, gamma-muurolene, gamma-cadinene; *Leaves/Twigs CT Alpha-Pinene*—Terpinolene, delta-3-carene, myrcene, cuminyl alcohol; *Leaves CT Limonene*—(Z)-3-hexenal, alpha-fenchene, myrcene, allo-ocimene, gamma-muurolene, delta-cadinene; *Leaves CT Myrcene*—Camphene, alpha-terpinene, para-cymene, delta-3-carene, gamma-terpinene, terpinolene, alpha-terpineol, beta-caryophyllene, bicyclogermacrene, delta-cadinene, spathulenol; *Leaves CT Sabinene*—Alpha-phellandrene, terpinolene, 1-terpineol, alpha-terpineol, bicyclogermacrene; *Leaves CT Camphor*—Delta-3-carene, delta-4-carene, alpha-copaene, alpha-humulene, gamma-muurolene, T-cadinol, alpha-cadinol; *Inflorescences*—Beta-pinene, myrcene; *Infructescences CT Myrcene*—Alpha-thujene, camphene, alpha-phellandrene, para-cymene, terpinolene, 1,3,8-para-menthatriene, beta-pinene oxide, alpha-terpineol, alpha-copaene, alpha-humulene, bicyclogermacrene, delta-cadinene; *Infructescences CT Sabinene*—Alpha-thujene, octanal, alpha-phellandrene, delta-2-carene, para-cymene, terpinolene, 1-terpineol, alpha-terpineol, beta-caryophyllene, bicyclogermacrene

PREFERRED COMPOSITION FOR CLINICAL USE:

Constituent	*Limonene CT*	*Alpha-Pinene CT*
Limonene	35%–55%	10%–25%
(Z)-Beta-Ocimene	15%–30%	5%–20%
Alpha-Pinene	5%–15%	35%–55%
(E)-Beta-Ocimene	2%–10%	0.1%–7%
Terpinolene	2%–10%	2%–10%
(E)-2-Hexenal	0.5%–10%	–
Beta-Pinene	0.5%–5%	2%–12%

REPORTED THERAPEUTIC PROPERTIES: Reduces the appearance of acne and blemishes, soothes skin irritation and itching, wound healing, relieves hemorrhoids, antioxidant, anticancer, antiseptic, antimicrobial, antiviral, liver protective, anti-inflammatory, analgesic, reduces fever, increase bile flow, relieves diarrhea, heals stomach ulcers, relieves baldness, aids respiratory function, antidepressant, relieves anxiety, combats anger and negative thoughts, relaxing

CAUTIONS:

Limonene, Sabinene, and Alpha-Pinene CTs

- None currently known.

Myrcene CTs

- May weakly interfere with the enzymes responsible for metabolizing medications (NSAIDs, proton-pump inhibitors, acetaminophen, antiepileptics, immune modulators, blood-sugar medications, blood pressure medications, antidepressants, antipsychotics, diabetic medications, antihistamines, antibiotics, and anesthetics), based on myrcene content.[3787,3788]

Camphor CT

- Avoid with epilepsy and Parkinson's disease due to the potential of camphor and 1,8-cineole to exacerbate or cause seizures/convulsions or reduce seizure-medication efficacy.[3789,3790,3791,3792]
- The significant camphor content may negatively impact red blood cells and increase the risk of jaundice in children with glucose-6-phosphate dehydrogenase deficiency (G6PD).[3793,3794]

SELECTED EVIDENCE:

- Smoketree leaves/twigs essential oil (limonene CT) demonstrated slightly higher activity against most bacteria (higher: *B. cereus, B. subtilis, S. aureus, S. epidermidis, S. faecalis*; equal: *E. coli, M. flavus, M. luteus*; lower: *P. mirabilis, S. typhimurium*) when compared to streptomycin in the disc-diffusion method and slightly lower activity in the microdilution method.[3795] The same researchers reported antifungal (*A. niger, A. ochraceus, A. versicolor, A. flavus, A. fumigatus, P. ochrochloron, P. funiculosum*) activity, but the activity was lower than bifonazole. The 2-bornanone CT of smoketree essential oil inhibited *E. coli*.[3796] Another study showed that the leavers/twigs essential oil (CT limonene) inhibited gram-positive bacteria (*B. cereus, S. aureus, S. epidermidis*) and fungi (*C. albicans, C. glabrata, C. tropicalis*).[3797]
- The camphor CT of smoketree essential oil exhibited weak antioxidant activity in the DPPH assay and good activity in the ABTS assay.[3798]

SMOKY TEA TREE (Blue-Green Tea Tree)

Leptospermum glaucescens Schauer.

FAMILY: Myrtaceae
NOTE: Middle
AROMA INTENSITY: Medium
AROMA: Sweet, fruity, herbaceous, medicinal
COMMON EXTRACTION METHOD: Steam distilled from the leaves
POSSIBLE SUBSTITUTE OILS: Kanuka, kunzea, tea tree
BLENDS WELL WITH: Bergamot, blue cypress, camphor, cedarwood, coriander, cypress, frankincense, juniper, lavender, lemon, lime, marjoram, orange, rosemary, sandalwood, tea tree
SUBCELLULAR LOCALIZATION | EPIGENETIC INFLUENCE: Currently unknown | Currently unknown
RECOMMENDED DILUTION RANGE: 3%–Neat

PRIMARY CONSTITUENTS:[3799,3800]

(E)-Beta-Ocimene CT

Constituent	%
Unidentified Sesquiterpene Alcohols	16.7%
(E)-Beta-Ocimene	12.7%
Unknown Compounds	10.0%
Globulol	9.4%
Viridiflorene	8.4%
Viridiflorol	5.5%
Spathulenol	3.9%
Gamma-Terpinene	3.5%

Alpha-Pinene CT

Constituent	%
Alpha-Pinene	53.1%
Unidentified Sesquiterpene Alcohol	16.0%
Para-Cymene	3.5%

OTHER CONSTITUENTS: *Alpha-Pinene CT*—Beta-pinene, limonene, alpha-terpineol, carvone, beta-elemene, gamma-eudesmol; *(E)-Beta-Ocimene CT*—Alpha-pinene, alpha-phellandrene, myrcene, beta-phellandrene, para-cymene, terpinolene, beta-elemene, beta-caryophyllene, terpinen-4-ol, aromadendrene, alloaromadendrene, humulene, alpha-muurolene, beta-selinene, alpha-selinene, bicyclogermacrene, delta-cadinene, palustrol, (E)-nerolidol, alpha-cadinol

PREFERRED COMPOSITION FOR CLINICAL USE*:

Constituent	
Alpha-Pinene	50%–80%
Viridiflorol	1%–8%
Viridiflorene (Ledene)	0.1%–5%
Globulol	0.1%–5%

* Very preliminary based on small sample size.

REPORTED THERAPEUTIC PROPERTIES: Antibacterial, antifungal, analgesic, anti-inflammatory, relieves headache, aids circulation, **supports respiratory function**, eases cough, reduces the appearance of blemishes, acne, and wrinkles, relieves insect bites, relaxing, relieves anxiety, combats anger

CAUTIONS:

- None currently known.

Selected Evidence:

- None found.

SNOWBUSH (African Rosemary, Wild Rosemary, Kapokbossie, Cape Snowbush)

Eriocephalus africanus L., *E. africanus* L. var. *africanus*, *E. africanus* L. var. *paniculatus*

Family: Asteraceae
Note: Middle
Aroma Intensity: Strong
Aroma: Green, spicy, bitter
Common Extraction Method: Steam distilled or hydrodistilled from the aerial parts or stems/leaves
Possible Substitute Oils: African bluegrass CT artemisia ketone, mugwort, sweet nancy, wild chamomile, kunzea, peta, yarrow
Blends Well With: Basil, bergamot, cedarwood, eucalyptus, fennel, frankincense, geranium, ginger, lemon, may chang, palmarosa, sandalwood, tangerine, vetiver, ylang ylang
Subcellular Localization | Epigenetic Influence: Currently unknown | Currently unknown
Recommended Dilution Range: 3%–33%; 50% for some conditions

Primary Constituents:[3801,3802,3803]

Artemisia Ketone CT	
Artemisia Ketone	11.8%–56.6%
Intermedeol	0.0%–11.6%
Gamma-Eudesmol	0.0%–5.6%
Alpha-Pinene	0.8%–4.1%
Yomogi Alcohol	2.2%–3.4%

Spathulenol CT	
Spathulenol	9.5%–40.0%
Piperitone	0.0%–17.2%
Bicyclogermacrene	0.0%–13.6%
Caryophyllene Oxide	0.0%–10.1%
Sabinene	0.0%–8.4%
Alpha-Cadinol or T-Muurolol	0.0%–8.4%
Para-Cymene	0.5%–7.4%
Camphor	0.0%–7.1%
Neryl Acetate	0.0%–5.7%
Beta-Caryophyllene	0.0%–5.1%
Terpinen-4-ol	0.6%–5.0%
1,8-Cineole	0.7%–4.6%
Alpha-Longipinene	0.0%–3.7%
Alpha-Pinene	0.0%–3.1%

Santolina Alcohol CT	
Santolina Alcohol	29.9%
Camphor	12.1%
Borneol	10.0%
Delta-Cadinene	9.0%

1,8-Cineole CT	
1,8-Cineole	23.6%
Chrysanthenone	18.5%
Alpha-Pinene	7.5%
Terpinen-4-ol	7.3%
Artemisia Ketone	4.5%
Para-Cymene	3.4%

Note: Other references also report a linalyl acetate CT and a 1,8-cineole CT with davanone.[3804,3805]

OTHER CONSTITUENTS: *Artemisia Ketone CT*—Beta-pinene, para-cymene, 1,8-cineole, artemisia alcohol, trans-pinocarveol, borneol, alpha-longipinene, pinocarvone, artemisia ketone isomer, myrtenal, beta-caryophyllene, bicyclogermacrene, kessane, liguloxide, elemol, spathulenol, caryophylla-4,8-dien-5-ol, T-muurolol, alpha-eudesmol, alpha-cadinol, en-in-dicycloether; *Spathulenol CT*—Borneol, chamazulene; *Santolina Alcohol CT*—Alpha-Pinene, yomogi alcohol, para-cymene, 1,8-cineole, terpinen-4-ol, bicyclogermacrene, spathulenol, alpha-cadinol, T-muurolol; *1,8-Cineole CT*—Sabinene, borneol, beta-caryophyllene, spathulenol, caryophyllene oxide

PREFERRED COMPOSITION FOR CLINICAL USE:
Too great of variance in data/samples to determine.

REPORTED THERAPEUTIC PROPERTIES: Antiseptic, anti-infectious, antifungal, antibacterial, antiviral, decongestant, stops excess bleeding, anti-inflammatory, antispasmodic, analgesic, relieves gout, supports respiratory function, eases cough, aids digestion, expels excess gas, diuretic, antidepressant, uplifting, relaxing, reduces anger, relieves anxiety

CAUTIONS:

1,8-Cineole CT

■ Avoid with epilepsy and Parkinson's disease due to potentially moderate 1,8-cineole content. May exacerbate or cause seizures/convulsions or reduce seizure-medication efficacy.[3806,3807,3808,3809]

■ May weakly interfere with the enzymes responsible for metabolizing medications (NSAIDs, proton-pump inhibitors, acetaminophen, antiepileptics, immune modulators, blood-sugar medications, blood pressure medications, antidepressants, antipsychotics, diabetic medications, antihistamines, antibiotics, and anesthetics) due to 1,8-cineole content.[3810,3811,3812]

■ May interfere with pentobarbital and other barbiturates (medications for anxiety and insomnia), based on 1,8-cineole content.[3813,3814]

Spathulenol CT or Santolina Alcohol CT

■ Avoid with epilepsy and Parkinson's disease due to potentially moderate 1,8-cineole and camphor content. May exacerbate or cause seizures/convulsions or reduce seizure-medication efficacy.[3815,3816,3817,3818]

Artemisia Ketone CT

■ None currently known.

SELECTED EVIDENCE:

- □ *In vitro* research demonstrated that snowbush essential oil (various CTs) inhibited *C. neoformans, B. cereus, B, subtilis, C. albicans, S. aureus, P. aeruginosa, S. enteritidis, E. faecalis, K. pneumoniae, E. coli, Y. enterocolitica, S. typhimurium, S. odorifera,* and MRSA.[3819]
- □ Snowbush essential oil CT artemisia ketone prevented the growth of weeds (*Amaranthus hybridus, Portulaca oleracea*) of Mediterranean summer crops.[3820]

SOUTHERNWOOD
(Lad's Love, Southern Wormwood)

Artemisia abrotanum L.

FAMILY: Asteraceae (Compositae)
NOTE: Top-Middle
AROMA INTENSITY: Medium
AROMA: Herbaceous, camphoraceous, slightly sweet
COMMON EXTRACTION METHOD: Steam distilled or hydrodistilled from the aerial parts
POSSIBLE SUBSTITUTE OILS: PSO
BLENDS WELL WITH: Bergamot, camphor, copaiba, eucalyptus, fennel, frankincense, grapefruit, juniper, lavender, lavandin, ledum, lemon, lime, myrtle, orange, peppermint, petitgrain, rosemary, Spanish sage, spike lavender, neroli, tangerine, wintergreen
SUBCELLULAR LOCALIZATION | EPIGENETIC INFLUENCE: Currently unknown | Currently unknown
RECOMMENDED DILUTION RANGE: 1.5%–20%

PRIMARY CONSTITUENTS:[3821,3822,3823,3824]

Camphor/1,8-Cineole CT

Camphor	20.3%–44.6%
1,8-Cineole	15.5%–33.2%
Camphene	2.4%–7.2%
Alpha-Eudesmol	0.2%–4.9%
Alpha-Bisabolol	0.0%–4.6%
2(10)-Pinen-3-one	0.0%–3.8%
Terpinen-4-ol	2.3%–3.7%
Sabinene	0.3%–3.6%
Borneol	2.1%–3.3%
Verbenol	0.0%–3.2%

1,8-Cineole/Bisabolol Oxide CT

1,8-Cineole	34.7%
Bisabolol Oxide	18.4%
Ascaridole	16.0%
Para-Cymene	7.9%
(E)-Nerolidol	4.2%
Spathulenol	4.0%
Bornyl Acetate	3.3%
Carvacrol	3.1%

Davanone CT

Davanone Derivative	22.5%
4-Methyl-pent-2-enolid	15.7%
Unknown (RI=1685)	12.4%
Unknown (RI=1711)	8.7%

1,8-Cineole/Silphiperfol CT

1,8-Cineole	24.5%
Silphiperfol-5-en-3-one A	18.9%
Borneol	9.3%
Unknown (RI=1513)	5.7%
Silphiperfol-5-en-3-one B-isomer	5.1%
Camphor	3.5%
Germacrene D	3.4%
Unknown (RI=1595)	3.0%

Trans-Sabinyl Acetate CT

Trans-Sabinyl Acetate	33.4%
Alpha-Terpineol	8.2%
Terpinen-4-ol	5.9%
Trans-Sabinol	5.1%
Sabinene	4.4%
Unknown (RI=1628)	4.4%
Linalool	4.2%
Unknown (RI=1636)	3.6%

Davana Ethers	5.5%
Cis-Davanone	5.2%
Unknown (RI=1688)	4.0%
Nordavanone	3.0%

Note: Other studies report 1,8-cineole/Silphiperfol-5-en-3-one A and 1,8-cinole as the major constituents of southernwood when extracted with solvents.[3825,3826] A congress abstract also reported the composition of hydrodistilled southernwood as 32.6% 1,8-cineole, 13.5% borneol, 10.2% presilphiperfolan-9α-ol and 8.0% p-cymene.[3827]

Other Constituents: *Camphor/1,8-Cineole CT*—Beta-pinene, alpha-terpinene, para-cymene, limonene, gamma-terpinene, trans-sabinene hydrate, cis-sabinene hydrate, p-menth-1-en-8-ol, pinocarvone (< 1.4%), alpha-terpineol, menth-1,5-dien-7-ol, 3(10)-caren-2ol, 1,4-dimethyl-4-propyl-2-on-1(2)-cyclohexene, trans-carvylacetate, methyl eugenol (<1.7%), geranyl isobutyrate, germacrene D, agarospirol, guaiol, trans-alpha-bisabolene epoxide; *1,8-Cineole/Silphiperfol CT*—Camphene, alpha-terpinene, terpinen-4-ol, trans-chrysanthenyl acetate, bornyl acetate, silphiperfol-5-ene, silphiperfolen-isomers, 7β-H-silphiperfol-5-ene, beta-caryophyllene, alpha-humulene; *1,8-Cineole/Bisabolol Oxide CT*—Alpha-phellandrene, alpha-terpinene, gamma-terpinene, terpinen-4-ol, myrtenol, pinocarvone (<1.9%); *Davanone CT*—Lavender lactone, cis-arbusculone, trans-arbusculone, 1-terpineol, cis-beta-terpineol, estragole (< 1.0%), piperitone, artedouglasia oxide C, artedouglasia oxide A, artedouglasia oxide B, artedouglasia oxide D, phytol isomer; *Trans-Sabinyl Acetate CT*—Alpha-thujene, alpha-pinene, camphene, myrcene, alpha-terpinene, gamma-terpinene, terpinolene, chrysanthenone, borneol, bornyl acetate, beta-caryophyllene, alpha-dehydro-ar-himachalene, caryophyllene oxide

Preferred Composition for Clinical Use:
Too great of variance in samples/data to determine.

Reported Therapeutic Properties: Antiallergic, aids respiratory function, antibacterial, antifungal, antiseptic, antiparasitic, relieves diarrhea, anti-inflammatory, antispasmodic, analgesic, balances menstruation and relieves painful menstruation, aids circulation, insect repellent, combats, jealousy and anger, stimulating, aids cognition and memory function

Cautions:

Camphor/1,8-Cineole CT

■ Avoid with children under six due to significant camphor content and moderate amounts of 1,8-cineole. Several cases of camphor poisoning and/or seizures from ingestion and topical application have been reported in children.[3828,3829] Ingestion of camphor-containing products has been lethal in children under age two.[3830] Children five years and up may use camphor-containing essential oils topically in dilutions no stronger than 5%. Camphor ingestion by infants and young children may cause cough, vomiting, seizure, burning sensation in the mucous membranes and eyes, or lack of voluntary coordination of muscle movements.[3831]

■ Caution is warranted during pregnancy and while lactating, due to significant camphor content. Ingestion of essential oils with significant levels of camphor may lead to abortion because fetuses lack the enzymes to process it.[3832] Camphor ingestion by infants and young children may cause cough, vomiting, seizure, burning sensation in the mucous membranes and eyes, or lack of voluntary coordination of muscle movements.[3833]

■ Avoid oral consumption due to significant camphor content. Essential oils with significant levels of camphor may be toxic when taken orally, usually single doses exceeding 2 mL, although the lethal dose for humans is estimated to be greater than 5 mL in a single dose.[3834,3835,3836]

■ Avoid with epilepsy and Parkinson's disease due to camphor and 1,8-cineole content, which have the potential to exacerbate or cause seizures/convulsions or reduce seizure-medication efficacy.[3837,3838,3839,3840]

■ Avoid with those who have a compromised liver or kidneys, due to the risk of increased liver enzymes and liver or kidney damage based on camphor.[3841,3842,3843,3844] This would also suggest that those taking medications that could cause liver damage should also use essential oils with significant camphor levels very cautiously or avoid them entirely.
■ The significant camphor content may negatively impact red blood cells and increase the risk of jaundice in children with glucose-6-phosphate dehydrogenase deficiency (G6PD).[3845,3846]
■ May weakly interfere with the enzymes responsible for metabolizing medications (NSAIDs, proton-pump inhibitors, acetaminophen, antiepileptics, immune modulators, blood-sugar medications, blood pressure medications, antidepressants, antipsychotics, diabetic medications, antihistamines, antibiotics, and anesthetics) due to 1,8-cineole content.[3847,3848,3849]
■ May interfere with pentobarbital and other barbiturates (medications for anxiety and insomnia), based on 1,8-cineole content.[3850,3851]

1,8-Cineole/Bisabolol Oxide and 1,8-Cineole/Silphiperfol CTs
■ Avoid with epilepsy and Parkinson's disease due to 1,8-cineole content. May exacerbate or cause seizures/convulsions or reduce seizure-medication efficacy based on 1,8-cineole content.[3852,3853,3854,3855]
■ May weakly interfere with the enzymes responsible for metabolizing medications (NSAIDs, proton-pump inhibitors, acetaminophen, antiepileptics, immune modulators, blood-sugar medications, blood pressure medications, antidepressants, antipsychotics, diabetic medications, antihistamines, antibiotics, and anesthetics) due to 1,8-cineole content.[3856,3857,3858]
■ May interfere with pentobarbital and other barbiturates (medications for anxiety and insomnia), based on 1,8-cineole content.[3859,3860]
■ *1,8-Cineole/Bisabolol Oxide CT*: Avoid use in children due to toxicity concerns and known poisoning and fatalities with ascaridole-containing essential oils in humans due to ascaridole content.[3861]
■ *1,8-Cineole/Bisabolol Oxide CT*: Avoid during pregnancy and lactation due to ascaridole content. Hydro-alcohol extracts of boldo (containing ascaridole) have reportedly caused abortions, fetal abnormalities, and low birth weight in animal studies when administered at 800 mg/kg/day.[3862,3863] Boldo leaf teas (containing ascaridole) are reportedly used to induce illegal abortions in humans.[3864]
■ *1,8-Cineole/Bisabolol Oxide CT*: Avoid oral consumption due to ascaridole content. Oral use of essential oils with significant ascaridole content can cause respiratory depression, overstimulation of the CNS (exaggerated reflexes, lack of coordination, convulsions, paralysis of motor and sensory nerves, and death by respiratory arrest), liver toxicity, and anaphylactic reactions.[3865,3866,3867] Human toxicity and fatal poisoning have been reported in children and adults with chenopodium essential oil (also high in ascaridole).[3868]
■ *1,8-Cineole/Bisabolol Oxide CT*: Avoid topical application, skin irritation or sensitization—ascoridole may cause skin sensitization.[3869,3870,3871]

Trans-Sabinyl Acetate CT
■ Avoid in children due to the toxicity of trans-sabinyl acetate.[3872]
■ Avoid during pregnancy. Essential oils rich in sabinyl acetate are toxic to developing embryos, cause fetal malformations, and prevent implantation during days 0–4 of gestation according to animal research.[3873,3874,3875,3876] Other research using a Spanish sage essential oil (containing 50% trans-sabinyl acetate) demonstrated trans-sabinyl acetate can cause abortions and maternal toxicity (which indicated a greater susceptibility to trans-sabinyl acetate toxicity during pregnancy) in animals.[3877] The researchers concluded that trans-sabinyl acetate possesses significant antifertility activity during the preimplantation period, and attribute the fetotoxic and abortifacient activity to this constituent.
■ Avoid oral consumption. Essential oils rich in sabinyl acetate irritate the mucosal lining of the gastrointestinal tract, damage the kidneys, cause blood in the urine, obstruct blood flow in the abdominal organs, and cause abnormally heavy menstrual bleeding, which the study authors attributed to the combined

effects of sabinene, sabinol, and trans-sabinyl acetate.[3878] However, more recent research suggests it is the trans-sabinyl acetate that causes toxicity and other adverse effects.

- Avoid with epilepsy and Parkinson's disease due to thujone and trans-sabinyl acetate content.[3879]
- Avoid in those who have a compromised liver due to the increased risk of liver toxicity and liver damage due to trans-sabinyl acetate content.[3880] This would also suggest that those taking medications that could cause liver damage should also use it very cautiously or avoid it entirely.

Davanone CT

- None currently known.

SELECTED EVIDENCE:

- ☐ Southernwood essential oil with 1,8-cineole and borneol as the major constituents repelled the *Aedes aegypti* mosquito.[3881]

SPANISH CEDAR (Cedrela, West Indian Cedar, Red Cedar, Cuban Cedar)

Cedrela odorata L., *C. mexicana* M. J. Roem

FAMILY: Meliaceae
NOTE: Middle
AROMA INTENSITY: Medium
AROMA: Woody, earthy, slightly sweet
COMMON EXTRACTION METHOD: Steam distilled from the wood (bark); may also be distilled from the leaves or stems
POSSIBLE SUBSTITUTE OILS: Copaiba, myrrh, cedarwood, hinoki
BLENDS WELL WITH: Agarwood, balsam fir, bergamot, cedarwood, clary sage, copaiba, cypress, frankincense, jasmine, juniper, lavender, myrrh, neroli, orange, pine, petitgrain, rose, rosemary, sandalwood, silver fir, spearmint, vetiver, white fir, ylang ylang
SUBCELLULAR LOCALIZATION | EPIGENETIC INFLUENCE: Currently unknown | Currently unknown
RECOMMENDED DILUTION RANGE: 3%–Neat

PRIMARY CONSTITUENTS:[3882,3883,3884,3885,3886,3887,3888]

Costa Rican		*Brazilian*	
Beta-Elemene	20.3%	Alpha-Copaene	15.6%
Germacrene D	15.4%	Delta-Cadinene	11.7%
1,10-di-epi-Cubenol	7.2%	Alpha-Cubebene	8.0%
Juniperol + Cedrol	7.1%	(E)-Nerolidol	8.0%
Beta-Acoradiene	7.0%	Beta-Curcumene	6.8%
Alpha-Cadinol	3.6%	Calarene	3.4%
Beta-Alaskene	3.5%	Beta-Cubebene	3.2%
Premnaspirodiene	3.2%	Germacrene D	2.6%
T-Cadinol	3.0%		

(E)-Cadina-1(6),4-Diene	2.9%
Beta-Caryophyllene	2.6%
Alpha-Acoradiene	2.5%
Germacrene B	2.5%

São Tomé and Príncipe

Alpha-Copaene	14.4%
Alpha-Cadinol	11.2%
Beta-Eudesmol	9.4%
Delta-Cadinene	9.2%
T-Muurolol	4.3%
Alpha-Cubebene	4.2%
Gamma-Eudesmol	3.4%
Hexadecanoic Acid	2.7%
(E)-Alpha-Bergamotene	2.6%
(E)-Calamenene	2.6%

Brazilian (Stems)

Caryophyllene Oxide	17.9%
Beta-Caryophyllene	17.2%
Cis-4(14),5-Muuroladiene	10.5%
(Z)-Beta-Caryophyllene	9.1%
Alpha-Humulene	7.7%
Beta-Elemene	5.4%
Humulene Epoxide	2.8%
Alpha-Copaene	2.6%
Selina-11-en-4-Alpha-ol	2.5%

Brazilian (Leaves)

Germacrene A	0.0%–22.6%
Beta-Elemene	6.8%–19.3%
Alpha-Santalene	0.0%–9.5%
Bicyclogermacrene	0.0%–7.6%
Beta-Acoradiene	0.0%–7.1%
Cembrane A	0.0%–6.5%
(Z)-Alpha-Bergamotene	0.0%–6.0%
Caryophyllene Oxide	0.8%–6.0%
Cedrane Diol	0.0%–6.0%
Alpha-Humulene	2.9%–4.4%
Isocaryophyllene	0.0%–4.2%
Beta-Bisabolene	0.0%–3.8%
Beta-Alaskene	0.0%–3.6%
Delta-Cadinene	2.2%–3.3%
Beta-Selinene	0.0%–3.3%
Beta-Caryophyllene	2.6%–2.9%
Alpha-Coapene	2.0%–2.9%

OTHER CONSTITUENTS: *Costa Rican*—Alpha-pinene, delta-elemene, alpha-copaene, beta-bourbonene, 6,9-guiadiene, alpha-humulene, beta-selinene, gamma-cadinene, delta-cadinene, hedycaryol, elemol; *São Tomé and Príncipe*—Thymol, beta-elemene, beta-caryophyllene, alloaromadendrene, alpha-selinene, alpha-muurolene, gamma-cadinene, cadina-1,4-diene, alpha-calacorene, elemol, spathulenol, caryophyllene oxide, globulol, guaiol, alpha-muurolol, T-cadinol, alpha-eudesmol, selin-11-en-4alpha-ol, 14-hydroxy-alpha-muurolene, benzyl benzoate, cryptomeridiol, alpha-bisabolol; *Brazilian*—Alpha-humulene, beta-farnesene, epi-cubenol, caryophyllene, alpha-muurolene, cadina-1,4-diene; *Brazilian (Leaves)*— Alpha-pinene, alpha-ylangne, beta-bourbonene, (Z)-caryophyllene, aromadendrene, alpha-guaiene, alpha-himachalene, beta-chamigrene, germacrene D, alpha-muurolene, gamma-cadinene, germacrene B, spathulenol, caryophyllene oxide, viridiflorol, 1-epi-cubenol, selin-11-en-4-alpha-ol, beta-funebrene, epi-beta-santalene, iso-bicyclogermacrene, (E)-alpha-bergamotene, gamma-humulene, delta-cadinene, humulene oxide, 14-oxy-2-muurolene; *Brazilian (Stems)*—Alpha-guaiene, gamma-himachalene, beta-selinene, alpha-selinene, germacrene A, gamma-cadinene, spathulenol, T-muurolol

PREFERRED COMPOSITION FOR CLINICAL USE:
Insufficient samples/data to determine.

REPORTED THERAPEUTIC PROPERTIES: Antibacterial, antifungal, antiparasitic, disinfectant, aids respiratory function, expectorant, astringent, wound healing, relieves boils and blisters, soothes chronic skin conditions, relieves fever, analgesic, anti-inflammatory, antispasmodic, relieves headache, balances blood-sugar levels, insect repellent, relieves anxiety, stress management, encourages a positive outlook

CAUTIONS:

Costa Rican, Brazilian (Leaves)

■ Avoid during pregnancy and lactation (very low risk). May contain moderate amounts of beta-elemene, which may prevent the formation of blood vessels necessary for fetal development according to animal research.[3889]

São Tomé and Príncipe, Brazilian, Brazilian (Stems)

■ None currently known.

SELECTED EVIDENCE:

- □ Costa Rican Spanish cedar essential oil inhibited the gram-positive bacterium *B. cereus* and the mold *A. niger*.[3890]

ST. JOHN'S WORT (Common St. Johnswort)

Hypericum perforatum L.

FAMILY: Hypericaceae
NOTE: Middle-Strong
AROMA INTENSITY: Medium
AROMA: Herbaceous, balsamic, sweet, earthy
COMMON EXTRACTION METHOD: Steam distilled from the flowering aerial parts
POSSIBLE SUBSTITUTE OILS: Helichrysum, melissa, lavender, German chamomile, ylang ylang, copaiba
BLENDS WELL WITH: Bergamot, calendula, ginger, grapefruit, helichrysum, lavender, lemon, lime, mandarin, Peru balsam, orange, tangerine, ylang ylang
SUBCELLULAR LOCALIZATION | EPIGENETIC INFLUENCE: Currently unknown | Currently unknown
RECOMMENDED DILUTION RANGE: 3%–33%; 50% for some conditions

PRIMARY CONSTITUENTS:[3891,3892,3893,3894,3895,3896,3897,3898,3899,3900,3901,3902,3903,3904,3905,3906,3907,3908,3909]

Alpha-Pinene CT		*Germacrene D CT*	
Alpha-Pinene	8.1%–36.7%	Germacrene D	12.0%–37.3%
Germacrene D	0.0%–13.7%	Beta-Caryophyllene	2.2%–28.4%
Alpha-Amorphene	0.0%–12.1%	Caryophyllene Oxide	0.5%–13.1%
Caryophyllene Oxide	0.1%–11.4%	ar-Curcumene	0.0%–13.0%
2-Methyloctane	0.0%–11.3%	Beta-Eudesma-4(15), 7-Dien-1-ol	0.0%–11.7%
Beta-Selinene	0.0%–10.3%	Gamma-Muurolene	0.0%–11.1%
Beta-Pinene	1.7%–8.3%	2-Methyloctane	0.0%–9.5%
Alpha-Cadalene	0.0%–7.8%	1-Tetradecanol	0.0%–9.4%
Beta-Caryophyllene	0.5%–7.6%	(Z)-Beta-Farnesene	0.0%–9.0%
1-Tetradecanol	0.0%–7.6%	Alpha-Pinene	0.0%–8.6%
n-Undecane	0.6%–7.0%	Spathulenol	0.3%–7.5%
n-Nonane	0.0%–7.0%	(E)-Beta-Farnesene	0.0%–7.1%
Thymol	0.0%–6.8%		

Alpha-Selinene	0.0%–6.2%
Alpha-Campholene Aldehyde	0.0%–6.1%
(E)-Anethole	0.0%–5.8%
Globulol	0.0%–5.5%
1-Dodecanol	0.0%–5.2%
T-Cadinol	0.0%–5.0%
Spathulenol	0.0%–4.7%
(E)-Beta-Ocimene	0.0%–4.4%
5-Methyl-Undecane	0.0%–4.2%
Delta-Cadinene	0.0%–4.2%
2-Methyldecane	0.2%–4.0%
(Z,Z)-Alpha-Farnesene	0.0%–4.0%
(Z)-Beta-Farnesene	0.0%–3.8%
Aromadendrene	0.0%–3.8%
Cumene	0.0%–3.7%
4-Isopropyl-1,6-dimethyl-decahydronaphthalene	0.0%–3.3%
Verbenol	0.0%–3.2%
Amorpha-4,9-Dien-2-ol	0.0%–3.0%

Caryophyllene Oxide CT

Caryophyllene Oxide	11.9%–18.5%
Beta-Caryophyllene	7.3%–16.8%
Spathulenol	0.0%–12.7%
1-Tetradecanol	7.0%–10.5%
1-Dodecanol	0.0%–4.1%
Caryophylladienol	3.5%–3.7%

Caryophyllene Oxide/β-Selinene CT

Caryophyllene Oxide	10.0%–13.3%
Beta-Selinene	5.5%–11.4%
Alpha-Selinene	5.0%–10.6%
Beta-Caryophyllene	4.1%–10.2%
Gamma-Muurolene	0.0%–9.0%
(E)-Beta-Ionone	0.0%–6.1%
Gamma-Curcumene	0.0%–5.5%
Spathulenol	0.0%–5.4%
Delta-Cadinene	1.1%–4.9%
Germacrene D	0.0%–4.9%
Alpha-Bisabolol	0.0%–4.8%
Dodecanol	0.0%–4.8%
Gamma-Cadinene	0.6%–4.3%
(E)-Beta-Farnesene	0.0%–3.9%
T-Cadinol	0.0%–3.8%
(E)-Nerolidol	0.9%–3.7%
Funebrol	0.0%–6.7%
n-Undecane	0.0%–6.3%
Alpha-Cadinol	0.0%–6.2%
Dodecanol	0.0%–6.1%
Beta-Selinene	0.0%–6.0%
Bicyclogermacrene	0.0%–5.8%
Delta-Cadinene	tr–4.9%
Gamma-Cadinene	0.3%–4.6%
(E)-Beta-Ocimene	0.0%–4.6%
Alpha-Muurolol	0.0%–4.3%
Beta-Pinene	0.0%–3.8%
(E)-Phytol	0.0%–3.5%
Terpinen-4-ol	0.0%–3.3%
Pentadecanol	0.0%–3.3%
Alloaromadendrene	0.0%–3.0%

Caryophyllene Oxide/Manool CT

Caryophyllene Oxide	15.9%–34.0%
Beta-Caryophyllene	1.6%–14.2%
Manool	tr–12.7%
m/z=149-Phthalide	tr–11.8%
1-Tetradecanol	0.9%–11.2%
Spathulenol	5.6%–10.6%
n-Eicosane	0.0%–10.1%
Tetradecanal	0.0%–8.9%
Viridiflorol	0.0%–6.9%
Alpha-Cadinol	0.0%–5.7%
n-Dodecanol	0.2%–5.6%
Khusinol	3.0%–5.5%
(E)-Nerolidol	2.3%–5.3%
Beta-Eudesma-4(15), 7-Dien-1-ol	tr–5.0%
Carotol	0.2%–4.5%
Caryophylla-4(14), 8(15)-dien-5-ol	tr–3.9%
n-Octadecane	0.0%–3.6%
Humulene Epoxide II	0.5%–3.1%
Germacrene D	tr–3.1%

Beta-Caryophyllene CT

Beta-Caryophyllene	14.2%–26.1%
Alpha-Selinene	0.0%–15.5%
2-Methyloctane	0.0%–13.1%
Caryophyllene Oxide	0.0%–15.9%
Gamma-Muurolene	3.5%–8.7%
2-Methyldecane	0.1%–7.9%

Selina-11-en-6-α-ol	0.0%–3.5%
Curzerene	0.0%–3.1%
Isopentyl Benzoate	0.0%–3.0%

Beta-Selinene CT

Beta-Selinene	10.2%–19.0%
Caryophyllene Oxide	2.9%–11.8%
Beta-Caryophyllene	4.4%–11.7%
Gamma-Muurolene	4.1%–10.7%
Alpha-Selinene	6.0%–10.6%
(E)-Beta-Farnesene	0.0%–7.7%
(E)-Beta-Ionone	0.0%–5.9%
Spathulenol	2.3%–5.5%
Bisabolol Oxide A	0.0%–5.4%
Gamma-Cadinene	2.8%–4.2%
Gamma-Curcumene	0.0%–3.4%
Germacrene D	0.0%–3.1%

Spathulenol CT

Spathulenol	21.1%–21.5%
Caryophyllene Oxide	4.4%–18.4%
Beta-Caryophyllene	0.2%–13.3%
1-Tetradecanol	2.3%–9.1%
(E,E)-Alpha-Farnesene	0.0%–8.4%
Gamma-Muurolene	6.9%–7.7%
(E)-Nerolidol	1.2%–6.5%
2-Methyldodecane	1.7%–4.0%
Delta-Cadinene	2.1%–3.2%

Delta-Cadinene CT

Delta-Cadinene	10.6%–22.6%
2,6-Dimethylheptane	7.4%–15.0%
Alpha-Pinene	5.6%–14.4%
Beta-Caryophyllene	2.1%–12.2%
Alpha-Humulene	5.3%–11.3%
n-Nonane	2.3%–11.0%
2-Methylnonane	2.0%–8.1%
Beta-Funebrene	0.0%–6.7%
Alpha-Bisabolol	0.4%–3.5%

α-Pinene/2,6-Dimethyl-Heptane CT

Alpha-Pinene	21.9%–26.0%
2,6-Dimethyl-Heptane	9.1%–15.2%
Beta-Funebrene	0.0%–12.4%
Beta-Pinene	0.0%–11.6%
Delta-Cadinene	0.4%–10.2%
n-Nonane	5.1%–9.8%
n-Pentadecanol	0.0%–6.9%
Dodecanol	0.0%–6.7%
Gamma-Amorphene	0.0%–6.5%
Spathulenol	0.5%–6.3%
Germacrene D	0.0%–6.3%
Delta-Cadinene	0.0%–5.4%
Alpha-Muurolene	0.0%–4.8%
2-Methyldodecane	0.0%–4.5%
Humulene Oxide II	0.2%–4.2%
1-Tetradecanol	0.0%–4.1%
(E)-Beta-Ionone	0.0%–4.1%
Gamma-Cadinene	0.0%–4.0%
n-Undecane	0.1%–3.9%
(E)-Beta-Farnesene	0.8%–3.6%
Beta-Funebrene	0.0%–3.2%
Alpha-Pinene	0.0%–3.1%

Manool CT

Manool	27.6%
Viridiflorol	11.1%
m/z=149-Phthalide	9.7%
Caryophyllene Oxide	7.7%
Spathulenol	4.5%
Beta-Caryophyllene	4.2%
Humulene Epoxide II	3.8%

Gamma-Cadinene CT

Gamma-Cadinene	16.9%
2,6-Dimethyl-Heptane	6.3%
1-Tetradecanol	6.3%
Alpha-Pinene	6.2%
(E,E)-Alpha-Farnesene	5.3%
3-Methylnonane	3.9%
Alpha-Acoradiene	3.2%

2,6-Dimethyl-Heptane CT

2,6-Dimethyl-Heptane	36.1%
Alpha-Pinene	23.6%
3-Methylnonane	10.1%
n-Nonane	4.7%

1-Tetradecanol CT

1-Tetradecanol	10.2%
Beta-Selinene	8.9%
Alpha-Pinene	5.4%
Alpha-Selinene	5.0%

Gamma-Cadinene	0.0%–8.7%
n-Dodecanol	0.6%–6.8%
Gamma-Himachalene	0.9%–6.0%
3-Methyl-Nonane	0.4%–5.7%
n-Undecane	tr–4.7%
Gamma-Elemene	0.6%–4.6%
Beta-Caryophyllene	0.0%–3.5%
Alpha-Humulene	0.2%–3.2%

2-Methyloctane CT

2-Methyloctane	35.7%
Alpha-Pinene	25.6%
2-Methylnonane	7.2%
2-Methyldecane	4.8%
Alpha-Cedrene	4.4%
Caryophyllene Oxide	4.2%
n-Nonane	3.3%

Alpha-Cadinol	3.6%

1-Tetradecanol/CO/Dodecanol CT

1-Tetradecanol	0.2%–15.0%
Caryophyllene Oxide	8.0%–14.4%
Dodecanol	0.8%–12.0%
Beta-Selinene	3.6%–8.5%
Beta-Caryophyllene	3.4%–7.5%
Alpha-Selinene	2.4%–7.5%
T-Muurolol	0.3%–6.3%
Gamma-Muurolene	2.5%–4.7%
(E)-2-Hexenol	0.5%–4.7%
Spathulenol	1.0%–4.6%
Gamma-Cadinene	1.6%–3.0%

OTHER CONSTITUENTS: *Alpha-Pinene CT*—Alpha-thujene, sabinene, 3-methylnonane, myrcene, alpha-terpinene, para-cymene, limonene, (Z)-beta-ocimene, gamma-terpinene, terpinolene, terpinen-4-ol, alpha-terpineol, beta-copaene, gamma-amorphene, gamma-muurolene, beta-guaiene, ledol, delta-cadinol, T-muurolol, (Z,E)-alpha-farnesol, benzyl benzoate, p-menth-1,8-dien-8-ol, p-cymen-8-ol, myrtenal, trans-carveol, tridecane, alpha-cubebene, alpha-copaene, pentadecane, calarene, 3-methylnonane, alpha-phellandrene, pulegone <1.0%), alpha-cedrene, phytol; *Germacrene D CT*—n-Nonane, para-cymene, linalool, campholenal, cis-carveol, aromadendrene, alpha-himachalene, alpha-humulene, humulene epoxide, 1-epi-cubenol; *Caryophyllene Oxide CT*—p-Cymen-8-ol, trans-carveol, 2-methyloctane, alpha-copaene, alpha-longipinene, beta-bourbonene, beta-longipinene, beta-copaene, alpha-himachalene, alpha-humulene, khusimene, (E)-beta-farnesene, beta-acoradene, germacrene D, ar-curcumene, gamma-himachalene, gamma-amorphene, beta-himachalene, gamma-cadinene, endo-bourbonanol, hexanoic acid, 2-methyldecane, alpha-cadinene, (Z)-3-hexenyl benzoate, nerolidol, lauric acid, salvialenone, funebrol, humulene oxide II, cadalene, 3-methyltetradecanol, beta-acorenol, myristic acid; *Caryophyllene Oxide/β-Selinene CT*—Cumene, beta-pinene, 2-methyldecane, n-undecane, beta-elemene, alpha-humulene, neo-intermedeol, 2-pentadecanone, n-hexadecanol, methyl hexadecanoate, hexadecanoic acid; *Caryophyllene Oxide/Manool CT*—Dimethyl heptane, 2-methyloctane, alpha-pinene, isoborneol, (Z)-beta-farnesene, alpha-humulene, trans-muurola-4(14),5-diene, delta-cadinene, alpha-calacorene, alpha-muurolol; *Beta-Selinene CT*—Alpha-pinene, 3-methylnonane, 2-methyldecane, n-undecane, beta-elemene, 4,5-di-epi-aristolochene, 7-epi-alpha-selinene, delta-cadinene, (E)-nerolidol, alpha-cadinol, n-tetradecanol, 2-pentadecanone, hexadecanoic acid, heneicosane, alpha-copaene, alpha-cedrene, alloaromadendrene; *Beta-Caryophyllene CT*—Beta-pinene, alpha-cubebene, alpha-ylangene, longicyclene, beta-cubebene, alpha-cedrene, beta-gurjunene, aromadendrene, alpha-humulene, beta-acoradiene, alloaromadendrene, (Z,E)-alpha-farnesene, n-tridecanal, (Z)-calamenene, (E)-nerolidol, methyltridecanol, cubenol, (E)-alpha-cadinol, cadina-1,4-dien-3-ol, (Z)-alpha-cadinol, caryophylla-3(15),7-dienol, cumene, 3-methylnonane, alpha-copaene, (Z)-beta-farnesene, n-hexadecanol, hexadecanoic acid; *Spathulenol CT*—Linalool, octanol, pinocarveol, p-mentha-1,5-dien-8-ol, alpha-terpineol, delta-butyl benzoate, isobutyl isobutyrate, alpha-copaene, isoamyl benzoate, alpha-himachalene, alpha-humulene, (E)-beta-farnesene, dodecanol, ar-curcumene, alpha-selinene, cis-calamenene, calacorene, hexenyl benzoate, humulene oxide II, T-cadinol, alpha-cadinol, benzyl benzoate, hexahydrofarnesylacetone; *Delta-Cadinene CT*—Beta-pinene, myrcene, allyl hexanoate, n-undecane, 2-

methyl-1-octanol, beta-elemene, dodecanal, gamma-elemene, (E)-beta-farnesene, alpha-acoradiene, n-dodecanol, gamma-himachalene, beta-selinene, delta-selinene, alpha-selinene, (E,E)-alpha-farnesene, (E)-nerolidol, caryophyllene oxide, dehydro-eudesmol, selina-11-en4-alpha-ol, 5-isocedranol, 1-tetracosane; *α-Pinene/2,6-Dimethyl-Heptane CT*—1,3-Dimethyl-benzene, myrcene, para-cymene, limonene, allyl hexanoate, n-tridecane, dodecanol, (E)-beta-farnesene, alpha-acoradiene, beta-selinene, delta-selinene, alpha-selinene, (E,E)-alpha-farnesene, (E)-nerolidol, caryophyllene oxide, dehydro-eudesmol, selina-11-en4-alpha-ol, beta-bisabolol, 1-tetradecanol, 5-isocedranol, n-pentadecanol, phytol; *2-Methyloctane CT*—Artemisia ketone, undecane, beta-thujone (< 1.0%), trans-pinocarveol, alpha-terpineol, verbenone, 2-methyldodecane, tridecane, beta-bourbonene, beta-caryophyllene, (E)-alpha-bergamotene, alpha-humulene, alloaromadendrene, germacrene D, 1-octadecene; *Manool CT*—2-methyloctane, isoborneol, (Z)-beta-farnesene, germacrene D, trans-muurola-4(14),5-diene, delta-cadinene, (E)-nerolidol, carotol, tetradecanol, alpha-muurolol, alpha-cadinol, 1-tetradecanol, khusinol, beta-eudesma-4(15),7-dien-1ol; *Gamma-Cadinene CT*—1,3-Dimethyl-benzene, n-nonane, n-undecane, n-tridecane, alpha-longipinene, dodecanal, beta-funebrene, alpha-humulene, n-dodecanol, gamma-himachalene, delta-cadinene, caryophyllene oxide, dehydro-eudesmol, 5-isocedranol, ethyl tetradecanoate; *2,6-Dimethyl-Heptane CT*—Myrcene, allyl hexanoate, n-undecane, 2-methyl-1-octanol, beta-elemene, gamma-elemene, beta-selinene, caryophyllene oxide, beta-bisabolol, 1-tetradecanol, n-tetracosane; *1-Tetradecanol CT*—(E)-3-Hexanal, n-nonane, beta-pinene, myrcene, para-cymene, n-undecane, alpha-campholenal, alpha-terpineol, myrtenol, trans-carveol, alpha-cubebene, alpha-ylangene, alpha-copaene, sativene, beta-caryophyllene, alpha-himachalene, trans-cadinene, delta-cadinene, alpha-cadinene, (E)-nerolidol, spathulenol, caryophyllene oxide, humulene epoxide II, beta-oplopenone, T-cadinol, T-muurolol, cadalene, (E)-sesquilavandulyl acetate, aristolone; *1-Tetradecanol/CO/Dodecanol CT*—(E)-2-Hexenal, n-nonane, sabinene, 6-methyl-5-hepten-2-one, cis-linalool oxide, n-undecane, linalool, n-nonanal, alpha-terpineol, (E)-2-decenal, alpha-ylangene, beta-cedrene, beta-gurjunene, aromadendrene, isopentyl benzoate, alpha-himachalene, (E)-beta-farnesene, alloaromadendrene, cis-muurola-4(14),5-diene, drima-7.9(11)-diene, gamma-amorphene, alpha-muurolene, cis-gamma-cadinene, alpha-cadinene, alpha-calacorene, (E)-nerolidol, dodecanoic acid, viridiflorol, humulene epoxide II, beta-himachalene oxide II, humulane-1.6-dien-3-ol, cis-cadin-4-en-7-ol, T-cadinol, cubenol, T-muurolol, alpha-muurolol, alpha-cadinol, himachalol, Selina-11-en-6-alpha-ol, cadalene, alpha-bisabolol, benzyl benzoate, trimethylpentadecanone, n-hexadecanol, n-eicosane, n-pentacosane

PREFERRED COMPOSITION FOR CLINICAL USE:

Constituent	*2-Methyloctane CT*
2-Methyloctane	20%–50%
Alpha-Pinene	8%–22%
Beta-Caryophyllene	2%–12%
Beta-Pinene	3%–10%
Germacrene D	1%–10%
n-Nonane	1%–8%
3-Methylnonane	0.5%–7%

REPORTED THERAPEUTIC PROPERTIES: Anti-inflammatory, antineuralgic, nervine, analgesic, wound healing, relieves itching and chronic skin conditions, restores dry, damaged, or aged skin, reduces the appearance of scars, heals bruises, blisters, and hemorrhoids, astringent, relieves burns, reduces the appearance of varicose veins, alleviates jet lag, antiallergic, antioxidant, antiviral, antibacterial, antifungal, antidepressant, relieves anxiety, promotes balance of mind, emotions, body, and spirit, improves mental clarity, stress management

CAUTIONS:

Caryophyllene Oxide/Manool CT

■ May interact with aspirin, blood pressure, antiplatelet, and anticoagulant medications, and increase the risk of bleeding among people with bleeding disorders, due to caryophyllene-oxide content.[3910]

Manool CT

■ May interact with aspirin, blood pressure, antiplatelet, and anticoagulant medications, and increase the risk of bleeding among people with bleeding disorders based on manool content.[3911]

All Other CTs

■ None currently known.

SELECTED EVIDENCE:

- □ *In vitro* research reported that the alpha-pinene CT of St. John's wort essential oil moderately to weakly inhibited *B. creus, L. monocytogenes, P. aeruginosa,* and *S. typhimurium*.[3912] The alpha-pinene CT inhibited *S. aureus* better than the antibiotic gentamicin and also inhibited *E. coli*.[3913]
- □ The alpha-pinene CT of St. John's wort essential oil exhibited weak antioxidant activity in the DPPH assay.[3914] Another study found that the alpha-pinene CT (also rich in alpha-amorphene) was a good antioxidant, especially according to the DPPH assay.[3915]
- □ St. John's wort essential oil CT 1-tetradecanol was larvicidal to *C. pipiens* mosquitoes.[3916]

SUGAR ROOT (Daily River Satinash, Sugargum, Kyaman)

Cleistocalyx operculatus DC., *C. nitidus* (Korth.) Blume, *Eucalyptus cladocalyx* F. Muell., *Jambosa nitida* Korth., *Syzygium cleistocalyx* (Merr.) P.S. Ashton

FAMILY: Myrtaceae
NOTE: Middle-Top
AROMA INTENSITY: Medium
AROMA: Herbaceous, fruity, earthy, spicy, peppery
COMMON EXTRACTION METHOD: Steam distilled or hydrodistilled from the leaves or buds
POSSIBLE SUBSTITUTE OILS: Hemp, hops, tagetes, lavender
BLENDS WELL WITH: Bergamot, copaiba, frankincense, grapefruit, hemp, hops, hyssop, lavender, lemon, lime, marjoram, orange, patchouli, tangerine, vetiver, ylang ylang
SUBCELLULAR LOCALIZATION | EPIGENETIC INFLUENCE: Currently unknown | Currently unknown
RECOMMENDED DILUTION RANGE: 3%–33%; neat for some conditions

PRIMARY CONSTITUENTS:[3917,3918,3919]

Buds		*Leaves*	
Gamma-Terpinene	5.8%	Myrcene	24.6%–69.7%
Globulol	5.6%	(Z)-Beta-Ocimene	4.8%–32.1%
Cis-Linalool Oxide	5.2%	Beta-Caryophyllene	1.1%–14.5%
Acorenol	5.1%	(E)-Beta-Ocimene	9.4%–12.2%
Camphene	4.1%	Linalool	0.5%–4.1%
Trans-Carveol	3.9%	Alpha-Pinene	0.0%–3.7%
Beta-Himachalol	3.8%		
Cyclobazzanene	3.1%		
Beta-Pinene	3.1%		
2,3-Dehydro-1,4-Cineole	3.0%		

OTHER CONSTITUENTS: *Leaves*—Alpha-thujene, beta-pinene, alpha-phellandrene, para-cymene, beta-phellandrene, gamma-terpinene, (E)-nerolidol, allo-ocimene, geranyl acetate, alpha-humulene, ledene, delta-cadinene, caryophyllene oxide; *Buds*—Myrcene, alpha-terpinene, alpha-pinene epoxide I, alpha-pinene epoxide II, cis-verbenol, isomenthone, terpinen-4-ol, nerol, lavandulyl acetate, cis-carvyl acetate, geranyl acetate, methyl jasmonate, presilphiperfol-1-ene, alpha-copaene, 6-epi-beta-cubebene, aromadendr-9-ene, gamma-muurolene, gamma-amorpehene, delta-cadinene, 1,8-oxidocadin-4-ene, marsupellol, 6alpha-hydroxygermacra-1(1),4-diene, trans-nuciferol, alpha-santalol acetate, falcarinol, rimuene, 1-hexanol, 2-heptanene, 5,7-dimethylbutyl isobutyrate 8, nanonic acid, benzyl benzoate

PREFERRED COMPOSITION FOR CLINICAL USE:
Insufficient data/samples to determine.

REPORTED THERAPEUTIC PROPERTIES: Aids digestion, relieves stomachache, relieves diarrhea, aids liver function, antiseptic, wound healing, heals burns, relieves skin itching and chronic skin conditions, antibacterial, antifungal, antioxidant, anticancer, balances blood-sugar levels, aids respiratory function, eases cough, anti-inflammatory, antispasmodic, analgesic, promotes restful sleep, relieves anxiety, stress management, boosts self-esteem, promotes creativity, combats negative thoughts, antidepressant

CAUTIONS:

Leaves

■ Caution is advised during pregnancy and lactation due to high beta-myrcene content. Extremely high doses (500 mg/kg) of beta-myrcene have been toxic to fetuses according to animal research.[3920,3921]

■ May interact with barbiturates (medications for anxiety and insomnia), antihistamines, benzodiazepines, tricyclic antidepressants, or other central nervous system depressant drugs, increasing their depressant effects.[3922,3923,3924]

■ May weakly interfere with the enzymes responsible for metabolizing medications (NSAIDs, proton-pump inhibitors, acetaminophen, antiepileptics, immune modulators, blood-sugar medications, blood pressure medications, antidepressants, antipsychotics, diabetic medications, antihistamines, antibiotics, and anesthetics), based on myrcene content.[3925,3926]

Buds

■ None currently known.

Selected Evidence:

- Sugar root leaf essential oil (69.7% myrcene) accelerated burn-wound healing in mice by reducing inflammation and wound infection.[3927] Wound healing (complete epidermal structure, thick and neatly arranged fibers) was 100% at day twenty with the sugar root essential oil, compared to 79% with tamanu oil and 71% in mice that were treated only with saline.
- Sugar root bud essential oil significantly reduced inflammation by suppressing the secretion of proinflammatory cytokines (TNF-α, IL-1β).[3928]
- *In vitro* research demonstrated that sugar root leaf essential oil (myrcene CT) is a weak antimicrobial against *A. niger, S. aureus, B. cereus,* and *C. albicans.*[3929] The bud oil inhibited a large panel of food spoilage, foodborne, and skin pathogens, including drug-resistant strains.[3930] Another study reported that the bud oil inhibited pathogenic *Xanthomonas* species.[3931]
- The bud essential oil exhibited good antioxidant activity in the DPPH assay.[3932]

SURINAM CHERRY (Pitanga, Pitangueira, Brazilian Cherry, Cayenne Cherry)

Eugenia uniflora L., *E. michelii* Lam., *Stenocalyx michelii* Berg, *S. brasiliensis* Berg.

Family: Myrtaceae
Note: Middle
Aroma Intensity: Medium
Aroma: Earthy, woody, spicy, green
Common Extraction Method: Steam distilled from the leaves or fruits
Possible Substitute Oils: Myrrh
Blends Well With: Camphor, cedarwood, frankincense, juniper, lavender, myrrh, myrtle, palo santo, sandalwood, spikenard
Subcellular Localization | Epigenetic Influence: Currently unknown | Currently unknown
Recommended Dilution Range: 5%–Neat

Primary Constituents:[3933,3934,3935,3936,3937,3938,3939,3940,3941,3942,3943,3944,3945,3946,3947,3948]

Selina-1,3,7(11)-Trien-8-One CT

Selina-1,3,7(11)-Trien-8-One	21.2%–48.5%
Selina-1,3,7(11)-Trien-8-One Oxide	17.0%–32.8%
Atractylone	0.0%–11.7%
Germacrene B	0.0%–10.5%
Spathulenol	0.0%–10.0%
Beta-Caryophyllene	0.0%–6.5%
Curzerene	0.0%–5.5%
Bicyclogermacrene	0.0%–4.8%
Germacrone	0.0%–4.3%
Beta-Elemene	0.0%–4.1%
Caryophyllene Oxide	0.0%–3.6%
Delta-Selinene	0.0%–3.3%

Curzerene CT

Curzerene	20.5%–48.1%
Germacrene B	0.5%–21.6%
Germacrone	1.2%–17.3%
Gamma-Elemene	0.3%–14.3%
Atractylone	0.0%–11.8%
(E)-Beta-Elemenone	0.0%–11.5%
Germacrene D	1.2%–8.8%
Beta-Caryophyllene	3.2%–8.7%
Germacrene A	0.0%–7.4%
Beta-Elemene	0.0%–5.5%
Viridiflorol	0.0%–3.2%

Atractylone CT

Atractylone	26.8%
Curzurene	18.0%
Germacrene B	9.3%
Zerumbone	4.2%
Gamma-Muurolene	3.6%
Viridiflorol	3.1%

Furanodiene CT

Furanodiene (Furanoelemene)	50.2%
Beta-Elemene	5.9%
Alpha-Cadinol	4.7%
Spathulenol	3.8%
Gamma-Elemene	3.5%
Germacrene B	3.5%
Globulol	3.1%

Isofurane-Germacrene CT

Isofurane-Germacrene	65.8%
Germacra-3,7,9-Trien-6-One	16.2%
Beta-Elemenone	4.5%
Gamma-Elemene	4.0%

Curzerene/Selina-1,3,7(11)-Trien-8-One CT

Curzerene	19.7%
Selina-1,3,7(11)-Trien-8-One	17.8%
Atractylone	16.9%
Furanodiene	9.6%
Oxidoselina-1,3,7(11)-Trien-8-One	5.9%
Germacrene B	5.8%
Beta-Caryophyllene	3.9%

Spathulenol CT

Spathulenol	15.8%
Alpha-Copaene	11.0%
Muurola-4,10-dien-3-ol	9.3%
Caryophyllene Oxide	8.9%
Alloaromadendrene	5.5%
Nootkatone	5.2%

Germacrene B CT

Germacrene B	21.2%
Selina-1,3,7(11)-Trien-8-One Oxide	19.3%
Beta-Caryophyllene	12.6%
Germacrene A	11.6%
Germacrene D	11.4%
Selina-1,3,7(11)-Trien-8-One	9.7%
Curzerene	3.8%

Germacrone CT

Germacrone	27.5%
Selina-1,3,7(11)-Trien-8-One	19.2%
Curzerene	11.3%
Oxidoselina-1,3,7(11)-Trien-8-One	11.0%
Atractylone	4.5%
Germacrene B	4.0%
(E)-Beta-Ocimene	3.9%

OTHER CONSTITUENTS: *Selina-1,3,7(11)-Trien-8-One CT*—Thujopsan-2α-ol, selina-3,11-dien-6α-ol, pogostol, 14-hydroxy-9-epi-(E)-caryophyllene, occidenol, nootkatol, 2α-hydroxyamorpha-4,7(11)-diene, (E)-beta-ocimene, gamma-elemene, delta-elemene, germacrene D, beta-selinene, gamma-muurolene, germacrene A, 4-acetoxy-germacra-l,8(11)-dien-9-one, alpha-cadinene, selina-3,7(11)-diene, globulol, guaiol, 7-epi-gamma-eudesmol; *Atractylone CT*—(E)-beta-ocimene, (Z)-beta-ocimene, alpha-bisabolene, delta-cadinene, caryophyllene alcohol, loniborneol, T-muurolol, cubenol; *Curzerene C*—Myrcene, limonene, linalool, delta-elemene, alpha-cubebene, alpha-humulene, beta-chamigrene, beta-selinene, spathulenol, beta-pinene, gamma-terpinene, delta-elemene, aromadendrene, alloaromadendrene, caryophyllene oxide, gamma-muurolene, globulol; *Furanodiene CT*—Delta-elemene, trans-beta-caryophyllene, germacrene D, ledol, viridiflorol, 1,10-di-epi-cubenol, 10-epi-beta-eudesmol, 1-epi-cubenol, gamma-eudesmol, T-cadinol, T-muurolol, beta-eudesmol, atractylone, alpha-bisabolol,(E,E)-germacrone, (E)-nerolidol acetate; *Spathulenol CT*—Para-cymene, gamma-terpinen-7-al, beta-caryophyllene, delta-guaiene, gamma-cadinene, beta-cadinene, germacrene B, viridiflorol, alpha-cadinol, alpha-santalol, amorpha-4,9-dien-2-ol, cedr-8-en-13-ol, germacrone, selina-1,3,7(11)-trien-8-one epoxide, nootkatol, trans-phytol; *Germacrene B CT*—Not listed; *Isofurane-Germacrene CT*—Germacrene B, (Z)-beta-elemene; *Curzerene/Selina-1,3,7(11)-Trien-8-One CT*—Beta-elemene, gamma-elemene, germacrene D, beta-selinene, bicyclogermacrene, beta-cadinene, spathulenol, guaiol, trans-beta-elemenone, germacrone; *Germacrone CT*—Myrcene, (Z)-beta-ocimene, beta-elemene, beta-caryophyllene, gamma-selinene, germacrene D, beta-selinene, ledene, 4,7(11)-selinadiene, guaiol, trans-beta-elemenone, T-cadinol

PREFERRED COMPOSITION FOR CLINICAL USE:
Too great of variance in composition to determine.

REPORTED THERAPEUTIC PROPERTIES: Analgesic, anti-inflammatory, antirheumatic, antiarthritic, relieves gout (reduces uric acid), reduces fever, aids digestion, relieves stomachache, expels excess gas, lowers high blood pressure, reduces cholesterol, **anticancer**, antibacterial, antifungal, antiviral, anti-infectious, antiparasitic, relieves diarrhea, balances blood-sugar levels, diuretic, enhances meditation, improves focus and concentration, promotes creativity, relaxing, reduces irritability

CAUTIONS:

■ Certain chemotypes may interact with (synergize or antagonize) the activity of antibiotics and antifungals.[3949,3950]

■ *Furanodiene CT*: Avoid during pregnancy and lactation (very low risk). Surinam cherry essential oil CT furanodiene contains significant amounts of furanodiene and moderate amounts of beta-elemene, which may prevent the formation of blood vessels necessary for fetal development according to animal research.[3951,3952,3953]

■ *Germacrone CT*: May interfere with the enzymes responsible for metabolizing medications (NSAIDs, proton-pump inhibitors, acetaminophen, antiepileptics, immune modulators, blood-sugar medications, blood pressure medications, antidepressants, antipsychotics, diabetic medications, antihistamines, antibiotics, and anesthetics) due to germacrone content.[3954]

SELECTED EVIDENCE:

- □ The spathulenol CT of Surinam cherry essential oil moderately killed breast cancer cells with good selectivity.[3955] Another study reported that the fruit and leaf essential oils with curzerene and selina-1,3,7(11)-trien-8-one or germacrone as the major constituents completely inhibited the growth of one breast cancer cell line (Hs 578T) and displayed excellent cytotoxicity against liver and prostate cancer cells.[3956]
- □ Oral administration of the curzerene CT of Surinam cherry essential oil reduced pain (acetic acid-induced abdominal constrictions; increased the latency time in hot plate test) and diminished body temperature.[3957] An unidentified CT of Surinam cherry essential oil demonstrated moderate anti-inflammatory effects in rats.[3958]
- □ A safety assessment of the selina-1,3,7(11)-trien-8-one CT of Surinam cherry concluded that it was nontoxic, nonmutagenic, nonphotosensitizing, and did not cause skin sensitization or irritation when tested at 1.5% in human volunteers.[3959]
- □ The curzerene CT of Surinam cherry essential oil exhibited antileishmanial activity against *Leishmania amazonensis* by activating macrophages (as evidenced by increased phagocytic capacity and lysosomal activity).[3960]
- □ *In vitro* research shows that Surinam cherry essential oil (atractylone CT) inhibits both bacteria and yeasts: *S. equi, S. epidermidis, C. dubliniensis, C. tropicalis, C. albicans, C. glabrata, C. parapsilosis, C. grubii* (serotype A), *C. gattii* (serotype C), *C. gattii* (serotype B), *C. neoformans,* and *S. cerevisiae*.[3961] A Surinam cherry essential oil with an unidentified compound as the primary constituent inhibited the gram-negative bacteria *Enterobacter aerogenes* and *Salmonella typhimurium*.[3962] The selina-1,3,7(11)-trien-8-one CT weakly inhibited *C. albicans, C. tropicalis,*

and *C. krusei* and reduced the virulence of *C. albicans* and *C. tropicalis*.[3963] Interestingly, the study authors reported that the oil modified the antimicrobial activity of fluconazole, including decreased activity against *C. albicans* and *C. krusei*. Another study found that Surinam cherry essential oil with isoflurane-germacrene as the major constituent inhibited *S. aureus* and *E. coli*, and improved the efficacy of antibiotics against the organisms.[3964] The spathulenol CT exhibited good activity against *C. albicans* and *C. parapsilosis,* moderate activity against *B. licheniformis* and weak activity against *B. subtilis*, but was inactive against the other tested gram-positive and gram-negative bacteria.[3965] Surinam cherry essential oil CT selina-1,3,7(11)-trien-8-one weakly inhibited *S. enteritidis, S. typhimurium, Salmonella* sp., and *P. aeruginosa*, with better inhibition of *S. aureus,* MRSA, MSSA, and *E. coli*.[3966] The germacrene B CT of Surinam cherry essential oil inhibited *L. monocytogens, S. aureus, P. aeruginosa, E. coli,* and *A. hidrophylla*.[3967] The selina-1,3,7(11)-trien-8-one CT inhibited *S. aureus* and *E. coli*.[3968] The furanodiene CT inhibited 80% of the dermatophytes (*Trichophyton rubrum, Trichophyton mentagraphytes, Microsporum canis*, and *Epidermophyton floccosum*) isolated from people.[3969] Surinam cherry essential oil (curzerene and selina-1,3,7(11)-trien-8-one CT) displayed varying degrees of antimicrobial activity against *B. cereus, S. aureus, E. coli,* and *P. aeruginosa*.[3970]

- □ Surinam cherry essential oil CT Germacrene B exhibited antioxidant activity in the DPPH, ABTS, and FRAP assays and reduced lipid peroxidation in the kidney of mice.[3971]
- □ Surinam cherry essential oil (spathulenol CT) showed promising activity against the parasite *T. brucei* (a cause of trypanosomiasis).[3972]
- □ Surinam cherry (Curzerene CT) essential oil was insecticidal to the common fruit fly (*Drosophila melanogaster*) by inducing oxidative stress.[3973]
- □ The curzerene CT of Surinam cherry essential oil was larvicidal to mosquitoes (*Anopheles subpictus, Aedes albopictus*, and *Culex tritaeniorhynchus*) and showed low toxicity against nontarget aquatic predators (larvivorous fishes, backswimmers, and waterbugs).[3974] The study authors concluded that curzerene was the major constituent responsible for the larvicidal effects.

SWEET ACACIA | SWEET ACACIA ABSOLUTE (Cassie, Huisache, Needle Bush)

Vachellia farnesiana (L.) Wight et Arn., *Acacia farnesiana* (L.) Wight et Arn.

FAMILY: Fabaceae
NOTE: Middle-Base
AROMA INTENSITY: Medium
AROMA: Sweet, green, fresh, honey, spicy, woody, slightly floral and fruity
COMMON EXTRACTION METHOD: Steam distilled or hydrodistilled from the flowers
POSSIBLE SUBSTITUTE OILS: Ylang ylang, calendula; mixtures of moringa, wintergreen, and ylang ylang
BLENDS WELL WITH: Amyris, Peru balsam, bergamot, clary sage, clove, davana, jasmine, lavender, lemon, neroli, orange, osmanthus, rose, sandalwood, spikenard, tangerine, vetiver, ylang ylang
SUBCELLULAR LOCALIZATION | EPIGENETIC INFLUENCE: Currently unknown | Currently unknown
RECOMMENDED DILUTION RANGE: *Essential Oil*—1.5%–20%, 50% for some conditions; *Absolute*—1%–3%

PRIMARY CONSTITUENTS:[3975,]

Essential Oil		*Absolute*	
Tricosane	12.2%–23.8%	Oleic Acid	7.8%–48.2%
Methyl Salicylate	4.4%–20.1%	Nonadecane	7.5%–28.4%
Nonadecane	11.0%–16.8%	Alpha-Amyrin	0.0%–17.5%
Pentacosane	6.7%–15.7%	Heneicosane	7.0%–17.1%
Heneicosane	10.8%–14.3%	Tricosane	4.2%–16.4%
Linoleic Acid	0.0%–11.6%	Hexadecanoic Acid	tr–12.1%
Hexahydrofarnesyl	2.2%–8.2%	Methyl Salicylate	1.9%–8.1%
Heptacosane	1.3%–5.7%	Ethyl Oleate	0.0%–8.0%
Kaurene	3.0%–5.3%	Benzyl Alcohol	2.2%–6.5%
Oleic Acid	0.0%–4.2%	Pentacosane	1.8%–5.9%
(E)-Beta-Ionone	0.8%–3.9%	Beta-Amyrin	0.0%–4.9%
Tetracosane	1.0%–3.0%	Lanosterol	0.0%–4.5%

OTHER CONSTITUENTS: *Essential Oil*—Linalool, benzyl acetate, decanal, geraniol, neryl acetate, geranyl acetate, (E)-alpha-ionone, dehydro-beta-ionone, geranyl acetone, 4-methylpentadecane, spathulenol, (2E,6Z)-farnesal, (2E,6E)-farnesyl acetate, sandaracopimara-8(14),15-diene, hexadecanoic acid, eicosane, stearic acid, docosane, hexacosane, nonacosane, 9,17-octadecadienal; *Absolute*—Benzaldehyde, chavicol, geraniol, methyl vanillin, vanillyl acetone, neophytadiene, hexahydrofarnesyl acetate, ethyl hexadecanoate, eicosane, kaurene, ethyl linoleate, docosane, tetracosane, hexadecanoic acid, phenylmethyl ester, squalene, heptacosane, nonacosane

PREFERRED COMPOSITION FOR CLINICAL USE:
Insufficient data to determine.

REPORTED THERAPEUTIC PROPERTIES: Nervine, reduces erectile dysfunction, eases cough, promotes balanced menstruation, soothes sore throat, analgesic, relieves headache, antispasmodic, anti-inflammatory, reduces fever, astringent, moisturizes and softens the skin, reduces skin inflammation and itching, wound healing, cleanses the urinary tract, promotes a feeling of general well-being, banishes negative thoughts, antidepressant, anti-anxiety, stress management, aphrodisiac

CAUTIONS:
Essential Oil

- Avoid with children under age 12. The high salicylate content in sweet acacia may increase the risk of a very serious and life-threatening illness called Reye syndrome in children under 12. Reye syndrome is sudden brain damage and liver dysfunction that most commonly occurs in children ages 4 to 12. It has an unknown cause, but it typically occurs in children who were given aspirin when they are experiencing a fever or recovering from flu-like symptoms or chicken pox.[3976]
- Do not use during pregnancy or lactation. May cause congenital abnormalities and fetal malformations.[3977,3978]
- Oral caution—As little as 2.5 mL can cause toxicity and 4 mL may be fatal in children. As little as 5 mL could result in methyl salicylate poisoning in adults, and more than 5 mL can be fatal.[3979,3980,3981,3982,3983] Do not exceed 5 drops per day orally for an adult.
- Avoid with epilepsy and Parkinson's disease due to methyl salicylate content.[3984,3985]

■ May interact with aspirin, blood pressure, antiplatelet, and anticoagulant medications, and increase the risk of bleeding among people with bleeding disorders both topically and orally.[3986,3987,3988,3989]
■ Avoid in individuals allergic to aspirin, methyl salicylate, or other NSAIDs; or those with salicylate sensitivities. Methyl salicylate may be metabolized to the known NSAID, salicylic acid, by the liver.[3990]

Absolute
■ Avoid oral consumption. Absolutes contain solvent residue and should not be taken orally. In addition, ingestion of benzyl derivatives (benzoic acids, benzoic salts, benzaldehyde, benzyl acetate, benzyl benzoate, and benzyl alcohol) may cause gastrointestinal irritation, vomiting, and diarrhea. The Joint Expert Committee for Food Additives (JECFA) has established an acceptable daily limit for the aforementioned constituents of 0–5 mg/kg body weight for adults to avoid GI discomfort. Based on this limit, it is suggested to limit the oral consumption of orange bird berry essential oil rich in benzyl derivatives to no more than 10 drops daily.
■ May interact with aspirin, blood pressure, antiplatelet, and anticoagulant medications, and increase the risk of bleeding among people with bleeding disorders, due to methyl salicylate, alpha-amyrin, and beta-amyrin content.[3991,3992,3993,3994] Beta-amyrin is a more potent inhibitor of platelet aggregation than aspirin and mixtures of the triterpenes alpha- and beta-amyrin significantly inhibit human platelet aggregation.[3995,3996]
■ Avoid in individuals allergic to aspirin, methyl salicylate, or other NSAIDs; or those with salicylate sensitivities. Methyl salicylate may be metabolized to the known NSAID, salicylic acid, by the liver.[3997]

SELECTED EVIDENCE:

- None found.

SWEET CICELY (Sweet Chervil, Garden Myrrh, Smooth Cicely, Great Chervil, British Myrrh, Anise)

Myrrhis odorata (L.) Scop.

FAMILY: Umbelliferae
NOTE: Middle
AROMA INTENSITY: Medium
AROMA: Sweet, licorice-like, woody, earthy, slightly floral
COMMON EXTRACTION METHOD: Steam distilled or hydrodistilled from the leaves
POSSIBLE SUBSTITUTE OILS: Fennel, anise
BLENDS WELL WITH: Anise, balsam fir, basil, bay laurel, bergamot, black pepper, camphor, cardamom, carrot seed, cinnamon, cypress, eucalyptus, fennel, geranium, grapefruit, juniper, lavender, lemon, lemon verbena, marjoram, myrrh, niaouli, orange, patchouli, pine, rose, rosemary, sandalwood, tangerine, ylang ylang
SUBCELLULAR LOCALIZATION | EPIGENETIC INFLUENCE: Currently unknown | Currently unknown
RECOMMENDED DILUTION RANGE: 3%–33%; 50% for some conditions

PRIMARY CONSTITUENTS:[3998,3999,4000,4001,4002,4003]

Trans-Anethole CT

Trans-Anethole	48.1%–85.0%
Methyl Eugenol	0.0%–14.3%
(E)-Nerolidol	0.0%–12.0%
Beta-Caryophyllene	1.0%–9.7%
Germacrene D	1.1%–5.0%
Methyl Chavicol	0.0%–4.1%

Para-Cymene CT

Para-Cymene	62.0%
Alpha-Terpinene	8.9%
Delta-Cadinene	5.3%
Patchoulene	4.0%
Alpha-Selinene	3.1%
Pyrazine	3.0%

(E)-Nerolidol CT

(E)-Nerolidol	27.1%
Cis-Anethole	15.9%
Germacrene D	6.5%
Beta-Caryophyllene	6.4%
Neophytadiene	5.3%
Beta-Bisabolene	5.1%
Caryophyllene Oxide	4.0%
(E)-Phytol	3.8%

OTHER CONSTITUENTS: *Trans-Anethole CT*—Myrcene, para-cymene, (E)-beta-ocimene, gamma-terpinene, terpinen-4-ol, longifolene, pinocarveol, gamma-muurolene, beta-selinene, beta-bisabolene, alpha-farnesene, delta-cadinene, myristicin, elemol, alpha-bergamotene, alpha-humulene, spathulenol, hexadecanol, (E)-phytol; *Para-Cymene CT*—Camphene, beta-pinene, limonene, alpha-gurjunene, beta-caryophyllene, alpha-patchoulene, seychellene, calamenene (isomer not identified); *(E)-Nerolidol CT*—Methyl chavicol (< 0.8%), thymol, alpha-humulene, bicyclogermacrene, spathulenol, methyl linoleate

PREFERRED COMPOSITION FOR CLINICAL USE:

Constituent	
Trans-Anethole	60%–85%
Beta-Caryophyllene	2%–12%
Germacrene D	0.5%–7%

REPORTED THERAPEUTIC PROPERTIES: Aids digestion, enhances appetite, expels excess gas, promotes red blood cell production, relieves anemia, purifies the blood, antiseptic, wound healing, relieves snakebites, antibacterial, antifungal, anti-infectious, disinfectant, antispasmodic, analgesic, antirheumatic, soothes gout, relieves sore throat, aids cardiovascular function, supports respiratory functions, relieves cough, expels excess mucus, lowers blood pressure, promotes balanced menstruation, diuretic, supports urinary and bladder health, promotes restful sleep, balances the emotions, relieves anxiety

CAUTIONS:

Trans-Anethole CT

- Use with caution and highly diluted for children under 6 due to methyl chavicol (estragole) and trans-anethole content (methyl chavicol and trans-anethole CTs).[4004]
- Long-term oral use of teas rich in trans-anethole (a weak phytoestrogen) may cause premature breast development in very young girls—usually under age 2.[4005] Long-term oral use is not recommended for children under age 12.
- Avoid during pregnancy and lactation. Large doses of fennel may be toxic to fetal cells based on animal research.[4006,4007] Nephrotoxicity (kidney toxicity) has been reported in breastfed infants whose mothers drank 2 liters of an herbal tea daily that contained fennel, licorice, anise, and goat's rue (ingredients that

contain trans-anethole). Symptoms improved within 36 hours after the mothers discontinued drinking the tea.[4008] It is also worth noting that trans-anethole may negatively affect fertility. Animal research suggests it decreases fertility by preventing the implantation of a fertilized egg in the uterus.[4009]

■ Avoid with epilepsy and Parkinson's due to fenchone content, which may exacerbate or trigger seizures/convulsions or reduce medication efficacy.[4010,4011]

■ May interact with aspirin, blood pressure, antiplatelet, and anticoagulant medications, and increase the risk of bleeding among people with bleeding disorders based on trans-anethole and methyl chavicol content.[4012,4013,4014,4015,4016]

■ May interact with ibuprofen and enhance its anti-inflammatory activity due to trans-anethole content.[4017]

■ May interact with diabetes medications and cause low blood sugar due to trans-anethole content.[4018]

■ May interact with antibiotics or antifungal drugs and possibly enhance or decrease their effectiveness based on trans-anethole and estragole content.[4019,4020,4021,4022]

■ May interfere with the enzymes responsible for metabolizing medications based on research with an herbal extract (NSAIDs, proton-pump inhibitors, acetaminophen, antiepileptics, immune modulators, blood-sugar medications, blood pressure medications, antidepressants, antipsychotics, diabetic medications, antihistamines, antibiotics, and anesthetics) due to trans-anethole, methyl eugenol, and methyl chavicol content.[4023,4024,4025,4026]

(E)-Nerolidol CT

■ May interfere with antibiotics and increase their effectiveness due to (E)-nerolidol content. (E)-nerolidol sensitizes bacteria to antibiotics).[4027]

■ May interfere with enzymes (CYP1A, CYP2B, CYP3A) responsible for metabolizing medications based on alpha-bisabolol content (NSAIDs, proton-pump inhibitors, acetaminophen, antiepileptics, immune modulators, blood-sugar medications, blood pressure medications, antidepressants, antipsychotics, diabetic medications, antihistamines, antibiotics, and anesthetics) due to (E)-nerolidol content.[4028]

Para-Cymene CT

■ None currently known.

SELECTED EVIDENCE:

- ☐ Sweet cicely CT para-cymene essential oil showed good antifungal activity against *A. niger, A. flavus, C. cladosporioides. P. funiculsum,* and *T. viride*, which exceeded the comparison drug bifonazole.[4029] The same research reported that its antibacterial activity was weak.

SWEET FRANKINCENSE (Frankincense Rivae, Ogaden)

Boswellia rivae Engl.

FAMILY: Burseraceae
NOTE: Base
AROMA INTENSITY: Medium
AROMA: Fresh, rich, woody, elegant, spicy, camphoraceous
COMMON EXTRACTION METHOD: Steam distilled or hydrodistilled from the oleogum resin
POSSIBLE SUBSTITUTE OILS: Dakara frankincense, frankincense, myrrh

BLENDS WELL WITH: Agarwood, basil, bay laurel, bergamot, black pepper, camphor, cassia, cinnamon, coriander, cypress, frankincense, geranium, grapefruit, lavender, lemon, myrrh, neroli, orange, palo santo, petitgrain, pine, ravensara, ravintsara, rose, sandalwood, spikenard, tangerine, vetiver, ylang ylang
SUBCELLULAR LOCALIZATION | EPIGENETIC INFLUENCE: Currently unknown | Currently unknown
RECOMMENDED DILUTION RANGE: 3%–50%; neat for some conditions

PRIMARY CONSTITUENTS:[4030,4031,4032,4033,4034,4035]

Limonene CT

Limonene	13.3%–28.0%
Delta-3-Carene	9.6%–17.3%
Alpha-Pinene	5.3%–16.7%
Para-Cymene	0.0%–7.1%
Trans-Verbenol	0.0%–6.8%
Cis-1,2-Limonene Epoxide	0.0%–4.6%
Verbenone	0.0%–4.3%
Trans-p-Mentha-2,8-Diene-1-ol	0.0%–3.9%
O-Cymene	0.8%–3.9%
m-Cymen-8-ol	0.0%–3.1%

Alpha-Pinene CT

Alpha-Pinene	32.5%–76.7%
Para-Cymene	5.7%–21.1%
Limonene	0.0%–19.6%
Alpha-Thujene	0.0%–10.0%
Delta-3-Carene	0.0%–6.7%
O-Cymene	0.0%–5.6%

OTHER CONSTITUENTS: *Alpha-Pinene CT*—Alpha-Campholene aldehyde, trans-verbenol, alpha-santalene; *Limonene CT*—Alpha-thujene, camphene, beta-pinene, myrcene, alpha-phellandrene, alpha-terpinene, cara-2,4-diene, sabinene, gamma-terpinene, beta-thujone (< 0.8%), trans-1,2-limonene peroxide, trans-sabinene hydrate, trans-pinocarveol, alpha-campholene aldehyde, cis-sabinene hydrate, 8,9-limonene epoxide II, pinocarvone, terpinen-4-ol, thymol, myrtenal, eucarvone, cis-verbenol, trans-pinocarveol, cis-p-mentha-2,8-diene-1-ol, alpha-terpineol, alpha-terpinyl acetate, carvone, myrtenol, trans-carveol, p-cymen-8-ol, cis-carveol, piperitenone, n-octyl acetate

PREFERRED COMPOSITION FOR CLINICAL USE:

Constituent	*Alpha-Pinene CT*
Alpha-Pinene	35%–60%
Delta-3-Carene	7%–18%
Beta-Pinene	7%–15%
Limonene	3%–13%
Para-Cymene	2%–12%
Sabinene	0.5%–5%
Alpha-Thujene	0.1%–5%

REPORTED THERAPEUTIC PROPERTIES: Analgesic, anti-inflammatory, antiarthritic, antirheumatic, anti-inflammatory, antispasmodic, aids bone health, antimicrobial, antifungal, antioxidant, astringent, aids digestion, expels excess gas, diuretic, supports respiratory function, expels excess mucus, immune modulation, would healing, nourishes the skin, restores dry, damaged, and mature skin, reduces the appearance of blemishes and wrinkles, clarifies thoughts, stress management, antidepressant, harmonizes the emotions

CAUTIONS:

- None currently known.

SELECTED EVIDENCE:

- *In vitro* research demonstrated that sweet frankincense CT limonene essential oil inhibited gram-positive bacteria (*S. aureus, S. epidermidis*, MRSA), gram-negative bacteria (*E. coli, P. aeruginosa*), and fungi (*C. albicans, C. tropicalis*).[4036] Remarkably, the oil inhibited *E. coli* better than the standard drug (Amikacin), buts activity against *P. aeruginosa* and *C. tropicalis* was rather weak compared to the standard drugs (Amphotericin B, Amikacin, and Vancomycin) and other frankincense species. Another study reported that dakara frankincense (unidentified CT) inhibited *S. aureus, B. cereus, P. aeruginosa, E. coli, C. albicans*, and *C. neoformans*, with synergistic and additive benefits observed when combined with opoponax.[4037] Sweet frankincense CT limonene essential oil significantly inhibited preformed *C. albicans* biofilms and prevented the formation of *C. albicans* biofilms at sub-MIC (minimum inhibitory concentration) levels.[4038]

SWEET VERNALGRASS (Flouve)

Anthoxanthum odoratum L.

FAMILY: Poaceae
NOTE: Top
AROMA INTENSITY: Medium
AROMA: Coumarinic, hay-like, herbaceous, hint of vanilla
COMMON EXTRACTION METHOD: Steam distilled or hydrodistilled from the aerial parts or leaves (grass tops); may also be solvent extracted
POSSIBLE SUBSTITUTE OILS: Cassia, Padang cassia
BLENDS WELL WITH: Chamomile (German, Roman), frankincense, galbanum, lavender, moringa, rosemary, sage, Spanish sage, tansy, thyme, vetiver
SUBCELLULAR LOCALIZATION | EPIGENETIC INFLUENCE: Currently unknown | Currently unknown
RECOMMENDED DILUTION RANGE: < 10%

PRIMARY CONSTITUENTS:[4039,4040,4041]

Aerial Parts	
Coumarin	16.2%–46.8%
Methyl Hexadecanoate	0.0%–9.9%
Octanol	0.0%–3.4%
Methyl Linoleate	0.0%–3.0%

Absolute (Leaves)	
Coumarin	68.0%
Carvacrol	9.7%
(E)-Beta-Ionone	3.1%
2-Phenethyl Alcohol	3.1%
Ethyl Methylmaleimide	3.0%

Leaves	
Coumarin	96.3%

OTHER CONSTITUENTS: *Aerial Parts (Hay)*—Valeraldehyde, methyl butyrate, methyl hexanoate, methyl octanoate, methyl nonanoate, p-vinylguaiacol, p-vinylguaiacol methyl ether, methyl dodecanoate, methyl tridecanoate, methyl tetradecanoate, 6,10,14-trimethyl-2-pentadecanone, diisobutyl phthalate, phytol, methyl linoleate, methyl oleate, tricosane; *Leaves*—Methyl linoleate; *Absolute*: Benzyl alcohol, dihydroactinodiolide, geranyl acetone, neryl acetone, phenol, thymol, eugenol, gamma-nonalactone, ethyl tetradecanoate

PREFERRED COMPOSITION FOR CLINICAL USE:
Insufficient samples/data to determine.

Reported Therapeutic Properties: Stimulates immune function, reduces fever, antispasmodic, analgesic, anti-inflammatory, antirheumatic, antiallergic, promotes restful sleep, reduces physical and mental fatigue, stimulating, uplifting, relieves anxiety

Cautions:

■ Use cautiously orally—maximum of 5 drops daily for adults due to potentially significant coumarin content. Animal research suggests that coumarin may be toxic to the liver and lungs when taken orally in very high doses.[4042] The total safe daily intake was established at up to 0.1 mg/kg body weight daily by the European Food Safety Authority.[4043] In addition, the coumarin content may interact with aspirin, blood pressure, antiplatelet, anticoagulant, and antidepressant medications, and increase the risk of bleeding among people with bleeding disorders.[4044]

Selected Evidence:

- □ None found.

SWEETFERN
(Sweet Fern)

Comptonia peregrina (L.) J.M. Coult., *C peregrina* (L.) J.M. Coult. var. *aspleniifolia* (L.) Fernald, *Liquidambar aspleniifolia* (L.) L., *L. peregrina* L., *Myrica aspleniifolia* L., *M. aspleniifolia* L. var. *tomentosa* (A. Chev) Gleason, *M. peregrina* (L.) Kuntze

Family: Myricaceae
Note: Middle
Aroma Intensity: Medium-Strong
Aroma: Earthy, sweet, herbaceous, woody, slightly medicinal
Common Extraction Method: Steam distilled or hydrodistilled from the aerial parts or leaves
Possible Substitute Oils: Copaiba, hemp, hops, guava leaf
Blends Well With: Balsam fir, bergamot, eucalyptus, frankincense, grapefruit, lavender, lemon, lime, myrtle, orange, palo santo, spruce (black), rose, silver fir, white fir
Subcellular Localization | Epigenetic Influence: Currently unknown | Currently unknown
Recommended Dilution Range: 3%–33%; neat for some conditions

Primary Constituents:[4045,4046,4047,4048]

Beta-Caryophyllene CT (Eastern Canada)		*Leaves (Eastern Canada)*	
Beta-Caryophyllene	27.0%–36.2%	Beta-Caryophyllene	15.2%–23.7%
Myrcene	7.3%–11.1%	Pentacosane	1.5%–18.8%
1,8-Cineole	9.7%–11.0%	Myrcene	0.2%–12.6%
(E)-Beta-Ocimene	2.5%–7.7%	Alpha-Humulene	7.4%–9.7%
Terpinen-4-ol	0.7%–7.2%	Linalool	0.5%–5.4%
Alpha-Pinene	3.1%–5.5%	Delta-Cadinene	4.3%–4.5%
Gamma-Terpinene	tr–4.2%	Caryophyllene Oxide	3.9%–4.1%
Para-Cymene	1.2%–3.4%	Germacrone	0.9%–3.9%
Linalool	0.0%–3.4%	Germacrene B	3.1%–3.7%

1,8-Cineole CT (Eastern USA)

1,8-Cineole	17.8%
Gamma-Terpinene	14.7%
Myrcene	9.9%
Beta-Caryophyllene	8.9%
Alpha-Pinene	7.0%
(E)-Beta-Ocimene	3.9%
(E)-Nerolidol	3.7%
(Z)-Beta-Ocimene	3.5%
Para-Cymene	3.4%

(E)-2-Hexenal CT (Eastern Canada)

(E)-2-Hexenal	16.8%–19.5%
Beta-Caryophyllene	10.7%–13.3%
1,8-Cineole	9.4%–10.1%
Myrcene + Alpha-Phellandrene	5.0%–7.0%
(E)-Nerolidol	2.7%–5.6%
Linalool	4.1%–5.4%
Eremophilene	0.8%–4.7%
Caryophyllene Oxide	2.0%–3.6%
(Z)-Hex-3-en-1-ol	2.2%–3.3%
Calamenene	0.7%–3.1%
(E)-Nerolidol	2.1%–3.4%
1,8-Cineole	0.0%–3.4%
Gamma-Cadinene	2.9%–3.0%

OTHER CONSTITUENTS: *Beta-Caryophyllene CT (Eastern Canada)*—Beta-pinene, alpha-terpinene, limonene, terpinolene, alpha-terpineol, beta-elemene, beta-chamigrene, unidentified sesquiterpene A, gamma-curcumene, ar-curcumene, beta-selinene, alpha-selinene, gamma-gurjunene, alpha-muurolene, 2-tridecanone, (E,E)-alpha-farnesene, gamma-cadinene, delta-cadinene, zonarene, (E)-gamma-bisabolene, Selina-3,7(11)-diene, germacrene B, (E)-nerolidol, T-cadinol, caryophyllene oxide, pentacosane; *1,8-Cineole CT (Eastern USA)*—Beta-pinene, alpha-terpinene, limonene, terpinolene, terpinen-4-ol, alpha-terpineol, beta-selinene, alpha-selinene, gamma-cadinene, delta-cadinene; *(E)-2-Hexenal CT (Eastern Canada)*—Alpha-pinene, hexanal, limonene, (Z)-beta-ocimene, gamma-terpinene, para-cymene, benzaldehyde, terpinen-4-ol, (E)-beta-elemene, alpha-humulene, alpha-terpineol, delta-selinene, delta-cadinene, gamma-cadinene, ethylacetophenone, benzyl alcohol, T-cadinol; *Leaves (Eastern Canada)*—Alpha-pinene, sabinene, (Z)-beta-ocimene, (E)-beta-ocimene, gamma-terpinene, terpinen-4-ol, alpha-terpineol, safrole, (E)-alpha-bergamotene, gamma-muurolene, gamma-curcumene, beta-selinene, ar-curcumene, alpha-selinene, alpha-muurolene, (E)-cadina-1,4-diene, (E)-gamma-bisabolene, alpha-cadinene, Selina-3,7-(11)-diene, humulene epoxide II, eremoligenol, caryophylla-4(14),8(15)-dien-5-ol, T-cadinol, T-muurolol, alpha-cadinol, unidentified oxygenated sesquiterpene, tricosane, tetracosane

PREFERRED COMPOSITION FOR CLINICAL USE:

Constituent	
Beta-Caryophyllene	24%–40%
Myrcene	5%–15%
1,8-Cineole	5%–15%
Alpha-Humulene	1%–12%
(E)-Beta-Ocimene	1.5%–10%

Alpha-Pinene	2%–8%
Terpinen-4-ol	0.5%–8%
Para-Cymene	0.5%–5%
(Z)-Beta-Ocimene	0.5%–5%
Beta-Pinene	0.1%–5%
Gamma-Terpinene	tr–5%
Linalool	tr–5%

Reported Therapeutic Properties: Astringent, **anticancer**, antibacterial, antifungal, relieves skin itching and chronic skin conditions, aids respiratory function, expels excess mucus, relieves diarrhea, anti-inflammatory, analgesic, antirheumatic, antispasmodic, relieves sprains and strains, insect repellent, enhances creativity, connects one to the universe, enhances meditation

Cautions:

- None currently known. Avoid with epilepsy if 1,8-cineole level is > 19.9%.

Selected Evidence:

- A fraction of sweetfern leaf essential oil obtained at 30–60 minutes of distillation killed human lung carcinoma and human colon adenocarcinoma cells.[4049]

SWEETGALE (Bog Myrtle, Wax Myrtle)

Myrica gale L., *M. gale* var. *subarctica* J. Rousseau, *M. gale* var. *subglabra* (A. Chev) Fernald, *M. gale* var. *tomentosa* C. DC., *Gale palustris* A. Chev.

Family: Myricaceae
Note: Middle-Top
Aroma Intensity: Medium
Aroma: Green, woody, citrusy, floral, medicinal
Common Extraction Method: Steam distilled or hydrodistilled from the leaves, fruits, or aerial parts; may also be distilled from the flowers
Possible Substitute Oils: Myrtle, hops, hemp, black pepper
Blends Well With: Basil, black pepper, cardamom, cassia, clove, copaiba, coriander, eucalyptus, frankincense, hemp, hops, lavender, lemon, marjoram, myrrh, myrtle, nutmeg, palo santo, patchouli, pine, ravensara, rosemary, sandalwood, ylang ylang
Subcellular Localization | Epigenetic Influence: Currently unknown | Currently unknown
Recommended Dilution Range: 3%–33%; neat for some conditions

Primary Constituents:[4050,4051,4052,4053,4054,4055,4056,4057,4058,4059,4060,4061,4062,4063]

α-Pinene/1,8-Cineole/Germacrene B CT (Aerial Parts)

Alpha-Pinene	15.7%–32.2%
1,8-Cineole	3.7%–23.9%
Germacrene B	1.4%–16.1%

Alpha-Pinene CT (Aerial Parts or Leaves)

Alpha-Pinene	17.8%–41.4%
1,8-Cineole	7.1%–20.0%
Delta/Gamma-Cadinene	0.0%–12.9%

Germacrone	0.0%–15.6%
Limonene	3.6%–10.8%
(E)-Nerolidol	0.7%–9.0%
Alpha-Phellandrene	0.0%–6.9%
Myrcene	0.0%–4.6%
Alpha-Terpinene	0.0%–4.2%
Para-Cymene	0.0%–3.7%
Beta-Pinene	2.5%–3.5%
Alpha-Terpineol	0.0%–3.1%
Gamma-Elemene	0.0%–3.0%

γ-Cadinene/α-Pinene CT (Aerial Parts)

UI Constituents	10.0%–22.9%
Gamma-Cadinene	8.4%–21.0%
1,8-Cineole	5.9%–18.7%
Alpha-Pinene	10.3%–17.4%
Limonene	5.2%–11.4%
Alpha-Terpinene	3.0%–7.2%
Alpha-Terpineol	1.4%–4.7%

Myrcene CT (Leaves)

Myrcene	12.1%–29.1%
Limonene	6.8%–15.3%
Selina-11-en-4-ol	0.0%–14.6%
Beta-Caryophyllene	tr–11.2%
Para-Cymene	2.4%–10.3%
Alpha-Phellandrene	0.7%–9.9%
Caryophyllene Oxide	0.0%–9.9%
Eremoligenol	0.0%–6.7%
Selina-4,11-Diene	0.0%–6.0%
Alpha-Terpinene	0.0%–5.6%
Alpha-Bisabolol	0.0%–5.0%
Alpha-Pinene	1.5%–4.7%
(E)-Beta-Ocimene	0.9%–4.5%
(Z)-Beta-Ocimene	0.9%–4.3%
Alpha-Humulene	0.0%–3.7%
Eudesmol (Isomer NI)	0.0%–3.6%
Beta-Bisabolene	0.0%–3.0%

Beta-Caryophyllene CT

Beta-Caryophyllene	18.6%
Limonene	10.4%
Selina-11-en-4-ol	8.1%
Alpha-Humulene	7.8%
Myrcene	7.7%
Para-Cymene	6.7%
Delta-Cadinene	5.3%
Alpha-Gurjunene	3.5%
Caryophyllene Oxide	3.5%

Limonene	5.0%–10.0%
Nerol	0.0%–7.7%
Alpha-Terpinyl Acetate	0.6%–7.0%
Myrcene	2.8%–6.4%
Para-Cymene	0.0%–6.0%
(E)-Nerolidol	0.0%–5.9%
Calamenene	0.0%–5.5%
Alpha-Humulene	0.0%–3.5%
Alpha-Terpinene	0.5%–3.3%
Cadinol (Isomer NI)	0.0%–3.3%
Beta-Phellandrene	0.0%–3.2%

Alpha-Pinene CT (Fruits)

Alpha-Pinene	22.6%
1,8-Cineole	18.9%
Germacrone	14.2%

Germacrone CT (Leaves)

Germacrone	25.1%
Alpha-Pinene	12.2%
Limonene	8.1%
Beta-Phellandrene	8.0%
1,8-Cineole	5.3%

(E,E)-Germacrone CT (Aerial Parts)

(E,E)-Germacrone	13.5%
Germacrene B	7.5%
Beta-Caryophyllene	5.6%
UI Constituents	4.8%
Limonene	4.1%
Alpha-Phellandrene	3.8%

Eudesm-11-en-4-ol CT (Leaves)

Eudesm-11-en-4-ol	11.5%
Myrcene	11.3%
Beta-Caryophyllene	8.4%
Alpha-Phellandrene	7.1%
Limonene	5.3%
Germacrone	4.3%
Para-Cymene	3.3%

Beta-Elemenone CT (Leaves)

Beta-Elemenone	14.3%
Alpha-Pinene	12.3%
Germacrene B	11.6%
1,8-Cineole	10.6%
Isobarbatene Ketone	4.4%

Germacrene B CT (Aerial Parts)

Germacrene B	23.4%
Germacrone	15.9%
(E)-Nerolidol	12.7%
1,8-Cineole	9.5%
Limonene	5.1%
Alpha-Pinene	4.7%
Gamma-Elemene	4.1%
Alpha-Phellandrene	3.4%

Flowers

Alpha-Pinene	28.1%–38.9%
1,8-Cineole	6.0%–14.7%
Gamma-Cadinene	0.4%–12.8%
UI Constituents	11.1%–12.3%
Alpha-Terpinene	3.8%–6.2%
Limonene	3.4%–5.3%
Beta-Elemenone	3.0%–4.3%
Beta-Pinene	1.8%–3.9%
(E)-Nerolidol	4.0%
Beta-Cadinene	3.9%

Beta-Elemenone CT (Leaves, var. tomentosa)

UI Constituents	14.5%–20.2%
Beta-Elemenone	13.3%–16.1%
1,8-Cineole	7.7%–15.9%
Para-Cymene	10.6%–13.1%
Limonene	9.8%–11.0%
Selina-3,7(11)-Diene	3.7%–11.0%
Beta-Pinene	5.0%–5.8%
Alpha-Pinene	1.9%–3.8%

UI=Unidentified; NI=Not identified

OTHER CONSTITUENTS: *α-Pinene/1,8-Cineole/Germacrene B CT (Aerial Parts)*—Beta-ocimene (isomer not identified), alpha-terpineol, beta-elemenone; *γ-Cadinene/α-Pinene CT (Aerial Parts)*—Beta-pinene, beta-ocimene (isomer not identified), beta-elemenone, (E)-nerolidol; *Alpha-Pinene CT (Aerial Parts or Leaves)*—Camphene, beta-pinene, alpha-phellandrene, (Z)-beta-ocimene, terpinolene, gamma-terpinene, terpinene-4-ol, citronellol, beta-caryophyllene, gamma-elemene, alpha-selinene, beta-selinene, valencene, beta-elemenone, juniper camphor; *Alpha-Pinene CT (Fruits)*—Beta-phellandrene, para-cymene, zonarene, selina-3,7(11)-diene, alpha-calacorene; *Myrcene CT (Leaves)*—Alpha-thujene, beta-phellandrene, delta-elemene, geranyl acetate, (Z)-alpha-bergamotene, beta-gurjunene, (E)-cadina-1(6),4-diene, gamma-gurjunene, gamma-curcumene, ar-curcumene, (E,E)-alpha-farnesene, beta-curcumene, delta-cadinene, (E)-nerolidol, humulene epoxide II, pentadecanal, linalool, citronellol, eremophilene, alpha-selinene, beta-selinene, alpha-terpinyl acetate, valencene, delta-selinene, trans-cadina-1(2),4-diene, Selina-3,7(11)-diene, alpha-bisabolol; *Beta-Caryophyllene CT*—Alpha-pinene, alpha-phellandrene, beta-phellandrene, (E)-beta-ocimene, (Z)-beta-ocimene, alpha-copaene, gamma-muurolene, (E)-nerolidol, alpha-bisabolol; *Germacrone CT (Leaves)*—Para-cymene, cubenol-1,10-di-epi, benzyl benzoate; *(E,E)-Germacrone CT (Aerial Parts)*—Alpha-pinene, para-cymene, beta-phellandrene, 1,8-cineole, (E)-beta-ocimene, (Z)-beta-ocimene, alpha-terpinyl acetate, gamma-elemene, alpha-humulene, drima-7,9(11)-diene, trans-cadina-1(6),4-diene, beta-chamigrene, gamma-muurolene, gamma-curcumene, beta-selinene, ar-curcumene, alapha-selinene, alpha-zingiberene, (E,E)-alpha-farnesene, beta-curcumene, Selina-4(15),7(11)-diene, Selina-3,7(11)-diene, beta-vetivenene, caryophyllene oxide, selina-11-en-4-ol, alpha-bisabolol; *Eudesm-11-en-4-ol CT (Leaves)*—Alpha-pinene, (Z)-beta-ocimene, (E)-beta-ocimene, gamma-terpinene, beta-phellandrene, alpha-terpinyl acetate, germacrene D, gamma-gurjunene, beta-selinene, alpha-selinene, (E,E)-alpha-farnesene, delta-cadinene, trans-cadina-1(2),4-diene, germacrene B, (E)-nerolidol, caryophyllene oxide, humulene epoxide II, alpha-bisabolol; *Germacrene B CT (Aerial Parts)*—Beta-pinene, myrcene, para-cymene; *Flowers*—Alpha-terpineol, (E)-nerolidol; *Beta-Elemenone CT (Leaves)*—Beta-pinene, myrcene, limonene, para-cymene; *Beta-Elemenone CT (Leaves, var. tomentosa)*—Alpha-phellandrene, alpha-terpinene, gamma-terpinene, isoamyl isovalerate, linalool, beta-elemene, terpinene-4-ol, beta-guaiene, gamma-elemene, alpha-terpinyl acetate, beta-selinene, alpha-selinene, eremophilene, lavandulol acetate, nerol, benzyl isovalerate, 3,7-dimethyl-10-(1-methylethylidene),3,7-cyclodecadien-1-one

PREFERRED COMPOSITION FOR CLINICAL USE:
Too great of variance to determine; however, a myrcene CT rich in sesquiterpenes may be preferred.

REPORTED THERAPEUTIC PROPERTIES: Anticancer, antibacterial, antifungal, antiparasitic, antiseptic, astringent, wound healing, reduces the appearance of acne and other blemishes, aids digestion, balances menstruation, analgesic, anti-inflammatory, antiarthritic, aids circulation, insect repellent, fosters courage, reduces mental fatigue, aids cognition and focus, stress management

CAUTIONS:
α-Pinene/1,8-Cineole/Germacrene B CT (Aerial Parts), Alpha-Pinene CT (Aerial Parts or Leaves), γ-Cadinene/α-Pinene CT (Aerial Parts), Alpha-Pinene CT (Fruits), Beta-Elemenone CT (Leaves, var. tomentosa), and Flowers
■ Avoid with epilepsy and Parkinson's disease due to 1,8-cineole content. May exacerbate or cause seizures/convulsions or reduce seizure-medication efficacy based on 1,8-cineole content.[4064,4065,4066,4067]

Myrcene CT (Leaves)
■ May weakly interfere with the enzymes responsible for metabolizing medications (NSAIDs, proton-pump inhibitors, acetaminophen, antiepileptics, immune modulators, blood-sugar medications, blood pressure medications, antidepressants, antipsychotics, diabetic medications, antihistamines, antibiotics, and anesthetics), based on myrcene content.[4068,4069]

(E,E)-Germacrone CT (Aerial Parts) and Germacrone CT (Leaves)
■ May interfere with the enzymes responsible for metabolizing medications (NSAIDs, proton-pump inhibitors, acetaminophen, antiepileptics, immune modulators, blood-sugar medications, blood pressure medications, antidepressants, antipsychotics, diabetic medications, antihistamines, antibiotics, and anesthetics) due to germacrone content.[4070]

Beta-Caryophyllene CT, Germacrene B CT (Aerial Parts), Eudesm-11-en-4-ol CT (Leaves), and Beta-Elemenone CT (Leaves)
■ None currently known.

SELECTED EVIDENCE:
- □ Sweetgale essential oil CT myrcene (two separate fractions tested individually) displayed anticancer activity against human lung and colon cancer cells.[4071] The oil with a richer sesquiterpene (eremoligenol, caryophyllene oxide, beta-caryophyllene, etc.) composition was more active.
- □ *In vitro* research demonstrated that sweetgale (var. *tomentosa*, CT beta-elemenone) essential oil inhibited *B. subtilis*.[4072] The fruit essential oil (alpha-pinene CT) inhibited foodborne fungi (*Cladosporium cladosporioides, Byssochlamys fulva, Paecilomyces variotii, Penicillum expansum, Penicillium corylophilum, Aspergilus flavus*).[4073]
- □ Sweetgale (var. *tomentosa*, CT beta-elemenone) essential oil exhibited very weak antioxidant activity in the DPPH, SOD, and lipid-peroxidation assays.[4074]
- □ Sweetgale absolute (alpha-pinene as the major constituent) repelled *Aedes aegypti* mosquitoes.[4075] Another study reported that sweetgale absolute (1,8-cineole, myrcene, and limonene as major constituents) repelled the castor bean tick (*Ixodes ricinus*)—50% repellency—when used at a 10% concentration.[4076]

TANA
(Issa)

Rhus taratana (Baker) H. Perrier

FAMILY: Anacardiaceae
NOTE: Top-Middle
AROMA INTENSITY: Medium
AROMA: Fresh, woody, spicy, balsamic, slightly fruity
COMMON EXTRACTION METHOD: Steam distilled or hydrodistilled from the leaves
POSSIBLE SUBSTITUTE OILS: Ferula, rabbitbrush, black pepper CT limonene, galbanum
BLENDS WELL WITH: Basil, bay laurel, bergamot, black pepper, camphor, cassia, cinnamon, coriander, cypress, geranium, grapefruit, lavender, lemon, myrrh, neroli, orange, pine, petitgrain, palo santo, ravensara, ravintsara, rose, sandalwood, spikenard, tangerine, vetiver, ylang ylang
SUBCELLULAR LOCALIZATION | EPIGENETIC INFLUENCE: Currently unknown | Currently unknown
RECOMMENDED DILUTION RANGE: 3%–50%; neat for some conditions

PRIMARY CONSTITUENTS:[4077,4078]

Limonene	25.0%–30.0%
Alpha-Pinene	16.0%–23.0%
1,8-Cineole + Beta-Phellandrene	16.0%–22.0%
Myrcene + Alpha-Phellandrene	12.0%–15.0%
Bea-Pinene	3.0%–5.0%
Camphene	2.0%–4.0%
Sabinene	1.5%–3.0%
Gamma-Terpinene	1.5%–3.0%
Alpha-Terpinene	1.0%–3.0%

Commercial Oil	
Alpha-Pinene	35.4%
Limonene + 1,8-Cineole + Beta-Phellandrene	22.5%
Beta-Pinene	17.9%
Alpha-Phellandrene	4.7%
Myrcene	3.9%

OTHER CONSTITUENTS: Para-cymene, terpinolene; *Commercial Oil*—Camphene, sabinene, alpha-terpinene, gamma-terpinene, terpinolene, terpinen-4-ol, alpha-terpineol, bornyl acetate, alpha-copaene, germacrene D, delta-cadinene

PREFERRED COMPOSITION FOR CLINICAL USE:
Insufficient data/samples to determine.

REPORTED THERAPEUTIC PROPERTIES: Aids digestion, heals stomach ulcers, antiviral, antibacterial, antifungal, improves male sexual function, aids circulation and lymph flow, relieves varicose veins, invigorating, stimulating, relieves mental fatigue, stress management

CAUTIONS:
- None currently known.

SELECTED EVIDENCE:
- ☐ None found.

TAURUS CEDAR (Lebanese Cedar, Lebanon Cedar, Cedar of Lebanon)

Cedrus libani A. Rich

FAMILY: Pinaceae
NOTE: Base-Middle
AROMA INTENSITY: Medium
AROMA: Woody, rich, earthy, balsamic, grassy
COMMON EXTRACTION METHOD: Hydrodistilled or steam distilled from the heartwood, wood, root, needles (leaves), shoots, cones, cone seeds, or oleoresin
POSSIBLE SUBSTITUTE OILS: Atlas cedarwood, Himalayan cedarwood
BLENDS WELL WITH: Balsam fir, bay laurel, bergamot, blue cypress, blue spruce, carrot seed, cedarwood, cinnamon, copaiba, cypress, myrtle, jasmine, juniper, lemon, neroli, palmarosa, patchouli, petitgrain, pine, rose, rosemary, ravensara, sandalwood, silver fir, Spanish sage, spruce, thyme, tsuga, vetiver, white fir
SUBCELLULAR LOCALIZATION | EPIGENETIC INFLUENCE: Currently unknown | Currently unknown
RECOMMENDED DILUTION RANGE: 5%–Neat

PRIMARY CONSTITUENTS:[4079,4080,4081,4082,4083,4084,4085,4086]

Wood (Himachalenes/Atlantones CT)	
Beta-Himachalene	8.1%–38.2%
(E)-Alpha-Atlantone	3.9%–32.2%
Himachalol	0.5%–19.7%
Alpha-Himachalene	2.6%–12.8%
(E)-10,11-Dihdroatlantone	0.0%–7.8%
Gamma-Himachalene	1.7%–7.6%
6,7-Epoxihimachalene	0.0%–7.5%
(Z)-Gamma-Atlantone	%–5.0%
(Z)-Alpha-Atlantone	0.0%–4.7%
Allohimachalol	0.0%–4.1%
(E)-Beta-Atlantone	0.0%–4.0%
(Z)-Beta-Atlantone	0.0%–3.2%

Needles	
Germacrene D	29.4%
1-Epi-Cubenol	6.3%
Trans-Alpha-Bisabolene	5.9%
Beta-Caryophyllene	5.6%
Delta-Cadinene	5.6%
Gamma-Muurolene	4.8%
4(14)-Salvialen-1-one	4.0%
T-Cadinol	3.4%

Wood (Himachalol CT)	
Himachalol	22.5%
Beta-Himachalene	21.9%
Alpha-Himachalene	10.5%
Gamma-Himachalene	9.1%
Allohimachalol	3.2%

Roots	
Beta-Himachalene	27.5%
(E)-Alpha-Atlantone	23.7%
Alpha-Himachalene	10.3%
Himachalol	9.7%
Gamma-Himachalene	7.1%

Shoots	
Alpha-Pinene	45.4%–46.8%
Myrcene	10.1%–22.6%
Limonene	13.6%–19.4%
Beta-Pinene	11.7%–16.3%
Beta-Caryophyllene	5.1%–5.6%

Oleoresin	
Alpha-Pinene	24.8%
Abieta-7,13-Diene	16.7%

Cone Seeds	
Alpha-Pinene	27.4%–33.5%
Beta-Pinene	17.8%–22.8%
Beta-Phellandrene	0.0%–20.0%
Cyclohexene	1.1%–8.4%
1-Hexen-3-yne	4.3%–7.9%
Bicyclo[2,2,1]Heptan-2-ol	3.9%–7.7%
Bicyclo[3,1,1]Hept-2-ene-2-Carboxaldehyde	3.5%–6.5%
Bornyl Acetate	0.0%–6.4%
3-Caren-10-al	0.0%–6.3%
Bicyclo[3,1,1]Heptan-3-ol	4.2%–6.0%
2(10)-Pinen-3-one	3.5%–5.8%
Bicyclo[3,1,1]Hept-3-en-2-one	0.0%–5.8%
Myrtenal	2.9%–5.7%
Trans-Pinocarveol	3.9%–5.6%
3-Cyclohexene-1-Methanol	0.0%–5.5%
Bicyclo[3,1,1]Hept-3-en-2-ol	0.0%–3.9%
Isopinocarveol	0.9%–3.8%
Limonene	2.4%–3.1%

Abieta-8,11,13-Triene	6.9%
Manool	5.8%
Terpinen-4-ol	3.7%
Alpha-Terpineol	3.4%

Cones (Alpha-Pinene CT)	
Alpha-Pinene	51.0%
Myrcene	13.0%
7,13-Abietadiene	3.2%
Terpinolene	3.1%

Cones (Limonene CT)	
Limonene	17.7%
Abieta-8,11,13-Triene	17.0%
Alpha-Pinene	12.3%
Beta-Pinene	8.3%
7,13-Abietadiene	8.3%
Palustradiene	7.1%
Myrcene	4.9%
(Z)-Alpha-Bisabolene	4.7%

Note: The Taurus cedar is an endangered and protected species with the most threatened populations being those in Lebanon and Syria. It is native to the mountains of the Mediterranean region and found in Lebanon, Turkey, and Syria.

OTHER CONSTITUENTS: *Wood (Himachalenes/Atlantones CT)*—1-Methyl-4-acetylcyclohex-1-ene, longifolene, beta-cedrene, 8,9-dehydroneoisolongifolene, alpha-ar-himachalene, beta-ar-himachalene, libanone, oxidohimachalene, (Z)-10,11-dihydroatlantone, (E)-nerolidol, isohimachalone, longiborneol, (Z)-gamma-atlantone, (E)-gamma-atlantone, allohimachalene, beta-himachalene oxide, beta-acorenol, himachalene oxide, gamma-dehydro-ar-himachalene, cis-cadina-1,4-diene, vestitenone; *Wood (Himachalol CT)*—Longiborneol, (Z)-gamma-atlantone, (E)-gamma-atlantone, (Z)-alpha-atlantone, (E)-alpha-atlantone, manool; *Roots*—Beta-cedrene, longiborneol, (Z)-beta-atlantone, (E)-beta-atlantone, (E)-gamma-atlantone, allohimachalol, (Z)-alpha-atlantone; *Needles*—Alpha-pinene, beta-pinene, myrcene, terpineol, alpha-cubebene, alpha-copaene, beta-bourbonene, beta-copaene, 4beta-10,-1(5),6-guaiadiene, alpha-himachalene, beta-humulene, gamma-amorphene, alpha-muurolene, alpha-cadinene, dodecanoic acid; *Shoots*—Camphene, alpha-humulene; *Oleoresin*—Camphene, beta-pinene, myrcene, delta-3-carene, para-cymene, limonene, terpinolene, para-cymenene, fenchol, borneol, methyl chavicol (< 1.8%), longifolene, beta-himachalene, diterpenes MG 272 (m2 134), manoyl oxide, diterpenes MG 272 (me/257); *Cones (Alpha-Pinene CT)*—Camphene, beta-phellandrene, delta-3-carene, alpha-terpinene, para-cymene, gamma-terpinene, alpha-terpineol, (E)-beta-farnesene, (E)-alpha-bisabolene, abietatriene; *Cones (Limonene CT)*—Para-cymene, neoabietadiee, 7-oxo-abieta-8,11,13-triene; *Cone Seeds*—Myrcene, 3-cyclohexen-1-ol, p-menth-1-en-8-ol, 3-diazo-1,7,7-trimethylbicyclo[2,2,1]heptan-2-one, trans-2-caren-4-ol, isobornyl propionate

PREFERRED COMPOSITION FOR CLINICAL USE:
Insufficient and too great of variance in samples/data to determine.

REPORTED THERAPEUTIC PROPERTIES: Wound healing, reduces skin irritation and itching, soothes chronic skin conditions, antibacterial, antifungal, antimicrobial, antiviral, antiparasitic, antiseptic, cleanses the urinary tract, aids respiratory function, antiallergic, modulates immune activity, heals stomach ulcers, anticancer, antioxidant, anti-inflammatory, analgesic, antispasmodic, balances blood-sugar levels, insecticidal, promotes restful sleep, calming, stress management, reduces anxiety

CAUTIONS:

- None currently known.

SELECTED EVIDENCE:

- Taurus cedar wood essential oil demonstrated antiproliferative activity against human chronic myelogenous leukemia cells.[4087] The same study reported that the oil stimulated erythroid differentiation (the process whereby red blood cells mature). Another study reported that the wood oil (Himachalenes/Atlantones CT) was cytotoxic to multidrug-resistant leukemia cells.[4088]
- Taurus cedar wood CT himachalol essential oil inhibited alpha-amylase activity.[4089] Alpha-amylase is the primary enzyme responsible for the breakdown of starch and sugar (dextrin, maltriose, maltose, and glucose) and inhibition of this enzyme is believed to improve glucose tolerance in people with diabetes.
- Animal research demonstrated that topical application of Taurus cedar cone essential oil (once daily for nine days; composition not reported) remarkably reduces inflammation and accelerates wound healing activities.[4090]
- The wood essential oil of Taurus cedar (himachalol CT) inhibited Phytochemical analysis and in vitro evaluation of the biological activity against Herpes simplex virus type-1 (HSV-1).[4091]
- Taurus cedar seed essential (composition not reported) was larvicidal to the *C. pipiens* mosquito.[4092]

TEA LEAF ABSOLUTE
(Tea Shrub)

Camellia sinensis (L.) Kuntze, *Thea sinensis* L.

FAMILY: Theaceae
NOTE: Note
AROMA INTENSITY: Medium
AROMA: Herbaceous, sweet, floral, woody, earthy
COMMON EXTRACTION METHOD: Steam distilled, hydrodistilled, or solvent extracted from the leaves
POSSIBLE SUBSTITUTE OILS: PSO
BLENDS WELL WITH: Anise, bergamot, cedarwood, clary sage, grapefruit, jasmine, labdanum, lavender, lemon, lime, nutmeg, palmarosa, rose, sage, tangerine
SUBCELLULAR LOCALIZATION | EPIGENETIC INFLUENCE: Currently unknown | Currently unknown
RECOMMENDED DILUTION RANGE: < 0.4%

PRIMARY CONSTITUENTS:[4093,4094]

Simultaneous HD/Solvent Extraction		*HS-SPME (Longjing Tea)*	
Beta-Pinene	10.4%–50.1%	Linalool	4.7%–22.5%
Alpha-Pinene	9.6%–30.3%	Unidentifieds	4.4%–20.3%
Alpha-Cadinol	0.0%–19.6%	(Z)-3-Hexenyl Hexanoate	3.5%–14.6%
Muurol-5-en-4-Alpha-ol	0.0%–17.9%	Geraniol	3.8%–10.9%

Muurol-5-en-4-Beta-ol	0.0%–12.6%	Myrcene	1.2%–5.8%
Beta-Ionone	0.0%–5.3%	(Z)-3-Hexenyl Butyrate	1.1%–5.5%
Delta-Muurolene	0.0%–4.5%	(E)-Nerolidol	0.9%–3.4%
Citronellal	0.0%–3.4%	Cedrene	0.3%–3.1%
Hexanal	0.0%–3.1%	Beta-Farnesene	0.2%–3.1%

Top Ten Volatiles in Various Types of Tea[4095]

White Tea	Green Tea	Oolong Tea	Black Tea	Pu-erh Tea
Hexanal	Geraniol	(E)-Nerolidol	(E)-2-Hexenal	1,2,3-Trimethoxybenzene
(E)-2-Hexenal	Linalool	Indole	Hexanal	Linalool Oxide II
Cis-Linalool Oxide II	(E)-Nerolidol	Linalool	Geraniol	Hexanal
Linalool	(Z)-Hexanoic Acid 3-Hexenyl Ester	Benzeneacetaldehyde	Linalool	Linalool Oxide I
Geraniol	Nonanal	Phytol	Linalool Oxide II	Benzaldehyde
Phenylethyl alcohol	3,7-Dimethyl-1,5,7-Octatrien-3-ol	Alpha-Farnesene	Benzeneacetaldehyde	Benzeneacetaldehyde
Benzaldehyde	(E)-2-Hexenal	Linalool Oxide I	Linalool Oxide I	Alpha-Terpineol
Linalool Oxide I	Phytol	3,7-Dimethyl-1,5,7-Octatrien-3-ol	3,7-Dimethyl-1,5,7-Octatrien-3-ol	Beta-Ionone
Benzeneacetaldehyde	Heptanal	Benzyl Nitrile	Benzaldehyde	Cedrol
(Z)-3-Hexen-1-ol	(Z)-Jasmone	Hexanal	Methyl Salicylate	Linalool

OTHER CONSTITUENTS: *Simultaneous HD/Solvent Extraction*—Hexanal, benzaldehyde, benzyl alcohol, linalool, nerol, para-cymene, methyl chavicol (< 1.5%), myrtenol, alpha-muurolene, nopinone, myrtenal, cubenol, globulol, myrcene, beta-phellandrene, limonene; *HS-SPME*—1-Octen-3-ol, limonene, ocimene, phenylethyl alcohol, naphthalene, methyl salicylate (< 2.9%), safranal, decanal, hexanoic acid anhydride, indole, tridecane, alpha-cubebene, copaene, hexyl hexanoate, (E)-2-hexenyl hexanoate, jasmone, tetradecane, longifolene, beta-caryophyllene, alpha-ionone, geranyl acetate, alkane, beta-ionone, pentadecane, alpha-farnesene, gamma-cadinene, delta-cadinene, cadina-1,4-diene, hexadecane, caffeine (< 3.0%)

PREFERRED COMPOSITION FOR CLINICAL USE:
Insufficient data\samples to determine.

REPORTED THERAPEUTIC PROPERTIES: Fragrance, reduces high cholesterol, helps maintain normal blood pressure, wound healing, antibacterial, antifungal, anti-infectious, promotes mental alertness

CAUTIONS:

- Avoid oral consumption. Absolutes contain solvent residue and should not be taken orally.
- Tea leaf extracts are strongly irritating to the skin and should not be used in concentrations higher than 0.3% to avoid sensitization.[4096,4097]

SELECTED EVIDENCE:

- None found.

THREELEAF SHRUBVERBENA
(Shrub Verbena)

Lantana trifolia L., *L mearnsii* Moldenke,
L. *mearnsii* var. *latibracteolata* Moldenke

FAMILY: Verbenaceae
NOTE: Middle
AROMA INTENSITY: Medium
AROMA: Herbaceous, floral, green, slightly sweet and woody
COMMON EXTRACTION METHOD: Hydrodistilled or steam distilled from the leaves, flowers, or aerial parts
POSSIBLE SUBSTITUTE OILS: Goldenrod, ylang ylang,
BLENDS WELL WITH: Bergamot, black pepper, black spruce, blue spruce, cedarwood, copaiba, cypress, ginger, goldenrod, lavender, lemon, neroli, orange, pine, sandalwood, tangerine, vetiver, ylang ylang
SUBCELLULAR LOCALIZATION | EPIGENETIC INFLUENCE: Currently unknown | Currently unknown
RECOMMENDED DILUTION RANGE: 3%–33%; 50% for some conditions

PRIMARY CONSTITUENTS:[4098,4099,4100,4101,4102]

Leaves	
Germacrene D	20.3%–45.1%
Tetradecane	0.0%–17.5%
(E)-Caryophyllene	6.3%–15.3%
(Z)-Beta-Ocimene	0.0%–15.1%
Bicyclogermacrene	0.0%–12.7%
Limonene	0.1%–12.1%
(E)-Nerolidol	0.0%–5.3%
(Z,E)-Alpha-Farnesene	0.0%–4.5%
Alpha-Humulene	0.0%–4.4%
Germacrene-D-4-ol	0.0%–4.2%
Linalool	0.0%–3.9%
Alpha-Copaene	0.0%–3.8%
Delta-Cadinene	1.3%–3.6%
(E)-Beta-Ocimene	0.9%–3.0%

Flowers	
Germacrene D	24.9%–38.1%
(E)-Caryophyllene	11.9%–16.1%
Bicyclogermacrene	0.8%–11.0%
Sabinene	tr–7.9%
Delta-Cadinene	1.2%–4.9%
Alpha-Humulene	1.5%–4.7%
Trans-p-Menth-2-en-ol	0.0%–3.8%
Germacrene-D-4-ol	0.0%–3.7%
1-Octen-3-ol	0.0%–3.5%
Spathulenol	0.0%–3.3%
(Z)-Beta-Ocimene	0.0%–3.3%
Alpha-Copaene	1.8%–3.0%

Aerial Parts	
Germacrene D	16.9%
Beta-Caryophyllene	13.5%
1-Octen-3-ol	6.7%
(Z)-Beta-Ocimene	5.7%
(E)-Nerolidol	3.4%
Alpha-Humulene	3.3%

OTHER CONSTITUENTS: *Leaves*—Alpha-pinene, camphene, myrcene, para-cymene, beta-phellandrene, 3-hexenol, 3-octanol, 1-terpineol, cis-piperitol, trans-piperitol, piperitone, bornyl acetate, beta-elemene, alloaromadendrene, alpha-muurolene, bicyclogermacrene, germacrene A, caryophyllene oxide, beta-bourbonene, epi-bicyclosphellandrene, beta-selinene, cis-p-menth-2-en-1-ol, isocaryophyllene, trans-p-menth-2-en-1-ol, aromadendrene, alpha-cadinene, germacrene B, humulene epoxide II, alpha-cadinol, spathulenol, dodecane, hexadecane, phytol; *Flowers*—Alpha-pinene, camphene, para-cymene, limonene, (Z)-beta-ocimene, 3-octanol, linalool, terpinen-4-ol, 1-terpineol, cis-piperitol, trans-piperitol, piperitone, bornyl acetate, beta-elemene, cis-p-menth-2-en-1-ol, isocaryophyllene, aromadendrene, alpha-terpineol, bicyclogermacrene, (Z,E)-alpha-farnesene, trans-piperitol, alloaromadendrene, epi-bicyclophellandrene, alpha-muurolene, (E,E)-alpha-farnesene, germacrene A, alpha-cadinene, germacrene B, (E)-nerolidol, humulene epoxide II, alpha-cadinol, T-cadinol, T-muurolol, alpha-muurolol, tetradecane, phenyl-3-methylbutanoate, phytol; *Aerial Parts*—(E)-Beta-ocimene, 3-hexenol, alpha-copaene, linalool, cis-p-menth-2-en-1-ol, trans-p-menth-2-en-1-ol, aromadendrene, bicyclogermacrene, (Z,E)-alpha-farnesene, trans-piperitol, (E,E)-alpha-farnesene, delta-cadinene, caryophyllene oxide, spathulenol, T-cadinol, alpha-muurolol

PREFERRED COMPOSITION FOR CLINICAL USE:
Insufficient samples/data to determine.

REPORTED THERAPEUTIC PROPERTIES: Aids liver health, relieves fever, analgesic, anti-inflammatory, antirheumatic, relieves painful menstruation, enhances male virility, soothes toothache, relieves headache, aids digestion, antibacterial, antifungal, anti-infectious, supports respiratory and lung function, eases cough, relieves snakebites, aids cardiovascular function, improves circulation, promotes oral hygiene, relieves dizziness, soothes skin irritation and rash, wound healing, antiallergic, combats hysteria, relaxing/calming, combats anger

CAUTIONS:
- None currently known.

SELECTED EVIDENCE:
- *In vitro* research demonstrates that threeleaf shrub verbena leaf essential oil (CT Germacrene D) possesses antimycobacterial properties.[4103]

THYME BERRY (Ehrenberg's Marjoram, Lebanese Oregano, Zaayter)

Origanum ehrenbergii Boiss.

FAMILY: Lamiaceae (Labiatae)
NOTE: Middle
AROMA INTENSITY: Strong
AROMA: Herbaceous, spicy, medicinal, fresh
COMMON EXTRACTION METHOD: Steam distilled or hydrodistilled from the aerial parts
POSSIBLE SUBSTITUTE OILS: Oregano, thyme, summer savory, winter savory, ajowan
BLENDS WELL WITH: Bay laurel, bergamot, camphor, cedarwood, chamomile (German, Roman), cinnamon, clove, cypress, eucalyptus, frankincense, ginger, lavender, lemon, marjoram, melaleuca (tea tree), myrrh, oregano, orange, pine, rosemary, Spanish sage, thyme
SUBCELLULAR LOCALIZATION | EPIGENETIC INFLUENCE: Currently unknown | Currently unknown
RECOMMENDED DILUTION RANGE: 1.5%–20%; 50% for some conditions

Primary Constituents:[4104,4105,4106,4107,4108]

Carvacrol CT

Carvacrol	71.4%–79.0%
Para-Cymene	4.4%–7.0%
Gamma-Terpinene	2.6%–6.1%

Carvacrol/Thymoquinone CT

Carvacrol	43.6%
Thymoquinone	37.0%
Thymol Methyl Ether	5.3%
Para-Cymene	4.7%

Thymol/Carvacrol CT

Thymol	19.6%–38.3%
Carvacrol	6.7%–36.8%
Para-Cymene	0.2%–16.1%
2-Isopropyl-1-Methoxy-4-Methylbenzene	0.0%–14.9%
Gamma-Terpinene	11.8%–14.5%
Carvacrol Methyl Ester	0.0%–7.2%
Carvacrol Methyl Ether	0.2%–3.8%
Myrcene	1.6%–3.2%

Note: Thyme berry is considered a threatened species.

Thymol CT

Thymol	38.4%–58.8%
Gamma-Terpinene	11.1%–17.1%
Para-Cymene	8.5%–12.9%
Carvacrol	1.6%–10.0%
Alpha-Terpinene	1.8%–3.5%
Carvacrol Methyl Ether	0.0%–3.4%

Para-Cymene CT

Para-Cymene	33.1%
Thymol	28.5%
Gamma-Terpinene	10.6%
Carvacrol Methyl Ether	4.2%

Other Constituents: *Carvacrol CT*—Alpha-thujene, alpha-pinene, 1-octen-3-ol, myrcene, alpha-terpinene, carvacrol methyl ether, thymol, beta-caryophyllene, beta-bisabolene; *Carvacrol/Thymoquinone CT*—Myrcene, beta-caryophyllene, aromadendrene, beta-bisabolene, delta-cadinene; *Thymol/Carvacrol CT*—Alpha-thujene, alpha-pinene, sabinene, 1-octren-3-ol, 3-octanol, alpha-terpinene, cis-sabinene hydrate, linalool, beta-caryophyllene, beta-bisabolene, thymol methyl ether, alpha-humulene, trans-beta-farnesene; *Thymol CT*—Alpha-thujene, alpha-pinene, 1-octen-3-ol, myrcene, 3-octanol, cis-sabinene hydrate, beta-caryophyllene, beta-bisabolene; *Para-Cymene CT*—Alpha-thujene, alpha-pinene, 1-octen-3-ol, 3-octanol, alpha-terpinene, cis-sabinene hydrate, terpinen-4-ol, carvacrol, beta-bisabolene

Preferred Composition for Clinical Use:

Constituent	*Carvacrol CT*	*Thymol CT*
Carvacrol	60%–80%	1%–12%
Para-Cymene	3%–12%	7%–20%
Gamma-Terpinene	2%–10%	10%–25%
Thymol	0.5%–5%	40%–60%
Myrcene	0.5%–4%	0.5%–5%

Reported Therapeutic Properties: Antimicrobial, antibacterial, antifungal, antiparasitic, antiviral, anti-infectious, antioxidant, aids respiratory function, relieves cough, expels excess mucus, antiarthritic, analgesic, anti-inflammatory, aids digestion, relieves stomachache, promotes localized circulation, preserves brain and memory function, antidepressant, relieves mental fatigue, promotes self-confidence

CAUTIONS:

Carvacrol CTs

■ Avoid during pregnancy and lactation. Oregano (also rich in carvacrol) may negatively affect embryonic development and encourage fetal cell death according to animal research.[4109] In addition, CTs with significant thymoquinone would require a caution because thymoquinone prevents the formation of new blood vessels (angiogenesis), which occurs during pregnancy beginning twenty-one days after conception.[4110,4111,4112] The growth of a dense network of blood vessels within the placenta is necessary for the exchange of respiratory gases, nutrients, and wastes between the mother and fetus. If angiogenesis is inhibited during pregnancy, it may lead to insufficient blood supply to the placenta and pregnancy complications (fetal growth restriction, preeclampsia, and fetal death). Thymoquinone also negatively affects embryonic development if administered during the second trimester (fetal malformations).[4113]

■ May interact with diabetic medications and cause low blood-sugar levels due to carvacrol content—carvacrol reduces blood glucose and enhances the antidiabetic activities of rosiglitazone in animals.[4114,4115,4116,4117]

■ May interact with aspirin, blood pressure, antiplatelet, and anticoagulant medications, and increase the risk of bleeding among people with bleeding disorders, due to carvacrol and thymol content.[4118,4119,4120,4121,4122]

■ May interact with anticholinergic (drugs used for asthma, incontinence, gastrointestinal cramps, muscular spasms, depression, and sleep disorders) and cholinergic medications (drugs used to reduce urinary retention, diagnose myasthenia gravis, and for glaucoma), based on carvacrol, thymol, or thymoquinone content.[4123,4124]

■ May interact with antibiotics or antifungals and possibly enhance or decrease their effects due to carvacrol or thymol content.[4125,4126,4127,4128,4129]

■ *With thymoquinone*: May interact with diabetes medications and cause low blood sugar based on thymoquinone content.[4130,4131,4132,4133,4134,4135,4136] Higher thymoquinone content will increase the effect on blood-sugar levels.[4137]

■ *With thymoquinone*: May interact with aspirin, blood pressure, antiplatelet, and anticoagulant medications, and increase the risk of bleeding among people with bleeding disorders, due to thymol, carvacrol, trans-anethole, and thymoquinone content.[4138,4139,4140,4141] Thymoquinone is known to cause platelet apoptosis (cell death).[4142,4143]

■ *With thymoquinone*: May interact with anticholinergic (drugs used for asthma, incontinence, gastrointestinal cramps, muscular spasms, depression, and sleep disorders) and cholinergic medications (drugs used to reduce urinary retention, diagnose myasthenia gravis, and for glaucoma), based on thymoquinone, thymol, and carvacrol content.[4144]

■ *With thymoquinone*: May interfere with the enzymes responsible for metabolizing medications based on trans-anethole and thymoquinone content (NSAIDs, proton-pump inhibitors, acetaminophen, antiepileptics, immune modulators, blood-sugar medications, blood pressure medications, antidepressants, antipsychotics, diabetic medications, antihistamines, antibiotics, and anesthetics).[4145,4146,4147,4148]

■ *With thymoquinone*: May interact with antibiotics and possibly enhance their effects due to thymol, thymoquinone, or carvacrol content.[4149,4150]

■ *With thymoquinone*: May interfere with the antiseizure medication valproate (possibly other antiseizure drugs) and enhance its anticonvulsive activity due to thymoquinone content.[4151]

■ May irritate mucous membranes (eyes, mouth, nasal passages, vagina, rectum).

Thymol CTs

■ Oral caution. Thymol has a longer half-life (the time it takes for half of the medication to metabolize or excrete half of the dosage) than most essential oil constituents and should not be administered orally for long periods of time.[4152] Thymol is a monoterpene phenol. Reports of fatalities in children who consumed 50 to 200 mg of phenols has been reported.[4153] It is recommended that chemotypes with high levels of thymol be limited to 10 drops per day orally for adults with a two- to seven-day break after twenty-one days of use; or limit ingestion to five or six days weekly.

■ May interact with aspirin, blood pressure, antiplatelet, and anticoagulant medications, and increase the risk of bleeding among people with bleeding disorders, due to thymol and carvacrol content.[4154,4155]

■ May interact with anticholinergic (drugs used for asthma, incontinence, gastrointestinal cramps, muscular spasms, depression, and sleep disorders) and cholinergic medications (drugs used to reduce urinary retention, diagnose myasthenia gravis, and for glaucoma), based on thymol and carvacrol content.[4156,4157]

■ May interact with antibiotics or antifungals and possibly enhance their effects based on thymol and carvacrol content.[4158,4159,4160,4161,4162,4163,4164,4165]

■ May irritate mucous membranes (eyes, mouth, nasal passages, vagina, rectum).

Para-Cymene CT

■ May interact with aspirin, blood pressure, antiplatelet, and anticoagulant medications, and increase the risk of bleeding among people with bleeding disorders, due to thymol content.[4166,4167]

■ May interact with anticholinergic (drugs used for asthma, incontinence, gastrointestinal cramps, muscular spasms, depression, and sleep disorders) and cholinergic medications (drugs used to reduce urinary retention, diagnose myasthenia gravis, and for glaucoma), based on thymol content.[4168,4169]

■ May interact with antibiotics or antifungals and possibly enhance their effects based on thymol content.[4170,4171,4172,4173,4174,4175,4176,4177]

■ May irritate mucous membranes (eyes, mouth, nasal passages, vagina, rectum).

SELECTED EVIDENCE:

- □ Thyme berry CT thymol/carvacrol essential oil showed anti-inflammatory activity through the inhibition of induced nitric oxide (NO) production.[4178]
- □ The thymol/carvacrol CT of thyme berry essential oil significantly inhibited the activity of both acetylcholinesterase (AChE) and butyrylcholinesterase (BChE).[4179] Inhibiting both AChE and BChE can help restore acetylcholinesterase levels in the brain and possibly relieve symptoms of Alzheimer's disease and dementia.
- □ Thyme berry CT carvacrol/thymoquinone essential oil moderately inhibited tyrosinase activity—tyrosinase is an enzyme involved in skin darkening.[4180] Inhibition of tyrosinase may lighten the skin. Interestingly, the researchers found that carvacrol was an excellent inhibitor of tyrosinase, but its activity was reduced when thymoquinone was present (as is the case with this CT of thyme berry essential oil).
- □ *In vitro* research demonstrated that thyme berry essential oil significantly inhibits *E. coli, P. aeruginosa, E. faecalis, S. enteritidis, S. aureus, B. subtilis* subsp. *spizienii*, and *C. albicans*.[4181]
- □ Thyme berry CT thymol/carvacrol essential oil exhibited good antioxidant activity in the DPPH and beta-carotene bleaching assays.[4182]

TOBACCO ABSOLUTE | TOBACCO

Nicotiana tabacum L.

FAMILY: Solanaceae
NOTE: Base
AROMA INTENSITY: Strong

AROMA: Herbaceous, pungent, woody, hay-like, smoky, slightly floral and spicy
COMMON EXTRACTION METHOD: Steam distilled, hydrodistilled, supercritical, or solvent extracted from the leaves or tobacco waste (midribs, leaf waste and dust)
POSSIBLE SUBSTITUTE OILS: Aspen, creeping thyme, helichrysum, lavender
BLENDS WELL WITH: Angelica, bergamot, cassia, cedarwood, chamomile (German, Roman), champaca, cinnamon, cistus, clary sage, clove, frangipani, frankincense, helichrysum, lavender, lemon, lime, patchouli, Peru balsam, rose, sandalwood, tangerine, vetiver, ylang ylang
SUBCELLULAR LOCALIZATION | EPIGENETIC INFLUENCE: Currently unknown | Currently unknown
RECOMMENDED DILUTION RANGE: *Essential Oil*—2%–20%; 33% for some conditions; *Absolute*—< 5%

PRIMARY CONSTITUENTS:[4183,4184,4185,4186,4187,4188,4189]

Neophytadiene/Solanone CT

Neophytadiene	1.6%–42.2%
Solanone	9.5%–29.5%
Pentadecanal	0.0%–7.9%
Megastigmatrienone-4	0.0%–7.0%
Epoxylabdenol II	0.0%–6.3%
Tetradecanoic Acid	0.0%–4.7%
(Z)-Beta-Damascenone	0.4%–4.1%
6,10,14-Trimethyl-2-Pentadecanone	0.0%–3.3%

(Z)-11-Tetradecenyl Acetate CT

(Z)-11-Tetradecenyl Acetate	53.0%–74.7%
Methyl eugenol	12.8%–38.1%
Solanone	4.9%–7.9%

Phytol CT

Phytol	16.4%
P-Xylene	12.4%
1,2,3-Trimethyl-Benzene	8.7%
Cyclohexane	8.4%
Nonane	4.4%
Decane	3.6%
Farnesol	3.1%
1-Naphthalenepropanol	3.1%
9-Methyl-Nonadecane	3.0%

Absolute

Neophytadiene	2.9%–5.8%
Stigmasterol	1.5%–4.1%
Cembra-2,7,11-Trien-4,6-diol (isomer)	2.0%–3.8%

Supercritical

Nonacosane	8.2%–28.4%
Neophytadiene	4.0%–15.0%
Nicotine	tr–12.0%
Di-n-Butylphthalate	3.3%–11.7%
Octasosane	4.3%–9.5%
Hentriacontane	tr–8.8%
Epoxylabdenol II	tr–8.2%
Eicosane	0.0%–8.2%
Tricosane	4.1%–7.9%
Solanesol	tr–7.4%
Heptacosane	2.8%–6.9%
Hexacosane	2.8%–6.3%
Sclareolide	0.0%–5.9%
Di-(2-Ethylhexyl) Phthalate	tr–4.8%
Dotriacontane	tr–3.9%

Note: Distillation and solvent extraction appears to be nicotine-free based on the available studies; however, supercritical extraction may also pull nicotine from the leaves, usually less than 2%, but as high as 12%.

OTHER CONSTITUENTS: *Neophytadiene/Solanone CT*—Camphor, (Z)-beta-damascone, geranyl acetate, geranyl acetone, gamma-gurjunene, pentadecanoic acid, germacrene D, spathulenol, 1-teradecene, ethyl linoleate, heptadecanal, farnesyl acetate, farnesyl acetone, cembrene, hexadecanoic acid, thunbergol, phytol, duvatriendiol, epoxylabdenol I, epoxylabdenol III; *(Z)-11-Tetradecenyl Acetate CT*—Geranyl acetone, 3-decen-1-yne, 5-ethyl-3-heptanone, n-octane, 4,6,8-trimethylnonane; *Phytol CT*—3-methyl-octane, 1-ethyl-3-methylcylohexane, 3,5-dimethyl-3-heptane, (1-methylethyl)-cyclohexane, 2,6-dimethyl-octane, propyl benzene, 1-ethyl-3-methyl-benzene, 1-methyl-2-propyl-cyclohexane, 1,2,5-trimrthyl-benzene, 4-methyl-decane, buthyl-cyclohexane, decahydro- Naphthalene, 1-methyl-2-(1-methylethyl)-benzene, (E)-2-tetradecaen-1-ol, 1-ethyl-2,4-dimethyl-benzene, undecane, 6,8-nonadien-2-one, isocaryophyllene, 2,6-dimethyl-heptadecane, tridecanal, thunbergol, 9-dodecyltradecahydro-anthracene, docosane; *Supercritical*—Hexadecanoic acid, epoxylabdenol I, epoxylabdenol III, heneicosane, pentacosane, tritriacontane, docosane; *Absolute*—Myristic acid, hexahydrofarnesyl acetate, palmitic acid, cembrene (14-isoprenyl-3,7,11-trimethylcyclotetradeca-1,3,6,10-tetraene), cembra-2,7,11-trien-4,6-diol, linoleic acid, 8,12-epoxy-13-hydroxy-labd-14-ene (8,12-epoxylabd-14-en-13-ol), oleic acid, sclareol, linolenic acid (9,12,15-18:3), ethyl ester, beta-levantenolide, squalene, cholesterol, methylsqualene, beta-sitosterol, fucosterol, beta-amyrin, 9,19 cyclolanost-24-en-3-ol (3-beta), alpha-tocopherol (isomer), alpha1-sitosterol, stigmast-8(14)en-3-beta-ol, stigmast-22-en-3-one, 4-methyl-, (4-alpha), stigmastane-3,6-dione (5-alpha)

PREFERRED COMPOSITION FOR CLINICAL USE:

Constituent	
Neophytadiene	20%–43%
Solanone	15%–30%
Epoxylabdenol II	2%–8%
(Z)-Beta-Damascenone	0.5%–5%
Epoxylabdenol I & III	1%–5%
6,10,14-Trimethyl-2-Pentadecanone	0.5%–5%

REPORTED THERAPEUTIC PROPERTIES: Perfumery, antimicrobial, wound healing, soothes irritated skin, heals bruises, antirheumatic, analgesic, relieves headache, supports hair and scalp health, antidepressant

CAUTIONS:

Neophytadiene/Solanone CT

■ None currently known, although research regarding the safety of the two primary constituents neophytadiene and solanone is lacking. A stinging nettle extract with neophytadiene (27.0%) as the primary constituent was nontoxic in rats up to the tested concentration of 800 mg/kg body weight.[4190]

Phytol CT

■ Avoid in individuals with Refsum disease (a neurological disorder caused by the overaccumulation of phytanic acid in cells and tissues) or who are AMACR-deficient (alphamethylacyl-CoA racemase-deficient) due to (E)-phytol content. It is believed that humans can convert phytol to phytanic acid. Animal research suggests that phytol can cause liver inflammation, liver failure, and death in AMCAR-deficient mice. AMACR is an enzyme found in the mitochondria that plays a role in the breakdown of fatty acids. AMACR deficiency causes neurological problems (seizures, cognitive decline, migraines, encephalopathy, loss of sensation due to nerve damage, muscle stiffness, lack of coordination, and vision problems) that begin in adulthood and progressively get worse. Phytol may also weakly promote tumor formation according to animal research.

(Z)-11-Tetradecenyl Acetate CT

■ Not recommended for therapy. (Z)-11-Tetradecenyl Acetate is a chemical pesticide, and high levels of methyl eugenol make this oil undesirable for safety reasons.

Absolute

■ Avoid oral consumption. Absolutes contain solvent residue and should not be taken orally.

Supercritical

- Not recommended due to the presence of nicotine.

Selected Evidence:

- □ *In vitro* research (three studies by the same group of researchers) demonstrated that tobacco essential oil CT neophytadiene/solanone inhibited *E. coli, S. aureus,* and *P. aeruginosa.*[4191,4192,4193]

TONKA ABSOLUTE (Cumaru, Kumaru, Tonquin)

Dipteryx odorata (Aubl.) Willd., *Coumarouna odora* Aubl., *C. odorata* Aubl.

Family: Fabaceae (Leguminosae)
Note: Base-Middle
Aroma Intensity: Medium
Aroma: Sweet, nutty, herbaceous, hay-like, tobacco-like, caramel-like, hint of vanilla
Common Extraction Method: Solvent extracted from the beans (seeds); may also be extracted from the flowers
Possible Substitute Oils: Peru balsam, cassia, ylang ylang
Blends Well With: Bergamot, cassia, cinnamon, clary sage, clove, grapefruit, helichrysum, jasmine, lavandin, lavender, neroli, patchouli, rose, sandalwood, vetiver, ylang ylang
Subcellular Localization | Epigenetic Influence: Currently unknown | Currently unknown
Recommended Dilution Range: < 10%

Primary Constituents:[4194,4195,4196]

Beans	
Coumarin	38.7%–58.7%
Ethyl Melilotate	3.9%–4.3%

Flowers	
Germacrene D	34.2%
Bicyclogermacrene	14.3%
Spathulenol	12.4%
Alpha-Cadinol	4.3%

Commercial (Beans)	
Benzyl Benzoate	20.0%–30.0%
Tetramethyl Acetyloctahydronaphthalenes	2.0%–5.0%
Methyldihydrojasmonate	2.0%–5.0%
Ethylene Brassylate	2.0%–5.0%

Other Constituents: *Beans*—Melilotc acid; *Commercial (Beans)*—Hexamethylindanopyran, coumarin, butylphenyl methylpropional; *Flowers*—Alpha-pinene, (E)-beta-ocimene, nonanal, alpha-copaene, beta-elemene, beta-caryophyllene (E)-beta-farnesene, delta-cadinene, globulol, cubenol, alpha-muurolol, (E,E)-farnesol, palmitic acid

Preferred Composition for Clinical Use:
Too great of variance in data to determine.

REPORTED THERAPEUTIC PROPERTIES: **Perfumery**, deodorant, antiseptic, expectorant, reduces blood pressure, insecticidal, aphrodisiac, promotes feelings of peace and tranquility, heals traumatic memories

CAUTIONS:

- Use cautiously orally—maximum of 5 drops daily for adults due to high coumarin content possible in bean absolute (coumarin content increases during drying). Animal research suggests that coumarin may be toxic to the liver and lungs when taken orally in high doses.[4197] The total safe daily intake was established at up to 0.1 mg/kg body weight daily by the European Food Safety Authority.[4198]
- May interact with antidepressants (MAOIs and SSRIs), based on inhibition of MAO by coumarin.[4199]
- Some texts warn against the use of coumarin-containing oils with blood-thinning drugs, but according to the available research, only coumarin derivatives inhibit platelet aggregation or act as anticoagulants, not coumarin itself.[4200,4201]

SELECTED EVIDENCE:

- None found.

TOOTHACHE TREE
(Southern Prickly Ash, Hercules' Club)

Zanthoxylum clava-herculis L., *Z. macrophyllum* Nutt.

FAMILY: Rutaceae
NOTE: Top
AROMA INTENSITY: Medium
AROMA: Citrusy, fruity, orange-like
COMMON EXTRACTION METHOD: Steam distilled from the leaves
POSSIBLE SUBSTITUTE OILS: Orange, guava leaf, clove, cape yellowwood
BLENDS WELL WITH: Basil (sweet), bergamot, cassia, cinnamon, citronella, clove, frankincense, grapefruit, lavender, lemon, lemongrass, lime, melissa, nutmeg, orange, tangerine
SUBCELLULAR LOCALIZATION | EPIGENETIC INFLUENCE: Currently unknown | Currently unknown
RECOMMENDED DILUTION RANGE: 3%–50%; neat for some conditions

PRIMARY CONSTITUENTS:[4202]

Limonene	43.6%–73.0%
1,8-Cineole	12.9%–43.3%
Linalool	0.0%–11.3%
Alpha-Thujene	tr–5.2%
Gamma-Terpinene	0.6%–4.2%
Alpha-Pinene	0.0%–3.8%
Alpha-Terpineol	tr–3.5%

OTHER CONSTITUENTS: Cis-3-hexenol, sabinene, myrcene, nonanal, cis-sabinene hydrate acetate, alpha-terpinyl acetate, beta-caryophyllene

PREFERRED COMPOSITION FOR CLINICAL USE:
Insufficient samples/data to determine.

Reported Therapeutic Properties: Antirheumatic, anti-inflammatory, analgesic, antispasmodic, soothes toothache, relieves stomachache, expels excess gas, reduces diarrhea, supports kidney function, aids cardiovascular function, anti-infectious, eases cough, aids respiratory function, reduces fever, aids circulation, diuretic, antidepressant, uplifting

Cautions:

■ Avoid with epilepsy and Parkinson's disease due to 1,8-cineole content. May exacerbate or cause seizures/convulsions or reduce seizure-medication efficacy based on 1,8-cineole content.[4203,4204,4205,4206]

■ May weakly interfere with the enzymes responsible for metabolizing medications (NSAIDs, proton-pump inhibitors, acetaminophen, antiepileptics, immune modulators, blood-sugar medications, blood pressure medications, antidepressants, antipsychotics, diabetic medications, antihistamines, antibiotics, and anesthetics) due to 1,8-cineole content.[4207,4208,4209]

■ May interfere with pentobarbital and other barbiturates (medications for anxiety and insomnia), based on 1,8-cineole content.[4210,4211]

Selected Evidence:

- □ None found.

TREEMOSS ABSOLUTE (Cedar Moss)

Pseudevernia furfuracea (L.) Zopf, *Lichen furfuraceus* L., *Borrera furfuracea* (L.) Ach., *Evernia furfuracea* (L.) W. Mann, *Tenorea furfuracea* (L.) Tornab., *Physcia furfuracea* (L.) DC., *Parmelia furfuracea* (L.) Ach.

Family: Parmeliaceae
Note: Base
Aroma Intensity: Medium
Aroma: Woody, herbaceous, green, seaweed-like, slightly sweet
Common Extraction Method: Solvent extracted from the lichens
Possible Substitute Oils: Oak moss
Blends Well With: Anise, bay laurel, cedarwood, clary sage, cypress, eucalyptus, geranium, ginger, jasmine, lavender, lemon, lime, neroli, orange, palmarosa, patchouli, petitgrain, rose, tea tree (melaleuca), vetiver, ylang ylang
Subcellular Localization | Epigenetic Influence: Currently unknown | Currently unknown
Recommended Dilution Range: < 0.2%

Primary Constituents:[4212,]

Traditional Absolute	
Beta-Sitosterol + Stigmasterol	4.1%
Dehydroabietic acid	4.1%
Methyl Beta-Orinolcarboxylate	3.7%
Physodone	2.4%

Resin Acid Free Absolute	
Beta-Sitosterol + Stigmasterol	8.0%
Physodone	4.7%
Methyl Beta-Orinolcarboxylate	3.6%

OTHER CONSTITUENTS: *Traditional Absolute*—2′-O-methylphysodone; *Resin Acid Free Absolute*—Ethyl chlorohaematommate, 2′-O-methylphysodone

PREFERRED COMPOSITION FOR CLINICAL USE:
Insufficient data/samples to determine.

REPORTED THERAPEUTIC PROPERTIES: Perfumery, antimicrobial, analgesic, anti-inflammatory, reduces high cholesterol, stress management, releases traumatic memories and emotions, relieves anxiety

CAUTIONS:
- Avoid oral consumption. Absolutes contain solvent residue and should not be taken orally.
- Treemoss absolute can be irritating to the skin and poses a high risk of sensitization—sensitization to oxidized resin acids and common constituents in treemoss absolute are documented.[4213,4214]

SELECTED EVIDENCE:
- ☐ None found.

TROPICAL BLACK SAGE (Black Sage, Wild Sage, Erva-baleeira)

Varronia curassavica Jacq., *Cordia curassavica* (Jacq.) Roem. & Schult.

FAMILY: Boraginaceae
NOTE: Middle
AROMA INTENSITY: Medium-Strong
AROMA: Herbaceous, earthy, woody
COMMON EXTRACTION METHOD: Steam distilled or hydrodistilled from the leaves
POSSIBLE SUBSTITUTE OILS: Copaiba, sandalwood, black pepper, melissa, turmeric, ginger
BLENDS WELL WITH: Angelica root, black pepper, caraway, chamomile (German, Roman), copaiba, coriander, cumin, ginger, lavender, nutmeg, perilla, rosemary, sage, turmeric
SUBCELLULAR LOCALIZATION | EPIGENETIC INFLUENCE: Currently unknown | Currently unknown
RECOMMENDED DILUTION RANGE: 3%–50%; neat for some conditions

PRIMARY CONSTITUENTS:[4215,4216,4217,4218,4219,4220,4221,4222]

Shyobunol CT	
Shyobunol	14.3%–28.4%
Germacrene D-4-ol	2.8%–12.6%
7-Cyclodecen-1-ona,7-Methylene-10-(1-propyl)	0.0%–11.5%
Beta-Caryophyllene	2.6%–9.0%
Alpha-Turmerone	0.0%–8.2%
Bicyclogermacrene	3.4%–5.7%
Alpha-Cadinol	1.1%–5.6%
(2E,6E)-Methyl Farnesoate	0.0%–4.6%
Ledol	0.0%–4.5%

Caryophyllene Oxide CT	
Caryophyllene Oxide	15.0%–23.1%
Beta-Caryophyllene	7.7%–14.7%
Alpha-Pinene	1.1%–14.4%
Aromadendrene	0.0%–6.8%
9-Epi-Beta-Caryophyllene	0.4%–6.4%
Humulene Epoxide II	3.1%–5.0%
Delta-Cadinene	2.5%–4.1%
Spathulenol	0.9%–3.6%
Cis-4-Thujanol Acetate	0.0%–3.5%

(2E,6E)-Farnesyl Acetate	0.0%–4.3%
T-Muurolol	0.0%–4.2%
Shyobunone IV	0.0%–3.0%

Shyobunones CT

Shyobunone II	12.0%–14.8%
Shyobunone IV	11.8%–14.0%
1-Epi-Cubenol	0.0%–13.3%
7-Cyclodecen-1-ona.7-Methylene-10-(1-propyl)	10.7%–12.9%
Shyobunone I	3.4%–5.9%
Shyobunol	2.6%–5.1%
Germacrene D	1.4%–4.9%
Beta-Caryophyllene	4.0%–4.1%
Bicyclogermacrene	1.9%–3.4%
Spathulenol	1.5%–3.1%

Beta-Caryophyllene/Turmerone CT

Beta-Caryophyllene	12.8%–31.1%
Beta-Phellandrene	0.0%–21.4%
Bicyclogermacrene	2.2%–12.1%
Alpha-Turmerone	0.0%–18.6%
7-Cyclodecen-1-ona,7-Methyl-3-Methylene-10-(1-Propyl)	0.0%–11.4%
Alpha-Humulene	3.3%–10.3%
ar-Turmerone	0.0%–9.3%
Delta-Cadinene	0.0%–7.8%
Caryophyllene Oxide	0.6%–7.2%
Germacrene D	1.2%–6.8%
Shyobunol	0.0%–6.5%
Alpha-Gurjunene	0.0%–6.4%
Alpha-Pinene	0.8%–6.1%
Alpha-Zingiberene	0.0%–5.7%
Spathulenol	1.5%–5.2%
Alpha-Cadinol	0.0%–4.9%
Germacrene D-4-ol	0.0%–4.8%
Para-Cymene	0.0%–4.1%
T-Muurolol	0.0%–4.0%
Globulol	0.0%–3.8%
Shyobunone IV	0.0%–3.4%
Beta-Sesquiphellandrene	0.0%–3.0%

Elemol CT

Elemol	37.3%
Beta-Eudesmol	19.2%
Spathulenol	11.3%
Cadina4(5),10(14)diene	7.9%
Beta-Selinene	3.8%

Beta-Caryophyllene CT

Beta-Caryophyllene	22.6%–27.4%
Gamma-Elemene	14.9%–17.2%
Gamma-Muurolene	9.1%–9.8%
Alpha-Pinene	6.0%–9.5%
Alpha-Humulene	3.3%–4.7%
Aristolene	0.0%–4.6%
Spathulenol	2.1%–4.3%
Beta-Pinene	3.1%–4.0%
Beta-Elemene	2.8%–3.5%

Beta-Caryophyllene/Epiglobulol CT

Beta-Caryophyllene	19.9%
Epiglobulol	15.3%
7-Cyclodecen-1-ona.7-Methylene-10-(1-propyl)	8.8%
Alpha-Humulene	6.2%
Tricyclene	6.2%
Camphene	4.7%
Bicyclogermacrene	4.2%
(2E,6E)-Methyl Farnesoate	3.9%
Alpha-Pinene	3.6%
Alpha-Cedrene Epoxide	3.0%

Alpha-Pinene CT

Alpha-Pinene	17.4%–42.4%
Beta-Caryophyllene	9.3%–19.2%
Alpha-Gurjunene	0.0%–11.7%
(Z)-Alpha-Bergamotene	0.0%–10.0%
Alpha-Santalene	0.0%–9.2%
Gamma-Terpinene	0.0%–8.8%
Alloaromadendrene	0.5%–8.2%
Beta-Pinene	0.2%–5.6%
Bicyclogermacrene	0.0%–5.0%
Delta-Elemene	0.0%–4.9%
Alpha-Santalol	0.0%–4.6%
Alpha-Humulene	3.1%–4.5%
Alpha-Bergamotol	0.0%–3.9%
Limonene	0.0%–3.7%
Alpha-Bisabolone Oxide A	0.0%–3.7%
Cubenol	0.0%–3.4%
Helifolenol B	0.0%–3.1%

Alpha-Santalene CT

Alpha-Santalene	20.2%–20.4%
(E)-Alpha-Santalol	6.0%–8.7%
(E,E)-Alpha-Bergamotol	1.3%–8.1%

Calamenene	3.7%
Androst-1-ene-3,17-diol,1-methyl-, (3beta,5alpha,17beta)-	3.1%

Germacrene D-4-ol

Germacrene D-4-ol	42.0%
Syobunol	9.8%
Alpha-Cadinol	5.0%
Beta-Caryophyllene	4.5%
Bicyclogermacrene	4.5%
T-Muurolol	4.3%

Sabinene CT

Sabinene	38.9%
Beta-Caryophyllene	7.9%
Terpinen-4-ol	5.2%
Germacrene D	4.0%
Bicyclogermacrene	3.5%
Gamma-Terpinene	3.3%
Alpha-Pinene	3.2%
Alpha-Gurjunene	3.0%

7-Cyclodecen-1-ona,7-Methyl-3-Methylene-10-(1-Propyl) CT

7-Cyclodecen-1-ona,7-Methyl-3-Methylene-10-(1-Propyl)	16.7%–50.2%
ar-Turmerone	0.0%–14.7%
Syobunone IV	1.8%–12.0%
Shyobunol	0.3%–10.7%
Beta-Caryophyllene	2.3%–10.4%
Alpha-Turmerone	0.0%–9.8%
Dehydroxy Isocalamendiol	0.0%–7.0%
Kessane	0.0%–6.8%
Spathulenol	0.0%–6.2%
Germacrene D	0.0%–5.1%
Delta-Cadinene	0.0%–5.0%
Bicyclogermacrene	0.0%–4.5%
Beta-Bisabolene	0.0%–4.0%
T-Muurolol	0.0%–3.6%
Beta-Phellandrnee	0.0%–3.5%
Germacrene D-4-ol	0.0%–3.4%
Alpha-Cadinol	0.0%–3.3%
(2E,6E)-Methyl Farnesoate	0.0%–3.3%
Shyobunone II	0.0%–3.2%

Sesquisabinene	5.1%–6.4%
Beta-Bisabolene	3.8%–4.9%
Beta-Atlantol	0.0%–3.7%
Delta-Cadinene	2.2%–3.2%
Di-Epi-Alpha-Cedrene	0.0%–3.1%

Tricyclene CT

Tricyclene	22.2%–36.0%
Camphene	16.6%–27.4%
Beta-Caryophyllene	6.1%–14.3%
Viridiflorol	0.0%–9.0%
Alpha-Pinene	3.5%–5.6%
Alpha-Humulene	2.6%–4.6%
Bicyclogermacrene	1.5%–4.2%
Sesquisabinene	0.0%–4.1%

ar-Turmerone CT

ar-Turmerone	15.4%
Beta-Caryophyllene	10.4%
Alpha-Turmerone	10.3%
Spathulenol	4.8%
Alpha-Humulene	3.8%

Alpha-Zingiberene CT

Alpha-Zingiberene	20.8%–35.8%
Beta-Caryophyllene	5.2%–20.2%
Beta-Sesquiphellandrene	10.7%–6.2%
Bicyclogermacrene	2.6%–10.7%
7-Cyclodecen-1-ona,7-Methyl-3-Methylene-10-(1-Propyl)	0.0%–8.8%
Alpha-Pinene	1.1%–7.7%
Beta-Phellandrene	0.0%–6.8%
Alpha-Humulene	1.8%–5.6%
Beta-Bisabolene	0.0%–4.5%
Germacrene D	1.0%–4.4%
Spathulenol	0.8%–4.4%
Beta-Pinene	0.0%–4.0%
ar-Curcumene	1.6%–3.8%
Shyobunol	0.0%–3.4%
Alpha-Cadinol	0.0%–3.0%

OTHER CONSTITUENTS: *Shyobunol CT*—Beta-elemene, alpha-humulene, alloaromadendrene, caryophyllene oxide, cubenol; *Shyobunones CT*—Delta-elemene, beta-bourbonene, beta-elemene, alpha-gurjunene, alpha-humulene, alloaromadendrene, germacrene D-4-ol, caryophyllene oxide, viridiflorol, alpha-cadinol; *Caryophyllene Oxide CT*—Beta-pinene, 1,8-cineole, beta-thujone (< 2.1%), camphenol, alpha-campholenal, trans-verbenol, cis-beta-terpineol, verbenone, trans-carveol, isobornyl acetate, neo-iso-verbenol acetate, delta-elemene, verbenol acetate, citronellyl acetate, beta-elemene, 1.7-di-epi-alpha-cedrene, alloaromadendrene, alpha-humulene, gamma-curcumene, germacrene D, pepizonarene, germacrene A, valerianol, 14-hydroxy-9-epi-beta-caryophyllene, ar-turmerone, beta-bisabolol, (Z)-alpha-santalol, acorenone, 14-hydroxy-alpha-humulene, iso-longifolol, pentadecanol; *Beta-Caryophyllene CT*—Sabinene, delta-3-carene, beta-bourbonene, caryophyllene oxide; *Beta-Caryophyllene/Epiglobulol CT*—Beta-pinene, trans-murola-3.5-diene, alloaromadendrene, beta-selinene, gamma-cadinene, cubebol, delta-cadinene, trans-cadina-1.4-diene, spathulenol, caryophyllene oxide, humulene epoxide II, alpha-cadinol, shyobunol; *Beta-Caryophyllene/Turmerone CT*—Tricyclene, camphene, camphene, beta-pinene, limonene, gamma-terpinene, delta-elemene, beta-bourbonene, beta-elemene, (E)-alpha-bergamotene, alloaromadendrene, ar-curcumene, beta-bisabolene, ar-turmerol, viridiflorol, ledol, 1-epi-cubenol, dehydroxy isocalamendiol, alpha-acorenol, (Z)-gamma-atlantone, curlone, (Z)-alpha-atlantone(2E,6E)-methyl farnesoate; *Alpha-Pinene CT*—Alpha-thujene, sabinene, beta-phellandrene, 1,8-cineole, beta-elemene, sesquithujene, beta-gurjunene, gamma-muurolene, bicyclogermacrene, beta-bisabolene, gamma-cadinene, caryophyllene oxide, (E)-beta-farnesene; *Elemol CT*—4-Isopropyl-3,7-dimethyl-3a,3b,4,5,6,7-hexahydro-1-H-cyclopenten [1,3] cyclopropa [1,2] benzene, isocaryophyllene, 5-(2,3-dimethyltricyclo 2.2.1.02, 6-hept-3-y1)-2-methyl-2-penten-1-ol, hexahydro-2,5,5-trimethyl-2H-2,4a-ethanonaphthalen-8(5H)-one; *Alpha-Santalene CT*—Alpha-pinene, bornyl acetate, delta-elemene, alpha-copaene, sesquithujene, (Z)-alpha-bergamotene, trans-beta-farnesene, alpha-humulene, trans-muurola-4(15),5-diene, alpha-bulnesene, valencene, germacrene A, cis-gamma-bisabolene, cubebol, cis-sesquisabinene hydrate, caryophyllene oxide, trans-sesquisabinene hydrate, beta-eudesmol, selin-11-en-4-alpha-ol, trans-epi-beta-santalol, trans-alpha-santalol acetate; *Germacrene D-4-ol*—Delta-elemene, beta-elemene, alpha-humulene, alloaromadendrene, germacrene D, gamma-cadinene, spathulenol, caryophyllene oxide, ledol; *Sabinene CT*—Camphene, alpha-terpinene, limonene, cis-sabinene hydrate, trans-sabinene hydrate, alpha-cubebene, alpha-copaene, beta-elemene, trans-murola-3.5-diene, alpha-humulene, beta-selinene, delta-cadinene, trans-cadina-1.4-diene, caryophyllene oxide, ledol, cubenol; *Tricyclene CT*—Limonene, beta-elemene, germacrene D, beta-bisabolene, delta-cadinene, 1,8-cineole, delta-elemene, beta-bourbonene, spathulenol, caryophyllene oxide, ledol, gamma-elemene; *ar-Turmerone CT*—Tricyclene, camphene, (E)-alpha-bergamotene, gamma-elemene, alpha-zingiberene, bicyclogermacrene, beta-sesquiphellandrene, alpha-turmerol, caryophyllene oxide, alpha-acorenol, curlone, (Z)-alpha-atlantone; *Alpha-Zingiberene CT*—Sabinene, para-cymene, 1,8-cineole, gamma-terpinene, beta-elemene, (E)-alpha-bergamotene, gamma-elemene, alloaromadendrene, shyobunone IV, globulol, germacrene D-4-ol, caryophyllene oxide, viridiflorol, germacra-4, curlone, (2E,6E)-methyl farnesoate; *7-Cyclodecen-1-ona,7-Methyl-3-Methylene-10-(1-Propyl) CT*—Tricyclene, camphene, para-cymene, gamma-terpinene, delta-elemene, beta-bourbonene, beta-elemene, alpha-gurjunene, (E)-alpha-bergamotene, alpha-humulene, alloaromadendrene, shyobunone I, caryophyllene oxide, alpha-acorenol, curlone, (Z)-alpha-atlantone

Preferred Composition for Clinical Use:
Too great of variance in samples to determine, although types rich in beta-caryophyllene and alpha-humulene are preferred.

Reported Therapeutic Properties: **Anti-inflammatory**, analgesic, antirheumatic, antiarthritic, eases fever, soothes tendonitis, wound healing, relieves chronic skin conditions and itching, aids respiratory function, antimicrobial, anti-infectious, antiparasitic, digestive aid, relieves stomachache, heals stomach ulcers, anticancer, aids memory function, relieves anxiety, stress management

CAUTIONS:

Alpha-Zingiberene CT

■ May interfere with the enzymes responsible for metabolizing medications (NSAIDs, proton-pump inhibitors, acetaminophen, antiepileptics, immune modulators, blood-sugar medications, blood pressure medications, antidepressants, antipsychotics, diabetic medications, antihistamines, antibiotics, and anesthetics) due to alpha-zingiberene content.[4223]

All Other CTs

■ None currently known.

SELECTED EVIDENCE:

- □ *In vitro* research demonstrated that tropical black sage CT elemol essential oil inhibited *Artemia salina, Staphylococcus aureus, Staphylococcus epidermidis, Bacillus subtilis, Sarcina lutea, Vibrio cholera, Vibrio cholera, Vibrio cholera (two strains), Escherichia coli, Aspergillus niger, Trichophyton mentagrophytes, Fusarium sporotrichum, Fusarium moniliforme*, and *Rhyzoctonia solan*.[4224]
- □ Tropical black sage essential oil (various CTs) demonstrated antiprotozoal activity against the parasite responsible for white spot disease in fish—*Ichthyophthirius multifiliis*.[4225] The germacrene D-4-ol CT was the most effective.

TUBEROSE ABSOLUTE (Polyanthus Lily)

Polianthes tuberosa L., *P. tuberosa* L. cv. Mahallati, *P. tuberosa* L. cv. Double

FAMILY: Agavaceae

NOTE: Middle-Base

AROMA INTENSITY: Strong

AROMA: Sweet, floral, honey-like

COMMON EXTRACTION METHOD: Solvent extracted from the flower petals

POSSIBLE SUBSTITUTE OILS: Jasmine, rose

BLENDS WELL WITH: Bergamot, chamomile (German, Roman), clary sage, clove, coriander, frankincense, geranium, ginger, grapefruit, jasmine, lavender, lemon, lime, Mediterranean mandarin, neroli, orange, palmarosa, patchouli, petitgrain, rose, rosewood, sandalwood, tangerine, vetiver, ylang ylang

SUBCELLULAR LOCALIZATION | EPIGENETIC INFLUENCE: Currently unknown | Currently unknown

RECOMMENDED DILUTION RANGE: 1%–3%

PRIMARY CONSTITUENTS:[4226,4227,4228,4229]

cv. Double (Hexane I)

Methyl Benzoate	12.7%–15.4%
Benzyl Benzoate	8.9%–12.9%
Methyl Anthranilate	4.4%–8.7%
Indole	1.4%–5.6%
Geraniol	1.2%–5.5%
Benzyl Salicylate	2.3%–4.7%
1,8-Cineole	3.5%–4.6%
Farnesol	1.7%–4.5%

cv. Double (Hexane II)

Benzyl Benzoate	24.3%
Pentacosane	19.2%
Delta-Decalactone	15.0%
1-Hexadecene	8.9%
2,4,-Di-Tert-Butylphenol	8.4%
Ecosanol	5.0%
Heptacosane	4.6%
(Z)-3-Hexenyl 2-Oxopropanoate	3.7%

cv. Double (Petroleum Ether)

Pentacosane	29.4%
Delta-Decalactone	18.1%
Heptacosane	12.5%
Benzyl Benzoate	10.3%
(E)-Methyl Isoeugenol	6.1%
Methyl Isoeugenol	4.9%
Methyl Anthranilate	4.2%
Alpha-Farnesol	4.1%
Tricosane	3.7%

cv. Mahallati (Hydrodistilled + Toluene)

Methyl Benzoate	1.1%–40.5%
n-Docosane	0.7%–14.9%
Ethyl Benzene	3.7%–13.3%
p-Xylene	3.4%–13.3%
n-Eicosane	1.8%–13.2%
Methyl Anthranilate	1.2%–11.5%
1,8-Cineole	2.7%–11.3%
(E)-Methyl Isoeugenol	0.3%–9.6%
2(E)-6-(E)-Methyl Farnesoate	tr–7.6%
n-Nonadecane	1.6%–7.0%
Benzyl Benzaote	1.0%–6.7%
n-Heneicosane	tr–6.6%
2-(Z)-6-(E)-Farnesol	1.5%–5.0%
n-Octadecane	1.5%–4.9%
Indole	0.7%–4.4%
Methyl Salicylate	0.1%–4.0%
Alpha-Terpineol	0.5%–3.3%

cv. Mahallati (Accelerated Hexane)

Benzyl Benzoate	tr–18.3%
Methyl Benzoate	tr–7.5%
(E)-Methyl Isoeugenol	tr–6.1%
1,8-Cineole	tr–3.7%

cv. Double (Enfleurage)

Methyl Benzoate	30.2%–44.9%
(Z)-3-Hexenyl 2-Oxopropanoate	0.0%–27.4%
Benzyl Benzoate	3.8%–23.6%
Delta-Decalactone	0.0%–13.3%
Methyl Salicylate	7.2%–12.1%
(E)-Methyl Isoeugenol	1.8%–8.8%
Methyl Anthranilate	0.0%–4.5%
1-Tetradecene	0.0%–4.2%
(Z)-Nerolidol	0.0%–3.4%

cv. Mahallati (Hexane)

Benzyl Salicylate	10.0%–26.8%
Methyl Linoleate	1.0%–20.8%
Nerylacetone	tr–18.6%
Ethyl Benzoate	7.8%–16.9%
n-Nonane	1.0%–16.9%
Benzyl Benzoate	tr–16.3%
Ethyl Hexadecanoate	1.8%–8.7%
Ethyl Isovalerate	0.8%–7.5%
(E)-Nerolidol Acetate	tr–7.3%
n-Octadecanol	tr–5.3%
n-Heptanol	0.6%–4.8%
Butyl Acetate	2.5%–3.2%
Linalool	tr–3.2%
Delta-Decalactone	tr–3.2%

OTHER CONSTITUENTS: *cv. Double (Hexane I)*—benzaldehyde, Eugenol, beta-pinene, benzoic acid, nerol; *cv. Double (Hexane II)*—Methyl eugenol, methyl isoeugenol, (E)-methyl isoeugenol, alpha-farnesol, benzyl salicylate, tricosane; *cv. Double (Hexane II; Petroleum Ether)*—1-Tetradecene, methyl eugenol, (Z)-methyl isoeugenol, 1-hexadecene, benzyl salicylate; *cv. Double (Enfleurage)*—Indole, geranial, methyl eugenol, (Z)-beta-farnesene, 2,4-decadien-1-al, (E)-methyl isoeugenol, 2-heptadecanone; *cv. Mahallati (Hydrodistilled + Toluene)*—(Z)-beta-farnesene, germacrene D, n-hexadecane, 2-trans-6-trans-Farnesyl acetate; *cv. Mahallati (Accelerated Hexane)*—Alpha-terpineol, methyl linoleate; *cv. Mahallati (Hexane)*—n-Hexanol, (E)-methyl isoeugenol

PREFERRED COMPOSITION FOR CLINICAL USE:
Insufficient and too great a variance in samples/data to determine.

REPORTED THERAPEUTIC PROPERTIES: Anti-inflammatory, antispasmodic, analgesic, improves circulation, nourishes and moisturizes the skin, reduces fine lines, blemishes, and wrinkles, deodorant, centers the mind, provides emotional strength, improves emotional depth, promotes serenity, sedating, relieves anxiety, aphrodisiac

CAUTIONS:

■ Avoid with children under age 12. The possibility of moderate methyl salicylate levels in tuberose absolute may increase the risk of a very serious and life-threatening illness called Reye syndrome in children under 12. Reye syndrome is sudden brain damage and liver dysfunction that most commonly occurs in children ages 4 to 12. It has an unknown cause, but it typically occurs in children who were given aspirin when they are experiencing a fever or recovering from flu-like symptoms or chicken pox.[4230]

■ Do not use during pregnancy or lactation. Salicylates cross the placental barrier and ingestion of 3 g of salicylate has caused abortion of a 33-week-old fetus eight days after ingestion.[4231] Methyl salicylate may also cause congenital abnormalities and fetal malformations according to animal research.[4232,4233] The American Academy of Pediatrics recommends salicylates be used cautiously during breastfeeding.[4234]

■ Avoid oral consumption. Absolutes contain solvent residue and should not be taken orally. In addition, ingestion of benzoic acids, benzoic salts (calcium, potassium, and sodium), benzaldehyde, benzyl acetate, benzyl benzoate, and benzyl alcohol may cause gastrointestinal irritation, vomiting, and diarrhea. The Joint Expert Committee for Food Additives (JECFA) has established an acceptable daily limit for the aforementioned constituents of 0–5 mg/kg body weight for adults to avoid GI discomfort. Based on this limit, it is suggested to limit the oral consumption to no more than 5 drops daily.

■ Avoid with epilepsy and Parkinson's disease due to potential moderate methyl salicylate and 1,8-cineole levels.[4235,4236,4237,4238,4239]

■ May interact with aspirin, blood pressure, antiplatelet, and anticoagulant medications, and increase the risk of bleeding among people with bleeding disorders both topically and orally.[4240,4241,4242,4243]

■ Avoid in individuals allergic to aspirin, methyl salicylate, or other NSAIDs; or those with salicylate sensitivities (methyl salicylate and benzyl salicylate). Methyl salicylate may be metabolized to the known NSAID, salicylic acid, by the liver.[4244]

SELECTED EVIDENCE:

- □ Inhalation of tuberose absolute (from a handkerchief) for fifteen to twenty minutes reduced test anxiety in seventh-grade students.[4245]

TURKISH THYME (Kekik)

Thymus revolutus Célak

FAMILY: Lamiaceae (Labiatae)
NOTE: Middle
AROMA INTENSITY: Strong
AROMA: Herbaceous, spicy, medicinal
COMMON EXTRACTION METHOD: Steam distilled or hydrodistilled from the aerial parts
POSSIBLE SUBSTITUTE OILS: Oregano, summer savory, mountain savory

BLENDS WELL WITH: Bergamot, grapefruit, lavandin, lavender, lemon, lime, mountain savory, orange, oregano, pine, rosemary, tangerine, tea tree, summer savory, thyme
SUBCELLULAR LOCALIZATION | EPIGENETIC INFLUENCE: Currently unknown | Currently unknown
RECOMMENDED DILUTION RANGE: 1.5%–20%; 50% for some conditions

PRIMARY CONSTITUENTS:[4246,4247]

Carvacrol	24.5%–43.1%
Gamma-Terpinene	0.7%–20.9%
Para-Cymene	13.9%–15.5%
Thymol	4.6%–15.4%
Borneol	0.3%–14.7%
Beta-Caryophyllene	3.6%–5.1%
Caryophyllene Oxide	1.4%–4.2%
Pulegone	0.0%–4.1%
Linalool	1.2%–3.0%

OTHER CONSTITUENTS: Alpha-thujene, alpha-pinene, camphene, 7-octen-4-ol, 1-octen-3-ol, myrcene, alpha-terpinene, limonene, alpha-thujone (< 0.7%), trans-thujon-4-ol, camphor, terpinen-4-ol, sabineol, linalyl propionate, p-cymen-8-ol

PREFERRED COMPOSITION FOR CLINICAL USE:
Insufficient samples/data to determine.

REPORTED THERAPEUTIC PROPERTIES: Antimicrobial, **antibacterial**, **antifungal**, antioxidant, anticancer, antispasmodic, anti-inflammatory, aids digestion, expels excess gas, supports respiratory function, eases cough, expels excess mucus, insect repellent, antidepressant, calming

CAUTIONS:

- Avoid during pregnancy and lactation. Oregano (also rich in carvacrol) may negatively affect embryonic development and encourage fetal cell death according to animal research.[4248]
- May interact with diabetic medications and cause low blood-sugar levels due to carvacrol content—carvacrol reduces blood glucose and enhances the antidiabetic activities of rosiglitazone in animals.[4249,4250,4251,4252]
- May interact with aspirin, blood pressure, antiplatelet, and anticoagulant medications, and increase the risk of bleeding among people with bleeding disorders, due to carvacrol and thymol content.[4253,4254,4255,4256,4257]
- May interact with anticholinergic (drugs used for asthma, incontinence, gastrointestinal cramps, muscular spasms, depression, and sleep disorders) and cholinergic medications (drugs used to reduce urinary retention, diagnose myasthenia gravis, and for glaucoma), based on thymol and carvacrol content.[4258,4259]
- May interact with antibiotics or antifungals and possibly enhance or decrease their effects due to carvacrol or thymol content.[4260,4261,4262,4263,4264]
- May irritate mucous membranes (eyes, mouth, nasal passages, vagina, rectum).

SELECTED EVIDENCE:

- Turkish thyme CT gamma-terpinene (full composition not reported) was toxic to lung (both parent line and drug-resistant), alveolar basal epithelial, and squamous carcinoma cancer cells.[4265]

- *In vitro* research demonstrated that Turkish thyme inhibits gram-positive (*B. brevis, B. cereus, B. subtilis, B subtilis* var. *niger, M. smegmatis. S. aureus*) and gram-negative bacteria (*A. hydrophila, E. coli, K. pneumoniae, M. morganii, P. mirabilis, P. aeruginosa, Y. enterocolitica*), and yeasts (*C. albicans, S. cerevisiae*).[4266]
- Turkish thyme exhibited moderate antioxidant activity in the DPPH assay.[4267]
- Turkish thyme (composition not reported) repelled the saltmarsh mosquito (*Ochlerotatus caspius*).[4268]

VIETNAMESE BALM
(Crested Latesummer Mint, Kinh Giới)

Elsholtzia ciliata (Thunb.) Hyl., *E. cristata* Willd., *E. patrinii* Kuntze, *Sideritis ciliata* Thunb.

FAMILY: Lamiaceae (Labiatae)
NOTE: Middle
AROMA INTENSITY: Medium
AROMA: Fresh, spicy to fresh, herbaceous, lemony
COMMON EXTRACTION METHOD: Steam distilled or hydrodistilled from the aerial parts or leaves
POSSIBLE SUBSTITUTE OILS: Tagetes, clove, melissa,
BLENDS WELL WITH: Bergamot, black pepper, cassia, cinnamon, clove, lavender, lemon, lemongrass, melissa, nutmeg, orange, patchouli, tangerine, vetiver, ylang ylang
SUBCELLULAR LOCALIZATION | EPIGENETIC INFLUENCE: Currently unknown | Currently unknown
RECOMMENDED DILUTION RANGE: 3%–20%; 50% for some conditions

PRIMARY CONSTITUENTS:[4269,4270,4271,4272,4273,4274,4275,4276,4277]

Dehydroelsholtzia Ketone CT	
Dehydroelsholtzia Ketone	63.3%–85.0%
Elsholtzia Ketone	5.7%–33.6%
Alpha-Dehydro-Elsholtzione	0.0%–5.7%
Para-Cymene	0.0%–4.3%
Perillene	0.0%–3.9%
Alpha-Humulene	0.0%–3.8%
Camphor	0.0%–3.6%

Citral CT	
Geranial	19.5%–26.5%
Neral	15.2%–20.5%
Limonene	4.1%–14.2%
(E)-Beta-Ocimene	%–14.0%
(Z)-Beta-Farnesene	6.2%–11.7%
Linalool	%–8.3%
1-Octen-3-ol	4.8%–7.1%
Nerol	3.5%–6.2%
Beta-Caryophyllene	3.0%–4.9%
Geraniol	0.0%–3.8%

Elsholtzia Ketone CT	
Elsholtzia Ketone	26.4%–86.0%
Dehydroelsholtzia Ketone	0.0%–30.1%
Unidentified Lactone	0.0%–25.8%
Citronellol	0.0%–7.7%
Alpha-Humulene	0.0%–7.1%
3-Octanol	0.0%–4.9%
1,8-Cineole	tr–3.5%
Linalool	0.0%–3.5%
Acetyl-Methylfuran	0.0%–3.1%

Rosefuran CT	
Rosefuran (Alpha-Naginatene)	84.8%
Benzeneacetatealdehyde	4.6%

OTHER CONSTITUENTS: *Dehydroelsholtzia Ketone CT*—Beta-Bourbonene, (E)-3-hexen-1-ol, alpha-pinene, sabinene, 1-octen-3-ol, 3-octanol, 1,8-cineole, linalool, p-cymen-7-ol, acetophenone, beta-thujone (< 1.5%), geranyl acetate, beta-caryophyllene, germacrene D, caryophyllene oxide, n-methyl-2-formylpyrrole, beta-bourbonene; *Elsholtzia Ketone CT*—Para-cymene, beta-naginatel, camphor, 3-octyl-acetate, geranyl acetate, beta-caryophyllene; *Citral CT*—Sabinene, 3-octanone, 1,8-cineole, (Z)-beta-ocimene, 5-hepten-2-one, 6-methyl-, methyl chavicol (< 0.9%), neryl acetate, alpha-humulene, beta-bisabolene, citronellol, (Z)-nerolidol, (E)-nerolidol, caryophyllene oxide, acetophenone, (Z)-3-hexen-1-ol, 1-octen-1-yl acetate; *Rosefuran CT*—1-Octen-3-ol, para-cymene, beta-caryophyllene, caryophyllene oxide

PREFERRED COMPOSITION FOR CLINICAL USE:

Constituent	
Dehydroelsholtzia Ketone	55%–81%
Elsholtzia Ketone	5%–30%
Alpha-Humulene	tr–5%

REPORTED THERAPEUTIC PROPERTIES: Antiviral, antibacterial, antifungal, anti-infectious, antioxidant, relieves headache, soothes sore throat, reduces fever, anti-inflammatory, analgesic, antirheumatic, anticancer, aids digestion, relieves diarrhea, supports kidney function, aids circulation, diuretic, weight management, astringent, relieves hangover symptoms, reduces symptoms of drug withdrawal, relieves anxiety, antidepressant, promotes happiness and contentment

CAUTIONS:

Dehydroelsholtzia Ketone CT and Elsholtzia Ketone CT

▪ None currently known.

Citral CT

▪ Caution is advised during pregnancy and lactation due to high citral content. Large doses of citral may negatively affect fetal development according to animal studies.[4278]

▪ There is a moderate risk that when taken orally, it may interfere with enzymes responsible for metabolizing medications (NSAIDs, proton-pump inhibitors, acetaminophen, antiepileptics, immune modulators, blood-sugar medications, blood pressure medications, antidepressants, antipsychotics, diabetic medications, antihistamines, antibiotics, and anesthetics) due to the citral content.[4279,4280,4281,4282]

▪ Use cautiously with diabetic medications. Large oral doses of citral may improve insulin sensitivity and lower blood-glucose levels according to animal research.[4283,4284]

▪ May interact with antibiotics and possibly enhance their effects due to citral content.[4285,4286]

▪ Dilution is recommended for topical application due to high citral content and the risk for skin irritation or sensitivity.[4287,4288] May also irritate mucous membranes.

Rosefuran CT

▪ Rosefuran safety has not been studied as of this writing but many furans are toxic and therefore this CT is not recommended until safety can be assessed.

SELECTED EVIDENCE:

- □ *In vitro* research shows that Vietnamese balm CT dehydroelsholtzia ketone essential oil reduces the proliferation of human glioblastoma (U87), pancreatic cancer (Panc-1), and triple negative breast cancer (MDA-MB231) cells.[4289] The oil was selective in nature based on the significantly higher viability of fibroblast cells exposed to the same concentrations of essential oil.
- □ Drug abuse modifies the brains dopamine system (specifically the dopamine transporter (DAT) protein undergoes adaptive neurobiological changes that reduces its activity/expression and the

reuptake of dopamine by neurons). This modification triggers withdrawal symptoms and increases the likelihood of relapse during drug addiction treatment. Vietnamese balm CT elsholtzia ketone upregulated expression of DAT (by inducing hyperphosphorylation of cyclic-AMP response element-binding protein (CREB), MAPK, and Akt), suggesting it may rehabilitate brain dopamine function in former drug abusers.[4290]

- □ The dehydroelsholtzia ketone CT of Vietnamese balm inhibited antibiotic-susceptible and -resistant strains of *S. aureus, S. pneumoniae, S. enteritidis,* and *S. typhimurium*.[4291]
- □ The citral CT of Vietnamese balm essential oil demonstrated moderate antileishmanial activity.[4292]

VIETNAMESE CILANTRO (Vietnamese Coriander, Hot Mint, Vietnamese Mint, Solomon's Seal, Rau Răm)

Persicaria odorata (Lour.), Polygonum odoratum (Mill.) Druce, P. officinale All.

FAMILY: Polygonaceae
NOTE: Middle
AROMA INTENSITY: Strong
AROMA: Green, pungent, herbaceous, citrusy, slightly floral
COMMON EXTRACTION METHOD: Steam distilled or hydrodistilled from the leaves
POSSIBLE SUBSTITUTE OILS: Kesum, cinnamon
BLENDS WELL WITH: Bergamot, chamomile (German, Roman), cinnamon, lavender, lemon, marjoram, orange, oregano, tea tree, thyme, vetiver
SUBCELLULAR LOCALIZATION | EPIGENETIC INFLUENCE: Currently unknown | Currently unknown
RECOMMENDED DILUTION RANGE: 3%–33%; 50% for some conditions

PRIMARY CONSTITUENTS:[4293,4294,4295]

Dodecanal	44.1%–55.5%
Decanal	11.6%–27.7%
Decanol	1.1%–10.9%
Pentacosane	0.0%–7.3%
P-Anisaldehyde	0.0%–6.4%
Beta-Caryophyllene	0.0%–3.8%
Undecanal	0.1%–3.6%
Alpha-Humulene	1.5%–3.4%
Dodecanol	1.3%–3.3%

Note: Some consider Vietnamese coriander (*Persicaria odorata*) and kesum (*Persicaria minor*) to be the same plant.

OTHER CONSTITUENTS: Undecane, dodecanol, undecanol, naphthalene-d8, caryophyllene oxide, farnesol isomer, (E)-beta-farnesene, methyl vanillin, alloaromadendrene epoxide, humulene epoxide II, zierone, neo-isomenthol, 2E-dodecenal, 2E-tridecenal, unknown sesquiterpenes

Preferred Composition for Clinical Use:

Constituent	
Dodecanal	40%–58%
Decanal	10%–30%
Decanol	1%–12%
Alpha-Humulene	0.5%–5%
Undecanal	0.1%–5%
Dodecanol	0.1%–5%

Reported Therapeutic Properties: Antimicrobial, **antibacterial**, antifungal, anticancer, aids digestion, heals stomach ulcers, relieves diarrhea, nourishes the scalp and hair, analgesic, anti-inflammatory, antioxidant, antiaging, insect repellent, relaxing, stress management

Cautions:

- None currently known.

Selected Evidence:

- Vietnamese cilantro demonstrates significant antibacterial activity against *S. choleraesuis*.[4296] Its two primary constituents (dodecanal, decanal) and two minor constituents (dodecanol, 2E-dodecenal) were also active against *S. choleraesuis*.

WAMPEE
(Wampi)

Clausena lansium (Lour.) Skeels, *C. wampi* (Blanco) Oliver., *C. punctata* (Sonn.) Rehd. & E.H. Wils

Family: Rutaceae
Note: Middle; Base
Aroma Intensity: Medium
Aroma: Anise-like, sweet; Woody, herbaceous, earthy; Sweet, fruity
Common Extraction Method: Steam distilled or hydrodistilled from the leaves, wood, fruits, pericarp, sarcocarp, flowers, or seeds; may also be headspace (extraction of volatile molecules from materials by changing it from a solid/liquid matrix to vapor phase by heating) extracted
Possible Substitute Oils: Pink pepper, betony, helichrysum, sandalwood, copaiba, black pepper, German chamomile, ginger
Blends Well With: Amyris, bergamot, black pepper, cedarwood, cardamom, carrot, clary sage, clove, frankincense, geranium, ginger, grapefruit, lavandin, lavender, lemon, lime, marjoram, may chang, myrrh, neroli, nutmeg, orange, palo santo, patchouli, rosemary, sage, sandalwood, Spanish sage, tangerine, tea tree (melaleuca), vetiver, yarrow, ylang ylang
Subcellular Localization | Epigenetic Influence: Currently unknown | Currently unknown
Recommended Dilution Range: 3%–50%; neat for some conditions

PRIMARY CONSTITUENTS:[4297,4298,4299,4300,4301,4302]

Wood

Beta-Phellandrene	55.0%
Alpha-Pinene	4.0%

Leaves (Delta-Cadinene CT)

Delta-Cadinene	47.2%
Beta-Bisabolene	20.0%
UI Sesquiterpenes	16.7%

Leaves (Sabinene CT)

Sabinene	14.9%
Beta-Bisabolene	9.9%
Butabal	8.6%
Beta-Caryophyllene	7.7%
Alpha-Zingiberene	6.5%
2-Ethylfuran	4.6%
2-Nonanone	3.4%
2-Propanone	3.0%

Leaves (Beta-Caryophyllene CT)

Beta-Caryophyllene	13.7%–33.3%
Beta-Sesquiphellandrene	3.5%–21.8%
Beta-Bisabolene	3.0%–12.2%
(E)-Beta-Farnesene	1.1%–11.7%
Trans-Alpha-Bergamotene	0.0%–10.6%
(Z)-Alpha-Santalol	0.0%–10.4%
(Z)-Alpha-Bisabolene	0.0%–8.0%
Alpha-Phellandrene	0.0%–7.5%
Methyl Palmitate	0.1%–7.1%
Alpha-Sinensal	0.0%–5.9%
Isophytol	0.0%–5.2%
Alpha-Longipinene	0.0%–4.9%
Caryophyllene Oxide	0.4%–4.5%
Alpha-Humulene	2.2%–3.9%
Sabinene	0.1%–3.5%
Alloaromadendrene	0.0%–3.5%
6-Methyl-5-Hepten-2-one	tr–3.3%

Leaves (Mixed CT)

Trans-Alpha-Bergamotene	7.9%–21.4%
Beta-Bisabolene	10.5%–17.5%
(Z)-Alpha-Santalol	6.0%–13.9%
Beta-Caryophyllene	5.0%–11.4%
Elixine	3.2%–10.9%
(E)-Beta-Farnesene	4.9%–7.1%
6-Methyl-5-Hepten-2-one	4.4%–5.7%
Alpha-Phellandrene	0.1%–5.6%
Spathulenol	2.8%–4.5%

Leaves ((Z)-Alpha-Santalol CT)

(Z)-Alpha-Santalol	14.7%–21.3%
Santalol	12.2%–17.3%
Beta-Caryophyllene	5.5%–12.7%
Beta-Bisabolene	5.3%–9.1%
Caryophyllene Oxide	3.2%–7.8%
Spathulenol	2.3%–6.4%
Alpha-Farnesene	1.3%–5.6%
Alpha-Longipinene	3.8%–5.3%
Beta-Bisabolene	2.5%–3.6%
Beta-Sesquiphellandrene	1.1%–3.1%

Leaves (Santalol/Santalene CT)

(Z)-Alpha-Santalol	31.7%
Alpha-Santalene	19.5%
Beta-Caryophyllene	8.3%
(E)-Nerolidol	6.0%
Trans-Alpha-Bergamotene	5.0%
Farnesol	3.6%
Beta-Bisabolol	3.1%
Beta-Bisabolene	3.0%

Leaves ((Z)-Beta-Santalol CT)

(Z)-Beta-Santalol	35.2%
Beta-Bisabolol	13.7%
Methyl Santalol	6.9%
Ledol	6.5%
Alpha-Sinensal	5.6%
(E)-Nerolidol	5.0%
Trans-Alpha-Bergamotol	4.4%
9-Octadecenamide	3.8%

Fruit (Sabinene CT)

Sabinene	50.6%–69.1%
3-Cyclohexen-1-ol	0.3%–15.2%
Alpha-Phellandrene	5.0%–10.6%
Alpha-Pinene	2.1%–9.4%
Cyclohexene	0.2%–6.5%
1,4-Cyclohexadiene	0.4%–6.2%
4-Carene	0.4%–4.0%
Myrcene	1.7%–3.2%

Pericarp (Mixed CT)

Beta-Bisabolene	8.6%–15.3%
Beta-Sesquiphellandrene	2.5%–15.3%
Elixine	2.9%–11.9%
Alpha-Phellandrene	2.0%–11.2%
(Z)-Alpha-Santalol	0.2%–9.4%
Beta-Caryophyllene	1.6%–9.0%

Fruit (Beta-Phellandrene CT)

Beta-Phellandrene	37.0%
Terpinen-4-ol	17.6%
Juniperene	13.8%
Gamma-Terpinene	5.2%
Alpha-Pinene	4.8%
Alpha-Phellandrene	4.3%
4-Carene	3.0%

Pericarp (Beta-Phellandrene CT)

Beta-Phellandrene	23.2%–45.2%
Beta-Sesquiphellandrene	0.7%–10.9%
Beta-Bisabolene	3.9%–10.2%
(Z)-Alpha-Santalol	0.0%–7.7%
Trans-Alpha-Bergamotene	0.2%–7.1%
Trans-(Z)-Alpha-Bisabolene Epoxide	0.0%–5.8%
Alpha-Pinene	0.0%–5.6%
Beta-Caryophyllene	2.8%–5.0%
Alpha-Phellandrene	0.0%–4.8%
Santalol	0.0%–4.4%
(E)-Beta-Farnesene	1.0%–4.0%
Elixine	0.0%–3.9%
Spathulenol	0.1%–3.6%
Trans-Alpha-Bergamotene	0.0%–8.6%
(E)-Beta-Farnesene	4.0%–8.0%
Alpha-Humulene	0.7%–3.8%
Alpha-Himachalene	0.9%–3.4%

Sarcocarp

(Z)-Beta-Santalol	52.0%
Alpha-Santalol	15.5%
UI Compound	8.3%
Farnesol	5.2%
Alpha-Sinensal	4.0%
Hexadecanoic Acid	3.9%
UI Compound	3.5%

Flower

(Z)-Beta-Santalol	50.6%
9-Octadecenamide	17.2%
UI Compound	6.7%
Alpha-Sinensal	4.1%
Trans-Alpha-Bergamotol	3.2%

Seeds (Sabinene CT)

Sabinene	17.8%–83.6%
Beta-Sesquiphellandrene	tr–22.3%
Beta-Caryophyllene	0.6%–16.6%
Beta-Bisabolene	0.2%–15.2%
Alpha-Phellandrene	0.7%–12.7%
Trans-Alpha-Bergamotene	0.0%–7.0%
Alpha-Pinene	0.5%–4.3%
Myrcene	2.9%–4.1%

Seeds (Beta-Phellandrene CT)

Beta-Phellandrene	54.8%
Limonene	23.6%
p-menth-1-en-4-ol	7.5%

OTHER CONSTITUENTS: *Wood*—Sabinene, myrcene, alpha-phellandrene, para-cymene, (E)-beta-ocimene, linalool, trans-p-menth-2-en-ol, methyl methanthanilate, beta-cubebene, beta-caryophyllene, gamma-muurolene, alpha-cadinene, delta-cadinene, spathulenol, levomenol; *Leaves (Delta-Cadinene CT)*—Beta-caryophyllene, elemene (isomer NI), alpha-humulene, beta-farnesene, gamma-elemene, beta-sinensal, phytol; *Leaves (Sabinene CT)*—Propanal, 2-mthylfuran, 1-pentene, acetic acid, hexanal, 2-hexenal, alpha-pinene, camphene, benzaldehyde, 6-methyl-5-hepten-2-one, alpha-phellandrene, myrcene, 1,3,6-octatriene, linalool, (E)-4,8-dimethyl-1,3,7-nonatriene, alpha-bergamotene, ar-curcumene, beta-sesquiphellandrene, ; *Leaves (Beta-Caryophyllene CT)*—Benzene acetaldehyde, trans-alpha-bergamotene, alpha-longipinene, beta-santalene, beta-himachalene, ar-curcumene, gamma-elemene, delta-cadinene, (E)-nerolidol, dendrolasin, spathulenol, gamma-gurjunene, alpha-bisabolol, trans-(Z)-alpha-bisabolene epoxide; *Leaves ((Z)-Alpha-Santalol CT)*—Sabinene, beta-phellandrene, elixine, alpha-funebrene, 7,11-dimethyl-3-methylene-1,6,10-dodecatriene, (E)-beta-farnesene, alpha-humulene, beta-himachalene, gamma-cadinene, ar-curcumene, alpha-bergamotene, acoradiene, (E)-nerolidol, cubenol, alpha-santalol, trans-(Z)-alpha-bisabolene epoxide phytone, phytol; *Leaves (Santalol/Santalene CT)*—Alpha-cedrene, alloaromadendrene, alpha-farnesene, myristicin, caryophyllene oxide, alpha-trans-bergamotol, hexadecanoic acid, alpha-sinensal; *Leaves ((Z)-Beta-Santalol CT)*—2-Hexenal, beta-caryophyllene, alpha-farnesene, alpha-longipinene, epiglobulol, denderalasin, spathulenol, caryophyllene oxide, farnesol; *Leaves (Mixed CT)*—Benzaldehyde, sabinene, melonal, alpha-longipinene, alloaromadendrene, alpha-humulene, alpha-himachalene, ar-curcumene, alpha-farnesene, beta-sesquiphellandrene, (E)-nerolidol, caryophyllene oxide, (Z)-alpha-bisabolene, cedrol, (E)-nuciferol, gamma-gurjunene, alpha-bisabolol; *Fruit (Sabinene CT)*—Ethanol, acetic acid, hexanal, beta-fenchyl alcohol, ar-curcumene, alpha-farnesene; *Fruit (Beta-Phellandrene CT)*—Beta-pinene, myrcene, para-cymene, delta-2-carene, 1-methyl-4-(1-methyl ethyl-2-cyclohexene-1-ol, unidentified ketone, alpha-terpineol, beta-caryophyllene, limonene; *Pericarp (Beta-Phellandrene CT)*—Sabinene, para-cymene, terpinene-4-ol, cryptone, phellandral, alpha-cubebene, alpha-copaene, 7,11-dimethyl-3-methylene-1,6,10-dodecatriene, alpha-himachalene, beta-himachalene, gamma-cadinene, gamma-elemene, alpha-amorphene, acoradiene, alpha-eudesmol, cubenol, (E)-nuciferol, alpha-cadinol, beta-bisabolene, alpha-santalol, ledol, cedr-8-en-13-ol, lanceol, farnesyl acetate, phytol; *Pericarp (Mixed CT)*—2-Hexenal, alpha-pinene, benzaldehyde, sabinene, 6-methyl-hepten-2-one, myrcene, beta-phellandrene, alpha-terpinene, para-cymene, benzene acetaldehyde, melonal, gamma-terpinene, alpha-longipinene, alloaromadendrene, beta-santalene, isoledene, ionene, beta-cadinene, gamma-muurolene, beta-himachalene, ar-curcumene, (Z)-beta-farnesene, cubebene, beta-selinene, gamma-elemene, epizonarene, beta-vatirenee, delta-cadinene, cubinene, alpha-bisabolene, (E)-nerolidol, spathulenol, caryophyllene oxide, alpha-bisabolol, (Z)-alpha-bisabolene; *Sarcocarp*—Caryophyllene oxide, phytol, stearic acid; *Flower*—(E)-Nerolidol, ledol, (Z)-alpha-santalol, lanceol, hexadecanoic acid, palmitamide, stearamide; *Seeds (Sabinene CT)*—Alpha-thujene, 4-carene, gamma-terpinene, beta-caryophyllene; *Seeds (Beta-Phellandrene CT)*—Beta-pinene, para-cymene, gamma-terpinene, alpha-terpineol, methyl isopropenyl-cyclohexen-1-ol, p-menth-1-en-8-ol, beta-caryophyllene, trans-alpha-bergamotol

PREFERRED COMPOSITION FOR CLINICAL USE:
Too great of variance in samples/data to determine.

REPORTED THERAPEUTIC PROPERTIES: Aids digestion, relieves stomach upset, heals stomach ulcers, expels excess gas, eases cough, aids respiratory function, wound healing, relieves skin conditions and itching, relieves dandruff, supports liver health, antioxidant, anticancer, balances blood sugar levels, analgesic, anti-inflammatory, antirheumatic, relieves fever, brain protective, insecticidal, insect repellent, aids memory function, increase focus and concentration, improves mental sharpness

CAUTIONS:
Wood, Fruit (Beta-Phellandrene CT), Pericarp (Beta-Phellandrene CT), and Seeds (Beta-Phellandrene CT)
■ May interact with anticholinergic (drugs used for asthma, incontinence, gastrointestinal cramps, muscular spasms, depression, and sleep disorders) and cholinergic medications (drugs used to reduce urinary retention, diagnose myasthenia gravis, and for glaucoma) based on AChE inhibitory activity of beta-phellandrene.[4303]

All Other CTs
■ None currently known.

SELECTED EVIDENCE:
- □ Wampee fruit CT beta-phellandrene essential oil showed inhibition towards mouse pancreatic cancer tumors.[4304]

- *In vitro* research showed that wampee pericarp essential oil is more active against *Candida* species than the leaf oil (beta-caryophyllene and (Z)-alpha-santalol CTs).[4305] The study authors attributed this to the higher contents of beta-phellandrene, beta-sesquiphellandrene, and beta-caryophyllene in the pericarp oil. Fruit oil CT beta-phellandrene inhibited *S. aureus, B. subtilis,* and *E. coli.*[4306]
- The beta-phellandrene CT of wampee fruit essential oil exhibited antioxidant activity in the hydroxyl radical scavenging assay.[4307]
- Wampee leaf CT (Z)-alpha-santalol/alpha-santalene essential oil repelled and was toxic to the booklouse (*Liposcelis bostrychophila*).[4308]

WATER PEPPER
(Marshpepper Knotweed, Laska Plant)

Persicaria hydropiper L., *Polygonum hydropiper* L.

FAMILY: Polygonaceae
NOTE: Middle
AROMA INTENSITY: Medium-strong
AROMA: Pungent, herbaceous, tobacco-like; or green, fresh, fruity
COMMON EXTRACTION METHOD: Steam distilled or hydrodistilled from the leaves, flowers, or aerial parts
POSSIBLE SUBSTITUTE OILS: Lavender, German chamomile
BLENDS WELL WITH: Cedarwood, clary sage, lavandin, lavender, mugwort, patchouli, pine, rosemary, sage, sandalwood, Spanish sage
SUBCELLULAR LOCALIZATION | EPIGENETIC INFLUENCE: Currently unknown | Currently unknown
RECOMMENDED DILUTION RANGE: 3%–50%

PRIMARY CONSTITUENTS:[4309,4310]

Leaves		*Aerial Parts*	
Polygodial	44.3%	(E)-2-Hexenal	34.5%
Neophytadiene Isomer I	21.0%	(Z)-3-Hexen-1-ol	31.1%
(E)-Beta-Farnesene	7.8%	Decanal	10.9%
Neophytadiene Isomer II	7.7%	Hexanal	5.1%
Alpha-Humulene	6.2%	Dodecanal	3.0%
Beta-Caryophyllene	6.1%		
Neophytadiene Isomer III	3.3%		

Flowers	
Polygodial	76.7%
Beta-Caryophyllene	6.7%
Alpha-Humulene	5.0%
(E)-Beta-Farnesene	3.9%

OTHER CONSTITUENTS: *Leaves*—Isodremin, (E)-nerolidol, apha-bisabolol; *Flowers*—Isodrimenin, drimenin, alpha-bisabolol, alpha-muurolene, aristolone, neophytadiene isomer I; *Aerial Parts*—Beta-caryophyllene, alpha-humulene, (Z)-2-penten-1-ol, (E)-2-hexen-1-ol, 6-methyl-5-hepten-2-ol, 1-decanol, (Z)-nerolidol, octadecanal, tert-butyl propanoate, hexyl formate

PREFERRED COMPOSITION FOR CLINICAL USE:
Insufficient data/samples to determine.

REPORTED THERAPEUTIC PROPERTIES: Antibacterial, antifungal, antiparasitic, anti-inflammatory, analgesic, anticancer, antioxidant, astringent, wound healing, heals hemorrhoids, reduces excess bleeding, reduces blood pressure, diuretic, balances menstruation, eases painful menstruation, aids digestion, expels excess gas, relieves diarrhea, soothes skin itching, insect repellent, mentally stimulating, aids memory function

CAUTIONS:
Polygodial CTs
■ May enhance the effectiveness of antifungals based on polygodial content. Polygodial enhances the transport of other molecules inside cells and improves the antifungal activity of actinomycin D.[4311]
■ May interact with aspirin, blood pressure, antiplatelet, and anticoagulant medications, and increase the risk of bleeding among people with bleeding disorders, due to potent human-platelet-aggregation inhibition by polygodial.[4312]

Aerial Parts
■ None currently known.

SELECTED EVIDENCE:
- □ Water pepper essential oil protected against stomach ulcers by inhibiting urease activity.[4313] Ureases are enzymes that catabolize urea, which causes an increased production of ammonia and carbamic acid. A high concentration of ammonia can disturb the stomach mucosal lining and alkalize the stomach, allowing *H. pylori* to survive, increasing the risk of ulcers.
- □ Water pepper leaf (38.3% decahydronaphthalene, 36.3% bicyclo[2.2.2]oct-2-ene,1,2,3,6-tetramethyl) and flower (41.4% caryophyllene oxide, 18.2% beta-caryophyllene epoxide, 16.1% humulene epoxide, 4.8% β-elemene cis-1,3-diisopropenyl-trans-4-vinyl-4-methylcyclohexane, 3.8% 3,5-diisopropenyl-1,1,2-trimethylcyclohexane, 1.8% limonene) essential oils with inhibited acetylcholinesterase activity, suggesting it may be beneficial for dementia and Alzheimer's disease.[4314] The leaf oil was more effective than the flower oil. Both oils also exhibited antioxidant activity in the DPPH and ABTS assays.
- □ Solvent extracted essential oils exhibited good mosquito repellent activity.[4315,4316]

WHISTLING PINE
(Beach Sheoak, Australian Pine, Bull-Oak, Common Ironwood, Paina, Beefwood)

Casuarina equisetifolia L., *C. litorea* L. ex Fosberg & Sachet

FAMILY: Casuarinaceae
NOTE: Middle
AROMA INTENSITY: Medium
AROMA: Fresh, woody, spicy, green, waxy, slightly citrusy
COMMON EXTRACTION METHOD: Steam distilled or hydrodistilled from the leaves (needles); may also be distilled from the stem bark or fruits
POSSIBLE SUBSTITUTE OILS: Moringa, melissa, dill, pink pepper, elemi

BLENDS WELL WITH: Amyris, bergamot, black pepper, cardamom, carrot, cedarwood, clary sage, clove, frankincense, geranium, ginger, grapefruit, lavandin, lavender, lemon, lime, marjoram, may chang, myrrh, neroli, nutmeg, orange, palo santo, patchouli, rosemary, sage, sandalwood, Spanish sage, tangerine, tea tree (melaleuca), vetiver, yarrow, ylang ylang
SUBCELLULAR LOCALIZATION | EPIGENETIC INFLUENCE: Currently unknown | Currently unknown
RECOMMENDED DILUTION RANGE: 3%–50%; neat for some conditions

PRIMARY CONSTITUENTS:[4317,4318,4319]

Leaves CT Pentadecanal	
Pentadecanal	32.0%
1,8-Cineole	13.1%
Dill Apiole	7.2%
Alpha-Phellandrene	7.0%
Delta-Terpinene	6.9%
Tetradecanal	4.7%
Para-Cymene	3.7%

Fruits	
Caryophyllene Oxide	11.7%
Trans-Linalool Oxide	11.5%
1,8-Cineole	9.7%
Alpha-Terpineol	8.8%
Alpha-Pinene	8.5%
Guaiol	6.4%
Alpha-Terpinyl Acetate	6.6%
Alpha-Campholenal	5.2%
Terpinen-4-ol	4.6%
Spathulenol	4.4%
Benzaldehyde	3.4%

Leaves CT Alpha-Phellandrene	
Alpha-Phellandrene	40.6%
Para-Cymene	15.7%
1,8-Cineole	14.1%
Terpinolene	8.4%
Alpha-Pinene	3.7%

Stem Bark	
n-Hexadecanoic Acid	18.7%
(Z)-13-Octadecanoic Acid	17.8%
Tridecane	11.8%
Undecane	10.5%
Henatriacontane	8.9%
Nonanal	8.6%
Methyl Ricinoleate	5.0%

OTHER CONSTITUENTS: *Leaves CT Pentadecanal*—Heptanal, alpha-thujene, alpha-pinene, camphene, sabinene, beta-pinene, myrcene, limonene, alpha-terpinene, linalool, nonanal, cis-ascaridole (< 0.8%), dodecanal, beta-caryophyllene, geranyl acetone, (E)-beta-ionone, tridecanal, beta-sesquiphellandrene, hexahydrofarnesylacetone; *Leaves CT Alpha-Phellandrene*—(2E)-Hexenal, beta-pinene, alpha-terpineol, methyl salicylate (< 0.6%), beta-cyclocitral, beta-elemene, beta-caryophyllene, geranyl acetone, aromadendrene, germacrene D, (E)-beta-ionone, (E)-nerolidol, (3Z)-hexenyl benzoate, carotol, dill apiole (< 2.1%); *Fruits*—N/A; *Stem Bark*—Oxirane, nonanoic acid, 1H-indene, 1,4-methano-1H-indene, dodecanoic acid, tetradecanoic acid, hexadecanoic acid, 9,12-octadecadienoic acid, 9-octadecenoic acid

PREFERRED COMPOSITION FOR CLINICAL USE:
Insufficient data/samples to determine.

REPORTED THERAPEUTIC PROPERTIES: Antibacterial, antifungal, antiseptic, antiallergic, liver protective, analgesic, relieves diarrhea, astringent, wound healing, reduces the appearance of blemishes, antioxidant, insecticidal, stress management, relieves anxiety, aids cognition

CAUTIONS:

■ None currently known, although a GC-MS should be reviewed to determine levels of dill apiole and 1,8-cineole. If dill apiole is greater than or equal to 10%, cautions for children, during pregnancy, orally, and in those who have a compromised liver would be necessary.

SELECTED EVIDENCE:

- □ *In vitro* evidence demonstrated that whistling pine essential oil CT alpha-phellandrene inhibits *B. cereus, S. aureus, A. niger, C. albicans, E. coli*, and *P. aeruginosa*, with greatest activity against *B. cereus* and *S. aureus*.[4320]
- □ Whistling pine stem bark essential oil was significantly mosquitocidal against *Ae. aegypti* and *An. gambiae* mosquitoes (carriers of malaria and arboviruses).[4321]

WHITEBRUSH
(Brazilian Lavender, Bee Brush)

Aloysia gratissima (Gillies & Hook.) Troncoso, *Verbena gratissima* Gill et Hook, *Aloysia lycioides* Cham, *Lippia lycioides* (Cham) Steudel

FAMILY: Verbenaceae
NOTE: Middle
AROMA INTENSITY: Strong
AROMA: Fresh, woody, balsamic, spicy, hint of vanilla
COMMON EXTRACTION METHOD: Hydrodistilled or steam distilled from the leaves; may also be distilled from the aerial parts or flowers (inflorescences)
POSSIBLE SUBSTITUTE OILS: Ylang ylang, goldenrod, marjoram, nutmeg, plai, blue cypress
BLENDS WELL WITH: Blue cypress, cassia, cedarwood, cypress, eucalyptus, frankincense, lavender, pine, rosemary, sage, sandalwood, tea tree, ylang ylang
SUBCELLULAR LOCALIZATION | EPIGENETIC INFLUENCE: Currently unknown | Currently unknown
RECOMMENDED DILUTION RANGE: Pinocamphones CTs—$\leq$ 0.5%; CTs with Thujones—1.5%–20%; Other CTs—3%–50%

PRIMARY CONSTITUENTS:[4322,4323,4324,4325,4326,4327,4328,4329,4330]

Leaves (Germacrene D CT)

Constituent	%
Germacrene D	13.1%
1,8-Cineole	12.4%
Beta-Caryophyllene	12.4%
Beta-Pinene	11.2%
Limonene	8.7%
Bicyclogermacrene	8.3%
Alpha-Pinene	3.7%
Viridiflorol	3.3%
Gamma-Terpinene	3.1%

Leaves (Pinocamphones CT)

Constituent	%
Trans-Pinocarveol Acetate	8.2%–17.6%
Pinocamphone	16.1%–16.3%
Beta-Pinene	0.0%–12.0%

Aerial Parts (1,8-Cineole CT)

Constituent	%
1,8-Cineole	17.6%
Guaiol	10.3%
Germacrene D	6.2%
Bicyclogermacrene	5.6%
Beta-Caryophyllene	5.0%
Delta-Elemene	5.0%
Bulnesol	4.5%
Sabinene	4.3%
Limonene	4.1%
Gamma-Elemene	3.3%

Aerial Parts (Pinocamphones CT)

Constituent	%
Isopinocamphone	25.4%
Limonene	15.1%

Guaiol	8.5%–11.5%
Alpha-Humulene	1.1%–7.2%
Germacrene B	0.0%–5.6%
Isopinocamphone	4.9%–6.0%
Gamma-Muurolene	3.8%–4.6%
Trans-Sabinol	0.0%–4.5%
Bicyclogermacrene	2.8%–4.2%
Myrtenol	1.8%–3.7%
Bulnesol	3.1%–3.7%
Caryophyllene Oxide	2.6%–3.3%
Beta-Caryophyllene	0.0%–3.1%
Trans-Pinocarveol	0.0%–3.0%

Leaves (Sabinene CT)

Sabinene	30.0%–35.3%
Beta-Pinene	7.8%–9.7%
Unidentified Hydrocarbon	7.4%–7.9%
Beta-Bisabolene	2.1%–5.8%
Copaenol	4.2%–4.3%
Copaenone	3.6%–4.1%
Unidentified Sesquiterpenol	1.5%–3.9%
Globulol or Ledol	3.1%–3.5%
Alpha-Cedrene	2.9%–3.4%
Alpha-Pinene	2.7%–3.0%

Aerial Parts (Beta-Elemene CT)

Beta-Elemene	23.6%–35.7%
10-Epi-Cubenol	0.4%–13.4%
Viridiflorol	0.0%–10.0%
Humulene Epoxide II	0.0%–9.7%
Germacrene D	3.8%–8.3%
Aromadendrene Epoxy	2.7%–6.9%
Unidentified Constituent	0.0%–6.6%
Bicyclogermacrene	0.0%–5.9%
Para-Cymene	1.4%–5.8%
(E)-Nerolidol	0.0%–4.5%
Sesquilavandulol	0.0%–4.5%
Beta-Caryophyllene	2.8%–3.8%
Germacrene A	0.7%–3.6%
Elemol Acetate	0.0%–3.0%
Unidentified Constituent	0.0%–3.0%

Flowers (Guaiol CT)

Guaiol	19.5%
Germacrene B	10.5%
Bulnesol	10.0%
Beta-Caryophyllene	8.9%
Caryophyllene Oxide	8.2%
Trans-Pinocarveol Acetate	6.2%
Bicyclogermacrene	4.2%
Spathulenol	3.8%

Guaiol	12.7%
Cis-Pinocarveol Acetate	8.3%
Pinocamphone	7.2%
Beta-Caryophyllene	5.7%
Acorenone	4.9%
Germacrene D	3.8%
Bulnesol	3.7%

Aerial Parts (Beta-Caryophyllene CT)

Beta-Caryophyllene	15.2%–28.0%
Alpha-Thujone	0.0%–17.5%
Linalool	0.0%–13.9%
Bicyclogermacrene	3.8%–12.6%
(E)-Nerolidol	0.0%–11.6%
Germacrene D	3.5%–10.1%
Caryophyllene Oxide	0.0%–10.1%
Spathulenol	0.0%–7.7%
Terpinen-4-yl Acetate	0.0%–6.2%
Viridiflorol	0.0%–6.2%
Gamma-Elemene	0.0%–4.9%
Alpha-Humulene	1.5%–4.3%
Germacrene B	0.0%–3.8%
Beta-Bisabolene	0.0%–3.0%
Trans-Sesquisabinene Hydrate	0.0%–3.0%

Aerial Parts (Viridiflorol CT)

Viridiflorol	33.6%
Beta-Elemene	14.4%
Bicyclogermacrene	12.8%
Globulol	5.4%

Flowers (Sabinene CT)

Sabinene	18.6%
Beta-Pinene	10.7%
Globulol	8.7%
Beta-Bisabolene	7.7%
Beta-Caryophyllene	6.8%
Unidentified Hydrocarbon	6.5%
Alpha-Curcumene	3.6%

Flowers (Globulol CT)

Globulol	39.7%
Sabinene	9.5%
Beta-Caryophyllene	5.2%
Caryophyllene Oxide	4.9%
Beta-Pinene	4.8%
1,8-Cineole	4.1%
Spathulenol	3.6%
Sesquiterpene Alcohol	3.4%

OTHER CONSTITUENTS: *Leaves (Germacrene D CT)*—Alpha-thujene, sabinene, myrcene, para-cymene, (E)-beta-ocimene, alpha-thujone (< 1.9%), terpinen-4-ol, delta-elemene, alpha-humulene, guaiol; *Leaves (Pinocamphones CT)*—Trans-sabinene hydrate, linalool, cis-pinal-2-ol, trans-verbenol, terpinen-4-ol, para-cymen-8-ol, trans-carveol, bornyl acetate, beta-bourbonene, beta-elemene, alpha-bulnesol, cubebol, spathulenol, 10-epi-gamma-eudesmol; *Leaves (Sabinene CT)*—Alpha-terpinene, limonene, 1,8-cineole, para-cymene, terpinolene, alpha-elemene, pinocamphone (< 2.1%), isopinocamphone (< 2.2%), cis-sabinene hydrate, isocaryophyllene, beta-caryophyllene, beta-cadinene, alpha-curcumene, calamenene, caryophyllene oxide, spathulenol; *Aerial Parts (1,8-Cineole CT)*—Cis-sabinene hydrate, borneol, alpha-terpineol, beta-elemene, beta-copaene, aromadendrene, alpha-humulene, elemol, spathulenol, caryophyllene oxide, germacrene B, humulene epoxide II, rosifoliol, 1-epi-cubenol, hinesol, beta-eudesmol, 7-epi-alpha-eudesmol, nootkatol; *Flowers (Guaiol CT)*—Trans-Sabinol, trans-verbenol, pinocamphone (< 2.2%), isopinocamphone (< 0.8%), myrtenol, beta-elemene, gamma-elemene, alpha-humulene, gamma-muurolene, alpha-bulnesene, cubebol, 10-epi-gamma-eudemol; *Aerial Parts (Pinocamphones CT)*—Myrcene, delta-3-carene, alpha-humulene, elemol, beta-eudesmol; *Aerial Parts (Beta-Caryophyllene CT)*—Alpha-pinene, limonene, beta-thujone < 1.5%), 3-thujanol, trans-piperityl acetate, alpha-longipinene, alpha-terpinyl acetate, beta-bourbonene, beta-elemene, beta-elemene isomer, beta-cubebene, alpha-gurjunene, cis-3,5-muuroladiene, dehydro aromadendrene, alloaromadendrene, germacrene A, delta-cadinene, aromadendrene epoxy, cedrol, epoxy-alloaromadendrene, cedr-8(15)-en-10-α-ol, eremoligenol, selin-11-en-4alpha-ol, caryophylla-4(14),8(15)-dien-5-β-ol, 14-hydroxy-9-epi-(E)-caryophyllene, elemol acetate, eudesma-4(15),7-dien-1-β-ol, (E)-bisabolol-11-ol; *Aerial Parts (Beta-Elemene CT)*—Alpha-pinene, beta-pinene, gamma-terpinene, delta-elemene, alpha-copaene, beta-elemene isomer, gamma-elemene, alloaromadendrene, alpha-humulene, beta-selinene, germacrene B, cis-muurol-5-en-4-α-ol, germacrene D-4-ol, caryophyllene oxide, globulol, cedr-8(15)-en-4-α-ol, selin-11-en-4-α-ol, alpha-muurolol, 14-hydroxy-9-epi-(E)-caryophyllene, selin-3,11-dien-6-α-ol, 3-thujopsanone, eudesma-4(15),7-dien-1-β-ol; *Aerial Parts (Viridiflorol CT)*—Para-cymene, delta-elemene, beta-elemene isomer, beta-caryophyllene, gamma-elemene, alloaromadendrene, alpha-humulene, germacrene D, beta-selinene, 10-epi-cubebol, germacrene B, humulene epoxide II, epoxy-alloaromadendrene, 14-hydroxy-9-epi-(E)-caryophyllene, selina-3,11-dien-6-α-ol, 3-thujopsanone; *Flowers (Globulol CT)*—Alpha-Pinene, alpha-terpinene, limonene, pinocamphone (< 2.3%), isopinocamphone (< 2.0%), beta-cadinene, beta-bisabolene, alpha-curcumene, calamenene, copaenone, copaenol; *Flowers (Sabinene CT)*—Alpha-Pinene, alpha-terpinene, limonene, 1,8-cineole, para-cymene, alpha-elemene, isopinocamphone (< 2.5%), pinocamphone (< 2.2%), alpha-cedrene, beta-cadinene, calamenene, caryophyllene oxide, copaenone, spathulenol, copaenol

PREFERRED COMPOSITION FOR CLINICAL USE:
Insufficient samples/data to determine.

REPORTED THERAPEUTIC PROPERTIES: Analgesic, anti-inflammatory, antispasmodic, relieves headache, aids respiratory function, eases cough, digestive aid, eases stomachache, bladder disorder treatment, antibacterial, antifungal, antiviral, antiparasitic, insecticidal, relaxing/calming, relieves anxiety, aphrodisiac, antidepressant

CAUTIONS:

Leaves (Germacrene D CT) and Aerial Parts (1,8-Cineole CT)

■ Caution is advised for people with epilepsy and other convulsive disorders, due to the possible presence of moderate amounts of 1,8-cineole and small amounts of thujone. These constituents may cause or exacerbate seizures or reduce medication efficacy.[4331,4332,4333,4334,4335]

Pinocamphones CTs

■ Avoid during pregnancy and lactation due to the neurotoxicity of pinocamphone and isopinocamphone.

■ Avoid use in children under twelve due to the neurotoxicity of pinocamphone and isopinocamphone.

■ Avoid oral consumption. Pinocamphone and isopinocamphone are highly neurotoxic, and pinocamphone is lethal at doses above 0.05 mL/kg body weight.[4336]

■ Avoid with epilepsy and other seizure disorders, due to pinocamphone and isopinocamphone content. The isomers pinocamphone and isopinocamphone are neurotoxic, and high doses cause convulsions in humans and animals. It only required 0.05 g/kg of pinocamphone to cause convulsion in animals, and as low as 3 mg/kg of oils rich in pinocamphone can cause convulsions in humans.[4337,4338] These constituents may cause or exacerbate seizures or reduce medication efficacy.

Aerial Parts (Beta-Caryophyllene CT)

■ Avoid with children under 6 due to thujone content. Cases of seizure have been reported in young children due to exposure to sage essential oil (rich in thujones) exposure.[4339,4340]

■ Avoid during pregnancy and while lactating, due to thujones content. Essential oils high in thujone content may cause abortion.[4341,4342]

■ Avoid oral consumption. Ingesting plants or products high in thujones can cause vomiting, stomachache, diarrhea, and gastroenteritis followed by absorption disorders, headache, nervous agitation, and chronic convulsions, and symptoms of liver and kidney toxicity extending to yellow liver atrophy (extensive and rapid death of liver cells—parenchymal, possibly by fatty degeneration), arrhythmia, and myocardial bleeding.[2852] Thujone is considered significantly neurotoxic and may damage the liver, and it is estimated that as little as 15 mg orally may negatively impact the central nervous system.[4343,4344,4345] Taking 20 drops of thuja oil (also with significant thujones content) twice daily for five days caused an adult woman to have a tonic seizure.[2033] As few as 12 drops of sage oil (also with significant thujones content) orally have reportedly caused a seizure and short coma in an adult.[4346]

■ Avoid with epilepsy and Parkinson's disease due to thujone and 1,8-cineole content.[4347,4348,4349,4350,4351]

■ Avoid with those who have a compromised liver or kidneys, due to the risk of kidney toxicity, kidney failure, or liver toxicity.[4352,4353,4354] Consuming extremely large amounts of thujone-rich essential oils (10 mL) has caused kidney failure.[4355] This would also suggest that those taking medications that could cause liver damage should also use tansy very cautiously or avoid it entirely.

Aerial Parts (Sabinene CT), Flowers (Guaiol CT), Flowers (Sabinene CT), and Flowers (Globulol CT)

■ Caution is advised for people with epilepsy and other convulsive disorders, due to the possible presence of small amounts of pinocamphones. The isomers pinocamphone and isopinocamphone are neurotoxic, and high doses cause convulsions in humans and animals. It only required 0.05 g/kg of pinocamphone to cause convulsion in animals, and as low as 3 mg/kg of oils rich in pinocamphone can cause convulsions in humans.[4356,4357] These constituents may cause or exacerbate seizures or reduce medication efficacy.

Aerial Parts (Beta-Elemene CT)

■ Avoid during pregnancy and lactation (very low risk) due to moderate beta-elemene content. Beta-elemene may prevent the formation of blood vessels necessary for fetal development according to animal research.[4358]

■ May interact with aspirin, blood pressure, antiplatelet, and anticoagulant medications, and increase the risk of bleeding among people with bleeding disorders based on beta-element content. Beta-elemene inhibits platelet aggregation and increases coagulation time.[4359,4360]

Aerial Parts (Viridiflorol CT)

■ None currently known.

SELECTED EVIDENCE:

- *In vitro* research reported that the germacrene D CT of whitebrush essential oil inhibited *B. cereus* and *T. mentagrophytes*.[4361] The pinocamphones CT (leaf) oil inhibited *P. aeruginosa* and *S. pneumoniae*, whereas the flowers oil CT guaiol inhibited *P. aeruginosa, S. pneumoniae*, and *C. albicans*.[4362] The pinocamphones CT (leaf oil) of whitebrush essential oil inhibited oral pathogens (*C. albicans, F. nucleatum, P. gingivalis, S. sanguis, S. mitis*) but was only effective against some of the pathogens in biofilm prevention.[4363]
- An unknown CT of whitebrush essential oil inhibited the herpes simplex virus type 1 (HSV-1) and the Junin virus (JUNV) *in vitro*.[4364]
- Whitebrush CT germacrene D essential oil exhibited weak antioxidant activity in the DPPH assay.[4365]
- The 1,8-cineole CT of whitebrush essential oil killed promastigotes and intracellular amastigotes of *L. amazonensis* without harming macrophages.[4366] Isolated guaiol was also effective.

WHITE CEDAR
(Cedar of Goa, Mexican White Cypress, Cedro Blanco, Portuguese Cypress)

Cupressus lusitanica Mill.

FAMILY: Cupressaceae
NOTE: Middle
AROMA INTENSITY: Medium
AROMA: Fruity, berry-like, fresh, evergreen, slightly woody
COMMON EXTRACTION METHOD: Hydrodistilled or steam distilled from the needles (leaves), twigs, branchlets, bark, or needles and twigs
POSSIBLE SUBSTITUTE OILS: Pine, cypress, juniper berry
BLENDS WELL WITH: Bergamot, clary sage, fennel, grapefruit, juniper, lavender, lemon, lime, mandarin, orange, pine, rosemary, tangerine, yuzu
SUBCELLULAR LOCALIZATION | EPIGENETIC INFLUENCE: Currently unknown | Currently unknown
RECOMMENDED DILUTION RANGE: 3%–50%; neat for some conditions

PRIMARY CONSTITUENTS:[4367,4368,4369,4370,4371]

Needles, Twigs, Branchlets	
Alpha-Pinene	18.0%
Beta-Pinene + Sabinene	13.2%
Delta-3-Carene + Myrcene	8.2%
Limonene	6.1%
Gamma-Terpinene	5.6%
Terpinen-4-ol	4.1%
Beta-Phellandrene	3.5%
Alpha-Terpinene	3.1%

Needles (Alpha-Cadinol CT)	
Alpha-Cadinol	6.6%
Alpha-Terpinyl Acetate	4.9%

Needles (Alpha-Pinene CT)	
Alpha-Pinene	39.9%–82.3%
Limonene	4.2%–17.6%
Isobornyl Acetate	0.2%–9.6%
Cis-Muurola-4(14),5-Diene	0.3%–6.7%
Nezukol	0.0%–4.0%

Needles (Germacrene D CT)	
Germacrene D	18.5%
Epizonarene	8.2%
Cis-Calamenene	8.2%
Terpinen-4-ol	6.3%
Linalool	6.0%

Alpha-Gurjunene	4.3%
Totarol	3.7%
Sclareol	3.5%

Needles (Umbellulone CT)

Umbellulone	18.4%
Alpha-Pinene	10.0%
Sabinene	8.2%
Limonene	7.9%
Delta-3-Carene	6.9%
Terpinen-4-ol	6.1%
O-Cymene	5.8%
Linalool	3.9%
(E)-Muurola-4(14),5-Diene	3.4%

Umbellulone	6.0%
Di-Epi-Alpha-Cedrene	4.9%
ar-Curcumene	4.1%

Bark

Androstan-17-One-3-Ethyl-3-Hydroxy	51.6%
Totarol	18.7%

OTHER CONSTITUENTS: *Needles, Twigs, Branchlets*—Para-cymene, terpinolene, linalool, bornyl acetate, terpinen-4-yl acetate, beta-caryophyllene, alpha-terpineol, alpha-terpinyl acetate, cardinene isomer; *Needles (Alpha-Pinene CT)*—(3Z)-Hexenol, camphene, sabinene, beta-pinene, myrcene, alpha-terpinene, gamma-terpinene, terpinolene, camphor, umbellulone (< 0.8%), alpha-terpinyl acetate, terpinen-4-ol, beta-caryophyllene, cis-muurola-3,5-diene, epizonarene, delta-cadinene, alpha-cadinol; *Needles (Alpha-Cadinol CT)*—Caryophyllene oxide, beta-cubebene; *Needles (Umbellulone CT)*—Alpha-phellandrene, alpha-fenchene, myrcene, delta-2-carene, para-cymenene, p-(Z)-menth-2-en-1-ol, camphor, [α,α],4-Trimethyl-benzenemethanol, piperitone, terpinolene, (E)-muurola-3,5-diene, epizonarene, cis-calamenene, (Z)-14-nor-Muurol-5-en-4-one, nezukol; *Needles (Germacrene D CT)*—Alpha-pinene, delta-3-carene, para-cymene, limonene, 1,8-cineole, cis-linalool oxide, p-cymen-8-ol, linalyl acetate, beta-caryophyllene, alpha-humulene, alpha-amorphene, alpha-muurolene, alpha-calacorene, cedrol, caryophyllene oxide, 14-norcadin-5-en-4-one, naphthalene; *Bark*—Not reported

PREFERRED COMPOSITION FOR CLINICAL USE:
Insufficient samples/data to determine.

REPORTED THERAPEUTIC PROPERTIES: Nervine, anti-inflammatory, analgesic, antirheumatic, soothes hemorrhoids, reduces oily skin, reduces excess sweating, heals varicose veins, astringent, wound healing, combats chronic skin conditions, promotes balanced menstruation and relieves painful menstruation, aids circulation, stops excess bleeding, antibacterial, antifungal, insecticidal, insect repellent, grounding, antidepressant, relieves anxiety, promotes feelings of peace and joy

CAUTIONS:

Alpha-Pinene CTs, Bark

■ None currently known. Androstan-17-One-3-Ethyl-3-Hydroxy (3-Ethyl-3-hydroxyandrostan-17-one) is not well studied and little is known about this rare 21-carbon constituent.

Umbellulone CT, Needles (Germacrene D CT)

■ Not recommended for essential oil therapy due to moderate umbellulone (a monoterpene ketone) content. The headache tree (*Umbellularia californica*) contains significant amounts of umbellulone. Inhalation of umbellulone has been associated with cluster headaches or migraines by diffusing to nerve terminals and triggering a cascade of chemical events that activate the Transient receptor potential ankyrin (TRPA1) and the trigeminovascular system.[4372,4373] While the germacrene D CT has lower amounts of umbellulone, it should still be used cautiously and a complete composition analysis be reviewed to ensure the umbellulone content doesn't exceed 9.9% before use.

Alpha-Cadinol CT

■ The alpha-cadinol CT contains the phytoestrogen sclareol. Theoretically, sclareol could increase estradiol levels and potentiate the action of estrogen-replacement drugs. Because sclareol exerts much weaker estrogenic effects than synthetic hormones administered during hormone-replacement therapy, it is highly unlikely that it will interact with hormone-replacement therapy. In general, phytoestrogens are balancing in nature, but women who are undergoing hormone-replacement therapy should have their estradiol levels checked regularly and consult with a physician before using the alpha-cadinol CT.

SELECTED EVIDENCE:

- □ White cedar CT alpha-pinene essential oil demonstrated antibacterial activity against *Bacillus cereus* and antifungal activity against *Aspergillus niger in vitro*.[4374] The germacrene D chemotype inhibited gram-negative bacteria (*E. coli., K. pneumonia, P. mirabilis, P. aeruginosa, S. typhi, S. flexerni*), gram-positive bacteria (*E. faecalis, S. aureus*), and fungi (*C. albicans, C. glabrata, C. krusei, C. lusitaniae, C. parapsilosis, C. tropicalis*), with greatest activity against *E. faecalis, P. mirabilis*, and *C. albicans*.[4375]
- □ The umbellulone CT of white cedar essential oil was toxic to stored-grain insects (*Acanthoscelides obtectus, Sitotroga cerealella*) and repelled another stored-grain pest (*Tribolium castaneum*).[4376]

WHITE CYPRESS (Emerald Cypress, Northern Cypress-Pine)

Callitris columellaris F. Muell.

FAMILY: Cupressaceae
NOTE: Top-Middle
AROMA INTENSITY: Medium
AROMA: Spicy, woody, fresh, piney, fruity
COMMON EXTRACTION METHOD: Steam distilled from the wood; may also be steam distilled from the leaves (needles) and twigs
POSSIBLE SUBSTITUTE OILS: Blue cypress; blue spruce, black spruce
BLENDS WELL WITH: Balsam fir, bergamot, black pepper, blue cypress, cardamom, cedarwood, cistus, frankincense, juniper berry, lavandin, lavender, lemon, lime, myrtle, orange, pine, sandalwood, silver fir, white fir, ylang ylang
SUBCELLULAR LOCALIZATION | EPIGENETIC INFLUENCE: Currently unknown | Currently unknown
RECOMMENDED DILUTION RANGE: 5%–Neat

PRIMARY CONSTITUENTS:[4377,4378]

Leaves (Needles)		*Citronellic Acid CT (Wood)*	
Limonene	17.7%–37.0%	Citronellic Acid	25.7%
Bornyl Acetate	0.8%–27.1%	Guaiol	14.6%
Alpha-Pinene	13.9%–17.2%	Beta-Eudesmol	7.5%
Alpha-Fenchyl Acetate	0.1%–5.7%	Alpha-Eudesmol	7.1%
Myrcene	3.6%–5.5%	Columellarin	6.7%
Bicyclogermacrene	0.0%–3.6%	Gamma-Eudesmol	3.7%
Germacrene D	0.0%–3.5%	Dihydrocolumellarin	3.4%
Camphor	0.0%–3.4%		

Note: Another study reported that white cypress essential oil was predominantly dihydrocolumellarin but did not provide chemical composition.[4379]

Other Constituents: *Citronellic Acid CT (Wood)*—Methyl citronellate, selina-4,11-diene, beta-selinene, alpha-selinene, bulnesol, cadalene, (–)-gamma-costol, eudesmanolide isomer, callitrisin, germacranolide, dihydrocallitrisin; *Leaves (Needles)*—Tricyclene, alpha-fenchene, beta-pinene, para-cymene, citronellal, borneol, terpinen-4-ol, alpha-terpineol, pulegone (< 1.9%), geraniol, citronellyl acetate, geranyl acetate, beta-caryophyllene, alpha-humulene, spathulenol

Preferred Composition for Clinical Use:
Insufficient samples/data to determine.

Reported Therapeutic Properties: Analgesic, anti-inflammatory, anti-infectious, eases cough, aids respiratory function, antiallergic, antioxidant, lowers high blood pressure, reduces the appearance of blemishes, wound healing, relieves chronic skin conditions, insecticidal, grounding, enhances meditation, stress management, relieves anxiety

Cautions:
- None currently known.

Selected Evidence:
- □ None found.

WHITE WORMWOOD
(Armoise, Wild Mugwort, White Mugwort)

Artemisia herba-alba Asso

Family: Asteraceae
Note: Middle
Aroma Intensity: Medium
Aroma: Camphoraceous, herbaceous, spicy, fresh
Common Extraction Method: Hydrodistilled or steam distilled from the aerial parts; may also be distilled from individual or combinations of aerial parts (stems, leaves, flowers) or roots
Possible Substitute Oils: Sage, tansy, wormwood, mugwort, Spanish sage, camphor, tagetes
Blends Well With: Bay laurel, bergamot, cajeput, camphor, chamomile (German, Roman), coriander, eucalyptus, lavender, lemon, lime, orange, neroli, palo santo, peppermint, petitgrain, ravensara, rosemary, sage, Spanish sage, spikenard, tangerine, tansy, thyme, turmeric
Subcellular Localization | Epigenetic Influence: Currently unknown | Currently unknown
Recommended Dilution Range: 1.5%–20%

Primary Constituents:[4380,4381,4382,4383,4384,4385,4386,4387,4388,4389,4390,4391,4392,4393,4394,4395,4396,4397,4398,4399,4400]

Alpha-Thujone CT		*Camphor CT*	
Alpha-Thujone	11.4%–45.5%	Camphor	13.3%–43.1%
Camphor	0.0%–22.9%	Trans-Pinocarveol	0.0%–16.9%
Trans-Sabinyl Acetate	0.0%–22.5%	Alpha-Thujone	0.0%–16.5%
Beta-Thujone	2.2%–22.4%	Chrysanthenone	0.0%–15.8%
Chrysanthenone	0.0%–19.0%	Beta-Thujone	1.7%–15.0%
Germacrene D	0.0%–14.5%	Trans-Sabinyl Acetate	0.0%–12.1%
Cis-Chrysanthenyl Acetate	0.0%–13.0%	1,8-Cineole	tr–11.2%
		Terpinen-4-ol	0.6%–8.7%
1,8-Cineole	1.1%–9.8%	Camphene	tr–7.2%

Davana Ether Isomers	0.0%–8.7%
Davanone	0.0%–8.0%
Gamma-Muurolene	0.0%–7.3%
Bicyclogermacrene	0.0%–6.2%
(+,-)-Lepidozene	0.0%–5.7%
Camphene	0.0%–5.6%
3-Hydroxyisodavanone	0.0%–4.3%
Terpinen-4-ol	0.0%–3.4%
Borneol	0.0%–3.4%
Piperitone	0.0%–3.4%

Beta-Thujone CT

Beta-Thujone	24.1%–43.0%
Alpha-Thujone	1.3%–23.2%
Camphor	1.4%–16.7%
Artemisia Alcohol	0.0%–10.9%
1,8-Cineole	0.0%–10.2%
Alpha-Phellandrene	0.0%–10.0%
Cis-Sabinene Hydrate	0.0%–5.6%
Borneol	0.0%–4.1%
Dihydro Myrcenol	0.0%–3.2%

Chrysanthenone CT

Chrysanthenone	16.2%–30.6%
Camphor	15.6%–24.4%
Beta-Thujone	2.7%–14.2%
Alpha-Thujone	2.4%–9.1%
1,8-Cineole	3.0%–8.9%
Piperitenone	0.0%–7.6%
Camphene	2.8%–4.5%
Filifolone	2.0%–4.3%

1,8-Cineole CT

1,8-Cineole	18.4%–41.1%
Camphor	5.6%–17.1%
Para-Cymene	0.0%–15.5%
Myrcene	0.0%–14.0%
Davanone	0.0%–13.4%
Chrysanthenone	0.3%–13.3%
Borneol	0.3%–10.8%
Yomogi Alcohol	0.0%–9.5%
Bornyl Acetate	tr–8.6%
Alpha-Pinene	0.0%–8.2%
Gamma-Terpinene	0.5%–5.9%
Beta-Thujone	0.0%–5.7%
Cis-Chrysanthenyl Acetate	0.0%–5.6%
Filifolone	0.0%–7.0%
Cis-Chrysanthenyl Acetate	0.0%–5.8%
Germacrene D	0.0%–5.3%
Borneol	tr–5.2%
Bornyl Acetate	0.0%–3.8%

Artemisia Alcohol CT

Artemisia Alcohol	26.4%–31.8%
6-Camphenone	15.9%–27.9%
Beta-Thujone	10.9%–14.2%
Cis-Sabinene Hydrate Acetate	2.4%–13.6%
Trans-Sabinene Hydrate	6.2%–8.7%
Chrysanthenone	0.8%–5.1%
Cis-Sabinene Hydrate	2.8%–4.2%
Alpha-Phellandrene	1.2%–3.2%

Cis-Chrysanthenol CT

Cis-Chrysanthenol	13.8%–27.87%
1,8-Cineole	4.8%–18.8%
Limonene	0.0%–12.6%
Xanthoxylin	0.0%–9.1%
Alpha-Terpineol	0.0%–7.0%
Para-Cymene	0.0%–6.7%
Cis-Chrysanthenyl Acetate	0.5%–6.4%
Gamma-Muurolene	0.0%–4.5%
Chrysanthenone	0.6%–4.4%
Lavandulol	0.0%–3.8%
Alpha-Terpineol	0.0%–3.4%
Alpha-Terpinyl Acetate	0.0%–3.2%
Isobornyl Acetate	0.0%–3.2%
Alpha-Pinene	0.0%–3.0%
Myrcene	0.0%–3.0%

Alpha-Pinene CT

Alpha-Pinene	45.9%
Borneol	11.3%
Alpha-Terpineol	6.5%
Camphene	3.9%
Gamma-Terpinene	3.2%

Nordavanone CT

Nordavanaone	27.1%
Artemisia Alcohol	22.3%
Alpha-Thujone	21.1%

Alpha-Terpineol	0.0%–4.7%
Alpha-Thujone	0.0%–4.6%
Trans-Sabinyl Acetate	0.0%–4.2%
Terpinen-4-ol	2.0%–3.9%
Santolina Alcohol	0.0%–3.7%
Camphene	1.2%–3.2%
Germacrene D	0.0%–3.1%
Artemisia Alcohol	0.0%–3.0%

Cis-Chrysanthenyl Acetate CT

Cis-Chrysanthenyl Acetate	18.4%–25.0%
Alpha-Pinene	0.0%–14.1%
1,8-Cineole	7.4%–12.2%
Para-Cymene	1.2%–8.9%
Cis-Chrysanthenol	4.9%–7.1%
Myrcene	0.0%–5.8%
Cis-Jasmone	0.0%–4.0%

Davanone CT

Davanone	13.4%–45.7%
Davanone Ether Isomers	0.3%–19.2%
Para-Cymene	0.0%–18.4%
Chrysanthenone	0.0%–17.4%
Trans-Sabinyl Acetate	0.0%–11.1%
Cis-Chrysanthenyl Acetate	0.0%–9.1%
Myrcene	0.0%–8.4%
Gamma-Terpinene	0.0%–5.9%
3-Hydroxyisodavanone	0.0%–5.0%
Davana Ether	0.5%–4.4%
Germacrene D	0.0%–4.1%
Camphor	0.0%–4.0%
Isodavanone	0.0%–3.6%
Unidentified Compound	0.0%–3.4%
Cis-Arbusculone	0.0%–3.0%
Isochrysanthone	0.0%–3.0%

Davanone Ether Isomers CT

Davanone Ether Isomers	10.0%–23.2%
Chrysanthenone	1.1%–11.7%
Davanone	7.1%–11.0%
Cis-Chrysanthenyl Acetate	0.0%–9.5%
Trans-Sabinyl Acetate	0.0%–6.5%
Trans-Pinene Hydrate	9.0%
Iso-3-Thujanol	7.8%

Para-Cymene CT

Para-Cymene	19.9%
Alpha-Pinene	17.2%
Myrcene	10.9%
1,8-Cineole	8.6%
Camphor	8.5%
Terpinen-4-ol	3.9%

Piperitone CT

Piperitone	26.5%
Ethyl Cinnamate	9.5%
Camphor	8.7%
Hexadecanoic Acid	6.9%
Hexahydrofarnesyl Acetone	5.7%
4-Methylphenylethanol	5.0%
4-Methylisopulegone	3.5%

Trans-Sabinyl Acetate CT

Trans-Sabinyl Acetate	72.8%
Alpha-Thujone	6.7%
Germacrene D	4.1%
Camphor	3.6%
Trans-Pinocarveol	3.5%

Balanced CT

Cis-Chrysanthenyl Acetate	10.6%
Trans-Sabinyl Acetate	9.1%
Alpha-Thujone	8.7%
Davana Ether Isomers	7.8%
Chrysanthenone	4.9%
Cis-Arbusculone	3.6%
Spathulenol	3.6%
Trans-Arbusculone	3.0%
1,8-Cineole	3.0%

Roots

Alpha-Thujone	18.2%
Camphor	14.6%
Curcumen-15-al	14.3%

Davanone Ether	2.5%–6.2%	Beta-Thujone	5.7%
Trans-Pinocarveol	0.6%–5.4%	Trans-Pinocarveol	3.6%
3-Hydroxyisodavanone	4.8%–5.4%	Piperitone	3.3%
Germacrene D	2.4%–4.8%	1,8-Cineole	3.0%
Cis-Arbusculone	2.8%–4.6%		
Unidentified Compound	1.6%–4.2%		
Isochrysthanenone	0.0%–4.0%		
Trans-Arbusculone	2.4%–3.6%		
Bicyclogermacrene	1.7%–3.0%		

OTHER CONSTITUENTS: *Alpha-Thujone CT*—Alpha-pinene, alpha-phellandrene, sabinene, para-cymene, 1-nitro-cyclohex-1-ene, bornyl acetate, alpha-copaene, beta-caryophyllene, delta-cadinene, gamma-epoxy-elemene, 1,2,3-trimethyl benzene, santolina alcohol, trans-sabinene hydrate, filifolene, trans-p-menth-1-en-1-ol, trans-pinocarveol, trans-verbenol, pinocarvone, spathulenol, 2-hexenal, santolina triene, alpha-terpinene, cis-arbusculone, trans-arbusculone, myrcene, cis-piperitol, trans-piperitol, carvacrol, jasmone, nordavanone, myrtenol, alpha-humulene, ethyl cinnamate, beta-selinene, epiglobulol, rosifolol, davana ether, T-cadinol, alpha-acorenol, beta-savanone-2-ol, phytol, xanthoxylintrans-dihydrofarnesol; *Beta-Thujone CT*—Tricyclene, myrcene, trans-pinene hydrate, iso-3-thujanol, nordavanone, iso-3-thujanol acetate, chrysanthenone, pinocarvone, myrtenal, myrtenol, piperitone oxide, bornyl acetate, jasmone, 2-hexanal, alpha-pinene, camphene, para-cymene, bicyclogermacrene, spathulenol; *Artemisia Alcohol CT*—Alpha-pinene, sabinene, artemiseole, alpha-terpinene, dehydro myrcenol, 6-camphenol, alpha-campholenal, camphor, neoiso-3-thujanol, trans-beta-terpineol, trans-chrysanthenyl acetate, trans-sabinene hydrate acetate, iso-3-thujanol acetate, neo-3-thujnanol acetate, myrtenyl acetate, delta-elemene, beta-caryophyllene; *Chrysanthenone CT*—(2,3,3-Trimethyl-2-oxiranyl)methanol, mesitylene, trans-pinocarveol, pinocarvone, borneol, verbenone, (E)-beta-ocimene, trans-chrysanthenyl acetate, isopiperitenone, piperitenone oxide; *Camphor CT*—Sabinene, delta-2-carene, gamma-terpinene, cis-sabinene hydrate, isochrysanthenone, pinocarvone, alpha-terpineol, myrtenol, trans-piperitol, cuminaldehyde, piperitone oxide, myrtenyl acetate, jasmone, ethyl cinnamate, bicyclogermacrene, spathulenol, epiglobulol, alpha-acorenol, T-cadinol, alpha-cadinol, santolina triene, trans-verbenol, terpinolene, alpha-pinene, para-cymene, sabina ketone, thymol, verbenone, germacrene B, filifolone, carvacrol, germacrene D, 1-epi-cubenol, caryophyllene oxide, (E)-ethyl cinnamate, alpha-ylangene, lavandulyl acetate; *Cis-Chrysanthenol CT*—Terpinen-4-ol, trans-pinocarveol, pinocarvone, beta-cubebene, gamma-elemene, alpha-copaene, davanone, jasmone, selin-11-en-4alpha-ol, spathulenol; *Alpha-Pinene CT*—Alpha-terpinene, para-cymene, (+)-4-carene, hotrienol, terpinolene, methyl cinnamate, jasmone, germacrene D, limonene, rosefuran, (E)-nerolidol, caryophyllene oxide; *1,8-Cineole CT*—Santolina triene, 1-octen-3-ol, gamma-terpinene, artemisia ketone, cis-p-menth-2-en-ol, pinocarvone, lavandulol, myrtenal, myrtenol, octyl acetate, carvone, piperitone, cis-chrysanthenyl acetate, ethyl cinnamate, spathulenol, xanthoxylin, isochrysanthenone, cis-p-menth-2-en-1-ol, trans-p-menth-2-en-ol, alpha-copaene, gamma-muurolene, davana ether, caryophyllene oxide; *Cis-Chrysanthenyl Acetate CT*—Camphene, beta-pinene, alpha-terpinene, gamma-terpinene, limonene, terpinen-4-ol, alpha-terpineol, p-cymen-8-ol, camphor, lavandulol, chrysanthenone, trans-pinocarveol, gamma-muurolene, spathulenol, davanone; *Nordavanone CT*—Alpha-pinene, alpha-terpinene, beta-thujone (< 1.7%), camphor, neoiso-3-thujanol, trans-sabinene hydrate acetate, delta-elemene, neo-3-thujanol acetate; *Davanone CT*—2.5-Diethenyl-2-methyltetrahydrofuran, sabinene, 1,8-cineole, trans-arbusculone, trans-pinocarveol, delta-verbenol, terpinen-4-ol, nordavanone, cuminaldehyde, piperitone, cis-chrysanthenyl acetate, jasmone, bicyclogermacrene, spathulenol, alpha-cadinol, (E)-dihydrofarnesol, beta-savanone-2-ol, 3-methyl-but-2-

enoic acid,1.7.7-trimethylbicyclo[2.2.1]hept-2-yl ester; *Davanone Ether Isomers CT*—2.5-Diethenyl-2-methyltetrahydrofuran, 5.5-dimethyl-2(5H)-furanone, myrcene, para-cymene, cis-sabinol, pinocarvone, borneol, terpinen-4-ol, myrtenol, trans-piperitol, nordavanone, cuminaldehyde, piperitone oxide, jasmone, alpha-humulene, beta-selinene, lyratyl isovalerate, (E)-nerolidol, spathulenol, globulol, caryophyllene oxide, beta-copaen-4-alpha-ol, alpha-acorenol, alpha-cadinol, selin-11-en-4-alpha-ol; *Para-Cymene CT*—Camphene, beta-pinene, alpha-terpinene, gamma-terpinene, isochrysanthenone, trans-pinocarveol, borneol, lavandulol, alpha-terpineol, bornyl acetate, alpha-copaene, gamma-muurolene, spathulenol, davanone; *Piperitone CT*—Borneol, carvacrol, 2-hydroxypiperitone, 4-oxoisophorone, thymol, tetradecanoic acid, decanoic acid, dodecanoic acid, 2-methoxy-4-vinylpohenol, dihydroactinidiolide, coumaran, benzyl alcohol; *Trans-Sabinyl Acetate CT*—Para-cymene, 1,8-cineole, cis-p-menth-2,8-dien-1-ol, pinocamphone (< 0.6%), terpinen-4-ol, piperitone, bicyclogermacrene; *Balanced CT*—Alpha-pinene, camphene, alpha-terpinene, gamma-terpinene, para-cymene, filifolone, beta-thujone (< 2.6%), iso-pinocarveol, camphor, borneol, cis-piperitol, cuminal, nor-chrysanthemic acid methyl ester, cis-jasmone, 1-methoxy-4-nitro benzene, bicyclogermacrene, (E)-nerolidol, viridiflorol, alpha-cadinol; *Roots*—Camphene, sabinene, trans-sabinene hydrate, cis-p-menth-2-en-ol, chrysanthenone, isoborneol, pinocarvone, terpinen-4-ol, naphthalene, myrtenal, trans-piperitol, carvone, trans-sabinyl acetate, spathulenol, caryophyllene oxide, globulol, alpha-cadinol, valeranone, mayurone

PREFERRED COMPOSITION FOR CLINICAL USE:
Too great a variability in composition to determine.

REPORTED THERAPEUTIC PROPERTIES: Antiparasitic, antibacterial, antifungal, **antimicrobial**, anti-infectious, diuretic, antiseptic, promotes menstruation, relieves menstrual cramps, antispasmodic, analgesic, anti-inflammatory, antineuralgic, nervine, expectorant, opens the airways, supports respiratory health, eases cough, relieves stomachache, relieves diarrhea, balances blood sugar, wound healing, insect repellent, promotes restful sleep, aids memory and cognition, reduces mental exhaustion

CAUTIONS:
▪ *Given the great variance in chemical composition and potential to contain several constituents that require cautions, white wormwood essential oil should not be used in essential oil therapy without a complete composition analysis and approval from a qualified health professional or essential oil therapist. The following are general cautions based on composition and should be observed when meaningful amounts of the listed constituents—or combinations of the constituents—are present, or avoided completely in more vulnerable people (pregnancy, lactation, children, epileptics). See other monographs with similar composition for more details:*

- Thujones: Children, pregnancy, lactation, liver disorders, kidney disorders, epilepsy, oral use, drugs metabolized by the CYP450 enzyme pathway
- Trans-Sabinyl Acetate: Children, pregnancy, lactation, oral use, epilepsy, live disorders, kidney disorders
- Camphor: Children, pregnancy, lactation, epilepsy, oral use, broken skin, G6PD, compromised liver
- 1,8-Cineole: Epilepsy, oral use, barbiturates, drugs metabolized by the CYP450 enzyme pathway
- Piperitone: Antibiotics

▪ Multiple chemotypes have shown antiacetylcholinesterase activity suggesting the oil may interact with anticholinergic (drugs used for asthma, incontinence, gastrointestinal cramps, muscular spasms, depression, and sleep disorders) and cholinergic medications (drugs used to reduce urinary retention, diagnose myasthenia gravis, and for glaucoma).[4401,4402]

Selected Evidence:

- Leaf (22.5% 2,5-octadecadiynoic acid, methyl ester; 20.4% 1,8-cineole; 15.8% alpha-bulnesene) essential oil and capitulum (19.6% (+-)-2,3,6,7-tetramethyl-1,4,4aα,5,8,8aβ,9β,9aα,10,10aα-decahydroanthracen-9-ol; 11.1% beta-guaiene; 9.5% 2,5-octadecadiynoic acid, methyl ester) essential oil were cytotoxic to rodent mastocytoma and kidney cancer cells.[4403] An interesting CT or aerial parts with verbenol as the major constituent was less effective.
- White wormwood CT chrysanthenone essential oil reduced intestinal spasms in isolated rat jejunum by influencing calcium channels via muscarinic receptors.[4404]
- The beta-thujone CT of white wormwood essential oil significantly inhibited nitric oxide (NO) production triggered by LPS without being toxic to normal cells.[4405]
- Various chemotypes (trans-sabinyl acetate, alpha-thujone, and camphor) of white wormwood essential oil demonstrated the ability to inhibit acetylcholinesterase activity, with the camphor CT being the highest inhibition and the alpha-thujone CT the lowest.[4406] Another study reported the balanced CT inhibited AChE activity.[4407] Inhibition of AChE prevents the breakdown of acetylcholine, which is essential for memory and thinking. People with neurodegenerative diseases make less acetylcholine, and the diseases often break it down at a faster rate, leading to acetylcholine deficits.
- *In vitro* research demonstrates that white wormwood essential oil (beta-thujone, nordavanaone, and Artemisia alcohol CTs) inhibit *E. coli, B. subtilis, M, luteus, S. aureus, A. niger, P. expansum, P. digitatum, C. versicolor, C. puteana, P. plancenta,* and *G. trabeum* depending on the concentration used of each CT.[4408] The camphor CT of white wormwood essential oil inhibited *L. monocytogenes* (greatest activity), *N. meningitidis, H, influenzae, K. pneumoniae, Pantoea* sp., *E. coli, S. aureus,* and *S. pneumoniae*.[4409] Another study reported that the camphor CT inhibited *S. aureus, E. coli, K. pneumoniae,* and *P. aeruginosa*, with maximum activity against *S. aureus*.[4410] White wormwood CT alpha-pinene demonstrated good antimicrobial activity against *E. coli, S. enteritidis, P. aeruginosa, S. aureus, E. faecalis, C. albicans,* and *A. funigatus*, with antimicrobial activity frequently exceeding that of a group of antibiotics (oxacillin, ticarcillin, carbenicillin, colistrin, piperacillin, erythromycin, tetracycline, and nystatin).[4411] The camphor CT exhibited antibiofilm activity against *S. epidermidis*.[4412] Interestingly, the same researchers found that the camphor CT antagonized the antimicrobial activity of chlorhexidine. A thujone-rich (70%, isomer not identified) CT of white wormwood essential oil inhibited *S. typhimurium* and *E. coli*.[4413] When tested against eight strains of *L. pneumophila*, white wormwood essential oil only inhibited some of the strains.[4414] The beta-thujone CT inhibited the fungi *T. rubrum* and *E. floccosum*.[4415] A camphor CT exhibited low antimicrobial activity against eight bacterial strains (*B. subtilis, S. aureus, S. epidermidis, S. faecalis, E. coli, K. pneumoniae, P. vulgaris, P. aeruginosa*).[4416] The balanced CT exhibited strong antimicrobial activity against fungi (*F. solani, Fusarium* sp., *A. oxysporum*) and bacteria (*E. coli, P. aeruginosa, K. pneumoniae, B. cereus, E. faecalis, and S, typhimurium*).[4417] An unidentified CT of white wormwood essential oil inhibited *St. hemolyticus, Sh. Sonnei, Sal. typhosa, S. aureus,* and *E. coli*.[4418]
- Oral administration of 400 mg/kg body weight of white wormwood CT camphor essential oil improved antioxidant activity of lamb muscles but did not protect against lipid oxidation in meat stored for seven days.[4419] The oil also possessed good antioxidant properties. Another study found that the addition of 400 ppm of white wormwood essential oil to lamb food improved fatty acid composition and healthy properties of lamb meat obtained from supplemented lambs.[4420]
- White wormwood CT alpha-thujone essential oil exhibits concentration dependent antioxidant activity that varies from weak to moderate in the DPPPH and reducing power assays.[4421] It performed better during the beta-carotene bleaching method assay, providing antioxidant activity near equivalent to BHT. Various chemotypes (trans-sabinyl acetate, alpha-thujone, and camphor) of white wormwood essential oil exhibited good antioxidant activity in the ferric ion reducing

assay, with alpha-thujone CTs demonstrating the best antioxidant activity and the camphor CT the lowest.[4422] A thujone-rich (70%, isomer not identified) CT of white wormwood essential oil exhibited good antioxidant activity.[4423] The balanced CT showed interesting antioxidant activity as evaluated by the 2,2-diphenyl-1-picrylhydrazyl and the β-carotene bleaching methods.[4424]

- White wormwood CT camphor essential oil was active against *Leishmania infantum* promastigotes by triggering cell apoptosis and cell cycle arrest.[4425] Other research reported that white wormwood essential oil (unknown composition) is active against *Leishmania tropica* and *Leishmania major*.[4426]
- The piperitone CT of white wormwood essential oil repelled tick (*Ixodes riinus*) nymphs.[4427]
- White wormwood CT camphor essential oil was toxic to two flour beetles (*Tribolium castaneum* and *Tribolium confusum*) in contact and fumigant tests.[4428]
- Topical application of white wormwood essential oil CT camphor was nonirritating when applied to rabbit skin at up to 30% concentration.[4429] Irritation occurred above 30%.

WILD CELERY (Mountain Celery, Kelus, Karafs-e-koohi)

Kelussia odoratissima Mozaff.

FAMILY: Apiaceae (Umbelliferae)
NOTE: Middle
AROMA INTENSITY: Medium
AROMA: Freshm green, sweet, spicy
COMMON EXTRACTION METHOD: Steam distilled or hydrodistilled from the aerial parts; may also be distilled from the fruits, roots, stems, or leaves
POSSIBLE SUBSTITUTE OILS: Lovage, Himalayan angelica
BLENDS WELL WITH: Basil (sweet), bay laurel, cassia, cinnamon, clove, frankincense, hairy basil, holy basil, lavender, lovage, myrrh, oakmoss, ocotea, patchouli, opoponax, rose, tamala
SUBCELLULAR LOCALIZATION | EPIGENETIC INFLUENCE: Currently unknown | Currently unknown
RECOMMENDED DILUTION RANGE: 3%–20%; 50% for some conditions

PRIMARY CONSTITUENTS:[4430,4431,4432,4433,4434,4435,4436,4437]

(Z)-Ligustilide CT	
(Z)-Ligustilide	34.5%–85.9%
(E)-Ligustilide	0.0%–11.8%
(Z)-3-Butylidenephthalide	0.4%–8.8%
2-Octen-1-ol Acetate	0.0%–7.2%
Dec-9-en-1-ol	0.0%–5.9%
Thymol	0.0%–4.7%
Piperitone Oxide	0.0%–3.1%

Neophyadiene CT	
Neophyadiene	51.8%–64.0%
3-n-Butyl Phthalide	15.5%–32.6%

3-n-Butyl Phthalide CT	
3-n-Butyl Phthalide	40.1%
Neophyadiene	25.8%
Butylidene Phthalide	9.4%
6-Butyl-1.4-Cycloheptadin	4.4%
E3-Tetradeceneylacetate	4.1%

Fruits	
(Z)-Ligustilide	29.2%
Germacrene B	15.9%
Germacrene D	15.5%
Beta-Phellandrene	9.2%

Butylidene Phthalide*	2.1%–6.3%
6-Butyl-1.4-Cycloheptadin	1.8%–5.0%
E3-Tetradeceneylacetate	1.5%–4.7%
Hexadecanoic Acid	1.7%–4.5%

Leaves (3NBP CT)	
3-n-Butyl Phthalide	25.1%
Pentyl Benzoate	15.1%
3-Octen-1-ol Acetate	9.6%
Cuparene	6.8%
3-Butyldine Phthalide*	6.7%
Alpha-Copaene	5.3%
Delta-Cadinene	4.1%

Delta-Cadinene	5.8%
(Z)-Beta-Ocimene	3.8%
Gamma-Elemene	3.4%

Leaves (Beta-Terpinene CT)	
Beta-Terpinene	23.0%
Sabinene	9.0%
Alpha-Thujene	8.4%
Delta-Cadinene	4.1%
Limonene	3.3%
Borneol	3.0%

Flowers	
1,8-Cineole	22.0%
Camphor	20.0%
Alpha-Pinene	19.0%
Camphene	12.0%
Bornyl Acetate	5.8%
Borneol	4.8%
Sabinene	3.9%

Stems	
Borneol	36.9%
Bornyl Acetate	14.0%
1,8-Cineole	13.6%
Camphor	9.5%
Artemisia Alcohol	4.1%

* Correct isomer not identified.

OTHER CONSTITUENTS: *(Z)-Ligustilide CT*—(E)-beta-ocimene, eugenol, alpha-copaene, beta-elemene, alpha-guaiene, beta-selinene, delta-selinene, alpha-muurolene, trans-beta-guaiene, germacrene A, cadina-1,4-diene, alpha-cadinene, delta-cadinene, alpha-cubebene, beta-caryophyllene, beta-himachalene, (Z)-gamma-bisabolene, 5-pentyl cyclohexa-1,3-diene, p-vinylguaiacol, 4-hydroxybenzaldehyde, alpha-acoradiene, 10-epi-beta-acoradiene, germacrene D, kessane, (Z)-beta-santalol acetate; *3-n-Butyl Phthalide CT*—Hexadecanoic acid; *Neophyadiene CT*—Not listed in study; *Fruits*—Alpha-pinene, alpha-copaene, beta-elemene, alpha-guaiene, alpha-muurolene, delta-selinene, germacrene A, cadina-1,4-diene, alpha-cadinene, (Z)-3-butylidenephthalide; *Leaves (3NBP CT)*—n-Propyl benzene, alpha-cubebene, beta-caryophyllene, (E)-beta-farnesene, beta-acoradiene, beta-himachalene; *Leaves (Beta-Terpinene CT)*—Alpha-pinene, alpha-copaene; *Flowers*—Myrcene, limonene, alpha-terpineol, alpha-humulene, alpha-amorphene, delta-cadinene, beta-caryophyllene, calarene; *Stems*—Camphene, limonene, terpinolene, pinocarvone, germacrene D, delta-cadinene, n-decanone

PREFERRED COMPOSITION FOR CLINICAL USE:

Constituent	
(Z)-Ligustilide	60%–88%
(E)-Ligustilide	tr–10%
(Z)-3-Butylidenephthalide	0.1%–8%

REPORTED THERAPEUTIC PROPERTIES: Anti-inflammatory, antirheumatic, analgesic, antispasmodic, neuroprotective, aids digestion, heals gastrointestinal ulcers, anticancer, reduces high blood pressure, supports cardiovascular function, reduces high cholesterol, stimulates glutathione production, antioxidant, antiparasitic, antimicrobial, insecticidal, sedating, aids cognitive function and memory

CAUTIONS:

(Z)-Ligustilide CTs

■ Avoid during pregnancy and lactation due to the estrogenic activity of (Z)-ligustilide and other essential oils with significant (Z)-ligustilide content. (Z)-ligustilide may prolong menstruation and stimulate or relax the uterus.[4438,4439,4440]

■ May interact with diabetic medications and cause low blood-sugar levels based on significant (Z)-ligustilide content.[4441]

■ May interact with aspirin, blood pressure, antiplatelet, and anticoagulant medications, and increase the risk of bleeding among people with bleeding disorders, due to ligustilide content. (Z)-ligustilide inhibits platelet aggregation without affecting coagulation time of peripheral blood.[4442]

■ May interact with levodopa (L-dopa) and increase its cytotoxic effects (causes neuronal cell death) due to (Z)-ligustilide content.[4443]

3-n-Butyl Phthalide and Neophyadiene CTs

■ May interact with aspirin, blood pressure, antiplatelet, and anticoagulant medications, and increase the risk of bleeding among people with bleeding disorders, due to 3-n-Butyl Phthalide. 3-n-Butyl Phthalide is a potent inhibitor or platelet aggregation in rats and platelet activation in human platelets.[4444,4445] However, a clinical trial found that it had little influence on blood coagulation in humans.[4446]

■ May interact with diabetic medications and cause low blood-sugar levels based on significant 3-n-Butyl Phthalide content. 3-n-Butyl Phthalide reduces blood-glucose levels in animals.[4447] However, a clinical trial found that it had very little influence on blood-glucose levels in humans.[4448]

■ May weakly interfere with the enzymes responsible for metabolizing medications (NSAIDs, proton-pump inhibitors, acetaminophen, antiepileptics, immune modulators, blood-sugar medications, blood pressure medications, antidepressants, antipsychotics, diabetic medications, antihistamines, antibiotics, and anesthetics) due to 3-n-Butyl Phthalide. High concentrations of 3-n-Butyl Phthalide inhibit CYP2C19 *in vitro*.[4449]

■ May interfere with pentobarbital and other barbiturates (medications for anxiety and insomnia) because 3-n-Butyl Phthalide prolongs the effects of pentobarbital in mice.[4450]

Borneol CT

■ May interact with aspirin, blood pressure, antiplatelet, and anticoagulant medications, and increase the risk of bleeding among people with bleeding disorders, due to the anticoagulant properties of borneol.[4451,4452]

■ May interfere with enzymes responsible for metabolizing medications (NSAIDs, proton-pump inhibitors, acetaminophen, antiepileptics, immune modulators, blood-sugar medications, blood pressure medications, antidepressants, antipsychotics, diabetic medications, antihistamines, antibiotics, and anesthetics), based on borneol content.[4453,4454,4455,4456,4457]

1,8-Cineole CT

■ Avoid with epilepsy and Parkinson's disease due to 1,8-cineole and camphor content. May exacerbate or cause seizures/convulsions or reduce seizure-medication efficacy based on 1,8-cineole and camphor content.[4458,4459,4460,4461,4462,4463]

■ The camphor content in wild celery may negatively impact red blood cells and increase the risk of jaundice in children with glucose-6-phosphate dehydrogenase deficiency (G6PD).[4464,4465]

■ Avoid with children under six due to moderate camphor content. Several cases of camphor poisoning and/or seizures from ingestion and topical application have been reported in children.[4466,4467] Ingestion of camphor-containing products has been lethal in children under age two.[4468] Children five years and up may use camphor-containing essential oils topically in dilutions no stronger than 5%. Camphor ingestion by infants and young children may cause cough, vomiting, seizure, burning sensation in the mucous membranes and eyes, or lack of voluntary coordination of muscle movements.[4469]

■ Caution is warranted during pregnancy and while lactating, due to moderate camphor content. Ingestion of essential oils with significant levels of camphor may lead to abortion because fetuses lack the enzymes to process it.[4470] Camphor ingestion by infants and young children may cause cough, vomiting, seizure, burning sensation in the mucous membranes and eyes, or lack of voluntary coordination of muscle movements.[4471]

Beta-Terpinene CT

■ None currently known.

SELECTED EVIDENCE:

- □ Repeated injection of wild celery CT ligustilide essential oil (5 and 10 mg/kg body weight) for four days significantly reduced morphine withdrawal symptoms, which suggests the oil reversed neuronal adaptation that takes place during morphine use.[4472]
- □ Injection of wild celery CT (Z)-ligustilide essential oil one hour prior to stroke reduced brain injury (necrotic tissue, edema, inflammation, and oxidative stress).[4473] (Z)-ligustilide produced similar results at lower doses than the whole oil, which made the researchers determine it may be responsible for the neuroprotective properties of wild celery essential oil.
- □ The (Z)-ligustilide CT of wild celery caused a sedating (cumulative reduction in locomotor activity) effect in mice when administered at 50 mg/kg body weight.[4474]
- □ Wild celery CT (Z)-Ligustilide essential oil was active against promastigotes of *Leishmania major*.[4475]
- □ Wild celery CT (Z)-ligustilide was an effective larvicide against *Anopheles stephensi* mosquitoes.[4476] Another study determined that the (Z)-ligustilide CT was larvicidal to both *Anopheles stephensi* and *Culex pipiens*.[4477]

WILD GARLIC

Allium vineale L.

FAMILY: Liliaceae
NOTE: Top
AROMA INTENSITY: Strong
AROMA: Pungent, sulfur-like, garlicky
COMMON EXTRACTION METHOD: Hydrodistilled from the leaves and bulbs
POSSIBLE SUBSTITUTE OILS: Garlic
BLENDS WELL WITH: Basil (sweet), cumin, lavender, onion, peppermint, rosemary
SUBCELLULAR LOCALIZATION | EPIGENETIC INFLUENCE: Currently unknown | Currently unknown
RECOMMENDED DILUTION RANGE: 1.5%–5.0%

PRIMARY CONSTITUENTS:[4478]

Dimethyl Trisulfide	3.8%–17.4%
Benzaldehyde	0.1%–16.4%
Allyl Methyl Trisulfide	7.9%–13.2%
Allyl (E)-1-Propenyl Disulfide	7.9%–12.5%

Methyl (Z)-1-Propenyl Disulfide	2.6%–12.5%
Diallyl Disulfide	4.4%–12.2%
Diallyl Trisulfide	2.8%–10.5%
p-Vinylguaiacol	5.2%–6.5%
Allyl Methyl Disulfide	2.3%–6.1%
5-Methyl-1,2,3,4-Tetrathiane	0.1%–6.1%
Allyl (Z)-1-Propenyl Disulfide	2.8%–4.3%
Dimethyl Tetrasulfide	0.8%–4.0%
Methyl (Z)-1-Propenyl Disulfide	1.1%–3.1%
Allyl (Z)-1-Propenyl Trisulfide	0.1%–3.0%

OTHER CONSTITUENTS: 2-Furaldehydem (2E)-hexenal, (3Z)-hexenol, 2,4-diemthylthiophene, 1-propenyl propyl disulfide, methyl mehtylthiomethyl disulfide, 4-methyl-1,2,3-trithiolane, methyl propyl trisulfide, methyl (Z)-1-propenyl trisulfide, methyl (E)-1-propenyl trisulfide, allyl methylthiomethyl disulfide, allyl propyl trisulfide, methyl methylthiomethyl trisulfide, allyl methyl tetrasulfide, allyl methylthiolmethyl trisulfide, 4-methyl-1,2,3,5,6-pentathiepane

PREFERRED COMPOSITION FOR CLINICAL USE:
Insufficient samples/data to determine.

REPORTED THERAPEUTIC PROPERTIES: Antibacterial, antimicrobial, antifungal, antiparasitic, analgesic, antirheumatic, anti-inflammatory, anticancer, antioxidant, **supports cardiovascular function**, supports normal blood pressure, reduces high cholesterol, relieves diarrhea, aids respiratory function, expels excess mucus, boosts immune function, removes warts and corns, relieves chronic skin conditions, wound healing (especially deep), reduces the appearance of acne and blemishes, promotes hair growth, aids circulation, relieves male impotence, weight management, insect repellent, insecticide, combats anger and rage, releases traumatic memories and emotions, increases awareness, mentally stimulating

CAUTIONS:

■ Caution is advised during pregnancy and nursing. Garlic odor was detected in the amniotic fluid of women who consumed a single capsule of ginger herb, and a review concluded that garlic herb has the potential to cause abortion, trigger menstruation, or stimulate the uterus.[4479,4480] However, no adverse outcomes were reported in pregnant women who took 800 mg of garlic herb daily during the third trimester of pregnancy, and diallyl sulfide prevents embryo-feto toxicity.[4481,4482,4483] Animal research also reports that garlic herbal oil—containing diallyl trisulfide—reduces pregnancy complications (decreased newborn body weight, abortion, and musculoskeletal abnormalities) due to its ability to reduce cholesterol in pregnant rats with high cholesterol.[4484] Based on conflicting information, it is best to use garlic cautiously during pregnancy until definitive research determines its safety. In addition, ingestion of garlic by lactating mothers reportedly altered the flavor of breast milk because garlic constituents are secreted in the breast milk and made infants prone to extended nursing.[4485,4486]

■ May interact with aspirin, blood pressure, antiplatelet, and anticoagulant medications, and increase the risk of bleeding among people with bleeding disorders. Garlic herbal oil, essential oil, and volatile components inhibit platelet aggregation.[4487,4488,4489,4490,4491,4492,4493,4494,4495,4496,4497] The study authors attributed this effect to the presence of essential oil or volatile organosulfur constituents (diallyl trisulfide, methyl allyl trisulfide).

■ May interfere with enzymes responsible for metabolizing medications (NSAIDs, proton-pump inhibitors, acetaminophen, antiepileptics, immune modulators, blood-sugar medications, blood pressure medications, antidepressants, antipsychotics, diabetic medications, antihistamines, antibiotics, and anesthetics) due to diallyl disulfide, diallyl trisulfide, allyl methyl sulfide, and allyl diallyl sulfide content.[4498,4499,4500,4501,4502] Diallyl disulfide and diallyl trisulfide also interfere with protein transporters that move drugs across cell membranes, which can increase or decrease drug effectiveness.[4503]

- May interact with antifungal medications (ketoconazole) and increase their effectiveness based on presence of sulfur-containing compounds similar to garlic essential oil.[4504]
- May be photosensitizing due to reports that diallyl disulfide causes photo contact dermatitis.[4505,4506] Avoid sun exposure to area of application for at least twelve hours after topical application.
- Topical application may cause skin irritation or skin sensitivity reactions. Cases of contact dermatitis or irritation have been reported after using garlic herbal oil, with diallyl disulfide being a major cause of this irritation.[4507,4508] Dilutions under 3% are strongly recommended.

SELECTED EVIDENCE:

- ☐ None found.

WILD PANSY
(Johnny Jumpup, Heart's Ease)

Viola tricolor L.

FAMILY: Violaceae
NOTE: Middle
AROMA INTENSITY: Strong
AROMA: Herbaceous, sweet, floral
COMMON EXTRACTION METHOD: Steam distilled from the flowering aerial parts
POSSIBLE SUBSTITUTE OILS: German chamomile, Roman chamomile
BLENDS WELL WITH: Bergamot, cassia, cedarwood, cinnamon, eucalyptus, geranium, German chamomile, lavender, neroli, patchouli, pine, ravintsara, Roman chamomile, rose, ylang ylang
SUBCELLULAR LOCALIZATION | EPIGENETIC INFLUENCE: Currently unknown | Currently unknown
RECOMMENDED DILUTION RANGE: 5%–Neat

PRIMARY CONSTITUENTS:[4509]

Fresh		*Dried*	
Bisabolone Oxide	43.3%	Palmitic Acid	21.6%
Palmitic Acid	12.6%	(E)-Phytol	7.3%
Bisabolol Oxide	7.8%	Hexahydrofarnesyl Acetone	4.1%
(E)-Phytol	6.5%	n-Tricosane	3.1%
2-Methyl Benzyl Alcohol	5.6%		
n-Pentacosane	4.5%		
(E)-Beta-Farnesene	4.0%		

OTHER CONSTITUENTS: *Fresh*—Methyl salicylate (< 1.5%), spathulenol, bisabolol oxide B, hexahydrofarnesyl acetone, benzyl salicylate (< 0.7%), linoleic acid, n-tricosane; *Dried*—2-Pentyl-furan, n-nonanal, methyl salicylate (< 1.3%), geranyl acetone, beta-ionone, lauric acid, myristic acid, farnesyl acetone C, n-pentacosane

PREFERRED COMPOSITION FOR CLINICAL USE:
Insufficient samples/data to determine.

Reported Therapeutic Properties: Aids cardiovascular function, anti-infectious, antibacterial, relieves fever, anti-inflammatory, anticonvulsive, antispasmodic, wound healing, relieves chronic skin conditions, antiallergic, aids respiratory function, eases cough, expels excess mucus, cleanses the blood, stimulates lymph flow, promotes restful sleep, relieves anxiety, combats anger and frustration

Cautions:

- Use cautiously in people with salicylate sensitivities due to the possible presence of methyl salicylate and benzyl salicylate.

Selected Evidence:

- *In vitro* research demonstrates that wild pansy inhibits *S. aureus, S. epidermidis, K. pneumonia,* and *E. coli.*[4510]

WIRE GRASS
(Wild Lemongrass, Agri Palma)

Elionurus muticus (Spreng.) Kuntze

Family: Poaceae (Gramineae)
Note: Middle
Aroma Intensity: Medium-Strong
Aroma: Fresh, lemony, grassy to sweet, woody, earthy, and slightly fruity
Common Extraction Method: Steam distilled from the leaves or aerial parts
Possible Substitute Oils: Lemongrass, melissa, citronella, copaiba, hemp
Blends Well With: Basil, black pepper, cassia, cedarwood, cinnamon, clary sage, geranium, lavandin, lavender, lemon, lemongrass, marjoram, melissa, ocotea, oregano, patchouli, rosemary, sandalwood, spikenard, tamala, tangerine, tea tree (melaleuca), ylang ylang
Subcellular Localization | Epigenetic Influence: Currently unknown | Currently unknown
Recommended Dilution Range: 1.5%–20%; 50% for some conditions

Primary Constituents:[4511,4512,4513,4514,4515]

Citral CT	
Geranial	35.0%–47.3%
Neral	31.5%–45.8%
Linalool	1.7%–11.8%
Beta-Pinene	0.0%–11.3%
Camphene	0.0%–5.0%
Borneol	0.0%–3.7%
ar-Curcumene	0.3%–3.6%
1,8-Cineole	0.4%–3.0%

Beta-Caryophyllene/Spathulenol CT	
Beta-Caryophyllene	17.9%–19.8%
Spathulenol	7.4%–18.6%
Caryophyllene Oxide	0.0%–13.4%

Citral/Caryophyllene CT	
Geranial	41.8%
Neral	29.0%
Caryophyllene	11.0%
Caryophyllene Oxide	10.6%
Neryl Acetate	7.6%

Myrcene CT	
Myrcene	11.6%–35.6%
Geraniol	1.7%–17.1%
Delta-Cadinene	0.0%–12.8%
Gamma-Cadinene	0.0%–12.3%
(Z)-Beta-Ocimene	6.0%–11.5%
Germacrene D-4-ol	0.8%–9.4%

Camphene	0.0%–11.5%
Bicyclogermacrene	0.0%–10.5%
Trans-Sesquisabinene Hydrate	0.0%–4.8%
Myrcene	0.3%–4.5%
(Z)-Beta-Farnesene	0.0%–4.5%
Alpha-Humulene	3.0%–4.0%
Limonene	0.2%–3.4%
(Z)-Trimenal	0.0%–3.0%

Bicyclogermacrene CT

Bicyclogermacrene	19.5%–33.3%
Beta-Caryophyllene	14.1%–18.5%
Myrcene	0.5%–7.1%
Spathulenol	5.6%–6.6%
Caryophyllene Oxide	2.4%–5.1%
Limonene	3.5%–4.6%
Alpha-Humulene	3.0%–3.5%
Viridiflorol	2.1%–3.1%

Borneol CT

Borneol	27.7%
Bisabolol	17.2%
Bornyl Acetate	11.7%
Pinocarvone	10.0%
Beta-Pinene	4.2%
Camphor	3.5%
Camphene	2.6%–8.3%
Borneol	2.4%–6.6%
Alpha-Pinene	2.3%–6.6%
Gamma-Terpinene	0.0%–5.7%
(E)-Beta-Ocimene	tr–5.3%
Alpha-Cadinol	1.0%–5.2%
Geranyl Acetate	0.1%–4.5%
Linalool	tr–4.5%
T-Muurolol	0.0%–3.8%
Germacrene A	0.0%–3.6%
Alpha-Muurolene	0.0%–3.4%
Bornyl Acetate	tr–3.1%
Tricyclene	0.7%–3.0%

Neryl Acetate CT

Neryl Acetate	12.1%
Geranyl Butyrate	11.6%
Beta-Eudesmol	9.7%
(E)-Nerolidol	9.5%
Linalool	8.7%
Nonanal	7.3%
Caryophyllene Oxide	3.6%

Other Constituents: *Citral CT*—Alpha-pinene, myrcene, 6-Methyl-5-hepten-2-one, limonene, (Z)-beta-ocimene, cis-cis-photocitral, trans-trans-photocitral, geraniol, myrtenal, piperitone, geranyl acetate, beta-elemene, beta-caryophyllene, 7-alpha-epi-selinene, geranyl butyrate; *Citral/Caryophyllene CT*—Not listed; *Beta-Caryophyllene/Spathulenol CT*—Unidentified constituents (21.3%–39.0%), alpha-thujene, alpha-pinene, myrcene, epi-beta-santalene, alpha-patchoulene, ar-curcumene, ledol, viridiflorol, acorenone, cedrene-8-en-9alpha-ol; *Bicyclogermacrene CT*—Beta-elemene, epi-beta-santalene, alpha-patchoulene, 9-epi-(E)-caryophyllene, ar-curcumene, trans-beta-guaiene, ledol; *Myrcene CT*—Limonene, beta-bourbonene, undecan-4-one, beta-caryophyllene, elemicin, alpha-humulene, neral, germacrene D, alpha-cadinene, elemol, beta-cadinol, torreyol, dill apiole (< 1.9%), apiole (< 1.5%); *Borneol CT*—Alpha-pinene, alpha-phellandrene, limonene, 1,8-cineole, nonanal, trans-pinocarveol, piperitone, neryl acetate, caryophyllene oxide; *Neral Acetate CT*—Alpha-pinene, beta-pinene, camphor, borneol, pinocarvone, bornyl acetate, ar-curcumene, valencene, spathulenol, isospathulenol, alpha-cadinol

Preferred Composition for Clinical Use:
Insufficient samples/data to determine.

Reported Therapeutic Properties: Antibacterial, antifungal, antimicrobial, antiviral, antiparasitic, analgesic, anti-inflammatory, antirheumatic, antispasmodic, nervine, antioxidant, relieves toothache, relieves stomachache, aids digestion, relieves diarrhea, reduces fever, insect repellent, insecticidal, antidepressant, aids mental clarity, aids memory function

CAUTIONS:

Citral CTs

■ Caution is advised during pregnancy and lactation due to high citral content. Large doses of citral may negatively affect fetal development according to animal studies.[4516]

■ There is a moderate risk that when wire grass is taken orally, it may interfere with enzymes responsible for metabolizing medications (NSAIDs, proton-pump inhibitors, acetaminophen, antiepileptics, immune modulators, blood-sugar medications, blood pressure medications, antidepressants, antipsychotics, diabetic medications, antihistamines, antibiotics, and anesthetics) due to the citral content.[4517,4518,4519,4520]

■ Use cautiously with diabetic medications. Large oral doses of citral may improve insulin sensitivity and lower blood-glucose levels according to animal research.[4521,4522]

■ May interact with antibiotics and possibly enhance their effects due to citral content.[4523,4524]

■ There is a moderate risk that wire grass may interfere with pentobarbital and other barbiturates (medications for anxiety and insomnia), based on citral content.[4525]

■ Dilution is recommended for topical application due to high citral content and the risk for skin irritation or sensitivity.[4526,4527] May also irritate mucous membranes.

Myrcene CT

■ May weakly interfere with the enzymes responsible for metabolizing medications (NSAIDs, proton-pump inhibitors, acetaminophen, antiepileptics, immune modulators, blood-sugar medications, blood pressure medications, antidepressants, antipsychotics, diabetic medications, antihistamines, antibiotics, and anesthetics), based on myrcene content.[4528]

Borneol CT

■ May interact with aspirin, blood pressure, antiplatelet, and anticoagulant medications, and increase the risk of bleeding among people with bleeding disorders, due to the anticoagulant properties of borneol.[4529,4530]

■ May interfere with enzymes responsible for metabolizing medications (NSAIDs, proton-pump inhibitors, acetaminophen, antiepileptics, immune modulators, blood-sugar medications, blood pressure medications, antidepressants, antipsychotics, diabetic medications, antihistamines, antibiotics, and anesthetics), based on borneol content.[4531,4532,4533,4534,4535]

Beta-Caryophyllene/Spathulenol CT and Neryl Acetate CTs

■ None currently known.

SELECTED EVIDENCE:

- □ Both the beta-caryophyllene/spathulenol and bicyclogermacrene CTs of wire grass essential oil were active against *Bacillus cereus, Pseudomonas aeruginosa*, and *Staphylococcus aureus*, but were inactive against *Escherichia coli.*[4536]
- □ The citral CT of wire grass (also rich in caryophyllene) showed contact toxicity on the stored-grain pest *Sitophilus oryzae* and repelled both *Tribolium castaneum* and *Sitophilus oryzae.*[4537]

WOOLY CUMIN
(Kamune es Sufi, Moudrayga)

Ammodaucus leucotrichus (Coss. & Dur.) Coss. & Dur., *A. leucotrichus* ssp. *leucotrichus* Coss. & Dur.

FAMILY: Apiaceae
NOTE: Top-Middle
AROMA INTENSITY: Medium
AROMA: Spicy, green, fruity, herbaceous, woody
COMMON EXTRACTION METHOD: Steam distilled or hydrodistilled from the seeds (fruits)
POSSIBLE SUBSTITUTE OILS: Perilla, nutmeg
BLENDS WELL WITH: Angelica root, black pepper, caraway, chamomile (German, Roman), coriander, cumin, lavender, nutmeg, perilla, rosemary, turmeric
SUBCELLULAR LOCALIZATION | EPIGENETIC INFLUENCE: Currently unknown | Currently unknown
RECOMMENDED DILUTION RANGE: 1%–10%

PRIMARY CONSTITUENTS:[4538,4539,4540,4541,4542,4543,4544]

Perillaldehyde	37.5%–88.7%
Limonene	1.7%–29.2%
Alpha-Pinene	0.0%–7.1%
Perilla Alcohol	0.0%–7.0%
Shybinol	0.0%–6.9%
Methyl Perillate	0.0%–6.4%
Delta-3-Carene	0.0%–4.4%
Beta-Caryophyllene	0.0%–3.6%
Caryophyllene Acetate	0.0%–3.0%

OTHER CONSTITUENTS: Allyl isovalerate, beta-pinene, camphene, myrcene, alpha-phellandrene, delta-3-carene, delta-cadinene, 11-acetoxyeudesman-4-a-ol, cis-limonene oxide, cuminal, beta-cubebene, cuminaldehyde, alpha-terpinyl-7-al, gamma-terpinen-7-al, 3-thujanone, acetic acid, (2-isopropenylcyclopentylidene)-, methyl ester

PREFERRED COMPOSITION FOR CLINICAL USE:

Constituent	
Perillaldehyde	58%–80%
Limonene	7%–25%
Alpha-Pinene	1.0%–8%

REPORTED THERAPEUTIC PROPERTIES: Anti-infectious, antimicrobial, antibacterial, antifungal, antioxidant, anticancer, anti-inflammatory, reduces fever, aids digestion, decreases colitis symptoms, relieves nausea/vomiting, antiallergic, balances menstruation, aids thyroid function, relieves snakebites, wound healing, antiseptic, eases sore throat, antidepressant, stress management, aphrodisiac

CAUTIONS:

■ Dilution strongly recommended and avoidance by people with sensitive skin is recommended. Perilla workers (also rich in perillaldehyde) are highly prone to skin irritation by perilla and those with dermatitis are highly likely to experience allergic reactions.[4545]

SELECTED EVIDENCE:

- □ Wooly cumin essential oil demonstrated strong anti-inflammatory activity when compared to dexamethasone (a steroid used to treat inflammation) and inhibited the secretion of inflammatory cytokines.[4546]
- □ Wooly cumin essential oil demonstrated antidepressant and neurological benefits based on its ability to selectively inhibit butyrylcholinesterase (BChE) (while not influencing AChE) and monoamine oxidase (MAO) enzyme activities.[4547] Inhibition of BChE has been shown to reduce Aß-amyloid levels (a hallmark of dementia and Alzheimer's disease). The oil inhibited MAO-A and strongly inhibited MAO-B, which was primarily attributed to perillaldehyde—although the whole essential oil was most effective. MAO-A inhibition leads to increased levels of serotonin, norepinephrine, and dopamine in the brain, which leads to reduced depression. In addition, inhibition of MAO-B preserves dopamine levels in the brain, which helps relieve symptoms of Parkinson's disease.
- □ Wooly cumin was inactive against breast cancer cells.[4548]
- □ *In vitro* research demonstrated that wooly cumin essential oil was most active against *Micrococcus luteus* and moderately active against *Klebsiella pneumonia, Staphylococcus aureus,* and *Bacillus subtilis.*[4549] Another study reported that wooly cumin inhibited gram-positive bacteria (*B. subtilis*), gram-negative bacteria (*E. coli*), yeasts (*S. cerevisiae, C. albicans*), and fungi (*P. expansum, A. flavus*), with strong inhibition of *S. cerevisiae.*[4550] Wooly cumin was effective against *S. aureus, E. coli*, and *P. aeruginosa*, *S. aureus* being the most sensitive bacteria.[4551] A wooly cumin with higher limonene content (16.6%) was more active against microorganisms (*E. coli, S. aureus, B. subtilis, M. luteus, A. niger, P. expansum, P. digitatum, C. puteana, C. versicolor, P. placenta, G. trabeum*) than a wooly cumin with high perillaldehyde (88.7%) and less limonene (8.7%). Wooly cumin also displayed significant activity against phytopathogenic fungi that cause apples rot (*B. cinerea, P. expansum*, and *R. stolonifer*).[4552] High antibacterial activity was observed against *Escherichia coli, S. aureus, E. cloaceae, B. cereus*, and *S. typhimurium*, and good antifungal activity against *F. oxysporum* and *A. flavus*.[4553] Wooly cumin possesses antimycotoxin activity against *A. flavus* and *A. ochraceus*.[4554] Wooly cumin tends to have stronger activity against gram-positive bacteria compared to gram-negative bacteria.[4555]
- □ Wooly cumin essential oil exhibited antioxidant activity similar to BHT in the β-carotene bleaching test.[4556] Another study found that wooly cumin is only a weak to moderate antioxidant according to the DDPH and β-carotene/linoleic acid assays.[4557]

XYLOPIA
(Meiú, Pindaíba)

Xylopia laevigata (Mart.) R .E. Fr.

Family: Annonaceae
Note: Middle-Base; Top
Aroma Intensity: Medium
Aroma: Herbaceous, green, floral, woody; fruity, sweet, woody
Common Extraction Method: Steam distilled or hydrodistilled from the leaves or fruits
Possible Substitute Oils: Ylang ylang, goldenrod, helichrysum, copaiba; frankincense, elemi, pink pepper
Blends Well With: Bay laurel, bergamot, cajeput, cardamom, chamomile (German, Roman), clary sage, clove, copaiba, coriander, cypress, eucalyptus, geranium, ginger, juniper, lavender, melissa, neroli, orange, palmarosa, patchouli, rose, sandalwood, turmeric, vetiver, ylang ylang
Subcellular Localization | Epigenetic Influence: Currently unknown | Currently unknown
Recommended Dilution Range: 3%–33%; 50% for some conditions

Primary Constituents:[4558,4559,4560,4561,4562]

Leaves (Germacrene D CT)

Germacrene D	18.9%–60.4%
Bicyclogermacrene	11.7%–18.4%
Beta-Elemene	0.0%–9.5%
Delta-Selinene	0.0%–9.2%
Beta-Caryophyllene	6.4%–8.6%
Gamma-Muurolene	0.6%–8.6%
Germacrene B	3.7%–7.3%
Delta-Cadinene	1.2%–6.8%
Alpha-Copaene	2.5%–4.8%
Delta-Elemene	1.6%–4.5%

Fruits

Limonene	56.2%
Alpha-Pinene	28.0%
Beta-Pinene	5.5%
Germacrene D	3.9%
Camphene	3.0%

Leaves (Gamma-Muurolene CT)

Gamma-Muurolene	17.8%–18.0%
Delta-Cadinene	12.2%–13.5%
Germacrene D	6.5%–9.1%
Bicyclogermacrene	7.0%–7.8%
Alpha-Copaene	6.0%–7.2%
Beta-Caryophyllene	5.4%–5.9%
Gamma-Cadinene	4.5%–4.7%
Aromadendrene	4.0%–4.7%
Gamma-Amorphene	3.9%–4.4%
Limonene	1.4%–3.4%
Germacrene B	2.9%–3.2%
Alpha-Cubebene	2.5%–3.0%

Other Constituents: *Leaves (Germacrene D CT)*—Alpha-pinene, beta-pinene, (Z)-beta-ocimene, limonene, beta-bourbonene, alpha-ylangene, beta-cubebene, gamma-elemene, (Z)-alpha-bisabolene, beta-copaene, aromadendrene, alpha-humulene, trans-muurola-4(14),5-diene, delta-amorphene, germacrene A, alpha-cadinene, gamma-cadinene, spathulenol, neo-intermedeol; *Leaves (Gamma-Muurolene CT)*—Alpha-pinene, (Z)-beta-ocimene, delta-elemene, beta-cubebene, alpha-ylangene, alpha-humulene, alpha-muurolene, delta-amorphene, trans-cadina-1,4-diene, alpha-cadinene, alpha-calacorene, spathulenol, alpha-muurolol, delta-cadinol, alpha-cadinene, caryophyllene oxide; *Fruits*—Myrcene, alpha-humulene

PREFERRED COMPOSITION FOR CLINICAL USE:
Insufficient samples/data to determine.

REPORTED THERAPEUTIC PROPERTIES: **Anti-inflammatory**, analgesic, reduces fever, eases cough, aids respiratory function, aids digestion, expels excess gas, relieves nausea, soothes skin infections, antimicrobial, antiparasitic, **anticancer**, antioxidant, lowers high blood pressure, supports cardiovascular function, insecticidal, insect repellent, improves self-confidence, relaxing, relieves anxiety

CAUTIONS:
- None currently known.

SELECTED EVIDENCE:
- Xylopia leaf (both the germacrene D and gamma-muurolene CTs) essential oil displayed anticancer activity against promyelocytic leukemia, ovarian cancer, glioblastoma, and colon cancer cells *in vitro*.[4563] Another study reported potent antiproliferative activity by the gamma-muurolene CT of the leaf oil against malignant melanoma, ovarian, and lung cancer cells.[4564] The scientists also reported that the oil prevented tumor growth in mice. The fruit oil was not toxic to promyelocytic leukemia, liver, melanoma, or myelogenous leukemia cancer cells.[4565]
- The leaf essential oil (gamma-muurolene CT) demonstrated significant anti-inflammatory, antinociceptive, and antioxidant activity in mice.[4566] Another study reported an ORAC (oxygen radical absorbance capacity) score of 1468.96 μmol of TE/g.[4567]
- Xylopia leaf essential oil CT germacrene D exhibited trypanocidal activity against epimastigote and trypomastigote forms of *T. cruzi* (a parasite that can cause fever, anorexia, myocarditis, swollen lymph nodes, and the simultaneous enlargement of the spleen and liver).[4568] The gamma-muurolene CT also exhibited trypanocidal activity.[4569]
- *In vitro* research showed that xylopia leaf (gamma-muurolene CT) essential oil inhibited *S. aureus, S. epidermidis,* and *C. albicans*.[4570]
- Xylopia leaf CT germacrene D essential oil provided weak protection against mosquitoes.[4571] It also displayed moderate larvicidal activity against *Aedes aegypti* mosquito larvae.[4572]

YANDEE
(York Gum)

Eucalyptus loxophleba Benth., *E. loxophleba* subsp. *gratiae* Brooker, E. *loxophleba* subsp. *lissophloia* L. A.S.Johnson & K.D.Hill, E. *loxophleba* subsp. *loxophleba* Benth., E. *loxophleba* subsp. *supralaevis* L.A.S.Johnson & K.D.Hill

FAMILY: Myrtaceae
NOTE: Top
AROMA INTENSITY: Strong
AROMA: Fresh, medicinal, slightly woody
COMMON EXTRACTION METHOD: Steam distilled or hydrodistilled from the
POSSIBLE SUBSTITUTE OILS: Eucalyptus, cajeput, ravintsara CT 1,8-cineole, myrtle

Blends Well With: Bay laurel, bergamot, blue cypress, cajeput, camphor, cedarwood, chamomile (German, Roman), eucalyptus, geranium, ginger, grapefruit, lavender, lemon, juniper, lemongrass, myrtle, peppermint, petitgrain, pine, ravensara, ravintsara, rosemary, sage, tea tree, tangerine, thyme
Subcellular Localization | Epigenetic Influence: Currently unknown | Currently unknown
Recommended Dilution Range: 3%–33%; neat for some conditions

Primary Constituents:[4573,4574,4575,4576,4577]

1,8-Cineole	24.2%–78.8%
4-Methyl-2-Pentyl Acetate	0.0%–19.8%
Alpha-Pinene	0.0%–15.2%
Aromadendrene	0.3%–10.0%
Spathulenol	0.0%–8.3%
Beta-Phellandrene	0.0%–6.2%
Viridiflorol	0.0%–6.0%
Bicyclogermacrene	0.0%–5.1%
Trans-Pinocarveol	0.0%–4.7%
Para-Cymene	1.0%–4.4%
Limonene	0.0%–3.7%
Caryophyllene Oxide	0.0%–3.0%
Globulol	0.0%–3.0%

Other Constituents: Myrcene, beta-pinene, alpha-phellandrene, terpinene-1-ol, alpha-gurjunene, terpinene-4-ol, alloaromadendrene, alpha-terpineol, ledene, epiglobulol, guaiol, beta-eudesmol, alpha-eudesmol, 2,2-dimethyl-3-heptanol, isoamyl acetate, pinocarvone, neoalloocimene, beta-selinene, 3-methylbutanal, beta-caryophyllene

Preferred Composition for Clinical Use:
Insufficient and too great of variance in samples/data to determine.

Reported Therapeutic Properties: Analgesic, antibacterial, antifungal, antirheumatic, antiseptic, antispasmodic, anesthetic, antiviral, antineuralgic, nervine, **decongestant**, diuretic, expectorant, eases cough, **supports respiratory function**, reduces fevers, promotes oral cleanliness, helps balance blood-sugar levels, wound healing, insect repellent, invigorating, aids concentration and memory, combats negative emotions, relieves mental exhaustion

Cautions:

- Avoid with children under age 3, particularly around the nose and mouth and use cautiously in children under age 5 due to significant 1,8-cineole content. 1,8-cineole may cause seizures, central nervous system problems, or respiratory distress in young children.[4578,4579,4580]
- Caution is warranted orally due to the significant amounts of 1,8-cineole. Limit it to small doses internally (adults—maximum 10 drops daily). Toxicity has been reported when eucalyptus (also high in 1,8-cineole) was ingested in large doses, and as few as 4 to 5 drops may cause problems in very sensitive individuals.[4581,4582,4583,4584,4585,4586]
- Avoid with epilepsy and Parkinson's disease due to 1,8-cineole content. May exacerbate or cause seizures/convulsions or reduce seizure-medication efficacy based on 1,8-cineole content.[4587,4588,4589,4590]
- May interact with diabetic medications due to significant 1,8-cineole content. 1,8-cineole significantly inhibits alpha-glucosidase activity, particularly when synergized with other constituents (alpha-pinene,

beta-pinene, and para-cymene).[4591,4592,4593] Alpha-glucosidase is an enzyme that breaks down carbohydrates by chemical reaction with water. Inhibiting its activity postpones glucose absorption and therefore the impact of carbohydrates on blood-sugar levels. Studies suggest that oils rich in 1,8-cineole may significantly lower glucose in animals, which may increase the action of diabetic medications.[4594,4595]

■ May interact with aspirin, blood pressure, antiplatelet, and anticoagulant medications, and increase the risk of bleeding among people with bleeding disorders (low risk) due to significant 1,8-cineole content.[4596] 1,8-cineole is a weak inhibitor of platelet aggregation.[4597]

■ May weakly interfere with the enzymes responsible for metabolizing medications (NSAIDs, proton-pump inhibitors, acetaminophen, antiepileptics, immune modulators, blood-sugar medications, blood pressure medications, antidepressants, antipsychotics, diabetic medications, antihistamines, antibiotics, and anesthetics) due to 1,8-cineole content.[4598,4599,4600]

■ May interfere with pentobarbital and other barbiturates (medications for anxiety and insomnia), based on 1,8-cineole content.[4601,4602]

SELECTED EVIDENCE:

- □ *In vitro* evidence demonstrated that yandee essential oil inhibits *S. aureus, E. coli, E. faecalis, S. epidermidis, C. albicans, S. typhimurium, P. aeruginosa,* and *A. bauimanii.*[4603]

YELLOW EVERLASTING DAISY (Bessmertnik Osh)

Helichrysum aucheri Boiss., *H. arenarium* subsp. *aucheri, H. arenarium* subsp. *ponticum*

FAMILY: Asteraceae (Compositae)
NOTE: Middle
AROMA INTENSITY: Medium
AROMA: Earthy, floral, woody
COMMON EXTRACTION METHOD: Steam distilled, hydrodistilled, or headspace microextracted from the aerial parts or flowers
POSSIBLE SUBSTITUTE OILS: Helichrysum, dwarf everlast, black pepper, copaiba, melissa
BLENDS WELL WITH: Bergamot, chamomile (German, Roman), clary sage, clove, cypress, geranium, grapefruit, helichrysum, lavender, lemon, lime, melaleuca (tea tree), neroli, orange, oregano, rosemary, tangerine, vetiver, ylang ylang
SUBCELLULAR LOCALIZATION | EPIGENETIC INFLUENCE: Currently unknown | Currently unknown
RECOMMENDED DILUTION RANGE: 3%–50%; neat for some conditions

PRIMARY CONSTITUENTS:[4604,4605]

Aerial Parts (HD)		*Flowers (HD)*	
Alpha-Pinene	39.6%	Alpha-Pinene	38.5%
1,8-Cineole	19.7%	Beta-Caryophyllene	23.0%
Beta-Caryophyllene	7.3%	1,8-Cineole	12.0%
Isoborneol	3.0%	Alpha-Humulene	6.6%
		Beta-Pinene	5.4%

Flowers (SFME)		*Flowers (HD-SPME)*	
Isoborneol	28.2%	Carvacrol	37.7%
Beta-Caryophyllene	12.9%	Alpha-Pinene	19.7%
Delta-Cadinene	6.3%	Beta-Caryophyllene	8.5%
Bornyl Acetate	6.0%	1,8-Cineole	7.4%
Limonene	5.9%	Isoborneol	3.7%
Caryophyllene Oxide	5.2%		
Camphene	4.4%		
Germacrene D	4.3%		
Alpha-Pinene	4.2%		
Alpha-Humulene	4.1%		
Alpha-Terpinyl Acetate	3.3%		
Beta-Cubebene	3.1%		

HD=Hydrodistillation; HD-SPME=Headspace microextraction; SFME=Hydrodistillation, solvent-free microwave extraction

OTHER CONSTITUENTS: *Aerial Parts*—Camphene, beta-pinene, limonene, fenchone, alpha-campholenal, camphor, alpha-terpineol, alpha-humulene, beta-selinene, caryolan-8-ol, caryophyllene oxide, caryophylla-4(14),8(15)dien-5.β-ol; *Flowers (HD)*—Camphene, limonene, gamma-terpinene, alpha-trans-bergamotene, beta-selinene, beta-bisabolene, (E)-nerolidol, caryolan-8-ol, palmitic acid; *Flowers (HD-SPME)*—Camphene, beta-pinene, myrcene, limonene, n-nonanal, alpha-terpineol, n-decanal, thymoquinone, geraniol, alpha-terpinyl acetate, n-tetradecane, (Z)-beta-farnesene, beta-selinene, delta-cadinene, beta-bisabolene, caryola-8-ol, caryophyllene oxide; *Flowers (SFME)*—Beta-pinene, beta-selinene, beta-bisabolene

PREFERRED COMPOSITION FOR CLINICAL USE:
Insufficient samples/data to determine.

REPORTED THERAPEUTIC PROPERTIES: Antibacterial, antifungal, antiviral, anti-inflammatory, antirheumatic, antiarthritic, analgesic, antioxidant, relieves burns/chronic skin conditions, wound healing, antiseptic, heals boils and blisters, antispasmodic, supports respiratory function, aids mental clarity, combats anger, stress management

CAUTIONS:
Carvacrol CT

■ Avoid during pregnancy and lactation. Oregano (also rich in carvacrol) may negatively affect embryonic development and encourage fetal cell death according to animal research.[4606]

■ May interact with diabetic medications and cause low blood-sugar levels due to carvacrol content—carvacrol reduces blood glucose and enhances the antidiabetic activities of rosiglitazone in animals.[4607,4608,4609,4610]

■ May interact with aspirin, blood pressure, antiplatelet, and anticoagulant medications, and increase the risk of bleeding among people with bleeding disorders, due to carvacrol and thymol content.[4611,4612,4613,4614,4615]

■ May interact with anticholinergic (drugs used for asthma, incontinence, gastrointestinal cramps, muscular spasms, depression, and sleep disorders) and cholinergic medications (drugs used to reduce urinary retention, diagnose myasthenia gravis, and for glaucoma), based on thymol and carvacrol content.[4616,4617]

■ May interact with antibiotics or antifungals and possibly enhance or decrease their effects due to carvacrol or thymol content.[4618,4619,4620,4621,4622]

■ May irritate mucous membranes (eyes, mouth, nasal passages, vagina, rectum).

All other CTs

■ None currently known, although caution should be exercised with epilepsy and convulsive disorders, due to the possible presence of moderate 1,8-cineole levels.

SELECTED EVIDENCE:

- □ None found.

ZIZIPHORA (Chopot, Kora-Dare)

Ziziphora tenuior L.

FAMILY: Lamiaceae
NOTE: Middle
AROMA INTENSITY: Strong
AROMA: Minty, herbaceous, fresh
COMMON EXTRACTION METHOD: Steam distilled or hydrodistilled from the aerial parts
POSSIBLE SUBSTITUTE OILS: Pennyroyal, applemint CT pulegone, Japanese catnip, lesser calamint
BLENDS WELL WITH: Applemint, citronella, geranium, lavender, pennyroyal, peppermint, rosemary, sage, Spanish sage, spearmint
SUBCELLULAR LOCALIZATION | EPIGENETIC INFLUENCE: Currently unknown | Currently unknown
RECOMMENDED DILUTION RANGE: 1.0%–5.0%

PRIMARY CONSTITUENTS:[4623,4624,4625,4626,4627,4628,4629,4630,4631]

Pulegone CT	
Pulegone	46.8%–90.1%
P-Menth-3-en-8-ol	0.0%–12.5%
Limonene	0.5%–8.3%
Isomenthone	0.0%–6.6%
8-Hydroxymenthone	0.0%–6.2%
Caryophyllene Oxide	0.0%–5.3%
Thymol	0.0%–4.3%
1,8-Cineole	0.0%–4.2%
Alpha-Pinene	tr–3.4%
Terpinolene	0.0%–3.1%

Alpha-Terpineol CT	
Alpha-Terpineol	16.2%–19.5%
Thymol	10.4%–13.9%
Benzene Methyl-1-Methylethyl	0.0%–8.8%
Para-Cymene	0.0%–6.1%
Geraniol	4.4%–5.8%
Geranyl Acetate	0.0%–5.2%
1,8-Cineole	4.1%–5.1%
Linalool	0.0%–3.8%

1,8-Cineole CT	
1,8-Cineole	19.6%
Spathulenol	12.2%
Myrcene	7.6%
Beta-Caryophyllene	7.5%
Beta-Pinene	5.4%
Germacrene D	4.9%
Alpha-Pinene	4.3%
Beta-Bourbonene	4.0%
Alpha-Copaene	3.3%

SFE	
P-Mentha-3-en-8-ol	54.0%
Pulegone	38.5%

Beta-Caryophyllene	3.4%–3.5%
Borneol	2.1%–3.3%
(Z)-Beta-Ocimene	2.8%–3.2%
Alpha-Pinene	3.1%–3.2%

OTHER CONSTITUENTS: *Pulegone CT*—Alpha-thujene, sabinene, camphene, beta-pinene, myrcene, gamma-terpinene, isopulegol, menthol, camphor, trans-pulegone oxide, 8-hydroxyisomenthone, carvacrol, beta-caryophyllene, isopulegone-I, 3,7-dimehtyloct-1,5,7-trien-3-ol, isopiperitenone, benzenepropanoic acid, beta-cubebene, germacrene D, gamma-cadinene; *Alpha-Terpineol CT*—Alpha-phellandrene, camphene, sabinene, beta-pinene, myrcene, alpha-terpinene, gamma-terpinene, camphor, terpinen-4-ol, carvone, neral, carvacrol methyl ether, carvacrol, ethylene methacrylate, beta-cubebene, alpha-amorphene, spathulenol, caryophyllene oxide, T-cadinol; *1,8-Cineole CT*—Alpha-thujene, camphene, sabinene, 1-ocen-3-ol, para-cymene, (E)-beta-ocimene, gamma-terpinene, linalool, terpinen-4-ol, pulegone, aromadendrene, bicyclogermacrene, gamma-cadinene, delta-cadinene, isospathulenol, oplopenone, hexadecanoic acid, tricosane; *SFE*—Alpha-thujene, beta-pinene, p-mentha-3,8-diene, 4aα,7α,7aα-nepetalactone

PREFERRED COMPOSITION FOR CLINICAL USE:
Should only be used if the benefits outweigh the risks as determined by a qualified professional.

Constituent	
Pulegone	65%–85%
Limonene	0.5%–10%
1,8-Cineole	tr–5%

REPORTED THERAPEUTIC PROPERTIES: Antifungal, antibacterial, antimicrobial, antiseptic, anti-inflammatory, reduces fever, analgesic, relieves headache and migraine, relieves diarrhea, expels excess gas, eases cough, relieves uterine and vaginal infections, stimulates menstruation, wound healing, soothes hemorrhoids, relieves edema, reduces high blood pressure, promotes restful sleep, insecticidal, calms hysteria, stress management, relieves anxiety, antidepressant

CAUTIONS:
Pulegone CT and SFE

■ Avoid with children under age twelve. Multiple organ failure, liver failure, brain swelling (cerebral edema), severe epileptic encephalopathy (severe brain disorder with seizures), organ cell death (necrosis), and death has been reported in infants who were administered mint tea containing pennyroyal essential oil (also rich in pulegone).[4632] Pulegone and its metabolite menthofuran (or metabolites of menthofuran) are believed to cause the toxic effects of pennyroyal, which include liver and kidney toxicity, bronchiolar cell destruction, depletion of liver glutathione levels (the primary protective antioxidant of the liver), and direct cellular damage.[4633,4634,4635] Children are more likely to experience these effects.[4636]

■ Avoid during pregnancy and lactation due to significant amounts of pulegone and potential toxicity to multiple organs.[4637,4638] Pennyroyal (also rich in pulegone) is believed to trigger uterine contractions by irritating the genito-urinary system. Pennyroyal is commonly used to induce abortions.[4639,4640] While pennyroyal herb is considered abortifacient, the essential oil is not.[4641] When pennyroyal essential oil is used alone to induce abortion, it generally is unsuccessful but causes serious health consequences to the mother. Carachipita is an abortant sold over the counter in South America that contains pennyroyal, yerba de la perdiz, oregano, and marsh rosemary. The literature reports successful use of Carachipita to induce abortion, but not without serious consequences to the mother (multiple organ failure, vaginal bleeding, destruction of red blood cells, numbness, dizziness, and death).[4642,4643,4644] It is prudent to avoid Japanese catnip oil while pregnant or nursing due to the toxicity of pulegone alone, despite a lack of evidence that it is a direct abortifacient.

■ Avoid oral consumption due to the toxic potential of pulegone and its metabolites to multiple organs. Menthofuran is generally a very minor constituent (or absent) in ziziphora essential oil, but it is considered more toxic than pulegone.[4645] In addition, pulegone is metabolized by human liver cytochrome P-450s (CYP2E1, CYP1A2, and CYP2C19) to menthofuran after administration, which actually increases its toxic effect to the liver.[4646,4647] Pulegone is also metabolized to p-cresol, which depletes glutathione levels and can increase liver damage.[4648] Animal research suggests oral administration of pulegone is significantly toxic to the liver and causes death at 300 mg/kg (over 23 mL for a human adult equivalent dose) for two weeks.[4649] The same research concluded that lower doses of 75 and 150 mg/kg for three months resulted in liver, kidney, bone marrow, heart, stomach, lung, ovarian, and thymus damage, and one death in animals. A two-year animal study reported urinary cancer and papillomas, liver cancer, osteosarcoma, and multiple organ damage or enlargement (liver, kidney, olfactory epithelium, lung, bile duct).[4650] Ingestion of pulegone-rich essential oils can cause abdominal pain and tenderness, nausea, bloody vomiting, burning of the throat, fever, lethargy alternating with agitation, confusion, delirium, restlessness, seizures, dizziness, auditory and visual hallucinations, increased blood pressure, rapid heartbeat, swollen blood vessels in the lungs and filling of the alveoli with blood, excessive acidity of the blood, blood clots, abortion, liver failure, kidney failure, respiratory failure, shock, coma, and death.[4651] Neurological damage has been reported in adults who consumed 2.5 to 5.0 mL of pennyroyal essential oil, moderate to severe poisoning with single doses of 10 mL, and deaths occurred at over 15 mL of ingestion.[4652,4653,4654,4655,4656,4657,4658,4659,4660,4661,4662]
■ Avoid with epilepsy and Parkinson's disease due to pulegone content, which may exacerbate or cause seizures or convulsions.[4663,4664,4665]
■ Avoid in people with compromised liver or kidney due to significant potential to cause liver and kidney toxicity or damage caused by pulegone.[4666,4667,4668,4669,4670] This warning includes use with medications known to harm the liver.

1,8-Cineole CT
■ Avoid with epilepsy and Parkinson's disease due to 1,8-cineole and camphor content. May exacerbate or cause seizures/convulsions or reduce seizure-medication efficacy based on 1,8-cineole content.[4671,4672,4673,4674]

Alpha-Terpineol CT
■ None currently known.

Selected Evidence:

- □ Ziziphora CT pulegone essential oil displayed anti-inflammatory activity (decreased nitric oxide production) at nontoxic concentrations.[4675]
- □ *In vitro* research demonstrates that ziziphora CT alpha-terpineol essential oil significantly inhibits *A. flavus*, *A. parasiticus*, *A. niger* (better than the standard drug amphotericin), moderately inhibits *C. albicans, S. aureus, S., saprophyticus,* and *E. coli,* and inhibits *E. faecalis, B. cereus, S. typhimurium,* and *P. aeruginosa*.[4676] The pulegone CT inhibited multiple dermatophytes, yeasts, and *Aspergillus* strains by inhibiting germ tube formation with a filamentation inhibition rate over 80% (both important virulence factors).[4677]
- □ The addition of 0.2% ziziphora essential oil (unknown CT) to a yogurt-based drink inhibited *S. aureus*.[4678]
- □ Injection of 0.07 mL/kg body weight of ziziphora essential oil (assumed by the authors to be pulegone type) produced a central nervous system depressant effect in mice.[4679]
- □ Ziziphora CT 1,8-cineole essential oil exhibited considerable antioxidant capacity in the DPPH assay, particularly its polar fraction.[4680]

REFERENCES:

[1] Ekor M. The growing use of herbal medicines: issues relating to adverse reactions and challenges in monitoring safety. *Front Pharmacol.* 2013;4:177.

[2] Wachtel-Galor S, Benzie FF. Herbal Medicine: Biomolecular and Clinical Aspects. 2nd edition. Chapter 1, Herbal Medicine. CRC Press: Boca Raton, FL, 2011.

[3] Barnes PM, Bloom B, Nahin R. Complementary and alternative medicine use among adults and children: United States, 2007. CDC National Health Statistics Report #12. 2008. Available at: www.cdc.gov/nchs/data/nhsr/nhsr012.pdf.

[4] Committee on the Use of Complementary, and Alternative Medicine by the American Public Board on Health Promotion Disease Prevention Institute of Medicine. (2005). Complementary and Alternative Medicine in the United States. Washington, DC: The National Academies Press

[5] Braun LA, Tiralongo E, Wilkinson JM, et al. Perceptions, use and attitudes of pharmacy customers on complementary medicines and pharmacy practice. *BMC Complement Altern Med.* 2010 Jul 20; 10():38.

[6] Anquez-Traxler C. The legal and regulatory framework of herbal medicinal products in the European Union: a focus on the traditional herbal medicines category. *Drug Inf J.* 2011;45:15–23

[7] Calapai G. European legislation on herbal medicines: a look into the future. *Drug Saf.* 2008;31(5):428-31.

[8] Subhose V, Srinivas P, Narayana A. Basic principles of pharmaceutical science in Ayurvĕda. *Bulletin Indian Institute History Medicine.* 2005;35(2):83–92.

[9] Pandey MM, Rastogi S, Rawat AKS. Indian Traditional Ayurvedic System of Medicine and Nutritional Supplementation. *Evidence Based Complemetn Alternat Med.*2013;2013:376327.

[10] Xutian S, Zhang J, Louise W. New exploration and understanding of traditional Chinese medicine. *Am J Chin Med.* 2009;37:411–26.

[11] Kendall D. Dao of Chinese Medicine, Understanding an Ancient Healing Art. Oxford University Press: Great Clarendon St, Oxford, 2002, pages 30-55.

[12] Song BK, Won JH, Kim S. Historical Medical Value of Donguibogam. J Pharmacopuncture. 2016 Mar;19(1):16–20.

[13] Zeinalian M, Eshaghi M, Naji H, et al. Iranian-Islamic traditional medicine: An ancient comprehensive personalized medicine. *Adv Biomed Res.* 2015; 4: 191.

[14] Hatami H. Public health and preventive medicine according to Avicenna. *Sci J Islam Repub Iran Med Counc.* 2000;18:223–38.

[15] Burt S. Essential oils: their antibacterial properties and potential applications in foods--a review. *Int J Food Microbiol.* 2004 Aug 1;94(3):223-53.

[16] Plant J, Stephens B. Effects of Essential Oils on Transcriptional Regulation in Human Cells. *Med Aromat Plants.* 2015;4:200.

[17] Bhat KV, Balasundaran M, Balagopalan M. Identification of Santalum album and Osyris lanceolata through morphological and biochemical characteristics and molecular markers to check adulteration. Kerala Forest Research Institute.

[18] Mwang'ingo PL, Teklehaimanot Z, Hall JB, et al. African sandalwood (Osyris lanceolata): Resource assessment and quality variation among populations in Tanzania. *South African Forestry J.* 2003 Nov;199:77-88.

[19] Kreipl AT, Konig WA. Sesquiterpenes from the east African sandalwood Osyris tenuifolia. *Phytochem.* 2004;65:2045-9.

[20] Adeogun OO, Maroyi A, Afolayan AJ. Variation in the chemical composition of essential oils from Artemisia afra (Jacq) ex-Wild leaf obtained by different methods and the effect of oil extracts on Artemia salina L. *Trop J Pharm Res.* 2018 Mar;17(3):519-28.

[21] Chagonda LS, Makanda C, Chalchat JC. The essential oil of cultivated Artemisia afra (Jacq.) from Zimbabwe. *Flavour Frag J.* 1999;14:140-2.

[22] Vagionas K, Graikou K, Chinou IB. Chemical Analysis and Antimicrobial Activity of Essential Oils from the Aromatic Plants Artemisia Afra Jacq. and Leonotis Ocymifolia (Burm. F.) lwarsson var. Raineriana (Vision1) lwarsson Growing In Tanzania. *J Essent Oil Res.* 2007 Jul/Aug;19:301-305.

[23] Mangena T, Muyima NYO. Comparative evaluation of the antimicrobial activities of essential oils of Artemisia afra, Pteronia incana and Rosmarinus officinals on selected bacteria and yeast strains. *Lett Appl Microbiol.* 1999;28:291-6.

[24] Graven EH, Deans SG, Svoboda KP, et al. Antimicrobial and Antioxidative Properties of the Volatile (Essential) Oil of Artemisia afra Jacq. *Flavour Fragr J.* 1992;7:121-3.

[25] Oyedemi SO, Pirochenva G, Mabinya LV, et al. Compositions and comparisons of antimicrobial potencies of some essential oils and antibiotics against selected bacteria. *African J Biotechnol.* 2008 Nov;7(22):4140-6.

[26] Asfaw N, Demissew S. Essential oil composition of four Artemisia species from Ethiopia. *Bull Chem Soc Ethip.* 2015;29(1):123-8.

[27] Burits M, Asres K, Bucar F. The Antioxidant Activity of the Essential Oils of Artemisia afra, Artemisia abyssinica and Juniperus procera. *Phytother Res.* 2001;15:103-8.

[28] Mwangi JW, Achola KJ, Sinei KA, et al. Essential Oil Constituents of Artemisia afra Willd. *J Essent Oil Res.* 1995;7(1):97-99.

[29] Graven EH, Webber L, Venter M, et al. The Development of Artemisia afra (Jacq.) as a New Essential Oil Crop. *J Essent Oil Res.* 1990;2(5):215-20.

[30] Oyedeji AO, Afolayan AJ, Hutchings A. Compositional variation of the essential oils of Artemisia afra Jacq. from three provinces in South Africa--a case study of its safety. *Nat Prod Commun.* 2009 Jun;4(6):849-52.

[31] Vagionas K, Graikou K, Chinou IB. Chemical Analysis and Antimicrobial Activity of Essential Oils from the Aromatic Plants Artemisia Afra Jacq. and Leonotis Ocymifolia (Burm. F.) lwarsson var. Raineriana (Vision1) lwarsson Growing In Tanzania. *J Essent Oil Res.* 2007 Jul/Aug;19:301-305.

[32] Mangena T, Muyima NYO. Comparative evaluation of the antimicrobial activities of essential oils of Artemisia afra, Pteronia incana and Rosmarinus officinals on selected bacteria and yeast strains. *Lett Appl Microbiol.* 1999;28:291-6.

[33] Graven EH, Deans SG, Svoboda KP, et al. Antimicrobial and Antioxidative Properties of the Volatile (Essential) Oil of Artemisia afra Jacq. *Flavour Fragr J.* 1992;7:121-3.

[34] Oyedemi SO, Pirochenva G, Mabinya LV, et al. Compositions and comparisons of antimicrobial potencies of some essential oils and antibiotics against selected bacteria. *African J Biotechnol.* 2008 Nov;7(22):4140-6.

[35] Gundidza M. Antifungal activity of essential oil from Artemisia afra Jacq. *Cent Afr J Med.* 1993 Jul;39(7):140-2.

[36] Graven EH, Deans SG, Svoboda KP, et al. Antimicrobial and Antioxidative Properties of the Volatile (Essential) Oil of Artemisia afra Jacq. *Flavour Fragr J.* 1992;7:121-3.

[37] Burits M, Asres K, Bucar F. The Antioxidant Activity of the Essential Oils of Artemisia afra, Artemisia abyssinica and Juniperus procera. *Phytother Res.* 2001;15:103-8.

[38] Ramdani M, Lograda T, Chalard P, et al. Chemical and antimicrobial properties of essential oils of Abies numidica, endemic species of Algeria. *Int J Phytopharmacol.* 2014;5(6):432-40.

[39] Tlili Ait-Kaki Y, S Bennadja, Chefrour A. Revalorisation d'une essence endémique, le sapin de Numidie (Abies numidica). Fl Medit. 2013;23:123-29.

[40] Tlili Ait Laki Y, Bennadja S, Abdelghani D. The Therapeutic Importance of Products Extracted from the Fir Tree of Numidia (Abies numidica) and Research on its Antibacterial Activity. *Kastamonu Univ J Forestry Faculty*. 2012;Special Issue:279-82.
[41] Ramdani M, Lograda T, Chalard P, et al. Chemical and antimicrobial properties of essential oils of Abies numidica, endemic species of Algeria. *Int J Phytopharmacol*. 2014;5(6):432-40.
[42] Mouloud G, Samir M, Rabah B, et al. Antioxidant and antimicrobial activities of endemic tree Abies numidica growing in Babor Mountains from Algeria. *Global J Res Med Plants & Indigen Med.* 2016;5(1):19-28.
[43] Mouloud G, Samir M, Rabah B, et al. Antioxidant and antimicrobial activities of endemic tree Abies numidica growing in Babor Mountains from Algeria. *Global J Res Med Plants & Indigen Med.* 2016;5(1):19-28.
[44] Chalchat JC, Garry RP, Mathieu JP. Composition of the Volatile Fraction from Honduras Styrax, Liquidambar styraciflua L. *J Essent Oil Res.* 1994;6(1):73-5.
[45] Fernandez X, Lizzani-Cuvelier L, Loiseau AM, et al. Chemical composition of the essential oils from Turkish and Honduras Styrax. *Flavour Fragr J.* 2005;20:70-3.
[46] El-Readi MZ, Eid HH, Ashour ML, et al. Variations of the chemical composition and bioactivity of essential oils from leaves and stems of Liquidambar styraciflua (Altingiaceae). *J Pharmacy Pharmacol.* 2013;65:1653-63.
[47] Tattje DHE, Bos R, Bruins AP. Constituents of essential oil from leaves of Liquidambar styraciflua L. *J Med Plant Res.* 1980;38;79-85.
[48] European Commission. Opinion on Liquidambar spp. Balsma Extracts and Oils (Storax), sesitization only. Available at: https://ec.europa.eu/health/ph_risk/committees/04_sccp/docs/sccp_o_025a.pdf.
[49] Agency for Toxic Substances & Disease Registry. Toxic Substances Portal – Styrene. Available at: https://www.atsdr.cdc.gov/phs/phs.asp?id=419&tid=74.
[50] Occupational Safety and Health Administration. Styrene. Available at: https://www.osha.gov/SLTC/styrene/hazards.html.
[51] European Commission. Opinion on Liquidambar spp. Balsma Extracts and Oils (Storax), sesitization only. Available at: https://ec.europa.eu/health/ph_risk/committees/04_sccp/docs/sccp_o_025a.pdf.
[52] El-Readi MZ, Eid HH, Ashour ML, et al. Variations of the chemical composition and bioactivity of essential oils from leaves and stems of Liquidambar styraciflua (Altingiaceae). *J Pharmacy Pharmacol.* 2013;65:1653-63.
[53] El-Readi MZ, Eid HH, Ashour ML, et al. Variations of the chemical composition and bioactivity of essential oils from leaves and stems of Liquidambar styraciflua (Altingiaceae). *J Pharmacy Pharmacol.* 2013;65:1653-63.
[54] El-Readi MZ, Eid HH, Ashour ML, et al. Variations of the chemical composition and bioactivity of essential oils from leaves and stems of Liquidambar styraciflua (Altingiaceae). *J Pharmacy Pharmacol.* 2013;65:1653-63.
[55] Godinho WM, Farnezi MM, Pereira IM, et al. Volatile constituents from leaves of Blepharocalyx salicifolius (Kunth) O. Berg (Myrtaceae). *Bol Latinoam Caribe Plant Med Aromat.* 2014;13(3):249–53.
[56] Moreira JJS, Pereira CS, Ethur EM, et al. Volatile Constituents Composition of Blepharocalyx salicifolius Leaf Oil. *J Essent Oil Res.* 1999 Jan/Feb;11:45-8.
[57] Hernandez JJ, Ragone MI, Bonazzola P, et al. Antitussive, antispasmodic, bronchodilating and cardiac inotropic effects of the essential oil from Blepharocalyx salicifolius leaves. *J Ethnopharmacol.* 2018;210:107-117.
[58] Bueon da Costa O, Soares Del Menezzi CH, Celino Benedito LE, et al. Essential Oil Constituents and Yields from Leaves of Blepharocalyx salicifolius (Kunt) O. Berg and Myracrodruon urundeuva (Allemão) Collected during Daytime. *Int J Forestry Res.* 2014;2014:982576.
[59] Limberger RP, Sobral MEG, Zuanazzi JAS, et al. Biological Activities and Essential Oil Composition of Leaves of Blepharocalyx salicifolius. *Pharm Biol.* 2001;39(4):308-11.
[60] Furtado FB, Borges BC, Teixeira TL, et al. Chemical Composition and Bioactivity of Essential Oil from Blepharocalyx salicifolius. *Int J Mol Sci.* 2018;19(33):1-13.
[61] Castelo AVM, del Menezzi CHS, Resck IS, et al. Seasonal variation in the yield and chemical composition of essential oils from two Brazillian native Arbustrive species. *J Appl Sci.* 2012;12(8):753-60.
[62] Dellcassa E, Lorenzo D, Modello L, et al. Uruguayan Essential Oils. Part IX. Composition of Leaf Oil of Blepharocalyx tweediei (Hook, et Arn.) Berg var. tweediei (Myrtaceae). *J Essent Oil Res.* 1998 Nov/Dec;9:673-6.
[63] Furlan R, Zacchino S, Gattuso M, et al. Constituents of the Essential Oil from Leaves and Seeds of Blepharocalyx tweediei (Hook, et Arn.) Berg and B. gigantea Lillo. *J Essent Oil Res.* 2002 May/Jun;14:175-8.
[64] Tucker AO, Maciarello MJ. Volatile leaf oils of American Myrtaceae. I. Belpharocalyx cruckshanksii (Hook. & Arn.) Niedenzu of Chile and B. salicifolius (Humb., Bonpl. &Kunth) Berg of Argentina. *J Essent Oil Res.* 1993 May/Jun;5:333-5.
[65] Craig JO. Poisoning by the volatile oils in childhood. *Arch Dis Child.* 1953;28:259-67.
[66] Melis K. Bochner A, Janssens G. Accidental nasal eucalyptol and menthol instillation. *Eur J Pediatr.* 1989 Aug;148(8)786-7.
[67] Day LM, Ozanne-Smith J, Parsons BJ, et al. Eucalyptus oil poisoning among young children: mechanisms of access and the potential prevention. *Aust N Z J Public Health.* 1997 Jun;21(3):297-302.
[68] Day LM, Ozanne-Smith J, Parsons BJ, et al. Eucalyptus oil poisoning among young children: mechanisms of access and the potential prevention. *Aust N Z J Public Health.* 1997 Jun;21(3):297-302.
[69] Myott E. Case of eucalyptus poisoning. *Brit M J.* 1906;1:558.
[70] Hindle RC. Eucalyptus oil ingestion. *N Z Med J.* 1994 May 11;107(977)185-6.
[71] Tibballs J. Clinical effects and management of eucalyptus oil ingestion in infants and small children. *Med J Aust.* 1995 Aug;163(4):177-80.
[72] Waldman W, Barwina M, Sein Anand J. Accidental intoxication with eucalyptus oil—a case report. *Przeql Lek.* 2011;68(8):555-6.
[73] Day LM, Ozanne-Smith J, Parsons BJ, et al. Eucalyptus oil poisoning among young children: mechanisms of access and the potential prevention. *Aust N Z J Public Health.* 1997 Jun;21(3):297-302.
[74] Burkhard PR, Burkhardt K, Haenggeli CA, et al. Plant-induced seizures: reappearance of an old problem. *J Neurol.* 1999 Aug;246(8):667-70.
[75] Waldman N. Seizure caused by dermal application of over-the-counter eucalyptus oil head lice preparation. *Clin Toxicol (Phila).* 2011 Oct;49(8):750-1.
[76] Craig JO. Poisoning by the volatile oils in childhood. *Arch Dis Child.* 1953;28:259-67.
[77] Mathew T, Kamath V, Kumar RS, et al. Eucalyptus oil inhalation-induced seizure: A novel, underrecognized, preventable cause of acute symptomatic seizure. *Epilepsia Open.* 2017 Jul 4;2(3):350-354.
[78] Unger M, Frank A. Simultaneous determination of the inhibitory potency of herbal extracts on the activity of six major cytochrome P450 enzymes using liquid chromatography/mass spectrometry and automated online extraction. *Rapid Commun Mass Spectrom.* 2004;18(19):2273-81.
[79] Jori A, Bianchetti A, Prestini PE, et al. Effect of eucalyptol (1,8-cineole) on the metabolism of other drugs in rats and in man. *Eur J Pharmacol.* 1970;9(3):362-66.

[80] Kim NH, Hyun SH, Jin CH, et al. Pretreatment with 1,8-cineole potentiates thioacetamide-induced hepatotoxicity and immunosuppression. *Arch Pharm Res*. 2004 Jul;27(7):781-9.
[81] Jori A, Bianchetti A, Prestini PE, et al. Effect of eucalyptol (1,8-cineole) on the metabolism of other drugs in rats and in man. *Eur J Pharmacol.* 1970;9(3):362-66.
[82] de Sousa DP, Raphael E, Brocksom U, et al. Sedative effect of monoterpene alcohols in mice: A preliminary screening. *Verlag der Zeitschrift fur Naturforschung*. 2007;62c:563-66.
[83] Burkhard PR, Burkhardt K, Haenggeli CA, et al. Plant-induced seizures: reappearance of an old problem. *J Neurol.* 1999 Aug;246(8):667-70.
[84] Waldman N. Seizure caused by dermal application of over-the-counter eucalyptus oil head lice preparation. *Clin Toxicol (Phila)*. 2011 Oct;49(8):750-1.
[85] Craig JO. Poisoning by the volatile oils in childhood. *Arch Dis Child.* 1953;28:259-67.
[86] Mathew T, Kamath V, Kumar RS, et al. Eucalyptus oil inhalation-induced seizure: A novel, underrecognized, preventable cause of acute symptomatic seizure. *Epilepsia Open*. 2017 Jul 4;2(3):350-354.
[87] Unger M, Frank A. Simultaneous determination of the inhibitory potency of herbal extracts on the activity of six major cytochrome P450 enzymes using liquid chromatography/mass spectrometry and automated online extraction. *Rapid Commun Mass Spectrom.* 2004;18(19):2273-81.
[88] Jori A, Bianchetti A, Prestini PE, et al. Effect of eucalyptol (1,8-cineole) on the metabolism of other drugs in rats and in man. *Eur J Pharmacol.* 1970;9(3):362-66.
[89] Kim NH, Hyun SH, Jin CH, et al. Pretreatment with 1,8-cineole potentiates thioacetamide-induced hepatotoxicity and immunosuppression. *Arch Pharm Res*. 2004 Jul;27(7):781-9.
[90] Jori A, Bianchetti A, Prestini PE, et al. Effect of eucalyptol (1,8-cineole) on the metabolism of other drugs in rats and in man. *Eur J Pharmacol.* 1970;9(3):362-66.
[91] de Sousa DP, Raphael E, Brocksom U, et al. Sedative effect of monoterpene alcohols in mice: A preliminary screening. *Verlag der Zeitschrift fur Naturforschung*. 2007;62c:563-66.
[92] Furtado FB, Borges BC, Teixeira TL, et al. Chemical Composition and Bioactivity of Essential Oil from Blepharocalyx salicifolius. *Int J Mol Sci*. 2018;19(33):1-13.
[93] Limberger RP, Sobral MEG, Zuanazzi JAS, et al. Biological Activities and Essential Oil Composition of Leaves of Blepharocalyx salicifolius. *Pharm Biol*. 2001;39(4):308-11.
[94] Furtado FB, Borges BC, Teixeira TL, et al. Chemical Composition and Bioactivity of Essential Oil from Blepharocalyx salicifolius. *Int J Mol Sci*. 2018;19(33):1-13.
[95] Andrade MA, Azevedo CD, Motta FN, et al. Essential oils: in vitro activity against Leishmania amazonensis, cytotoxicity and chemical composition. *BMC Complement Altern Med*. 2016 Nov 8;16(1):444.
[96] Andrade MA, das Gracas Cardoso M, de Andrade J, et al. Chemical Composition and Antioxidant Activity of Essential Oils from Cinnamodendron dinisii Schwacke and Siparuna guianensis Aublet. *Antioxidants*. 2013;2:384-97.
[97] Zoghbi MDGB, Andrada EHA, Santos AS, et al. Essential Oils of Siparuna guianensis Aubl. *J Essent Oil Res.* 1998 Sep-Oct;10:543-6.
[98] Fernandes Machado SM, Adauto Ribeiro VAF, Militao JSLT, et al. Seasonal Variation of (E)-Verolidol in Siparuna guianensis Aublet and 13C-NMR Spectral Assignments of (E)- and (Z)-Nerolidol. *J Essent Oil Res.* 2001 Mar-Apr;13:130-1.
[99] Souza Aguiar RW, dos Santos SF, da Silva Morgado F, et al. Insecticidal and Repellent Activity of Siparuna guianensis Aubl. (Negramina) against Aedes aegypti and Culex quinquefasciatus. *PLoS One*. 2015 Feb 3;10(w)e0116765.
[100] Ferreira TP, Ponki Kuckelhaus JG, Tschoeke PH, et al. Influence of seasonality on the yield and composition of the essential oil of Siparuna guianensis Aublet. *African J Biotech*. 2017 Jul 19;16(29):1611-8.
[101] Fischer DCH, Limberger RP, Henriques AT, et al. Essential oils from fruits and leaves of Siparuna guianensis (Aubl.) Tulasne from southeastern Brazil. *J Essent Oil Res*. 2005 Jan-Feb;17:101-2.
[102] Ferreira TP, Oliveira EE, Tschoeke PH, et al. Potential use of Negramina (Siparuna guianensis Aubl.) essential oil to control wax moths and its selectivity in relation to honey bees. *Industrial Crops Prod*. 2017 Dec;109:151-57.
[103] Faria de Bessa NG, Bezerra Periera MA, Ferraz V, et al. Antimicrobial activity and medicinal biomass of Siparuna guianensis in Brazilian Cerrado forest, a global hotspot. *J Medicinal Plants Res*. 2015 Oct;9(37):968-80.
[104] Franke Portella AC, Munaro M, Donizete Ascencio S, et al. Physical and chemical characterization of essential oil of Siparuna guianensis Aublet. *Quim Nova*. 2014 jun;37(5):0100-4042.
[105] Andrade EHA, Zoghbi MGB, Machado LB. Seasonal variation of germacrone and atractylone in the essential oils Siparuna guianensis Aublet. *Rev Bras Pl Med*. 2004;6(3):62-4.
[106] Delgado IF, Carvalho RR, Nogueira, et al. Study on embryofetotoxicity of b-myrcene in the rat. *Food and Chemical Toxicology*. 1993;31(1):31-5.
[107] Paumgartten FJ, De-Carvalho RR, Souza CA, et al. Study of the effects of beta-myrcene on rat fertility and general reproductive performance. *Braz J Med Biol Res*. 1998 Jul;31(7):955-65.
[108] do Vale TG, Furtado EC, Santos JG Jr, et al. Central effects of citral, myrcene and limonene, constituents of essential oil chemotypes from Lippia alba (Mill.) n.e. Brown. *Phytomedicine*. 2002 Dec;9(8):709-14.
[109] Segelman AB, Sofia RD, Segelman FP, et al. Cannabis sativa L. (marijuana). V. Pharmacological evaluation of marijuana aqueous extract and volatile oil. *J Pharm Sci*. 1974 Jun;63(6):962-4.
[110] Freitas JC1, Presgrave OA, Fingola FF, et al. Effect of beta-myrcene on pentobarbital sleeping time. *Braz J Med Biol Res*. 1993 May;26(5):519-23.
[111] De Olivera AC, Ribeiro-Pinto LF, Paumgartten JR. In vitro inhibition of CYP2B1 monoxygenase by beta-myrcene and other monoterpenoid constituents. *Br J Nutr*. 1999;81:289-95.
[112] De-Oliveira AC1, Ribeiro-Pinto LF, Otto SS, et al. Induction of liver monooxygenases by beta-myrcene. *Toxicology*. 1997 Dec 26;124(2):135-40.
[113] De Olivera AC, Ribeiro-Pinto LF, Paumgartten JR. In vitro inhibition of CYP2B1 monoxygenase by beta-myrcene and other monoterpenoid constituents. *Br J Nutr*. 1999;81:289-95.
[114] De-Oliveira AC1, Ribeiro-Pinto LF, Otto SS, et al. Induction of liver monooxygenases by beta-myrcene. *Toxicology*. 1997 Dec 26;124(2):135-40.
[115] Burkhard PR, Burkhardt K, Haenggeli CA, et al. Plant-induced seizures: reappearance of an old problem. *J Neurol.* 1999 Aug;246(8):667-70.
[116] Waldman N. Seizure caused by dermal application of over-the-counter eucalyptus oil head lice preparation. *Clin Toxicol (Phila)*. 2011 Oct;49(8):750-1.
[117] Craig JO. Poisoning by the volatile oils in childhood. *Arch Dis Child.* 1953;28:259-67.

[118] Brehm-Stecher BF, Johnson EA. Sensitization of Staphylococcus aureus and Escherichia coli to antibiotics by the sesquiterpenoids nerolidol, farnesol, bisabolol, and apritone. *Antimicrob Agents Chemother.* 2003 Oct;47(10):3357-60.
[119] Špičáková A, Szotáková B, Dimunová D, et al. Nerolidol and Farnesol Inhibit Some Cytochrome P450 Activities but Did Not Affect Other Xenobiotic-Metabolizing Enzymes in Rat and Human Hepatic Subcellular Fractions. *Molecules.* 2017 Mar 24;22(4). pii: E509.
[120] Pimkaew P1, Küblbeck J, Petsalo A, et al. Interactions of sesquiterpenes zederone and germacrone with the human cytochrome P450 system. *Toxicol In Vitro.* 2013 Sep;27(6):2005-12.
[121] Mendes Ferreira da Costa MDC, da Silva AG, Sant'Anna da Silva AP, et al. Essential Oils from Leaves of Medicinal Plants of Brazilian Flora: Chemical Composition and Activity against Candida Species. *Medicines (Basel).* 2017 Jun; 4(2): 27.
[122] Andrade MA, Cardoso MDS, Domes MDS, et al. Biological activity of the essential oils from Cinnamodendron dinisii and Siparuna guianensis. *Braz J Microbiol.* 2015 Jan-Mar;46(1):e1678-4405.
[123] de Melo DC, Dantas Miranda ML, Ferreira Jr. WG, et al. Anticariogenic and Antimycobacterial Activities of the Essential Oil of Siparuna guianensis Aublet (Siparunaceae). *Orbital Electronic J Chem.* 207 Jan-Mar;9(1):55-60.
[124] Faria de Bessa NG, Bezerra Periera MA, Ferraz V, et al. Antimicrobial activity and medicinal biomass of Siparuna guianensis in Brazilian Cerrado forest, a global hotspot. *J Medicinal Plants Res.* 2015 Oct;9(37):968-80.
[125] Andrade MA, Cardoso MDS, Domes MDS, et al. Biological activity of the essential oils from Cinnamodendron dinisii and Siparuna guianensis. *Braz J Microbiol.* 2015 Jan-Mar;46(1):e1678-4405.
[126] Andrade MA, dos Santos Azevedo C, Motta FM, et al. Essential oils: in vitro activity against Leishmania amazonensis, cytotoxicity and chemical composition. *BMC Complement Altern Med.* 2016 Nov;16:444.
[127] Andrade MA, das Gracas Cardoso M, de Andrade J, et al. Chemical Composition and Antioxidant Activity of Essential Oils from Cinnamodendron dinisii Schwacke and Siparuna guianensis Aublet. *Antioxidants.* 2013;2:384-97.
[128] Souza Aguiar RW, dos Santos SF, da Silva Morgado F, et al. Insecticidal and Repellent Activity of Siparuna guianensis Aubl. (Negramina) against Aedes aegypti and Culex quinquefasciatus. *PLoS One.* 2015 Feb 3;10(w)e0116765.
[129] Dudai N, Shachter A, Satyal P, et al. Chemical Composition and Monoterpenoid Enantiomeric Distribution of the Essential Oils from Apharsemon (Commiphora gileadensis). *Medicines.* 2017;4(66):1-9.
[130] Amiel E, Ofir R, Dudai N, et al. β-Caryophyllene, a Compound Isolated from the Biblical Balm of Gilead (Commiphora gileadensis), Is a Selective Apoptosis Inducer for Tumor Cell Lines. *Evidence-Based Compl Alt Med.* 2012;2012:872394.
[131] Amiel E, Ofir R, Dudai N, et al. β-Caryophyllene, a Compound Isolated from the Biblical Balm of Gilead (Commiphora gileadensis), Is a Selective Apoptosis Inducer for Tumor Cell Lines. *Evidence-Based Compl Alt Med.* 2012;2012:872394.
[132] Dehsheikh AB, Sourestani MM, Dehsheikh PB, et al. A Comparative Study of Essential Oil Constituents and Phenolic Compounds of Arabian Lilac (Vitex trifolia var. Purpurea): An Evidence of Season Effects. *Foods.* 2019;8(52):1-14.
[133] Devi WR, Singh CB. Chemical composition, anti-dermatophytic activity, antioxidant and total phenolic content within the leaves essential oil of Vitex trifolia. *Int J Phytocosmetics Nat Ingred.* 2014;1(5):1-7.
[134] Craig JO. Poisoning by the volatile oils in childhood. *Arch Dis Child.* 1953;28:259-67.
[135] Melis K. Bochner A, Janssens G. Accidental nasal eucalyptol and menthol instillation. *Eur J Pediatr.* 1989 Aug;148(8)786-7.
[136] Day LM, Ozanne-Smith J, Parsons BJ, et al. Eucalyptus oil poisoning among young children: mechanisms of access and the potential prevention. *Aust N Z J Public Health.* 1997 Jun;21(3):297-302.
[137] Day LM, Ozanne-Smith J, Parsons BJ, et al. Eucalyptus oil poisoning among young children: mechanisms of access and the potential prevention. *Aust N Z J Public Health.* 1997 Jun;21(3):297-302.
[138] Myott E. Case of eucalyptus poisoning. *Brit M J.* 1906;1:558.
[139] Hindle RC. Eucalyptus oil ingestion. *N Z Med J.* 1994 May 11;107(977)185-6.
[140] Tibballs J. Clinical effects and management of eucalyptus oil ingestion in infants and small children. *Med J Aust.* 1995 Aug;163(4):177-80.
[141] Waldman W, Barwina M, Sein Anand J. Accidental intoxication with eucalyptus oil—a case report. *Przeql Lek.* 2011;68(8):555-6.
[142] Day LM, Ozanne-Smith J, Parsons BJ, et al. Eucalyptus oil poisoning among young children: mechanisms of access and the potential prevention. *Aust N Z J Public Health.* 1997 Jun;21(3):297-302.
[143] Burkhard PR, Burkhardt K, Haenggeli CA, et al. Plant-induced seizures: reappearance of an old problem. *J Neurol.* 1999 Aug;246(8):667-70.
[144] Waldman N. Seizure caused by dermal application of over-the-counter eucalyptus oil head lice preparation. *Clin Toxicol (Phila).* 2011 Oct;49(8):750-1.
[145] Craig JO. Poisoning by the volatile oils in childhood. *Arch Dis Child.* 1953;28:259-67.
[146] Mathew T, Kamath V, Kumar RS, et al. Eucalyptus oil inhalation-induced seizure: A novel, underrecognized, preventable cause of acute symptomatic seizure. *Epilepsia Open.* 2017 Jul 4;2(3):350-354.
[147] Unger M, Frank A. Simultaneous determination of the inhibitory potency of herbal extracts on the activity of six major cytochrome P450 enzymes using liquid chromatography/mass spectrometry and automated online extraction. *Rapid Commun Mass Spectrom.* 2004;18(19):2273-81.
[148] Jori A, Bianchetti A, Prestini PE, et al. Effect of eucalyptol (1,8-cineole) on the metabolism of other drugs in rats and in man. *Eur J Pharmacol.* 1970;9(3):362-66.
[149] Kim NH, Hyun SH, Jin CH, et al. Pretreatment with 1,8-cineole potentiates thioacetamide-induced hepatotoxicity and immunosuppression. *Arch Pharm Res.* 2004 Jul;27(7):781-9.
[150] Jori A, Bianchetti A, Prestini PE, et al. Effect of eucalyptol (1,8-cineole) on the metabolism of other drugs in rats and in man. *Eur J Pharmacol.* 1970;9(3):362-66.
[151] de Sousa DP, Raphael E, Brocksom U, et al. Sedative effect of monoterpene alcohols in mice: A preliminary screening. *Verlag der Zeitschrift fur Naturforschung.* 2007;62c:563-66.
[152] Devi WR, Singh CB. Chemical composition, anti-dermatophytic activity, antioxidant and total phenolic content within the leaves essential oil of Vitex trifolia. *Int J Phytocosmetics Nat Ingred.* 2014;1(5):1-7.
[153] Devi WR, Singh CB. Chemical composition, anti-dermatophytic activity, antioxidant and total phenolic content within the leaves essential oil of Vitex trifolia. *Int J Phytocosmetics Nat Ingred.* 2014;1(5):1-7.
[154] Tandon S, Mittal AK, Pant AK. Insect growth regulatory activity of Vitex trifolia and Vitex agnus-castus essential oils against Spilosoma obliqua. *Fitoterapia.* 2008 Jun;79(4):283-6.
[155] Chandrasekaran T, Thyagarajan A, Santhakumari PG, et al. Larvicidal activity of essential oil from Vitex negundo and Vitex trifolia on dengue vector mosquito Aedes aegypti. *Rev Soc Bras Med Trop.* 2019 Jul 29;52:e20180459.
[156] de Macedo DG, Souza MMA, Norais-Braga MFB, et al. Effect of seasonality on chemical profile and antifungal activity of essential oil isolated from leaves Psidium salutare (Kunth) O. Berg. *PeerJ.* 2018:1-19.
[157] Pino JA, Marbot R, Bello A. Volatile Compounds of Psidium salutare (H.B.K.) Berg. Fruit. *J Agric Food Chem.* 2002;50:5146-5148.

[158] Pino JA, Bello A, Urquiola A, et al. Leaf Oil of Psidium salutare (HBK) Berg. from Cuba. *J Essent Oil Res.* 2003 Jan/Feb;15:19-20.
[159] de Macedo DG, Souza MMA, Norais-Braga MFB, et al. Effect of seasonality on chemical profile and antifungal activity of essential oil isolated from leaves Psidium salutare (Kunth) O. Berg. *PeerJ.* 2018:1-19.
[160] Lin WY, Kuo YH, Chang YL, et al. Anti-platelet aggregation and chemical constituents from the rhizome of Gynura japonica. *Planta Med.* 2003 Aug;69(8):757-64.
[161] de Macedo DG, Souza MMA, Norais-Braga MFB, et al. Effect of seasonality on chemical profile and antifungal activity of essential oil isolated from leaves Psidium salutare (Kunth) O. Berg. *PeerJ.* 2018:1-19.
[162] Ramanoelina PAR, Rasoarahona RE. Chemical composition of the leaf oil of Psiadia altissima (Compositeae). *J Essent Oil Res.* 1994 Nov/Dec;6:565-70.
[163] Liberty Natural. Certificate Of Analysis Sheet ARINA (IARY) ESSENTIAL OIL - Product No: 1222 PSIADIA ALTISSIMA. Available at: https://www.libertynatural.com/coa/1222.htm.
[164] Ramanoelina AR, Terrom GP, Bianchini JP, et al. [Antibacterial action of essential oils extracted from Madagascar plants]. *Arch Inst Pasteur Madagascar.* 1987;53(1):217-26.
[165] Okinczyc P, Szumny A, Szperlik J, et al. Profile of Polyphenolic and Essential Oil Composition of Polish Propolis, Black Poplar and Aspens Buds. *Molecules.* 2018;23:1262.
[166] Politano VT, Diener RM, Christian MS, et al. Oral and dermal developmental toxicity studies of phenylethyl alcohol in rats. *Int J Toxicol.* 2013 Jan-Feb;32(1):32-8.
[167] Liu Q, Zhang S, Yang X, et al. Differentiation of essential oils in Atractylodes lancea and Atractylodes koreana by gas chromatography with mass spectrometry. J Sep Sci. 2016;39(24):4773-4780.
[168] Zhou J, Fang L, Wang X, et al. Comparison of the volatile compounds of crude and processed Atractylodis rhizome analyzed by GC-MS. *African J Pharmacy Parmacol.* 2012 Jul;6(28):2155-60.
[169] Bruns K, Dolhaine H, Weber U. Über die Zusammensetzung des atherischel Ols aus Atractylis lancea DC (Compositae). *Parfume und Kosmetik.* 1982;63:237-9.
[170] Ouyang Z, Yang L, Su SL, et al. [Fingerprint of volatile oil of Atractylodes lancea by GC-MS]. *Yao Xue Xue Bao.* 2007 Sep;42(9):968-72.
[171] Tsuneki H, Ma EL, Kobayashi S, et al. Antiangiogenic activity of beta-eudesmol in vitro and in vivo. *Eur J Pharmacol.* 2005 Apr 11;512(2-3):105-15.
[172] Arora CK, Arora RB, Mesta CK, et al. Hypotensive activity of beta-eudesmol and some related sesquiterpenes. *Ind J Med Res.* 1967 May;55(5):463-72.
[173] Wang BG, Hong X, Li L, Zhou J, et al. Chemical constituents of two Chinese Magnoliaceae plants, Tsoongiodendron odorum and Manglietiastrum sinicum, and their inhibition of platelet aggregation. *Planta Med.* 2000 Aug;66(6):511-5.
[174] Tsuneki H, Ma EL, Kobayashi S, et al. Antiangiogenic activity of beta-eudesmol in vitro and in vivo. *Eur J Pharmacol.* 2005 Apr 11;512(2-3):105-15.
[175] Arora CK, Arora RB, Mesta CK, et al. Hypotensive activity of beta-eudesmol and some related sesquiterpenes. *Ind J Med Res.* 1967 May;55(5):463-72.
[176] Wang BG, Hong X, Li L, Zhou J, et al. Chemical constituents of two Chinese Magnoliaceae plants, Tsoongiodendron odorum and Manglietiastrum sinicum, and their inhibition of platelet aggregation. *Planta Med.* 2000 Aug;66(6):511-5.
[177] Morita M, Chin M, Tomita M, et al. Blood platelet aggregation inhibitors from Atractylodes lancea (an oriental drug). *Jpn Kokai Tokkyo Koho.* 1992;JP04173733:223-229.
[178] Zhang H, Han T, Sun LN, et al. Regulative effects of essential oil from Atractylodes lancea on delayed gastric emptying in stress-induced rats. *Phytomedicine.* 2008 Aug;15(8):602-11.
[179] Wang Y, Dai CC, Chen Y. [Antimicrobial activity of volatile oil from Atractylodes lancea against three species of endophytic fungi and seven species of exogenous fungi]. *Ying Yong Sheng Tai Xue Bao.* 2009 Nov;20(11):2778-84.
[180] Ogunbinu AO, Ogunwande IA, Flamini G, et al. Volatile Compounds of Persea americana Mill from Nigeria. *JEOBP.* 2007;10(2):133-8.
[181] Larijani K, Rustaiyan A, Azar PA, et al. Composition of essential oil of leaves of Persea Americana cultivated in Iran. *Chem Nat Compounds.* 2010;46(3):489-90
[182] Sagrero-Nieves L. Volatile Components of Avocado Leaves (Persea americana Mill) from the Mexican Race. *J Agric Food Sci.* 1995;67:49-51.
[183] Granados-Echegoyen C, Perez-Pacheco R, Alonso-Hernandez N, et al. Chemical characterization and mosquito larvicidal activity of essential oil from leaves of Persea americana Mill (Lauraceae) against Culex quinquefasciatus (Say). *Asian Pac J Trop Dis.* 2015;5(5):463-7.
[184] Pino JA, Rosado A, Aguero J. Volatile Components of Avocado (Persea americana Mill.) Fruits. *JEssent Oil Res.* 200 May/Jun;12:377-8.
[185] Brehm-Stecher BF, Johnson EA. Sensitization of Staphylococcus aureus and Escherichia coli to antibiotics by the sesquiterpenoids nerolidol, farnesol, bisabolol, and apritone. *Antimicrob Agents Chemother.* 2003 Oct;47(10):3357-60.
[186] Špičáková A, Szotáková B, Dimunová D, et al. Nerolidol and Farnesol Inhibit Some Cytochrome P450 Activities but Did Not Affect Other Xenobiotic-Metabolizing Enzymes in Rat and Human Hepatic Subcellular Fractions. *Molecules.* 2017 Mar 24;22(4). pii: E509.
[187] Granados-Echegoyen C, Perez-Pacheco R, Alonso-Hernandez N, et al. Chemical characterization and mosquito larvicidal activity of essential oil from leaves of Persea americana Mill (Lauraceae) against Culex quinquefasciatus (Say). *Asian Pac J Trop Dis.* 2015;5(5):463-7.
[188] Memarzadeh SM, Pirbalouti AG, AdibMejad M. Chemical Composition and Yield of Essential Oils From Bakhtiari Savory (Satureja Bachtiarica Bunge.) Under Different Extraction Methods. *Ind Crops Prod.* 2015;76:809-16.
[189] Sefidkon F, Jamzad Z. Essential Oil of Satureja bachtiarica Bunge. *J Essent Oil Res.* 2000 Sep/Oct;12(5):545-6.
[190] Bakhtiarpoor M, Setorki M, Kaffashian MR. Effects of Essential Oil of Satureja Bachtiarica Bunge in a Rat Model of Reserpine-Induced Depression. *Iran J Med Sci.* 2018 Jul;43(4):409-15.
[191] Falsafi T, Moradi P, Mahboubi M, et al. Chemical composition and anti-Helicobacter pylori effect of Satureja bachtiarica Bunge essential oil. *Phytomed.* 2014;24:173-7.
[192] Moeil M, Karami F, Tavallali H, et al. Chemical Composition of the Essential oil of Satureja bachtiarica Bunge. from Iran. *Iranian J Pharm Sci.* 2012 Autumn;8(4):277-81.
[193] Babadi E, Pirbalouti AG, Nourafcan H, et al. Bioactivity of essential oil of Bakhtiari savory (Lamiaeae). *Electronic J Biol.* 2012;8(4):73-8.
[194] Abdolrahimzadeh H, Bolandparvaz S, Abbasi HR, et al. Antimicrobial Survey of Local Herbal Drugs against Acinetobacter baumannii Isolated from Patients Admitted to a Level-I Trauma Center. *Bull Emerg Trauma.* 2018;6(4):355-62.
[195] Hashemi MB, Niakousari M, Saharkhiz MJ. Antioxidant activity of Satureja bachtiarica Bunge essential oil in rapeseed oil irradiated with UV rays. *European J Lipid Sci Tech.* 2011 Sep;113(9):1132-7.
[196] Domaracky M, Rehak P, Juhas S, et al. Effects of selected plant essential oils on the growth and development of mouse preimplantation embryos in vivo. *Physiol Res.* 2007;56(1):97-104.

[197] Lemhadri A, Zeggwagh NA, Maghrani M, et al. Anti-hypoglycaemic activity of the aqueous extract of Origanum vulgare growing wild in Tafilalet region. *J Ethnopharmacol.* 2004 Jun;92(2-3):251-6.
[198] Bayramoglu G, Senturk H, Bayramoglu A, et al. Carvacrol partially reverses symptoms of diabetes in STZ-induced diabetic rats. *Cytotechnology.* 2014 Mar;66(2):251–257.
[199] Ezhumalai M, Radhiga T, Pugalendi KV. Antihyperglycemic effect of carvacrol in combination with rosiglitazone in high-fat diet-induced type 2 diabetic C57BL/6J mice. *Mol Cell Biochem.* 2014 Jan;385(1-2):23-31.
[200] Umasankar K, Ramya DR, Nambikkairaj B. Antidiabetic and antidyslipidemic nature of carvacrol, a monoterpenic phenol studied in high-fat - fed and low-dose streptozotocin-induced experimental diabetic rats. *J Pharm Res.* 2015;9(8):484-90.
[201] Tognolini M, Barocelli E, Ballabeni V, et al. Comparative screening of plant essential oils; phenylpropanoid moiety as basic core for antiplatelet activity. *Life Sci.* 2006 Feb 23;78(13):1419-32.
[202] Tsai HH, Lin HW, Chen YL, et al. A review of potential harmful interactions between anticoagulant/antiplatelet agents and Chinese herbal medicines. *PLoS One.* 2013 May 9;8(5):e64255.
[203] Karkabounas S, Kostoula OK, Daskalou T, et al. Anticarcinogenic and antiplatelet effects of carvacrol. *Exp Oncol.* 2006 Jun;28(2):121-5.
[204] Baser KH. Biological and pharmacological activities of carvacrol and carvacrol bearing essential oils. *Curr Pharm Des.* 2008;14(29):3106-19.
[205] Okazaki K, Kawazoe K, Takaishi Y. Human platelet aggregation inhibitors from thyme (Thymus vulgaris L.). *Phytother Res.* 2002 Jun;16(4):398-9.
[206] Jukic M, Politeo O, Maksimmovic M, et al. In vitro acetylcholinesterase inhibitory properties of thymol, carvacrol and their derivatives thymoquinone and thymohydroquinone. *Phytother Res.* 2007;21(3):259-61.
[207] Askin H, Yildiz M, Ayar A. Effects of Thymol and Carvacrol on Acetylcholinesterase from Drosophila melanogaster. *Acta Physica Polinica A.* 2017;132(3):720-22.
[208] Medeiros Barreto H, Cerqueira Fontinele F, Pereira de Oliveira A, et al. Phytochemical prospection and modulation of antibiotic activity in vitro by Lippia origanoides H.B.K. in methicillin resistant Staphylococcus aureus. *Biomed Res Int.* 2014;305610:1–7.
[209] Aelenei P, Miron A, Trifan A, et al. Essential Oils and Their Components as Modulators of Antibiotic Activity against Gram-Negative Bacteria. *Medicines.* 2016;3(19):2-34.
[210] Magi G, Marini E, Facinelli B. Antimicrobial activity of essential oils and carvacrol, and synergy of carvacrol and erythromycin, against clinical, erythromycin-resistant Group A Streptococci. *Front Microbiol.* 2015;6:165.
[211] Veras HNH, Rodrigues FFG, Colares AV, et al. Synergistic antibiotic activity of volatile compounds from the essential oil of Lippia sidoides and thymol. *Fitoterapia.* 2012 Apr;83(3):508-12.
[212] Langeveld WT, Veldhuizen EJ, Burt SA. Synergy between essential oil components and antibiotics: A review. *Crit Rev Microbiol.* 2014 Feb;40(1):76-94.
[213] Kohlert C, Schindler G, Marz RW, et al. Systemic availability and pharmacokinetics of thymol in humans. *J Clin Pharmacol.* 2002 Jul;42(7):731-7.
[214] Toxicology Data Network, National Library of Medicine. Thymol. Available at: http://toxnet.nlm.nih.gov/cgibin/sis/search/a?dbs+hsdb:@term+@DOCNO+866.
[215] Bakhtiarpoor M, Setorki M, Kaffashian MR. Effects of Essential Oil of Satureja Bachtiarica Bunge in a Rat Model of Reserpine-Induced Depression. *Iran J Med Sci.* 2018 Jul;43(4):409-15.
[216] Falsafi T, Moradi P, Mahboubi M, et al. Chemical composition and anti-Helicobacter pylori effect of Satureja bachtiarica Bunge essential oil. *Phytomed.* 2014;24:173-7.
[217] Babadi E, Pirbalouti AG, Nourafcan H, et al. Bioactivity of essential oil of Bakhtiari savory (Lamiaeae). *Electronic J Biol.* 2012;8(4):73-8.
[218] Abdolrahimzadeh H, Bolandparvaz S, Abbasi HR, et al. Antimicrobial Survey of Local Herbal Drugs against Acinetobacter baumannii Isolated from Patients Admitted to a Level-I Trauma Center. *Bull Emerg Trauma.* 2018;6(4):355-62.
[219] Hashemi MB, Niakousari M, Saharkhiz MJ. Antioxidant activity of Satureja bachtiarica Bunge essential oil in rapeseed oil irradiated with UV rays. *European J Lipid Sci Tech.* 2011 Sep;113(9):1132-7.
[220] Arambewela L, Kumaratunga KGA, Dias K. Studies on Piper betle of Sri Lanka. *J Natn Sci Foundation Sri lanka.* 2005;33(2):133-9.
[221] Muhammed Arif M. Isolation, structure elucidation and properties of secondary metabolites in plants. Chapter III. Thesis. Department of Chemistry, University of Calicut, 2007. Available at: http://shodhganga.inflibnet.ac.in/bitstream/10603/20288/9/09_chapter%203.pdf.
[222] Mohottalage S, Tabacchi R, Guerin PM. Components from Sri Lankan Piper betle L. leaf oil and their analogues showing toxicity against the housefly, Musca domestica. *Flavour Fragr J.* 2007;22:130-8.
[223] Rawat AKS, Tripathi RD, Khan AJ. Essentoial Oil Components as Markers for Identification of Piper betle L. Cultivars. *Biochem Systematics Ecol.* 1989;17(1):35-8.
[224] Preethy TT, Elsy CR, Beena C. Profiling of essential oil in Tirur vettilai (Piper betle L.), a group of unique betle vine land races from Malappuram district, Kerala. *J Pharmacognosy Phytochem.* 2017;6(3):774-8.
[225] Caburian AB, Osi MO. Characterization and Evaluation of Antimicrobial Activity of the Essential Oil from the Leaves of Piper betle L. *E-Int Sci Res J.* 2010;2(1):2094-1749.
[226] Sugumaran M, Suresh GM, Sankarnarayanan K, et al. Chemical composition and antimicrobial activity of vellaikodi variety of Piper betle Linn Leaf oil against dental pathogens. *Int J PharmTech Res.* 2011 Oct-Dec;3(4):2135-9.
[227] Misra P, Kumar K, Khare P, et al. Pro-apoptotic effect of the landrace Bangla Mahoba of Piper betle on Leishmania donovani may be due to the high content of eugenol. *J Med Microbiol.* 2009;58:1058-66.
[228] Jantan IB, Ahmad AR, Ahmad AS, et al. A Comparative Study of the Essential Oils of Five piper Species from Peninsular Malaysia. *Flavour Frag J.* 1994;9:339-42.
[229] Rimando AM, Han BH, Park JH, et al. Studies on the Constituents of Philippine Piper betle Leaves. *Arch Pharm Res.* 1986;9(2):93-7.
[230] Astuti EJ, Basuki SA, Sabrian A, et al. Steam and Water Distillation of Piper Betle, Ocimum Basilicum, Cymbopogon Winterianus, and Citrus Hystrix Leaves for Activity of Insect Repellent against Mosquito. *Adv Health Sci Res.* 2017;2:240-6.
[231] Jirovetz L, Puschmann C, Buchbauer G, et al. Analysis of the essential oil of Piper betle L. Leaves from south- India using GC/FID, GC/MS and olfactometry. *Sci Pharm.* 1999 Dec;67(4):305-12.
[232] Basak S, Guha P. Modelling the effect of essential oil of betle leaf (Piper betle L.) on germination, growth, and apparent lag time of Penicillium expansum on semi-synthetic media. *Int J Food Microbiology.* 2015 Dec;215:171-8.
[233] Suryasnata D, Sandeep IS, Parida R, et al. Variation in Volatile Constituents and Eugenol Content of Five Important Betlevine (Piper betle L.) Landraces Exported from Eastern India. *J Essent Oil Bearing Plants.* 2016;19(7):1788-93.
[234] Tanh L, Dung NX, Bighelli A, et al. Combination of capillary GC, GC/MS and 13C-NMR for the characterization of the rhizome oil of Piper betle L. (Piperaceae) from Vietnam. *Spectroscopy.* 1996/1997;13:131-6.

[235] Periyanayagam K, Mubeen M, Saleem MM, et al. Phytochemical studies and GC/MS analysis on the isolated essential oil from the leaves of Piper betle L.var.sirugamani1(SGM1). *J Pharm Res.* 2011;4(7):2411-3.
[236] Prakash B, Shukla R, Singh P, et al. Efficacy of chemically characterized Piper betle L. essential oil against fungal and aflatoxin contamination of some edible commodities and its antioxidant activity. *Int J Food Microbiol.* 2010 Aug 15;142(1-2):114-9.
[237] Vaghasiya Y., Nair R. and Chanda S. Investigation of some piper species for antibacterial and anti-inflammatory property. *Int J Pharmacol.* 2007;3(5):400-405.
[238] Perera WH, Bizzo HR, Gama PE, et al. Essential oil constituents from high altitude Brazilian species with antimicrobial activity: Baccharis parvidentata Malag., Hyptis monticola Mart. ex Benth. and Lippia origanoides Kunth. *J Essent Oil Res.* 2016;2:109-16.
[239] Vesselinovitch SD, Rao KVN, Mihalovich N. Transplacental and Lactational Carcinogenesis by Safrole. *Cancer Res.* 1979 Nov;39:4378-80.
[240] Lu LJ, Disher RM, Randerath K. Differences in the covalent binding of benzo[a]pyrene, safrole, 1'-hydroxysafrole, and 4-aminobiphenyl to DNA of pregnant and non-pregnant mice. *Cancer Lett.* 1986 Apr;31(1):43-52.
[241] NIH. Report on carcinogenesis, fourteenth edition. Safrole. Available at: https://ntp.niehs.nih.gov/ntp/roc/content/profiles/safrole.pdf.
[242] Tsai HH, Lin HW, Chen YL, et al. A review of potential harmful interactions between anticoagulant/antiplatelet agents and Chinese herbal medicines. *PLoS One.* 2013 May 9;8(5):e64255.
[243] Ioannides C, Delaforge M, Parke DV. Interactions of safrole and isosafrole and their metabolites with cytochromes P-450. *Chem Biol Interact.* 1985 May;53(3):303-11.
[244] Human and Environmental Risk Assessment on ingredients of Household Cleaning Products. Isoeugenol. 4-Hydroxy-3-methoxy-1-propen-1-yl benzene. CAS 97-54-1. Available at: https://www.heraproject.com/files/19-f-05-hera%20isoeugenol%20(corrections%20may%202005).pdf.
[245] European Chemicals Agency. CLH Report. Proposal for Harmonised Classification and Labelling. Available at: https://echa.europa.eu/documents/10162/13626/clh_report_isoeugenol_en.pdf.
[246] Chen R, Chen J, Cheng S, et al. Assessment of embryotoxicity of constituents in cosmetics by the embryonic stem cell tes. *Toxicol Mech Methods.* 2010 Mar;20(3):112-18.
[247] Domaracky M, Rehak P, Juhas S, et al. Effects of selected plant essential oils on the growth and development of mouse preimplantation embryos in vivo. *Physiol Res.* 2007;56(1):97-104.
[248] Vrskova D, Modra H. Evaluation of the developmental toxicity of 2-phenoxyethanol and clove oil anesthetics using the Frog Embryo Teratogenesis Assay: Xenopus (FETAX). *Veterinarami Medicina.* 2012;57(5):245-50.
[249] Amini A, Cheraghi E, Safaee MR, et al. The role of eugenol in the reduction of teratogenic effects of retinoic acid on skeletal morphology of mice embryo. *Yakhteh Medical Journal.* 2003;4:195-200.
[250] Chen SJ, Wang MH, Chen IJ. Antiplatelet and calcium inhibitory properties of eugenol and sodium eugenol acetate. *Gen Pharmacol.* 1996 Jun;27(4):629-33.
[251] Tognolini M, Barocelli E, Ballabeni V, et al. Comparative screening of plant essential oils; phenylpropanoid moiety as basic core for antiplatelet activity. *Life Sci.* 2006 Feb 23;78(13):1419-32.
[252] Heck AM, DeWitt BA, Lukes AL. Potential interactions between alternative therapies and warfarin. *Am J Health Syst Pharm.* 2000;57(13):1221-1227.
[253] Saaeed SA, Gilani AH. Antithrombotic activity of clove oil. *J Pak Med Assoc.* 1994;44(5):112-15.
[254] Kamatou GP, Vermaak I, Viljoen AM. Eugenol—from the remote Maluku Islands to the international market place: A review of a remarkable and versatile molecule. *Molecules.* 2012;17:6953-81.
[255] Cochrane ML. Inhibition of Cytochrome P450 2C9 by essential oils. Available at: https://libres.uncg.edu/ir/uncg/listing.aspx?id=18102.
[256] Kumaravelu P, Dakshinamoorthy DP, Subramaniam S, et al. Effect of eugenol on drug-metabolizing enzymes of carbon tetrachloride-intoxicated rat liver. *Biochem Pharmacol.* 1995 May;49(11):1703-7.
[257] Dohi S, Terasaki M, Makino M. Acetylcholinesterase inhibitory activity and chemical composition of commercial essential oils. *J Agric Food Chem.* 2009 May 27;57(10):4313-8.
[258] Langeveld WT, Veldhuizen EJ, Burt SA. Synergy between essential oil constituents and antibiotics. *Crit Rev Microbiol.* 2014 Feb;40(1);76-94.
[259] Sarrami N, Pemberton MN, Thornhill MH, et al. Adverse reactions associated with the use of eugenol in dentistry. *British Dental J.*2002;193:253-55.
[260] Tammannavar P, Pushpalatha C, Jain S, et al. An unexpected positive hypersensitive reaction to eugenol. *BMJ Case Rep.* 2013; 2013:bcr2013009464.
[261] Human and Environmental Risk Assessment on ingredients of Household Cleaning Products. Isoeugenol. 4-Hydroxy-3-methoxy-1-propen-1-yl benzene. CAS 97-54-1. Available at: https://www.heraproject.com/files/19-f-05-hera%20isoeugenol%20(corrections%20may%202005).pdf.
[262] European Chemicals Agency. CLH Report. Proposal for Harmonised Classification and Labelling. Available at: https://echa.europa.eu/documents/10162/13626/clh_report_isoeugenol_en.pdf.
[263] European Medicines Agency. Public statement on the use of herbal medicinal products containing estragole. Available at: http://www.ema.europa.eu/docs/en_GB/document_library/Scientific_guideline/2010/04/WC500089960.pdf.
[264] Ding W, Levy DD, Bishop ME, et al. In vivo genotoxicity of estragole in male F344 rats. *Environ Mol Mutagen.* 2015 May;56(4):356-65.
[265] Eisenmann SW, Poulev A, Struwe L, et al. Qualitative variation of anti-diabetic constituents in different tarragon (Artemisia dracunculus L.) cytotypes. *Fitoterapia.* 2011 Oct;82(7):1062–74.
[266] European Medicines Agency. Public statement on the use of herbal medicinal products containing estragole. Available at: http://www.ema.europa.eu/docs/en_GB/document_library/Scientific_guideline/2010/04/WC500089960.pdf.
[267] European Medicines Agency. Public statement on the use of herbal medicinal products containing estragole. Available at: http://www.ema.europa.eu/docs/en_GB/document_library/Scientific_guideline/2010/04/WC500089960.pdf.
[268] European Commission. Opinion of the Scientific Committee on Food on estragole (1-Allyl-4-methoxybenzene). Available at: http://ec.europa.eu/food/fs/sc/scf/out104_en.pdf.
[269] California Environmental Protection Agency. Evidence of the carcinogenicity of estragole. Available at: http://oehha.ca.gov/prop65/pdf/estragf.pdf.
[270] Drinkwater NR, Miller EC, Miller JA, et al. Hepatocarcinogenicity of estragole (1-allyl-4-methoxybenzene) and 1'-hydroxyestragole in the mouse and mutagenicity of 1'-acetoxyestragole in bacteria. *J Natl Cancer Inst.* 1976 Dec;57(6):1323-31.
[271] Zeller A, Horst K, Rychlik M. Study of the metabolism of estragole in humans consuming fennel tea. *Chem Res Toxicol.* 2009 Dec;22(12):1929-37.
[272] Chen R, Chen J, Cheng S, et al. Assessment of embryotoxicity of constituents in cosmetics by the embryonic stem cell tes. *Toxicol Mech Methods.* 2010 Mar;20(3):112-18.
[273] Domaracky M, Rehak P, Juhas S, et al. Effects of selected plant essential oils on the growth and development of mouse preimplantation embryos in vivo. *Physiol Res.* 2007;56(1):97-104.

[274] Vrskova D, Modra H. Evaluation of the developmental toxicity of 2-phenoxyethanol and clove oil anesthetics using the Frog Embryo Teratogenesis Assay: Xenopus (FETAX). *Veterinarami Medicina*. 2012;57(5):245-50.
[275] Amini A, Cheraghi E, Safaee MR, et al. The role of eugenol in the reduction of teratogenic effects of retinoic acid on skeletal morphology of mice embryo. *Yakhteh Medical Journal*. 2003;4:195-200.
[276] Chang MC, Uang BJ, Tsai CY, et al. Hydroxychavicol, a novel betle leaf component, inhibits platelet aggregation by suppression of cyclooxygenase, thromboxane production and calcium mobilization. *Br J Pharmacol*. 2007 Sep;152(1):73-82.
[277] Jesonbabu J, Spandana N, Reddy M, et al. A bioactive compound from piper betel with anticoagulant activity. *Int J Pharm Pharmaceutical Sci*. 2012 Jan;4:109-12.
[278] Janssens J, Laekeman GM, Pieters LA, et al. Nutmeg oil: identification and quantitation of its most active constituents as inhibitors of platelet aggregation. *J Ethnopharmacol*. 1990 May;29(2):179-88.
[279] Cho JH, Lee CH, Son DJ, et al. Antiplatelet activity of phenylpropanoids isolated from Eugenia caryophyllata leaf oil. *Food Sci Biotech*. 2004 Jun;13(3):315-7.
[280] Xiang CP. Han JX, Li XC, et al. Chemical Composition and Acetylcholinesterase Inhibitory Activity of Essential Oils from Piper Species. *J Agric Food Chem*. 2017;65:3702-10.
[281] Human and Environmental Risk Assessment on ingredients of Household Cleaning Products. Isoeugenol. 4-Hydroxy-3-methoxy-1-propen-1-yl benzene. CAS 97-54-1. Available at: https://www.heraproject.com/files/19-f-05-hera%20isoeugenol%20(corrections%20may%202005).pdf.
[282] European Chemicals Agency. CLH Report. Proposal for Harmonised Classification and Labelling. Available at: https://echa.europa.eu/documents/10162/13626/clh_report_isoeugenol_en.pdf.
[283] Sanubol A, Chaveerach A, Tanee T, et al. Pre-clinical evaluation of extracts and essential oils from betle-like scent Piper species identified potential cancer treatment. *Afr J Tradit Complement Altern Med*. 2017;14(1):89–102.
[284] Sucipto TH, Aisyah N, Lestari P, et al. Betle leaf essential oil for the hemophiliac patients and its antibacterial effets on Mycobacterium tuberculosis. *Indonesian J Trop Infect Disease*. 2016 Dec;6(3):63-7.
[285] Sugumaran M, Suresh GM, Sankarnarayanan K, et al. Chemical composition and antimicrobial activity of vellaikodi variety of Piper betle Linn Leaf oil against dental pathogens. *Int J PharmTech Res*. 2011 Oct-Dec;3(4):2135-9.
[286] Paridhi B, Ashita US, Swati P, et al. An Invitro Study Of Determination Of Anti-Bacterial, Antioxidant, Anti- Inflammatory Potential Of Piper Betle Essential Oil. *NJIRM*. 2015;6(2):37-44.
[287] Sucipto TH, Aisyah N, Lestari P, et al. Betle leaf essential oil for the hemophiliac patients and its antibacterial effets on Mycobacterium tuberculosis. *Indonesian J Trop Infect Disease*. 2016 Dec;6(3):63-7.
[288] Basak S, Guha P. Modelling the effect of essential oil of betle leaf (Piper betle L.) on germination, growth, and apparent lag time of Penicillium expansum on semi-synthetic media. *Int J Food Microbiology*. 2015 Dec;215:171-8.
[289] Arambewela LSR, Arawwawala LDAM, Kumaratunga KG, et al. Investigations on Piper betle grown in Sri Lanka. *Pharmacogn Rev*. 2011 Jul-Dec; 5(10): 159–163.
[290] Prakash B, Shukla R, Singh P, et al. Efficacy of chemically characterized Piper betle L. essential oil against fungal and aflatoxin contamination of some edible commodities and its antioxidant activity. *Int J Food Microbiol*. 2010 Aug 15;142(1-2):114-9.
[291] Xiang CP. Han JX, Li XC, et al. Chemical Composition and Acetylcholinesterase Inhibitory Activity of Essential Oils from Piper Species. *J Agric Food Chem*. 2017;65:3702-10.
[292] Mohottalage S, Tabacchi R, Guerin PM. Components from Sri Lankan Piper betle L. leaf oil and their analogues showing toxicity against the housefly, Musca domestica. *Flavour Fragr J*. 2007;22:130-8.
[293] Astuti EJ, Basuki SA, Sabrian A, et al. Steam and Water Distillation of Piper Betle, Ocimum Basilicum, Cymbopogon Winterianus, and Citrus Hystrix Leaves for Activity of Insect Repellent against Mosquito. *Adv Health Sci Res*. 2017;2:240-6.
[294] Vasantha-Srinivasan P, Senthil-Nathan S, Ponsankar A, et al. Comparative analysis of mosquito (Diptera: Culicidae: Aedes aegypti Liston) responses to the insecticide Temephos and plant derived essential oil derived from Piper betle L. *Ecotoxicol Environ Saf*. 2017 May;139:439-446.
[295] Arambewela LSR, Arawwawala LDAM, Kumaratunga KG, et al. Investigations on Piper betle grown in Sri Lanka. *Pharmacogn Rev*. 2011 Jul-Dec; 5(10): 159–163.
[296] Kumarasinghe SP, Karunaweera ND, Ihalamulla RL, et al. Larvicidal effects of mineral turpentine, low aromatic white spirits, aqueous extracts of Cassia alata, and aqueous extracts, ethanolic extracts and essential oil of betle leaf (Piper betle) on Chrysomya megacephala. *Int J Dermatol*. 2002 Dec;41(12):877-80.
[297] Vasantha-Srinivasan P, Senthil-Nathan S, Thanigaivel A, et al. Developmental response of Spodoptera litura Fab. to treatments of crude volatile oil from Piper betle L. and evaluation of toxicity to earthworm, Eudrilus eugeniae Kinb. *Chemosphere*. 2016 Jul;155:336-347.
[298] Vasantha-Srinivasan P, Senthil-Nathan S, Ponsankar A, et al. Acute toxicity of chemical pesticides and plant-derived essential oil on the behavior and development of earthworms, Eudrilus eugeniae (Kinberg) and Eisenia fetida (Savigny). *Environ Sci Pollut Res Int*. 2018 Apr;25(11):10371-10382.
[299] dos Santos ERQ, Maia GSF, Fontes Junior EA, et al. Linalool-rich essential oils from the Amazon display antidepressant-type effect in rodents. *J Ethnopharm*. 2018;212:43-9.
[300] Sarrazin SLF, Oliveira RB, Maia JGS, et al. Antibacterial Activity of the Rosewood (Aniba rosaeodora and A. parviflora) Linalool-rich Oils from the Amazon. *Eur J Med Plants*. 2016;12(2):1-9.
[301] da Silva JK, Maia JGS, Dosoky NS, et al. Antioxidant, Antimicrobial, and Cytotoxic Properties of Aniba parviflora Essential Oils from the Amazon. *Nat Prod Commun*. 2016;11(7):1025-8.
[302] Zoghbi MDGB, Ohashi ST, Salomao RDP, et al. Chemical Variability of Aniba rosaeodora Oils. *Global J Sci Frontier Res*. 2015;10(5):2249-4626.
[303] Stillpoint Aromatics. Rosewood Essential Oil (Aniba parviflora). Available at: https://www.stillpointaromatics.com/rosewood-essential-oil-aromatherapy.
[304] Rosato A, Piarulli M, Corbo F, et al. In vitro synergistic antibacterial action of certain combination of gentamicin and essential oils. *Curr Med Chem*. 2010;17(28):3289–95.
[305] Cardoso NN, Alviano CS, Blank AF, et al. Synergism Effect of the Essential Oil from Ocimum basilicum var. Maria Bonita and Its Major Components with Fluconazole and Its Influence on Ergosterol Biosynthesis. *Evid Based Complement Alternat Med*. 2016;2016:5647182.
[306] Bonesi M, Menichini F, Tundis R, et al. Acetylcholinesterase and butyrylcholinesterase inhibitory activity of Pinus species essential oils and their constituents. *J Enzyme Inhib Med Chem*. 2010 Oct;25(5):622-8.
[307] da Silva JK, Maia JGS, Dosoky NS, et al. Antioxidant, Antimicrobial, and Cytotoxic Properties of Aniba parviflora Essential Oils from the Amazon. *Nat Prod Commun*. 2016;11(7):1025-8.

[308] dos Santos ERQ, Maia GSF, Fontes Junior EA, et al. Linalool-rich essential oils from the Amazon display antidepressant-type effect in rodents. *J Ethnopharm*. 2018;212:43-9.
[309] Sarrazin SLF, Oliveira RB, Maia JGS, et al. Antibacterial Activity of the Rosewood (Aniba rosaeodora and A. parviflora) Linalool-rich Oils from the Amazon. *Eur J Med Plants*. 2016;12(2):1-9.
[310] da Silva JK, Maia JGS, Dosoky NS, et al. Antioxidant, Antimicrobial, and Cytotoxic Properties of Aniba parviflora Essential Oils from the Amazon. *Nat Prod Commun*. 2016;11(7):1025-8.
[311] da Silva JK, Maia JGS, Dosoky NS, et al. Antioxidant, Antimicrobial, and Cytotoxic Properties of Aniba parviflora Essential Oils from the Amazon. *Nat Prod Commun*. 2016;11(7):1025-8.
[312] Baby S, Dan M, Thaha ARM, et al. High content of zerumbone in volatile oils of Zingiber zerumbet from southern India and Malaysia. *Flavour Frag J*. 2009 Nov;24(6):301-8.
[313] Srivastava AK, Srivastava SK, Shah NC. Essential Oil Composition of Zingiber zerumbet (L.) Sm. from India. *J Essent Oil Res*. 2000 Sep/Oct;12:595-7.
[314] Chane-Ming J, Vera R, Chalchat JC, et al. Chemical Composition of the Essential Oil from Rhizomes, Leaves and Flowers of Zingiber zerumbef Smith from Reunion Island. *J Essent Oil Res*. 2003 May/Jun;15:202-5.
[315] Rana VS, Verdeguer M, Blazquez MA. Chemical Composition of the Essential Oil of Zingiber zerumbet var. darcyi. *Nat Prod Commun*. 2012;7(10):1369-70.
[316] Bhuiyan MNI, Chowdhury JU, Begum J. Chemical investigation of the leaf and rhizome essential oils of Zingiber zerumbet (L.) Smith from Bangladesh. *Bangladesh J Pharmacol*. 2009;4:9-12.
[317] Singh CB, Chanu SB, Kh L, et al. Chemical composition and biological activity of the essential oil of rhizome of Zingiber zerumbet (L.) Smith. *J Pharmacognosy Phytochem*. 2014;3(3):130-3.
[318] Malek SNA, Ibrahim H, Lai HS, et al. Essential oils of Zingiber ottesii Valet. And ZIngiber zerumbet (L.) Sm. from Sabah, Malaysia. *Malaysian J Sci*. 2005;24:49-58.
[319] Worananthakij W, Doungtadum K, Mingkwan N, et al. The Effects of Soil Parameters on Efficiency of Essential Oil from Zingiber zerumbet (L.) Smith in Thailand, *Int J Agric Biosys Eng*. 2016;10(7):425-9.
[320] Rana VS, Ahluwalia V, Shakil NA, et al. Essential oil composition, antifungal, and seedling growth inhibitory effects of zerumbone from Zingiber zerumbet Smith. *J Essent Oil Res*. 2017;29(4):320-9.
[321] Jantan I, Raweh SM, Sirat HM, et al. Inhibitory effect of compounds from Zingiberaceae species on human platelet aggregation. *Phytomed*. 2008;15:306-9.
[322] Al-Saffar FJ. Effects on Renal Histology, Renal Metabolizing Enzymes and Oxidative Stress Biomarkers in Rats Administered Zerumbone Against Osteoarthritis. *Asian J Cell Biol*.2015;10(3):66-79.
[323] Burkhard PR, Burkhardt K, Haenggeli CA, et al. Plant-induced seizures: reappearance of an old problem. *J Neurol*. 1999 Aug;246(8):667–70.
[324] Millet Y, Jouglard J, Steinmetz MD, et al. Toxicity of some essential plant oils. Clinical and experimental study. *Clin Toxicol*. 1981 Dec;18(12):1485–98.
[325] Charanraj KJ, G MV, S M. Camphor poisoning in a child. *Natl Med J India*. 2013 Jan-Feb;26(1):60.
[326] Narayan S, Singh N. Camphor poisoning-An unusual cause of seizure. *Med J Armed Forces India*. 2012 Jul;68(3):252–53.
[327] Olowe SA, Ransome-Kuti O. The risk of jaundice in glucose-6-phosphate dehydrogenase deficient babies exposed to menthol. *Acta Paediatr Scand*. 1980 May;69(3):341-5.
[328] Dillon Remy M, Manning Alleyne P, Bratt DE, et al. Neonatal jaundice at Port-of-Spain General Hospital abstract. *West Indian Med J*. 1987;36(Suppl):28.
[329] Brehm-Stecher BF, Johnson EA. Sensitization of Staphylococcus aureus and Escherichia coli to antibiotics by the sesquiterpenoids nerolidol, farnesol, bisabolol, and apritone. *Antimicrob Agents Chemother*. 2003 Oct;47(10):3357-60.
[330] Khalid MH, Akhtar MN, Mohamad AS, et al. Antinociceptive effect of the essential oil of Zingiber zerumbet in mice: possible mechanisms. *J Ethnopharmacol*. 2011 Sep 1;137(1):345-51.
[331] Sulaiman MR, Tengku Mohamad TA, Shaik Mossadeq WM, et al. Antinociceptive activity of the essential oil of Zingiber zerumbet. *Planta Med*. 2010 Feb;76(2):107-12.
[332] Pinho FVS de Arujo, Coelho-de-Souza AN, Morais SM, et al. Antinociceptive effects of the essential oil of Alpinia zerumbet on mice. *Phytomed*. 2005;12:482-6.
[333] Zakaria ZA, Mohamad AS, Ahmad MS, et al. Preliminary analysis of the anti-inflammatory activity of essential oils of Zingiber zerumbet. *Biol Res Nurs*. 2011 Oct;13(4):425-32.
[334] Singh CB, Chanu SB, Kh L, et al. Chemical composition and biological activity of the essential oil of rhizome of Zingiber zerumbet (L.) Smith. *J Pharmacognosy Phytochem*. 2014;3(3):130-3.
[335] Singh CB, Chanu SB, Kh L, et al. Chemical composition and biological activity of the essential oil of rhizome of Zingiber zerumbet (L.) Smith. *J Pharmacognosy Phytochem*. 2014;3(3):130-3.
[336] Soonwera M, Wongnet O, Sittichok S. Ovicidal effect of essential oils from Zingiberaceae plants and Eucalytus globulus on eggs of head lice, Pediculus humanus capitis De Geer. *Phytomedicine*. 2018 Aug 1;47:93-104.
[337] Batubara I, Suparto IH, Sadiah S, et al. Effect of Zingiber zerumbet essential oils and zerumbone inhalation on body weight of Sprague Dawley rat. *Pak J Biol Sci*. 2013 Oct 1;16(19):1028-33.
[338] Singh CB, Chanu SB, Kh L, et al. Chemical composition and biological activity of the essential oil of rhizome of Zingiber zerumbet (L.) Smith. *J Pharmacognosy Phytochem*. 2014;3(3):130-3.
[339] Le TB, Beaufay C, Nghiem DT, et al. In Vitro Anti-Leishmanial Activity of Essential Oils Extracted from Vietnamese Plants. *Molecules*. 2017 Jun 27;22(7).
[340] Singh CB, Chanu SB, Kh L, et al. Chemical composition and biological activity of the essential oil of rhizome of Zingiber zerumbet (L.) Smith. *J Pharmacognosy Phytochem*. 2014;3(3):130-3.
[341] Worananthakij W, Doungtadum K, Mingkwan N, et al. The Effects of Soil Parameters on Efficiency of Essential Oil from Zingiber zerumbet (L.) Smith in Thailand, *Int J Agric Biosys Eng*. 2016;10(7):425-9.
[342] Phukerd U, Soonwera M. Repellency of essential oils extracted from Thai native plants against Aedes aegypti (Linn.) and Culex quinquefasciatus (Say). *Parasitol Res*. 2014 Sep;113(9):3333-40.
[343] Sutthanont N, Choochote W, Tuetun B, et al. Chemical composition and larvicidal activity of edible plant-derived essential oils against the pyrethroid-susceptible and -resistant strains of Aedes aegypti (Diptera: Culicidae). *J Vector Ecol*. 2010 Jun;35(1):106-15.
[344] Wu Y, Guo SS, Huang DY, et al. Contact and Repellant Activities of Zerumbone and Its Analogues from the Essential Oil of Zingiber zerumbet (L.) Smith against Lasioderma serricorne. *J Oleo Sci*. 2017 Apr 3;66(4):399-405.

[345] Suthisut D, Fields PG, Chandrapatya A. Contact Toxicity, Feeding Reduction, and Repellency of Essential Oils From Three Plants From the Ginger Family (Zingiberaceae) and Their Major Components Against Sitophilus zeamais and Tribolium castaneum. *J Economic Entomol.* 2011;104(4):1445-54.
[346] Aswalam EF, Hassanali A. Constituents of the essential oil of Vernonia amygdalina AS Maize weevil protectants. *Tropical Subtropical Agroecosystems.* 2006;6:95-102.
[347] Ogunbinu AO, Flamini G, Cioni PL, et al. Essential Oil Constituents of Eclipta prostrata (L.) L. and Vernonia amygdalina Delile. *Nat Prod Commun.* 2009;4(3):421-4.
[348] Sonibare OO, Sonibare MA, Adesanya EO, et al. Essential Oil Composition of Vernonia amygdalina Del. from Southwestern Nigeria. *J Essental Oil Bearing Plants.* 2009;12(1):55-8.
[349] Aswalam EF, Emosairue SO, Hassanali A. Contribution of different constituents to the toxicity of the essential oil constituents of Vernonia amygdalina (Compositae) and Xylopia aetiopica (Annonaceae) on maize weevil, Sitophilus zeamais Motschulsky (Coleoptera: Curculionidae). *African J Botech.* 2008 Aug;7(16):2957-62.
[350] Ocheng F, Bwanga F, Joloba M, et al. Essential Oils from Ugandan Aromatic Medicinal Plants: Chemical Composition and Growth Inhibitory Effects on Oral Pathogens. Evid Based Complement Alternat Med. 2015;2015:230832.
[351] Burkhard PR, Burkhardt K, Haenggeli CA, et al. Plant-induced seizures: reappearance of an old problem. *J Neurol.* 1999 Aug;246(8):667-70.
[352] Waldman N. Seizure caused by dermal application of over-the-counter eucalyptus oil head lice preparation. *Clin Toxicol (Phila).* 2011 Oct;49(8):750-1.
[353] Craig JO. Poisoning by the volatile oils in childhood. *Arch Dis Child.* 1953;28:259-67.
[354] Mathew T, Kamath V, Kumar RS, et al. Eucalyptus oil inhalation-induced seizure: A novel, underrecognized, preventable cause of acute symptomatic seizure. *Epilepsia Open.* 2017 Jul 4;2(3):350-354.
[355] Unger M, Frank A. Simultaneous determination of the inhibitory potency of herbal extracts on the activity of six major cytochrome P450 enzymes using liquid chromatography/mass spectrometry and automated online extraction. *Rapid Commun Mass Spectrom.* 2004;18(19):2273-81.
[356] Jori A, Bianchetti A, Prestini PE, et al. Effect of eucalyptol (1,8-cineole) on the metabolism of other drugs in rats and in man. *Eur J Pharmacol.* 1970;9(3):362-66.
[357] Kim NH, Hyun SH, Jin CH, et al. Pretreatment with 1,8-cineole potentiates thioacetamide-induced hepatotoxicity and immunosuppression. *Arch Pharm Res.* 2004 Jul;27(7):781-9.
[358] Jori A, Bianchetti A, Prestini PE, et al. Effect of eucalyptol (1,8-cineole) on the metabolism of other drugs in rats and in man. *Eur J Pharmacol.* 1970;9(3):362-66.
[359] de Sousa DP, Raphael E, Brocksom U, et al. Sedative effect of monoterpene alcohols in mice: A preliminary screening. *Verlag der Zeitschrift fur Naturforschung.* 2007;62c:563-66.
[360] Tognolini M, Barocelli E, Ballabeni V, et al. Comparative screening of plant essential oils; phenylpropanoid moiety as basic core for antiplatelet activity. *Life Sci.* 2006 Feb 23;78(13):1419-32.
[361] Okazaki K, Kawazoe K, Takaishi Y. Human platelet aggregation inhibitors from thyme (Thymus vulgaris L.). *Phytother Res.* 2002 Jun;16(4):398-9.
[362] Medeiros Barreto H, Cerqueira Fontinele F, Pereira de Oliveira A, et al. Phytochemical prospection and modulation of antibiotic activity in vitro by Lippia origanoides H.B.K. in methicillin resistant Staphylococcus aureus. *Biomed Res Int.* 2014;305610:1–7.
[363] Aelenei P, Miron A, Trifan A, et al. Essential Oils and Their Components as Modulators of Antibiotic Activity against Gram-Negative Bacteria. *Medicines.* 2016;3(19):2-34.
[364] Veras HNH, Rodrigues FFG, Colares AV, et al. Synergistic antibiotic activity of volatile compounds from the essential oil of Lippia sidoides and thymol. *Fitoterapia.* 2012 Apr;83(3):508-12.
[365] Langeveld WT, Veldhuizen EJ, Burt SA. Synergy between essential oil components and antibiotics: A review. *Crit Rev Microbiol.* 2014 Feb;40(1):76-94.
[366] Miladinovic DL, Ilic BS, Kocic BD, et al. Antibacterial investigation of thyme essential oil and its main constituents on combination with tetracycline. *J Med Food.* 2015 Aug;18(8):935-7.
[367] Shin S, Lim S. Antifungal effects of herbal essential oils alone and in combination with ketoconazole against Trichophyton spp. *J Appl Micribiol.* 2004;97:1289-96.
[368] Kissels W, Wu X, Santos RR, et al. Short communication: Interaction of the isomers carvacrol and thymol with the antibiotics doxycycline and tilmicosin: In vitro effects against pathogenic bacteria commonly found in the respiratory tract of calves. *J Dairy Sci.* 2017 Feb;100(2):970-974.
[369] van den Brink DM, van Miert JNI, Dacremont G, et al. Characterization of the Final Step in the Conversion of Phytol into Phytanic Acid. *J Biol Chem.* 2005 Jul 22;280(29):26838-44.
[370] Selkälä EM, Nair RR, Schmitz W, et al. Phytol is lethal for Amacr-deficient mice. *Biochim Biophys Acta.* 2015 Oct;1851(10):1394-405.
[371] McGinty D, Letizia CS, Api AM. Fragrance material review on phytol. *Food Chem Toxicol.* 2010;48:S59–S63.
[372] National Institutes of Health. U.S. National Library of Medicine. Alpha-methylacyl-CoA racemase deficiency. Available at: https://ghr.nlm.nih.gov/condition/alpha-methylacyl-coa-racemase-deficiency.
[373] Kagoura M, Matsui C, Morohashi M. Phytol is a novel tumor promoter on ICR mouse skin. *Jpn J Cancer Res.* 1999 Apr;90(4):377-84.
[374] Ocheng F, Bwanga F, Joloba M, et al. Essential Oils from Ugandan Aromatic Medicinal Plants: Chemical Composition and Growth Inhibitory Effects on Oral Pathogens. *Evid Based Complement Alternat Med.* 2015;2015:230832.
[375] Aswalam EF, Hassanali A. Constituents of the essential oil of Vernonia amygdalina AS Maize weevil protectants. *Tropical Subtropical Agroecosystems.* 2006;6:95-102.
[376] Aswalam EF, Hassanali A. Constituents of the essential oil of Vernonia amygdalina AS Maize weevil protectants. *Tropical Subtropical Agroecosystems.* 2006;6:95-102.
[377] Orav A, Kailas T, Muurisepp M. Composition of blackcurrant aroma isolated from leaves, buds, and berries of Ribes nigrum L. *Proc Estonian Acad Sci Chem.* 2002;51(4):225-34.
[378] The Scientific Division, Camilli, Albert & Laloue. The Composition of Blackcurrant Absolute (Ribes Nigrum). *Perf Flavorist.* 1984/1985 Dec-Jan;9:39-42.
[379] Dvaranauskaite A, Venskutonis PR, Raynaud C, et al. Aroma composition of blackcurrant bud extracts isolated by simultaneous distillation/extraction. *Foocbalt.* 2008:84-8.
[380] Dvaranauskaite A, Venskutonis PR, Raynaud C, et al. Variations in the essential oil composition in buds of six blackcurrant (Ribes nigrum L.) cultivars at various development phases. *Food Chem.* 2009;114:671-9.
[381] Stevic T, Savikin K, Ristic M, et al. Composition and antimicrobial activity of the essential oil of the leaves of black currant (Ribes nigrum L.) cultivar Čačanska crna. *J Serb Chem Soc.* 2010;75(1):35-43.

[382] Kilic CS, Koyuncu M, Ozek T, et al. Essential Oil of the Leaves of Ribes nigrum L. from Turkey. *J Essent Oil Res*. 2008;20(6):512-4.
[383] Oprea E, Radulescu V, Balotescu C, et al. Chemical and biological studies of Ribes nigrum L. buds essential oil. *BioFactors*. 2008;34:3-12.
[384] Dordevic BS, Pljevljakusic DS, Savikin KP, et al. Essential oil from black currant buds as chemotaxonomy marker and antimicrobial agent. *Chem Biodivers*. 2014;11:1228-40.
[385] Nursten HE, Williams AA. Volatile constituents of the black currant, Ribes nigrum L. I. A commercial black-currant distillate. *J Sci FD Agric*. 1969 Feb;20:91-8.
[386] Dvaranauskaite A, Venskutonis PR, Raynaud C, et al. Characterization of Steam Volatiles in the Essential Oil of Black Currant Buds and the Antioxidant Properties of Different Bud Extracts. *J Agric Food Chem*. 2008;56:3279-86.
[387] Le Quere JL, Latrasse A. Composition of the Essential Oils of Blackcurrant Buds (Ribes nigrum L.). *J Agric Food Chem*. 1990;38:3-10.
[388] Aazza S, Lyoussi B, Miguel MG. Antioxidant and antiacetylcholinesterase activities of some commercial essential oils and their major compounds. *Molecules*. 2011 Sep 7;16(9):7672-90.
[389] Miyazawa M, Yamafuji C. Inhibition of acetylcholinesterase activity by bicyclic monoterpenoids. *J Agric Food Chem*. 2005 Mar 9;53(5):1765-8.
[390] Oprea E, Radulescu V, Balotescu C, et al. Chemical and biological studies of Ribes nigrum L. buds essential oil. *BioFactors*. 2008;34:3-12.
[391] Dordevic BS, Pljevljakusic DS, Savikin KP, et al. Essential oil from black currant buds as chemotaxonomy marker and antimicrobial agent. *Chem Biodivers*. 2014;11:1228-40.
[392] Stevic T, Savikin K, Ristic M, et al. Composition and antimicrobial activity of the essential oil of the leaves of black currant (Ribes nigrum L.) cultivar Čačanska crna. *J Serb Chem Soc*. 2010;75(1):35-43.
[393] Jerkovic I, Mastelic J. Volatile compounds from leaf-buds of Populus nigra L. (Salicaceae). *Phytochem*. 2003;63:109-13.
[394] Okinczyc P, Szumny A, Szperlik J, et al. Profile of Polyphenolic and Essential Oil Composition of Polish Propolis, Black Poplar and Aspens Buds. *Molecules*. 2018;23:1262.
[395] Tsuneki H, Ma EL, Kobayashi S, et al. Antiangiogenic activity of beta-eudesmol in vitro and in vivo. *Eur J Pharmacol*. 2005 Apr 11;512(2-3):105-15.
[396] Arora CK, Arora RB, Mesta CK, et al. Hypotensive activity of beta-eudesmol and some related sesquiterpenes. *Ind J Med Res*. 1967 May;55(5):463-72.
[397] Wang BG, Hong X, Li L, Zhou J, et al. Chemical constituents of two Chinese Magnoliaceae plants, Tsoongiodendron odorum and Manglietiastrum sinicum, and their inhibition of platelet aggregation. *Planta Med*. 2000 Aug;66(6):511-5.
[398] Kozan E, Ilhan M, Tümen I, et al. The scolicidal activity of the essential oil obtained from the needles of Pinus nigra Arn. subsp. pallasiana (Lamb.) Holmboe on hydatid cyst. *J Ethnopharmacol*. 2019 Feb 11;235:243-247.
[399] Almarie A, Mama A, Rukunudin I. Chemical composition and herbicidal effects of Melaleuca bracteata F.Muell. essential oil against some weedy species. *Int J Sci Eng Res*. 2016 Jan;7(1):507-14.
[400] Goswami P, Verma SK, Chauhan A, et al. Chemical Composition and Antibacterial Activity of Melaleuca bracteata Essential Oil from India: A Natural Source of Methyl Eugenol. *Nat Prod Commun*. 2017;12(6):965-8.
[401] Li Y, Ye Z, Wang W, et al. Composition analysis of essential oil from Melaleuca bracteata leaves using ultrasound assisted extraction and its antioxidative and antimicrobial activities. *BioResources*. 2018;13(4):8488-8504.
[402] Brophy JJ, Goldsack RJ, Doran JC, et al. A Comparison of the Leaf Oils of Melaleuca squamophloia with Those of Its Close Relatives, M. styphelioides and M. bracteata. *J Essent Oil Res*. 1999 May/Jun;11(3):327-32.
[403] Siddique S, Parveen Z, Firadaus-e-Bareen, et al. Chemical composition, antibacterial and antioxidant activities of essential oils from leaves of three Melaleuca species of Pakistani flora. *Arabian J Chem*. 2017 Feb. [Epub ahead of print].
[404] Aboutabl EA, El Tohamy SF, de Pooter HL, et al. A Comparative Study of the Essential Oils from Three Melaleuca Species Growing in Egypt. *Flavour Frag J*. 1991;6:139-41.
[405] Siddique S, Parveen Z, Firdaus-e-Bareen, et al. Chemical Composition and Insecticidal Activities of Essential Oils of Myrtaceae against Tribolium castaneum (Coleoptera: Tenebrionidae). *Pol J Environ Stud*. 2017;26(4):1653-62.
[406] Johnson JD, Ryan MJ, Toft JD II, et al. Two-year toxicity and carcinogenicity study of methyl eugenol in F344/N rats and B6C3F(1) mice. *J Agric Food Chem*. 2000 Aug;48(8):3620-32.
[407] European Commission. Opinion of the Scientific Committee on Food on methyl eugenol (4-Allyl-1,2- dimethoxybenzene). Available at: http://ec.europa.eu/food/fs/sc/scf/out102_en.pdf.
[408] National Toxicology Program. NTP Toxicology and Carcinogenesis Studies of Methyl eugenol (CAS NO. 93-15-2) in F344/N Rats and B6C3F1 Mice (Gavage Studies). *Natl Toxicol Program Tech Rep Ser*. 2000 Jul;491:1-412.
[409] National Toxicology Program. Carcinogenesis Studies of Eugenol (CAS No. 97-53-0) in F344/N Rats and B6C3F1 Mice (Feed Studies). Technical Report Series No. 223. NIH Publication No. 84-1779. 1983. U.S. DHHS, PHS, NIH, NTP, Research Triangle Park, NC.
[410] Johnson JD, Ryan MJ, Toft JD II, et al. Two-year toxicity and carcinogenicity study of methyl eugenol in F344/N rats and B6C3F(1) mice. *J Agric Food Chem*. 2000 Aug;48(8):3620-32.
[411] National Toxicology Program. NTP Toxicology and Carcinogenesis Studies of Methyl eugenol (CAS NO. 93-15-2) in F344/N Rats and B6C3F1 Mice (Gavage Studies). *Natl Toxicol Program Tech Rep Ser*. 2000 Jul;491:1-412.
[412] Kerckaert GA, Brauninger R, LeBoeuf RA, et al. Use of the Syrian hamster embryo cell transformation assay for carcinogenicity prediction of chemicals currently being tested by the National Toxicology Program in rodent bioassays. *Environ Health Perspect*. 1996;104:1075-84.
[413] National Toxicology Program. Carcinogenesis Studies of Eugenol (CAS No. 97-53-0) in F344/N Rats and B6C3F1 Mice (Feed Studies). Technical Report Series No. 223. NIH Publication No. 84-1779. 1983. U.S. DHHS, PHS, NIH, NTP, Research Triangle Park, NC.
[414] Johnson JD, Ryan MJ, Toft JD II, et al. Two-year toxicity and carcinogenicity study of methyl eugenol in F344/N rats and B6C3F(1) mice. *J Agric Food Chem*. 2000 Aug;48(8):3620-32.
[415] European Commission. Opinion of the Scientific Committee on Food on methyl eugenol (4-Allyl-1,2- dimethoxybenzene). Available at: http://ec.europa.eu/food/fs/sc/scf/out102_en.pdf.
[416] National Toxicology Program. NTP Toxicology and Carcinogenesis Studies of Methyl eugenol (CAS NO. 93-15-2) in F344/N Rats and B6C3F1 Mice (Gavage Studies). *Natl Toxicol Program Tech Rep Ser*. 2000 Jul;491:1-412.
[417] National Toxicology Program. Carcinogenesis Studies of Eugenol (CAS No. 97-53-0) in F344/N Rats and B6C3F1 Mice (Feed Studies). Technical Report Series No. 223. NIH Publication No. 84-1779. 1983. U.S. DHHS, PHS, NIH, NTP, Research Triangle Park, NC.
[418] Ahmad A, Khan A, Ahmad Khan L, et al. In vitro synergy of eugenol and methyl eugenol with fluconazole against clinical Candida isolates. *J Med Microbiol*. 2010;59:1178-84.
[419] Lee SE, Lee BH, Choi WS, et al. Fumigant toxicity of volatile natural products from Korean spices and medicinal plants towards the rice weevil, Sitophilus oryzae (L). *Pest Manag Sci*. 2001;57:548-53.

[420] Grice ID, Rogers KL, Griffiths LR. Isolation of Bioactive Compounds That Relate to the Anti-Platelet Activity of Cymbopogon ambiguous. *Evidence-Based Complement Alternat Med.* 2011;2011:467134.
[421] Fajemiroye JO, Galdino PM, De Paula JA, et al. Anxiolytic and antidepressant like effects of natural food flavour (E)-methyl isoeugenol. *Food Funct.* 2014 Aug;5(8):1819-28.
[422] Goswami P, Verma SK, Chauhan A, et al. Chemical Composition and Antibacterial Activity of Melaleuca bracteata Essential Oil from India: A Natural Source of Methyl Eugenol. *Nat Prod Commun.* 2017;12(6):965-8.
[423] Li Y, Ye Z, Wang W, et al. Composition analysis of essential oil from Melaleuca bracteata leaves using ultrasound assisted extraction and its antioxidative and antimicrobial activities. *BioResources.* 2018;13(4):8488-8504.
[424] Siddique S, Parveen Z, Firadaus-e-Bareen, et al. Chemical composition, antibacterial and antioxidant activities of essential oils from leaves of three Melaleuca species of Pakistani flora. *Arabian J Chem.* 2017 Feb. [Epub ahead of print].
[425] Li Y, Ye Z, Wang W, et al. Composition analysis of essential oil from Melaleuca bracteata leaves using ultrasound assisted extraction and its antioxidative and antimicrobial activities. *BioResources.* 2018;13(4):8488-8504.
[426] Siddique S, Parveen Z, Firadaus-e-Bareen, et al. Chemical composition, antibacterial and antioxidant activities of essential oils from leaves of three Melaleuca species of Pakistani flora. *Arabian J Chem.* 2017 Feb. [Epub ahead of print].
[427] Siddique S, Parveen Z, Firdaus-e-Bareen, et al. Chemical Composition and Insecticidal Activities of Essential Oils of Myrtaceae against Tribolium castaneum (Coleoptera: Tenebrionidae). *Pol J Environ Stud.* 2017;26(4):1653-62.
[428] Pandey A, Chowdhury A. Volatile constituents of the rhizome oil of Curcuma caesia Roxb. from central India. *Flavour Frag J.* 2003;18(5):463-5.
[429] Angel GR, Menon N, Vimala B, et al. Essential oil composition of eight starchy Curcuma species. *Ind Crops Prod.* 2014;60:233-8.
[430] Mukunthan KS, Anil Kumar NV, Balaji S, et al. Analysis of Essential Oil Constituents in Rhizome of Curcuma caesia Roxb. from South India. *J Essent Oil Bear Plants.* 2014;17(4):647-51.
[431] Behura S, Srivastava VK. Essential Oils of Leaves of Curcuma Species. *J Essent Oil Res.* 2004 Mar/Apr;16:109-10.
[432] Dayma S, Jain, N, Pathak AK. Comparative Study of Volatile oil of Two Different Species Curcuma caesia and Curcuma longa Rhizome. *Curr Res Pharm Sci.* 2013;01:35-8.
[433] Sastri BN. The Wealth of India. Raw Materials. CSIR. New Delhi. 1962.
[434] Banerjee AK, Kaul VK, Nigam SS. Chemical examination of the essential oil of Curcuma caesia Roxb. *Essenze e Derivati Agrumari.* 1984;54:117-21.
[435] Buddhasukh D, Smith J, Ternai B. Essential oil of Curcuma caesia Roxb. J Sci Faculty Chiang Mai University. 1995;1-22(1-2):14-16.
[436] Khine H, Weiss D, Graber N, et al. A cluster of children with seizures caused by camphor poisoning. *Pediatrics.* 2009 May;123(5):1269-72.
[437] Michiels EA, Mazor SS. Toddler with seizures due to ingesting camphor at an Indian celebration. *Pediatr Emerg Care.* 2010 Aug;26(8):574-75.
[438] Koren G. Medications which can kill a toddler with one tablet or teaspoonful. *J Toxicol Clin Toxicol. 1*993;31(3):407-13.
[439] Flaman Z, Pellechia-Clarke S, Bailey B, et al. Unintentional exposure of young children to camphor and eucalyptus oils. *Paediatr Child Health.* 2001 Feb;6(2):80-83.
[440] Rabl W, Katzgraber F, Steinlechner M. Camphor ingestion for abortion (case report). *Forensic Sci Int.* 1997 Sep 19;89(1-2):137-40.
[441] Flaman Z, Pellechia-Clarke S, Bailey B, et al. Unintentional exposure of young children to camphor and eucalyptus oils. *Paediatr Child Health.* 2001 Feb;6(2):80-83.
[442] Manoguerra AS, Erdman AR, Wax PM, et al. Camphor poisoning: an evidence-based practice guideline for out-of- hospital management. *Clin Toxicol (Phila).*2006;44(4):357-70.
[443] Gibson DE, Moore GP, Pfaff JA. Camphor ingestion. *Am J Emerg Med.* 1989 Jan;7(1):41-43.
[444] Koppel C, Martends F, Schirop T, et al. Hemoperfusion in acute camphor poisoning. *Intensive Care Med.* 1988;14(4):431-33.
[445] Burkhard PR, Burkhardt K, Haenggeli CA, et al. Plant-induced seizures: reappearance of an old problem. *J Neurol.* 1999 Aug;246(8):667–70.
[446] Millet Y, Jouglard J, Steinmetz MD, et al. Toxicity of some essential plant oils. Clinical and experimental study. *Clin Toxicol.* 1981 Dec;18(12):1485–98.
[447] Charanraj KJ, G MV, S M. Camphor poisoning in a child. *Natl Med J India.* 2013 Jan-Feb;26(1):60.
[448] Narayan S, Singh N. Camphor poisoning-An unusual cause of seizure. *Med J Armed Forces India.* 2012 Jul;68(3):252–53.
[449] Olowe SA, Ransome-Kuti O. The risk of jaundice in glucose-6-phosphate dehydrogenase deficient babies exposed to menthol. *Acta Paediatr Scand.* 1980 May;69(3):341-5.
[450] Dillon Remy M, Manning Alleyne P, Bratt DE, et al. Neonatal jaundice at Port-of-Spain General Hospital abstract. *West Indian Med J.* 1987;36(Suppl):28.
[451] Burkhard PR, Burkhardt K, Haenggeli CA, et al. Plant-induced seizures: reappearance of an old problem. *J Neurol.* 1999 Aug;246(8):667-70.
[452] Waldman N. Seizure caused by dermal application of over-the-counter eucalyptus oil head lice preparation. *Clin Toxicol (Phila).* 2011 Oct;49(8):750-1.
[453] Craig JO. Poisoning by the volatile oils in childhood. *Arch Dis Child.* 1953;28:259-67.
[454] Mathew T, Kamath V, Kumar RS, et al. Eucalyptus oil inhalation-induced seizure: A novel, underrecognized, preventable cause of acute symptomatic seizure. *Epilepsia Open.* 2017 Jul 4;2(3):350-354.
[455] Unger M, Frank A. Simultaneous determination of the inhibitory potency of herbal extracts on the activity of six major cytochrome P450 enzymes using liquid chromatography/mass spectrometry and automated online extraction. *Rapid Commun Mass Spectrom.* 2004;18(19):2273-81.
[456] Jori A, Bianchetti A, Prestini PE, et al. Effect of eucalyptol (1,8-cineole) on the metabolism of other drugs in rats and in man. *Eur J Pharmacol.* 1970;9(3):362-66.
[457] Kim NH, Hyun SH, Jin CH, et al. Pretreatment with 1,8-cineole potentiates thioacetamide-induced hepatotoxicity and immunosuppression. *Arch Pharm Res.* 2004 Jul;27(7):781-9.
[458] Jori A, Bianchetti A, Prestini PE, et al. Effect of eucalyptol (1,8-cineole) on the metabolism of other drugs in rats and in man. *Eur J Pharmacol.* 1970;9(3):362-66.
[459] de Sousa DP, Raphael E, Brocksom U, et al. Sedative effect of monoterpene alcohols in mice: A preliminary screening. *Verlag der Zeitschrift fur Naturforschung.* 2007;62c:563-66.
[460] Garg SC, Jain RK. Antimicrobial Efficacy of Essential Oil from Curcuma caesia. *Indian J Microbiol.* 1998;38(3):169-70.
[461] Banerjee A, Nigam SS. Antifungal activity of the essential oil of Curcuma caesia Roxb. *Indian J Med Res.* 1976 Sep;64(9):1318-21.
[462] Rustaiyan A, Mojob F, Salsali M, et al. Composition of the Essential Oil of Tanacetum polycephalum Schultz. Bip. *J Essent Oil Res.* 1999;11(4):497-8.
[463] Mojarrad M, Sarghein SH, Sonboli A. Chemical diversity of the essential oils of twenty populations of Tanacetum polycephalum Sch. Bip. from Iran. *Nat Prod Res.* 2018;1478-6427.

[464] Rezazadeh F, Mahdavi M, Motavalizadehkakhky A, et al. Essential Oil Composition of Three Population of Tanacetum polycephalum from Iran and their Antimicrobial Activity. *J Essent Oil Res.* 2014;17(2):317-30.
[465] Mirahmadi SF, Sefidkon F, Aalifar M, et al. Essential Oil Composition of Tanacetum polycephalum subsp. duderanum (Boiss) Podl., A Plant Endemic from Iran. *J Essent Oil Bearing Plants.* 2011;14(6):742-5.
[466] Javidnia K, Miri R, Soltani M, et al. Composition of the essential oil of Tanacetum polycephalum Schultz Bip. subsp. farsicum Podl. from Iran. *J Essent Oil Res.* 2008 May/June;20:209-11.
[467] Nori-Shargh D, Norouzi-Arasi H, Mirza M, et al. Chemical composition of the essential oil of Tanacetum polycephalum (Schultz Bip. ssp. heterophyllum). *Flavour Frag J.* 1999;14:105-6.
[468] Nezhadali A, Roudi BS, Akbarpour M. Chemical composition of the essential oils from the flower of Tanacetum polycephalum subsp. duderanum as a herbal plant in Iran. *Der Pharma Chemica.* 2009;1(2):27-31.
[469] Mahdavi M, Jouri MH, Rezazadeh F. Identification of chemical compositions in the essential oil of Tanacetum polycephalum Scultz-Bip. Iran. *Int J Med Arom Plants.* 2013 sep;3(3):395-400.
[470] Mahdavi M, Jouri MH, Mahmoudi J, et al. Investigating the altitude effect on the quantity and quality of the essential oil in Tanacetum polycephalum Sch.-Bip. polycephalum in the Baladeh region of Nour, Iran. *Chinese J Nat Med.* 2013;11(5):553-9.
[471] Azadeh CS, Whiteon F. In field conditions, the yield and composition of essential oils, leaves, stems and flower buds Tanacetum polycephalum Schultz. Bip. Subsp. polycephalum. J Res Iranian Herbs Med Herbs. 1996;33(4):67.
[472] Rezaee MB, Jaimand K, Naderi M. Chemical Composition of the Essential Oil of Tanacetum polycephalum subsp. polycephalum from Different Locations of Azerbaijan Province, Iran. *J Med Plants By-products.* 2012;1:71-4.
[473] Amiri H. Chemical composition, antibacterial and antioxidant activity of the essential oil of Tanacetum polycephalum Schultz. Bip. *Int J Botany.* 2007;3(3):321-4.
[474] Haliciglu O, Astarcioglu G, Yaprak I, et al. Toxicity of Salvia officinalis in a newborn and a child: an alarming report. *Pediatr Neurol.* 2011 Oct;45(4):259-60.
[475] Lachenmeier DW, Walch SG. Epileptic seizure caused by accidental ingestion of sage (Salvia officinalis L.) oil in children: a rare, exceptional case or a threat to public health. *Pediatr Neurol.* 2012 Mar;46(3):201.
[476] da Silva Costa KC, Bexerra SB, Norte CM, et al. Medicinal plants with teratogenic potential: current considerations. *Braz J Pharm Sci.* 2012;48(3):427-33.
[477] Brinker F. Herb Contraindications and Drug Interactions. 2nd ed. Sandy, OR. Eclectic Medical Publications. 1998.
[478] Frohne D. Giftpflanzen: Cupressaceae. Stuttgart: Wissenschaftliche Verlagsgesellschaft mbH; 1997. pp. 153–6.
[479] European Medicines Agency. Public statement on the use of herbal medicinal products containing thujone. Available at: http://www.ema.europa.eu/docs/en_GB/document_library/Public_statement/2011/02/WC500102294.pdf.
[480] Millet Y, Jouglard J, Steinmetz MD, et al. Toxicity of some essential plant oils. Clinical and experimental study. *Clin Toxicol.* 1981 Dec;18(12):1485-98.
[481] Cristovao L, Carvalho F, Bastos MDL, et al. Hepatotoxicity of an essential oil of Salvia officinalis L.: an in vitro study using freshly isolated rat hepatocytes. *Congress Biomarkers.* 2001 Sep:165.
[482] Millet Y, Jouglard J, Steinmetz MD, et al. Toxicity of some essential plant oils. Clinical and experimental study. *Clin Toxicol.* 1981 Dec;18(12):1485-98.
[483] Burkhard PR, Burkhardt K, Haenggeli CA, et al. Plant-induced seizures: reappearance of an old problem. *J Neurol.* 1999 Aug;246(8):667-70.
[484] Burkhard PR, Burkhardt K, Haenggeli CA, et al. Plant-induced seizures: reappearance of an old problem. *J Neurol.* 1999 Aug;246(8):667-70.
[485] Hold KM, Sirisoma NS, Ikeda T, et al. α-Thujone (the active component of absinthe): γ-Aminobutyric acid type A receptor modulation and metabolic detoxification. *Proc Natl Acad Sci USA.* 2000 Apr 11;97(8):3826–3831.
[486] Waldman N. Seizure caused by dermal application of over-the-counter eucalyptus oil head lice preparation. *Clin Toxicol (Phila).* 2011 Oct;49(8):750-1.
[487] Craig JO. Poisoning by the volatile oils in childhood. *Arch Dis Child.* 1953;28:259-67.
[488] Mathew T, Kamath V, Kumar RS, et al. Eucalyptus oil inhalation-induced seizure: A novel, underrecognized, preventable cause of acute symptomatic seizure. *Epilepsia Open.* 2017 Jul 4;2(3):350-354.
[489] Uc A, Bishop WP, Sanders KD. Camphor hepatotoxicity. *South Med J.* 2000;93:596-98.
[490] Frohne D. Giftpflanzen: Cupressaceae. Stuttgart: Wissenschaftliche Verlagsgesellschaft mbH; 1997. pp. 153–6.
[491] Dolan LC, Matulka RA, Burdock GA. Naturally Occurring Food Toxins. *Toxins (Basel).* 2010 Sep;2(9):2289–2332.
[492] United States National Toxicology Program (NTP). Alpha-Thujone. Dec 10, 1997. Available at: http://ntp.niehs.nih.gov/index.cfm?objectid=03DB8C36-E7A1-9889-3BDF8436F2A8C51F.
[493] Unger M, Frank A. Simultaneous determination of the inhibitory potency of herbal extracts on the activity of six major cytochrome P450 enzymes using liquid chromatography/mass spectrometry and automated online extraction. *Rapid Commun Mass Spectrom.* 2004;18(19):2273-81.
[494] Jori A, Bianchetti A, Prestini PE, et al. Effect of eucalyptol (1,8-cineole) on the metabolism of other drugs in rats and in man. *Eur J Pharmacol.* 1970;9(3):362-66.
[495] Abass K, Reponen P, Mattila S, et al. Metabolism of α-thujone in human hepatic preparations in vitro. *Xenobiotica.* 2011 Feb;41(2):101-11.
[496] Kim NH, Hyun SH, Jin CH, et al. Pretreatment with 1,8-cineole potentiates thioacetamide-induced hepatotoxicity and immunosuppression. *Arch Pharm Res.* 2004 Jul;27(7):781-9.
[497] Khine H, Weiss D, Graber N, et al. A cluster of children with seizures caused by camphor poisoning. *Pediatrics.* 2009 May;123(5):1269-72.
[498] Michiels EA, Mazor SS. Toddler with seizures due to ingesting camphor at an Indian celebration. *Pediatr Emerg Care.* 2010 Aug;26(8):574-75.
[499] Koren G. Medications which can kill a toddler with one tablet or teaspoonful. *J Toxicol Clin Toxicol. 1*993;31(3):407-13.
[500] Flaman Z, Pellechia-Clarke S, Bailey B, et al. Unintentional exposure of young children to camphor and eucalyptus oils. *Paediatr Child Health.* 2001 Feb;6(2):80-83.
[501] Haliciglu O, Astarcioglu G, Yaprak I, et al. Toxicity of Salvia officinalis in a newborn and a child: an alarming report. *Pediatr Neurol.* 2011 Oct;45(4):259-60.
[502] Lachenmeier DW, Walch SG. Epileptic seizure caused by accidental ingestion of sage (Salvia officinalis L.) oil in children: a rare, exceptional case or a threat to public health. *Pediatr Neurol.* 2012 Mar;46(3):201.
[503] Rabl W, Katzgraber F, Steinlechner M. Camphor ingestion for abortion (case report). *Forensic Sci Int.* 1997 Sep 19;89(1-2):137-40.
[504] Flaman Z, Pellechia-Clarke S, Bailey B, et al. Unintentional exposure of young children to camphor and eucalyptus oils. *Paediatr Child Health.* 2001 Feb;6(2):80-83.
[505] da Silva Costa KC, Bexerra SB, Norte CM, et al. Medicinal plants with teratogenic potential: current considerations. *Braz J Pharm Sci.* 2012;48(3):427-33.

[506] Brinker F. Herb Contraindications and Drug Interactions. 2nd ed. Sandy, OR. Eclectic Medical Publications. 1998.
[507] Manoguerra AS, Erdman AR, Wax PM, et al. Camphor poisoning: an evidence-based practice guideline for out-of- hospital management. *Clin Toxicol (Phila).*2006;44(4):357-70.
[508] Gibson DE, Moore GP, Pfaff JA. Camphor ingestion. *Am J Emerg Med.* 1989 Jan;7(1):41-43.
[509] Koppel C, Martends F, Schirop T, et al. Hemoperfusion in acute camphor poisoning. *Intensive Care Med.* 1988;14(4):431-33.
[510] Burkhard PR, Burkhardt K, Haenggeli CA, et al. Plant-induced seizures: reappearance of an old problem. *J Neurol.* 1999 Aug;246(8):667–70.
[511] Millet Y, Jouglard J, Steinmetz MD, et al. Toxicity of some essential plant oils. Clinical and experimental study. *Clin Toxicol.* 1981 Dec;18(12):1485–98.
[512] Charanraj KJ, G MV, S M. Camphor poisoning in a child. *Natl Med J India.* 2013 Jan-Feb;26(1):60.
[513] Narayan S, Singh N. Camphor poisoning-An unusual cause of seizure. *Med J Armed Forces India.* 2012 Jul;68(3):252–53.
[514] Uc A, Bishop WP, Sanders KD. Camphor hepatotoxicity. *South Med J.* 2000;93:596-98.
[515] Uc A, Bishop WP, Sanders KD. Camphor hepatotoxicity. *South Med J.* 2000;93:596-98.
[516] Frohne D. Giftpflanzen: Cupressaceae. Stuttgart: Wissenschaftliche Verlagsgesellschaft mbH; 1997. pp. 153–6.
[517] Dolan LC, Matulka RA, Burdock GA. Naturally Occurring Food Toxins. *Toxins (Basel).* 2010 Sep;2(9):2289–2332.
[518] Olowe SA, Ransome-Kuti O. The risk of jaundice in glucose-6-phosphate dehydrogenase deficient babies exposed to menthol. *Acta Paediatr Scand.* 1980 May;69(3):341-5.
[519] Dillon Remy M, Manning Alleyne P, Bratt DE, et al. Neonatal jaundice at Port-of-Spain General Hospital abstract. *West Indian Med J.* 1987;36(Suppl):28.
[520] Covington TR, et al. Handbook of Nonprescription Drugs. 11th ed. Washington, D.C.: American Pharmaceutical Association. 1996.
[521] Craig JO. Poisoning by the volatile oils in childhood. *Arch Dis Child.* 1953;28:259-67.
[522] Melis K. Bochner A, Janssens G. Accidental nasal eucalyptol and menthol instillation. *Eur J Pediatr.* 1989 Aug;148(8)786-7.
[523] Day LM, Ozanne-Smith J, Parsons BJ, et al. Eucalyptus oil poisoning among young children: mechanisms of access and the potential prevention. *Aust N Z J Public Health.* 1997 Jun;21(3):297-302.
[524] Day LM, Ozanne-Smith J, Parsons BJ, et al. Eucalyptus oil poisoning among young children: mechanisms of access and the potential prevention. *Aust N Z J Public Health.* 1997 Jun;21(3):297-302.
[525] Myott E. Case of eucalyptus poisoning. *Brit M J.* 1906;1:558.
[526] Hindle RC. Eucalyptus oil ingestion. *N Z Med J.* 1994 May 11;107(977)185-6.
[527] Tibballs J. Clinical effects and management of eucalyptus oil ingestion in infants and small children. *Med J Aust.* 1995 Aug;163(4):177-80.
[528] Waldman W, Barwina M, Sein Anand J. Accidental intoxication with eucalyptus oil—a case report. *Przeql Lek.* 2011;68(8):555-6.
[529] Day LM, Ozanne-Smith J, Parsons BJ, et al. Eucalyptus oil poisoning among young children: mechanisms of access and the potential prevention. *Aust N Z J Public Health.* 1997 Jun;21(3):297-302.
[530] Burkhard PR, Burkhardt K, Haenggeli CA, et al. Plant-induced seizures: reappearance of an old problem. *J Neurol.* 1999 Aug;246(8):667-70.
[531] Waldman N. Seizure caused by dermal application of over-the-counter eucalyptus oil head lice preparation. *Clin Toxicol (Phila).* 2011 Oct;49(8):750-1.
[532] Craig JO. Poisoning by the volatile oils in childhood. *Arch Dis Child.* 1953;28:259-67.
[533] Mathew T, Kamath V, Kumar RS, et al. Eucalyptus oil inhalation-induced seizure: A novel, underrecognized, preventable cause of acute symptomatic seizure. *Epilepsia Open.* 2017 Jul 4;2(3):350-354.
[534] Gray AM, Flatt PR. Antihyperglycemic actions of Eucalyptus globulus (Eucalyptus) are associated with pancreatic and extra pancreatic effects in mice. *J Nutr.* 1998 Dec;128(12):2319–23.
[535] Basak SS, Candan F. Effect of Laurus nobilis L. Essential Oil and its Main Components on α-glucosidase and Reactive Oxygen Species Scavenging Activity. *Iran J Pharm Res.* 2013 Spring;12(2):367-79.
[536] Fathiazad F, Mazandarani M, Hamedeyazdan. Phytochemical analysis and antioxidant activity of Hyssopus officinalis L. from Iran. *Adv Pharm Bull.* 2011 Dec;1(2):63–67.
[537] Gray AM, Flatt PR. Antihyperglycemic actions of Eucalyptus globulus (Eucalyptus) are associated with pancreatic and extra pancreatic effects in mice. *J Nutr.* 1998 Dec;128(12):2319-23.
[538] Dey B. Chemo-profiling of eucalyptus and study of its hypoglycemic potential. *World J Diabetes.* 2013 Oct 15;4(5):170–76.
[539] Tsai HH, Lin HW, Chen YL, et al. A review of potential harmful interactions between anticoagulant/antiplatelet agents and Chinese herbal medicines. *PLoS One.* 2013 May 9;8(5):e64255.
[540] Moharam BA, Jantan I, bin Ahmad F, et al. Antiplatelet Aggregation and Platelet Activating Factor (PAF) Receptor Antagonistic Activities of the Essential Oils of Five Goniothalamus Species. *Molecules.* 2010;15:5124-38.
[541] Unger M, Frank A. Simultaneous determination of the inhibitory potency of herbal extracts on the activity of six major cytochrome P450 enzymes using liquid chromatography/mass spectrometry and automated online extraction. *Rapid Commun Mass Spectrom.* 2004;18(19):2273-81.
[542] Jori A, Bianchetti A, Prestini PE, et al. Effect of eucalyptol (1,8-cineole) on the metabolism of other drugs in rats and in man. *Eur J Pharmacol.* 1970;9(3):362-66.
[543] Kim NH, Hyun SH, Jin CH, et al. Pretreatment with 1,8-cineole potentiates thioacetamide-induced hepatotoxicity and immunosuppression. *Arch Pharm Res.* 2004 Jul;27(7):781-9.
[544] Jori A, Bianchetti A, Prestini PE, et al. Effect of eucalyptol (1,8-cineole) on the metabolism of other drugs in rats and in man. *Eur J Pharmacol.* 1970;9(3):362-66.
[545] de Sousa DP, Raphael E, Brocksom U, et al. Sedative effect of monoterpene alcohols in mice: A preliminary screening. *Verlag der Zeitschrift fur Naturforschung.* 2007;62c:563-66.
[546] Rabl W, Katzgraber F, Steinlechner M. Camphor ingestion for abortion (case report). *Forensic Sci Int.* 1997 Sep 19;89(1-2):137-40.
[547] Flaman Z, Pellechia-Clarke S, Bailey B, et al. Unintentional exposure of young children to camphor and eucalyptus oils. *Paediatr Child Health.* 2001 Feb;6(2):80-83.
[548] da Silva Costa KC, Bexerra SB, Norte CM, et al. Medicinal plants with teratogenic potential: current considerations. *Braz J Pharm Sci.* 2012;48(3):427-33.
[549] Brinker F. Herb Contraindications and Drug Interactions. 2nd ed. Sandy, OR. Eclectic Medical Publications. 1998.
[550] Burkhard PR, Burkhardt K, Haenggeli CA, et al. Plant-induced seizures: reappearance of an old problem. *J Neurol.* 1999 Aug;246(8):667–70.
[551] Millet Y, Jouglard J, Steinmetz MD, et al. Toxicity of some essential plant oils. Clinical and experimental study. *Clin Toxicol.* 1981 Dec;18(12):1485–98.
[552] Charanraj KJ, G MV, S M. Camphor poisoning in a child. *Natl Med J India.* 2013 Jan-Feb;26(1):60.
[553] Narayan S, Singh N. Camphor poisoning-An unusual cause of seizure. *Med J Armed Forces India.* 2012 Jul;68(3):252–53.

554 Olowe SA, Ransome-Kuti O. The risk of jaundice in glucose-6-phosphate dehydrogenase deficient babies exposed to menthol. *Acta Paediatr Scand.* 1980 May;69(3):341-5.
555 Dillon Remy M, Manning Alleyne P, Bratt DE, et al. Neonatal jaundice at Port-of-Spain General Hospital abstract. *West Indian Med J.* 1987;36(Suppl):28.
556 Burkhard PR, Burkhardt K, Haenggeli CA, et al. Plant-induced seizures: reappearance of an old problem. *J Neurol.* 1999 Aug;246(8):667-70.
557 Waldman N. Seizure caused by dermal application of over-the-counter eucalyptus oil head lice preparation. *Clin Toxicol (Phila).* 2011 Oct;49(8):750-1.
558 Craig JO. Poisoning by the volatile oils in childhood. *Arch Dis Child.* 1953;28:259-67.
559 Mathew T, Kamath V, Kumar RS, et al. Eucalyptus oil inhalation-induced seizure: A novel, underrecognized, preventable cause of acute symptomatic seizure. *Epilepsia Open.* 2017 Jul 4;2(3):350-354.
560 Khine H, Weiss D, Graber N, et al. A cluster of children with seizures caused by camphor poisoning. *Pediatrics.* 2009 May;123(5):1269-72.
561 Michiels EA, Mazor SS. Toddler with seizures due to ingesting camphor at an Indian celebration. *Pediatr Emerg Care.* 2010 Aug;26(8):574-75.
562 Koren G. Medications which can kill a toddler with one tablet or teaspoonful. *J Toxicol Clin Toxicol. 1*993;31(3):407-13.
563 Flaman Z, Pellechia-Clarke S, Bailey B, et al. Unintentional exposure of young children to camphor and eucalyptus oils. *Paediatr Child Health.* 2001 Feb;6(2):80-83.
564 Rabl W, Katzgraber F, Steinlechner M. Camphor ingestion for abortion (case report). *Forensic Sci Int.* 1997 Sep 19;89(1-2):137-40.
565 Flaman Z, Pellechia-Clarke S, Bailey B, et al. Unintentional exposure of young children to camphor and eucalyptus oils. *Paediatr Child Health.* 2001 Feb;6(2):80-83.
566 Manoguerra AS, Erdman AR, Wax PM, et al. Camphor poisoning: an evidence-based practice guideline for out-of- hospital management. *Clin Toxicol (Phila).*2006;44(4):357-70.
567 Gibson DE, Moore GP, Pfaff JA. Camphor ingestion. *Am J Emerg Med.* 1989 Jan;7(1):41-43.
568 Koppel C, Martends F, Schirop T, et al. Hemoperfusion in acute camphor poisoning. *Intensive Care Med.* 1988;14(4):431-33.
569 Burkhard PR, Burkhardt K, Haenggeli CA, et al. Plant-induced seizures: reappearance of an old problem. *J Neurol.* 1999 Aug;246(8):667–70.
570 Millet Y, Jouglard J, Steinmetz MD, et al. Toxicity of some essential plant oils. Clinical and experimental study. *Clin Toxicol.* 1981 Dec;18(12):1485–98.
571 Charanraj KJ, G MV, S M. Camphor poisoning in a child. *Natl Med J India.* 2013 Jan-Feb;26(1):60.
572 Narayan S, Singh N. Camphor poisoning-An unusual cause of seizure. *Med J Armed Forces India.* 2012 Jul;68(3):252–53.
573 Olowe SA, Ransome-Kuti O. The risk of jaundice in glucose-6-phosphate dehydrogenase deficient babies exposed to menthol. *Acta Paediatr Scand.* 1980 May;69(3):341-5.
574 Dillon Remy M, Manning Alleyne P, Bratt DE, et al. Neonatal jaundice at Port-of-Spain General Hospital abstract. *West Indian Med J.* 1987;36(Suppl):28.
575 Burkhard PR, Burkhardt K, Haenggeli CA, et al. Plant-induced seizures: reappearance of an old problem. *J Neurol.* 1999 Aug;246(8):667–70.
576 Millet Y, Jouglard J, Steinmetz MD, et al. Toxicity of some essential plant oils. Clinical and experimental study. *Clin Toxicol.* 1981 Dec;18(12):1485–98.
577 Charanraj KJ, G MV, S M. Camphor poisoning in a child. *Natl Med J India.* 2013 Jan-Feb;26(1):60.
578 Narayan S, Singh N. Camphor poisoning-An unusual cause of seizure. *Med J Armed Forces India.* 2012 Jul;68(3):252–53.
579 Li Y, Shao X, Xu J, et al. Effects and possible mechanism of tea tree oil against Botrytis cinerea and Penicillium expansum in vitro and in vivo test. *Can J Microbiol.* 2017 Mar;63(3):219-227.
580 Li YH, Sun XP, Zhang YQ, et al. The antithrombotic effect of borneol related to its anticoagulant property. *Am J Chin Med.* 2008;36(4):719-27.
581 Hiroi T, Miyazaki Y, Kobayashi Y, et al. Induction of hepatic P450s in rat by essential wood and leaf oils. *Xenobiotica.* 1995 May;25(5):457-67.
582 Chen JY, Wang JJ, Meng MR, et al. [Borneol is an inducer of rat hepatic CYP2D activity in vivo]. *Yao Xue Xue Bao.* 2015 Apr;50(4):459-63.
583 Zhang R1, Mi SQ, Wang NS. Effect of borneol on cytochrome P450 3A enzyme and midazolam pharmacokinetics in rats. *Eur J Drug Metab Pharmacokinet.* 2013 Sep;38(3):159-69.
584 Kim NH, Hyun SH, Jin CH, et al. Pretreatment with 1,8-cineole potentiates thioacetamide-induced hepatotoxicity and immunosuppression. *Arch Pharm Res.* 2004 Jul;27(7):781-9.
585 Marín I, Sayas-Barberá E, Viuda-Martos M, et al. Chemical Composition, Antioxidant and Antimicrobial Activity of Essential Oils from Organic Fennel, Parsley, and Lavender from Spain. *Foods.* 2016 Mar 4;5(1).
586 de Sousa DP, Raphael E, Brocksom U, et al. Sedative effect of monoterpene alcohols in mice: A preliminary screening. Verlag der Zeitschrift fur Naturforschung. 2007;62c:563-66.
587 Rezazadeh F, Mahdavi M, Motavalizadehkakhky A, et al. Essential Oil Composition of Three Population of Tanacetum polycephalum from Iran and their Antimicrobial Activity. *J Essent Oil Res.* 2014;17(2):317-30.
588 Amiri H. Chemical composition, antibacterial and antioxidant activity of the essential oil of Tanacetum polycephalum Schultz. Bip. *Int J Botany.* 2007;3(3):321-4.
589 Amiri H. Chemical composition, antibacterial and antioxidant activity of the essential oil of Tanacetum polycephalum Schultz. Bip. *Int J Botany.* 2007;3(3):321-4.
590 Le TX, Ho ASH, Mah SH, et al. Determination of borneol and other chemical compounds of essential oil of Dryobalanops aromatica exudate from Malaysia. *Trop J Pharm Res.* 2016 Jun;15(6):1293-7.
591 Kamariah AS, Ozek T, Demirci B, et al. Chemical Composition of Leaf and Seed Oils of Dryobalanops aromatica Gaertn. (Dipterocarpaceae). *Asean J Sci Technol Dev.* 2012;29(2):105-14.
592 Tisserand R, Young R. Essential Oil Safety, Second Edition. Churchill Livingstone: 2014.
593 Fujiwara Y, Ito M. Synergistic effect of fragrant herbs in Japanese scent sachets. *Planta Med.* 2015 Feb;81(3):193-9.
594 Weyerstahl P, Marschall H, Bork WR, et al. Constituents of the Absolute of Boronia megastigma Nees. from Tasmania. *Flavour Frag J.* 1995;10:297-311.
595 Farag RS, Shalaby AS, El-Baroty GA, et al. Chemical and biological evaluation of the essential oils of different Melaleuca species. *Phytother Res.* 2004;18(1):30-35.
596 Aboutabl E, Tohamy S, De Footer H, et al. A comparative study of the essential oils from three Melaleuca species growing in Egypt. Flavour Fragr J. 1991;6(2):139-141.
597 Amri I, Mancini E, de Martino L, et al. Chemical Composition and Biological Activities of the Essential Oils from Three Melaleuca Species Grown in Tunisia. *Int J Mol Sci.* 2012;13:16580-91.

[598] Chabir N, Romdhane M, Valentin A, et al. Chemical Study and Antimalarial, Antioxidant, and Anticancer Activities of Melaleuca armillaris (Sol Ex Gateau) Sm Essential Oil. *J Med Food.* 2011;14(11):1383-8.

[599] Hayouni EA, Bouix M, Abedrabba M, et al. Mechanism of action of Melaleuca armillaris (Sol. Ex Gaertu) Sm. essential oil on six LAB strains as assessed by multiparametric flow cytometry and automated microtiter-based assay. *Food Chem.* 2008;111:707-8.

[600] Siddique S, Parveen Z, Firadaus-e-Bareen, et al. The essential oil of Melaleuca armillaris (Sol. ex Gaertn.) Sm. leaves from Pakistan: A potential source of eugenol methyl ether. *Ind Crops Prod.* 2017;109:912-7.

[601] Craig JO. Poisoning by the volatile oils in childhood. *Arch Dis Child.* 1953;28:259-67.

[602] Melis K. Bochner A, Janssens G. Accidental nasal eucalyptol and menthol instillation. *Eur J Pediatr.* 1989 Aug;148(8)786-7.

[603] Day LM, Ozanne-Smith J, Parsons BJ, et al. Eucalyptus oil poisoning among young children: mechanisms of access and the potential prevention. *Aust N Z J Public Health.* 1997 Jun;21(3):297-302.

[604] Day LM, Ozanne-Smith J, Parsons BJ, et al. Eucalyptus oil poisoning among young children: mechanisms of access and the potential prevention. *Aust N Z J Public Health.* 1997 Jun;21(3):297-302.

[605] Myott E. Case of eucalyptus poisoning. *Brit M J.* 1906;1:558.

[606] Hindle RC. Eucalyptus oil ingestion. *N Z Med J.* 1994 May 11;107(977)185-6.

[607] Tibballs J. Clinical effects and management of eucalyptus oil ingestion in infants and small children. *Med J Aust.* 1995 Aug;163(4):177-80.

[608] Waldman W, Barwina M, Sein Anand J. Accidental intoxication with eucalyptus oil—a case report. *Przeql Lek.* 2011;68(8):555-6.

[609] Day LM, Ozanne-Smith J, Parsons BJ, et al. Eucalyptus oil poisoning among young children: mechanisms of access and the potential prevention. *Aust N Z J Public Health.* 1997 Jun;21(3):297-302.

[610] Burkhard PR, Burkhardt K, Haenggeli CA, et al. Plant-induced seizures: reappearance of an old problem. *J Neurol.* 1999 Aug;246(8):667-70.

[611] Waldman N. Seizure caused by dermal application of over-the-counter eucalyptus oil head lice preparation. *Clin Toxicol (Phila).* 2011 Oct;49(8):750-1.

[612] Craig JO. Poisoning by the volatile oils in childhood. *Arch Dis Child.* 1953;28:259-67.

[613] Mathew T, Kamath V, Kumar RS, et al. Eucalyptus oil inhalation-induced seizure: A novel, underrecognized, preventable cause of acute symptomatic seizure. *Epilepsia Open.* 2017 Jul 4;2(3):350-354.

[614] Unger M, Frank A. Simultaneous determination of the inhibitory potency of herbal extracts on the activity of six major cytochrome P450 enzymes using liquid chromatography/mass spectrometry and automated online extraction. *Rapid Commun Mass Spectrom.* 2004;18(19):2273-81.

[615] Jori A, Bianchetti A, Prestini PE, et al. Effect of eucalyptol (1,8-cineole) on the metabolism of other drugs in rats and in man. *Eur J Pharmacol.* 1970;9(3):362-66.

[616] Kim NH, Hyun SH, Jin CH, et al. Pretreatment with 1,8-cineole potentiates thioacetamide-induced hepatotoxicity and immunosuppression. *Arch Pharm Res.* 2004 Jul;27(7):781-9.

[617] Jori A, Bianchetti A, Prestini PE, et al. Effect of eucalyptol (1,8-cineole) on the metabolism of other drugs in rats and in man. *Eur J Pharmacol.* 1970;9(3):362-66.

[618] de Sousa DP, Raphael E, Brocksom U, et al. Sedative effect of monoterpene alcohols in mice: A preliminary screening. *Verlag der Zeitschrift fur Naturforschung.* 2007;62c:563-66.

[619] Craig JO. Poisoning by the volatile oils in childhood. *Arch Dis Child.* 1953;28:259-67.

[620] Melis K. Bochner A, Janssens G. Accidental nasal eucalyptol and menthol instillation. *Eur J Pediatr.* 1989 Aug;148(8)786-7.

[621] Day LM, Ozanne-Smith J, Parsons BJ, et al. Eucalyptus oil poisoning among young children: mechanisms of access and the potential prevention. *Aust N Z J Public Health.* 1997 Jun;21(3):297-302.

[622] Day LM, Ozanne-Smith J, Parsons BJ, et al. Eucalyptus oil poisoning among young children: mechanisms of access and the potential prevention. *Aust N Z J Public Health.* 1997 Jun;21(3):297-302.

[623] Myott E. Case of eucalyptus poisoning. *Brit M J.* 1906;1:558.

[624] Hindle RC. Eucalyptus oil ingestion. *N Z Med J.* 1994 May 11;107(977)185-6.

[625] Tibballs J. Clinical effects and management of eucalyptus oil ingestion in infants and small children. *Med J Aust.* 1995 Aug;163(4):177-80.

[626] Waldman W, Barwina M, Sein Anand J. Accidental intoxication with eucalyptus oil—a case report. *Przeql Lek.* 2011;68(8):555-6.

[627] Day LM, Ozanne-Smith J, Parsons BJ, et al. Eucalyptus oil poisoning among young children: mechanisms of access and the potential prevention. *Aust N Z J Public Health.* 1997 Jun;21(3):297-302.

[628] Burkhard PR, Burkhardt K, Haenggeli CA, et al. Plant-induced seizures: reappearance of an old problem. *J Neurol.* 1999 Aug;246(8):667-70.

[629] Waldman N. Seizure caused by dermal application of over-the-counter eucalyptus oil head lice preparation. *Clin Toxicol (Phila).* 2011 Oct;49(8):750-1.

[630] Craig JO. Poisoning by the volatile oils in childhood. *Arch Dis Child.* 1953;28:259-67.

[631] Mathew T, Kamath V, Kumar RS, et al. Eucalyptus oil inhalation-induced seizure: A novel, underrecognized, preventable cause of acute symptomatic seizure. *Epilepsia Open.* 2017 Jul 4;2(3):350-354.

[632] Gray AM, Flatt PR. Antihyperglycemic actions of Eucalyptus globulus (Eucalyptus) are associated with pancreatic and extra pancreatic effects in mice. *J Nutr.* 1998 Dec;128(12):2319–23.

[633] Basak SS, Candan F. Effect of Laurus nobilis L. Essential Oil and its Main Components on α-glucosidase and Reactive Oxygen Species Scavenging Activity. *Iran J Pharm Res.* 2013 Spring;12(2):367-79.

[634] Fathiazad F, Mazandarani M, Hamedeyazdan. Phytochemical analysis and antioxidant activity of Hyssopus officinalis L. from Iran. *Adv Pharm Bull.* 2011 Dec;1(2):63–67.

[635] Gray AM, Flatt PR. Antihyperglycemic actions of Eucalyptus globulus (Eucalyptus) are associated with pancreatic and extra pancreatic effects in mice. *J Nutr.* 1998 Dec;128(12):2319-23.

[636] Dey B. Chemo-profiling of eucalyptus and study of its hypoglycemic potential. *World J Diabetes.* 2013 Oct 15;4(5):170–76.

[637] Tsai HH, Lin HW, Chen YL, et al. A review of potential harmful interactions between anticoagulant/antiplatelet agents and Chinese herbal medicines. *PLoS One.* 2013 May 9;8(5):e64255.

[638] Moharam BA, Jantan I, bin Ahmad F, et al. Antiplatelet Aggregation and Platelet Activating Factor (PAF) Receptor Antagonistic Activities of the Essential Oils of Five Goniothalamus Species. *Molecules.* 2010;15:5124-38.

[639] Unger M, Frank A. Simultaneous determination of the inhibitory potency of herbal extracts on the activity of six major cytochrome P450 enzymes using liquid chromatography/mass spectrometry and automated online extraction. *Rapid Commun Mass Spectrom.* 2004;18(19):2273-81.

[640] Jori A, Bianchetti A, Prestini PE, et al. Effect of eucalyptol (1,8-cineole) on the metabolism of other drugs in rats and in man. *Eur J Pharmacol.* 1970;9(3):362-66.

[641] Kim NH, Hyun SH, Jin CH, et al. Pretreatment with 1,8-cineole potentiates thioacetamide-induced hepatotoxicity and immunosuppression. *Arch Pharm Res.* 2004 Jul;27(7):781-9.

[642] Jori A, Bianchetti A, Prestini PE, et al. Effect of eucalyptol (1,8-cineole) on the metabolism of other drugs in rats and in man. *Eur J Pharmacol.* 1970;9(3):362-66.
[643] de Sousa DP, Raphael E, Brocksom U, et al. Sedative effect of monoterpene alcohols in mice: A preliminary screening. *Verlag der Zeitschrift fur Naturforschung*. 2007;62c:563-66.
[644] Johnson JD, Ryan MJ, Toft JD II, et al. Two-year toxicity and carcinogenicity study of methyl eugenol in F344/N rats and B6C3F(1) mice. *J Agric Food Chem.* 2000 Aug;48(8):3620-32.
[645] European Commission. Opinion of the Scientific Committee on Food on methyl eugenol (4-Allyl-1,2- dimethoxybenzene). Available at: http://ec.europa.eu/food/fs/sc/scf/out102_en.pdf.
[646] National Toxicology Program. NTP Toxicology and Carcinogenesis Studies of Methyl eugenol (CAS NO. 93-15-2) in F344/N Rats and B6C3F1 Mice (Gavage Studies). *Natl Toxicol Program Tech Rep Ser.* 2000 Jul;491:1-412.
[647] National Toxicology Program. Carcinogenesis Studies of Eugenol (CAS No. 97-53-0) in F344/N Rats and B6C3F1 Mice (Feed Studies). Technical Report Series No. 223. NIH Publication No. 84-1779. 1983. U.S. DHHS, PHS, NIH, NTP, Research Triangle Park, NC.
[648] Johnson JD, Ryan MJ, Toft JD II, et al. Two-year toxicity and carcinogenicity study of methyl eugenol in F344/N rats and B6C3F(1) mice. *J Agric Food Chem.* 2000 Aug;48(8):3620-32.
[649] National Toxicology Program. NTP Toxicology and Carcinogenesis Studies of Methyl eugenol (CAS NO. 93-15-2) in F344/N Rats and B6C3F1 Mice (Gavage Studies). *Natl Toxicol Program Tech Rep Ser.* 2000 Jul;491:1-412.
[650] Kerckaert GA, Brauninger R, LeBoeuf RA, et al. Use of the Syrian hamster embryo cell transformation assay for carcinogenicity prediction of chemicals currently being tested by the National Toxicology Program in rodent bioassays. *Environ Health Perspect.* 1996;104:1075-84.
[651] National Toxicology Program. Carcinogenesis Studies of Eugenol (CAS No. 97-53-0) in F344/N Rats and B6C3F1 Mice (Feed Studies). Technical Report Series No. 223. NIH Publication No. 84-1779. 1983. U.S. DHHS, PHS, NIH, NTP, Research Triangle Park, NC.
[652] Johnson JD, Ryan MJ, Toft JD II, et al. Two-year toxicity and carcinogenicity study of methyl eugenol in F344/N rats and B6C3F(1) mice. *J Agric Food Chem.* 2000 Aug;48(8):3620-32.
[653] European Commission. Opinion of the Scientific Committee on Food on methyl eugenol (4-Allyl-1,2- dimethoxybenzene). Available at: http://ec.europa.eu/food/fs/sc/scf/out102_en.pdf.
[654] National Toxicology Program. NTP Toxicology and Carcinogenesis Studies of Methyl eugenol (CAS NO. 93-15-2) in F344/N Rats and B6C3F1 Mice (Gavage Studies). *Natl Toxicol Program Tech Rep Ser.* 2000 Jul;491:1-412.
[655] National Toxicology Program. Carcinogenesis Studies of Eugenol (CAS No. 97-53-0) in F344/N Rats and B6C3F1 Mice (Feed Studies). Technical Report Series No. 223. NIH Publication No. 84-1779. 1983. U.S. DHHS, PHS, NIH, NTP, Research Triangle Park, NC.
[656] Ahmad A, Khan A, Ahmad Khan L, et al. In vitro synergy of eugenol and methyl eugenol with fluconazole against clinical Candida isolates. *J Med Microbiol.* 2010;59:1178-84.
[657] Lee SE, Lee BH, Choi WS, et al. Fumigant toxicity of volatile natural products from Korean spices and medicinal plants towards the rice weevil, Sitophilus oryzae (L). *Pest Manag Sci.* 2001;57:548-53.
[658] Chabir N, Romdhane M, Valentin A, et al. Chemical Study and Antimalarial, Antioxidant, and Anticancer Activities of Melaleuca armillaris (Sol Ex Gateau) Sm Essential Oil. *J Med Food.* 2011;14(11):1383-8.
[659] Yvon Y, Raoelison EG, Razafindrazaka R, et al. Relation between chemical composition or antioxidant activity and antihypertensive activity for six essential oils. *J Food Sci.* 2012 Aug;77(8):H184-91.
[660] Farag RS, Shalaby AS, El-Baroty GA, et al. Chemical and biological evaluation of the essential oils of different Melaleuca species. *Phytother Res.* 2004;18(1):30-35.
[661] Siddique S, Parveen Z, Firadaus-e-Bareen, et al. The essential oil of Melaleuca armillaris (Sol. ex Gaertn.) Sm. leaves from Pakistan: A potential source of eugenol methyl ether. *Ind Crops Prod.* 2017;109:912-7.
[662] Rizk M, Ibrahim N, El-Rigal N. Comparative in vivo antioxidant levels in Schistosoma mansoni infected mice treated with praziquantel or the essential oil of Melaleuca armillaris leaves. *Pak J Biol Sci.* 2012 Oct 15;15(20):971-8.
[663] Chabir N, Romdhane M, Valentin A, et al. Chemical Study and Antimalarial, Antioxidant, and Anticancer Activities of Melaleuca armillaris (Sol Ex Gateau) Sm Essential Oil. *J Med Food.* 2011;14(11):1383-8.
[664] Farag RS, Shalaby AS, El-Baroty GA, et al. Chemical and biological evaluation of the essential oils of different Melaleuca species. *Phytother Res.* 2004;18(1):30-35.
[665] Farag RS, Shalaby AS, El-Baroty GA, et al. Chemical and biological evaluation of the essential oils of different Melaleuca species. *Phytother Res.* 2004;18(1):30-35.
[666] Chabir N, Romdhane M, Valentin A, et al. Chemical Study and Antimalarial, Antioxidant, and Anticancer Activities of Melaleuca armillaris (Sol Ex Gateau) Sm Essential Oil. *J Med Food.* 2011;14(11):1383-8.
[667] Yvon Y, Raoelison EG, Razafindrazaka R, et al. Relation between chemical composition or antioxidant activity and antihypertensive activity for six essential oils. *J Food Sci.* 2012 Aug;77(8):H184-91.
[668] Siddique S, Parveen Z, Firadaus-e-Bareen, et al. The essential oil of Melaleuca armillaris (Sol. ex Gaertn.) Sm. leaves from Pakistan: A potential source of eugenol methyl ether. *Ind Crops Prod.* 2017;109:912-7.
[669] Apel MA, Sobral M, Zuanazzi JA, et al. Essential oil composition of four Plinia species (Myrtaceae). *Flavour Fragr J.* 2006;21:565-567.
[670] Lago JHG, Souza ED, Mariane B, et al. Chemical and Biological Evaluation of Essential Oils from Two Species of Myrtaceae — Eugenia uniflora L. and Plinia trunciflora (O. Berg) Kausel. *Molecules.* 2011;16:9827-37.
[671] Lowenstein L, Ballew DH. Fatal acute hemolytic anemia, thrombocytopenic purpura, nephrosis and hepatitis resulting from ingestion of a constituent containing apiol. *Can Med Assoc J.* 1958 Feb 1;78(3):195-8.
[672] Basso U. [Study of death caused by apiol and experimental study of its toxic action]. *Arch Patol Clin Med.* 1957;34(4):308-22.
[673] Barni B, Barni I. [The apiol poisoning]. *Rassegna casistica.* 1967;3(2):197-221.
[674] Newall CA, Anderson LA, Philpson JD. Herbal Medicine: A Guide for Healthcare Professionals. London, UK: The Pharmaceutical Press, 1996.
[675] Colalillo R. Apiol poisoning in 7 women and 2 children. *Rivista di Tossicologia Sperimentale e Clinica.* 1988;18(2):125-130.
[676] Mumolo M. [On a case of acute and fatal poisoning caused by apiol]. *Recenti Prog Med.* 1964 Feb;36:139-51.
[677] Lowenberg K. Cerbral damage in a case of fatal poisoning due to a compound if ergot and apiol (ergoapiol). *JAMA.* 1938;110(8):573-5.
[678] Hermann K, Le Roux A, Fiddes FS. Death from apiol used as abortifacient. *Lancet.* 1956 Jun 16;270(6929):937-9.
[679] Ciganda C, Laborde A. Herbal infusions used for induced abortion. *J Toxicol Clin Toxicol.* 2003;41:235-239.
[680] Lowenstein L, Ballew DH. Fatal acute hemolytic anemia, thrombocytopenic purpura, nephrosis and hepatitis resulting from ingestion of a constituent containing apiol. *Can Med Assoc J.* 1958 Feb 1;78(3):195-8.
[681] Basso U. [Study of death caused by apiol and experimental study of its toxic action]. *Arch Patol Clin Med.* 1957;34(4):308-22.
[682] Barni B, Barni I. [The apiol poisoning]. *Rassegna casistica.* 1967;3(2):197-221.

[683] D'Aprile F. [Clinical study-Experimental apiol poisoning]. *Annali di Ostetricia Ginecologia.* 1928;50:1204-27.

[684] Peralta-Bohorquez AF, Parada F, Quijano CE, et al. Analysis of Volatile Compounds of Sour Guava (Psidium guineense Swartz) Fruit. *J Essent Oil Res.* 2010 Nov-Dec;22:493-8.

[685] do Nascimento KF, Figueira Moreira FM, Santos JA, et al. Antioxidant, anti-inflammatory, antiproliferative and antimycobacterial activities of the essential oil of Psidium guineense Sw. and spathulenol. *J Ethnopharmacol.* 2017 Aug 24;210:351-358.

[686] Da Silva JD, Luz AIR, da Silva MHL, et al. Essential oils of the leaves and stems of four Psidium spp. *Flavour Frag J.* 2003;18:240-3.

[687] Tucker AO, Maciarello MJ. Volatile leaf oils of American Myrtaceae. III. Psidium cattleianum Sabine, P. friedrichsthalianum (Berg) Niedenzu, P. guajava L., P. guineense Sw., and P. sartorianum (Berg) Niedenzu. *J Essent Oil Res.* 1995 Mar/Apr;7:187-90.

[688] do Nascimento KF, Figueira Moreira FM, Santos JA, et al. Antioxidant, anti-inflammatory, antiproliferative and antimycobacterial activities of the essential oil of Psidium guineense Sw. and spathulenol. *J Ethnopharmacol.* 2017 Aug 24;210:351-358.

[689] Santos FA, Rao V, Silveira E. Anti-inflammatory and analgesic activities of the essential oil of Psidium guianense. *Fitoterapia.* 1997 Jan;68(1):65-8.

[690] Marques AM, Lima CHP, Alviano DS, et al. Traditional use, chemical composition and antimicrobial activity of Pectis brevipedunculata essential oil: A correlated lemongrass species in Brazil. *Emir J Food Agric.* 2013;25(10):798-808.

[691] Pereira SL, Marques AM, Sudo RT, et al. Vasodilator Activity of the Essential Oil from Aerial Parts of Pectis brevipedunculata and Its Main Constituent Citral in Rat Aorta. *Molecules.* 2013;18:3072-85.

[692] Nogueira AC, Carvalho RR, Souza CA, et al. Study on the embryofeto-toxicity of citral in the rat. *Toxicology.* 1995;96(2):105-13.

[693] Seo KA, Kim H, Ku HY, et al. The monoterpenoids citral and geraniol are moderate inhibitors of the CYP2B6 hydroxylase activity. *Chem Biol Interact.* 2008;174:141-46.

[694] Sheweita SA, Newairy AA, Mansour HA, et al. Effect of some hypoglycemic herbs on the activity of phase I and II drug-metabolizing enzymes in alloxan-induced diabetic rats. *Toxicology.* 2002 May 24;17(2):131-39.

[695] Raner GM, Vaz AD, Coon MJ. Metabolism of all-trans, 9-cis, and 13-cis isomers of retinal by purified isozymes of microsomal cytochrome P450 and mechanism-based inhibition of retinoid oxidation by citral. *Mol Pharmacol.* 1996;49(3):515-22.

[696] De Olivera AC, Ribeiro-Pinto LF, Paumgartten JR. In vitro inhibition of CYP2B1 monoxygenase by beta-myrcene and other monoterpenoid constituents. Br J Nutr. 1999;81:289-95.

[697] Modak T, Mukhopadhaya A. Effects of citral, a naturally occurring antiadipogenic molecule, on an energy-intense diet model of obesity. *Indian J Pharmacol.* 2011 May-Jun;43(3):300-05.

[698] Adeneye AA, Agbaje EO. Hypoglycemic and hypolipidemic effects of fresh leaf aqueous extract of Cymbopogon citratus Stapf. in rats. *J Ethnopharmacol.* 2007 Jul 25;112(3):440-4.

[699] Choi JY, Damte D, Seung-Jin L, et al. Antimicrobial activity of lemongrass and oregano essential oil against standard antibiotic resistant Staphylococcus aureus and field isolates from chronic mastitis cow. *International Journal of Phytomedicine.* 2012;4(1):134-39.

[700] Zanini SF, Silva-Angulo AB, Rosenthal A, et al. Effect of citral and carvacrol on the susceptibility of Listeria monocytogenes and Listeria innocua to antibiotics. *Lett Appl Microbiol.* 2014 May;58(5):486-92.

[701] do Vale TG, Furtado EC, Santos JG Jr, et al. Central effects of citral, myrcene and limonene, constituents of essential oil chemotypes from Lippia alba (Mill.) n.e. Brown. *Phytomedicine.* 2002 Dec;9(8):709-14.

[702] Kuhn GO, McCampbell P, Singmaster G, et al. Application of microencapsulation technology to improve the stability of citral in rodent diets. *Fundam Appl Toxicol.* 1991 Oct;17(3):635-40.

[703] De Mozzi P, Johnston GA. An outbreak of allergic contact dermatitis caused by citral in beauticians working in a health spa. *Contact Dermatitis.* 2014 Jun;70(6):377-9.

[704] Pereira SL, Marques AM, Sudo RT, et al. Vasodilator Activity of the Essential Oil from Aerial Parts of Pectis brevipedunculata and Its Main Constituent Citral in Rat Aorta. *Molecules.* 2013;18:3072-85.

[705] Marques AM, Lima CHP, Alviano DS, et al. Traditional use, chemical composition and antimicrobial activity of Pectis brevipedunculata essential oil: A correlated lemongrass species in Brazil. *Emir J Food Agric.* 2013;25(10):798-808.

[706] Ghasemi Y, Abedtash H, Morowvat MH, et al. Essential oil composition and bioinformatic analysis of Spanish broom (Spartium junceum L.). *Trends Pharm Sci.* 2015;1(2):97-104.

[707] Lawrence BM. Progress in Essential Oils. *Perf Flavorist.* 2006;31:44-55.

[708] Nadaf M, Halimi M, Mortazavi M. Identification of Nonpolar Chemical Composition Spartium junceum flower Growing in Iran by GC-MS. *Middle-East J Sci Res.* 2012;11(2):221-4.

[709] Miraldi E, Ferri S, Giorgi G. Identification of Volatile Constituents from the Flower Oil of Spartium junceum. *J Essent Oil Res.* 2004;156(6):568-70.

[710] Bezic N, Dunkic V, Radonic A. Anatomical and chemical adaptation of Spartium juneceum in arid habitat. *Acta Biol Cracoviensai.* 2003;45(3):43-7.

[711] Lawrence BM. Essential oils 1988-1991. Allured Publishing, Wheaton. 1995.

[712] Burkhard PR, Burkhardt K, Haenggeli CA, et al. Plant-induced seizures: reappearance of an old problem. *J Neurol.* 1999 Aug;246(8):667–70.

[713] Millet Y, Jouglard J, Steinmetz MD, et al. Toxicity of some essential plant oils. Clinical and experimental study. *Clin Toxicol.* 1981 Dec;18(12):1485–98.

[714] Charanraj KJ, G MV, S M. Camphor poisoning in a child. *Natl Med J India.* 2013 Jan-Feb;26(1):60.

[715] Narayan S, Singh N. Camphor poisoning-An unusual cause of seizure. *Med J Armed Forces India.* 2012 Jul;68(3):252–53.

[716] Olowe SA, Ransome-Kuti O. The risk of jaundice in glucose-6-phosphate dehydrogenase deficient babies exposed to menthol. *Acta Paediatr Scand.* 1980 May;69(3):341-5.

[717] Dillon Remy M, Manning Alleyne P, Bratt DE, et al. Neonatal jaundice at Port-of-Spain General Hospital abstract. *West Indian Med J.* 1987;36(Suppl):28.

[718] Opdyke DLJ. Monographs on fragrance raw materials. *Food Cosmet Toxicol.* 1976;14(Suppl):779.

[719] Paroul N, Dalla Rosa RL, Piazza SP, et al. Chemical Composition and Antioxidant Activity of Baccharis trimera Pers and Baccharis dracunculifolia DC (Asteraceae). *Perspectiva Erchim.* 2016 Sep;40(151):55-64.

[720] Marostica Jr. MB, Daugsch A, Moraes CS, et al. Comparison of volatile and polyphenolic compounds in Brazilian green propolis and its botanical origin Baccharis dracunculifolia. *Ciênc Tecnol Aliment, Campinas.* 2008 Jan-Mar;28(1):178-81.

[721] Queiroga CL, Bastos JK, de Sousa JPB, et al. Comparison of the Chemical Composition of the Essential Oil and the Water Soluble Oil of Baccharis dracunculifolia DC. (Asteraceae). *J Essent Oil Res.* 2008 Mar/Apr;20:111-3.

[722] Frizzo CD, Anti-Serafini L, Laguna SE, et al. Essential oil variability in Baccharis uncinella DC and Baccharis dracunculifolia DC growing wild in southern Brazil, Bolivia and Uruguay. *Flavour Frag J.* 2008;23:99-106.

[723] de Assis Lage TC, Montanari RM, Fernandes SA, et al. Chemical composition and acaricidal activity of the essential oil of Baccharis dracunculifolia De Candole (1836) and its constituents nerolidol and limonene on larvae and engorged females of Rhipicephalus microplus (Acari: Ixodidae). *Exp Parasitol.* 2015;148:24-9.

[724] Bufalo MC, Candeias JMG, Sousa JPG, et al. In vitro cytotoxic activity of Baccharis dracunculifolia and propolis against HEp-2 cells. *Nat Prod Res.* 2010 Nov;24(10):1710-8.

[725] Klopell FC, Lemos M, Barreto Sousa JP, et al. Nerolidol, an Antiulcer Constituent from the Essential Oil of Baccharis dracunculifolia DC (Asteraceae). *Z Natuforsch.* 2007:537-42.

[726] Chaaban A, Martins CEN, Bretanha LC, et al. Insecticide activity of Baccharis dracunculifolia essential oil against Cochliomyia macellaria (Diptera: Calliphoridae). *Nat Prod Res.* 2017 Oct 20:1-5.

[727] Parreira NA, Magalhães LG, Morais DR, et al. Antiprotozoal, schistosomicidal, and antimicrobial activities of the essential oil from the leaves of Baccharis dracunculifolia. *Chem Biodivers.* 2010 Apr;7(4):993-1001.

[728] Brehm-Stecher BF, Johnson EA. Sensitization of Staphylococcus aureus and Escherichia coli to antibiotics by the sesquiterpenoids nerolidol, farnesol, bisabolol, and apritone. *Antimicrob Agents Chemother.* 2003 Oct;47(10):3357-60.

[729] Špičáková A, Szotáková B, Dimunová D, et al. Nerolidol and Farnesol Inhibit Some Cytochrome P450 Activities but Did Not Affect Other Xenobiotic-Metabolizing Enzymes in Rat and Human Hepatic Subcellular Fractions. *Molecules.* 2017 Mar 24;22(4). pii: E509.

[730] Bufalo MC, Candeias JMG, Sousa JPG, et al. In vitro cytotoxic activity of Baccharis dracunculifolia and propolis against HEp-2 cells. Nat Prod Res. 2010 Nov;24(10):1710-8.

[731] Klopell FC, Lemos M, Barreto Sousa JP, et al. Nerolidol, an Antiulcer Constituent from the Essential Oil of Baccharis dracunculifolia DC (Asteraceae). *Z Natuforsch.* 2007:537-42.

[732] Massignani JJ, Lemos M, Maistro EL, et al. Antiulcerogenic activity of the essential oil of Baccharis dracunculifolia on different experimental models in rats. *Phytother Res.* 2009 Oct;23(10):1355-60.

[733] Silva NC, Barbosa L, Seito LN, et al. Antimicrobial activity and phytochemical analysis of crude extracts and essential oils from medicinal plants. *Nat Prod Res.* 2012;26(16):1510-4.

[734] Pereira CA, da Costa AC, Machado AK, et al. Enzymatic activity, sensitivity to antifungal drugs and Baccharis dracunculifolia essential oil by Candida strains isolated from the oral cavities of breastfeeding infants and in their mothers' mouths and nipples. *Mycopathologia.* 2011 Feb;171(2):103-9.

[735] Búfalo MC, Figueiredo AS, de Sousa JP, et al. Anti-poliovirus activity of Baccharis dracunculifolia and propolis by cell viability determination and real-time PCR. *J Appl Microbiol.* 2009 Nov;107(5):1669-80.

[736] Pereira CA, Costa AC, Liporoni PC, et al. Antibacterial activity of Baccharis dracunculifolia in planktonic cultures and biofilms of Streptococcus mutans. *J Infect Public Health.* 2016 May-Jun;9(3):324-30.

[737] Pedrazzi V, Leite MF, Tavares RC, et al. Herbal mouthwash containing extracts of Baccharis dracunculifolia as agent for the control of biofilm: clinical evaluation in humans. *ScientificWorldJournal.* 2015;2015:712683.

[738] Parreira NA, Magalhães LG, Morais DR, et al. Antiprotozoal, schistosomicidal, and antimicrobial activities of the essential oil from the leaves of Baccharis dracunculifolia. *Chem Biodivers.* 2010 Apr;7(4):993-1001.

[739] Paroul N, Dalla Rosa RL, Piazza SP, et al. Chemical Composition and Antioxidant Activity of Baccharis trimera Pers and Baccharis dracunculifolia DC (Asteraceae). *Perspectiva Erchim.* 2016 Sep;40(151):55-64.

[740] de Assis Lage TC, Montanari RM, Fernandes SA, et al. Chemical composition and acaricidal activity of the essential oil of Baccharis dracunculifolia De Candole (1836) and its constituents nerolidol and limonene on larvae and engorged females of Rhipicephalus microplus (Acari: Ixodidae). *Exp Parasitol.* 2015;148:24-9.

[741] Chaaban A, Martins CEN, Bretanha LC, et al. Insecticide activity of Baccharis dracunculifolia essential oil against Cochliomyia macellaria (Diptera: Calliphoridae). *Nat Prod Res.* 2017 Oct 20:1-5.

[742] Mothana RA, Alsaid MS, Hasoon SS, et al. Antimicrobial and antioxidant activities and gas chromatography mass spectrometry (GC/MS) analysis of the essential oils of Ajuga bracteosa Wall. ex Benth. and Lavandula dentata L. growing wild in Yemen. *J Med Plants Res.* 2012 Apr;6(15):3066-71.

[743] Vohra A, Kaur H. Chemical investigation of medicinal plant Ajuga bracteosa. *J Nat Prod Plant Resour.* 2011;1(1):37-45.

[744] Vohra A, Kaur H. Chemical investigation of medicinal plant Ajuga bracteosa. *J Nat Prod Plant Resour.* 2011;1(1):37-45.

[745] Mothana RA, Alsaid MS, Hasoon SS, et al. Antimicrobial and antioxidant activities and gas chromatography mass spectrometry (GC/MS) analysis of the essential oils of Ajuga bracteosa Wall. ex Benth. and Lavandula dentata L. growing wild in Yemen. *J Med Plants Res.* 2012 Apr;6(15):3066-71.

[746] Mothana RA, Alsaid MS, Hasoon SS, et al. Antimicrobial and antioxidant activities and gas chromatography mass spectrometry (GC/MS) analysis of the essential oils of Ajuga bracteosa Wall. ex Benth. and Lavandula dentata L. growing wild in Yemen. *J Med Plants Res.* 2012 Apr;6(15):3066-71.

[747] Chizzola R. Composition of the Essential Oils from Anthriscus cerefolium var. trichocarpa and A. caucalis Growing Wild in the Urban Area of Vienna (Austria). *Nat Prod Commun.* 2011;6(8):1147-50.

[748] Lai P, Rao H, Gao Y. Chemical Composition, Cytotoxic, Antimicrobial and Antioxidant Activities of Essential oil from Anthriscus caucalis M. Bieb Grown in China. *Rec Nat Prod.* 20187;12(3):290-4.

[749] Lai P, Rao H, Gao Y. Chemical Composition, Cytotoxic, Antimicrobial and Antioxidant Activities of Essential oil from Anthriscus caucalis M. Bieb Grown in China. *Rec Nat Prod.* 20187;12(3):290-4.

[750] Lai P, Rao H, Gao Y. Chemical Composition, Cytotoxic, Antimicrobial and Antioxidant Activities of Essential oil from Anthriscus caucalis M. Bieb Grown in China. *Rec Nat Prod.* 20187;12(3):290-4.

[751] Lai P, Rao H, Gao Y. Chemical Composition, Cytotoxic, Antimicrobial and Antioxidant Activities of Essential oil from Anthriscus caucalis M. Bieb Grown in China. *Rec Nat Prod.* 20187;12(3):290-4.

[752] Wanner JKR, Bail S, Jirovetz L, et al. Chemical composition, olfactory analysis and antibacterial activity of selected essential oils (mainly used in perfumery products). Conference: International Symposium on Essential Oils. 2009 Sep.

[753] Brehm-Stecher BF, Johnson EA. Sensitization of Staphylococcus aureus and Escherichia coli to antibiotics by the sesquiterpenoids nerolidol, farnesol, bisabolol, and apritone. *Antimicrob Agents Chemother.* 2003 Oct;47(10):3357-60.

[754] Wanner J, Schmidt E, Bail S, et al. Chemical composition and antibacterial activity of selected essential oils and some of their main compounds. *Nat Prod Commun.* 2010 Sep;5(9):1359-64.

[755] Hajdari A, Mustafa B, Gashi V, et al. Chemical composition of the essential oils of ripe berries of Juniperus oxycedrus L., growing wild in Kosovo. *Biochem System Ecol.* 2014;57:90-94.

[756] Barrero AF, Oltra JE, Altarejos J, et al. Minor Components in the Essential Oil of Juniperus oxycedrus L. Wood. *Flavour Frag J.* 1993;8:185-9.

[757] Medini H, Marzouki H, Chemli R, et al. Comparison of the antimicrobial activity and the essential oil composition of Juniperus oxycedrus subsp. rufescens obtained by hydrodistillation and supercritical carbon dioxide extraction methods. *Chem Mat Compounds.* 2009;45(5):739-41.

[758] Hanene M, Ameur E, Larbi KM, et al. Chemical composition of the essential oils of the berries of Juniperus oxycedrus L. ssp. rufescens (L. K.) and Juniperus oxycedrus L. ssp. macrocarpa (S. & m.) Ball. and their antioxidant activities. *Nat Prod Res*. 2012;26(9):810-20.
[759] Angioni A, Barra A, Russo MT, et al. Chemical Composition of the Essential Oils of Juniperus from Ripe and Unripe Berries and Leaves and Their Antimicrobial Activity. *J Agric Food Chem*. 2003;51:3073-8.
[760] Loizzo MR, Tundis R, Conforti F, et al. Comparative chemical composition, antioxidant and hypoglycaemic activities of Juniperus oxycedrus ssp. oxycedrus L. berry and wood oils from Lebanon. *Food Chem*. 2007;105:572-8.
[761] Hayta S, Bagci E. Essential oil constituents of the leaves, bark and cones of Juniperus oxycedrus subsp. oxycedrus L. from Turkey. *Acta Botanica Gallica: Botany Letters*. 2014;161(2):201-7.
[762] Boti JB, Bighelli A, Cavaleiro C, et al. Chemical variability of Juniperus oxycedrus ssp. oxycedrus berry and leaf oils from Corsica, analysed by combination of GC, GC–MS and 13C-NMR. *Flavour Frag J*. 2006;212:268-73.
[763] Sela F, Karapandzova M, Stefkiv G, et al. Chemical composition and antimicrobial activity of berry essential oil of Juniperus oxycedrus L. (Cupressaceae) grown wild in Republic of Macedonia. *Macedonian Pharm Bull*. 2013;49(1,2):41-8.
[764] Medini H, Manongiu B, Aicha N, et al. Chemical and Antibacterial Polymorphism of Juniperus oxycedrus ssp. oxycedrus and Juniperus oxycedrus ssp. macrocarpa (Cupressaceae) Leaf Essential Oils from Tunisia. *J Chem*. 2013;2013:389252.
[765] Rajouani N, Benyamma A, Romane A, et al. Chemical Composition, Antioxidant and Antibacterial Activity of Juniperus oxycedrus subsp. oxycedrus L. berry, essential oil from Morocco. *J Appl Chem Envir Prot*. 2015;1(1):9-19.
[766] Alan S, Kurkcuglu M, Sener G. Composition of the Essential Oils of Juniperus oxycedrus L. subsp. oxycedrus Growing in Turkey. *Turk J Pharm Sci*. 2016;13(3):300-3.
[767] Llorens-Molina JA, Vacas S, Sabater J. Essential oil composition of berries and leaves of Juniperus oxycedrus ssp. oxycedras L. from two typical substrates of Valencia (Spain). *Nat Volatiles & Essent Oils*. 2016;3(1):23-30.
[768] Sezik E, Kocakulak E, Baser KHC, et al. Composition of the essential oils of Juniperus oxycedrus subsp. macrocarpa from Turkey. *Chem Nat Comp*. 2005 May;3:1-3.
[769] Fahed L, Khoury M, Stien D, et al. Essential Oils Composition and Antimicrobial Activity of Six Conifers Harvested in Lebanon. *Chem Biodiversity*. 2017;14(2):1-7.
[770] Medini H, Elaissi A, Khouja M, et al. Leaf Essential Oil of Juniperus oxycedrus L. (Cupressaceae) Harvested in Northern Tunisia: Composition and Intra-Specific Variability. *Chem Biodiversity*. 2010;7(5):1254-1266.
[771] Salido S, Altarejos J, Nogueras M, et al. Chemical studies of essential oils of Juniperus oxycedrus ssp. badia. *J Ethnopharm*. 2002;81:129-34.
[772] Budavari S. The Merck Index, 11th edition. Merck, New Jersey. 1989.
[773] Hardonniere K, Saunier E, Lemaire A, et al. The environmental carcinogen benzo[a]pyrene induces a Warburg-like metabolic reprogramming dependent on NHE1 and associated with cell survival. *Sci Reports*. 2016;6:30776.
[774] Phillips DH1, Schoket B, Hewer A, et al. DNA adduct formation in human and mouse skin by mixtures of polycyclicaromatic hydrocarbons. *IARC Sci Publ*. 1990;(104):223-9.
[775] Hughes NC, Pfau W, Hewer A, et al. Covalent binding of polycyclic aromatic hydrocarbon components of coal tar to DNA in mouse skin. *Carcinogenesis*. 1993 Jan;14(1):135-44.
[776] Schoket B, Horkay I, Kósa A, et al. Formation of DNA adducts in the skin of psoriasis patients, in human skin in organ culture, and in mouse skin and lung following topical application of coal-tar and juniper tar. *J Invest Dermatol*. 1990 Feb;94(2):241-6.
[777] Takizawa Y, Noda M, Kawamura T, et al. Survey on mutagenicity of natural food additives II. Summary and Conclusion. *Mutat Res*. 1985;147:275.
[778] Delgado IF, Carvalho RR, Nogueira, et al. Study on embryofetotoxicity of b-myrcene in the rat. *Food and Chemical Toxicology*. 1993;31(1):31-5.
[779] Paumgartten FJ, De-Carvalho RR, Souza CA, et al. Study of the effects of beta-myrcene on rat fertility and general reproductive performance. *Braz J Med Biol Res*. 1998 Jul;31(7):955-65.
[780] do Vale TG, Furtado EC, Santos JG Jr, et al. Central effects of citral, myrcene and limonene, constituents of essential oil chemotypes from Lippia alba (Mill.) n.e. Brown. *Phytomedicine*. 2002 Dec;9(8):709-14.
[781] Segelman AB, Sofia RD, Segelman FP, et al. Cannabis sativa L. (marijuana). V. Pharmacological evaluation of marijuana aqueous extract and volatile oil. *J Pharm Sci*. 1974 Jun;63(6):962-4.
[782] Freitas JC1, Presgrave OA, Fingola FF, et al. Effect of beta-myrcene on pentobarbital sleeping time. *Braz J Med Biol Res*. 1993 May;26(5):519-23.
[783] De Olivera AC, Ribeiro-Pinto LF, Paumgartten JR. In vitro inhibition of CYP2B1 monoxygenase by beta-myrcene and other monoterpenoid constituents. *Br J Nutr*. 1999;81:289-95.
[784] De-Oliveira AC1, Ribeiro-Pinto LF, Otto SS, et al. Induction of liver monooxygenases by beta-myrcene. *Toxicology*. 1997 Dec 26;124(2):135-40.
[785] Loizzo MR, Tundis R, Conforti F, et al. Comparative chemical composition, antioxidant and hypoglycaemic activities of Juniperus oxycedrus ssp. oxycedrus L. berry and wood oils from Lebanon. *Food Chem*. 2007;105:572-8.
[786] Neves A, Rosa S, Gonçalves J, et al. Screening of five essential oils for identification of potential inhibitors of IL-1-induced Nf-kappaB activation and NO production in human chondrocytes: characterization of the inhibitory activity of alpha-pinene. *Planta Med*. 2010 Feb;76(3):303-8.
[787] Loizzo MR, Saab AM, Tundis R, et al. Phytochemical analysis and in vitro antiviral activities of the essential oils of seven Lebanon species. *Chem Biodivers*. 2008 Mar;5(3):461-70.
[788] Medini H, Marzouki H, Chemli R, et al. Comparison of the antimicrobial activity and the essential oil composition of Juniperus oxycedrus subsp. rufescens obtained by hydrodistillation and supercritical carbon dioxide extraction methods. *Chem Mat Compounds*. 2009;45(5):739-41.
[789] Angioni A, Barra A, Russo MT, et al. Chemical Composition of the Essential Oils of Juniperus from Ripe and Unripe Berries and Leaves and Their Antimicrobial Activity. *J Agric Food Chem*. 2003;51:3073-8.
[790] Sela F, Karapandzova M, Stefkiv G, et al. Chemical composition and antimicrobial activity of berry essential oil of Juniperus oxycedrus L. (Cupressaceae) grown wild in Republic of Macedonia. *Macedonian Pharm Bull*. 2013;49(1,2):41-8.
[791] Medini H, Manongiu B, Aicha N, et al. Chemical and Antibacterial Polymorphism of Juniperus oxycedrus ssp. oxycedrus and Juniperus oxycedrus ssp. macrocarpa (Cupressaceae) Leaf Essential Oils from Tunisia. *J Chem*. 2013;2013:389252.
[792] Rajouani N, Benyamma A, Romane A, et al. Chemical Composition, Antioxidant and Antibacterial Activity of Juniperus oxycedrus subsp. oxycedrus L. berry, essential oil from Morocco. *J Appl Chem Envir Prot*. 2015;1(1):9-19.
[793] Fahed L, Khoury M, Stien D, et al. Essential Oils Composition and Antimicrobial Activity of Six Conifers Harvested in Lebanon. *Chem Biodiversity*. 2017;14(2):1-7.
[794] Costa S, Cavadas C, Cavaleiro C, et al. In vitro susceptibility of Trypanosoma brucei brucei to selected essential oils and their major components. *Exp Parasitol*. 2018 Jul;190:34-40.

[795] Medini H, Marzouki H, Chemli R, et al. Comparison of the antimicrobial activity and the essential oil composition of Juniperus oxycedrus subsp. rufescens obtained by hydrodistillation and supercritical carbon dioxide extraction methods. *Chem Mat Compounds*. 2009;45(5):739-41.
[796] Rajouani N, Benyamma A, Romane A, et al. Chemical Composition, Antioxidant and Antibacterial Activity of Juniperus oxycedrus subsp. oxycedrus L. berry, essential oil from Morocco. *J Appl Chem Envir Prot*. 2015;1(1):9-19.
[797] Athanassiou CG, Kavallieratos NC, Evergetis E, et al. Insecticidal efficacy of silica gel with Juniperus oxycedrus ssp. oxycedrus (Pinales: Cupressaceae) essential oil against Sitophilus oryzae (Coleoptera: Curculionidae) and Tribolium confusum (Coleoptera: Tenebrionidae). *J Econ Entomol*. 2013 Aug;106(4):1902-10.
[798] Apel MA, Sobral M, Zuanazzi JA, et al. Essential oil composition of four Plinia species (Myrtaceae). *Flavour Fragr J*. 2006;21:565-567.
[799] Miguel G, Simoes M, Figeiredo AC, et al. Composition and antioxidant activities of the essential oils of Thymus caespititius, Thymus camphoratus and Thymus mastichina. *Food Chem*. 2004;86:183-8.
[800] Zuzarte M, Alves-Silva JM, Alves M, et al. New insights on the anti-inflammatory potential and safety profile of Thymus carnosus and Thymus camphoratus essential oils and their main compounds.
[801] Miguel MG, Costa LA, Figueiredo AC, et al. Assessment of the antioxidant ability of Thymus albicans, Th. mastichina, Th. camphoratus and Th. carnosus essential oils by TBARS and Micellar Model systems. *Nat Prod Commun*. 2007 Jan;2(4):399-406.
[802] Faleiro ML, Miguel MG, Ladeiro F, et al. Antimicrobial activity of essential oils isolated from Portuguese endemic species of Thymus. *Lett Appl Microbiol*. 2003;36(1):35-40.
[803] Burkhard PR, Burkhardt K, Haenggeli CA, et al. Plant-induced seizures: reappearance of an old problem. *J Neurol*. 1999 Aug;246(8):667-70.
[804] Waldman N. Seizure caused by dermal application of over-the-counter eucalyptus oil head lice preparation. *Clin Toxicol (Phila)*. 2011 Oct;49(8):750-1.
[805] Craig JO. Poisoning by the volatile oils in childhood. *Arch Dis Child*. 1953;28:259-67.
[806] Mathew T, Kamath V, Kumar RS, et al. Eucalyptus oil inhalation-induced seizure: A novel, underrecognized, preventable cause of acute symptomatic seizure. *Epilepsia Open*. 2017 Jul 4;2(3):350-354.
[807] Unger M, Frank A. Simultaneous determination of the inhibitory potency of herbal extracts on the activity of six major cytochrome P450 enzymes using liquid chromatography/mass spectrometry and automated online extraction. *Rapid Commun Mass Spectrom*. 2004;18(19):2273-81.
[808] Jori A, Bianchetti A, Prestini PE, et al. Effect of eucalyptol (1,8-cineole) on the metabolism of other drugs in rats and in man. *Eur J Pharmacol*. 1970;9(3):362-66.
[809] Kim NH, Hyun SH, Jin CH, et al. Pretreatment with 1,8-cineole potentiates thioacetamide-induced hepatotoxicity and immunosuppression. *Arch Pharm Res*. 2004 Jul;27(7):781-9.
[810] Jori A, Bianchetti A, Prestini PE, et al. Effect of eucalyptol (1,8-cineole) on the metabolism of other drugs in rats and in man. *Eur J Pharmacol*. 1970;9(3):362-66.
[811] de Sousa DP, Raphael E, Brocksom U, et al. Sedative effect of monoterpene alcohols in mice: A preliminary screening. *Verlag der Zeitschrift fur Naturforschung*. 2007;62c:563-66.
[812] Craig JO. Poisoning by the volatile oils in childhood. *Arch Dis Child*. 1953;28:259-67.
[813] Melis K. Bochner A, Janssens G. Accidental nasal eucalyptol and menthol instillation. *Eur J Pediatr*. 1989 Aug;148(8)786-7.
[814] Day LM, Ozanne-Smith J, Parsons BJ, et al. Eucalyptus oil poisoning among young children: mechanisms of access and the potential prevention. *Aust N Z J Public Health*. 1997 Jun;21(3):297-302.
[815] Day LM, Ozanne-Smith J, Parsons BJ, et al. Eucalyptus oil poisoning among young children: mechanisms of access and the potential prevention. *Aust N Z J Public Health*. 1997 Jun;21(3):297-302.
[816] Myott E. Case of eucalyptus poisoning. *Brit M J*. 1906;1:558.
[817] Hindle RC. Eucalyptus oil ingestion. *N Z Med J*. 1994 May 11;107(977)185-6.
[818] Tibballs J. Clinical effects and management of eucalyptus oil ingestion in infants and small children. *Med J Aust*. 1995 Aug;163(4):177-80.
[819] Waldman W, Barwina M, Sein Anand J. Accidental intoxication with eucalyptus oil—a case report. *Przeql Lek*. 2011;68(8):555-6.
[820] Day LM, Ozanne-Smith J, Parsons BJ, et al. Eucalyptus oil poisoning among young children: mechanisms of access and the potential prevention. *Aust N Z J Public Health*. 1997 Jun;21(3):297-302.
[821] Burkhard PR, Burkhardt K, Haenggeli CA, et al. Plant-induced seizures: reappearance of an old problem. *J Neurol*. 1999 Aug;246(8):667-70.
[822] Waldman N. Seizure caused by dermal application of over-the-counter eucalyptus oil head lice preparation. *Clin Toxicol (Phila)*. 2011 Oct;49(8):750-1.
[823] Craig JO. Poisoning by the volatile oils in childhood. *Arch Dis Child*. 1953;28:259-67.
[824] Mathew T, Kamath V, Kumar RS, et al. Eucalyptus oil inhalation-induced seizure: A novel, underrecognized, preventable cause of acute symptomatic seizure. *Epilepsia Open*. 2017 Jul 4;2(3):350-354.
[825] Unger M, Frank A. Simultaneous determination of the inhibitory potency of herbal extracts on the activity of six major cytochrome P450 enzymes using liquid chromatography/mass spectrometry and automated online extraction. *Rapid Commun Mass Spectrom*. 2004;18(19):2273-81.
[826] Jori A, Bianchetti A, Prestini PE, et al. Effect of eucalyptol (1,8-cineole) on the metabolism of other drugs in rats and in man. *Eur J Pharmacol*. 1970;9(3):362-66.
[827] Kim NH, Hyun SH, Jin CH, et al. Pretreatment with 1,8-cineole potentiates thioacetamide-induced hepatotoxicity and immunosuppression. *Arch Pharm Res*. 2004 Jul;27(7):781-9.
[828] Jori A, Bianchetti A, Prestini PE, et al. Effect of eucalyptol (1,8-cineole) on the metabolism of other drugs in rats and in man. *Eur J Pharmacol*. 1970;9(3):362-66.
[829] de Sousa DP, Raphael E, Brocksom U, et al. Sedative effect of monoterpene alcohols in mice: A preliminary screening. *Verlag der Zeitschrift fur Naturforschung*. 2007;62c:563-66.
[830] Zuzarte M, Alves-Silva JM, Alves M, et al. New insights on the anti-inflammatory potential and safety profile of Thymus carnosus and Thymus camphoratus essential oils and their main compounds.
[831] Faleiro ML, Miguel MG, Ladeiro F, et al. Antimicrobial activity of essential oils isolated from Portuguese endemic species of Thymus. *Lett Appl Microbiol*. 2003;36(1):35-40.
[832] Miguel G, Simoes M, Figeiredo AC, et al. Composition and antioxidant activities of the essential oils of Thymus caespititius, Thymus camphoratus and Thymus mastichina. *Food Chem*. 2004;86:183-8.
[833] Miguel MG, Costa LA, Figueiredo AC, et al. Assessment of the antioxidant ability of Thymus albicans, Th. mastichina, Th. camphoratus and Th. carnosus essential oils by TBARS and Micellar Model systems. *Nat Prod Commun*. 2007 Jan;2(4):399-406.
[834] Baser KHC, Demirci B, Ozek T, et al. Composition of the Essential Oils of Five Coleonema Species from South Africa. *J Essent Oil Res*. 2006;18:26-29.
[835] Berger RG, Akkan Z, Drawert F. The Essential Oil of Coleonema album (Rutaceae) and of a Photomixotrophic Cell Culture Derived thereof. *Z Naturforsch*. 1990;452:187-95.

[836] Fajinmi OO, Gruz J, Trakowski P, et al. Antifungal and antioxidant activities of Coleonema album and C. pulchellum against skin diseases. *Pharmaceutical Biol.* 2017;55(1):1249-55.
[837] Fajinmi OO, Gruz J, Trakowski P, et al. Antifungal and antioxidant activities of Coleonema album and C. pulchellum against skin diseases. *Pharmaceutical Biol.* 2017;55(1):1249-55.
[838] Joy MT, Verghese J, Retamar JA, et al. Essential oil of Xanthoxylum rhetsa. *Flav Fragrance J.* 1986;1:165-8.
[839] Rana VS, Blazquez MA. Volatile Constituents of the Seed Coat of Zanthoxylum rhetsa (Roxb.) DC. *J Essent Oil res.* 2010 Sep/Oct;22:430-2.
[840] Wongkattiya N, Akekawatchai C, Sanguansermsri P, et al. Chemical compositions and biological properties of essential oils from Zanthoxylum rhetsa (Roxb.) DC and Zanthoxylum limonella Alston. *Afr J Tradit Complement Altern Med.* 2018;15(2):12-18.
[841] Naik RR, Shakya AK, Khalaf NA, et al. GC-MS Analysis and Biological Evaluation of Essential Oil of Zanthoxylum Rhesta (Roxb.) DC Pericarp. *Jordan J Pharm Sci.* 2015;8(3):181-92.
[842] Jirovetz L, Buchbauer G, Shafi MP, et al. Analysis of the aroma compounds of the essential oil of seeds of the spice plant Zanthoxylum rhetsa from southern India. *Z Lebensm Unters Forsch A.* 1998;206:228-9.
[843] Shafi PM, Jose B, Radhamani KT, et al. Influence of pH on essential oil composition of Zanthoxylum rhetsa seeds obtained by steam distillation. *Flavour Frag J.* 2006 Mar;21(2):317-18.
[844] Shafi PM, Saidutty A, Clery RA. Volatile Constituents of Zanthoxylum rhetsa Leaves and Seeds. *J Essent Oil Res.* 2000;12(2):179-82.
[845] Tangjitjaroenkun J, Chavasiri W, Thunyaharn S, et al. Bactericidal effects and time–kill studies of the essential oil from the fruits of Zanthoxylum limonella on multi-drug resistant bacteria. *J Essent Oil Res.* 2012 Aug;24(4):363-70.
[846] Mertas A, Garbusińska A, Szliszka E, et al. The influence of tea tree oil (Melaleuca alternifolia) on fluconazole activity against fluconazoleresistant Candida albicans strains. *Biomed Res Int.* 2015;2015:590470.
[847] Wongkattiya N, Akekawatchai C, Sanguansermsri P, et al. Chemical compositions and biological properties of essential oils from Zanthoxylum rhetsa (Roxb.) DC and Zanthoxylum limonella Alston. *Afr J Tradit Complement Altern Med.* 2018;15(2):12-18.
[848] Naik RR, Shakya AK, Khalaf NA, et al. GC-MS Analysis and Biological Evaluation of Essential Oil of Zanthoxylum Rhesta (Roxb.) DC Pericarp. *Jordan J Pharm Sci.* 2015;8(3):181-92.
[849] Itthipanichpong C, Ruangrungsi N, Pattanaautsahakit C. Chemical compositions and pharmacological effects of essential oil from the fruit of Zanthoxylum limonella. *J Med Assoc Thai.* 2002 Jun;85 Suppl 1:S344-54.
[850] Naik RR, Shakya AK, Khalaf NA, et al. GC-MS Analysis and Biological Evaluation of Essential Oil of Zanthoxylum Rhesta (Roxb.) DC Pericarp. *Jordan J Pharm Sci.* 2015;8(3):181-92.
[851] Wongkattiya N, Akekawatchai C, Sanguansermsri P, et al. Chemical compositions and biological properties of essential oils from Zanthoxylum rhetsa (Roxb.) DC and Zanthoxylum limonella Alston. *Afr J Tradit Complement Altern Med.* 2018;15(2):12-18.
[852] Naik RR, Shakya AK, Khalaf NA, et al. GC-MS Analysis and Biological Evaluation of Essential Oil of Zanthoxylum Rhesta (Roxb.) DC Pericarp. *Jordan J Pharm Sci.* 2015;8(3):181-92.
[853] Tangjitjaroenkun J, Chavasiri W, Thunyaharn S, et al. Bactericidal effects and time–kill studies of the essential oil from the fruits of Zanthoxylum limonella on multi-drug resistant bacteria. *J Essent Oil Res.* 2012 Aug;24(4):363-70.
[854] Wannissorn B, Jarikasem S, Siriwangchai T, et al. Antibacterial properties of essential oils from Thai medicinal plants. *Fitoterapia.* 2005;76:233-236
[855] Nanasombat S, Wimuttigosol P. Antimicrobial and antioxidant activity of spice essential oils. *Food Sci Biotechnol.* 2011;20(1):45-53.
[856] Wongkattiya N, Akekawatchai C, Sanguansermsri P, et al. Chemical compositions and biological properties of essential oils from Zanthoxylum rhetsa (Roxb.) DC and Zanthoxylum limonella Alston. *Afr J Tradit Complement Altern Med.* 2018;15(2):12-18.
[857] Naik RR, Shakya AK, Khalaf NA, et al. GC-MS Analysis and Biological Evaluation of Essential Oil of Zanthoxylum Rhesta (Roxb.) DC Pericarp. *Jordan J Pharm Sci.* 2015;8(3):181-92.
[858] Das NG, Dhiman S, Talukdar PK, et al. Synergistic mosquito-repellent activity of Curcuma longa, Pogostemon heyneanus and Zanthoxylum limonella essential oils. *J Infect Public Health.* 2015 Jul-Aug;8(4):323-8.
[859] Sunil D, Bipul R, Pronobesh C, Das NG, et al. Field evaluation of repellency of a polyherbal essential oil against blackflies and its dermal toxicity using rat model. *Trop Biomed.* 2012 Sep;29(3):391-7.
[860] Trongtokit Y, Rongsriyam Y, Komalamisra N, et al. Comparative repellency of 38 essential oils against mosquito bites. *Phytother Res.* 2005 Apr;19(4):303-9.
[861] Trongtokit Y, Rongsriyam Y, Komalamisra N, et al. Laboratory and field trial of developing medicinal local Thai plant products against four species of mosquito vectors. *Southeast Asian J Trop Med Public Health.* 2004 Jun;35(2):325-33.
[862] Das NG, Baruah I, Talukdar PK, et al. Evaluation of botanicals as repellents against mosquitoes. *J Vector Borne Dis.* 2003 Mar-Jun;40(1-2):49-53.
[863] Pitasawat B, Champakaew D, Choochote W, et al. Aromatic plant-derived essential oil: an alternative larvicide for mosquito control. *Fitoterapia.* 2007 Apr;78(3):205-10.
[864] Mohamed EI. Agrochemical Studies on Dianthus caryophyllus L. Grown in Egypt. *Int J PharmTech Res.* 2016;9(4):113-7.
[865] El-Ghorab AH, Mahgoub MH, Bekheta M. Effect of Some Bioregulators on the Chemical Composition of Essential Oil and its Antioxidant Activity of Egyptian Carnation (Dianthus caryophyllus L.). *J Essent Oil Bearing Plants.* 2006;9(3):214-22.
[866] Al-Sanfi AE. Chemical contents and medical importance of Dianthus caryophyllus- A review. *IOSR J Pharm.* 2017 Mar;7(3):61-71.
[867] Kimbaris A, Koliopoulos G, Michaelakis A, et al. Bioactivity of Dianthus caryophyllus, Lepidium sativum, Pimpinella anisum, and Illicium verum essential oils and their major components against the West Nile vector Culex pipiens. *Parasitology Res.* 2012;111(6):2403-10.
[868] Golestani MR, Rad M, Bassami M, et al. Analysis and evaluation of antibacterial effects of new herbal formulas, AP-001 and AP-002, against Escherichia coli O157:H7. *Life Sci.* 2015 Aug 15;135:22-6.
[869] Domaracky M, Rehak P, Juhas S, et al. Effects of selected plant essential oils on the growth and development of mouse preimplantation embryos in vivo. *Physiol Res.* 2007;56(1):97-104.
[870] Vrskova D, Modra H. Evaluation of the developmental toxicity of 2-phenoxyethanol and clove oil anesthetics using the Frog Embryo Teratogenesis Assay: Xenopus (FETAX). *Veterinarami Medicina.* 2012;57(5):245-50.
[871] Amini A, Cheraghi E, Safaee MR, et al. The role of eugenol in the reduction of teratogenic effects of retinoic acid on skeletal morphology of mice embryo. *Yakhteh Medical Journal.* 2003;4:195-200.
[872] Liu H, Ren C, Liu W, et al. Embryotoxicity estimation of commonly used compounds with embryonic stem cell test. *Mol Med Rep.* 2017 Jul; 16(1): 263–271.
[873] Liu H, Ren C, Liu W, et al. Embryotoxicity estimation of commonly used compounds with embryonic stem cell test. *Mol Med Rep.* 2017 Jul; 16(1): 263–271.
[874] Chen R, Chen J, Cheng S, et al. Assessment of embryotoxicity of constituents in cosmetics by the embryonic stem cell tes. *Toxicol Mech Methods.* 2010 Mar;20(3):112-18.

[875] Chen SJ, Wang MH, Chen IJ. Antiplatelet and calcium inhibitory properties of eugenol and sodium eugenol acetate. *Gen Pharmacol.* 1996 Jun;27(4):629-33.
[876] Tognolini M, Barocelli E, Ballabeni V, et al. Comparative screening of plant essential oils; phenylpropanoid moiety as basic core for antiplatelet activity. *Life Sci.* 2006 Feb 23;78(13):1419-32.
[877] Heck AM, DeWitt BA, Lukes AL. Potential interactions between alternative therapies and warfarin. *Am J Health Syst Pharm.* 2000;57(13):1221-1227.
[878] Saaeed SA, Gilani AH. Antithrombotic activity of clove oil. *J Pak Med Assoc.* 1994;44(5):112-15.
[879] Kamatou GP, Vermaak I, Viljoen AM. Eugenol—from the remote Maluku Islands to the international market place: A review of a remarkable and versatile molecule. *Molecules.* 2012;17:6953-81.
[880] Langeveld WT, Veldhuizen EJ, Burt SA. Synergy between essential oil constituents and antibiotics. *Crit Rev Microbiol.* 2014 Feb;40(1);76-94.
[881] Cochrane ML. Inhibition of Cytochrome P450 2C9 by essential oils. Available at: https://libres.uncg.edu/ir/uncg/listing.aspx?id=18102.
[882] Sarrami N, Pemberton MN, Thornhill MH, et al. Adverse reactions associated with the use of eugenol in dentistry. *British Dental J.* 2002;193:253-55.
[883] Tammannavar P, Pushpalatha C, Jain S, et al. An unexpected positive hypersensitive reaction to eugenol. *BMJ Case Rep.* 2013; 2013:bcr2013009464.
[884] Domaracky M, Rehak P, Juhas S, et al. Effects of selected plant essential oils on the growth and development of mouse preimplantation embryos in vivo. *Physiol Res.* 2007;56(1):97-104.
[885] Vrskova D, Modra H. Evaluation of the developmental toxicity of 2-phenoxyethanol and clove oil anesthetics using the Frog Embryo Teratogenesis Assay: Xenopus (FETAX). *Veterinarami Medicina.* 2012;57(5):245-50.
[886] Amini A, Cheraghi E, Safaee MR, et al. The role of eugenol in the reduction of teratogenic effects of retinoic acid on skeletal morphology of mice embryo. *Yakhteh Medical Journal.* 2003;4:195-200.
[887] Liu H, Ren C, Liu W, et al. Embryotoxicity estimation of commonly used compounds with embryonic stem cell test. *Mol Med Rep.* 2017 Jul; 16(1): 263–271.
[888] Liu H, Ren C, Liu W, et al. Embryotoxicity estimation of commonly used compounds with embryonic stem cell test. *Mol Med Rep.* 2017 Jul; 16(1): 263–271.
[889] Chen R, Chen J, Cheng S, et al. Assessment of embryotoxicity of constituents in cosmetics by the embryonic stem cell tes. *Toxicol Mech Methods.* 2010 Mar;20(3):112-18.
[890] Chen SJ, Wang MH, Chen IJ. Antiplatelet and calcium inhibitory properties of eugenol and sodium eugenol acetate. *Gen Pharmacol.* 1996 Jun;27(4):629-33.
[891] Tognolini M, Barocelli E, Ballabeni V, et al. Comparative screening of plant essential oils; phenylpropanoid moiety as basic core for antiplatelet activity. *Life Sci.* 2006 Feb 23;78(13):1419-32.
[892] Heck AM, DeWitt BA, Lukes AL. Potential interactions between alternative therapies and warfarin. *Am J Health Syst Pharm.* 2000;57(13):1221-1227.
[893] Saaeed SA, Gilani AH. Antithrombotic activity of clove oil. *J Pak Med Assoc.* 1994;44(5):112-15.
[894] Kamatou GP, Vermaak I, Viljoen AM. Eugenol—from the remote Maluku Islands to the international market place: A review of a remarkable and versatile molecule. *Molecules.* 2012;17:6953-81.
[895] Langeveld WT, Veldhuizen EJ, Burt SA. Synergy between essential oil constituents and antibiotics. *Crit Rev Microbiol.* 2014 Feb;40(1);76-94.
[896] Cochrane ML. Inhibition of Cytochrome P450 2C9 by essential oils. Available at: https://libres.uncg.edu/ir/uncg/listing.aspx?id=18102.
[897] Dohi S, Terasaki M, Makino M. Acetylcholinesterase inhibitory activity and chemical composition of commercial essential oils. *J Agric Food Chem.* 2009 May 27;57(10):4313-8.
[898] Sarrami N, Pemberton MN, Thornhill MH, et al. Adverse reactions associated with the use of eugenol in dentistry. *British Dental J.* 2002;193:253-55.
[899] Tammannavar P, Pushpalatha C, Jain S, et al. An unexpected positive hypersensitive reaction to eugenol. *BMJ Case Rep.* 2013; 2013:bcr2013009464.
[900] Bulut C, Altiok E, Bayraktar O, et al. Antioxidative and antimicrobial screening of 19 commercial essential oils in Turkey. *Acta Hort.* 2009;826(14):111-6.
[901] Golestani MR, Rad M, Bassami M, et al. Analysis and evaluation of antibacterial effects of new herbal formulas, AP-001 and AP-002, against Escherichia coli O157:H7. *Life Sci.* 2015 Aug 15;135:22-6.
[902] Bulut C, Altiok E, Bayraktar O, et al. Antioxidative and antimicrobial screening of 19 commercial essential oils in Turkey. *Acta Hort.* 2009;826(14):111-6.
[903] Kimbaris A, Koliopoulos G, Michaelakis A, et al. Bioactivity of Dianthus caryophyllus, Lepidium sativum, Pimpinella anisum, and Illicium verum essential oils and their major components against the West Nile vector Culex pipiens. *Parasitology Res.* 2012;111(6):2403-10.
[904] Hagedorn ML, Brown SM. The constituents of cascarilla oil (croton eluteria bennett). *Flavour Frag J.* 1991 Dep;6(3):193-204.
[905] Djabou N, Andreani S, Varesi L, et al. Analysis of the volatile fraction of Teucrium marum L. *Flavour Frag J.* 2013;28:14-24
[906] Sanz J, Mus M, Rossello JA. Volatile components variation in the Teucrium marurm complex (Lamiaceae) from the Balearic Islands. *Botanical J Linnean Soc.* 2000;132:253-61.
[907] Ricci D, Fraternale D, Giamperi L, et al. Chemical composition, antimicrobial and antioxidant activity of the essential oil of Teucrium marum (Lamiaceae). *J Ethnopharmacol.* 2005;98:195-200.
[908] Ricci D, Fraternale D, Giamperi L, et al. Chemical composition, antimicrobial and antioxidant activity of the essential oil of Teucrium marum (Lamiaceae). *J Ethnopharmacol.* 2005;98:195-200.
[909] Ricci D, Fraternale D, Giamperi L, et al. Chemical composition, antimicrobial and antioxidant activity of the essential oil of Teucrium marum (Lamiaceae). *J Ethnopharmacol.* 2005;98:195-200.
[910] dos Santos ERQ, Maia GSF, Fontes Junior EA, et al. Linalool-rich essential oils from the Amazon display antidepressant-type effect in rodents. *J Ethnopharm.* 2018;212:43-9.
[911] Tucker AO, Maciarello MJ, Alkire BH. Essential Oil of Aeollanthus suaveolens Mart, ex Spreng. (Lamiaceae). *J Essent Oil Res.* 2001 May/Jun;13:198-9.
[912] Maia JGS, Zoghbi MDGB, Andrade HA. Essential oils of Aeollanthus suaveolens Matt. ex Spreng. *J Essent Oil Res.* 2003 Mar/Apr;15:86-7.
[913] Martins RL, Simoes RC, Rabelo EDM, et al. Chemical Composition, an Antioxidant, Cytotoxic and Microbiological Activity of the Essential Oil from the Leaves of Aeollanthus suaveolens Mart. ex Spreng. *PLOS One.* 2016 Dec: 0166684.
[914] Simionatto E, Porto C, Stuker CZ, et al. Chemical composition and antimicrobial activity of the essential oil from Aeolanthus suaveolens Mart. Ex Spreng. *Quim Nova.* 2007;30(8):1923-5.
[915] Martins RL. ESTUDO QUÍMICO E AVALIAÇÃO DAS ATIVIDADES ANTIOXIDANTE, MICROBIOLÓGICA, DE CITOTOXICIDADE E INSETICIDA DO ÓLEO ESSENCIAL DE Aeollanthus suaveolens Mart. ex Spreng (LAMIACEAE). 2016. Available at: http://www2.unifap.br/ppgcf/files/2015/04/Dissertacao_Rosany-Lopes-Martins_16-09-2016.pdf.

[916] dos Santos ERQ, Maia GSF, Fontes Junior EA, et al. Linalool-rich essential oils from the Amazon display antidepressant-type effect in rodents. *J Ethnopharm*. 2018;212:43-9.
[917] Simionatto E, Porto C, Stuker CZ, et al. Chemical composition and antimicrobial activity of the essential oil from Aeolanthus suaveolens Mart. Ex Spreng. *Quim Nova*. 2007;30(8):1923-5.
[918] Martins RL, Simoes RC, Rabelo EDM, et al. Chemical Composition, an Antioxidant, Cytotoxic and Microbiological Activity of the Essential Oil from the Leaves of Aeollanthus suaveolens Mart. ex Spreng. *PLOS One*. 2016 Dec: 0166684.
[919] Martins RL, Simoes RC, Rabelo EDM, et al. Chemical Composition, an Antioxidant, Cytotoxic and Microbiological Activity of the Essential Oil from the Leaves of Aeollanthus suaveolens Mart. ex Spreng. *PLOS One*. 2016 Dec: 0166684.
[920] Su YC, Ho CL, Wang EIC. Analysis of leaf essential oils from the indigenous five conifers of Taiwan. *Flavour Fragr J*. 2006 May/Jun;21(3):447-52.
[921] Chung MJ, Cheng SS, Lin CY, et al. Profiling of volatile compounds from five interior decoration timbers in Taiwan using TD/GC–MS/FID. *J Wood Sci*. 2018;64:823-35.
[922] Ho CL, Yang SS, Chang TM, et al. Composition, Antioxidant, Antimicrobial and Anti-wood-decay Fungal Activities of the Twig Essential Oil of Taiwania cryptomerioides from Taiwan. *Nat Prod Commun*. 2012;7(2):261-4.
[923] Chang ST, Cheng SS, Wang SY. Antitermitic activity of essential oils and components from Taiwania (Taiwania cryptomerioides). *J Chem Ecology*. 2001;27(4):717-24.
[924] Ho CL, Yang SS, Chang TM, et al. Composition, Antioxidant, Antimicrobial and Anti-wood-decay Fungal Activities of the Twig Essential Oil of Taiwania cryptomerioides from Taiwan. *Nat Prod Commun*. 2012;7(2):261-4.
[925] Chung MJ, Cheng SS, Lin CY, et al. Profiling of volatile compounds from five interior decoration timbers in Taiwan using TD/GC–MS/FID. *J Wood Sci*. 2018;64:823-35.
[926] Ho CL, Yang SS, Chang TM, et al. Composition, Antioxidant, Antimicrobial and Anti-wood-decay Fungal Activities of the Twig Essential Oil of Taiwania cryptomerioides from Taiwan. *Nat Prod Commun*. 2012;7(2):261-4.
[927] Chang ST, Cheng SS, Wang SY. Antitermitic activity of essential oils and components from Taiwania (Taiwania cryptomerioides). *J Chem Ecology*. 2001;27(4):717-24.
[928] Chang ST, Chen PF, Wang SY, et al. Antimite activity of essential oils and their constituents from Taiwania cryptomerioides. *J Med Entomol*. 2001 May;38(3):455-7.
[929] Su LC, Huang CG, Chang ST, et al. An improved bioassay facilitates the screening of repellents against cat flea, Ctenocephalides felis (Siphonaptera: Pulicidae). *Pest Manag Sci*. 2014 Feb;70(2):264-70.
[930] Chen HC, Chi HS, Lin LY. Headspace Solid-Phase Microextraction Analysis of Volatile Components in Narcissus tazetta var. chinensis Roem. *Molecules*. 2013;18:13723-34.
[931] Bruno S, Laurentis ND, Amico A, et al. Chemical Investigation and Cytologic Localization of Essential Oils in the Flowers of Narcissus tazetta. *Int J Pharmacog*. 1999;32(4):357-61.
[932] Shikhiev AS, Goryaev MI, Sharipova FS, et al. An investigation of the chemical composition of the absolute oil of Narcissus tazetta florepleno. *Plenum Publishing Corporation*. 1974:409-10.
[933] National Library of Medicine. TOXNET. Benzyl acetate. Available at: https://toxnet.nlm.nih.gov/cgibin/sis/search/a?dbs+hsdb:@term+@DOCNO+2851.
[934] JECFA. Summary of evaluations performed by the Joint FAO/WHO Expert Committee on Food Additives. 2001. Available at: https://toxnet.nlm.nih.gov/cgi-bin/sis/search/a?dbs+hsdb:@term+@DOCNO+2851.
[935] Miyazawa M, Fujita T, Yamafuji C, et al. Chemical composition of volatile oil from the roots of Periploca sepium. *J Oleo Sci*. 2004;53(10):511-3.
[936] Chu SS, Jiang GH, Liu WL, et al. Insecticidal activity of the root bark essential oil of Periploca sepium Bunge and its main component. *Nat Prod Res*. 2012 May;26(10):926-32.
[937] Wang J, Liu H, Zhao J, et al. Antimicrobial and Antioxidant Activities of the Root Bark Essential Oil of Periploca sepium and Its Main Component 2-Hydroxy-4-methoxybenzaldehyde. *Molecules*. 2010;15:5807-17.
[938] Wang J, Liu H, Zhao J, et al. Antimicrobial and Antioxidant Activities of the Root Bark Essential Oil of Periploca sepium and Its Main Component 2-Hydroxy-4-methoxybenzaldehyde. *Molecules*. 2010;15:5807-17.
[939] Wang J, Liu H, Zhao J, et al. Antimicrobial and Antioxidant Activities of the Root Bark Essential Oil of Periploca sepium and Its Main Component 2-Hydroxy-4-methoxybenzaldehyde. *Molecules*. 2010;15:5807-17.
[940] Chu SS, Jiang GH, Liu WL, et al. Insecticidal activity of the root bark essential oil of Periploca sepium Bunge and its main component. *Nat Prod Res*. 2012 May;26(10):926-32.
[941] Kim DY, Won KJ, Hwang DI, et al. Chemical composition, antioxidant and anti-melanogenic activities of essential oils from Chrysanthemum boreale Makino at different harvesting stages. *Chem Biodiversity*. 2018;15(2):1-13.
[942] Kim BS, Park SJ, Kim MK, et al. Inhibitory Effects of Chrysanthemum boreale Essential Oil on Biofilm Formation and Virulence Factor Expression of Streptococcus mutans. *Evidence-Based Complement Altern Med*. 2015;2015:1-12.
[943] Hong CU. Essential Oil Composition of Chrysanthemum boreale and Chrysanthemum indicum. *Appl Biological Chem*. 2002;45(2):108-13.
[944] Khine H, Weiss D, Graber N, et al. A cluster of children with seizures caused by camphor poisoning. *Pediatrics*. 2009 May;123(5):1269-72.
[945] Michiels EA, Mazor SS. Toddler with seizures due to ingesting camphor at an Indian celebration. *Pediatr Emerg Care*. 2010 Aug;26(8):574-75.
[946] Koren G. Medications which can kill a toddler with one tablet or teaspoonful. *J toxicol Clin Toxicol*. 1993;31(3):407- 13.
[947] Flaman Z, Pellechia-Clarke S, Bailey B, et al. Unintentional exposure of young children to camphor and eucalyptus oils. *Paediatr Child Health*. 2001 Feb;6(2):80-83.
[948] Hindle RC. Eucalyptus oil ingestion. *N Z Med J*. 1994 May 11;107(977)185-6.
[949] Tibballs J. Clinical effects and management of eucalyptus oil ingestion in infants and small children. *Med J Aust*. 1995 Aug;163(4):177-80.
[950] Halicioglu O, Astarcioglu G, Yaprak I, et al. Toxicity of Salvia officinalis in a newborn and a child: an alarming report. *Pediatr Neurol*. 2011 Oct;45(4):259-60.
[951] Rabl W, Katzgraber F, Steinlechner M. Camphor ingestion for abortion (case report). *Forensic Sci Int*. 1997 Sep 19;89(1-2):137-40.
[952] da Silva Costa KC, Bexerra SB, Norte CM, et al. Medicinal plants with teratogenic potential: current considerations. *Braz J Pharm Sci*. 2012;48(3):427-33.
[953] Brinker F. Herb Contraindications and Drug Interactions. 2nd ed. Sandy, OR. Eclectic Medical Publications. 1998.
[954] Flaman Z, Pellechia-Clarke S, Bailey B, et al. Unintentional exposure of young children to camphor and eucalyptus oils. *Paediatr Child Health*. 2001 Feb;6(2):80-83.
[955] Manoguerra AS, Erdman AR, Wax PM, et al. Camphor poisoning: an evidence-based practice guideline for out-of- hospital management. *Clin Toxicol (Phila)*.2006;44(4):357-70.
[956] Gibson DE, Moore GP, Pfaff JA. Camphor ingestion. *Am J Emerg Med*. 1989 Jan;7(1):41-43.

[957] Koppel C, Martends F, Schirop T, et al. Hemoperfusion in acute camphor poisoning. *Intensive Care Med.* 1988;14(4):431-33.
[958] European Medicines Agency. Public statement on the use of herbal medicinal products containing thujone. Available at: http://www.ema.europa.eu/docs/en_GB/document_library/Public_statement/2011/02/WC500102294.pdf.
[959] Millet Y, Jouglard J, Steinmetz MD, et al. Toxicity of some essential plant oils. Clinical and experimental study. *Clin Toxicol.* 1981 Dec;18(12):1485-98.
[960] Cristovao L, Carvalho F, Bastos MDL, et al. Hepatotoxicity of an essential oil of Salvia officinalis L.: an in vitro study using freshly isolated rat hepatocytes. *Congress Biomarkers.* 2001 Sep:165.
[961] Burkhard PR, Burkhardt K, Haenggeli CA, et al. Plant-induced seizures: reappearance of an old problem. *J Neurol.* 1999 Aug;246(8):667-70.
[962] Burkhard PR, Burkhardt K, Haenggeli CA, et al. Plant-induced seizures: reappearance of an old problem. *J Neurol.* 1999 Aug;246(8):667-70.
[963] Narayan S, Singh N. Camphor poisoning-An unusual cause of seizure. *Med J Armed Forces India.* 2012 Jul;68(3):252-53.
[964] Chanaranaj KJ, G MV, S M. Camphor poisoning in a child. *Natl Med J India.* 2013 Jan-Feb;26(1):60.
[965] Perry NB, Anderson RE, Brennan NJ, et al. Essential oils from dalmatian sage (Salvia officinalis L.) variations among individuals, plant parts, seasons, and sites. *J Agric Food Chem.* 1999;47:2048-54.
[966] Millet Y, Jouglard J, Steinmetz MD, et al. Toxicity of some essential plant oils. Clinical and experimental study. *Clin Toxicol.* 1981 Dec;18(12):1485-98.
[967] Olowe SA, Ransome-Kuti O. The risk of jaundice in glucose-6-phosphate dehydrogenase deficient babies exposed to menthol. *Acta Paediatr Scand.* 1980 May;69(3):341-5.
[968] Dillon Remy M, Manning Alleyne P, Bratt DE, et al. Neonatal jaundice at Port-of-Spain General Hospital abstract. *West Indian Med J.* 1987;36(Suppl):28.
[969] Kim DY, Won KJ, Hwang DI, et al. Chemical composition, antioxidant and anti-melanogenic activities of essential oils from Chrysanthemum boreale Makino at different harvesting stages. *Chem Biodiversity.* 2018;15(2):1-13.
[970] Kim BS, Park SJ, Kim MK, et al. Inhibitory Effects of Chrysanthemum boreale Essential Oil on Biofilm Formation and Virulence Factor Expression of Streptococcus mutans. *Evidence-Based Complement Altern Med.* 2015;2015:1-12.
[971] Kim DY, Won KJ, Yoon MS, et al. Chrysanthemum boreale Makino essential oil induces keratinocyte proliferation and skin regeneration. *Nat Prod Res.* 2015;29(6):562-4.
[972] Kim KJ, Kim YH, Yu HH, et al. Antibacterial activity and chemical composition of essential oil of Chrysanthemum boreale. *Planta Med.* 2003 Mar;69(3):274-7.
[973] Cosmetic Ingredient Review. Safety Assessment of Citrus-Derived Peel Oils as Used in Cosmetics. 2014. Available at: https://www.cir-safety.org/sites/default/files/cpeelo092014FR.pdf.
[974] Thi Nguyen TT, Thi Tran TT, Hua TM, et al. Investigation of peel and leaf essential oils of Citrus clementina Hort. ex Tan. growing in the south of Vietnam. *J Essent Oil Res.* 2015;28(2):96-103.
[975] Germana MA, Palazzolo E, Chiancone B, et al. Characterization of Leaf Essential Oil Composition of Homozygous and Heterozygous Citrus clementina Hort. Extan. and its Ancestors. *J Essent Oil Bearing Plants.* 2013;16(1):92-101.
[976] Miguel MG, Dandlen S, Figueiredo AC, et al. Essential oils of flowers of Citrus sinensis and Citrus clementina cultivated in Algarve, Portugal. *Acta Hort.* 2008;773:89-94.
[977] Boudries H, Loupassaki S, Ladjal Ettoumi Y, et al. Chemical profile, antimicrobial and antioxidant activities of Citrus reticulate and Citrus clementina (L.) essential oil. *Int Food Res J.* 2017 Aug;24(4):1782-92.
[978] Dugo G, Cotroneo A, Trozzi A, et al. On the genuineness of citrus essential oils. part xxv. the essential oil from 'wenzhou honey' oranges. *Flavour Frag J.* 1988;3(4):161-166.
[979] Ruberto G, Biondi D, Piattelli M, et al. Essential Oil of the New Citrus Hybrid, Citrus clementina x C. limon. *J Essent Oil Res.* 1994;6(1):1-8.
[980] Cosmetic Ingredient Review. Safety Assessment of Citrus-Derived Peel Oils as Used in Cosmetics. 2014. Available at: https://www.cir-safety.org/sites/default/files/cpeelo092014FR.pdf.
[981] Chen W, Lu Y, Gao J, et al. Beta-elemene inhibits melanoma growth and metastasis via suppressing vascular endothelial growth factormediated angiogenesis. *Cancer Chemo Pharmacol.* 2011 Apr;67(4):799-808.
[982] Thi Nguyen TT, Thi Tran TT, Hua TM, et al. Investigation of peel and leaf essential oils of Citrus clementina Hort. ex Tan. growing in the south of Vietnam. *J Essent Oil Res.* 2015;28(2):96-103.
[983] Boudries H, Loupassaki S, Ladjal Ettoumi Y, et al. Chemical profile, antimicrobial and antioxidant activities of Citrus reticulate and Citrus clementina (L.) essential oil. *Int Food Res J.* 2017 Aug;24(4):1782-92.
[984] Thi Nguyen TT, Thi Tran TT, Hua TM, et al. Investigation of peel and leaf essential oils of Citrus clementina Hort. ex Tan. growing in the south of Vietnam. *J Essent Oil Res.* 2015;28(2):96-103.
[985] Boudries H, Loupassaki S, Ladjal Ettoumi Y, et al. Chemical profile, antimicrobial and antioxidant activities of Citrus reticulate and Citrus clementina (L.) essential oil. *Int Food Res J.* 2017 Aug;24(4):1782-92.
[986] Bakkour Y, Makhoul S, El-Nakat H, et al. Chemical analysis of the essential oils from Punica granatum, Vitis vinifera and Cucurbita maxima seeds growing in Lebanon by GC/MS. *J Nat Prod.* 2011;4:71-74.
[987] Yalcin H, Kavuncuoglu H, Ekici L, et al. Determination of fatty acid composition, volatile componenets, physico-chemical, and bioactive properties of grape (Vitis vinifera) seed and seed oil. *J food Proc Preserv.* 2017 Apr;41(2):e12854.
[988] Boido E, LLoret A, Medina K, et al. Aroma Composition of Vitis vinifera Cv. Tannat: the Typical Red Wine from Uruguay. *J Agric Food Chem.* 2003;51:5408-13.
[989] Edens Garden. Cognac Essential Oil. Available at: https://cdn.shopify.com/s/files/1/0380/8537/files/Cognac_7_17.pdf?2259714687948636368.
[990] Miracle Botanicals. GC/MS Batch Number: 170525E. Available at: https://www.dropbox.com/s/wemly69p40r6ur6/Green%20Cognac%20GCMS%20Miracle.pdf?dl=0.
[991] Apel MA, Sobral M, Zuanazzi JA, et al. Essential oil composition of four Plinia species (Myrtaceae). *Flavour Fragr J.* 2006;21:565-567.
[992] Khodakov G, Kotikov I. Component composition of essential oil from Artemisia taurica. Chem Nat Compounds. 2008;44(2):261-262.
[993] Bayramoglu M, Candan F. Antioxidant properties of volatile oils obtained from Artemisia taurica Willd. and Salvia kronenburgii Rech. Fil. plants and their effects on xanthine oxidase. *African J Biotech.* 2014 jan;13(5):683-92.
[994] Haliciglu O, Astarcioglu G, Yaprak I, et al. Toxicity of Salvia officinalis in a newborn and a child: an alarming report. *Pediatr Neurol.* 2011 Oct;45(4):259-60.
[995] Lachenmeier DW, Walch SG. Epileptic seizure caused by accidental ingestion of sage (Salvia officinalis L.) oil in children: a rare, exceptional case or a threat to public health. *Pediatr Neurol.* 2012 Mar;46(3):201.
[996] da Silva Costa KC, Bexerra SB, Norte CM, et al. Medicinal plants with teratogenic potential: current considerations. *Braz J Pharm Sci.* 2012;48(3):427-33.

[997] Brinker F. Herb Contraindications and Drug Interactions. 2nd ed. Sandy, OR. Eclectic Medical Publications. 1998.
[998] Frohne D. Giftpflanzen: Cupressaceae. Stuttgart: Wissenschaftliche Verlagsgesellschaft mbH; 1997. pp. 153–6.
[999] European Medicines Agency. Public statement on the use of herbal medicinal products containing thujone. Available at: http://www.ema.europa.eu/docs/en_GB/document_library/Public_statement/2011/02/WC500102294.pdf.
[1000] Millet Y, Jouglard J, Steinmetz MD, et al. Toxicity of some essential plant oils. Clinical and experimental study. *Clin Toxicol.* 1981 Dec;18(12):1485-98.
[1001] Cristovao L, Carvalho F, Bastos MDL, et al. Hepatotoxicity of an essential oil of Salvia officinalis L.: an in vitro study using freshly isolated rat hepatocytes. *Congress Biomarkers.* 2001 Sep:165.
[1002] Millet Y, Jouglard J, Steinmetz MD, et al. Toxicity of some essential plant oils. Clinical and experimental study. *Clin Toxicol.* 1981 Dec;18(12):1485-98.
[1003] Burkhard PR, Burkhardt K, Haenggeli CA, et al. Plant-induced seizures: reappearance of an old problem. *J Neurol.* 1999 Aug;246(8):667-70.
[1004] Burkhard PR, Burkhardt K, Haenggeli CA, et al. Plant-induced seizures: reappearance of an old problem. *J Neurol.* 1999 Aug;246(8):667-70.
[1005] Hold KM, Sirisoma NS, Ikeda T, et al. α-Thujone (the active component of absinthe): γ-Aminobutyric acid type A receptor modulation and metabolic detoxification. *Proc Natl Acad Sci USA.* 2000 Apr 11;97(8):3826–3831.
[1006] Waldman N. Seizure caused by dermal application of over-the-counter eucalyptus oil head lice preparation. *Clin Toxicol (Phila).* 2011 Oct;49(8):750-1.
[1007] Craig JO. Poisoning by the volatile oils in childhood. *Arch Dis Child.* 1953;28:259-67.
[1008] Mathew T, Kamath V, Kumar RS, et al. Eucalyptus oil inhalation-induced seizure: A novel, underrecognized, preventable cause of acute symptomatic seizure. *Epilepsia Open.* 2017 Jul 4;2(3):350-354.
[1009] Uc A, Bishop WP, Sanders KD. Camphor hepatotoxicity. *South Med J.* 2000;93:596-98.
[1010] Frohne D. Giftpflanzen: Cupressaceae. Stuttgart: Wissenschaftliche Verlagsgesellschaft mbH; 1997. pp. 153–6.
[1011] Dolan LC, Matulka RA, Burdock GA. Naturally Occurring Food Toxins. *Toxins (Basel).* 2010 Sep;2(9):2289–2332.
[1012] United States National Toxicology Program (NTP). Alpha-Thujone. Dec 10, 1997. Available at: http://ntp.niehs.nih.gov/index.cfm?objectid=03DB8C36-E7A1-9889-3BDF8436F2A8C51F.
[1013] Unger M, Frank A. Simultaneous determination of the inhibitory potency of herbal extracts on the activity of six major cytochrome P450 enzymes using liquid chromatography/mass spectrometry and automated online extraction. *Rapid Commun Mass Spectrom.* 2004;18(19):2273-81.
[1014] Jori A, Bianchetti A, Prestini PE, et al. Effect of eucalyptol (1,8-cineole) on the metabolism of other drugs in rats and in man. *Eur J Pharmacol.* 1970;9(3):362-66.
[1015] Abass K, Reponen P, Mattila S, et al. Metabolism of α-thujone in human hepatic preparations in vitro. *Xenobiotica.* 2011 Feb;41(2):101-11.
[1016] Kim NH, Hyun SH, Jin CH, et al. Pretreatment with 1,8-cineole potentiates thioacetamide-induced hepatotoxicity and immunosuppression. *Arch Pharm Res.* 2004 Jul;27(7):781-9.
[1017] Khine H, Weiss D, Graber N, et al. A cluster of children with seizures caused by camphor poisoning. *Pediatrics.* 2009 May;123(5):1269-72.
[1018] Michiels EA, Mazor SS. Toddler with seizures due to ingesting camphor at an Indian celebration. *Pediatr Emerg Care.* 2010 Aug;26(8):574-75.
[1019] Koren G. Medications which can kill a toddler with one tablet or teaspoonful. *J Toxicol Clin Toxicol. 1*993;31(3):407-13.
[1020] Flaman Z, Pellechia-Clarke S, Bailey B, et al. Unintentional exposure of young children to camphor and eucalyptus oils. *Paediatr Child Health.* 2001 Feb;6(2):80-83.
[1021] Rabl W, Katzgraber F, Steinlechner M. Camphor ingestion for abortion (case report). *Forensic Sci Int.* 1997 Sep 19;89(1-2):137-40.
[1022] Flaman Z, Pellechia-Clarke S, Bailey B, et al. Unintentional exposure of young children to camphor and eucalyptus oils. *Paediatr Child Health.* 2001 Feb;6(2):80-83.
[1023] Manoguerra AS, Erdman AR, Wax PM, et al. Camphor poisoning: an evidence-based practice guideline for out-of- hospital management. *Clin Toxicol (Phila).*2006;44(4):357-70.
[1024] Gibson DE, Moore GP, Pfaff JA. Camphor ingestion. *Am J Emerg Med.* 1989 Jan;7(1):41-43.
[1025] Koppel C, Martends F, Schirop T, et al. Hemoperfusion in acute camphor poisoning. *Intensive Care Med.* 1988;14(4):431-33.
[1026] Burkhard PR, Burkhardt K, Haenggeli CA, et al. Plant-induced seizures: reappearance of an old problem. *J Neurol.* 1999 Aug;246(8):667–70.
[1027] Millet Y, Jouglard J, Steinmetz MD, et al. Toxicity of some essential plant oils. Clinical and experimental study. *Clin Toxicol.* 1981 Dec;18(12):1485–98.
[1028] Charanraj KJ, G MV, S M. Camphor poisoning in a child. *Natl Med J India.* 2013 Jan-Feb;26(1):60.
[1029] Narayan S, Singh N. Camphor poisoning-An unusual cause of seizure. *Med J Armed Forces India.* 2012 Jul;68(3):252–53.
[1030] Uc A, Bishop WP, Sanders KD. Camphor hepatotoxicity. *South Med J.* 2000;93:596-98.
[1031] Unger M, Frank A. Simultaneous determination of the inhibitory potency of herbal extracts on the activity of six major cytochrome P450 enzymes using liquid chromatography/mass spectrometry and automated online extraction. *Rapid Commun Mass Spectrom.* 2004;18(19):2273-81.
[1032] Jori A, Bianchetti A, Prestini PE, et al. Effect of eucalyptol (1,8-cineole) on the metabolism of other drugs in rats and in man. *Eur J Pharmacol.* 1970;9(3):362-66.
[1033] Kim NH, Hyun SH, Jin CH, et al. Pretreatment with 1,8-cineole potentiates thioacetamide-induced hepatotoxicity and immunosuppression. *Arch Pharm Res.* 2004 Jul;27(7):781-9.
[1034] Jori A, Bianchetti A, Prestini PE, et al. Effect of eucalyptol (1,8-cineole) on the metabolism of other drugs in rats and in man. *Eur J Pharmacol.* 1970;9(3):362-66.
[1035] de Sousa DP, Raphael E, Brocksom U, et al. Sedative effect of monoterpene alcohols in mice: A preliminary screening. *Verlag der Zeitschrift fur Naturforschung.* 2007;62c:563-66.
[1036] Olowe SA, Ransome-Kuti O. The risk of jaundice in glucose-6-phosphate dehydrogenase deficient babies exposed to menthol. *Acta Paediatr Scand.* 1980 May;69(3):341-5.
[1037] Dillon Remy M, Manning Alleyne P, Bratt DE, et al. Neonatal jaundice at Port-of-Spain General Hospital abstract. *West Indian Med J.* 1987;36(Suppl):28.
[1038] Covington TR, et al. Handbook of Nonprescription Drugs. 11th ed. Washington, D.C.: American Pharmaceutical Association. 1996.
[1039] Bayramoglu M, Candan F. Antioxidant properties of volatile oils obtained from Artemisia taurica Willd. and Salvia kronenburgii Rech. Fil. plants and their effects on xanthine oxidase. *African J Biotech.* 2014 jan;13(5):683-92.
[1040] Veiga Jr VF, Rosas EC, Carvalho MV, et al. Chemical composition and anti-inflammatory activity of copaiba oils from Copaifera cearensis Huber ex Ducke, Copaifera reticulata Ducke and Copaifera multijuga Hayne—A comparative study. *J Ethnopharm.* 2007;112:248-54.
[1041] Zoghbi MDGB, Andrade EHA, Martins-da-Silva RCV, et al. Chemical Variation in the Volatiles of Copaifera reticulata Ducke (Leguminosae) Growing Wild in the States of Pará and Amapá, Brazil. *J Essent Oil Res.* 2009 Nov/Dec;21:501-3.

[1042] Lameira OA, Martins-da-Silva RCV. Seasonal Variation in the Volatiles of Copaifera duckei Dwyer Growing Wild in the State of Para Brazil. *J Essent Oil Res.* 2009 Mar/Apr;21:105-7.
[1043] Zoghbi MDGB, Martins-da-Silva RCV, Trigo JR. Volatiles of Oleoresins of Copaifera paupera (Herzog) Dwyer, C. piresii Dwyer and C. pubiflora Benth. (Leguminosae). *J Essent Oil Res.* 2009 Sep/Oct;21:403-4.
[1044] Santos AO, Ueda-Nakamura T, Dias Filho BP, et al. Effect of Brazilian copaiba oils on Leishmania amazonensis. *J Ethnopharmacol.* 2008;120:204-8.
[1045] Sant'Anna BMP, Fontes SP, Pinto AC, et al. Characterization of woody odorant contributors in copaiba oil (Copaifera multijuga Hayne). *J Braz Chem Soc.* 2007;18(5):984-9.
[1046] Barbosa PCS, Medeiros RS, Sampaio PTB, et al. Influence of abiotic factors on the chemical composition of copaiba oil (Copaifera multijuga Hayne): Soil composition, seasonality and diameter at breast height. *J Braz Chem Soc.* 2012;23(10:1823-33.
[1047] Cascon V, Gilbert B. Characterization of the chemical composition of oleoresins of Copaifera guianensis Desf., Copaifera duckei Dwyer and Copaifera multijuga Hayne. *Phytochem.* 2000;55:773-8.
[1048] Herrero-Jauregui C, Casado MA, Zoghbi MDGB, et al. Chemical Variability of Copaifera reticulata Ducke Oleoresin. *Chem Biodivers.* 2011;8:674-85.
[1049] Guimaraes-Santos A, Santos DS, Santos IR, et al. Copaiba oil-resin treatment is neuroprotective and reduces neutrophil recruitment and microglia activation after motor cortex excitotoxic injury. *Evidence-Based Complement Altern Med.* 2012;2012:918174.
[1050] Bardaji DKR, da Silva JJM, Bianchi TC, et al. Copaifera reticulata oleoresin: Chemical characterization and antibacterial properties against oral pathogens. *Anaerobe.* 2016;40:18-27.
[1051] Souza Barbosa P, Moreira Wiedemann L, da Silva Medeiros R, et al. Phytochemical Fingerprints of Copaiba Oils (Copaifera multijuga Hayne) Determined by Multivariate Analysis. *Chem Biodiversity.* 2013;10(7):1350-60.
[1052] Zoghbi MDGB, Lameira OA, Oliveira ECP. Seasonal Variation of Oleoresin and Volatiles from Copaifera martii Hayne Growing Wild in the State of Pará, Brazil. *J Essent Oil Res.* 2007;19(6):504-6.
[1053] Silva MT, Borges LL, de Sousa Fiuza T, et al. Viscosity of the Oil-resins and Chemical Composition of the Essential Oils from Oils-resins of Copaifera multijuga Hayne Growing in the National Forest Saracá-Taquera Brazil. 2017;20(5):1226-34.
[1054] Lin WY, Kuo YH, Chang YL, et al. Anti-platelet aggregation and chemical constituents from the rhizome of Gynura japonica. *Planta Med.* 2003 Aug;69(8):757-64.
[1055] Nascimento AM, Brandão MG, Oliveira GB, et al. Synergistic bactericidal activity of Eremanthus erythropappus oil or beta-bisabolene with ampicillin against Staphylococcus aureus. *Antonie Van Leeuwenhoek.* 2007 Jul;92(1):95-100.
[1056] Veiga Jr VF, Rosas EC, Carvalho MV, et al. Chemical composition and anti-inflammatory activity of copaiba oils from Copaifera cearensis Huber ex Ducke, Copaifera reticulata Ducke and Copaifera multijuga Hayne—A comparative study. *J Ethnopharm.* 2007;112:248-54.
[1057] Lima SRM, Veiga Jr. VF, Christo HB, et al. In vivo and in vitro studies on the anticancer activity of Copaifera multijuga Hayne and its fractions. *Phytother Res.* 2003;17:1048–53.
[1058] Kobayashi C, Fontanive TO, Enzweiler BG, et al. Pharmacological evaluation of Copaifera multijuga oil in rats. *Pharm Biol.* 2011;49(3):306-13.
[1059] Guimaraes-Santos A, Santos DS, Santos IR, et al. Copaiba oil-resin treatment is neuroprotective and reduces neutrophil recruitment and microglia activation after motor cortex excitotoxic injury. *Evidence-Based Complement Altern Med.* 2012;2012:918174.
[1060] Gonçalves Dias FG, Jorge AT, de Freitas Pereira L, et al. Use of Copaifera multijuga for acute corneal repair after chemical injury: A clinical, histopathological and toxicogenetic study. *Biomed Pharmacother.* 2017 Dec;96:1193-1198.
[1061] Bardaji DKR, da Silva JJM, Bianchi TC, et al. Copaifera reticulata oleoresin: Chemical characterization and antibacterial properties against oral pathogens. *Anaerobe.* 2016;40:18-27.
[1062] Santos AO, Ueda-Nakamura T, Dias Filho BP, et al. Effect of Brazilian copaiba oils on Leishmania amazonensis. *J Ethnopharmacol.* 2008;120:204-8.
[1063] Rondon FC, Bevilaqua CM, Accioly MP, et al. In vitro efficacy of Coriandrum sativum, Lippia sidoides and Copaifera reticulata against Leishmania chagasi. *Rev Bras Parasitol Vet.* 2012 Jul-Sep;21(3):185-91.
[1064] Kobayashi C, Fontanive TO, Enzweiler BG, et al. Pharmacological evaluation of Copaifera multijuga oil in rats. *Pharm Biol.* 2011;49(3):306-13.
[1065] Bylaite E, Venskutonis R, Roozen JP, et al. Composition of Essential Oil of Costmary [Balsamita major (L.) Desf.] at Different Growth Phases. *J Agric Food Chem.* 2000;48:2409-14.
[1066] Derakhshani Z, Hassani A, Sefidkon F. Growth Characters, Essential Oil Content and Terpene Composition of Costmary (Chrysanthemum balsamita L.) as Influenced by Zinc Nutrition and Different Moisture Regimes. *J Essent Oil Bearing Plants.* 2014;17(5):1046-56.
[1067] Hassanpouraghdam MB, Tabatabaie SJ, Nazemiyeh H, et al. Essential Oil Composition of Hydroponically Grown Chrysanthemum balsamita L. *J Essent Oil Bearing Plants.* 2008;11(6):649-54.
[1068] Hassanpouraghdam MB. Flowerhead volatile oil composition of soilless culture-grown Chrysanthemum balsamita L. *Nat Prod Res.* 2009 Mar;23(7):672-7.
[1069] Haliciglu O, Astarcioglu G, Yaprak I, et al. Toxicity of Salvia officinalis in a newborn and a child: an alarming report. *Pediatr Neurol.* 2011 Oct;45(4):259-60.
[1070] Lachenmeier DW, Walch SG. Epileptic seizure caused by accidental ingestion of sage (Salvia officinalis L.) oil in children: a rare, exceptional case or a threat to public health. *Pediatr Neurol.* 2012 Mar;46(3):201.
[1071] da Silva Costa KC, Bexerra SB, Norte CM, et al. Medicinal plants with teratogenic potential: current considerations. *Braz J Pharm Sci.* 2012;48(3):427-33.
[1072] Brinker F. Herb Contraindications and Drug Interactions. 2nd ed. Sandy, OR. Eclectic Medical Publications. 1998.
[1073] Frohne D. Giftpflanzen: Cupressaceae. Stuttgart: Wissenschaftliche Verlagsgesellschaft mbH; 1997. pp. 153–6.
[1074] European Medicines Agency. Public statement on the use of herbal medicinal products containing thujone. Available at: http://www.ema.europa.eu/docs/en_GB/document_library/Public_statement/2011/02/WC500102294.pdf.
[1075] Millet Y, Jouglard J, Steinmetz MD, et al. Toxicity of some essential plant oils. Clinical and experimental study. *Clin Toxicol.* 1981 Dec;18(12):1485-98.
[1076] Cristovao L, Carvalho F, Bastos MDL, et al. Hepatotoxicity of an essential oil of Salvia officinalis L.: an in vitro study using freshly isolated rat hepatocytes. *Congress Biomarkers.* 2001 Sep:165.
[1077] Millet Y, Jouglard J, Steinmetz MD, et al. Toxicity of some essential plant oils. Clinical and experimental study. *Clin Toxicol.* 1981 Dec;18(12):1485-98.
[1078] Burkhard PR, Burkhardt K, Haenggeli CA, et al. Plant-induced seizures: reappearance of an old problem. *J Neurol.* 1999 Aug;246(8):667-70.
[1079] Burkhard PR, Burkhardt K, Haenggeli CA, et al. Plant-induced seizures: reappearance of an old problem. *J Neurol.* 1999 Aug;246(8):667-70.

[1080] Hold KM, Sirisoma NS, Ikeda T, et al. α-Thujone (the active component of absinthe): γ-Aminobutyric acid type A receptor modulation and metabolic detoxification. *Proc Natl Acad Sci USA*. 2000 Apr 11;97(8):3826–3831.
[1081] Waldman N. Seizure caused by dermal application of over-the-counter eucalyptus oil head lice preparation. *Clin Toxicol (Phila)*. 2011 Oct;49(8):750-1.
[1082] Craig JO. Poisoning by the volatile oils in childhood. *Arch Dis Child.* 1953;28:259-67.
[1083] Mathew T, Kamath V, Kumar RS, et al. Eucalyptus oil inhalation-induced seizure: A novel, underrecognized, preventable cause of acute symptomatic seizure. *Epilepsia Open*. 2017 Jul 4;2(3):350-354.
[1084] Uc A, Bishop WP, Sanders KD. Camphor hepatotoxicity. *South Med J.* 2000;93:596-98.
[1085] Frohne D. Giftpflanzen: Cupressaceae. Stuttgart: Wissenschaftliche Verlagsgesellschaft mbH; 1997. pp. 153–6.
[1086] Dolan LC, Matulka RA, Burdock GA. Naturally Occurring Food Toxins. *Toxins (Basel)*. 2010 Sep;2(9):2289–2332.
[1087] United States National Toxicology Program (NTP). Alpha-Thujone. Dec 10, 1997. Available at: http://ntp.niehs.nih.gov/index.cfm?objectid=03DB8C36-E7A1-9889-3BDF8436F2A8C51F.
[1088] Unger M, Frank A. Simultaneous determination of the inhibitory potency of herbal extracts on the activity of six major cytochrome P450 enzymes using liquid chromatography/mass spectrometry and automated online extraction. *Rapid Commun Mass Spectrom*. 2004;18(19):2273-81.
[1089] Jori A, Bianchetti A, Prestini PE, et al. Effect of eucalyptol (1,8-cineole) on the metabolism of other drugs in rats and in man. *Eur J Pharmacol.* 1970;9(3):362-66.
[1090] Abass K, Reponen P, Mattila S, et al. Metabolism of α-thujone in human hepatic preparations in vitro. *Xenobiotica*. 2011 Feb;41(2):101-11.
[1091] Kim NH, Hyun SH, Jin CH, et al. Pretreatment with 1,8-cineole potentiates thioacetamide-induced hepatotoxicity and immunosuppression. *Arch Pharm Res*. 2004 Jul;27(7):781-9.
[1092] Muruganathan U, Srinivasan S, Indumathi D. Antihyperglycemic effect of carvone: Effect on the levels of glycoprotein components in streptozotocin-induced diabetic rats. *J Acute Disease.* 2013;2(4):310-15.
[1093] Souza FV, da Rocha MB, de Souza DP, et al. (-)-Carvone: antispasmodic effect and mode of action. *Fitoterapia.* 2013 Mar;85:20-24.
[1094] Rajeshwari T, Raja B. Antihypertensive, antihyperlipidemicand antioxidantinfluence of D-carvone in L-NAME induced hypertensive rats. *Int J Pharmaceutical Biol Arch.* 2014;5(4):82-88.
[1095] de Sousa DP, Farias Nobrega FF, de Almeida RN. Influence of the chirality of (R)-(-)- and (S)-(+)-carvone in the central nervous system: a comparative study. *Chirality*. 2007 May;19(4):264-268.
[1096] Akdogan M, Gultekin F, Yontem M. Effect of Mentha piperita (Labiatae) and Mentha spicata (Labiatae) on iron absorption in rats. *Toxicol Ind Health.* 2004 Sep;20(6-10):119-22.
[1097] Brophy JM, Lassak EV. The volatile leaf oils of Melaleuca armillaris, M. dissitiflora and M. trichostachya. *J Proc Royal Soc NSW.* 1983;116:7-10.
[1098] Williams LR, Lusunzi I. Essential oil from Melaleuca dissitiflora: a potential source of high quality tea tree oil. *Ind Crops Prod.* 1994;2:211-7.
[1099] Park HM, Kim J, Chang KS, et al. Larvicidal activity of Myrtaceae essential oils and their components against Aedes aegypti, acute toxicity on Daphnia magna, and aqueous residue. *J Med Entomol.* 2011 Mar;48(2):405-10.
[1100] Mertas A, Garbusińska A, Szliszka E, et al. The influence of tea tree oil (Melaleuca alternifolia) on fluconazole activity against fluconazoleresistant Candida albicans strains. *Biomed Res Int.* 2015;2015:590470.
[1101] Craig JO. Poisoning by the volatile oils in childhood. *Arch Dis Child.* 1953;28:259-67.
[1102] Melis K. Bochner A, Janssens G. Accidental nasal eucalyptol and menthol instillation. *Eur J Pediatr.* 1989 Aug;148(8)786-7.
[1103] Day LM, Ozanne-Smith J, Parsons BJ, et al. Eucalyptus oil poisoning among young children: mechanisms of access and the potential prevention. *Aust N Z J Public Health.* 1997 Jun;21(3):297-302.
[1104] Day LM, Ozanne-Smith J, Parsons BJ, et al. Eucalyptus oil poisoning among young children: mechanisms of access and the potential prevention. *Aust N Z J Public Health.* 1997 Jun;21(3):297-302.
[1105] Myott E. Case of eucalyptus poisoning. *Brit M J.* 1906;1:558.
[1106] Hindle RC. Eucalyptus oil ingestion. *N Z Med J.* 1994 May 11;107(977)185-6.
[1107] Tibballs J. Clinical effects and management of eucalyptus oil ingestion in infants and small children. *Med J Aust.* 1995 Aug;163(4):177-80.
[1108] Waldman W, Barwina M, Sein Anand J. Accidental intoxication with eucalyptus oil—a case report. *Przeql Lek.* 2011;68(8):555-6.
[1109] Day LM, Ozanne-Smith J, Parsons BJ, et al. Eucalyptus oil poisoning among young children: mechanisms of access and the potential prevention. *Aust N Z J Public Health.* 1997 Jun;21(3):297-302.
[1110] Burkhard PR, Burkhardt K, Haenggeli CA, et al. Plant-induced seizures: reappearance of an old problem. *J Neurol.* 1999 Aug;246(8):667-70.
[1111] Waldman N. Seizure caused by dermal application of over-the-counter eucalyptus oil head lice preparation. *Clin Toxicol (Phila)*. 2011 Oct;49(8):750-1.
[1112] Craig JO. Poisoning by the volatile oils in childhood. *Arch Dis Child.* 1953;28:259-67.
[1113] Mathew T, Kamath V, Kumar RS, et al. Eucalyptus oil inhalation-induced seizure: A novel, underrecognized, preventable cause of acute symptomatic seizure. *Epilepsia Open*. 2017 Jul 4;2(3):350-354.
[1114] Unger M, Frank A. Simultaneous determination of the inhibitory potency of herbal extracts on the activity of six major cytochrome P450 enzymes using liquid chromatography/mass spectrometry and automated online extraction. *Rapid Commun Mass Spectrom.* 2004;18(19):2273-81.
[1115] Jori A, Bianchetti A, Prestini PE, et al. Effect of eucalyptol (1,8-cineole) on the metabolism of other drugs in rats and in man. *Eur J Pharmacol.* 1970;9(3):362-66.
[1116] Kim NH, Hyun SH, Jin CH, et al. Pretreatment with 1,8-cineole potentiates thioacetamide-induced hepatotoxicity and immunosuppression. *Arch Pharm Res*. 2004 Jul;27(7):781-9.
[1117] Jori A, Bianchetti A, Prestini PE, et al. Effect of eucalyptol (1,8-cineole) on the metabolism of other drugs in rats and in man. *Eur J Pharmacol.* 1970;9(3):362-66.
[1118] de Sousa DP, Raphael E, Brocksom U, et al. Sedative effect of monoterpene alcohols in mice: A preliminary screening. *Verlag der Zeitschrift fur Naturforschung*. 2007;62c:563-66.
[1119] Park HM, Kim J, Chang KS, et al. Larvicidal activity of Myrtaceae essential oils and their components against Aedes aegypti, acute toxicity on Daphnia magna, and aqueous residue. *J Med Entomol.* 2011 Mar;48(2):405-10.
[1120] Park IK, Shin SC. Fumigant activity of plant essential oils and components from garlic (Allium sativum) and clove bud (Eugenia caryophyllata) oils against the Japanese termite (Reticulitermes speratus Kolbe). *J Agric Food Chem*. 2005 Jun 1;53(11):4388-92.
[1121] Paaver U, Orav A, Arak E, et al. Phytochemical analysis of the essential oil of Thymus serpyllum L. growing wild in Estonia. *Nat Prod Res.* 2008 Jan;22(2):108-15.
[1122] Kirilov V, Stikhareva T, Mukanov B, et al. Composition of the Essential Oil of Thymus serpyllum L. from Northern Kazakhstan. *J Essent Oil Bearing Plants.* 2016'19(1):212-22.

[1123] Raal A, Paaver U, Arak E, et al. Content and composition of the essential oil of Thymus serpyllum L. growing wild in Estonia. *Medicina (Kaunas)*. 2004;40(8):795-800.
[1124] Sefidkon F, Dabiri M, Mirmostafa SA. The Composition of Thymus serpyllum L. Oil. *J Essen Oil Res*. 2004 May/Jun;16:184-5.
[1125] Nardoni S, Giovanelli S, Pistelli L, et al. In Vitro Activity of Twenty Commercially Available, Plant-Derived Essential Oils against Selected Dermatophyte Species. *Nat Prod Commun*. 2015 Aug;10(8):1473-8.
[1126] Jarić S, Mitrović M, Pavlović P, et al. Review of Ethnobotanical, Phytochemical, and Pharmacological Study of Thymus serpyllum L. *Evid Based Complement Alternat Med*. 2015;2015:101978.
[1127] Cavara S, Maksimović M, Vidic D. The essential oil of Thymus aureopunctatus (Beck) K. Malý. *Nat Prod Commun*. 2009 Mar;4(3):415-20.
[1128] Domaracky M, Rehak P, Juhas S, et al. Effects of selected plant essential oils on the growth and development of mouse preimplantation embryos in vivo. *Physiol Res*. 2007;56(1):97-104.
[1129] Lemhadri A, Zeggwagh NA, Maghrani M, et al. Anti-hypoglycaemic activity of the aqueous extract of Origanum vulgare growing wild in Tafilalet region. *J Ethnopharmacol*. 2004 Jun;92(2-3):251-6.
[1130] Bayramoglu G, Senturk H, Bayramoglu A, et al. Carvacrol partially reverses symptoms of diabetes in STZ-induced diabetic rats. *Cytotechnology*. 2014 Mar;66(2):251–257.
[1131] Ezhumalai M, Radhiga T, Pugalendi KV. Antihyperglycemic effect of carvacrol in combination with rosiglitazone in high-fat diet-induced type 2 diabetic C57BL/6J mice. *Mol Cell Biochem*. 2014 Jan;385(1-2):23-31.
[1132] Umasankar K, Ramya DR, Nambikkairaj B. Antidiabetic and antidyslipidemic nature of carvacrol, a monoterpenic phenol studied in high-fat - fed and low-dose streptozotocin-induced experimental diabetic rats. *J Pharm Res*. 2015;9(8):484-90.
[1133] Tognolini M, Barocelli E, Ballabeni V, et al. Comparative screening of plant essential oils; phenylpropanoid moiety as basic core for antiplatelet activity. *Life Sci*. 2006 Feb 23;78(13):1419-32.
[1134] Tsai HH, Lin HW, Chen YL, et al. A review of potential harmful interactions between anticoagulant/antiplatelet agents and Chinese herbal medicines. *PLoS One*. 2013 May 9;8(5):e64255.
[1135] Karkabounas S, Kostoula OK, Daskalou T, et al. Anticarcinogenic and antiplatelet effects of carvacrol. *Exp Oncol*. 2006 Jun;28(2):121-5.
[1136] Baser KH. Biological and pharmacological activities of carvacrol and carvacrol bearing essential oils. *Curr Pharm Des*. 2008;14(29):3106-19.
[1137] Okazaki K, Kawazoe K, Takaishi Y. Human platelet aggregation inhibitors from thyme (Thymus vulgaris L.). *Phytother Res*. 2002 Jun;16(4):398-9.
[1138] Jukic M, Politeo O, Maksimmovic M, et al. In vitro acetylcholinesterase inhibitory properties of thymol, carvacrol and their derivatives thymoquinone and thymohydroquinone. *Phytother Res*. 2007;21(3):259-61.
[1139] Askin H, Yildiz M, Ayar A. Effects of Thymol and Carvacrol on Acetylcholinesterase from Drosophila melanogaster. *Acta Physica Polinica A*. 2017;132(3):720-22.
[1140] Medeiros Barreto H, Cerqueira Fontinele F, Pereira de Oliveira A, et al. Phytochemical prospection and modulation of antibiotic activity in vitro by Lippia origanoides H.B.K. in methicillin resistant Staphylococcus aureus. *Biomed Res Int*. 2014;305610:1–7.
[1141] Aelenei P, Miron A, Trifan A, et al. Essential Oils and Their Components as Modulators of Antibiotic Activity against Gram-Negative Bacteria. *Medicines*. 2016;3(19):2-34.
[1142] Magi G, Marini E, Facinelli B. Antimicrobial activity of essential oils and carvacrol, and synergy of carvacrol and erythromycin, against clinical, erythromycin-resistant Group A Streptococci. *Front Microbiol*. 2015;6:165.
[1143] Veras HNH, Rodrigues FFG, Colares AV, et al. Synergistic antibiotic activity of volatile compounds from the essential oil of Lippia sidoides and thymol. *Fitoterapia*. 2012 Apr;83(3):508-12.
[1144] Langeveld WT, Veldhuizen EJ, Burt SA. Synergy between essential oil components and antibiotics: A review. *Crit Rev Microbiol*. 2014 Feb;40(1):76-94.
[1145] Kohlert C, Schindler G, Marz RW, et al. Systemic availability and pharmacokinetics of thymol in humans. *J Clin Pharmacol*. 2002 Jul;42(7):731-7.
[1146] Toxicology Data Network, National Library of Medicine. Thymol. Available at: http://toxnet.nlm.nih.gov/cgibin/sis/search/a?dbs+hsdb:@term+@DOCNO+866.
[1147] Tognolini M, Barocelli E, Ballabeni V, et al. Comparative screening of plant essential oils; phenylpropanoid moiety as basic core for antiplatelet activity. *Life Sci*. 2006 Feb 23;78(13):1419-32.
[1148] Okazaki K, Kawazoe K, Takaishi Y. Human platelet aggregation inhibitors from thyme (Thymus vulgaris L.). *Phytother Res*. 2002 Jun;16(4):398-9.
[1149] Jukic M, Politeo O, Maksimmovic M, et al. In vitro acetylcholinesterase inhibitory properties of thymol, carvacrol and their derivatives thymoquinone and thymohydroquinone. *Phytother Res*. 2007;21(3):259-61.
[1150] Askin H, Yildiz M, Ayar A. Effects of Thymol and Carvacrol on Acetylcholinesterase from Drosophila melanogaster. *Acta Physica Polinica A*. 2017;132(3):720-22.
[1151] Medeiros Barreto H, Cerqueira Fontinele F, Pereira de Oliveira A, et al. Phytochemical prospection and modulation of antibiotic activity in vitro by Lippia origanoides H.B.K. in methicillin resistant Staphylococcus aureus. *Biomed Res Int*. 2014;305610:1–7.
[1152] Aelenei P, Miron A, Trifan A, et al. Essential Oils and Their Components as Modulators of Antibiotic Activity against Gram-Negative Bacteria. *Medicines*. 2016;3(19):2-34.
[1153] Magi G, Marini E, Facinelli B. Antimicrobial activity of essential oils and carvacrol, and synergy of carvacrol and erythromycin, against clinical, erythromycin-resistant Group A Streptococci. *Front Microbiol*. 2015;6:165.
[1154] Veras HNH, Rodrigues FFG, Colares AV, et al. Synergistic antibiotic activity of volatile compounds from the essential oil of Lippia sidoides and thymol. *Fitoterapia*. 2012 Apr;83(3):508-12.
[1155] Langeveld WT, Veldhuizen EJ, Burt SA. Synergy between essential oil components and antibiotics: A review. *Crit Rev Microbiol*. 2014 Feb;40(1):76-94.
[1156] Miladinovic DL, Ilic BS, Kocic BD, et al. Antibacterial investigation of thyme essential oil and its main constituents on combination with tetracycline. *J Med Food*. 2015 Aug;18(8):935-7.
[1157] Shin S, Lim S. Antifungal effects of herbal essential oils alone and in combination with ketoconazole against Trichophyton spp. *J Appl Micribiol*. 2004;97:1289-96.
[1158] Kissels W, Wu X, Santos RR, et al. Short communication: Interaction of the isomers carvacrol and thymol with the antibiotics doxycycline and tilmicosin: In vitro effects against pathogenic bacteria commonly found in the respiratory tract of calves. *J Dairy Sci*. 2017 Feb;100(2):970-974.
[1159] Seo KA, Kim H, Ku HY, et al. The monoterpenoids citral and geraniol are moderate inhibitors of the CYP2B6 hydroxylase activity. *Chem Biol Interact*. 2008;174:141-46.

[1160] Ibrahim SM, El-Denshary ES, Abdallah DM. Geraniol, Alone and in Combination with Pioglitazone, Ameliorates Fructose-Induced Metabolic Syndrome in Rats via the Modulation of Both Inflammatory and Oxidative Stress Status. *PLoS One.* 2015 Feb 13;10(2):e0117516.
[1161] Babukumar S, Vinothkumar V, Sankaranarayanan C, et al. Geraniol, a natural monoterpene, ameliorates hyperglycemia by attenuating the key enzymes of carbohydrate metabolism in streptozotocin-induced diabetic rats. *Pharm Biol.* 2017 Dec;55(1):1442-1449.
[1162] Shin S, Lim S. Antifungal effects of herbal essential oils alone and in combination with ketoconazole against Trichophyton spp. *J Appl Micribiol.* 2004;97:1289-96.
[1163] Lorenzi V, Muselli A, Bernardini AF, et al. Geraniol restores antibiotic activities against multidrug-resistant isolates from gram-negative species. *Antimicrob Agents Chemother.* 2009 May;53(5):2209-11.
[1164] Li L, Li ZW, Yin ZQ, et al. Antibacterial activity of leaf essential oil and its constituents from Cinnamomum longepaniculatum. *Int J Clin Exp Med.* 2014;7(7):1721-7.
[1165] Hagvall L. Cytochrome P450-mediated activation of the fragrance constituent geraniol forms potent contact allergens. *Tox Appl Pharm.* 2008 Dec;233(2):308-13.
[1166] Olowe SA, Ransome-Kuti O. The risk of jaundice in glucose-6-phosphate dehydrogenase deficient babies exposed to menthol. *Acta Paediatr Scand.* 1980 May;69(3):341-5.
[1167] Dillon Remy M, Manning Alleyne P, Bratt DE, et al. Neonatal jaundice at Port-of-Spain General Hospital abstract. *West Indian Med J* 1987;36(Suppl):28.
[1168] Brehm-Stecher BF, Johnson EA. Sensitization of Staphylococcus aureus and Escherichia coli to antibiotics by the sesquiterpenoids nerolidol, farnesol, bisabolol, and apritone. *Antimicrob Agents Chemother.* 2003 Oct;47(10):3357-60.
[1169] Špičáková A, Szotáková B, Dimunová D, et al. Nerolidol and Farnesol Inhibit Some Cytochrome P450 Activities but Did Not Affect Other Xenobiotic-Metabolizing Enzymes in Rat and Human Hepatic Subcellular Fractions. *Molecules.* 2017 Mar 24;22(4). pii: E509.
[1170] Burkhard PR, Burkhardt K, Haenggeli CA, et al. Plant-induced seizures: reappearance of an old problem. *J Neurol.* 1999 Aug;246(8):667-70.
[1171] Waldman N. Seizure caused by dermal application of over-the-counter eucalyptus oil head lice preparation. *Clin Toxicol (Phila).* 2011 Oct;49(8):750-1.
[1172] Craig JO. Poisoning by the volatile oils in childhood. *Arch Dis Child.* 1953;28:259-67.
[1173] Mathew T, Kamath V, Kumar RS, et al. Eucalyptus oil inhalation-induced seizure: A novel, underrecognized, preventable cause of acute symptomatic seizure. *Epilepsia Open.* 2017 Jul 4;2(3):350-354.
[1174] Narayan S, Singh N. Camphor poisoning-An unusual cause of seizure. *Med J Armed Forces India.* 2012 Jul;68(3):252-53.
[1175] Chanaranaj KJ, G MV, S M. Camphor poisoning in a child. *Natl Med J India.* 2013 Jan-Feb;26(1):60.
[1176] Lin WY, Kuo YH, Chang YL, et al. Anti-platelet aggregation and chemical constituents from the rhizome of Gynura japonica. *Planta Med.* 2003 Aug;69(8):757-64.
[1177] Nardoni S, Giovanelli S, Pistelli L, et al. In Vitro Activity of Twenty Commercially Available, Plant-Derived Essential Oils against Selected Dermatophyte Species. *Nat Prod Commun.* 2015 Aug;10(8):1473-8.
[1178] Sokolić-Mihalak D, Frece J, Slavica A, et al. The effects of wild thyme (Thymus serpyllum L.) essential oil components against ochratoxin-producing Aspergilli. *Arh Hig Rada Toksikol.* 2012 Dec;63(4):457-62.
[1179] Abu-Darwish MS, Al-Ramamneh EA, Kyslychenko VS, et al. The antimicrobial activity of essential oils and extracts of some medicinal plants grown in Ash-shoubak region - South of Jordan. *Pak J Pharm Sci.* 2012 Jan;25(1):239-46.
[1180] Oussalah M, Caillet S, Saucier L, et al. Antimicrobial effects of selected plant essential oils on the growth of a Pseudomonas putida strain isolated from meat. *Meat Sci.* 2006 Jun;73(2):236-44.
[1181] Rasooli I, Mirmostafa SA. Antibacterial properties of Thymus pubescens and Thymus serpyllum essential oils. *Fitoterapia.* 2002 Jun;73(3):244-50.
[1182] Knezevic P, Aleksic Sabo V, Simin N, et al. A colorimetric broth microdilution method for assessment of Helicobacter pylori sensitivity to antimicrobial agents. *J Pharm Biomed Anal.* 2018 Apr 15;152:271-278.
[1183] Topal U, Sasaki M, Goto M, et al. Chemical compositions and antioxidant properties of essential oils from nine species of Turkish plants obtained by supercritical carbon dioxide extraction and steam distillation. *Int J Food Sci Nutr.* 2008 Nov-Dec;59(7-8):619-34.
[1184] Kulisić T, Krisko A, Dragović-Uzelac V, et al. The effects of essential oils and aqueous tea infusions of oregano (Origanum vulgare L. spp. hirtum), thyme (Thymus vulgaris L.) and wild thyme (Thymus serpyllum L.) on the copper-induced oxidation of human low-density lipoproteins. *Int J Food Sci Nutr.* 2007 Mar;58(2):87-93.
[1185] Štefanidesová K, Škultéty Ľ, Sparagano OAE, et al. The repellent efficacy of eleven essential oils against adult Dermacentor reticulatus ticks. *Ticks Tick Borne Dis.* 2017 Aug;8(5):780-786.
[1186] Mothana RA, Al-Said MS, Raishi M, et al. Chemical composition, anti-inflammatory and antioxidant activities of the essential oil of Piper cubeba L. *Romanian Biotech Lett.* 2017'22(2):12366-76.
[1187] Esperandim VR, da Silva Ferreiea D, Sousa Rezende KC, et al. In Vitro Antiparasitic Activity and Chemical Composition of the Essential Oil Obtained from the Fruits of Piper cubeba. *Planta Med.* 2013;79"1653-5.
[1188] Singh G, Marimuthu P, de Heluani CS, et al. Chemical Constituents, Antioxidative and Antimicrobial Activities of Essential Oil and Oleoresin of Tailed Pepper (Piper Cubeba L). *Int J Food Engineering.* 2007;3(6):1-22.
[1189] Bos R, Woerdenbag HJ, Kayser O, et al. Essential Oil Constituents of Piper cubeba L. fils. from Indonesia. *J Essent Oil Res.* 2007 Jan/Feb;19:14-17.
[1190] Shankaracharya NB, Jaganmohan Rao L, Nagalakshmi S, et al. Studies on the chemical composition of Cubeb (Piper cubeba Linn.). *Pafai J.* 1995 Mar;33-9.
[1191] Sumathykutty M, Rao J, Padmakumari K, et al. Essential oil constituents of some Piper species. Flavour Frag J. 1999;14(5):279-282.
[1192] Terhune SJ, Hogg JW, Lawrence BM. Bicyclosesquiphellandrene and 1-epibicyclosesquiphellandrene: Two new dienes based on the cadalene skeleton. *Phytochem.* 1974;13:1183-5. In: Bos R, Woerdenbag HJ, Kayser O, et al. Essential Oil Constituents of Piper cubeba L. fils. from Indonesia. *J Essent Oil Res.* 2007 Jan/Feb;19:14-17.
[1193] Lawrence BM. Progress in essential oils. *Perfum Flavor.* 1980;5(5):27-32. In: Bos R, Woerdenbag HJ, Kayser O, et al. Essential Oil Constituents of Piper cubeba L. fils. from Indonesia. *J Essent Oil Res.* 2007 Jan/Feb;19:14-17.
[1194] Magalhães LG, de Souza JM, Wakabayashi KA, et al. In vitro efficacy of the essential oil of Piper cubeba L. (Piperaceae) against Schistosoma mansoni. Parasitol Res. 2012 May;110(5):1747-54.
[1195] Esperandim VR, da Silva Ferreiea D, Sousa Rezende KC, et al. In Vitro Antiparasitic Activity and Chemical Composition of the Essential Oil Obtained from the Fruits of Piper cubeba. *Planta Med.* 2013;79"1653-5.
[1196] Mothana R, Alsaid M, Khaled JM, et al. Assessment of antinociceptive, antipyretic and antimicrobial activity of Piper cubeba L. essential oil in animal models. *Pak J Pharm Sci.* 2016 Mar;29(2 Suppl):671-7.

[1197] Singh G, Marimuthu P, de Heluani CS, et al. Chemical Constituents, Antioxidative and Antimicrobial Activities of Essential Oil and Oleoresin of Tailed Pepper (Piper Cubeba L). *Int J Food Engineering.* 2007;3(6):1-22.
[1198] Alharbi NS, Khaled JM, Alzaharni KE, et al. Effects of Piper cubeba L. essential oil on methicillin-resistant Staphylococcus aureus: an AFM and TEM study.
[1199] Mothana RA, Al-Said MS, Raishi M, et al. Chemical composition, anti-inflammatory and antioxidant activities of the essential oil of Piper cubeba L. *Romanian Biotech Lett.* 2017'22(2):12366-76.
[1200] Magalhães LG, de Souza JM, Wakabayashi KA, et al. In vitro efficacy of the essential oil of Piper cubeba L. (Piperaceae) against Schistosoma mansoni. Parasitol Res. 2012 May;110(5):1747-54.
[1201] Esperandim VR, da Silva Ferreiea D, Sousa Rezende KC, et al. In Vitro Antiparasitic Activity and Chemical Composition of the Essential Oil Obtained from the Fruits of Piper cubeba. *Planta Med.* 2013;79"1653-5.
[1202] Singh G, Marimuthu P, de Heluani CS, et al. Chemical Constituents, Antioxidative and Antimicrobial Activities of Essential Oil and Oleoresin of Tailed Pepper (Piper Cubeba L). *Int J Food Engineering.* 2007;3(6):1-22.
[1203] Chaubey MK. Biological activities of Zingiber officinale (Zingiberaceae) and Piper cubeba (Piperaceae) essential oils against pulse beetle, Callosobruchus chinensis (Coleoptera: Bruchidae). *Pak J Biol Sci.* 2013 Jun 1;16(11):517-23.
[1204] Villa-Ruano N, Pacheco-Hernandez Y, Arate-Reyes JA, et al. Volatile composition and biological activities of the leaf essential oil from Zanthoxylum limoncello grown in Oaxaca, México. *Chem Biodiversity.* 2018 Dec 4. [Epub ahead of print]
[1205] Villa-Ruano N, Pacheco-Hernandez Y, Arate-Reyes JA, et al. Volatile composition and biological activities of the leaf essential oil from Zanthoxylum limoncello grown in Oaxaca, México. *Chem Biodiversity.* 2018 Dec 4. [Epub ahead of print]
[1206] Villa-Ruano N, Pacheco-Hernandez Y, Arate-Reyes JA, et al. Volatile composition and biological activities of the leaf essential oil from Zanthoxylum limoncello grown in Oaxaca, México. *Chem Biodiversity.* 2018 Dec 4. [Epub ahead of print]
[1207] Tripathi YC, Anjum N, Rana A. Chemical Composition and In vitro Antifungal and Antioxidant Activities of Essential Oil from Murraya koenigii (L.) Spreng. Leaves. *Asian J Biomed Pharm Sci.* 2018;8(65):6-13.
[1208] Senthilkumar A, Gopalakrishnan B, Jayaraman M, et al. Chemical composition and antibacterial activity of essential oil from the leaves of Murraya koenigii (L.) Spreng. *J Exp Sci.* 2014;5:1-4.
[1209] Chowdhury JU, Bhuiyan NI, Yusuf M. Chemical composition of the leaf essential oils of Murraya koenigii (L.) Spreng and Murraya paniculata (L.) Jack. *Bangladesh J Pharmacol.* 2008;3:59-63.
[1210] Thi Nguyen TT, Diep TT, Hoang V, et al. Investigation of Curry Leaf Essential Oils of Murraya koenigii Spreng. Growing in the South of Vietnam. *J Essent Oil Bear Plants.* 2012;15(6):1021-9.
[1211] Syamasundar KV, Srinivasulu B, Ananda PLG, et al. Chemo variations of wild curry leaf (Murraya koenigii Spreng.) from western Ghats of India. *J Pharmacog.* 2012;3(2):126-30.
[1212] Rao BRR, Rajput DK, Mallavarapu GR. Chemical diversity in curry leaf (Murraya koenigii) essential oils. *Food Chem.* 2011;126:989-94.
[1213] Verma RS, Chauhan A, Padalia RC, et al. Phytochemical Diversity of Murraya koenigii (L.) Spreng. from Western Himalaya. *Chem Biodivers.* 2013;10:628-40.
[1214] Aslam I, Aslam F, Qamar S, et al. GC-MS analysis and in vitro antibacterial potential of volatile chemical constituents from leaves of Murraya koenigii L. Spreng. *Ann Phytomed.* 2017;6(1):143-7.
[1215] Jamil R, Nasir NN, Ramli H, et al. Extraction of essential oil from Murraya koenigii leaves: Potential study for application as natural insect repellent. *J Eng Appl Sci.* 2016 Feb;11(4):2248-52.
[1216] Rana VS, Juyal JP, Rashmi, et al. Chemical constituents of the volatile oil of Murraya koenigii leaves. *Int J Aromatherapy.* 2004;14:23-5.
[1217] Wong KC, Tie DY. The essential oil of the leaves of Murraya koenigii Spreng. *J Essent Oil Res.* 1993 Jul/Aug;5:371-4.
[1218] Ganesh PS, Vittal RR. In vitro antibiofilm activity of Murraya koenigii essential oil extracted using supercritical fluid CO_2 method against Pseudomonas aeruginosa PAO1. *Nat Prod Res.* 2015;29(24):2295-8.
[1219] Rajendran MP, Pallaiyan BB, Selvaraj N, et al. Chemical composition, antibacterial and antioxidant profile of essential oil from Murraya koenigii (L.) leaves. *Avicenna J Phytomed.* 2014 May;4(3):200-14.
[1220] Erkan N, Tao Z, Rupasinghe HP, et al. Antibacterial activities of essential oils extracted from leaves of Murraya koenigii by solvent-free microwave extraction and hydro-distillation. *Nat Prod Commun.* 2012 Jan;7(1):121-4.
[1221] Nagappan T, Ramasamy P, Wahid ME, et al. Biological activity of carbazole alkaloids and essential oil of Murraya koenigii against antibiotic resistant microbes and cancer cell lines. *Molecules.* 2011 Nov 21;16(11):9651-64.
[1222] Bonesi M, Menichini F, Tundis R, et al. Acetylcholinesterase and butyrylcholinesterase inhibitory activity of Pinus species essential oils and their constituents. *J Enzyme Inhib Med Chem.* 2010 Oct;25(5):622-8.
[1223] Aazza S, Lyoussi B, Miguel MG. Antioxidant and antiacetylcholinesterase activities of some commercial essential oils and their major compounds. *Molecules.* 2011 Sep 7;16(9):7672-90.
[1224] Miyazawa M, Yamafuji C. Inhibition of acetylcholinesterase activity by bicyclic monoterpenoids. *J Agric Food Chem.* 2005 Mar 9;53(5):1765-8.
[1225] Nagappan T, Ramasamy P, Wahid ME, et al. Biological activity of carbazole alkaloids and essential oil of Murraya koenigii against antibiotic resistant microbes and cancer cell lines. *Molecules.* 2011 Nov 21;16(11):9651-64.
[1226] Kaur I, Bhatia S, Bhati Y, et al. Augmented primary humoral immune response and decreased cell-mediated immunity by Murraya koenigii in rats. *J Basic Clin Physiol Pharmacol.* 2014 May 1;25(2):211-5.
[1227] Tripathi YC, Anjum N, Rana A. Chemical Composition and In vitro Antifungal and Antioxidant Activities of Essential Oil from Murraya koenigii (L.) Spreng. Leaves. *Asian J Biomed Pharm Sci.* 2018;8(65):6-13.
[1228] Senthilkumar A, Gopalakrishnan B, Jayaraman M, et al. Chemical composition and antibacterial activity of essential oil from the leaves of Murraya koenigii (L.) Spreng. *J Exp Sci.* 2014;5:1-4.
[1229] Aslam I, Aslam F, Qamar S, et al. GC-MS analysis and in vitro antibacterial potential of volatile chemical constituents from leaves of Murraya koenigii L. Spreng. *Ann Phytomed.* 2017;6(1):143-7.
[1230] Ganesh PS, Rai RV. Inhibition of quorum-sensing-controlled virulence factors of Pseudomonas aeruginosa by Murraya koenigii essential oil: a study in a Caenorhabditis elegans infectious model. *J Med Microbiol.* 2016 Dec;65(12):1528-1535.
[1231] Ganesh PS, Vittal RR. In vitro antibiofilm activity of Murraya koenigii essential oil extracted using supercritical fluid CO_2 method against Pseudomonas aeruginosa PAO1. *Nat Prod Res.* 2015;29(24):2295-8.
[1232] Rajendran MP, Pallaiyan BB, Selvaraj N, et al. Chemical composition, antibacterial and antioxidant profile of essential oil from Murraya koenigii (L.) leaves. *Avicenna J Phytomed.* 2014 May;4(3):200-14.
[1233] Erkan N, Tao Z, Rupasinghe HP, et al. Antibacterial activities of essential oils extracted from leaves of Murraya koenigii by solvent-free microwave extraction and hydro-distillation. *Nat Prod Commun.* 2012 Jan;7(1):121-4.

[1234] Nagappan T, Ramasamy P, Wahid ME, et al. Biological activity of carbazole alkaloids and essential oil of Murraya koenigii against antibiotic resistant microbes and cancer cell lines. *Molecules.* 2011 Nov 21;16(11):9651-64.
[1235] Tripathi YC, Anjum N, Rana A. Chemical Composition and In vitro Antifungal and Antioxidant Activities of Essential Oil from Murraya koenigii (L.) Spreng. Leaves. *Asian J Biomed Pharm Sci.* 2018;8(65):6-13.
[1236] Rajendran MP, Pallaiyan BB, Selvaraj N, et al. Chemical composition, antibacterial and antioxidant profile of essential oil from Murraya koenigii (L.) leaves. *Avicenna J Phytomed.* 2014 May;4(3):200-14.
[1237] Erkan N, Tao Z, Rupasinghe HP, et al. Antibacterial activities of essential oils extracted from leaves of Murraya koenigii by solvent-free microwave extraction and hydro-distillation. *Nat Prod Commun.* 2012 Jan;7(1):121-4.
[1238] Jamil R, Nasir NN, Ramli H, et al. Extraction of essential oil from Murraya koenigii leaves: Potential study for application as natural insect repellent. *J Eng Appl Sci.* 2016 Feb;11(4):2248-52.
[1239] Reddy SG, Kirti Dolma S, Koundal R, et al. Chemical composition and insecticidal activities of essential oils against diamondback moth, Plutella xylostella (L.) (Lepidoptera: Yponomeutidae). *Nat Prod Res.* 2016 Aug;30(16):1834-8.
[1240] Clery RA, Cason JRL, Zelenay V. Constituents of Cypriol Oil (Cyperus scariosus R.Br.): N-Containing Molecules and Key Aroma Components. *J Agric Food Chem.* 2016;64:4566-73.
[1241] Srivastava RK, Singh A, Srivastava GP, et al. Chemical Constituents and Biological Activities of Promising Aromatic Plant Nagarmotha (Cyperus scariosus R.Br.): A Review. *Proc Indian Natn Sci Acad.* 2014 Sep;80(3):525-36.
[1242] Kumar A, Chahal KK, Kataria D. A review on phytochemistry and pharmacological activities of Cyperus scariosus. *J Pharmacognosy Phytochem.* 2017;6(1):510-17.
[1243] Kumar A, Niranjan A, Lehri A, et al. Effect of Geographical Climatic Conditions on Yield, Chemical Composition and Carbon Isotope Composition of Nagarmotha (Cyperus scariosus R. Br.) Essential Oil. *J Essent Oil Bearing Plants.* 2016;19(2):368-73.
[1244] Pandey AK, Chowdhury AR. Essential oil of Cyperus scariosus R. Br. tubers from Central India. *Indian Perf.* 2002;46(4):325-8.
[1245] Gupta SK, Sharma RC, Aggarwal OP, et al. Anti-inflammatory activity of the oil isolated from Cyperus scariosus (R. Br.). *Indian J Exp Biol.* 1972 Jan;10(1):41-2.
[1246] Bekana D, Kebede T, Assefa M, et al. Comparative Phytochemical Analyses of Resins of Boswellia Species (B. papyrifera (Del.) Hochst., B. neglecta S. Moore, and B. rivae Engl.) from Northwestern, Southern, and Southeastern Ethiopia. *ISRN Analytical Chem.* 2014;2014:374678.
[1247] Baser KHC, Demirci B, Dekebo A, et al. John Wiley & Sons, Ltd. Essential oils of some Boswellia spp., Myrrh and Opopanax. *Flavour Frag J.* 2003;18:153-6.
[1248] Asefa M, Shimelis B, Kassa H, et al. Chemical investigations on frankincense from Boswellia trees to improve production, handling and grading practices for the export market in Ethiopia. *G J P&A Tech.* 2012;02(5):15-30.
[1249] Provan GJ, Gray AI, Waterman PG. Monoterpene-rich Resins from some Kenyan Burseraceae. *Flavour Frag J.* 1987;2:115-8.
[1250] Basar S. Phytochemical investigations of Boswellia species: Comparative studies on the essential oils, pyrolysates and boswellic acids of Boswellia carterii Birdw., Boswellia serrata Roxb., Boswellia frereana Birdw., Boswellia neglecta S. Moore and Boswellia rivae Engl. 2005 thesis.
[1251] Van Vuuren SF, Kamatou GPP, Viljoen AM. Volatile composition and antimicrobial activity of twenty commercial frankincense essential oil samples. *South African J Botany.* 2010;76:686-91.
[1252] Van Vuuren SF, Kamatou GPP, Viljoen AM. Volatile composition and antimicrobial activity of twenty commercial frankincense essential oil samples. *South African J Botany.* 2010;76:686-91.
[1253] de Rapper S, Van Vuuren SF, Kamatou GP, et al. The additive and synergistic antimicrobial effects of select frankincense and myrrh oils--a combination from the pharaonic pharmacopoeia. *Lett Appl Microbiol.* 2012 Apr;54(4):352-8.
[1254] DeCarlo A, Johnson S, Okeke-Aula KI, et al. Compositional analysis of the essential oil of Boswellia dalzielii frankincense from West Africa reveals two major chemotypes. *Phytochem.* 2019;164:24-32.
[1255] DeCarlo A, JohnsonS, Ouedraogo A, et al. Chemical Composition of the Oleogum Resin Essential Oils of Boswellia dalzielii from Burkina Faso. *Plants.* 2019;8:223.
[1256] Kohoude MJ, Gbaguidi F, Agbani P, et al. Chemical composition and biological activities of extracts and essential oil of Boswellia dalzielii leaves. *Pharm Biol.* 2017;55:33–42.
[1257] Kubmarawa D, Ogunwande IA, Okorie DA, et al. Constituents of the essential oils of Boswellia dalzielii Hutch. from Nigeria. *J Essent Oil Res.* 2006;18:119–120.
[1258] Frank A, Unger M. Analysis of frankincense from various Boswellia species with inhibitory activity on human drug metabolizing cytochrome P450 enzymes using liquid chromatography mass spectrometry after automated on-line extraction. *J Chromatogra A.* 2006 Apr 21;1112(1-2):255-62.
[1259] Jardat N, Adwan L, K'aibni S, et al. Chemical composition, anthelmintic, antibacterial and antioxidant effects of Thymus bovei essential oil. *Complement Altern Med.* 2016;16:418.
[1260] Hassan STS, Berchova-Bimova K, Sudomova M, et al. In Vitro Study of Multi-Therapeutic Properties of Thymus bovei Benth. Essential Oil and Its Main Component for Promoting Their Use in Clinical Practice. *J Clin Med.* 2018;7:283.
[1261] Seo KA, Kim H, Ku HY, et al. The monoterpenoids citral and geraniol are moderate inhibitors of the CYP2B6 hydroxylase activity. *Chem Biol Interact.* 2008;174:141-46.
[1262] Ibrahim SM, El-Denshary ES, Abdallah DM. Geraniol, Alone and in Combination with Pioglitazone, Ameliorates Fructose-Induced Metabolic Syndrome in Rats via the Modulation of Both Inflammatory and Oxidative Stress Status. *PLoS One.* 2015 Feb 13;10(2):e0117516.
[1263] Srinivasan S, Muruganathan U. Antidiabetic efficacy of citronellol, a citrus monoterpene by ameliorating the hepatic key enzymes of carbohydrate metabolism in streptozotocin-induced diabetic rats. *Chem Biol Interact.* 2016 Apr 25;250:38-46.
[1264] Shin S, Lim S. Antifungal effects of herbal essential oils alone and in combination with ketoconazole against Trichophyton spp. *J Appl Micribiol.* 2004;97:1289-96.
[1265] Hagvall L. Cytochrome P450-mediated activation of the fragrance constituent geraniol forms potent contact allergens. *Tox Appl Pharm.* 2008 Dec;233(2):308-13.
[1266] Hassan STS, Berchova-Bimova K, Sudomova M, et al. In Vitro Study of Multi-Therapeutic Properties of Thymus bovei Benth. Essential Oil and Its Main Component for Promoting Their Use in Clinical Practice. *J Clin Med.* 2018;7:283.
[1267] Hassan STS, Berchova-Bimova K, Sudomova M, et al. In Vitro Study of Multi-Therapeutic Properties of Thymus bovei Benth. Essential Oil and Its Main Component for Promoting Their Use in Clinical Practice. *J Clin Med.* 2018;7:283.
[1268] Hassan STS, Berchova-Bimova K, Sudomova M, et al. In Vitro Study of Multi-Therapeutic Properties of Thymus bovei Benth. Essential Oil and Its Main Component for Promoting Their Use in Clinical Practice. *J Clin Med.* 2018;7:283.
[1269] Jardat N, Adwan L, K'aibni S, et al. Chemical composition, anthelmintic, antibacterial and antioxidant effects of Thymus bovei essential oil. *Complement Altern Med.* 2016;16:418.

[1270] Jardat N, Adwan L, K'aibni S, et al. Chemical composition, anthelmintic, antibacterial and antioxidant effects of Thymus bovei essential oil. *Complement Altern Med.* 2016;16:418.

[1271] Jardat N, Adwan L, K'aibni S, et al. Chemical composition, anthelmintic, antibacterial and antioxidant effects of Thymus bovei essential oil. *Complement Altern Med.* 2016;16:418.

[1272] Behbahani BA, Yazdi FT, Vasiee A, et al. Oliveria decumbens essential oil: Chemical compositions and antimicrobial activity against the growth of some clinical and standard strains causing infection. *Microbial Pathogen.* 2018;114:449-52.

[1273] Sajjadi SE, Hoseini SA. Essential Oil Constituents of Oliveria decumbens Vent. *J Essent Oil Res.* 2002 May/Jun;14:220-1.

[1274] Hajimehdipoor H, Samadi N, Mozaffarian V, et al. Chemical Composition and Antimicrobial Activity of Oliveria decumbens Volatile Oil from West of Iran. *J Med Plants.* 2010 Winter;9(Sup6):39-44.

[1275] Mahboubi M, Mohammadi-Yeganeh S, Bokaee S, et al. Antimicrobial activity of essential oil from Oliveria decumbens and its synergy with vancomycin against Staphylococcus aureus. *Herba Polonica.* 2007;53(4):69-76.

[1276] Khajehie N, Golmakani MT, Eblaghi M, et al. Evaluating the Effects of Microwave-Assisted Hydrodistillation on Antifungal and Radical Scavenging Activities of Oliveria decumbens and Chaerophyllum macropodum Essential Oils. *J Food Protect.* 2017;80(5):783-91.

[1277] Amin G, Sourmaghi MHS, Zahedi M, et al. Essential oil composition and antimicrobial activity of Oliveria decumbens. *Fitoterapia.* 2005;76:704-7.

[1278] Saidi M. Antioxidant Activities and Chemical Composition of Essential Oils from Satureja khuzestanica, Oliveria decumbens and Thymus daenensis. *J Essent Oil Bearing Plants.* 2014;17(3):513-21.

[1279] Sereshti H, Izadmanesh Y, Samadi S. Optimized ultrasonic assisted extraction-dispersive liquid-liquid microextraction coupled with gas chromatography for determination of essential oil of Oliveria decumbens Vent. *J Chromatogr A.* 2011 Jul 22;1218(29):4593-8.

[1280] Kohlert C, Schindler G, Marz RW, et al. Systemic availability and pharmacokinetics of thymol in humans. *J Clin Pharmacol.* 2002 Jul;42(7):731-7.

[1281] Toxicology Data Network, National Library of Medicine. Thymol. Available at: http://toxnet.nlm.nih.gov/cgibin/sis/search/a?dbs+hsdb:@term+@DOCNO+866.

[1282] Tognolini M, Barocelli E, Ballabeni V, et al. Comparative screening of plant essential oils; phenylpropanoid moiety as basic core for antiplatelet activity. *Life Sci.* 2006 Feb 23;78(13):1419-32.

[1283] Okazaki K, Kawazoe K, Takaishi Y. Human platelet aggregation inhibitors from thyme (Thymus vulgaris L.). *Phytother Res.* 2002 Jun;16(4):398-9.

[1284] Jukic M, Politeo O, Maksimmovic M, et al. In vitro acetylcholinesterase inhibitory properties of thymol, carvacrol and their derivatives thymoquinone and thymohydroquinone. *Phytother Res.* 2007;21(3):259-61.

[1285] Askin H, Yildiz M, Ayar A. Effects of Thymol and Carvacrol on Acetylcholinesterase from Drosophila melanogaster. *Acta Physica Polinica A.* 2017;132(3):720-22.

[1286] Medeiros Barreto H, Cerqueira Fontinele F, Pereira de Oliveira A, et al. Phytochemical prospection and modulation of antibiotic activity in vitro by Lippia origanoides H.B.K. in methicillin resistant Staphylococcus aureus. *Biomed Res Int.* 2014;305610:1–7.

[1287] Aelenei P, Miron A, Trifan A, et al. Essential Oils and Their Components as Modulators of Antibiotic Activity against Gram-Negative Bacteria. *Medicines.* 2016;3(19):2-34.

[1288] Magi G, Marini E, Facinelli B. Antimicrobial activity of essential oils and carvacrol, and synergy of carvacrol and erythromycin, against clinical, erythromycin-resistant Group A Streptococci. *Front Microbiol.* 2015;6:165.

[1289] Veras HNH, Rodrigues FFG, Colares AV, et al. Synergistic antibiotic activity of volatile compounds from the essential oil of Lippia sidoides and thymol. *Fitoterapia.* 2012 Apr;83(3):508-12.

[1290] Langeveld WT, Veldhuizen EJ, Burt SA. Synergy between essential oil components and antibiotics: A review. *Crit Rev Microbiol.* 2014 Feb;40(1):76-94.

[1291] Miladinovic DL, Ilic BS, Kocic BD, et al. Antibacterial investigation of thyme essential oil and its main constituents on combination with tetracycline. *J Med Food.* 2015 Aug;18(8):935-7.

[1292] Shin S, Lim S. Antifungal effects of herbal essential oils alone and in combination with ketoconazole against Trichophyton spp. *J Appl Micribiol.* 2004;97:1289-96.

[1293] Kissels W, Wu X, Santos RR, et al. Short communication: Interaction of the isomers carvacrol and thymol with the antibiotics doxycycline and tilmicosin: In vitro effects against pathogenic bacteria commonly found in the respiratory tract of calves. *J Dairy Sci.* 2017 Feb;100(2):970-974.

[1294] Mahboubi M, Taghizadeh M, Khamechian T, et al. The Wound Healing Effects of Herbal Cream Containing Oliveria Decumbens and Pelargonium Graveolens Essential Oils in Diabetic Foot Ulcer Model. *World J Plast Surg.* 2018 Jan;7(1):45–50.

[1295] Behbahani BA, Yazdi FT, Vasiee A, et al. Oliveria decumbens essential oil: Chemical compositions and antimicrobial activity against the growth of some clinical and standard strains causing infection. *Microbial Pathogen.* 2018;114:449-52.

[1296] Mahboubi M, Mohammadi-Yeganeh S, Bokaee S, et al. Antimicrobial activity of essential oil from Oliveria decumbens and its synergy with vancomycin against Staphylococcus aureus. *Herba Polonica.* 2007;53(4):69-76.

[1297] Hajimehdipoor H, Samadi N, Mozaffarian V, et al. Chemical Composition and Antimicrobial Activity of Oliveria decumbens Volatile Oil from West of Iran. *J Med Plants.* 2010 Winter;9(Sup6):39-44.

[1298] Mahboubi M, Kazempour N, Mahboubi A. The efficacy of essential oils as natural preservatives in vegetable oil. *J Diet Suppl.* 2014 Dec;11(4):334-46.

[1299] Khajehie N, Golmakani MT, Eblaghi M, et al. Evaluating the Effects of Microwave-Assisted Hydrodistillation on Antifungal and Radical Scavenging Activities of Oliveria decumbens and Chaerophyllum macropodum Essential Oils. *J Food Protect.* 2017;80(5):783-91.

[1300] Khajehie N, Golmakani MT, Eblaghi M, et al. Evaluating the Effects of Microwave-Assisted Hydrodistillation on Antifungal and Radical Scavenging Activities of Oliveria decumbens and Chaerophyllum macropodum Essential Oils. *J Food Protect.* 2017;80(5):783-91.

[1301] Saidi M. Antioxidant Activities and Chemical Composition of Essential Oils from Satureja khuzestanica, Oliveria decumbens and Thymus daenensis. *J Essent Oil Bearing Plants.* 2014;17(3):513-21.

[1302] Samadi N, Masoum S, Mehrara B, et al. Application of linear multivariate calibration techniques to identify the peaks responsible for the antioxidant activity of Satureja hortensis L. and Oliveria decumbens Vent. essential oils by gas chromatography-mass spectrometry. *J Chromatogr B Analyt Technol Biomed Life Sci.* 2015 Sep 15;1001:75-81.

[1303] Giatropoulos A, Kimbaris A, Michaelakis A, et al. Chemical composition and assessment of larvicidal and repellent capacity of 14 Lamiaceae essential oils against Aedes albopictus. *Parasitol Res.* 2018 Jun;117(6):1953-1964.

[1304] Mitropoulou G, Fitsiou E, Stavropoulou E, et al. Composition, antimicrobial, antioxidant, and antiproliferative activity of Origanum dictamnus (dittany) essential oil. *Microbial Ecol Health Dis.* 2015;26:26543.
[1305] Stavropoulou A, Loulakakis K, Magan N, et al. Origanum dictamnus Oil Vapour Suppresses the Development of Grey Mould in Eggplant Fruit In Vitro. *BioMed Res Int.* 2014;2014:562679.
[1306] Argyropoulou C, Papadatou M, Grigoriadou C, et al. Evaluation of the Essential Oil Content of Cretan Dittany Cultivated in Northern Greece. *Med Aromat Plants.* 2014;3(2):1000157.
[1307] Karousou R, Kokkini S. The genus Origanum (Labiatae) in Crete: distribution and essential oils. *Bocconea.* 2003;16(2):717-21.
[1308] Chorianopoulos N, Kalpoutzakis E, Aligiannis N, et al. Essential Oils of Satureja, Origanum, and Thymus Species: Chemical Composition and Antibacterial Activities Against Foodborne Pathogens. *J Agric Food Chem.* 2004;52:8261-7.
[1309] Sivropoulou A, Papanikolaou E, Nikolaou C, et al. Antimicrobial and Cytotoxic Activities of Origanum Essential Oils. *J Agric Food Chem.* 1996;44:1202-5.
[1310] Economakis C, Demetzos C, Anastassaki T, et al. Volatile Constituents of Bracts and Leaves of Wild and Cultivated Origanum dictamnus. *Planta Med.* 1999;65:189-91.
[1311] Economakis C, Karioti A, Skaltsa H, et al. Effect of solution conductivity on the volatile constituents of Origanum dictamnus L. in nutrient film culture. *J Agric Food Chem.* 2005 Mar 9;53(5):1656-60.
[1312] Marrelli M, Conforti F, Formisano C, et al. Composition, antibacterial, antioxidant and antiproliferative activities of essential oils from three Origanum species growing wild in Lebanon and Greece. *Nat Prod Res.* 2016;30(6):735-9.
[1313] Economakis C, Skaltsa H, Demetzos C, et al. Effect of phosphorus concentration of the nutrient solution on the volatile constituents of leaves and bracts of Origanum dictamnus. *J Agric Food Chem.* 2002;50(22):6276-6280.
[1314] Economakis C, Karioti A, Skaltsa H, et al. Effect of solution conductivity on the volatile constituents of Origanum dictamnus L. in nutrient film culture. *J Agric Food Chem.* 2005 Mar 9;53(5):1656-60.
[1315] Economakis C, Skaltsa H, Demetzos C, et al. Effect of phosphorus concentration of the nutrient solution on the volatile constituents of leaves and bracts of Origanum dictamnus. *J Agric Food Chem.* 2002;50(22):6276-6280.
[1316] Domaracky M, Rehak P, Juhas S, et al. Effects of selected plant essential oils on the growth and development of mouse preimplantation embryos in vivo. *Physiol Res.* 2007;56(1):97-104.
[1317] Lemhadri A, Zeggwagh NA, Maghrani M, et al. Anti-hypoglycaemic activity of the aqueous extract of Origanum vulgare growing wild in Tafilalet region. *J Ethnopharmacol.* 2004 Jun;92(2-3):251-6.
[1318] Bayramoglu G, Senturk H, Bayramoglu A, et al. Carvacrol partially reverses symptoms of diabetes in STZ-induced diabetic rats. *Cytotechnology.* 2014 Mar;66(2):251–257.
[1319] Ezhumalai M, Radhiga T, Pugalendi KV. Antihyperglycemic effect of carvacrol in combination with rosiglitazone in high-fat diet-induced type 2 diabetic C57BL/6J mice. *Mol Cell Biochem.* 2014 Jan;385(1-2):23-31.
[1320] Umasankar K, Ramya DR, Nambikkairaj B. Antidiabetic and antidyslipidemic nature of carvacrol, a monoterpenic phenol studied in high-fat - fed and low-dose streptozotocin-induced experimental diabetic rats. *J Pharm Res.* 2015;9(8):484-90.
[1321] Tognolini M, Barocelli E, Ballabeni V, et al. Comparative screening of plant essential oils; phenylpropanoid moiety as basic core for antiplatelet activity. *Life Sci.* 2006 Feb 23;78(13):1419-32.
[1322] Tsai HH, Lin HW, Chen YL, et al. A review of potential harmful interactions between anticoagulant/antiplatelet agents and Chinese herbal medicines. *PLoS One.* 2013 May 9;8(5):e64255.
[1323] Karkabounas S, Kostoula OK, Daskalou T, et al. Anticarcinogenic and antiplatelet effects of carvacrol. *Exp Oncol.* 2006 Jun;28(2):121-5.
[1324] Baser KH. Biological and pharmacological activities of carvacrol and carvacrol bearing essential oils. *Curr Pharm Des.* 2008;14(29):3106-19.
[1325] Okazaki K, Kawazoe K, Takaishi Y. Human platelet aggregation inhibitors from thyme (Thymus vulgaris L.). *Phytother Res.* 2002 Jun;16(4):398-9.
[1326] Jukic M, Politeo O, Maksimmovic M, et al. In vitro acetylcholinesterase inhibitory properties of thymol, carvacrol and their derivatives thymoquinone and thymohydroquinone. *Phytother Res.* 2007;21(3):259-61.
[1327] Askin H, Yildiz M, Ayar A. Effects of Thymol and Carvacrol on Acetylcholinesterase from Drosophila melanogaster. *Acta Physica Polinica A.* 2017;132(3):720-22.
[1328] Medeiros Barreto H, Cerqueira Fontinele F, Pereira de Oliveira A, et al. Phytochemical prospection and modulation of antibiotic activity in vitro by Lippia origanoides H.B.K. in methicillin resistant Staphylococcus aureus. *Biomed Res Int.* 2014;305610:1–7.
[1329] Aelenei P, Miron A, Trifan A, et al. Essential Oils and Their Components as Modulators of Antibiotic Activity against Gram-Negative Bacteria. *Medicines.* 2016;3(19):2-34.
[1330] Magi G, Marini E, Facinelli B. Antimicrobial activity of essential oils and carvacrol, and synergy of carvacrol and erythromycin, against clinical, erythromycin-resistant Group A Streptococci. *Front Microbiol.* 2015;6:165.
[1331] Veras HNH, Rodrigues FFG, Colares AV, et al. Synergistic antibiotic activity of volatile compounds from the essential oil of Lippia sidoides and thymol. *Fitoterapia.* 2012 Apr;83(3):508-12.
[1332] Langeveld WT, Veldhuizen EJ, Burt SA. Synergy between essential oil components and antibiotics: A review. *Crit Rev Microbiol.* 2014 Feb;40(1):76-94.
[1333] Mitropoulou G, Fitsiou E, Stavropoulou E, et al. Composition, antimicrobial, antioxidant, and antiproliferative activity of Origanum dictamnus (dittany) essential oil. *Microbial Ecol Health Dis.* 2015;26:26543.
[1334] Marrelli M, Conforti F, Formisano C, et al. Composition, antibacterial, antioxidant and antiproliferative activities of essential oils from three Origanum species growing wild in Lebanon and Greece. *Nat Prod Res.* 2016;30(6):735-9.
[1335] Koutroumanidou E, Kimbaris A, Kortsaris A, et al. Increased seizure latency and decreased severity of pentylenetetrazol-induced seizures in mice after essential oil administration. *Epilepsy Res Treat.* 2013;2013:532657.
[1336] Mitropoulou G, Fitsiou E, Stavropoulou E, et al. Composition, antimicrobial, antioxidant, and antiproliferative activity of Origanum dictamnus (dittany) essential oil. *Microbial Ecol Health Dis.* 2015;26:26543.
[1337] Chorianopoulos N, Kalpoutzakis E, Aligiannis N, et al. Essential Oils of Satureja, Origanum, and Thymus Species: Chemical Composition and Antibacterial Activities Against Foodborne Pathogens. *J Agric Food Chem.* 2004;52:8261-7.
[1338] Marrelli M, Conforti F, Formisano C, et al. Composition, antibacterial, antioxidant and antiproliferative activities of essential oils from three Origanum species growing wild in Lebanon and Greece. *Nat Prod Res.* 2016;30(6):735-9.
[1339] Stavropoulou A, Loulakakis K, Magan N, et al. Origanum dictamnus Oil Vapour Suppresses the Development of Grey Mould in Eggplant Fruit In Vitro. *BioMed Res Int.* 2014;2014:562679.
[1340] Mitropoulou G, Fitsiou E, Stavropoulou E, et al. Composition, antimicrobial, antioxidant, and antiproliferative activity of Origanum dictamnus (dittany) essential oil. *Microbial Ecol Health Dis.* 2015;26:26543.

[1341] Marrelli M, Conforti F, Formisano C, et al. Composition, antibacterial, antioxidant and antiproliferative activities of essential oils from three Origanum species growing wild in Lebanon and Greece. *Nat Prod Res*. 2016;30(6):735-9.
[1342] Ntalli NG, Ferrari F, Giannakou I, et al. Phytochemistry and nematicidal activity of the essential oils from 8 Greek Lamiaceae aromatic plants and 13 terpene components. *J Agric Food Chem*. 2010 Jul 14;58(13):7856-63.
[1343] Giatropoulos A, Kimbaris A, Michaelakis A, et al. Chemical composition and assessment of larvicidal and repellent capacity of 14 Lamiaceae essential oils against Aedes albopictus. *Parasitol Res*. 2018 Jun;117(6):1953-1964.
[1344] Said-Al Ahl HAH, Sabra AS, El Gendy ANG, et al. Changes in content and chemical composition of Dracocephalum moldavica L. essential oil at different harvest dates. *J Med Plants Studies*. 2015;3(2):61-4.
[1345] Golpravar AR, Hadipanah A, Gheisari MM, et al. Chemical constituents of essential oil of Dracocephalum moldavica L. and Dracocephalum kotschyi Boiss. from Iran. *Acta Agriculturae Slovenica*. 2016 Apr;107(1):25-31.
[1346] Aziz EE, Hussein MS, Wahba HE, et al. Essential Oil Constituents of Dracocephalum moldavica L. Grown under Salt Stress and Different Sources of Soil Amendment. *Middle-East J Sci Res*. 2013;16(5):706-13.
[1347] Yousefzadeh S, Modarres-Sanavy S, Sefidkon F, et al. Effects of Azocompost and urea on the herbage yield and contents and compositions of essential oils from two genotypes of dragonhead (Dracocephalum moldavica L.) in two regions of Iran. *Food Chem*. 2013;138(2):1407-13.
[1348] Kakasy A, Lemberkovics B, Simándi B, et al. Comparative study of traditional essential oil and supercritical fluid extracts of Moldavian dragonhead (Dracocephalum moldavica L.). Flav Fragrance J. 2006;21(4):598-603.
[1349] Nogueira AC, Carvalho RR, Souza CA, et al. Study on the embryofeto-toxicity of citral in the rat. *Toxicology*. 1995;96(2):105-13.
[1350] Seo KA, Kim H, Ku HY, et al. The monoterpenoids citral and geraniol are moderate inhibitors of the CYP2B6 hydroxylase activity. *Chem Biol Interact*. 2008;174:141-46.
[1351] Seo KA, Kim H, Ku HY, et al. The monoterpenoids citral and geraniol are moderate inhibitors of the CYP2B6 hydroxylase activity. *Chem Biol Interact*. 2008;174:141-46.
[1352] Sheweita SA, Newairy AA, Mansour HA, et al. Effect of some hypoglycemic herbs on the activity of phase I and II drug-metabolizing enzymes in alloxan-induced diabetic rats. *Toxicology*. 2002 May 24;17(2):131-39.
[1353] Raner GM, Vaz AD, Coon MJ. Metabolism of all-trans, 9-cis, and 13-cis isomers of retinal by purified isozymes of microsomal cytochrome P450 and mechanism-based inhibition of retinoid oxidation by citral. *Mol Pharmacol*. 1996;49(3):515-22.
[1354] De Olivera AC, Ribeiro-Pinto LF, Paumgartten JR. In vitro inhibition of CYP2B1 monoxygenase by beta-myrcene and other monoterpenoid constituents. Br J Nutr. 1999;81:289-95.
[1355] Ibrahim SM, El-Denshary ES, Abdallah DM. Geraniol, Alone and in Combination with Pioglitazone, Ameliorates Fructose-Induced Metabolic Syndrome in Rats via the Modulation of Both Inflammatory and Oxidative Stress Status. *PLoS One*. 2015 Feb 13;10(2):e0117516.
[1356] Babukumar S, Vinothkumar V, Sankaranarayanan C, et al. Geraniol, a natural monoterpene, ameliorates hyperglycemia by attenuating the key enzymes of carbohydrate metabolism in streptozotocin-induced diabetic rats. *Pharm Biol*. 2017 Dec;55(1):1442-1449.
[1357] Modak T, Mukhopadhaya A. Effects of citral, a naturally occurring antiadipogenic molecule, on an energy-intense diet model of obesity. *Indian J Pharmacol*. 2011 May-Jun;43(3):300-05.
[1358] Adeneye AA, Agbaje EO. Hypoglycemic and hypolipidemic effects of fresh leaf aqueous extract of Cymbopogon citratus Stapf. in rats. *J Ethnopharmacol*. 2007 Jul 25;112(3):440-4.
[1359] Shin S, Lim S. Antifungal effects of herbal essential oils alone and in combination with ketoconazole against Trichophyton spp. *J Appl Micribiol*. 2004;97:1289-96.
[1360] Lorenzi V, Muselli A, Bernardini AF, et al. Geraniol restores antibiotic activities against multidrug-resistant isolates from gram-negative species. *Antimicrob Agents Chemother*. 2009 May;53(5):2209-11.
[1361] Li L, Li ZW, Yin ZQ, et al. Antibacterial activity of leaf essential oil and its constituents from Cinnamomum longepaniculatum. *Int J Clin Exp Med*. 2014;7(7):1721-7.
[1362] do Vale TG, Furtado EC, Santos JG Jr, et al. Central effects of citral, myrcene and limonene, constituents of essential oil chemotypes from Lippia alba (Mill.) n.e. Brown. *Phytomedicine*. 2002 Dec;9(8):709-14.
[1363] Hagvall L. Cytochrome P450-mediated activation of the fragrance constituent geraniol forms potent contact allergens. *Tox Appl Pharm*. 2008 Dec;233(2):308-13.
[1364] Kuhn GO, McCampbell P, Singmaster G, et al. Application of microencapsulation technology to improve the stability of citral in rodent diets. *Fundam Appl Toxicol*. 1991 Oct;17(3):635-40.
[1365] De Mozzi P, Johnston GA. An outbreak of allergic contact dermatitis caused by citral in beauticians working in a health spa. *Contact Dermatitis*. 2014 Jun;70(6):377-9.
[1366] do Vale TG1, Furtado EC, Santos JG Jr, Vet al. Central effects of citral, myrcene and limonene, constituents of essential oil chemotypes from Lippia alba (Mill.) n.e. Brown. *Phytomedicine*. 2002 Dec;9(8):709-14.
[1367] Maham M, Akbari H, Delazar A. Chemical Composition and Antinociceptive Effect of the Essential Oil of Dracocephalum moldavica L. *Pharmaceutical Sci*. 2013;18(4):187-92.
[1368] Judzentiene A, Butkiene R. Chemical Composition of the Essential Oils of Wild Helichrysum arenarium (L.) with Differently Colored Inflorescences from Eastern Lithuania. *J Essent Oil Res*. 2006;18(1):80-83.
[1369] Lemberkovics E, Czinner E, Balazs A, et al. New data on composition of essential oil from inflorescence of everlasting (Helichrysum arenarium (L.) Moench). *Acta Pharm Hungarica*. 2001 Aug;71:187-91.
[1370] Radusiene J, Judzentiene A. Volatile composition of Helichrysum arenarium field accessions with differently coloured inflorescences. *Biologija*. 2008;54(2):116-20.
[1371] Rancic A, Sokovic M, Vukojevic J, et al. Chemical Composition and Antimicrobial Activities of Essential Oils of Myrrhis odorata (L.) Scop, Hypericum perforatum L and Helichrysum arenarium (L.) Moench. *J Essent Oil Res*. 2005 May/Jun;17:341-5.
[1372] Noori M, Poodineh M, Hakimzadeh V, et al. Composition of Helichrysum arenarium essential oil and antimicrobial activity against some food-borne pathogens. *Biotechnol*. 2015;11(4):121-5.
[1373] Bihatsi-Karsai E, Vitanyi G, Lelik L. Composition of the Essential Oil from the Inflorescence of Helichrysum arenarium(L.) Moench. *J Essent Oil Res*. 2000 Nov/Dec;12:728-30.
[1374] Rancic A, Sokovic M, Vukojevic J, et al. Chemical Composition and Antimicrobial Activities of Essential Oils of Myrrhis odorata (L.) Scop, Hypericum perforatum L and Helichrysum arenarium (L.) Moench. *J Essent Oil Res*. 2005 May/Jun;17:341-5.
[1375] Noori M, Poodineh M, Hakimzadeh V, et al. Composition of Helichrysum arenarium essential oil and antimicrobial activity against some food-borne pathogens. *Biotechnol*. 2015;11(4):121-5.
[1376] Sani AM. Inhibitory Effect of Helichrysum arenarium Essential Oil on the Growth of Food Contaminated Microorganisms. *Int J Agric Biosys Eng*. 2014;8(8):839-43.

[1377] Gao X, Wei J, Hong L, et al. Comparative Analysis of Chemical Composition, Anti-Inflammatory Activity and Antitumor Activity in Essential Oils from Siegesbeckia orientalis, S. glabrescens and S. pubescens with an ITS Sequence Analysis. *Molecules*. 2018;23:2185.
[1378] Gao X, Wei J, Hong L, et al. Comparative Analysis of Chemical Composition, Anti-Inflammatory Activity and Antitumor Activity in Essential Oils from Siegesbeckia orientalis, S. glabrescens and S. pubescens with an ITS Sequence Analysis. *Molecules*. 2018;23:2185.
[1379] Gao X, Wei J, Hong L, et al. Comparative Analysis of Chemical Composition, Anti-Inflammatory Activity and Antitumor Activity in Essential Oils from Siegesbeckia orientalis, S. glabrescens and S. pubescens with an ITS Sequence Analysis. *Molecules*. 2018;23:2185.
[1380] Nyalambisa M, Oyemitan IA, Matewu R, et al. Volatile constituents and biological activities of the leaf and root of Echinacea species from South Africa. *Saudi Pharm J*. 2017;25:381-6.
[1381] Diraz E, Karaman S, Koca N. Fatty Acıd and Essentıal Oıl Composıtıon of Echınacea Purpurea (L.) Moench, Growıng in Kahramanmaras-Turkey. *Int Conf Environ Biol Sci*. 2016 Dec;21-22:35-37.
[1382] Hudaib M, Cavrini V, Bellardi MG, et al. Characterization of the Essential Oils of Healthy and Virus Infected Echinacea purpurea (L.) Moench Plants. *J Essent Oil Res*. 2002;14(6):427-30.
[1383] Nyalambisa M, Oyemitan IA, Matewu R, et al. Volatile constituents and biological activities of the leaf and root of Echinacea species from South Africa. *Saudi Pharm J*. 2017;25:381-6.
[1384] Vaverkova S, Mikulasova M., Haban M, et al. Variability of the essential oil from three sorts of Echinacea MOENCH genus during ontogenesis. *Ceska A Slovenska Farmacie*. 2007;56(c.3):121-4.
[1385] Delgado IF, Carvalho RR, Nogueira, et al. Study on embryofetotoxicity of b-myrcene in the rat. *Food and Chemical Toxicology*. 1993;31(1):31-5.
[1386] Paumgartten FJ, De-Carvalho RR, Souza CA, et al. Study of the effects of beta-myrcene on rat fertility and general reproductive performance. *Braz J Med Biol Res*. 1998 Jul;31(7):955-65.
[1387] do Vale TG, Furtado EC, Santos JG Jr, et al. Central effects of citral, myrcene and limonene, constituents of essential oil chemotypes from Lippia alba (Mill.) n.e. Brown. *Phytomedicine*. 2002 Dec;9(8):709-14.
[1388] Segelman AB, Sofia RD, Segelman FP, et al. Cannabis sativa L. (marijuana). V. Pharmacological evaluation of marijuana aqueous extract and volatile oil. *J Pharm Sci*. 1974 Jun;63(6):962-4.
[1389] Freitas JC1, Presgrave OA, Fingola FF, et al. Effect of beta-myrcene on pentobarbital sleeping time. *Braz J Med Biol Res*. 1993 May;26(5):519-23.
[1390] De Olivera AC, Ribeiro-Pinto LF, Paumgartten JR. In vitro inhibition of CYP2B1 monoxygenase by beta-myrcene and other monoterpenoid constituents. *Br J Nutr*. 1999;81:289-95.
[1391] De-Oliveira AC1, Ribeiro-Pinto LF, Otto SS, et al. Induction of liver monooxygenases by beta-myrcene. *Toxicology*. 1997 Dec 26;124(2):135-40.
[1392] Yu D, Yuan Y, Jiang L, et al. Anti-inflammatory effects of essential oil in Echinacea purpurea L. *Pak J Pharm Sci*. 2013 Mar;26(2):403-8.
[1393] Tucker AO, Maciarello MJ, Brown RC. Essential Oils from the Oleo-gum-resins of Elephant Tree or Torote (Bursera microphylla A. Gray, Burseraceae) from Arizona. *J Essent Oil Res*. 2009 Jan/Feb;21:57-58.
[1394] Abdallah IB, Baatour O, Mechrgui K, et al. Essential oil composition of walnut tree (Juglans regia L.)' leaves from Tunisia. *J Essent Oil Res*. 2016;6(28):545-50.
[1395] Okan K, Aydin S, Apaydin E, et al. Antimicrobial Activity of Essential Oils from Juglans regia L. (Juglandaceae) Leaves Grown in the West Anatolian Area. *ProEnviornment*. 2018;11:32-6.
[1396] Rather M, Dar B, Dar M, et al. Chemical composition, antioxidant and antibacterial activities of the leaf essential oil of Juglans regia L. and its constituents. *Phytomedicine*. 2012;19(13):1185-1190.
[1397] Verma RS, Padalia RC, Chauhan A, et al. Phytochemical analysis of the leaf volatile oil of walnut tree (Juglans regia L.) from western Himalaya. *Ind Crops Prod*. 2013;42:195-201.
[1398] Paudel P, Satyal P, Dosoky NS, et al. Juglans regia and J. nigra, Two Trees Important in Traditional Medicine: A Comparison of Leaf Essential Oil Compositions and Biological Activities. *Nat Prod Commun*. 2013;8(10):1481-6.
[1399] Koroglu A, Baldemir A, Ozek G, et al. 11-Hydroxy-2,4-cycloeudesmane from the Leaf Oil of Juglans regia and Evaluation of its Larvicidal Activity. *Nat Prod Commun*. 2016;11(10):1421-4.
[1400] Karabulut AK, Ulger H, Pratten MK. Protection by free oxygen radical scavenging enzymes against salicylate- induced embryonic malformation in vitro. *Toxicol In Vitro*. 2000 Aug;14(4):297-307.
[1401] Overman DO, White JA. Comparative teratogenic effects of methyl salicylate applied orally or topically to hamsters. *Teratology*. 1983 Dec;28(3):421-6.
[1402] Chen R, Chen J, Cheng S, et al. Assessment of embryotoxicity of constituents in cosmetics by the embryonic stem cell tes. *Toxicol Mech Methods*. 2010 Mar;20(3):112-18.
[1403] Domaracky M, Rehak P, Juhas S, et al. Effects of selected plant essential oils on the growth and development of mouse preimplantation embryos in vivo. *Physiol Res*. 2007;56(1):97-104.
[1404] Vrskova D, Modra H. Evaluation of the developmental toxicity of 2-phenoxyethanol and clove oil anesthetics using the Frog Embryo Teratogenesis Assay: Xenopus (FETAX). *Veterinarami Medicina*. 2012;57(5):245-50.
[1405] Amini A, Cheraghi E, Safaee MR, et al. The role of eugenol in the reduction of teratogenic effects of retinoic acid on skeletal morphology of mice embryo. *Yakhteh Medical Journal*. 2003;4:195-200.
[1406] National Institutes of Health, Medline Plus. Reye Syndrome. Available at: http://www.nlm.nih.gov/medlineplus/ency/article/001565.htm.
[1407] Chan TY. Potential dangers from topical preparations containing methyl salicylate. *Hum Exp Toxicol*. 1996 Sep;15(9):747-5.
[1408] Botma M, Colquhoun-Flannery W, Leighton S. Laryngeal oedema caused by accidental ingestion of Oil of Wintergreen. *Int J Pediatr Otorhinolaryngol*. 2001 May 11;58(3):229-32.
[1409] Chyka PA. et al. Salicylate poisoning: an evidence-based consensus guideline for out-of-hospital management. *Clin Toxicol (Phila)*. 2007;45(2):95-13.
[1410] Johnson PN. Methyl salicylate/aspirin (salicylate) equivalence: who do you trust? *Vet Hum Toxicol*. 1984 Aug; 26(4):317-318
[1411] Howrie DL, Moriarty R, Breit R. Candy flavoring as a source of salicylate poisoning. *Pediatrics*. 1985 May 1;75(5):869-71.
[1412] Chen SJ, Wang MH, Chen IJ. Antiplatelet and calcium inhibitory properties of eugenol and sodium eugenol acetate. *Gen Pharmacol*. 1996 Jun;27(4):629-33.
[1413] Tognolini M, Barocelli E, Ballabeni V, et al. Comparative screening of plant essential oils; phenylpropanoid moiety as basic core for antiplatelet activity. *Life Sci*. 2006 Feb 23;78(13):1419-32.
[1414] Heck AM, DeWitt BA, Lukes AL. Potential interactions between alternative therapies and warfarin. *Am J Health Syst Pharm*. 2000;57(13):1221-1227.
[1415] Saaeed SA, Gilani AH. Antithrombotic activity of clove oil. *J Pak Med Assoc*. 1994;44(5):112-15.
[1416] Tanen DA, Danish DC, Reardon JM, et al. Comparison of oral aspirin versus topical applied methyl salicylate for platelet inhibition. *Ann Pharmacother*. 2008 Oct;42(10):1396-401.

[1417] Le Bourhis B, Soenen AM. Recherches sur l'action psychotrope de quelques substances aromatiques utilisees en alimentation. *Food Cosmet Toxicol.* 1973;11:1-9.
[1418] Joss JD, LeBlond RF. Potentiation of warfarin anticoagulation associated with topical methyl salicylate. *Ann Pharmacother.* 2000 Jun;34(6):729-33.
[1419] Tisserand R, Young R. Essential Oil Safety (Second Edition). 2014 Churchill Livingstone Elsevier.
[1420] Burkhard PR, Burkhardt K, Haenggeli CA, et al. Plant-induced seizures: reappearance of an old problem. *J Neurol.* 1999 Aug;246(8):667-70.
[1421] Kamatou GP, Vermaak I, Viljoen AM. Eugenol—from the remote Maluku Islands to the international market place: A review of a remarkable and versatile molecule. *Molecules.* 2012;17:6953-81.
[1422] Cochrane ML. Inhibition of Cytochrome P450 2C9 by essential oils. Available at: https://libres.uncg.edu/ir/uncg/listing.aspx?id=18102.
[1423] Kumaravelu P, Dakshinamoorthy DP, Subramaniam S, et al. Effect of eugenol on drug-metabolizing enzymes of carbon tetrachloride-intoxicated rat liver. *Biochem Pharmacol.* 1995 May;49(11):1703-7.
[1424] Langeveld WT, Veldhuizen EJ, Burt SA. Synergy between essential oil constituents and antibiotics. *Crit Rev Microbiol.* 2014 Feb;40(1);76-94.
[1425] Orra P, Bartle WR, Walker SE, et al. Serum concentrations of salicylic acid following topically applied salicylate derivatives. *Ann Pharmacother.* 1996 Sep;30(9):935-40.
[1426] Parker D, Martinez C, Stanley C, et al. The analysis of methyl salicylate and salicylic acid from Chinese herbal medicine ingestion. *J Analytical Tox.* 2004 Apr;28:214-16.
[1427] Sarrami N, Pemberton MN, Thornhill MH, et al. Adverse reactions associated with the use of eugenol in dentistry. *British Dental J.*2002;193:253-55.
[1428] Tammannavar P, Pushpalatha C, Jain S, et al. An unexpected positive hypersensitive reaction to eugenol. *BMJ Case Rep.* 2013; 2013:bcr2013009464.
[1429] Human and Environmental Risk Assessment on ingredients of Household Cleaning Products. Isoeugenol. 4-Hydroxy-3-methoxy-1-propen-1-yl benzene. CAS 97-54-1. Available at: https://www.heraproject.com/files/19-f-05-hera%20isoeugenol%20(corrections%20may%202005).pdf.
[1430] European Chemicals Agency. CLH Report. Proposal for Harmonised Classification and Labelling. Available at: https://echa.europa.eu/documents/10162/13626/clh_report_isoeugenol_en.pdf.
[1431] Tognolini M, Barocelli E, Ballabeni V, et al. Comparative screening of plant essential oils; phenylpropanoid moiety as basic core for antiplatelet activity. *Life Sci.* 2006 Feb 23;78(13):1419-32.
[1432] Okazaki K, Kawazoe K, Takaishi Y. Human platelet aggregation inhibitors from thyme (Thymus vulgaris L.). *Phytother Res.* 2002 Jun;16(4):398-9.
[1433] Medeiros Barreto H, Cerqueira Fontinele F, Pereira de Oliveira A, et al. Phytochemical prospection and modulation of antibiotic activity in vitro by Lippia origanoides H.B.K. in methicillin resistant Staphylococcus aureus. *Biomed Res Int.* 2014;305610:1–7.
[1434] Aelenei P, Miron A, Trifan A, et al. Essential Oils and Their Components as Modulators of Antibiotic Activity against Gram-Negative Bacteria. *Medicines.* 2016;3(19):2-34.
[1435] Veras HNH, Rodrigues FFG, Colares AV, et al. Synergistic antibiotic activity of volatile compounds from the essential oil of Lippia sidoides and thymol. *Fitoterapia.* 2012 Apr;83(3):508-12.
[1436] Langeveld WT, Veldhuizen EJ, Burt SA. Synergy between essential oil components and antibiotics: A review. *Crit Rev Microbiol.* 2014 Feb;40(1):76-94.
[1437] Miladinovic DL, Ilic BS, Kocic BD, et al. Antibacterial investigation of thyme essential oil and its main constituents on combination with tetracycline. *J Med Food.* 2015 Aug;18(8):935-7.
[1438] Shin S, Lim S. Antifungal effects of herbal essential oils alone and in combination with ketoconazole against Trichophyton spp. *J Appl Micribiol.* 2004;97:1289-96.
[1439] van den Brink DM, van Miert JNI, Dacremont G, et al. Characterization of the Final Step in the Conversion of Phytol into Phytanic Acid. *J Biol Chem.* 2005 Jul 22;280(29):26838-44.
[1440] Selkälä EM, Nair RR, Schmitz W, et al. Phytol is lethal for Amacr-deficient mice. *Biochim Biophys Acta.* 2015 Oct;1851(10):1394-405.
[1441] McGinty D, Letizia CS, Api AM. Fragrance material review on phytol. *Food Chem Toxicol.* 2010;48:S59–S63.
[1442] National Institutes of Health. U.S. National Library of Medicine. Alpha-methylacyl-CoA racemase deficiency. Available at: https://ghr.nlm.nih.gov/condition/alpha-methylacyl-coa-racemase-deficiency.
[1443] Kagoura M, Matsui C, Morohashi M. Phytol is a novel tumor promoter on ICR mouse skin. *Jpn J Cancer Res.* 1999 Apr;90(4):377-84.
[1444] Lin WY, Kuo YH, Chang YL, et al. Anti-platelet aggregation and chemical constituents from the rhizome of Gynura japonica. *Planta Med.* 2003 Aug;69(8):757-64.
[1445] Okan K, Aydin S, Apaydin E, et al. Antimicrobial Activity of Essential Oils from Juglans regia L. (Juglandaceae) Leaves Grown in the West Anatolian Area. *ProEnviornment.* 2018;11:32-6.
[1446] Rather M, Dar B, Dar M, et al. Chemical composition, antioxidant and antibacterial activities of the leaf essential oil of Juglans regia L. and its constituents. *Phytomedicine.* 2012;19(13):1185-1190.
[1447] Rather M, Dar B, Dar M, et al. Chemical composition, antioxidant and antibacterial activities of the leaf essential oil of Juglans regia L. and its constituents. *Phytomedicine.* 2012;19(13):1185-1190.
[1448] Stefanovic M, Ristic M, Popovic Z, et al. Chemical Composition and Interpopulation Variability of Essential Oils of Taxus baccata L. from Serbia. *Chem Biodiveristy.* 2016;13:943-53.
[1449] Radulovic N, Blagojevic P, Palic R. Chemical Composition of the Essential Oil Hydrodistilled from Serbian Taxus baccata L. *J Essent Oil Res.* 2010 Sep/Oct;22:458-61.
[1450] Yasar S. Volatile Constituents of Taxus baccata L. Leaves from Western and Southern Turkey. *Asian J Chem.* 2013;25(16):9123-5.
[1451] Cope RB. The dangers of yew ingestion. *Veterinary Medicine Toxicology Brief.* 2005 Sep:646-50.
[1452] Tranca S, Petrisor CL. A fatal case of Taxus poisoning. *Clujul Med.* 2013;86(3):279–281.
[1453] Chograni H, Zaouali Y, Rajeb C, et al. Essential Oil Variation among Natural Populations of Lavandula multifida L. (Lamiaceae). *Chem Biodivers.* 2010;7(4):933-942.
[1454] Soro NDK, Majdouli K, Moutaouakil K, et al. Chemical composition and Antibacterial power of Lavandula multifida L. essential oil against multiresistant strains of Escherichia coli, Pseudomonas aeruginosa and Klebsiella pneumoniae isolated in hospital. *Int J Innovation Sci Res.* 2014 Sep;9(2):464-72.
[1455] Msaada K, Salem N, Tammar S, et al. Essential Oil Composition of Lavandula dentata, L. stoechas and L. multifida Cultivated in Tunisia. *J Essent Oil Bear Plants.* 2012;15(6):1030-9.
[1456] Sellam K, Ramchoun M, Alem C, et al. Biological investigations of antioxidant-antimicrobial properties and chemical composition of essential oil from L avandula multifida. *Oxid Antioxid Med Sci.* 2013;2(3):211-6.

[1457] Saadi A, Brada M, Kouidri M, et al. Chemical Composition and Content of Essential Oil of Lavandula multifida from Algeria. *Chem Nat Compounds*. 2016 Jan;52(1):162-4.
[1458] Khadir A, Bendahou M, Benbelaid F, et al. Chemical Composition and Anti-MRSA Activity of Essential Oil and Ethanol Extract of Lavandula multifida L. from Algeria. *J Essent Oil Bear Plants*. 2016;19(3):712-8.
[1459] Douhri B, Douhri H, Farah A, et al. Phytochemical Analysis and Antibacterial Activity of Essential Oil Of Lavandula multifida L. *Int J Innovation Sci Res*. 2014 May;1(2):116-26.
[1460] Messaoud C, Chograni H, Boussaid M. Chemical composition and antioxidant activities of essential oils and methanol extracts of three wild Lavandula L. species. *Nat Prod Res*. 2012 Nov;26(21):1976-84.
[1461] Rhaffari LE, Ismaili-Alaoui M, Belkamel J, et al. Chemical composition and antibacterial properties of the essential oil of Lavandula multifida L. *Int J Essent Oil Ther*. 2007;1:1-4.
[1462] Domaracky M, Rehak P, Juhas S, et al. Effects of selected plant essential oils on the growth and development of mouse preimplantation embryos in vivo. *Physiol Res*. 2007;56(1):97-104.
[1463] Lemhadri A, Zeggwagh NA, Maghrani M, et al. Anti-hypoglycaemic activity of the aqueous extract of Origanum vulgare growing wild in Tafilalet region. *J Ethnopharmacol*. 2004 Jun;92(2-3):251-6.
[1464] Bayramoglu G, Senturk H, Bayramoglu A, et al. Carvacrol partially reverses symptoms of diabetes in STZ-induced diabetic rats. *Cytotechnology*. 2014 Mar;66(2):251–257.
[1465] Ezhumalai M, Radhiga T, Pugalendi KV. Antihyperglycemic effect of carvacrol in combination with rosiglitazone in high-fat diet-induced type 2 diabetic C57BL/6J mice. *Mol Cell Biochem*. 2014 Jan;385(1-2):23-31.
[1466] Umasankar K, Ramya DR, Nambikkairaj B. Antidiabetic and antidyslipidemic nature of carvacrol, a monoterpenic phenol studied in high-fat - fed and low-dose streptozotocin-induced experimental diabetic rats. *J Pharm Res*. 2015;9(8):484-90.
[1467] Tognolini M, Barocelli E, Ballabeni V, et al. Comparative screening of plant essential oils; phenylpropanoid moiety as basic core for antiplatelet activity. *Life Sci*. 2006 Feb 23;78(13):1419-32.
[1468] Tsai HH, Lin HW, Chen YL, et al. A review of potential harmful interactions between anticoagulant/antiplatelet agents and Chinese herbal medicines. *PLoS One*. 2013 May 9;8(5):e64255.
[1469] Karkabounas S, Kostoula OK, Daskalou T, et al. Anticarcinogenic and antiplatelet effects of carvacrol. *Exp Oncol*. 2006 Jun;28(2):121-5.
[1470] Baser KH. Biological and pharmacological activities of carvacrol and carvacrol bearing essential oils. *Curr Pharm Des*. 2008;14(29):3106-19.
[1471] Okazaki K, Kawazoe K, Takaishi Y. Human platelet aggregation inhibitors from thyme (Thymus vulgaris L.). *Phytother Res*. 2002 Jun;16(4):398-9.
[1472] Jukic M, Politeo O, Maksimmovic M, et al. In vitro acetylcholinesterase inhibitory properties of thymol, carvacrol and their derivatives thymoquinone and thymohydroquinone. *Phytother Res*. 2007;21(3):259-61.
[1473] Askin H, Yildiz M, Ayar A. Effects of Thymol and Carvacrol on Acetylcholinesterase from Drosophila melanogaster. *Acta Physica Polinica A*. 2017;132(3):720-22.
[1474] Medeiros Barreto H, Cerqueira Fontinele F, Pereira de Oliveira A, et al. Phytochemical prospection and modulation of antibiotic activity in vitro by Lippia origanoides H.B.K. in methicillin resistant Staphylococcus aureus. *Biomed Res Int*. 2014;305610:1–7.
[1475] Aelenei P, Miron A, Trifan A, et al. Essential Oils and Their Components as Modulators of Antibiotic Activity against Gram-Negative Bacteria. *Medicines*. 2016;3(19):2-34.
[1476] Magi G, Marini E, Facinelli B. Antimicrobial activity of essential oils and carvacrol, and synergy of carvacrol and erythromycin, against clinical, erythromycin-resistant Group A Streptococci. *Front Microbiol*. 2015;6:165.
[1477] Veras HNH, Rodrigues FFG, Colares AV, et al. Synergistic antibiotic activity of volatile compounds from the essential oil of Lippia sidoides and thymol. *Fitoterapia*. 2012 Apr;83(3):508-12.
[1478] Langeveld WT, Veldhuizen EJ, Burt SA. Synergy between essential oil components and antibiotics: A review. *Crit Rev Microbiol*. 2014 Feb;40(1):76-94.
[1479] Kohlert C, Schindler G, Marz RW, et al. Systemic availability and pharmacokinetics of thymol in humans. *J Clin Pharmacol*. 2002 Jul;42(7):731-7.
[1480] Toxicology Data Network, National Library of Medicine. Thymol. Available at: http://toxnet.nlm.nih.gov/cgibin/sis/search/a?dbs+hsdb:@term+@DOCNO+866.
[1481] Tognolini M, Barocelli E, Ballabeni V, et al. Comparative screening of plant essential oils; phenylpropanoid moiety as basic core for antiplatelet activity. *Life Sci*. 2006 Feb 23;78(13):1419-32.
[1482] Okazaki K, Kawazoe K, Takaishi Y. Human platelet aggregation inhibitors from thyme (Thymus vulgaris L.). *Phytother Res*. 2002 Jun;16(4):398-9.
[1483] Jukic M, Politeo O, Maksimmovic M, et al. In vitro acetylcholinesterase inhibitory properties of thymol, carvacrol and their derivatives thymoquinone and thymohydroquinone. *Phytother Res*. 2007;21(3):259-61.
[1484] Askin H, Yildiz M, Ayar A. Effects of Thymol and Carvacrol on Acetylcholinesterase from Drosophila melanogaster. *Acta Physica Polinica A*. 2017;132(3):720-22.
[1485] Medeiros Barreto H, Cerqueira Fontinele F, Pereira de Oliveira A, et al. Phytochemical prospection and modulation of antibiotic activity in vitro by Lippia origanoides H.B.K. in methicillin resistant Staphylococcus aureus. *Biomed Res Int*. 2014;305610:1–7.
[1486] Aelenei P, Miron A, Trifan A, et al. Essential Oils and Their Components as Modulators of Antibiotic Activity against Gram-Negative Bacteria. *Medicines*. 2016;3(19):2-34.
[1487] Magi G, Marini E, Facinelli B. Antimicrobial activity of essential oils and carvacrol, and synergy of carvacrol and erythromycin, against clinical, erythromycin-resistant Group A Streptococci. *Front Microbiol*. 2015;6:165.
[1488] Veras HNH, Rodrigues FFG, Colares AV, et al. Synergistic antibiotic activity of volatile compounds from the essential oil of Lippia sidoides and thymol. *Fitoterapia*. 2012 Apr;83(3):508-12.
[1489] Langeveld WT, Veldhuizen EJ, Burt SA. Synergy between essential oil components and antibiotics: A review. *Crit Rev Microbiol*. 2014 Feb;40(1):76-94.
[1490] Miladinovic DL, Ilic BS, Kocic BD, et al. Antibacterial investigation of thyme essential oil and its main constituents on combination with tetracycline. *J Med Food*. 2015 Aug;18(8):935-7.
[1491] Shin S, Lim S. Antifungal effects of herbal essential oils alone and in combination with ketoconazole against Trichophyton spp. *J Appl Micribiol*. 2004;97:1289-96.
[1492] Kissels W, Wu X, Santos RR, et al. Short communication: Interaction of the isomers carvacrol and thymol with the antibiotics doxycycline and tilmicosin: In vitro effects against pathogenic bacteria commonly found in the respiratory tract of calves. *J Dairy Sci*. 2017 Feb;100(2):970-974.

[1493] Sellam K, Ramchoun M, Alem C, et al. Biological investigations of antioxidant-antimicrobial properties and chemical composition of essential oil from L avandula multifida. *Oxid Antioxid Med Sci.* 2013;2(3):211-6.
[1494] Khadir A, Bendahou M, Benbelaid F, et al. Chemical Composition and Anti-MRSA Activity of Essential Oil and Ethanol Extract of Lavandula multifida L. from Algeria. *J Essent Oil Bear Plants.* 2016;19(3):712-8.
[1495] Douhri B, Douhri H, Farah A, et al. Phytochemical Analysis and Antibacterial Activity of Essential Oil Of Lavandula multifida L. *Int J Innovation Sci Res.* 2014 May;1(2):116-26.
[1496] Zuzarte M, Vale-Silva L, Gonçalves MJ, et al. Antifungal activity of phenolic-rich Lavandula multifida L. essential oil. *Eur J Clin Microbiol Infect Dis.* 2012 Jul;31(7):1359-66.
[1497] Soro NDK, Majdouli K, Moutaouakil K, et al. Chemical composition and Antibacterial power of Lavandula multifida L. essential oil against multiresistant strains of Escherichia coli, Pseudomonas aeruginosa and Klebsiella pneumoniae isolated in hospital. *Int J Innovation Sci Res.* 2014 Sep;9(2):464-72.
[1498] Benbelaid F, Bendahou M, Khadir A, et al. Antimicrobial activity of essential oil of Lavandula multifida L. *J Microbiol Biotech Res.* 2012;2(2):244-7.
[1499] Rhaffari LE, Ismaili-Alaoui M, Belkamel J, et al. Chemical composition and antibacterial properties of the essential oil of Lavandula multifida L. *Int J Essent Oil Ther.* 2007;1:1-4.
[1500] Zuzarte M, Vale-Silva L, Gonçalves MJ, et al. Antifungal activity of phenolic-rich Lavandula multifida L. essential oil. *Eur J Clin Microbiol Infect Dis.* 2012 Jul;31(7):1359-66.
[1501] Messaoud C, Chograni H, Boussaid M. Chemical composition and antioxidant activities of essential oils and methanol extracts of three wild Lavandula L. species. *Nat Prod Res.* 2012 Nov;26(21):1976-84.
[1502] Li J, Tian YZ, Sun BY, et al. Analysis on Volatile Constituents in Leaves and Fruits of Ficus carica by GC-MS. *Chinese Herbal Med.* 2011;4(1):63-9.
[1503] Ayoub N, Singab AN, Mostafa N, et al. Volatile Constituents of Leaves of Ficus carica Linn. Grown in Egypt. *J Essent Oil Bearing Plants.* 2010;13(3):316-21.
[1504] Kaiser R. 1986. New natural products of structural and olfactory interest identified in fig leaf absolute (Ficus carica). In: Progress in Essential Oil Research. Berlin, New York: Walter de Gruyter & Co.
[1505] Ivanov I, Dincheva I, Badjakov I, et al. GC-MS analysis of unpolar fraction from Ficus carica L. (fig) leaves. *Int Food Res J.* 2018 Feb;25(1):282-6.
[1506] Russo F, Caproraso N, Paduano A, et al. Characterisation of volatile compounds in Cilento (Italy) figs (Ficus carica L.) cv. Dottato as affected by the drying process. *Int J Food Prop.* 2017;20(S2):S1366-76.
[1507] Johnson S. Medicinal Essential Oils: The Science and Practice of Evidence-Based Essential Oil Therapy. 2017. Scott A Johnson Professional Writing Services, LLC: Orem, Utah.
[1508] Opdyke DLJ, Letizia C. Monographs on fragrance raw materials. *Food Cosmet Toxicol.* 1982;20(Suppl):691-2.
[1509] Fournier G, Paris M, Dumitresco SM, et al. Contribution to the study of Plectranthus fruticosus leaf essential oil. *Planta Med,* 1986 Dec;(6):486-88.
[1510] Casares R. Juniperus sabina. Food Cosmet Toxicol. 1964;2:680-81. Found in U.S. Food and Drug Administration. FDA Poisonous Plant Database. Available at: http://www.accessdata.fda.gov/scripts/plantox/detail.cfm?id=5133.
[1511] Pages N, Salazar M, Chamorro G, et al. Teratological evaluation of Plectranthus fruticosus leaf essential oil. *Planta Med.* 1988 Aug;54(4):296-8.
[1512] Chamorro G, Salazar M, Fournier G, et al. Anti-implantation effects of various extracts of Plectranthus fruticosus on pregnant rats. *Planta Med.* 1991 Feb;57(1):81.
[1513] Pages N, Fournier G, Chamorro G, et al. Teratological evaluation of Juniperus sabina essential oil in mice. *Planta Med.* 1989 Apr;55(2):144-46.
[1514] Pages N, Fournier G, Baudel C, et al. Sabinyl Acetate, the Main Component of Juniperus sabina L'Hérit. Essential oil, is Responsible for Antiimplantation Effect. *Phytother Res.* 1996 Aug;10(5):438-440.
[1515] Pages N, Fournier G, Velut V, et al. Potential teratogenicity in mice of the essential oil of Salvia lavandulifolia Vahl. Study of a fraction rich in sabinyl acetate. *Phytother Res.* 1992 Mar;6(2):80-83.
[1516] Casares R. Juniperus sabina. Food Cosmet Toxicol. 1964;2:680-81. Found in U.S. Food and Drug Administration. FDA Poisonous Plant Database. Available at: http://www.accessdata.fda.gov/scripts/plantox/detail.cfm?id=5133.
[1517] Burkhard PR, Burkhardt K, Haenggeli CA, et al. Plant-induced seizures: reappearance of an old problem. *J Neurol.* 1999 Aug;246(8):667-70.
[1518] Pages N, Fournier G, Chamorro G, et al. Teratological evaluation of Juniperus sabina essential oil in mice. *Planta Med.* 1989 Apr;55(2):144-46.
[1519] Kerntopf MR, de Alburquerque RL, Machado MIL, et al. Essential Oils from Leaves, Stems and Roots of Plectranthus barbatus Andr. (Labiatae) Grown in Brazil. *J Essent Oil Res.* 2002;14(2):
[1520] Bandeira JM, Barbosa FF, Barbosa LMP, et al. Composição do óleo essencial de quatro espécies do gênero Plectranthus. *Rev Bras P. Med.* 2011;13(2):157-64.
[1521] Rodrigues FFG, Costa JGM, Rodriguez FFG, et al. Study of the Interference between Plectranthus Species Essential Oils from Brazil and Aminoglycosides. *Evid Based Complement Alternat Med.* 2013;2013:724161.
[1522] Marques AM, Lima MCHP, Costa Fo HA, et al. Chemical profile analysis of three different species of Plectranthus. *Emir J Food Agric.* 2012;24(2):137-41.
[1523] Misra L, Tyagi BR, Ahmad A, et al. Variability in the Chemical Composition of the Essential Oil of Coleus forskohlii Genotypes. *J Essent Oil Res.* 1994 May;6(3):243-7.
[1524] Govindarajan M, Rajeswary M, Hoti S, et al. Eugenol, α-pinene and β-caryophyllene from Plectranthus barbatus essential oil as eco-friendly larvicides against malaria, dengue and Japanese encephalitis mosquito vectors. *Parasitol Res.* 2015 Oct;115(2):807-15.
[1525] Mota L., Figueiredo AC, Pedro LG, et al. Volatile-Oils Composition, and Bioactivity of the Essential Oils of Plectranthus barbatus, P. neochilus, and P. ornatus Grown in Portugal. *Chem Biodiversity.* 2014 May;11(5):719-32.
[1526] Singh R, Saklani S, Kumar A, et al. Chemical Composition of the Essential Oil from Coleus forskohlii. *Int J Nat Prod Res.* 2011;1(3):38-43.
[1527] Santos NO, Mariane B, Lago JH, et al. Assessing the Chemical Composition and Antimicrobial Activity of Essential Oils from Brazilian Plants-Eremanthus erythropappus (Asteraceae), Plectrantuns barbatus, and P. amboinicus (Lamiaceae). *Molecules.* 2015 May 11;20(5):8440-52.
[1528] Takshak S, Agrawal SB. The role of supplemental ultraviolet-B radiation in altering the metabolite profile, essential oil content and composition, and free radical scavenging activities of Coleus forskohlii, an indigenous medicinal plant. *Environ Sci Pollut Res Int.* 2016 Apr;23(8):7324-37.
[1529] Gelmini F, Squillace P, Testa C, et al. GC-MS characterisation and biological activity of essential oils from different vegetative organs of Plectranthus barbatus and Plectranthus caninus cultivated in north Italy. *Nat Prod Res.* 2015;29(11):993-8.
[1530] Domaracky M, Rehak P, Juhas S, et al. Effects of selected plant essential oils on the growth and development of mouse preimplantation embryos in vivo. *Physiol Res.* 2007;56(1):97-104.
[1531] Vrskova D, Modra H. Evaluation of the developmental toxicity of 2-phenoxyethanol and clove oil anesthetics using the Frog Embryo Teratogenesis Assay: Xenopus (FETAX). *Veterinarami Medicina.* 2012;57(5):245-50.

[1532] Amini A, Cheraghi E, Safaee MR, et al. The role of eugenol in the reduction of teratogenic effects of retinoic acid on skeletal morphology of mice embryo. *Yakhteh Medical Journal*. 2003;4:195-200.
[1533] Liu H, Ren C, Liu W, et al. Embryotoxicity estimation of commonly used compounds with embryonic stem cell test. *Mol Med Rep*. 2017 Jul; 16(1): 263–271.
[1534] Liu H, Ren C, Liu W, et al. Embryotoxicity estimation of commonly used compounds with embryonic stem cell test. *Mol Med Rep*. 2017 Jul; 16(1): 263–271.
[1535] Chen R, Chen J, Cheng S, et al. Assessment of embryotoxicity of constituents in cosmetics by the embryonic stem cell tes. *Toxicol Mech Methods*. 2010 Mar;20(3):112-18.
[1536] Chen SJ, Wang MH, Chen IJ. Antiplatelet and calcium inhibitory properties of eugenol and sodium eugenol acetate. *Gen Pharmacol*. 1996 Jun;27(4):629-33.
[1537] Tognolini M, Barocelli E, Ballabeni V, et al. Comparative screening of plant essential oils; phenylpropanoid moiety as basic core for antiplatelet activity. *Life Sci*. 2006 Feb 23;78(13):1419-32.
[1538] Heck AM, DeWitt BA, Lukes AL. Potential interactions between alternative therapies and warfarin. *Am J Health Syst Pharm*. 2000;57(13):1221-1227.
[1539] Saaeed SA, Gilani AH. Antithrombotic activity of clove oil. *J Pak Med Assoc*. 1994;44(5):112-15.
[1540] Kamatou GP, Vermaak I, Viljoen AM. Eugenol—from the remote Maluku Islands to the international market place: A review of a remarkable and versatile molecule. *Molecules*. 2012;17:6953-81.
[1541] Langeveld WT, Veldhuizen EJ, Burt SA. Synergy between essential oil constituents and antibiotics. *Crit Rev Microbiol*. 2014 Feb;40(1);76-94.
[1542] Rodrigues FFG, Costa JGM, Rodriguez FFG, et al. Study of the Interference between Plectranthus Species Essential Oils from Brazil and Aminoglycosides. *Evid Based Complement Alternat Med*. 2013;2013:724161.
[1543] Cochrane ML. Inhibition of Cytochrome P450 2C9 by essential oils. Available at: https://libres.uncg.edu/ir/uncg/listing.aspx?id=18102.
[1544] Sarrami N, Pemberton MN, Thornhill MH, et al. Adverse reactions associated with the use of eugenol in dentistry. *British Dental J*. 2002;193:253-55.
[1545] Tammannavar P, Pushpalatha C, Jain S, et al. An unexpected positive hypersensitive reaction to eugenol. *BMJ Case Rep*. 2013; 2013:bcr2013009464.
[1546] Bonesi M, Menichini F, Tundis R, et al. Acetylcholinesterase and butyrylcholinesterase inhibitory activity of Pinus species essential oils and their constituents. *J Enzyme Inhib Med Chem*. 2010 Oct;25(5):622-8.
[1547] Camara CC, Nascimento NRF, Macedo-Filho CL, et al. Antispasmodic Effect of the Essential Oil of Plectranthus barbatus and some Major Constituents on the Guinea-Pig Ileum. *Planta Med*. 2003;69(12):1080-1085.
[1548] Mota L., Figueiredo AC, Pedro LG, et al. Volatile-Oils Composition, and Bioactivity of the Essential Oils of Plectranthus barbatus, P. neochilus, and P. ornatus Grown in Portugal. *Chem Biodiversity*. 2014 May;11(5):719-32.
[1549] Rodrigues FFG, Costa JGM, Rodriguez FFG, et al. Study of the Interference between Plectranthus Species Essential Oils from Brazil and Aminoglycosides. *Evid Based Complement Alternat Med*. 2013;2013:724161.
[1550] Santos NO, Mariane B, Lago JH, et al. Assessing the Chemical Composition and Antimicrobial Activity of Essential Oils from Brazilian Plants-Eremanthus erythropappus (Asteraceae), Plectrantuns barbatus, and P. amboinicus (Lamiaceae). *Molecules*. 2015 May 11;20(5):8440-52.
[1551] Gelmini F, Squillace P, Testa C, et al. GC-MS characterisation and biological activity of essential oils from different vegetative organs of Plectranthus barbatus and Plectranthus caninus cultivated in north Italy. *Nat Prod Res*. 2015;29(11):993-8.
[1552] Mota L., Figueiredo AC, Pedro LG, et al. Volatile-Oils Composition, and Bioactivity of the Essential Oils of Plectranthus barbatus, P. neochilus, and P. ornatus Grown in Portugal. *Chem Biodiversity*. 2014 May;11(5):719-32.
[1553] Gelmini F, Squillace P, Testa C, et al. GC-MS characterisation and biological activity of essential oils from different vegetative organs of Plectranthus barbatus and Plectranthus caninus cultivated in north Italy. *Nat Prod Res*. 2015;29(11):993-8.
[1554] Govindarajan M, Rajeswary M, Hoti S, et al. Eugenol, α-pinene and β-caryophyllene from Plectranthus barbatus essential oil as eco-friendly larvicides against malaria, dengue and Japanese encephalitis mosquito vectors. *Parasitol Res*. 2015 Oct;115(2):807-15.
[1555] Msaada K, Salem N, Tammar S, et al. Essential Oil Composition of Lavandula dentata, L. stoechas and L. multifida Cultivated in Tunisia. *JEOBP*. 2012;15(6):1030-9.
[1556] Touati B, Chograni H, Hassen I, et al. Chemical Composition of the Leaf and Flower Essential Oils of Tunisian Lavandula dentata L. (Lamiaceae). *Chem Biodiversity*. 2011;8:1560-9.
[1557] Dris D, Tine-Djebbar F, Soltani N. Lavandula dentata essential oils: chemical composition and larvicidal activity against Culiseta longiareolata and Culex pipiens (Diptera: Culicidae). *African Entomol*. 2017;25(2):387-94.
[1558] Bousmaha L, Boti JB, Bekkara FA, et al. Infraspecific chemical variability of the essential oil of Lavandula dentata L. from Algeria. *Flavour Frag J*. 2006;21:368-72.
[1559] Imelouane B, Elbachiri A, Ankit M, et al. Physico-Chemical Compositions and Antimicrobial Activity of Essential Oil of Eastern Moroccan Lavandula dentata. *Int J Agric Biol*. 2009;11(2):113-8.
[1560] Sudria C, Pinol MT, Palazon J, et al. Influence of plant growth regulators on the growth and essential oil content of cultured Lavandula dentata plantlets. *Plant Cell Tissue Organ Culture*. 1999;58:177-84.
[1561] Masetto MAM, Descahamps C, Mogor AF, et al. Yield and composition of essential oil from inflorescences and leaves of lavender (Lavandula dentata L.) in different flower development stages and harvest times. *Rev Bras Pl Med Botucatu*. 2011;13(4):413-21.
[1562] Mothana RA, Alsaid MS, Hasoon SS, et al. Antimicrobial and antioxidant activities and gas chromatography mass spectrometry (GC/MS) analysis of the essential oils of Ajuga bracteosa Wall. ex Benth. and Lavandula dentata L. growing wild in Yemen. *J Med Plant Res*. 2012 Apr;6(15):3066-71.
[1563] Al-Sarar AS, Hussein HI, Abobakr Y, et al. Fumigant toxicity and antiacetylcholinesterase activity of Saudi Mentha longifolia and Lavandula dentata Species against Callosobruchus maculatus (F.) (Coleoptera: Bruchidae). *Turk Entonol Derg*. 2014;38(1):11-18.
[1564] Abdel-Hady NM, Abdallah GM, Idris NF. Phytochemical studies and in vitro antioxidant activity of two Lavandula species (Lamiaceae) against streptozotocin induced oxidative stress in albino rats. *J Biomed Pharm Res*. 2014;3(4):30-40.
[1565] Khine H, Weiss D, Graber N, et al. A cluster of children with seizures caused by camphor poisoning. *Pediatrics*. 2009 May;123(5):1269-72.
[1566] Michiels EA, Mazor SS. Toddler with seizures due to ingesting camphor at an Indian celebration. *Pediatr Emerg Care*. 2010 Aug;26(8):574-75.
[1567] Koren G. Medications which can kill a toddler with one tablet or teaspoonful. *J Toxicol Clin Toxicol*. *1*993;31(3):407-13.
[1568] Flaman Z, Pellechia-Clarke S, Bailey B, et al. Unintentional exposure of young children to camphor and eucalyptus oils. *Paediatr Child Health*. 2001 Feb;6(2):80-83.
[1569] Rabl W, Katzgraber F, Steinlechner M. Camphor ingestion for abortion (case report). *Forensic Sci Int*. 1997 Sep 19;89(1-2):137-40.

[1570] Flaman Z, Pellechia-Clarke S, Bailey B, et al. Unintentional exposure of young children to camphor and eucalyptus oils. *Paediatr Child Health.* 2001 Feb;6(2):80-83.
[1571] Manoguerra AS, Erdman AR, Wax PM, et al. Camphor poisoning: an evidence-based practice guideline for out-of- hospital management. *Clin Toxicol (Phila).*2006;44(4):357-70.
[1572] Gibson DE, Moore GP, Pfaff JA. Camphor ingestion. *Am J Emerg Med.* 1989 Jan;7(1):41-43.
[1573] Koppel C, Martends F, Schirop T, et al. Hemoperfusion in acute camphor poisoning. *Intensive Care Med.* 1988;14(4):431-33.
[1574] Burkhard PR, Burkhardt K, Haenggeli CA, et al. Plant-induced seizures: reappearance of an old problem. *J Neurol.* 1999 Aug;246(8):667–70.
[1575] Millet Y, Jouglard J, Steinmetz MD, et al. Toxicity of some essential plant oils. Clinical and experimental study. *Clin Toxicol.* 1981 Dec;18(12):1485–98.
[1576] Charanraj KJ, G MV, S M. Camphor poisoning in a child. *Natl Med J India.* 2013 Jan-Feb;26(1):60.
[1577] Narayan S, Singh N. Camphor poisoning-An unusual cause of seizure. *Med J Armed Forces India.* 2012 Jul;68(3):252–53.
[1578] Waldman N. Seizure caused by dermal application of over-the-counter eucalyptus oil head lice preparation. *Clin Toxicol (Phila).* 2011 Oct;49(8):750-1.
[1579] Craig JO. Poisoning by the volatile oils in childhood. *Arch Dis Child.* 1953;28:259-67.
[1580] Mathew T, Kamath V, Kumar RS, et al. Eucalyptus oil inhalation-induced seizure: A novel, underrecognized, preventable cause of acute symptomatic seizure. *Epilepsia Open.* 2017 Jul 4;2(3):350-354.
[1581] Olowe SA, Ransome-Kuti O. The risk of jaundice in glucose-6-phosphate dehydrogenase deficient babies exposed to menthol. *Acta Paediatr Scand.* 1980 May;69(3):341-5.
[1582] Dillon Remy M, Manning Alleyne P, Bratt DE, et al. Neonatal jaundice at Port-of-Spain General Hospital abstract. *West Indian Med J.* 1987;36(Suppl):28.
[1583] Burkhard PR, Burkhardt K, Haenggeli CA, et al. Plant-induced seizures: reappearance of an old problem. *J Neurol.* 1999 Aug;246(8):667–70.
[1584] Millet Y, Jouglard J, Steinmetz MD, et al. Toxicity of some essential plant oils. Clinical and experimental study. *Clin Toxicol.* 1981 Dec;18(12):1485–98.
[1585] Charanraj KJ, G MV, S M. Camphor poisoning in a child. *Natl Med J India.* 2013 Jan-Feb;26(1):60.
[1586] Narayan S, Singh N. Camphor poisoning-An unusual cause of seizure. *Med J Armed Forces India.* 2012 Jul;68(3):252–53.
[1587] Olowe SA, Ransome-Kuti O. The risk of jaundice in glucose-6-phosphate dehydrogenase deficient babies exposed to menthol. *Acta Paediatr Scand.* 1980 May;69(3):341-5.
[1588] Dillon Remy M, Manning Alleyne P, Bratt DE, et al. Neonatal jaundice at Port-of-Spain General Hospital abstract. *West Indian Med J.* 1987;36(Suppl):28.
[1589] Day LM, Ozanne-Smith J, Parsons BJ, et al. Eucalyptus oil poisoning among young children: mechanisms of access and the potential prevention. *Aust N Z J Public Health.* 1997 Jun;21(3):297-302.
[1590] Myott E. Case of eucalyptus poisoning. *Brit M J.* 1906;1:558.
[1591] Hindle RC. Eucalyptus oil ingestion. *N Z Med J.* 1994 May 11;107(977)185-6.
[1592] Tibballs J. Clinical effects and management of eucalyptus oil ingestion in infants and small children. *Med J Aust.* 1995 Aug;163(4):177-80.
[1593] Waldman W, Barwina M, Sein Anand J. Accidental intoxication with eucalyptus oil—a case report. *Przeql Lek.* 2011;68(8):555-6.
[1594] Day LM, Ozanne-Smith J, Parsons BJ, et al. Eucalyptus oil poisoning among young children: mechanisms of access and the potential prevention. *Aust N Z J Public Health.* 1997 Jun;21(3):297-302.
[1595] Burkhard PR, Burkhardt K, Haenggeli CA, et al. Plant-induced seizures: reappearance of an old problem. *J Neurol.* 1999 Aug;246(8):667-70.
[1596] Waldman N. Seizure caused by dermal application of over-the-counter eucalyptus oil head lice preparation. *Clin Toxicol (Phila).* 2011 Oct;49(8):750-1.
[1597] Craig JO. Poisoning by the volatile oils in childhood. *Arch Dis Child.* 1953;28:259-67.
[1598] Mathew T, Kamath V, Kumar RS, et al. Eucalyptus oil inhalation-induced seizure: A novel, underrecognized, preventable cause of acute symptomatic seizure. *Epilepsia Open.* 2017 Jul 4;2(3):350-354.
[1599] Unger M, Frank A. Simultaneous determination of the inhibitory potency of herbal extracts on the activity of six major cytochrome P450 enzymes using liquid chromatography/mass spectrometry and automated online extraction. *Rapid Commun Mass Spectrom.* 2004;18(19):2273-81.
[1600] Jori A, Bianchetti A, Prestini PE, et al. Effect of eucalyptol (1,8-cineole) on the metabolism of other drugs in rats and in man. *Eur J Pharmacol.* 1970;9(3):362-66.
[1601] Kim NH, Hyun SH, Jin CH, et al. Pretreatment with 1,8-cineole potentiates thioacetamide-induced hepatotoxicity and immunosuppression. *Arch Pharm Res.* 2004 Jul;27(7):781-9.
[1602] Jori A, Bianchetti A, Prestini PE, et al. Effect of eucalyptol (1,8-cineole) on the metabolism of other drugs in rats and in man. *Eur J Pharmacol.* 1970;9(3):362-66.
[1603] de Sousa DP, Raphael E, Brocksom U, et al. Sedative effect of monoterpene alcohols in mice: A preliminary screening. *Verlag der Zeitschrift fur Naturforschung.* 2007;62c:563-66.
[1604] Burkhard PR, Burkhardt K, Haenggeli CA, et al. Plant-induced seizures: reappearance of an old problem. *J Neurol.* 1999 Aug;246(8):667–70.
[1605] Millet Y, Jouglard J, Steinmetz MD, et al. Toxicity of some essential plant oils. Clinical and experimental study. *Clin Toxicol.* 1981 Dec;18(12):1485–98.
[1606] Charanraj KJ, G MV, S M. Camphor poisoning in a child. *Natl Med J India.* 2013 Jan-Feb;26(1):60.
[1607] Narayan S, Singh N. Camphor poisoning-An unusual cause of seizure. *Med J Armed Forces India.* 2012 Jul;68(3):252–53.
[1608] Olowe SA, Ransome-Kuti O. The risk of jaundice in glucose-6-phosphate dehydrogenase deficient babies exposed to menthol. *Acta Paediatr Scand.* 1980 May;69(3):341-5.
[1609] Dillon Remy M, Manning Alleyne P, Bratt DE, et al. Neonatal jaundice at Port-of-Spain General Hospital abstract. *West Indian Med J.* 1987;36(Suppl):28.
[1610] Khine H, Weiss D, Graber N, et al. A cluster of children with seizures caused by camphor poisoning. *Pediatrics.* 2009 May;123(5):1269-72.
[1611] Michiels EA, Mazor SS. Toddler with seizures due to ingesting camphor at an Indian celebration. *Pediatr Emerg Care.* 2010 Aug;26(8):574-75.
[1612] Koren G. Medications which can kill a toddler with one tablet or teaspoonful. *J Toxicol Clin Toxicol. 1*993;31(3):407-13.
[1613] Flaman Z, Pellechia-Clarke S, Bailey B, et al. Unintentional exposure of young children to camphor and eucalyptus oils. *Paediatr Child Health.* 2001 Feb;6(2):80-83.
[1614] Rabl W, Katzgraber F, Steinlechner M. Camphor ingestion for abortion (case report). *Forensic Sci Int.* 1997 Sep 19;89(1-2):137-40.
[1615] Flaman Z, Pellechia-Clarke S, Bailey B, et al. Unintentional exposure of young children to camphor and eucalyptus oils. *Paediatr Child Health.* 2001 Feb;6(2):80-83.

[1616] Manoguerra AS, Erdman AR, Wax PM, et al. Camphor poisoning: an evidence-based practice guideline for out-of- hospital management. *Clin Toxicol (Phila).*2006;44(4):357-70.
[1617] Gibson DE, Moore GP, Pfaff JA. Camphor ingestion. *Am J Emerg Med.* 1989 Jan;7(1):41-43.
[1618] Koppel C, Martends F, Schirop T, et al. Hemoperfusion in acute camphor poisoning. *Intensive Care Med.* 1988;14(4):431-33.
[1619] Burkhard PR, Burkhardt K, Haenggeli CA, et al. Plant-induced seizures: reappearance of an old problem. *J Neurol.* 1999 Aug;246(8):667–70.
[1620] Millet Y, Jouglard J, Steinmetz MD, et al. Toxicity of some essential plant oils. Clinical and experimental study. *Clin Toxicol.* 1981 Dec;18(12):1485–98.
[1621] Charanraj KJ, G MV, S M. Camphor poisoning in a child. *Natl Med J India.* 2013 Jan-Feb;26(1):60.
[1622] Narayan S, Singh N. Camphor poisoning-An unusual cause of seizure. *Med J Armed Forces India.* 2012 Jul;68(3):252–53.
[1623] Uc A, Bishop WP, Sanders KD. Camphor hepatotoxicity. *South Med J.* 2000;93:596-98.
[1624] Olowe SA, Ransome-Kuti O. The risk of jaundice in glucose-6-phosphate dehydrogenase deficient babies exposed to menthol. *Acta Paediatr Scand.* 1980 May;69(3):341-5.
[1625] Dillon Remy M, Manning Alleyne P, Bratt DE, et al. Neonatal jaundice at Port-of-Spain General Hospital abstract. *West Indian Med J.* 1987;36(Suppl):28.
[1626] Covington TR, et al. Handbook of Nonprescription Drugs. 11th ed. Washington, D.C.: American Pharmaceutical Association. 1996.
[1627] Khine H, Weiss D, Graber N, et al. A cluster of children with seizures caused by camphor poisoning. *Pediatrics.* 2009 May;123(5):1269-72.
[1628] Michiels EA, Mazor SS. Toddler with seizures due to ingesting camphor at an Indian celebration. *Pediatr Emerg Care.* 2010 Aug;26(8):574-75.
[1629] Koren G. Medications which can kill a toddler with one tablet or teaspoonful. *J Toxicol Clin Toxicol. 1*993;31(3):407-13.
[1630] Flaman Z, Pellechia-Clarke S, Bailey B, et al. Unintentional exposure of young children to camphor and eucalyptus oils. *Paediatr Child Health.* 2001 Feb;6(2):80-83.
[1631] Rabl W, Katzgraber F, Steinlechner M. Camphor ingestion for abortion (case report). *Forensic Sci Int.* 1997 Sep 19;89(1-2):137-40.
[1632] Flaman Z, Pellechia-Clarke S, Bailey B, et al. Unintentional exposure of young children to camphor and eucalyptus oils. *Paediatr Child Health.* 2001 Feb;6(2):80-83.
[1633] Manoguerra AS, Erdman AR, Wax PM, et al. Camphor poisoning: an evidence-based practice guideline for out-of- hospital management. *Clin Toxicol (Phila).*2006;44(4):357-70.
[1634] Gibson DE, Moore GP, Pfaff JA. Camphor ingestion. *Am J Emerg Med.* 1989 Jan;7(1):41-43.
[1635] Koppel C, Martends F, Schirop T, et al. Hemoperfusion in acute camphor poisoning. *Intensive Care Med.* 1988;14(4):431-33.
[1636] Burkhard PR, Burkhardt K, Haenggeli CA, et al. Plant-induced seizures: reappearance of an old problem. *J Neurol.* 1999 Aug;246(8):667–70.
[1637] Millet Y, Jouglard J, Steinmetz MD, et al. Toxicity of some essential plant oils. Clinical and experimental study. *Clin Toxicol.* 1981 Dec;18(12):1485–98.
[1638] Charanraj KJ, G MV, S M. Camphor poisoning in a child. *Natl Med J India.* 2013 Jan-Feb;26(1):60.
[1639] Narayan S, Singh N. Camphor poisoning-An unusual cause of seizure. *Med J Armed Forces India.* 2012 Jul;68(3):252–53.
[1640] Waldman N. Seizure caused by dermal application of over-the-counter eucalyptus oil head lice preparation. *Clin Toxicol (Phila).* 2011 Oct;49(8):750-1.
[1641] Craig JO. Poisoning by the volatile oils in childhood. *Arch Dis Child.* 1953;28:259-67.
[1642] Mathew T, Kamath V, Kumar RS, et al. Eucalyptus oil inhalation-induced seizure: A novel, underrecognized, preventable cause of acute symptomatic seizure. *Epilepsia Open.* 2017 Jul 4;2(3):350-354.
[1643] Uc A, Bishop WP, Sanders KD. Camphor hepatotoxicity. *South Med J.* 2000;93:596-98.
[1644] Olowe SA, Ransome-Kuti O. The risk of jaundice in glucose-6-phosphate dehydrogenase deficient babies exposed to menthol. *Acta Paediatr Scand.* 1980 May;69(3):341-5.
[1645] Dillon Remy M, Manning Alleyne P, Bratt DE, et al. Neonatal jaundice at Port-of-Spain General Hospital abstract. *West Indian Med J.* 1987;36(Suppl):28.
[1646] Unger M, Frank A. Simultaneous determination of the inhibitory potency of herbal extracts on the activity of six major cytochrome P450 enzymes using liquid chromatography/mass spectrometry and automated online extraction. *Rapid Commun Mass Spectrom.* 2004;18(19):2273-81.
[1647] Jori A, Bianchetti A, Prestini PE, et al. Effect of eucalyptol (1,8-cineole) on the metabolism of other drugs in rats and in man. *Eur J Pharmacol.* 1970;9(3):362-66.
[1648] Kim NH, Hyun SH, Jin CH, et al. Pretreatment with 1,8-cineole potentiates thioacetamide-induced hepatotoxicity and immunosuppression. *Arch Pharm Res.* 2004 Jul;27(7):781-9.
[1649] Jori A, Bianchetti A, Prestini PE, et al. Effect of eucalyptol (1,8-cineole) on the metabolism of other drugs in rats and in man. *Eur J Pharmacol.* 1970;9(3):362-66.
[1650] de Sousa DP, Raphael E, Brocksom U, et al. Sedative effect of monoterpene alcohols in mice: A preliminary screening. *Verlag der Zeitschrift fur Naturforschung.* 2007,62c:563-66.
[1651] Covington TR, et al. Handbook of Nonprescription Drugs. 11th ed. Washington, D.C.: American Pharmaceutical Association. 1996.
[1652] Qneibi M, Jaradat N, Hawash M, et al. The Neuroprotective Role of Origanum syriacum L. and Lavandula dentata L. Essential Oils through Their Effects on AMPA Receptors. *Biomed Res Int.* 2019 Mar 11;2019:5640173.
[1653] Imelouane B, Elbachiri A, Ankit M, et al. Physico-Chemical Compositions and Antimicrobial Activity of Essential Oil of Eastern Moroccan Lavandula dentata. *Int J Agric Biol.* 2009;11(2):113-8.
[1654] Mothana RA, Alsaid MS, Hasoon SS, et al. Antimicrobial and antioxidant activities and gas chromatography mass spectrometry (GC/MS) analysis of the essential oils of Ajuga bracteosa Wall. ex Benth. and Lavandula dentata L. growing wild in Yemen. *J Med Plant Res.* 2012 Apr;6(15):3066-71.
[1655] Abdel-Hady NM, Abdallah GM, Idris NF. Phytochemical studies and in vitro antioxidant activity of two Lavandula species (Lamiaceae) against streptozotocin induced oxidative stress in albino rats. *J Biomed Pharm Res.* 2014;3(4):30-40.
[1656] Mothana RA, Alsaid MS, Hasoon SS, et al. Antimicrobial and antioxidant activities and gas chromatography mass spectrometry (GC/MS) analysis of the essential oils of Ajuga bracteosa Wall. ex Benth. and Lavandula dentata L. growing wild in Yemen. *J Med Plant Res.* 2012 Apr;6(15):3066-71.
[1657] Dris D, Tine-Djebbar F, Soltani N. Lavandula dentata essential oils: chemical composition and larvicidal activity against Culiseta longiareolata and Culex pipiens (Diptera: Culicidae). *African Entomol.* 2017;25(2):387-94.
[1658] Al-Sarar AS, Hussein HI, Abobakr Y, et al. Fumigant toxicity and antiacetylcholinesterase activity of Saudi Mentha longifolia and Lavandula dentata Species against Callosobruchus maculatus (F.) (Coleoptera: Bruchidae). *Turk Entonol Derg.* 2014;38(1):11-18.
[1659] Cossetin LF, Santi EMT, Cossetin JF, et al. In vitro Safety and Efficacy of Lavender Essential Oil (Lamiales: Lamiaceae) as an Insecticide Against Houseflies (Diptera: Muscidae) and Blowflies (Diptera: Calliphoridae). *J Econ Entomol.* 2018 May 26. [Epub ahead of print]

[1660] Djebir S, Ksouri S, Trigui M, et al. Chemical Composition and Acaricidal Activity of the Essential Oils of Some Plant Species of Lamiaceae and Myrtaceae against the Vector of Tropical Bovine Theileriosis: Hyalomma scupense (syn. Hyalomma detritum). *Biomed Res Int.* 2019 Feb 7;2019:7805467.
[1661] Hattori R, Muraki S, Yoshida T. Chemical composition of the absolute from gardenia flower. *Agric Biol Chem.* 1978;42(7):1351-6.
[1662] Chaichana J, Niwatananun W, Vejabhikul S, et al. Volatile constituents and biological activities of Gadenia jasminoides. *J Health Res.* 2009;23(3):141-5.
[1663] Obuzor GU, Nwaokolo MI. Composition of Essential Oil of Nigerian (Niger Delta) Grown Gardenia Jasminoide Flower. *Int Arch Appl Sci Technol.* 2010 Dec;1(2):32-6.
[1664] Sukatta U, Dilokkunanant U, Vajrodaya S, et al. Essential Oil and Scents Extraction from Gardenia jasminoides J. Ellis Flowers. 2011.
[1665] Kanlayavattankul M, Lourith N. Volatile profile and sensory property of Gardenia jasminoides aroma extracts. *J Cosmet Sci.* 2015 Nov/Dec;66:371-7.
[1666] National Library of Medicine. TOXNET. Benzyl acetate. Available at: https://toxnet.nlm.nih.gov/cgibin/sis/search/a?dbs+hsdb:@term+@DOCNO+2851.
[1667] JECFA. Summary of evaluations performed by the Joint FAO/WHO Expert Committee on Food Additives. 2001. Available at: https://toxnet.nlm.nih.gov/cgi-bin/sis/search/a?dbs+hsdb:@term+@DOCNO+2851.
[1668] Chaichana J, Niwatananun W, Vejabhikul S, et al. Volatile constituents and biological activities of Gadenia jasminoides. *J Health Res.* 2009;23(3):141-5.
[1669] Chaichana J, Niwatananun W, Vejabhikul S, et al. Volatile constituents and biological activities of Gadenia jasminoides. *J Health Res.* 2009;23(3):141-5.
[1670] Baser KHC, Ermin N, Demircakmak B. The Essential Oil of Anthriscus cerefolium (L.) Hoffm. (Chervil) Growing Wild in Turkey. *J Essent Oil Res.* 1998 Jul/Aug;10:463-4.
[1671] Rigand J, Sarris J. Les constituants volatils de l'arome du cerfeuil. *Sci Aliment.* 1982;2:163-72.
[1672] Simandi B, Oszagyan M, Lemberkovics E, et al. Comparison of the Volatile Composition of Chervil Oil Obtained by Hydrodistillation and Supercritical Fluid Extraction. *J Essent Oil Res.* 1996 May;8(3):305-6.
[1673] Kilic O. Essential Oil Composition of Aerial Parts of Two Anthriscus Pers. Species From Turkey. *TEOP.* 2017;20(2):591-6.
[1674] Lemberkovics E, Petri G, Vitanyi G, et al. Essential Oil Composition of Chervil Growing Wild in Hungary. Part 1. *J Essent Oil Res.* 1994 Jul/Aug;6:421-2.
[1675] Chizzola R. Composition of the Essential Oils from Anthriscus cerefolium var. trichocarpa and A. caucalis Growing Wild in the Urban Area of Vienna (Austria). *Nat Prod Commun.* 2011;6(8):1147-50.
[1676] European Medicines Agency. Public statement on the use of herbal medicinal products containing estragole. Available at: http://www.ema.europa.eu/docs/en_GB/document_library/Scientific_guideline/2010/04/WC500089960.pdf.
[1677] European Medicines Agency. Public statement on the use of herbal medicinal products containing estragole. Available at: http://www.ema.europa.eu/docs/en_GB/document_library/Scientific_guideline/2010/04/WC500089960.pdf.
[1678] Ding W, Levy DD, Bishop ME, et al. In vivo genotoxicity of estragole in male F344 rats. *Environ Mol Mutagen.* 2015 May;56(4):356-65.
[1679] Eisenmann SW, Poulev A, Struwe L, et al. Qualitative variation of anti-diabetic constituents in different tarragon (Artemisia dracunculus L.) cytotypes. *Fitoterapia.* 2011 Oct;82(7):1062–74.
[1680] European Medicines Agency. Public statement on the use of herbal medicinal products containing estragole. Available at: http://www.ema.europa.eu/docs/en_GB/document_library/Scientific_guideline/2010/04/WC500089960.pdf.
[1681] European Commission. Opinion of the Scientific Committee on Food on methyl eugenol (4-Allyl-1,2- dimethoxybenzene). Available at: http://ec.europa.eu/food/fs/sc/scf/out102_en.pdf.
[1682] European Commission. Opinion of the Scientific Committee on Food on estragole (1-Allyl-4-methoxybenzene). Available at: http://ec.europa.eu/food/fs/sc/scf/out104_en.pdf.
[1683] California Environmental Protection Agency. Evidence of the carcinogenicity of estragole. Available at: http://oehha.ca.gov/prop65/pdf/estragf.pdf.
[1684] Drinkwater NR, Miller EC, Miller JA, et al. Hepatocarcinogenicity of estragole (1-allyl-4-methoxybenzene) and 1'-hydroxyestragole in the mouse and mutagenicity of 1'-acetoxyestragole in bacteria. *J Natl Cancer Inst.* 1976 Dec;57(6):1323-31.
[1685] Zeller A, Horst K, Rychlik M. Study of the metabolism of estragole in humans consuming fennel tea. *Chem Res Toxicol.* 2009 Dec;22(12):1929-37.
[1686] Lee HS. Anticoagulant properties of constituents derived from fennel (Foeniculum vulgare Gaertner) fruits. *Food Sci Biotech.* 2006 Oct;15(5):763-67.
[1687] Khan MM. Masters Thesis: Inhibition of Cytochrome P450 2E1 and Cytochrome P450 2A6 by Essential Oils: Tarragon (Artemisia dracunculus) and Basil (Ocimum basilicum). 2014. Available at: https://libres.uncg.edu/ir/uncg/f/Khan_uncg_0154M_11587.pdf.
[1688] Sayyah M, Nadjafnia L, Kamalinejad M. Anticonvulsant activity and chemical composition of Artemisia dracunculus L. essential oil. *J Ethnopharmacol.* 2004 Oct;94(2-3):283-7.
[1689] Shin S, Pyun MS. Anti-Candida effects of estragole in combination with ketoconazole or amphotericin B. *Phytother Res.* 2004 Oct;18(10):827-30.
[1690] Shin S. Essential oil compounds from Agastache rugosa as antifungal agents against Trichophyton species. *Arch Pharm Res.* 2004 Mar;27(3):295-9.
[1691] Shin S, Kang CA. Antifungal activity of the essential oil of Agastache rugosa Kuntze and its synergism with ketoconazole. *Lett Appl Microbiol.* 2003;36(2):111-5.
[1692] Wanner JKR, Bail S, Jirovetz L, et al. Chemical composition, olfactory analysis and antibacterial activity of selected essential oils (mainly used in perfumery products). Conference: International Symposium on Essential Oils. 2009 Sep.
[1693] Burkhard PR, Burkhardt K, Haenggeli CA, et al. Plant-induced seizures: reappearance of an old problem. *J Neurol.* 1999 Aug;246(8):667-70.
[1694] Waldman N. Seizure caused by dermal application of over-the-counter eucalyptus oil head lice preparation. *Clin Toxicol (Phila).* 2011 Oct;49(8):750-1.
[1695] Craig JO. Poisoning by the volatile oils in childhood. *Arch Dis Child.* 1953;28:259-67.
[1696] Mathew T, Kamath V, Kumar RS, et al. Eucalyptus oil inhalation-induced seizure: A novel, underrecognized, preventable cause of acute symptomatic seizure. *Epilepsia Open.* 2017 Jul 4;2(3):350-354.
[1697] Unger M, Frank A. Simultaneous determination of the inhibitory potency of herbal extracts on the activity of six major cytochrome P450 enzymes using liquid chromatography/mass spectrometry and automated online extraction. *Rapid Commun Mass Spectrom.* 2004;18(19):2273-81.
[1698] Jori A, Bianchetti A, Prestini PE, et al. Effect of eucalyptol (1,8-cineole) on the metabolism of other drugs in rats and in man. *Eur J Pharmacol.* 1970;9(3):362-66.

[1699] Kim NH, Hyun SH, Jin CH, et al. Pretreatment with 1,8-cineole potentiates thioacetamide-induced hepatotoxicity and immunosuppression. *Arch Pharm Res*. 2004 Jul;27(7):781-9.
[1700] Jori A, Bianchetti A, Prestini PE, et al. Effect of eucalyptol (1,8-cineole) on the metabolism of other drugs in rats and in man. *Eur J Pharmacol*. 1970;9(3):362-66.
[1701] de Sousa DP, Raphael E, Brocksom U, et al. Sedative effect of monoterpene alcohols in mice: A preliminary screening. *Verlag der Zeitschrift fur Naturforschung*. 2007;62c:563-66.
[1702] Bahrami S, Jalai R, Ramezani Z, et al. In vitro Scolicidal Effect of Lepidium sativum Essential Oil. *J Ardabil University Med Sci*. 2016 Winter;15(4):395-403.
[1703] Jaradat N, Al-lahham S, Abualhasan MN, et al. Chemical Constituents, Antioxidant, Cyclooxygenase Inhibitor, and Cytotoxic Activities of Teucrium pruinosum Boiss. Essential Oil. *BioMed Res Int*. 2018;2018:4034689.
[1704] Hammami S, el Mokni R, Faidi K, et al. Chemical composition and antioxidant activity of essential oil from aerial parts of Teucrium flavum L. subsp. flavum growing spontaneously in Tunisia. *Nat Prod Comm*. 2015;29(24):2336-40.
[1705] Presti ML, Crupi ML, Costa R, et al. Seasonal Variations of Teucrium flavum L. Essential Oil. *J Essent Oil Res*. 2010 May/Jun;22:211-6.
[1706] Bellomaria B, Arnold N, Valentini G, et al. Essential Oil of Teucrium flavum subsp. hellenicum from Greece. *J essent Oil Res*. 1998 Mar/Apr;10:131-3.
[1707] Cozzani S, Muselloi A, Desjobert JM, et al. Chemical composition of essential oil of Teucrium polium subsp. capitatum (L.) from Corsica. *Flavour Frag J*. 2005;20:436-41.
[1708] Menichini F, Conforti F, Rigano D, et al. Phytochemical composition, anti-inflammatory and antitumour activities of four Teucrium essential oils from Greece. *Food Chem*. 2009;115:679-86.
[1709] Aburjai T, Hudaib M, Cavrini V. Composition of the Essential Oil from Jordanian Germander (Teucrium polium L.). *J Essent Oil Res*. 2006 Jan/Feb;18:97-9.
[1710] Eikani MH, Goodarzina I, Mizra M. Comparison between the Essential Oil and Supercritical Carbon Dioxide Extract of Teucrium polium L. *J Essent Oil Res*. 1999 Jul/Aug;11:470-2.
[1711] Othman MB, Salah-Fatnassi KBH, Ncibi S, et al. Antimicrobial activity of essential oil and aqueous and ethanol extracts of Teucrium polium L. subsp. gabesianum (L.H.) from Tunisia. *Physiol Mol Biol Plants*. 2017 Jul-Sep;23(3):723-9.
[1712] Lograda T, Ramdani M, Chalard P, et al. Chemical Analysis and Antimicrobial Activity of Teucrium polium L. Essential Oil from Eastern Algeria. *Am J Adv Drug Delivery*. 2014;2(6):697-710.
[1713] Kabouche A, Kabouche Z, Ghannadi A, et al. Analysis of the Essential Oil of Teucrium polium ssp. aurasiacum from Algeria. *J Essent Oil Res*. 2007 Jan/Feb;19:44-6.
[1714] Vahdani M, Faridi P, Zarshenas MM, et al. Major Compounds and Antimicrobial Activity of Essential Oils from Five Iranian Endemic Medicinal Plants. *Pharmacog J*. 2011 Jun;3(22):48-53.
[1715] Sadeghi H, Jamalpoor S, Shirzadi MH. Variability in essential oil of Teucrium polium L. of different latitudinal populations. *Ind Crops Prod*. 2014;54:130-4.
[1716] Boulila A, Béjaoui A, Messaoud C, et al. Variation of volatiles in Tunisian populations of Teucrium polium L. (Lamiaceae). *Chem Biodivers*. 2008 Jul;5(7):1389-400.
[1717] Jaradat N, Al-lahham S, Abualhasan MN, et al. Chemical Constituents, Antioxidant, Cyclooxygenase Inhibitor, and Cytotoxic Activities of Teucrium pruinosum Boiss. Essential Oil. *BioMed Res Int*. 2018;2018:4034689.
[1718] Menichini F, Conforti F, Rigano D, et al. Phytochemical composition, anti-inflammatory and antitumour activities of four Teucrium essential oils from Greece. *Food Chem*. 2009;115:679-86.
[1719] Jaradat N, Al-lahham S, Abualhasan MN, et al. Chemical Constituents, Antioxidant, Cyclooxygenase Inhibitor, and Cytotoxic Activities of Teucrium pruinosum Boiss. Essential Oil. *BioMed Res Int*. 2018;2018:4034689.
[1720] Menichini F, Conforti F, Rigano D, et al. Phytochemical composition, anti-inflammatory and antitumour activities of four Teucrium essential oils from Greece. *Food Chem*. 2009;115:679-86.
[1721] Abdollahi M, Karimpour H, Monsef-Esfehani HR. Antinociceptive effects of Teucrium polium L total extract and essential oil in mouse writhing test. *Pharmacol Res*. 2003 Jul;48(1):31-5.
[1722] Othman MB, Salah-Fatnassi KBH, Ncibi S, et al. Antimicrobial activity of essential oil and aqueous and ethanol extracts of Teucrium polium L. subsp. gabesianum (L.H.) from Tunisia. *Physiol Mol Biol Plants*. 2017 Jul-Sep;23(3):723-9.
[1723] Lograda T, Ramdani M, Chalard P, et al. Chemical Analysis and Antimicrobial Activity of Teucrium polium L. Essential Oil from Eastern Algeria. *Am J Adv Drug Delivery*. 2014;2(6):697-710.
[1724] Vahdani M, Faridi P, Zarshenas MM, et al. Major Compounds and Antimicrobial Activity of Essential Oils from Five Iranian Endemic Medicinal Plants. *Pharmacog J*. 2011 Jun;3(22):48-53.
[1725] Ibrahim SR, Abdallah HM, Mohamed GA, et al. Volatile oil profile of some lamiaceous plants growing in Saudi Arabia and their biological activities. *Z Naturforsch C*. 2017 Jan 1;72(1-2):35-41.
[1726] Jaradat N, Al-lahham S, Abualhasan MN, et al. Chemical Constituents, Antioxidant, Cyclooxygenase Inhibitor, and Cytotoxic Activities of Teucrium pruinosum Boiss. Essential Oil. *BioMed Res Int*. 2018;2018:4034689.
[1727] Hammami S, el Mokni R, Faidi K, et al. Chemical composition and antioxidant activity of essential oil from aerial parts of Teucrium flavum L. subsp. flavum growing spontaneously in Tunisia. *Nat Prod Comm*. 2015;29(24):2336-40.
[1728] Sayyad R, Farahmandfar R. Influence of Teucrium polium L. essential oil on the oxidative stability of canola oil during storage. *J Food Sci Technol*. 2017 Sep;54(10):3073-3081.
[1729] Khani A, Heydarian M. Fumigant and repellent properties of sesquiterpene-rich essential oil from Teucrium polium subsp. capitatum (L.). *Asian Pac J Trop Med*. 2014 Dec;7(12):956-61.
[1730] Adams RP, Kauffmann M, Callahan F. The leaf essential oil of Abies grandis (Doug. ex D. Don) Lindl. (Pinaceae): revisited 38 years later. *Phytologia*. 2015 Jan;97(1):1-5.
[1731] Von Rudloff E. Chemosystematic studies in the genus Abies. II. Leaf oil analysis of grand fir. *Can J Bot*. 1976;54:1926-31.
[1732] Cavalli JF, Tomi F, Bernardini AF, et al. Chemical variability of the essential oil of Helichrysum faradifani Sc. Ell. from Madagascar. *Flavour Frag J*. 2006;21:111-4.
[1733] Ralijerson LB, Rabehaja DJR, Rajaonarison JF, et al. Comparison Between the Fresh and Dry Essential Oil of Helichrysum faradifani Scott Elliot from Madagascar. *J Esent Oil Res*. 2005 Nov/Dec;17:597-600.
[1734] Benelli G, Pvela R, Rakotosaona R, et al. Chemical composition and insecticidal activity of the essential oil from Helichrysum faradifani endemic to Madagascar. *Nat Prod Res*. 2018 Jul;32(14):1690-1698.

[1735] Cavalli JF, Ranarivelo L, Ratsimbason M, et al. Constituents of the essential oil of six Helichrysum species from Madagascar. *Flavour Frag J.* 2001;16:253-6.
[1736] Benelli G, Pvela R, Rakotosaona R, et al. Chemical composition and insecticidal activity of the essential oil from Helichrysum faradifani endemic to Madagascar. *Nat Prod Res.* 2018 Jul;32(14):1690-1698.
[1737] Cvetkovikj I, Stefkov G, Karapandzova M, et al. Essential oil composition of Salvia fruticosa Mill. populations from Balkan Peninsula. *Macedonian Pharm Bull.* 2015;61(1):19-26.
[1738] Putievsky E, Ravid U, Dudai N. The essential oil and yield components from various plant parts of Salvia fruticosa. *J Nat Prod.* 1986 Nov-Dec;49(6):1015-17.
[1739] Karousou R, Vokou D, Kokkini S. Variation of Salvia fruticose essential oils on the island of Crete (Greece). *Bot Acta.* 1998;111:250-54.
[1740] Askun T, Baser KHC, Tumen G, et al. Characterization of essential oils of some Salvia species and their antimycobacterial activities. *Turk J Biol.* 2010;34:89-95.
[1741] Elmann A, Mordechay S, Rindner M, et al. Protective Effects of the Essential Oil of Salvia fruticosa and Its Constituents on Astrocytic Susceptibility to Hydrogen Peroxide-Induced Cell Death. *J Agric Food Chem.* 2009 Aug 12;57(15):6636-41.
[1742] Papageorgiou V, Gardeli C, Mallouchos A, et al. Variation of the chemical profile and antioxidant behavior of Rosmarinus officinalis L. and Salvia fruticosa Miller grown in Greece. *J Agric Food Chem.* 2008 Aug 27;56(16):7254-64.
[1743] Karioti A, Skaltsa H, Demetzos C, et al. Effect of Nitrogen Concentration of the Nutrient Solution on the Volatile Constituents of Leaves of Salvia fruticosa Mill. in Solution Culture. *J Agric Food Chem.* 2003;51:6505-8.
[1744] Haliciglu O, Astarcioglu G, Yaprak I, et al. Toxicity of Salvia officinalis in a newborn and a child: an alarming report. *Pediatr Neurol.* 2011 Oct;45(4):259-60.
[1745] Lachenmeier DW, Walch SG. Epileptic seizure caused by accidental ingestion of sage (Salvia officinalis L.) oil in children: a rare, exceptional case or a threat to public health. *Pediatr Neurol.* 2012 Mar;46(3):201.
[1746] Khine H, Weiss D, Graber N, et al. A cluster of children with seizures caused by camphor poisoning. *Pediatrics.* 2009 May;123(5):1269-72.
[1747] Michiels EA, Mazor SS. Toddler with seizures due to ingesting camphor at an Indian celebration. *Pediatr Emerg Care.* 2010 Aug;26(8):574-75.
[1748] Koren G. Medications which can kill a toddler with one tablet or teaspoonful. *J Toxicol Clin Toxicol. 1*993;31(3):407-13.
[1749] Flaman Z, Pellechia-Clarke S, Bailey B, et al. Unintentional exposure of young children to camphor and eucalyptus oils. *Paediatr Child Health.* 2001 Feb;6(2):80-83.
[1750] Craig JO. Poisoning by the volatile oils in childhood. *Arch Dis Child.* 1953;28:259-67.
[1751] Melis K. Bochner A, Janssens G. Accidental nasal eucalyptol and menthol instillation. *Eur J Pediatr.* 1989 Aug;148(8)786-7.
[1752] Day LM, Ozanne-Smith J, Parsons BJ, et al. Eucalyptus oil poisoning among young children: mechanisms of access and the potential prevention. *Aust N Z J Public Health.* 1997 Jun;21(3):297-302.
[1753] da Silva Costa KC, Bexerra SB, Norte CM, et al. Medicinal plants with teratogenic potential: current considerations. *Braz J Pharm Sci.* 2012;48(3):427-33.
[1754] Brinker F. Herb Contraindications and Drug Interactions. 2nd ed. Sandy, OR. Eclectic Medical Publications. 1998.
[1755] Rabl W, Katzgraber F, Steinlechner M. Camphor ingestion for abortion (case report). *Forensic Sci Int.* 1997 Sep 19;89(1-2):137-40.
[1756] Flaman Z, Pellechia-Clarke S, Bailey B, et al. Unintentional exposure of young children to camphor and eucalyptus oils. *Paediatr Child Health.* 2001 Feb;6(2):80-83.
[1757] Frohne D. Giftpflanzen: Cupressaceae. Stuttgart: Wissenschaftliche Verlagsgesellschaft mbH; 1997. pp. 153–6.
[1758] European Medicines Agency. Public statement on the use of herbal medicinal products containing thujone. Available at: http://www.ema.europa.eu/docs/en_GB/document_library/Public_statement/2011/02/WC500102294.pdf.
[1759] Millet Y, Jouglard J, Steinmetz MD, et al. Toxicity of some essential plant oils. Clinical and experimental study. *Clin Toxicol.* 1981 Dec;18(12):1485-98.
[1760] Cristovao L, Carvalho F, Bastos MDL, et al. Hepatotoxicity of an essential oil of Salvia officinalis L.: an in vitro study using freshly isolated rat hepatocytes. *Congress Biomarkers.* 2001 Sep:165.
[1761] Millet Y, Jouglard J, Steinmetz MD, et al. Toxicity of some essential plant oils. Clinical and experimental study. *Clin Toxicol.* 1981 Dec;18(12):1485-98.
[1762] Burkhard PR, Burkhardt K, Haenggeli CA, et al. Plant-induced seizures: reappearance of an old problem. *J Neurol.* 1999 Aug;246(8):667-70.
[1763] Manoguerra AS, Erdman AR, Wax PM, et al. Camphor poisoning: an evidence-based practice guideline for out-of- hospital management. *Clin Toxicol (Phila).*2006;44(4):357-70.
[1764] Gibson DE, Moore GP, Pfaff JA. Camphor ingestion. *Am J Emerg Med.* 1989 Jan;7(1):41-43.
[1765] Koppel C, Martends F, Schirop T, et al. Hemoperfusion in acute camphor poisoning. *Intensive Care Med.* 1988;14(4):431-33.
[1766] Burkhard PR, Burkhardt K, Haenggeli CA, et al. Plant-induced seizures: reappearance of an old problem. *J Neurol.* 1999 Aug;246(8):667-70.
[1767] Hold KM, Sirisoma NS, Ikeda T, et al. α-Thujone (the active component of absinthe): γ-Aminobutyric acid type A receptor modulation and metabolic detoxification. *Proc Natl Acad Sci USA.* 2000 Apr 11;97(8):3826–3831.
[1768] Waldman N. Seizure caused by dermal application of over-the-counter eucalyptus oil head lice preparation. *Clin Toxicol (Phila).* 2011 Oct;49(8):750-1.
[1769] Craig JO. Poisoning by the volatile oils in childhood. *Arch Dis Child.* 1953;28:259-67.
[1770] Mathew T, Kamath V, Kumar RS, et al. Eucalyptus oil inhalation-induced seizure: A novel, underrecognized, preventable cause of acute symptomatic seizure. *Epilepsia Open.* 2017 Jul 4;2(3):350-354.
[1771] Unger M, Frank A. Simultaneous determination of the inhibitory potency of herbal extracts on the activity of six major cytochrome P450 enzymes using liquid chromatography/mass spectrometry and automated online extraction. *Rapid Commun Mass Spectrom.* 2004;18(19):2273-81.
[1772] Jori A, Bianchetti A, Prestini PE, et al. Effect of eucalyptol (1,8-cineole) on the metabolism of other drugs in rats and in man. *Eur J Pharmacol.* 1970;9(3):362-66.
[1773] Abass K, Reponen P, Mattila S, et al. Metabolism of α-thujone in human hepatic preparations in vitro. *Xenobiotica.* 2011 Feb;41(2):101-11.
[1774] Kim NH, Hyun SH, Jin CH, et al. Pretreatment with 1,8-cineole potentiates thioacetamide-induced hepatotoxicity and immunosuppression. *Arch Pharm Res.* 2004 Jul;27(7):781-9.
[1775] Burkhard PR, Burkhardt K, Haenggeli CA, et al. Plant-induced seizures: reappearance of an old problem. *J Neurol.* 1999 Aug;246(8):667-70.
[1776] Waldman N. Seizure caused by dermal application of over-the-counter eucalyptus oil head lice preparation. *Clin Toxicol (Phila).* 2011 Oct;49(8):750-1.
[1777] Craig JO. Poisoning by the volatile oils in childhood. *Arch Dis Child.* 1953;28:259-67.
[1778] Mathew T, Kamath V, Kumar RS, et al. Eucalyptus oil inhalation-induced seizure: A novel, underrecognized, preventable cause of acute symptomatic seizure. *Epilepsia Open.* 2017 Jul 4;2(3):350-354.

[1779] Day LM, Ozanne-Smith J, Parsons BJ, et al. Eucalyptus oil poisoning among young children: mechanisms of access and the potential prevention. *Aust N Z J Public Health.* 1997 Jun;21(3):297-302.
[1780] Myott E. Case of eucalyptus poisoning. *Brit M J.* 1906;1:558.
[1781] Hindle RC. Eucalyptus oil ingestion. *N Z Med J.* 1994 May 11;107(977)185-6.
[1782] Tibballs J. Clinical effects and management of eucalyptus oil ingestion in infants and small children. *Med J Aust.* 1995 Aug;163(4):177-80.
[1783] Waldman W, Barwina M, Sein Anand J. Accidental intoxication with eucalyptus oil—a case report. *Przeql Lek.* 2011;68(8):555-6.
[1784] Day LM, Ozanne-Smith J, Parsons BJ, et al. Eucalyptus oil poisoning among young children: mechanisms of access and the potential prevention. *Aust N Z J Public Health.* 1997 Jun;21(3):297-302.
[1785] Unger M, Frank A. Simultaneous determination of the inhibitory potency of herbal extracts on the activity of six major cytochrome P450 enzymes using liquid chromatography/mass spectrometry and automated online extraction. *Rapid Commun Mass Spectrom.* 2004;18(19):2273-81.
[1786] Jori A, Bianchetti A, Prestini PE, et al. Effect of eucalyptol (1,8-cineole) on the metabolism of other drugs in rats and in man. *Eur J Pharmacol.* 1970;9(3):362-66.
[1787] Kim NH, Hyun SH, Jin CH, et al. Pretreatment with 1,8-cineole potentiates thioacetamide-induced hepatotoxicity and immunosuppression. *Arch Pharm Res.* 2004 Jul;27(7):781-9.
[1788] Jori A, Bianchetti A, Prestini PE, et al. Effect of eucalyptol (1,8-cineole) on the metabolism of other drugs in rats and in man. *Eur J Pharmacol.* 1970;9(3):362-66.
[1789] de Sousa DP, Raphael E, Brocksom U, et al. Sedative effect of monoterpene alcohols in mice: A preliminary screening. *Verlag der Zeitschrift fur Naturforschung.* 2007;62c:563-66.
[1790] Elmann A, Mordechay S, Rindner M, et al. Protective Effects of the Essential Oil of Salvia fruticosa and Its Constituents on Astrocytic Susceptibility to Hydrogen Peroxide-Induced Cell Death. *J Agric Food Chem.* 2009 Aug 12;57(15):6636-41.
[1791] Sarac N, Ugur A. The in vitro antimicrobial activities of the essential oils of some Lamiaceae species from Turkey. *J Med Food.* 2009 Aug;12(4):902-7.
[1792] Chovanová R, Mezovská J, Vaverková Š, et al. The inhibition the Tet(K) efflux pump of tetracycline resistant Staphylococcus epidermidis by essential oils from three Salvia species. *Lett Appl Microbiol.* 2015 Jul;61(1):58-62.
[1793] Askun T, Baser KHC, Tumen G, et al. Characterization of essential oils of some Salvia species and their antimycobacterial activities. *Turk J Biol.* 2010;34:89-95.
[1794] Soković M, Tzakou O, Pitarokili D, et al. Antifungal activities of selected aromatic plants growing wild in Greece. *Nahrung.* 2002 Oct;46(5):317-20.
[1795] Daferera DJ, Ziogas BN, Polissiou MG. GC-MS analysis of essential oils from some Greek aromatic plants and their fungitoxicity on Penicillium digitatum. *J Agric Food Chem.* 2000 Jun;48(6):2576-81.
[1796] Zeidán-Chuliá F, Rybarczyk-Filho JL, Gursoy M, et al. Bioinformatical and in vitro approaches to essential oil-induced matrix metalloproteinase inhibition. *Pharm Biol.* 2012 Jun;50(6):675-86.
[1797] Papageorgiou V, Gardeli C, Mallouchos A, et al. Variation of the chemical profile and antioxidant behavior of Rosmarinus officinalis L. and Salvia fruticosa Miller grown in Greece. *J Agric Food Chem.* 2008 Aug 27;56(16):7254-64.
[1798] Limberger RP, Aboy AL, Bassani VL, et al. Essential Oils from Four Mikania Species (Asteraceae). *J Essent Oil Res.* 2001 Julk/Aug;13:225-8.
[1799] Rehder VLG, Sartoratto A, Rodrigues MVN. Essential oils composition from leaves, inflorescences and seeds of Mikania laevigata Schultz Bip. ex Baker and Mikania glomerata Sprengel. *Rev Bras Pl Med, Botucatu.* 2006;8(esp):116-118.
[1800] Limberger R, Suyenaga ES, Henriques AT, et al. Chemical Composition of Essential Oils from Three Southern Brazilian Species of Mikania (Asteraceae). *J Essent Oil Res.* 1998 Jul/Aug;10:363-7.
[1801] Duarte MCT, Figueira GM, Satoratto A, et al. Anti-Candida activity of Brazilian medicinal plants. *J Ethnopharmacol.* 2005;97:305-11.
[1802] Lin WY, Kuo YH, Chang YL, et al. Anti-platelet aggregation and chemical constituents from the rhizome of Gynura japonica. *Planta Med.* 2003 Aug;69(8):757-64.
[1803] Duarte MCT, Figueira GM, Satoratto A, et al. Anti-Candida activity of Brazilian medicinal plants. *J Ethnopharmacol.* 2005;97:305-11.
[1804] Priestap HA, Bennett BC, Quirke JME. Investigation of the essential oils of Bidens pilosa var minor, Bidens alba, and Flaveria linearis. *J Essent Oil Res.* 2008 Sep/Oct;20:396-402.
[1805] Goudoum A, Ngamo Tinkeu LS, Ngassoum MB, et al. Insecticidal and Antifungal Properties of Essential Oil of Bidens Pilosa Linn. Var. Radita (Asteraceae) Towards Stored Bambara Groundnut Insect and Fungi Pests. *Asian J Agric Food Sci.* 2016 Apr;04(2):66-72.
[1806] Goudoum A, Abdou AB, Ngamo LST, et al. Antioxidant activities of essential oil of Bidens pilosa (Linn. Var. Radita) used for the preservation of food qualities in North Cameroon. *Food Sci Nutr.* 2016;4(5):671-8.
[1807] Deba F, Xuan TD, Yasuda M, et al. Chemical composition and antioxidant, antibacterial and antifungal activities of the essential oils from Bidens pilosa Linn. var. Radiata. *Food Control.* 2008;19:346-52.
[1808] Ocheng F, Bwanga F, Joloba M, et al. Essential Oils from Ugandan Aromatic Medicinal Plants: Chemical Composition and Growth Inhibitory Effects on Oral Pathogens. *Evidence-Based Complement Alternat Med.* 2015;2015:230832.
[1809] Wat CK, Biswas RK, Graham EA, et al. Ultraviolet-mediated cytotoxic activity of phenylheptatriyne from Bidens pilosa L. *J Nat Prod.* 1979 Jan-Feb;42(1):103-11.
[1810] Gong HH, Kagan J, Seitz R, et al. The phototoxicity of phenylheptatriyne: oxygen-dependent hemolysis of human erythrocytes and inactivation of Escherichia coli. *Photochem Photobiol.* 1988 Jan;47(1):55-63.
[1811] Ketel DH. Chemistry and biology of naturally occurring acetylenes and related compounds. *Flavour Frag J.* 1990 Mar;5(1):69-69.
[1812] Deba F, Xuan TD, Yasuda M, et al. Chemical composition and antioxidant, antibacterial and antifungal activities of the essential oils from Bidens pilosa Linn. var. Radiata. *Food Control.* 2008;19:346-52.
[1813] Ocheng F, Bwanga F, Joloba M, et al. Essential Oils from Ugandan Aromatic Medicinal Plants: Chemical Composition and Growth Inhibitory Effects on Oral Pathogens. *Evidence-Based Complement Alternat Med.* 2015;2015:230832.
[1814] Goudoum A, Ngamo Tinkeu LS, Ngassoum MB, et al. Insecticidal and Antifungal Properties of Essential Oil of Bidens Pilosa Linn. Var. Radita (Asteraceae) Towards Stored Bambara Groundnut Insect and Fungi Pests. *Asian J Agric Food Sci.* 2016 Apr;04(2):66-72.
[1815] Goudoum A, Abdou AB, Ngamo LST, et al. Antioxidant activities of essential oil of Bidens pilosa (Linn. Var. Radita) used for the preservation of food qualities in North Cameroon. *Food Sci Nutr.* 2016;4(5):671-8.
[1816] Falchero Luca, Coppa M, Fossi A, et al. Essential oil composition of lady's mantle (Alchemilla xanthochlora Rothm.) growing wild in Alpine pastures. *Nat Prod Res.* 2009;23:15:1367-72.
[1817] Tbanca N, Avonto C, Wang M, et al. Comparative investigation of Umbellularia californica and Laurus nobilis leaf essential oils and identification of constituents active against Aedes aegypti. *J Agric Food Chem.* 2013;61:12283-91.

[1818] Kelsey RG, McCuistion O, Karchesy J. Bark and leaf essential oil of Umbellularia Californica, California bay laurel, from Oregon. *Nat Prod Commun.* 2007 Jan;2(7):779-80.
[1819] Nassini R, Materazzi S, Vriens J, et al. The 'headache tree' via umbellulone and TRPA1 activates the trigeminovascular system. *Brain.* 2012 Feb;135(Pt 2):376-90.
[1820] Benemei S, Appendino G, Geppetti P. Pleasant natural scent with unpleasant effects: cluster headache-like attacks triggered by Umbellularia californica. *Cephalalgia.* 2010 Jun;30(6):744-6.
[1821] Wagstaff HM, Maman SR, Tufte MJ, et al. The Effect of Umbellularia californica Essential Oil on Blood Vessel Diameter in Frogs. *Am J Undergraduate Res.* 2016 Jan;13(1):59-64.
[1822] Satyal P. Private communication. December 2018.
[1823] Mehmeti V. Essential Oils of Japanese and Austrian Liverworts and of Selected Japanese Higher Plants. University thesis. Available at: https://core.ac.uk/download/pdf/11588573.pdf.
[1824] Nagaki M, Goto Y, Yamanouchi K, et al. Phytochemical profile and antioxidant activity of the leaf of Aomori Hiba (Thujopsis dolabrata SIEB, et ZUCC. Var hondai MAKINO). *J Hirosaki Univ Health Welfare.* 2018;9(1):1-7.
[1825] Jeong HU, Kwon SS, Kong TY, et al. Inhibitory effects of cedrol, β-cedrene, and thujopsene on cytochrome P450 enzyme activities in human liver microsomes. *J Toxicol Environ Health A.* 2014;77(22-24):1522-32.
[1826] Fujita M, Aoki T. Allergic contact dermatitis to pyridoxine ester and hinokitiol. *Conbtact Dermatitis.* 1982 Feb;9(1):61-5.
[1827] Hosono K, Wakita M. Two Cases of Allergic Contact Dermatitis to Hinokitiol and Pantothenylethylether in Hair Growth Promoter. *J Stage.* 1997 Jan;39(1):42-7.
[1828] Nagata T, Fujino Y, Toume K, et al. Anti-cancer Effect in Volatile Components of Hiba Essential Oil (Thujopsis dolabrata). *J Clin Exper Pharmacol.* 2016;6(4):214.
[1829] Nam KW, Noh JK, Kim SK, et al. Essential Oil of Thujopsis dolobrata Suppresses Atopic Dermatitis-Like Skin Lesions in NC/Nga Mice. *Biomol Ther.* 2011;19(1):102-8.
[1830] Matsuura T, Yamaguchi T, Zaike Y, et al. Reduction of the chronic stress response by inhalation of hiba (Thujopsis dolabrata) essential oil in rats. *Biosc Biotech Biochem.* 2014;78(7):1135-9.
[1831] Nagaki M, Goto Y, Yamanouchi K, et al. Phytochemical profile and antioxidant activity of the leaf of Aomori Hiba (Thujopsis dolabrata SIEB, et ZUCC. Var hondai MAKINO). *J Hirosaki Univ Health Welfare.* 2018;9(1):1-7.
[1832] Thappa RK, Kaul P, Christi AM, et al. Variability in the Essential Oil of Angelica glauca Edgew of Different Geographical Regions. *J Essent Oil Res.* 2005 Jul;17(4):361-3.
[1833] Purohit VK, Andola HC, Haider SZ, et al. Essential Oil Constituents of Angelica glauca Edgew. Roots: An Endangered Species from Uttarakhand Himalaya (India). *Natl Acad Sci Lett.* 2015 Sep/Oct;38(5):445-7.
[1834] Agnihotri VK, Thappa RK, Meena B, et al. Essential oil composition of aerial parts of Angelica glauca growing wild in North-West Himalaya (India). *Phytochem.* 2004;65:2411-3.
[1835] Irshad M, Aziz S, Habib-ur-Rehman, et al. GC-MS Analysis and Antifungal Activity of Essential oils of Angelica glauca, Plectranthus rugosus, and Valeriana wallichii. *J Essential Oil Bearing Plants.* 2012;15(1):15-21.
[1836] Chauhan RS, Nautiyal MC, Bahuguna YM, et al. Volatile Composition of Underground Parts of Angelica glauca Edgew. from Two Distant Populations of India. *J Essent Oil Bearing Plants.* 2017;20(3):851-4.
[1837] Kaul PN, Mallavarapu GR, Chamoli RP. The essential oil composition of Angelica glauca roots. *Planta Medica.* 1996 Mar;62(1):80-1.
[1838] Circosta C, Pasquale RD, Palumbo DR, et al. Estrogenic activity of standardized extract of Angelica sinensis. *Phytother Res.* 2006 Aug;20(8):665–69.
[1839] Shi M, Chang L, He G. [Stimulating action of Carthamus tinctorius L., Angelica sinensis (Oliv.) Diels and Leonurus sibiricus L. on the uterus]. *Zhongguo Zhong Yao Za Zhi.* 1995 Mar;20(3):173-5, 192.
[1840] Shi M, Chang L, He G. [Stimulating action of Carthamus tinctorius L., Angelica sinensis (Oliv.) Diels and Leonurus sibiricus L. on the uterus]. *Zhongguo Zhong Yao Za Zhi.* 1995 Mar;20(3):173-5, 192.
[1841] Brindis F, Rodríguez R, Bye R, et al. (Z)-3-butylidenephthalide from Ligusticum porteri, an α-glucosidase inhibitor. *J Nat Prod.* 2011 Mar 25;74(3):314-20.
[1842] Zhang L, Du JR, Wang J, et al. Z-ligustilide extracted from Radix Angelica Sinensis decreased platelet aggregation induced by ADP ex vivo and arterio-venous shunt thrombosis in vivo in rats. *Yakugaku Zasshi.* 2009 Jul;129(7):855-9.
[1843] Qi H, Zhao J, Han Y, et al. Z-ligustilide potentiates the cytotoxicity of dopamine in rat dopaminergic PC12 cells. *Neurotox Res.* 2012 Nov;22(4):345-54.
[1844] Sharma S, Rasal VP, Patil PA, et al. Effect of Angelica glauca essential oil on allergic airway changes induced by histamine and ovalbumin in experimental animals. *Indian J Pharmacol.* 2017 Jan-Feb;49(1):55-59.
[1845] Lohani H, Gwari G, Andola HC, et al. α-Pinene Rich Volatile Constituents of Cupressus torulosa D. Don from Uttarakhand Himalaya. *Indian J Pharm Sci.* 2012 May-Jun; 74(3): 278–280.
[1846] Marchuk NY. The research of essential oil's chemical composition in species of genus Cupressus L., growing in Crimean south coast. *Ukranian Bioorganica Acta.* 2014;1:40-43.
[1847] Giatropoulos A, Pitarokili D, Papaioannou F, et al. Essential oil composition, adult repellency and larvicidal activity of eight Cupressaceae species from Greece against Aedes albopictus (Diptera: Culicidae). *Parasitology Research.* 2013;112(3):1113-23.
[1848] Padalia RC, Verma RS, Chauhan A, et al. Essential oil compositions of branchlets and cones of Cupressus torulosa D. Don. *J Essent Oil Res.* 2013;25(4):251-6.
[1849] Lohani H, Kumar A, Bhandari U, et al. Effect of Seasonal and Tree Girth Size Variation on Cupressus torulosa D. Don Leaves Essential Oil Composition Growing in Uttarakhand. *J Essent Oil Bear Plants.* 2014;17(6):1257-67.
[1850] Molly, ZOils. Private communication. August 2018.
[1851] Nassini R, Materazzi S, Vriens J, et al. The 'headache tree' via umbellulone and TRPA1 activates the trigeminovascular system. *Brain.* 2012 Feb;135(Pt 2):376-90.
[1852] Benemei S, Appendino G, Geppetti P. Pleasant natural scent with unpleasant effects: cluster headache-like attacks triggered by Umbellularia californica. *Cephalalgia.* 2010 Jun;30(6):744-6.
[1853] Gupta S, Bhagat M, Sudan R, et al. Analysis of chemical composition of Cupressus torulosa (D.Don) essential oil and bioautography guided evaluation of its antimicrobial fraction. *Infian J Exp Biol.*
[1854] Giatropoulos A, Pitarokili D, Papaioannou F, et al. Essential oil composition, adult repellency and larvicidal activity of eight Cupressaceae species from Greece against Aedes albopictus (Diptera: Culicidae). *Parasitology Research.* 2013;112(3):1113-23.
[1855] Satyal P, Setzer WN. Chemical composition and enantiomeric distribution of monoterpenoids of the essential oil of Abies spectabilis from Nepal. *Am J Essent Oils Nat Prod.* 2017;5(1):22-6.
[1856] Verma RS, Padalia RC, Goswami P, et al. Assessing productivity and essential oil quality of Himalayan thyme (Thymus linearis Benth.) in the subtropical region of north India. *Ind Crops Prod.* 2016;94:557-61.

[1857] Verma RS, Padalia RC, Chanotiya CS, et al. Chemical investigation of the essential oil of Thymus linearis (Benth. ex Benth) from western Himalaya, India. *Nat Prod Res.* 2010 Dec 15;24(20):1890-6.
[1858] Chandra M, Prakash O, Bachheti RK, et al. Essential Oil Composition, Phenolic Constituents, Antioxidant and Pharmacological Activities of Thymus linearis Benth. Collected from Uttarakhand Region of India. *J Essent Oil Bearing Plants.* 2016;19(2):277-89.
[1859] Kohlert C, Schindler G, Marz RW, et al. Systemic availability and pharmacokinetics of thymol in humans. *J Clin Pharmacol.* 2002 Jul;42(7):731-7.
[1860] Toxicology Data Network, National Library of Medicine. Thymol. Available at: http://toxnet.nlm.nih.gov/cgibin/sis/search/a?dbs+hsdb:@term+@DOCNO+866.
[1861] Tognolini M, Barocelli E, Ballabeni V, et al. Comparative screening of plant essential oils; phenylpropanoid moiety as basic core for antiplatelet activity. *Life Sci.* 2006 Feb 23;78(13):1419-32.
[1862] Okazaki K, Kawazoe K, Takaishi Y. Human platelet aggregation inhibitors from thyme (Thymus vulgaris L.). *Phytother Res.* 2002 Jun;16(4):398-9.
[1863] Jukic M, Politeo O, Maksimmovic M, et al. In vitro acetylcholinesterase inhibitory properties of thymol, carvacrol and their derivatives thymoquinone and thymohydroquinone. *Phytother Res.* 2007;21(3):259-61.
[1864] Askin H, Yildiz M, Ayar A. Effects of Thymol and Carvacrol on Acetylcholinesterase from Drosophila melanogaster. *Acta Physica Polinica A.* 2017;132(3):720-22.
[1865] Medeiros Barreto H, Cerqueira Fontinele F, Pereira de Oliveira A, et al. Phytochemical prospection and modulation of antibiotic activity in vitro by Lippia origanoides H.B.K. in methicillin resistant Staphylococcus aureus. *Biomed Res Int.* 2014;305610:1–7.
[1866] Aelenei P, Miron A, Trifan A, et al. Essential Oils and Their Components as Modulators of Antibiotic Activity against Gram-Negative Bacteria. *Medicines.* 2016;3(19):2-34.
[1867] Magi G, Marini E, Facinelli B. Antimicrobial activity of essential oils and carvacrol, and synergy of carvacrol and erythromycin, against clinical, erythromycin-resistant Group A Streptococci. *Front Microbiol.* 2015;6:165.
[1868] Veras HNH, Rodrigues FFG, Colares AV, et al. Synergistic antibiotic activity of volatile compounds from the essential oil of Lippia sidoides and thymol. *Fitoterapia.* 2012 Apr;83(3):508-12.
[1869] Langeveld WT, Veldhuizen EJ, Burt SA. Synergy between essential oil components and antibiotics: A review. *Crit Rev Microbiol.* 2014 Feb;40(1):76-94.
[1870] Miladinovic DL, Ilic BS, Kocic BD, et al. Antibacterial investigation of thyme essential oil and its main constituents on combination with tetracycline. *J Med Food.* 2015 Aug;18(8):935-7.
[1871] Shin S, Lim S. Antifungal effects of herbal essential oils alone and in combination with ketoconazole against Trichophyton spp. *J Appl Micribiol.* 2004;97:1289-96.
[1872] Kissels W, Wu X, Santos RR, et al. Short communication: Interaction of the isomers carvacrol and thymol with the antibiotics doxycycline and tilmicosin: In vitro effects against pathogenic bacteria commonly found in the respiratory tract of calves. *J Dairy Sci.* 2017 Feb;100(2):970-974.
[1873] Joulain D. Study of the fragrance given off by certain springtime flowers. 1986. In: Brunke EJ. Progress in essential oil research. De Gruyter, pp. 57-67.
[1874] Radulovic N, Blagojevic P, Palic R. Composition of Diethyl Ether Flower Extracts of Lonicera fragrantissima Lindl. & Paxton (Caprifoliaceae). *Nat Prod Commun.* 2009;4(0):1-4.
[1875] Ilies DC, Radulescu V, Dutu L. Volatile constituents from the flowers of two species of Honeysuckle (Lonicera japonica and Lonicera caprifolium). *Farmacia.* 2014;62(1):194-201.
[1876] Schlotzhauer WS, Pair SD, Horvat RJ. Volatile Constituents from the Flowers of Japanese Honeysuckle (Lonicera japonica). *J Agric Food Chem.* 1996;44:206-9.
[1877] Vukovic N, Kacaniova M, Hleba L, et al. Chemical composition of the essential oils from the flower, leaf and stem of Lonicera japonica. *Nat Prod Commun.* 2012;7(5):641-44.
[1878] Rahman A, Kang SC. In vitro control of food-borne and food spoilage bacteria by essential oil and ethanol extracts of Lonicera japonica Thunb. *Food Chem.* 2009;116:670-5.
[1879] Wu C, Wang F, Liu J, et al. A comparison of volatile fractions obtained from Lonicera macranthoides via different extraction processes: ultrasound, microwave, Soxhlet extraction, hydrodistillation, and cold maceration. *Integr Med Res.* 2015;4:171-7.
[1880] Špičáková A, Szotáková B, Dimunová D, et al. Nerolidol and Farnesol Inhibit Some Cytochrome P450 Activities but Did Not Affect Other Xenobiotic-Metabolizing Enzymes in Rat and Human Hepatic Subcellular Fractions. *Molecules.* 2017 Mar 24;22(4). pii: E509.
[1881] Ganzera M, Schneider P, Stuppner H. Inhibitory effects of the essential oil of chamomile (Matricaria recutita L.) and its major constituents on human cytochrome P450 enzymes. *Life Sci.* 2006 Jan 18;78(8):856-61.
[1882] Yang SP, Raner Gm. Cytochrome P450 expression and activities in human tongue cells and their modulation by green tea Extract. *Toxicol Appl Pharmacol.* 2005;202:140-150.
[1883] Johnson S. Supercritical Essential Oils. 2017. Scott A Johnson Professional Writing Services, LLC: Orem, Utah.
[1884] EWG. Dibutyl Phthalate. Available at: https://www.ewg.org/skindeep/ingredient/701929/DIBUTYL_PHTHALATE/#.
[1885] Rahman A, Kang SC. In vitro control of food-borne and food spoilage bacteria by essential oil and ethanol extracts of Lonicera japonica Thunb. *Food Chem.* 2009;116:670-5.
[1886] Rahman A, Al-Reza SM, Siddiqui SA, et al. Antifungal potential of essential oil and ethanol extracts of Lonicera japonica Thunb. against dermatophytes. *EXCLI J.* 2014 Apr 28;13:427-36.
[1887] Tomsone L, Kruma Z, Galoburda R. Composition of Volatile Compounds of Horseradish Roots (Armoracia rusticana L.) Depending on the Genotype. *Proc Latv Univ Agr.* 2013 Jul;29(324):1-10.
[1888] Nhat NM. Study On Horseradish (Armoracia Rusticana) Essential Oil And Comparison With The Related Species – Debreceni Horseradish (Armoracia Macrocarpa). Doctoral thesis. Available at: https://dea.lib.unideb.hu/dea/bitstream/handle/2437/231292/Finalized_thesis.pdf?sequence=1.
[1889] Chen H, Akinkurolere RO, Zhang H. Fumigant activity of plant essential oil from Armoracia rusticana (L.) on Plodia interpunctella (Lepidoptera: Pyralidae) and Sitophilus zeamais (Coleoptera: Curculionidae). *African J Biotech.* 2011 Feb;10(7):1200-5.
[1890] Gilbert J, Nursten H. Volatile constituents of horseradish roots. *J Sci Food Agric.* 1972;23(4):527-539.
[1891] Lawrence BM. Essential Oils 1981-1987. Allured Publishing, Wheaton.
[1892] Park IK, Choi KS, Kim DH, et al. Fumigant activity of plant essential oils and components from horseradish (Armoracia rusticana), anise (Pimpinella anisum) and garlic (Allium sativum) oils against Lycoriella ingenua (Diptera: Sciaridae). *Pest Manag Sci.* 2006;62:723-8.
[1893] European Food Safety Association. Scientific Opinion on the safety of allyl isothiocyanate for the proposed uses as a food additivel EFSA Panel on Food Additives and Nutrient Sources added to Food (ANS). *EFSA J.* 2010;8(12):1943.

[1894] European Food Safety Association. Scientific Opinion on the safety of allyl isothiocyanate for the proposed uses as a food additivel EFSA Panel on Food Additives and Nutrient Sources added to Food (ANS). *EFSA J.* 2010;8(12):1943.
[1895] Lee DS, Kim TH, Jung YS. Inhibitory effect of allyl isothiocyanate on platelet aggregation. *J Agric Food Chem.* 2014 Jul 23;62(29):7131-9.
[1896] Mori N, Kurata M, Yamazaki H, et al. Intragastric administration of allyl isothiocyanate reduces hyperglycemia in intraperitoneal glucose tolerance test (IPGTT) by enhancing blood glucose consumption in mice. *J Nutr Sci Vitaminol (Tokyo).* 2013;59(1):56-63.
[1897] Mori N, Kurata M, Yamazaki H, et al. Allyl isothiocyanate increases carbohydrate oxidation through enhancing insulin secretion by TRPV1. *Biosci Biotechnol Biochem.* 2018 Apr;82(4):698-708.
[1898] Grant, W.M. Toxicology of the Eye. 3rd ed. Springfield, IL: Charles C. Thomas Publisher, 1986., p. 70.
[1899] Gosselin, R.E., R.P. Smith, H.C. Hodge. Clinical Toxicology of Commercial Products. 5th ed. Baltimore: Williams and Wilkins, 1984., p. II-352.
[1900] Lerbaek A, Rastogi SC, Menné T. Allergic contact dermatitis from allyl isothiocyanate in a Danish cohort of 259 selected patients. *Contact Dermatitis.* 2004 Aug;51(2):79-83.
[1901] IARC. Monographs on the Evaluation of the Carcinogenic Risk of Chemicals to Humans. Geneva: World Health Organization, International Agency for Research on Cancer, 1972-PRESENT. (Multivolume work). Available at: http://monographs.iarc.fr/ENG/Classification/index.php p. V73 40 (1999).
[1902] Brand G, Jacquot L. Sensitization and desensitization to allyl isothiocyanate (mustard oil) in the nasal cavity. *Chem Senses.* 2002 Sep;27(7):593-8.
[1903] Kloucek P, Smid J, Flesar J, et al. In vitro inhibitory activity of essential oil vapors against Ascosphaera apis. *Nat Prod Commun.* 2012 Feb;7(2):253-6.
[1904] Chen H, Akinkurolere RO, Zhang H. Fumigant activity of plant essential oil from Armoracia rusticana (L.) on Plodia interpunctella (Lepidoptera: Pyralidae) and Sitophilus zeamais (Coleoptera: Curculionidae). *African J Biotech.* 2011 Feb;10(7):1200-5.
[1905] Park IK, Choi KS, Kim DH, et al. Fumigant activity of plant essential oils and components from horseradish (Armoracia rusticana), anise (Pimpinella anisum) and garlic (Allium sativum) oils against Lycoriella ingenua (Diptera: Sciaridae). *Pest Manag Sci.* 2006;62:723-8.
[1906] Yun YK, Kim HK, Kim JR, et al. Contact and fumigant toxicity of Armoracia rusticana essential oil, allyl isothiocyanate and related compounds to Dermatophagoides farinae. *Pest Manag Sci.* 2012 May;68(5):788-94.
[1907] Guenther E. 1949-1952. The essential oils. Vols. 1–6. Van Nostrand, New York.
[1908] IARC Monographs – 101. Methyl eugenol.
[1909] Brophy JJ, Goldsack RJ. Chemistry of the Australian Gymnosperms—Part 5: Leaf essential oils of some endemic Tasmanian Gymnosperms: Diselma archeri, Lagarostrobos franklinii, Microcachrys tetragona and Phyllocladus aspleniifolius. *J Essen Oil Res.* 2003 Jul/Aug;15:217-20.
[1910] Johnson JD, Ryan MJ, Toft JD II, et al. Two-year toxicity and carcinogenicity study of methyl eugenol in F344/N rats and B6C3F(1) mice. *J Agric Food Chem.* 2000 Aug;48(8):3620-32.
[1911] European Commission. Opinion of the Scientific Committee on Food on methyl eugenol (4-Allyl-1,2- dimethoxybenzene). Available at: http://ec.europa.eu/food/fs/sc/scf/out102_en.pdf.
[1912] National Toxicology Program. NTP Toxicology and Carcinogenesis Studies of Methyl eugenol (CAS NO. 93-15-2) in F344/N Rats and B6C3F1 Mice (Gavage Studies). *Natl Toxicol Program Tech Rep Ser.* 2000 Jul;491:1-412.
[1913] National Toxicology Program. Carcinogenesis Studies of Eugenol (CAS No. 97-53-0) in F344/N Rats and B6C3F1 Mice (Feed Studies). Technical Report Series No. 223. NIH Publication No. 84-1779. 1983. U.S. DHHS, PHS, NIH, NTP, Research Triangle Park, NC.
[1914] Johnson JD, Ryan MJ, Toft JD II, et al. Two-year toxicity and carcinogenicity study of methyl eugenol in F344/N rats and B6C3F(1) mice. *J Agric Food Chem.* 2000 Aug;48(8):3620-32.
[1915] National Toxicology Program. NTP Toxicology and Carcinogenesis Studies of Methyl eugenol (CAS NO. 93-15-2) in F344/N Rats and B6C3F1 Mice (Gavage Studies). *Natl Toxicol Program Tech Rep Ser.* 2000 Jul;491:1-412.
[1916] Kerckaert GA, Brauninger R, LeBoeuf RA, et al. Use of the Syrian hamster embryo cell transformation assay for carcinogenicity prediction of chemicals currently being tested by the National Toxicology Program in rodent bioassays. *Environ Health Perspect.* 1996;104:1075-84.
[1917] National Toxicology Program. Carcinogenesis Studies of Eugenol (CAS No. 97-53-0) in F344/N Rats and B6C3F1 Mice (Feed Studies). Technical Report Series No. 223. NIH Publication No. 84-1779. 1983. U.S. DHHS, PHS, NIH, NTP, Research Triangle Park, NC.
[1918] Johnson JD, Ryan MJ, Toft JD II, et al. Two-year toxicity and carcinogenicity study of methyl eugenol in F344/N rats and B6C3F(1) mice. *J Agric Food Chem.* 2000 Aug;48(8):3620-32.
[1919] European Commission. Opinion of the Scientific Committee on Food on methyl eugenol (4-Allyl-1,2- dimethoxybenzene). Available at: http://ec.europa.eu/food/fs/sc/scf/out102_en.pdf.
[1920] National Toxicology Program. NTP Toxicology and Carcinogenesis Studies of Methyl eugenol (CAS NO. 93-15-2) in F344/N Rats and B6C3F1 Mice (Gavage Studies). *Natl Toxicol Program Tech Rep Ser.* 2000 Jul;491:1-412.
[1921] National Toxicology Program. Carcinogenesis Studies of Eugenol (CAS No. 97-53-0) in F344/N Rats and B6C3F1 Mice (Feed Studies). Technical Report Series No. 223. NIH Publication No. 84-1779. 1983. U.S. DHHS, PHS, NIH, NTP, Research Triangle Park, NC.
[1922] Ahmad A, Khan A, Ahmad Khan L, et al. In vitro synergy of eugenol and methyl eugenol with fluconazole against clinical Candida isolates. *J Med Microbiol.* 2010;59:1178-84.
[1923] Lee SE, Lee BH, Choi WS, et al. Fumigant toxicity of volatile natural products from Korean spices and medicinal plants towards the rice weevil, Sitophilus oryzae (L). *Pest Manag Sci.* 2001;57:548-53.
[1924] Hosokawa K, Fukunaga Y. Production of essential oils by flowers of Hyacinthus orientalis L. regenerated in vitro. *Plant Cell Reports.* 19995;14:575-9.
[1925] Atahacoba T, Hehob H, Лазаров A, et al. Low temperature extraction of essential oil bearing plants by liquefied gases. 3. Hyacinth flowers (Hyacinthus orientalis L.). *НАУЧНИ ТРУДОВЕ НА РУСЕНСКИЯ УНИВЕРСИТЕТ.* 2009 Jan;48(9):131-5.
[1926] Brunke EJ, Hammerschmidt FJ, Schmaus G. Headsapce analysis of hyacinth flowers. *Flavour Frag J.* 1994;9:59-69.
[1927] Dulgheru. Research on the composition of volatile oils produced by some species of flowers in Romania. Doctoral Thesis. Available at: https://www.usamv.ro/images/Programe_de_studii/Doctorat/Teze_de_doctorat/Arhiva_2013/Dulgheru_EN.pdf.
[1928] Veluthoor S, Kelsey RG, Gonzalez-Hernandez MP, et al. Composition of the heartwood essential oil of incense cedar (Calocedrus decurrens Torr.). *Holzforschung.* 2011;65:333-6.
[1929] Ibrahim TA, El-Hela AA, El-Hefnaway HM, et al. Chemical Composition and Antimicrobial Activities of Essential Oils of Some Coniferous Plants Cultivated in Egypt. *Iranian J Pharm Res.* 2017;16(1):328-37.
[1930] Garcia G, Tissandie L, Filippi JJ, et al. New Pinane Derivatives Found in Essential Oils of Calocedrus decurrens. *Molecules.* 2017;22:921.
[1931] Rudloff EV. The leaf oil terpene composition of incense cedar and coast redwood.*Can J Chem.* 1981;59:285-7.
[1932] Adams RP, Nguyen S, Hsieh CF, et al. The lead essential oils of the Genus Calocedrus. *J Essent Oil Res.* 2006 Nov/Dec;18:654-8.
[1933] Domaracky M, Rehak P, Juhas S, et al. Effects of selected plant essential oils on the growth and development of mouse preimplantation embryos in vivo. *Physiol Res.* 2007;56(1):97-104.

[1934] Paramasivam A, Kalaimangai M, Sambantham S, et al. Anti-angiogenic activity of thymoquinone by the down-regulation of VEGF using zebrafish (Danio rerio) model. *Biomed Prevent Nutr*. 2012 Jul-Sep;2(3):169-73.
[1935] Yi T, Cho SG, Yi A, et al. Thymoquinone inhibits tumor angiogenesis and tumor growth through suppressing AKT and ERK signaling pathways. *Mol Cancer Ther*. 2008 Jul;7(7):1789-96.
[1936] Peng L, Liu A, Shen Y, et al. Antitumor and anti-angiogenesis effects of thymoquinone on osteosarcoma through the NF-κB pathway. *Oncol Rep*. 2013 Feb;29(2):571-8.
[1937] AbuKhader MM, Khater SH, Al-Matubsi HY. Acute effects of thymoquinone on the pregnant rat and embryo-fetal development. *Drug Chem Toxicol*. 2013 Jan;36(1):27-34.
[1938] Lemhadri A, Zeggwagh NA, Maghrani M, et al. Anti-hypoglycaemic activity of the aqueous extract of Origanum vulgare growing wild in Tafilalet region. *J Ethnopharmacol*. 2004 Jun;92(2-3):251-6.
[1939] Bayramoglu G, Senturk H, Bayramoglu A, et al. Carvacrol partially reverses symptoms of diabetes in STZ-induced diabetic rats. *Cytotechnology*. 2014 Mar;66(2):251–257.
[1940] Ezhumalai M, Radhiga T, Pugalendi KV. Antihyperglycemic effect of carvacrol in combination with rosiglitazone in high-fat diet-induced type 2 diabetic C57BL/6J mice. *Mol Cell Biochem*. 2014 Jan;385(1-2):23-31.
[1941] Umasankar K, Ramya DR, Nambikkairaj B. Antidiabetic and antidyslipidemic nature of carvacrol, a monoterpenic phenol studied in high-fat - fed and low-dose streptozotocin-induced experimental diabetic rats. *J Pharm Res*. 2015;9(8):484-90.
[1942] Bamosa AO. A review on the hypoglycemic effect of nigella sativa and thymoquinone. *Rev Articles*. 2015;3(1):2-7.
[1943] Ali Sangi SM, Sulaiman MI, Abd El-wahab MF, et al. Antihyperglycemic effect of thymoquinone and oleuropein, on streptozotocin-induced diabetes mellitus in experimental animals. *Pharmacogn Mad*. 2015 Oct;11(Suppl 2):S251-57.
[1944] El-Dakhakhny M, Mady N, Lembert N, et al. The hypoglycemic effect of Nigella sativa oil is mediated by extrapancreatic actions. *Planta Med*. 2002 May;68(5):465-6.
[1945] Fararh KM, Shimizu Y, Shiina T, et al. Thymoquinone reduces hepatic glucose production in diabetic hamsters. *Res Vet Sci*. 2005 Dec;79(3):219-23.
[1946] Razavi BM, Hosseinzadeh H. A review of the effects of Nigella sativa L. and its constituent, thymoquinone, in metabolic syndrome. *J Endocrinol Invest*. 2014 Nov;37(11):1031-40.
[1947] Sheikh BA, Pari L, Rathinham A, et al. Trans-anethole, a terpenoid ameliorates hyperglycemia by regulating key enzymes of carbohydrate metabolism in streptozotocin induced diabetic rats. *Biochimie*. 2015 May;112:57-65.
[1948] Pari L, Sheikh BA. Antihyperglycemic effect of trans-anethole in streptozotocin induced diabetic rats with special reference to glycoprotein components. *Int J Adv Res Biol Sci*. 2015;2(5):28-34.
[1949] MM. Thymoquinone: a promising antidiabetic agent. *Int J Diabetes Dev Countries*. 2012 Jun;32(2):65-68.
[1950] Jukic M, Politeo O, Maksimmovic M, et al. In vitro acetylcholinesterase inhibitory properties of thymol, carvacrol and their derivatives thymoquinone and thymohydroquinone. *Phytother Res*. 2007;21(3):259-61.
[1951] Tognolini M, Barocelli E, Ballabeni V, et al. Comparative screening of plant essential oils; phenylpropanoid moiety as basic core for antiplatelet activity. *Life Sci*. 2006 Feb 23;78(13):1419-32.
[1952] Okazaki K, Kawazoe K, Takaishi Y. Human platelet aggregation inhibitors from thyme (Thymus vulgaris L.). *Phytother Res*. 2002 Jun;16(4):398-9.
[1953] Lee HS. Anticoagulant properties of constituents derived from fennel (Foeniculum vulgare Gaertner) fruits. *Food Sci Biotech*. 2006 Oct;15(5):763-67.
[1954] Thushara RM, Hemshekhar M, Santhosh MS, et al. Differential action of phytochemicals on platelet apoptosis: a biological overview. *Curr Med Chem*. 2013;20(8):1018-27.
[1955] Towhid ST, Schmidt EM, Schmid E, et al. Thymoquinone-induced platelet apoptosis. *J Cell Biochem*. 2011 Nov;112(11):3112-21.
[1956] Jukic M, Politeo O, Maksimmovic M, et al. In vitro acetylcholinesterase inhibitory properties of thymol, carvacrol and their derivatives thymoquinone and thymohydroquinone. *Phytother Res*. 2007;21(3):259-61.
[1957] Subehan UT, Iwata H, Kadota S, et al. Mechanism-based inhibition of CYP3A4 and CYP2D6 by Indonesian medicinal plants. *J Ethnopharmacol*. 2006 May;105(3):449-55.
[1958] Al-Jenoobi FI, Al-Thukair AA, Abbas FA, et al. Effect of black seed on dextromethorphan o- and n-demethylation in human liver microsomes and healthy human subjects. *Drug Metabolism Letters*. 2010;4(1):51–55.
[1959] Liu X, Park JH, Abd El-Aty AM, et al. Isolation of volatiles from Nigella sativa seeds using microwave-assisted extraction: effect of whole extracts on canine and murine CYP1A. *Biomed Chromatogr*. 2013 Jul;27(7):938-45.
[1960] Ahmad A, Khan RM, Alkharfy KM, et al. Effects of Thymoquinone on the Pharmacokinetics and Pharmacodynamics of Glibenclamide in a Rat Model. *Nat Prod Commun*. 2015 Aug;10(8):1395-8.
[1961] Langeveld WT, Veldhuizen EJ, Burt SA. Synergy between essential oil constituents and antibiotics. *Crit Rev Microbiol*. 2014 Feb;40(1);76-94.
[1962] Kouidhi B, Zmantar T, Jrah H, et al. Antibacterial and resistance-modifying activities of thymoquinone against oral pathogens. *Ann Clin Microbiol Antimicrob*. 2011 Jun 27;10:29.
[1963] Raza M, Alghasham AA, Alorainy MS, et al. Potentiation of Valproate-induced Anticonvulsant Response by Nigella sativa Seed Constituents: The Role of GABA Receptors. *Int J Health Sci (Qassim)*. 2008 Jan;2(1):15-25.
[1964] Ibrahim TA, El-Hela AA, El-Hefnaway HM, et al. Chemical Composition and Antimicrobial Activities of Essential Oils of Some Coniferous Plants Cultivated in Egypt. *Iranian J Pharm Res*. 2017;16(1):328-37.
[1965] Singh S. Chemical Constituents of Essential Oil from Anethum Sowa Kurz. Seed. *J Chem Pharm Res*. 2012;4(9):4156-60.
[1966] Vats N, Pandey R. Analysis of oil extracted from Anethum sowa. *Int J Chem Sci*. 2012;10(4):2010-14.
[1967] Saleh-e-IN MM, Sultana A, Husain M, et al. Chemical Constituents of Essential Oil from Anethum sowa L. Herb (Leaf and Stem) Growing in Bangladesh. *Bangladesh J Sci Res*. 2010;45(2):173-6.
[1968] Saleh-e-in MM, Sultana A, Husain , et al. Chemical composition of essential oil of Anethum sowa L. stem. *Indian Perf.* 2008 Oct-Dec;52:64-6.
[1969] Saleh-e-In MM, sultana N, Rahim MM, et al. Chemical composition and pharmacological significance of Anethum Sowa L. Root. *BMC Complement Altern Med*. 2017;17:127.
[1970] Lawrence BM. Essential oils 1988-1991. Allured Publishing, Wheaton. 1995.
[1971] Shankaracharya NB, Mohan Rao LJ, Puranaik J, et al. Studies on chemical and technological aspects of Indian dill seed (Anethum sowa Roxb.). J Food Sci Technol. 2000;37:368-372.
[1972] Nautiyal OP, Tiwari KK. Extraction of Dill Seed Oil (Anethum sowa) Using Supercritical Carbon Dioxide and Comparison with Hydrodistillation. *Ind Eng Chem Res*. 2011;50:5723-6.

[1973] No author listed. Anethum sowa. Review of the literature. Available at: http://shodhganga.inflibnet.ac.in/bitstream/10603/31971/9/09_chapter3.pdf.
[1974] Lowenstein L, Ballew DH. Fatal acute hemolytic anemia, thrombocytopenic purpura, nephrosis and hepatitis resulting from ingestion of a constituent containing apiol. *Can Med Assoc J.* 1958 Feb 1;78(3):195-8.
[1975] Basso U. [Study of death caused by apiol and experimental study of its toxic action]. *Arch Patol Clin Med.* 1957;34(4):308-22.
[1976] Barni B, Barni I. [The apiol poisoning]. *Rassegna casistica.* 1967;3(2):197-221.
[1977] Newall CA, Anderson LA, Philpson JD. Herbal Medicine: A Guide for Healthcare Professionals. London, UK: The Pharmaceutical Press, 1996.
[1978] Colalillo R. Apiol poisoning in 7 women and 2 children. *Rivista di Tossicologia Sperimentale e Clinica.* 1988;18(2):125-130.
[1979] Mumolo M. [On a case of acute and fatal poisoning caused by apiol]. *Recenti Prog Med.* 1964 Feb;36:139-51.
[1980] Lowenberg K. Cerbral damage in a case of fatal poisoning due to a compound if ergot and apiol (ergoapiol). *JAMA.* 1938;110(8):573-5.
[1981] Hermann K, Le Roux A, Fiddes FS. Death from apiol used as abortifacient. *Lancet.* 1956 Jun 16;270(6929):937-9.
[1982] Ciganda C, Laborde A. Herbal infusions used for induced abortion. *J Toxicol Clin Toxicol.* 2003;41:235-239.
[1983] Lowenstein L, Ballew DH. Fatal acute hemolytic anemia, thrombocytopenic purpura, nephrosis and hepatitis resulting from ingestion of a constituent containing apiol. *Can Med Assoc J.* 1958 Feb 1;78(3):195-8.
[1984] Basso U. [Study of death caused by apiol and experimental study of its toxic action]. *Arch Patol Clin Med.* 1957;34(4):308-22.
[1985] Barni B, Barni I. [The apiol poisoning]. *Rassegna casistica.* 1967;3(2):197-221.
[1986] D'Aprile F. [Clinical study-Experimental apiol poisoning]. *Annali di Ostetricia Ginecologia.* 1928;50:1204-27.
[1987] Hermann K, Le Roux A, Fiddes FS. Death from apiol used as abortifacient. *Lancet.* 1956 Jun 16;270(6929):937-9.
[1988] Lowenstein L, Ballew DH. Fatal acute hemolytic anemia, thrombocytopenic purpura, nephrosis and hepatitis resulting from ingestion of a constituent containing apiol. *Can Med Assoc J.* 1958 Feb 1;78(3):195-8.
[1989] Basso U. [Study of death caused by apiol and experimental study of its toxic action]. *Arch Patol Clin Med.* 1957;34(4):308-22.
[1990] Mumolo M. [On a case of acute and fatal poisoning caused by apiol]. *Recenti Prog Med.* 1964 Feb;36:139-51.
[1991] Barni B, Barni I. [The apiol poisoning]. *Rassegna Casistica.* 1967;3(2):197-221.
[1992] D'Aprile F. [Clinical study-Experimental apiol poisoning]. *Annali di Ostetricia Ginecologia.* 1928;50:1204-27.
[1993] Muruganathan U, Srinivasan S, Indumathi D. Antihyperglycemic effect of carvone: Effect on the levels of glycoprotein components in streptozotocin-induced diabetic rats. *J Acute Disease.* 2013;2(4):310-15.
[1994] Souza FV, da Rocha MB, de Souza DP, et al. (-)-Carvone: antispasmodic effect and mode of action. *Fitoterapia.* 2013 Mar;85:20-24.
[1995] Rajeshwari T, Raja B. Antihypertensive, antihyperlipidemicand antioxidantinfluence of D-carvone in L-NAME induced hypertensive rats. *Int J Pharmaceutical Biol Arch.* 2014;5(4):82-88.
[1996] de Sousa DP, Farias Nobrega FF, de Almeida RN. Influence of the chirality of (R)-(-)- and (S)-(+)-carvone in the central nervous system: a comparative study. *Chirality.* 2007 May;19(4):264-268.
[1997] Lowenstein L, Ballew DH. Fatal acute hemolytic anemia, thrombocytopenic purpura, nephrosis and hepatitis resulting from ingestion of a constituent containing apiol. *Can Med Assoc J.* 1958 Feb 1;78(3):195-8.
[1998] Basso U. [Study of death caused by apiol and experimental study of its toxic action]. *Arch Patol Clin Med.* 1957;34(4):308-22.
[1999] Barni B, Barni I. [The apiol poisoning]. *Rassegna casistica.* 1967;3(2):197-221.
[2000] Newall CA, Anderson LA, Philpson JD. Herbal Medicine: A Guide for Healthcare Professionals. London, UK: The Pharmaceutical Press, 1996.
[2001] Colalillo R. Apiol poisoning in 7 women and 2 children. *Rivista di Tossicologia Sperimentale e Clinica.* 1988;18(2):125-130.
[2002] Sarbhoy AK, Varshney JL, Maheshwari ML, et al. Efficacy of some essential oils and their constituents on few ubiquitous molds. *Zentralbl Bakteriol Naturwiss.* 1978;133(7-8):723-5.
[2003] Keawsa-ard S, Kongtaweelert S. Antioxidant, Antibacterial, Anticancer Activities and Chemical Constituents of the Essential Oil from Mesua ferrea Leaves. Chiang Mai J Sci. 2012;39(3):455-63.
[2004] Choudhury S, Ahmed R, Barthel A, et al. Volatile Oils of Mesua ferrea (L.) from Assam, India. *J Essent Oil Res.* 1998 Sep/Oct;10:497-501.
[2005] Choudhury S, Ahmed R, Barthel A, et al. Volatile Oils of Mesua ferrea (L.) from Assam, India. *J Essent Oil Res.* 1998 Sep/Oct;10:497-501.
[2006] Keawsa-ard S, Kongtaweelert S. Antioxidant, Antibacterial, Anticancer Activities and Chemical Constituents of the Essential Oil from Mesua ferrea Leaves. Chiang Mai J Sci. 2012;39(3):455-63.
[2007] Keawsa-ard S, Kongtaweelert S. Antioxidant, Antibacterial, Anticancer Activities and Chemical Constituents of the Essential Oil from Mesua ferrea Leaves. Chiang Mai J Sci. 2012;39(3):455-63.
[2008] Keawsa-ard S, Kongtaweelert S. Antioxidant, Antibacterial, Anticancer Activities and Chemical Constituents of the Essential Oil from Mesua ferrea Leaves. Chiang Mai J Sci. 2012;39(3):455-63.
[2009] Craveiro AA, Andrade CHS, Matos FJA, et al. Essential oils from Brazilian Rutaceae. I. Genus Pilocarpus. *J Nat Prod.* 2004 Jul;42(60):669-71.
[2010] Taveira FSN, Andrade EHA, Lima WN, et al. Seasonal variation in the essential oil of Pilocarpus microphyllus Stapf. *Anais da Acad Brasil de Ciencias.* 2003;75(1):27-31.
[2011] Andrade-Neto M, de A. Cunha U, Mafezoli J, et al. Volatile constituents of Pilocarpus trachyllophus Holmes and Pilocarpus jaborandi Holmes (Rutaceae) from northeast of brazil. *J Essent Oil Res.* 2000 Nov/Dec;12(6):769-74.
[2012] Tisserand R, Young R. Essential Oil Safety (Second Edition), p. 395-396. 2014 Churchill Livingstone Elsevier.
[2013] Plant Therapy. GC/MS Batch Number: J40100. Jack Pine. Available at: https://www.planttherapy.com/test_reports/Jack%20Pine%20J40100.pdf.
[2014] Plant Therapy. GC/MS Batch Number: J40101. Jack Pine. Available at: https://www.planttherapy.com/test_reports/Jack%20Pine%20J40101.pdf
[2015] Sadgrove NJ, Jones GL. Medicinal compounds, chemically and biologically characterized from extracts of Australian Callitris endlicheri and C. glaucophylla (Cupressaceae): Used traditionally in Aboriginal and colonial pharmacopoeia. *J Ethnopharmacol.* 2014 May 14;153(3):872-83.
[2016] Roupetz R. Phytochemistry and pharmacology of volatile components of Callitris glaucophylla wood. University Thesis. Available at: http://othes.univie.ac.at/5397/1/2009-06-04_0206219.pdf.
[2017] Doimo L. Azulenes, Costols and Γ-Lactones from Cypress-Pines (Callitris columellaris, C. glaucophylla and C. intratropica) Distilled Oils and Methanol Extracts. *J Essent Oil Res.* 2001 Jan/Feb;13:25-29.
[2018] Roupetz R. Phytochemistry and pharmacology of volatile components of Callitris glaucophylla wood. University Thesis. Available at: http://othes.univie.ac.at/5397/1/2009-06-04_0206219.pdf.
[2019] Mahindru SN. Indian plant perfumes. Metropolatin, 1992.
[2020] Mishra PK, Shukla R, Singh P, et al. Antifungal, anti-aflatoxigenic, and antioxidant efficacy of Jamrosa essential oil for preservation of herbal raw materials. *Int Biodeterioration Biodegradation.* 2012;74:11-16.
[2021] Akbar N, Saxena BK. Isolation of Geraniol content from various essential oils. *Asian J Exp Chem.* 2009 June & Dec;4(1&2):14-17.

[2022] Kak SN, Bhan MK, Rekha K. Development of Improved Clones of Jamrosa [Cymbopogon nardus (L.) Rendle var. Confertiflorus (Steud.) Bor. x C. jwarancusa (Jones) Schult.] through Induced Mutations. *J Essent Oil Res*. 2000;12(1):108-10.
[2023] Nayak S, Debata BK, Srivastava VK, et al. Evaluation of agronomically useful somaclonal variants in Jamrosa (a hybrid Cymbopogon) and detection of genetic changes through RAPD. *Plant Sci*. 2003;164:1029-35.
[2024] Bharat CS, Praveen D. Evaluation of in vitro antimicrobial potential and phytochemical analysis of spruce, cajeput and jamrosa essential oil against clinical isolates. *Int J Green Pharm*. 2016 Jan-Mar;10(1):27-32.
[2025] Seo KA, Kim H, Ku HY, et al. The monoterpenoids citral and geraniol are moderate inhibitors of the CYP2B6 hydroxylase activity. *Chem Biol Interact*. 2008;174:141-46.
[2026] Ibrahim SM, El-Denshary ES, Abdallah DM. Geraniol, Alone and in Combination with Pioglitazone, Ameliorates Fructose-Induced Metabolic Syndrome in Rats via the Modulation of Both Inflammatory and Oxidative Stress Status. *PLoS One*. 2015 Feb 13;10(2):e0117516.
[2027] Babukumar S, Vinothkumar V, Sankaranarayanan C, et al. Geraniol, a natural monoterpene, ameliorates hyperglycemia by attenuating the key enzymes of carbohydrate metabolism in streptozotocin-induced diabetic rats. *Pharm Biol*. 2017 Dec;55(1):1442-1449.
[2028] Shin S, Lim S. Antifungal effects of herbal essential oils alone and in combination with ketoconazole against Trichophyton spp. *J Appl Micribiol*. 2004;97:1289-96.
[2029] Lorenzi V, Muselli A, Bernardini AF, et al. Geraniol restores antibiotic activities against multidrug-resistant isolates from gram-negative species. *Antimicrob Agents Chemother*. 2009 May;53(5):2209-11.
[2030] Li L, Li ZW, Yin ZQ, et al. Antibacterial activity of leaf essential oil and its constituents from Cinnamomum longepaniculatum. *Int J Clin Exp Med*. 2014;7(7):1721-7.
[2031] Hagvall L. Cytochrome P450-mediated activation of the fragrance constituent geraniol forms potent contact allergens. *Tox Appl Pharm*. 2008 Dec;233(2):308-13.
[2032] Nogueira AC, Carvalho RR, Souza CA, et al. Study on the embryofeto-toxicity of citral in the rat. *Toxicology*. 1995;96(2):105-13.
[2033] Seo KA, Kim H, Ku HY, et al. The monoterpenoids citral and geraniol are moderate inhibitors of the CYP2B6 hydroxylase activity. *Chem Biol Interact*. 2008;174:141-46.
[2034] Sheweita SA, Newairy AA, Mansour HA, et al. Effect of some hypoglycemic herbs on the activity of phase I and II drug-metabolizing enzymes in alloxan-induced diabetic rats. *Toxicology*. 2002 May 24;17(2):131-39.
[2035] Raner GM, Vaz AD, Coon MJ. Metabolism of all-trans, 9-cis, and 13-cis isomers of retinal by purified isozymes of microsomal cytochrome P450 and mechanism-based inhibition of retinoid oxidation by citral. *Mol Pharmacol*. 1996;49(3):515-22.
[2036] Modak T, Mukhopadhaya A. Effects of citral, a naturally occurring antiadipogenic molecule, on an energy-intense diet model of obesity. *Indian J Pharmacol*. 2011 May-Jun;43(3):300-05.
[2037] Adeneye AA, Agbaje EO. Hypoglycemic and hypolipidemic effects of fresh leaf aqueous extract of Cymbopogon citratus Stapf. in rats. *J Ethnopharmacol*. 2007 Jul 25;112(3):440-4.
[2038] Choi JY, Damte D, Seung-Jin L, et al. Antimicrobial activity of lemongrass and oregano essential oil against standard antibiotic resistant Staphylococcus aureus and field isolates from chronic mastitis cow. *Int J Phytomedicine*. 2012;4(1):134-39.
[2039] Zanini SF, Silva-Angulo AB, Rosenthal A, et al. Effect of citral and carvacrol on the susceptibility of Listeria monocytogenes and Listeria innocua to antibiotics. *Lett Appl Microbiol*. 2014 May;58(5):486-92.
[2040] De Mozzi P, Johnston GA. An outbreak of allergic contact dermatitis caused by citral in beauticians working in a health spa. *Contact Dermatitis*. 2014 Jun;70(6):377-9.
[2041] Mishra PK, Prakash B, Kedia A, et al. Efficacy of Caesulia axillaris, Cymbopogon khasans and Cymbopogon martinii essential oils as plant based preservatives against degradation of raw materials of Andrographis paniculata by fungal spoilage. *Int Biodeterioratoin Bidegeneration*. 2015;104:244-50.
[2042] Mishra PK, Shukla R, Singh P, et al. Antifungal, anti-aflatoxigenic, and antioxidant efficacy of Jamrosa essential oil for preservation of herbal raw materials. *Int Biodeterioration Biodegradation*. 2012;74:11-16.
[2043] Mishra PK, Shukla R, Singh P, et al. Antifungal, anti-aflatoxigenic, and antioxidant efficacy of Jamrosa essential oil for preservation of herbal raw materials. *Int Biodeterioration Biodegradation*. 2012;74:11-16.
[2044] Zhou Y, Tang N, Huang L, et al. Effects of Salt Stress on Plant Growth, Antioxidant Capacity, Glandular Trichome Density, and Volatile Exudates of Schizonepeta tenuifolia Briq. *Int J Mol Sci*. 2018;19(1):252.
[2045] Lee YH, Wang CM, Liu PY, et al. Volatile Oils of Nepeta tenuifolia (Jing Jie) as an Alternative Medicine against Multidrug-Resistant Pathogenic Microbes. *Can J Infectious Dis Med Microbiol*. 2018;20187:8347403.
[2046] Yu S, Chen Y, Zhang L, et al. Quantitative Comparative Analysis of the Bio-Active and Toxic Constituents of Leaves and Spikes of Schizonepeta tenuifolia at Different Harvesting Times. *Int J Mol Sci*. 2011;12(10):6335-44.
[2047] Park I, Kim L, Choi I, et al. Fumigant Activity of Plant Essential Oils and Components from Schizonepeta tenuifolia Against Lycoriella ingenua (Diptera: Sciaridae). *J Economic Entomol*. 2006;99(5):1717-1721.
[2048] Chun MH, Kim EK, Lee KR, et al. Quality control of Schizonepeta tenuifolia Briq by solid-phase microextraction gas chromatography/mass spectrometry and principal component analysis. *Microchemical J*. 2010;95:25-31.
[2049] Liu C, Srividya N, Parrish AN, et al. Morphology of glandular trichomes of Japanese catnip (Schizonepeta tenuifolia Briquet) and developmental dynamics of their secretory activity. *Phytochem*. 2018;150:23-30.
[2050] Yann JY, Lee HS. Changes in acaricidal potency by introducing functional radicals and an acaricidal constituent isolated from Schizonepeta tenuifolia. *J Agric Food Chem*. 2013;61:11511-6.
[2051] Guo SC, Yu S, Qian Y, et al. Correlation of antioxidant activity and volatile oil chemical components from Schizonepeta tenuifolia herbs by chemometric methods. *Int J Food Prop*. 2017;20(1):S1082-92.
[2052] Bakerink JA, Gospe SM Jr, Dimand RJ, et al. Multiple organ failure after ingestion of pennyroyal oil from herbal tea in two infants. *Pediatrics*. 1996;98(5):944-47.
[2053] Sudekum M, Poppenga RH, Raju N, et al. Pennyroyal oil toxicosis in a dog. *J Am Vet Med Assoc*. 1992;200:817-8.
[2054] Anderson IB, Mullen WH, Meeker JE, et al. Pennyroyal toxicity: measurement of toxic metabolite levels in two cases and review of the literature. *Ann Intern Med*. 1996;124:726-34.
[2055] Seeff L, Stickel F, Navarro VJ. Hepatotoxicity of herbals and dietary supplements. In, Kaplowitz N, DeLeve LD, eds. Drug-induced liver disease. 3rd ed. Amsterdam: Elsevier, 2013, pp. 631-58.
[2056] Mullen W, Anderson I, Oishii S, et al. Accidental pennyroyal oil ingestion in a toddler with the first human serum metabolite detection. *Vet Hum Toxicol*. 1994;36:342.
[2057] Barnes J, Anderson LA, Philpson JD. Herbal Medicine: A Guide for Healthcare Professionals. London, UK: The Pharmaceutical Press, 1996.

[2058] Zimmerman HJ. Unconventional drugs. Miscellaneous drugs and diagnostic chemicals. In, Zimmerman, HJ. Hepatotoxicity: the adverse effects of drugs and other chemicals on the liver. 2nd ed. Philadelphia: Lippincott, 1999: pp. 731-34.
[2059] Anderson IB, Mullen WH, Meeker JE, et al. Pennyroyal toxicity: measurement of toxic metabolite levels in two cases and review of the literature. *Ann Intern Med.* 1996;124:726-34.
[2060] Centers for Disease Control. Fatality and illness associated with consumption of pennyroyal oil - Colorado. *MMWR.* 1978;27:511-13.
[2061] Soares PM, Assreuy AM, Souza EP, et al. Inhibitory effects of the essential oil of Mentha pulegium on the isolated rat myometrium. *Planta Med.* 2005 Mar;71(3):214-8.
[2062] Ciganda C, Laborde A. Herbal infusions used for induced abortion. *J Toxicol Clin Toxicol.* 2003;41(3):235-9.
[2063] Valance WB. Pennyroyal poisoning: a fatal case. *Lancet.* 1955;266 (6895):850-51.
[2064] Gold J, Cates W Jr. Herbal abortifacients. *JAMA.* 1980;243:1365-66.
[2065] Gordon WP, Huitric AC, Seth CL, et al. The metabolism of the abortifacient terpene, (R)-(+)-pulegone, to a proximate toxin, menthofuran. *Drug Metab Dispos.* 1987;15:589–94.
[2066] Khojasteh-Bakht SC, Chen W, Koenigs LL, et al. Metabolism of (R)-(+)-pulegone and (R)-(+)-menthofuran by human liver cytochrome P-450s: evidence for formation of a furan epoxide. *Drug Metab Dispos.* 1999 May;27(5):574-80.
[2067] Gordon P, Khojasteh SC. A decades-long investigation of acute metabolism-based hepatotoxicity by herbal constituents: a case study of pennyroyal oil. *Drug Metab Rev.* 2015 Feb;47(1):12-20.
[2068] Zhou SF, Xue CC, Yu XQ, et al. Metabolic activation of herbal and dietary constituents and its clinical and toxicological implications: an update. *Curr Drug Metab.* 2007;8:526–553.
[2069] National Toxicology Program. Toxicology and carcinogenesis studies of pulegone (CAS No. 89-82-7) in F344/N rats and B6C3F1 mice (gavage studies). *Natl Toxicol Program Tech Rep Ser.* 2011 Aug;(563):1-201.
[2070] National Toxicology Program. Toxicology and carcinogenesis studies of pulegone (CAS No. 89-82-7) in F344/N rats and B6C3F1 mice (gavage studies). *Natl Toxicol Program Tech Rep Ser.* 2011 Aug;(563):1-201.
[2071] Sullivan JB Jr, Rumack BH, Thomas H Jr, et al. Pennyroyal oil poisoning and hepatotoxicity. *JAMA.* 1979;242:2873-4.
[2072] The Review of Natural Products by Facts and Comparisons. St. Louis, MO: Wolters Kluwer Co., 1999.
[2073] Foster S, Tyler VE. Tyler's Honest Herbal: A Sensible Guide to the Use of Herbs and Related Remedies. 3rd ed., Binghamton, NY: Haworth Herbal Press, 1993.
[2074] Martindale W. Martindale the Extra Pharmacopoeia. Pharmaceutical Press, 1999.
[2075] McGuffin M, Hobbs C, Upton R, Goldberg A, eds. American Herbal Products Association's Botanical Safety Handbook. Boca Raton, FL: CRC Press, LLC, 1997.
[2076] Sullivan JB Jr, Rumack BH, Thomas H Jr, et al. Pennyroyal oil poisoning and hepatotoxicity. *JAMA.* 1979;242:2873-4.
[2077] Anderson IB, Mullen WH, Meeker JE, et al. Pennyroyal toxicity: measurement of toxic metabolite levels in two cases and review of the literature. *Ann Intern Med.* 1996;124:726-34.
[2078] Allen WT. Note on a case of supposed poisoning by pennyroyal. *Lancet.* 1897;1(3841):1022-23.
[2079] Girling J. Poisoning by pennyroyal. *Br Med J.* 1887;1:1214.
[2080] Braithwaite PF. A case of poisoning by pennyroyal: recovery. *Br Med J.* 1906;2:865.
[2081] Buechel DW, Haverlah VC, Gardner ME. Pennyroyal oil ingestion: report of a case. *J Am Osteopath Assoc.* 1983;82:793-4.
[2082] Flynn EF. Poisoning by essence of pennyroyal. *Br Med J.* 1893;2:1270.
[2083] Burkhard PR, Burkhardt K, Haenggeli CA, et al. Plant-induced seizures: reappearance of an old problem. *J Neurol.* 1999 Aug;246(8):667-70.
[2084] Early DF. Pennyroyal: a rare cause of epilepsy. *Lancet.* 1961; 278 (7202): 580-81.
[2085] Sánchez-Borzone ME, Marin LD, García DA. Effects of Insecticidal Ketones Present in Mint Plants on GABAA Receptor from Mammalian Neurons. *Pharmacogn Mag.* 2017 Jan-Mar;13(49):114-117.
[2086] Sullivan JB Jr, Rumack BH, Thomas H Jr, et al. Pennyroyal oil poisoning and hepatotoxicity. *JAMA.* 1979;242:2873-4.
[2087] Sudekum M, Poppenga RH, Raju N, et al. Pennyroyal oil toxicosis in a dog. *J Am Vet Med Assoc.* 1992;200:817-8.
[2088] Anderson IB, Mullen WH, Meeker JE, et al. Pennyroyal toxicity: measurement of toxic metabolite levels in two cases and review of the literature. *Ann Intern Med.* 1996;124:726-34.
[2089] Zimmerman HJ. Unconventional drugs. Miscellaneous drugs and diagnostic chemicals. In, Zimmerman, HJ. Hepatotoxicity: the adverse effects of drugs and other chemicals on the liver. 2nd ed. Philadelphia: Lippincott,1999: pp. 731-34.
[2090] Seeff L, Stickel F, Navarro VJ. Hepatotoxicity of herbals and dietary supplements. In, Kaplowitz N, DeLeve LD, eds. Drug-induced liver disease. 3rd ed. Amsterdam: Elsevier, 2013, pp. 631-58.
[2091] Shan MQ, Qian Y, Yu S, et al. Anti-inflammatory effect of volatile oil from Schizonepeta tenuifolia on carrageenin-induced pleurisy in rats and its application to study of appropriate harvesting time coupled with multi-attribute comprehensive index method. *J Ethnopharmacol.* 2016 Dec 24;194:580-586.
[2092] Ge WH, Guo JY, Shen YJ, et al. [Effects of volatie oil of Schizonepeta tenuifolia Briq herb and Saposhnikovia divaricata Schischke root on proinflammatory cytokine expression and regulation]. *Zhongguo Zhong Yao Za Zhi.* 2007 Sep;32(17):1777-9.
[2093] Lee SO, Park IK, Choi GJ, et al. Fumigant activity of essential oils and components of Illicium verum and Schizonepeta tenuifolia against Botrytis cinerea and Colletotrichum gloeosporioides. *J Microbiol Biotechnol.* 2007 Sep;17(9):1568-72.
[2094] Lee YH, Wang CM, Liu PY, et al. Volatile Oils of Nepeta tenuifolia (Jing Jie) as an Alternative Medicine against Multidrug-Resistant Pathogenic Microbes. *Can J Infect Dis Med Microbiol.* 2018 Apr 11;2018:8347403.
[2095] Guo SC, Yu S, Qian Y, et al. Correlation of antioxidant activity and volatile oil chemical components from Schizonepeta tenuifolia herbs by chemometric methods. *Int J Food Prop.* 2017;20(1):S1082-92.
[2096] Park I, Kim L, Choi I, et al. Fumigant Activity of Plant Essential Oils and Components from Schizonepeta tenuifolia Against Lycoriella ingenua (Diptera: Sciaridae). *J Economic Entomol.* 2006;99(5):1717-1721.
[2097] Yann JY, Lee HS. Changes in acaricidal potency by introducing functional radicals and an acaricidal constituent isolated from Schizonepeta tenuifolia. *J Agric Food Chem.* 2013;61:11511-6.
[2098] Liang Y, Li JL, Xu S, et al. Evaluation of repellency of some Chinese medicinal herbs essential oils against Liposcelis bostrychophila (Psocoptera: Liposcelidae) and Tribolium castaneum (Coleoptera: Tenebrionidae). *J Econ Entomol.* 2013 Feb;106(1):513-9.
[2099] Rustaiyan A, Mazloomifar H, Masoudi S, et al. Volatile Oils of Ducrosia assadii Alava. and Prangos acaulis (DC.) Bornm. from Iran. *J Essent Oil Res.* 2006 Nov/Dec;18:682-4.
[2100] Meshkatalsadat MH, Mirzaei HH. Chemical compositions of the essential oils of stems, leaves and flowers of Prangos acaulis (DC) Bornm. *Pakistan J Biol Sci.* 2007;10(16):2775-7.
[2101] Meshkataksadat MH, Bamoniri A, Batooli H. The bioactive and volatile constituents of Prangos acaulis (DC) Bronm extracted using hydrodistillation and nanoscale injection techniques. *Digest J Nanomaterials Biostructures.* 2010 Mar;5(1):263-6.

2102 Mirzaei HH, Meshkataksadat MH, Soheilivand S. Determination of essential oil composition and Prangos acaulis (DC) Bornm obtained by hydrodistillation and supercritical fluid extraction methods. *J Appl Sci.* 2007;7(17):2535-8.
2103 Murugan R, Mallavarapu GR, Padmashree KV, et al. Volatile oil composition of Pogostemon heyneanus and comparison of its composition with patchouli oil. *Nat Prod Commun.* 2010 Dec;5(12):1961-4.
2104 Dharmadasa RM, Rathnayake RMDH, Abeysinghe DC, et al. Screening of Local and Introduced Varieties of Pogostemon heyneanus Benth. (Lamiaceae), for Superior Quality Physical, Chemical and Biological Parameters. *World J Agric Res.* 2014;2(6):261-6.
2105 Blank AF, Pergentino Sant'ana TC, Dantos PS, et al. Chemical characterization of the essential oil from patchouli accessions harvested over four seasons. *Ind Crops Prod.* 2011 Jul;34(1):831-7.
2106 Rubini D, Banu SF, Nisha P, et al. Essential oils from unexplored aromatic plants quench biofilm formation and virulence of Methicillin resistant Staphylococcus aureus. *Microb Pathog.* 2018 Sep;122:162-173.
2107 Das NG, Dhiman S, Talukdar PK, et al. Synergistic mosquito-repellent activity of Curcuma longa, Pogostemon heyneanus and Zanthoxylum limonella essential oils. *J Infect Public Health.* 2015 Jul-Aug;8(4):323-8.
2108 Azzaz NAE, El-Khateeb AY, Farag AA. Chemical composition and biological activity of the essential oils of Cyperus articulatus. *Int J Acad Res.* 2014 Sep;6(5):265-9.
2109 Zoghbi MDGB, Andrade EHA, Oliveira J, et al. Yield and Chemical Composition of the Essential Oil of the Stems and Rhizomes of Cyperus articulatus L. Cultivated in the State of Para', Brazil. *J Essent Oil Res.* 2006 Jan/Feb;18:10-2.
2110 Olawore NO, Usman LA, Ogunwande IA, et al. Constituents of Rhizome Essential Oils of Two Types of Cyperus articulatus L. Grown in Nigeria. *J Essent Oil Res.* 2006 Nov/Dec;18:604-6.
2111 Teixeira Duarte MC, Figuueira GM, Sartoratto A, et al. Anti-Candida activity of Brazilian medicinal plants. *J Ethnopharmacol.* 2005;97:305-11.
2112 N'gola V, Tchiyamba PL, Otoungou FVF. Composition of the essential oils of three Cyperus species from Congo. *African J Plant Sci.* 2014 Nov;8(11):473-81.
2113 Hassanein HD, Nazij NM, Shahat A, et al. Chemical Diversity of Essential Oils from Cyperus articulatus, Cyperus esculentus and Cyperus papyrus. *J Essent Oil Bear Plants.* 2014;17(2):251-64.
2114 Bersan SM, Galvão LC, Goes VF, et al. Action of essential oils from Brazilian native and exotic medicinal species on oral biofilms. BMC Complement Altern Med. 2014 Nov 18;14:451.
2115 Zoghbi MDGB, Andrade EHA, Carreira LMM, et al. Comparison of the Main Components of the Essential Oils of "priprioca": Cyperus articulatus var. articulatus L., C. articulatus var. nodosus L., C. prolixus Kunth and C. rotundus L. *J Essent Oil Res.* 2008 Jan/Feb;20:42-6.
2116 Lin WY, Kuo YH, Chang YL, et al. Anti-platelet aggregation and chemical constituents from the rhizome of Gynura japonica. *Planta Med.* 2003 Aug;69(8):757-64.
2117 Azzaz NAE, El-Khateeb AY, Farag AA. Chemical composition and biological activity of the essential oils of Cyperus articulatus. *Int J Acad Res.* 2014 Sep;6(5):265-9.
2118 Teixeira Duarte MC, Figuueira GM, Sartoratto A, et al. Anti-Candida activity of Brazilian medicinal plants. *J Ethnopharmacol.* 2005;97:305-11.
2119 Bersan SM, Galvão LC, Goes VF, et al. Action of essential oils from Brazilian native and exotic medicinal species on oral biofilms. BMC Complement Altern Med. 2014 Nov 18;14:451.
2120 Joulain D, Laurent R. The absolute from Karo-Karounde flowers. *Dev Food Sci.* 1988;18:607-25.
2121 Ford RA, Api AM, Letizia CS. Karo karounde absolute. *Food Chem Toxicol.* 1992;30(SUPPL):61S.
2122 Rusdi NA, Goh HH, Baharum SN. GC-MS/Olfactometric characterisation and aroma extraction dilution analysis of aroma active compounds in Polygonum minus essential oil. *POJ.* 2016;9(4):289-94.
2123 Baharum SN, Bunwan H, Ghani MA, et al. Analysis of the Chemical Composition of the Essential Oil of Polygonum minus Huds. Using Two-Dimensional Gas Chromatography-Time-of-Flight Mass Spectrometry (GC-TOF MS). *Molecules.* 2015;15:7006-15.
2124 Ullah H, Wilfred CD, Shaharun MS. Ionic liquid based ultrasonication-assisted extraction of essential oil from the leaves of Persicaria minor and conductor-like screening model for realistic solvents study. *Se Pu.* 2017 Jun 8;35(6):656-664.
2125 Yaacob KB. Kesom Oil—A Natural Source of Aliphatic Aldehydes. *Perfume Flavorist.* 1987 Oct/Nov;12:27-30.
2126 Lin WY, Kuo YH, Chang YL, et al. Anti-platelet aggregation and chemical constituents from the rhizome of Gynura japonica. *Planta Med.* 2003 Aug;69(8):757-64.
2127 Raina VK, Kumar A, Srivastava SK, et al. Essential oil composition of 'kewda' (Pandanus odoratissimus) from India. *Flavour Fragr J.* 2004;19:434-6.
2128 Nasim N, Behera JK, Sandeep IS, et al. Phytochemical analysis of flower from Pandanus odorifer (Forssk.) Kuntze for industrial application. *Nat Prod Commun.* 2018;32(20):2494-7.
2129 Maheshwari ML. Composition of essential oil from flowers of keora (Pandanus odoratissimus Linn.) by capillary gas chromatography. *Indian Perfumer.* 1995;39:45–48.
2130 Misra R, Dash PK, Rao YR. Chemical Composition of the Essential Oils of Kewda and Ketaki. *J Essent Oil Res.* 2000 Mar/Apr;12:175-8.
2131 Nasim N, Ray A, Sing S, et al. Characterization of Kewda volatile components by comprehensive two-dimensional gas chromatography time-of-flight mass spectrometry. *Nat Prod Res.* 2017 Apr;31(7):853-856.
2132 Tahvilian R, Moradi R, Zhaleh H, et al. Chemical composition and screening of antibacterial activity of essential oil of Pistacia khinjuk against two selected pathogenic bacteria. *Ann Trop Med Public Health.* 2017;10(5):1159-64.
2133 Azimi M, Sharifan A, Ghiasi Tarzi B. The Use of Pistacia khinjuk Essential Oil to Modulate Shelf-Life and Organoleptic Traits of Mechanically Deboned Chicken Meat. *J Food Process Preservation.* 2017;41(2):n/a-n/a.
2134 Pirbalouti AG, Aghaee K. Chemical Composition of Essential Oil of Pistacia khinjuk Stocks Grown in Bakhtiari Zagross Mountains, Iran. *Electronic J Biol.* 2011;7(4):67-9.
2135 Tahvilian R, Moradi R, Zhale H, et al. Ethnomedicinal Plants: In vitro antibacterial effect of essential oil of Pistacia khinjuk. *Int J Sci Eng Res.* 2016 Oct;7(1):437-47.
2136 Tahvilian R, Moradi R, Zhale H, et al. Ethnomedicinal Plants: Study on Antifungal Activity of Essential Oil of Pistacia khinjuk (Combined with the Dominance γ-Terpinene) Against Candida albicans. *Int J Pharm Clin Res.* 2016;8(10)1369-73.
2137 Taran M, Azizi E, Shikhvaisi A, et al. The Anthelmintic Effect of Pistacia khinjuk Against Protoscoleces of Echinococcus granulosus. *World J Zoology.* 2009;4(4):291-5.
2138 Pooter HLD, Schamp NM, Aboutabl EA, et al. Essential Oils from the Leaves of Three Pistacia Species Grown in Egypt. *Flavour Frag J.* 1991;6:229-32.
2139 Taran M, Azizi E, Shikhvaisi A, et al. The Anthelmintic Effect of Pistacia khinjuk Against Protoscoleces of Echinococcus granulosus. *World J Zoology.* 2009;4(4):291-5.
2140 Tahvilian R, Moradi R, Zhale H, et al. Chemical composition and screening of antibacterial activity of essential oil of Pistacia khinjuk against two selected pathogenic bacteria. *Ann Trop Med Public Health.* 2017;10(5):1159-64.

[2141] Tahvilian R, Moradi R, Zhale H, et al. Ethnomedicinal Plants: In vitro antibacterial effect of essential oil of Pistacia khinjuk. *Int J Sci Eng Res*. 2016 Oct;7(1):437-47.
[2142] Tahvilian R, Moradi R, Zhale H, et al. Ethnomedicinal Plants: Study on Antifungal Activity of Essential Oil of Pistacia khinjuk (Combined with the Dominance γ-Terpinene) Against Candida albicans. *Int J Pharm Clin Res*. 2016;8(10)1369-73.
[2143] Tahvilian R, Moradi R, Hajialiani M, et al. Evaluation of Antibacterial properties of essential oil of Pistacia khinjuk. *Int Res Pharm Sci*. 2017;7(1):1-6.
[2144] Azimi M, Sharifan A, Ghiasi Tarzi B. The Use of Pistacia khinjuk Essential Oil to Modulate Shelf-Life and Organoleptic Traits of Mechanically Deboned Chicken Meat. *J Food Process Preservation*. 2017;41(2):n/a-n/a.
[2145] Montanari RM, Barbosa LCA, Demuner AJ, et al. Chemical composition and antibacterial activity of essential oils from Verbenaceae species: Alternative sources of (E)-caryophyllene and germacrene-D. *Quim Nova*. 2011;34(9):1550-5.
[2146] Bendera MM. Isolation and characterization of essential oils from Ocimum americanum, Lantana camara, Lantana trifolia and Tephrosia vogelii. Egerton University Thesis. October 2007. Available at: http://ir-library.egerton.ac.ke/jspui/bitstream/123456789/287/1/Isolation%20and%20characterisation%20of%20essential%20oils%20from%20Ocium%20americunum%2CLantana%20camara%2Clantana%20trifolia%20and%20Tephrosia%20vogelii.PDF.
[2147] Satyal P, Crouch RA, Monzote L, et al. The Chemical Diversity of Lantana camara: Analyses of Essential Oil Samples from Cuba, Nepal, and Yemen. *Chem Biodiverity*. 2016;13:336-42.
[2148] Chowhury JU, Nandi NC, Bhuiyan NI. Chemical composition of leaf essential oil of Lantana camara L. from Bangladesh. *Bangladesh J Bot*. 2007 Dec;36(2):193-4.
[2149] Sundufu AJ, Shoushan H. Chemical composition of the essential oils of Lantana camara L. occurring in south China. *Flavour Frag J*. 2004;19:229-32.
[2150] Randrianalijaona JA, Ramanoelina PAR, Rasoarahona JRE, et al. Seasonal and chemotype influences on the chemical composition of Lantana camara L. Essential oils from Madagascar. *Analytica Chimica Acta*. 2005;545:46-52.
[2151] Khan MK, Srivastaca SK, Jain N, et al. Chemical composition of fruit and stem essential oils of Lantana camara from northern India. *Flavour Fragr J*. 2003;18:376-9.
[2152] Seth R, Mohan M, Singh P, et al. Chemical composition and antibacterial properties of the essential oil and extracts of Lantana camara Linn. from Uttarakhand (India). *Asian Pac J Trop Biomed*. 2012:S1207-11.
[2153] Barros LM, Duarte AE, Morais-Braga MFB, et al. Chemical Characterization and Trypanocidal, Leishmanicidal and Cytotoxicity Potential of Lantana camara L. (Verbenaceae) Essential Oil. *Molecules*. 2016;21(209):1-9.
[2154] de Sena Filho JG, Rabbani ARC, dos Santos Silva TR, et al. Chemical and molecular characterization of fifteen species from Lantana (Verbenaceae) genus. *Biochem Systematics Ecology*. 2012 Dec;45:130-7.
[2155] Rana VS, Prasad D, Blazquez MA. Chemical Composition of the Leaf Oil of Lantana camara. *J Essent Oil Res*. 2005 Mar/Apr;17:198-200.
[2156] Khan M, Srivastava SK, Syamasundar KV, et al. Chemical composition of leaf and flower essential oil of Lantana camara from India. *Flavour Fragr J*. 2002;17:75-7.
[2157] Macedo IT, de Oliveira LM, Camurça-Vasconcelos AL, et al. In vitro effects of Coriandrum sativum, Tagetes minuta, Alpinia zerumbet and Lantana camara essential oils on Haemonchus contortus. *Rev Bras Parasitol Vet*. 2013 Oct-Dec;22(4):463-9.
[2158] Passos JL, Barbosa LC, Demuner AJ, et al. Chemical characterization of volatile compounds of Lantana camara L. and L. radula Sw. and their antifungal activity. *Molecules*. 2012 Sep 27;17(10):11447-55.
[2159] Tesch NR, Mora F, Rojas L, et al. Chemical composition and antibacterial activity of the essential oil of Lantana camara var. moritziana. *Nat Prod Commun*. 2011 Jul;6(7):1031-4.
[2160] Sousa EO, Silva NF, Rodrigues FF, et al. Chemical composition and resistance-modifying effect of the essential oil of Lantana camara Linn. Pharmacogn Mag. 2010 Apr;6(22):79-82.
[2161] Sousa EO, Silva NF, Rodrigues FF, et al. Chemical composition and resistance-modifying effect of the essential oil of Lantana camara Linn. *Pharmacogn Mag*. 2010 Apr;6(22):79-82.
[2162] Lin WY, Kuo YH, Chang YL, et al. Anti-platelet aggregation and chemical constituents from the rhizome of Gynura japonica. *Planta Med*. 2003 Aug;69(8):757-64.
[2163] Seth R, Mohan M, Singh P, et al. Chemical composition and antibacterial properties of the essential oil and extracts of Lantana camara Linn. from Uttarakhand (India). *Asian Pac J Trop Biomed*. 2012:S1207-11.
[2164] Passos JL, Barbosa LC, Demuner AJ, et al. Chemical characterization of volatile compounds of Lantana camara L. and L. radula Sw. and their antifungal activity. *Molecules*. 2012 Sep 27;17(10):11447-55.
[2165] Salada JT, Balala LM, Vasquez EA. Phytochemical and Antibacterial Studies of Lantana camara L. Leaf Fraction and Essential Oil. *Int J Sci Res Pub*. 2015 Mar;5(3):1-5.
[2166] Tesch NR, Mora F, Rojas L, et al. Chemical composition and antibacterial activity of the essential oil of Lantana camara var. moritziana. *Nat Prod Commun*. 2011 Jul;6(7):1031-4.
[2167] Sousa EO, Silva NF, Rodrigues FF, et al. Chemical composition and resistance-modifying effect of the essential oil of Lantana camara Linn. *Pharmacogn Mag*. 2010 Apr;6(22):79-82.
[2168] Kurade NP, Jaitak V, Kaul VK, et al. Chemical composition and antibacterial activity of essential oils of Lantana camara, Ageratum houstonianum and Eupatorium adenophorum. *Pharm Biol*. 2010 May;48(5):539-44.
[2169] Deena MJ, Thoppil JE. Antimicrobial activity of the essential oil of Lantana camara. *Fitoterapia*. 2000 Aug;71(4):453-5.
[2170] Zoubiri S, Baaliouamer A. Larvicidal activity of two Algerian Verbenaceae essential oils against Culex pipiens. *Vet Parasitol*. 2011 Sep 27;181(2-4):370-3.
[2171] Dua VK, Pandey AC, Dash AP. Adulticidal activity of essential oil of Lantana camara leaves against mosquitoes. *Indian J Med Res*. 2010 Mar;131:434-9.
[2172] Bouda H, Tapondjou LA, Fontem DA, et al. Effect of essential oils from leaves of Ageratum conyzoides, Lantana camara and Chromolaena odorata on the mortality of Sitophilus zeamais (Coleoptera, Curculionidae). *J Stored Prod Res*. 2001 Apr;37(2):103-109.
[2173] Zoubiri S, Baaliouamer A. GC and GC/MS analyses of the Algerian Lantana camara leaf essential oil: Effect against Sitophilus granarius adults.*J Saudi Chemical Soc*. 2012;16:91-7.
[2174] Barros LM, Duarte AE, Morais-Braga MFB, et al. Chemical Characterization and Trypanocidal, Leishmanicidal and Cytotoxicity Potential of Lantana camara L. (Verbenaceae) Essential Oil. *Molecules*. 2016;21(209):1-9.
[2175] Macedo IT, de Oliveira LM, Camurça-Vasconcelos AL, et al. In vitro effects of Coriandrum sativum, Tagetes minuta, Alpinia zerumbet and Lantana camara essential oils on Haemonchus contortus. *Rev Bras Parasitol Vet*. 2013 Oct-Dec;22(4):463-9.
[2176] Joshi R, Sharma P, Sharma V, et al. Analysis of the essential oil of large cardamom (Amomum subulatum Roxb.) growing in different agro-climatic zones of Himachal Pradesh, India. *J Sci Food Agric*. 2013 Apr;93(6):1303-9.

[2177] Rout PK, Sahoo D, Jena KS, et al. Analysis of the Oil of Large Cardamom (Amomum subulatum Roxb.) Growing in Sikkim. *J Essent Oil Res.* 2003 Jul/Aug;265-6.
[2178] Adegoke GO, Jagan MRL, Shankaracharya NB. A comparison of the essential oils of Aframomum daniellii (Hook. F.) K. Schum. and Amomum subulatum Roxb. *Flavour Frag J.* 1998 Sep 1;13(5):349-52.
[2179] Pura Naik J, Jagan Mohan Rao L, Mohan Kumar T, et al. Chemical composition of the volatile oil from the pericarp (husk) of large cardamom (Amomum subulatum Roxb.). *Flavour Frag J.* 2004;19(5):441-4.
[2180] Vijayan AK, Leela NK, Dhanalakshmi M, et al. Volatile oil composition of four popular varieties of large cardamom (Amomum subulatum Roxb.). *J Spices Aromatic Crops.* 2017;26(1):37-43.
[2181] Satyal P, Dosoky NS, Kincer BL, et al. Chemical Compositions and Biological Activities of Amomum subulatum Essential Oils from Nepal. *Nat Prod Commun.* 2012;7(9):1233-6.
[2182] Bhandrari AK, Baunthiyal M, Bisht VK, et al. Assessment of Intra-Specific Variation in Essential Oil Composition in Amomum subulatum Roxb. Cultivated in Uttarakhand, India. *Asian J Chem Pharm Sci.* 2017;2(1):1-5.
[2183] Govindarajan M, Rajeswary M, Senthilmurugan S, et al. Larvicidal activity of the essential oil from Amomum subulatum Roxb. (Zingiberaceae) against Anopheles subpictus, Aedes albopictus and Culex tritaeniorhynchus (Diptera: Culicidae), and non-target impact on four mosquito natural enemies. *Phys Mol Plant Pathol.* 2018 Jan;101:219-24.
[2184] Craig JO. Poisoning by the volatile oils in childhood. *Arch Dis Child.* 1953;28:259-67.
[2185] Melis K. Bochner A, Janssens G. Accidental nasal eucalyptol and menthol instillation. *Eur J Pediatr.* 1989 Aug;148(8)786-7.
[2186] Day LM, Ozanne-Smith J, Parsons BJ, et al. Eucalyptus oil poisoning among young children: mechanisms of access and the potential prevention. *Aust N Z J Public Health.* 1997 Jun;21(3):297-302.
[2187] Day LM, Ozanne-Smith J, Parsons BJ, et al. Eucalyptus oil poisoning among young children: mechanisms of access and the potential prevention. *Aust N Z J Public Health.* 1997 Jun;21(3):297-302.
[2188] Myott E. Case of eucalyptus poisoning. *Brit M J.* 1906;1:558.
[2189] Hindle RC. Eucalyptus oil ingestion. *N Z Med J.* 1994 May 11;107(977)185-6.
[2190] Tibballs J. Clinical effects and management of eucalyptus oil ingestion in infants and small children. *Med J Aust.* 1995 Aug;163(4):177-80.
[2191] Waldman W, Barwina M, Sein Anand J. Accidental intoxication with eucalyptus oil—a case report. *Przeql Lek.* 2011;68(8):555-6.
[2192] Day LM, Ozanne-Smith J, Parsons BJ, et al. Eucalyptus oil poisoning among young children: mechanisms of access and the potential prevention. *Aust N Z J Public Health.* 1997 Jun;21(3):297-302.
[2193] Burkhard PR, Burkhardt K, Haenggeli CA, et al. Plant-induced seizures: reappearance of an old problem. *J Neurol.* 1999 Aug;246(8):667-70.
[2194] Waldman N. Seizure caused by dermal application of over-the-counter eucalyptus oil head lice preparation. *Clin Toxicol (Phila).* 2011 Oct;49(8):750-1.
[2195] Craig JO. Poisoning by the volatile oils in childhood. *Arch Dis Child.* 1953;28:259-67.
[2196] Mathew T, Kamath V, Kumar RS, et al. Eucalyptus oil inhalation-induced seizure: A novel, underrecognized, preventable cause of acute symptomatic seizure. *Epilepsia Open.* 2017 Jul 4;2(3):350-354.
[2197] Gray AM, Flatt PR. Antihyperglycemic actions of Eucalyptus globulus (Eucalyptus) are associated with pancreatic and extra pancreatic effects in mice. *J Nutr.* 1998 Dec;128(12):2319–23.
[2198] Basak SS, Candan F. Effect of Laurus nobilis L. Essential Oil and its Main Components on α-glucosidase and Reactive Oxygen Species Scavenging Activity. *Iran J Pharm Res.* 2013 Spring;12(2):367-79.
[2199] Fathiazad F, Mazandarani M, Hamedeyazdan. Phytochemical analysis and antioxidant activity of Hyssopus officinalis L. from Iran. *Adv Pharm Bull.* 2011 Dec;1(2):63–67.
[2200] Gray AM, Flatt PR. Antihyperglycemic actions of Eucalyptus globulus (Eucalyptus) are associated with pancreatic and extra pancreatic effects in mice. *J Nutr.* 1998 Dec;128(12):2319-23.
[2201] Dey B. Chemo-profiling of eucalyptus and study of its hypoglycemic potential. *World J Diabetes.* 2013 Oct 15;4(5):170–76.
[2202] Moharam BA, Jantan I, bin Ahmad F, et al. Antiplatelet Aggregation and Platelet Activating Factor (PAF) Receptor Antagonistic Activities of the Essential Oils of Five Goniothalamus Species. *Molecules.* 2010;15:5124-38.
[2203] Unger M, Frank A. Simultaneous determination of the inhibitory potency of herbal extracts on the activity of six major cytochrome P450 enzymes using liquid chromatography/mass spectrometry and automated online extraction. *Rapid Commun Mass Spectrom.* 2004;18(19):2273-81.
[2204] Jori A, Bianchetti A, Prestini PE, et al. Effect of eucalyptol (1,8-cineole) on the metabolism of other drugs in rats and in man. *Eur J Pharmacol.* 1970;9(3):362-66.
[2205] Kim NH, Hyun SH, Jin CH, et al. Pretreatment with 1,8-cineole potentiates thioacetamide-induced hepatotoxicity and immunosuppression. *Arch Pharm Res.* 2004 Jul;27(7):781-9.
[2206] Jori A, Bianchetti A, Prestini PE, et al. Effect of eucalyptol (1,8-cineole) on the metabolism of other drugs in rats and in man. *Eur J Pharmacol.* 1970;9(3):362-66.
[2207] de Sousa DP, Raphael E, Brocksom U, et al. Sedative effect of monoterpene alcohols in mice: A preliminary screening. *Verlag der Zeitschrift fur Naturforschung.* 2007;62c:563-66.
[2208] Satyal P, Dosoky NS, Kincer BL, et al. Chemical Compositions and Biological Activities of Amomum subulatum Essential Oils from Nepal. *Nat Prod Commun.* 2012;7(9):1233-6.
[2209] Jafri MA, Farah, Javed K, et al. Evaluation of the gastric antiulcerogenic effect of large cardamom (fruits of Amomum subulatum Roxb). *J Ethnopharmacol.* 2001 May;75(2-3):89-94.
[2210] Agnihotri SAA, Wakode SR, Ali M. Chemical composition, antimicrobial and topical anti-inflammatory activity of essential oil of Amomum subulatum fruits. *Acta Poloniae Pharmaceutica Drug Res.* 2012;69(6):1177-81.
[2211] Naveed R, Hussain I, Tawab A, et al. Antimicrobial activity of the bioactive components of essential oils from Pakistani spices against Salmonella and other multi-drug resistant bacteria. *BMC Complement Altern Med.* 213;13:365.
[2212] Agnihotri SAA, Wakode SR, Ali M. Chemical composition, antimicrobial and topical anti-inflammatory activity of essential oil of Amomum subulatum fruits. *Acta Poloniae Pharmaceutica Drug Res.* 2012;69(6):1177-81.
[2213] Agnihotri S, Wakode S. Antimicrobial activity of essential oil and various extracts of fruits of greater cardamom. *Indian J Pharm Sci.* 2010 Sep;72(5):657-9.
[2214] Kapoor IPS, Singh B, Sing G. Essential oil and oleoresins of Cardamom (Amomum subulatum Roxb.) as natural food preservatives for sweet orange (Citrus sinensis) juice. *J Food Process Eng.* 2011;34:1101-13.
[2215] Satyal P, Dosoky NS, Kincer BL, et al. Chemical Compositions and Biological Activities of Amomum subulatum Essential Oils from Nepal. *Nat Prod Commun.* 2012;7(9):1233-6.

[2216] Govindarajan M, Rajeswary M, Senthilmurugan S, et al. Larvicidal activity of the essential oil from Amomum subulatum Roxb. (Zingiberaceae) against Anopheles subpictus, Aedes albopictus and Culex tritaeniorhynchus (Diptera: Culicidae), and non-target impact on four mosquito natural enemies. *Phys Mol Plant Pathol.* 2018 Jan;101:219-24.
[2217] Satyal P, Dosoky NS, Kincer BL, et al. Chemical Compositions and Biological Activities of Amomum subulatum Essential Oils from Nepal. *Nat Prod Commun.* 2012;7(9):1233-6.
[2218] Derbesy M, Touche J, Zola A. The Essential Oil of Santolina chamaecyparissus L. *J Essent Oil Res.* 1989;1(6):269-75.
[2219] Demirci B, Ozek T, Baser KHC. Chemical Composition of Santolina chamaecyparissus L. Essential Oil. *J Essent Oil Res.* 2000 Sep-Oct;12:625-7.
[2220] Giner RM, Manez S, Rios JL. Seasonal variations in the essential oil of Santolina chamaecyparisus L. *Sci Pharm.* 1993;61:169-73.
[2221] Lawrence BM, Santolina oil. In: Essential Oils, 1992-1994. pp 57-59, Allured Publ. Corp., Carol Stream, IL, USA (1995).
[2222] Vernin G. Volatile Constituents of the Essential Oil of Santolina chamaecyparissus L. *J Essent Oil Res.* 1991 Jan-Feb;3:49-53.
[2223] Perez-Alonso MJ, Velasco-Negueruela A. Essential Oil Components of Santolina chamaecyparissus L. *Flavour Frag J.* 1992;7:37-41.
[2224] Salah-Fatnassi KBH, Hassayoun F, Cheraif I, et al. Chemical composition, antibacterial and antifungal activities of flowerhead and root essential oils of Santolina chamaecyparissus L., growing wild in Tunisia. *Saudi J Biol Sci.* 2017;24:875-82.
[2225] Djeddi S, Djebile K, Hadjbourega G, et al. In vitro Antimicrobial Properties and Chemical Composition of Santolina chamaecyparissus Essential Oil from Algeria. *Nat Prod Comm.* 2012;7(7):937-40.
[2226] Khubeiz MJ, Mansour G. In Vitro Antifungal, Antimicrobial Properties and Chemical Composition of Santolina chamaecyparissus Essential Oil in Syria. *Int J Toxicol Pharmacol Res.* 2016;8(5):372-8.
[2227] Misra LN, Siddiqui MS, Srivastava SK, et al. Gas chromatographic examination of an essential oil of Santolina chamaecyparissus L. *Perf Flavorist.* 1981 Apr-May;6:25-27.
[2228] Elsharkawy ER. Antitcancer effect and Seasonal variation in oil constituents of Santolina chamaecyparissus. *Chem Mat Res.* 2014;6(3):85-91.
[2229] Garg SN, Gupta D, Mehta VK, et al. Volatile Constituents of the Essential Oil of Santolina chamaecyparissus Linn, from the Southern Hills of India. *J Essent Oil Res.* 2001 Jul-Aug;13(4):234-5.
[2230] El-Sahhar KF, Nassar DM, Farag HM. Morphological and Anatomical Studies of Santolina chamaecyparissus L. (Asteraceae) Ii. Anatomical Characteristics and Volatile Oil. *Res J Agric Biol Sci.* 2011;7(6):413-22.
[2231] Ruiz-Navajas Y, Viuda-Martos M, Perez-Alvarez J, et al. Chemical Characterization and Antibacterial Activity of Two Aromatic Herbs (Santolina chamaecyparissus and Sideritis angustifolia) Widely Used in the Folk Medicine. *J Food Safety.* 2012;32(4):426-34.
[2232] Ahmed N, Tahar D, Soumia K, et al. Chemical composition, antioxidant and antimicrobial activities of the essential oil of Santolina chamaecyparissus L. of Algeria. *J Coastal Life Led.* 2015;3(3):220-7.
[2233] Ortiz de Elguea-Culebras G, Sánchez-Vioque R, Berruga MI, et al. Biocidal Potential and Chemical Composition of Industrial Essential Oils from Hyssopus officinalis, Lavandula × intermedia var. Super, and Santolina chamaecyparissus. *Chem Biodivers.* 2018 Jan;15(1).
[2234] Grosso C. Coelho JA, Urieta JS, et al. Herbicidal Activity of Volatiles from Coriander, Winter Savory, Cotton Lavender, and Thyme Isolated by Hydrodistillation and Supercritical Fluid Extraction. *J Agric Food Chem.* 2010;58:1007-13.
[2235] Grosso C. Coelho JA, Urieta JS, et al. Herbicidal Activity of Volatiles from Coriander, Winter Savory, Cotton Lavender, and Thyme Isolated by Hydrodistillation and Supercritical Fluid Extraction. *J Agric Food Chem.* 2010;58:1007-13.
[2236] Ahuja A, Bakshi S, Sharma S, et al. Production of volatile terpenes by proliferating shoots and micropropagated plants of Santolina chamaecyparissus L. (cotton lavender). *Flavour Frag J.* 2005;20(4):403-406.
[2237] Suresh B, Sriram S, Dhanaraj SA, et al. Anticandidal activity of Santolina chamaecyparissus volatile oil. *J Ethnopharm.* 1997;55:151-9.
[2238] Olowe SA, Ransome-Kuti O. The risk of jaundice in glucose-6-phosphate dehydrogenase deficient babies exposed to menthol. *Acta Paediatr Scand.* 1980 May;69(3):341-5.
[2239] Dillon Remy M, Manning Alleyne P, Bratt DE, et al. Neonatal jaundice at Port-of-Spain General Hospital abstract. *West Indian Med J.* 1987;36(Suppl):28.
[2240] Suresh B, Sriram S, Dhanaraj SA, et al. Anticandidal activity of Santolina chamaecyparissus volatile oil. *J Ethnopharm.* 1997;55:151-9.
[2241] Burkhard PR, Burkhardt K, Haenggeli CA, et al. Plant-induced seizures: reappearance of an old problem. *J Neurol.* 1999 Aug;246(8):667-70.
[2242] Waldman N. Seizure caused by dermal application of over-the-counter eucalyptus oil head lice preparation. *Clin Toxicol (Phila).* 2011 Oct;49(8):750-1.
[2243] Craig JO. Poisoning by the volatile oils in childhood. *Arch Dis Child.* 1953;28:259-67.
[2244] Mathew T, Kamath V, Kumar RS, et al. Eucalyptus oil inhalation-induced seizure: A novel, underrecognized, preventable cause of acute symptomatic seizure. *Epilepsia Open.* 2017 Jul 4;2(3):350-354.
[2245] Suresh B, Sriram S, Dhanaraj SA, et al. Anticandidal activity of Santolina chamaecyparissus volatile oil. *J Ethnopharm.* 1997;55:151-9.
[2246] Haliciglu O, Astarcioglu G, Yaprak I, et al. Toxicity of Salvia officinalis in a newborn and a child: an alarming report. *Pediatr Neurol.* 2011 Oct;45(4):259-60.
[2247] Lachenmeier DW, Walch SG. Epileptic seizure caused by accidental ingestion of sage (Salvia officinalis L.) oil in children: a rare, exceptional case or a threat to public health. *Pediatr Neurol.* 2012 Mar;46(3):201.
[2248] da Silva Costa KC, Bexerra SB, Norte CM, et al. Medicinal plants with teratogenic potential: current considerations. *Braz J Pharm Sci.* 2012;48(3):427-33.
[2249] Brinker F. Herb Contraindications and Drug Interactions. 2nd ed. Sandy, OR. Eclectic Medical Publications. 1998.
[2250] Frohne D. Giftpflanzen: Cupressaceae. Stuttgart: Wissenschaftliche Verlagsgesellschaft mbH; 1997. pp. 153–6.
[2251] European Medicines Agency. Public statement on the use of herbal medicinal products containing thujone. Available at: http://www.ema.europa.eu/docs/en_GB/document_library/Public_statement/2011/02/WC500102294.pdf.
[2252] Millet Y, Jouglard J, Steinmetz MD, et al. Toxicity of some essential plant oils. Clinical and experimental study. *Clin Toxicol.* 1981 Dec;18(12):1485-98.
[2253] Cristovao L, Carvalho F, Bastos MDL, et al. Hepatotoxicity of an essential oil of Salvia officinalis L.: an in vitro study using freshly isolated rat hepatocytes. *Congress Biomarkers.* 2001 Sep:165.
[2254] Millet Y, Jouglard J, Steinmetz MD, et al. Toxicity of some essential plant oils. Clinical and experimental study. *Clin Toxicol.* 1981 Dec;18(12):1485-98.
[2255] Burkhard PR, Burkhardt K, Haenggeli CA, et al. Plant-induced seizures: reappearance of an old problem. *J Neurol.* 1999 Aug;246(8):667-70.
[2256] Burkhard PR, Burkhardt K, Haenggeli CA, et al. Plant-induced seizures: reappearance of an old problem. *J Neurol.* 1999 Aug;246(8):667-70.
[2257] Hold KM, Sirisoma NS, Ikeda T, et al. α-Thujone (the active component of absinthe): γ-Aminobutyric acid type A receptor modulation and metabolic detoxification. *Proc Natl Acad Sci USA.* 2000 Apr 11;97(8):3826–3831.
[2258] Waldman N. Seizure caused by dermal application of over-the-counter eucalyptus oil head lice preparation. *Clin Toxicol (Phila).* 2011 Oct;49(8):750-1.

[2259] Craig JO. Poisoning by the volatile oils in childhood. *Arch Dis Child.* 1953;28:259-67.
[2260] Mathew T, Kamath V, Kumar RS, et al. Eucalyptus oil inhalation-induced seizure: A novel, underrecognized, preventable cause of acute symptomatic seizure. *Epilepsia Open.* 2017 Jul 4;2(3):350-354.
[2261] Uc A, Bishop WP, Sanders KD. Camphor hepatotoxicity. *South Med J.* 2000;93:596-98.
[2262] Frohne D. Giftpflanzen: Cupressaceae. Stuttgart: Wissenschaftliche Verlagsgesellschaft mbH; 1997. pp. 153–6.
[2263] Dolan LC, Matulka RA, Burdock GA. Naturally Occurring Food Toxins. *Toxins (Basel).* 2010 Sep;2(9):2289–2332.
[2264] United States National Toxicology Program (NTP). Alpha-Thujone. Dec 10, 1997. Available at:
http://ntp.niehs.nih.gov/index.cfm?objectid=03DB8C36-E7A1-9889-3BDF8436F2A8C51F.
[2265] Suresh B, Sriram S, Dhanaraj SA, et al. Anticandidal activity of Santolina chamaecyparissus volatile oil. *J Ethnopharm.* 1997;55:151-9.
[2266] Khine H, Weiss D, Graber N, et al. A cluster of children with seizures caused by camphor poisoning. *Pediatrics.* 2009 May;123(5):1269-72.
[2267] Michiels EA, Mazor SS. Toddler with seizures due to ingesting camphor at an Indian celebration. *Pediatr Emerg Care.* 2010 Aug;26(8):574-75.
[2268] Koren G. Medications which can kill a toddler with one tablet or teaspoonful. *J Toxicol Clin Toxicol. 1*993;31(3):407-13.
[2269] Flaman Z, Pellechia-Clarke S, Bailey B, et al. Unintentional exposure of young children to camphor and eucalyptus oils. *Paediatr Child Health.* 2001 Feb;6(2):80-83.
[2270] Rabl W, Katzgraber F, Steinlechner M. Camphor ingestion for abortion (case report). *Forensic Sci Int.* 1997 Sep 19;89(1-2):137-40.
[2271] Flaman Z, Pellechia-Clarke S, Bailey B, et al. Unintentional exposure of young children to camphor and eucalyptus oils. *Paediatr Child Health.* 2001 Feb;6(2):80-83.
[2272] Manoguerra AS, Erdman AR, Wax PM, et al. Camphor poisoning: an evidence-based practice guideline for out-of- hospital management. *Clin Toxicol (Phila).*2006;44(4):357-70.
[2273] Gibson DE, Moore GP, Pfaff JA. Camphor ingestion. *Am J Emerg Med.* 1989 Jan;7(1):41-43.
[2274] Koppel C, Martends F, Schirop T, et al. Hemoperfusion in acute camphor poisoning. *Intensive Care Med.* 1988;14(4):431-33.
[2275] Burkhard PR, Burkhardt K, Haenggeli CA, et al. Plant-induced seizures: reappearance of an old problem. *J Neurol.* 1999 Aug;246(8):667–70.
[2276] Millet Y, Jouglard J, Steinmetz MD, et al. Toxicity of some essential plant oils. Clinical and experimental study. *Clin Toxicol.* 1981 Dec;18(12):1485–98.
[2277] Charanraj KJ, G MV, S M. Camphor poisoning in a child. *Natl Med J India.* 2013 Jan-Feb;26(1):60.
[2278] Narayan S, Singh N. Camphor poisoning-An unusual cause of seizure. *Med J Armed Forces India.* 2012 Jul;68(3):252–53.
[2279] Uc A, Bishop WP, Sanders KD. Camphor hepatotoxicity. *South Med J.* 2000;93:596-98.
[2280] Uc A, Bishop WP, Sanders KD. Camphor hepatotoxicity. *South Med J.* 2000;93:596-98.
[2281] Frohne D. Giftpflanzen: Cupressaceae. Stuttgart: Wissenschaftliche Verlagsgesellschaft mbH; 1997. pp. 153–6.
[2282] Dolan LC, Matulka RA, Burdock GA. Naturally Occurring Food Toxins. *Toxins (Basel).* 2010 Sep;2(9):2289–2332.
[2283] Olowe SA, Ransome-Kuti O. The risk of jaundice in glucose-6-phosphate dehydrogenase deficient babies exposed to menthol. *Acta Paediatr Scand.* 1980 May;69(3):341-5.
[2284] Dillon Remy M, Manning Alleyne P, Bratt DE, et al. Neonatal jaundice at Port-of-Spain General Hospital abstract. *West Indian Med J.* 1987;36(Suppl):28.
[2285] Suresh B, Sriram S, Dhanaraj SA, et al. Anticandidal activity of Santolina chamaecyparissus volatile oil. *J Ethnopharm.* 1997;55:151-9.
[2286] Khine H, Weiss D, Graber N, et al. A cluster of children with seizures caused by camphor poisoning. *Pediatrics.* 2009 May;123(5):1269-72.
[2287] Michiels EA, Mazor SS. Toddler with seizures due to ingesting camphor at an Indian celebration. *Pediatr Emerg Care.* 2010 Aug;26(8):574-75.
[2288] Koren G. Medications which can kill a toddler with one tablet or teaspoonful. *J Toxicol Clin Toxicol. 1*993;31(3):407-13.
[2289] Flaman Z, Pellechia-Clarke S, Bailey B, et al. Unintentional exposure of young children to camphor and eucalyptus oils. *Paediatr Child Health.* 2001 Feb;6(2):80-83.
[2290] Rabl W, Katzgraber F, Steinlechner M. Camphor ingestion for abortion (case report). *Forensic Sci Int.* 1997 Sep 19;89(1-2):137-40.
[2291] Flaman Z, Pellechia-Clarke S, Bailey B, et al. Unintentional exposure of young children to camphor and eucalyptus oils. *Paediatr Child Health.* 2001 Feb;6(2):80-83.
[2292] Manoguerra AS, Erdman AR, Wax PM, et al. Camphor poisoning: an evidence-based practice guideline for out-of- hospital management. *Clin Toxicol (Phila).*2006;44(4):357-70.
[2293] Gibson DE, Moore GP, Pfaff JA. Camphor ingestion. *Am J Emerg Med.* 1989 Jan;7(1):41-43.
[2294] Koppel C, Martends F, Schirop T, et al. Hemoperfusion in acute camphor poisoning. *Intensive Care Med.* 1988;14(4):431-33.
[2295] Burkhard PR, Burkhardt K, Haenggeli CA, et al. Plant-induced seizures: reappearance of an old problem. *J Neurol.* 1999 Aug;246(8):667–70.
[2296] Millet Y, Jouglard J, Steinmetz MD, et al. Toxicity of some essential plant oils. Clinical and experimental study. *Clin Toxicol.* 1981 Dec;18(12):1485–98.
[2297] Charanraj KJ, G MV, S M. Camphor poisoning in a child. *Natl Med J India.* 2013 Jan-Feb;26(1):60.
[2298] Narayan S, Singh N. Camphor poisoning-An unusual cause of seizure. *Med J Armed Forces India.* 2012 Jul;68(3):252–53.
[2299] Olowe SA, Ransome-Kuti O. The risk of jaundice in glucose-6-phosphate dehydrogenase deficient babies exposed to menthol. *Acta Paediatr Scand.* 1980 May;69(3):341-5.
[2300] Dillon Remy M, Manning Alleyne P, Bratt DE, et al. Neonatal jaundice at Port-of-Spain General Hospital abstract. *West Indian Med J.* 1987;36(Suppl):28.
[2301] Li Y, Shao X, Xu J, et al. Effects and possible mechanism of tea tree oil against Botrytis cinerea and Penicillium expansum in vitro and in vivo test. *Can J Microbiol.* 2017 Mar;63(3):219-227.
[2302] Li YH, Sun XP, Zhang YQ, et al. The antithrombotic effect of borneol related to its anticoagulant property. *Am J Chin Med.* 2008;36(4):719-27.
[2303] Hiroi T, Miyazaki Y, Kobayashi Y, et al. Induction of hepatic P450s in rat by essential wood and leaf oils. *Xenobiotica.* 1995 May;25(5):457-67.
[2304] Chen JY, Wang JJ, Meng MR, et al. [Borneol is an inducer of rat hepatic CYP2D activity in vivo]. *Yao Xue Xue Bao.* 2015 Apr;50(4):459-63.
[2305] Zhang R1, Mi SQ, Wang NS. Effect of borneol on cytochrome P450 3A enzyme and midazolam pharmacokinetics in rats. *Eur J Drug Metab Pharmacokinet.* 2013 Sep;38(3):159-69.
[2306] Kim NH, Hyun SH, Jin CH, et al. Pretreatment with 1,8-cineole potentiates thioacetamide-induced hepatotoxicity and immunosuppression. *Arch Pharm Res.* 2004 Jul;27(7):781-9.
[2307] Marín I, Sayas-Barberá E, Viuda-Martos M, et al. Chemical Composition, Antioxidant and Antimicrobial Activity of Essential Oils from Organic Fennel, Parsley, and Lavender from Spain. *Foods.* 2016 Mar 4;5(1).
[2308] Burkhard PR, Burkhardt K, Haenggeli CA, et al. Plant-induced seizures: reappearance of an old problem. *J Neurol.* 1999 Aug;246(8):667–70.
[2309] Millet Y, Jouglard J, Steinmetz MD, et al. Toxicity of some essential plant oils. Clinical and experimental study. *Clin Toxicol.* 1981 Dec;18(12):1485–98.
[2310] Charanraj KJ, G MV, S M. Camphor poisoning in a child. *Natl Med J India.* 2013 Jan-Feb;26(1):60.

[2311] Narayan S, Singh N. Camphor poisoning-An unusual cause of seizure. *Med J Armed Forces India.* 2012 Jul;68(3):252–53.
[2312] Unger M, Frank A. Simultaneous determination of the inhibitory potency of herbal extracts on the activity of six major cytochrome P450 enzymes using liquid chromatography/mass spectrometry and automated online extraction. *Rapid Commun Mass Spectrom.* 2004;18(19):2273-81.
[2313] Jori A, Bianchetti A, Prestini PE, et al. Effect of eucalyptol (1,8-cineole) on the metabolism of other drugs in rats and in man. *Eur J Pharmacol.* 1970;9(3):362-66.
[2314] Kim NH, Hyun SH, Jin CH, et al. Pretreatment with 1,8-cineole potentiates thioacetamide-induced hepatotoxicity and immunosuppression. *Arch Pharm Res.* 2004 Jul;27(7):781-9.
[2315] Jori A, Bianchetti A, Prestini PE, et al. Effect of eucalyptol (1,8-cineole) on the metabolism of other drugs in rats and in man. *Eur J Pharmacol.* 1970;9(3):362-66.
[2316] de Sousa DP, Raphael E, Brocksom U, et al. Sedative effect of monoterpene alcohols in mice: A preliminary screening. *Verlag der Zeitschrift fur Naturforschung.* 2007;62c:563-66.
[2317] Suresh B, Sriram S, Dhanaraj SA, et al. Anticandidal activity of Santolina chamaecyparissus volatile oil. *J Ethnopharm.* 1997;55:151-9.
[2318] Elsharkawy ER. Antitcancer effect and Seasonal variation in oil constituents of Santolina chamaecyparissus. *Chem Mat Res.* 2014;6(3):85-91.
[2319] Suresh B, Sriram S, Dhanaraj SA, et al. Anticandidal activity of Santolina chamaecyparissus volatile oil. *J Ethnopharm.* 1997;55:151-9.
[2320] Djeddi S, Djebile K, Hadjbourega G, et al. In vitro Antimicrobial Properties and Chemical Composition of Santolina chamaecyparissus Essential Oil from Algeria. *Nat Prod Comm.* 2012;7(7):937-40.
[2321] Khubeiz MJ, Mansour G. In Vitro Antifungal, Antimicrobial Properties and Chemical Composition of Santolina chamaecyparissus Essential Oil in Syria. *Int J Toxicol Pharmacol Res.* 2016;8(5):372-8.
[2322] Ruiz-Navajas Y, Viuda-Martos M, Perez-Alvarez J, et al. Chemical Characterization and Antibacterial Activity of Two Aromatic Herbs (Santolina chamaecyparissus and Sideritis angustifolia) Widely Used in the Folk Medicine. *J Food Safety.* 2012;32(4):426-34.
[2323] Salah-Fatnassi KBH, Hassayoun F, Cheraif I, et al. Chemical composition, antibacterial and antifungal activities of flowerhead and root essential oils of Santolina chamaecyparissus L., growing wild in Tunisia. *Saudi J Biol Sci.* 2017;24:875-82.
[2324] Ahmed N, Tahar D, Soumia K, et al. Chemical composition, antioxidant and antimicrobial activities of the essential oil of Santolina chamaecyparissus L. of Algeria. *J Coastal Life Led.* 2015;3(3):220-7.
[2325] Ruiz-Navajas Y, Viuda-Martos M, Sendra E, et al. In vitro antioxidant and antifungal properties of essential oils obtained from aromatic herbs endemic to the southeast of Spain. *J Food Prot.* 2013 Jul;76(7):1218-25.
[2326] Diánez F, Santos M1, Parra C, et al. Screening of antifungal activity of 12 essential oils against eight pathogenic fungi of vegetables and mushroom. *Lett Appl Microbiol.* 2018 Oct;67(4):400-410.
[2327] Ahmed N, Tahar D, Soumia K, et al. Chemical composition, antioxidant and antimicrobial activities of the essential oil of Santolina chamaecyparissus L. of Algeria. *J Coastal Life Led.* 2015;3(3):220-7.
[2328] Ruiz-Navajas Y, Viuda-Martos M, Sendra E, et al. In vitro antioxidant and antifungal properties of essential oils obtained from aromatic herbs endemic to the southeast of Spain. *J Food Prot.* 2013 Jul;76(7):1218-25.
[2329] Ortiz de Elguea-Culebras G, Sánchez-Vioque R, Berruga MI, et al. Biocidal Potential and Chemical Composition of Industrial Essential Oils from Hyssopus officinalis, Lavandula × intermedia var. Super, and Santolina chamaecyparissus. *Chem Biodivers.* 2018 Jan;15(1).
[2330] Suresh B, Sriram S, Balasubramanian S, et al. Santolina volatile oil, a potential antifeedant for the control of the brown plant hopper, milaparvatha lugens. *Anc Sci Life.* 1996 Jan;15(3):166-8.
[2331] Seo SM, Kim J, Kang J, et al. Fumigant toxicity and acetylcholinesterase inhibitory activity of 4 Asteraceae plant essential oils and their constituents against Japanese termite (Reticulitermes speratus Kolbe). *Pestic Biochem Physiol.* 2014 Jul;113:55-61.
[2332] Grosso C. Coelho JA, Urieta JS, et al. Herbicidal Activity of Volatiles from Coriander, Winter Savory, Cotton Lavender, and Thyme Isolated by Hydrodistillation and Supercritical Fluid Extraction. *J Agric Food Chem.* 2010;58:1007-13.
[2333] Bareemizadeh F, Karimi N, Ghasempour HR, et al. Essential oil composition of Allium ampeloprasum L. var. atroviolaceum and Allium iranicum. *Int J Biosciences.* 2014;4(1):372-7.
[2334] Selim YA, Sakeran MI. Effect of Time Distillation on Chemical Constituents and Anti-Diabetic Activity of the Essential Oil from Dark Green Parts of Egyptian Allium ampeloprasum L. *J Essent Oil Bearing Plants.* 2014 Dec;17(5):838-46.
[2335] Mnayer D, Fabiano-Tixier AS, Petitcolas E, et al. Chemical Composition, Antibacterial and Antioxidant Activities of Six Essentials Oils from the Alliaceae Family. *Molecules.* 2014;19:20034-53.
[2336] Zaman R, Parvez M, Ali MS, et al. Evaluation of Antifertility Effect of Methanolic Bulb Extract of Allium cepa on Swiss Albino Male and Teratogenic Effect on Female Mice. *Adv Biol Res.* 2015;9(2):128-32.
[2337] Farnsworth NR, Bingel AS, Cordell GA, et al. Potential value of plants as sources of new antifertility agents I. *J Pharm Sci.* 1975 Apr;64(4):535-98.
[2338] Mennella JA, Johnson A, Beauchamp GK. Garlic ingestion by pregnant women alters the odor of amniotic fluid. *Chem Senses.* 1995;20:207-9.
[2339] Ziaei S, Hantoshzadeh S, Rezasoltani P, et al. The effect of garlic tablet on plasma lipids and platelet aggregation in nulliparous pregnants at high risk of preeclampsia. *Eur J Obstet Gynecol Reprod Biol.* 2001;99:201-6.
[2340] Soltani PR. Preeclampisia is an important complication of pregnancy which can result in morbidity and mortality in mother, fetus and the neonate. *J Med Counc Islamic Repub Iran.* 2005;23(3):319.
[2341] Kim SH, Lee IC, Baek HS, et al. Induction of cytochrome P450 3A1 expression by diallyl disulfide: protective effects against cyclophosphamideinduced embryo-fetal developmental toxicity. *Food Chem Toxicol.* 2014 Jul;69:312-9.
[2342] El Sayyad HI, Abou_El-Naga AM, Gadallah AA, et al. Protective effects of Allium sativum against defects of hypercholesterolemia on pregnant rats and their offspring. *Int J Clin Exp Med.* 2010; 3(2): 152–163.
[2343] Mennella JA, Beauchamp GK. The effects of repeated exposure to garlic-flavored milk on the nursling's behavior. *Pediatr Res.* 1993;34:805-8.
[2344] Mennella JA, Beauchamp GK. Maternal diet alters the sensory qualities of human milk and the nursling's behavior. *Pediatrics.* 1991;88:737-44.
[2345] Goldman IL, Kopelberg M, Debaene JE, et al. Antiplatelet activity in onion (Allium cepa) is sulfur dependent. *Thromb Haemost.* 1996 Sep;76(3):450-2.
[2346] [No authors listed]. The effect of essential oil of garlic on hyperlipemia and platelet aggregation--an analysis of 308 cases. Cooperative Group for Essential Oil of Garlic. *J Tradit Chin Med.* 1986 Jun;6(2):117-20.
[2347] Barrie SA, Wright JV, Pizzorno JE. Effects of Garlic Oil on Platelet Aggregation, Serum Lipids and Blood Pressure in Humans. *J Orthomolecular Med.* 1987;2(1):15-21.
[2348] Makheja AN, Vanderhoek JY, Bailey JM. Inhibition of platelet aggregation and thromboxane synthesis by onion and garlic. *Lancet.* 1979 Apr 7;1(8119):781.
[2349] Boullin DJ. Garlic as a platelet inhibitor. *Lancet.* 1981 Apr 4;1(8223):776-7.
[2350] Fenwick GR, Hanley AB. The genus Allium--Part 3. *Crit Rev Food Sci Nutr.* 1985;23(1):1-73.

[2351] Lawson LD, Hughes BG. Characterization of the formation of allicin and other thiosulfinates from garlic. *Planta Med.* 1992 Aug;58(4):345-50.
[2352] Bordia A, Verma SK, Srivastava KC. Effect of garlic (Allium sativum) on blood lipids, blood sugar, fibrinogen and fibrinolytic activity in patients with coronary artery disease. *Prostaglandins Leukot Essent Fatty Acids.* 1998 Apr;58(4):257-63.
[2353] Bordia A. Effect of garlic on human platelet aggregation in vitro. *Atherosclerosis.* 1978 Aug;30(4):355-60.
[2354] Qi R, Liao F, Inoue K, et al. Inhibition by diallyl trisulfide, a garlic component, of intracellular Ca(2+) mobilization without affecting inositol-1,4,5-trisphosphate (IP(3)) formation in activated platelets. *Biochem Pharmacol.* 2000 Nov 15;60(10):1475-83.
[2355] Ariga T, Takeda A, Teramoto S, et al. Inhibition Site of Methylallyl Trisulfide: A Volatile Oil Component of Garlic, in the Platelet Arachidonic Acid Cascade. *Food Factors for Cancer Prevention.* 1997:231-34.
[2356] Lawson LD, Ransom DK, Hughes BG. Inhibition of whole blood platelet-aggregation by compounds in garlic clove extracts and commercial garlic products. *Thromb Res.* 1992 Jan 15;65(2):141-56.
[2357] Siess MH, Le Bon AM, Canivenc-Lavier MC, et al. Modification of hepatic drug metabolising enzymes in rats treated with alkyl sulfides. *Cancer Lett.* 1997;120:195–201.
[2358] Lai YS, Chen WC, Ho CT, et al. Garlic essential oil protects against obesity-triggered nonalcoholic fatty liver disease through modulation of lipid metabolism and oxidative stress. *J Agric Food Chem.* 2014 Jun 25;62(25):5897-906.
[2359] Taubert D, Glöckner R, Müller D, et al. The garlic ingredient diallyl sulfide inhibits cytochrome P450 2E1 dependent bioactivation of acrylamide to glycidamide. *Toxicol Lett.* 2006 Jun 20;164(1):1-5.
[2360] Davenport DM, Wargovich MJ. Modulation of cytochrome P450 enzymes by organosulfur constituents from garlic. *Food Chem Toxicol.* 2005 Dec;43(12):1753-62.
[2361] Fisher CD, Augustine LM, Maher JM, et al. Induction of drug-metabolizing enzymes by garlic and allyl sulfide compounds via activation of constitutive androstane receptor and nuclear factor E2-related factor 2. *Drug Metab Dispos.* 2007 Jun;35(6):995-1000.
[2362] Lii CK, Tsai CW, Wu CC. Garlic allyl sulfides display differential modulation of rat cytochrome P450 2B1 and the placental form glutathione S transferase in various organs. *J Agric Food Chem.* 2006 Jul 12;54(14):5191-6.
[2363] Johnson S. Supercritical Essential Oils: A Companion Resource to Medicinal Essential Oils. 2017. Scott A Johnson Professional Writing Services, LLC: Orem, Utah.
[2364] Pyun MS, Shin S. Antifungal effects of the volatile oils from Allium plants against Trichophyton species and synergism of the oils with ketoconazole. *Phytomedicine.* 2006 Jun;13(6):394-400.
[2365] Benezra C, Ducombs G, Sell Y, Foussereau J. Plant contact dermatitis. Philadelphia: B.C. DECKER; 1985. p.240-3.
[2366] Coz CJL, Ducombs G. Plants and Plant Products. Chapter 41. Available at: http://eknygos.lsmuni.lt/springer/99/751-800.pdf.
[2367] Navarro JA, del Pozo MD, Gastaminza G, et al. Allium cepa seeds: A new occupational allergen. *J Allergy Clinical Immunol.* 1995 Nov; 96(15):690-93.
[2368] Lee TY, Lam TH. Contact dermatitis due to topical treatment with garlic in Hong Kong. *Contact Dermatitis.* 1991 Mar;24(3):193-6.
[2369] Cabanillas M, Fernandez-Redondo V, Toribio J. Allergic contact dermatitis to plants in a Spanish dermatology department: a 7-year review. *Contact Dermatitis.* 2006 Aug;55(2):84-91.
[2370] Selim YA, Sakeran MI. Effect of Time Distillation on Chemical Constituents and Anti-Diabetic Activity of the Essential Oil from Dark Green Parts of Egyptian Allium ampeloprasum L. *J Essent Oil Bearing Plants.* 2014 Dec;17(5):838-46.
[2371] Nguansangiam S, Angsubhakorn S, Bhamarapravati S, et al. Effects of elephant garlic volatile oil (Allium ampeloprasum) and T-2 toxin on murine skin. *Southeast Asian J Trop Med Public Health.* 2003 Dec;34(4):899-905.
[2372] Kocic-Tanackov SD, Dimic GR, Tepic AN, et al. Influence of Allium amperloprasum L. and Allium cepa L. essential oils on the growth of some moulds. *Zbornik Matice srpske za prirodne nauke.* 2009 Jan;116:121-30.
[2373] Maciel mv, Morais SM, Bevilaqua CML, et al. Chemical composition of Eucalyptus spp. essential oils and their insecticidal effects on Lutzomyia longipalpis. *Vet Parasitol.* 2010;167:1-7.
[2374] Correia Ribeiro WL, Freitas Macedo IT, Leite dos Santos JM, et al. Activity of chitosan-encapsulated Eucalyptus staigeriana essential oil on Haemonchus contortus. *Exp Parasitol.* 2013 Sep;135(1):24-9.
[2375] Correia Ribeiro WL, Camurca-Vasconcelos ALF, Freitas Macedo IT, er al. In vitro effects of Eucalyptus staigeriana nanoemulsion on Haemonchus contortus and toxicity in rodents. *Vet Parasitol.* 2015 Sep 15;212(3-4):444-7.
[2376] Correia Ribeiro WL, Freitas Macedo IT, Leite dos Santos JM, et al. Activity of chitosan-encapsulated Eucalyptus staigeriana essential oil on Haemonchus contortus. *Exp Parasitol.* 2013 Sep;135(1):24-9.
[2377] de Aquino Mesquita M, E Silva Júnior JB, Panassol AM, et al. Anthelmintic activity of Eucalyptus staigeriana encapsulated oil on sheep gastrointestinal nematodes. *Parasitol Res.* 2013 Sep;112(9):3161-5.
[2378] Macedo IT, Bevilaqua CM, de Oliveira LM, et al. Anthelmintic effect of Eucalyptus staigeriana essential oil against goat gastrointestinal nematodes. *Vet Parasitol.* 2010 Oct 11;173(1-2):93-8.
[2379] Maciel mv, Morais SM, Bevilaqua CML, et al. Chemical composition of Eucalyptus spp. essential oils and their insecticidal effects on Lutzomyia longipalpis. *Vet Parasitol.* 2010;167:1-7.
[2380] Cruz GS, Wanderley-Teixeira V, Oliveira JV, et al. Sublethal Effects of Essential Oils From Eucalyptus staigeriana (Myrtales: Myrtaceae), Ocimum gratissimum (Lamiales: Laminaceae), and Foeniculum vulgare (Apiales: Apiaceae) on the Biology of Spodoptera frugiperda (Lepidoptera: Noctuidae). *J Econ Entomol.* 2016 Apr;109(2):660-6.
[2381] Fraternale D, Giamperi L, Ricci D, et al. Composition of Essential Oil as a Taxonomic Marker for Calamintha nepeta (L.) Savi ssp. nepeta. *J Essent Oil Res.* 1998 Sep/Oct;10:568-70.
[2382] Demirci B, Temel HE, Portakal T, et al. Inhibitory effect of Calamintha nepeta subsp. glandulosa essential oil on lipoxygenase. *Türk Biyokimya Dergisi.* 2011;36(4):290-5.
[2383] Ristorcelli D, Tomi F, Cassanova J. Essential oils of Calamintha subsp. nepeta and subsp. glandulosa from Corsica (France). *J Essent Oil Res.* 1996 Jul/Aug;8:363-6.
[2384] Kitic D, Jovanovic T, Ristic M, et al. Chemical Composition and Antimicrobial Activity of the Essential Oil of Calaminfha nepeta (L.) Savi ssp glandulosa (Req.) P. W. Ball from Montenegro. *J Essent Oil Res.* 2002 Mar/Apr;14:150-2.
[2385] Sarer E, Pancali SS. Composition of the Essential Oil from Calamintha nepeta (L.) Savi ssp. glandulosa (Req.) P. W. Ball. *Flavour Frag J.* 1998;13:31-2.
[2386] Alan S, Kurkcuoglu M, Baser KHC. Composition of Essential Oils of Calamintha nepeta (L.) Savi Subsp. nepeta and Calamintha nepeta (L.) Savi Subsp. glandulosa (Req.) P.W. Ball. *Asian J Chem.* 2011;23(6):2357-60.
[2387] Bozovic M, Ragno R. Calamintha nepeta (L.) Savi and its Main Essential Oil Constituent Pulegone: Biological Activities and Chemistry. *Molecules.* 2017;22(2):290.

[2388] Riela S, Bruno M, Formisano C, et al. Effects of solvent-free microwave extraction on the chemical composition of essential oil of Calamintha nepeta (L.) Savi compared with the conventional production method. *J Sep Sci.* 2008;31:1110-17.
[2389] Kirimer N, Baser KHC. Composition of the essential oil of Calamintha nepeta subsp. glandulosa. *J Essent Oil Res.* 1992 Mar/Apr;4:189-90.
[2390] Baldovini N, Ristorcelli D, Tomi F, et al. Infraspecific variability of the essential oil of Calamintha nepeta from Corsica (France). *Flavour Frag J.* 2000;15:50-4.
[2391] Vieira M, Bessa L, Martins M, et al. Chemical Composition, Antibacterial, Antibiofilm and Synergistic Properties of Essential Oils from Eucalyptus globulus Labill. and Seven Mediterranean Aromatic Plants. *Chem Biodivers.* 2017;14(6):e1700006.
[2392] Božović M, Garzoli S, Sabatino M, et al. Essential Oil Extraction, Chemical Analysis and Anti-Candida Activity of Calamintha nepeta (L.) Savi subsp. glandulosa (Req.) Ball-New Approaches. *Molecules.* 2017 Jan 26;22(2).
[2393] Arantes S, Piçarra A, Candeias F, et al. Antioxidant activity and cholinesterase inhibition studies of four flavouring herbs from Alentejo. *Nat Prod Res.* 2017 Sep;31(18):2183-2187.
[2394] Gormez A, Bozari S, Yanmis D, et al. Chemical Composition and Antibacterial Activity of Essential Oils of Two Species of Lamiaceae against Phytopathogenic Bacteria. *Pol J Microbiol.* 2015;64(2):121-7.
[2395] Colombo RP, Martínez A, Fernández di Pardo A, et al. Differential effects of two strains of Rhizophagus intraradices on dry biomass and essential oil yield and composition in Calamintha nepeta. *Rev Argent Microbiol.* 2013 Apr-Jun;45(2):114-8.
[2396] Araniti F, Lupini A, Sorgonà A, et al. Phytotoxic activity of foliar volatiles and essential oils of Calamintha nepeta (L.) Savi. *Nat Prod Res.* 2013;27(18):1651-6.
[2397] Miladinović DL, Ilić BS, Mihajilov-Krstev TM, et al. Investigation of the chemical composition-antibacterial activity relationship of essential oils by chemometric methods. *Anal Bioanal Chem.* 2012 May;403(4):1007-18.
[2398] Kitic D, Stojanovic G, Palic R, et al. Chemical Composition and Microbial Activity of the Essential Oil of Calamintha nepeta (L.) Savi ssp. nepeta var. subisodonda (Borb.) Hayek from Serbia. *J Essent Oil Res.* 2005 Nov/Dec;17:701-3.
[2399] Flamini G, Cioni P, Puleio R, et al. Antimicrobial activity of the essential oil of Calamintha nepeta and its constituent pulegone against bacteria and fungi. *Phytother Res.* 1999;13(4):349-51.
[2400] Panizzi L, Flamini G, Cioni PL, et al. Compositoin and antimicrobial properties of essential oils of four Mediterranean Lamiaceae. *J Ethnopharm.* 1993;39:167-70.
[2401] Negro C, Notarnicola S, de Bellis L, et al. Intraspecific variability of the essential oil of Calamintha nepeta subsp. nepeta from Southern Italy (Apulia). *Nat Prod Res.* 2013 Mar;27(4-5):331-9.
[2402] Bakerink JA, Gospe SM Jr, Dimand RJ, et al. Multiple organ failure after ingestion of pennyroyal oil from herbal tea in two infants. *Pediatrics.* 1996;98(5):944-47.
[2403] Sudekum M, Poppenga RH, Raju N, et al. Pennyroyal oil toxicosis in a dog. *J Am Vet Med Assoc.* 1992;200:817-8.
[2404] Anderson IB, Mullen WH, Meeker JE, et al. Pennyroyal toxicity: measurement of toxic metabolite levels in two cases and review of the literature. *Ann Intern Med.* 1996;124:726-34.
[2405] Seeff L, Stickel F, Navarro VJ. Hepatotoxicity of herbals and dietary supplements. In, Kaplowitz N, DeLeve LD, eds. Drug-induced liver disease. 3rd ed. Amsterdam: Elsevier, 2013, pp. 631-58.
[2406] Mullen W, Anderson I, Oishii S, et al. Accidental pennyroyal oil ingestion in a toddler with the first human serum metabolite detection. *Vet Hum Toxicol.* 1994;36:342.
[2407] Barnes J, Anderson LA, Philpson JD. Herbal Medicine: A Guide for Healthcare Professionals. London, UK: The Pharmaceutical Press, 1996.
[2408] Zimmerman HJ. Unconventional drugs. Miscellaneous drugs and diagnostic chemicals. In, Zimmerman, HJ. Hepatotoxicity: the adverse effects of drugs and other chemicals on the liver. 2nd ed. Philadelphia: Lippincott,1999: pp. 731-34.
[2409] Anderson IB, Mullen WH, Meeker JE, et al. Pennyroyal toxicity: measurement of toxic metabolite levels in two cases and review of the literature. *Ann Intern Med.* 1996;124:726-34.
[2410] Centers for Disease Control. Fatality and illness associated with consumption of pennyroyal oil - Colorado. *MMWR.* 1978;27:511-13.
[2411] Soares PM, Assreuy AM, Souza EP, et al. Inhibitory effects of the essential oil of Mentha pulegium on the isolated rat myometrium. *Planta Med.* 2005 Mar;71(3):214-8.
[2412] Ciganda C, Laborde A. Herbal infusions used for induced abortion. *J Toxicol Clin Toxicol.* 2003;41(3):235-9.
[2413] Valance WB. Pennyroyal poisoning: a fatal case. *Lancet.* 1955;266 (6895):850-51.
[2414] Gold J, Cates W Jr. Herbal abortifacients. *JAMA.* 1980;243:1365-66.
[2415] Gordon WP, Huitric AC, Seth CL, et al. The metabolism of the abortifacient terpene, (R)-(+)-pulegone, to a proximate toxin, menthofuran. *Drug Metab Dispos.* 1987;15:589–94.
[2416] Khojasteh-Bakht SC, Chen W, Koenigs LL, et al. Metabolism of (R)-(+)-pulegone and (R)-(+)-menthofuran by human liver cytochrome P-450s: evidence for formation of a furan epoxide. *Drug Metab Dispos.* 1999 May;27(5):574-80.
[2417] Gordon P, Khojasteh SC. A decades-long investigation of acute metabolism-based hepatotoxicity by herbal constituents: a case study of pennyroyal oil. *Drug Metab Rev.* 2015 Feb;47(1):12-20.
[2418] Zhou SF, Xue CC, Yu XQ, et al. Metabolic activation of herbal and dietary constituents and its clinical and toxicological implications: an update. *Curr Drug Metab.* 2007;8:526–553.
[2419] National Toxicology Program. Toxicology and carcinogenesis studies of pulegone (CAS No. 89-82-7) in F344/N rats and B6C3F1 mice (gavage studies). *Natl Toxicol Program Tech Rep Ser.* 2011 Aug;(563):1-201.
[2420] National Toxicology Program. Toxicology and carcinogenesis studies of pulegone (CAS No. 89-82-7) in F344/N rats and B6C3F1 mice (gavage studies). *Natl Toxicol Program Tech Rep Ser.* 2011 Aug;(563):1-201.
[2421] Sullivan JB Jr, Rumack BH, Thomas H Jr, et al. Pennyroyal oil poisoning and hepatotoxicity. *JAMA.* 1979;242:2873-4.
[2422] The Review of Natural Products by Facts and Comparisons. St. Louis, MO: Wolters Kluwer Co., 1999.
[2423] Foster S, Tyler VE. Tyler's Honest Herbal: A Sensible Guide to the Use of Herbs and Related Remedies. 3rd ed., Binghamton, NY: Haworth Herbal Press, 1993.
[2424] Martindale W. Martindale the Extra Pharmacopoeia. Pharmaceutical Press, 1999.
[2425] McGuffin M, Hobbs C, Upton R, Goldberg A, eds. American Herbal Products Association's Botanical Safety Handbook. Boca Raton, FL: CRC Press, LLC, 1997.
[2426] Sullivan JB Jr, Rumack BH, Thomas H Jr, et al. Pennyroyal oil poisoning and hepatotoxicity. *JAMA.* 1979;242:2873-4.
[2427] Anderson IB, Mullen WH, Meeker JE, et al. Pennyroyal toxicity: measurement of toxic metabolite levels in two cases and review of the literature. *Ann Intern Med.* 1996;124:726-34.
[2428] Allen WT. Note on a case of supposed poisoning by pennyroyal. *Lancet.* 1897;1(3841):1022-23.
[2429] Girling J. Poisoning by pennyroyal. *Br Med J.* 1887;1:1214.
[2430] Braithwaite PF. A case of poisoning by pennyroyal: recovery. *Br Med J.* 1906;2:865.

[2431] Buechel DW, Haverlah VC, Gardner ME. Pennyroyal oil ingestion: report of a case. *J Am Osteopath Assoc.* 1983;82:793-4.
[2432] Flynn EF. Poisoning by essence of pennyroyal. *Br Med J.* 1893;2:1270.
[2433] Burkhard PR, Burkhardt K, Haenggeli CA, et al. Plant-induced seizures: reappearance of an old problem. *J Neurol.* 1999 Aug;246(8):667-70.
[2434] Early DF. Pennyroyal: a rare cause of epilepsy. *Lancet.* 1961; 278 (7202): 580-81.
[2435] Sánchez-Borzone ME, Marin LD, García DA. Effects of Insecticidal Ketones Present in Mint Plants on GABAA Receptor from Mammalian Neurons. *Pharmacogn Mag.* 2017 Jan-Mar;13(49):114-117.
[2436] Sullivan JB Jr, Rumack BH, Thomas H Jr, et al. Pennyroyal oil poisoning and hepatotoxicity. *JAMA.* 1979;242:2873-4.
[2437] Sudekum M, Poppenga RH, Raju N, et al. Pennyroyal oil toxicosis in a dog. *J Am Vet Med Assoc.* 1992;200:817-8.
[2438] Anderson IB, Mullen WH, Meeker JE, et al. Pennyroyal toxicity: measurement of toxic metabolite levels in two cases and review of the literature. *Ann Intern Med.* 1996;124:726-34.
[2439] Zimmerman HJ. Unconventional drugs. Miscellaneous drugs and diagnostic chemicals. In, Zimmerman, HJ. Hepatotoxicity: the adverse effects of drugs and other chemicals on the liver. 2nd ed. Philadelphia: Lippincott,1999: pp. 731-34.
[2440] Seeff L, Stickel F, Navarro VJ. Hepatotoxicity of herbals and dietary supplements. In, Kaplowitz N, DeLeve LD, eds. Drug-induced liver disease. 3rd ed. Amsterdam: Elsevier, 2013, pp. 631-58.
[2441] Murayama M, Kumaroo KK. Inhibitors of ex vivo aggregation of human platelets induced by decompression, during reduced barometric pressure. *Thromb Res.* 1986 May 15;42(4):511-6.
[2442] Sessa R, Di Pietro M, De Santis F, et al. Effects of Mentha suaveolens essential oil on Chlamydia trachomatis. *Biomed Res Int.* 2015;2015:508071.
[2443] Stringaro A, Vavala E, Colone M, et al. Effects of Mentha suaveolens Essential Oil Alone or in Combination with Other Drugs in Candida albicans. *Evid Based Complement Alternat Med.* 2014;2014:125904.
[2444] Civitelli L, Panella S, Marcocci ME, et al. In vitro inhibition of herpes simplex virus type 1 replication by Mentha suaveolens essential oil and its main component piperitenone oxide. *Phytomedicine.* 2014 May 15;21(6):857-65.
[2445] Shahverdi AR, Mirzaie S, Rafii F, et al. Monoterpenes as nitrofurantoin resistance modulating agents: minimal structural requirements, molecular dynamics simulations, and the effect of piperitone on the emergence of nitrofurantoin resistance in Enterobacteriaceae. *J Mol Model.* 2015 Aug;21(8):198.
[2446] Shahverdi AR, Rafii F, Tavassoli F, et al. Piperitone from Mentha longifolia var. chorodictya Rech F. reduces the nitrofurantoin resistance of strains of enterobacteriaceae. *Phytother Res.* 2004 Nov;18(11):911-4.
[2447] Scazzocchio F, Garzoli S, Conti C, et al. Properties and limits of some essential oils: chemical characterisation, antimicrobial activity, interaction with antibiotics and cytotoxicity. Nat Prod Res. 2016 Sep;30(17):1909-18.
[2448] Bhatia SP, McGinty D, Letizia CS, et al. Fragrance material review on isopulegol. *Food Chem Toxicol.* 2008;46:S185–S189.
[2449] Api AM, Belsito D, Bhatia S, et al. RIFM fragrance ingredient safety assessment, Isopulegol, CAS Registry Number 89-79-2. *Food Chem Toxicol.* 2016 Nov;97S:S129-S135.
[2450] Demirci B, Temel HE, Portakal T, et al. Inhibitory effect of Calamintha nepeta subsp. glandulosa essential oil on lipoxygenase. *Türk Biyokimya Dergisi.* 2011;36(4):290-5.
[2451] Arantes S, Piçarra A, Candeias F, et al. Antioxidant activity and cholinesterase inhibition studies of four flavouring herbs from Alentejo. *Nat Prod Res.* 2017 Sep;31(18):2183-2187.
[2452] Kitic D, Jovanovic T, Ristic M, et al. Chemical Composition and Antimicrobial Activity of the Essential Oil of Calaminfha nepeta (L.) Savi ssp glandulosa (Req.) P. W. Ball from Montenegro. *J Essent Oil Res.* 2002 Mar/Apr;14:150-2.
[2453] Vieira M, Bessa L, Martins M, et al. Chemical Composition, Antibacterial, Antibiofilm and Synergistic Properties of Essential Oils from Eucalyptus globulus Labill. and Seven Mediterranean Aromatic Plants. *Chem Biodivers.* 2017;14(6):e1700006.
[2454] Božović M, Garzoli S, Sabatino M, et al. Essential Oil Extraction, Chemical Analysis and Anti-Candida Activity of Calamintha nepeta (L.) Savi subsp. glandulosa (Req.) Ball-New Approaches. *Molecules.* 2017 Jan 26;22(2).
[2455] Miladinović DL, Ilić BS, Mihajilov-Krstev TM, et al. Investigation of the chemical composition-antibacterial activity relationship of essential oils by chemometric methods. *Anal Bioanal Chem.* 2012 May;403(4):1007-18.
[2456] Kitic D, Stojanovic G, Palic R, et al. Chemical Composition and Microbial Activity of the Essential Oil of Calamintha nepeta (L.) Savi ssp. nepeta var. subisodonda (Borb.) Hayek from Serbia. *J Essent Oil Res.* 2005 Nov/Dec;17:701-3.
[2457] Flamini G, Cioni P, Puleio R, et al. Antimicrobial activity of the essential oil of Calamintha nepeta and its constituent pulegone against bacteria and fungi. *Phytother Res.* 1999;13(4):349-51.
[2458] Panizzi L, Flamini G, Cioni PL, et al. Compositoin and antimicrobial properties of essential oils of four Mediterranean Lamiaceae. *J Ethnopharm.* 1993;39:167-70.
[2459] Gormez A, Bozari S, Yanmis D, et al. Chemical Composition and Antibacterial Activity of Essential Oils of Two Species of Lamiaceae against Phytopathogenic Bacteria. *Pol J Microbiol.* 2015;64(2):121-7.
[2460] Shafiq MI, Ahmed M, Rasul A, et al. Chemical Composition of the Essential Oils of Nutmeg and Mace by GC-FID/MS Indigenous to Pakistan and Evaluation of their Biological Activities. *Latin Am J Pharm.* 2016;35(10):2176-84.
[2461] Choo LC, Wong SM, Liew KY. Essential oil of nutmeg pericarp. *J Sci Food Agric.* 1999(79):1954-7.
[2462] Singh G, Marimuthu P, de Helijani CS, et al. Antimicrobial and Antioxidant Potentials of Essential Oil and Acetone Extract ofMyristica fragrans Houtt. (Aril Part). *Food Microbiol Safety.* 2005;70(2):141-8.
[2463] Lawrence BM. Progress in essential oils, numbers 1-6. *Perfume & Flavorist.* 2000;25(3):6-68.
[2464] Lawrence BM. Essential oils 1988-1991. 1995, Allured Publishing: Wheaton.
[2465] Forrest JE, Heacock RA. Nutmeg and mace, the psychotropic spices from Myristica fragrans. *Lloydia.* 1972 Dec;35(4):440-9.
[2466] Ehrenpreis JE, DesLauriers, Lank P, et al. Nutmeg poisonings: A retrospective review of 10 years experience from the Illinois Poison Control center, 2001-2011. *J Med Toxicol.* 2014 Jun;10(2):148-51.
[2467] Sivathanu S, Sampath S, Suresh David H, et al. Myristicin and phenytoin toxicity in an infant. *BMJ Case Rep.* 2014 Jun;2014.
[2468] Carstairs SD, Cantrell FL. The spice of life: an analysis of nutmeg exposures in California. *Clin Toxicol (Phila).* 2011 Mar;49(3):177-80.
[2469] Randerath K, Putman KL, Randerath E. Flavor constituents in cola drinks induce hepatic DNA adducts in adult and fetal mice. *Biochem Biophys Res Commun.* 1993 Apr 15;192(1):61-8.
[2470] Vesselinovitch SD, Rao KVN, Mihailovich N. Transplacental and lactational carcinogenesis by safrole. *Cancer Res.* 1979 Nov;39(11):4378-80.
[2471] NIH. Report on carcinogenesis, fourteenth edition. Safrole. Available at: https://ntp.niehs.nih.gov/ntp/roc/content/profiles/safrole.pdf.
[2472] Murray M. Toxicological actions of plant-derived and anthropogenic methylenedioxyphenyl-substituted chemicals in mammals and insects. *J Toxicol Environ Health B Crit Rev.* 2012;15(6):365-95.

[2473] Jin M, Kijima A, Hibi D, et al. In vivo genotoxicity of methyl eugenol in gpt delta transgenic rats following medium-term exposure. *Toxicol Sci.* 2013 Feb;131(2):387-94.
[2474] Jin M, Kijima A, Suzuki Y, et al. Comprehensive toxicity study of safrole using medium-term animal model with gpt delta rats. *Toxicology.* 2011 Dec 18;290(2-3):312-21.
[2475] Hallstrom H, Thuvander A. Toxicological evaluation of myristicin. *Natural Toxins.* 1997;5:186-92.
[2476] Brenner N, Frank OS, Knight E. Chronic nutmeg psychosis. *J R Soc Med.* 1993;86:179-80.
[2477] Sangalli BC, Sangalli B, Chiang W. Toxicology of nutmeg abuse. *Clin Toxicol.* 2000;38(6):671-78.
[2478] Tsai HH, Lin HW, Chen YL, et al. A review of potential harmful interactions between anticoagulant/antiplatelet agents and Chinese herbal medicines. *PLoS One.* 2013 May 9;8(5):e64255.
[2479] Forrest JE, Heacock RA. Nutmeg and mace, the psychotropic spices from Myristica fragrans. *Lloydia.* 1972 Dec;35(4):440-9.
[2480] Leiter E1, Hitchcock G, Godwin S, et al. Evaluation of the anxiolytic properties of myristicin, a component of nutmeg, in the male Sprague-Dawley rat. *AANA J.* 2011 Apr;79(2):109-14.
[2481] Truitt EB Jr, Duritz G, Ebersberger EM. Evidence of monoamine oxidase inhibition by myristicin and nutmeg. *Proc Soc Exp Biol Med.* 1963 Mar;112:647-50.
[2482] Banjeree S, Sharma R, Kale RK, et al. Influence of certain essential oils on carcinogen-metabolizing enzyme soluble sulfhydryls in mouse liver. *Nutr Cancer.* 1994;21(3):263-69.
[2483] Kimura Y, Ito H, Hatano T. Effects of mace and nutmeg on human cytochrome P450 3A4 and 2C9 activity. *Biol Pharm Bull.* 2012;33(12):1977-87.
[2484] Jeong HG, Yun CH. Induction of rat hepatic cytochrome P450 enzymes by myristicin. *Biochem Biophys Res Commun.* 1995 Dec26;217(3):966-71.
[2485] Zhao R, Wang W, Zhao L, et al. Effect of volatile oil from nutmeg on liver licrosomal cytochrome P450 in mice.
Zhonqquo Zhonq Yao Za Zhi. 2009Feb;34(4):447-49.
[2486] Zhao R. [Effect of volatile oil from nutmeg on liver microsomal cytochrome P450 in mice.] *Zhongguo Zhong Yao Za Zhi.* 2009 Feb;34(4):447-9.
[2487] Yang AH, He X, Chen JX, et al. Identification and characterization of reactive metabolites in myristicin-mediated mechanism-based inhibition of CYP1A2. *Chem Biol Interact.* 2015 Jul 25;237:133-40.
[2488] Ioannides C, Delaforge M, Parke DV. Interactions of safrole and isosafrole and their metabolites with cytochromes P-450. *Chem Biol Interact.* 1985 May;53(3):303-11.
[2489] Ehrenpreis JE, DesLauriers, Lank P, et al. Nutmeg poisonings: A retrospective review of 10 years experience from the Illinois Poison Control center, 2001-2011. *J Med Toxicol.* 2014 Jun;10(2):148-51.
[2490] Sivathanu S, Sampath S, Suresh David H, et al. Myristicin and phenytoin toxicity in an infant. *BMJ Case Rep.* 2014 Jun;2014.
[2491] Carstairs SD, Cantrell FL. The spice of life: an analysis of nutmeg exposures in California. *Clin Toxicol (Phila).* 2011 Mar;49(3):177-80.
[2492] Hallstrom H, Thuvander A. Toxicological evaluation of myristicin. *Natural Toxins.* 1997;5:186-92.
[2493] Brenner N, Frank OS, Knight E. Chronic nutmeg psychosis. *J R Soc Med.* 1993;86:179-80.
[2494] Sangalli BC, Sangalli B, Chiang W. Toxicology of nutmeg abuse. *Clin Toxicol.* 2000;38(6):671-78.
[2495] Forrest JE, Heacock RA. Nutmeg and mace, the psychotropic spices from Myristica fragrans. *Lloydia.* 1972 Dec;35(4):440-9.
[2496] Leiter E1, Hitchcock G, Godwin S, et al. Evaluation of the anxiolytic properties of myristicin, a component of nutmeg, in the male Sprague-Dawley rat. *AANA J.* 2011 Apr;79(2):109-14.
[2497] Truitt EB Jr, Duritz G, Ebersberger EM. Evidence of monoamine oxidase inhibition by myristicin and nutmeg. *Proc Soc Exp Biol Med.* 1963 Mar;112:647-50.
[2498] Banjeree S, Sharma R, Kale RK, et al. Influence of certain essential oils on carcinogen-metabolizing enzyme soluble sulfhydryls in mouse liver. *Nutr Cancer.* 1994;21(3):263-69.
[2499] Kimura Y, Ito H, Hatano T. Effects of mace and nutmeg on human cytochrome P450 3A4 and 2C9 activity. *Biol Pharm Bull.* 2012;33(12):1977-87.
[2500] Jeong HG, Yun CH. Induction of rat hepatic cytochrome P450 enzymes by myristicin. *Biochem Biophys Res Commun.* 1995 Dec26;217(3):966-71.
[2501] Zhao R, Wang W, Zhao L, et al. Effect of volatile oil from nutmeg on liver licrosomal cytochrome P450 in mice.
Zhonqquo Zhonq Yao Za Zhi. 2009Feb;34(4):447-49.
[2502] Zhao R. [Effect of volatile oil from nutmeg on liver microsomal cytochrome P450 in mice.] *Zhongguo Zhong Yao Za Zhi.* 2009 Feb;34(4):447-9.
[2503] Yang AH, He X, Chen JX, et al. Identification and characterization of reactive metabolites in myristicin-mediated mechanism-based inhibition of CYP1A2. *Chem Biol Interact.* 2015 Jul 25;237:133-40.
[2504] Ioannides C, Delaforge M, Parke DV. Interactions of safrole and isosafrole and their metabolites with cytochromes P-450. *Chem Biol Interact.* 1985 May;53(3):303-11.
[2505] Singh G, Marimuthu P, de Helijani CS, et al. Antimicrobial and Antioxidant Potentials of Essential Oil and Acetone Extract ofMyristica fragrans Houtt. (Aril Part). *Food Microbiol Safety.* 2005;70(2):141-8.
[2506] Shafiq MI, Ahmed M, Rasul A, et al. Chemical Composition of the Essential Oils of Nutmeg and Mace by GC-FID/MS Indigenous to Pakistan and Evaluation of their Biological Activities. *Latin Am J Pharm.* 2016;35(10):2176-84.
[2507] Misharina TA, Terenina MB, Krikunova NI. [Antioxidant properties of essential oils]. Prikl Biokhim Mikrobiol. 2009 Nov-Dec;45(6):710-6.
[2508] Barton AFM, Tjamdra J, Nicholas PG. Chemical Evaluation of Volatile Oils in Eucalyptus Species. *J Agric Food Chem.* 1989;37:1253-7.
[2509] Wildy DT, Pate JS, Bartle JR. Variations in composition and yield of leaf oils from alley-farmed oil mallees (Eucalyptus spp.) at a range of contrasting sites in the Western Australian wheatbelt. *Forect Ecol Manag.* 2000;134:205-17.
[2510] Mendham D, Drew D, Abadi A, et al. Potential cineole cost and availability from mallee eucalypt shelterbelt
plantations in Western Australia. *CSIRO Food Nutr.* 2015:1-17.
[2511] Aldoghaim FS, Flematti GR, Hammer KA, et al. Antimicrobial Activity of Several Cineole-Rich Western Australian Eucalyptus Essential Oils. *Microorganisms.* 2018 Dec 3;6(4).
[2512] Craig JO. Poisoning by the volatile oils in childhood. *Arch Dis Child.* 1953;28:259-67.
[2513] Melis K. Bochner A, Janssens G. Accidental nasal eucalyptol and menthol instillation. *Eur J Pediatr.* 1989 Aug;148(8)786-7.
[2514] Day LM, Ozanne-Smith J, Parsons BJ, et al. Eucalyptus oil poisoning among young children: mechanisms of access and the potential prevention. *Aust N Z J Public Health.* 1997 Jun;21(3):297-302.
[2515] Day LM, Ozanne-Smith J, Parsons BJ, et al. Eucalyptus oil poisoning among young children: mechanisms of access and the potential prevention. *Aust N Z J Public Health.* 1997 Jun;21(3):297-302.
[2516] Myott E. Case of eucalyptus poisoning. *Brit M J.* 1906;1:558.
[2517] Hindle RC. Eucalyptus oil ingestion. *N Z Med J.* 1994 May 11;107(977)185-6.
[2518] Tibballs J. Clinical effects and management of eucalyptus oil ingestion in infants and small children. *Med J Aust.* 1995 Aug;163(4):177-80.
[2519] Waldman W, Barwina M, Sein Anand J. Accidental intoxication with eucalyptus oil—a case report. *Przeql Lek.* 2011;68(8):555-6.
[2520] Day LM, Ozanne-Smith J, Parsons BJ, et al. Eucalyptus oil poisoning among young children: mechanisms of access and the potential prevention. *Aust N Z J Public Health.* 1997 Jun;21(3):297-302.

[2521] Burkhard PR, Burkhardt K, Haenggeli CA, et al. Plant-induced seizures: reappearance of an old problem. *J Neurol.* 1999 Aug;246(8):667-70.
[2522] Waldman N. Seizure caused by dermal application of over-the-counter eucalyptus oil head lice preparation. *Clin Toxicol (Phila).* 2011 Oct;49(8):750-1.
[2523] Craig JO. Poisoning by the volatile oils in childhood. *Arch Dis Child.* 1953;28:259-67.
[2524] Mathew T, Kamath V, Kumar RS, et al. Eucalyptus oil inhalation-induced seizure: A novel, underrecognized, preventable cause of acute symptomatic seizure. *Epilepsia Open.* 2017 Jul 4;2(3):350-354.
[2525] Gray AM, Flatt PR. Antihyperglycemic actions of Eucalyptus globulus (Eucalyptus) are associated with pancreatic and extra pancreatic effects in mice. *J Nutr.* 1998 Dec;128(12):2319–23.
[2526] Basak SS, Candan F. Effect of Laurus nobilis L. Essential Oil and its Main Components on α-glucosidase and Reactive Oxygen Species Scavenging Activity. *Iran J Pharm Res.* 2013 Spring;12(2):367-79.
[2527] Fathiazad F, Mazandarani M, Hamedeyazdan. Phytochemical analysis and antioxidant activity of Hyssopus officinalis L. from Iran. *Adv Pharm Bull.* 2011 Dec;1(2):63–67.
[2528] Gray AM, Flatt PR. Antihyperglycemic actions of Eucalyptus globulus (Eucalyptus) are associated with pancreatic and extra pancreatic effects in mice. *J Nutr.* 1998 Dec;128(12):2319-23.
[2529] Dey B. Chemo-profiling of eucalyptus and study of its hypoglycemic potential. *World J Diabetes.* 2013 Oct 15;4(5):170–76.
[2530] Tsai HH, Lin HW, Chen YL, et al. A review of potential harmful interactions between anticoagulant/antiplatelet agents and Chinese herbal medicines. *PLoS One.* 2013 May 9;8(5):e64255.
[2531] Moharam BA, Jantan I, bin Ahmad F, et al. Antiplatelet Aggregation and Platelet Activating Factor (PAF) Receptor Antagonistic Activities of the Essential Oils of Five Goniothalamus Species. *Molecules.* 2010;15:5124-38.
[2532] Unger M, Frank A. Simultaneous determination of the inhibitory potency of herbal extracts on the activity of six major cytochrome P450 enzymes using liquid chromatography/mass spectrometry and automated online extraction. *Rapid Commun Mass Spectrom.* 2004;18(19):2273-81.
[2533] Jori A, Bianchetti A, Prestini PE, et al. Effect of eucalyptol (1,8-cineole) on the metabolism of other drugs in rats and in man. *Eur J Pharmacol.* 1970;9(3):362-66.
[2534] Kim NH, Hyun SH, Jin CH, et al. Pretreatment with 1,8-cineole potentiates thioacetamide-induced hepatotoxicity and immunosuppression. *Arch Pharm Res.* 2004 Jul;27(7):781-9.
[2535] Jori A, Bianchetti A, Prestini PE, et al. Effect of eucalyptol (1,8-cineole) on the metabolism of other drugs in rats and in man. *Eur J Pharmacol.* 1970;9(3):362-66.
[2536] de Sousa DP, Raphael E, Brocksom U, et al. Sedative effect of monoterpene alcohols in mice: A preliminary screening. *Verlag der Zeitschrift fur Naturforschung.* 2007;62c:563-66.
[2537] Aldoghaim FS, Flematti GR, Hammer KA, et al. Antimicrobial Activity of Several Cineole-Rich Western Australian Eucalyptus Essential Oils. *Microorganisms.* 2018 Dec 3;6(4).
[2538] Vishnupriya M, Nishaa S, Sasikumar JM, et al. Chemical composition and antioxidant activity of essential oil from Curcuma Amada Roxb. *Int Res J Pharm.* 2012;3(6):99-103.
[2539] Padalia RC, Verma RS, Sundaresan V, et al. Volatile terpenoid compositions of leaf and rhizome of Curcuma amada Roxb. from Northern India. *J Essent Oil Res.* 2013;25(1):17-22.
[2540] Singh G, Singh OP, de Lampasona MP, et al. Curcuma amada Roxb – Chemical composition of rhizome oil. *Indian Perf.* 2003 Apr/Jun;47(2):143-6.
[2541] George M, Britto SJ, Arulappan MT, et al. Phytochemical, antioxidant and antibacterial studies on the essential oil of the rhizome of Curcuma amada Roxb. *Int J Current Res.* 2015 Jul;7(07):18098-104.
[2542] Tamta A, Prakash O, Punetha H, et al. Chemical composition and in vitro antioxidant potential of essential oil and rhizome extracts of Curcuma amada Roxb. *Cogent Chemistry* 2016;2:1168067.
[2543] Choudhury SN, Rabha LC, Kanjilal PB, et al. Essential oil of Curcuma amada Roxb. From Northeastern India. *J Essent Oil Res.* 1996;8:79-80.
[2544] Dutt S, Tayal JN. Chemical composition of the essential oil derived from the rhizomes of Curcuma amada Roxb. *Indian Soap J.* 1941;7:200-5.
[2545] Srivastava AK, Srivastava SK, Shah NC. Constituents of the rhizome essential oil of Curcuma amada Roxb. from India. *J Essent Oil Res.* 2001;13:63-4.
[2546] Mustafa A, Ali M, Khan NZ. Volatile Oil Constituents of the Fresh Rhizomes of Curcuma amada Roxb. *J Essent Oil Res.* 2005;17(5):490-1.
[2547] Delgado IF, Carvalho RR, Nogueira, et al. Study on embryofetotoxicity of b-myrcene in the rat. *Food and Chemical Toxicology.* 1993;31(1):31-5.
[2548] Paumgartten FJ, De-Carvalho RR, Souza CA, et al. Study of the effects of beta-myrcene on rat fertility and general reproductive performance. *Braz J Med Biol Res.* 1998 Jul;31(7):955-65.
[2549] do Vale TG, Furtado EC, Santos JG Jr, et al. Central effects of citral, myrcene and limonene, constituents of essential oil chemotypes from Lippia alba (Mill.) n.e. Brown. *Phytomedicine.* 2002 Dec;9(8):709-14.
[2550] Segelman AB, Sofia RD, Segelman FP, et al. Cannabis sativa L. (marijuana). V. Pharmacological evaluation of marijuana aqueous extract and volatile oil. *J Pharm Sci.* 1974 Jun;63(6):962-4.
[2551] Freitas JC1, Presgrave OA, Fingola FF, et al. Effect of beta-myrcene on pentobarbital sleeping time. *Braz J Med Biol Res.* 1993 May;26(5):519-23.
[2552] De Olivera AC, Ribeiro-Pinto LF, Paumgartten JR. In vitro inhibition of CYP2B1 monoxygenase by beta-myrcene and other monoterpenoid constituents. *Br J Nutr.* 1999;81:289-95.
[2553] De-Oliveira AC1, Ribeiro-Pinto LF, Otto SS, et al. Induction of liver monooxygenases by beta-myrcene. *Toxicology.* 1997 Dec 26;124(2):135-40.
[2554] Kerckaert GA, Brauninger R, LeBoeuf RA, et al. Use of the Syrian hamster embryo cell transformation assay for carcinogenicity prediction of chemicals currently being tested by the National Toxicology Program in rodent bioassays. *Environ Health Perspect.* 1996;104:1075-84.
[2555] López ML, Hernández A, Chamorro G, et al. alpha-Asarone toxicity in long-term cultures of adult rat hepatocytes. *Planta Med.* 1993 Apr;59(2):115-20.
[2556] Chamorro G, Salazar M, Salazar S, et al. [Pharmacology and toxicology of Guatteria gaumeri and alpha-asarone.] *Rev Invest Clin.* 1993;45(6):597-604.
[2557] European Medicines Agency. Public statement on the use of herbal medicinal products containing asarone. Available at: http://www.ema.europa.eu/docs/en_GB/document_library/Scientific_guideline/2010/04/WC500089956.pdf.
[2558] Salazar M1, Salazar S, Ulloa V, et al. [Teratogenic action of alpha-asarone in the mouse]. *J Toxicol Clin Exp.* 1992 Aug;12(3):149-54.
[2559] Yabiku HK. Calamus oil - Toxicological aspects and their control in alcoholic beverages. M. S. Thesis, Sao Paulo, Brazil, Submitted to FAO/WHO. 1980.
[2560] Unger P, Melzig MF. Comparative Study of the Cytotoxicity and Genotoxicity of Alpha- and Beta-Asarone. *Sci Pharm.* 2012 Jul-Sep; 80(3):663–668.

[2561] Hagan EC, Hansen WH, Fitzhugh OG, et al. Food flavourings and constituents of related structure. II. Subacute and chronic toxicity. *Food Cosmet Toxicol.* 1967 Apr;5(2):141-57.
[2562] Taylor JM. (Food and Drug Administration) Personal communication to the World Health Organization concerning unpublished studies on beta-asarone and calamus oils. 1981. Available at: http://www.inchem.org/documents/jecfa/jecmono/v16je04.htm.
[2563] Taylor JM, et al. Toxicity of oil of calamus (Jammus variety). *Toxicol Exptl Pharmacol.* 1967;10:405.
[2564] Weinberg M. Studies conducted with Calamus. Unpublished report from Foster D. Snell, Inc. submitted to the World Health Organization by Comitato Per Lo Studio Delle Bevande Alcooliche Aromatizzate. 1969. Available at: http://www.inchem.org/documents/jecfa/jecmono/v16je04.htm.
[2565] Zuba D, Bogumila B. Alpha- and beta-asarone in herbal medicinal products. A case study. *Forensic Sci Int.* 2012 Nov 30;223(1-3):e5-9.
[2566] Meyer F, Meyer E. Percutaneous absorption of essential oils and their constituents. *Arzneimittel-Forsch.* 1965;9:516.
[2567] Zuba D, Bogumila B. Alpha- and beta-asarone in herbal medicinal products. A case study. *Forensic Sci Int.* 2012 Nov 30;223(1-3):e5-9.
[2568] Dandiya PC, Cullumbine H. Studies on Acorus calamus. III. Some pharmacological actions of the volatile oil. *J Pharmacol Exptl Therap.* 1959;125:353-59.
[2569] Menon MK, Dandiya PC. The mechanism of the tranquillizing action of asarone from Acorus calamus Linn. *J Pharm Pharmacol.* 1967 Mar;19(3):170-5.
[2570] Baxter RM, Dandiya PC, Kandel SI, et al. Separation of the hypnotic-potentiating principles from the essential oil of Acorus calamus L. of Indian origin by liquid-gas chromatograph. *Nature.* 1960 Feb;185:466-67.
[2571] Dandiya PC, Baxter RM, Walker GC, et al. Studies on Acorns calamus. II. Investigation of volatile oil. *J Pharm Pharmacol.* 1959;11:163-68.
[2572] Unger P, Melzig MF. Comparative Study of the Cytotoxicity and Genotoxicity of Alpha- and Beta-Asarone. *Sci Pharm.* 2012 Jul-Sep; 80(3):663–668.
[2573] Hagan EC, Hansen WH, Fitzhugh OG, et al. Food flavourings and constituents of related structure. II. Subacute and chronic toxicity. *Food Cosmet Toxicol.* 1967 Apr;5(2):141-57.
[2574] Taylor JM. (Food and Drug Administration) Personal communication to the World Health Organization concerning unpublished studies on beta-asarone and calamus oils. 1981. Available at: http://www.inchem.org/documents/jecfa/jecmono/v16je04.htm.
[2575] Taylor JM, et al. Toxicity of oil of calamus (Jammus variety). *Toxicol Exptl Pharmacol.* 1967;10:405.
[2576] Weinberg M. Studies conducted with Calamus. Unpublished report from Foster D. Snell, Inc. submitted to the World Health Organization by Comitato Per Lo Studio Delle Bevande Alcooliche Aromatizzate. 1969. Available at: http://www.inchem.org/documents/jecfa/jecmono/v16je04.htm.
[2577] Burkhard PR, Burkhardt K, Haenggeli CA, et al. Plant-induced seizures: reappearance of an old problem. *J Neurol.* 1999 Aug;246(8):667–70.
[2578] Millet Y, Jouglard J, Steinmetz MD, et al. Toxicity of some essential plant oils. Clinical and experimental study. *Clin Toxicol.* 1981 Dec;18(12):1485–98.
[2579] Charanraj KJ, G MV, S M. Camphor poisoning in a child. *Natl Med J India.* 2013 Jan-Feb;26(1):60.
[2580] Narayan S, Singh N. Camphor poisoning-An unusual cause of seizure. *Med J Armed Forces India.* 2012 Jul;68(3):252–53.
[2581] Olowe SA, Ransome-Kuti O. The risk of jaundice in glucose-6-phosphate dehydrogenase deficient babies exposed to menthol. *Acta Paediatr Scand.* 1980 May;69(3):341-5.
[2582] Dillon Remy M, Manning Alleyne P, Bratt DE, et al. Neonatal jaundice at Port-of-Spain General Hospital abstract. *West Indian Med J.* 1987;36(Suppl):28.
[2583] George M, Britto SJ, Arulappan MT, et al. Phytochemical, antioxidant and antibacterial studies on the essential oil of the rhizome of Curcuma amada Roxb. *Int J Current Res.* 2015 Jul;7(07):18098-104.
[2584] Tamta A, Prakash O, Punetha H, et al. Phytochemical Profiling and Antifungal Activity of Essential Oil and Rhizome Extracts of Curcuma Amada Roxb. *Organic Med Chem.* 2017;4(1):001-005.
[2585] Vishnupriya M, Nishaa S, Sasikumar JM, et al. Chemical composition and antioxidant activity of essential oil from Curcuma Amada Roxb. *Int Res J Pharm.* 2012;3(6):99-103.
[2586] George M, Britto SJ, Arulappan MT, et al. Phytochemical, antioxidant and antibacterial studies on the essential oil of the rhizome of Curcuma amada Roxb. *Int J Current Res.* 2015 Jul;7(07):18098-104.
[2587] Singh D, Singh AK. Repellent and insecticidal properties of essential oils against housefly (Musca domestica L). *Insect Sci Applic.* 1991;12:487–492.
[2588] Rali T, Wossa SW, Leach DN. Comparative Chemical Analysis of the Essential Oil Constituents in the Bark, Heartwood and Fruits of Cryptocarya massoy (Oken) Kosterm. (Lauraceae) from Papua New Guinea. *Molecules.* 2007;12:149-54.
[2589] Pratiwi SUT, Hertiani T. Efficacy of massoia oil in combination with some Indonesian medicinal plants oils as anti-biofilm agent towards Candida albicans. *IJPSR.* 2017;8(5):2013-25.
[2590] Rolli E, Marieschi M, Maietti S, et al. Phytotoxic Effects and Phytochemical Fingerprinting of Hydrodistilled Oil, Enriched Fractions, and Isolated Compounds Obtained from Cryptocarya massoy (Oken) Kosterm. Bark. *Chem Biodivers.* 2016 Jan;13(1):66-76.
[2591] National Library of Medicine. TOXNET. Benzyl acetate. Available at: https://toxnet.nlm.nih.gov/cgibin/ sis/search/a?dbs+hsdb:@term+@DOCNO+2851.
[2592] JECFA. Summary of evaluations performed by the Joint FAO/WHO Expert Committee on Food Additives. 2001. Available at: https://toxnet.nlm.nih.gov/cgi-bin/sis/search/a?dbs+hsdb:@term+@DOCNO+2851.
[2593] Permanasari P, Hertiani T, Yuswanto A. In vitro Evaluation of Massoia Bark Essential Oil and C-10 Massoialactone Potency as Immunomodulator. *J Essent Oil Bearing Plants.* 2017;20(2):459-67.
[2594] Hertiani T, Pratiwi S, Yuswanto A, et al. Potency of Massoia Bark in Combating Immunosuppressed-related Infection. *Pharmacogn Mag.* 2016 May;12(Suppl 3):S363–S370.
[2595] Pratiwi SUT, Hertiani T. Efficacy of massoia oil in combination with some Indonesian medicinal plants oils as anti-biofilm agent towards Candida albicans. *IJPSR.* 2017;8(5):2013-25.
[2596] Hertiani T, Pratiwi S, Yuswanto A, et al. Potency of Massoia Bark in Combating Immunosuppressed-related Infection. *Pharmacogn Mag.* 2016 May;12(Suppl 3):S363–S370.
[2597] Rivero-Cruz I, Duarte G, Navarrete A, et al. Chemical Composition and Antimicrobial and Spasmolytic Properties of Poliomintha longiflora and Lippia graveolens Essential Oils. *J Food Sci.* 2011;76(2):C309-17.
[2598] Calvo-Irabien LM, Parra-Tabla V, Acosta-Arriola V, et al. Phytochemical Diversity of the Essential Oils of Mexican Oregano (Lippia graveolens Kunth) Populations along an Edapho-Climatic Gradient. *Chem Biodivers.* 2014;11:1010-20.
[2599] Calvo-Irabien LM, Yam-Puc JA, Dzib G, et al. Effect of Postharvest Drying on the Composition of Mexican Oregano (Lippia graveolens) Essential Oil. *J Herbs Spices Med Plants.* 2009;15:281-7.

[2600] Soto-Armenta LC, Sacramento-Rivero JC, Acereto-Escoffie PO, et al. Extraction Yield of Essential Oil from Lippia graveolens Leaves by Steam Distillation at Laboratory and Pilot Scales. *J Essent Oil Res.* 2017;20(3):610-21.
[2601] Peschiutta ML, Arena JS, Ramirez Sanchez A, et al. Effectiveness of Mexican oregano essential oil from the Dominican Republic (Lippia graveolens) against maize pests (Sitophilus zeamais and Fusarium verticillioides). *Agriscientia.* 2016;33(2):89-97.
[2602] Arana-Sanchez A, Estarron-Espinosa M, Obledo-Vazquez EN, et al. Antimicrobial and antioxidant activities of Mexican oregano essential oils (Lippia graveolens H. B. K.) with different composition when microencapsulated in b-cyclodextrin. *J Appl Microbiol.* 2010;50:585-90.
[2603] Hernandez T, Canales M, Avila JG, et al. Composition and antibacterial activity of essential oil of Lippia graveolens H.B.K. (Verbenaceae). *Boletín Latinoamericano y del Caribe de Plantas Medicinales y Aromáticas.* 2009;8(4):295-300.
[2604] Pilau MR, Alves SH, Weiblen R, et al. Antiviral activity of the Lippia graveolens (Mexican oregano) essential oil and its main compound carvacrol against human and animal viruses. *Braz J Mirobiol.* 2011;42:1616-24.
[2605] Vazquez RS, Dunford NT. Bioactive Components of Mexican Oregano Oil as Affected by Moisture and Plant Maturity. *J Essent Oil Res.* 2005 Nov/Dec;17-668-71.
[2606] Vernin G, Lageot C, Gaydou EM, et al. Analysis of the essential oil of Lippia graveolens HBK from El Salvador. *Flavour Fragr J.* 2001;16:219-26.
[2607] Domaracky M, Rehak P, Juhas S, et al. Effects of selected plant essential oils on the growth and development of mouse preimplantation embryos in vivo. *Physiol Res.* 2007;56(1):97-104.
[2608] Lemhadri A, Zeggwagh NA, Maghrani M, et al. Anti-hypoglycaemic activity of the aqueous extract of Origanum vulgare growing wild in Tafilalet region. *J Ethnopharmacol.* 2004 Jun;92(2-3):251-6.
[2609] Bayramoglu G, Senturk H, Bayramoglu A, et al. Carvacrol partially reverses symptoms of diabetes in STZ-induced diabetic rats. *Cytotechnology.* 2014 Mar;66(2):251–257.
[2610] Ezhumalai M, Radhiga T, Pugalendi KV. Antihyperglycemic effect of carvacrol in combination with rosiglitazone in high-fat diet-induced type 2 diabetic C57BL/6J mice. *Mol Cell Biochem.* 2014 Jan;385(1-2):23-31.
[2611] Umasankar K, Ramya DR, Nambikkairaj B. Antidiabetic and antidyslipidemic nature of carvacrol, a monoterpenic phenol studied in high-fat - fed and low-dose streptozotocin-induced experimental diabetic rats. *J Pharm Res.* 2015;9(8):484-90.
[2612] Tognolini M, Barocelli E, Ballabeni V, et al. Comparative screening of plant essential oils; phenylpropanoid moiety as basic core for antiplatelet activity. *Life Sci.* 2006 Feb 23;78(13):1419-32.
[2613] Tsai HH, Lin HW, Chen YL, et al. A review of potential harmful interactions between anticoagulant/antiplatelet agents and Chinese herbal medicines. *PLoS One.* 2013 May 9;8(5):e64255.
[2614] Karkabounas S, Kostoula OK, Daskalou T, et al. Anticarcinogenic and antiplatelet effects of carvacrol. *Exp Oncol.* 2006 Jun;28(2):121-5.
[2615] Baser KH. Biological and pharmacological activities of carvacrol and carvacrol bearing essential oils. *Curr Pharm Des.* 2008;14(29):3106-19.
[2616] Okazaki K, Kawazoe K, Takaishi Y. Human platelet aggregation inhibitors from thyme (Thymus vulgaris L.). *Phytother Res.* 2002 Jun;16(4):398-9.
[2617] Jukic M, Politeo O, Maksimmovic M, et al. In vitro acetylcholinesterase inhibitory properties of thymol, carvacrol and their derivatives thymoquinone and thymohydroquinone. *Phytother Res.* 2007;21(3):259-61.
[2618] Askin H, Yildiz M, Ayar A. Effects of Thymol and Carvacrol on Acetylcholinesterase from Drosophila melanogaster. *Acta Physica Polinica A.* 2017;132(3):720-22.
[2619] Medeiros Barreto H, Cerqueira Fontinele F, Pereira de Oliveira A, et al. Phytochemical prospection and modulation of antibiotic activity in vitro by Lippia origanoides H.B.K. in methicillin resistant Staphylococcus aureus. *Biomed Res Int.* 2014;305610:1–7.
[2620] Aelenei P, Miron A, Trifan A, et al. Essential Oils and Their Components as Modulators of Antibiotic Activity against Gram-Negative Bacteria. *Medicines.* 2016;3(19):2-34.
[2621] Magi G, Marini E, Facinelli B. Antimicrobial activity of essential oils and carvacrol, and synergy of carvacrol and erythromycin, against clinical, erythromycin-resistant Group A Streptococci. *Front Microbiol.* 2015;6:165.
[2622] Veras HNH, Rodrigues FFG, Colares AV, et al. Synergistic antibiotic activity of volatile compounds from the essential oil of Lippia sidoides and thymol. *Fitoterapia.* 2012 Apr;83(3):508-12.
[2623] Langeveld WT, Veldhuizen EJ, Burt SA. Synergy between essential oil components and antibiotics: A review. *Crit Rev Microbiol.* 2014 Feb;40(1):76-94.
[2624] Kohlert C, Schindler G, Marz RW, et al. Systemic availability and pharmacokinetics of thymol in humans. *J Clin Pharmacol.* 2002 Jul;42(7):731-7.
[2625] Toxicology Data Network, National Library of Medicine. Thymol. Available at: http://toxnet.nlm.nih.gov/cgibin/sis/search/a?dbs+hsdb:@term+@DOCNO+866.
[2626] Burkhard PR, Burkhardt K, Haenggeli CA, et al. Plant-induced seizures: reappearance of an old problem. *J Neurol.* 1999 Aug;246(8):667-70.
[2627] Waldman N. Seizure caused by dermal application of over-the-counter eucalyptus oil head lice preparation. *Clin Toxicol (Phila).* 2011 Oct;49(8):750-1.
[2628] Craig JO. Poisoning by the volatile oils in childhood. *Arch Dis Child.* 1953;28:259-67.
[2629] Mathew T, Kamath V, Kumar RS, et al. Eucalyptus oil inhalation-induced seizure: A novel, underrecognized, preventable cause of acute symptomatic seizure. *Epilepsia Open.* 2017 Jul 4;2(3):350-354.
[2630] Tsuneki H, Ma EL, Kobayashi S, et al. Antiangiogenic activity of beta-eudesmol in vitro and in vivo. *Eur J Pharmacol.* 2005 Apr 11;512(2-3):105-15.
[2631] Arora CK, Arora RB, Mesta CK, et al. Hypotensive activity of beta-eudesmol and some related sesquiterpenes. *Ind J Med Res.* 1967 May;55(5):463-72.
[2632] Wang BG, Hong X, Li L, Zhou J, et al. Chemical constituents of two Chinese Magnoliaceae plants, Tsoongiodendron odorum and Manglietiastrum sinicum, and their inhibition of platelet aggregation. *Planta Med.* 2000 Aug;66(6):511-5.
[2633] Ruiz NQ, Campo YC, Stashenko EE, et al. Antigenotoxic Effect Against Ultraviolet Radiation-induced DNA Damage of the Essential Oils from Lippia Species. *Photochem Photobiol.* 2017;93:1063-72.
[2634] Machado M, Dinis AM, Salgueiro L, et al. Anti-Giardia activity of phenolic-rich essential oils: effects of Thymbra capitata, Origanum virens, Thymus zygis subsp. sylvestris, and Lippia graveolens on trophozoites growth, viability, adherence, and ultrastructure. *Parasitol Res.* 2010 Apr;106(5):1205-15.
[2635] Rivero-Cruz I, Duarte G, Navarrete A, et al. Chemical Composition and Antimicrobial and Spasmolytic Properties of Poliomintha longiflora and Lippia graveolens Essential Oils. *J Food Sci.* 2011;76(2):C309-17.

[2636] Machado M, Sousa Mdo C, Salgueiro L, et al. Effects of essential oils on the growth of Giardia lamblia trophozoites. *Nat Prod Commun.* 2010 Jan;5(1):137-41.
[2637] Rivero-Cruz I, Duarte G, Navarrete A, et al. Chemical Composition and Antimicrobial and Spasmolytic Properties of Poliomintha longiflora and Lippia graveolens Essential Oils. *J Food Sci.* 2011;76(2):C309-17.
[2638] Peschiutta ML, Arena JS, Ramirez Sanchez A, et al. Effectiveness of Mexican oregano essential oil from the Dominican Republic (Lippia graveolens) against maize pests (Sitophilus zeamais and Fusarium verticillioides). *Agriscientia.* 2016;33(2):89-97.
[2639] Arana-Sanchez A, Estarron-Espinosa M, Obledo-Vazquez EN, et al. Antimicrobial and antioxidant activities of Mexican oregano essential oils (Lippia graveolens H. B. K.) with different composition when microencapsulated in b-cyclodextrin. *J Appl Microbiol.* 2010;50:585-90.
[2640] Hernandez T, Canales M, Avila JG, et al. Composition and antibacterial activity of essential oil of Lippia graveolens H.B.K. (Verbenaceae). *Boletín Latinoamericano y del Caribe de Plantas Medicinales y Aromáticas.* 2009;8(4):295-300.
[2641] Soares IH, Loreto ÉS, Rossato L, et al. In vitro activity of essential oils extracted from condiments against fluconazole-resistant and -sensitive Candida glabrata. *J Mycol Med.* 2015 Sep;25(3):213-7.
[2642] Miller AB, Cates RG, Lawrence M, et al. The antibacterial and antifungal activity of essential oils extracted from Guatemalan medicinal plants. *Pharm Biol.* 2015 Apr;53(4):548-54.
[2643] Pilau MR, Alves SH, Weiblen R, et al. Antiviral activity of the Lippia graveolens (Mexican oregano) essential oil and its main compound carvacrol against human and animal viruses. *Braz J Mirobiol.* 2011;42:1616-24.
[2644] Arana-Sanchez A, Estarron-Espinosa M, Obledo-Vazquez EN, et al. Antimicrobial and antioxidant activities of Mexican oregano essential oils (Lippia graveolens H. B. K.) with different composition when microencapsulated in b-cyclodextrin. *J Appl Microbiol.* 2010;50:585-90.
[2645] Martinez-Velazquez M, Rosario-Cruz R, Castillo-Herrera G, et al. Acaricidal effect of essential oils from Lippia graveolens (Lamiales: Verbenaceae), Rosmarinus officinalis (Lamiales: Lamiaceae), and Allium sativum (Liliales: Liliaceae) against Rhipicephalus (Boophilus) microplus (Acari: Ixodidae). *J Med Entomol.* 2011 Jul;48(4):822-7.
[2646] Peschiutta ML, Arena JS, Ramirez Sanchez A, et al. Effectiveness of Mexican oregano essential oil from the Dominican Republic (Lippia graveolens) against maize pests (Sitophilus zeamais and Fusarium verticillioides). *Agriscientia.* 2016;33(2):89-97.
[2647] Kotoky R, Bordoloi M, Yadav A, et al. Chemical composition and antibacterial activity of essential oil from fruit of Micromelum integerrimum (Buch.-Ham. ex DC.) Wight & Arn. ex M. Roem. *Nat Prod Res.* 2018 Jun. [Epub ahead of print]
[2648] Biqiang C. Preliminary study on Micromelum integerrimum. *Flavour Frag Cosmetics.* 1999.
[2649] Kotoky R, Bordoloi M, Yadav A, et al. Chemical composition and antibacterial activity of essential oil from fruit of Micromelum integerrimum (Buch.-Ham. ex DC.) Wight & Arn. ex M. Roem. *Nat Prod Res.* 2018 Jun. [Epub ahead of print]
[2650] Perriot R, Breme K, Meierhenrich UJ, et al. Chemical Composition of French Mimosa Absolute Oil. *J Agric Food Chem.* 2010;58:1844-9.
[2651] Johnson S, DeCarlo A, Satyal P, et al. Organic Certification is Not Enough: The Case of the Methoxydecane Frankincense. *Plants.* 2019;8(4):88.
[2652] de Abreu Matos FJ, Lacerda Machado MI, Craveiro AA, et al. Essential Oil of Mentha x villosa Huds. from Northeastern Brazil. *J Essent Oil Res.* 1999 Jan/Feb;11:41-44.
[2653] Lawrence BM. Essential Oils (1979-1980). Allured Publishing Corp., Carol Stream, IL (1981).
[2654] Kokkini S. Doctoral dissertation: Taxonomic studies in the genus Mentha in Greece. Institute of Systemic Botany, Univeristy of Thessaloniki, Thessalonik, Greece (1983).
[2655] Lima TC, da Silva TKM, Silva FL, et al. Larvicidal activity of Mentha x villosa Hudson essential oil, rotundifolone and derivatives. *Chemosphere.* 2014;104:37-43.
[2656] Fogaca RTH, Cavalcante ADA, Serpa AKL, et al. The Effects of Essential Oil of Mentha x villosa on Skeletal Muscle of the Toad. *Phytother Res.* 1997;11:552-7.
[2657] Teles S, Pereira JA, Santos CHB, et al. Effect of geographical origin on the essential oil content and composition of fresh and dried Mentha ×villosa Hudson leaves. *Ind Crops Prod.* 2013;46:1-7.
[2658] Pavela R, Kaffkova K, Kumsta M. Chemical Composition and Larvicidal Activity of Essential Oils from Different Mentha L. and Pulegium Species against Culex quinquefasciatus Say (Diptera: Culicidae). *Plant Protect Sci.* 2014;50(1):36-42.
[2659] Gracindo LAMB, Grisi MCM, Silva DB, et al. Chemical characterization of mint (Mentha spp.) germplasm at Federal District, Brazil. *Rev Bras Pl Med Botucatu.* 2006;8(esp.):5-9.
[2660] Alves LF, Bonfim FPG, de Oliveira SG, et al. Biomass, content, yield and chemical composition of mint (Mentha x villosa Huds.) essential oil in response to withholding irrigation. *AJCS.* 2018;12(04):519-23.
[2661] Lahlou S, Carneiro-Leao RFL, Leal-Cardoso JH. Cardiovascular effects of the essential oil of Mentha x villosa in DOCA-salt-hypertensive rats. *Phytomed.* 20025;9:715-20.
[2662] da Cruz Almeida ET, de Medeiros Barbosa I, Tavares JF, et al. Inactivation of Spoilage Yeasts by Mentha spicata L. and M. × villosa Huds. Essential Oils in Cashew, Guava, Mango, and Pineapple Juices. *Front Microbiol.* 2018 May 25;9:1111.
[2663] da Silva Bezerra Guerra KS1, Silva RL, Souza Maia MB, et al. Embryo and fetal toxicity of Mentha x villosa essential oil in Wistar rats. *Pharm Biol.* 2012 Jul;50(7):871-7.
[2664] Fogaca RTH, Cavalcante ADA, Serpa AKL, et al. The Effects of Essential Oil of Mentha x villosa on Skeletal Muscle of the Toad. *Phytother Res.* 1997;11:552-7.
[2665] Guedes DN, Silva DF, Barbosa-Filho JM, et al. Endothelium-dependent hypotensive and vasorelaxant effects of the essential oil from aerial parts of Mentha x villosa in rats. *Phytomed.* 2004;11:490-7.
[2666] Guedes DN, Silva DF, Barbosa-Filho JM, et al. Endothelium-dependent hypotensive and vasorelaxant effects of the essential oil from aerial parts of Mentha x villosa in rats. *Phytomedicine.* 2004 Sep;11(6):490-7.
[2667] Lahlou S, Magalhães PJ, Carneiro-Leão RF, et al. Involvement of nitric oxide in the mediation of the hypotensive action of the essential oil of Mentha x villosa in normotensive conscious rats. *Planta Med.* 2002 Aug;68(8):694-9.
[2668] Guedes DN, Silva DF, Barbosa-Filho JM, et al. Calcium antagonism and the vasorelaxation of the rat aorta induced by rotundifolone. *Braz J Med Biol Res.* 2004 Dec;37(12):1881-7.
[2669] Sessa R, Di Pietro M, De Santis F, et al. Effects of Mentha suaveolens essential oil on Chlamydia trachomatis. *Biomed Res Int.* 2015;2015:508071.
[2670] Stringaro A, Vavala E, Colone M, et al. Effects of Mentha suaveolens Essential Oil Alone or in Combination with Other Drugs in Candida albicans. *Evid Based Complement Alternat Med.* 2014;2014:125904.
[2671] Civitelli L, Panella S, Marcocci ME, et al. In vitro inhibition of herpes simplex virus type 1 replication by Mentha suaveolens essential oil and its main component piperitenone oxide. *Phytomedicine.* 2014 May 15;21(6):857-65.

[2672] Scazzocchio F, Garzoli S, Conti C, et al. Properties and limits of some essential oils: chemical characterisation, antimicrobial activity, interaction with antibiotics and cytotoxicity. Nat Prod Res. 2016 Sep;30(17):1909-18.
[2673] Muruganathan U, Srinivasan S, Indumathi D. Antihyperglycemic effect of carvone: Effect on the levels of glycoprotein components in streptozotocin-induced diabetic rats. *J Acute Disease*. 2013;2(4):310-15.
[2674] Souza FV, da Rocha MB, de Souza DP, et al. (-)-Carvone: antispasmodic effect and mode of action. *Fitoterapia*. 2013 Mar;85:20-24.
[2675] Rajeshwari T, Raja B. Antihypertensive, antihyperlipidemicand antioxidantinfluence of D-carvone in L-NAME induced hypertensive rats. *Int J Pharmaceutical Biol Arch*. 2014;5(4):82-88.
[2676] de Sousa DP, Farias Nobrega FF, de Almeida RN. Influence of the chirality of (R)-(-)- and (S)-(+)-carvone in the central nervous system: a comparative study. *Chirality*. 2007 May;19(4):264-268.
[2677] Amaral RG, Fonseca CS, Silva TK, et al. Evaluation of the cytotoxic and antitumour effects of the essential oil from Mentha x villosa and its main compound, rotundifolone. *J Pharm Pharmacol*. 2015 Aug;67(8):1100-6.
[2678] Sousa PJC, Linard CFBM, Azevedo-Batista D, et al. Antinociceptive effects of the essential oil of Mentha x villosa leaf and its major constituent piperitenone oxide in mice. *Braz J Med Biol Res*. 2009;42:655-9.
[2679] Fogaca RTH, Cavalcante ADA, Serpa AKL, et al. The Effects of Essential Oil of Mentha x villosa on Skeletal Muscle of the Toad. *Phytother Res*. 1997;11:552-7.
[2680] Lahlou S, Carneiro-Leao RFL, Leal-Cardoso JH. Cardiovascular effects of the essential oil of Mentha x villosa in DOCA-salt-hypertensive rats. *Phytomed*. 20025;9:715-20.
[2681] Guedes DN, Silva DF, Barbosa-Filho JM, et al. Endothelium-dependent hypotensive and vasorelaxant effects of the essential oil from aerial parts of Mentha x villosa in rats. *Phytomed*. 2004;11:490-7.
[2682] Guedes DN, Silva DF, Barbosa-Filho JM, et al. Endothelium-dependent hypotensive and vasorelaxant effects of the essential oil from aerial parts of Mentha x villosa in rats. *Phytomedicine*. 2004 Sep;11(6):490-7.
[2683] Lahlou S, Magalhães PJ, Carneiro-Leão RF, et al. Involvement of nitric oxide in the mediation of the hypotensive action of the essential oil of Mentha x villosa in normotensive conscious rats. *Planta Med*. 2002 Aug;68(8):694-9.
[2684] Silva DF, Araújo IG, Albuquerque JG, et al. Rotundifolone-induced relaxation is mediated by BK(Ca) channel activation and Ca(v) channel inactivation. *Basic Clin Pharmacol Toxicol*. 2011 Dec;109(6):465-75.
[2685] Guedes DN, Silva DF, Barbosa-Filho JM, et al. Calcium antagonism and the vasorelaxation of the rat aorta induced by rotundifolone. *Braz J Med Biol Res*. 2004 Dec;37(12):1881-7.
[2686] Matos-Rocha TJ, Cavalcanti MG, Veras DL, et al. Ultrastructural changes in Schistosoma mansoni male worms after in vitro incubation with essential oil of Mentha x villosa Huds. *Rev Inst Med Trop Sao Paulo*. 2016;58:4.
[2687] Matos-Rocha TJ, dos Santos Cavalcanti MG, Barbosa-Filho JM, et al. In vitro evaluation of schistosomicidal activity of essential oil of Mentha x villosa and some of its chemical constituents in adult worms of Schistosoma mansoni. *Planta Med*. 2013 Sep;79(14):1307-12.
[2688] Arruda TA, Antunes RMP, Catao RMR, et al. Preliminary study of the antimicrobial activity of Mentha x villosa Hudson essential oil, rotundifolone and its analogues. *Rev Brasil Farmacog*. 2006 Jul/Sep;16(3):307-11.
[2689] da Cruz Almeida ET, de Medeiros Barbosa I, Tavares JF, et al. Inactivation of Spoilage Yeasts by Mentha spicata L. and M. × villosa Huds. Essential Oils in Cashew, Guava, Mango, and Pineapple Juices. *Front Microbiol*. 2018 May 25;9:1111.
[2690] Guerra ICD, de Oliveira PDL, de Souza Pontes AL, et al. Coatings comprising chitosan and Mentha piperita L. or Mentha × villosa Huds essential oils to prevent common postharvest mold infections and maintain the quality of cherry tomato fruit. *Int J Food Microbiol*. 2015 Dec 2;214:168-178.
[2691] Lima TC, da Silva TKM, Silva FL, et al. Larvicidal activity of Mentha x villosa Hudson essential oil, rotundifolone and derivatives. *Chemosphere*. 2014;104:37-43.
[2692] Pavela R, Kaffkova K, Kumsta M. Chemical Composition and Larvicidal Activity of Essential Oils from Different Mentha L. and Pulegium Species against Culex quinquefasciatus Say (Diptera: Culicidae). *Plant Protect Sci*. 2014;50(1):36-42.
[2693] Lee NH, Lee SM, Lee TM, et al. GC-MS Analyses of the Essential Oils Obtained from Pinaceae Leaves in Korea. *J Essent Oil Bearing Plants*. 2015;18(3):538-42.
[2694] Satou T, Matsuura M, Takahashiu M, et al. Components of Essential Oils Extracted from Leaves and Shoots of Abies Species in Japan. *Chem Biodivers*. 2011;8:1132-8.
[2695] Aazza S, Lyoussi B, Miguel MG. Antioxidant and antiacetylcholinesterase activities of some commercial essential oils and their major compounds. *Molecules*. 2011 Sep 7;16(9):7672-90.
[2696] Miyazawa M, Yamafuji C. Inhibition of acetylcholinesterase activity by bicyclic monoterpenoids. *J Agric Food Chem*. 2005 Mar 9;53(5):1765-8.
[2697] Fahed L, Khoury M, Stien D, et al. Essential Oils Composition and Antimicrobial Activity of Six Conifers Harvested in Lebanon. *Chem Biodiversity*. 2017 Feb;14(2):e1600235.
[2698] Saad AM, Mohammed MMD, Ghareeb MA, et al. Chemical Composition and Antimicrobial Activity of the Essential Oil of the Leaves of Cupressus macrocarpa Hartweg. ex Gordon. *J Appl Pharm Sci*. 2017 Sep;7(9):207-12.
[2699] Marchuk NY. The research of essential oil's chemical composition in species of genus Cupressus L., growing in Crimean south coast. *Ukrainica Bioa=organica Acta*. 2014;1:40-43.
[2700] El-Ghorab AH, El-Massry KF, Shaaban HA. Effect of Drying on the Chemical Composition of the Egyptian Cupressus macrocarpa (Hartw.ex Gordon) Essential Oils and their Biological Characteristics. *Jeobp*. 2007:10(5):399-411.
[2701] Salem MZM, Elansary HO, Ali HM, et al. Bioactivity of essential oils extracted from Cupressus macrocarpa branchlets and Corymbia citriodora leaves grown in Egypt. *BMC Complement Altern Med*. 2018; 18: 23.
[2702] Giatropoulos A, Pitarokili, Papaioannou F, et al. Essential oil composition, adult repellency and larvicidal activity of eight Cupressaceae species from Greece against Aedes albopictus (Diptera: Culicidae). *Parasitol Res*. 2013;112:1113-23.
[2703] Nogueira AC, Carvalho RR, Souza CA, et al. Study on the embryofeto-toxicity of citral in the rat. *Toxicology*. 1995;96(2):105-13.
[2704] Seo KA, Kim H, Ku HY, et al. The monoterpenoids citral and geraniol are moderate inhibitors of the CYP2B6 hydroxylase activity. *Chem Biol Interact*. 2008;174:141-46.
[2705] Sheweita SA, Newairy AA, Mansour HA, et al. Effect of some hypoglycemic herbs on the activity of phase I and II drug-metabolizing enzymes in alloxan-induced diabetic rats. *Toxicology*. 2002 May 24;17(2):131-39.
[2706] Raner GM, Vaz AD, Coon MJ. Metabolism of all-trans, 9-cis, and 13-cis isomers of retinal by purified isozymes of microsomal cytochrome P450 and mechanism-based inhibition of retinoid oxidation by citral. *Mol Pharmacol*. 1996;49(3):515-22.
[2707] Modak T, Mukhopadhaya A. Effects of citral, a naturally occurring antiadipogenic molecule, on an energy-intense diet model of obesity. *Indian J Pharmacol*. 2011 May-Jun;43(3):300-05.

[2708] Adeneye AA, Agbaje EO. Hypoglycemic and hypolipidemic effects of fresh leaf aqueous extract of Cymbopogon citratus Stapf. in rats. *J Ethnopharmacol.* 2007 Jul 25;112(3):440-4.
[2709] Choi JY, Damte D, Seung-Jin L, et al. Antimicrobial activity of lemongrass and oregano essential oil against standard antibiotic resistant Staphylococcus aureus and field isolates from chronic mastitis cow. *Int J Phytomedicine*. 2012;4(1):134-39.
[2710] Zanini SF, Silva-Angulo AB, Rosenthal A, et al. Effect of citral and carvacrol on the susceptibility of Listeria monocytogenes and Listeria innocua to antibiotics. *Lett Appl Microbiol.* 2014 May;58(5):486-92.
[2711] De Mozzi P, Johnston GA. An outbreak of allergic contact dermatitis caused by citral in beauticians working in a health spa. *Contact Dermatitis.* 2014 Jun;70(6):377-9.
[2712] Saad AM, Mohammed MMD, Ghareeb MA, et al. Chemical Composition and Antimicrobial Activity of the Essential Oil of the Leaves of Cupressus macrocarpa Hartweg. ex Gordon. *J Appl Pharm Sci.* 2017 Sep;7(9):207-12.
[2713] Fahed L, Khoury M, Stien D, et al. Essential Oils Composition and Antimicrobial Activity of Six Conifers Harvested in Lebanon. *Chem Biodiversity.* 2017 Feb;14(2):e1600235.
[2714] El-Ghorab AH, El-Massry KF, Shaaban HA. Effect of Drying on the Chemical Composition of the Egyptian Cupressus macrocarpa (Hartw.ex Gordon) Essential Oils and their Biological Characteristics. *Jeobp.* 2007:10(5):399-411.
[2715] Salem MZM, Elansary HO, Ali HM, et al. Bioactivity of essential oils extracted from Cupressus macrocarpa branchlets and Corymbia citriodora leaves grown in Egypt. *BMC Complement Altern Med.* 2018; 18: 23.
[2716] Manivannan R, Kumar MS, Jawahar N, et al. A comparative antimicrobial study on the essential oil of the leaves of various species of cupressus. *Anc Sci Life.* 2005 Jan;24(3):131-3.
[2717] El-Ghorab AH, El-Massry KF, Shaaban HA. Effect of Drying on the Chemical Composition of the Egyptian Cupressus macrocarpa (Hartw.ex Gordon) Essential Oils and their Biological Characteristics. *Jeobp.* 2007:10(5):399-411.
[2718] Giatropoulos A, Pitarokili, Papaioannou F, et al. Essential oil composition, adult repellency and larvicidal activity of eight Cupressaceae species from Greece against Aedes albopictus (Diptera: Culicidae). *Parasitol Res.* 2013;112:1113-23.
[2719] Zoghbi MDGB, Andrade EHA, Carreira LMM, et al. Comparison of the Main Components of the Essential Oils of "priprioca": Cyperus articulatus var. articulatus L., C. articulatus var. nodosus L., C. prolixus Kunth and C. rotundus L. *J Essent Oil Res.* 2008 Jan/Feb;20:42-6.
[2720] Lin WY, Kuo YH, Chang YL, et al. Anti-platelet aggregation and chemical constituents from the rhizome of Gynura japonica. *Planta Med.* 2003 Aug;69(8):757-64.
[2721] Bruce TJA, Birkett MA, Blande J, et al. Response of economically important aphids to components of Hemizygia petiolata essential oil. *Pest Manag Sci.* 2005;61:1115-21.
[2722] Jiang ZT, Li R, Zuo YM. Composition of Essential Oil of Brassica juncea (L.) Coss. From China. *J Essent Oil Res.* 1999;11(4):503-6.
[2723] Yu JC, Jiang ZT, Li R, et al. Chemical Composition of the Essential Oils of Brassica juncea (L.) Coss. Grown in Different Regions, Hebei, Shaanxi and Shandong, of China. *J Food Drug Analysis.* 2003;11(1):22-6.
[2724] Peng C, Zhao SQ, Zhang J, et al. Chemical composition, antimicrobial property and microencapsulation of Mustard (Sinapis alba) seed essential oil by complex coacervation. *Food Chem.* 2014;165:560-8.
[2725] Reyes-Jurado F, López-Malo A, Palou E, et al. Antimicrobial Activity of Individual and Combined Essential Oils against Foodborne Pathogenic Bacteria. *J Food Prot.* 2016 Feb;79(2):309-15.
[2726] Mejía-Garibay B, Palou E, López-Malo A. Composition, diffusion, and antifungal activity of black mustard (Brassica nigra) essential oil when applied by direct addition or vapor phase contact. *J Food Prot.* 2015 Apr;78(4):843-8.
[2727] Singh S, Das SS, singh G, et al. Comparison of chemical composition, antioxidant and antimicrobial potentials of essential oils and oleoresins obtained from seeds of Brassica juncea and sinapis Alba. *MOJ Food Process Technol.* 2017;4(4):113-20.
[2728] European Food Safety Association. Scientific Opinion on the safety of allyl isothiocyanate for the proposed uses as a food additive1 EFSA Panel on Food Additives and Nutrient Sources added to Food (ANS). *EFSA J.* 2010;8(12):1943.
[2729] European Food Safety Association. Scientific Opinion on the safety of allyl isothiocyanate for the proposed uses as a food additive1 EFSA Panel on Food Additives and Nutrient Sources added to Food (ANS). *EFSA J.* 2010;8(12):1943.
[2730] Lee DS, Kim TH, Jung YS. Inhibitory effect of allyl isothiocyanate on platelet aggregation. *J Agric Food Chem.* 2014 Jul 23;62(29):7131-9.
[2731] Mori N, Kurata M, Yamazaki H, et al. Intragastric administration of allyl isothiocyanate reduces hyperglycemia in intraperitoneal glucose tolerance test (IPGTT) by enhancing blood glucose consumption in mice. *J Nutr Sci Vitaminol (Tokyo).* 2013;59(1):56-63.
[2732] Mori N, Kurata M, Yamazaki H, et al. Allyl isothiocyanate increases carbohydrate oxidation through enhancing insulin secretion by TRPV1. *Biosci Biotechnol Biochem.* 2018 Apr;82(4):698-708.
[2733] Von Skramlik EV. Uber die Giftigkeit und Verträglichkeit von ätherischen Olen. *Pharmazie.* 1959;14:435-45.
[2734] Brand G, Jacquot L. Sensitization and desensitization to allyl isothiocyanate (mustard oil) in the nasal cavity. *Chem Senses.* 2002 Sep;27(7):593-8.
[2735] Leung AY, Foster S. Encyclopedia of common natural ingredients used in food, drugs, and cosmetics, 2nd edition. 2003; John Wiley, New Jersey.
[2736] Evans FJ, Schmidt RJ. Plants and plant products that induce contact dermatitis. *Planta Med.* 1980;38:289-316.
[2737] Lerbaek A, Rastogi SC, Menné T. Allergic contact dermatitis from allyl isothiocyanate in a Danish cohort of 259 selected patients. *Contact Dermatitis.* 2004 Aug;51(2):79-83.
[2738] Brand G, Jacquot L. Sensitization and desensitization to allyl isothiocyanate (mustard oil) in the nasal cavity. *Chem Senses.* 2002 Sep;27(7):593-8.
[2739] Ramzy AG, Lammintausta K, Matura M, et al. Isothiocyanates are important as haptens in contact allergy to chloroprene rubber. *Br J Dermatol.* 2017 Aug;177(2):522-530.
[2740] Angelini G, Vena GA, Filotico R, et al. Allergic contact dermatitis from Capparis spinosa L. applied as wet compresses. *Contact Dermatitis.* 1991 may;24:382.
[2741] Rycroft RJG, Menne T, Frosch PJ, et al. Textbook of Contact Dermatitis. 2001; Springer, New York.
[2742] Singh S, Das SS, singh G, et al. Comparison of chemical composition, antioxidant and antimicrobial potentials of essential oils and oleoresins obtained from seeds of Brassica juncea and sinapis Alba. *MOJ Food Process Technol.* 2017;4(4):113-20.
[2743] Peng C, Zhao SQ, Zhang J, et al. Chemical composition, antimicrobial property and microencapsulation of Mustard (Sinapis alba) seed essential oil by complex coacervation. *Food Chem.* 2014;165:560-8.
[2744] Reyes-Jurado F, López-Malo A, Palou E, et al. Antimicrobial Activity of Individual and Combined Essential Oils against Foodborne Pathogenic Bacteria. *J Food Prot.* 2016 Feb;79(2):309-15.
[2745] Singh S, Das SS, singh G, et al. Comparison of chemical composition, antioxidant and antimicrobial potentials of essential oils and oleoresins obtained from seeds of Brassica juncea and sinapis Alba. *MOJ Food Process Technol.* 2017;4(4):113-20.

[2746] Plastina P, Apriantini A, Meijerink J, et al. In Vitro Anti-Inflammatory and Radical Scavenging Properties of Chinotto (Citrus myrtifolia Raf.) Essential Oils. *Nutrients*. 2018;10:783.
[2747] Lota ML, Serra DR, Jacquemond C, et al. Chemical variability of peel and leaf essential oils of sour orange. *Flavour Frag J*. 2001;16:89-96.
[2748] Huang Y, Pu Z. The Chemical Composition of the Leaf Essential Oils from 110 Citrus Species, Cultivars, Hybrids and Varieties of Chinese Origin. *Perf Flavorist*. 2000 Jan/Feb;25:53-66.
[2749] Chivala F, Doglia G, et al. Essential Oil Constituents of Chinotto (Citrus aurantium L. var. myrtifolia Guill.)
[2750] Scordino M, Sabatino L, Belligno A, et al. Flavonoids and furocoumarins distribution of unripe chinotto (Citrus × myrtifolia Rafinesque) fruit: beverage processing homogenate and juice characterization. *Eur Food Res Technol*. 2011 Nov;233:759.
[2751] Plastina P, Apriantini A, Meijerink J, et al. In Vitro Anti-Inflammatory and Radical Scavenging Properties of Chinotto (Citrus myrtifolia Raf.) Essential Oils. *Nutrients*. 2018;10:783.
[2752] Plastina P, Apriantini A, Meijerink J, et al. In Vitro Anti-Inflammatory and Radical Scavenging Properties of Chinotto (Citrus myrtifolia Raf.) Essential Oils. *Nutrients*. 2018;10:783.
[2753] Benyelles B, Allali H, Fekih N, et al. Chemical Composition of the Volatile Components of Tropaeolum majus L. (Garden Nasturtium) from North Western Algeria. *PhytoChem BioSub J*. 2015;9(3):92-7.
[2754] Butnariu M, Bostan C. Antimicrobial and anti-inflammatory activities of the volatile oil compounds from Tropaeolum majus L. (Nasturtium). *African J Biotechnol*. 2011 Jun;10(31):5900-9.
[2755] Benyelles B, Allali H, Touaibia M, et al. Antioxidant and Anti-Inflammatory Activities of the Volatile Components of Tropaeolum Majus L. (Garden Nasturtium) From North Western Algeria. *J Pharm Chem Biol Sci*. 2016 Sep-Nov;4(3):431-9.
[2756] Adebiyi A, Adaikan PG, Prasad Rn. Pregnancy outcomes following pre- and post-implanatation exposure of Sprague-Dawley rats to benzyl isothiocyanate. *Food Chem Toxicol*. 2004;42:715-20.
[2757] Mennicke WH, Gorler K, Krumbiegel G, et al. Studies on the metabolism and excretion of benzyl isothiocyanate in man. *Xenobiotica*. 1988 Apr;18(4):441-7.
[2758] Wattenburg LW. Inhibition of carcinogenic effects of polycyclic hydrocarbons by benzyl isothiocyanate and related compounds. *J Natl Cancer Inst*. 1977 Feb;58(2):395-8.
[2759] Rao CV. Benzyl Isothiocyanate: Double Trouble for Breast Cancer Cells. *Cancer Prev Res (Phila)*. 2013 Aug;6(8):760-3.
[2760] Moreno RL, Kent UM, Hodge K, et al. Inactivation of cytochrome P450 2E1 by benzyl isothiocyanate. *Chem Res Toxicol*. 1999 Jul;12(7):582-7.
[2761] Goosen TC, Kent UM, Brand L, et al. Inactivation of cytochrome P450 2B1 by benzyl isothiocyanate, a chemopreventative agent from cruciferous vegetables. *Chem Res Toxicol*. 2000 Dec;13(12):1349-59.
[2762] Ramzy AG, Lammintausta K, Matura M, et al. Isothiocyanates are important as haptens in contact allergy to chloroprene rubber. *Br J Dermatol*. 2017 Aug;177(2):522-530.
[2763] Angelini G, Vena GA, Filotico R, et al. Allergic contact dermatitis from Capparis spinosa L. applied as wet compresses. *Contact Dermatitis*. 1991 may;24:382.
[2764] Rycroft RJG, Menne T, Frosch PJ, et al. Textbook of Contact Dermatitis. 2001; Springer, New York.
[2765] Leung AY, Foster S. Encyclopedia of common natural ingredients used in food, drugs, and cosmetics, 2nd edition. 2003; John Wiley, New Jersey.
[2766] Evans FJ, Schmidt RJ. Plants and plant products that induce contact dermatitis. *Planta Med*. 1980;38:289-316.
[2767] Lerbaek A, Rastogi SC, Menné T. Allergic contact dermatitis from allyl isothiocyanate in a Danish cohort of 259 selected patients. *Contact Dermatitis*. 2004 Aug;51(2):79-83.
[2768] Brand G, Jacquot L. Sensitization and desensitization to allyl isothiocyanate (mustard oil) in the nasal cavity. *Chem Senses*. 2002 Sep;27(7):593-8.
[2769] Von Skramlik EV. Uber die Giftigkeit und Verträglichkeit von ätherischen Olen. *Pharmazie*. 1959;14:435-45.
[2770] Brand G, Jacquot L. Sensitization and desensitization to allyl isothiocyanate (mustard oil) in the nasal cavity. *Chem Senses*. 2002 Sep;27(7):593-8.
[2771] Benyelles B, Allali H, Touaibia M, et al. Antioxidant and Anti-Inflammatory Activities of the Volatile Components of Tropaeolum Majus L. (Garden Nasturtium) From North Western Algeria. *J Pharm Chem Biol Sci*. 2016 Sep-Nov;4(3):431-9.
[2772] Benyelles B, Allali H, Touaibia M, et al. Antioxidant and Anti-Inflammatory Activities of the Volatile Components of Tropaeolum Majus L. (Garden Nasturtium) From North Western Algeria. *J Pharm Chem Biol Sci*. 2016 Sep-Nov;4(3):431-9.
[2773] Benyelles B, Allali H, Touaibia M, et al. Antioxidant and Anti-Inflammatory Activities of the Volatile Components of Tropaeolum Majus L. (Garden Nasturtium) From North Western Algeria. *J Pharm Chem Biol Sci*. 2016 Sep-Nov;4(3):431-9.
[2774] Benyelles B, Allali H, Touaibia M, et al. Antioxidant and Anti-Inflammatory Activities of the Volatile Components of Tropaeolum Majus L. (Garden Nasturtium) From North Western Algeria. *J Pharm Chem Biol Sci*. 2016 Sep-Nov;4(3):431-9.
[2775] Palá-Paúl J, Usano-Alemany J, Granda E, et al. Chemical composition, antifungal and antibacterial activity of the essential oil of Chamaecyparis nootkatensis from Spain. *Nat Prod Commun*. 2009 Jul;4(7):1007-10.
[2776] Liu X. Analysis and Comparison of Essential Oil Components Extracted from the Heartwoods of Leyland Cypress, Alaska Yellow Cedar, and Monterey Cypress. Master's Thesis.
[2777] Khasawnwh MA, Ziong Y, Peralta-Cruz J, et al. Biologically Important Eremophilane Sesquiterpenes from Alaska Cedar Heartwood Essential Oil and Their Semi-Synthetic Derivatives. *Molecules*. 2011;16:4775-85.
[2778] Domaracky M, Rehak P, Juhas S, et al. Effects of selected plant essential oils on the growth and development of mouse preimplantation embryos in vivo. *Physiol Res*. 2007;56(1):97-104.
[2779] Lemhadri A, Zeggwagh NA, Maghrani M, et al. Anti-hypoglycaemic activity of the aqueous extract of Origanum vulgare growing wild in Tafilalet region. *J Ethnopharmacol*. 2004 Jun;92(2-3):251-6.
[2780] Bayramoglu G, Senturk H, Bayramoglu A, et al. Carvacrol partially reverses symptoms of diabetes in STZ-induced diabetic rats. *Cytotechnology*. 2014 Mar;66(2):251–257.
[2781] Ezhumalai M, Radhiga T, Pugalendi KV. Antihyperglycemic effect of carvacrol in combination with rosiglitazone in high-fat diet-induced type 2 diabetic C57BL/6J mice. *Mol Cell Biochem*. 2014 Jan;385(1-2):23-31.
[2782] Umasankar K, Ramya DR, Nambikkairaj B. Antidiabetic and antidyslipidemic nature of carvacrol, a monoterpenic phenol studied in high-fat - fed and low-dose streptozotocin-induced experimental diabetic rats. *J Pharm Res*. 2015;9(8):484-90.
[2783] Tognolini M, Barocelli E, Ballabeni V, et al. Comparative screening of plant essential oils; phenylpropanoid moiety as basic core for antiplatelet activity. *Life Sci*. 2006 Feb 23;78(13):1419-32.

[2784] Tsai HH, Lin HW, Chen YL, et al. A review of potential harmful interactions between anticoagulant/antiplatelet agents and Chinese herbal medicines. *PLoS One*. 2013 May 9;8(5):e64255.

[2785] Karkabounas S, Kostoula OK, Daskalou T, et al. Anticarcinogenic and antiplatelet effects of carvacrol. *Exp Oncol*. 2006 Jun;28(2):121-5.

[2786] Baser KH. Biological and pharmacological activities of carvacrol and carvacrol bearing essential oils. *Curr Pharm Des*. 2008;14(29):3106-19.

[2787] Okazaki K, Kawazoe K, Takaishi Y. Human platelet aggregation inhibitors from thyme (Thymus vulgaris L.). *Phytother Res*. 2002 Jun;16(4):398-9.

[2788] Medeiros Barreto H, Cerqueira Fontinele F, Pereira de Oliveira A, et al. Phytochemical prospection and modulation of antibiotic activity in vitro by Lippia origanoides H.B.K. in methicillin resistant Staphylococcus aureus. *Biomed Res Int*. 2014;305610:1–7.

[2789] Aelenei P, Miron A, Trifan A, et al. Essential Oils and Their Components as Modulators of Antibiotic Activity against Gram-Negative Bacteria. *Medicines*. 2016;3(19):2-34.

[2790] Magi G, Marini E, Facinelli B. Antimicrobial activity of essential oils and carvacrol, and synergy of carvacrol and erythromycin, against clinical, erythromycin-resistant Group A Streptococci. *Front Microbiol*. 2015;6:165.

[2791] Veras HNH, Rodrigues FFG, Colares AV, et al. Synergistic antibiotic activity of volatile compounds from the essential oil of Lippia sidoides and thymol. *Fitoterapia*. 2012 Apr;83(3):508-12.

[2792] Langeveld WT, Veldhuizen EJ, Burt SA. Synergy between essential oil components and antibiotics: A review. *Crit Rev Microbiol*. 2014 Feb;40(1):76-94.

[2793] Palá-Paúl J, Usano-Alemany J, Granda E, et al. Chemical composition, antifungal and antibacterial activity of the essential oil of Chamaecyparis nootkatensis from Spain. *Nat Prod Commun*. 2009 Jul;4(7):1007-10.

[2794] Johnston WH, Karchesy JJ, Constantine GH, et al. Antimicrobial activity of some Pacific Northwest woods against anaerobic bacteria and yeast. *Phytother Res*. 2001 Nov;15(7):586-8.

[2795] Addesso KM, Oliver JB, O'Neal PA, et al. Efficacy of Nootka Oil as a Biopesticide for Management of Imported Fire Ants (Hymenoptera: Formicidae). *J Econ Entomol*. 2017 Aug 1;110(4):1547-1555.

[2796] Ucar M, Ucar G, Ozdemir H. Composition of Essential Oils from Fir (Abies) Wood Species Grown in Turkey. *Chem Nat Compounds*. 2015;51(2):356-358.

[2797] Tumen I, Hafizoglu H, Kilic A, et al. Yields and Constituents of Essential Oil from Cones of Pinaceae spp. Natively Grown in Turkey. *Molecules*. 2010;15:5797-5806.

[2798] Ucar G, Ucar MB, Ozdemir H, et al. Chemical Characterization of Volatile Needle Oils from Anatolian Fir Species: Abies nordmanniana (Stev.) Mattf., A. bornmülleriana Mattf., A. equitrojani Aschers et Sint. and A. cilicica Carr. *J Essent Oil Res*. 2010;22(6):548-54.

[2799] Rezaeieh KAP, Yildirim B, Kumlay AM, et al. Chemical Characterization of Needle Volatile Oil of Firwood (Abies nordmanniana subsp. bornmuelleriana Mattf.) from Eastern Turkey. *Int Sci Res J*. 2016 Aug;72(8):363-7.

[2800] Tumen I, Akkol EK, Süntar I, et al. Wound repair and anti-inflammatory potential of essential oils from cones of Pinaceae: preclinical experimental research in animal models. *J Ethnopharmacol*. 2011 Oct 11;137(3):1215-20.

[2801] Radulescu V, Saviuc C, Chifiriuc C, et al. Chemical Composition and Antimicrobial Activity of Essential Oil from Shoots Spruce (Picea abies L). *Rev Chim (Bucharest)*. 2011;62(1): 69-74.

[2802] Korica A, Polis O, Spalvis K, et al. Quantitative and qualitative seasonal changes of Scots Pine and Norway Spruce Foliage essential oils in Latvia, and the extraction dynamics thereof. *Baltic Forestry*. 2015;21(1):51-8.

[2803] Kartnig T, Still F, Reinthaler F. Antimicrobial activity of the essential oil of young pine shoots (Picea abies L.). *J Ethnopharmacol*. 1991;35:155-7.

[2804] Lee NH, Lee SM, Lee TM, et al. GC-MS Analyses of the Essential Oils Obtained from Pinaceae Leaves in Korea. 2015;18:3:538-42.

[2805] Zule J, Tisler V, Zurej A, et al. Isolation and characterization of essential oils from the cones of Norway spruce (Picea abies Karst.), European larch (Larix decidua Mill.) and Scots pine (Pinus sylvestric L.). *Zbornik gozdarstva in lesarstva*. 2003;71:159-172.

[2806] Radulescu V, Saviuc C, Chifiriuc C, et al. Chemical Composition and Antimicrobial Activity of Essential Oil from Shoots Spruce (Picea abies L). *Rev Chim (Bucharest)*. 2011;62(1): 69-74.

[2807] Kartnig T, Still F, Reinthaler F. Antimicrobial activity of the essential oil of young pine shoots (Picea abies L.). *J Ethnopharmacol*. 1991;35:155-7.

[2808] Yagi S, Babiker R, Tzanova T, et al. Chemical composition, antiproliferative, antioxidant and antibacterial activities of essential oils from aromatic plants growing in Sudan. *J Essent Oil Bearing Plants*. 2014;17(3):522-32.

[2809] Al-Massarani S, Al-Enzi F, Al-Tamimi M, et al. Composition & biological activity of Cyperus rotundus L. tuber volatiles from Saudi Arabia. *Nat Volatiles & Essent Oils*. 2016;3(2):26-34.

[2810] Hu QP, Cao XM, Hao DL, et al. Chemical Composition, Antioxidant, DNA Damage Protective, Cytotoxic and Antibacterial Activities of Cyperus rotundus Rhizomes Essential Oil against Foodborne Pathogens. *Asian Pac J Trop Med*. 2016 Aug;9(8):763-70.

[2811] Richa T, Suneet K. Chemical Constituents of the Essential oil of Cyperus rotundus Linn. *Int J Drug Dev & Res*. 2014 April-Jun;6(2):57-60.

[2812] Aghassi A, Naeemy A, Feizbakhsh A. Chemical Composition of the Essential Oil of Cyperus rotundus L. from Iran. *J Essent Oil Bearing Plants*. 2013;16(3):382-86.

[2813] Zoghbi MDGB, Andrade EHA, Carreira LMM, et al. Comparison of the Main Components of the Essential Oils of "priprioca": Cyperus articulatus var. articulatus L., C. articulatus var. nodosus L., C. prolixus Kunth and C. rotundus L. *J Essent Oil Res*. 2008 Jan/Feb;20:42-6.

[2814] Essaidi I, Koubaier HBH, Snoussi A, et al. Chemical Composition of Cyperus rotundus L. Tubers Essential Oil from the South of Tunisia, Antioxidant Potentiality and Antibacterial Activity against Foodborne Pathogens. *J Essent Oil Bearing Plants*. 2014;17(3):522-32.

[2815] Lawal OA, Oyedeji AO. Chemical Composition of the Essential Oils of Cyperus rotundus L. from South Africa. *Molecules*. 2009;14:2909-17.

[2816] Kilani S, Ledauphin J, Bouhlel I, et al. Comparative Study of Cyperus rotundus Essential Oil by a Modified GC/MS Analysis Method. Evaluation of Its Antioxidant, Cytotoxic, and Apoptotic Effects. *Chem Biodivers*. 2008;5:729-42.

[2817] Ekundayo O, Oderinde R, Ogundeyin M. Essential Oil Constituents of Cypevus tuberosus Rottb. Rhizomes. *Flavour Fragr J*. 1991;6:261-4.

[2818] Kilani S, Abdelwahed A, Ammar RB, et al. Chemical Composition, Antibacterial and Antimutagenic Activities of Essential Oil from (Tunisian) Cyperus rotundus. *J Essent Oil Res*. 2005 Nov/Dec;17:695-700.

[2819] Jirovetz L, Wobus A, Buchbauer G, et al. Comparative Analysis of the Essential Oil and SPME-Headspace Aroma Compounds of Cyperus rotundus L. Roots/Tubers from South-India Using GC, GC-MS and Olfactometry. *J Essent Oil Bearing Plants*. 2004;7(2):100-6.

[2820] Al-Massarani S, Al-Enzi F, Al-Tamimi M, et al. Composition & biological activity of Cyperus rotundus L. tuber volatiles from Saudi Arabia. *Nat Volatiles & Essent Oils*. 2016;3(2):26-34.

[2821] Hu QP, Cao XM, Hao DL, et al. Chemical Composition, Antioxidant, DNA Damage Protective, Cytotoxic and Antibacterial Activities of Cyperus rotundus Rhizomes Essential Oil against Foodborne Pathogens. *Asian Pac J Trop Med*. 2016 Aug;9(8):763-70.

[2822] Yagi S, Babiker R, Tzanova T, et al. Chemical composition, antiproliferative, antioxidant and antibacterial activities of essential oils from aromatic plants growing in Sudan. *J Essent Oil Bearing Plants*. 2014;17(3):522-32.

[2823] Kilani S, Ledauphin J, Bouhlel I, et al. Comparative Study of Cyperus rotundus Essential Oil by a Modified GC/MS
Analysis Method. Evaluation of Its Antioxidant, Cytotoxic, and Apoptotic Effects. *Chem Biodivers*. 2008;5:729-42.
[2824] Mohammed GF. Topical Cyperus rotundus oil: a new therapeutic modality with comparable efficacy to Alexandrite laser photo-epilation. *Aesthet Surg J*. 2014 Feb;34(2):298-305.
[2825] Farouk G, El-Kaream ABD. Role of Cyperus rotundus oil in decreasing hair growth. *J Intercult Ethnopharmacol*. 2012;1(2):111-118.
[2826] Chen Y, Zhao YY, Wang XY, et al. [GC-MS analysis and analgesic activity of essential oil from fresh rhizoma of Cyperus rotundus]. *Zhong Yao Cai*. 2011 Aug;34(8):1225-9.
[2827] Essaidi I, Koubaier HBH, Snoussi A, et al. Chemical Composition of Cyperus rotundus L. Tubers Essential Oil from the South of Tunisia, Antioxidant Potentiality and Antibacterial Activity against Foodborne Pathogens. *J Essent Oil Bearing Plants*. 2014;17(3):522-32.
[2828] Al-Massarani S, Al-Enzi F, Al-Tamimi M, et al. Composition & biological activity of Cyperus rotundus L. tuber volatiles from Saudi Arabia. *Nat Volatiles & Essent Oils*. 2016;3(2):26-34.
[2829] Hu QP, Cao XM, Hao DL, et al. Chemical Composition, Antioxidant, DNA Damage Protective, Cytotoxic and Antibacterial Activities of Cyperus rotundus Rhizomes Essential Oil against Foodborne Pathogens. *Asian Pac J Trop Med*. 2016 Aug;9(8):763-70.
[2830] Yagi S, Babiker R, Tzanova T, et al. Chemical composition, antiproliferative, antioxidant and antibacterial activities of essential oils from aromatic plants growing in Sudan. *J Essent Oil Bearing Plants*. 2014;17(3):522-32.
[2831] Duarte MC, Figueira GM, Sartoratto A, et al. Anti-Candida activity of Brazilian medicinal plants. *J Ethnopharmacol*. 2005 Feb 28;97(2):305-11.
[2832] Essaidi I, Koubaier HBH, Snoussi A, et al. Chemical Composition of Cyperus rotundus L. Tubers Essential Oil from the South of Tunisia, Antioxidant Potentiality and Antibacterial Activity against Foodborne Pathogens. *J Essent Oil Bearing Plants*. 2014;17(3):522-32.
[2833] Hu QP, Cao XM, Hao DL, et al. Chemical Composition, Antioxidant, DNA Damage Protective, Cytotoxic and Antibacterial Activities of Cyperus rotundus Rhizomes Essential Oil against Foodborne Pathogens. *Asian Pac J Trop Med*. 2016 Aug;9(8):763-70.
[2834] Yagi S, Babiker R, Tzanova T, et al. Chemical composition, antiproliferative, antioxidant and antibacterial activities of essential oils from aromatic plants growing in Sudan. *J Essent Oil Bearing Plants*. 2014;17(3):522-32.
[2835] Kilani S, Ledauphin J, Bouhlel I, et al. Comparative Study of Cyperus rotundus Essential Oil by a Modified GC/MS
Analysis Method. Evaluation of Its Antioxidant, Cytotoxic, and Apoptotic Effects. *Chem Biodivers*. 2008;5:729-42.
[2836] Al-Massarani S, Al-Enzi F, Al-Tamimi M, et al. Composition & biological activity of Cyperus rotundus L. tuber volatiles from Saudi Arabia. *Nat Volatiles & Essent Oils*. 2016;3(2):26-34.
[2837] Chansang A, Champakaew D, Junkum A, et al. Synergy in the adulticidal efficacy of essential oils for the improvement of permethrin toxicity against Aedes aegypti L. (Diptera: Culicidae). *Parasit Vectors*. 2018 Jul 13;11(1):417.
[2838] Menut C, Bessiere JM, Hassani MS, et al. Chemical and biological studies of Ocotea comoriensis bark essential oil. *Flavour Frag J*. 2002;17:459-61.
[2839] Menut C, Bessiere JM, Hassani MS, et al. Chemical and biological studies of Ocotea comoriensis bark essential oil. *Flavour Frag J*. 2002;17:459-61.
[2840] Chaverri C, Ciccio JF. Essential oil of trees of the genus Ocotea (Lauraceae) in Costa Rica. I. Ocotea brenesii. *Rev Biol Trop*. 2005 Sep-Dec;53(3-4):431-6.
[2841] Rigobert A, Rahanira R, Onja A, et al. Chemical Composition, Antioxidant and Antibacterial Properties of Essential Oil from Ocotea Auriculiformis Kost. (Lauraceae) Leaves, Endemic of Madagascar. *Eur Sci J*. 2017 Nov;13(33):364-77.
[2842] Rigobert A, Rahanira R, Onja A, et al. Chemical Composition, Antioxidant and Antibacterial Properties of Essential Oil from Ocotea Auriculiformis Kost. (Lauraceae) Leaves, Endemic of Madagascar. *Eur Sci J*. 2017 Nov;13(33):364-77.
[2843] Rigobert A, Rahanira R, Onja A, et al. Chemical Composition, Antioxidant and Antibacterial Properties of Essential Oil from Ocotea Auriculiformis Kost. (Lauraceae) Leaves, Endemic of Madagascar. *Eur Sci J*. 2017 Nov;13(33):364-77.
[2844] Abdel-Lateef E, Rabia I, Abdel-Gawad M, et al. In vitro antischistosomal activity of Allium cepa L. (Red onion_ extracts and identification of the essential oil by GC-MS. *J Microbiol Biotech Food Sci*. 2014;7(4):421-5.
[2845] El-Wakil EA, El-Sayed MM, Abdel-Lateef EES. GC-MS Investigation of Essential oil and antioxidant activity of Egyptian White Onion (Allium cepa L.). *Int J Pharm Sci Res*. 2015 Mar;6(3):537-43.
[2846] Kocic-Tanackov S, Dimic G, Levic J, et al. Effects of Onion (Allium cepa L.) and Garlic (Allium sativum L.) Essential Oils on the Aspergillus versicolor Growth and Sterigmatocystin Production. *Food Microbiol Safety*. 2012;00(0):M1-M7.
[2847] Ehiabhi OS, Edet UU, Walker TM, et al. Constituents of Essential Oils of Apium graveolens L., Allium cepa L., and Voacanga africana Staph. From Nigeria. *J Essent Oil Bear Plants*. 2006;9(2):126-32.
[2848] Mnayer D, Fabiano-Tixier AS, Petitcolas E, et al. Chemical Composition, Antibacterial and Antioxidant Activities
of Six Essentials Oils from the Alliaceae Family. *Molecules*. 2014;19:20034-53.
[2849] Ndoye Foe FM, Tchinang TFK, Nyegue AM, et al. Chemical composition, in vitro antioxidant and anti-inflammatory properties of essential oils of four dietary and medicinal plants from Cameroon. *BMC Complement Altern Med*. 2016 Apr 7;16:117.
[2850] Zaman R, Parvez M, Ali MS, et al. Evaluation of Antifertility Effect of Methanolic Bulb Extract of Allium cepa on Swiss Albino Male and Teratogenic Effect on Female Mice. *Adv Biol Res*. 2015;9(2):128-32.
[2851] Farnsworth NR, Bingel AS, Cordell GA, et al. Potential value of plants as sources of new antifertility agents I. *J Pharm Sci*. 1975 Apr;64(4):535-98.
[2852] Mennella JA, Johnson A, Beauchamp GK. Garlic ingestion by pregnant women alters the odor of amniotic fluid. *Chem Senses*. 1995;20:207-9.
[2853] Ziaei S, Hantoshzadeh S, Rezasoltani P, et al. The effect of garlic tablet on plasma lipids and platelet aggregation in nulliparous pregnants at high risk of preeclampsia. *Eur J Obstet Gynecol Reprod Biol*. 2001;99:201-6.
[2854] Soltani PR. Preeclampsia is an important complication of pregnancy which can result in morbidity and mortality in mother, fetus and the neonate. *J Med Counc Islamic Repub Iran*. 2005;23(3):319.
[2855] Kim SH, Lee IC, Baek HS, et al. Induction of cytochrome P450 3A1 expression by diallyl disulfide: protective effects against cyclophosphamideinduced embryo-fetal developmental toxicity. *Food Chem Toxicol*. 2014 Jul;69:312-9.
[2856] El Sayyad HI, Abou_El-Naga AM, Gadallah AA, et al. Protective effects of Allium sativum against defects of hypercholesterolemia on pregnant rats and their offspring. *Int J Clin Exp Med*. 2010; 3(2): 152–163.
[2857] Mennella JA, Beauchamp GK. The effects of repeated exposure to garlic-flavored milk on the nursling's behavior. *Pediatr Res*. 1993;34:805-8.
[2858] Mennella JA, Beauchamp GK. Maternal diet alters the sensory qualities of human milk and the nursling's behavior. *Pediatrics*. 1991;88:737-44.
[2859] Goldman IL, Kopelberg M, Debaene JE, et al. Antiplatelet activity in onion (Allium cepa) is sulfur dependent. *Thromb Haemost*. 1996 Sep;76(3):450-2.
[2860] [No authors listed]. The effect of essential oil of garlic on hyperlipemia and platelet aggregation--an analysis of 308 cases. Cooperative Group for Essential Oil of Garlic. *J Tradit Chin Med*. 1986 Jun;6(2):117-20.
[2861] Barrie SA, Wright JV, Pizzorno JE. Effects of Garlic Oil on Platelet Aggregation, Serum Lipids and Blood Pressure in Humans. *J Orthomolecular Med*. 1987;2(1):15-21.

[2862] Makheja AN, Vanderhoek JY, Bailey JM. Inhibition of platelet aggregation and thromboxane synthesis by onion and garlic. *Lancet.* 1979 Apr 7;1(8119):781.
[2863] Boullin DJ. Garlic as a platelet inhibitor. *Lancet.* 1981 Apr 4;1(8223):776-7.
[2864] Fenwick GR, Hanley AB. The genus Allium--Part 3. *Crit Rev Food Sci Nutr.* 1985;23(1):1-73.
[2865] Lawson LD, Hughes BG. Characterization of the formation of allicin and other thiosulfinates from garlic. *Planta Med.* 1992 Aug;58(4):345-50.
[2866] Bordia A, Verma SK, Srivastava KC. Effect of garlic (Allium sativum) on blood lipids, blood sugar, fibrinogen and fibrinolytic activity in patients with coronary artery disease. *Prostaglandins Leukot Essent Fatty Acids.* 1998 Apr;58(4):257-63.
[2867] Bordia A. Effect of garlic on human platelet aggregation in vitro. *Atherosclerosis.* 1978 Aug;30(4):355-60.
[2868] Qi R, Liao F, Inoue K, et al. Inhibition by diallyl trisulfide, a garlic component, of intracellular Ca(2+) mobilization without affecting inositol-1,4,5-trisphosphate (IP(3)) formation in activated platelets. *Biochem Pharmacol.* 2000 Nov 15;60(10):1475-83.
[2869] Ariga T, Takeda A, Teramoto S, et al. Inhibition Site of Methylallyl Trisulfide: A Volatile Oil Component of Garlic, in the Platelet Arachidonic Acid Cascade. *Food Factors for Cancer Prevention.* 1997:231-34.
[2870] Lawson LD, Ransom DK, Hughes BG. Inhibition of whole blood platelet-aggregation by compounds in garlic clove extracts and commercial garlic products. *Thromb Res.* 1992 Jan 15;65(2):141-56.
[2871] Siess MH, Le Bon AM, Canivenc-Lavier MC, et al. Modification of hepatic drug metabolising enzymes in rats treated with alkyl sulfides. *Cancer Lett.* 1997;120:195–201.
[2872] Lai YS, Chen WC, Ho CT, et al. Garlic essential oil protects against obesity-triggered nonalcoholic fatty liver disease through modulation of lipid metabolism and oxidative stress. *J Agric Food Chem.* 2014 Jun 25;62(25):5897-906.
[2873] Taubert D, Glöckner R, Müller D, et al. The garlic ingredient diallyl sulfide inhibits cytochrome P450 2E1 dependent bioactivation of acrylamide to glycidamide. *Toxicol Lett.* 2006 Jun 20;164(1):1-5.
[2874] Davenport DM, Wargovich MJ. Modulation of cytochrome P450 enzymes by organosulfur constituents from garlic. *Food Chem Toxicol.* 2005 Dec;43(12):1753-62.
[2875] Fisher CD, Augustine LM, Maher JM, et al. Induction of drug-metabolizing enzymes by garlic and allyl sulfide compounds via activation of constitutive androstane receptor and nuclear factor E2-related factor 2. *Drug Metab Dispos.* 2007 Jun;35(6):995-1000.
[2876] Lii CK, Tsai CW, Wu CC. Garlic allyl sulfides display differential modulation of rat cytochrome P450 2B1 and the placental form glutathione S transferase in various organs. *J Agric Food Chem.* 2006 Jul 12;54(14):5191-6.
[2877] Johnson S. Supercritical Essential Oils: A Companion Resource to Medicinal Essential Oils. 2017. Scott A Johnson Professional Writing Services, LLC: Orem, Utah.
[2878] Pyun MS, Shin S. Antifungal effects of the volatile oils from Allium plants against Trichophyton species and synergism of the oils with ketoconazole. *Phytomedicine.* 2006 Jun;13(6):394-400.
[2879] Scheman A, Gupta S. Photoallergic contact dermatitis from diallyl disulfide. *Contact Dermatitis.* 2001 Sep;45(3):179.
[2880] Alvarez MS, Jacobs S, Jiang SB, Brancaccio RR, et al. Photocontact allergy to diallyl disulfide. *Am J Contact Dermat.* 2003 Sep;14(3):161-5.
[2881] Benezra C, Ducombs G, Sell Y, Foussereau J. Plant contact dermatitis. Philadelphia: B.C. DECKER; 1985. p.240-3.
[2882] Coz CJL, Ducombs G. Plants and Plant Products. Chapter 41. Available at: http://eknygos.lsmuni.lt/springer/99/751-800.pdf.
[2883] Navarro JA, del Pozo MD, Gastaminza G, et al. Allium cepa seeds: A new occupational allergen. *J Allergy Clinical Immunol.* 1995 Nov; 96(15):690-93.
[2884] Lee TY, Lam TH. Contact dermatitis due to topical treatment with garlic in Hong Kong. *Contact Dermatitis.* 1991 Mar;24(3):193-6.
[2885] Cabanillas M, Fernandez-Redondo V, Toribio J. Allergic contact dermatitis to plants in a Spanish dermatology department: a 7-year review. *Contact Dermatitis.* 2006 Aug;55(2):84-91.
[2886] Ndoye Foe FM, Tchinang TFK, Nyegue AM, et al. Chemical composition, in vitro antioxidant and anti-inflammatory properties of essential oils of four dietary and medicinal plants from Cameroon. *BMC Complement Altern Med.* 2016 Apr 7;16:117.
[2887] Abdel-Lateef E, Rabia I, Abdel-Gawad M, et al. In vitro antischistosomal activity of Allium cepa L. (Red onion_ extracts and identification of the essential oil by GC-MS. *J Microbiol Biotech Food Sci.* 2014;7(4):421-5.
[2888] Kocic-Tanackov S, Dimic G, Levic J, et al. Effects of Onion (Allium cepa L.) and Garlic (Allium sativum L.) Essential Oils on the Aspergillus versicolor Growth and Sterigmatocystin Production. *Food Microbiol Safety.* 2012;00(0):M1-M7.
[2889] Kocic-Tanackov SD, Dimic GR, Tepic AN, et al. Influence of Allium ampeloprasum L. and Allium cepa L.
essential oils on the growth of some yeasts and moulds. Zbornik Matice srpske za prirodne nauke. 2009 Jan:121-129.
[2890] Golestani MR, Rad M, Bassami M, et al. Analysis and evaluation of antibacterial effects of new herbal formulas, AP-001 and AP-002, against Escherichia coli O157:H7. *Life Sci.* 2015 Aug 15;135:22-6.
[2891] El-Wakil EA, El-Sayed MM, Abdel-Lateef EES. GC-MS Investigation of Essential oil and antioxidant activity of Egyptian White Onion (Allium cepa L.). *Int J Pharm Sci Res.* 2015 Mar;6(3):537-43.
[2892] Ndoye Foe FM, Tchinang TFK, Nyegue AM, et al. Chemical composition, in vitro antioxidant and anti-inflammatory properties of essential oils of four dietary and medicinal plants from Cameroon. *BMC Complement Altern Med.* 2016 Apr 7;16:117.
[2893] Tonzibo ZF, Koffi AM, Wognin EL, et al. Intraspecific Chemical Variability of Essential Oil of Hoslundia opposita. *American J Analytical Chem.* 2013;4:77-82.
[2894] Akolade JO, Muhammad NO, Usman L, et al. Effects of Essential Oil Extracted from the Leaves of Hoslundia opposita V. on Selected Biochemical Indices in Rats. *British J Pharmacol Toxicol.* 2014;5(3):109-14.
[2895] Chagonda LS, Chalchat JC. The essential oil of wild and cultivated Hoslundia opposita Vahl. from Zimbabwe. *Flavour Frag J.* 2005;20:193-5.
[2896] Tonzibo ZF, Coffy AA, Chalchat JC, et al. Chemical composition of essential oils of Hoslundia opposita Vahl. from Ivory Coast. *Flavour Frag J.* 2006;21:789-91.
[2897] Usman LA, Zubair MF, Adebayo SA, et al. Chemical Composition of Leaf and Fruit Essential Oils of Hoslundia opposita Vahl Grown in Nigeria. *American-Eurasian J Agric Environ Sci.* 2010;8(1):40-3.
[2898] Ocheng F, Bwanga F, Joloba M, et al. Essential Oils from Ugandan Aromatic Medicinal Plants: Chemical Composition and Growth Inhibitory Effects on Oral Pathogens. *Evid Based Complement Alternat Med.* 2015;2015:230832.
[2899] Babarinde SA, Pitan OOR, Ajala MO, et al. Insectifugal and insecticidal potentials of two tropical botanical essential oils against cowpea seed bruchid. *Environ Sci Pollut Res Int.* 2017 Aug;24(24):19785-19794.
[2900] Craig JO. Poisoning by the volatile oils in childhood. *Arch Dis Child.* 1953;28:259-67.
[2901] Melis K. Bochner A, Janssens G. Accidental nasal eucalyptol and menthol instillation. *Eur J Pediatr.* 1989 Aug;148(8)786-7.
[2902] Day LM, Ozanne-Smith J, Parsons BJ, et al. Eucalyptus oil poisoning among young children: mechanisms of access and the potential prevention. *Aust N Z J Public Health.* 1997 Jun;21(3):297-302.
[2903] Day LM, Ozanne-Smith J, Parsons BJ, et al. Eucalyptus oil poisoning among young children: mechanisms of access and the potential prevention. *Aust N Z J Public Health.* 1997 Jun;21(3):297-302.

[2904] Myott E. Case of eucalyptus poisoning. *Brit M J.* 1906;1:558.
[2905] Hindle RC. Eucalyptus oil ingestion. *N Z Med J.* 1994 May 11;107(977)185-6.
[2906] Tibballs J. Clinical effects and management of eucalyptus oil ingestion in infants and small children. *Med J Aust.* 1995 Aug;163(4):177-80.
[2907] Waldman W, Barwina M, Sein Anand J. Accidental intoxication with eucalyptus oil—a case report. *Przeql Lek.* 2011;68(8):555-6.
[2908] Day LM, Ozanne-Smith J, Parsons BJ, et al. Eucalyptus oil poisoning among young children: mechanisms of access and the potential prevention. *Aust N Z J Public Health.* 1997 Jun;21(3):297-302.
[2909] Burkhard PR, Burkhardt K, Haenggeli CA, et al. Plant-induced seizures: reappearance of an old problem. *J Neurol.* 1999 Aug;246(8):667-70.
[2910] Waldman N. Seizure caused by dermal application of over-the-counter eucalyptus oil head lice preparation. *Clin Toxicol (Phila).* 2011 Oct;49(8):750-1.
[2911] Craig JO. Poisoning by the volatile oils in childhood. *Arch Dis Child.* 1953;28:259-67.
[2912] Mathew T, Kamath V, Kumar RS, et al. Eucalyptus oil inhalation-induced seizure: A novel, underrecognized, preventable cause of acute symptomatic seizure. *Epilepsia Open.* 2017 Jul 4;2(3):350-354.
[2913] Unger M, Frank A. Simultaneous determination of the inhibitory potency of herbal extracts on the activity of six major cytochrome P450 enzymes using liquid chromatography/mass spectrometry and automated online extraction. *Rapid Commun Mass Spectrom.* 2004;18(19):2273-81.
[2914] Jori A, Bianchetti A, Prestini PE, et al. Effect of eucalyptol (1,8-cineole) on the metabolism of other drugs in rats and in man. *Eur J Pharmacol.* 1970;9(3):362-66.
[2915] Kim NH, Hyun SH, Jin CH, et al. Pretreatment with 1,8-cineole potentiates thioacetamide-induced hepatotoxicity and immunosuppression. *Arch Pharm Res.* 2004 Jul;27(7):781-9.
[2916] Jori A, Bianchetti A, Prestini PE, et al. Effect of eucalyptol (1,8-cineole) on the metabolism of other drugs in rats and in man. *Eur J Pharmacol.* 1970;9(3):362-66.
[2917] de Sousa DP, Raphael E, Brocksom U, et al. Sedative effect of monoterpene alcohols in mice: A preliminary screening. *Verlag der Zeitschrift fur Naturforschung.* 2007;62c:563-66.
[2918] Chen R, Chen J, Cheng S, et al. Assessment of embryotoxicity of constituents in cosmetics by the embryonic stem cell tes. *Toxicol Mech Methods.* 2010 Mar;20(3):112-18.
[2919] Domaracky M, Rehak P, Juhas S, et al. Effects of selected plant essential oils on the growth and development of mouse preimplantation embryos in vivo. *Physiol Res.* 2007;56(1):97-104.
[2920] Vrskova D, Modra H. Evaluation of the developmental toxicity of 2-phenoxyethanol and clove oil anesthetics using the Frog Embryo Teratogenesis Assay: Xenopus (FETAX). *Veterinarami Medicina.* 2012;57(5):245-50.
[2921] Amini A, Cheraghi E, Safaee MR, et al. The role of eugenol in the reduction of teratogenic effects of retinoic acid on skeletal morphology of mice embryo. *Yakhteh Medical Journal.* 2003;4:195-200.
[2922] Chen SJ, Wang MH, Chen IJ. Antiplatelet and calcium inhibitory properties of eugenol and sodium eugenol acetate. *Gen Pharmacol.* 1996 Jun;27(4):629-33.
[2923] Tognolini M, Barocelli E, Ballabeni V, et al. Comparative screening of plant essential oils; phenylpropanoid moiety as basic core for antiplatelet activity. *Life Sci.* 2006 Feb 23;78(13):1419-32.
[2924] Heck AM, DeWitt BA, Lukes AL. Potential interactions between alternative therapies and warfarin. *Am J Health Syst Pharm.* 2000;57(13):1221-1227.
[2925] Saaeed SA, Gilani AH. Antithrombotic activity of clove oil. *J Pak Med Assoc.* 1994;44(5):112-15.
[2926] Kamatou GP, Vermaak I, Viljoen AM. Eugenol—from the remote Maluku Islands to the international market place: A review of a remarkable and versatile molecule. *Molecules.* 2012;17:6953-81.
[2927] Cochrane ML. Inhibition of Cytochrome P450 2C9 by essential oils. Available at: https://libres.uncg.edu/ir/uncg/listing.aspx?id=18102.
[2928] Kumaravelu P, Dakshinamoorthy DP, Subramaniam S, et al. Effect of eugenol on drug-metabolizing enzymes of carbon tetrachloride-intoxicated rat liver. *Biochem Pharmacol.* 1995 May;49(11):1703-7.
[2929] Dohi S, Terasaki M, Makino M. Acetylcholinesterase inhibitory activity and chemical composition of commercial essential oils. *J Agric Food Chem.* 2009 May 27;57(10):4313-8.
[2930] Langeveld WT, Veldhuizen EJ, Burt SA. Synergy between essential oil constituents and antibiotics. *Crit Rev Microbiol.* 2014 Feb;40(1);76-94.
[2931] Sarrami N, Pemberton MN, Thornhill MH, et al. Adverse reactions associated with the use of eugenol in dentistry. *British Dental J.*2002;193:253-55.
[2932] Tammannavar P, Pushpalatha C, Jain S, et al. An unexpected positive hypersensitive reaction to eugenol. *BMJ Case Rep.* 2013; 2013:bcr2013009464.
[2933] National Library of Medicine. TOXNET. Benzyl acetate. Available at: https://toxnet.nlm.nih.gov/cgibin/ sis/search/a?dbs+hsdb:@term+@DOCNO+2851.
[2934] JECFA. Summary of evaluations performed by the Joint FAO/WHO Expert Committee on Food Additives. 2001. Available at: https://toxnet.nlm.nih.gov/cgi-bin/sis/search/a?dbs+hsdb:@term+@DOCNO+2851.
[2935] van den Brink DM, van Miert JNI, Dacremont G, et al. Characterization of the Final Step in the Conversion of Phytol into Phytanic Acid. *J Biol Chem.* 2005 Jul 22;280(29):26838-44.
[2936] Selkälä EM, Nair RR, Schmitz W, et al. Phytol is lethal for Amacr-deficient mice. *Biochim Biophys Acta.* 2015 Oct;1851(10):1394-405.
[2937] McGinty D, Letizia CS, Api AM. Fragrance material review on phytol. *Food Chem Toxicol.* 2010;48:S59–S63.
[2938] National Institutes of Health. U.S. National Library of Medicine. Alpha-methylacyl-CoA racemase deficiency. Available at: https://ghr.nlm.nih.gov/condition/alpha-methylacyl-coa-racemase-deficiency.
[2939] Selkälä EM, Nair RR, Schmitz W, et al. Phytol is lethal for Amacr-deficient mice. *Biochim Biophys Acta.* 2015 Oct;1851(10):1394-405.
[2940] Khine H, Weiss D, Graber N, et al. A cluster of children with seizures caused by camphor poisoning. *Pediatrics.* 2009 May;123(5):1269-72.
[2941] Michiels EA, Mazor SS. Toddler with seizures due to ingesting camphor at an Indian celebration. *Pediatr Emerg Care.* 2010 Aug;26(8):574-75.
[2942] Koren G. Medications which can kill a toddler with one tablet or teaspoonful. *J Toxicol Clin Toxicol. 1*993;31(3):407-13.
[2943] Flaman Z, Pellechia-Clarke S, Bailey B, et al. Unintentional exposure of young children to camphor and eucalyptus oils. *Paediatr Child Health.* 2001 Feb;6(2):80-83.
[2944] Rabl W, Katzgraber F, Steinlechner M. Camphor ingestion for abortion (case report). *Forensic Sci Int.* 1997 Sep 19;89(1-2):137-40.
[2945] Flaman Z, Pellechia-Clarke S, Bailey B, et al. Unintentional exposure of young children to camphor and eucalyptus oils. *Paediatr Child Health.* 2001 Feb;6(2):80-83.
[2946] Manoguerra AS, Erdman AR, Wax PM, et al. Camphor poisoning: an evidence-based practice guideline for out-of- hospital management. *Clin Toxicol (Phila).*2006;44(4):357-70.
[2947] Gibson DE, Moore GP, Pfaff JA. Camphor ingestion. *Am J Emerg Med.* 1989 Jan;7(1):41-43.

[2948] Koppel C, Martends F, Schirop T, et al. Hemoperfusion in acute camphor poisoning. *Intensive Care Med.* 1988;14(4):431-33.
[2949] Burkhard PR, Burkhardt K, Haenggeli CA, et al. Plant-induced seizures: reappearance of an old problem. *J Neurol.* 1999 Aug;246(8):667–70.
[2950] Charanraj KJ, G MV, S M. Camphor poisoning in a child. *Natl Med J India.* 2013 Jan-Feb;26(1):60.
[2951] Narayan S, Singh N. Camphor poisoning-An unusual cause of seizure. *Med J Armed Forces India.* 2012 Jul;68(3):252–53.
[2952] Uc A, Bishop WP, Sanders KD. Camphor hepatotoxicity. *South Med J.* 2000;93:596-98.
[2953] Olowe SA, Ransome-Kuti O. The risk of jaundice in glucose-6-phosphate dehydrogenase deficient babies exposed to menthol. *Acta Paediatr Scand.* 1980 May;69(3):341-5.
[2954] Dillon Remy M, Manning Alleyne P, Bratt DE, et al. Neonatal jaundice at Port-of-Spain General Hospital abstract. *West Indian Med J.* 1987;36(Suppl):28.
[2955] Covington TR, et al. Handbook of Nonprescription Drugs. 11th ed. Washington, D.C.: American Pharmaceutical Association. 1996.
[2956] Lin WY, Kuo YH, Chang YL, et al. Anti-platelet aggregation and chemical constituents from the rhizome of Gynura japonica. *Planta Med.* 2003 Aug;69(8):757-64.
[2957] Akolade JO, Muhammad NO, Usman LA, et al. Antidyslipidemic effect of leaf essential oil of Hoslundia opposita Vahl in alloxan-induced diabetic rats. *Int J Tropical Med Public Health.* 2011;1(1):6-11.
[2958] Akolade JO, Usman LA, Okereke OE, et al. Antidiabetic potentials of essential oil extracted from the leaves of Hoslundia opposita Vahl. *J Med Food.* 2014 Oct;17(10):1122-8.
[2959] Muhammad NO, Akolade JO, Usman L, et al. Haematological parameters of alloxan-induced diabetic rats treated with leaf essential oil of Hoslundia opposita (Vahl). *EXCLI J.* 2012;11:670-6.
[2960] Ocheng F, Bwanga F, Joloba M, et al. Essential Oils from Ugandan Aromatic Medicinal Plants: Chemical Composition and Growth Inhibitory Effects on Oral Pathogens. *Evid Based Complement Alternat Med.* 2015;2015:230832.
[2961] Babarinde SA, Pitan OOR, Ajala MO, et al. Insectifugal and insecticidal potentials of two tropical botanical essential oils against cowpea seed bruchid. *Environ Sci Pollut Res Int.* 2017 Aug;24(24):19785-19794.
[2962] Babarinde SA, Akinyemi AO, Usman LA, et al. Toxicity and repellency of Hoslundia opposita Vahl (Lamiaceae) leaves' essential oil against rust-red flour beetle, Tribolium castaneum Herbst (Coleoptera: Tenebrionidae). *Nat Prod Res.* 2014;28(6):365-71.
[2963] Fernandez X, Lizzani-Cuvelier L, Loiseau AM, et al. Chemical composition of the essential oils from Turkish and Honduras Styrax. *Flavour Fragr J.* 2005;20:70-3.
[2964] Gurbuz I, Yesilada E, Demirci B, et al. Characterization of volatiles and anti-ulcerogenic effect of Turkish sweetgum balsam (Styrax liquidus). *J Ethnopharmacology.* 2013 Jun;148(1):332-6.
[2965] Altop A, Erener G, Duru ME, et al. Effects of essential oils from Liquidambar orientalis Mill. leaves on growth performance, carcass and some organ traits, some blood metabolites and intestinal microbiota in broilers. *British Poultry Sci.* 2018;59(1):121-7.
[2966] Park IK. Fumigant Toxicity of Oriental Sweetgum (Liquidambar orientalis) and Valerian (Valeriana wallichii) Essential Oils and Their Components, Including Their Acetylcholinesterase Inhibitory Activity, against Japanese Termites (Reticulitermes speratus). *Molecules.* 2014;19:12547-8.
[2967] Kim J, Seo SM, Lee SG, et al. Nematicidal activity of plant essential oils and components from coriander (Coriandrum sativum), oriental sweetgum (Liquidambar orientalis), and valerian (Valeriana wallichii) essential oils against pine wood nematode (Bursaphelenchus xylophilus). *J Agric Food Chem.* 2008;56:7316-20.
[2968] Duru ME. Cakir A, Harmandar M. Composition of the volatile oils isolated from the leaves of Liquidambar orientalis Mill. var. orientalis and L. orientalis var. integriloba from Turkey. *Flavour Fragr J.* 2002;17:95-8.
[2969] Hafizoglu H, Reunanen M, Istek A. Chemical constituents of balsam from liquidambar orientalis. *Holzforcehung.* 1996;50:116–117.
[2970] European Commission. Opinion on Liquidambar spp. Balsma Extracts and Oils (Storax), sesitization only. Available at: https://ec.europa.eu/health/ph_risk/committees/04_sccp/docs/sccp_o_025a.pdf.
[2971] Agency for Toxic Substances & Disease Registry. Toxic Substances Portal – Styrene. Available at: https://www.atsdr.cdc.gov/phs/phs.asp?id=419&tid=74.
[2972] Occupational Safety and Health Administration. Styrene. Available at: https://www.osha.gov/SLTC/styrene/hazards.html.
[2973] Mertas A, Garbusińska A, Szliszka E, et al. The influence of tea tree oil (Melaleuca alternifolia) on fluconazole activity against fluconazoleresistant Candida albicans strains. *Biomed Res Int.* 2015;2015:590470.
[2974] Gurbuz I, Yesilada E, Demirci B, et al. Characterization of volatiles and anti-ulcerogenic effect of Turkish sweetgum balsam (Styrax liquidus). *J Ethnopharmacology.* 2013 Jun;148(1):332-6.
[2975] Altop A, Erener G, Duru ME, et al. Effects of essential oils from Liquidambar orientalis Mill. leaves on growth performance, carcass and some organ traits, some blood metabolites and intestinal microbiota in broilers. *British Poultry Sci.* 2018;59(1):121-7.
[2976] Park IK. Fumigant Toxicity of Oriental Sweetgum (Liquidambar orientalis) and Valerian (Valeriana wallichii) Essential Oils and Their Components, Including Their Acetylcholinesterase Inhibitory Activity, against Japanese Termites (Reticulitermes speratus). *Molecules.* 2014;19:12547-8.
[2977] Kim J, Seo SM, Lee SG, et al. Nematicidal activity of plant essential oils and components from coriander (Coriandrum sativum), oriental sweetgum (Liquidambar orientalis), and valerian (Valeriana wallichii) essential oils against pine wood nematode (Bursaphelenchus xylophilus). *J Agric Food Chem.* 2008;56:7316-20.
[2978] Mojab F, Nickavar B, Hooshdar Tehrani H. Essential oil analysis of Nepeta crispa and N. menthoides from Iran. *Int J Pharm Sci.* 2009;5(1):43-46.
[2979] Kahkeshani N, Hadjiakhoondi A, Navidpour L, et al. Chemodiversity of Nepeta menthoides Boiss. & Bohse. essential oil from Iran and antimicrobial, acetylcholinesterase inhibitory and cytotoxic properties of 1,8-cineole chemotype. *Nat Prod Res.* 2017 Oct. [Epub ahead of print]
[2980] Sonboli A, Gholipour A, Yousefzadi M, et al. Antibacterial activity and composition of the essential oil of Nepeta menthoides from Iran. *Nat Prod Commun.* 2009;4(2):283-6.
[2981] Kahkeshani N, Razzaghirad Y, Ostad SN, et al. Cytotoxic, Acetylcholinesterase Inhibitor and Antioxidant Activity of Nepeta menthoides Boiss & Buhse essential oil. *J Essent Oil Bearing Plants.* 2014;17(4):544-52.
[2982] Nazemiyeh H, Razavi SM, Asnaashari S, et al. Chemical composition of the essential oil of Nepeta menthoides Boiss & Buhse. *Pharm Sci.* 2009;14(9)-283-9.
[2983] Baiazandeh MM. Essential Oil Composition of Nepeta menthoides Boiss. et Bushe from Iran. *J Essent Oil Res.* 2006 Mar/Apr;18:144-5.
[2984] Kahkeshani N, Hadjiakhoondi A, Navidpour L, et al. Chemodiversity of Nepeta menthoides Boiss. & Bohse. essential oil from Iran and antimicrobial, acetylcholinesterase inhibitory and cytotoxic properties of 1,8-cineole chemotype. *Nat Prod Commun.* 2017 Oct 15:1-4.
[2985] Craig JO. Poisoning by the volatile oils in childhood. *Arch Dis Child.* 1953;28:259-67.
[2986] Melis K. Bochner A, Janssens G. Accidental nasal eucalyptol and menthol instillation. *Eur J Pediatr.* 1989 Aug;148(8)786-7.
[2987] Day LM, Ozanne-Smith J, Parsons BJ, et al. Eucalyptus oil poisoning among young children: mechanisms of access and the potential prevention. *Aust N Z J Public Health.* 1997 Jun;21(3):297-302.

[2988] Day LM, Ozanne-Smith J, Parsons BJ, et al. Eucalyptus oil poisoning among young children: mechanisms of access and the potential prevention. *Aust N Z J Public Health.* 1997 Jun;21(3):297-302.
[2989] Myott E. Case of eucalyptus poisoning. *Brit M J.* 1906;1:558.
[2990] Hindle RC. Eucalyptus oil ingestion. *N Z Med J.* 1994 May 11;107(977)185-6.
[2991] Tibballs J. Clinical effects and management of eucalyptus oil ingestion in infants and small children. *Med J Aust.* 1995 Aug;163(4):177-80.
[2992] Waldman W, Barwina M, Sein Anand J. Accidental intoxication with eucalyptus oil—a case report. *Przeql Lek.* 2011;68(8):555-6.
[2993] Day LM, Ozanne-Smith J, Parsons BJ, et al. Eucalyptus oil poisoning among young children: mechanisms of access and the potential prevention. *Aust N Z J Public Health.* 1997 Jun;21(3):297-302.
[2994] Burkhard PR, Burkhardt K, Haenggeli CA, et al. Plant-induced seizures: reappearance of an old problem. *J Neurol.* 1999 Aug;246(8):667-70.
[2995] Waldman N. Seizure caused by dermal application of over-the-counter eucalyptus oil head lice preparation. *Clin Toxicol (Phila).* 2011 Oct;49(8):750-1.
[2996] Craig JO. Poisoning by the volatile oils in childhood. *Arch Dis Child.* 1953;28:259-67.
[2997] Mathew T, Kamath V, Kumar RS, et al. Eucalyptus oil inhalation-induced seizure: A novel, underrecognized, preventable cause of acute symptomatic seizure. *Epilepsia Open.* 2017 Jul 4;2(3):350-354.
[2998] Kahkeshani N, Hadjiakhoondi A, Navidpour L, et al. Chemodiversity of Nepeta menthoides Boiss. & Bohse. essential oil from Iran and antimicrobial, acetylcholinesterase inhibitory and cytotoxic properties of 1,8-cineole chemotype. *Nat Prod Res.* 2017 Oct. [Epub ahead of print]
[2999] Kahkeshani N, Razzaghirad Y, Ostad SN, et al. Cytotoxic, Acetylcholinesterase Inhibitor and Antioxidant Activity of Nepeta menthoides Boiss & Buhse essential oil. *J Essent Oil Bearing Plants.* 2014;17(4):544-52.
[3000] Unger M, Frank A. Simultaneous determination of the inhibitory potency of herbal extracts on the activity of six major cytochrome P450 enzymes using liquid chromatography/mass spectrometry and automated online extraction. *Rapid Commun Mass Spectrom.* 2004;18(19):2273-81.
[3001] Jori A, Bianchetti A, Prestini PE, et al. Effect of eucalyptol (1,8-cineole) on the metabolism of other drugs in rats and in man. *Eur J Pharmacol.* 1970;9(3):362-66.
[3002] Kim NH, Hyun SH, Jin CH, et al. Pretreatment with 1,8-cineole potentiates thioacetamide-induced hepatotoxicity and immunosuppression. *Arch Pharm Res.* 2004 Jul;27(7):781-9.
[3003] Jori A, Bianchetti A, Prestini PE, et al. Effect of eucalyptol (1,8-cineole) on the metabolism of other drugs in rats and in man. *Eur J Pharmacol.* 1970;9(3):362-66.
[3004] de Sousa DP, Raphael E, Brocksom U, et al. Sedative effect of monoterpene alcohols in mice: A preliminary screening. *Verlag der Zeitschrift fur Naturforschung.* 2007;62c:563-66.
[3005] Burkhard PR, Burkhardt K, Haenggeli CA, et al. Plant-induced seizures: reappearance of an old problem. *J Neurol.* 1999 Aug;246(8):667-70.
[3006] Waldman N. Seizure caused by dermal application of over-the-counter eucalyptus oil head lice preparation. *Clin Toxicol (Phila).* 2011 Oct;49(8):750-1.
[3007] Craig JO. Poisoning by the volatile oils in childhood. *Arch Dis Child.* 1953;28:259-67.
[3008] Mathew T, Kamath V, Kumar RS, et al. Eucalyptus oil inhalation-induced seizure: A novel, underrecognized, preventable cause of acute symptomatic seizure. *Epilepsia Open.* 2017 Jul 4;2(3):350-354.
[3009] Kahkeshani N, Hadjiakhoondi A, Navidpour L, et al. Chemodiversity of Nepeta menthoides Boiss. & Bohse. essential oil from Iran and antimicrobial, acetylcholinesterase inhibitory and cytotoxic properties of 1,8-cineole chemotype. *Nat Prod Res.* 2017 Oct. [Epub ahead of print]
[3010] Kahkeshani N, Razzaghirad Y, Ostad SN, et al. Cytotoxic, Acetylcholinesterase Inhibitor and Antioxidant Activity of Nepeta menthoides Boiss & Buhse essential oil. *J Essent Oil Bearing Plants.* 2014;17(4):544-52.
[3011] Unger M, Frank A. Simultaneous determination of the inhibitory potency of herbal extracts on the activity of six major cytochrome P450 enzymes using liquid chromatography/mass spectrometry and automated online extraction. *Rapid Commun Mass Spectrom.* 2004;18(19):2273-81.
[3012] Jori A, Bianchetti A, Prestini PE, et al. Effect of eucalyptol (1,8-cineole) on the metabolism of other drugs in rats and in man. *Eur J Pharmacol.* 1970;9(3):362-66.
[3013] Kim NH, Hyun SH, Jin CH, et al. Pretreatment with 1,8-cineole potentiates thioacetamide-induced hepatotoxicity and immunosuppression. *Arch Pharm Res.* 2004 Jul;27(7):781-9.
[3014] Jori A, Bianchetti A, Prestini PE, et al. Effect of eucalyptol (1,8-cineole) on the metabolism of other drugs in rats and in man. *Eur J Pharmacol.* 1970;9(3):362-66.
[3015] de Sousa DP, Raphael E, Brocksom U, et al. Sedative effect of monoterpene alcohols in mice: A preliminary screening. *Verlag der Zeitschrift fur Naturforschung.* 2007;62c:563-66.
[3016] Kahkeshani N, Hadjiakhoondi A, Navidpour L, et al. Chemodiversity of Nepeta menthoides Boiss. & Bohse. essential oil from Iran and antimicrobial, acetylcholinesterase inhibitory and cytotoxic properties of 1,8-cineole chemotype. *Nat Prod Res.* 2017 Oct. [Epub ahead of print]
[3017] Kahkeshani N, Razzaghirad Y, Ostad SN, et al. Cytotoxic, Acetylcholinesterase Inhibitor and Antioxidant Activity of Nepeta menthoides Boiss & Buhse essential oil. *J Essent Oil Bearing Plants.* 2014;17(4):544-52.
[3018] Kahkeshani N, Hadjiakhoondi A, Navidpour L, et al. Chemodiversity of Nepeta menthoides Boiss. & Bohse. essential oil from Iran and antimicrobial, acetylcholinesterase inhibitory and cytotoxic properties of 1,8-cineole chemotype. *Nat Prod Res.* 2017 Oct. [Epub ahead of print]
[3019] Kahkeshani N, Razzaghirad Y, Ostad SN, et al. Cytotoxic, Acetylcholinesterase Inhibitor and Antioxidant Activity of Nepeta menthoides Boiss & Buhse essential oil. *J Essent Oil Bearing Plants.* 2014;17(4):544-52.
[3020] Kahkeshani N, Hadjiakhoondi A, Navidpour L, et al. Chemodiversity of Nepeta menthoides Boiss. & Bohse. essential oil from Iran and antimicrobial, acetylcholinesterase inhibitory and cytotoxic properties of 1,8-cineole chemotype. *Nat Prod Res.* 2017 Oct. [Epub ahead of print]
[3021] Sonboli A, Gholipour A, Yousefzadi M, et al. Antibacterial activity and composition of the essential oil of Nepeta menthoides from Iran. *Nat Prod Commun.* 2009;4(2):283-6.
[3022] Pratiwi SUT, Lagendijk EL, de Weert S, et al. Effect of Cinnamomum burmannii Nees ex Bl. And Massoia aromatica Becc. Essential Oils on Planktonic Growth and Biofilm formation of Pseudomonas aeruginosa and Staphylococcus aureus In Vitro. *Int J App Res Nat Prod.* 2015 Mar;8(2):1-13.
[3023] Chen L, Su J, Li L, et al. A new source of natural D-borneol and its characteristic. *J Med Plants Res.* 2011 Aug;5(15):3440-7.
[3024] Ji XD, Pu QL, Garraffo HM, et al. Essential Oils of the Leaf, Bark and Branch of Cinnamomum burmannii Blume. *J Essent Oil Res.* 1991 Sep/Oct;3:373-5.
[3025] Domaracky M, Rehak P, Juhas S, et al. Effects of selected plant essential oils on the growth and development of mouse preimplantation embryos in vivo. *Physiol Res.* 2007;56(1):97-104.
[3026] Forschmidt P. Teratogenic activity of flavor additives. *Teratology.* 1979;19:26A.
[3027] Mantovani A, Stazi AV, Macri C, et al. Pre-natal (segment II) toxicity study of cinnamic aldehyde in the Sprague-Dawley rat. *Food Chem Toxicol.* 1989;27:781-86.

[3028] Huang J, Wang S, Luo X, et al. Cinnamaldehyde reduction of platelet aggregation and thrombosis in rodents. *Thromb Res.* 2007;119(3):337-42.
[3029] Kim SY, Koo YK, Koo JY, et al. Platelet anti-aggregation activities of constituents from Cinnamomum cassia. *J Med Food.* 2010 Oct;13(5):1069-74.
[3030] Takenaga M, Hirai A, Terano T, et al. In vitro effect of cinnamic aldehyde, a main component of Cinnamomi Cortex, on human platelet aggregation and arachidonic acid metabolism. *J Pharmacobiodyn.* 1987 May;10(5):201-208.
[3031] Mack TS, Raner GM. Regulation of cytochrome P450 2A6 and phase II enzymes by unsaturated aldehydes. University of North Carolina at Greensboro. Available at: http://libres.uncg.edu/ir/uncg/listing.aspx?id=2343.
[3032] Wickramasinghe RH, Muller G, Norpoth K. Spectral evidence of interaction of spice constituents with hepatic microsomal cytochrome P-450. *Cytobios.* 1980;29(113):25-27.
[3033] Singh G, Maurya S, deLampasona MP, et al. A comparison of chemical antioxidant and antimicrobial studies of cinnamon leaf and bark volatile oils, oleoresins and their constituents. *Food Chem Toxicol.* 2007;45:1650–61.
[3034] Verspohl EJ, Bauer K, Neddermann E. Antidiabetic effect of cinnamomum cassia and cinamomum zeylanicum in vivo and in vitro. *Phytother Res.* 2005 Mar;19(3):203-06.
[3035] Hua P, Guijun Z, Guixing R. Antidiabetic effects of cinnamon oil in diabetic KK-A mice. *Food Chem Toxicol.* 2010 Aug-Sep;48(8-9):2344-9.
[3036] Sung HK, Sun HH, Choung SY. Anti-diabetic effect of cinnamon extract on blood glucose in db/db mice. *J Ethnopharmacol.* 2006 Mar 8;104(1-2):119-23.
[3037] Langeveld WT, Veldhuizen EJ, Burt SA. Synergy between essential oil constituents and antibiotics. *Crit Rev Microbiol.* 2014 Feb;40(1);76-94.
[3038] Palaniappan K, Holley RA. Use of natural antimicrobials to increase antibiotic susceptibility of drug resistant bacteria. *In J Food Microbiol.* 2010 Jun 15;140(2-3):164-8.
[3039] National Toxicology Program. Cinnamaldehyde. 1989 Dec. Available at:
https://ntp.niehs.nih.gov/ntp/htdocs/chem_background/exsumpdf/cinnamaldehyde_508.pdf.
[3040] National Toxicology Program. Cinnamaldehyde. 1989 Dec. Available at:
https://ntp.niehs.nih.gov/ntp/htdocs/chem_background/exsumpdf/cinnamaldehyde_508.pdf.
[3041] Felter SP, Vasallo JD, Carlton BD, et al. A safety assessment of coumarin taking into account species specific toxicokinetics. *Food Chem Toxicol.* 2006 Apr;44(4):462-75.
[3042] European Food Safety Authority. Coumarin in flacourings and other food ingredients with flavouring properties: Scientific opinion of the panel on food additives, flavourings, processing aids and materials in contact with food (AFC). Available at: http://www.efsa.europa.eu/sites/default/files/scientific_output/files/main_documents/793.pdf.
[3043] Kim SY, Koo YK, Koo JY, et al. Platelet anti-aggregation activities of constituents from Cinnamomum cassia. *J Med Food.* 2010 Oct;13(5):1069-74.
[3044] Craig JO. Poisoning by the volatile oils in childhood. *Arch Dis Child.* 1953;28:259-67.
[3045] Melis K. Bochner A, Janssens G. Accidental nasal eucalyptol and menthol instillation. *Eur J Pediatr.* 1989 Aug;148(8)786-7.
[3046] Day LM, Ozanne-Smith J, Parsons BJ, et al. Eucalyptus oil poisoning among young children: mechanisms of access and the potential prevention. *Aust N Z J Public Health.* 1997 Jun;21(3):297-302.
[3047] Day LM, Ozanne-Smith J, Parsons BJ, et al. Eucalyptus oil poisoning among young children: mechanisms of access and the potential prevention. *Aust N Z J Public Health.* 1997 Jun;21(3):297-302.
[3048] Myott E. Case of eucalyptus poisoning. *Brit M J.* 1906;1:558.
[3049] Hindle RC. Eucalyptus oil ingestion. *N Z Med J.* 1994 May 11;107(977)185-6.
[3050] Tibballs J. Clinical effects and management of eucalyptus oil ingestion in infants and small children. *Med J Aust.* 1995 Aug;163(4):177-80.
[3051] Waldman W, Barwina M, Sein Anand J. Accidental intoxication with eucalyptus oil—a case report. *Przeql Lek.* 2011;68(8):555-6.
[3052] Day LM, Ozanne-Smith J, Parsons BJ, et al. Eucalyptus oil poisoning among young children: mechanisms of access and the potential prevention. *Aust N Z J Public Health.* 1997 Jun;21(3):297-302.
[3053] Burkhard PR, Burkhardt K, Haenggeli CA, et al. Plant-induced seizures: reappearance of an old problem. *J Neurol.* 1999 Aug;246(8):667-70.
[3054] Waldman N. Seizure caused by dermal application of over-the-counter eucalyptus oil head lice preparation. *Clin Toxicol (Phila).* 2011 Oct;49(8):750-1.
[3055] Craig JO. Poisoning by the volatile oils in childhood. *Arch Dis Child.* 1953;28:259-67.
[3056] Mathew T, Kamath V, Kumar RS, et al. Eucalyptus oil inhalation-induced seizure: A novel, underrecognized, preventable cause of acute symptomatic seizure. *Epilepsia Open.* 2017 Jul 4;2(3):350-354.
[3057] Unger M, Frank A. Simultaneous determination of the inhibitory potency of herbal extracts on the activity of six major cytochrome P450 enzymes using liquid chromatography/mass spectrometry and automated online extraction. *Rapid Commun Mass Spectrom.* 2004;18(19):2273-81.
[3058] Jori A, Bianchetti A, Prestini PE, et al. Effect of eucalyptol (1,8-cineole) on the metabolism of other drugs in rats and in man. *Eur J Pharmacol.* 1970;9(3):362-66.
[3059] Kim NH, Hyun SH, Jin CH, et al. Pretreatment with 1,8-cineole potentiates thioacetamide-induced hepatotoxicity and immunosuppression. *Arch Pharm Res.* 2004 Jul;27(7):781-9.
[3060] Jori A, Bianchetti A, Prestini PE, et al. Effect of eucalyptol (1,8-cineole) on the metabolism of other drugs in rats and in man. *Eur J Pharmacol.* 1970;9(3):362-66.
[3061] de Sousa DP, Raphael E, Brocksom U, et al. Sedative effect of monoterpene alcohols in mice: A preliminary screening. *Verlag der Zeitschrift fur Naturforschung.* 2007;62c:563-66.
[3062] Li Y, Shao X, Xu J, et al. Effects and possible mechanism of tea tree oil against Botrytis cinerea and Penicillium expansum in vitro and in vivo test. *Can J Microbiol.* 2017 Mar;63(3):219-227.
[3063] Li YH, Sun XP, Zhang YQ, et al. The antithrombotic effect of borneol related to its anticoagulant property. *Am J Chin Med.* 2008;36(4):719-27.
[3064] Hiroi T, Miyazaki Y, Kobayashi Y, et al. Induction of hepatic P450s in rat by essential wood and leaf oils. *Xenobiotica.* 1995 May;25(5):457-67.
[3065] Chen JY, Wang JJ, Meng MR, et al. [Borneol is an inducer of rat hepatic CYP2D activity in vivo]. *Yao Xue Xue Bao.* 2015 Apr;50(4):459-63.
[3066] Zhang R1, Mi SQ, Wang NS. Effect of borneol on cytochrome P450 3A enzyme and midazolam pharmacokinetics in rats. *Eur J Drug Metab Pharmacokinet.* 2013 Sep;38(3):159-69.
[3067] Kim NH, Hyun SH, Jin CH, et al. Pretreatment with 1,8-cineole potentiates thioacetamide-induced hepatotoxicity and immunosuppression. *Arch Pharm Res.* 2004 Jul;27(7):781-9.
[3068] Marín I, Sayas-Barberá E, Viuda-Martos M, et al. Chemical Composition, Antioxidant and Antimicrobial Activity of Essential Oils from Organic Fennel, Parsley, and Lavender from Spain. *Foods.* 2016 Mar 4;5(1).
[3069] Pratiwi SUT, Lagendijk EL, de Weert S, et al. Effect of Cinnamomum burmannii Nees ex Bl. And Massoia aromatica Becc. Essential Oils on Planktonic Growth and Biofilm formation of Pseudomonas aeruginosa and Staphylococcus aureus In Vitro. *Int J App Res Nat Prod.* 2015 Mar;8(2):1-13.

[3070] Awang AFIBm Susanti D, Taher M. Antimicrobial Activity and Synergic Effect of Cinnamomum burmannii's Essential Oil & its isolated compound (Cinnamaldehyde). *Int Conf Chem Agric Med Sci.* 2013 Dec:26-29.

[3071] Kurkcuoglu M, Baser KHC, Vural M. Composition of the essential oil of Pastinaca sativa L. subsp. urens (Req. ex Gordon) Celak. *Chem Nat Compounds.* 2006;42(1):114-15.

[3072] Kubeczka KH, Stahl E. [On the essential oils of the Apiaceae (Umbelliferae). I. The root oil of Pastinaca sativa]. *Planta Med.* 1975 Jun;27(3):235-41.

[3073] Stahl-Biskup E, Wichtmann E. Composition of the essential oils from roots of some apiaceae in relation to the development of their oil duct systems. *Flavour Frag J.* 1991;6(4):249-255.

[3074] Stahl E. Variation of myristicin content in cultivated parsnip roots (Pastinaca sativa ssp. sativa var. hortensis). *J Agric Food Chem.* 1981 Jul;29(4):890-2.

[3075] Kapetanos C, Karioti A, Bojovic S, et al. Chemical and Principal-Component Analyses of the Essential Oils of Apioideae Taxa (Apiaceae) from Central Balkan. *Chem Biodivers.* 2008;5:101-19.

[3076] Stahl-Biskup E, Wichtmann E. Composition of the essential oils from roots of some apiaceae in relation to the development of their oil duct systems. *Flavour Frag J.* 1991;6(4):249-255.

[3077] Randerath K, Putman KL, Randerath E. Flavor constituents in cola drinks induce hepatic DNA adducts in adult and fetal mice. *Biochem Biophys Res Commun.* 1993 Apr 15;192(1):61-8.

[3078] Vesselinovitch SD, Rao KVN, Mihailovich N. Transplacental and lactational carcinogenesis by safrole. *Cancer Res.* 1979 Nov;39(11):4378-80.

[3079] Forrest JE, Heacock RA. Nutmeg and mace, the psychotropic spices from Myristica fragrans. *Lloydia.* 1972 Dec;35(4):440-9.

[3080] Leiter E1, Hitchcock G, Godwin S, et al. Evaluation of the anxiolytic properties of myristicin, a component of nutmeg, in the male Sprague-Dawley rat. *AANA J.* 2011 Apr;79(2):109-14.

[3081] Truitt EB Jr, Duritz G, Ebersberger EM. Evidence of monoamine oxidase inhibition by myristicin and nutmeg. *Proc Soc Exp Biol Med.* 1963 Mar;112:647-50.

[3082] Jeong HG, Yun CH. Induction of rat hepatic cytochrome P450 enzymes by myristicin. *Biochem Biophys Res Commun.* 1995 Dec26;217(3):966-71.

[3083] Yang AH, He X, Chen JX, et al. Identification and characterization of reactive metabolites in myristicin-mediated mechanism-based inhibition of CYP1A2. *Chem Biol Interact.* 2015 Jul 25;237:133-40.

[3084] Matejic JS, DZamic AM, Mihajilov-Krstev T, et al. Antimicrobial potential of essential oil from Pastinaca sativa L. *Biologica Nyssana.* 2014 Sep;5(1):31-5.

[3085] Bumblauskiene L, Jakstas V, Janulis V, et al. Preliminary analysis on essential oil composition of Perilla L. cultivated in Lithuania. *Acta Poloniea Pharm Drug Res.* 2009;66(4):409-13.

[3086] Baser KHC, Demirci B, Donmez AA. Composition of the essential oil of Perilla frutescens (L.) britton from Turkey. *Flavour Fragr J.* 2003;18:122-3.

[3087] Gwari G, Lohani H, Haider SZ, et al. Chemical analysis and antioxidant activityin Perilla frutescens collected from Uttahakhand. *Int J Biol Pharm Res.* 2015;6(6):473-7.

[3088] Tabanca N, Demirci B, Ali A, et al. Essential oils of green and red Perilla frutescens as potential sources ofcompounds for mosquito management. *Ind Crops Prod.* 2015;65:36-44.

[3089] Sharma S, Arunachalam K, Phondani P. Variation in Yield and Composition of Perilla frutescens Essential Oil Across Altitudinal Gradients in the Indian Himalayan Region. *Iranian J Sci Tech Trans Sci.* 2017:1-10.

[3090] Zhu L. Li Y, Li B, et al. Aromatic Plants and Essential Constituents (Supplement 1). SATAS Library and Publishing House.

[3091] Verma RS, Padalia RC, Chauhan A. Volatile oil composition of Indian Perilla [Perilla frutescens (L.) Britton] collected at different phenophases. *J Essent Oil Res.* 2013;25(2):92-6.

[3092] Kim SJ, Kang EY, Won SE, et al. Chemical Composition and Comparison of Essential Oil Contents of Perilla frutescens Britton var. japonica HARA Leaves. *Korean J Med Crop Sci.* 2008;16(4):242-54.

[3093] Lin LY, Peng CC, Wang HE, et al. Active Volatile Constituents in Perilla frutescens Essential Oils and Improvement of Antimicrobial and Anti-Inflammatory Bioactivity by Fractionation. *J Essent Oil Bear Plants.* 2016;19(8):1957-83.

[3094] Kim SJ, Kang EY, Won SE, et al. Chemical Composition and Comparison of Essential Oil Contents of Perilla frutescens Britton var. japonica HARA Leaves. *Korean J Med Crop Sci.* 2008;16(4):242-54.

[3095] Ford RA, Letizia C, Api AM. Monographs on Fragrance Raw Materials. *Food Chem Toxicol.* 1988;26(Suppl.):397-8.

[3096] Garst JE, Wilson WC, Kristensen NC, et al. Species susceptibility to the pulmonary toxicity of 3-furyl isoamyl ketone (perilla ketone): in vivo support for involvement of the lung monooxygenase system. *J Anim Sci.* 1985 Jan;60(1):248-57.

[3097] Linnabary RD, Wilson BJ, Garst JE. Acute bovine pulmonary emphysema (ABPE): Perilla ketone as another cause. *Vet Hum Toxicol.* 1978;20:325-326

[3098] Breeze RG, Legreid WW, Bayly WM, et al. Perilla ketone toxicity: a chemical model for the study of equine restrictive lung disease. *Equine Vet J.* 1984 May;16(3):180-4.

[3099] Wilson BJ, Garst JE, Linnabary RD, et al. Perilla ketone: a potent lung toxin from the mint plant, Perilla frutescens Britton. *Science.* 1977 Aug 5;197(4303):573-4.

[3100] Garst JE, Wilson WC, Kristensen NC, et al. Species susceptibility to the pulmonary toxicity of 3-furyl isoamyl ketone (perilla ketone): in vivo support for involvement of the lung monooxygenase system. *J Anim Sci.* 1985 Jan;60(1):248-57.

[3101] Linnabary RD, Wilson BJ, Garst JE. Acute bovine pulmonary emphysema (ABPE): Perilla ketone as another cause. *Vet Hum Toxicol.* 1978;20:325-326

[3102] Breeze RG, Legreid WW, Bayly WM, et al. Perilla ketone toxicity: a chemical model for the study of equine restrictive lung disease. *Equine Vet J.* 1984 May;16(3):180-4.

[3103] Wilson BJ, Garst JE, Linnabary RD, et al. Perilla ketone: a potent lung toxin from the mint plant, Perilla frutescens Britton. *Science.* 1977 Aug 5;197(4303):573-4.

[3104] Randerath K, Putman KL, Randerath E. Flavor constituents in cola drinks induce hepatic DNA adducts in adult and fetal mice. *Biochem Biophys Res Commun.* 1993 Apr 15;192(1):61-8.

[3105] Vesselinovitch SD, Rao KVN, Mihailovich N. Transplacental and lactational carcinogenesis by safrole. *Cancer Res.* 1979 Nov;39(11):4378-80.

[3106] Forrest JE, Heacock RA. Nutmeg and mace, the psychotropic spices from Myristica fragrans. *Lloydia.* 1972 Dec;35(4):440-9.

[3107] Leiter E1, Hitchcock G, Godwin S, et al. Evaluation of the anxiolytic properties of myristicin, a component of nutmeg, in the male Sprague-Dawley rat. *AANA J.* 2011 Apr;79(2):109-14.

3108 Truitt EB Jr, Duritz G, Ebersberger EM. Evidence of monoamine oxidase inhibition by myristicin and nutmeg. *Proc Soc Exp Biol Med.* 1963 Mar;112:647-50.
3109 Jeong HG, Yun CH. Induction of rat hepatic cytochrome P450 enzymes by myristicin. *Biochem Biophys Res Commun.* 1995 Dec26;217(3):966-71.
3110 Yang AH, He X, Chen JX, et al. Identification and characterization of reactive metabolites in myristicin-mediated mechanism-based inhibition of CYP1A2. *Chem Biol Interact.* 2015 Jul 25;237:133-40.
3111 Ji WW, Li RP, Li M, et al. Antidepressant-like effect of essential oil of Perilla frutescens in a chronic, unpredictable, mild stress-induced depression model mice. *Chin J Nat Med.* 2014 Oct;12(10):753-9.
3112 Ji WW, Wang SY, Ma ZQ, et al. Effects of perillaldehyde on alternations in serum cytokines and depressive-like behavior in mice after lipopolysaccharide administration. *Pharmacol Biochem Behav.* 2014 Jan;116:1-8.
3113 Yi LT, Li J, Geng D, Liu BB, et al. Essential oil of Perilla frutescens-induced change in hippocampal expression of brain-derived neurotrophic factor in chronic unpredictable mild stress in mice. *J Ethnopharmacol.* 2013 May 2;147(1):245-53.
3114 Qiu J, Zhang X, Luo M, et al. Subinhibitory concentrations of perilla oil affect the expression of secreted virulence factor genes in Staphylococcus aureus. *PLoS One.* 2011 Jan 19;6(1):e16160.
3115 Li N, Zhang ZJ, Li XJ, et al. Microcapsules biologically prepared using Perilla frutescens (L.) Britt. essential oil and their use for extension of fruit shelf life. *J Sci Food Agric.* 2018 Feb;98(3):1033-1041.
3116 Gwari G, Lohani H, Haider SZ, et al. Chemical analysis and antioxidant activityin Perilla frutescens collected from Uttahakhand. *Int J Biol Pharm Res.* 2015;6(6):473-7.
3117 Tabanca N, Demirci B, Ali A, et al. Essential oils of green and red Perilla frutescens as potential sources ofcompounds for mosquito management. *Ind Crops Prod.* 2015;65:36-44.
3118 Dong ZX, Wang YW, Liu QZ, et al. Laboratory Screening of 26 Essential Oils Against Cacopsylla chinensis (Hemiptera: Psyllidae) and Field Confirmation of the Top Performer, Perilla frutescens (Lamiales: Lamiaceae). *J Econ Entomol.* 2019 Jan 11. [Epub ahead of print]
3119 Lawrence BM. Progress in essential oils. *Perf Flavorist.* 1995;6:47-48.
3120 Weyerstahl P, Wahlburg H, Splittgerber U, et al. Volatile constituents of brazilian phoebe oil. *Flavour Fragr J.* 1994;9(4):179-186.
3121 Srivilai J, Phimnuan P, Jaisabai J, et al. Curcuma aeruginosa Roxb. essential oil slows hair-growth and lightens skin in axillae; a randomised, double blinded trial. *Phytomedicine.* 2017 Feb 15;25:29-38.
3122 Akarchariya N, Sirilun S, Julsrigival J, et al. Chemical profiling and antimicrobial activity of essential oil from Curcuma aeruginosa Roxb., Curcuma glans K. Larsen & J. Mood and Curcuma cf. xanthorrhiza Roxb. collected in Thailand. *Asian Pac J Tropical Biomed.* 2017 Sep 20. Epub Ahead of Print.
3123 Kamazeri TSAT, Samah OA, Taher M, et al. Antimicrobial activity and essential oils of Curcuma aeruginosa, Curcuma mangga, and Zingiber cassumunar from Malaysia. *Asian Pac J Tropical Biomed.* 2012 Mar;5(3):202-9.
3124 Zwaving J, Bos R. Analysis of the Essential Oils of Five Curcuma Species. *Flavour Frag J.* 1992;7(1):19-22.
3125 Theanphong O, Mingvanish W, Kirdmanee C. Chemical constituents and biological activities of essential oil form Curcuma aeruginosa Roxb. rhizome. *BHST.* 2015;13(1):6-16.
3126 Angel GR, Menon N, Vimala B, et al. Essential oil composition of eight starchy Curcuma species. *Ind Crops Products.* 2014;60:233-8.
3127 Sirat HM, Jamil S, Hussain J. Essential Oil of Curcuma aeruginosa Roxb. from Malaysia. *J Essent Oil Res.* 1998;10(4):453-8.
3128 Khine H, Weiss D, Graber N, et al. A cluster of children with seizures caused by camphor poisoning. *Pediatrics.* 2009 May;123(5):1269-72.
3129 Michiels EA, Mazor SS. Toddler with seizures due to ingesting camphor at an Indian celebration. *Pediatr Emerg Care.* 2010 Aug;26(8):574-75.
3130 Koren G. Medications which can kill a toddler with one tablet or teaspoonful. *J Toxicol Clin Toxicol. 1*993;31(3):407-13.
3131 Flaman Z, Pellechia-Clarke S, Bailey B, et al. Unintentional exposure of young children to camphor and eucalyptus oils. *Paediatr Child Health.* 2001 Feb;6(2):80-83.
3132 Rabl W, Katzgraber F, Steinlechner M. Camphor ingestion for abortion (case report). *Forensic Sci Int.* 1997 Sep 19;89(1-2):137-40.
3133 Flaman Z, Pellechia-Clarke S, Bailey B, et al. Unintentional exposure of young children to camphor and eucalyptus oils. *Paediatr Child Health.* 2001 Feb;6(2):80-83.
3134 Manoguerra AS, Erdman AR, Wax PM, et al. Camphor poisoning: an evidence-based practice guideline for out-of- hospital management. *Clin Toxicol (Phila).*2006;44(4):357-70.
3135 Gibson DE, Moore GP, Pfaff JA. Camphor ingestion. *Am J Emerg Med.* 1989 Jan;7(1):41-43.
3136 Koppel C, Martends F, Schirop T, et al. Hemoperfusion in acute camphor poisoning. *Intensive Care Med.* 1988;14(4):431-33.
3137 Burkhard PR, Burkhardt K, Haenggeli CA, et al. Plant-induced seizures: reappearance of an old problem. *J Neurol.* 1999 Aug;246(8):667–70.
3138 Millet Y, Jouglard J, Steinmetz MD, et al. Toxicity of some essential plant oils. Clinical and experimental study. *Clin Toxicol.* 1981 Dec;18(12):1485–98.
3139 Charanraj KJ, G MV, S M. Camphor poisoning in a child. *Natl Med J India.* 2013 Jan-Feb;26(1):60.
3140 Narayan S, Singh N. Camphor poisoning-An unusual cause of seizure. *Med J Armed Forces India.* 2012 Jul;68(3):252–53.
3141 Uc A, Bishop WP, Sanders KD. Camphor hepatotoxicity. *South Med J.* 2000;93:596-98.
3142 Olowe SA, Ransome-Kuti O. The risk of jaundice in glucose-6-phosphate dehydrogenase deficient babies exposed to menthol. *Acta Paediatr Scand.* 1980 May;69(3):341-5.
3143 Dillon Remy M, Manning Alleyne P, Bratt DE, et al. Neonatal jaundice at Port-of-Spain General Hospital abstract. *West Indian Med J.* 1987;36(Suppl):28.
3144 Covington TR, et al. Handbook of Nonprescription Drugs. 11th ed. Washington, D.C.: American Pharmaceutical Association. 1996.
3145 Pimkaew P1, Küblbeck J, Petsalo A, et al. Interactions of sesquiterpenes zederone and germacrone with the human cytochrome P450 system. *Toxicol In Vitro.* 2013 Sep;27(6):2005-12.
3146 Srivilai J, Phimnuan P, Jaisabai J, et al. Curcuma aeruginosa Roxb. essential oil slows hair-growth and lightens skin in axillae; a randomised, double blinded trial. *Phytomedicine.* 2017 Feb 15;25:29-38.
3147 Akarchariya N, Sirilun S, Julsrigival J, et al. Chemical profiling and antimicrobial activity of essential oil from Curcuma aeruginosa Roxb., Curcuma glans K. Larsen & J. Mood and Curcuma cf. xanthorrhiza Roxb. collected in Thailand. Asian Pac J Tropical Biomed. 2017 Sep 20. Epub Ahead of Print.
3148 Kamazeri TSAT, Samah OA, Taher M, et al. Antimicrobial activity and essential oils of Curcuma aeruginosa, Curcuma mangga, and Zingiber cassumunar from Malaysia. *Asian Pac J Tropical Biomed.* 2012 Mar;5(3):202-9.
3149 Theanphong O, Mingvanish W, Kirdmanee C. Chemical constituents and biological activities of essential oil form Curcuma aeruginosa Roxb. rhizome. *BHST.* 2015;13(1):6-16.
3150 Skoula M, Grayer RJ. Volatile oils of Coridothymus capitatus, Satureja thymbra, Satureja spinosa and Thymbra calostachya (Lamiaceae) from Crete. *Flavour Frag J.* 2005;20:573-6.

[3151] Hafi MA, Beyrouthy ME, Ouaini N, et al. Chemical Composition and Antimicrobial Activity of Satureja, Thymus, and Thymbra Species Grown in Lebanon. *Chem Biodivers*. 2017 May;14(5):n/a-n/a.
[3152] Öztürk M. Anticholinesterase and antioxidant activities of Savoury (Satureja thymbra L.) with identified major terpenes of the essential oil. *Food Chem*. 2012;134(1):48-54.
[3153] Azaz AD, Kurkcuoglu M, Satil F, et al. In vitro antimicrobial activity and chemical composition of some Satureja essential oils. *Flavour Fragr J*. 2005;20(6):587-91.
[3154] Giatropoulos A, Kimbaris A, Michaelakis A, eta l. Chemical composition and assessment of larvicidal and repellent capacity of 14 Lamiaceae essential oils against Aedes albopictus. *Parasitol Res*. 2018;117(6):1953-1964.
[3155] El Beyrouthy M, Arnold-Apostolides N, Cazier F, et al. Chemical Composition of the Essential Oil of Aerial Parts of Satureja thymbra L. Growing Wild in Lebanon. *Acta Hort*. 2012;997:59-66.
[3156] Chorianopolous N, Evergets E, Mallouchos A, et al. Characterization of the Essential Oil Volatiles of Satureja thymbra and Satureja parnassica: Influence of Harvesting Time and Antimicrobial Activity. *J Agric Food Chem*. 2006;54:3139-45.
[3157] Economou G, Panagopoulos G, Tarantilis P. et al. Variability in essential oil content and composition of Origanum hirtum L., Origanum onites L., Coridothymus capitatus (L.) and Satureja thymbra L. populations from the Greek island Ikaria. *Ind Crops Prod*. 2011;33:236-41.
[3158] Markovic T, Chatopoulou P, Siljegovic J, et al. Chemical analysis and antimicrobial activities of the essential oils of Satureja thymbra L. and Thymbra spicata L. and their main components. *Arch Biol Sci Belgrade*. 2011;63(2):457-64.
[3159] Giweli A, Dzamic AM, Sokovic M, et al. Antimicrobial and Antioxidant Activities of Essential Oils of Satureja thymbra Growing Wild in Libya. *Molecules*. 2012;17:4836-50.
[3160] Karabay-Yavasoglu N, Baykan S, Ozturk B, et al. Evaluation of the Antinociceptive and Anti-inflammatory Activities of Satureja thymbra L. Essential Oil. *Pharm Biol*. 2006;44(8):585-91.
[3161] Goren AC, Topcu G, Bislel G, et al. Analysis of essential oil of Satureja thymbra by hydrodistillation, thermal desorber, and headspace GC/MS techniques and its antimicrobial activity. *Nat Prod Res*. 2004;18:189-95.
[3162] Mueller-Riebau, Nerger F, Yegen O. Chemical composition and fungitoxic properties to phytopathogenic fungi of essential oils of selected aromatic plants growing wild in Turkey. *J Agric Chem*. 1995;43:2262-6.
[3163] Chorianopolous N, Evergets E, Mallouchos A, et al. Essential oils of Satureja, Origanum, and Thymus species: Chemical composition and antibacterial activities against foodborne pathogens. *J Agric Food Chem*. 2005 Jan;52(26):8261-7.
[3164] Michaelakis A, Theotokatos SA, Koliopoulos G, et al. Essential oils of Satureja species: Insecticidal effect on Culex pipiens larvae (Diptera: Culicade). *Molecules*. 2007;12:2567-78.
[3165] Karousou R, Koureas DN, Kokkini S. Essential oil composition is related to the natural habitats: Coridothymus capitatus and Satureja thymbra in NATURA 2000 sites of Crete. *Phytochemistry*. 2005 Nov;66(22):2668-73.
[3166] Domaracky M, Rehak P, Juhas S, et al. Effects of selected plant essential oils on the growth and development of mouse preimplantation embryos in vivo. *Physiol Res*. 2007;56(1):97-104.
[3167] Lemhadri A, Zeggwagh NA, Maghrani M, et al. Anti-hypoglycaemic activity of the aqueous extract of Origanum vulgare growing wild in Tafilalet region. *J Ethnopharmacol*. 2004 Jun;92(2-3):251-6.
[3168] Bayramoglu G, Senturk H, Bayramoglu A, et al. Carvacrol partially reverses symptoms of diabetes in STZ-induced diabetic rats. *Cytotechnology*. 2014 Mar;66(2):251–257.
[3169] Ezhumalai M, Radhiga T, Pugalendi KV. Antihyperglycemic effect of carvacrol in combination with rosiglitazone in high-fat diet-induced type 2 diabetic C57BL/6J mice. *Mol Cell Biochem*. 2014 Jan;385(1-2):23-31.
[3170] Umasankar K, Ramya DR, Nambikkairaj B. Antidiabetic and antidyslipidemic nature of carvacrol, a monoterpenic phenol studied in high-fat - fed and low-dose streptozotocin-induced experimental diabetic rats. *J Pharm Res*. 2015;9(8):484-90.
[3171] Tognolini M, Barocelli E, Ballabeni V, et al. Comparative screening of plant essential oils; phenylpropanoid moiety as basic core for antiplatelet activity. *Life Sci*. 2006 Feb 23;78(13):1419-32.
[3172] Tsai HH, Lin HW, Chen YL, et al. A review of potential harmful interactions between anticoagulant/antiplatelet agents and Chinese herbal medicines. *PLoS One*. 2013 May 9;8(5):e64255.
[3173] Karkabounas S, Kostoula OK, Daskalou T, et al. Anticarcinogenic and antiplatelet effects of carvacrol. *Exp Oncol*. 2006 Jun;28(2):121-5.
[3174] Baser KH. Biological and pharmacological activities of carvacrol and carvacrol bearing essential oils. *Curr Pharm Des*. 2008;14(29):3106-19.
[3175] Okazaki K, Kawazoe K, Takaishi Y. Human platelet aggregation inhibitors from thyme (Thymus vulgaris L.). *Phytother Res*. 2002 Jun;16(4):398-9.
[3176] Jukic M, Politeo O, Maksimmovic M, et al. In vitro acetylcholinesterase inhibitory properties of thymol, carvacrol and their derivatives thymoquinone and thymohydroquinone. *Phytother Res*. 2007;21(3):259-61.
[3177] Askin H, Yildiz M, Ayar A. Effects of Thymol and Carvacrol on Acetylcholinesterase from Drosophila melanogaster. *Acta Physica Polinica A*. 2017;132(3):720-22.
[3178] Medeiros Barreto H, Cerqueira Fontinele F, Pereira de Oliveira A, et al. Phytochemical prospection and modulation of antibiotic activity in vitro by Lippia origanoides H.B.K. in methicillin resistant Staphylococcus aureus. *Biomed Res Int*. 2014;305610:1–7.
[3179] Aelenei P, Miron A, Trifan A, et al. Essential Oils and Their Components as Modulators of Antibiotic Activity against Gram-Negative Bacteria. *Medicines*. 2016;3(19):2-34.
[3180] Magi G, Marini E, Facinelli B. Antimicrobial activity of essential oils and carvacrol, and synergy of carvacrol and erythromycin, against clinical, erythromycin-resistant Group A Streptococci. *Front Microbiol*. 2015;6:165.
[3181] Veras HNH, Rodrigues FFG, Colares AV, et al. Synergistic antibiotic activity of volatile compounds from the essential oil of Lippia sidoides and thymol. *Fitoterapia*. 2012 Apr;83(3):508-12.
[3182] Langeveld WT, Veldhuizen EJ, Burt SA. Synergy between essential oil components and antibiotics: A review. *Crit Rev Microbiol*. 2014 Feb;40(1):76-94.
[3183] Kohlert C, Schindler G, Marz RW, et al. Systemic availability and pharmacokinetics of thymol in humans. *J Clin Pharmacol*. 2002 Jul;42(7):731-7.
[3184] Toxicology Data Network, National Library of Medicine. Thymol. Available at: http://toxnet.nlm.nih.gov/cgibin/ sis/search/a?dbs+hsdb:@term+@DOCNO+866.
[3185] Tognolini M, Barocelli E, Ballabeni V, et al. Comparative screening of plant essential oils; phenylpropanoid moiety as basic core for antiplatelet activity. *Life Sci*. 2006 Feb 23;78(13):1419-32.
[3186] Okazaki K, Kawazoe K, Takaishi Y. Human platelet aggregation inhibitors from thyme (Thymus vulgaris L.). *Phytother Res*. 2002 Jun;16(4):398-9.

[3187] Jukic M, Politeo O, Maksimmovic M, et al. In vitro acetylcholinesterase inhibitory properties of thymol, carvacrol and their derivatives thymoquinone and thymohydroquinone. *Phytother Res.* 2007;21(3):259-61.
[3188] Askin H, Yildiz M, Ayar A. Effects of Thymol and Carvacrol on Acetylcholinesterase from Drosophila melanogaster. *Acta Physica Polinica A.* 2017;132(3):720-22.
[3189] Medeiros Barreto H, Cerqueira Fontinele F, Pereira de Oliveira A, et al. Phytochemical prospection and modulation of antibiotic activity in vitro by Lippia origanoides H.B.K. in methicillin resistant Staphylococcus aureus. *Biomed Res Int.* 2014;305610:1–7.
[3190] Aelenei P, Miron A, Trifan A, et al. Essential Oils and Their Components as Modulators of Antibiotic Activity against Gram-Negative Bacteria. *Medicines.* 2016;3(19):2-34.
[3191] Magi G, Marini E, Facinelli B. Antimicrobial activity of essential oils and carvacrol, and synergy of carvacrol and erythromycin, against clinical, erythromycin-resistant Group A Streptococci. *Front Microbiol.* 2015;6:165.
[3192] Veras HNH, Rodrigues FFG, Colares AV, et al. Synergistic antibiotic activity of volatile compounds from the essential oil of Lippia sidoides and thymol. *Fitoterapia.* 2012 Apr;83(3):508-12.
[3193] Langeveld WT, Veldhuizen EJ, Burt SA. Synergy between essential oil components and antibiotics: A review. *Crit Rev Microbiol.* 2014 Feb;40(1):76-94.
[3194] Miladinovic DL, Ilic BS, Kocic BD, et al. Antibacterial investigation of thyme essential oil and its main constituents on combination with tetracycline. *J Med Food.* 2015 Aug;18(8):935-7.
[3195] Shin S, Lim S. Antifungal effects of herbal essential oils alone and in combination with ketoconazole against Trichophyton spp. *J Appl Micribiol.* 2004;97:1289-96.
[3196] Kissels W, Wu X, Santos RR, et al. Short communication: Interaction of the isomers carvacrol and thymol with the antibiotics doxycycline and tilmicosin: In vitro effects against pathogenic bacteria commonly found in the respiratory tract of calves. *J Dairy Sci.* 2017 Feb;100(2):970-974.
[3197] Fitsiou E, Anestopoulos I, Chlichlia K, et al. Antioxidant and Antiproliferative Properties of the Essential Oils of Satureja thymbra and Satureja parnassica and their Major Constituents. *Anticancer Res.* 2016 Nov;36(11):5757-5763.
[3198] Öztürk M. Anticholinesterase and antioxidant activities of Savoury (Satureja thymbra L.) with identified major terpenes of the essential oil. *Food Chem.* 2012;134(1):48-54.
[3199] Karabay-Yavasoglu N, Baykan S, Ozturk B, et al. Evaluation of the Antinociceptive and Anti-inflammatory Activities of Satureja thymbra L. Essential Oil. *Pharm Biol.* 2006;44(8):585-91.
[3200] Azaz AD, Kurkcuoglu M, Satil F, et al. In vitro antimicrobial activity and chemical composition of some Satureja essential oils. *Flavour Fragr J.* 2005;20(6):587-91.
[3201] Gören AC, Topçu G, Bilsel G, et al. Analysis of essential oil of Satureja thymbra by hydrodistillation, thermal desorber, and headspace GC/MS techniques and its antimicrobial activity. *Nat Prod Res.* 2004 Apr;18(2):189-95.
[3202] Chorianopoulos N, Giaouris E, Skandamis P, et al. Disinfectant test against monoculture and mixed-culture biofilms composed of technological, spoilage and pathogenic bacteria: bactericidal effect of essential oil and hydrosol of Satureja thymbra and comparison with standard acid–base sanitizers. J App Microbiology. 2008;104(6):1586-96.
[3203] Lebert I, Leroy S, Talon R. Effect of industrial and natural biocides on spoilage, pathogenic and technological strains grown in biofilm. *Food Microbiol.* 2007 May;24(3):281-7.
[3204] Markovic T, Chatopoulou P, Siljegovic J, et al. Chemical analysis and antimicrobial activities of the essential oils of Satureja thymbra L. and Thymbra spicata L. and their main components. *Arch Biol Sci Belgrade.* 2011;63(2):457-64.
[3205] Giweli A, Dzamic AM, Sokovic M, et al. Antimicrobial and Antioxidant Activities of Essential Oils of Satureja thymbra Growing Wild in Libya. *Molecules.* 2012;17:4836-50.
[3206] Chorianopolous N, Evergets E, Mallouchos A, et al. Essential oils of Satureja, Origanum, and Thymus species: Chemical composition and antibacterial activities against foodborne pathogens. *J Agric Food Chem.* 2005 Jan;52(26):8261-7.
[3207] Piras A, Cocco V, Falconieri D, et al. Isolation of the volatile oil from Satureja thymbra by supercritical carbon dioxide extraction: chemical composition and biological activity. *Nat Prod Commun.* 2011 Oct;6(10):1523-6.
[3208] Boziaris IS, Chorianopoulos NG, Haroutounian SA, et al. Effect of Satureja thymbra essential oil on growth-no growth interfaces of Listeria monocytogenes Scott A and Salmonella Enteritidis PT4, at various temperatures, pH, and water activities. *J Food Prot.* 2011 Jan;74(1):45-54.
[3209] Sarac N, Ugur A. Antimicrobial activities of the essential oils of Origanum onites L., Origanum vulgare L. subspecies hirtum (Link) Ietswaart, Satureja thymbra L., and Thymus cilicicus Boiss. & Bal. growing wild in Turkey. *J Med Food.* 2008 Sep;11(3):568-73.
[3210] Dell'Agli M, Sanna C, Rubiolo P, et al. Anti-plasmodial and insecticidal activities of the essential oils of aromatic plants growing in the Mediterranean area. *Malar J.* 2012 Jul 2;11:219.
[3211] Öztürk M. Anticholinesterase and antioxidant activities of Savoury (Satureja thymbra L.) with identified major terpenes of the essential oil. *Food Chem.* 2012;134(1):48-54.
[3212] Fitsiou E, Anestopoulos I, Chlichlia K, et al. Antioxidant and Antiproliferative Properties of the Essential Oils of Satureja thymbra and Satureja parnassica and their Major Constituents. *Anticancer Res.* 2016 Nov;36(11):5757-5763.
[3213] Giatropoulos A, Kimbaris A, Michaelakis A, eta l. Chemical composition and assessment of larvicidal and repellent capacity of 14 Lamiaceae essential oils against Aedes albopictus. *Parasitol Res.* 2018;117(6):1953-1964.
[3214] Michaelakis A, Theotokatos SA, Koliopoulos G, et al. Essential oils of Satureja species: Insecticidal effect on Culex pipiens larvae (Diptera: Culicade). *Molecules.* 2007;12:2567-78.
[3215] Karamaouna F, Kimbaris A, Michaelakis A, et al. Insecticidal activity of plant essential oils against the vine mealybug, Planococcus ficus. *J Insect Sci.* 2013;13:142.
[3216] Dell'Agli M, Sanna C, Rubiolo P, et al. Anti-plasmodial and insecticidal activities of the essential oils of aromatic plants growing in the Mediterranean area. *Malar J.* 2012 Jul 2;11:219.
[3217] Ayvaz A, Sagdic O, Karaborklu S, et al. Insecticidal activity of the essential oils from different plants against three stored-product insects. *J Insect Sci.* 2010;10:21.
[3218] Cetin H, Cilek JE, Oz E, Aydin L, et al. Acaricidal activity of Satureja thymbra L. essential oil and its major components, carvacrol and gamma-terpinene against adult Hyalomma marginatum (Acari: Ixodidae). *Vet Parasitol.* 2010 Jun 24;170(3-4):287-90.
[3219] Figueredo AC, Barroso JG, Pedro LG, et al. Composition of the Essential Oil of Lavandula pinnata L. fil. var. pinnata Grown on Madeira. *Flavour Frag J.* 1995;10:93-6.
[3220] Argentieri MP, Lucia BD, Cristiano G, et al. Compositional Analysis of Lavandula pinnata Essential Oils. *Nat Prod Commun.* 2016;11(3):287-90.
[3221] Lis-Balchin M. (2002) Chemical composition of essential oils from different species, hybrids and cultivars of Lavandula. In Lavender – The genus Lavandula. Lis-Balchin M (Ed). Taylor & Francis, London, UK, 251- 262. Available at: http://www.vuzllib.su/books/8954-Lavender/28--24_Chemical_composition_of_essentialoils_from _different_species,_hybrids_and_cultivars_of_Lavandu.

[3222] Bonesi M, Menichini F, Tundis R, et al. Acetylcholinesterase and butyrylcholinesterase inhibitory activity of Pinus species essential oils and their constituents. *J Enzyme Inhib Med Chem*. 2010 Oct;25(5):622-8.
[3223] Domaracky M, Rehak P, Juhas S, et al. Effects of selected plant essential oils on the growth and development of mouse preimplantation embryos in vivo. *Physiol Res*. 2007;56(1):97-104.
[3224] Lemhadri A, Zeggwagh NA, Maghrani M, et al. Anti-hypoglycaemic activity of the aqueous extract of Origanum vulgare growing wild in Tafilalet region. *J Ethnopharmacol*. 2004 Jun;92(2-3):251-6.
[3225] Bayramoglu G, Senturk H, Bayramoglu A, et al. Carvacrol partially reverses symptoms of diabetes in STZ-induced diabetic rats. *Cytotechnology*. 2014 Mar;66(2):251–257.
[3226] Ezhumalai M, Radhiga T, Pugalendi KV. Antihyperglycemic effect of carvacrol in combination with rosiglitazone in high-fat diet-induced type 2 diabetic C57BL/6J mice. *Mol Cell Biochem*. 2014 Jan;385(1-2):23-31.
[3227] Umasankar K, Ramya DR, Nambikkairaj B. Antidiabetic and antidyslipidemic nature of carvacrol, a monoterpenic phenol studied in high-fat - fed and low-dose streptozotocin-induced experimental diabetic rats. *J Pharm Res*. 2015;9(8):484-90.
[3228] Tognolini M, Barocelli E, Ballabeni V, et al. Comparative screening of plant essential oils; phenylpropanoid moiety as basic core for antiplatelet activity. *Life Sci*. 2006 Feb 23;78(13):1419-32.
[3229] Tsai HH, Lin HW, Chen YL, et al. A review of potential harmful interactions between anticoagulant/antiplatelet agents and Chinese herbal medicines. *PLoS One*. 2013 May 9;8(5):e64255.
[3230] Karkabounas S, Kostoula OK, Daskalou T, et al. Anticarcinogenic and antiplatelet effects of carvacrol. *Exp Oncol*. 2006 Jun;28(2):121-5.
[3231] Baser KH. Biological and pharmacological activities of carvacrol and carvacrol bearing essential oils. *Curr Pharm Des*. 2008;14(29):3106-19.
[3232] Okazaki K, Kawazoe K, Takaishi Y. Human platelet aggregation inhibitors from thyme (Thymus vulgaris L.). *Phytother Res*. 2002 Jun;16(4):398-9.
[3233] Jukic M, Politeo O, Maksimmovic M, et al. In vitro acetylcholinesterase inhibitory properties of thymol, carvacrol and their derivatives thymoquinone and thymohydroquinone. *Phytother Res*. 2007;21(3):259-61.
[3234] Askin H, Yildiz M, Ayar A. Effects of Thymol and Carvacrol on Acetylcholinesterase from Drosophila melanogaster. *Acta Physica Polinica A*. 2017;132(3):720-22.
[3235] Medeiros Barreto H, Cerqueira Fontinele F, Pereira de Oliveira A, et al. Phytochemical prospection and modulation of antibiotic activity in vitro by Lippia origanoides H.B.K. in methicillin resistant Staphylococcus aureus. *Biomed Res Int*. 2014;305610:1–7.
[3236] Aelenei P, Miron A, Trifan A, et al. Essential Oils and Their Components as Modulators of Antibiotic Activity against Gram-Negative Bacteria. *Medicines*. 2016;3(19):2-34.
[3237] Magi G, Marini E, Facinelli B. Antimicrobial activity of essential oils and carvacrol, and synergy of carvacrol and erythromycin, against clinical, erythromycin-resistant Group A Streptococci. *Front Microbiol*. 2015;6:165.
[3238] Veras HNH, Rodrigues FFG, Colares AV, et al. Synergistic antibiotic activity of volatile compounds from the essential oil of Lippia sidoides and thymol. *Fitoterapia*. 2012 Apr;83(3):508-12.
[3239] Langeveld WT, Veldhuizen EJ, Burt SA. Synergy between essential oil components and antibiotics: A review. *Crit Rev Microbiol*. 2014 Feb;40(1):76-94.
[3240] Lawrence BM. Essential Oils 1988-1991. Allured Publishing, Wheaton: Carol Stream, IN, 1995.
[3241] Ehret C, Maupetit P, Petrzilka M. New organoleptically important constituents of narcissus absolute (Narcissus poeticus L.). *J Essent Oil Res*. 1992 Jan/Feb;4:41-47.
[3242] Okello EJ, Dimaki C, Howes MJR, et al. In vitro inhibition of human acetyl- and butyrylcholinesterase by Narcissus poeticus L. (Amaryllidaceae) flower absolute. *Int J Essential Oil Therapeutics*. 2008;2:105-10.
[3243] Okello EJ, Dimaki C, Howes MJR, et al. In vitro inhibition of human acetyl- and butyrylcholinesterase by Narcissus poeticus L. (Amaryllidaceae) flower absolute. *Int J Essential Oil Therapeutics*. 2008;2:105-10.
[3244] Rameshkumar KB, Sudheesh N, George V, et al. Volatile Constituents of the Roots of Cyperus compressus Linn. *J Essent Oil Res*. 2011 May/Jun;23:39-41.
[3245] Lin WY, Kuo YH, Chang YL, et al. Anti-platelet aggregation and chemical constituents from the rhizome of Gynura japonica. *Planta Med*. 2003 Aug;69(8):757-64.
[3246] Tucker AO, Maciarello MJ, Karchesy JJ. Commercial "Rose of Cedar" Oil, the Wood Oil of Port Orford Cedar, Chamaecyparis lawsoniana (A. Murray) Parl. (Cupressaceae). *J Essent Oil Res*. 2000 Jan/Feb;12:24-6.
[3247] Duringer JM, Swan LR, Walker DB, et al. Acute aquatic toxicity of western juniper (Juniperus occidentalis) foliage and Port Orford cedar (Chamaecyparis lawsoniana) heartwood oils. *Environ Monit Assess*. 2010;170:585-98.
[3248] Kritchevsky G, Anderson AB. Composition of the volatile Oil from Chamaecyparis lawsoniana Stump Heartwood. *J American Pharm Assoc*. 1955 Sep;44(9):535-38.
[3249] Enami SA, Massoomi H, Moghadam MS, et al. Identification of Volatile Oil Components from Aerial Parts of Chamaecyparis lawsoniana by GC-MS and 13CNMR Methods. *JEOBP*. 2009;12(6):661-5.
[3250] Giatropoulos A, Pitarokili D, Papaioannou F, et al. Essential oil composition, adult repellency and larvicidal activity of eight Cupressaceae species from Greece against Aedes albopictus (Diptera: Culicidae). *Parasitol Res*. 2013;112:1113-23.
[3251] Pala-Paul J, Usano-Alemany J, Granda E, et al. Antifungal and Antibacterial Activity of the Essential oil of Chamaecyparis lawsoniana from Spain. *Nat Prod Commun*. 2012;7(1):1383-6.
[3252] Burkhard PR, Burkhardt K, Haenggeli CA, et al. Plant-induced seizures: reappearance of an old problem. *J Neurol*. 1999 Aug;246(8):667–70.
[3253] Millet Y, Jouglard J, Steinmetz MD, et al. Toxicity of some essential plant oils. Clinical and experimental study. *Clin Toxicol*. 1981 Dec;18(12):1485–98.
[3254] Charanraj KJ, G MV, S M. Camphor poisoning in a child. *Natl Med J India*. 2013 Jan-Feb;26(1):60.
[3255] Narayan S, Singh N. Camphor poisoning-An unusual cause of seizure. *Med J Armed Forces India*. 2012 Jul;68(3):252–53.
[3256] Pala-Paul J, Usano-Alemany J, Granda E, et al. Antifungal and Antibacterial Activity of the Essential oil of Chamaecyparis lawsoniana from Spain. *Nat Prod Commun*. 2012;7(1):1383-6.
[3257] Giatropoulos A, Pitarokili D, Papaioannou F, et al. Essential oil composition, adult repellency and larvicidal activity of eight Cupressaceae species from Greece against Aedes albopictus (Diptera: Culicidae). *Parasitol Res*. 2013;112:1113-23.
[3258] Dolan MC, Dietrich G, Panella NA, et al. Biocidal activity of three wood essential oils against Ixodes scapularis (Acari: Ixodidae), Xenopsylla cheopis (Siphonaptera: Pulicidae), and Aedes aegypti (Diptera: Culicidae). *J Econ Entomol*. 2007 Apr;100(2):622-5.
[3259] Salgueiro L, Vila R, Tomas X, et al. Chemical polymorphism of the essential oil of Thymus carnosus from Portugal. *Phytochem*. 1995;38(2):391-6.
[3260] Sefidkon F, Askari F, Mirmostafa SA. The Essential Oil of Thymus carnosus Boiss. From Iran. *J Essent Oil Res*. 2001;13(3):192-3.
[3261] Marhuenda E, Menendez M, de La Lastra CA. Constituents of essential oil of Thymus carnosus Boiss. *J Chromatrography*. 1988;436:103-6.

[3262] Miguel MG, Costa LA, Figueiredo AC, et al. Assessment of the antioxidant ability of Thymus albicans, Th. mastichina, Th. camphoratus and Th. carnosus essential oils by TBARS and Micellar Model systems. *Nat Prod Commun.* 2007 Jan;2(4):399-406.

[3263] Zuzarte M, Alves-Silva JM, Alves M, et al. New insights on the anti-inflammatory potential and safety profile of Thymus carnosus and Thymus camphoratus essential oils and their main compounds. *J Ethnopharmacol.* 2018;225:10-7.

[3264] Aazza S, El-Guendouz S, Miguel MG, et al. Antioxidant, Anti-inflammatory and Anti-hyperglycaemic Activities of Essential Oils from Thymbra capitata, Thymus albicans, Thymus caespititius, Thymus carnosus, Thymus lotocephalus and Thymus mastichina from Portugal. *Nat Prod Commun.* 2016 Jul;11(7):1029-1038.

[3265] Li Y, Shao X, Xu J, et al. Effects and possible mechanism of tea tree oil against Botrytis cinerea and Penicillium expansum in vitro and in vivo test. *Can J Microbiol.* 2017 Mar;63(3):219-227.

[3266] Li YH, Sun XP, Zhang YQ, et al. The antithrombotic effect of borneol related to its anticoagulant property. *Am J Chin Med.* 2008;36(4):719-27.

[3267] Hiroi T, Miyazaki Y, Kobayashi Y, et al. Induction of hepatic P450s in rat by essential wood and leaf oils. *Xenobiotica.* 1995 May;25(5):457-67.

[3268] Chen JY, Wang JJ, Meng MR, et al. [Borneol is an inducer of rat hepatic CYP2D activity in vivo]. *Yao Xue Xue Bao.* 2015 Apr;50(4):459-63.

[3269] Zhang R1, Mi SQ, Wang NS. Effect of borneol on cytochrome P450 3A enzyme and midazolam pharmacokinetics in rats. *Eur J Drug Metab Pharmacokinet.* 2013 Sep;38(3):159-69.

[3270] Kim NH, Hyun SH, Jin CH, et al. Pretreatment with 1,8-cineole potentiates thioacetamide-induced hepatotoxicity and immunosuppression. *Arch Pharm Res.* 2004 Jul;27(7):781-9.

[3271] Marín I, Sayas-Barberá E, Viuda-Martos M, et al. Chemical Composition, Antioxidant and Antimicrobial Activity of Essential Oils from Organic Fennel, Parsley, and Lavender from Spain. *Foods.* 2016 Mar 4;5(1).

[3272] Tognolini M, Barocelli E, Ballabeni V, et al. Comparative screening of plant essential oils; phenylpropanoid moiety as basic core for antiplatelet activity. *Life Sci.* 2006 Feb 23;78(13):1419-32.

[3273] Okazaki K, Kawazoe K, Takaishi Y. Human platelet aggregation inhibitors from thyme (Thymus vulgaris L.). *Phytother Res.* 2002 Jun;16(4):398-9.

[3274] Jukic M, Politeo O, Maksimmovic M, et al. In vitro acetylcholinesterase inhibitory properties of thymol, carvacrol and their derivatives thymoquinone and thymohydroquinone. *Phytother Res.* 2007;21(3):259-61.

[3275] Askin H, Yildiz M, Ayar A. Effects of Thymol and Carvacrol on Acetylcholinesterase from Drosophila melanogaster. *Acta Physica Polinica A.* 2017;132(3):720-22.

[3276] Medeiros Barreto H, Cerqueira Fontinele F, Pereira de Oliveira A, et al. Phytochemical prospection and modulation of antibiotic activity in vitro by Lippia origanoides H.B.K. in methicillin resistant Staphylococcus aureus. *Biomed Res Int.* 2014;305610:1–7.

[3277] Aelenei P, Miron A, Trifan A, et al. Essential Oils and Their Components as Modulators of Antibiotic Activity against Gram-Negative Bacteria. *Medicines.* 2016;3(19):2-34.

[3278] Magi G, Marini E, Facinelli B. Antimicrobial activity of essential oils and carvacrol, and synergy of carvacrol and erythromycin, against clinical, erythromycin-resistant Group A Streptococci. *Front Microbiol.* 2015;6:165.

[3279] Veras HNH, Rodrigues FFG, Colares AV, et al. Synergistic antibiotic activity of volatile compounds from the essential oil of Lippia sidoides and thymol. *Fitoterapia.* 2012 Apr;83(3):508-12.

[3280] Langeveld WT, Veldhuizen EJ, Burt SA. Synergy between essential oil components and antibiotics: A review. *Crit Rev Microbiol.* 2014 Feb;40(1):76-94.

[3281] Miladinovic DL, Ilic BS, Kocic BD, et al. Antibacterial investigation of thyme essential oil and its main constituents on combination with tetracycline. *J Med Food.* 2015 Aug;18(8):935-7.

[3282] Shin S, Lim S. Antifungal effects of herbal essential oils alone and in combination with ketoconazole against Trichophyton spp. *J Appl Micribiol.* 2004;97:1289-96.

[3283] Kissels W, Wu X, Santos RR, et al. Short communication: Interaction of the isomers carvacrol and thymol with the antibiotics doxycycline and tilmicosin: In vitro effects against pathogenic bacteria commonly found in the respiratory tract of calves. *J Dairy Sci.* 2017 Feb;100(2):970-974.

[3284] Zuzarte M, Alves-Silva JM, Alves M, et al. New insights on the anti-inflammatory potential and safety profile of Thymus carnosus and Thymus camphoratus essential oils and their main compounds. *J Ethnopharmacol.* 2018;225:10-7.

[3285] Aazza S, El-Guendouz S, Miguel MG, et al. Antioxidant, Anti-inflammatory and Anti-hyperglycaemic Activities of Essential Oils from Thymbra capitata, Thymus albicans, Thymus caespititius, Thymus carnosus, Thymus lotocephalus and Thymus mastichina from Portugal. *Nat Prod Commun.* 2016 Jul;11(7):1029-1038.

[3286] Miguel MG, Costa LA, Figueiredo AC, et al. Assessment of the antioxidant ability of Thymus albicans, Th. mastichina, Th. camphoratus and Th. carnosus essential oils by TBARS and Micellar Model systems. *Nat Prod Commun.* 2007 Jan;2(4):399-406.

[3287] Miguel MG, Figueiredo AC, Costa MM, et al. Effect of the volatile constituents isolated from Thymus albicans, Th. mastichina, Th. carnosus and Thymbra capitata in sunflower oil. *Nahrung.* 2003 Dec;47(6):397-402.

[3288] Aazza S, El-Guendouz S, Miguel MG, et al. Antioxidant, Anti-inflammatory and Anti-hyperglycaemic Activities of Essential Oils from Thymbra capitata, Thymus albicans, Thymus caespititius, Thymus carnosus, Thymus lotocephalus and Thymus mastichina from Portugal. *Nat Prod Commun.* 2016 Jul;11(7):1029-1038.

[3289] Satyal P, Powers CN, V. Parducci R, et al. Chemical Composition, Enantiomeric Distribution, and Antifungal Activity of the Oleoresin Essential Oil of Protium amazonicum from Ecuador. *Medicines.* 2014;4(70):1-11.

[3290] Satyal P, Powers CN, V. Parducci R, et al. Chemical Composition, Enantiomeric Distribution, and Antifungal Activity of the Oleoresin Essential Oil of Protium amazonicum from Ecuador. *Medicines.* 2014;4(70):1-11.

[3291] Mayekiso B, Magwa ML, Coopoosamy RM. Variation in the essential oil constituents of Pteronia incana (Asteraceae). *African J Biotechnol.* 2006 Jun;5(12):1220-6.

[3292] Bruns K, Meiertoberens M. Volatile Constituents of Pteronia incana (Compositae). *Flavour Frag J.* 1987;2:157-62.

[3293] Hulley IM, Viljoen AM, Tilney PM, et al. The ethnobotany, leaf anatomy, essential oil variation and biological activity of Pteronia incana (Asteraceae). *South African J Botany.* 2010;76:668-75.

[3294] Mangena T, Muyima NYO. Comparative evaluation of the antimicrobial activities of essential oils of Artemisia afra, Pteronia incana and Rosmarinus officinals on selected bacteria and yeast strains. *Lett Appl Microbiol.* 1999;28:291-6.

[3295] Oyedemi SO, Pirochenva G, Mabinya LV, et al. Compositions and comparisons of antimicrobial potencies of some essential oils and antibiotics against selected bacteria. *African J Biotechnol.* 2008 Nov;7(22):4140-6.

[3296] Burkhard PR, Burkhardt K, Haenggeli CA, et al. Plant-induced seizures: reappearance of an old problem. *J Neurol.* 1999 Aug;246(8):667-70.

[3297] Waldman N. Seizure caused by dermal application of over-the-counter eucalyptus oil head lice preparation. *Clin Toxicol (Phila).* 2011 Oct;49(8):750-1.

[3298] Craig JO. Poisoning by the volatile oils in childhood. *Arch Dis Child.* 1953;28:259-67.

[3299] Mathew T, Kamath V, Kumar RS, et al. Eucalyptus oil inhalation-induced seizure: A novel, underrecognized, preventable cause of acute symptomatic seizure. *Epilepsia Open*. 2017 Jul 4;2(3):350-354.
[3300] Unger M, Frank A. Simultaneous determination of the inhibitory potency of herbal extracts on the activity of six major cytochrome P450 enzymes using liquid chromatography/mass spectrometry and automated online extraction. *Rapid Commun Mass Spectrom*. 2004;18(19):2273-81.
[3301] Jori A, Bianchetti A, Prestini PE, et al. Effect of eucalyptol (1,8-cineole) on the metabolism of other drugs in rats and in man. *Eur J Pharmacol*. 1970;9(3):362-66.
[3302] Kim NH, Hyun SH, Jin CH, et al. Pretreatment with 1,8-cineole potentiates thioacetamide-induced hepatotoxicity and immunosuppression. *Arch Pharm Res*. 2004 Jul;27(7):781-9.
[3303] Jori A, Bianchetti A, Prestini PE, et al. Effect of eucalyptol (1,8-cineole) on the metabolism of other drugs in rats and in man. *Eur J Pharmacol*. 1970;9(3):362-66.
[3304] de Sousa DP, Raphael E, Brocksom U, et al. Sedative effect of monoterpene alcohols in mice: A preliminary screening. *Verlag der Zeitschrift fur Naturforschung*. 2007;62c:563-66.
[3305] Hulley IM, Viljoen AM, Tilney PM, et al. The ethnobotany, leaf anatomy, essential oil variation and biological activity of Pteronia incana (Asteraceae). *South African J Botany*. 2010;76:668-75.
[3306] Mangena T, Muyima NYO. Comparative evaluation of the antimicrobial activities of essential oils of Artemisia afra, Pteronia incana and Rosmarinus officinals on selected bacteria and yeast strains. *Lett Appl Microbiol*. 1999;28:291-6.
[3307] Oyedemi SO, Pirochenva G, Mabinya LV, et al. Compositions and comparisons of antimicrobial potencies of some essential oils and antibiotics against selected bacteria. *African J Biotechnol*. 2008 Nov;7(22):4140-6.
[3308] Perfect Potion. Sandalwood Monograph (May 2016). Available at: https://www.perfectpotion.com.au/news/monographs/sandalwood-monograph/.
[3309] James Cook University. Development of high quality sandalwood oil. Available at: http://www.thesummitvanuatu.com/wp-content/uploads/2016/01/Tony-Page-Paper-on-Vanuatu-swood.pdf.
[3310] Mendes Ferreira da Costa MDC, da Silva AG, Sant'Anna da Silva AP, et al. Essential Oils from Leaves of Medicinal Plants of Brazilian Flora: Chemical Composition and Activity against Candida Species. *Medicines (Basel)*. 2017 Jun; 4(2): 27.
[3311] Andrade MA, dos Santos Azevedo C, Nader Motta F, et al. Essential oils: in vitro activity against Leishmania amazonensis, cytotoxicity and chemical composition. *BMC Complement Altern Med*. 2016;16:444.
[3312] Pereira R, Pereira AL, Ferreira MM, et al. Evaluation of the antimicrobial and antioxidant activity of 7-hydroxy-4', 6-dimethoxy-isoflavone and essential oil from Myroxylon peruiferum L.f. *An Acad Bras Cienc*. 2019;91(2):e20180204.
[3313] Mendes Ferreira da Costa MDC, da Silva AG, Sant'Anna da Silva AP, et al. Essential Oils from Leaves of Medicinal Plants of Brazilian Flora: Chemical Composition and Activity against Candida Species. *Medicines (Basel)*. 2017 Jun; 4(2): 27.
[3314] Andrade MA, dos Santos Azevedo C, Nader Motta F, et al. Essential oils: in vitro activity against Leishmania amazonensis, cytotoxicity and chemical composition. *BMC Complement Altern Med*. 2016;16:444.
[3315] Pereira R, Pereira AL, Ferreira MM, et al. Evaluation of the antimicrobial and antioxidant activity of 7-hydroxy-4', 6-dimethoxy-isoflavone and essential oil from Myroxylon peruiferum L.f. *An Acad Bras Cienc*. 2019;91(2):e20180204.
[3316] Shrestha S, Poudel A, Satyal P, et al. Chemical composition and biological activity of the leaf essential oil of Callistemon citrinus from Nepal. *Am J Essential Oils Nat Prod*. 2015;2(5):29-33.
[3317] Andola HC, Haider SZ, Singh P, et al. Composition of the Essential Oil of Callistemon citrinus (Curtis) Skeels from Uttarakhand (India). *Nat Acad Sci Lett*. 2014;40:389-92.
[3318] Gupta S, Kuman A, Srivastava K, et al. Antimicrobial Activity and Chemical Composition of Callistemon comboynensis and C. citrinus Leaf Essential Oils from the Northern Plains of India. *Nat Prod Commun*. 2008;3(11):1931-4.
[3319] Zandi-Sohani N, Hojjati M, Antonio A, et al. Volatile Composition of the Essential Oil of Callistemon citrinus Leaves from Iran. *J Essent Oil Bearing Plants*. 2012;15(5):703-7.
[3320] Oyedeji OO, Lawal OA, Shode FO, et al. Chemical Composition and Antibacterial Activity of the Essential Oils of Callistemon citrinus and Callistemon viminalis from South Africa. *Molecules*. 2009;14:1990-8.
[3321] Kumar D, Sukapaka M, Babu GDK, et al. Chemical Composition and In Vitro Cytotoxicity of Essential Oils from Leaves and Flowers of Callistemon citrinus from Western Himalayas. *PLOS One*. 2015 Aug 15:1-18.
[3322] Rosas-Burgos EC, El-Zaeddi H, Burgos-Hernandez A, et al. Antifungal activity of Callistemon citrinus essential oils against phyto-pathogenic fungi. *J Food Process Technol*. 2015;6(5):98.
[3323] Larayetan RA, Okoh OO, Sadimenko A, et al. Terpene constituents of the aerial parts, phenolic content, antibacterial potential, free radical scavenging and antioxidant activity of Callistemon citrinus (Curtis) Skeels (Myrtaceae) from Eastern Cape Province of South Africa. *BMC Complement Altern Med*. 2017 Jun 5;17(1):292.
[3324] Craig JO. Poisoning by the volatile oils in childhood. *Arch Dis Child*. 1953;28:259-67.
[3325] Melis K. Bochner A, Janssens G. Accidental nasal eucalyptol and menthol instillation. *Eur J Pediatr*. 1989 Aug;148(8)786-7.
[3326] Day LM, Ozanne-Smith J, Parsons BJ, et al. Eucalyptus oil poisoning among young children: mechanisms of access and the potential prevention. *Aust N Z J Public Health*. 1997 Jun;21(3):297-302.
[3327] Day LM, Ozanne-Smith J, Parsons BJ, et al. Eucalyptus oil poisoning among young children: mechanisms of access and the potential prevention. *Aust N Z J Public Health*. 1997 Jun;21(3):297-302.
[3328] Myott E. Case of eucalyptus poisoning. *Brit M J*. 1906;1:558.
[3329] Hindle RC. Eucalyptus oil ingestion. *N Z Med J*. 1994 May 11;107(977)185-6.
[3330] Tibballs J. Clinical effects and management of eucalyptus oil ingestion in infants and small children. *Med J Aust*. 1995 Aug;163(4):177-80.
[3331] Waldman W, Barwina M, Sein Anand J. Accidental intoxication with eucalyptus oil—a case report. *Przeql Lek*. 2011;68(8):555-6.
[3332] Day LM, Ozanne-Smith J, Parsons BJ, et al. Eucalyptus oil poisoning among young children: mechanisms of access and the potential prevention. *Aust N Z J Public Health*. 1997 Jun;21(3):297-302.
[3333] Burkhard PR, Burkhardt K, Haenggeli CA, et al. Plant-induced seizures: reappearance of an old problem. *J Neurol*. 1999 Aug;246(8):667-70.
[3334] Waldman N. Seizure caused by dermal application of over-the-counter eucalyptus oil head lice preparation. *Clin Toxicol (Phila)*. 2011 Oct;49(8):750-1.
[3335] Craig JO. Poisoning by the volatile oils in childhood. *Arch Dis Child*. 1953;28:259-67.
[3336] Mathew T, Kamath V, Kumar RS, et al. Eucalyptus oil inhalation-induced seizure: A novel, underrecognized, preventable cause of acute symptomatic seizure. *Epilepsia Open*. 2017 Jul 4;2(3):350-354.
[3337] Gray AM, Flatt PR. Antihyperglycemic actions of Eucalyptus globulus (Eucalyptus) are associated with pancreatic and extra pancreatic effects in mice. *J Nutr*. 1998 Dec;128(12):2319–23.

[3338] Basak SS, Candan F. Effect of Laurus nobilis L. Essential Oil and its Main Components on α-glucosidase and Reactive Oxygen Species Scavenging Activity. *Iran J Pharm Res.* 2013 Spring;12(2):367-79.
[3339] Fathiazad F, Mazandarani M, Hamedeyazdan. Phytochemical analysis and antioxidant activity of Hyssopus officinalis L. from Iran. *Adv Pharm Bull.* 2011 Dec;1(2):63–67.
[3340] Gray AM, Flatt PR. Antihyperglycemic actions of Eucalyptus globulus (Eucalyptus) are associated with pancreatic and extra pancreatic effects in mice. *J Nutr.* 1998 Dec;128(12):2319-23.
[3341] Dey B. Chemo-profiling of eucalyptus and study of its hypoglycemic potential. *World J Diabetes.* 2013 Oct 15;4(5):170–76.
[3342] Tsai HH, Lin HW, Chen YL, et al. A review of potential harmful interactions between anticoagulant/antiplatelet agents and Chinese herbal medicines. *PLoS One.* 2013 May 9;8(5):e64255.
[3343] Moharam BA, Jantan I, bin Ahmad F, et al. Antiplatelet Aggregation and Platelet Activating Factor (PAF) Receptor Antagonistic Activities of the Essential Oils of Five Goniothalamus Species. *Molecules.* 2010;15:5124-38.
[3344] Unger M, Frank A. Simultaneous determination of the inhibitory potency of herbal extracts on the activity of six major cytochrome P450 enzymes using liquid chromatography/mass spectrometry and automated online extraction. *Rapid Commun Mass Spectrom.* 2004;18(19):2273-81.
[3345] Jori A, Bianchetti A, Prestini PE, et al. Effect of eucalyptol (1,8-cineole) on the metabolism of other drugs in rats and in man. *Eur J Pharmacol.* 1970;9(3):362-66.
[3346] Kim NH, Hyun SH, Jin CH, et al. Pretreatment with 1,8-cineole potentiates thioacetamide-induced hepatotoxicity and immunosuppression. *Arch Pharm Res.* 2004 Jul;27(7):781-9.
[3347] Jori A, Bianchetti A, Prestini PE, et al. Effect of eucalyptol (1,8-cineole) on the metabolism of other drugs in rats and in man. *Eur J Pharmacol.* 1970;9(3):362-66.
[3348] de Sousa DP, Raphael E, Brocksom U, et al. Sedative effect of monoterpene alcohols in mice: A preliminary screening. *Verlag der Zeitschrift fur Naturforschung.* 2007;62c:563-66.
[3349] Kumar D, Sukapaka M, Babu GDK, et al. Chemical Composition and In Vitro Cytotoxicity of Essential Oils from Leaves and Flowers of Callistemon citrinus from Western Himalayas. *PLOS One.* 2015 Aug 15:1-18.
[3350] Adesina AB, Josephine OO. Inhibitory effects of the volatile oils of Callistemon citrinus (Curtis) Skeels and Eucalyptus citriodora Hook (Myrtaceae) on the acetylcholine induced contraction of isolated rat ileum. *Pak J Pharm Sci.* 2012 Apr;25(2):435-9.
[3351] Gupta S, Kuman A, Srivastava K, et al. Antimicrobial Activity and Chemical Composition of Callistemon comboynensis and C. citrinus Leaf Essential Oils from the Northern Plains of India. *Nat Prod Commun.* 2008;3(11):1931-4.
[3352] Oyedeji OO, Lawal OA, Shode FO, et al. Chemical Composition and Antibacterial Activity of the Essential Oils of Callistemon citrinus and Callistemon viminalis from South Africa. *Molecules.* 2009;14:1990-8.
[3353] Rosas-Burgos EC, El-Zaeddi H, Burgos-Hernandez A, et al. Antifungal activity of Callistemon citrinus essential oils against phyto-pathogenic fungi. *J Food Process Technol.* 2015;6(5):98.
[3354] Larayetan RA, Okoh OO, Sadimenko A, et al. Terpene constituents of the aerial parts, phenolic content, antibacterial potential, free radical scavenging and antioxidant activity of Callistemon citrinus (Curtis) Skeels (Myrtaceae) from Eastern Cape Province of South Africa. *BMC Complement Altern Med.* 2017 Jun 5;17(1):292.
[3355] Larayetan RA, Okoh OO, Sadimenko A, et al. Terpene constituents of the aerial parts, phenolic content, antibacterial potential, free radical scavenging and antioxidant activity of Callistemon citrinus (Curtis) Skeels (Myrtaceae) from Eastern Cape Province of South Africa. *BMC Complement Altern Med.* 2017 Jun 5;17(1):292.
[3356] Shrestha S, Poudel A, Satyal P, et al. Chemical composition and biological activity of the leaf essential oil of Callistemon citrinus from Nepal. *Am J Essential Oils Nat Prod.* 2015;2(5):29-33.
[3357] Zandi-Sohani N, Hojjati M, Carbonell-Barrachina ÁA. Insecticidal and repellent activities of the essential oil of Callistemon citrinus (Myrtaceae) against Callosobruchus maculatus (F.) (Coleoptera: Bruchidae). *Neotrop Entomol.* 2013 Feb;42(1):89-94.
[3358] Thappa RK, Agarwal SG, Kalia NK, et al. Essential Oils from Exotic Eucalyptus: Leaf Oils of Eucalyptus Youmonii, Eucalyptus Macrorhyncha, Eucalyptus Macarthuri, and Eucalyptus Cinera from Northwest Himalays (India). *J Wood Chem Technol.* 1990;10(4):543-9.
[3359] Brophy JJ, Lassak EV, Win S, et al. The volatile oils of Eucalyptus youmanii and Eucalyptus macrorhyncha. *J Sci Soc Thailand.* 1982;8:137-45.
[3360] Tsuneki H, Ma EL, Kobayashi S, et al. Antiangiogenic activity of beta-eudesmol in vitro and in vivo. *Eur J Pharmacol.* 2005 Apr 11;512(2-3):105-15.
[3361] Craig JO. Poisoning by the volatile oils in childhood. *Arch Dis Child.* 1953;28:259-67.
[3362] Melis K. Bochner A, Janssens G. Accidental nasal eucalyptol and menthol instillation. *Eur J Pediatr.* 1989 Aug;148(8)786-7.
[3363] Day LM, Ozanne-Smith J, Parsons BJ, et al. Eucalyptus oil poisoning among young children: mechanisms of access and the potential prevention. *Aust N Z J Public Health.* 1997 Jun;21(3):297-302.
[3364] Day LM, Ozanne-Smith J, Parsons BJ, et al. Eucalyptus oil poisoning among young children: mechanisms of access and the potential prevention. *Aust N Z J Public Health.* 1997 Jun;21(3):297-302.
[3365] Myott E. Case of eucalyptus poisoning. *Brit M J.* 1906;1:558.
[3366] Hindle RC. Eucalyptus oil ingestion. *N Z Med J.* 1994 May 11;107(977)185-6.
[3367] Tibballs J. Clinical effects and management of eucalyptus oil ingestion in infants and small children. *Med J Aust.* 1995 Aug;163(4):177-80.
[3368] Waldman W, Barwina M, Sein Anand J. Accidental intoxication with eucalyptus oil—a case report. *Przeql Lek.* 2011;68(8):555-6.
[3369] Day LM, Ozanne-Smith J, Parsons BJ, et al. Eucalyptus oil poisoning among young children: mechanisms of access and the potential prevention. *Aust N Z J Public Health.* 1997 Jun;21(3):297-302.
[3370] Burkhard PR, Burkhardt K, Haenggeli CA, et al. Plant-induced seizures: reappearance of an old problem. *J Neurol.* 1999 Aug;246(8):667-70.
[3371] Waldman N. Seizure caused by dermal application of over-the-counter eucalyptus oil head lice preparation. *Clin Toxicol (Phila).* 2011 Oct;49(8):750-1.
[3372] Craig JO. Poisoning by the volatile oils in childhood. *Arch Dis Child.* 1953;28:259-67.
[3373] Mathew T, Kamath V, Kumar RS, et al. Eucalyptus oil inhalation-induced seizure: A novel, underrecognized, preventable cause of acute symptomatic seizure. *Epilepsia Open.* 2017 Jul 4;2(3):350-354.
[3374] Unger M, Frank A. Simultaneous determination of the inhibitory potency of herbal extracts on the activity of six major cytochrome P450 enzymes using liquid chromatography/mass spectrometry and automated online extraction. *Rapid Commun Mass Spectrom.* 2004;18(19):2273-81.
[3375] Jori A, Bianchetti A, Prestini PE, et al. Effect of eucalyptol (1,8-cineole) on the metabolism of other drugs in rats and in man. *Eur J Pharmacol.* 1970;9(3):362-66.
[3376] Kim NH, Hyun SH, Jin CH, et al. Pretreatment with 1,8-cineole potentiates thioacetamide-induced hepatotoxicity and immunosuppression. *Arch Pharm Res.* 2004 Jul;27(7):781-9.
[3377] Jori A, Bianchetti A, Prestini PE, et al. Effect of eucalyptol (1,8-cineole) on the metabolism of other drugs in rats and in man. *Eur J Pharmacol.* 1970;9(3):362-66.

[3378] de Sousa DP, Raphael E, Brocksom U, et al. Sedative effect of monoterpene alcohols in mice: A preliminary screening. *Verlag der Zeitschrift fur Naturforschung*. 2007;62c:563-66.
[3379] von Rudloff E. The chemical composition of the volatile oil of the leaves of Picea rubens Sarg. and chemotaxonomic correlations with other North American spruce species : Gas-liquid chromatography of terpenes—XIV. *Phytochem*. 1966;5:331-41.
[3380] Kocak A, Kilic O. Identification of Essential Oil Composition of Four Picea Mill. (Pinaceae) Species from Canada. *J Agric Sci Technol*. 2014:209-14.
[3381] Owokotomo IA, Ekundayo O, Oladosu IA, et al. Analysis of the Essential Oils of Leaves and Stems of Crassocephalum crepidioides Growing in South Western Nigeria. *Int J Chem*. 2012 Apr;4(2):34-7.
[3382] Thankur S, Koundal R, Kumar D, et al. Volatile Composition and Cytotoxic Activity of Aerial Parts of Crassocephalum crepidioides growing in Western Himalaya, India. *Indian J Pharm Sci*. 2019;81(1):167-72.
[3383] Hung NH, Satyal P, Dai DN, et al. Chemical Compositions of Crassocephalum crepidioides Essential Oils and Larvicidal Activities Against Aedes aegypti, Aedes albopictus, and Culex quinquefasciatus. *Nat Prod Commun*. 2019 Jun:1-5.
[3384] Joshi RK. Study on essential oil composition of the roots of Crassocephalum crepidioides (Benth.) S. Moore. *J Chil Chem Soc*. 2014;59(1):2363-5.
[3385] Kohlert C, Schindler G, Marz RW, et al. Systemic availability and pharmacokinetics of thymol in humans. *J Clin Pharmacol*. 2002 Jul;42(7):731-7.
[3386] Toxicology Data Network, National Library of Medicine. Thymol. Available at: http://toxnet.nlm.nih.gov/cgibin/sis/search/a?dbs+hsdb:@term+@DOCNO+866.
[3387] Tognolini M, Barocelli E, Ballabeni V, et al. Comparative screening of plant essential oils; phenylpropanoid moiety as basic core for antiplatelet activity. *Life Sci*. 2006 Feb 23;78(13):1419-32.
[3388] Okazaki K, Kawazoe K, Takaishi Y. Human platelet aggregation inhibitors from thyme (Thymus vulgaris L.). *Phytother Res*. 2002 Jun;16(4):398-9.
[3389] Jukic M, Politeo O, Maksimmovic M, et al. In vitro acetylcholinesterase inhibitory properties of thymol, carvacrol and their derivatives thymoquinone and thymohydroquinone. *Phytother Res*. 2007;21(3):259-61.
[3390] Askin H, Yildiz M, Ayar A. Effects of Thymol and Carvacrol on Acetylcholinesterase from Drosophila melanogaster. *Acta Physica Polinica A*. 2017;132(3):720-22.
[3391] Medeiros Barreto H, Cerqueira Fontinele F, Pereira de Oliveira A, et al. Phytochemical prospection and modulation of antibiotic activity in vitro by Lippia origanoides H.B.K. in methicillin resistant Staphylococcus aureus. *Biomed Res Int*. 2014;305610:1–7.
[3392] Aelenei P, Miron A, Trifan A, et al. Essential Oils and Their Components as Modulators of Antibiotic Activity against Gram-Negative Bacteria. *Medicines*. 2016;3(19):2-34.
[3393] Magi G, Marini E, Facinelli B. Antimicrobial activity of essential oils and carvacrol, and synergy of carvacrol and erythromycin, against clinical, erythromycin-resistant Group A Streptococci. *Front Microbiol*. 2015;6:165.
[3394] Veras HNH, Rodrigues FFG, Colares AV, et al. Synergistic antibiotic activity of volatile compounds from the essential oil of Lippia sidoides and thymol. *Fitoterapia*. 2012 Apr;83(3):508-12.
[3395] Langeveld WT, Veldhuizen EJ, Burt SA. Synergy between essential oil components and antibiotics: A review. *Crit Rev Microbiol*. 2014 Feb;40(1):76-94.
[3396] Miladinovic DL, Ilic BS, Kocic BD, et al. Antibacterial investigation of thyme essential oil and its main constituents on combination with tetracycline. *J Med Food*. 2015 Aug;18(8):935-7.
[3397] Shin S, Lim S. Antifungal effects of herbal essential oils alone and in combination with ketoconazole against Trichophyton spp. *J Appl Micribiol*. 2004;97:1289-96.
[3398] Kissels W, Wu X, Santos RR, et al. Short communication: Interaction of the isomers carvacrol and thymol with the antibiotics doxycycline and tilmicosin: In vitro effects against pathogenic bacteria commonly found in the respiratory tract of calves. *J Dairy Sci*. 2017 Feb;100(2):970-974.
[3399] Delgado IF, Carvalho RR, Nogueira, et al. Study on embryofetotoxicity of b-myrcene in the rat. *Food and Chemical Toxicology*. 1993;31(1):31-5.
[3400] Paumgartten FJ, De-Carvalho RR, Souza CA, et al. Study of the effects of beta-myrcene on rat fertility and general reproductive performance. *Braz J Med Biol Res*. 1998 Jul;31(7):955-65.
[3401] do Vale TG, Furtado EC, Santos JG Jr, et al. Central effects of citral, myrcene and limonene, constituents of essential oil chemotypes from Lippia alba (Mill.) n.e. Brown. *Phytomedicine*. 2002 Dec;9(8):709-14.
[3402] Segelman AB, Sofia RD, Segelman FP, et al. Cannabis sativa L. (marijuana). V. Pharmacological evaluation of marijuana aqueous extract and volatile oil. *J Pharm Sci*. 1974 Jun;63(6):962-4.
[3403] Freitas JC1, Presgrave OA, Fingola FF, et al. Effect of beta-myrcene on pentobarbital sleeping time. *Braz J Med Biol Res*. 1993 May;26(5):519-23.
[3404] De Olivera AC, Ribeiro-Pinto LF, Paumgartten JR. In vitro inhibition of CYP2B1 monoxygenase by beta-myrcene and other monoterpenoid constituents. *Br J Nutr*. 1999;81:289-95.
[3405] De-Oliveira AC1, Ribeiro-Pinto LF, Otto SS, et al. Induction of liver monooxygenases by beta-myrcene. *Toxicology*. 1997 Dec 26;124(2):135-40.
[3406] Owokotomo IA, Ekundayo O, Abayomi TG, et al. In-vitro anti-cholinesterase activity of essential oil from four tropical medicinal plants. *Toxicology Reports*. 2015;2:850-7.
[3407] Owokotomo IA, Owokotomo EP. Anti-bacterial and brine shrimps lethality studies of the essential oils of Crassocephalum crepidioides (Benth S. More) grown in south west Nigeria. *African J Pure Appl Chem*. 2018 Jan;12(1):1-7.
[3408] Hung NH, Satyal P, Dai DN, et al. Chemical Compositions of Crassocephalum crepidioides Essential Oils and Larvicidal Activities Against Aedes aegypti, Aedes albopictus, and Culex quinquefasciatus. *Nat Prod Commun*. 2019 Jun:1-5.
[3409] Taha KF, Shakour ZTA. Chemical Composition and Antibacterial Activity of Volatile Oil of Sequoia sempervirens (Lamb.) Grown in Egypt. *Med Arom Plants*. 2016;5(3):100245.
[3410] Okamoto RA, Ellison BO, Kepner RE. Volatile Terpenes in Sequoia sempervirens Foliage. Changes in Composition during Maturation. *J Agric Food Chem*. 1981;29:324-6.
[3411] Taha KF, Shakour ZTA. Chemical Composition and Antibacterial Activity of Volatile Oil of Sequoia sempervirens (Lamb.) Grown in Egypt. *Med Arom Plants*. 2016;5(3):100245.
[3412] Taha KF, Shakour ZTA. Chemical Composition and Antibacterial Activity of Volatile Oil of Sequoia sempervirens (Lamb.) Grown in Egypt. *Med Arom Plants*. 2016;5(3):100245.
[3413] Safaeian R, Amin G, Azarnivand H. Phytochemistry of Prangos ferulacea (L.) Lindl. in one of the Habitats of Zagros Mountain. *Ecopersia*. 2013;1(1):63-74.
[3414] Safaeian R, Amin G, Azarnivand H. Study of Chemical Compositions of the Essential Oils of Aerial Parts of Prangos ferulacea (L.) Lindly in Growing Stage in the North of Fars Province. *Acta Hort*. 2012;964:71-6.

[3415] Sajjadi SE, Shokoohinia Y, Gholamzadeh S. Chemical composition of essential oil of Prangos ferulacea (L.) Lindl. Roots. *Chemija.* 2011;22(3):178-80.
[3416] Yousefi K, Hamedeyazdan S, Hodaei D, et al. An in vitro ethnopharmacological study on Prangos ferulacea: a wound healing agent. *Bioimpacts.* 2017;7(2):75-82.
[3417] Delnavazi MR, Soleimani M, Hadjiakhoondi A, et al. Isolation of Phenolic Derivatives and Essential Oil Analysis of Prangos ferulacea (L.) Lindl. Aerial Parts. *Iran J Pharm Res.* 2017 Winter;16(Suppl):207-215.
[3418] Amiri H. Essential oil variation of Pr. ferulacea (L.)Lindl. in different stage of plant growth. *J Med Plants.* 2007;21:36-41.
[3419] Sefidkon F. Study essential oils composition of aerial parts and seed of Pr. ferulacea. *J IMAPR.* 2000;5:47-60.
[3420] Aazza S, Lyoussi B, Miguel MG. Antioxidant and antiacetylcholinesterase activities of some commercial essential oils and their major compounds. *Molecules.* 2011 Sep 7;16(9):7672-90.
[3421] Miyazawa M, Yamafuji C. Inhibition of acetylcholinesterase activity by bicyclic monoterpenoids. *J Agric Food Chem.* 2005 Mar 9;53(5):1765-8.
[3422] Bonesi M, Menichini F, Tundis R, et al. Acetylcholinesterase and butyrylcholinesterase inhibitory activity of Pinus species essential oils and their constituents. *J Enzyme Inhib Med Chem.* 2010 Oct;25(5):622-8.
[3423] Yousefi K, Hamedeyazdan S, Hodaei D, et al. An in vitro ethnopharmacological study on Prangos ferulacea: a wound healing agent. *Bioimpacts.* 2017;7(2):75-82.
[3424] Nazemisalman B, Vahabi S, Yazdinejad A, et al. Comparison of antimicrobial effect of Ziziphora tenuior, Dracocephalum moldavica, Ferula gummosa, and Prangos ferulacea essential oil with chlorhexidine on Enterococcus faecalis: An in vitro study. *Dent Res J (Isfahan).* 2018 Mar-Apr;15(2):111-116.
[3425] Yousefi K, Hamedeyazdan S, Hodaei D, et al. An in vitro ethnopharmacological study on Prangos ferulacea: a wound healing agent. *Bioimpacts.* 2017;7(2):75-82.
[3426] Priestap HA, Bennett BC, Quirke JME. Investigation of the essential oils of Bidens pilosa var minor, Bidens alba, and Flaveria linearis. J Essent Oil Res. 2008 Sep/Oct;20:396-402.
[3427] Wat CK, Biswas RK, Graham EA, et al. Ultraviolet-mediated cytotoxic activity of phenylheptatriyne from Bidens pilosa L. *J Nat Prod.* 1979 Jan-Feb;42(1):103-11.
[3428] Gong HH, Kagan J, Seitz R, et al. The phototoxicity of phenylheptatriyne: oxygen-dependent hemolysis of human erythrocytes and inactivation of Escherichia coli. *Photochem Photobiol.* 1988 Jan;47(1):55-63.
[3429] Ketel DH. Chemistry and biology of naturally occurring acetylenes and related compounds. *Flavour Frag J.* 1990 Mar;5(1):69-69.
[3430] Dabiri M, Sefidkon F, Yousefi M, et al. Volatile Components of Pelargonium roseum R. Br. *J Essent Oil Bearing Plants.* 2011;14(1):114-7.
[3431] Azhdarzadeh F, Hojjati M, Didehban ST. Chemical composition and antimicrobial activity of Pelargonium roseum essential oil from southwest of Iran. *J Food Bioprocess Eng.* 2015:47-53.
[3432] Tabari MA, Moghaddam AH, Maggi F, et al. Anxiolytic and antidepressant activities of Pelargonium roseum essential oil on Swiss albino mice: Possible involvement of serotonergic transmission. *Phytother Res.* 2018 Jun;32(6):1014-1022.
[3433] Abouhosseini Tabari M, Youssefi MR. In vitro and in vivo evaluations of Pelargonium roseum essential oil activity against Trichomonas gallinae. *Avicenna J Phytomed.* 2018 Mar-Apr;8(2):136-142.
[3434] Ibrahim SM, El-Denshary ES, Abdallah DM. Geraniol, Alone and in Combination with Pioglitazone, Ameliorates Fructose-Induced Metabolic Syndrome in Rats via the Modulation of Both Inflammatory and Oxidative Stress Status. *PLoS One.* 2015 Feb 13;10(2):e0117516.
[3435] Srinivasan S, Muruganathan U. Antidiabetic efficacy of citronellol, a citrus monoterpene by ameliorating the hepatic key enzymes of carbohydrate metabolism in streptozotocin-induced diabetic rats. *Chem Biol Interact.* 2016 Apr 25;250:38-46.
[3436] Lim S, Shin S. Effects of citronellol and thymol on cell membrane composition of Candida albicans. *Korean J Pharmacognosy.* 2009 Dec;40(4):357-64.
[3437] Shin S, Lim S. Antifungal effects of herbal essential oils alone and in combination with ketoconazole against Trichophyton spp. *J Appl Micribiol.* 2004;97:1289-96.
[3438] Tabari MA, Moghaddam AH, Maggi F, et al. Anxiolytic and antidepressant activities of Pelargonium roseum essential oil on Swiss albino mice: Possible involvement of serotonergic transmission. *Phytother Res.* 2018 Jun;32(6):1014-1022.
[3439] Carmen G, Hancu G. Antimicrobial and Antifungal Activity of Pelargonium roseum Essential Oils. *Adv Pharm Bull.* 2014;4(Suppl 2):511-4.
[3440] Azhdarzadeh F, Hojjati M, Didehban ST. Chemical composition and antimicrobial activity of Pelargonium roseum essential oil from southwest of Iran. *J Food Bioprocess Eng.* 2015:47-53.
[3441] Abouhosseini Tabari M, Youssefi MR. In vitro and in vivo evaluations of Pelargonium roseum essential oil activity against Trichomonas gallinae. *Avicenna J Phytomed.* 2018 Mar-Apr;8(2):136-142.
[3442] Yarahmadi F, Rajabpour A, Zandi Sohani N, et al. Investigating contact toxicity of Geranium and Artemisia essential oils on Bemisia tabaci Gen. *Avicenna J Phytomed.* 2013 Spring;3(2):106-11.
[3443] Pirali-Kheirabadi K, Razzaghi-Abyaneh M, Halajian A. Acaricidal effect of Pelargonium roseum and Eucalyptus globulus essential oils against adult stage of Rhipicephalus (Boophilus) annulatus in vitro. *Vet Parasitol.* 2009 Jun 10;162(3-4):346-9.
[3444] Ueyama Y, Hashimoto S, Nii H, et al. The essential oil from the flowers of Rosa rugosa Thunb. var. plena Regel. *Flavour Fragr J.* 1990;5(4):219-222.
[3445] Zhu L, Li Y, Li B, et al. Aromatic plants and essential constituents. 1996, Hai Feng.
[3446] Ibrahim SM, El-Denshary ES, Abdallah DM. Geraniol, Alone and in Combination with Pioglitazone, Ameliorates Fructose-Induced Metabolic Syndrome in Rats via the Modulation of Both Inflammatory and Oxidative Stress Status. *PLoS One.* 2015 Feb 13;10(2):e0117516.
[3447] Srinivasan S, Muruganathan U. Antidiabetic efficacy of citronellol, a citrus monoterpene by ameliorating the hepatic key enzymes of carbohydrate metabolism in streptozotocin-induced diabetic rats. *Chem Biol Interact.* 2016 Apr 25;250:38-46.
[3448] Lim S, Shin S. Effects of citronellol and thymol on cell membrane composition of Candida albicans. *Korean J Pharmacognosy.* 2009 Dec;40(4):357-64.
[3449] Shin S, Lim S. Antifungal effects of herbal essential oils alone and in combination with ketoconazole against Trichophyton spp. *J Appl Micribiol.* 2004;97:1289-96.
[3450] Hagvall L. Cytochrome P450-mediated activation of the fragrance constituent geraniol forms potent contact allergens. *Tox Appl Pharm.* 2008 Dec;233(2):308-13.
[3451] Weyerstahl P, Kaul VK, Weirauch M, et al. Volatile constituents of Artemisia vestita oil. *Planta Med.* 1987 Feb;53(1):66-72.
[3452] Yang C, Hu DH, Feng Y. Essential oil of Artemisia vestita exhibits potent in vitro and in vivo antibacterial activity: Investigation of the effect of oil on biofilm formation, leakage of potassium ions and survival curve measurement. *Mol Med Reports.* 2015;12:5762-70.
[3453] Chu SS, Liu QR, Liu ZL. Insecticidal activity and chemical composition of the essential oil of Artemisia vestita from China against Sitophilus zeamais. *Biochem System Ecol.* 2010;38:489-92.

[3454] Rather MA, Dar BA, Shah WA, et al. Comprehensive GC–FID, GC–MS and FT-IR spectroscopic analysis of the volatile aroma constituents of Artemisia indica and Artemisia vestita essential oils. *Asian J Chem.* 2017 May;10(Sup2):S3798-803.
[3455] Khine H, Weiss D, Graber N, et al. A cluster of children with seizures caused by camphor poisoning. *Pediatrics.* 2009 May;123(5):1269-72.
[3456] Michiels EA, Mazor SS. Toddler with seizures due to ingesting camphor at an Indian celebration. *Pediatr Emerg Care.* 2010 Aug;26(8):574-75.
[3457] Koren G. Medications which can kill a toddler with one tablet or teaspoonful. *J Toxicol Clin Toxicol.* 1993;31(3):407-13.
[3458] Flaman Z, Pellechia-Clarke S, Bailey B, et al. Unintentional exposure of young children to camphor and eucalyptus oils. *Paediatr Child Health.* 2001 Feb;6(2):80-83.
[3459] Rabl W, Katzgraber F, Steinlechner M. Camphor ingestion for abortion (case report). *Forensic Sci Int.* 1997 Sep 19;89(1-2):137-40.
[3460] Flaman Z, Pellechia-Clarke S, Bailey B, et al. Unintentional exposure of young children to camphor and eucalyptus oils. *Paediatr Child Health.* 2001 Feb;6(2):80-83.
[3461] Manoguerra AS, Erdman AR, Wax PM, et al. Camphor poisoning: an evidence-based practice guideline for out-of- hospital management. *Clin Toxicol (Phila).*2006;44(4):357-70.
[3462] Gibson DE, Moore GP, Pfaff JA. Camphor ingestion. *Am J Emerg Med.* 1989 Jan;7(1):41-43.
[3463] Koppel C, Martends F, Schirop T, et al. Hemoperfusion in acute camphor poisoning. *Intensive Care Med.* 1988;14(4):431-33.
[3464] Burkhard PR, Burkhardt K, Haenggeli CA, et al. Plant-induced seizures: reappearance of an old problem. *J Neurol.* 1999 Aug;246(8):667-70.
[3465] Waldman N. Seizure caused by dermal application of over-the-counter eucalyptus oil head lice preparation. *Clin Toxicol (Phila).* 2011 Oct;49(8):750-1.
[3466] Craig JO. Poisoning by the volatile oils in childhood. *Arch Dis Child.* 1953;28:259-67.
[3467] Mathew T, Kamath V, Kumar RS, et al. Eucalyptus oil inhalation-induced seizure: A novel, underrecognized, preventable cause of acute symptomatic seizure. *Epilepsia Open.* 2017 Jul 4;2(3):350-354.
[3468] Narayan S, Singh N. Camphor poisoning-An unusual cause of seizure. *Med J Armed Forces India.* 2012 Jul;68(3):252-53.
[3469] Chanaranaj KJ, G MV, S M. Camphor poisoning in a child. *Natl Med J India.* 2013 Jan-Feb;26(1):60.
[3470] Olowe SA, Ransome-Kuti O. The risk of jaundice in glucose-6-phosphate dehydrogenase deficient babies exposed to menthol. *Acta Paediatr Scand.* 1980 May;69(3):341-5.
[3471] Dillon Remy M, Manning Alleyne P, Bratt DE, et al. Neonatal jaundice at Port-of-Spain General Hospital abstract. *West Indian Med J.* 1987;36(Suppl):28.
[3472] Haliciglu O, Astarcioglu G, Yaprak I, et al. Toxicity of Salvia officinalis in a newborn and a child: an alarming report. *Pediatr Neurol.* 2011 Oct;45(4):259-60.
[3473] Lachenmeier DW, Walch SG. Epileptic seizure caused by accidental ingestion of sage (Salvia officinalis L.) oil in children: a rare, exceptional case or a threat to public health. *Pediatr Neurol.* 2012 Mar;46(3):201.
[3474] Khine H, Weiss D, Graber N, et al. A cluster of children with seizures caused by camphor poisoning. *Pediatrics.* 2009 May;123(5):1269-72.
[3475] Michiels EA, Mazor SS. Toddler with seizures due to ingesting camphor at an Indian celebration. *Pediatr Emerg Care.* 2010 Aug;26(8):574-75.
[3476] Koren G. Medications which can kill a toddler with one tablet or teaspoonful. *J Toxicol Clin Toxicol.* 1993;31(3):407-13.
[3477] Craig JO. Poisoning by the volatile oils in childhood. *Arch Dis Child.* 1953;28:259-67.
[3478] Day LM, Ozanne-Smith J, Parsons BJ, et al. Eucalyptus oil poisoning among young children: mechanisms of access and the potential prevention. *Aust N Z J Public Health.* 1997 Jun;21(3):297-302.
[3479] Jori A, Bianchetti A, Prestini PE, et al. Effect of eucalyptol (1,8-cineole) on the metabolism of other drugs in rats and in man. *Eur J Pharmacol.* 1970;9(3):362-66.
[3480] Rabl W, Katzgraber F, Steinlechner M. Camphor ingestion for abortion (case report). Forensic Sci Int. 1997 Sep 19;89(1-2):137-40.
[3481] da Silva Costa KC, Bexerra SB, Norte CM, et al. Medicinal plants with teratogenic potential: current considerations. *Braz J Pharm Sci.* 2012;48(3):427-33.
[3482] Brinker F. Herb Contraindications and Drug Interactions. 2nd ed. Sandy, OR. Eclectic Medical Publications. 1998.
[3483] Flaman Z, Pellechia-Clarke S, Bailey B, et al. Unintentional exposure of young children to camphor and eucalyptus oils. *Paediatr Child Health.* 2001 Feb;6(2):80-83.
[3484] Newall CA, Anderson LA, Phillipson JD. "Herbal Medicines: A Guide for Health-care Professionals." London: The Pharmaceutical Press, 1996, 108.
[3485] European Medicines Agency. Community herbal monograph on Eucalyptus globulus Labill., Eucalyptus polybractea R.T. Baker and/or Eucalyptus smithii R.T. Baker, aetheroleum. 2013 Jun. Available at: http://www.ema.europa.eu/docs/en_GB/document_library/Herbal_-_Community_herbal_monograph/2013/07/WC500147008.pdf.
[3486] Day LM, Ozanne-Smith J, Parsons BJ, et al. Eucalyptus oil poisoning among young children: mechanisms of access and the potential prevention. *Aust N Z J Public Health.* 1997 Jun;21(3):297-302.
[3487] Myott E. Case of eucalyptus poisoning. *Brit M J.* 1906;1:558.
[3488] Hindle RC. Eucalyptus oil ingestion. *N Z Med J.* 1994 May 11;107(977)185-6.
[3489] Tibballs J. Clinical effects and management of eucalyptus oil ingestion in infants and small children. *Med J Aust.* 1995 Aug;163(4):177-80.
[3490] Waldman W, Barwina M, Sein Anand J. Accidental ontoxication with eucalyptus oil—a case report. *Przeql Lek.* 2011;68(8):555-6.
[3491] De Vincenzi M, Silano M, De Vincenzi A, et al. Constituents of aromatic plants: eucalyptol. *Fitoterapia.* 2002 Jun;73(3):269-75.
[3492] Burkhard PR, Burkhardt K, Haenggeli CA, et al. Plant-induced seizures: reappearance of an old problem. *J Neurol.* 1999 Aug;246(8):667-70.
[3493] Waldman N. Seizure caused by dermal application of over-the-counter eucalyptus oil head lice preparation. *Clin Toxicol (Phila).* 2011 Oct;49(8):750-1.
[3494] Craig JO. Poisoning by the volatile oils in childhood. *Arch Dis Child.* 1953;28:259-67.
[3495] Mathew T, Kamath V, Kumar RS, et al. Eucalyptus oil inhalation-induced seizure: A novel, underrecognized, preventable cause of acute symptomatic seizure. *Epilepsia Open.* 2017 Jul 4;2(3):350-354.
[3496] Unger M, Frank A. Simultaneous determination of the inhibitory potency of herbal extracts on the activity of six major cytochrome P450 enzymes using liquid chromatography/mass spectrometry and automated online extraction. *Rapid Commun Mass Spectrom.* 2004;18(19):2273-81.
[3497] Jori A, Bianchetti A, Prestini PE, et al. Effect of eucalyptol (1,8-cineole) on the metabolism of other drugs in rats and in man. *Eur J Pharmacol.* 1970;9(3):362-66.
[3498] Kim NH, Hyun SH, Jin CH, et al. Pretreatment with 1,8-cineole potentiates thioacetamide-induced hepatotoxicity and immunosuppression. *Arch Pharm Res.* 2004 Jul;27(7):781-9.
[3499] Jori A, Bianchetti A, Prestini PE, et al. Effect of eucalyptol (1,8-cineole) on the metabolism of other drugs in rats and in man. *Eur J Pharmacol.* 1970;9(3):362-66.

[3500] de Sousa DP, Raphael E, Brocksom U, et al. Sedative effect of monoterpene alcohols in mice: A preliminary screening. *Verlag der Zeitschrift fur Naturforschung*. 2007;62c:563-66.

[3501] Yang C, Hu DH, Feng Y. Essential oil of Artemisia vestita exhibits potent in vitro and in vivo antibacterial activity: Investigation of the effect of oil on biofilm formation, leakage of potassium ions and survival curve measurement. *Mol Med Reports*. 2015;12:5762-70.

[3502] Chu SS, Liu QR, Liu ZL. Insecticidal activity and chemical composition of the essential oil of Artemisia vestita from China against Sitophilus zeamais. *Biochem System Ecol*. 2010;38:489-92.

[3503] Laakso Y, Hiltunen R. The enantiomeric composition of monoterpene hydrocarbons as a chemotaxonomic marker in Abies sachalinensis (fr. schm.) mast. and A. mayriana miy. et kudo needle essential oils. Flavour Frag J. 1994;9(5):223-7.

[3504] Kawai K, Takahashi C, Miyamoto T, et al. Chemical differences between two populations of Abies sachalinensis. *Phytochem*. 1993;32(2):331-4.

[3505] Satou T, Matsuura M, Takahashi M, et al. Anxiolytic-like effect of essential oil extracted from Abies sachalinensis. *Flavour Frag J*. 2011;26:416-20.

[3506] Satou T, Matsuura M, Murakami S, et al. Composition and seasonal variation of the essential oil from Abies sachalinensis from Hokkaido, Japan. *Nat Prod Commun*. 2009 Jun;4(6):845-8.

[3507] Satou T, Matsuura M, Takahashi M, et al. Anxiolytic-like effect of essential oil extracted from Abies sachalinensis. *Flavour Frag J*. 2011;26:416-20.

[3508] Vega-Vela NE, Delgado-Avila WA, Chacon-Sanchez MI. Genetic structure and essential oil diversity of the aromatic shrub Lippia origanoides Kunth (Verbenaceae) in two populations from northern Colombia. *Agronomia Colombiana*. 2013;31(1):7-17.

[3509] Tozin LRS, Marques MOM, Rodrigues TM. Glandular trichome density and essential oil composition in leaves and inflorescences of Lippia origanoides Kunth (Verbenaceae) in the Brazilian Cerrado. *Anais da Academia Brasileira de Ciências*. 2015;87(2):943-53.

[3510] Teixeira ML, Cardoso MDG, Figueiredo ACS, et al. Essential Oils from Lippia origanoides Kunth. and Mentha spicata L.: Chemical Composition, Insecticidal and Antioxidant Activities. *Am J Plant Sci*. 2014;5:1181-90.

[3511] Sarrazin SLF, da Silva LA, de Assuncao APF, et al. Antimicrobial and Seasonal Evaluation of the Carvacrol-Chemotype Oil from Lippia origanoides Kunth. *Molecules*. 2015;20:1860-71.

[3512] Oliveira DR, Leitao GG, Bizzo HR, et al. Chemical and antimicrobial analyses of essential oil of Lippia origanoides H.B.K. *Food Chem*. 2007;101:236-40.

[3513] dos Santos FJB, Lopes JAD, Cito AMGL, et al. Composition and Biological Activity of Essential Oils from Lippia origanoides H.B.K. *J Essent Oil Res*. 2004 Sept/Oct;16:504-6.

[3514] Vicuna GG, Stashenko EE, Fuentes JL. Chemical composition of the Lippia origanoides essential oils and their antigenotoxicity against bleomycin-induced DNA damage. *Fitoterpia*. 2010 Jul;81(5):343-9.

[3515] Teles S, Pereira JA, de Oliveira LM, et al. Lippia origanoides H.B.K. essential oil production, composition, and antioxidant activity under organic and mineral fertilization: Effect of harvest moment. *Ind Crops Prod*. 2014;60:217-25.

[3516] Majolo C, da Rocha SIB, Chagas EC, et al. Chemical composition of Lippia spp. essential oil and antimicrobial activity against Aeromonas hydrophila. *Agric Res*. 2017;48:2380-7.

[3517] da Silva AP, Silva NF, Andrade EHA, et al. Tyrosinase inhibitory activity, molecular docking studies and antioxidant potential of chemotypes of Lippia origanoides (Verbenaceae) essential oils. *PLoS One*. 2017 May 1;12(5):e0175598.

[3518] Escobar P, Milena Leal S, Herrera LV, et al. Chemical composition and antiprotozoal activities of Colombian Lippia spp essential oils and their major components. *Mem Inst Oswaldo Cruz*. 2010 Mar;105(2):184-90.

[3519] Stashenko EE, Martínez JR, Ruíz CA, et al. Lippia origanoides chemotype differentiation based on essential oil GC-MS and principal component analysis. *J Sep Sci*. 2010 Jan;33(1):93-103.

[3520] Meneses R, Ocazionez RE, Martínez JR, et al. Inhibitory effect of essential oils obtained from plants grown in Colombia on yellow fever virus replication in vitro. *Ann Clin Microbiol Antimicrob*. 2009 Mar 6;8:8.

[3521] Perera WH, Bizzo HR, Gama PE, et al. Essential oil constituents from high altitude Brazilian species with antimicrobial activity: Baccharis parvidentata Malag., Hyptis monticola Mart. ex Benth. and Lippia origanoides Kunth. *J Essent Oil Res*. 2016;2:109-16.

[3522] Domaracky M, Rehak P, Juhas S, et al. Effects of selected plant essential oils on the growth and development of mouse preimplantation embryos in vivo. *Physiol Res*. 2007;56(1):97-104.

[3523] Lemhadri A, Zeggwagh NA, Maghrani M, et al. Anti-hypoglycaemic activity of the aqueous extract of Origanum vulgare growing wild in Tafilalet region. *J Ethnopharmacol*. 2004 Jun;92(2-3):251-6.

[3524] Bayramoglu G, Senturk H, Bayramoglu A, et al. Carvacrol partially reverses symptoms of diabetes in STZ-induced diabetic rats. *Cytotechnology*. 2014 Mar;66(2):251–257.

[3525] Ezhumalai M, Radhiga T, Pugalendi KV. Antihyperglycemic effect of carvacrol in combination with rosiglitazone in high-fat diet-induced type 2 diabetic C57BL/6J mice. *Mol Cell Biochem*. 2014 Jan;385(1-2):23-31.

[3526] Umasankar K, Ramya DR, Nambikkairaj B. Antidiabetic and antidyslipidemic nature of carvacrol, a monoterpenic phenol studied in high-fat - fed and low-dose streptozotocin-induced experimental diabetic rats. *J Pharm Res*. 2015;9(8):484-90.

[3527] Tognolini M, Barocelli E, Ballabeni V, et al. Comparative screening of plant essential oils; phenylpropanoid moiety as basic core for antiplatelet activity. *Life Sci*. 2006 Feb 23;78(13):1419-32.

[3528] Tsai HH, Lin HW, Chen YL, et al. A review of potential harmful interactions between anticoagulant/antiplatelet agents and Chinese herbal medicines. *PLoS One*. 2013 May 9;8(5):e64255.

[3529] Karkabounas S, Kostoula OK, Daskalou T, et al. Anticarcinogenic and antiplatelet effects of carvacrol. *Exp Oncol*. 2006 Jun;28(2):121-5.

[3530] Baser KH. Biological and pharmacological activities of carvacrol and carvacrol bearing essential oils. *Curr Pharm Des*. 2008;14(29):3106-19.

[3531] Okazaki K, Kawazoe K, Takaishi Y. Human platelet aggregation inhibitors from thyme (Thymus vulgaris L.). *Phytother Res*. 2002 Jun;16(4):398-9.

[3532] Jukic M, Politeo O, Maksimmovic M, et al. In vitro acetylcholinesterase inhibitory properties of thymol, carvacrol and their derivatives thymoquinone and thymohydroquinone. *Phytother Res*. 2007;21(3):259-61.

[3533] Askin H, Yildiz M, Ayar A. Effects of Thymol and Carvacrol on Acetylcholinesterase from Drosophila melanogaster. *Acta Physica Polinica A*. 2017;132(3):720-22.

[3534] Medeiros Barreto H, Cerqueira Fontinele F, Pereira de Oliveira A, et al. Phytochemical prospection and modulation of antibiotic activity in vitro by Lippia origanoides H.B.K. in methicillin resistant Staphylococcus aureus. *Biomed Res Int*. 2014;305610:1–7.

[3535] Aelenei P, Miron A, Trifan A, et al. Essential Oils and Their Components as Modulators of Antibiotic Activity against Gram-Negative Bacteria. *Medicines*. 2016;3(19):2-34.

[3536] Magi G, Marini E, Facinelli B. Antimicrobial activity of essential oils and carvacrol, and synergy of carvacrol and erythromycin, against clinical, erythromycin-resistant Group A Streptococci. *Front Microbiol*. 2015;6:165.

[3537] Veras HNH, Rodrigues FFG, Colares AV, et al. Synergistic antibiotic activity of volatile compounds from the essential oil of Lippia sidoides and thymol. *Fitoterapia.* 2012 Apr;83(3):508-12.
[3538] Langeveld WT, Veldhuizen EJ, Burt SA. Synergy between essential oil components and antibiotics: A review. *Crit Rev Microbiol.* 2014 Feb;40(1):76-94.
[3539] Kohlert C, Schindler G, Marz RW, et al. Systemic availability and pharmacokinetics of thymol in humans. *J Clin Pharmacol.* 2002 Jul;42(7):731-7.
[3540] Toxicology Data Network, National Library of Medicine. Thymol. Available at: http://toxnet.nlm.nih.gov/cgibin/sis/search/a?dbs+hsdb:@term+@DOCNO+866.
[3541] Tognolini M, Barocelli E, Ballabeni V, et al. Comparative screening of plant essential oils; phenylpropanoid moiety as basic core for antiplatelet activity. *Life Sci.* 2006 Feb 23;78(13):1419-32.
[3542] Okazaki K, Kawazoe K, Takaishi Y. Human platelet aggregation inhibitors from thyme (Thymus vulgaris L.). *Phytother Res.* 2002 Jun;16(4):398-9.
[3543] Jukic M, Politeo O, Maksimmovic M, et al. In vitro acetylcholinesterase inhibitory properties of thymol, carvacrol and their derivatives thymoquinone and thymohydroquinone. *Phytother Res.* 2007;21(3):259-61.
[3544] Askin H, Yildiz M, Ayar A. Effects of Thymol and Carvacrol on Acetylcholinesterase from Drosophila melanogaster. *Acta Physica Polinica A.* 2017;132(3):720-22.
[3545] Medeiros Barreto H, Cerqueira Fontinele F, Pereira de Oliveira A, et al. Phytochemical prospection and modulation of antibiotic activity in vitro by Lippia origanoides H.B.K. in methicillin resistant Staphylococcus aureus. *Biomed Res Int.* 2014;305610:1–7.
[3546] Aelenei P, Miron A, Trifan A, et al. Essential Oils and Their Components as Modulators of Antibiotic Activity against Gram-Negative Bacteria. *Medicines.* 2016;3(19):2-34.
[3547] Magi G, Marini E, Facinelli B. Antimicrobial activity of essential oils and carvacrol, and synergy of carvacrol and erythromycin, against clinical, erythromycin-resistant Group A Streptococci. *Front Microbiol.* 2015;6:165.
[3548] Veras HNH, Rodrigues FFG, Colares AV, et al. Synergistic antibiotic activity of volatile compounds from the essential oil of Lippia sidoides and thymol. *Fitoterapia.* 2012 Apr;83(3):508-12.
[3549] Langeveld WT, Veldhuizen EJ, Burt SA. Synergy between essential oil components and antibiotics: A review. *Crit Rev Microbiol.* 2014 Feb;40(1):76-94.
[3550] Miladinovic DL, Ilic BS, Kocic BD, et al. Antibacterial investigation of thyme essential oil and its main constituents on combination with tetracycline. *J Med Food.* 2015 Aug;18(8):935-7.
[3551] Shin S, Lim S. Antifungal effects of herbal essential oils alone and in combination with ketoconazole against Trichophyton spp. *J Appl Micribiol.* 2004;97:1289-96.
[3552] Kissels W, Wu X, Santos RR, et al. Short communication: Interaction of the isomers carvacrol and thymol with the antibiotics doxycycline and tilmicosin: In vitro effects against pathogenic bacteria commonly found in the respiratory tract of calves. *J Dairy Sci.* 2017 Feb;100(2):970-974.
[3553] Craig JO. Poisoning by the volatile oils in childhood. *Arch Dis Child.* 1953;28:259-67.
[3554] Melis K. Bochner A, Janssens G. Accidental nasal eucalyptol and menthol instillation. *Eur J Pediatr.* 1989 Aug;148(8)786-7.
[3555] Day LM, Ozanne-Smith J, Parsons BJ, et al. Eucalyptus oil poisoning among young children: mechanisms of access and the potential prevention. *Aust N Z J Public Health.* 1997 Jun;21(3):297-302.
[3556] Day LM, Ozanne-Smith J, Parsons BJ, et al. Eucalyptus oil poisoning among young children: mechanisms of access and the potential prevention. *Aust N Z J Public Health.* 1997 Jun;21(3):297-302.
[3557] Myott E. Case of eucalyptus poisoning. *Brit M J.* 1906;1:558.
[3558] Hindle RC. Eucalyptus oil ingestion. *N Z Med J.* 1994 May 11;107(977)185-6.
[3559] Tibballs J. Clinical effects and management of eucalyptus oil ingestion in infants and small children. *Med J Aust.* 1995 Aug;163(4):177-80.
[3560] Waldman W, Barwina M, Sein Anand J. Accidental intoxication with eucalyptus oil—a case report. *Przeql Lek.* 2011;68(8):555-6.
[3561] Day LM, Ozanne-Smith J, Parsons BJ, et al. Eucalyptus oil poisoning among young children: mechanisms of access and the potential prevention. *Aust N Z J Public Health.* 1997 Jun;21(3):297-302.
[3562] Burkhard PR, Burkhardt K, Haenggeli CA, et al. Plant-induced seizures: reappearance of an old problem. *J Neurol.* 1999 Aug;246(8):667-70.
[3563] Waldman N. Seizure caused by dermal application of over-the-counter eucalyptus oil head lice preparation. *Clin Toxicol (Phila).* 2011 Oct;49(8):750-1.
[3564] Craig JO. Poisoning by the volatile oils in childhood. *Arch Dis Child.* 1953;28:259-67.
[3565] Mathew T, Kamath V, Kumar RS, et al. Eucalyptus oil inhalation-induced seizure: A novel, underrecognized, preventable cause of acute symptomatic seizure. *Epilepsia Open.* 2017 Jul 4;2(3):350-354.
[3566] Unger M, Frank A. Simultaneous determination of the inhibitory potency of herbal extracts on the activity of six major cytochrome P450 enzymes using liquid chromatography/mass spectrometry and automated online extraction. *Rapid Commun Mass Spectrom.* 2004;18(19):2273-81.
[3567] Jori A, Bianchetti A, Prestini PE, et al. Effect of eucalyptol (1,8-cineole) on the metabolism of other drugs in rats and in man. *Eur J Pharmacol.* 1970;9(3):362-66.
[3568] Kim NH, Hyun SH, Jin CH, et al. Pretreatment with 1,8-cineole potentiates thioacetamide-induced hepatotoxicity and immunosuppression. *Arch Pharm Res.* 2004 Jul;27(7):781-9.
[3569] Jori A, Bianchetti A, Prestini PE, et al. Effect of eucalyptol (1,8-cineole) on the metabolism of other drugs in rats and in man. *Eur J Pharmacol.* 1970;9(3):362-66.
[3570] de Sousa DP, Raphael E, Brocksom U, et al. Sedative effect of monoterpene alcohols in mice: A preliminary screening. *Verlag der Zeitschrift fur Naturforschung.* 2007;62c:563-66.
[3571] Burkhard PR, Burkhardt K, Haenggeli CA, et al. Plant-induced seizures: reappearance of an old problem. *J Neurol.* 1999 Aug;246(8):667-70.
[3572] Waldman N. Seizure caused by dermal application of over-the-counter eucalyptus oil head lice preparation. *Clin Toxicol (Phila).* 2011 Oct;49(8):750-1.
[3573] Craig JO. Poisoning by the volatile oils in childhood. *Arch Dis Child.* 1953;28:259-67.
[3574] Mathew T, Kamath V, Kumar RS, et al. Eucalyptus oil inhalation-induced seizure: A novel, underrecognized, preventable cause of acute symptomatic seizure. *Epilepsia Open.* 2017 Jul 4;2(3):350-354.
[3575] Brehm-Stecher BF, Johnson EA. Sensitization of Staphylococcus aureus and Escherichia coli to antibiotics by the sesquiterpenoids nerolidol, farnesol, bisabolol, and apritone. *Antimicrob Agents Chemother.* 2003 Oct;47(10):3357-60.
[3576] Špičáková A, Szotáková B, Dimunová D, et al. Nerolidol and Farnesol Inhibit Some Cytochrome P450 Activities but Did Not Affect Other Xenobiotic-Metabolizing Enzymes in Rat and Human Hepatic Subcellular Fractions. *Molecules.* 2017 Mar 24;22(4). pii: E509.

[3577] Vicuna GG, Stashenko EE, Fuentes JL. Chemical composition of the Lippia origanoides essential oils and their antigenotoxicity against bleomycin-induced DNA damage. *Fitoterpia.* 2010 Jul;81(5):343-9.
[3578] Menezes PMN, Brito MC, de Paiva GO, et al. Relaxant effect of Lippia origanoides essential oil in guinea-pig trachea smooth muscle involves potassium channels and soluble guanylyl cyclase. *J Ethnopharmacol.* 2018 Jun 28;220:16-25.
[3579] da Silva AP, Silva NF, Andrade EHA, et al. Tyrosinase inhibitory activity, molecular docking studies and antioxidant potential of chemotypes of Lippia origanoides (Verbenaceae) essential oils. *PLoS One.* 2017 May 1;12(5):e0175598.
[3580] Meneses R, Ocazionez RE, Martínez JR, et al. Inhibitory effect of essential oils obtained from plants grown in Colombia on yellow fever virus replication in vitro. *Ann Clin Microbiol Antimicrob.* 2009 Mar 6;8:8.
[3581] Sarrazin SLF, da Silva LA, de Assuncao APF, et al. Antimicrobial and Seasonal Evaluation of the Carvacrol-Chemotype Oil from Lippia origanoides Kunth. *Molecules.* 2015;20:1860-71.
[3582] Oliveira DR, Leitao GG, Bizzo HR, et al. Chemical and antimicrobial analyses of essential oil of Lippia origanoides H.B.K. *Food Chem.* 2007;101:236-40.
[3583] Majolo C, da Rocha SIB, Chagas EC, et al. Chemical composition of Lippia spp. essential oil and antimicrobial activity against Aeromonas hydrophila. *Agric Res.* 2017;48:2380-7.
[3584] Hernandes C, Pina ES, Taleb-Contini SH, et al. Lippia origanoides essential oil: an efficient and safe alternative to preserve food, cosmetic and pharmaceutical products. *J Appl Microbiol.* 2017 Apr;122(4):900-910.
[3585] Perera WH, Bizzo HR, Gama PE, et al. Essential oil constituents from high altitude Brazilian species with antimicrobial activity: Baccharis parvidentata Malag., Hyptis monticola Mart. ex Benth. and Lippia origanoides Kunth. *J Essent Oil Res.* 2016;2:109-16.
[3586] Teixeira ML, Cardoso MDG, Figueiredo ACS, et al. Essential Oils from Lippia origanoides Kunth. and Mentha spicata L.: Chemical Composition, Insecticidal and Antioxidant Activities. *Am J Plant Sci.* 2014;5:1181-90.
[3587] Teles S, Pereira JA, de Oliveira LM, et al. Lippia origanoides H.B.K. essential oil production, composition, and antioxidant activity under organic and mineral fertilization: Effect of harvest moment. *Ind Crops Prod.* 2014;60:217-25.
[3588] Teixeira ML, Cardoso MDG, Figueiredo ACS, et al. Essential Oils from Lippia origanoides Kunth. and Mentha spicata L.: Chemical Composition, Insecticidal and Antioxidant Activities. *Am J Plant Sci.* 2014;5:1181-90.
[3589] Chagas AC, Oliveira MC, Giglioti R, et al. Efficacy of 11 Brazilian essential oils on lethality of the cattle tick Rhipicephalus (Boophilus) microplus. *Ticks Tick Borne Dis.* 2016 Apr;7(3):427-32.
[3590] Borges AR, Aires JR, Higino TM, et al. Trypanocidal and cytotoxic activities of essential oils from medicinal plants of Northeast of Brazil. *Exp Parasitol.* 2012 Oct;132(2):123-8.
[3591] Escobar P, Milena Leal S, Herrera LV, et al. Chemical composition and antiprotozoal activities of Colombian Lippia spp essential oils and their major components. *Mem Inst Oswaldo Cruz.* 2010 Mar;105(2):184-90.
[3592] Gowda VSV. Global Emerging Trends on Sustainable Production of Natural Sandalwood. Available at: http://www.fao.org/3/ap001e/ap001e15.pdf.
[3593] Brophy JJ, Goldsack RJ. Doran JC, et al. Heartwood Oils of Santalum macgregorii F. Muell. (PNG Sandalwood). 2009 May/Jun;21:249-53.
[3594] Howes MJR, Simmonds MSJ, Kite GC. Evaluation of the quality of sandalwood essential oils by gas chromatography–mass spectrometry. *J Chromatography A.* 2004;1028:307-12.
[3595] Butaud JF, Raharivelomanana P, Bianchini JP, et al. A New Chemotype of Sandalwood (Santalum insulare Bertero ex A. DC.) from Marquesas Islands. *J Essent Oil Res.* 2003 Sep/Oct;15:323-6.
[3596] Braun NA, Butaud JF, Bianchini JP, et al. Eastern Polynesian Sandalwood Oil (Santalum insulare Bertero ex A. DC.) – a Detailed Investigation. *Nat Prod Commun.* 2007;2(6):695-9.
[3597] Formacek K, Kubeckza RA. Essential Oils Analysis by Capillary Gas Chromatography and Carbon-13 NMR Spectroscopy. 1972. John Wiley, New York.
[3598] Zhu L, Li Y, Li B, et al. Aromatic plants and essential constituents. 1996, Hai Feng.
[3599] Simic A, Sokovic MD, Ristic M, et al. The Chemical Composition of some Lauraceae Essential Oils and Their Antifungal Activities. *Phytother Res.* 2004;18:713-7.
[3600] Zwaving JH, Bos R. Composition of the Essential Oil from the Root of Sassafras albidum (Nutt.) Nees. *J Essent Oil Res.* 1996;8(2):193-5.
[3601] Council of Europe's Committee of Experts on Cosmetic Products. Sassafras albidum essential oil, monograph no. 36. Active ingredients used in cosmetics: safety survey, Council of Europe Publishing. 2008; 347-350.
[3602] Mossi AJ, Zanella CA, Kubiak G, et al. Essential oil of Ocotea odorifera: An alternative against Sitophilus zeamais. *Renewable Agric Food Sys.* 2014 Jun;29(2):161-6.
[3603] Alcoba AET, de Melo DC, de Andrade PM, et al. Chemical composition and in vitro antileishmanial and cytotoxic activities of the essential oils of Ocotea dispersa (Nees) Mez and Ocotea odorifera (Vell) Rohwer (Lauraceae). *Nat Prod Res.* 2017 Oct 12:1-4. [Epub ahead of print]
[3604] Sukchareon O, Sirirote P, Thanaboripat D. Control of aflatoxigenic strains by Cinnamomum porrectum essential oil. *J Food Sci Technol.* 2017 Aug;54(9):2929-35.
[3605] Palanuvej C, Werawatganone P, Lipipun V, et al. Chemical composition and antimicrobial activity against Candida albicans of essential oil from leaves of Cinnamomum porrectum. *Thai J Health Res.* 2006 Jan;20(1):69-76.
[3606] Subki SYM, Jamal JA, Husain K, et al. Characterisation of leaf essential oils of three Cinnamomum species from Malaysia by gas chromatography and multivariate data analysis. *Pharmacognosy J.* 2013;5:22-9.
[3607] Vesselinovitch SD, Rao KVN, Mihalovich N. Transplacental and Lactational Carcinogenesis by Safrole. *Cancer Res.* 1979 Nov;39:4378-80.
[3608] Lu LJ, Disher RM, Randerath K. Differences in the covalent binding of benzo[a]pyrene, safrole, 1'-hydroxysafrole, and 4-aminobiphenyl to DNA of pregnant and non-pregnant mice. *Cancer Lett.* 1986 Apr;31(1):43-52.
[3609] NIH. Report on carcinogenesis, fourteenth edition. Safrole. Available at: https://ntp.niehs.nih.gov/ntp/roc/content/profiles/safrole.pdf.
[3610] Tsai HH, Lin HW, Chen YL, et al. A review of potential harmful interactions between anticoagulant/antiplatelet agents and Chinese herbal medicines. *PLoS One.* 2013 May 9;8(5):e64255.
[3611] Ioannides C, Delaforge M, Parke DV. Interactions of safrole and isosafrole and their metabolites with cytochromes P-450. *Chem Biol Interact.* 1985 May;53(3):303-11.
[3612] Alcoba AET, de Melo DC, de Andrade PM, et al. Chemical composition and in vitro antileishmanial and cytotoxic activities of the essential oils of Ocotea dispersa (Nees) Mez and Ocotea odorifera (Vell) Rohwer (Lauraceae). *Nat Prod Res.* 2017 Oct 12:1-4. [Epub ahead of print]
[3613] Simic A, Sokovic MD, Ristic M, et al. The Chemical Composition of some Lauraceae Essential Oils and Their Antifungal Activities. *Phytother Res.* 2004;18:713-7.
[3614] Sukchareon O, Sirirote P, Thanaboripat D. Control of aflatoxigenic strains by Cinnamomum porrectum essential oil. *J Food Sci Technol.* 2017 Aug;54(9):2929-35.
[3615] Palanuvej C, Werawatganone P, Lipipun V, et al. Chemical composition and antimicrobial activity against Candida albicans of essential oil from leaves of Cinnamomum porrectum. *Thai J Health Res.* 2006 Jan;20(1):69-76.

[3616] Phongpaichit S, Kummee S, Nilrat L, et al. Palanuvej C, Werawatganone P, Lipipun V, et al. Antimicrobial activity of the oil from the root of Cinnamomum porrectum. *Thai Herbs II*. 2006 Mar;29(Suppl. 1):11-16.
[3617] Mossi AJ, Zanella CA, Kubiak G, et al. Essential oil of Ocotea odorifera: An alternative against Sitophilus zeamais. *Renewable Agric Food Sys*. 2014 Jun;29(2):161-6.
[3618] Kwon SY, Kim KL. Aroma Characteristics of Essential Oil from Citrus unshiu Juice Concentrate By-products in Jeju. *Kor J Asthet Cosmetol*. 2015 Apr;13(2):251-9.
[3619] Tao NG, Liu YJ, Zhang JH, et al. Chemical composition of essential oil from the peel of Satsuma mandarin. *African J Biotech*. 2008 May 2;7(9):1261-4.
[3620] Kita Y, Nakatani Y, Kobayashi A, et al. Composition of Peel Oil from Citrus Unshiu. *Agr Biol Chem*. 1969;33(11):1559-65.
[3621] Yang XN, Kang SC. Chemical composition, antioxidant and antibacterial activities of essential oil from Korean Citrus unshiu peel. 2013;2(3):42-9.
[3622] Kekelidze NA, Dzhanikashvili MI, Tatarishvili AN, et al. Essential oils of the leaves of grafts of Citrus unshiu.
[3623] Yang E, Kim S, Oh T, et al. Peucedanum japonicum andCitrus unshiu essential oils inhibit the growth of antibiotic-resistant skin pathogens. *Annals Microbiol*. 2009;59(3):623-8.
[3624] Dugrand-Judek A, Olry A, Hehn A, et al. The Distribution of Coumarins and Furanocoumarins in Citrus Species Closely Matches Citrus Phylogeny and Reflects the Organization of Biosynthetic Pathways. *PLoS One*. 2015;10(11):e0142757.
[3625] Yang E, Kim S, Oh T, et al. Peucedanum japonicum andCitrus unshiu essential oils inhibit the growth of antibiotic-resistant skin pathogens. *Annals Microbiol*. 2009;59(3):623-8.
[3626] Yang XN, Kang SC. Chemical composition, antioxidant and antibacterial activities of essential oil from Korean Citrus unshiu peel. 2013;2(3):42-9.
[3627] Yang E, Kim S, Oh T, et al. Peucedanum japonicum andCitrus unshiu essential oils inhibit the growth of antibiotic-resistant skin pathogens. *Annals Microbiol*. 2009;59(3):623-8.
[3628] Yang XN, Kang SC. Chemical composition, antioxidant and antibacterial activities of essential oil from Korean Citrus unshiu peel. 2013;2(3):42-9.
[3629] Reis Simas DL, Merida-Reyes MS, Alejandro Munoz M, et al. Chemical Composition and Evaluation of Antinociceptive Activity of the Essential Oil of Stevia serrata Cav. from Guatemala. *Nat Prod Res*. 2017 Nov 13:1-3. [Epub ahead of print]
[3630] Reis Simas DL, Merida-Reyes MS, Alejandro Munoz M, et al. Chemical Composition and Evaluation of Antinociceptive Activity of the Essential Oil of Stevia serrata Cav. from Guatemala. *Nat Prod Res*. 2017 Nov 13:1-3. [Epub ahead of print]
[3631] Shah AJ, Gilani AH, Abbas K, et al. Studies on the Chemical Composition and Possible Mechanisms Underlying the Antispasmodic and Bronchodilatory Activities of the Essential Oil of Artemisia maritima L. *Arch Pharm Res*. 2011;34(8):1227-38.
[3632] Stappen I, Wanner J, Tabanca N, et al. Chemical Composition and Biological Effects of Artemisia maritima and Artemisia nilagirica Essential Oils from Wild Plants of Western Himalaya. *Planta Med*. 2014;80:1079-87.
[3633] Mohan M, Pandey AK, Singh P, et al. Evaluation of Artemisia maritima L. Essential Oil for its Chemical and Biological Properties against Some Foodborne Pathogens. *Analytical Chem Letters*. 2016;6(1):47-54.
[3634] Mathela CS, Kharkwal H, Shah GC. Essential Oil Composition of Some Himalayan Artemisia Species. *J Essent Oil Res*. 1994 Jul/Aug;6:345-8.
[3635] Sah S, Lohani H, Narayan O, et al. Volatile Constituents of Artemisia maritima Linn. Grown in Garhwal Himalaya. *J Essential Oil Bearing Plants*. 2010;13(5):603-6.
[3636] Khine H, Weiss D, Graber N, et al. A cluster of children with seizures caused by camphor poisoning. *Pediatrics*. 2009 May;123(5):1269-72.
[3637] Michiels EA, Mazor SS. Toddler with seizures due to ingesting camphor at an Indian celebration. *Pediatr Emerg Care*. 2010 Aug;26(8):574-75.
[3638] Koren G. Medications which can kill a toddler with one tablet or teaspoonful. *J Toxicol Clin Toxicol. 1*993;31(3):407-13.
[3639] Flaman Z, Pellechia-Clarke S, Bailey B, et al. Unintentional exposure of young children to camphor and eucalyptus oils. *Paediatr Child Health*. 2001 Feb;6(2):80-83.
[3640] Haliciglu O, Astarcioglu G, Yaprak I, et al. Toxicity of Salvia officinalis in a newborn and a child: an alarming report. *Pediatr Neurol*. 2011 Oct;45(4):259-60.
[3641] Lachenmeier DW, Walch SG. Epileptic seizure caused by accidental ingestion of sage (Salvia officinalis L.) oil in children: a rare, exceptional case or a threat to public health. *Pediatr Neurol*. 2012 Mar;46(3):201.
[3642] Rabl W, Katzgraber F, Steinlechner M. Camphor ingestion for abortion (case report). *Forensic Sci Int*. 1997 Sep 19;89(1-2):137-40.
[3643] Flaman Z, Pellechia-Clarke S, Bailey B, et al. Unintentional exposure of young children to camphor and eucalyptus oils. *Paediatr Child Health*. 2001 Feb;6(2):80-83.
[3644] da Silva Costa KC, Bexerra SB, Norte CM, et al. Medicinal plants with teratogenic potential: current considerations. *Braz J Pharm Sci*. 2012;48(3):427-33.
[3645] Brinker F. Herb Contraindications and Drug Interactions. 2nd ed. Sandy, OR. Eclectic Medical Publications. 1998.
[3646] Manoguerra AS, Erdman AR, Wax PM, et al. Camphor poisoning: an evidence-based practice guideline for out-of- hospital management. *Clin Toxicol (Phila)*.2006;44(4):357-70.
[3647] Gibson DE, Moore GP, Pfaff JA. Camphor ingestion. *Am J Emerg Med*. 1989 Jan;7(1):41-43.
[3648] Koppel C, Martends F, Schirop T, et al. Hemoperfusion in acute camphor poisoning. *Intensive Care Med*. 1988;14(4):431-33.
[3649] Burkhard PR, Burkhardt K, Haenggeli CA, et al. Plant-induced seizures: reappearance of an old problem. *J Neurol*. 1999 Aug;246(8):667–70.
[3650] Millet Y, Jouglard J, Steinmetz MD, et al. Toxicity of some essential plant oils. Clinical and experimental study. *Clin Toxicol*. 1981 Dec;18(12):1485–98.
[3651] Charanraj KJ, G MV, S M. Camphor poisoning in a child. *Natl Med J India*. 2013 Jan-Feb;26(1):60.
[3652] Narayan S, Singh N. Camphor poisoning-An unusual cause of seizure. *Med J Armed Forces India*. 2012 Jul;68(3):252–53.
[3653] Uc A, Bishop WP, Sanders KD. Camphor hepatotoxicity. *South Med J*. 2000;93:596-98.
[3654] Uc A, Bishop WP, Sanders KD. Camphor hepatotoxicity. *South Med J*. 2000;93:596-98.
[3655] Frohne D. Giftpflanzen: Cupressaceae. Stuttgart: Wissenschaftliche Verlagsgesellschaft mbH; 1997. pp. 153–6.
[3656] Dolan LC, Matulka RA, Burdock GA. Naturally Occurring Food Toxins. *Toxins (Basel)*. 2010 Sep;2(9):2289–2332.
[3657] Olowe SA, Ransome-Kuti O. The risk of jaundice in glucose-6-phosphate dehydrogenase deficient babies exposed to menthol. *Acta Paediatr Scand*. 1980 May;69(3):341-5.
[3658] Dillon Remy M, Manning Alleyne P, Bratt DE, et al. Neonatal jaundice at Port-of-Spain General Hospital abstract. *West Indian Med J*. 1987;36(Suppl):28.
[3659] Covington TR, et al. Handbook of Nonprescription Drugs. 11th ed. Washington, D.C.: American Pharmaceutical Association. 1996.
[3660] Burkhard PR, Burkhardt K, Haenggeli CA, et al. Plant-induced seizures: reappearance of an old problem. *J Neurol*. 1999 Aug;246(8):667-70.

[3661] Waldman N. Seizure caused by dermal application of over-the-counter eucalyptus oil head lice preparation. *Clin Toxicol (Phila)*. 2011 Oct;49(8):750-1.

[3662] Craig JO. Poisoning by the volatile oils in childhood. *Arch Dis Child.* 1953;28:259-67.

[3663] Mathew T, Kamath V, Kumar RS, et al. Eucalyptus oil inhalation-induced seizure: A novel, underrecognized, preventable cause of acute symptomatic seizure. *Epilepsia Open*. 2017 Jul 4;2(3):350-354.

[3664] Unger M, Frank A. Simultaneous determination of the inhibitory potency of herbal extracts on the activity of six major cytochrome P450 enzymes using liquid chromatography/mass spectrometry and automated online extraction. *Rapid Commun Mass Spectrom.* 2004;18(19):2273-81.

[3665] Jori A, Bianchetti A, Prestini PE, et al. Effect of eucalyptol (1,8-cineole) on the metabolism of other drugs in rats and in man. *Eur J Pharmacol.* 1970;9(3):362-66.

[3666] Kim NH, Hyun SH, Jin CH, et al. Pretreatment with 1,8-cineole potentiates thioacetamide-induced hepatotoxicity and immunosuppression. *Arch Pharm Res.* 2004 Jul;27(7):781-9.

[3667] Jori A, Bianchetti A, Prestini PE, et al. Effect of eucalyptol (1,8-cineole) on the metabolism of other drugs in rats and in man. *Eur J Pharmacol.* 1970;9(3):362-66.

[3668] de Sousa DP, Raphael E, Brocksom U, et al. Sedative effect of monoterpene alcohols in mice: A preliminary screening. *Verlag der Zeitschrift fur Naturforschung*. 2007;62c:563-66.

[3669] Haliciglu O, Astarcioglu G, Yaprak I, et al. Toxicity of Salvia officinalis in a newborn and a child: an alarming report. *Pediatr Neurol.* 2011 Oct;45(4):259-60.

[3670] Lachenmeier DW, Walch SG. Epileptic seizure caused by accidental ingestion of sage (Salvia officinalis L.) oil in children: a rare, exceptional case or a threat to public health. *Pediatr Neurol.* 2012 Mar;46(3):201.

[3671] da Silva Costa KC, Bexerra SB, Norte CM, et al. Medicinal plants with teratogenic potential: current considerations. *Braz J Pharm Sci.* 2012;48(3):427-33.

[3672] Brinker F. Herb Contraindications and Drug Interactions. 2nd ed. Sandy, OR. Eclectic Medical Publications. 1998.

[3673] Frohne D. Giftpflanzen: Cupressaceae. Stuttgart: Wissenschaftliche Verlagsgesellschaft mbH; 1997. pp. 153–6.

[3674] European Medicines Agency. Public statement on the use of herbal medicinal products containing thujone. Available at: http://www.ema.europa.eu/docs/en_GB/document_library/Public_statement/2011/02/WC500102294.pdf.

[3675] Millet Y, Jouglard J, Steinmetz MD, et al. Toxicity of some essential plant oils. Clinical and experimental study. *Clin Toxicol.* 1981 Dec;18(12):1485-98.

[3676] Cristovao L, Carvalho F, Bastos MDL, et al. Hepatotoxicity of an essential oil of Salvia officinalis L.: an in vitro study using freshly isolated rat hepatocytes. *Congress Biomarkers.* 2001 Sep:165.

[3677] Millet Y, Jouglard J, Steinmetz MD, et al. Toxicity of some essential plant oils. Clinical and experimental study. *Clin Toxicol.* 1981 Dec;18(12):1485-98.

[3678] Burkhard PR, Burkhardt K, Haenggeli CA, et al. Plant-induced seizures: reappearance of an old problem. *J Neurol.* 1999 Aug;246(8):667-70.

[3679] Burkhard PR, Burkhardt K, Haenggeli CA, et al. Plant-induced seizures: reappearance of an old problem. *J Neurol.* 1999 Aug;246(8):667-70.

[3680] Hold KM, Sirisoma NS, Ikeda T, et al. α-Thujone (the active component of absinthe): γ-Aminobutyric acid type A receptor modulation and metabolic detoxification. *Proc Natl Acad Sci USA.* 2000 Apr 11;97(8):3826–3831.

[3681] Waldman N. Seizure caused by dermal application of over-the-counter eucalyptus oil head lice preparation. *Clin Toxicol (Phila)*. 2011 Oct;49(8):750-1.

[3682] Craig JO. Poisoning by the volatile oils in childhood. *Arch Dis Child.* 1953;28:259-67.

[3683] Mathew T, Kamath V, Kumar RS, et al. Eucalyptus oil inhalation-induced seizure: A novel, underrecognized, preventable cause of acute symptomatic seizure. *Epilepsia Open*. 2017 Jul 4;2(3):350-354.

[3684] Uc A, Bishop WP, Sanders KD. Camphor hepatotoxicity. *South Med J.* 2000;93:596-98.

[3685] Frohne D. Giftpflanzen: Cupressaceae. Stuttgart: Wissenschaftliche Verlagsgesellschaft mbH; 1997. pp. 153–6.

[3686] Dolan LC, Matulka RA, Burdock GA. Naturally Occurring Food Toxins. *Toxins (Basel)*. 2010 Sep;2(9):2289–2332.

[3687] United States National Toxicology Program (NTP). Alpha-Thujone. Dec 10, 1997. Available at: http://ntp.niehs.nih.gov/index.cfm?objectid=03DB8C36-E7A1-9889-3BDF8436F2A8C51F.

[3688] Unger M, Frank A. Simultaneous determination of the inhibitory potency of herbal extracts on the activity of six major cytochrome P450 enzymes using liquid chromatography/mass spectrometry and automated online extraction. *Rapid Commun Mass Spectrom.* 2004;18(19):2273-81.

[3689] Jori A, Bianchetti A, Prestini PE, et al. Effect of eucalyptol (1,8-cineole) on the metabolism of other drugs in rats and in man. *Eur J Pharmacol.* 1970;9(3):362-66.

[3690] Abass K, Reponen P, Mattila S, et al. Metabolism of α-thujone in human hepatic preparations in vitro. *Xenobiotica.* 2011 Feb;41(2):101-11.

[3691] Kim NH, Hyun SH, Jin CH, et al. Pretreatment with 1,8-cineole potentiates thioacetamide-induced hepatotoxicity and immunosuppression. *Arch Pharm Res.* 2004 Jul;27(7):781-9.

[3692] Shah AJ, Gilani AH, Abbas K, et al. Studies on the Chemical Composition and Possible Mechanisms Underlying the Antispasmodic and Bronchodilatory Activities of the Essential Oil of Artemisia maritima L. *Arch Pharm Res.* 2011;34(8):1227-38.

[3693] Mohan M, Pandey AK, Singh P, et al. Evaluation of Artemisia maritima L. Essential Oil for its Chemical and Biological Properties against Some Foodborne Pathogens. *Analytical Chem Letters.* 2016;6(1):47-54.

[3694] Stappen I, Wanner J, Tabanca N, et al. Chemical Composition and Biological Effects of Artemisia maritima and Artemisia nilagirica Essential Oils from Wild Plants ofWestern Himalaya. *Planta Med.* 2014;80:1079-87.

[3695] Maxia A, Frau MA, Marongiu B, et al. Biological Assays and Chemical Composition of Volatile Oils of Bupleurum fruticosum L. (Apiaceae). *Rec Nat Prod.* 2011;5(2):92-99.

[3696] Liu K, Lota ML, Casanova J, et al. The Essential Oil of Bupleurum fruticosum L. from Corsica: A Comprehensive Study. *Chem Biodiversity.* 20096:2244-54.

[3697] Bertoli A, Pistelli L, Morelli I, et al. Volatile constituents of micropropagated plants of Bupleurum fruticosum L. *Plant Sci.* 2004;167:807-10.

[3698] Chizzola R. Composition of the fruit essential oil of Bupleurum fruiticosum grown in Southern France. *Chem Nat Compounds.* 2008;44(6):792-3.

[3699] Manunta A, Trillini B, Fraternale D. Secretory Tissues and Essential Oil Composition of Bupleurum fruticosum L. *J Essent Oil Res.* 1992;4(5)461-6.

[3700] Lorente I, Ocete MA, Zarzuelo A, et al. Bioactivity of the essential oil of Bupleurum fruticosum. *J Nat Prod.* 1989 Mar-Apr;52(2):267-72.

[3701] Maxia A, Frau MA, Marongiu B, et al. Biological Assays and Chemical Composition of Volatile Oils of Bupleurum fruticosum L. (Apiaceae). *Rec Nat Prod.* 2011;5(2):92-99.

[3702] Kim JH. Seasonal variations in the content and composition of essential oil from Zanthoxylum piperitum. *J EcolField Biol.* 2012;35(3):195-201.

[3703] Wu Y, Shimoda M, Osajima Y. Volatile aroma compounds in young leaves and green fruits of Japanese pepper. *Nippon Nogeikagaku Kaishi.* 1996;70(9):100-5.
[3704] Donald GR, Fernandes PD, Boylan F. Antinociceptive Activity of Zanthoxylum piperitum DC. Essential Oil. *Evid Based Complement Alternat Med.* 2016;2016:3840398.
[3705] Lee HS. Insecticidal Toxicities and Essential Oil Compositions of Zanthoxylum piperitum and Zanthoxylum schinifolium Fruits in Korea. *J Essent Oil Bearing Plants.* 2016;19(8):2065-71.
[3706] Sakai T, Yoshihara K, Hirose Y. Constituents of fruit oil from Japanese pepper. *Bull Chem Soc Japan.* 1968 Aug;41:1945-50.
[3707] Bonesi M, Menichini F, Tundis R, et al. Acetylcholinesterase and butyrylcholinesterase inhibitory activity of Pinus species essential oils and their constituents. *J Enzyme Inhib Med Chem.* 2010 Oct;25(5):622-8.
[3708] Aazza S, Lyoussi B, Miguel MG. Antioxidant and antiacetylcholinesterase activities of some commercial essential oils and their major compounds. *Molecules.* 2011 Sep 7;16(9):7672-90.
[3709] Miyazawa M, Yamafuji C. Inhibition of acetylcholinesterase activity by bicyclic monoterpenoids. *J Agric Food Chem.* 2005 Mar 9;53(5):1765-8.
[3710] Lee JHK, Chang KM, Kim GH. Anti-inflammatory Activities of Chopi (Zanthoxylum piperitum AP DC) Essential Oil: Suppression of the Inducible Nitric Oxide Synthase and Cellular Adhesion. *Food Sci Biotech.* 2009 Dec;18(6):1371-8.
[3711] Lee JH, Jang KM, Kim GH. Composition of components and anti-inflammatory activities of Zanthoxylum piperitum A.P. DC and Zanthoxylum schinifolium essential oils. 2008 Apr;22(Suppl. 2):
[3712] Donald GR, Fernandes PD, Boylan F. Antinociceptive Activity of Zanthoxylum piperitum DC. Essential Oil. *Evid Based Complement Alternat Med.* 2016;2016:3840398.
[3713] Lee JH, Jang M, Seo J, et al. Antibacterial effects of natural volatile essential oil from zanthoxylum piperitum A.P. DC. against foodborne pathogens. *J Food Biochem.* 2012 Dec;36(6):667-74.
[3714] Kamsuk K, Choochote W, Chaithong U, et al. Effectiveness of Zanthoxylum piperitum-derived essential oil as an alternative repellent under laboratory and field applications. *Parasitol Res.* 2007 Jan;100(2):339-45.
[3715] Choochote W, Chaithong U, Kamsuk K, et al. Repellent activity of selected essential oils against Aedes aegypti. *Fitoterapia.* 2007 Jul;78(5):359-64.
[3716] Park IK, Kim LS, Choi IH, et al. Fumigant activity of plant essential oils and components from Schizonepeta tenuifolia against Lycoriella ingenua (Diptera: Sciaridae). *J Econ Entomol.* 2006 Oct;99(5):1717-21.
[3717] Lee HS. Insecticidal Toxicities and Essential Oil Compositions of Zanthoxylum piperitum and Zanthoxylum schinifolium Fruits in Korea. *J Essent Oil Bearing Plants.* 2016;19(8):2065-71.
[3718] Brunke EJ, Hammerschmidt FJ, Schmaus G, et al. The Essential Oil of Rhus coriaria L. Fruits. *Flavour Fragr J.* 1993;8:209-14.
[3719] Kurucu S, Koyuncu M, Guvenc A. The essential oil of Rhus coriana L. (Sumac). 1993 Sep/Oct;5:481-6.
[3720] Morshedloo MR, Maggi F, Neko HT, et al. Sumac (Rhus coriaria L.) fruit: Essential oil variability in Iranian populations. *Industrial Crops Prod.* 2018;111:1-7.
[3721] Farag MA, Fayek MN, Reidah IA. Volatile profiling in Rhus coriaria fruit (sumac) from three different geographical origins and upon roasting as analyzed via solid-phase microextraction. *PeerJ.* 2018;6:e5121.
[3722] Giovanelli S, Giusti G, Cioni PL, et al. Aroma profile and essential oil composition of Rhus coriaria fruitsfrom four Sicilian sites of collection. *Industrial Crops Prod.* 2017;97:166-74.
[3723] Mojab F, Abdi K, Nazarnia A. Chemical components of the essential oil from fruits and leaves of Rhus coriaria. *J Med Plants.* 2012 Sep;11(44):55-61.
[3724] Burkhard PR, Burkhardt K, Haenggeli CA, et al. Plant-induced seizures: reappearance of an old problem. *J Neurol.* 1999 Aug;246(8):667-70.
[3725] Waldman N. Seizure caused by dermal application of over-the-counter eucalyptus oil head lice preparation. *Clin Toxicol (Phila).* 2011 Oct;49(8):750-1.
[3726] Craig JO. Poisoning by the volatile oils in childhood. *Arch Dis Child.* 1953;28:259-67.
[3727] Mathew T, Kamath V, Kumar RS, et al. Eucalyptus oil inhalation-induced seizure: A novel, underrecognized, preventable cause of acute symptomatic seizure. *Epilepsia Open.* 2017 Jul 4;2(3):350-354.
[3728] Narayan S, Singh N. Camphor poisoning-An unusual cause of seizure. *Med J Armed Forces India.* 2012 Jul;68(3):252-53.
[3729] Chanaranaj KJ, G MV, S M. Camphor poisoning in a child. *Natl Med J India.* 2013 Jan-Feb;26(1):60.
[3730] Zhaleh M, Sohrabi N, Zangeneh MM, et al. Chemical Composition and Antibacterial Effects of Essential Oil of Rhus coriaria Fruits in the West of Iran (Kermanshah). *J Essent Oil Bearing Plants.* 2018;21(2):493-501.
[3731] Collin G, St-Gelais A, Turcotte M, et al. Composition of the essential oil and of some extracts of the aerial parts of Artemisia ludoviciana var. latiloba Nutt. *American J Essential Oils Nat Prod.* 2016;4(4):28-38.
[3732] Baldemir A, Karaman U, Ilgun S, et al. Antiparasitic Efficacy of Artemisia ludoviciana Nutt. (Asteraceae) Essential Oil for Acanthamoeba castellanii, Leishmania infantum and Trichomonas vaginalis. *Indian J Pharm Ed Res.* 2018 Jul-Sep;52(3):416-25.
[3733] Lopes-Lutz D, Alviano DS, Aliano CS, et al. Screening of chemical composition, antimicrobial and antioxidant activities of Artemisia essential oils. *Phytochem.* 2008;69:1732-8.
[3734] Miracle Botanicals. Owyhee GCMS Batch Number: 46279. Available at: https://www.dropbox.com/s/pzscn8z2hw80l68/Owyhee%20GCMS%20Miracle.pdf?dl=0.
[3735] Aromaticplantproject.com. Owyhee – Artemisia ludoviciana: A new essential oil & hydrosol. The plant, the plant, the name. Available at: http://www.aromaticplantproject.com/articles_archive/Owyhee_Artemisia_Ludoviciana.html.
[3736] Khine H, Weiss D, Graber N, et al. A cluster of children with seizures caused by camphor poisoning. *Pediatrics.* 2009 May;123(5):1269-72.
[3737] Michiels EA, Mazor SS. Toddler with seizures due to ingesting camphor at an Indian celebration. *Pediatr Emerg Care.* 2010 Aug;26(8):574-75.
[3738] Koren G. Medications which can kill a toddler with one tablet or teaspoonful. *J Toxicol Clin Toxicol. 1993*;31(3):407-13.
[3739] Flaman Z, Pellechia-Clarke S, Bailey B, et al. Unintentional exposure of young children to camphor and eucalyptus oils. *Paediatr Child Health.* 2001 Feb;6(2):80-83.
[3740] Rabl W, Katzgraber F, Steinlechner M. Camphor ingestion for abortion (case report). *Forensic Sci Int.* 1997 Sep 19;89(1-2):137-40.
[3741] Flaman Z, Pellechia-Clarke S, Bailey B, et al. Unintentional exposure of young children to camphor and eucalyptus oils. *Paediatr Child Health.* 2001 Feb;6(2):80-83.
[3742] Manoguerra AS, Erdman AR, Wax PM, et al. Camphor poisoning: an evidence-based practice guideline for out-of- hospital management. *Clin Toxicol (Phila).*2006;44(4):357-70.
[3743] Gibson DE, Moore GP, Pfaff JA. Camphor ingestion. *Am J Emerg Med.* 1989 Jan;7(1):41-43.
[3744] Koppel C, Martends F, Schirop T, et al. Hemoperfusion in acute camphor poisoning. *Intensive Care Med.* 1988;14(4):431-33.

[3745] Burkhard PR, Burkhardt K, Haenggeli CA, et al. Plant-induced seizures: reappearance of an old problem. *J Neurol.* 1999 Aug;246(8):667–70.
[3746] Millet Y, Jouglard J, Steinmetz MD, et al. Toxicity of some essential plant oils. Clinical and experimental study. *Clin Toxicol.* 1981 Dec;18(12):1485–98.
[3747] Charanraj KJ, G MV, S M. Camphor poisoning in a child. *Natl Med J India.* 2013 Jan-Feb;26(1):60.
[3748] Narayan S, Singh N. Camphor poisoning-An unusual cause of seizure. *Med J Armed Forces India.* 2012 Jul;68(3):252–53.
[3749] Uc A, Bishop WP, Sanders KD. Camphor hepatotoxicity. *South Med J.* 2000;93:596-98.
[3750] Unger M, Frank A. Simultaneous determination of the inhibitory potency of herbal extracts on the activity of six major cytochrome P450 enzymes using liquid chromatography/mass spectrometry and automated online extraction. *Rapid Commun Mass Spectrom.* 2004;18(19):2273-81.
[3751] Jori A, Bianchetti A, Prestini PE, et al. Effect of eucalyptol (1,8-cineole) on the metabolism of other drugs in rats and in man. *Eur J Pharmacol.* 1970;9(3):362-66.
[3752] Kim NH, Hyun SH, Jin CH, et al. Pretreatment with 1,8-cineole potentiates thioacetamide-induced hepatotoxicity and immunosuppression. *Arch Pharm Res.* 2004 Jul;27(7):781-9.
[3753] Jori A, Bianchetti A, Prestini PE, et al. Effect of eucalyptol (1,8-cineole) on the metabolism of other drugs in rats and in man. *Eur J Pharmacol.* 1970;9(3):362-66.
[3754] de Sousa DP, Raphael E, Brocksom U, et al. Sedative effect of monoterpene alcohols in mice: A preliminary screening. *Verlag der Zeitschrift fur Naturforschung.* 2007;62c:563-66.
[3755] Olowe SA, Ransome-Kuti O. The risk of jaundice in glucose-6-phosphate dehydrogenase deficient babies exposed to menthol. *Acta Paediatr Scand.* 1980 May;69(3):341-5.
[3756] Dillon Remy M, Manning Alleyne P, Bratt DE, et al. Neonatal jaundice at Port-of-Spain General Hospital abstract. *West Indian Med J.* 1987;36(Suppl):28.
[3757] Covington TR, et al. Handbook of Nonprescription Drugs. 11th ed. Washington, D.C.: American Pharmaceutical Association. 1996.
[3758] Burkhard PR, Burkhardt K, Haenggeli CA, et al. Plant-induced seizures: reappearance of an old problem. *J Neurol.* 1999 Aug;246(8):667–70.
[3759] Millet Y, Jouglard J, Steinmetz MD, et al. Toxicity of some essential plant oils. Clinical and experimental study. *Clin Toxicol.* 1981 Dec;18(12):1485–98.
[3760] Charanraj KJ, G MV, S M. Camphor poisoning in a child. *Natl Med J India.* 2013 Jan-Feb;26(1):60.
[3761] Narayan S, Singh N. Camphor poisoning-An unusual cause of seizure. *Med J Armed Forces India.* 2012 Jul;68(3):252–53.
[3762] Anaya-Eugenio GD, Rivero-Cruz I1, Bye R, et al. Antinociceptive activity of the essential oil from Artemisia ludoviciana. *J Ethnopharmacol.* 2016 Feb 17;179:403-11.
[3763] Baldemir A, Karaman U, Ilgun S, et al. Antiparasitic Efficacy of Artemisia ludoviciana Nutt. (Asteraceae) Essential Oil for Acanthamoeba castellanii, Leishmania infantum and Trichomonas vaginalis. *Indian J Pharm Ed Res.* 2018 Jul-Sep;52(3):416-25.
[3764] Lopes-Lutz D, Alviano DS, Aliano CS, et al. Screening of chemical composition, antimicrobial and antioxidant activities of Artemisia essential oils. *Phytochem.* 2008;69:1732-8.
[3765] Lopes-Lutz D, Alviano DS, Aliano CS, et al. Screening of chemical composition, antimicrobial and antioxidant activities of Artemisia essential oils. *Phytochem.* 2008;69:1732-8.
[3766] Shah WA, Dar MY, Zagar MI, et al. Chemical composition and antimicrobial activity of the leaf essential oil of Skimmia laureola growing wild in Jammu and Kashmir, India. *Nat Prod Res.* 2013;27(11):1023-7.
[3767] Barkatullah, Ibrar M, Muhammad N, et al. Chemical Composition and Biological Activities of the Essential Oil of Skimmia laureola Leaves. *Molecules.* 2015;20:4735-45.
[3768] Dar MY. Prospecting of High Altitude Aromatic Plants for Essential Oil Chemistry and Bio Activity. Available at: http://shodhganga.inflibnet.ac.in/jspui/handle/10603/191929.
[3769] Shah WA, Dar MY, Kuratull-Ain, et al. Comparison of Terpene Composition of Skimmia laureola Using Hydrodistillation and HS-SPME Techniques. *J Essent Oil Bearing Plants.* 2012;15(1):116-21.
[3770] Pandey V, Chauhan A, Verma RS, et al. Chemical Investigation of Skimmia laureola (Rutaceae) Essential Oil for Characterization of New Constituents. *J Essent Oil Bear Plants.* 2015;18(4):791-7.
[3771] Mehmood F, Khan ZU, Manzoor F, et al. Analysis of Insect toxicity and repellent activity of Phytochemicals from "Skimmia laureola, Nair" against "Black garden ant, Lasius niger" of Pakistan. *Pak J Pharm Sci.* 2016 May;29(3):789-93.
[3772] Irshad M, Aziz S, Rehman H-U, et al. Antioxidant and antimicrobial activities of essential oil of Skimmea laureola growing wild in the State of Jammu and Kashmir. *J Med Plant Res.* 2012;6(9):1680-1684.
[3773] Gondwal M, Nee Pant GJ, Singh Gautam BP, et al. Chemical Constituents and Biological Activities of Genus "Skimmia". *Nat Prod J.* 2015;5:91-102.
[3774] Muhammad N, Barkatullah, Ibrar M, et al. In vivo screening of essential oils of Skimmia laureola leaves for antinociceptive and antipyretic activity. *Asian Pac J Trop Biomed.* 2013;3(3):202-6.
[3775] Barkatullah, Ibrar M, Muhammad N, et al. Chemical Composition and Biological Activities of the Essential Oil of Skimmia laureola Leaves. *Molecules.* 2015;20:4735-45.
[3776] Shah WA, Dar MY, Zagar MI, et al. Chemical composition and antimicrobial activity of the leaf essential oil of Skimmia laureola growing wild in Jammu and Kashmir, India. *Nat Prod Res.* 2013;27(11):1023-7.
[3777] Barkatullah, Ibrar M, Muhammad N, et al. Chemical Composition and Biological Activities of the Essential Oil of Skimmia laureola Leaves. *Molecules.* 2015;20:4735-45.
[3778] Irshad M, Aziz S, Rehman H-U, et al. Antioxidant and antimicrobial activities of essential oil of Skimmea laureola growing wild in the State of Jammu and Kashmir. *J Med Plant Res.* 2012;6(9):1680-1684.
[3779] Irshad M, Aziz S, Rehman H-U, et al. Antioxidant and antimicrobial activities of essential oil of Skimmea laureola growing wild in the State of Jammu and Kashmir. *J Med Plant Res.* 2012;6(9):1680-1684.
[3780] Mehmood F, Khan ZU, Manzoor F, et al. Analysis of Insect toxicity and repellent activity of Phytochemicals from "Skimmia laureola, Nair" against "Black garden ant, Lasius niger" of Pakistan. *Pak J Pharm Sci.* 2016 May;29(3):789-93.
[3781] Demirci B, Demirci F, Baser KHC. Composition of the essential oil of Cotinus coggygria Scop. from Turkey. *Flavour Frag J.* 2003;18:43-4.
[3782] Novakovic M, Vukkovic I, Janackovic P, et al. Chemical composition, antibacterial and antifungal activity of the essential oils of Cotinus coggygria from Serbia. *J Serb Chem Soc.* 2007;72(11):1045-51.
[3783] Shagun S, Sujata B, Manjul S. Chemical Profiling, Antioxidant and Antibacterial Properties of Cotinus coggygria Essential Oil from Western Himalaya. *Int J Pharmacog Phytochem Res.* 2016;8(7):1183-6.
[3784] Fraternale D, Ricci D. Chemical Composition and Antimicrobial Activity of the Essential oil of Cotinus coggygria Scoop. from Italy. *J Essent Oil Bearing Plants.* 2014;17(3):366-70.

[3785] Tzakou O, Bazos I, Yannitsaros A. Essential oils of leaves, inflorescences and infructescences of spontaneous Cotinus coggygria Scop. from Greece. *Flavour Fragr J.* 2005;20:531-3.
[3786] Tsankova ET, Dyulgerov AS, Milenkov BK. Chemical Composition of the Bulgarian Sumac Oil. *J Essent Oil Res.* 1993;5(2):205-7.
[3787] De Olivera AC, Ribeiro-Pinto LF, Paumgartten JR. In vitro inhibition of CYP2B1 monoxygenase by beta-myrcene and other monoterpenoid constituents. *Br J Nutr.* 1999;81:289-95.
[3788] De-Oliveira AC1, Ribeiro-Pinto LF, Otto SS, et al. Induction of liver monooxygenases by beta-myrcene. *Toxicology.* 1997 Dec 26;124(2):135-40.
[3789] Burkhard PR, Burkhardt K, Haenggeli CA, et al. Plant-induced seizures: reappearance of an old problem. *J Neurol.* 1999 Aug;246(8):667–70.
[3790] Millet Y, Jouglard J, Steinmetz MD, et al. Toxicity of some essential plant oils. Clinical and experimental study. *Clin Toxicol.* 1981 Dec;18(12):1485–98.
[3791] Charanraj KJ, G MV, S M. Camphor poisoning in a child. *Natl Med J India.* 2013 Jan-Feb;26(1):60.
[3792] Narayan S, Singh N. Camphor poisoning-An unusual cause of seizure. *Med J Armed Forces India.* 2012 Jul;68(3):252–53.
[3793] Olowe SA, Ransome-Kuti O. The risk of jaundice in glucose-6-phosphate dehydrogenase deficient babies exposed to menthol. *Acta Paediatr Scand.* 1980 May;69(3):341-5.
[3794] Dillon Remy M, Manning Alleyne P, Bratt DE, et al. Neonatal jaundice at Port-of-Spain General Hospital abstract. *West Indian Med J.* 1987;36(Suppl):28.
[3795] Novakovic M, Vukkovic I, Janackovic P, et al. Chemical composition, antibacterial and antifungal activity of the essential oils of Cotinus coggygria from Serbia. *J Serb Chem Soc.* 2007;72(11):1045-51.
[3796] Shagun S, Sujata B, Manjul S. Chemical Profiling, Antioxidant and Antibacterial Properties of Cotinus coggygria Essential Oil from Western Himalaya. *Int J Pharmacog Phytochem Res.* 2016;8(7):1183-6.
[3797] Fraternale D, Ricci D. Chemical Composition and Antimicrobial Activity of the Essential oil of Cotinus coggygria Scoop. from Italy. *J Essent Oil Bearing Plants.* 2014;17(3):366-70.
[3798] Shagun S, Sujata B, Manjul S. Chemical Profiling, Antioxidant and Antibacterial Properties of Cotinus coggygria Essential Oil from Western Himalaya. *Int J Pharmacog Phytochem Res.* 2016;8(7):1183-6.
[3799] Dragar VA. A preliminary survey of selected species of endemic plants to determine commercial cropping potential for essential oils. University of Tasmania, Masters Thesis. Available at: http://eprints.utas.edu.au/19544/1/whole_DragarValerieAnnette1985_thesis.pdf.
[3800] Brophy JJ, Goldsack RJ, Bean AR, et al. Leaf essential oils of the genus Leptospermum (Myrtaceae) in eastern Australia, Part 4.* Leptospermum deanei and allies. *Flavour Frag J.* 1999;14:92-7.
[3801] Merle H, Verdeguer M, Blazquez MA, et al. Chemical composition of the essential oils from Eriocephalus africanus L. var. africanus populations growing in Spain. *Flavour Frag J.* 2007;22:461-4.
[3802] Verdeguer M, Blazquez MA, Boira H. Phytotoxic effects of Lantana camara, Eucalyptus camaldulensis and Eriocephalus africanus essential oils in weeds of Mediterranean summer crops. *Bichem System Ecol.* 2009;37:362-9.
[3803] Njenga EW. The chemotaxanomy, phylogeny and biological activity of the genus Eriocephalus L. (Asteraceae). University of the WitwatersrandDoctoral Thesis. Available at: https://www.researchgate.net/profile/Elizabeth_Njenga/publication/277404829_The_chemotaxonomy_Phylogeny_and_Biological_activity_of_the_genus_Eriocephalus_L_Asteraceae_species/links/556a2f1d08aeab77722128a2/The-chemotaxonomy-Phylogeny-and-Biological-activity-of-the-genus-Eriocephalus-L-Asteraceae-species.pdf.
[3804] Tisserand R, Young R. Essential Oil Safety, Second Edition. Churchill Livingstone: 2014.
[3805] Wikipedia. Ericocephalus africanus. Available at: https://en.wikipedia.org/wiki/Eriocephalus_africanus.
[3806] Burkhard PR, Burkhardt K, Haenggeli CA, et al. Plant-induced seizures: reappearance of an old problem. *J Neurol.* 1999 Aug;246(8):667-70.
[3807] Waldman N. Seizure caused by dermal application of over-the-counter eucalyptus oil head lice preparation. *Clin Toxicol (Phila).* 2011 Oct;49(8):750-1.
[3808] Craig JO. Poisoning by the volatile oils in childhood. *Arch Dis Child.* 1953;28:259-67.
[3809] Mathew T, Kamath V, Kumar RS, et al. Eucalyptus oil inhalation-induced seizure: A novel, underrecognized, preventable cause of acute symptomatic seizure. *Epilepsia Open.* 2017 Jul 4;2(3):350-354.
[3810] Unger M, Frank A. Simultaneous determination of the inhibitory potency of herbal extracts on the activity of six major cytochrome P450 enzymes using liquid chromatography/mass spectrometry and automated online extraction. *Rapid Commun Mass Spectrom.* 2004;18(19):2273-81.
[3811] Jori A, Bianchetti A, Prestini PE, et al. Effect of eucalyptol (1,8-cineole) on the metabolism of other drugs in rats and in man. *Eur J Pharmacol.* 1970;9(3):362-66.
[3812] Kim NH, Hyun SH, Jin CH, et al. Pretreatment with 1,8-cineole potentiates thioacetamide-induced hepatotoxicity and immunosuppression. *Arch Pharm Res.* 2004 Jul;27(7):781-9.
[3813] Jori A, Bianchetti A, Prestini PE, et al. Effect of eucalyptol (1,8-cineole) on the metabolism of other drugs in rats and in man. *Eur J Pharmacol.* 1970;9(3):362-66.
[3814] de Sousa DP, Raphael E, Brocksom U, et al. Sedative effect of monoterpene alcohols in mice: A preliminary screening. *Verlag der Zeitschrift fur Naturforschung.* 2007;62c:563-66.
[3815] Burkhard PR, Burkhardt K, Haenggeli CA, et al. Plant-induced seizures: reappearance of an old problem. *J Neurol.* 1999 Aug;246(8):667-70.
[3816] Waldman N. Seizure caused by dermal application of over-the-counter eucalyptus oil head lice preparation. *Clin Toxicol (Phila).* 2011 Oct;49(8):750-1.
[3817] Craig JO. Poisoning by the volatile oils in childhood. *Arch Dis Child.* 1953;28:259-67.
[3818] Mathew T, Kamath V, Kumar RS, et al. Eucalyptus oil inhalation-induced seizure: A novel, underrecognized, preventable cause of acute symptomatic seizure. *Epilepsia Open.* 2017 Jul 4;2(3):350-354.
[3819] Njenga EW. The chemotaxanomy, phylogeny and biological activity of the genus Eriocephalus L. (Asteraceae). University of the WitwatersrandDoctoral Thesis. Available at: https://www.researchgate.net/profile/Elizabeth_Njenga/publication/277404829_The_chemotaxonomy_Phylogeny_and_Biological_activity_of_the_genus_Eriocephalus_L_Asteraceae_species/links/556a2f1d08aeab77722128a2/The-chemotaxonomy-Phylogeny-and-Biological-activity-of-the-genus-Eriocephalus-L-Asteraceae-species.pdf.
[3820] Verdeguer M, Blazquez MA, Boira H. Phytotoxic effects of Lantana camara, Eucalyptus camaldulensis and Eriocephalus africanus essential oils in weeds of Mediterranean summer crops. *Bichem System Ecol.* 2009;37:362-9.
[3821] Khodakov G, Kotikov I, Pankovetskii V. Component composition of essential oil from Artemisia abrotanum and A. dracunculus. *Chem Nat Compounds.* 2010;45(6):905-908.
[3822] Obistioiu D, Cristina RT, Schmerold I, et al. Chemical characterization by GC-MS and in vitro activity against Candida albicans of volatile fractions prepared from Artemisia dracunculus, Artemisia abrotanum, Artemisia absinthium and Artemisia vulgaris. *Chem Central J.* 2014;8:6.

[3823] Pino JA, Marbot R, Marti MP. Leaf Oil of Artemisia abrotanum L. Grown in Cuba. *J Essent Oil Res.* 2011 Jan/Feb;23:119-20.
[3824] Mucciarelli M, Caramiello R, Maffei M. Essential Oils from Some Artemisia Species Growing Spontaneously in North-West Italy. *Flavour Fragr J.* 1995;10:25-33.
[3825] Radulovic NS, Blagojevic PD, Zlatkovic BK, et al. A GC/MS Profile of the Volatile Constituents of the Aerial Parts of Artemisia abrotanum L. (Asteraceae) from Serbia. *S Afr J Chem.* 2009;62:30-32.
[3826] Vostrowsky O, Michaelis K, Ihm H, et al. Das ätherische Öl von Artemisia abrotanum L. *Z Lebensmitt Unters Forsch.* 1984;179:125–128.
[3827] Tabanca N, Demirci B, Blythe EK, et al. Composition of Artemisia abrotanum and A. pontica Essential Oils and Their Repellent Activity against Aedes aegypti. *Planta Med.* 2011;77:PE52.
[3828] Khine H, Weiss D, Graber N, et al. A cluster of children with seizures caused by camphor poisoning. *Pediatrics.* 2009 May;123(5):1269-72.
[3829] Michiels EA, Mazor SS. Toddler with seizures due to ingesting camphor at an Indian celebration. *Pediatr Emerg Care.* 2010 Aug;26(8):574-75.
[3830] Koren G. Medications which can kill a toddler with one tablet or teaspoonful. *J Toxicol Clin Toxicol.* 1993;31(3):407-13.
[3831] Flaman Z, Pellechia-Clarke S, Bailey B, et al. Unintentional exposure of young children to camphor and eucalyptus oils. *Paediatr Child Health.* 2001 Feb;6(2):80-83.
[3832] Rabl W, Katzgraber F, Steinlechner M. Camphor ingestion for abortion (case report). *Forensic Sci Int.* 1997 Sep 19;89(1-2):137-40.
[3833] Flaman Z, Pellechia-Clarke S, Bailey B, et al. Unintentional exposure of young children to camphor and eucalyptus oils. *Paediatr Child Health.* 2001 Feb;6(2):80-83.
[3834] Manoguerra AS, Erdman AR, Wax PM, et al. Camphor poisoning: an evidence-based practice guideline for out-of- hospital management. *Clin Toxicol (Phila).*2006;44(4):357-70.
[3835] Gibson DE, Moore GP, Pfaff JA. Camphor ingestion. *Am J Emerg Med.* 1989 Jan;7(1):41-43.
[3836] Koppel C, Martends F, Schirop T, et al. Hemoperfusion in acute camphor poisoning. *Intensive Care Med.* 1988;14(4):431-33.
[3837] Burkhard PR, Burkhardt K, Haenggeli CA, et al. Plant-induced seizures: reappearance of an old problem. *J Neurol.* 1999 Aug;246(8):667–70.
[3838] Millet Y, Jouglard J, Steinmetz MD, et al. Toxicity of some essential plant oils. Clinical and experimental study. *Clin Toxicol.* 1981 Dec;18(12):1485–98.
[3839] Charanraj KJ, G MV, S M. Camphor poisoning in a child. *Natl Med J India.* 2013 Jan-Feb;26(1):60.
[3840] Narayan S, Singh N. Camphor poisoning-An unusual cause of seizure. *Med J Armed Forces India.* 2012 Jul;68(3):252–53.
[3841] Uc A, Bishop WP, Sanders KD. Camphor hepatotoxicity. *South Med J.* 2000;93:596-98.
[3842] Uc A, Bishop WP, Sanders KD. Camphor hepatotoxicity. *South Med J.* 2000;93:596-98.
[3843] Frohne D. Giftpflanzen: Cupressaceae. Stuttgart: Wissenschaftliche Verlagsgesellschaft mbH; 1997. pp. 153–6.
[3844] Dolan LC, Matulka RA, Burdock GA. Naturally Occurring Food Toxins. *Toxins (Basel).* 2010 Sep;2(9):2289–2332.
[3845] Olowe SA, Ransome-Kuti O. The risk of jaundice in glucose-6-phosphate dehydrogenase deficient babies exposed to menthol. *Acta Paediatr Scand.* 1980 May;69(3):341-5.
[3846] Dillon Remy M, Manning Alleyne P, Bratt DE, et al. Neonatal jaundice at Port-of-Spain General Hospital abstract. *West Indian Med J.* 1987;36(Suppl):28.
[3847] Unger M, Frank A. Simultaneous determination of the inhibitory potency of herbal extracts on the activity of six major cytochrome P450 enzymes using liquid chromatography/mass spectrometry and automated online extraction. *Rapid Commun Mass Spectrom.* 2004;18(19):2273-81.
[3848] Jori A, Bianchetti A, Prestini PE, et al. Effect of eucalyptol (1,8-cineole) on the metabolism of other drugs in rats and in man. *Eur J Pharmacol.* 1970;9(3):362-66.
[3849] Kim NH, Hyun SH, Jin CH, et al. Pretreatment with 1,8-cineole potentiates thioacetamide-induced hepatotoxicity and immunosuppression. *Arch Pharm Res.* 2004 Jul;27(7):781-9.
[3850] Jori A, Bianchetti A, Prestini PE, et al. Effect of eucalyptol (1,8-cineole) on the metabolism of other drugs in rats and in man. *Eur J Pharmacol.* 1970;9(3):362-66.
[3851] de Sousa DP, Raphael E, Brocksom U, et al. Sedative effect of monoterpene alcohols in mice: A preliminary screening. *Verlag der Zeitschrift fur Naturforschung.* 2007;62c:563-66.
[3852] Burkhard PR, Burkhardt K, Haenggeli CA, et al. Plant-induced seizures: reappearance of an old problem. *J Neurol.* 1999 Aug;246(8):667-70.
[3853] Waldman N. Seizure caused by dermal application of over-the-counter eucalyptus oil head lice preparation. *Clin Toxicol (Phila).* 2011 Oct;49(8):750-1.
[3854] Craig JO. Poisoning by the volatile oils in childhood. *Arch Dis Child.* 1953;28:259-67.
[3855] Mathew T, Kamath V, Kumar RS, et al. Eucalyptus oil inhalation-induced seizure: A novel, underrecognized, preventable cause of acute symptomatic seizure. *Epilepsia Open.* 2017 Jul 4;2(3):350-354.
[3856] Unger M, Frank A. Simultaneous determination of the inhibitory potency of herbal extracts on the activity of six major cytochrome P450 enzymes using liquid chromatography/mass spectrometry and automated online extraction. *Rapid Commun Mass Spectrom.* 2004;18(19):2273-81.
[3857] Jori A, Bianchetti A, Prestini PE, et al. Effect of eucalyptol (1,8-cineole) on the metabolism of other drugs in rats and in man. *Eur J Pharmacol.* 1970;9(3):362-66.
[3858] Kim NH, Hyun SH, Jin CH, et al. Pretreatment with 1,8-cineole potentiates thioacetamide-induced hepatotoxicity and immunosuppression. *Arch Pharm Res.* 2004 Jul;27(7):781-9.
[3859] Jori A, Bianchetti A, Prestini PE, et al. Effect of eucalyptol (1,8-cineole) on the metabolism of other drugs in rats and in man. *Eur J Pharmacol.* 1970;9(3):362-66.
[3860] de Sousa DP, Raphael E, Brocksom U, et al. Sedative effect of monoterpene alcohols in mice: A preliminary screening. *Verlag der Zeitschrift fur Naturforschung.* 2007;62c:563-66.
[3861] European Medicines Agency. Assessment report on Peumus boldus Molina, folium. Available at: http://www.ema.europa.eu/docs/en_GB/document_library/Herbal_-_HMPC_assessment_report/2009/12/WC500018102.pdf.
[3862] European Medicines Agency. Assessment report on Peumus boldus Molina, folium. Available at: http://www.ema.europa.eu/docs/en_GB/document_library/Herbal_-_HMPC_assessment_report/2009/12/WC500018102.pdf.
[3863] Almeida ER, Melo AM, Xavier H. Toxicological evaluation of the hydro-alcohol extract of the dry leaves of Peumus boldus and boldine in rats. *Phytother Res.* 2000 Mar;14(2):99-102.
[3864] Mitchell EMH, Heumann S, Araujo A, et al. Brazilian adolescents' knowledge and beliefs about abortion methods: a school-based internet inquiry. *BMC Womens Health.* 2014;14:27.
[3865] Osol A, ed. The Dispensatory of the United States of America. 25th ed. Philadelphia, PA: Lippincott, 1955.
[3866] Piscaglia F, Leoni S, Venturi A, et al. Caution in the use of boldo in herbal laxatives: a case of hepatotoxicity. *Scand J Gastroenterol.* 2005;40:236-39.
[3867] Monzon S, Lezaun A, Saenz D, et al. Anaphylaxis to boldo infusion, a herbal remedy. *Allergy.* 2004;59:1019-20.

[3868] European Medicines Agency. Assessment report on Peumus boldus Molina, folium. Available at: http://www.ema.europa.eu/docs/en_GB/document_library/Herbal_-_HMPC_assessment_report/2009/12/WC500018102.pdf.
[3869] Chittiboyina AG, Avonto C, Khan IA. What Happens after Activation of Ascaridole? Reactive Constituents and Their Implications for Skin Sensitization. *Chem Res Toxicol.* 2016 Sep 19;29(9):1488-92.
[3870] Sciarrone D, Ragonese C, Carnovale C, et al. Evaluation of tea tree oil quality and ascaridole: a deep study by means of chiral and multi heart-cuts multidimensional gas chromatography system coupled to mass spectrometry detection. *J Chromatogr A.* 2010 Oct 8;1217(41):6422-7.
[3871] Krutz NL, Hennen J1, Korb C, et al. Activation of the Endoperoxide Ascaridole Modulates Its Sensitizing Capacity. *Toxicol Sci.* 2015 Oct;147(2):515-23.
[3872] Casares R. Juniperus sabina. Food Cosmet Toxicol. 1964;2:680-81. Found in U.S. Food and Drug Administration. FDA Poisonous Plant Database. Available at: http://www.accessdata.fda.gov/scripts/plantox/detail.cfm?id=5133.
[3873] Pages N, Salazar M, Chamorro G, et al. Teratological evaluation of Plectranthus fruticosus leaf essential oil. *Planta Med.* 1988 Aug;54(4):296-8.
[3874] Chamorro G, Salazar M, Fournier G, et al. Anti-implantation effects of various extracts of Plectranthus fruticosus on pregnant rats. *Planta Med.* 1991 Feb;57(1):81.
[3875] Pages N, Fournier G, Chamorro G, et al. Teratological evaluation of Juniperus sabina essential oil in mice. *Planta Med.* 1989 Apr;55(2):144-46.
[3876] Pages N, Fournier G, Baudel C, et al. Sabinyl Acetate, the Main Component of Juniperus sabina L'Hérit. Essential oil, is Responsible for Antiimplantation Effect. *Phytother Res.* 1996 Aug;10(5):438-440.
[3877] Pages N, Fournier G, Velut V, et al. Potential teratogenicity in mice of the essential oil of Salvia lavandulifolia Vahl. Study of a fraction rich in sabinyl acetate. *Phytother Res.* 1992 Mar;6(2):80-83.
[3878] Casares R. Juniperus sabina. Food Cosmet Toxicol. 1964;2:680-81. Found in U.S. Food and Drug Administration. FDA Poisonous Plant Database. Available at: http://www.accessdata.fda.gov/scripts/plantox/detail.cfm?id=5133.
[3879] Burkhard PR, Burkhardt K, Haenggeli CA, et al. Plant-induced seizures: reappearance of an old problem. *J Neurol.* 1999 Aug;246(8):667-70.
[3880] Pages N, Fournier G, Chamorro G, et al. Teratological evaluation of Juniperus sabina essential oil in mice. *Planta Med.* 1989 Apr;55(2):144-46.
[3881] Tabanca N, Demirci B, Blythe EK, et al. Composition of Artemisia abrotanum and A. pontica Essential Oils and Their Repellent Activity against Aedes aegypti. *Planta Med.* 2011;77:PE52.
[3882] Villanueva HE, Tuten JA, Haber WA, et al. Chemical composition and antimicrobial activity of the bark essential oil of Cedrela odorata from Monteverde, Costa Rica. *Der Pharma Chemica.* 2009;1(2):14-8.
[3883] Maia BHLNS, de Paula JR, Sant'Ana J, et al. Essential Oils of Toona and Cedrela Species (Meliaceae): Taxonomic and Ecological Implications. *J Braz Chem Soc.* 2000;11(6):629-39.
[3884] Martins AP, Salguiero LR, Proenca A, et al. Chemical Composition of the Bark Oil of Cedrela odorata from S. Tomé and Príncipe. *J Essent Oil Res.* 2003 Nov/Dec;15:422-4.
[3885] Lawrence BM. Progress in essential oils. *Perfum FLav.* 1987;12(3):58-60.
[3886] Asekun O, Ekundayo O. Constituents of the leaf essential oil of Cedrela odorata L. from Nigeria. Flavour Fragr J. 1999;14(6):390-2.
[3887] Hardt I, Rieck A, Fricke C, et al. Enantiomeric composition of sesquiterpene hydrocarbons of the essential oil of Cedrela odorata L. *Flavour Frag J.* 1995;10(3):165-71.
[3888] Motl O, Trka A. Composition of the Essential Oil from Cedrela odorata L. *J Soc Cosmetic Chemists.* 1973;24:747-51.
[3889] Chen W, Lu Y, Gao J, et al. Beta-elemene inhibits melanoma growth and metastasis via suppressing vascular endothelial growth factormediated angiogenesis. *Cancer Chemo Pharmacol.* 2011 Apr;67(4):799-808.
[3890] Villanueva HE, Tuten JA, Haber WA, et al. Chemical composition and antimicrobial activity of the bark essential oil of Cedrela odorata from Monteverde, Costa Rica. *Der Pharma Chemica.* 2009;1(2):14-8.
[3891] Cirak C, Bertoli A, Pistelli L, et al. Essential oil composition and variability of Hypericum perforatum from wild populations of northern Turkey. *Pharm Biol.* 2010;48(8):906-14.
[3892] Chauhan RS, Vashistha RK, Nautiyal MC. Essential Oil Composition of Hypericum perforatum L. from Cultivated Source. *J Essent Oil Res.* 2011 May/Jun;23:1-6.
[3893] Radusiene J, Judzentiene A, Bernotiene G. Essential oil composition and variability of Hypericum perforatum L. growing in Lithuania. *Biochem System Ecol.* 2005;33:113-24.
[3894] Pirbalouti AG, Fatahi-Vanani M, Craker L, et al. Chemical composition and bioactivity of essential oils of Hypericum helianthemoides, Hypericum perforatum and Hypericum scabrum. *Pharm Biol.* 2013 Sep;1-7.
[3895] Mockute D, Bernotiene G, Judzentiene A. The Essential Oils with Dominant Germacrene D of Hypericum perforatum L. Growing Wild in Lithuania. *J Essent Oil Res.* 2008 Mar/Apr;20:128-31.
[3896] Schwob I, Bessiere JM, Masotti V, et al. Changes in essential oil composition in Saint John's wort (Hypericum perforatum L.) aerial parts during its phenological cycle. *Biochem System Ecol.* 2004;32:735-45.
[3897] Đorđević AS, Chemical composition of Hypericum perforatum L. essential oil. *Adv Tech.* 2015;4(1):64-8.
[3898] Gudzic B, Dordevic S, Palic R, et al. Essential oils of Hypericum olympicum L. and Hypericum perforatum L. *Flavour Fragr J.* 2001;16:201-3.
[3899] Schwob I, Bessiere JM, Viano J. Composition of the essential oils of Hypericum perforatum L. from southeastern France. *CR Biolgies.* 2002;325:781-5.
[3900] Sharapov FS, Gulmurodov IS, Setzer WN. Essential oil composition of Hypericum perforatum L. and Hypericum scabrum L. growing wild in Tajikistan. *J Chem Pharm Res.* 2010;2(6):284-90.
[3901] Bardhi N, Stefkov G, Karapandzova M, et al. Essential oil composition of indigenous populations of Hypericum perforatum L. from southern Albania. *Macedonian J Chem Chemical Eng.* 2015;34(2):333-41.
[3902] Morshedloo MR, Ebadi A, Maggi F, et al. Chemical characterization of the essential oil compositions from Iranian populations of Hypericum perforatum L. *Ind Crops Prod.* 2015;565-73.
[3903] Hajdari A, Mustafa B, Nebija D, et al. Essential oil composition and variability of Hypericum perforatum L. from wild population in Kosovo. *Current Issues Pharm Med Sci.* 2014;27(1):51-4.
[3904] Hosni K, Msaada K, Ben Taarit M, et al. Volatile Constituents of Two Hypericum Species from Tunisia. *Nat Prod Commun.* 2011;6(0):1-4.
[3905] Rouis Z, Laamari A, Abid N, et al. Chemical composition and larvicidal activity of several essential oils from Hypericum species from Tunisia. *Parasitol Res.* 2013 Feb;112(2):699-705.
[3906] Marandino A, De Martino L, Mancini E, et al. Chemical composition and possible in vitro antigermination activity of three Hypericum essential oils. *Nat Prod Commun.* 2011 Nov;6(11):1735-8.
[3907] Akhbari M, Batooli H, Mozdianfard M. Comparative study of composition and biological activities of SDE prepared essential oils from flowers and fruits of two Hypericum species from central Iran. *Nat Prod Res.* 2012;26(3):193-202.

[3908] Bertoli A, Cirak C, Leonardi M, et al. Morphogenetic changes in essential oil composition of Hypericum perforatum during the course of ontogenesis. *Pharm Biol.* 2011 Jul;49(7):741-51.
[3909] Helmja K, Vaher M, Püssa T, et al. Variation in the composition of the essential oils, phenolic compounds and mineral elements of Hypericum perforatum L. growing in Estonia. *Nat Prod Res.* 2011 Mar;25(5):496-510.
[3910] Lin WY, Kuo YH, Chang YL, et al. Anti-platelet aggregation and chemical constituents from the rhizome of Gynura japonica. *Planta Med.* 2003 Aug;69(8):757-64.
[3911] Kurashov EA, Fedorova EV, Krylova JV, et al. Assessment of the Potential Biological Activity of Low Molecular Weight Metabolites of Freshwater Macrophytes with QSAR. *Scientifica (Cairo).* 2016;2016:1205680.
[3912] Pirbalouti AG, Fatahi-Vanani M, Craker L, et al. Chemical composition and bioactivity of essential oils of Hypericum helianthemoides, Hypericum perforatum and Hypericum scabrum. *Pharm Biol.* 2013 Sep;1-7.
[3913] Akhbari M, Batooli H, Mozdianfard M. Comparative study of composition and biological activities of SDE prepared essential oils from flowers and fruits of two Hypericum species from central Iran. *Nat Prod Res.* 2012;26(3):193-202.
[3914] Pirbalouti AG, Fatahi-Vanani M, Craker L, et al. Chemical composition and bioactivity of essential oils of Hypericum helianthemoides, Hypericum perforatum and Hypericum scabrum. *Pharm Biol.* 2013 Sep;1-7.
[3915] Akhbari M, Batooli H, Mozdianfard M. Comparative study of composition and biological activities of SDE prepared essential oils from flowers and fruits of two Hypericum species from central Iran. *Nat Prod Res.* 2012;26(3):193-202.
[3916] Rouis Z, Laamari A, Abid N, et al. Chemical composition and larvicidal activity of several essential oils from Hypericum species from Tunisia. *Parasitol Res.* 2013 Feb;112(2):699-705.
[3917] Dosoky NS, Pokharel SK, Setzer W. Leaf essential oil composition, antimicrobial and cytotoxic activities of Cleistocalyx operculatus from Hetauda, Nepal. *Am J Essental Oils Nat Prod.* 2015;2(5):34-7.
[3918] Dung NT, Kim JM, Kang SC. Chemical composition, antimicrobial and antioxidant activities of the essential oil and the ethanol extract of Cleistocalyx operculatus (Roxb.) Merr and Perry buds. *Food Chem Toxicol.* 2008;46:3632-9.
[3919] Dung NX, Luu HV, Khoi TT, et al. GC and GC/MS Analysis of the Leaf Oil of Cleistocalyx operculatus Roxb. Merr. et Perry (Syn. Eugenia operculata Roxb.; Syzygicum mervosum DC.). *J Essent Oil Res.* 1994 Nov/Dec;6:661-2.
[3920] Delgado IF, Carvalho RR, Nogueira, et al. Study on embryofetotoxicity of b-myrcene in the rat. *Food and Chemical Toxicology.* 1993;31(1):31-5.
[3921] Paumgartten FJ, De-Carvalho RR, Souza CA, et al. Study of the effects of beta-myrcene on rat fertility and general reproductive performance. *Braz J Med Biol Res.* 1998 Jul;31(7):955-65.
[3922] do Vale TG, Furtado EC, Santos JG Jr, et al. Central effects of citral, myrcene and limonene, constituents of essential oil chemotypes from Lippia alba (Mill.) n.e. Brown. *Phytomedicine.* 2002 Dec;9(8):709-14.
[3923] Segelman AB, Sofia RD, Segelman FP, et al. Cannabis sativa L. (marijuana). V. Pharmacological evaluation of marijuana aqueous extract and volatile oil. *J Pharm Sci.* 1974 Jun;63(6):962-4.
[3924] Freitas JC1, Presgrave OA, Fingola FF, et al. Effect of beta-myrcene on pentobarbital sleeping time. *Braz J Med Biol Res.* 1993 May;26(5):519-23.
[3925] De Olivera AC, Ribeiro-Pinto LF, Paumgartten JR. In vitro inhibition of CYP2B1 monoxygenase by beta-myrcene and other monoterpenoid constituents. *Br J Nutr.* 1999;81:289-95.
[3926] De-Oliveira AC1, Ribeiro-Pinto LF, Otto SS, et al. Induction of liver monooxygenases by beta-myrcene. *Toxicology.* 1997 Dec 26;124(2):135-40.
[3927] Tan GB, Le NT, Dam SM. Potential Use of Essential Oil Isolated from Cleistocalyx operculatus Leaves as a Topical Dermatological Agent for Treatment of Burn Wound. *Dermatol Res Practice.* 2018;2018:2730169.
[3928] Dung NT, Bajpai VK, Yoon JI, et al. Anti-inflammatory effects of essential oil isolated from the buds of Cleistocalyx operculatus (Roxb.) Merr and Perry. *Food Chem Toxicol.* 2009;47:449-53.
[3929] Dosoky NS, Pokharel SK, Setzer W. Leaf essential oil composition, antimicrobial and cytotoxic activities of Cleistocalyx operculatus from Hetauda, Nepal. *Am J Essental Oils Nat Prod.* 2015;2(5):34-7.
[3930] Dung NT, Kim JM, Kang SC. Chemical composition, antimicrobial and antioxidant activities of the essential oil and the ethanol extract of Cleistocalyx operculatus (Roxb.) Merr and Perry buds. *Food Chem Toxicol.* 2008;46:3632-9.
[3931] Bajpai V, Dung N, Suh H, et al. Antibacterial Activity of Essential Oil and Extracts of Cleistocalyx operculatus Buds Against the Bacteria of Xanthomonas spp. J American Oil Chemists Soc. 2010;87(11):1341-1349.
[3932] Dung NT, Kim JM, Kang SC. Chemical composition, antimicrobial and antioxidant activities of the essential oil and the ethanol extract of Cleistocalyx operculatus (Roxb.) Merr and Perry buds. *Food Chem Toxicol.* 2008;46:3632-9.
[3933] Costa DP, Santos SC, Seraphin JC, et al. Seasonal Variability of Essential Oils of Eugenia uniflora Leaves. *J Braz Chem Soc.* 2009;20(7):1287-93.
[3934] Lago JHG, Souza ED, Mariane B, et al. Chemical and Biological Evaluation of Essential Oils from Two Species of Myrtaceae — Eugenia uniflora L. and Plinia trunciflora (O. Berg) Kausel. *Molecules.* 2011;16:9827-37.
[3935] da Franca Rodrigues KA, Amorim LV, de Oliveira JMG, et al. Eugenia uniflora L. Essential Oil as a Potential Anti-Leishmania Agent: Effects on Leishmania amazonensis and Possible Mechanisms of Action. *Evidence-Based Complement Altern Med.* 2013;2013:279726.
[3936] Melo RM, Correa VFS, Amorim ACL, et al. Identification of Impact Aroma Compounds in Eugenia uniflora L. (Brazilian Pitanga) Leaf Essential Oil. *J Braz Chem Soc.* 2007;18(1):179-83.
[3937] Costa DP, Alves Filho EG, Silva LMA, et al. Influence of Fruit Biotypes on the Chemical Composition and Antifungal Activity of the Essential Oils of Eugenia uniflora Leaves. *J Braz Chem Soc.* 2010;21(5):851-8.
[3938] de Morais SM, Craveiro AA, Lacerda Machado MI, et al. Volatile Constituents of Eugenia uniflora Leaf Oil from Northeastern Brazil. *J Essent Oil Res.* 1996 Jul/Aug;8:449-51.
[3939] Gallucci S, Neto AP, Porto C, et al. Essential Oil of Eugenia uniflora L.: An Industrial Perfumery Approach. *J Essent Oil Res.* 2010 Mar/Apr;22:176-8.
[3940] Becker NA, Volcao LM., Camargo TM, et al. Biological properties of Eugenia uniflora L. essential oil: phytochemistry composition and antimicrobial activity against gram negative bacteria. 2017;29(1):22-30.
[3941] Kolb N, Ferreyra D, Kolb E, et al. Estudio comparative de los aceites esenciales de las mirtaceas de la provincial de misiones.
[3942] Dos Santos JFS, Rocha JE, Bezerra CF, et al. Chemical composition, antifungal activity and potential anti-virulence evaluation of the Eugenia uniflora essential oil against Candida spp. *Food Chem.* 2018 Sep 30;261:233-239.
[3943] Govindarajan M, Rajeswary M, Senthilmurugan S, et al. Curzerene, trans-β-elemenone, and γ-elemene as effective larvicides against Anopheles subpictus, Aedes albopictus, and Culex tritaeniorhynchus: toxicity on non-target aquatic predators. *Environ Sci Pollut Res Int.* 2018 Apr;25(11):10272-10282.
[3944] Sobeh M, Braun M, Krstin S, et al. Chemical Profiling of the Essential Oils of Syzygium aqueum, Syzygium samarangense and Eugenia uniflora and Their Discrimination Using Chemometric Analysis. *Chem Biodivers.* 2016; 13(11):1537-50.
[3945] Barbosa LN, Probst IDS, Teles Andrade BFM, et al. In vitro antibacterial and chemical properties of essential oils including native plants from Brazil against pathogenic and resistant bacteria. *J Oleo Sci.* 2015;64(3):289-98.

[3946] Victoria FN, Lenardão EJ, Savegnago L, et al. Essential oil of the leaves of Eugenia uniflora L.: antioxidant and antimicrobial properties. *Food Chem Toxicol.* 2012 Aug;50(8):2668-74.
[3947] Silva NCC, Barbosa L, Seito LN, et al. Antimicrobial activity and phytochemical analysis of crude extracts and essential oils from medicinal plants. *Nat Prod Res.* 2012;26(16):1510-4.
[3948] Ogunwande IA, Olawore NO, Ekundayo O, et al. Studies on the essential oils composition, antibacterial and cytotoxicity of Eugenia uniflora L. *Int J Aromatherapy.* 2005;15:147-52.
[3949] Dos Santos JFS, Rocha JE, Bezerra CF, et al. Chemical composition, antifungal activity and potential anti-virulence evaluation of the Eugenia uniflora essential oil against Candida spp. *Food Chem.* 2018 Sep 30;261:233-239.
[3950] Pereira NLF, Aquino PEA, Júnior JGAS, et al. In vitro evaluation of the antibacterial potential and modification of antibiotic activity of the Eugenia uniflora L. essential oil in association with led lights. *Microb Pathog.* 2017 Sep;110:512-518.
[3951] Zhou L, Zhang K, Li J, et al. Inhibition of vascular endothelial growth factor-mediated angiogenesis involved in reproductive toxicity induced by sesquiterpenoids of Curcuma zedoaria in rats. *Reprod Toxicol.* 2013 Jun;37:62-9.
[3952] Chaiworapongsa T, Romero R, Kusanovic JP, et al. Unexplained fetal death is associated with increased concentrations of anti-angiogenic factors in amniotic fluid. *J Matern Fetal Neonatal Med.* 2010 Aug;23(8):794-805.
[3953] Chen W, Lu Y, Gao J, et al. Beta-elemene inhibits melanoma growth and metastasis via suppressing vascular endothelial growth factormediated angiogenesis. *Cancer Chemo Pharmacol.* 2011 Apr;67(4):799-808.
[3954] Pimkaew P1, Küblbeck J, Petsalo A, et al. Interactions of sesquiterpenes zederone and germacrone with the human cytochrome P450 system. *Toxicol In Vitro.* 2013 Sep;27(6):2005-12.
[3955] Sobeh M, Braun M, Krstin S, et al. Chemical Profiling of the Essential Oils of Syzygium aqueum, Syzygium samarangense and Eugenia uniflora and Their Discrimination Using Chemometric Analysis. *Chem Biodivers.* 2016; 13(11):1537-50.
[3956] Ogunwande IA, Olawore NO, Ekundayo O, et al. Studies on the essential oils composition, antibacterial and cytotoxicity of Eugenia uniflora L. *Int J Aromatherapy.* 2005;15:147-52.
[3957] Amorim ACL, Lima CKF, Hovell AMC, et al. Antinociceptive andhypothermicevaluationoftheleafessentialoiland isolated terpenoidsfrom Eugenia uniflora L. (BrazilianPitanga). *Phytomedicine.* 2009;16:923-8.
[3958] Schapoval EES, Silveira SM, Miranda M, et al. Evaluation of some pharmacological activities of Eugenia uniflora L. *J Ethnopharmacology.* 1994;44:137-42.
[3959] Gallucci S, Neto AP, Porto C, et al. Essential Oil of Eugenia uniflora L.: An Industrial Perfumery Approach. *J Essent Oil Res.* 2010 Mar/Apr;22:176-8.
[3960] da Franca Rodrigues KA, Amorim LV, de Oliveira JMG, et al. Eugenia uniflora L. Essential Oil as a Potential Anti-Leishmania Agent: Effects on Leishmania amazonensis and Possible Mechanisms of Action. *Evidence-Based Complement Altern Med.* 2013;2013:279726.
[3961] Lago JHG, Souza ED, Mariane B, et al. Chemical and Biological Evaluation of Essential Oils from Two Species of Myrtaceae — Eugenia uniflora L. and Plinia trunciflora (O. Berg) Kausel. *Molecules.* 2011;16:9827-37.
[3962] Becker NA, Volcao LM., Camargo TM, et al. Biological properties of Eugenia uniflora L. essential oil: phytochemistry composition and antimicrobial activity against gram negative bacteria. 2017;29(1):22-30.
[3963] Dos Santos JFS, Rocha JE, Bezerra CF, et al. Chemical composition, antifungal activity and potential anti-virulence evaluation of the Eugenia uniflora essential oil against Candida spp. *Food Chem.* 2018 Sep 30;261:233-239.
[3964] Pereira NLF, Aquino PEA, Júnior JGAS, et al. In vitro evaluation of the antibacterial potential and modification of antibiotic activity of the Eugenia uniflora L. essential oil in association with led lights. *Microb Pathog.* 2017 Sep;110:512-518.
[3965] Sobeh M, Braun M, Krstin S, et al. Chemical Profiling of the Essential Oils of Syzygium aqueum, Syzygium samarangense and Eugenia uniflora and Their Discrimination Using Chemometric Analysis. *Chem Biodivers.* 2016; 13(11):1537-50.
[3966] Barbosa LN, Probst IDS, Teles Andrade BFM, et al. In vitro antibacterial and chemical properties of essential oils including native plants from Brazil against pathogenic and resistant bacteria. *J Oleo Sci.* 2015;64(3):289-98.
[3967] Victoria FN, Lenardão EJ, Savegnago L, et al. Essential oil of the leaves of Eugenia uniflora L.: antioxidant and antimicrobial properties. *Food Chem Toxicol.* 2012 Aug;50(8):2668-74.
[3968] Silva NCC, Barbosa L, Seito LN, et al. Antimicrobial activity and phytochemical analysis of crude extracts and essential oils from medicinal plants. *Nat Prod Res.* 2012;26(16):1510-4.
[3969] Lima EO1, Gompertz OF, Giesbrecht AM, et al. In vitro antifungal activity of essential oils obtained from officinal plants against dermatophytes. *Mycoses.* 1993 Sep-Oct;36(9-10):333-6.
[3970] Ogunwande IA, Olawore NO, Ekundayo O, et al. Studies on the essential oils composition, antibacterial and cytotoxicity of Eugenia uniflora L. *Int J Aromatherapy.* 2005;15:147-52.
[3971] Victoria FN, Lenardão EJ, Savegnago L, et al. Essential oil of the leaves of Eugenia uniflora L.: antioxidant and antimicrobial properties. *Food Chem Toxicol.* 2012 Aug;50(8):2668-74.
[3972] Sobeh M, Braun M, Krstin S, et al. Chemical Profiling of the Essential Oils of Syzygium aqueum, Syzygium samarangense and Eugenia uniflora and Their Discrimination Using Chemometric Analysis. *Chem Biodivers.* 2016; 13(11):1537-50.
[3973] da Cunha FAB, Wallau GL, Pinho AI, et al. Eugenia uniflora leaves essential oil induces toxicity in Drosophila melanogaster: involvement of oxidative stress mechanisms. *Toxicol Res.* 2015;3:1-32.
[3974] Govindarajan M, Rajeswary M, Senthilmurugan S, et al. Curzerene, trans-β-elemenone, and γ-elemene as effective larvicides against Anopheles subpictus, Aedes albopictus, and Culex tritaeniorhynchus: toxicity on non-target aquatic predators. *Environ Sci Pollut Res Int.* 2018 Apr;25(11):10272-10282.
[3975] Farag RS, Shalaby AS, El-Baroty GA, et al. Chemical and biological evaluation of the essential oils of different Melaleuca species. *Phytother Res.* 2004;18(1):30-35.
[3976] National Institutes of Health, Medline Plus. Reye Syndrome. Available at: http://www.nlm.nih.gov/medlineplus/ency/article/001565.htm.
[3977] Karabulut AK, Ulger H, Pratten MK. Protection by free oxygen radical scavenging enzymes against salicylate- induced embryonic malformation in vitro. *Toxicol In Vitro.* 2000 Aug;14(4):297-307.
[3978] Overman DO, White JA. Comparative teratogenic effects of methyl salicylate applied orally or topically to hamsters. *Teratology.* 1983 Dec;28(3):421-6.
[3979] Chan TY. Potential dangers from topical preparations containing methyl salicylate. *Hum Exp Toxicol.* 1996 Sep;15(9):747-5.
[3980] Botma M, Colquhoun-Flannery W, Leighton S. Laryngeal oedema caused by accidental ingestion of Oil of Wintergreen. *Int J Pediatr Otorhinolaryngol.* 2001 May 11;58(3):229-32.
[3981] Chyka PA. et al. Salicylate poisoning: an evidence-based consensus guideline for out-of-hospital management. *Clin Toxicol (Phila).* 2007;45(2):95-13.

[3982] Johnson PN. Methyl salicylate/aspirin (salicylate) equivalence: who do you trust? *Vet Hum Toxicol.* 1984 Aug; 26(4):317-318
[3983] Howrie DL, Moriarty R, Breit R. Candy flavoring as a source of salicylate poisoning. *Pediatrics.* 1985 May 1;75(5):869-71.
[3984] Tisserand R, Young R. Essential Oil Safety (Second Edition). 2014 Churchill Livingstone Elsevier.
[3985] Burkhard PR, Burkhardt K, Haenggeli CA, et al. Plant-induced seizures: reappearance of an old problem. *J Neurol.* 1999 Aug;246(8):667-70.
[3986] Tanen DA, Danish DC, Reardon JM, et al. Comparison of oral aspirin versus topical applied methyl salicylate for platelet inhibition. *Ann Pharmacother.* 2008 Oct;42(10):1396-401.
[3987] Le Bourhis B, Soenen AM. Recherches sur l'action psychotrope de quelques substances aromatiques utilisees en alimentation. *Food Cosmet Toxicol.* 1973;11:1-9.
[3988] Joss JD, LeBlond RF. Potentiation of warfarin anticoagulation associated with topical methyl salicylate. *Ann Pharmacother.* 2000 Jun;34(6):729-33.
[3989] Chan TY. Potential dangers from topical preparations containing methyl salicylate. *Hum Exp Toxicol.* 1996;15:747- 50.
[3990] Parker D, Martinez C, Stanley C, et al. The analysis of methyl salicylate and salicylic acid from Chinese herbal medicine ingestion. *J Analytical Tox.* 2004 Apr;28:214-16.
[3991] Tanen DA, Danish DC, Reardon JM, et al. Comparison of oral aspirin versus topical applied methyl salicylate for platelet inhibition. *Ann Pharmacother.* 2008 Oct;42(10):1396-401.
[3992] Le Bourhis B, Soenen AM. Recherches sur l'action psychotrope de quelques substances aromatiques utilisees en alimentation. *Food Cosmet Toxicol.* 1973;11:1-9.
[3993] Joss JD, LeBlond RF. Potentiation of warfarin anticoagulation associated with topical methyl salicylate. *Ann Pharmacother.* 2000 Jun;34(6):729-33.
[3994] Chan TY. Potential dangers from topical preparations containing methyl salicylate. *Hum Exp Toxicol.* 1996;15:747- 50.
[3995] Ching J, Chua T, Chin L, β-Amyrin from Ardisia elliptica Thunb. Is more potent than aspirin in inhibiting collegen-induced platelet aggregation. *Ind J Exp Biol.* 2010;48:275-9.
[3996] Aragao GF, Carneiro LMV, Junior APF, et al. Antiplatelet Activity of α.- and β.-Amyrin, Isomeric Mixture from Protium heptaphyllum. *Pharm Biol.* 2007;45(5):343-9.
[3997] Parker D, Martinez C, Stanley C, et al. The analysis of methyl salicylate and salicylic acid from Chinese herbal medicine ingestion. *J Analytical Tox.* 2004 Apr;28:214-16.
[3998] Dobravalskyte D. Le bioraffinage de myrrhis odorata, Tussilago farfara et Calamintha grandiflora pour la production D/Aromes et D'Antioxidants. Doctorat De L'Universite de Toulouse. 2013 Nov.
[3999] Dobravalskyte D, Venskutonis PR, Zebib B, et al. Essential oil composition of Myrrhis odorata (L.) Scop. leaves grown in Lithuania and France. *J Essent Oil Res.* 2013;25(1):44-8.
[4000] Uusitalo JS, Jalonen JE, Aflatuni A, et al. Essential Leaf Oil Composition of Myrrhis odorata (L.) Scop. Grown in Finland. *J Essent Oil Res.* 1999;11(4):423-5.
[4001] Tkachenko KG, Zenkevich IG. Chemical composition of the leaf oil of Myrrhis odorata (L.) Scop. *J Essent Oil Res.* 1993 May/Jun;5:329-31.
[4002] Rancic A, Sokovic M, Vukojevic J, et al. Chemical Composition and Antimicrobial Activities of Essential Oils of Myrrhis odorata (L.) Scop, Hypericum perforatum L and Helichrysum arenarium (L.) Moench. *J Essent Oil Res.* 2005 May/Jun;17(3):341-5.
[4003] Mickiene R, Bakutis B, Maruska A, et al. Effect of the volatile secondary metabolites of Monarda didyma L., Lamium album L. and Myrrhis odorata L. plants against Micromycetes of indoor environments of animals. *Veterinarija Ir Zootechnika (Vet Med Zoot).* 2014;68(90):48-54.
[4004] European Medicines Agency. Public statement on the use of herbal medicinal products containing estragole. Available at: http://www.ema.europa.eu/docs/en_GB/document_library/Scientific_guideline/2010/04/WC500089960.pdf.
[4005] Turkyilmaz Z, Karabulut R, Sonmez K, et al. A striking and frequent cause of premature thelarche in children: Foeniculum vulgare. *J Pediatr Surg.* 2008 Nov;43(11):2109-11.
[4006] Ostad SN, Khakinegard B, Sabzevari O. Evaluation of the teratogenicity of fennel essential oil (FEO) on the rat embryo limb buds culture. *Toxicol.* 2004 Oct;18(5):623-7.
[4007] European Medicines Agency. Public statement on the use of herbal medicinal products containing estragole. Available at: http://www.ema.europa.eu/docs/en_GB/document_library/Scientific_guideline/2010/04/WC500089960.pdf.
[4008] Rosti L, Nardini A, Bettini ME, et al. Toxic effects of a herbal tea mixture in two newborns. *Acta Paediatrica.* 1994;83:683.
[4009] Dhar SK. Antifertility activity and hormonal profile of trans-anethole in rats. *Indian J Physiol Pharmacol.* 1995;39(1):63-67.
[4010] Burkhard PR, Burkhardt K, Haenggeli CA, et al. Plant-induced seizures: reappearance of an old problem. *J Neurol.* 1999 Aug;246(8):667-70.
[4011] Skalli S, Bencheikh RS. Epileptic seizure induced by fennel essential oil. *Epileptic Discord.* 2011;13:345-347.
[4012] Tognolini M, Barocelli E, Ballabeni V, et al. Comparative screening of plant essential oils; phenylpropanoid moiety as basic core for antiplatelet activity. *Life Sci.* 2006 Feb 23;78(13):1419-32.
[4013] Yoshioka M, Tamada TT. Aromatic factors of anti-platelet aggregation in fennel oil. *Biogenic Amines.* 2005 Apr;19(2):89-96.
[4014] Rather MA, Dar BA, Sofi SN, et al. Foeniculum vulgare: A comprehensive review of its traditional use, phytochemistry, pharmacology, and safety. *Arabian J Chem.* 2012 Apr;XX:1-10.
[4015] Tognolini M, Ballabeni V, Bertoni S, et al. Protective effect of Foeniculum vulgare essential oil and anethole in an experimental model of theombosis. *Pharm Res.* 2007;56:254-60.
[4016] Lee HS. Anticoagulant properties of constituents derived from fennel (Foeniculum vulgare Gaertner) fruits. *Food Sci Biotech.* 2006 Oct;15(5):763-67.
[4017] Wisniewski-Rebecca ES, Rocha BA, Wiirzler LA, et al. Synergistic effects of anethole and ibuprofen in acute inflammatory response. *Chem Biol Interact.* 2015 Dec 5;242:247-53.
[4018] El-Soud N, El-Laithy N, El-Saeed G, et al. Antidiabetic activities of Foeniculum vulgare Mill. Essential oil in streptozotocin-induced diabetic rats. *Macedonian J Med.* 2011;173:1857-5773.
[4019] Shin S, Pyun MS. Anti-Candida effects of estragole in combination with ketoconazole or amphotericin B. *Phytother Res.* 2004 Oct;18(10):827-30.
[4020] Shin S. Essential oil compounds from Agastache rugosa as antifungal agents against Trichophyton species. *Arch Pharm Res.* 2004 Mar;27(3):295-9.
[4021] Shin S, Kang CA. Antifungal activity of the essential oil of Agastache rugosa Kuntze and its synergism with ketoconazole. *Lett Appl Microbiol.* 2003;36(2):111-5.
[4022] Kubo I, Fujita K. Naturally occurring anti-Salmonella agents. *J Agric Food Chem.* 2001 Dec;49(12):5750-4.
[4023] Subehan UT, Iwata H, Kadota S, et al. Mechanism-based inhibition of CYP3A4 and CYP2D6 by Indonesian medicinal plants. *J Ethnopharmacol.* 2006 May;105(3):449-55.

[4024] Subehan Z, Kadota SF, Tezuka Y. Inhibition on human liver cytochrome P450 3A4 by constituents of fennel (Foeniculum vulgare): Identification and characterization of a mechanism-based inactivator. *J Agric Food Chem*. 2007 Dec;55(25):10162-67.

[4025] Khan MM. Masters Thesis: Inhibition of Cytochrome P450 2E1 and Cytochrome P450 2A6 by Essential Oils: Tarragon (Artemisia dracunculus) and Basil (Ocimum basilicum). 2014. Available at: https://libres.uncg.edu/ir/uncg/f/Khan_uncg_0154M_11587.pdf.

[4026] Rompelberg CJ, Verhagen H, van Bladeren PJ. Effects of the naturally occurring alkenylbenzenes eugenol and trans-anethole on drugmetabolizing enzymes in the rat liver. *Food Chem Toxicol*. 1993 Sep;31(9):637-45.

[4027] Brehm-Stecher BF, Johnson EA. Sensitization of Staphylococcus aureus and Escherichia coli to antibiotics by the sesquiterpenoids nerolidol, farnesol, bisabolol, and apritone. *Antimicrob Agents Chemother*. 2003 Oct;47(10):3357-60.

[4028] Špičáková A, Szotáková B, Dimunová D, et al. Nerolidol and Farnesol Inhibit Some Cytochrome P450 Activities but Did Not Affect Other Xenobiotic-Metabolizing Enzymes in Rat and Human Hepatic Subcellular Fractions. *Molecules*. 2017 Mar 24;22(4). pii: E509.

[4029] Rancic A, Sokovic M, Vukojevic J, et al. Chemical Composition and Antimicrobial Activities of Essential Oils of Myrrhis odorata (L.) Scop, Hypericum perforatum L and Helichrysum arenarium (L.) Moench. *J Essent Oil Res*. 2005 May/Jun;17(3):341-5.

[4030] Bekana D, Kebede T, Assefa M, et al. Comparative Phytochemical Analyses of Resins of Boswellia Species (B. papyrifera (Del.) Hochst., B. neglecta S. Moore, and B. rivae Engl.) from Northwestern, Southern, and Southeastern Ethiopia. *ISRN Analytical Chem*. 2014;2014:374678.

[4031] Baser KHC, Demirci B, Dekebo A, et al. John Wiley & Sons, Ltd. Essential oils of some Boswellia spp., Myrrh and Opopanax. *Flavour Frag J*. 2003;18:153-6.

[4032] Asefa M, Shimelis B, Kassa H, et al. Chemical investigations on frankincense from Boswellia trees to improve production, handling and grading practices for the export market in Ethiopia. *G J P&A Tech*. 2012;02(5):15-30.

[4033] Basar S. Phytochemical investigations of Boswellia species: Comparative studies on the essential oils, pyrolysates and boswellic acids of Boswellia carterii Birdw., Boswellia serrata Roxb., Boswellia frereana Birdw., Boswellia neglecta S. Moore and Boswellia rivae Engl. 2005 thesis.

[4034] Camarda L, Dayton T, Stefano VD, et al. Chemical composition and antimicrobial activity of some oleogum resin essential oils from Boswellia spp (Burseraceae). *Annali di Chimica*. 2007;97;837-43.

[4035] Schillaci D, Arizza V, Dayton T, et al. In vitro anti-biofilm activity of Boswellia spp. oleogum resin essential oils. *Lett Appl Microbiol*. 2008 Nov;47(5):433-8.

[4036] Camarda L, Dayton T, Stefano VD, et al. Chemical composition and antimicrobial activity of some oleogum resin essential oils from Boswellia spp (Burseraceae). *Annali di Chimica*. 2007;97;837-43.

[4037] de Rapper S, Van Vuuren SF, Kamatou GP, et al. The additive and synergistic antimicrobial effects of select frankincense and myrrh oils--a combination from the pharaonic pharmacopoeia. *Lett Appl Microbiol*. 2012 Apr;54(4):352-8.

[4038] Schillaci D, Arizza V, Dayton T, et al. In vitro anti-biofilm activity of Boswellia spp. oleogum resin essential oils. *Lett Appl Microbiol*. 2008 Nov;47(5):433-8.

[4039] Tava A. Coumarin-Containing Grass: Volatiles from Sweet Vernalgrass (Anthoxanthum odoratum L.). *J Essent Oil Res*. 2001 Sep/Oct;13:367-70.

[4040] Lawrence BM. Progress in essential oils. *Perf Flavourist*. 2009;34:54-56.

[4041] Hildebrand T, Tufte M, Manwill P, et al. Antimicrobial Properties of Essential Oils Isolated from Anthoxanthum (Poaceae) Against Soil Bacteria. Botany 2013.

[4042] Felter SP, Vasallo JD, Carlton BD, et al. A safety assessment of coumarin taking into account species specific toxicokinetics. *Food Chem Toxicol*. 2006 Apr;44(4):462-75.

[4043] European Food Safety Authority. Coumarin in flacourings and other food ingredients with flavouring properties: Scientific opinion of the panel on food additives, flavourings, processing aids and materials in contact with food (AFC). Available at: http://www.efsa.europa.eu/sites/default/files/scientific_output/files/main_documents/793.pdf.

[4044] Kim SY, Koo YK, Koo JY, et al. Platelet anti-aggregation activities of constituents from Cinnamomum cassia. *J Med Food*. 2010 Oct;13(5):1069-74.

[4045] Collin G, Gagnon H. Chemical composition and stability of the hydrosol obtained during the production of essential oils. III. The case of Myrica gale L., Comptonia peregrina (L.) Coulter and Ledum groenlandicum Retzius. *Am J Essent Oil Nat Prod*. 2016;4(1):07-19.

[4046] Lawrence BM, Weaver KM. Essential oils and their constituents. XII. A Comparative chemical composition of the essential oils of Myrica gale and Comptonia peregrina. Planta Med. 1974;25(4):384-388.

[4047] Sylvestre M, Pichette A, Lavoie S, et al. Composition and cytotoxic activity of the leaf essential oil of Comptonia peregrina (L.) Coulter, Phytother Res. 2007;21(6):536-540.

[4048] Collin GJ, Hachey JM, Simard S. Extraction and GC-MS Analysis of the Essential Oil of Comptonia peregrina (L.) Coulter. *Flavour Frag J*. 1988;3:65-8.

[4049] Sylvestre M, Pichette A, Lavoie S, et al. Composition and cytotoxic activity of the leaf essential oil of Comptonia peregrina (L.) Coulter, Phytother Res. 2007;21(6):536-540.

[4050] Nakata M, Myoda T, Wakita Y, et al. Volatile components of essential oil from cultivates Myrica gale var, tomentosa and its antioxidant and antimicrobial activities. *J Oleo Sci*. 2013;62(9):755-62.

[4051] Collin G, Gagnon H. Chemical composition and stability of the hydrosol obtained during the production of essential oils. III. The case of Myrica gale L., Comptonia peregrina (L.) Coulter and Ledum groenlandicum Retzius. *Am J Essent Oils Nat Prod*. 2016;4(2):07-19.

[4052] Sylvestre M, Legault J, Dufour D, et al. Chemical composition and anticancer activity of leaf essential oil of Myrica gale L. *Phytomed*. 2005;12:299-304.

[4053] Sylvestre M, Legault J, Lavoie S, et al. Investigation of Leaf Essential Oil of Myrica gale L. from Quebec: Purification and Analysis of Oxygenated Fractions. *J Essent Oil Res*. 2006;18(1):38-41.

[4054] Popovici J, Bertrand C, Bagnarol E, et al. Chemical composition of essential oil and headspace-solid microextracts from fruits of Myrica gale L. and antifungal activity. *Nat Prod Commun*. 2008 Aug 15;22(12):1024-32.

[4055] Chang X, Martin P. N, P and K Fertilizers Alter Plant Growth, Essential Oil Yield and Gender of Sweet Gale (Myrica gale L. *Open Plant Sci J*. 2014;8:9-17.

[4056] Belanger A, Dextraze L, Isnardi MJ, et al. Chemical composition of essential oil and headspace of the Quebec myrique baumier wax myrtle (Myrica gale L.). *J Essent Oil Res*. 1997 Nov;9(6):657-62.

[4057] Von Schantz M, Kapetanidis I. Qualitative and quantitative study of the essential oils of Myrica gale (Myricaceae). *Pharmaceutica Acta Helvetiae*. 1971;46(10-11):649-656.

[4058] Tattje DHE, Bos R. Essential oils of Myrica gale. *Pharmaceutish Weekblad*. 1974;109(49):1189-1195.

[4059] Carlton RR, Waterman PG, Gray AI. Variation of leaf gland volatile oil within a population of sweet gale (Myrica gale) (Myricaceae). *Chemoecology*. 1992;3(1):45-54.

[4060] Negueruela VA, Alonso PMJ, Rico MM. Chemical composition of the essential oil of a relict Spanish population of Myrica gale L. *Anales de Bromatología*. 1982;34(2):231-238.
[4061] Svoboda KP, Inglis A, Hampson J, Galambosi B, et al. Biomass production, essential oil yield and composition of Myrica gale L. harvested from wild populations in Scotland and Finland. *Flavour Fragr J.* 1998;13:367-372.
[4062] Lawrence BM, Weaver KM. Essential oils and their constituents. XII. A Comparative chemical composition of the essential oils of Myrica gale and Comptonia peregrina. *Planta Med.* 1974;25(4):384-388.
[4063] Halim AF, Collins RP. Essential oil analysis of the Myricaceae of the Eastern United States. *Phytochem.* 1973;12:1077-1083.
[4064] Burkhard PR, Burkhardt K, Haenggeli CA, et al. Plant-induced seizures: reappearance of an old problem. *J Neurol.* 1999 Aug;246(8):667-70.
[4065] Waldman N. Seizure caused by dermal application of over-the-counter eucalyptus oil head lice preparation. *Clin Toxicol (Phila).* 2011 Oct;49(8):750-1.
[4066] Craig JO. Poisoning by the volatile oils in childhood. *Arch Dis Child.* 1953;28:259-67.
[4067] Mathew T, Kamath V, Kumar RS, et al. Eucalyptus oil inhalation-induced seizure: A novel, underrecognized, preventable cause of acute symptomatic seizure. *Epilepsia Open.* 2017 Jul 4;2(3):350-354.
[4068] De Olivera AC, Ribeiro-Pinto LF, Paumgartten JR. In vitro inhibition of CYP2B1 monoxygenase by beta-myrcene and other monoterpenoid constituents. *Br J Nutr.* 1999;81:289-95.
[4069] De-Oliveira AC1, Ribeiro-Pinto LF, Otto SS, et al. Induction of liver monooxygenases by beta-myrcene. *Toxicology.* 1997 Dec 26;124(2):135-40.
[4070] Pimkaew P1, Küblbeck J, Petsalo A, et al. Interactions of sesquiterpenes zederone and germacrone with the human cytochrome P450 system. *Toxicol In Vitro.* 2013 Sep;27(6):2005-12.
[4071] Sylvestre M, Legault J, Dufour D, et al. Chemical composition and anticancer activity of leaf essential oil of Myrica gale L. *Phytomed.* 2005;12:299-304.
[4072] Nakata M, Myoda T, Wakita Y, et al. Volatile components of essential oil from cultivates Myrica gale var, tomentosa and its antioxidant and antimicrobial activities. *J Oleo Sci.* 2013;62(9):755-62.
[4073] Popovici J, Bertrand C, Bagnarol E, et al. Chemical composition of essential oil and headspace-solid microextracts from fruits of Myrica gale L. and antifungal activity. *Nat Prod Commun.* 2008 Aug 15;22(12):1024-32.
[4074] Nakata M, Myoda T, Wakita Y, et al. Volatile components of essential oil from cultivates Myrica gale var, tomentosa and its antioxidant and antimicrobial activities. *J Oleo Sci.* 2013;62(9):755-62.
[4075] Jaenson TG, Pålsson K, Borg-Karlson AK. Evaluation of extracts and oils of mosquito (Diptera: Culicidae) repellent plants from Sweden and Guinea-Bissau. *J Med Entomol.* 2006 Jan;43(1):113-9.
[4076] Jaenson TG, Pålsson K, Borg-Karlson AK. Evaluation of extracts and oils of tick-repellent plants from Sweden. *Med Vet Entomol.* 2005 Dec;19(4):345-52.
[4077] Behra, private communication, 2003. In: Tisserand R, Young R. Essential Oil Safety, Second Edition. Churchill Livingstone: 2014. p. 434.
[4078] Oshadi, AYUS Essential Oils and Natural products Ltd. Private communication.
[4079] Saab A, Harb F, Koenig WA. Essential oils components in the leaves of Cedrus libani and Cedrus deodora. *Minerva Biotec.* 2009;21:201-5.
[4080] Yilmaz N, Alma MK, Nitz S, et al. Chemical compositions of the essential oils from oleoresin on cones of Cedrus libani. *Asian J Chem.* 2005;17(4):2300-6.
[4081] Canard D, Perru O, Tauzin V, et al. Terpene composition variations in diverse provenances of Cedrus libani (A.) Rich. and Cedrus atlantica Manet. *Trees.* 1997;11:504-10.
[4082] Baser KHC, Demircakmak B. The essential oil of Taurus cedar (Cedrus libani A. Rich): Recent results. *Chem Nat Comp.* 1995;31(1):16-21
[4083] Loizzo MR, Saab AM, Statti GA, et al. Composition and α-amylase inhibitory effect of essential oils from Cedrus libani. *Fitoterpia.* 2007;78:323-6.
[4084] Bilir N, Avci AB. Essential Oil in Taurus Cedar (Cedrus libani A. Rich) Seeds. *J Essent Oil Bearing Plants.* 2013;16(4):538-44.
[4085] Avci AB, Bilir N, Effects of growth and geographic characteristics on essential oil of Taurus cedar (Cedrus libani A, Rich.). *Fresenius Enviro Bullet.* 2017;26(6):3919-23.
[4086] Tumen I, Hafizoglu H, Kilic A, et al. Yields and constituents of essential oil from cones of Pinaceae spp. natively grown in Turkey. *Molecules.* 2010 Aug 24;15(8):5797-806.
[4087] Saab AM, Lampronti I, Borgatti M, et al. In vitro evaluation of the anti-proliferative activities of the wood essential oils of three Cedrus species against K562 human chronic myelogenous leukaemia cells. *Nat Prod Res.* 2012;26(23):2227-31.
[4088] Saab AM, Guerrini A, Sacchetti G, et al. Phytochemical analysis and cytotoxicity towards multidrug-resistant leukemia cells of essential oils derived from Lebanese medicinal plants. *Planta Med.* 2012 Dec;78(18):1927-31.
[4089] Loizzo MR, Saab AM, Statti GA, et al. Composition and α-amylase inhibitory effect of essential oils from Cedrus libani. *Fitoterpia.* 2007;78:323-6.
[4090] Tumen I, Akkol EK, Süntar I, et al. Wound repair and anti-inflammatory potential of essential oils from cones of Pinaceae: preclinical experimental research in animal models. *J Ethnopharmacol.* 2011 Oct 11;137(3):1215-20.
[4091] Loizzo MR, Saab A, Tundis R, et al. Phytochemical analysis and in vitro evaluation of the biological activity against herpes simplex virus type 1 (HSV-1) of Cedrus libani A. Rich. *Phytomedicine.* 2008 Jan;15(1-2):79-83.
[4092] Cetin H, Kurt Y, Isik K, et al. Larvicidal effect of Cedrus libani seed oils on mosquito Culex pipiens. *Pharm Biol.* 2009;47(8):665-8.
[4093] Rehman S, Bhatti HN, Iqbal Z, et al. Essential oil composition of commercial black tea (Camellia sinensis). *Int J Food Sci Technol.* 2008;43:346-50.
[4094] Lin J, Dai Y, Guo Y, et al. Volatile profile analysis and quality prediction of Longjing tea (Camellia sinensis) by HS-SPME/GC-MS. *J Zhejiang Univ-Sci B (Biomed & Biotechnol).* 2012;13(12):972-80.
[4095] Wang K, Liu F, Liu Z, et al. Comparison of catechins and volatile compounds among different types of tea using high performance liquid chromatograph and gas chromatograph mass spectrometer. *Int J Food Sci Technol.* 2011;46:1406–1412.
[4096] Cosmetic Ingredient Review. Safety Assessment of Camellia sinensis-Derived Ingredients as Used in Cosmetics. 2014. Available at: https://www.cir-safety.org/sites/default/files/camellia_2.pdf.
[4097] IFRA. Tea Leaf Absoute. Available at: file:///C:/Users/scott.johnson/Downloads/23273_STRS_2013_06_17_Tea_leaf_absolute%20(1).pdf.
[4098] Montanari RM, Barbosa LCA, Demuner AJ, et al. Chemical composition and antibacterial activity of essential oils from Verbenaceae species: Alternative sources of (E)-caryophyllene and germacrene-D. *Quim Nova.* 2011;34(9):1550-5.
[4099] Muhayimana A, Chalchat JC, Garry RP. Chemical Composition of Essential Oils of Lantana trifolia L. from Rwanda. *J Essent Oil Res.* 1998 Sep/Oct;10:547-9.
[4100] Bendera MM. Isolation and characterization of essential oils from Ocimum americanum, Lantana camara, Lantana trifolia and Tephrosia vogelii. Egerton University Thesis. October 2007. Available at: http://ir-library.egerton.ac.ke/jspui/bitstream/123456789/287/1/Isolation%20and%20characterisation%20of%20essential%20oils%20from%20Ocium%20americunum%2CLantana%20camara%2Clantana%20trifolia%20and%20Tephrosia%20vogelii.PDF.

[4101] Juliao LDS, Bizzo HR, Souza AM, et al. Essential Oils from two Lantana species with Antimycobacterial Activity. *Nat Prod Commun.* 2009;4(12):1733-6.
[4102] de Sena Filho JG, Rabbani ARC, dos Santos Silva TR, et al. Chemical and molecular characterization of fifteen species from Lantana (Verbenaceae) genus. *Biochem Systematics Ecology*. 2012 Dec;45:130-7.
[4103] Juliao LDS, Bizzo HR, Souza AM, et al. Essential Oils from two Lantana species with Antimycobacterial Activity. *Nat Prod Commun.* 2009;4(12):1733-6.
[4104] Hafi A, El Beyrouthy M, Ouaini N, et al. Chemical Composition and Antimicrobial Activity of Origanum libanoticum, Origanum ehrenbergii, and Origanum syriacum Growing Wild in Lebanon. *Chem Biodivers.* 2016 May 1;13(5):555-60.
[4105] American University of Beirut. Nature Conservation Center. Origanum ehrenbergii. Available at: https://website.aub.edu.lb/units/natureconservation/programs/medicinal/essential_oils/Pages/oe.aspx.
[4106] Figueredo G, Chalchat JC, Pasquier B. A Study of the Mediterranean Oregano Populations. Chemical Composition of Essential Oils of Origanum ehrenbergii Boiss. from Two Populations in Lebanon. *J Essent Oil Res.* 2005;17(6):593-6.
[4107] Loizzo MR, Menichini F, Conforti F, et al. Chemical analysis, antioxidant, antiinflammatory and anticholinesterase activities of Origanum ehrenbergii Boiss and Origanum syriacum L. essential oils. *Food Chem.* 2009;117:174-80.
[4108] El Khoury R, Jubeli RM, El Beyrouthy M, et al. Phytochemical screening and antityrosinase activity of carvacrol, thymoquinone, and four essential oils of Lebanese plants. *J Cosmetic Dermatol.* 2018 Oct 5. [Epub ahead of print]
[4109] Domaracky M, Rehak P, Juhas S, et al. Effects of selected plant essential oils on the growth and development of mouse preimplantation embryos in vivo. *Physiol Res.* 2007;56(1):97-104.
[4110] Paramasivam A, Kalaimangai M, Sambantham S, et al. Anti-angiogenic activity of thymoquinone by the down-regulation of VEGF using zebrafish (Danio rerio) model. *Biomed Prevent Nutr.* 2012 Jul-Sep;2(3):169-73.
[4111] Yi T, Cho SG, Yi A, et al. Thymoquinone inhibits tumor angiogenesis and tumor growth through suppressing AKT and ERK signaling pathways. *Mol Cancer Ther.* 2008 Jul;7(7):1789-96.
[4112] Peng L, Liu A, Shen Y, et al. Antitumor and anti-angiogenesis effects of thymoquinone on osteosarcoma through the NF-κB pathway. *Oncol Rep.* 2013 Feb;29(2):571-8.
[4113] AbuKhader MM, Khater SH, Al-Matubsi HY. Acute effects of thymoquinone on the pregnant rat and embryo-fetal development. *Drug Chem Toxicol.* 2013 Jan;36(1):27-34.
[4114] Lemhadri A, Zeggwagh NA, Maghrani M, et al. Anti-hypoglycaemic activity of the aqueous extract of Origanum vulgare growing wild in Tafilalet region. *J Ethnopharmacol.* 2004 Jun;92(2-3):251-6.
[4115] Bayramoglu G, Senturk H, Bayramoglu A, et al. Carvacrol partially reverses symptoms of diabetes in STZ-induced diabetic rats. *Cytotechnology*. 2014 Mar;66(2):251–257.
[4116] Ezhumalai M, Radhiga T, Pugalendi KV. Antihyperglycemic effect of carvacrol in combination with rosiglitazone in high-fat diet-induced type 2 diabetic C57BL/6J mice. *Mol Cell Biochem.* 2014 Jan;385(1-2):23-31.
[4117] Umasankar K, Ramya DR, Nambikkairaj B. Antidiabetic and antidyslipidemic nature of carvacrol, a monoterpenic phenol studied in high-fat - fed and low-dose streptozotocin-induced experimental diabetic rats. *J Pharm Res.* 2015;9(8):484-90.
[4118] Tognolini M, Barocelli E, Ballabeni V, et al. Comparative screening of plant essential oils; phenylpropanoid moiety as basic core for antiplatelet activity. *Life Sci.* 2006 Feb 23;78(13):1419-32.
[4119] Tsai HH, Lin HW, Chen YL, et al. A review of potential harmful interactions between anticoagulant/antiplatelet agents and Chinese herbal medicines. *PLoS One.* 2013 May 9;8(5):e64255.
[4120] Karkabounas S, Kostoula OK, Daskalou T, et al. Anticarcinogenic and antiplatelet effects of carvacrol. *Exp Oncol.* 2006 Jun;28(2):121-5.
[4121] Baser KH. Biological and pharmacological activities of carvacrol and carvacrol bearing essential oils. *Curr Pharm Des.* 2008;14(29):3106-19.
[4122] Okazaki K, Kawazoe K, Takaishi Y. Human platelet aggregation inhibitors from thyme (Thymus vulgaris L.). *Phytother Res.* 2002 Jun;16(4):398-9.
[4123] Jukic M, Politeo O, Maksimmovic M, et al. In vitro acetylcholinesterase inhibitory properties of thymol, carvacrol and their derivatives thymoquinone and thymohydroquinone. *Phytother Res.* 2007;21(3):259-61.
[4124] Askin H, Yildiz M, Ayar A. Effects of Thymol and Carvacrol on Acetylcholinesterase from Drosophila melanogaster. *Acta Physica Polinica A*. 2017;132(3):720-22.
[4125] Medeiros Barreto H, Cerqueira Fontinele F, Pereira de Oliveira A, et al. Phytochemical prospection and modulation of antibiotic activity in vitro by Lippia origanoides H.B.K. in methicillin resistant Staphylococcus aureus. *Biomed Res Int.* 2014;305610:1–7.
[4126] Aelenei P, Miron A, Trifan A, et al. Essential Oils and Their Components as Modulators of Antibiotic Activity against Gram-Negative Bacteria. *Medicines*. 2016;3(19):2-34.
[4127] Magi G, Marini E, Facinelli B. Antimicrobial activity of essential oils and carvacrol, and synergy of carvacrol and erythromycin, against clinical, erythromycin-resistant Group A Streptococci. *Front Microbiol.* 2015;6:165.
[4128] Veras HNH, Rodrigues FFG, Colares AV, et al. Synergistic antibiotic activity of volatile compounds from the essential oil of Lippia sidoides and thymol. *Fitoterapia*. 2012 Apr;83(3):508-12.
[4129] Langeveld WT, Veldhuizen EJ, Burt SA. Synergy between essential oil components and antibiotics: A review. *Crit Rev Microbiol.* 2014 Feb;40(1):76-94.
[4130] Bamosa AO. A review on the hypoglycemic effect of nigella sativa and thymoquinone. *Rev Articles.* 2015;3(1):2-7.
[4131] Ali Sangi SM, Sulaiman MI, Abd El-wahab MF, et al. Antihyperglycemic effect of thymoquinone and oleuropein, on streptozotocin-induced diabetes mellitus in experimental animals. *Pharmacogn Mad.* 2015 Oct;11(Suppl 2):S251-57.
[4132] El-Dakhakhny M, Mady N, Lembert N, et al. The hypoglycemic effect of Nigella sativa oil is mediated by extrapancreatic actions. *Planta Med.* 2002 May;68(5):465-6.
[4133] Fararh KM, Shimizu Y, Shiina T, et al. Thymoquinone reduces hepatic glucose production in diabetic hamsters. *Res Vet Sci.* 2005 Dec;79(3):219-23.
[4134] Razavi BM, Hosseinzadeh H. A review of the effects of Nigella sativa L. and its constituent, thymoquinone, in metabolic syndrome. *J Endocrinol Invest.* 2014 Nov;37(11):1031-40.
[4135] Sheikh BA, Pari L, Rathinham A, et al. Trans-anethole, a terpenoid ameliorates hyperglycemia by regulating key enzymes of carbohydrate metabolism in streptozotocin induced diabetic rats. *Biochimie*. 2015 May;112:57-65.
[4136] Pari L, Sheikh BA. Antihyperglycemic effect of trans-anethole in streptozotocin induced diabetic rats with special reference to glycoprotein components. *Int J Adv Res Biol Sci.* 2015;2(5):28-34.
[4137] MM. Thymoquinone: a promising antidiabetic agent. *Int J Diabetes Dev Countries.* 2012 Jun;32(2):65-68.

[4138] Jukic M, Politeo O, Maksimmovic M, et al. In vitro acetylcholinesterase inhibitory properties of thymol, carvacrol and their derivatives thymoquinone and thymohydroquinone. *Phytother Res.* 2007;21(3):259-61.
[4139] Tognolini M, Barocelli E, Ballabeni V, et al. Comparative screening of plant essential oils; phenylpropanoid moiety as basic core for antiplatelet activity. *Life Sci.* 2006 Feb 23;78(13):1419-32.
[4140] Okazaki K, Kawazoe K, Takaishi Y. Human platelet aggregation inhibitors from thyme (Thymus vulgaris L.). *Phytother Res.* 2002 Jun;16(4):398-9.
[4141] Lee HS. Anticoagulant properties of constituents derived from fennel (Foeniculum vulgare Gaertner) fruits. *Food Sci Biotech.* 2006 Oct;15(5):763-67.
[4142] Thushara RM, Hemshekhar M, Santhosh MS, et al. Differential action of phytochemicals on platelet apoptosis: a biological overview. *Curr Med Chem.* 2013;20(8):1018-27.
[4143] Towhid ST, Schmidt EM, Schmid E, et al. Thymoquinone-induced platelet apoptosis. *J Cell Biochem.* 2011 Nov;112(11):3112-21.
[4144] Jukic M, Politeo O, Maksimmovic M, et al. In vitro acetylcholinesterase inhibitory properties of thymol, carvacrol and their derivatives thymoquinone and thymohydroquinone. *Phytother Res.* 2007;21(3):259-61.
[4145] Subehan UT, Iwata H, Kadota S, et al. Mechanism-based inhibition of CYP3A4 and CYP2D6 by Indonesian medicinal plants. *J Ethnopharmacol.* 2006 May;105(3):449-55.
[4146] Al-Jenoobi FI, Al-Thukair AA, Abbas FA, et al. Effect of black seed on dextromethorphan o- and n-demethylation in human liver microsomes and healthy human subjects. *Drug Metabolism Letters.* 2010;4(1):51–55.
[4147] Liu X, Park JH, Abd El-Aty AM, et al. Isolation of volatiles from Nigella sativa seeds using microwave-assisted extraction: effect of whole extracts on canine and murine CYP1A. *Biomed Chromatogr.* 2013 Jul;27(7):938-45.
[4148] Ahmad A, Khan RM, Alkharfy KM, et al. Effects of Thymoquinone on the Pharmacokinetics and Pharmacodynamics of Glibenclamide in a Rat Model. *Nat Prod Commun.* 2015 Aug;10(8):1395-8.
[4149] Langeveld WT, Veldhuizen EJ, Burt SA. Synergy between essential oil constituents and antibiotics. *Crit Rev Microbiol.* 2014 Feb;40(1);76-94.
[4150] Kouidhi B, Zmantar T, Jrah H, et al. Antibacterial and resistance-modifying activities of thymoquinone against oral pathogens. *Ann Clin Microbiol Antimicrob.* 2011 Jun 27;10:29.
[4151] Raza M, Alghasham AA, Alorainy MS, et al. Potentiation of Valproate-induced Anticonvulsant Response by Nigella sativa Seed Constituents: The Role of GABA Receptors. *Int J Health Sci (Qassim).* 2008 Jan;2(1):15-25.
[4152] Kohlert C, Schindler G, Marz RW, et al. Systemic availability and pharmacokinetics of thymol in humans. *J Clin Pharmacol.* 2002 Jul;42(7):731-7.
[4153] Toxicology Data Network, National Library of Medicine. Thymol. Available at: http://toxnet.nlm.nih.gov/cgibin/ sis/search/a?dbs+hsdb:@term+@DOCNO+866.
[4154] Tognolini M, Barocelli E, Ballabeni V, et al. Comparative screening of plant essential oils; phenylpropanoid moiety as basic core for antiplatelet activity. *Life Sci.* 2006 Feb 23;78(13):1419-32.
[4155] Okazaki K, Kawazoe K, Takaishi Y. Human platelet aggregation inhibitors from thyme (Thymus vulgaris L.). *Phytother Res.* 2002 Jun;16(4):398-9.
[4156] Jukic M, Politeo O, Maksimmovic M, et al. In vitro acetylcholinesterase inhibitory properties of thymol, carvacrol and their derivatives thymoquinone and thymohydroquinone. *Phytother Res.* 2007;21(3):259-61.
[4157] Askin H, Yildiz M, Ayar A. Effects of Thymol and Carvacrol on Acetylcholinesterase from Drosophila melanogaster. *Acta Physica Polinica A.* 2017;132(3):720-22.
[4158] Medeiros Barreto H, Cerqueira Fontinele F, Pereira de Oliveira A, et al. Phytochemical prospection and modulation of antibiotic activity in vitro by Lippia origanoides H.B.K. in methicillin resistant Staphylococcus aureus. *Biomed Res Int.* 2014;305610:1–7.
[4159] Aelenei P, Miron A, Trifan A, et al. Essential Oils and Their Components as Modulators of Antibiotic Activity against Gram-Negative Bacteria. *Medicines.* 2016;3(19):2-34.
[4160] Magi G, Marini E, Facinelli B. Antimicrobial activity of essential oils and carvacrol, and synergy of carvacrol and erythromycin, against clinical, erythromycin-resistant Group A Streptococci. *Front Microbiol.* 2015;6:165.
[4161] Veras HNH, Rodrigues FFG, Colares AV, et al. Synergistic antibiotic activity of volatile compounds from the essential oil of Lippia sidoides and thymol. *Fitoterapia.* 2012 Apr;83(3):508-12.
[4162] Langeveld WT, Veldhuizen EJ, Burt SA. Synergy between essential oil components and antibiotics: A review. *Crit Rev Microbiol.* 2014 Feb;40(1):76-94.
[4163] Miladinovic DL, Ilic BS, Kocic BD, et al. Antibacterial investigation of thyme essential oil and its main constituents on combination with tetracycline. *J Med Food.* 2015 Aug;18(8):935-7.
[4164] Shin S, Lim S. Antifungal effects of herbal essential oils alone and in combination with ketoconazole against Trichophyton spp. *J Appl Micribiol.* 2004;97:1289-96.
[4165] Kissels W, Wu X, Santos RR, et al. Short communication: Interaction of the isomers carvacrol and thymol with the antibiotics doxycycline and tilmicosin: In vitro effects against pathogenic bacteria commonly found in the respiratory tract of calves. *J Dairy Sci.* 2017 Feb;100(2):970-974.
[4166] Tognolini M, Barocelli E, Ballabeni V, et al. Comparative screening of plant essential oils; phenylpropanoid moiety as basic core for antiplatelet activity. *Life Sci.* 2006 Feb 23;78(13):1419-32.
[4167] Okazaki K, Kawazoe K, Takaishi Y. Human platelet aggregation inhibitors from thyme (Thymus vulgaris L.). *Phytother Res.* 2002 Jun;16(4):398-9.
[4168] Jukic M, Politeo O, Maksimmovic M, et al. In vitro acetylcholinesterase inhibitory properties of thymol, carvacrol and their derivatives thymoquinone and thymohydroquinone. *Phytother Res.* 2007;21(3):259-61.
[4169] Askin H, Yildiz M, Ayar A. Effects of Thymol and Carvacrol on Acetylcholinesterase from Drosophila melanogaster. *Acta Physica Polinica A.* 2017;132(3):720-22.
[4170] Medeiros Barreto H, Cerqueira Fontinele F, Pereira de Oliveira A, et al. Phytochemical prospection and modulation of antibiotic activity in vitro by Lippia origanoides H.B.K. in methicillin resistant Staphylococcus aureus. *Biomed Res Int.* 2014;305610:1–7.
[4171] Aelenei P, Miron A, Trifan A, et al. Essential Oils and Their Components as Modulators of Antibiotic Activity against Gram-Negative Bacteria. *Medicines.* 2016;3(19):2-34.
[4172] Magi G, Marini E, Facinelli B. Antimicrobial activity of essential oils and carvacrol, and synergy of carvacrol and erythromycin, against clinical, erythromycin-resistant Group A Streptococci. *Front Microbiol.* 2015;6:165.
[4173] Veras HNH, Rodrigues FFG, Colares AV, et al. Synergistic antibiotic activity of volatile compounds from the essential oil of Lippia sidoides and thymol. *Fitoterapia.* 2012 Apr;83(3):508-12.
[4174] Langeveld WT, Veldhuizen EJ, Burt SA. Synergy between essential oil components and antibiotics: A review. *Crit Rev Microbiol.* 2014 Feb;40(1):76-94.

[4175] Miladinovic DL, Ilic BS, Kocic BD, et al. Antibacterial investigation of thyme essential oil and its main constituents on combination with tetracycline. *J Med Food.* 2015 Aug;18(8):935-7.
[4176] Shin S, Lim S. Antifungal effects of herbal essential oils alone and in combination with ketoconazole against Trichophyton spp. *J Appl Micribiol.* 2004;97:1289-96.
[4177] Kissels W, Wu X, Santos RR, et al. Short communication: Interaction of the isomers carvacrol and thymol with the antibiotics doxycycline and tilmicosin: In vitro effects against pathogenic bacteria commonly found in the respiratory tract of calves. *J Dairy Sci.* 2017 Feb;100(2):970-974.
[4178] Loizzo MR, Menichini F, Conforti F, et al. Chemical analysis, antioxidant, antiinflammatory and anticholinesterase activities of Origanum ehrenbergii Boiss and Origanum syriacum L. essential oils. *Food Chem.* 2009;117:174-80.
[4179] Loizzo MR, Menichini F, Conforti F, et al. Chemical analysis, antioxidant, antiinflammatory and anticholinesterase activities of Origanum ehrenbergii Boiss and Origanum syriacum L. essential oils. *Food Chem.* 2009;117:174-80.
[4180] El Khoury R, Jubeli RM, El Beyrouthy M, et al. Phytochemical screening and antityrosinase activity of carvacrol, thymoquinone, and four essential oils of Lebanese plants. *J Cosmetic Dermatol.* 2018 Oct 5. [Epub ahead of print]
[4181] Hafi A, El Beyrouthy M, Ouaini N, et al. Chemical Composition and Antimicrobial Activity of Origanum libanoticum, Origanum ehrenbergii, and Origanum syriacum Growing Wild in Lebanon. *Chem Biodivers.* 2016 May 1;13(5):555-60.
[4182] Loizzo MR, Menichini F, Conforti F, et al. Chemical analysis, antioxidant, antiinflammatory and anticholinesterase activities of Origanum ehrenbergii Boiss and Origanum syriacum L. essential oils. *Food Chem.* 2009;117:174-80.
[4183] Alagic S, Stancic I, Palic R, et al. Chemical Composition and Antimicrobial Activity of the Essential Oil of the Oriental Tobacco Yaka. *J Essent Oil Res.* 2002 May/Jun;14:230-2.
[4184] Wilson RA, Mookherjee BD, Vinals JF. A Comparative Analysis of the Volatile Constituents of Virginia, Burley, Turkish and Black Tobaccos. ACS Meeting, Kansas City. 1982.
[4185] Stojanovic G, Palic R, Alagic S, et al. Chemical composition and antimicrobial activity of the essential oil and CO2-extracts of semioriental tobacco. *Flavour Frag J.* 2000;15:335-8.
[4186] Nurnasari E, Subiyakto. Chemical Compound of Essential Oils from Three Different Area of Tobacco Leaves (Nicotiana tabaccum L.) in Indonesia. *J Life Sci Biomed.*, 2015 Nov;5(6):163-6.
[4187] Palic R, Stojanovic G, Alagic S, et al. Chemical composition and antimicrobial activity of the essential oil and CO2 extracts of semi-oriental tobacco, Otlja. *Flavour Frag J.* 2002;17:323-26.
[4188] Kidah MI. Chemical Constituents of the Essential Oil Extracted from Nicotiana tabacum Leaves. *Biotechno J Int.* 2018;21(1):1-4.
[4189] Cvetanovska L, Cvetanovska A, Krstikj M. Production of essential oil and concret in various types of Tobaco (Nicotiana tabacum L.). *Eur J Pharm Med Res.* 2018;5(03):57-60.
[4190] Singh R, Dar SA, Sharma P. Antibacterial activity and toxicological evaluation of semi purified hexane extract of Urtica dioica leaves. *Res J Med Plants.* 2012;6(2):123-35.
[4191] Alagic S, Stancic I, Palic R, et al. Chemical Composition and Antimicrobial Activity of the Essential Oil of the Oriental Tobacco Yaka. *J Essent Oil Res.* 2002 May/Jun;14:230-2.
[4192] Palic R, Stojanovic G, Alagic S, et al. Chemical composition and antimicrobial activity of the essential oil and CO2 extracts of semi-oriental tobacco, Otlja. *Flavour Frag J.* 2002;17:323-26.
[4193] Stojanovic G, Palic R, Alagic S, et al. Chemical composition and antimicrobial activity of the essential oil and CO2-extracts of semioriental tobacco. *Flavour Frag J.* 2000;15:335-8.
[4194] Andrade EHA, Zoghbi MDGB, Carreira LMM. Volatile constituents of the flowers of Dipteryx odorata (Audbl.) Willd. *J Essent Oil Res.* 2003 May/Jun;15:211-12.
[4195] Ehlers D, Pfister M, Bork WR, et al. HPLC analysis of tonka bean extracts. *Z Lebensm Unters Forsch.* 1995;201:278-82.
[4196] Crafter's Choice. Safety Data Sheet: Tonka Bean Noir Fragrance Oil. Available at: http://www.crafters-choice.com/PDFs/ProductDocs/10247-SDS.pdf.
[4197] Felter SP, Vasallo JD, Carlton BD, et al. A safety assessment of coumarin taking into account species specific toxicokinetics. *Food Chem Toxicol.* 2006 Apr;44(4):462-75.
[4198] European Food Safety Authority. Coumarin in flacourings and other food ingredients with flavouring properties: Scientific opinion of the panel on food additives, flavourings, processing aids and materials in contact with food (AFC). Available at: http://www.efsa.europa.eu/sites/default/files/scientific_output/files/main_documents/793.pdf.
[4199] Huong DTL, Jo YS, Lee MK, et al. Monoamine Oxidase Inhibitors from Cinnamomi Cortex. *Nat Prod Sci.* 2000;6:16-19.
[4200] Feuer G. The metabolism and biological activity of coumarins. *Progress Med Chem.* 1974;10:85-158.
[4201] Kontogiorgis C, Nicolotti O, Mangiatordi GF, et al. Studies on the antiplatelet and antithrombotic profile of anti-inflammatory coumarin derivatives. *J Enzyme Inhibition Med Chem.* 2015;30(6):925-33.
[4202] Eiter LC, Fadamiro H, Setzer WN. Seasonal Variation in the Leaf Essential Oil Composition of Zanthoxylum clava-herculis growing in Huntsville, Alabama. *Nat Prod Commun.* 2010;5(3):457-60.
[4203] Burkhard PR, Burkhardt K, Haenggeli CA, et al. Plant-induced seizures: reappearance of an old problem. *J Neurol.* 1999 Aug;246(8):667-70.
[4204] Waldman N. Seizure caused by dermal application of over-the-counter eucalyptus oil head lice preparation. *Clin Toxicol (Phila).* 2011 Oct;49(8):750-1.
[4205] Craig JO. Poisoning by the volatile oils in childhood. *Arch Dis Child.* 1953;28:259-67.
[4206] Mathew T, Kamath V, Kumar RS, et al. Eucalyptus oil inhalation-induced seizure: A novel, underrecognized, preventable cause of acute symptomatic seizure. *Epilepsia Open.* 2017 Jul 4;2(3):350-354.
[4207] Unger M, Frank A. Simultaneous determination of the inhibitory potency of herbal extracts on the activity of six major cytochrome P450 enzymes using liquid chromatography/mass spectrometry and automated online extraction. *Rapid Commun Mass Spectrom.* 2004;18(19):2273-81.
[4208] Jori A, Bianchetti A, Prestini PE, et al. Effect of eucalyptol (1,8-cineole) on the metabolism of other drugs in rats and in man. *Eur J Pharmacol.* 1970;9(3):362-66.
[4209] Kim NH, Hyun SH, Jin CH, et al. Pretreatment with 1,8-cineole potentiates thioacetamide-induced hepatotoxicity and immunosuppression. *Arch Pharm Res.* 2004 Jul;27(7):781-9.
[4210] Jori A, Bianchetti A, Prestini PE, et al. Effect of eucalyptol (1,8-cineole) on the metabolism of other drugs in rats and in man. *Eur J Pharmacol.* 1970;9(3):362-66.
[4211] de Sousa DP, Raphael E, Brocksom U, et al. Sedative effect of monoterpene alcohols in mice: A preliminary screening. *Verlag der Zeitschrift fur Naturforschung.* 2007;62c:563-66.
[4212] Joulain D, Tabacchi R. Lichen extracts as raw materials in perfumery. Part 2: treemoss. *Flavour Frag J.* 2009;24(3):105-16.

[4213] European Commision Scientific Committees. Opinion on Oak moss / tree moss (sensitization only). Available at: https://ec.europa.eu/health/ph_risk/committees/04_sccp/docs/sccp_o_131.pdf.
[4214] Uter W, Schmidt E, Lessmann H, et al. Contact sensitization to tree moss (Evernia furfuracea extract, INCI) is heterogeneous. *Contact Dermatitis*. 2012 Jul;67(1):36-41.
[4215] de Castro Nizio DA, Blank AF, Sampaio TS, et al. Distillation methods affect the chemical composition of Varronia curassavica Jacq. Essential oil. *Biosci J*. 2018 May/June;34(3):629-39.
[4216] Da S. Feijo EVR, de Oliveira RA, do B. Costa LC. Light affects Varronia curassavica essential oil yield by increasing trichomes frequency. *Rev Bras Farmacogn*. 2014;24:516-523.
[4217] Queiroz TB, Mendes ADR, Silva JCRL, et al. Content and chemical composition of the essential oil of 'erva-baleeira' (Varronia curassavica Jaqc.) as a function of harvesting times. *Rev Bras Plantas Med*. 2016; 18(Supl. 1):356-62.
[4218] Brandao DS, Costa KP, Vieira JTR, et al. Growing of tropical black sage and chemical composition of the essential oil. *Revista de Ciencias Agrarias*. 2017;40(4):823-9.
[4219] Hernandez T, Canales M, Teran B, et al. Antimicrobial activity of the essential oil and extracts of Cordia curassavica (Boraginaceae). *J Ethnopharmacology*. 2007;111:137-41.
[4220] Santos AV, Antunes e Defaveri AC, Bizzo HR, et al. In vitro propagation, histochemistry, and analysis of essential oil from conventionally propagated and in vitro-propagated plants of Varronia curassavica Jacq. *In Vitro Cell Dev Biol—Plant*. 2013;49:405–413.
[4221] de Castro Nizio DA, de Andrade Brito F, Sampaio TS, et al. Chemical diversity of native populations of Varronia curassavica Jacq. and antifungal activity against Lasiodoplodia theobromae. *Ind Crops Prod*. 2015 Dec 15;76:437-48.
[4222] de Castro Nizio DA, Fujimoto RY, Maria AN, et al. Essential oils of Varronia curassavica accessions have different activity against white spot disease in freshwater fish. *Parasitol Res*. 2018 Jan;117(1):97-105.
[4223] Jeena K, Liju VB, Viswanathan R, et al. Antimutagenic potential and modulation of carcinogen-metabolizing enzymes by ginger essential oil. *Phytother Res*. 2014 Jun;28(6):849-55.
[4224] Hernandez T, Canales M, Teran B, et al. Antimicrobial activity of the essential oil and extracts of Cordia curassavica (Boraginaceae). *J Ethnopharmacology*. 2007;111:137-41.
[4225] de Castro Nizio DA, Fujimoto RY, Maria AN, et al. Essential oils of Varronia curassavica accessions have different activity against white spot disease in freshwater fish. *Parasitol Res*. 2018 Jan;117(1):97-105.
[4226] Ahmadian M, Ahmadi N, Babaei A, et al. Comparison of volatile compounds at various developmental stages of tuberose (Polianthes tuberosa L. cv. Mahallati) flower with different extraction methods. *J Essent Oil Res*. 2018;30(3):197-206.
[4227] Selim SM, Matter FM, Hassanain MA, et al. Response of Growth, Flowering, Concrete Oil and its Component of Polianthes tuberosa L. cv. Double to Phosphorus Fertilizer and Gibberellic Acid. *Int J Curr Microbiol App Sci*. 2017;6(9):1639-52.
[4228] Rakthaworn P, Dilokkunanant U, Sukkatta U, et al. Tuberose Oil: Extractions and Its Chemical Components. Available at: http://kukr.lib.ku.ac.th/db/BKN_KAPI/search_detail/dowload_digital_file/11541/101191.
[4229] Extraction Methods for Tuberose Oil and Their Chemical Components. *Kasetsart J (Nat Sci)*. 2009;43:204-11.
[4230] National Institutes of Health, Medline Plus. Reye Syndrome. Available at: http://www.nlm.nih.gov/medlineplus/ency/article/001565.htm.
[4231] Ellenhorn MJ, Barceloux DG. Medical Toxicology - Diagnosis and Treatment of Human Poisoning. New York, NY: Elsevier Science Publishing Co., Inc. 1988., p. 564
[4232] Karabulut AK, Ulger H, Pratten MK. Protection by free oxygen radical scavenging enzymes against salicylate- induced embryonic malformation in vitro. *Toxicol In Vitro*. 2000 Aug;14(4):297-307.
[4233] Overman DO, White JA. Comparative teratogenic effects of methyl salicylate applied orally or topically to hamsters. *Teratology*. 1983 Dec;28(3):421-6.
[4234] Briggs GG, Freeman RK, Yaffe SJMD. Drugs in pregnancy and lactation. Baltimore, MD: Lippincott Williams & Wilkins; 1998.
[4235] Tisserand R, Young R. Essential Oil Safety (Second Edition). 2014 Churchill Livingstone Elsevier.
[4236] Burkhard PR, Burkhardt K, Haenggeli CA, et al. Plant-induced seizures: reappearance of an old problem. *J Neurol*. 1999 Aug;246(8):667-70.
[4237] Waldman N. Seizure caused by dermal application of over-the-counter eucalyptus oil head lice preparation. *Clin Toxicol (Phila)*. 2011 Oct;49(8):750-1.
[4238] Craig JO. Poisoning by the volatile oils in childhood. *Arch Dis Child*. 1953;28:259-67.
[4239] Mathew T, Kamath V, Kumar RS, et al. Eucalyptus oil inhalation-induced seizure: A novel, underrecognized, preventable cause of acute symptomatic seizure. *Epilepsia Open*. 2017 Jul 4;2(3):350-354.
[4240] Tanen DA, Danish DC, Reardon JM, et al. Comparison of oral aspirin versus topical applied methyl salicylate for platelet inhibition. *Ann Pharmacother*. 2008 Oct;42(10):1396-401.
[4241] Le Bourhis B, Soenen AM. Recherches sur l'action psychotrope de quelques substances aromatiques utilisees en alimentation. *Food Cosmet Toxicol*. 1973;11:1-9.
[4242] Joss JD, LeBlond RF. Potentiation of warfarin anticoagulation associated with topical methyl salicylate. *Ann Pharmacother*. 2000 Jun;34(6):729-33.
[4243] Chan TY. Potential dangers from topical preparations containing methyl salicylate. *Hum Exp Toxicol*. 1996;15:747- 50.
[4244] Parker D, Martinez C, Stanley C, et al. The analysis of methyl salicylate and salicylic acid from Chinese herbal medicine ingestion. *J Analytical Tox*. 2004 Apr;28:214-16.
[4245] Ghorat F, Shahrestani S, Tagabadi Z, et al. The Effect of Inhalation of Essential Oils of Polianthes Tuberosa on Test Anxiety in Students: A Clinical Trial. *Iran J Med Sci*. 2016 May;41(3 Suppl):S13.
[4246] Gokturk RS, Sagdic O, Oxkan G, et al. Essential Oil Compositions and Bioactivities of Thymus revolutus and Cyclotrichium origanifolium. *J Essent Oil Bear Plants*. 2013;16(6):795-805.
[4247] Karaman S, Digrak M, Ravid U, et al. Antibacterial and antifungal activity of the essential oils of Thymus reolutus Celak from Turkey. *J Ethnopharmacol*. 2001;76:183-6.
[4248] Domaracky M, Rehak P, Juhas S, et al. Effects of selected plant essential oils on the growth and development of mouse preimplantation embryos in vivo. *Physiol Res*. 2007;56(1):97-104.
[4249] Lemhadri A, Zeggwagh NA, Maghrani M, et al. Anti-hypoglycaemic activity of the aqueous extract of Origanum vulgare growing wild in Tafilalet region. *J Ethnopharmacol*. 2004 Jun;92(2-3):251-6.
[4250] Bayramoglu G, Senturk H, Bayramoglu A, et al. Carvacrol partially reverses symptoms of diabetes in STZ-induced diabetic rats. *Cytotechnology*. 2014 Mar;66(2):251–257.
[4251] Ezhumalai M, Radhiga T, Pugalendi KV. Antihyperglycemic effect of carvacrol in combination with rosiglitazone in high-fat diet-induced type 2 diabetic C57BL/6J mice. *Mol Cell Biochem*. 2014 Jan;385(1-2):23-31.

[4252] Umasankar K, Ramya DR, Nambikkairaj B. Antidiabetic and antidyslipidemic nature of carvacrol, a monoterpenic phenol studied in high-fat - fed and low-dose streptozotocin-induced experimental diabetic rats. *J Pharm Res.* 2015;9(8):484-90.
[4253] Tognolini M, Barocelli E, Ballabeni V, et al. Comparative screening of plant essential oils; phenylpropanoid moiety as basic core for antiplatelet activity. *Life Sci.* 2006 Feb 23;78(13):1419-32.
[4254] Tsai HH, Lin HW, Chen YL, et al. A review of potential harmful interactions between anticoagulant/antiplatelet agents and Chinese herbal medicines. *PLoS One.* 2013 May 9;8(5):e64255.
[4255] Karkabounas S, Kostoula OK, Daskalou T, et al. Anticarcinogenic and antiplatelet effects of carvacrol. *Exp Oncol.* 2006 Jun;28(2):121-5.
[4256] Baser KH. Biological and pharmacological activities of carvacrol and carvacrol bearing essential oils. *Curr Pharm Des.* 2008;14(29):3106-19.
[4257] Okazaki K, Kawazoe K, Takaishi Y. Human platelet aggregation inhibitors from thyme (Thymus vulgaris L.). *Phytother Res.* 2002 Jun;16(4):398-9.
[4258] Jukic M, Politeo O, Maksimmovic M, et al. In vitro acetylcholinesterase inhibitory properties of thymol, carvacrol and their derivatives thymoquinone and thymohydroquinone. *Phytother Res.* 2007;21(3):259-61.
[4259] Askin H, Yildiz M, Ayar A. Effects of Thymol and Carvacrol on Acetylcholinesterase from Drosophila melanogaster. *Acta Physica Polinica A.* 2017;132(3):720-22.
[4260] Medeiros Barreto H, Cerqueira Fontinele F, Pereira de Oliveira A, et al. Phytochemical prospection and modulation of antibiotic activity in vitro by Lippia origanoides H.B.K. in methicillin resistant Staphylococcus aureus. *Biomed Res Int.* 2014;305610:1–7.
[4261] Aelenei P, Miron A, Trifan A, et al. Essential Oils and Their Components as Modulators of Antibiotic Activity against Gram-Negative Bacteria. *Medicines.* 2016;3(19):2-34.
[4262] Magi G, Marini E, Facinelli B. Antimicrobial activity of essential oils and carvacrol, and synergy of carvacrol and erythromycin, against clinical, erythromycin-resistant Group A Streptococci. *Front Microbiol.* 2015;6:165.
[4263] Veras HNH, Rodrigues FFG, Colares AV, et al. Synergistic antibiotic activity of volatile compounds from the essential oil of Lippia sidoides and thymol. *Fitoterapia.* 2012 Apr;83(3):508-12.
[4264] Langeveld WT, Veldhuizen EJ, Burt SA. Synergy between essential oil components and antibiotics: A review. *Crit Rev Microbiol.* 2014 Feb;40(1):76-94.
[4265] Ozkan A, Erdogan A. Evaluation of cytotoxic, membrane, and DNA damaging effects of Thymus revolutus Célak essential oil on different cancer cells. *Turkish J Med.* 2017;47:702-14.
[4266] Gokturk RS, Sagdic O, Oxkan G, et al. Essential Oil Compositions and Bioactivities of Thymus revolutus and Cyclotrichium origanifolium. *J Essent Oil Bear Plants.* 2013;16(6):795-805.
[4267] Gokturk RS, Sagdic O, Oxkan G, et al. Essential Oil Compositions and Bioactivities of Thymus revolutus and Cyclotrichium origanifolium. *J Essent Oil Bear Plants.* 2013;16(6):795-805.
[4268] Koc S, Oz E, Cetin H. Repellent activities of some Labiatae plant essential oils against the saltmarsh mosquito Ochlerotatus caspius (Pallas, 1771) (Diptera: Culicidae). *Parasitol Res.* 2012 Jun;110(6):2205-9.
[4269] Pudziuvelyte L, Stankevicius M, Marusak A, et al. Chemical composition and anticancer activity of Elsholtzia ciliata essential oils and extracts prepared by different methods. *Ind Crops Prod.* 2017;107:90-6.
[4270] Korolyuk EA, Konig W, Tkachev AV. Composition of essential oils of Elsholtzia ciliata Thunb. Hyl. from the Novosibirsk region, Russia. *Химия растительного сырья.* 2002;1(C):31–6.
[4271] Bakova NI, Dmitriev LB, Mashanov VI, et al. Characterization of essential oils of different biotypes Elsholtzia ciliata (Thunb.) Hyl. *Izv Timiryazevsk S-kh Acad.* 1988;12:162–6.
[4272] Dung NX, Van Hac L, Le Huy Hai, et al. Composition of the essential oils from the aerial parts of Elsholtzia ciliata (Thunb.) Hyland. from Vietnam. *J Essent Oil Res.* 1996;8(1):107–9.
[4273] Thappa RK, Agarwal SG, Kapahi BK, Srivastava TN. Chemosystematics of the Himalayan Elsholtzia. *J Essent Oil Res.* 1999;11:97–103.
[4274] Fujuita Y, Tanaka Y, Iwamura J. Essential oil of Elsholtzia ciliata. *Nippon Kadaku Zasshi.* 1967;88(7):763–6.
[4275] Kharina TG, Kalinkina GI, Dembitsky AD, et al. Essential oil composition and morphological and biological characteristics of Elsholtzia ciliata (Thumb.). *Hyl Rastit Resur.* 1995;31(3):58–64.
[4276] Le TB, Beaufay C, Nghiem DT, et al. In Vitro Anti-Leishmanial Activity of Essential Oils Extracted from Vietnamese Plants. *Molecules.* 2017;22:1-13.
[4277] Shin S. Study on Activities of the Essential Oil from Elsholtzia ciliata against Some Antibiotic-Susceptible and -Resistant Strains of Pathogenic Bacteria. *Yakhak Hoeji.* 2010;54(2):122-5.
[4278] Nogueira AC, Carvalho RR, Souza CA, et al. Study on the embryofeto-toxicity of citral in the rat. *Toxicology.* 1995;96(2):105-13.
[4279] Seo KA, Kim H, Ku HY, et al. The monoterpenoids citral and geraniol are moderate inhibitors of the CYP2B6 hydroxylase activity. *Chem Biol Interact.* 2008;174:141-46.
[4280] Sheweita SA, Newairy AA, Mansour HA, et al. Effect of some hypoglycemic herbs on the activity of phase I and II drug-metabolizing enzymes in alloxan-induced diabetic rats. *Toxicology.* 2002 May 24;17(2):131-39.
[4281] Raner GM, Vaz AD, Coon MJ. Metabolism of all-trans, 9-cis, and 13-cis isomers of retinal by purified isozymes of microsomal cytochrome P450 and mechanism-based inhibition of retinoid oxidation by citral. *Mol Pharmacol.* 1996;49(3):515-22.
[4282] De Olivera AC, Ribeiro-Pinto LF, Paumgartten JR. In vitro inhibition of CYP2B1 monoxygenase by beta-myrcene and other monoterpenoid constituents. Br J Nutr. 1999;81:289-95.
[4283] Modak T, Mukhopadhaya A. Effects of citral, a naturally occurring antiadipogenic molecule, on an energy-intense diet model of obesity. *Indian J Pharmacol.* 2011 May-Jun;43(3):300-05.
[4284] Adeneye AA, Agbaje EO. Hypoglycemic and hypolipidemic effects of fresh leaf aqueous extract of Cymbopogon citratus Stapf. in rats. *J Ethnopharmacol.* 2007 Jul 25;112(3):440-4.
[4285] Choi JY, Damte D, Seung-Jin L, et al. Antimicrobial activity of lemongrass and oregano essential oil against standard antibiotic resistant Staphylococcus aureus and field isolates from chronic mastitis cow. *International Journal of Phytomedicine.* 2012;4(1):134-39.
[4286] Zanini SF, Silva-Angulo AB, Rosenthal A, et al. Effect of citral and carvacrol on the susceptibility of Listeria monocytogenes and Listeria innocua to antibiotics. *Lett Appl Microbiol.* 2014 May;58(5):486-92.
[4287] Kuhn GO, McCampbell P, Singmaster G, et al. Application of microencapsulation technology to improve the stability of citral in rodent diets. *Fundam Appl Toxicol.* 1991 Oct;17(3):635-40.
[4288] De Mozzi P, Johnston GA. An outbreak of allergic contact dermatitis caused by citral in beauticians working in a health spa. *Contact Dermatitis.* 2014 Jun;70(6):377-9.
[4289] Pudziuvelyte L, Stankevicius M, Marusak A, et al. Chemical composition and anticancer activity of Elsholtzia ciliata essential oils and extracts prepared by different methods. *Ind Crops Prod.* 2017;107:90-6.

4290 Choi MS, Choi BS, Kim SH, et al. Essential Oils from the Medicinal Herbs Upregulate Dopamine Transporter in Rat Pheochromocytoma Cells. *J Med Food.* 2015 Oct;18(10):1112-20.
4291 Shin S. Study on Activities of the Essential Oil from Elsholtzia ciliata against Some Antibiotic-Susceptible and -Resistant Strains of Pathogenic Bacteria. *Yakhak Hoeji.* 2010;54(2):122-5.
4292 Le TB, Beaufay C, Nghiem DT, et al. In Vitro Anti-Leishmanial Activity of Essential Oils Extracted from Vietnamese Plants. *Molecules.* 2017;22:1-13.
4293 Potter TL, Fagerson IS, Craker LE. Composition of Vietnamese coriander leaf oil. Acta Hortic. 1993;344:305-11.
4294 Fujita K, Chavasiri W, Kubo I. Anti-Salmonella Activity of Volatile Compounds of Vietnam Coriander. *Phytother Res.* 2015 Jul;29(7):1081-7.
4295 Hunter MV, Brophy JJ, Ralph BJ, et al. Composition of Polygonum odoratum Lour. from Southern Australia. *J Essent Oil Res.* 1997;9(5):603-4.
4296 Fujita K, Chavasiri W, Kubo I. Anti-Salmonella Activity of Volatile Compounds of Vietnam Coriander. *Phytother Res.* 2015 Jul;29(7):1081-7.
4297 Jantan I, Mohad Ali NA, Mat Ali R. The essential oils of Clausena lansium. *J Tropical Forest Sci.* 1992;5(4):512-517.
4298 Chokeprasert P, Charles AL, Sue KH, et al. Volatile components of the leaves, fruits and seeds of wampee [Clausena lansium (Lour.) Skeels]. *J Food Comp Analysis.* 2007;20:52-6.
4299 He Z, Zhang L, Chen J, et al. Correlation between Chemical Composition and Antifungal Activity of Clausena lansium Essential Oil against Candida spp. *Molecules.*2019;24:1394.
4300 Guo SS, Wang Y, Chen ZY, et al. Essential Oils from Clausena Species in China: Santalene Sesquiterpenes Resource and Toxicity against Liposcelis bostrychophila. *J Chem.* 2018:7813675.
4301 Liang Q, Zheng P, Xu R, et al. Properties and Exactions of Essential Oil from the Fruit of Clausena Lansium(Lour.) Skeels in Zhaoqing Area. *Int J Adv Res Chem Sci.* 2016 Jul;3(7):20-6.
4302 Zhao J, Nan P, Zhong Y. Chemical Composition of the Essential Oils of Clausena lansium from Hainan Island, China. *Z Naturforsch.* 2003 Nov;59c:153-6.
4303 Bonesi M, Menichini F, Tundis R, et al. Acetylcholinesterase and butyrylcholinesterase inhibitory activity of Pinus species essential oils and their constituents. *J Enzyme Inhib Med Chem.* 2010 Oct;25(5):622-8.
4304 Liang Q, Zheng P, Xu R, et al. Properties and Exactions of Essential Oil from the Fruit of Clausena Lansium(Lour.) Skeels in Zhaoqing Area. *Int J Adv Res Chem Sci.* 2016 Jul;3(7):20-6.
4305 He Z, Zhang L, Chen J, et al. Correlation between Chemical Composition and Antifungal Activity of Clausena lansium Essential Oil against Candida spp. *Molecules.*2019;24:1394.
4306 Liang Q, Zheng P, Xu R, et al. Properties and Exactions of Essential Oil from the Fruit of Clausena Lansium(Lour.) Skeels in Zhaoqing Area. *Int J Adv Res Chem Sci.* 2016 Jul;3(7):20-6.
4307 Liang Q, Zheng P, Xu R, et al. Properties and Exactions of Essential Oil from the Fruit of Clausena Lansium(Lour.) Skeels in Zhaoqing Area. *Int J Adv Res Chem Sci.* 2016 Jul;3(7):20-6.
4308 Guo SS, Wang Y, Chen ZY, et al. Essential Oils from Clausena Species in China: Santalene Sesquiterpenes Resource and Toxicity against Liposcelis bostrychophila. *J Chem.* 2018:7813675.
4309 Prota N, Mumm R, Bouwmeester HJ, et al. Comparison of the chemical composition of three species of smartweed (genus Persicaria) with a focus on drimane sesquiterpenoids. *Phytochem.* 2014;108:129-36.
4310 Jiang J. Volatile composition of the laksa plant (Polygonum hydropiper L.), a potential source of green note aroma compounds. *Flavour Fragr J.* 2005;20:455-9.
4311 Kubo I, Taniguchi M. Polygodial, an antifungal potentiator. *J Nat Prod.* 1988 Jan-Feb;51(1):22-9.
4312 Abe M, Ozawa Y, Uda Y, et al. A novel labdane-type trialdehyde from myoga (Zingiber mioga Roscoe) that potently inhibits human platelet aggregation and human 5-lipoxygenase. *Biosci Biotechnol Biochem.* 2006;10:2494–2500.
4313 Ayaz M,Junaid M, Ullah F, et al. GC-MS Analysis and Gastroprotective Evaluations of Crude Extracts, Isolated Saponins, and Essential Oil from Polygonum hydropiper L. *Frontiers Chem.* 2017 Aug;5(58):1-11.
4314 Ayaz M, Junaid M, Ullah F, et al. Comparative chemical profiling, cholinesterase inhibitions and anti-radicals properties of essential oils from Polygonum hydropiper L: A Preliminary anti- Alzheimer's study. *Lipids Health Disease.* 2015;14:141.
4315 Maheswaran R, Ignacimuthu S. Bioefficacy of essential oil from Polygonum hydropiper L. against mosquitoes, Anopheles stephensi and Culex quinquefasciatus. *Ecotoxicol Environ Saf.* 2013 Nov;97:26-31.
4316 Maheswaran R, Ignacimuthu S. Effect of Polygonum hydropiper L. against dengue vector mosquito Aedes albopictus L. *Parasitol Res.* 2014 Sep;113(9):3143-50.
4317 Essien EE, Newby JM, Walker TM, et al. Essential oil constituents, anticancer and antimicrobial activity of Ficus mucoso and Casuarina equisetifolia leaves. *Am J Essential Oils Nat Prod.* 2016;4(1):1-6.
4318 Ogunwande IA, Flamini G, Adefuye AE, et al. Chemical compositions of Casuarina equisetifolia L., Eucalyptus toreliana L. and Ficus elastica Roxb. ex Hornem cultivated in Nigeria. *South African J Botany.* 2011;77:645-9.
4319 Adeosun C, Bamidele I, Salau B, et al. Essential oil from the stem bark of Casuarina equisetifolia exhibits mosquitocidal potential. *J Phtyopathol Pest Manag.* 2016;3(3):53-62.
4320 Ogunwande IA, Flamini G, Adefuye AE, et al. Chemical compositions of Casuarina equisetifolia L., Eucalyptus toreliana L. and Ficus elastica Roxb. ex Hornem cultivated in Nigeria. *South African J Botany.* 2011;77:645-9.
4321 Adeosun C, Bamidele I, Salau B, et al. Essential oil from the stem bark of Casuarina equisetifolia exhibits mosquitocidal potential. *J Phtyopathol Pest Manag.* 2016;3(3):53-62.
4322 Santos TG, Laemmle J, Rebelo RA, et al. Chemical composition and antimicrobial activity of Aloysia gratissima (Verbenaceae) leaf essential oil. *J Essent Oil Res.* 2015;27(2):125-30.
4323 Santos FM, Pinto JEBP, Bertolucci SKV, et al. Chemical composition and antimicrobial activity of the essential oil from the leaves and flowers of Aloysia gratissima. *Rev Bras Pl Med (Campinas).* 2013;15(4):583-8.
4324 Trovati G, Chierice GO, Sanches EA, et al. Essential Oil Composition of Aloysia gratissima From Brazil. *J Essent Oil Res.* 2009 Jul/Aug;21:1-2.
4325 Soler E, Dellacassa E, Moyna P. Composition of Aloysia gratissima flower essential oil. *Planta Med.* 1986 Dec;(6):488-90.
4326 Ricciardi GAL, van Baren CM, Di Leo Lira P, et al. Volatile constituents from aerial parts of Aloysia gratissima (Gillies & Hook.) Tronc. var. gratissima growing in Corrientes, Argentina. *Flavour Frag J.* 2006;21:698-703.
4327 Trovati G, Sanches ES, Chierice GO. TLC Separation and Identification of the Essential Oil Constituents from Aloysia gratissima. *J Liquid Chrom Related Tech.* 2009;32:890-5.
4328 Soler E, Dellcassa E, Moyna P. Composition of Aloysia gratissima leaf essential oil. *Phytochem.* 1986;25(6):1343-5.
4329 Freitas Garcia MC, Sores DC, Santana RC, et al. The in vitro antileishmanial activity of essential oil from Aloysia gratissima and guaiol, its major sesquiterpene against Leishmania amazonensis. *Parasitol.* 2018 Jan 21:1-9.

[4330] Bersan SM, Galvão LC, Goes VF, et al. Action of essential oils from Brazilian native and exotic medicinal species on oral biofilms. *BMC Complement Altern Med.* 2014 Nov 18;14:451.
[4331] Burkhard PR, Burkhardt K, Haenggeli CA, et al. Plant-induced seizures: reappearance of an old problem. *J Neurol.* 1999 Aug;246(8):667-70.
[4332] Hold KM, Sirisoma NS, Ikeda T, et al. α-Thujone (the active component of absinthe): γ-Aminobutyric acid type A receptor modulation and metabolic detoxification. *Proc Natl Acad Sci USA.* 2000 Apr 11;97(8):3826–3831.
[4333] Waldman N. Seizure caused by dermal application of over-the-counter eucalyptus oil head lice preparation. *Clin Toxicol (Phila).* 2011 Oct;49(8):750-1.
[4334] Craig JO. Poisoning by the volatile oils in childhood. *Arch Dis Child.* 1953;28:259-67.
[4335] Mathew T, Kamath V, Kumar RS, et al. Eucalyptus oil inhalation-induced seizure: A novel, underrecognized, preventable cause of acute symptomatic seizure. *Epilepsia Open.* 2017 Jul 4;2(3):350-354.
[4336] Millet Y, Tognetti P, Lavaire-Perlovisi M, et al. Experimental study of the toxic convulsant properties of commercial preparations of essences of sage and hyssop. *Rev Electroencephalogr Neurophysiol.* Clin. 1979;9:12–8.
[4337] Millet Y, Jouglard J, Steinmetz MD, et al. Toxicity of some essential plant oils. Clinical and experimental study. *Clin Toxicol.* 1981 Dec;18(12):1485-98.
[4338] Millet Y, Tognetti P, Steinmetz MD, et al. Etude de la toxicite d'huiles essentielles des propertes toxiques convulsivantes des essences de sauge et d'hysope du commerce. Medicine Legale, Toxicologie. 1980;23:9-21.
[4339] Haliciglu O, Astarcioglu G, Yaprak I, et al. Toxicity of Salvia officinalis in a newborn and a child: an alarming report. *Pediatr Neurol.* 2011 Oct;45(4):259-60.
[4340] Lachenmeier DW, Walch SG. Epileptic seizure caused by accidental ingestion of sage (Salvia officinalis L.) oil in children: a rare, exceptional case or a threat to public health. *Pediatr Neurol.* 2012 Mar;46(3):201.
[4341] da Silva Costa KC, Bexerra SB, Norte CM, et al. Medicinal plants with teratogenic potential: current considerations. *Braz J Pharm Sci.* 2012;48(3):427-33.
[4342] Brinker F. Herb Contraindications and Drug Interactions. 2nd ed. Sandy, OR. Eclectic Medical Publications. 1998.
[4343] European Medicines Agency. Public statement on the use of herbal medicinal products containing thujone. Available at: http://www.ema.europa.eu/docs/en_GB/document_library/Public_statement/2011/02/WC500102294.pdf.
[4344] Millet Y, Jouglard J, Steinmetz MD, et al. Toxicity of some essential plant oils. Clinical and experimental study. *Clin Toxicol.* 1981 Dec;18(12):1485-98.
[4345] Cristovao L, Carvalho F, Bastos MDL, et al. Hepatotoxicity of an essential oil of Salvia officinalis L.: an in vitro study using freshly isolated rat hepatocytes. *Congress Biomarkers.* 2001 Sep:165.
[4346] Burkhard PR, Burkhardt K, Haenggeli CA, et al. Plant-induced seizures: reappearance of an old problem. *J Neurol.* 1999 Aug;246(8):667-70.
[4347] Burkhard PR, Burkhardt K, Haenggeli CA, et al. Plant-induced seizures: reappearance of an old problem. *J Neurol.* 1999 Aug;246(8):667-70.
[4348] Hold KM, Sirisoma NS, Ikeda T, et al. α-Thujone (the active component of absinthe): γ-Aminobutyric acid type A receptor modulation and metabolic detoxification. *Proc Natl Acad Sci USA.* 2000 Apr 11;97(8):3826–3831.
[4349] Waldman N. Seizure caused by dermal application of over-the-counter eucalyptus oil head lice preparation. *Clin Toxicol (Phila).* 2011 Oct;49(8):750-1.
[4350] Craig JO. Poisoning by the volatile oils in childhood. *Arch Dis Child.* 1953;28:259-67.
[4351] Mathew T, Kamath V, Kumar RS, et al. Eucalyptus oil inhalation-induced seizure: A novel, underrecognized, preventable cause of acute symptomatic seizure. *Epilepsia Open.* 2017 Jul 4;2(3):350-354.
[4352] Uc A, Bishop WP, Sanders KD. Camphor hepatotoxicity. *South Med J.* 2000;93:596-98.
[4353] Frohne D. Giftpflanzen: Cupressaceae. Stuttgart: Wissenschaftliche Verlagsgesellschaft mbH; 1997. pp. 153–6.
[4354] Dolan LC, Matulka RA, Burdock GA. Naturally Occurring Food Toxins. *Toxins (Basel).* 2010 Sep;2(9):2289–2332.
[4355] United States National Toxicology Program (NTP). Alpha-Thujone. Dec 10, 1997. Available at: http://ntp.niehs.nih.gov/index.cfm?objectid=03DB8C36-E7A1-9889-3BDF8436F2A8C51F.
[4356] Millet Y, Jouglard J, Steinmetz MD, et al. Toxicity of some essential plant oils. Clinical and experimental study. *Clin Toxicol.* 1981 Dec;18(12):1485-98.
[4357] Millet Y, Tognetti P, Steinmetz MD, et al. Etude de la toxicite d'huiles essentielles des propertes toxiques convulsivantes des essences de sauge et d'hysope du commerce. Medicine Legale, Toxicologie. 1980;23:9-21.
[4358] Chen W, Lu Y, Gao J, et al. Beta-elemene inhibits melanoma growth and metastasis via suppressing vascular endothelial growth factormediated angiogenesis. *Cancer Chemo Pharmacol.* 2011 Apr;67(4):799-808.
[4359] Mei-Lin Xie,Qun Lu,Zhen-Lun Gu. Inhibitory Effect of β-Elemene Emulsion on Platelet Aggregation and its Mechanism. *J Chinese Pham Sci.* 1998;7(2):112-114 .
[4360] Qun L, Lujia Z, Meilin X, et al. Inhibitory effect of β-elemene on thrombosis and its mechanism in rats. *Chinese J Mod Appl Pharm.* 1999-04.
[4361] Santos TG, Laemmle J, Rebelo RA, et al. Chemical composition and antimicrobial activity of Aloysia gratissima (Verbenaceae) leaf essential oil. *J Essent Oil Res.* 2015;27(2):125-30.
[4362] Santos FM, Pinto JEBP, Bertolucci SKV, et al. Chemical composition and antimicrobial activity of the essential oil from the leaves and flowers of Aloysia gratissima. *Rev Bras Pl Med (Campinas).* 2013;15(4):583-8.
[4363] Bersan SM, Galvão LC, Goes VF, et al. Action of essential oils from Brazilian native and exotic medicinal species on oral biofilms. *BMC Complement Altern Med.* 2014 Nov 18;14:451.
[4364] García CC, Talarico L, Almeida N, et al. Virucidal activity of essential oils from aromatic plants of San Luis, Argentina. *Phytother Res.* 2003 Nov;17(9):1073-5.
[4365] Santos TG, Laemmle J, Rebelo RA, et al. Chemical composition and antimicrobial activity of Aloysia gratissima (Verbenaceae) leaf essential oil. *J Essent Oil Res.* 2015;27(2):125-30.
[4366] Freitas Garcia MC, Sores DC, Santana RC, et al. The in vitro antileishmanial activity of essential oil from Aloysia gratissima and guaiol, its major sesquiterpene against Leishmania amazonensis. *Parasitol.* 2018 Jan 21:1-9.
[4367] Carmo M, Frazão S. The essential oil of Cupressus lusitanicus Mill. *Flavour Frag J.* 1989;4(4):185-6.
[4368] Hassanzadeh SL, Tuten JA, Vogler B, et al. The chemical composition and antimicrobial activity of the leaf oil of Cupressus lusitanica from Monteverde, Costa Rica. *Pharmacognosy Res.* 2010 Jan-Feb;2(1):19–21.
[4369] Tum PK, Mainya NO. Extraction and Analysis of Essential Oils from Cypress Species (Cupressus lusitanica and Taxodium distichum) from Selected Areas in Kenya. *Int J Sci Tech.* 2016 Aug;4(8):26-34.
[4370] Bett PK, Deng AL, Ogendo JO, et al. Chemical composition of Cupressus lusitanica and Eucalyptus saligna leaf essential oils and bioactivity against major insect pests of stored food grains. *Ind Crops Prod.* 2016;82:51-62.

[4371] Teke GN, Elisée KN, Roger KJ. Chemical composition, antimicrobial properties and toxicity evaluation of the essential oil of Cupressus lusitanica Mill. leaves from Cameroon. *BMC Complement Altern Med.* 2013 Jun 13;13:130.

[4372] Nassini R, Materazzi S, Vriens J, et al. The 'headache tree' via umbellulone and TRPA1 activates the trigeminovascular system. *Brain.* 2012 Feb;135(Pt 2):376-90.

[4373] Benemei S, Appendino G, Geppetti P. Pleasant natural scent with unpleasant effects: cluster headache-like attacks triggered by Umbellularia californica. *Cephalalgia.* 2010 Jun;30(6):744-6.

[4374] Hassanzadeh SL, Tuten JA, Vogler B, et al. The chemical composition and antimicrobial activity of the leaf oil of Cupressus lusitanica from Monteverde, Costa Rica. *Pharmacognosy Res.* 2010 Jan-Feb;2(1):19–21.

[4375] Teke GN, Elisée KN, Roger KJ. Chemical composition, antimicrobial properties and toxicity evaluation of the essential oil of Cupressus lusitanica Mill. leaves from Cameroon. *BMC Complement Altern Med.* 2013 Jun 13;13:130.

[4376] Bett PK, Deng AL, Ogendo JO, et al. Chemical composition of Cupressus lusitanica and Eucalyptus saligna leaf essential oils and bioactivity against major insect pests of stored food grains. *Ind Crops Prod.* 2016;82:51-62.

[4377] Doimo L. Azulenes, Costols and Γ-Lactones from Cypress-Pines (Callitris columellaris, C. glaucophylla and C. intratropica) Distilled Oils and Methanol Extracts. *J Essent Oil Res.* 2001 Jan/Feb;13:25-29.

[4378] Doimo L. Azulenes, Costols and Γ-Lactones from Cypress-Pines (Callitris columellaris, C. glaucophylla and C. intratropica) Distilled Oils and Methanol Extracts. *J Essent Oil Res.* 2001 Jan/Feb;13:25-29.

[4379] Oprava A, Leach D, Beattie K, et al. Chemical composition and biological activity of the essential oils from native Australian Callitris species. *Planta Med.* 2010;76:SL35.

[4380] Kadri A, Chobba IB, Zarai Z, et al. Chemical constituents and antioxidant activity of the essential oil from aerial parts of Artemisia herba-alba grown in Tunisian semi-arid region. *African J Biotech.* 2011 Apr;10(15):2923-9.

[4381] Btisam O, Mohamed G, Badr S, et al. Effect of harvest date on yield, chemical composition and antimicrobial activity of Artemisia herba-alba essential oil. *Int J Sci Engin Res.* 2016 Feb;7(2):600-6.

[4382] Rekkab S, Abaza I, Chibani S, et al. Chemical composition of the essential oil of aerial parts of Artemisia herba-alba Asso. from Oum El-Bouaghi (Algeria) and chemotaxonomic survey. *J Mater Environ Sci.* 2016;7(12):4383-90.

[4383] Imelouane B, Bachiri AE, Ankit M, et al. Essential oil composition and antimicrobial activity of Artemisia herba-alba Asso grown in Morocco. *Banat's J Biotech.* 2010;I(2):48-55.

[4384] Lakehal S, Meliana A, Benmimoune S, et al. Essential Oil Composition and Antimicrobial Activity of Artemisia herba–alba Asso Grown in Algeria. *Med Chem.* 2016;6(6):435-9.

[4385] Dob T, Benabdelkader T. Chemical Composition of the Essential Oil of Artemisia herba-alba Asso Grown in Algeria. *J Essent Oil Res.* 2006 Nov/Dec;18:685-90.

[4386] Belhattab R, Amor L, Barroso JG, et al. Essential oil from Artemisia herba-alba Asso grown wild in Algeria: Variability assessment and comparison with an updated literature survey. *Arabian J Chem.* 2014 Apr;7(2):243-51.

[4387] Al-Shuneigat J, Al-Sarayreh S, Al-Qudah M, et al. GC-MS Analysis and Antibacterial Activity of the Essential Oil Isolated from Wild Artemisia herba-alba Grown in South Jordan. *British J Medicine Med Res.* 2015;5(3):297-302.

[4388] Hatem N, Wakim LH, Nemer N, et al. Chemical composition and antimicrobial activity of the essential oil of Artemisia herba-alba Asso., Asteraceae. *Int J Eng Sci Res Tech.* 2018 Apr;7(4):258-66.

[4389] Mofsen H, Ali F. Essential Oil Composition of Artemisia herba-alba from Southern Tunisia. *Molecules.* 2009;14:1585-94.

[4390] Bellili S, Dhifi W, Khsif Al-Garni AB, et al. Essential oil composition and variability of Artemisia herba-alba Asso. growing in Tunisia: comparison and chemometric investigation of different plant organs. *J Appl Pharm Sci.* 2016 Jul;6(07):038-042.

[4391] Aziz M, Karim A, Ouariachi ELE, et al. Relaxant Effect of Essential Oil of Artemisia herba-alba Asso. on Rodent Jejunum Contractions. *Sci Pharm.* 2012 Apr-Jun;80(2):457-67.

[4392] Feuerstein I, Muller D, Hobert K, et al. The constitution of the essential oils from Artemisia herba-alba populations of Israel and Sinai. *Phytochem.* 1986;25(10:2343-7.

[4393] Salido S, Valenzuela LR, Altarejos J, et al. Composition and infraspecific variability of Artemisia herba-alba from southern Spain. *Biochem System Ecol.* 2004;32:265-77.

[4394] Alabdullatif M, Boujezza I, Mekni M, et al. Enhancing blood donor skin disinfection using natural oils. *Transfusion.* 2017 Dec;57(12):2920-2927.

[4395] El-Seedi HR, Azeem M, Khalil NS, et al. Essential oils of aromatic Egyptian plants repel nymphs of the tick Ixodes ricinus (Acari: Ixodidae). *Exp Appl Acarol.* 2017 Sep;73(1):139-157.

[4396] Younsi F, Mehdi S, Aissi O, et al. Essential Oil Variability in Natural Populations of Artemisia campestris (L.) and Artemisia herba-alba (Asso) and Incidence on Antiacetylcholinesterase and Antioxidant Activities. *Chem Biodivers.* 2017 Jul;14(7).

[4397] Rafiq R, Hayek SA, Anyanwu U, et al. Antibacterial and Antioxidant Activities of Essential Oils from Artemisia herba-alba Asso., Pelargonium capitatum × radens and Laurus nobilis L. *Foods.* 2016 Apr 11;5(2).

[4398] Amri I, De Martino L, Marandino A, et al. Chemical composition and biological activities of the essential oil from Artemisia herba-alba growing wild in Tunisia. *Nat Prod Commun.* 2013 Mar;8(3):407-10.

[4399] Zouri S, Zouri N, Fakhfakh N, et al. Chemical composition and biological activities of a new essential oil chemotype of Tunisian Artemisia herba alba Asso. *J Med Plants Res.* 2010 May;4(10):871-80.

[4400] Fleisher Z, Fleisher F, Nachbar RB. Chemovariation of Artemisia herba alba Asso. Aromatic Plants of the Holy Land and the Sinai. Part XVI. *J Essent Oil Res.* 2002 May/Jun;14:156-60

[4401] Younsi F, Mehdi S, Aissi O, et al. Essential Oil Variability in Natural Populations of Artemisia campestris (L.) and Artemisia herba-alba (Asso) and Incidence on Antiacetylcholinesterase and Antioxidant Activities. *Chem Biodivers.* 2017 Jul;14(7).

[4402] Zouri S, Zouri N, Fakhfakh N, et al. Chemical composition and biological activities of a new essential oil chemotype of Tunisian Artemisia herba alba Asso. *J Med Plants Res.* 2010 May;4(10):871-80.

[4403] Tilaoui M, Ait Mouse H, Jaafari, et al. Comparative Phytochemical Analysis of Essential Oils from Different Biological Parts of Artemisia herba alba and Their Cytotoxic Effect on Cancer Cells. *PLoS One.* 2015 Jul 21;10(7):e0131799.

[4404] Aziz M, Karim A, Ouariachi ELE, et al. Relaxant Effect of Essential Oil of Artemisia herba-alba Asso. on Rodent Jejunum Contractions. *Sci Pharm.* 2012 Apr-Jun;80(2):457-67.

[4405] Abu-Darwish MS, Cabral C, Gonçalves MJ, et al. Artemisia herba-alba essential oil from Buseirah (South Jordan): Chemical characterization and assessment of safe antifungal and anti-inflammatory doses. *J Ethnopharmacol.* 2015 Nov 4;174:153-60.

[4406] Younsi F, Mehdi S, Aissi O, et al. Essential Oil Variability in Natural Populations of Artemisia campestris (L.) and Artemisia herba-alba (Asso) and Incidence on Antiacetylcholinesterase and Antioxidant Activities. *Chem Biodivers.* 2017 Jul;14(7).

[4407] Zouri S, Zouri N, Fakhfakh N, et al. Chemical composition and biological activities of a new essential oil chemotype of Tunisian Artemisia herba alba Asso. *J Med Plants Res*. 2010 May;4(10):871-80.
[4408] Btisam O, Mohamed G, Badr S, et al. Effect of harvest date on yield, chemical composition and antimicrobial activity of Artemisia herba-alba essential oil. *Int J Sci Engin Res*. 2016 Feb;7(2):600-6.
[4409] Imelouane B, Bachiri AE, Ankit M, et al. Essential oil composition and antimicrobial activity of Artemisia herba-alba Asso grown in Morocco. *Banat's J Biotech*. 2010;I(2):48-55.
[4410] Lakehal S, Meliana A, Benmimoune S, et al. Essential Oil Composition and Antimicrobial Activity of Artemisia herba–alba Asso Grown in Algeria. *Med Chem*. 2016;6(6):435-9.
[4411] Hatem N, Wakim LH, Nemer N, et al. Chemical composition and antimicrobial activity of the essential oil of Artemisia herba-alba Asso., Asteraceae. *Int J Eng Sci Res Tech*. 2018 Apr;7(4):258-66.
[4412] Alabdullatif M, Boujezza I, Mekni M, et al. Enhancing blood donor skin disinfection using natural oils. *Transfusion*. 2017 Dec;57(12):2920-2927.
[4413] Rafiq R, Hayek SA, Anyanwu U, et al. Antibacterial and Antioxidant Activities of Essential Oils from Artemisia herba-alba Asso., Pelargonium capitatum × radens and Laurus nobilis L. *Foods*. 2016 Apr 11;5(2).
[4414] Chaftar N, Girardot M4, Quellard N, et al. Activity of Six Essential Oils Extracted from Tunisian Plants against Legionella pneumophila. *Chem Biodivers*. 2015 Oct;12(10):1565-74.
[4415] Abu-Darwish MS, Cabral C, Gonçalves MJ, et al. Artemisia herba-alba essential oil from Buseirah (South Jordan): Chemical characterization and assessment of safe antifungal and anti-inflammatory doses. *J Ethnopharmacol*. 2015 Nov 4;174:153-60.
[4416] Amri I, De Martino L, Marandino A, et al. Chemical composition and biological activities of the essential oil from Artemisia herba-alba growing wild in Tunisia. *Nat Prod Commun*. 2013 Mar;8(3):407-10.
[4417] Zouri S, Zouri N, Fakhfakh N, et al. Chemical composition and biological activities of a new essential oil chemotype of Tunisian Artemisia herba alba Asso. *J Med Plants Res*. 2010 May;4(10):871-80.
[4418] Yashphe J, Segal R, Breuer A, et al. Antibacterial activity of Artemisia herba-alba. *J Pharm Sci*. 1979 Jul;68(7):924-5.
[4419] Aouadi D, Luciano G, Vasta V, et al. The antioxidant status and oxidative stability of muscle from lambs receiving oral administration of Artemisia herba alba and Rosmarinus officinalis essential oils. *Meat Sci*. 2014 Jun;97(2):237-43.
[4420] Vasta V, Aouadi D, Brogna DM, et al. Effect of the dietary supplementation of essential oils from rosemary and artemisia on muscle fatty acids and volatile compound profiles in Barbarine lambs. *Meat Sci*. 2013 Oct;95(2):235-41.
[4421] Kadri A, Chobba IB, Zarai Z, et al. Chemical constituents and antioxidant activity of the essential oil from aerial parts of Artemisia herba-alba grown in Tunisian semi-arid region. *African J Biotech*. 2011 Apr;10(15):2923-9.
[4422] Younsi F, Mehdi S, Aissi O, et al. Essential Oil Variability in Natural Populations of Artemisia campestris (L.) and Artemisia herba-alba (Asso) and Incidence on Antiacetylcholinesterase and Antioxidant Activities. *Chem Biodivers*. 2017 Jul;14(7).
[4423] Rafiq R, Hayek SA, Anyanwu U, et al. Antibacterial and Antioxidant Activities of Essential Oils from Artemisia herba-alba Asso., Pelargonium capitatum × radens and Laurus nobilis L. *Foods*. 2016 Apr 11;5(2).
[4424] Zouri S, Zouri N, Fakhfakh N, et al. Chemical composition and biological activities of a new essential oil chemotype of Tunisian Artemisia herba alba Asso. *J Med Plants Res*. 2010 May;4(10):871-80.
[4425] Aloui Z, Messaoud C, Haoues M, et al. Asteraceae Artemisia campestris and Artemisia herba-alba Essential Oils Trigger Apoptosis and Cell Cycle Arrest in Leishmania infantum Promastigotes. *Evid Based Complement Alternat Med*. 2016;2016:9147096.
[4426] Hatimi S, Boudouma M, Bichichi M, et al. [In vitro evaluation of antileishmania activity of Artemisia herba alba Asso]. *Bull Soc Pathol Exot*. 2001 Mar;94(1):29-31.
[4427] El-Seedi HR, Azeem M, Khalil NS, et al. Essential oils of aromatic Egyptian plants repel nymphs of the tick Ixodes ricinus (Acari: Ixodidae). *Exp Appl Acarol*. 2017 Sep;73(1):139-157.
[4428] Abbad A, Kasrati A, Jamali CA, et al. Insecticidal properties and chemical composition of essential oils of some aromatic herbs from Morocco. *Nat Prod Res*. 2014;28(24):2338-41.
[4429] Alabdullatif M, Boujezza I, Mekni M, et al. Enhancing blood donor skin disinfection using natural oils. *Transfusion*. 2017 Dec;57(12):2920-2927.
[4430] Pirbalouti AG, Aghaee K, Kashi A, et al. Chemical composition of the essential oil of wild and cultivated plant populations of Kelussia odoratissima Mozaff.*J Med Plant Res*. 2012 Jan;6(3):449-54.
[4431] Sajjadi SE, Shokoohinia Y, Mehramiri P. Isolation and characterization of steroids, phthalide and essential oil of the fruits of Kelussia odoratissima Mozaff., an endemic mountain celery. *Res Pharm Sci*. 2013;8(1):35-41.
[4432] Kheirabadi KP, Dehkordi SS, Kheibari P. Effect of Kelussia odoratissima Mozaff essential oil on promastigot form of Leishmania major (in vitro). *J HerbMed Pharmacol*. 2015;4(1):10-14.
[4433] Rabbani M, Sajjadi SE, Sadeghi. Chemical composition of the essential oil from Kelussia odoratissima Mozaff. and the evaluation of its sedative and anxiolytic effects in mice. *Clinics*. 2011;66(5):843-8.
[4434] Osanloo M, Amani A, Sereshti H, et al. Extraction and chemical composition essential oil of Kelussia odoratissima and comparison its larvicidal activity with Z-ligustilide (Major Constituent) against Anopheles stephensi. *J Entomol Zoology Studies*. 2017;5(4):611-6.
[4435] Ghasemi M, Mirlohi A, Ayyari M, et al. Kelussia odoratissima Mozaff. a rich source of essential fatty acids and phthalides. *J HerbMed Parmacol*. 2015;4(4):115-20.
[4436] Omidbaigi R, Sefidkon F, Saeedi K. Essential Oil Content and Composition of Kelussia odoratissima Mozaff. as an Iranian Endemic Plant. *J Essent Oil Bear Plants*. 2008;11(6):594-7.
[4437] Esmaeili A, Sharafian S, Mirian S, et al. Chemical Composition of Essential Oil from Leaves, Stems and Flowers of Kelussia odoratissima Mozaff. Grown in Iran. *J Essent Oil Bear Plants*. 2011;14(5):643-6.
[4438] Circosta C, Pasquale RD, Palumbo DR, et al. Estrogenic activity of standardized extract of Angelica sinensis. *Phytother Res*. 2006 Aug;20(8):665–69.
[4439] Shi M, Chang L, He G. [Stimulating action of Carthamus tinctorius L., Angelica sinensis (Oliv.) Diels and Leonurus sibiricus L. on the uterus]. *Zhongguo Zhong Yao Za Zhi*. 1995 Mar;20(3):173-5, 192.
[4440] Shi M, Chang L, He G. [Stimulating action of Carthamus tinctorius L., Angelica sinensis (Oliv.) Diels and Leonurus sibiricus L. on the uterus]. *Zhongguo Zhong Yao Za Zhi*. 1995 Mar;20(3):173-5, 192.
[4441] Brindis F, Rodríguez R, Bye R, et al. (Z)-3-butylidenephthalide from Ligusticum porteri, an α-glucosidase inhibitor. *J Nat Prod*. 2011 Mar 25;74(3):314-20.
[4442] Zhang L, Du JR, Wang J, et al. Z-ligustilide extracted from Radix Angelica Sinensis decreased platelet aggregation induced by ADP ex vivo and arterio-venous shunt thrombosis in vivo in rats. *Yakugaku Zasshi*. 2009 Jul;129(7):855-9.
[4443] Qi H, Zhao J, Han Y, et al. Z-ligustilide potentiates the cytotoxicity of dopamine in rat dopaminergic PC12 cells. *Neurotox Res*. 2012 Nov;22(4):345-54.

[4444] Xu HL, Feng YP. [Effects of 3-n-butylphthalide on thrombosis formation and platelet function in rats]. *Yao Xue Xue Bao*. 2001 May;36(5):329-33.

[4445] Ye J, Zhai L, Zhang S, et al. DL-3-n-butylphthalide inhibits platelet activation via inhibition of cPLA2-mediated TXA2 synthesis and phosphodiesterase. *Platelets*. 2015;26(8):736-44.

[4446] Zhang CMM, Xhao SMB, Zang YMB, et al. The efficacy and safety of Dl-3n-butylphthalide on progressive cerebral infarction: A randomized controlled STROBE study. *Medicine*. 2017 Jul;96(30):e7257.

[4447] Wang F, Ma J, Han F, et al. DL-3-n-butylphthalide delays the onset and progression of diabetic cataract by inhibiting oxidative stress in rat diabetic model. *Sci Rep*. 2016;6:19396.

[4448] Zhang CMM, Xhao SMB, Zang YMB, et al. The efficacy and safety of Dl-3n-butylphthalide on progressive cerebral infarction: A randomized controlled STROBE study. *Medicine*. 2017 Jul;96(30):e7257.

[4449] Zhao Q, Hu JP, Jiang J, et al. [Interaction of butylphthalide with rat and human liver CYP450 isoenzymes]. *Yao Xue Xue Bao*. 2015 May;50(5):541-6.

[4450] Bjeldanes LF, Kim IS. Sedative activity of celery oil constituents. *J Food Sci*. 1978 Jan;43(1):143-4.

[4451] Li Y, Shao X, Xu J, et al. Effects and possible mechanism of tea tree oil against Botrytis cinerea and Penicillium expansum in vitro and in vivo test. *Can J Microbiol*. 2017 Mar;63(3):219-227.

[4452] Li YH, Sun XP, Zhang YQ, et al. The antithrombotic effect of borneol related to its anticoagulant property. *Am J Chin Med*. 2008;36(4):719-27.

[4453] Hiroi T, Miyazaki Y, Kobayashi Y, et al. Induction of hepatic P450s in rat by essential wood and leaf oils. *Xenobiotica*. 1995 May;25(5):457-67.

[4454] Chen JY, Wang JJ, Meng MR, et al. [Borneol is an inducer of rat hepatic CYP2D activity in vivo]. *Yao Xue Xue Bao*. 2015 Apr;50(4):459-63.

[4455] Zhang R1, Mi SQ, Wang NS. Effect of borneol on cytochrome P450 3A enzyme and midazolam pharmacokinetics in rats. *Eur J Drug Metab Pharmacokinet*. 2013 Sep;38(3):159-69.

[4456] Kim NH, Hyun SH, Jin CH, et al. Pretreatment with 1,8-cineole potentiates thioacetamide-induced hepatotoxicity and immunosuppression. *Arch Pharm Res*. 2004 Jul;27(7):781-9.

[4457] Marín I, Sayas-Barberá E, Viuda-Martos M, et al. Chemical Composition, Antioxidant and Antimicrobial Activity of Essential Oils from Organic Fennel, Parsley, and Lavender from Spain. *Foods*. 2016 Mar 4;5(1).

[4458] Burkhard PR, Burkhardt K, Haenggeli CA, et al. Plant-induced seizures: reappearance of an old problem. *J Neurol*. 1999 Aug;246(8):667-70.

[4459] Waldman N. Seizure caused by dermal application of over-the-counter eucalyptus oil head lice preparation. *Clin Toxicol (Phila)*. 2011 Oct;49(8):750-1.

[4460] Craig JO. Poisoning by the volatile oils in childhood. *Arch Dis Child*. 1953;28:259-67.

[4461] Mathew T, Kamath V, Kumar RS, et al. Eucalyptus oil inhalation-induced seizure: A novel, underrecognized, preventable cause of acute symptomatic seizure. *Epilepsia Open*. 2017 Jul 4;2(3):350-354.

[4462] Charanraj KJ, G MV, S M. Camphor poisoning in a child. *Natl Med J India*. 2013 Jan-Feb;26(1):60.

[4463] Narayan S, Singh N. Camphor poisoning-An unusual cause of seizure. *Med J Armed Forces India*. 2012 Jul;68(3):252–53.

[4464] Olowe SA, Ransome-Kuti O. The risk of jaundice in glucose-6-phosphate dehydrogenase deficient babies exposed to menthol. *Acta Paediatr Scand*. 1980 May;69(3):341-5.

[4465] Dillon Remy M, Manning Alleyne P, Bratt DE, et al. Neonatal jaundice at Port-of-Spain General Hospital abstract. *West Indian Med J*. 1987;36(Suppl):28.

[4466] Khine H, Weiss D, Graber N, et al. A cluster of children with seizures caused by camphor poisoning. *Pediatrics*. 2009 May;123(5):1269-72.

[4467] Michiels EA, Mazor SS. Toddler with seizures due to ingesting camphor at an Indian celebration. *Pediatr Emerg Care*. 2010 Aug;26(8):574-75.

[4468] Koren G. Medications which can kill a toddler with one tablet or teaspoonful. *J Toxicol Clin Toxicol. 1*993;31(3):407-13.

[4469] Flaman Z, Pellechia-Clarke S, Bailey B, et al. Unintentional exposure of young children to camphor and eucalyptus oils. *Paediatr Child Health*. 2001 Feb;6(2):80-83.

[4470] Rabl W, Katzgraber F, Steinlechner M. Camphor ingestion for abortion (case report). *Forensic Sci Int*. 1997 Sep 19;89(1-2):137-40.

[4471] Flaman Z, Pellechia-Clarke S, Bailey B, et al. Unintentional exposure of young children to camphor and eucalyptus oils. *Paediatr Child Health*. 2001 Feb;6(2):80-83.

[4472] Rabbani M, Sajjadi SE, Izadi A. Alleviation ofMorphineWithdrawal Signs but Not Tolerance by the Essential Oil of Kelussia odoratissima Mozaff. *Evidence-Based Complement Altern Med*. 2012;2012:847096.

[4473] Barzegar-Fallah N, Baery N, Soleymankhani M, et al. Kelussia odoratissima Mozaff attenuates thromboembolic brain injury, possibly due to its Z-ligustilide content. *Brain Inj*. 2016;30(7):926-32.

[4474] Rabbani M, Sajjadi SE, Sadeghi. Chemical composition of the essential oil from Kelussia odoratissima Mozaff. and the evaluation of its sedative and anxiolytic effects in mice. *Clinics*. 2011;66(5):843-8.

[4475] Kheirabadi KP, Dehkordi SS, Kheibari P. Effect of Kelussia odoratissima Mozaff essential oil on promastigot form of Leishmania major (in vitro). *J HerbMed Pharmacol*. 2015;4(1):10-14.

[4476] Osanloo M, Amani A, Sereshti H, et al. Extraction and chemical composition essential oil of Kelussia odoratissima and comparison its larvicidal activity with Z-ligustilide (Major Constituent) against Anopheles stephensi. *J Entomol Zoology Studies*. 2017;5(4):611-6.

[4477] Vatandoost H, Sanei Dehkordi A, Sadeghi SM, et al. Identification of chemical constituents and larvicidal activity of Kelussia odoratissima Mozaffarian essential oil against two mosquito vectors Anopheles stephensi and Culex pipiens (Diptera: Culicidae). *Exp Parasitol*. 2012 Dec;132(4):470-4.

[4478] Satyal P, Craft JD, Dosoky NS, et al. The Chemical Compositions of the Volatile Oils of Garlic (Allium sativum) and Wild Garlic (Allium vineale). *Foods*. 2017 Aug 5;6(8).

[4479] Farnsworth NR, Bingel AS, Cordell GA, et al. Potential value of plants as sources of new antifertility agents I. *J Pharm Sci*. 1975 Apr;64(4):535-98.

[4480] Mennella JA, Johnson A, Beauchamp GK. Garlic ingestion by pregnant women alters the odor of amniotic fluid. *Chem Senses*. 1995;20:207-9.

[4481] Ziaei S, Hantoshzadeh S, Rezasoltani P, et al. The effect of garlic tablet on plasma lipids and platelet aggregation in nulliparous pregnants at high risk of preeclampsia. *Eur J Obstet Gynecol Reprod Biol*. 2001;99:201-6.

[4482] Soltani PR. Preeclampisia is an important complication of pregnancy which can result in morbidity and mortality in mother, fetus and the neonate. *J Med Counc Islamic Repub Iran*. 2005;23(3):319.

[4483] Kim SH, Lee IC, Baek HS, et al. Induction of cytochrome P450 3A1 expression by diallyl disulfide: protective effects against cyclophosphamideinduced embryo-fetal developmental toxicity. *Food Chem Toxicol*. 2014 Jul;69:312-9.

[4484] El Sayyad HI, Abou_El-Naga AM, Gadallah AA, et al. Protective effects of Allium sativum against defects of hypercholesterolemia on pregnant rats and their offspring. *Int J Clin Exp Med*. 2010; 3(2): 152–163.

[4485] Mennella JA, Beauchamp GK. The effects of repeated exposure to garlic-flavored milk on the nursling's behavior. *Pediatr Res*. 1993;34:805-8.

4486 Mennella JA, Beauchamp GK. Maternal diet alters the sensory qualities of human milk and the nursling's behavior. *Pediatrics*. 1991;88:737-44.
4487 [No authors listed]. The effect of essential oil of garlic on hyperlipemia and platelet aggregation--an analysis of 308 cases. Cooperative Group for Essential Oil of Garlic. *J Tradit Chin Med*. 1986 Jun;6(2):117-20.
4488 Barrie SA, Wright JV, Pizzorno JE. Effects of Garlic Oil on Platelet Aggregation, Serum Lipids and Blood Pressure in Humans. *J Orthomolecular Med*. 1987;2(1):15-21.
4489 Makheja AN, Vanderhoek JY, Bailey JM. Inhibition of platelet aggregation and thromboxane synthesis by onion and garlic. *Lancet*. 1979 Apr 7;1(8119):781.
4490 Boullin DJ. Garlic as a platelet inhibitor. *Lancet*. 1981 Apr 4;1(8223):776-7.
4491 Fenwick GR, Hanley AB. The genus Allium--Part 3. *Crit Rev Food Sci Nutr*. 1985;23(1):1-73.
4492 Lawson LD, Hughes BG. Characterization of the formation of allicin and other thiosulfinates from garlic. *Planta Med*. 1992 Aug;58(4):345-50.
4493 Bordia A, Verma SK, Srivastava KC. Effect of garlic (Allium sativum) on blood lipids, blood sugar, fibrinogen and fibrinolytic activity in patients with coronary artery disease. *Prostaglandins Leukot Essent Fatty Acids*. 1998 Apr;58(4):257-63.
4494 Bordia A. Effect of garlic on human platelet aggregation in vitro. *Atherosclerosis*. 1978 Aug;30(4):355-60.
4495 Qi R, Liao F, Inoue K, et al. Inhibition by diallyl trisulfide, a garlic component, of intracellular Ca(2+) mobilization without affecting inositol-1,4,5-trisphosphate (IP(3)) formation in activated platelets. *Biochem Pharmacol*. 2000 Nov 15;60(10):1475-83.
4496 Ariga T, Takeda A, Teramoto S, et al. Inhibition Site of Methylallyl Trisulfide: A Volatile Oil Component of Garlic, in the Platelet Arachidonic Acid Cascade. *Food Factors for Cancer Prevention*. 1997:231-34.
4497 Lawson LD, Ransom DK, Hughes BG. Inhibition of whole blood platelet-aggregation by compounds in garlic clove extracts and commercial garlic products. *Thromb Res*. 1992 Jan 15;65(2):141-56.
4498 Lai YS, Chen WC, Ho CT, et al. Garlic essential oil protects against obesity-triggered nonalcoholic fatty liver disease through modulation of lipid metabolism and oxidative stress. *J Agric Food Chem*. 2014 Jun 25;62(25):5897-906.
4499 Taubert D, Glöckner R, Müller D, et al. The garlic ingredient diallyl sulfide inhibits cytochrome P450 2E1 dependent bioactivation of acrylamide to glycidamide. *Toxicol Lett*. 2006 Jun 20;164(1):1-5.
4500 Davenport DM, Wargovich MJ. Modulation of cytochrome P450 enzymes by organosulfur constituents from garlic. *Food Chem Toxicol*. 2005 Dec;43(12):1753-62.
4501 Fisher CD, Augustine LM, Maher JM, et al. Induction of drug-metabolizing enzymes by garlic and allyl sulfide compounds via activation of constitutive androstane receptor and nuclear factor E2-related factor 2. *Drug Metab Dispos*. 2007 Jun;35(6):995-1000.
4502 Lii CK, Tsai CW, Wu CC. Garlic allyl sulfides display differential modulation of rat cytochrome P450 2B1 and the placental form glutathione S transferase in various organs. *J Agric Food Chem*. 2006 Jul 12;54(14):5191-6.
4503 Johnson S. Supercritical Essential Oils: A Companion Resource to Medicinal Essential Oils. 2017. Scott A Johnson Professional Writing Services, LLC: Orem, Utah.
4504 Pyun MS, Shin S. Antifungal effects of the volatile oils from Allium plants against Trichophyton species and synergism of the oils with ketoconazole. *Phytomedicine*. 2006 Jun;13(6):394-400.
4505 Scheman A, Gupta S. Photoallergic contact dermatitis from diallyl disulfide. *Contact Dermatitis*. 2001 Sep;45(3):179.
4506 Alvarez MS, Jacobs S, Jiang SB, Brancaccio RR, et al. Photocontact allergy to diallyl disulfide. *Am J Contact Dermat*. 2003 Sep;14(3):161-5.
4507 Lee TY, Lam TH. Contact dermatitis due to topical treatment with garlic in Hong Kong. *Contact Dermatitis*. 1991 Mar;24(3):193-6.
4508 Cabanillas M, Fernandez-Redondo V, Toribio J. Allergic contact dermatitis to plants in a Spanish dermatology department: a 7-year review. *Contact Dermatitis*. 2006 Aug;55(2):84-91.
4509 Anca T, Philippe V, Ilioara O, et al. Composition of essential oils of Viola tricolor and V. arvensis from Romania. *Chem Nat Compounds*. 2009;45(1):91-2.
4510 Bulut C, Altiok E, Bayraktar O, et al. Antioxidative and antimicrobial screening of 19 commercial essential oils in Turkey. *Acta Hort*. 2009;826(14):111-6.
4511 Scramin S, Saito ML, Pott A, et al. Essential Oil of Elyonurus muticus (Sprengel) O.Kuntze (Gramineae). *J Essent Oil Res*. 2000 May/Jun;12:298-300.
4512 Fuller TN, Bertrand C, Antonello B, et al. Chemical diversity of wild populations of Elionurus muticus (Spreng.) and the allelopathic effect of its essential oil. *J Essent Oil Res*. 2017;29(6):499-506.
4513 Fuller TN, Bertrand C, Simon A, et al. Elionurus muticus as an alternative source of citral from Pampa biome, Brazil. *J Oleo Sci*. 2014;63(11):1109-16.
4514 Hess SC, Peres TLP, Batista AL, et al. Evaluation of seasonal changes in chemical composition and antibacterial activity of Elyonurus muticus 9Sprengel) O. Kuntze (Gramineae). *Quim Nova*. 2007;30(2):370-3.
4515 Chagonda L, Makanda C, Chalchat J. The essential oils of wild and cultivated Cymbopogon validus (Stapf) Stapf ex Burtt Davy and Elionurus muticus (Spreng.) Kunth from Zimbabwe. *Flav Fragrance J*. 2000;15(2):100-104.
4516 Nogueira AC, Carvalho RR, Souza CA, et al. Study on the embryofeto-toxicity of citral in the rat. *Toxicology*. 1995;96(2):105-13.
4517 Seo KA, Kim H, Ku HY, et al. The monoterpenoids citral and geraniol are moderate inhibitors of the CYP2B6 hydroxylase activity. *Chem Biol Interact*. 2008;174:141-46.
4518 Sheweita SA, Newairy AA, Mansour HA, et al. Effect of some hypoglycemic herbs on the activity of phase I and II drug-metabolizing enzymes in alloxan-induced diabetic rats. *Toxicology*. 2002 May 24;17(2):131-39.
4519 Raner GM, Vaz AD, Coon MJ. Metabolism of all-trans, 9-cis, and 13-cis isomers of retinal by purified isozymes of microsomal cytochrome P450 and mechanism-based inhibition of retinoid oxidation by citral. *Mol Pharmacol*. 1996;49(3):515-22.
4520 De Olivera AC, Ribeiro-Pinto LF, Paumgartten JR. In vitro inhibition of CYP2B1 monoxygenase by beta-myrcene and other monoterpenoid constituents. Br J Nutr. 1999;81:289-95.
4521 Modak T, Mukhopadhaya A. Effects of citral, a naturally occurring antiadipogenic molecule, on an energy-intense diet model of obesity. *Indian J Pharmacol*. 2011 May-Jun;43(3):300-05.
4522 Adeneye AA, Agbaje EO. Hypoglycemic and hypolipidemic effects of fresh leaf aqueous extract of Cymbopogon citratus Stapf. in rats. *J Ethnopharmacol*. 2007 Jul 25;112(3):440-4.
4523 Choi JY, Damte D, Seung-Jin L, et al. Antimicrobial activity of lemongrass and oregano essential oil against standard antibiotic resistant Staphylococcus aureus and field isolates from chronic mastitis cow. *International Journal of Phytomedicine*. 2012;4(1):134-39.
4524 Zanini SF, Silva-Angulo AB, Rosenthal A, et al. Effect of citral and carvacrol on the susceptibility of Listeria monocytogenes and Listeria innocua to antibiotics. *Lett Appl Microbiol*. 2014 May;58(5):486-92.
4525 do Vale TG, Furtado EC, Santos JG Jr, et al. Central effects of citral, myrcene and limonene, constituents of essential oil chemotypes from Lippia alba (Mill.) n.e. Brown. *Phytomedicine*. 2002 Dec;9(8):709-14.

[4526] Kuhn GO, McCampbell P, Singmaster G, et al. Application of microencapsulation technology to improve the stability of citral in rodent diets. *Fundam Appl Toxicol.* 1991 Oct;17(3):635-40.
[4527] De Mozzi P, Johnston GA. An outbreak of allergic contact dermatitis caused by citral in beauticians working in a health spa. *Contact Dermatitis.* 2014 Jun;70(6):377-9.
[4528] De-Oliveira AC1, Ribeiro-Pinto LF, Otto SS, et al. Induction of liver monooxygenases by beta-myrcene. *Toxicology.* 1997 Dec 26;124(2):135-40.
[4529] Li Y, Shao X, Xu J, et al. Effects and possible mechanism of tea tree oil against Botrytis cinerea and Penicillium expansum in vitro and in vivo test. *Can J Microbiol.* 2017 Mar;63(3):219-227.
[4530] Li YH, Sun XP, Zhang YQ, et al. The antithrombotic effect of borneol related to its anticoagulant property. *Am J Chin Med.* 2008;36(4):719-27.
[4531] Hiroi T, Miyazaki Y, Kobayashi Y, et al. Induction of hepatic P450s in rat by essential wood and leaf oils. *Xenobiotica.* 1995 May;25(5):457-67.
[4532] Chen JY, Wang JJ, Meng MR, et al. [Borneol is an inducer of rat hepatic CYP2D activity in vivo]. *Yao Xue Xue Bao.* 2015 Apr;50(4):459-63.
[4533] Zhang R1, Mi SQ, Wang NS. Effect of borneol on cytochrome P450 3A enzyme and midazolam pharmacokinetics in rats. *Eur J Drug Metab Pharmacokinet.* 2013 Sep;38(3):159-69.
[4534] Kim NH, Hyun SH, Jin CH, et al. Pretreatment with 1,8-cineole potentiates thioacetamide-induced hepatotoxicity and immunosuppression. *Arch Pharm Res.* 2004 Jul;27(7):781-9.
[4535] Marín I, Sayas-Barberá E, Viuda-Martos M, et al. Chemical Composition, Antioxidant and Antimicrobial Activity of Essential Oils from Organic Fennel, Parsley, and Lavender from Spain. *Foods.* 2016 Mar 4;5(1).
[4536] Hess SC, Peres TLP, Batista AL, et al. Evaluation of seasonal changes in chemical composition and antibacterial activity of Elyonurus muticus 9Sprengel) O. Kuntze (Gramineae). *Quim Nova.* 2007;30(2):370-3.
[4537] Stefanazzi N, Stadler T, Ferrero A. Composition and toxic, repellent and feeding deterrent activity of essential oils against the stored-grain pests Tribolium castaneum (Coleoptera: Tenebrionidae) and Sitophilus oryzae (Coleoptera: Curculionidae). *Pest Manag Sci.* 2011;67:639-46.
[4538] Zarga MHA, Al-Jaber HI, Amer ZYB, et al. Chemical Composition, Antimicrobial and Antitumor Activities of Essential Oil of Ammodaucus leucotrichus Growing in Algeria. *J Biological Active Prod Nature.* 2013;3(3):224-31.
[4539] Louail Z, Kameli A, Benabdelkader T, et al. Antimicrobial and antioxidant activity of essential oil of Ammodaucus leucotrichus Coss. & Dur. Seeds. *J Mater Environ Sci.* 2016;7(7):2328-34.
[4540] Halla N, Helemno SA, Costa P, et al. Chemical profile and bioactive properties of the essential oil isolated from Ammodaucus leucotrichus fruits growing in Sahara and its evaluation as a cosmeceutical ingredient. *Ind Crops Prod.* 2018;119:249-54.
[4541] Manssouri M, Znini M, Harrak AE, et al. Antifungal activity of essential oil from the fruits of Ammodaucus leucotrichus Coss. & Dur., in liquid and vapour phase against postharvest phytopathogenic fungi in apples. *J Appl Pharm Sci.* 2016 May;6(5):131-6.
[4542] Sadaoui N, Bec N, Barragan-Montero V, et al. The essential oil of Algerian Ammodaucus leucotrichus Coss. & Dur. and its effect on the cholinesterase and monoamine oxidase activities. *Fitoterapia.* 2018;130:1-5.
[4543] El-Haci IA, Bekhechi C, Atik-Bekkara F, et al. Antimicrobial Activity of Ammodaucus leucotrichus Fruit Oil from Algerian Sahara. *Nat Prod Commun.* 2014;9(5):711-12.
[4544] Dahmane D, Dob T, Krimat S, et al. Chemical composition, antioxidant and antibacterial activities of the essential oils of medicinal plant Ammodaucus leucotrichus from Algeria. *J Essent Oil Res.* 2017;29(1):48-55.
[4545] Ford RA, Letizia C, Api AM. Monographs on Fragrance Raw Materials. *Food Chem Toxicol.* 1988;26(Suppl.):397-8.
[4546] Halla N, Helemno SA, Costa P, et al. Chemical profile and bioactive properties of the essential oil isolated from Ammodaucus leucotrichus fruits growing in Sahara and its evaluation as a cosmeceutical ingredient. *Ind Crops Prod.* 2018;119:249-54.
[4547] Sadaoui N, Bec N, Barragan-Montero V, et al. The essential oil of Algerian Ammodaucus leucotrichus Coss. & Dur. and its effect on the cholinesterase and monoamine oxidase activities. *Fitoterapia.* 2018;130:1-5.
[4548] Zarga MHA, Al-Jaber HI, Amer ZYB, et al. Chemical Composition, Antimicrobial and Antitumor Activities of Essential Oil of Ammodaucus leucotrichus Growing in Algeria. *J Biological Active Prod Nature.* 2013;3(3):224-31.
[4549] Zarga MHA, Al-Jaber HI, Amer ZYB, et al. Chemical Composition, Antimicrobial and Antitumor Activities of Essential Oil of Ammodaucus leucotrichus Growing in Algeria. *J Biological Active Prod Nature.* 2013;3(3):224-31.
[4550] Zarga MHA, Al-Jaber HI, Amer ZYB, et al. Chemical Composition, Antimicrobial and Antitumor Activities of Essential Oil of Ammodaucus leucotrichus Growing in Algeria. *J Biological Active Prod Nature.* 2013;3(3):224-31.
[4551] Halla N, Helemno SA, Costa P, et al. Chemical profile and bioactive properties of the essential oil isolated from Ammodaucus leucotrichus fruits growing in Sahara and its evaluation as a cosmeceutical ingredient. *Ind Crops Prod.* 2018;119:249-54.
[4552] Manssouri M, Znini M, Harrak AE, et al. Antifungal activity of essential oil from the fruits of Ammodaucus leucotrichus Coss. & Dur., in liquid and vapour phase against postharvest phytopathogenic fungi in apples. *J Appl Pharm Sci.* 2016 May;6(5):131-6.
[4553] El-Haci IA, Bekhechi C, Atik-Bekkara F, et al. Antimicrobial Activity of Ammodaucus leucotrichus Fruit Oil from Algerian Sahara. *Nat Prod Commun.* 2014;9(5):711-12.
[4554] Khaldi A, Meddah B, Moussaoui A, et al. Anti-mycotoxin Effect and Antifungal Properties of Essential Oil from Ammodaucus leucotrichus Coss. & Dur. on Aspergillus flavus and Aspergillus ochraceus. *J Essent Oil Bearing Plants.* 2017;20(1):36-44.
[4555] Dahmane D, Dob T, Krimat S, et al. Chemical composition, antioxidant and antibacterial activities of the essential oils of medicinal plant Ammodaucus leucotrichus from Algeria. *J Essent Oil Res.* 2017;29(1):48-55.
[4556] Louail Z, Kameli A, Benabdelkader T, et al. Antimicrobial and antioxidant activity of essential oil of Ammodaucus leucotrichus Coss. & Dur. Seeds. *J Mater Environ Sci.* 2016;7(7):2328-34.
[4557] Dahmane D, Dob T, Krimat S, et al. Chemical composition, antioxidant and antibacterial activities of the essential oils of medicinal plant Ammodaucus leucotrichus from Algeria. *J Essent Oil Res.* 2017;29(1):48-55.
[4558] da Silva TB, Alencar Menezes LR, Chaves Sampaio MF, et al. Chemical Composition and Anti-Trypanosoma cruzi Activity of Essential Oils Obtained from Leaves of Xylopia frutescens and X. laevigata (Annonaceae). *Nat Prod Commun.* 2013;8(3):403-6.
[4559] Quintans JDSS, Soares BM, Ferraz RPC, et al. Chemical Constituents and Anticancer Effects of the Essential Oil from Leaves of Xylopia laevigata. *Planta Med.* 2013;79:123-30.
[4560] Costa EV, da Silva TB, D'Souza Costa CO, et al. Chemical Composition of the Essential Oil from the Fresh Fruits of Xylopia laevigata and its Cytotoxic Evaluation. *Nat Prod Commun.* 2016;11(3):417-18.
[4561] Queiroz JCC, Antoniolli AR, Quintans-Juniot LJ, et al. Evaluation of the Anti-Inflammatory and Antinociceptive Effects of the Essential Oil from Leaves of Xylopia laevigata in Experimental Models. *Sci World J.* 2014;2014:816450.
[4562] Pereira TS, Esquissato GNM, Costa EV, et al. Mutagenic and cytostatic activities of the Xylopia laevigata essential oil in human lymphocytes. *Nat Prod Res.* 2019 Jun 14:1-4.
[4563] Quintans JDSS, Soares BM, Ferraz RPC, et al. Chemical Constituents and Anticancer Effects of the Essential Oil from Leaves of Xylopia laevigata. *Planta Med.* 2013;79:123-30.

[4564] Costa EV,da Silva TB, Alencar Menezes LR, et al. Biological activities of the essential oil from the leaves of Xylopia laevigata (Annonaceae). *J Essent Oil Res*. 2013;25:179-85.
[4565] Costa EV, da Silva TB, D'Souza Costa CO, et al. Chemical Composition of the Essential Oil from the Fresh Fruits of Xylopia laevigata and its Cytotoxic Evaluation. Nat Prod Commun. 2016;11(3):417-18.
[4566] Queiroz JCC, Antoniolli AR, Quintans-Juniot LJ, et al. Evaluation of the Anti-Inflammatory and Antinociceptive Effects of the Essential Oil from Leaves of Xylopia laevigata in Experimental Models. *Sci World J*. 2014;2014:816450.
[4567] Costa EV,da Silva TB, Alencar Menezes LR, et al. Biological activities of the essential oil from the leaves of Xylopia laevigata (Annonaceae). *J Essent Oil Res*. 2013;25:179-85.
[4568] da Silva TB, Alencar Menezes LR, Chaves Sampaio MF, et al. Chemical Composition and Anti-Trypanosoma cruzi Activity of Essential Oils Obtained from Leaves of Xylopia frutescens and X. laevigata (Annonaceae). *Nat Prod Commun*. 2013;8(3):403-6.
[4569] Costa EV,da Silva TB, Alencar Menezes LR, et al. Biological activities of the essential oil from the leaves of Xylopia laevigata (Annonaceae). *J Essent Oil Res*. 2013;25:179-85.
[4570] Costa EV,da Silva TB, Alencar Menezes LR, et al. Biological activities of the essential oil from the leaves of Xylopia laevigata (Annonaceae). *J Essent Oil Res*. 2013;25:179-85.
[4571] Nascimento AM, Maia TD, Soares TE, et al. Repellency and Larvicidal Activity of Essential oils from Xylopia laevigata, Xylopia frutescens, Lippia pedunculosa, and Their Individual Compounds against Aedes aegypti Linnaeus. *Neotrop Entomol*. 2017 Apr;46(2):223-230.
[4572] Costa EV,da Silva TB, Alencar Menezes LR, et al. Biological activities of the essential oil from the leaves of Xylopia laevigata (Annonaceae). *J Essent Oil Res*. 2013;25:179-85.
[4573] Aldoghaim FS, Flematti GR, Hammer KA. Antimicrobial Activity of Several Cineole-Rich Western Australian Eucalyptus Essential Oils. *Microorganisms*. 2018;6:122.
[4574] Assareh MH, Jaimand K, Rezaee MB. Chemical composition of the essential oils of six eucalyptus species (Myrtaceae) from south west of Iran. *J Essent Res*. 2007 Jan/Feb;19:8-10.
[4575] Rahimi-Nasrabadi M, Batooli H. Chemical composition of essential oils of two cultivated Eucalyptus species from the central Iran. *J Plant Breed Crop Sci*. 2011 Nov 15;3(13):379-81.
[4576] Foley W, Lassak E. The potential of bioactive constituents of Eucalyptus foliage as non-wood products from plantations. 2004.
[4577] Rahimi-Nasrabadi M, Ahmadi F, Batooli H. Chemical composition of essential oil and in vitro antioxidant activities of the essential oil and methanol extracts of Eucalyptus loxophleba. *Nat Prod Res*. 2012;26(7):669-74.
[4578] Craig JO. Poisoning by the volatile oils in childhood. *Arch Dis Child*. 1953;28:259-67.
[4579] Melis K. Bochner A, Janssens G. Accidental nasal eucalyptol and menthol instillation. *Eur J Pediatr*. 1989 Aug;148(8)786-7.
[4580] Day LM, Ozanne-Smith J, Parsons BJ, et al. Eucalyptus oil poisoning among young children: mechanisms of access and the potential prevention. *Aust N Z J Public Health*. 1997 Jun;21(3):297-302.
[4581] Day LM, Ozanne-Smith J, Parsons BJ, et al. Eucalyptus oil poisoning among young children: mechanisms of access and the potential prevention. *Aust N Z J Public Health*. 1997 Jun;21(3):297-302.
[4582] Myott E. Case of eucalyptus poisoning. *Brit M J*. 1906;1:558.
[4583] Hindle RC. Eucalyptus oil ingestion. *N Z Med J*. 1994 May 11;107(977)185-6.
[4584] Tibballs J. Clinical effects and management of eucalyptus oil ingestion in infants and small children. *Med J Aust*. 1995 Aug;163(4):177-80.
[4585] Waldman W, Barwina M, Sein Anand J. Accidental intoxication with eucalyptus oil—a case report. *Przeql Lek*. 2011;68(8):555-6.
[4586] Day LM, Ozanne-Smith J, Parsons BJ, et al. Eucalyptus oil poisoning among young children: mechanisms of access and the potential prevention. *Aust N Z J Public Health*. 1997 Jun;21(3):297-302.
[4587] Burkhard PR, Burkhardt K, Haenggeli CA, et al. Plant-induced seizures: reappearance of an old problem. *J Neurol*. 1999 Aug;246(8):667-70.
[4588] Waldman N. Seizure caused by dermal application of over-the-counter eucalyptus oil head lice preparation. *Clin Toxicol (Phila)*. 2011 Oct;49(8):750-1.
[4589] Craig JO. Poisoning by the volatile oils in childhood. *Arch Dis Child*. 1953;28:259-67.
[4590] Mathew T, Kamath V, Kumar RS, et al. Eucalyptus oil inhalation-induced seizure: A novel, underrecognized, preventable cause of acute symptomatic seizure. *Epilepsia Open*. 2017 Jul 4;2(3):350-354.
[4591] Gray AM, Flatt PR. Antihyperglycemic actions of Eucalyptus globulus (Eucalyptus) are associated with pancreatic and extra pancreatic effects in mice. *J Nutr*. 1998 Dec;128(12):2319–23.
[4592] Basak SS, Candan F. Effect of Laurus nobilis L. Essential Oil and its Main Components on α-glucosidase and Reactive Oxygen Species Scavenging Activity. *Iran J Pharm Res*. 2013 Spring;12(2):367-79.
[4593] Fathiazad F, Mazandarani M, Hamedeyazdan. Phytochemical analysis and antioxidant activity of Hyssopus officinalis L. from Iran. *Adv Pharm Bull*. 2011 Dec;1(2):63–67.
[4594] Gray AM, Flatt PR. Antihyperglycemic actions of Eucalyptus globulus (Eucalyptus) are associated with pancreatic and extra pancreatic effects in mice. *J Nutr*. 1998 Dec;128(12):2319-23.
[4595] Dey B. Chemo-profiling of eucalyptus and study of its hypoglycemic potential. *World J Diabetes*. 2013 Oct 15;4(5):170–76.
[4596] Tsai HH, Lin HW, Chen YL, et al. A review of potential harmful interactions between anticoagulant/antiplatelet agents and Chinese herbal medicines. *PLoS One*. 2013 May 9;8(5):e64255.
[4597] Moharam BA, Jantan I, bin Ahmad F, et al. Antiplatelet Aggregation and Platelet Activating Factor (PAF) Receptor Antagonistic Activities of the Essential Oils of Five Goniothalamus Species. *Molecules*. 2010;15:5124-38.
[4598] Unger M, Frank A. Simultaneous determination of the inhibitory potency of herbal extracts on the activity of six major cytochrome P450 enzymes using liquid chromatography/mass spectrometry and automated online extraction. *Rapid Commun Mass Spectrom*. 2004;18(19):2273-81.
[4599] Jori A, Bianchetti A, Prestini PE, et al. Effect of eucalyptol (1,8-cineole) on the metabolism of other drugs in rats and in man. *Eur J Pharmacol*. 1970;9(3):362-66.
[4600] Kim NH, Hyun SH, Jin CH, et al. Pretreatment with 1,8-cineole potentiates thioacetamide-induced hepatotoxicity and immunosuppression. *Arch Pharm Res*. 2004 Jul;27(7):781-9.
[4601] Jori A, Bianchetti A, Prestini PE, et al. Effect of eucalyptol (1,8-cineole) on the metabolism of other drugs in rats and in man. *Eur J Pharmacol*. 1970;9(3):362-66.
[4602] de Sousa DP, Raphael E, Brocksom U, et al. Sedative effect of monoterpene alcohols in mice: A preliminary screening. *Verlag der Zeitschrift fur Naturforschung*. 2007;62c:563-66.
[4603] Aldoghaim FS, Flematti GR, Hammer KA. Antimicrobial Activity of Several Cineole-Rich Western Australian Eucalyptus Essential Oils. *Microorganisms*. 2018;6:122.

[4604] Torabbeigi M, Azar PA, Meibodi ZA. Chemical Composition of Essential Oils of Aerial Parts of Helichrysum aucheri from Iran. *Analyt Chem Lett.* 2011;1(5&6):393-6.
[4605] Azar PA, Torabbeigi M, Tehrani MS, et al. Hydrodistillation, Solvent Free Microwave Assisted Extraction and Headspace-Solid Phase Microextraction for Analysis of Essential Oil of Flowers of Helichrysum aucheri. *Asian J Chem.* 2011;23(3):1209-11.
[4606] Domaracky M, Rehak P, Juhas S, et al. Effects of selected plant essential oils on the growth and development of mouse preimplantation embryos in vivo. *Physiol Res.* 2007;56(1):97-104.
[4607] Lemhadri A, Zeggwagh NA, Maghrani M, et al. Anti-hypoglycaemic activity of the aqueous extract of Origanum vulgare growing wild in Tafilalet region. *J Ethnopharmacol.* 2004 Jun;92(2-3):251-6.
[4608] Bayramoglu G, Senturk H, Bayramoglu A, et al. Carvacrol partially reverses symptoms of diabetes in STZ-induced diabetic rats. *Cytotechnology.* 2014 Mar;66(2):251–257.
[4609] Ezhumalai M, Radhiga T, Pugalendi KV. Antihyperglycemic effect of carvacrol in combination with rosiglitazone in high-fat diet-induced type 2 diabetic C57BL/6J mice. *Mol Cell Biochem.* 2014 Jan;385(1-2):23-31.
[4610] Umasankar K, Ramya DR, Nambikkairaj B. Antidiabetic and antidyslipidemic nature of carvacrol, a monoterpenic phenol studied in high-fat - fed and low-dose streptozotocin-induced experimental diabetic rats. *J Pharm Res.* 2015;9(8):484-90.
[4611] Tognolini M, Barocelli E, Ballabeni V, et al. Comparative screening of plant essential oils; phenylpropanoid moiety as basic core for antiplatelet activity. *Life Sci.* 2006 Feb 23;78(13):1419-32.
[4612] Tsai HH, Lin HW, Chen YL, et al. A review of potential harmful interactions between anticoagulant/antiplatelet agents and Chinese herbal medicines. *PLoS One.* 2013 May 9;8(5):e64255.
[4613] Karkabounas S, Kostoula OK, Daskalou T, et al. Anticarcinogenic and antiplatelet effects of carvacrol. *Exp Oncol.* 2006 Jun;28(2):121-5.
[4614] Baser KH. Biological and pharmacological activities of carvacrol and carvacrol bearing essential oils. *Curr Pharm Des.* 2008;14(29):3106-19.
[4615] Okazaki K, Kawazoe K, Takaishi Y. Human platelet aggregation inhibitors from thyme (Thymus vulgaris L.). *Phytother Res.* 2002 Jun;16(4):398-9.
[4616] Jukic M, Politeo O, Maksimmovic M, et al. In vitro acetylcholinesterase inhibitory properties of thymol, carvacrol and their derivatives thymoquinone and thymohydroquinone. *Phytother Res.* 2007;21(3):259-61.
[4617] Askin H, Yildiz M, Ayar A. Effects of Thymol and Carvacrol on Acetylcholinesterase from Drosophila melanogaster. *Acta Physica Polinica A.* 2017;132(3):720-22.
[4618] Medeiros Barreto H, Cerqueira Fontinele F, Pereira de Oliveira A, et al. Phytochemical prospection and modulation of antibiotic activity in vitro by Lippia origanoides H.B.K. in methicillin resistant Staphylococcus aureus. *Biomed Res Int.* 2014;305610:1–7.
[4619] Aelenei P, Miron A, Trifan A, et al. Essential Oils and Their Components as Modulators of Antibiotic Activity against Gram-Negative Bacteria. *Medicines.* 2016;3(19):2-34.
[4620] Magi G, Marini E, Facinelli B. Antimicrobial activity of essential oils and carvacrol, and synergy of carvacrol and erythromycin, against clinical, erythromycin-resistant Group A Streptococci. *Front Microbiol.* 2015;6:165.
[4621] Veras HNH, Rodrigues FFG, Colares AV, et al. Synergistic antibiotic activity of volatile compounds from the essential oil of Lippia sidoides and thymol. *Fitoterapia.* 2012 Apr;83(3):508-12.
[4622] Langeveld WT, Veldhuizen EJ, Burt SA. Synergy between essential oil components and antibiotics: A review. *Crit Rev Microbiol.* 2014 Feb;40(1):76-94.
[4623] Pirbalouti AG, Amirkhosravi A, Bordbar F, et al. Diversity in the chemical composition of essential oils of Ziziphora tenuior as a potential source of pulegone. *Chemija.* 2013;24(3):234-9.
[4624] Sezik E, Tumen G, Baser KHC. Zizipkora tenuior L., a New Source of Pulegone. *Flavour Frag J.* 1991;6:101-3.
[4625] Mahboubi M, Kazempour N, Hosseini H. Chemical Composition, Antimicrobial Activity of Essential Oil from Ziziphora tenuir L. Aerial Parts. *J Essent Oil Bearing Plants.* 2012;15(4):545-9.
[4626] Ganjali A, Pourramezani Harati M, Kaykhaii M. Antimicrobial Effect of Essential Oil of Ziziphora tenuior on Water by Heterotrophic Plate Counts Method in Kerman (Southeast of Iran). *Int J Chem Eng Appl.* 2016 Apr;7(2):120-3.
[4627] Batooli H, Akhabari M, Hosseinizadeh SMJ. Effect of different distillation methods on quantity and quality of essential oil of two ziziphora L. species. *J Herbal Drugs.* 2012 Autumn;3(3):135-46.
[4628] Delnavazi MR, Baba-ali F, Soufiabadi S, et al. Essential Oil Composition, Antioxidant Activity and Total Phenolic Content of Some Lamiaceae Taxa Growing in Northwest of Iran. *Pharm Sci.* 2014;20(1):22-8.
[4629] Abu-Darwish MS, Cabral C, Gonçalves MJ, et al. Ziziphora tenuior L. essential oil from Dana Biosphere Reserve (Southern Jordan); Chemical characterization and assessment of biological activities. *J Ethnopharmacol.* 2016 Dec 24;194:963-970.
[4630] Piryaei M, Abolghasemi MM, Nazemiyeh H. Fast determination of Ziziphora tenuior L. essential oil by inorganic-organic hybrid material based on ZnO nanoparticles anchored to a composite made from polythiophene and hexagonally ordered silica. *Nat Prod Res.* 2015;29(9):833-7.
[4631] Mahboubi M, Kazempour N, Mahboubi A. The efficacy of essential oils as natural preservatives in vegetable oil. *J Diet Suppl.* 2014 Dec;11(4):334-46.
[4632] Bakerink JA, Gospe SM Jr, Dimand RJ, et al. Multiple organ failure after ingestion of pennyroyal oil from herbal tea in two infants. *Pediatrics.* 1996;98(5):944-47.
[4633] Sudekum M, Poppenga RH, Raju N, et al. Pennyroyal oil toxicosis in a dog. *J Am Vet Med Assoc.* 1992;200:817-8.
[4634] Anderson IB, Mullen WH, Meeker JE, et al. Pennyroyal toxicity: measurement of toxic metabolite levels in two cases and review of the literature. *Ann Intern Med.* 1996;124:726-34.
[4635] Seeff L, Stickel F, Navarro VJ. Hepatotoxicity of herbals and dietary supplements. In, Kaplowitz N, DeLeve LD, eds. Drug-induced liver disease. 3rd ed. Amsterdam: Elsevier, 2013, pp. 631-58.
[4636] Mullen W, Anderson I, Oishii S, et al. Accidental pennyroyal oil ingestion in a toddler with the first human serum metabolite detection. *Vet Hum Toxicol.* 1994;36:342.
[4637] Barnes J, Anderson LA, Philpson JD. Herbal Medicine: A Guide for Healthcare Professionals. London, UK: The Pharmaceutical Press, 1996.
[4638] Zimmerman HJ. Unconventional drugs. Miscellaneous drugs and diagnostic chemicals. In, Zimmerman, HJ. Hepatotoxicity: the adverse effects of drugs and other chemicals on the liver. 2nd ed. Philadelphia: Lippincott, 1999: pp. 731-34.
[4639] Anderson IB, Mullen WH, Meeker JE, et al. Pennyroyal toxicity: measurement of toxic metabolite levels in two cases and review of the literature. *Ann Intern Med.* 1996;124:726-34.
[4640] Centers for Disease Control. Fatality and illness associated with consumption of pennyroyal oil - Colorado. *MMWR.* 1978;27:511-13.
[4641] Soares PM, Assreuy AM, Souza EP, et al. Inhibitory effects of the essential oil of Mentha pulegium on the isolated rat myometrium. *Planta Med.* 2005 Mar;71(3):214-8.
[4642] Ciganda C, Laborde A. Herbal infusions used for induced abortion. *J Toxicol Clin Toxicol.* 2003;41(3):235-9.
[4643] Valance WB. Pennyroyal poisoning: a fatal case. *Lancet.* 1955;266 (6895):850-51.

[4644] Gold J, Cates W Jr. Herbal abortifacients. *JAMA*. 1980;243:1365-66.

[4645] Gordon WP, Huitric AC, Seth CL, et al. The metabolism of the abortifacient terpene, (R)-(+)-pulegone, to a proximate toxin, menthofuran. *Drug Metab Dispos*. 1987;15:589–94.

[4646] Khojasteh-Bakht SC, Chen W, Koenigs LL, et al. Metabolism of (R)-(+)-pulegone and (R)-(+)-menthofuran by human liver cytochrome P-450s: evidence for formation of a furan epoxide. *Drug Metab Dispos*. 1999 May;27(5):574-80.

[4647] Gordon P, Khojasteh SC. A decades-long investigation of acute metabolism-based hepatotoxicity by herbal constituents: a case study of pennyroyal oil. *Drug Metab Rev*. 2015 Feb;47(1):12-20.

[4648] Zhou SF, Xue CC, Yu XQ, et al. Metabolic activation of herbal and dietary constituents and its clinical and toxicological implications: an update. *Curr Drug Metab*. 2007;8:526–553.

[4649] National Toxicology Program. Toxicology and carcinogenesis studies of pulegone (CAS No. 89-82-7) in F344/N rats and B6C3F1 mice (gavage studies). *Natl Toxicol Program Tech Rep Ser*. 2011 Aug;(563):1-201.

[4650] National Toxicology Program. Toxicology and carcinogenesis studies of pulegone (CAS No. 89-82-7) in F344/N rats and B6C3F1 mice (gavage studies). *Natl Toxicol Program Tech Rep Ser*. 2011 Aug;(563):1-201.

[4651] Sullivan JB Jr, Rumack BH, Thomas H Jr, et al. Pennyroyal oil poisoning and hepatotoxicity. *JAMA*. 1979;242:2873-4.

[4652] The Review of Natural Products by Facts and Comparisons. St. Louis, MO: Wolters Kluwer Co., 1999.

[4653] Foster S, Tyler VE. Tyler's Honest Herbal: A Sensible Guide to the Use of Herbs and Related Remedies. 3rd ed., Binghamton, NY: Haworth Herbal Press, 1993.

[4654] Martindale W. Martindale the Extra Pharmacopoeia. Pharmaceutical Press, 1999.

[4655] McGuffin M, Hobbs C, Upton R, Goldberg A, eds. American Herbal Products Association's Botanical Safety Handbook. Boca Raton, FL: CRC Press, LLC, 1997.

[4656] Sullivan JB Jr, Rumack BH, Thomas H Jr, et al. Pennyroyal oil poisoning and hepatotoxicity. *JAMA*. 1979;242:2873-4.

[4657] Anderson IB, Mullen WH, Meeker JE, et al. Pennyroyal toxicity: measurement of toxic metabolite levels in two cases and review of the literature. *Ann Intern Med*. 1996;124:726-34.

[4658] Allen WT. Note on a case of supposed poisoning by pennyroyal. *Lancet*. 1897;1(3841):1022-23.

[4659] Girling J. Poisoning by pennyroyal. *Br Med J*. 1887;1:1214.

[4660] Braithwaite PF. A case of poisoning by pennyroyal: recovery. *Br Med J*. 1906;2:865.

[4661] Buechel DW, Haverlah VC, Gardner ME. Pennyroyal oil ingestion: report of a case. *J Am Osteopath Assoc*. 1983;82:793-4.

[4662] Flynn EF. Poisoning by essence of pennyroyal. *Br Med J*. 1893;2:1270.

[4663] Burkhard PR, Burkhardt K, Haenggeli CA, et al. Plant-induced seizures: reappearance of an old problem. *J Neurol*. 1999 Aug;246(8):667-70.

[4664] Early DF. Pennyroyal: a rare cause of epilepsy. *Lancet*. 1961; 278 (7202): 580-81.

[4665] Sánchez-Borzone ME, Marin LD, García DA. Effects of Insecticidal Ketones Present in Mint Plants on GABAA Receptor from Mammalian Neurons. *Pharmacogn Mag*. 2017 Jan-Mar;13(49):114-117.

[4666] Sullivan JB Jr, Rumack BH, Thomas H Jr, et al. Pennyroyal oil poisoning and hepatotoxicity. *JAMA*. 1979;242:2873-4.

[4667] Sudekum M, Poppenga RH, Raju N, et al. Pennyroyal oil toxicosis in a dog. *J Am Vet Med Assoc*. 1992;200:817-8.

[4668] Anderson IB, Mullen WH, Meeker JE, et al. Pennyroyal toxicity: measurement of toxic metabolite levels in two cases and review of the literature. *Ann Intern Med*. 1996;124:726-34.

[4669] Zimmerman HJ. Unconventional drugs. Miscellaneous drugs and diagnostic chemicals. In, Zimmerman, HJ. Hepatotoxicity: the adverse effects of drugs and other chemicals on the liver. 2nd ed. Philadelphia: Lippincott,1999: pp. 731-34.

[4670] Seeff L, Stickel F, Navarro VJ. Hepatotoxicity of herbals and dietary supplements. In, Kaplowitz N, DeLeve LD, eds. Drug-induced liver disease. 3rd ed. Amsterdam: Elsevier, 2013, pp. 631-58.

[4671] Burkhard PR, Burkhardt K, Haenggeli CA, et al. Plant-induced seizures: reappearance of an old problem. *J Neurol*. 1999 Aug;246(8):667-70.

[4672] Waldman N. Seizure caused by dermal application of over-the-counter eucalyptus oil head lice preparation. *Clin Toxicol (Phila)*. 2011 Oct;49(8):750-1.

[4673] Craig JO. Poisoning by the volatile oils in childhood. *Arch Dis Child*. 1953;28:259-67.

[4674] Mathew T, Kamath V, Kumar RS, et al. Eucalyptus oil inhalation-induced seizure: A novel, underrecognized, preventable cause of acute symptomatic seizure. *Epilepsia Open*. 2017 Jul 4;2(3):350-354.

[4675] Abu-Darwish MS, Cabral C, Gonçalves MJ, et al. Ziziphora tenuior L. essential oil from Dana Biosphere Reserve (Southern Jordan); Chemical characterization and assessment of biological activities. *J Ethnopharmacol*. 2016 Dec 24;194:963-970.

[4676] Mahboubi M, Kazempour N, Hosseini H. Chemical Composition, Antimicrobial Activity of Essential Oil from Ziziphora tenuir L. Aerial Parts. *J Essent Oil Bearing Plants*. 2012;15(4):545-9.

[4677] Abu-Darwish MS, Cabral C, Gonçalves MJ, et al. Ziziphora tenuior L. essential oil from Dana Biosphere Reserve (Southern Jordan); Chemical characterization and assessment of biological activities. *J Ethnopharmacol*. 2016 Dec 24;194:963-970.

[4678] Abdolshahi A, Naybandi-Atashi S, Heydari-Majd M, et al. Antibacterial activity of some Lamiaceae species against Staphylococcus aureus in yoghurt-based drink (Doogh). *Cell Mol Biol (Noisy-le-grand)*. 2018 Jun 25;64(8):71-77.

[4679] Oxtturk Y, Aydin S, Tecik B, et al. Effects of Essential Oils from Certain Ziziphora Species on Swimming Performance in Mice. *Phytother Res*. 1995;9:225-7.

[4680] Delnavazi MR, Baba-ali F, Soufiabadi S, et al. Essential Oil Composition, Antioxidant Activity and Total Phenolic Content of Some Lamiaceae Taxa Growing in Northwest of Iran. *Pharm Sci*. 2014;20(1):22-8.

INDEX

D

E

F

G

Q

R

S

T